W9-AQI-839

Applied Numerical Methods for Engineers and Scientists

Applied Numerical Methods for Engineers and Scientists

Singiresu S. Rao

University of Miami, Coral Gables, Florida

Prentice Hall
Upper Saddle River, NJ 07458

Library of Congress Cataloging-in-Publication Data

Cip Data on file

Vice President and Editorial Director: *Marcia J. Horton*
Publisher: *Tom Robbins*
Acquisitions Editor: *Eric Frank*
Editorial Assistant: *Jessica Romeo*
Vice President and Director of Production and Manufacturing, ESM: *David W. Riccardi*
Executive Managing Editor: *Vince O'Brien*
Managing Editor: *David A. George*
Production Editor: *Lakshmi Balasubramanian*
Director of Creative Services: *Paul Belfanti*
Art Director: *Jayne Conte*
Cover Designer: *Bruce Kenselaar*
Art Editor: *Gregory Dulles*
Manufacturing Manager: *Trudy Pisciotti*
Manufacturing Buyer: *Lisa McDowell*
Marketing Manager: *Holly Stark*
Marketing Assistant: *Karen Moon*

Prentice-Hall, Inc.
Upper Saddle River, New Jersey 07458

Printed in the United States of America

10 9 8 7 6 5 4

ISBN 013089480X

Pearson Education Ltd., *London*
Pearson Education Australia Pty. Ltd., *Sydney*
Pearson Education Singapore, Pte. Ltd.
Pearson Education North Asia Ltd., *Hong Kong*
Pearson Education Canada, Inc., *Toronto*
Pearson Educacíon de Mexico, S.A. de C.V.
Pearson Education—Japan, *Tokyo*
Pearson Education Malaysia, Pte. Ltd.

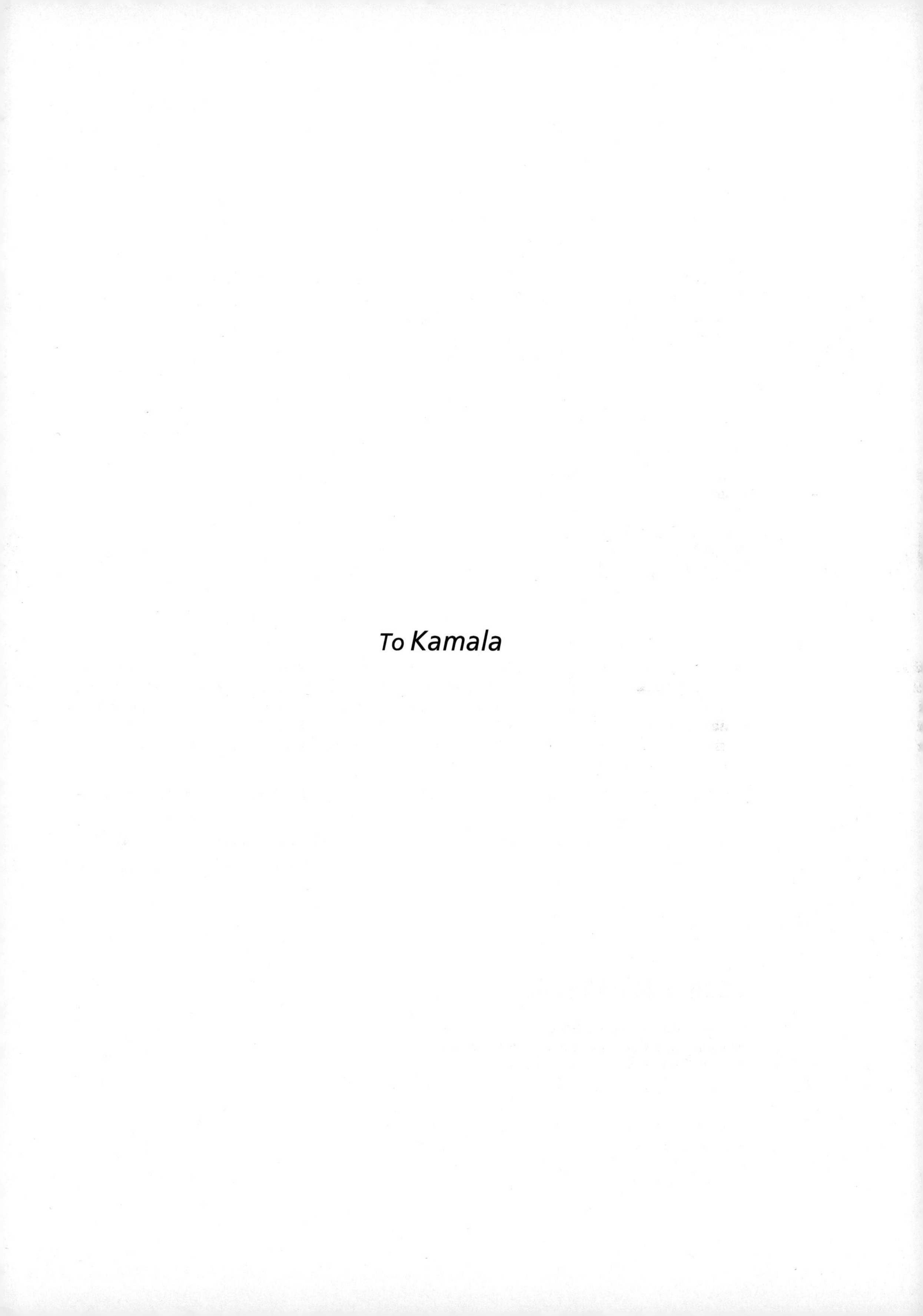

To Kamala

Contents

PREFACE

The use of numerical methods for the analysis, simulation, and design of engineering processes and systems has been increasing at a rapid rate in recent years. The availability of cheap high-speed computing power makes the numerical solution of even complex engineering problems economically feasible. In the face of ever increasing demands on engineering profession to perform better, the students who learn numerical methods in preparing to face the challenges of 21st century should learn not only the theory behind the methods, but also acquire skills to implement the methods for computer solution. In addition, the students should be aware of the many commercial software systems available and their use in the solution of engineering problems. Although a student may not learn all the numerical methods described in this book and use all the software systems available in any one course, he or she should be in a position to intelligently select and use suitable numerical methods and software systems as the need arises in practice.

The use of numerical methods in engineering can be considered partly science and partly art. Thus, a cookbook-type procedure will not be effective in learning the methods. A student should solve a problem using different approaches and a variety of software systems and experiment with the various parameters of the problem. The different results obtained through this process will form an experience base for selecting a suitable method and interpreting the results for a new problem. It is always desirable to compare and verify the results with other available solutions based on engineering judgment and intuition.

This book is intended for courses on numerical methods at the junior and senior level as well as at the beginning graduate level. The book also serves as a reference for numerical methods in engineering. Fortran and C programs, along with illustrative examples, are given in each chapter to implement many of the numerical methods discussed in that chapter. The use of commercial numerical softwares—MATLAB,[1] MAPLE[2] and MATHCAD[3]—in the solution of practical problems is demonstrated in every chapter. Even when a program from a software package is used, we need to understand the basic principles, purpose, and limitations of the program. Often, in many engineering applications, an available standard program cannot be

[1]MATLAB is a registered trademark of The MathWorks, Inc.

[2]MAPLE is a registered trademark of Waterloo Maple Software.

[3]MATHCAD is a registered trademark of MathSoft, Inc.

used directly; we need to adapt and modify it. This invariably requires a sound knowledge of the numerical method as well as some computational experience with the method. The book is aimed at presenting numerical methods along with their practical applications in a manner that helps students achieve the goals just outlined.

Organization

Applied Numerical Methods for Engineers is organized into 13 chapters and 6 appendices. Chapter 1 presents an overview of numerical methods, iterative processes, numerical errors, software available for numerical methods, programming languages, and the various aspects of computer program development. The methods of solving nonlinear equations are given in Chapter 2. The solution of sets of linear algebraic equations is presented in Chapter 3. Both direct and iterative methods are considered. The matrix eigenvalue problem is the topic of Chapter 4. Chapter 5 deals with the methods of curve fitting and interpolation. The probabilistic and statistical methods are considered in Chapter 6. The numerical differentiation and numerical integration are the topics of Chapters 7 and 8, respectively. The numerical solution of ordinary differential equations is considered in Chapters 9 and 10. While Chapter 9 presents the methods of solving initial-value problems, Chapter 10 deals with the solution of boundary-value problems. The numerical solution of partial differential equations is considered in Chapter 11. The optimization and the finite-element methods are presented in Chapters 12 and 13, respectively. Appendices A and B provide the basics of Fortran and C languages while Appendices C, D, and E summarize the basics of MAPLE, MATLAB, and MATHCAD, respectively. A review of matrix algebra is given in Appendix F. Finally, Appendix G presents tables of statistical distributions.

The material of the book provides flexible options for different types of numerical methods courses. A junior and senior level course may cover the basic techniques of Chapters 1, 2, 3, and 5 to 9. A first-level graduate course can cover Chapters 4, 10, 11, 12, and 13 as well. The prerequisites for using the text are elementary calculus, basic concepts of linear algebra, and an introduction to differential equations.

Each topic for *Applied Numerical Methods for Engineers* is self-contained. In derivations and developments, steps needed for continuity of understanding have been included to aid the reader at the introductory level. Representative engineering applications are given at the beginning of each chapter so that the reader can appreciate the practical use and application of the numerical methods presented in that chapter. Many sample problems are solved by using several methods, and the results are compared, discussed, and general conclusions are drawn. Most of the algorithms described in the book are implemented in the form of Fortran and C codes and are made available at the Web site of the book. The use of different commercial software systems, as well as the programs available at the Web site of the book, is illustrated in each chapter.

Features

The specific features of the book include

1. A variety of engineering applications at the beginning of each chapter to illustrate the practicality of the methods considered in that chapter.
2. The presentation of the material in a simple and user-friendly form. Illustrative examples follow the presentation of the topics.
3. A discussion of convergence rate, error, relative performance, and recommendations for the numerical methods.
4. Review questions to help students in reviewing and testing their understanding of the text material. These include multiple choice questions, questions with brief answers, true–false questions, questions involving matching of related descriptions, and fill-in-the-blank type questions. Answers to review questions can be found at the Web site of the book.
5. A summary of important algorithms in the instructor's manual.
6. Over 700 problems, with solutions in the instructor's manual.
7. The inclusion of several open ended, project type and design problems at the end of chapters.
8. Fortran and C programs for many of the methods presented in the book can be found at the Web site of the book.
9. The inclusion of examples and problems based on the use of MATLAB, MAPLE, and MATHCAD in every chapter.
10. References to lead the reader to specialized and advanced literature.
11. Brief biographical information and photographs of scientists and mathematicians who contributed to the development of numerical methods, found at the Web site of the book.

Web site of the book

The Fortran and C programs used in the book, answers to problems, solutions to review questions, and brief biographical information of scientists can be found at the web site of the book: http://www.prenhall.com/rao. Note that the programs and techniques presented in the book and at the web site are intended for use by students in learning the material. Although the material has been tested, no warranty is implied as to their accuracy. I would appreciate receiving any errors found in the book.

Acknowledgments

I would like to express my appreciation to the students who used the notes that led to the present text. I would like to thank Mr. Qiang Fu, Mr. Lingtao Cao and Ms. Qing Liu, graduate students at the University of Miami, for their help in solving some of the examples and problems using MATLAB, MAPLE, and MATHCAD. I wish to thank my family, wife Kamala, daughters Sridevi and Shobha, and grand daughter Siriveena, for their numerous intangible contributions to this work. In particular, I dedicate this book to my wife, Kamala, for providing me the inspiration and support in completing this book.

మా తెలుగు తల్లికి మల్లెపూ దండ

మా కన్న తల్లికి మంగళారతులు

SINGIRESU S. RAO
University of Miami
Coral Gables, FL

1

Introduction to Numerical Methods

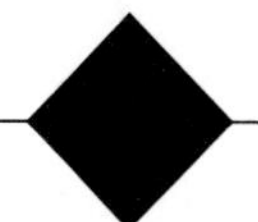

CHAPTER CONTENTS

1.1 Importance of Numerical Methods in Engineering

Most engineering analysis problems involve (1) the development of a mathematical model to represent all the important characteristics of the physical system; (2) the derivation of the governing equations of the model by applying physical laws, such as equilibrium equation, Newton's laws of motion, conservation of mass and conservation of energy; (3) solution of the governing equations; and (4) interpretation of the solution. Depending on the system being analyzed and the mathematical model used, the governing equations may be a set of linear or nonlinear algebraic equations, a set of transcendental equations, a set of ordinary or partial differential equations, a set of homogeneous equations leading to an eigenvalue problem, or an equation involving integrals or derivatives. We may or may not be able to find the solution of a governing equation analytically. If the solution can be represented in the form of a closed-form mathematical expression, it is called an analytical solution. Analytical solutions denote exact solutions that can be used to study the behavior of the system with varying parameters. Unfortunately, very few practical systems lead to analytical solutions, and hence analytical solutions are of limited use. In certain special types of problems, graphical solutions can be found to study the behavior of the system. However, graphical solutions usually are less accurate, awkward to use, can only be implemented if the dimensionality of the problem is less than or equal to three, and require more time. Numerical solutions are those that cannot be expressed in the form of mathematical expressions. They can be found using a suitable type of calculation-intensive process, known as a numerical method. For example, consider the integral

$$I_1 = \int_a^b x e^{-x^2}\, dx. \tag{1.1}$$

The value of this integral can be expressed analytically as

$$I_1 = \left(-\frac{1}{2} e^{-x^2} \right) \Bigg|_a^b = -\frac{1}{2} e^{-b^2} + \frac{1}{2} e^{-a^2} = \frac{1}{2} \left(e^{-a^2} - e^{-b^2} \right). \tag{1.2}$$

On the other hand, the integral

$$I_2 = \int_a^b f(x)\, dx = \int_a^b e^{-x^2}\, dx \tag{1.3}$$

does not have a closed-form (analytical) solution. This integral can only be evaluated numerically. Since the integral is the same as the area under the curve $f(x)$, its value can be estimated by breaking the area under the curve into small rectangles and adding the areas of the rectangles. (See Fig. 1.1.) Since numerical methods involve a large number of tedius arithmetic calculations, their use and popularity has been increasing with the development and availability of powerful and inexpensive computers. Numerical methods can be used to find solutions of even complex engineering problems. While analytical solutions usually require several simplifying assumptions of the physical system, numerical solutions do not require such assumptions. Although numerical solutions cannot provide an immediate insight into the behavior of the simplified physical system, they can be used to study the behavior of the true physical system.

1.2 Computers

1.2.1 Brief History

The abacus, developed in ancient China and Egypt about 3000 years ago, represents one of the earliest computers. The first systematic attempt to organize information processing resulted in the development of logarithmic and trigonometric tables in the 16th and 17th centuries. The slide rule, developed in 1654 by Robert Bissaker, was used for multiplications and divisions, as well as for the evaluation of

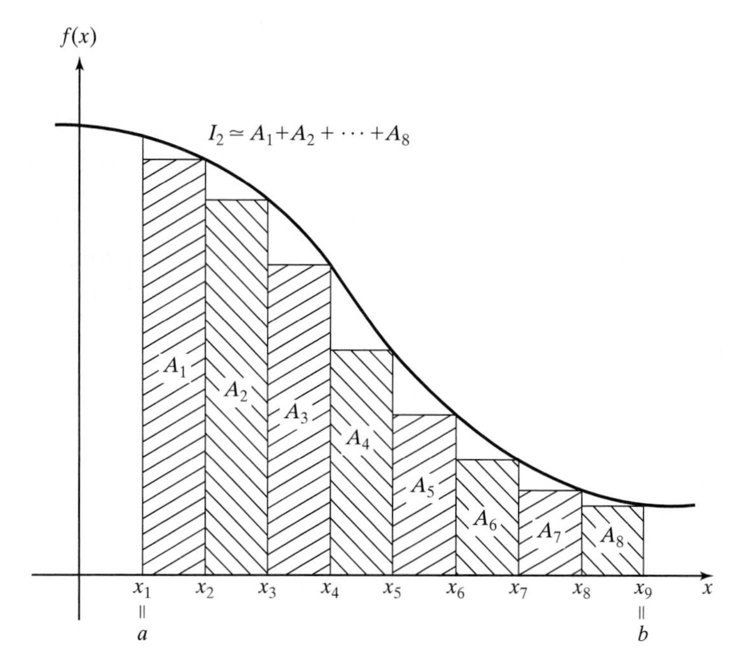

Figure 1.1 Numerical evaluation of the integral I_2.

square roots, logarithms, and trigonometric functions. Stimulated by the industrial revolution, the French philosopher and mathematician Blaise Pascal developed the first mechanical adding machine in 1642. Later, Gottfried Wilhelm von Leibniz, a German philosopher and mathematician, built a mechanical calculator in 1694. In 1804, Joseph Jacquard, a French loom designer, developed an automatic pattern loom whose sequence of operations was controlled by punched cards. The loom was used to produce intricate patterns and paved the way for the development of mechanical computers. The British mathematician Charles Babbage designed an automatic digital computer around 1833, but the machine, called the analytical engine, was never built.

The basic ideas of Babbage were implemented in the electromechanical Automatic-Sequence-Controlled Calculator (ASCC), also known as MARK I, which was developed as a joint project between Harvard University and IBM. The first entirely electronic universal calculator was built in 1945 at the University of Pennsylvania with the support of U.S. Army's Ballistic Research Laboratory. It used vacuum tubes and was called the Electronic Numerical Integrator And Computer (ENIAC). In 1950, there were approximately 20 automatic calculators and computers in the United States, with a total value of nearly $1 million. The first-generation computers, developed between 1950 and 1959, included machines such as UNIVAC I and 1103, IBM 701 and 704, and ERA 1101. The second-generation computers, produced between 1959 and 1963, were based on ferrite-core memories and transistors as circuit elements. CDC 1604 and 3600, IBM 1401, 1620, 7040 and 7094, and PDP 1 represent some of the second-generation computers. The most powerful computer systems, also known as supercomputers, represent the fifth-generation computers and were developed in 1980s. Although these mainframe computers were popular, they were too expensive for individual professionals to acquire. The development of integrated circuits consisting of thousands of transistors on tiny silicon chips lead to the invention of the personal computer (PC) or the microcomputer. The PC has dominated the computer industry and slowly became part of everyday life. As far back as January 1983, *Time* magazine selected the PC as its *Man of the Year*.

1.2.2 Hardware and Software

A digital computer functions through the interaction of the hardware (the physical component of the computer) and the software (the programs or instructions). The hardware is composed of the central processing unit (CPU) and the input/output–(I/O) units as shown in Fig. 1.2. The CPU comprises a memory unit, an arithmetic/logic unit (ALU), and a control unit. The instructions or data are entered by the user through the input device and are stored in the memory unit. The control unit decodes these data and causes the ALU to process the data and produce the output. The output is stored in the memory and is sent to the output unit.

The computer software is made up of system software, consisting of programs built into the computer, and user software, consisting of programs written by the user. The system software, stored in the memory of the computer, is in binary form

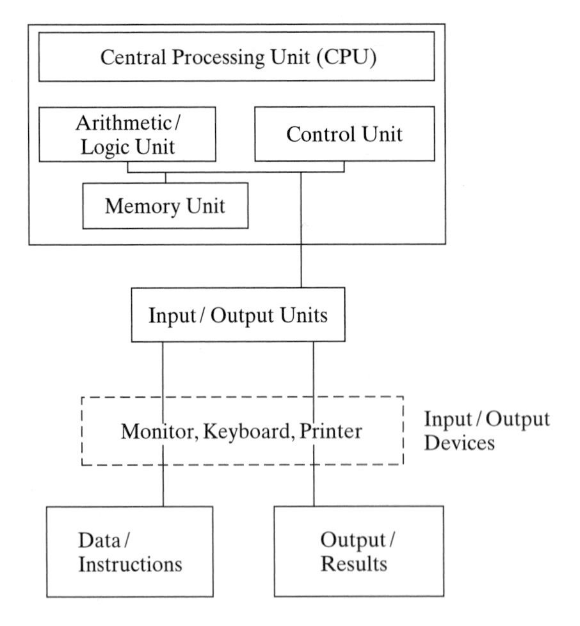

Figure 1.2 Central processing unit of a computer system.

and is written in a low-level (machine) language. If instructions are written in a low-level language, the necessary commands (program) will be lengthy and vary from machine to machine. Thus, the programs written in a low-level language are machine dependent and are not portable.

A major set of built-in programs (system software), called an operating system, controls the communications and operations with the system, schedules the tasks and interprets the user's instructions. Although the operating system used in a computer system depends on the manufacturer, the operating systems known as Unix and DOS (disk operating system) have been popular. User software used by engineers and scientists includes CAD and graphics tools, word processors, spread sheets, numerical analysis packages, symbolic manipulation routines, and simulation tools.

The user software is developed using a high-level language. The high-level languages permit the programmer to write instructions for the computer in a form that is similar to ordinary English and algebra. If instructions are written in a high-level language, the resulting program will be brief and independent of the machine. Thus, the programs written in high-level languages are portable.

The user programs, also called source codes, are not executed directly. A compiler, which is a part of the system software, translates the source code into a binary code (machine language), which is then loaded into computer memory and executed. If compilation of a source code is successful, it means that the translator understood the structure and syntax used in the source code. It does not imply that the instructions given are correct or can be executed. The binary code can be saved and used at a later time without a need to recompile. The binary code can be executed only on the specific type of computer on which it was compiled. On the other hand, source code written in a particular high-level language can be compiled and executed on any computer that has a compiler for that language.

1.3 Computer Programming Languages

Once the particular numerical method to be used for the solution of a given problem is selected, the method is to be transformed into computer code, or a program, for implementation on a specific computer using a high-level language. Some of the high-level languages are FORTRAN, COBOL, BASIC, Pascal, C, Ada, LISP, APL, and FORTH. FORTRAN (FORmula TRANslation) was the first high-level language developed primarily for use in engineering and science. It was developed by a committee sponsored by International Business Machines (IBM) and headed by John Backus in 1957. FORTRAN uses English-like commands and facilitates the easy development of even complex programs. It has undergone several modifications and improvements over the past several decades and has become one of the preferred computer languages for solving engineering and scientific problems. It was the first programming language to be standardized by the American National Standards Institute (ANSI) in 1966. The 1966 standards were designated as FORTRAN IV. Later, the standards were revised in 1977, and the new standard was designated as FORTRAN 77 (since it was completed in 1977). The most recent version standardized by ANSI, designated as FORTRAN 90, has some features similar to those used in the languages C and Pascal. FORTRAN 90 was followed in 1997 by a minor update and was designated as FORTRAN 95.

COBOL (COmmon Business Oriented Language) can be considered as the business equivalent of FORTRAN and is not popular in engineering applications. Due to its excellent input/output characteristics and ability to handle and process vast files of business information, it has been used widely in business applications.

BASIC (Beginner's All-purpose Symbolic Instruction Code) was developed in 1964 by John Kemeny and Thomas Kurtz of Dartmouth College. It is a widely used language for personal (micro) computers. It is much simpler to use compared to FORTRAN and is particularly useful for developing small programs. It does not have the versatility of FORTRAN in developing large and complex programs. Since BASIC is widely used on personal computers, several dialects and improved versions of the language were developed, especially due to the lack of any standardization. Some of the dialects include BetterBASIC, QuickBASIC, and TrueBASIC. It is not commonly used for engineering and science applications.

C is one of the most powerful and portable high-level languages that can be used to generate efficient codes for a variety of computers. It was developed by Dennis Ritchie of Bell Laboratories in 1974 and originally implemented on the Unix operating system. Although originally it was not intended to be a general purpose language, it has proved to be valuable for several applications such as systems programming, microprocessors, text processing, and a variety of application packages. In 1983, the ANSI formed a committee to standardize the C language based on the industry's *de facto* standard and issued a standard document in 1990.

Pascal was developed in 1971 by the Swiss computer scientist, Niklaus Wirth. The language was named after the French mathematician, Blaise Pascal, who in 1642 attempted to construct a mechanical device to perform simple calculations. Pascal is a high-level language that is powerful, easy to learn, and its syntax and organization tend to lead programmers to develop good programs. The deficiencies of Pascal include a primitive I/O system and the nonexistence of built-in interfaces to control the computers. To encourage compatibility between compilers, Pascal was standardized by ANSI in cooperation with International Standards Organization (ISO) and the Institute of Electrical and Electronic Engineers (IEEE).

Ada is named in honor of Augusta Ada Byron, the Countess of Lovelace and the daughter of the English poet Lord Byron. Ada was the assistant of the mathematician, Charles Babbage, who invented the calculating machine called the *Analytical Engine*. She wrote a computer program to compute Bernoulli numbers in 1830 on the *Analytical Engine*. Because of this effort, Ada may be considered the world's first computer programmer. Ada was developed under the initiative of the United States Department of Defense with the aim of standardizing military software. It is based on Pascal and was developed by a team of computer scientists lead by Jean Ichbian of CII Honeywell Bull in 1980. Ada enforces rules that lead to the development of more readable, portable, reliable, modular, maintainable, and efficient programs. Its popularity is increasing due to the availability of a variety of software developed for the military.

Appendices A and B present the basic concepts of Fortran and C languages, respectively.

1.4 Data Representation

The information to be stored in a computer may be in the form of numeric data, nonnumeric data, constants, and variables. In a digital computer, all information is stored in binary form. A *bit* is a binary digit (i.e., a zero or a one). A *byte* is a larger unit in which bits are organized. Usually, a byte consists of 8 bits. A *word* is a larger unit in which bytes are organized. For example, a 32-bit word consists of 4 bytes.

1.4.1 Numeric Data

Numeric data, in the form of numbers, are used to compute, count, and label. The numbers can be integers (fixed-point numbers) or real (floating-point numbers). Integer numbers denote whole numbers with no fractional part. In the decimal system, an integer I (base-10 number) is written as

$$I = a_{m-1}a_{m-2}\cdots a_2 a_1 a_0, \tag{1.4}$$

which is equivalent to

$$(I)_{10} = a_{m-1}(10)^{m-1} + a_{m-2}(10)^{m-2} + \cdots + a_2(10)^2 + a_1(10)^1 + a_0(10)^0. \tag{1.5}$$

In general, an m-digit integer (I_m) can be expressed in base b as

$$I_m = (a_{m-1}a_{m-2}\cdots a_1 a_0)_b; \quad a_j \in \{0, 1, 2, \ldots, b-1\}, \tag{1.6}$$

where the subscript b denotes the number base and the digits a_j can take any of the integer values, $0, 1, 2, \ldots, b-1$. Thus, the digits a_j can take values from 0 to 1 for binary numbers ($b = 2$) and from 0 to 9 for decimal numbers ($b = 10$). The decimal value of the number given by Eq. (1.6) is

$$I_m = \sum_{j=0}^{m-1} b^j (a_j). \tag{1.7}$$

For example,

$$I_m = (1\,0\,0\,1\,0)_2 = 2^4(1) + 2^3(0) + 2^2(0) + 2^1(1) + 2^0(0) = 16+0+0+2+0 = (18)_{10},$$

and

$$I_m = (1\,2\,3\,4)_{10} = 10^3(1) + 10^2(2) + 10^1(3) + 10^0(4) = (1\,2\,3\,4)_{10}.$$

1.4.2 Conversion of a Decimal Number to a Binary Number

A decimal number I can be represented in a binary form

$$(I)_{10} = (a_{m-1}a_{m-2}\cdots a_1 a_0)_2 = \sum_{j=0}^{m-1} 2^j (a_j)$$
$$= 2^{m-1}a_{m-1} + 2^{m-2}a_{m-2} + \cdots + 2^1 a_1 + 2^0 a_0 \tag{1.8}$$

using the procedure shown next. First, the number I is expressed as the sum of twice another number P_0 and a constant a_0:

$$I = 2P_0 + a_0 \quad (a_0 \text{ is 0 or 1}). \tag{1.9}$$

The number P_0 is then written as the sum of twice another number P_1 and a constant a_1:

$$P_0 = 2P_1 + a_1 \quad (a_1 \text{ is 0 or 1}). \tag{1.10}$$

This process is continued until the new number $P_{m-1} = 0$ is obtained. Thus, the procedure yields the sequences of numbers P_0, P_1, $P_2, \ldots, P_{m-1}$ and a_0, a_1, $a_2, \ldots, a_{m-1}$:

$$\begin{aligned} I &= 2P_0 + a_0; \\ P_0 &= 2P_1 + a_1; \\ P_1 &= 2P_2 + a_2; \\ &\vdots \\ P_{m-2} &= 2P_{m-1} + a_{m-1}(P_{m-1} = 0). \end{aligned} \tag{1.11}$$

The decimal number I can then be expressed in the binary form

$$(I)_{10} = (a_{m-1}a_{m-2}\cdots a_2 a_1 a_0)_2. \tag{1.12}$$

▶Example 1.1

Find the binary form of the number 193.

Solution

By using the procedure indicated by Eq. (1.11), we obtain

$$\begin{aligned} 193 &= 2 \times 96 + 1 \quad (N = 193,\ P_0 = 96,\ a_0 = 1); \\ 96 &= 2 \times 48 + 0 \quad (P_0 = 96,\ P_1 = 48,\ a_1 = 0); \\ 48 &= 2 \times 24 + 0 \quad (P_1 = 48,\ P_2 = 24,\ a_2 = 0); \\ 24 &= 2 \times 12 + 0 \quad (P_2 = 24,\ P_3 = 12,\ a_3 = 0); \\ 12 &= 2 \times 6 + 0 \quad (P_3 = 12,\ P_4 = 6,\ a_4 = 0); \end{aligned}$$

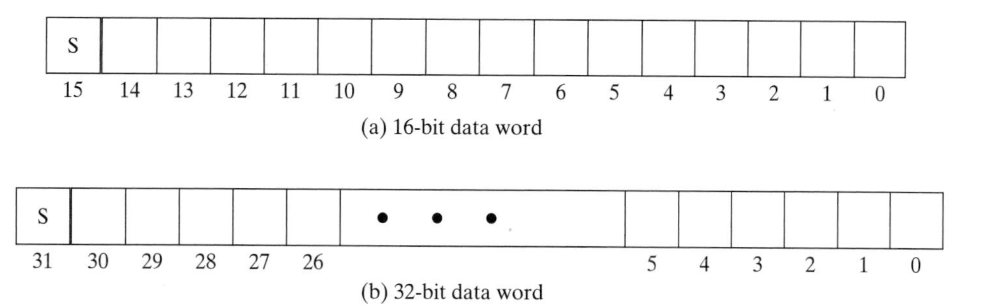

Figure 1.3 Internal representation of integers.

$$6 = 2 \times 3 + 0 \quad (P_4 = 6,\ P_5 = 3,\ a_5 = 0);$$

$$3 = 2 \times 1 + 1 \quad (P_5 = 3,\ P_6 = 1,\ a_6 = 1);$$

$$1 = 2 \times 0 + 1 \quad (P_6 = 1,\ P_7 = 0,\ a_7 = 1).$$ ◀

This gives the equivalence

$$193 = (11000001)_2.$$

In a computer, integers are stored as signed binary numbers as shown in Fig. 1.3, where the first digit denotes the sign ($S = 1$ if negative and $S = 0$ if positive). The range of an integer is $2^{15} - 1$ to -2^{15} (32767 to −32768) for a 16-bit word (number) and $2^{31} - 1$ to -2^{31} for a 32-bit word (number).

A real or floating-point number (R) contains an integer part and a fractional part. A fractional part, in decimal system, is given by the sum of negative powers of 10. Thus, a fractional number R, $0 < R < 1$, can be expressed as

$$R = b_1 \times 10^{-1} + b_2 \times 10^{-2} + \cdots + b_m \times 10^{-m} = 0.b_1 b_2 \cdots b_m, \tag{1.13}$$

where $b_1, b_2, \ldots, b_m$ are integers between 0 and 9, and m is the number of digits required to represent the number R. Often, infinitely many digits are required to represent a fraction in the decimal system. For example, the fraction $\frac{1}{6}$ is expressed as

$$\frac{1}{6} = 0.1666666\bar{6},$$

where the notation $\bar{6}$ indicates that the digit 6 is to be repeated infinite times for an exact representation of the fraction $\frac{1}{6}$. Similar to Eq. (1.13), a binary fraction is given by the sum of negative powers of 2. Thus, a fractional number R, $0 < R < 1$, can be expressed as

$$R = b_1 \times 2^{-1} + b_2 \times 2^{-2} + \cdots + b_m \times 2^{-m} = (0.b_1 b_2 \cdots b_m)_2, \tag{1.14}$$

where $b_1, b_2, \ldots, b_m$ are 0 or 1 and m is the number of digits required to represent the number R. For some numbers R, infinitely many digits are required for representation as a binary fraction. For example, the fraction $\frac{1}{6}$ can be expressed as a binary fraction as

$$\frac{1}{6} = (0.100110011\overline{1001}\ldots)_2,$$

where the group of four digits 1001 is repeated forever.

To convert a real number $R, 0 < R < 1$, into a binary fraction, we use Eq. (1.14). For this, the number R is doubled and separated into integer and fractional parts as

$$2R = b_1 + f_1 = \text{intg}\ (2R) + \text{frac}\ (2R), \tag{1.15}$$

where b_1 = integer part of $(2R)$ = intg $(2R)$ and f_1 = fractional part of $(2R)$ = frac $(2R)$. The fractional part is again doubled and the result expressed as a sum of an integer $b_2 = \text{intg}\ (2f_1)$ and a fraction $f_2 = \text{frac}\ (2f_1)$:

$$2f_1 = b_2 + f_2. \tag{1.16}$$

This process is continued until the fractional part f_i becomes zero. The process can be summarized as follows:

$$\begin{aligned}
2R &= \text{intg}\ (2R) + \text{frac}\ (2R) \equiv b_1 + f_1;\\
2f_1 &= \text{intg}\ (2f_1) + \text{frac}\ (2f_1) \equiv b_2 + f_2;\\
&\vdots\\
2f_i &= \text{intg}\ (2f_{i-1}) + \text{frac}\ (2f_{i-1}) \equiv b_i + f_i \quad (f_i = 0).
\end{aligned} \tag{1.17}$$

▶Example 1.2

Express the number $\frac{6}{10}$ as a binary number.

Solution

Noting that $R = 0.6$, the procedure indicated by Eq. (1.17) can be applied to obtain the following result:

$$\begin{aligned}
2R &= 1.2,\ b_1 = \text{intg}\ (2R) = 1, && f_1 = \text{frac}\ (2R) = 0.2;\\
2f_1 &= 0.4,\ b_2 = \text{intg}\ (2f_1) = 0, && f_2 = \text{frac}\ (2f_1) = 0.4;\\
2f_2 &= 0.8,\ b_3 = \text{intg}\ (2f_2) = 0, && f_3 = \text{frac}\ (2f_2) = 0.8;\\
2f_3 &= 1.6,\ b_4 = \text{intg}\ (2f_3) = 1, && f_4 = \text{frac}\ (2f_3) = 0.6;\\
2f_4 &= 1.2,\ b_5 = \text{intg}\ (2f_4) = 1, && f_5 = \text{frac}\ (2f_4) = 0.2;\\
2f_5 &= 0.4,\ b_6 = \text{intg}\ (2f_5) = 0, && f_6 = \text{frac}\ (2f_5) = 0.4;\\
2f_6 &= 0.8,\ b_7 = \text{intg}\ (2f_6) = 0, && f_7 = \text{frac}\ (2f_6) = 0.8;\\
2f_7 &= 1.6,\ b_8 = \text{intg}\ (2f_7) = 1, && f_8 = \text{frac}\ (2f_7) = 0.6.
\end{aligned}$$

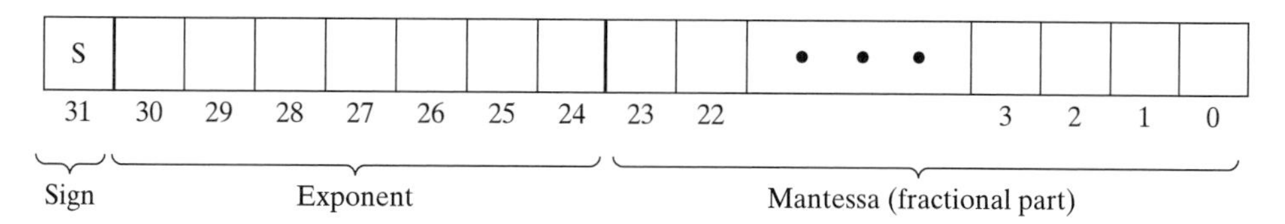

Figure 1.4 Internal representation of real numbers.

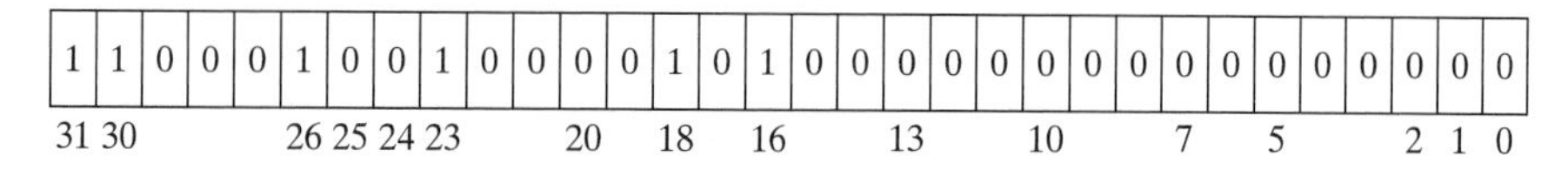

Figure 1.5 A real number.

This gives the desired relation

$$\frac{6}{10} = (0.1001\overline{1001}\ldots)_2.$$

A real number (R) is stored internally as a binary number as shown in Fig. 1.4, where bit 31 is used to store the sign ($S = 1$ if negative and $S = 0$ if positive), bits 24 through 30 are used to store the exponent of 16 increased by 64, and bits 0 through 23 are used to denote the magnitude or the fractional part. For example, the real number indicated in Fig. 1.5 can be converted to its decimal equivalent as follows:

(1) bit 31 gives the sign as $(-1)^1 = -1$.

(2) bits 24 through 30 give the exponent as

$$2^6(1) + 2^5(0) + 2^4(0) + 2^3(0) + 2^2(1) + 2^1(0) + 2^0(0) = 68.$$

(3) bits 0 through 23 give the mantessa (fractional part) as

$$\left(\frac{1}{2}\right)^1(1) + \left(\frac{1}{2}\right)^2(0) + \left(\frac{1}{2}\right)^3(0) + \left(\frac{1}{2}\right)^4(0) + \left(\frac{1}{2}\right)^5(0) + \left(\frac{1}{2}\right)^6(1)$$
$$+ \left(\frac{1}{2}\right)^7(0) + \left(\frac{1}{2}\right)^8(1) + \left(\frac{1}{2}\right)^9(0) + \cdots + \left(\frac{1}{2}\right)^{24}(0) = 0.51953125.$$

Thus, the decimal equivalent of the binary number is given by

$$-(0.51953125)(16^{68-64}) = -34048.0.$$

1.4.3 Nonnumeric Data

Nonnumeric, or character, data consist of one or more characters that include the letters 'A' through 'Z', the digits 0 through 9, symbols such as +, −, ∗, /, ·, (,), =, $,

Table 1.1 Typical constants in Fortran.

Type of constant	Representation in Fortran	External form
Integer number	1421	1421
Real number	-14.21 or -0.1421×10^2	-14.21, or $-0.1421\text{E}+02$
Double-precision number	1.421	$1.421\text{D}+00$
Complex number	(1.42, −2.25)	$1.42 - 2.35i$
Logical constant	.TRUE.	true
Character string	'NEWTON'	Newton

and the blank space. According to the Extended Binary Coded Decimal Interchange Code (EBCDIC) used by IBM, 8 bits are used to store a single character. In the American Standard Code for Information Interchange (ASCII) code used by most other computing systems, a single character is stored in 7 bits.

1.4.4 Constants and Variables

Data are represented as constants or variables in a high-level language such as Fortran. The value of a constant does not change during program execution. The constants include integers, real numbers, double-precision numbers, complex numbers, logical constants, and character strings as shown in Table 1.1. A variable, on the other hand, can be assigned different values during program execution. A variable name, to which a value is assigned, can be an integer, real, double precision, complex, logical variable or a character string. A variable name in Fortran can consist of one to six alphabetic or numeric characters, and the first character must be alphabetic.

1.5 Programming Structure

A programming structure denotes a scheme for processing data. A typical high-level language uses the following types of statements to process the data:

(1) Assignment

(2) Input/Output

(3) Control or Decision

(4) Specification

(5) Subprogram

An assignment statement is used to compute a quantity or assign a value to a variable. Examples of assignment statements are

(1) AREA = 4.15

(2) PI = 3.1416

DIA = 2.5

AREA = PI*DIA**2/4.0

The I/O statements denote instructions whereby information is transmitted to and from the computer. Control statements are used to direct the logical sequence of instructions in a computer program. Common types of control statements include "GO TO", "LOGICAL IF THEN ELSE", and the "DO LOOP."

A specification statement is used to establish a data type or data structure and to format an I/O record. A subprogram statement is used to implement a predefined procedure, known as a subprogram. A subprogram is used to execute a set of one or more statements that is repeated several times during the program. Instead of repeating the set of statements many times, they are written only once in the form of a subprogram, which is then invoked with a single subprogram statement whenever they are needed.

A computer program consists of different statements arranged in a logical sequence. A logical sequence implies an intelligent use of a programming structure.

1.6 Errors

1.6.1 Error and Relative Error

If $\tilde{x}$ is an approximation to x, the difference between the true value and the approximate value is called the error (E_x):

$$E_x = x - \tilde{x}. \tag{1.18}$$

The relative error, R_x, is defined as

$$R_x = \frac{x - \tilde{x}}{x}, \quad x \neq 0. \tag{1.19}$$

A number $\tilde{x}$ is considered to be an approximation to the true value x to d significant digits if d is the largest positive integer for which

$$\left| \frac{x - \tilde{x}}{x} \right| < \frac{1}{2} 10^{-d}. \tag{1.20}$$

▶Example 1.3

The number 3.1415927 is approximated as 3.1416. Find the following: (a) error, (b) relative error, and (c) number of significant digits of the approximation.

Solution

(a) The error (E_x) is given by Eq. (1.18):

$$E_x = x - \tilde{x} = 3.1415927 - 3.1416 = -0.0000073.$$

(b) The relative error (R_x) is given by Eq. (1.19):

$$R_x = \left|\frac{x - \tilde{x}}{x}\right| = \left|\frac{-0.0000073}{3.1415927}\right| = 0.0000023237.$$

(c) The relative error can be expressed as

$$R_x = 0.0000023237 < \frac{10^{-5}}{2},$$

and hence, $\tilde{x}$ approximates x to five significant digits. ◀

1.6.2 Sources of Error

In general, a numerical result is subject to the following types of errors:

(1) *Errors in mathematical modeling*

The simplifying approximations and assumptions made in representing a physical system by mathematical equations introduce error. For example, the finite element analysis is subject to discretization error. In such a case, the results of the mathematical model will be different from the measured or observed behavior of the physical system.

(2) *Blunders*

The programming errors, if undetected, introduce errors in the computed values. Thus, when a large program is written, it is a good practice to divide it into smaller subprograms and test each subprogram separately for accuracy.

(3) *Errors in input*

The input errors occur due to unavoidable reasons such as the errors in data transfer and the uncertainties associated with measurements.

(4) *Machine errors*

The floating-point representation of numbers involves rounding and chopping errors, as well as underflow and overflow errors. These errors are introduced at each arithmetic operation during the computations.

(5) *Truncation errors associated with the mathematical process*

The mathematical process used in the computation sometimes introduces a truncation error. For example, the approximate evaluation of an infinite series or an integral involving infinity as a limit of integration involve computational errors.

1.6.3 Propagation Error

The propagation error is the error in the output of a procedure due to the error in the input data. To find the propagation error, the output of a procedure (f) is considered as a function of the input parameters ($x_1, x_2, \ldots, x_n$):

$$f = f(x_1, x_2, \ldots, x_n) \equiv f(\vec{X}). \tag{1.21}$$

Here, $\vec{X} = \{x_1, x_2, \ldots, x_n\}^T$ is the vector of input parameters. If approximate values of the input parameters are used in the numerical computation, the value of f can be found using Taylor's series expansion about the approximate values $\bar{\vec{X}} = \{\bar{x}_1, \bar{x}_2, \ldots, \bar{x}_n\}^T$ as

$$f(x_1, x_2, \ldots, x_n) = f(\bar{x}_1, \bar{x}_2, \ldots, \bar{x}_n) + \frac{\partial f}{\partial x_1}(\bar{\vec{X}})(x_1 - \bar{x}_1) + \frac{\partial f}{\partial x_2}(\bar{\vec{X}})(x_2 - \bar{x}_2)$$

$$+ \cdots + \frac{\partial f}{\partial x_n}(\bar{\vec{X}})(x_n - \bar{x}_n) + \text{ higher order derivative terms.} \tag{1.22}$$

By neglecting the higher order derivative terms, the error in the output can be expressed as

$$\Delta f = f - \bar{f} \equiv f(x_1, x_2, \ldots, x_n) - f(\bar{x}_1, \bar{x}_2, \ldots, \bar{x}_n). \tag{1.23}$$

Denoting the errors in the input parameters as

$$\Delta x_i = x_i - \bar{x}_i, \quad i = 1, 2, \ldots, n, \tag{1.24}$$

we can estimate the propagation error (Δf) as

$$\Delta f \approx \sum_{i=1}^{n} \frac{\partial f}{\partial x_i}(\bar{\vec{X}})(x_i - \bar{x}_i). \tag{1.25}$$

If $f(x_1, x_2, \ldots, x_n) \neq 0$ and $x_i \neq 0$, the relative propagation error (ε_f) is given by

$$\varepsilon_f = \frac{\Delta f}{f} = \sum_{i=1}^{n} \left\{ \frac{x_i}{f(\vec{X})} \frac{\partial f}{\partial x_i}(\bar{\vec{X}}) \right\} \varepsilon_{x_i}, \tag{1.26}$$

where ε_{x_i} is the relative error in x_i:

$$\varepsilon_{x_i} = \frac{x_i - \bar{x}_i}{x_i}, \quad i = 1, 2, \ldots, n. \tag{1.27}$$

The quantity

$$c_i = \frac{x_i}{f(\vec{X})} \frac{\partial f}{\partial x_i}(\bar{\vec{X}}) \tag{1.28}$$

is called the amplification or the condition number of the relative input error ε_{x_i}.

The study of propagation error due to input errors is called the error analysis. The numerical problem or procedure is said to be well conditioned if the condition numbers (c_i) are reasonably bounded.

1.6.4 Truncation Error

Truncation error is the discrepancy introduced by the use of an approximate expression in place of an exact mathematical expression or formula. For example, consider the Taylor's series expansion of the function $\ln(1+x)$:

$$y(x) = \ln(1+x) = \sum_{i=1}^{\infty} \frac{(-1)^{i+1}}{i} x^i$$

$$= x - \frac{1}{2}x^2 + \frac{1}{3}x^3 - \frac{1}{4}x^4 + \frac{1}{5}x^5 - \frac{1}{6}x^6 + - \cdots; \quad |x| \le 1. \tag{1.29}$$

For simplicity, let the function $y(x)$ be approximated by the first four terms of the Taylor's series expansion. The resulting discrepancy between the exact function $y(x)$ and the approximate function, $\tilde{y}(x) = x - \frac{1}{2}x^2 + \frac{1}{3}x^3 - \frac{1}{4}x^4$, is called the truncation error.

▶Example 1.4

Find the truncation error in approximating the function $y(x) = \ln(1+x)$ by (a) $\tilde{y}_1(x) = x$, (b) $\tilde{y}_2(x) = x - \frac{1}{2}x^2$, (c) $\tilde{y}_3(x) = x - \frac{1}{2}x^2 + \frac{1}{3}x^3$, and $\tilde{y}_4(x) = x - \frac{1}{2}x^2 + \frac{1}{3}x^3 - \frac{1}{4}x^4$ over the range $0 \le x \le 1$.

Solution

We consider a representative value of x, $x = 0.5$, and find the truncation error. The exact value of the function is given by

$$y(x) = \ln(1+x) = \ln 1.5 = 0.405465108.$$

The approximate value of the function and the corresponding truncation error T_e can be computed as follows:

(a) $\tilde{y}_1(0.5) = 0.5;$

$$T_e = y(0.5) - \tilde{y}_1(0.5) = 0.405465108 - 0.5 = -0.094534892.$$

(b) $\tilde{y}_2(0.5) = 0.5 - \frac{1}{2}(0.5)^2 = 0.375;$

$$T_e = y(0.5) - \tilde{y}_2(0.5) = 0.405465108 - 0.375 = 0.030465108.$$

(c) $\tilde{y}_3(0.5) = 0.5 - \frac{1}{2}(0.5)^2 + \frac{1}{3}(0.5)^3 = 0.416666667;$

$$T_e = y(0.5) - \tilde{y}_3(0.5) = 0.405465108 - 0.416666667 = -0.011201559.$$

(d) $\tilde{y}_4(0.5) = 0.5 - \frac{1}{2}(0.5)^2 + \frac{1}{3}(0.5)^3 - \frac{1}{4}(0.5)^4 = 0.401041667;$

$$T_e = y(0.5) - \tilde{y}_4(0.5) = 0.405465108 - 0.401041667 = 0.004423441.$$

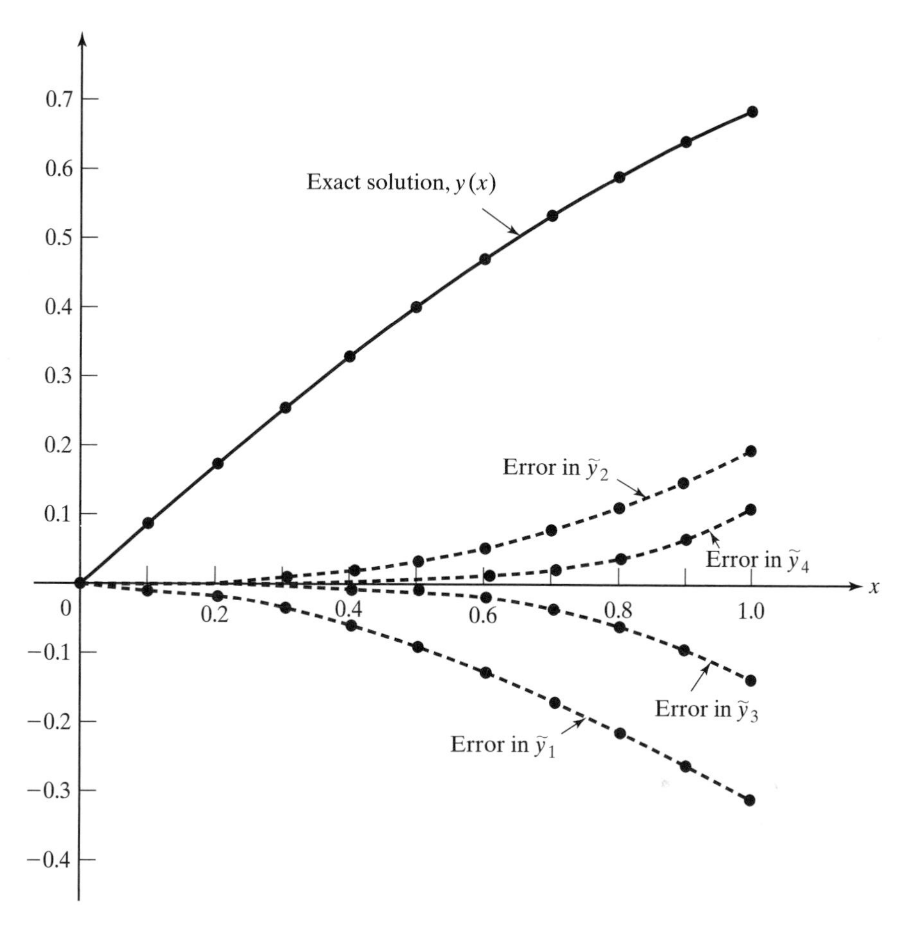

Figure 1.6 Truncation error.

The truncation errors involved by the approximations $\tilde{y}_1(x)$, $\tilde{y}_2(x)$, $\tilde{y}_3(x)$, and $\tilde{y}_4(x)$ for various values of x in the range $0 \leq x \leq 1$ are shown in Fig. 1.6. ◀

▶Example 1.5

Find the number of significant digits of the various approximations used in Example 1.4 at $x = 0.5$.

Solution

The relative errors of the various approximations can be expressed as

$$R_{x_1} = \left| \frac{y - \tilde{y}_1}{y} \right| = \left| \frac{-0.094534892}{0.405465108} \right| = 0.233151732 < \frac{10^{-0}}{2};$$

$$R_{x_2} = \left|\frac{y - \tilde{y}_2}{y}\right| = \left|\frac{0.030465108}{0.405465108}\right| = 0.075136201 < \frac{10^{-0}}{2};$$

$$R_{x_3} = \left|\frac{y - \tilde{y}_3}{y}\right| = \left|\frac{-0.011201559}{0.405465108}\right| = 0.027626444 < \frac{10^{-1}}{2};$$

$$R_{x_4} = \left|\frac{y - \tilde{y}_4}{y}\right| = \left|\frac{0.004423441}{0.405465108}\right| = 0.010909548 < \frac{10^{-1}}{2}.$$

Hence, the number of significant digits of the approximations $\tilde{y}_1$, $\tilde{y}_2$, $\tilde{y}_3$, and $\tilde{y}_4$ are given by 0, 0, 1, and 1, respectively. ◀

1.6.5 Round-off Error

Since only a finite number of digits are stored in a computer, the actual numbers may undergo chopping or rounding of the last digit. For example, let a number in decimal form be given by

$$x = 0.b_1\, b_2 \cdots b_i\, b_{i+1}\, b_{i+2}\, \cdots, \tag{1.30}$$

where $0 \le b_j \le 9$ for $j \ge 1$. If the maximum number of decimal digits used in the floating-point computations is i, then the chopped floating-point representation of x, x_{chop}, is given by

$$x_{\text{chop}} = 0.b_1\, b_2 \cdots b_i, \tag{1.31}$$

where the ith digit of x_{chop} is identical to the ith digit of x. On the other hand, the rounded floating-point representation of x, x_{round}, is given by

$$x_{\text{round}} = 0.b_1\, b_2 \cdots b_{i-1} d_i, \tag{1.32}$$

where $d_i (1 \le d_i \le 9)$ is obtained by rounding the number $d_i\; d_{i+1} d_{i+2} \cdots$, to the nearest integer. For example, the value of e is given by $e = 2.718281828459045\ldots$. The seven-digit representations of e by using chopping and rounding are given by

$$e_{\text{chop}} = 0.2718281 \times 10^1, \quad e_{\text{round}} = 0.2718282 \times 10^1.$$

1.6.6 Computational Error

The numerical solution of an engineering problem is found using a suitable algorithm. An algorithm is a finite set of precise instructions to be carried using the given initial data in a specified sequence in order to find the desired output. All the local computational errors involved in the various steps of an algorithm will accumulate to a computational error in the output. The local computational errors arise due to errors involved during arithmetic operations such as subtraction of numbers of near-equal magnitude and also when irrational numbers (such as $\sqrt{3}$ and π) are replaced by machine numbers with a finite number of digits.

Errors associated with arithmetic operations

When two numbers are used in an arithmetic operation, the numbers cannot be stored exactly by the floating-point representation. Let x and y be the exact numbers and $\bar{x}$ and $\bar{y}$ their approximate values. Then

$$x = \bar{x} + \varepsilon_x, \quad y = \bar{y} + \varepsilon_y, \tag{1.33}$$

where ε_x and ε_y denote the errors in x and y, respectively. For example, when a multiplication operation is used, the associated error (E) is given by

$$E = x\,y - \bar{x}\,\bar{y} = x\,y - (x - \varepsilon_x)(y - \varepsilon_y) = x\,\varepsilon_y + y\,\varepsilon_x - \varepsilon_x\,\varepsilon_y. \tag{1.34}$$

The relative error (R) is given by

$$\begin{aligned} R &= \frac{E}{x\,y} = \frac{\varepsilon_x}{x} + \frac{\varepsilon_y}{y} - \frac{\varepsilon_x}{x}\frac{\varepsilon_y}{y} \\ &= R_x + R_y - R_x\,R_y \approx R_x + R_y, \end{aligned} \tag{1.35}$$

where $|R_x| << 1$ and $|R_y| << 1$ with R_x and R_y denoting the relative errors in x and y, respectively, and the symbol $<<$ representing "much less than." Proceeding in a similar manner, the relative error associated with division operation, $\frac{x}{y}$, can be represented as

$$R \approx R_x - R_y. \tag{1.36}$$

▶Example 1.6

The values of $x = 2.71828183$ and $y = 2.71818283$ are represented as $\bar{x} = 2.7183$ and $\bar{y} = 2.7181$. Determine the error and the relative error associated with the operation $x - y$.

Solution

The errors in x and y can be expressed as

$$E_x = x - \bar{x} = -1.817 \times 10^{-5}, \quad R_x = -6.684 \times 10^{-6},$$

$$E_y = y - \bar{y} = 8.283 \times 10^{-5}, \quad R_y = 3.0473 \times 10^{-5}.$$

The error associated with subtraction can be found as

$$E = (x - y) - (\bar{x} - \bar{y}) = E_x - E_y = -10.1 \times 10^{-5},$$

$$R = \frac{E}{x - y} = \frac{-10.1 \times 10^{-5}}{0.000099} \approx -0.10202020.$$

It can be seen that although the error in $x - y$ is very small, the relative error is much larger than that of x and y alone. ◀

1.6.7 General Guidelines for Rounding of Numbers

The following procedure is used to round off numbers during numerical computations:

(1) If the round-off is done by retaining i digits, the last retained digit (the ith one) is increased by one if the first discarded digit is 6 or larger.

(2) If the last retained digit is odd and the first discarded digit is 5 or 5 followed by zeros, the last retained digit is increased by one.

(3) In all other cases, the last retained digit is unaltered.

▶ Example 1.7

Round off the following numbers to four significant digits: (a) 9.46932, (b) 201.72, (c) 200.550, (d) 200.650, and (e) 2.013501

Solution

By using the guidelines indicated just before, we can round off the numbers as follows:

(a) 9.469, (b) 201.7, (c) 200.6, (d) 200.6, and (e) 2.014. ◀

(4) During addition or subtraction, the rounding off of the final result is done such that the position of the last retained digit is the same as that of the most significant last retained digit in the original numbers that were added or subtracted.

▶ Example 1.8

Find the value of the following:

(a) $x = 3.3 - 2.868$.
(b) $y = 5.72 \times 10^{-7} + 4.6 \times 10^{-4} - 391 \times 10^{-4}$.

Solution

(a) $x = 3.3 - 2.868 = 0.432 = 0.4$.
(b) $y = 0.00572 \times 10^{-4} + 5.6 \times 10^{-4} - 3.91 \times 10^{-4} = 5.60572 \times 10^{-4} - 3.91 \times 10^{-4} = 1.69572 \times 10^{-4}$ Since the 6 in the middle number is the most significant last retained digit, the value of y is rounded as $y = 1.7 \times 10^{-4}$. ◀

(5) During multiplication or division, the round-off of the final result is done such that the number of significant digits is equal to the smallest number of significant digits used in the original numbers.

▶ Example 1.9

Find the value of (a) $x = 0.0839 \times 7.4$ and (b) $y = 932/0.48765$.

Solution

Since the smallest number of significant digits is two and three in the numbers involved with x and y, respectively, the answers are to be rounded to correspond to the same number of significant digits. This yields

(a) $x = 0.0839 \times 7.4 = 0.62086 \rightarrow 0.62.$

(b) $y = 932/0.48765 = 1911.20680816 \rightarrow 1910.$ ◀

(6) During multiple arithmetic operations, the operations are performed one at a time as indicated by the parentheses:

$$\text{(multiplication or division)} \pm \text{(multiplication or division)}$$

$$\text{(addition or subtraction)} \times /\text{(addition or subtraction)}.$$

In each step of the operation, the results are rounded as indicated in guidelines 4 and 5 before proceeding to the next operation, instead of only rounding the final result.

▶Example 1.10

Find the value of $x = (203.6 \times \{3.4 \times 10^{-4}\}) + (\{9.214 \times 10^{-4}\} \times 3.92)$.

Solution

The operations indicated in the parentheses are performed and the resulting numbers are rounded as

$$y = 203.6 \times \{3.4 \times 10^{-4}\} = 692.24 \times 10^{-4} = 69 \times 10^{-3};$$

$$z = \{9.214 \times 10^{-4}\} \times 3.92 = 36.11888 \times 10^{-4} = 3.62 \times 10^{-3}.$$

The addition of y and z yields $x = 72.62 \times 10^{-3}$, which can be rounded as $x = 73 \times 10^{-3}$. ◀

▶Example 1.11

Find the value of $x = \frac{7.210\times10^{-4} - 9.5\times10^{-6}}{3.456\times10^{2} + 6.9732}$.

Solution

The numbers in the numerator and denominator are first rewritten using the same exponent and then the subtraction or addition is performed:

$$y = 721.0 \times 10^{-6} - 9.5 \times 10^{-6} = 711.5 \times 10^{-6} \rightarrow 712 \times 10^{-6};$$

$$z = 3.456 \times 10^{2} + 0.069732 \times 10^{2} = 3.525732 \times 10^{2} \rightarrow 3.526 \times 10^{2}.$$

Finally, the division of y and z and subsequent rounding off gives

$$x = \frac{y}{z} = \frac{712 \times 10^{-6}}{3.526 \times 10^{2}} = 201.928530913 \times 10^{-8} \rightarrow 202 \times 10^{-8}.$$ ◀

1.7 Numerical Methods Considered

The behavior of any physical or engineering system can be described by one or more mathematical equation(s). If the mathematical equations are simple, the exact solution can be found in closed form. Although closed-form solutions are most desirable, for most engineering problems, the equations are quite complex for which the exact solution cannot be found. In such cases, numerical methods can be used to solve the mathematical equations using arithmetic operations in order to understand the behavior of the system. The various types of numerical methods discussed in this book are summarized next.

1.7.1 Solution of Nonlinear Equations

Many engineering problems involve the solution of one or more nonlinear equations. A nonlinear equation may be in the form of an algebraic, transcendental, or polynomial equation. For example, the determination of natural frequencies of a vibrating system, the temperature of a heated body from an energy balance, the friction factor corresponding to a turbulent fluid flow, and the transient current in an electrical circuit lead to different types of nonlinear equations. A simple nonlinear equation involves the determination of the root of the equation

$$f(x) = 0.$$

The typical iterative process used in the determination of the root, x^*, is shown in Fig. 1.7.

1.7.2 Simultaneous Linear Algebraic Equations

In engineering applications, a wide variety of mathematical problems are encountered. In areas such as solid mechanics, heat transfer, fluid mechanics, electrostatics, and combustion, the governing partial differential equations are usually solved using a finite element or finite difference technique. This converts the problem into a

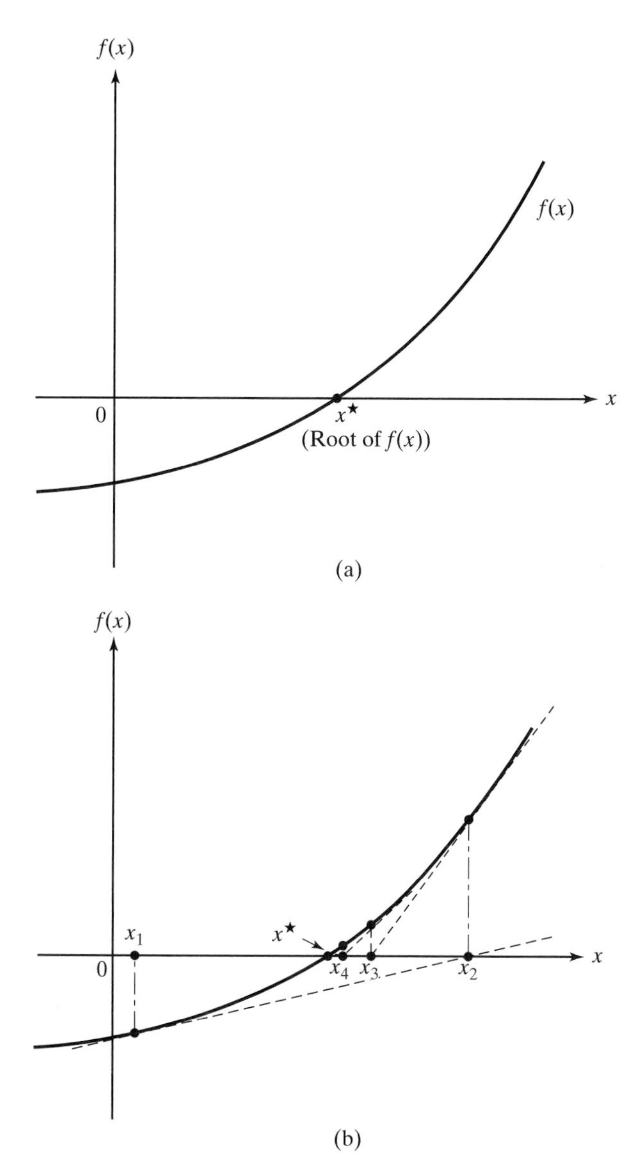

Figure 1.7 Solution of a nonlinear equation.

system of linear algebraic equations in terms of a set of unknown variables. A set of two linear equations can be stated as

$$a_{11}x_1 + a_{12}x_2 = b_1;$$

$$a_{21}x_1 + a_{22}x_2 = b_2.$$

The graphical interpretation of the solution of the two equations is shown in Fig. 1.8.

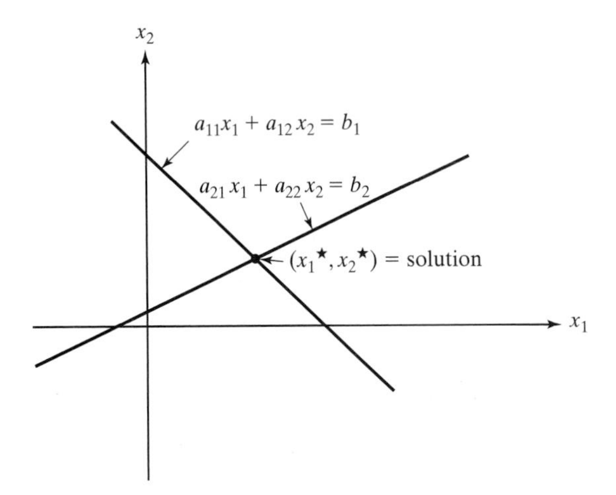

Figure 1.8 Solution of linear equations.

1.7.3 Solution of Matrix Eigenvalue Problem

The analysis of many engineering systems, such as the vibration of structures and machines, buckling of columns, and dynamic response of electrical systems, requires the solution of a set of homogeneous linear algebraic equations. In these problems, if the number of equations is n, there will be $n + 1$ unknowns. These problems are known as algebraic eigenvalue problems. For example, an eigenvalue problem involving two homogeneous equations can be stated as

$$(a_{11} - \omega)x_1 + a_{12}x_2 = 0,$$

and

$$a_{21}x_1 + (a_{22} - \omega)x_2 = 0,$$

where ω, called the eigenvalue, and $\vec{X} = \begin{Bmatrix} x_1 \\ x_2 \end{Bmatrix}$, known as the eigenvector, are the unknowns. The interpretations of the eigenvalue and the eigenvector are indicated in Fig. 1.9.

1.7.4 Curve Fitting and Interpolation

In certain physical problems, the values of a function may be available at a certain number of data points, and we may be required to estimate the function value at a missing data point. In such a case, a weighted average of the known function values at neighboring points can be used as an estimate of the missing functional value. Another approach is to first fit a curve using the available functional values at the data points and then estimate the missing functional value from the fitted curve. These approaches are known as interpolation and curve-fitting techniques. A typical curve-fitting problem can be stated as follows: Find a quadratic, $f(x) = a + bx + cx^2$,

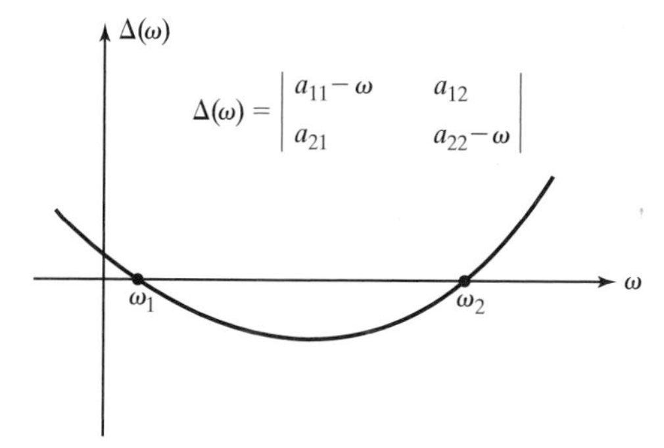

(a) Stage 1: Find value(s) of ω.

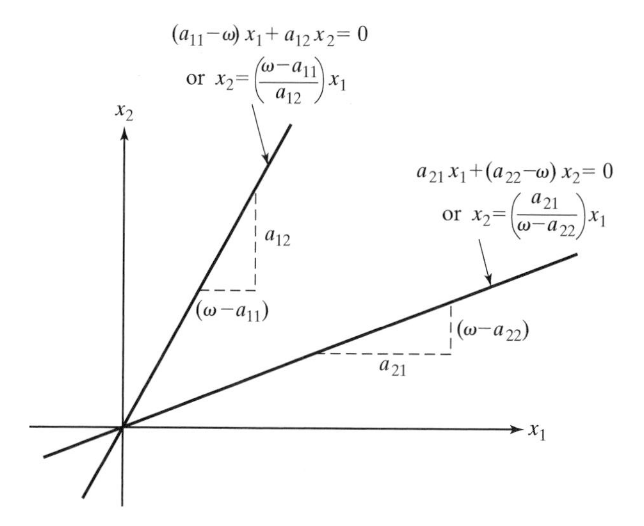

(b) Stage 2: Find relative values of x_1 and x_2 for known values of ω.

Figure 1.9 Solution of eigenvalue problem.

to fit the given data shown by dots in Fig. 1.10(a). Similarly, an interpolation problem involves finding an nth-degree polynomial that passes through $n + 1$ data points to estimate the value of the function in between the data points (Fig. 1.10(b)).

1.7.5 Statistical and Probability Methods

In many physical and engineering problems, numerical data are collected to understand physical phenomena. Then the principles of statistics and probability are used to analyze the data, develop models, and predict the behavior of the system. For example, if several values of an uncertain quantity, such as the wind load acting on a building, are measured as $x_1, x_2, \ldots$, they can be used to develop a histogram and a probability distribution of wind load as shown in Fig. 1.11.

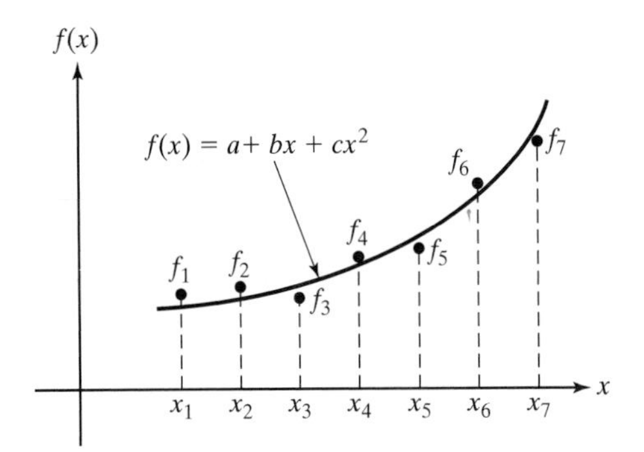

(a) Polynomial fit or nonlinear regression.

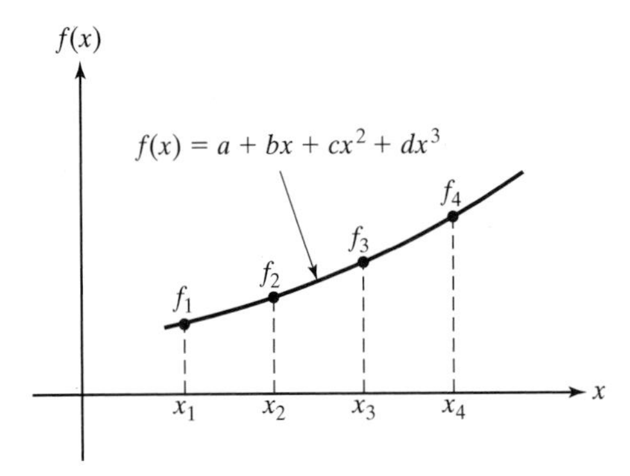

(b) Interpolation $f(x)$ passes through all data points.

Figure 1.10 Curve fitting and interpolation.

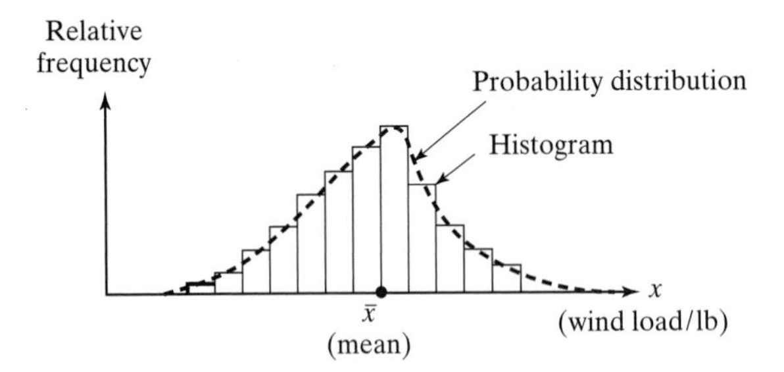

Figure 1.11 Probability distribution.

1.7.6 Numerical Differentiation

In certain physical problems, a function is to be differentiated without knowing the expression. In these cases, the values of the function are known only at a discrete set of points. Then we can use numerical differentiation. In this procedure, first a polynomial passing through all the data points is determined, and the resulting polynomial is then differentiated to find the approximate derivative of the unknown

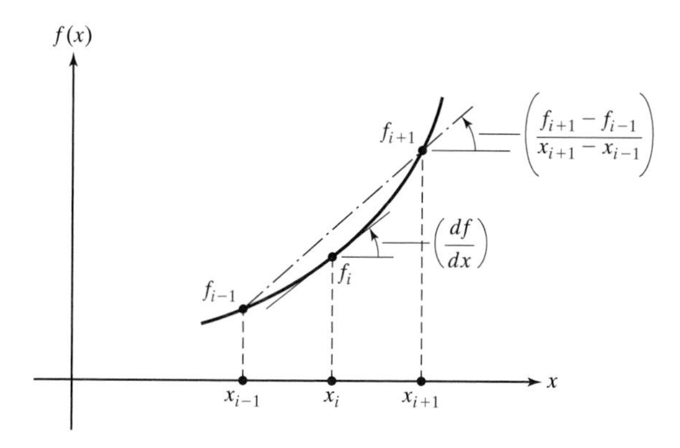

Figure 1.12 Numerical differentiation $\left(\frac{df}{dx}\text{ approximated by }\frac{f_{i+1}-f_{i-1}}{x_{i+1}-x_{i-1}}\right)$.

function. For example, the numerical derivative $\frac{df}{dx}$ can be found using a central difference formula as

$$\frac{df}{dx} \approx \frac{f_{i+1} - f_{i-1}}{x_{i+1} - x_{i-1}}.$$

This equation is shown graphically in Fig. 1.12.

1.7.7 Numerical Integration

The solution of many engineering problems requires the evaluation of an integral. If the function to be integrated is too complex or if the values of the function are known only at discrete values of the independent variable, numerical integration techniques are to be used. Basically, the variation of the function (to be integrated) is assumed to be a simple polynomial over a discrete interval, and then the integral is evaluated as the sum of the areas under the assumed polynomials over the various discrete intervals. For example, if the exact integral is indicated as in Eq. (1.3), the numerical integral can be evaluated as (Fig. 1.1)

$$I_2 \approx A_1 + A_2 + A_3 + \cdots + A_8.$$

1.7.8 Solution of Ordinary Differential Equations

Ordinary differential equations arise in the study of many physical phenomena such as dynamics, heat and mass transfer, current flow in electrical circuits, and chemical reactions. In some cases, partial differential equations can be transformed to ordinary differential equations. In all these cases, the solution of a set of one or more ordinary differential equations is required under specified initial or boundary conditions. For example, the solution of the first order differential equation

$$\frac{dy}{dx} = f(x, y)$$

can be found numerically by approximating the derivative as the slope of the function $y(x)$ at different values of x as indicated in Fig. 1.13.

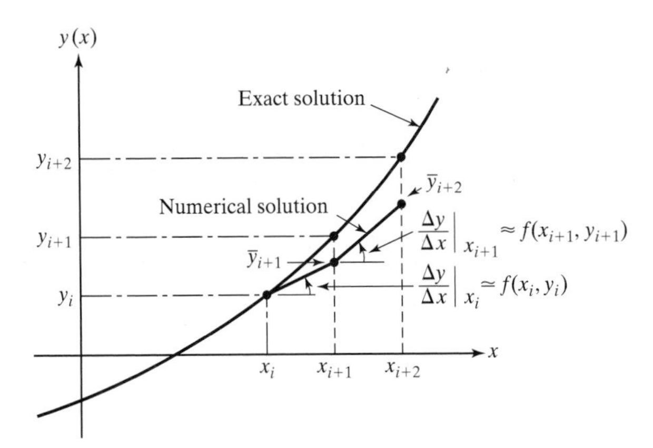

Figure 1.13 Solution of an ordinary differential equation.

1.7.9 Solution of Partial Differential Equations

The behavior of many physical systems are governed by differential equations involving two or more independent variables, known as partial differential equations. For example, the transient temperature distribution in a rod, the seepage (fluid) flow in soil, and the displacement of a plate under load require the solution of different types of partial differential equations. For example, a partial differential equation involving the spatial coordinate x and time t as independent variables is given by

$$c^2 \frac{\partial^2 w}{\partial x^2} = \frac{\partial^2 w}{\partial t^2},$$

where c is a constant and $w(x, t)$ is the unknown function. This equation can be approximated at various grid points of the solution domain using finite differences as (see Fig. 1.14)

$$\frac{c^2}{(\Delta x)^2}\left(w_{i-1,j} - 2w_{i,j} + w_{i+1,j}\right) = \frac{1}{(\Delta t)^2}\left(w_{i,j-1} - 2w_{i,j} + w_{i,j+1}\right); i, j = 1, 2, \ldots.$$

These equations represent a system of algebraic linear equations that can be solved easily.

1.7.10 Optimization

The analysis, design, and operation of many engineering systems involve the determination of certain variables so as to minimize an objective (cost) function while satisfying certain functional and economical constraints. The solution of such problems requires the use of analytical or numerical optimization techniques. If the equations for the objective and constraint functions are available in closed form and are simple, analytical methods of optimization can be used for the solution of the problem. On the other hand, if the objective and constraint equations are complex

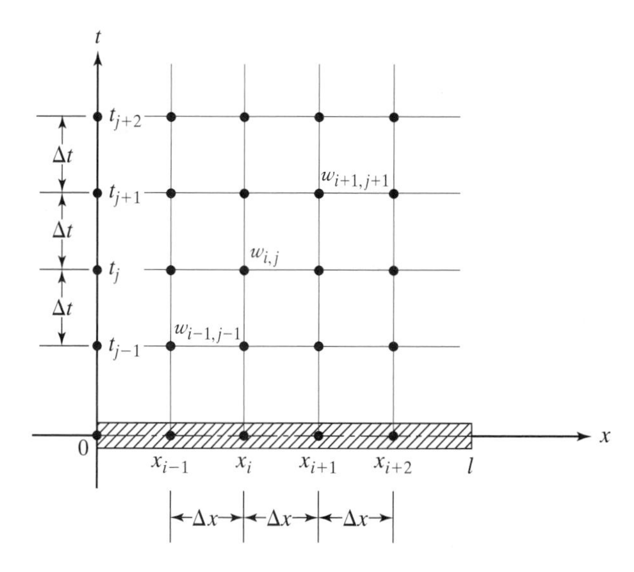

Figure 1.14 Approximation of partial differential equation.

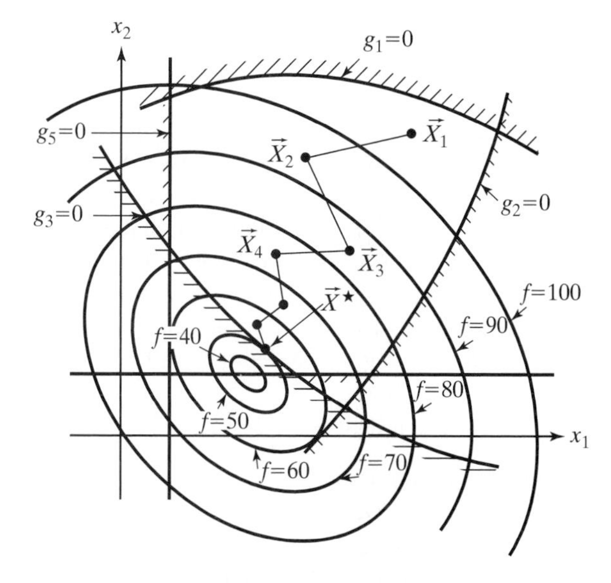

Figure 1.15 Optimization process.

or not available in closed form, numerical methods of optimization can be used for the solution of the problem. For example, to find the minimum of a function $f(\vec{X})$ subject to the constraints $g_j(\vec{X}) \leq 0, j = 1, 2, \ldots, m$, first a starting vector $\vec{X}_1$ is assumed. Then the vector is iteratively improved to find $\vec{X}_2, \vec{X}_3, \ldots$, and ultimately, the optimum vector, $\vec{X}^*$, as shown in Fig. 1.15.

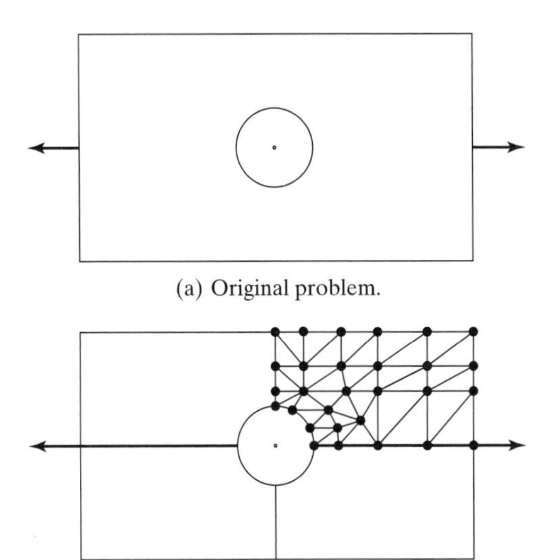

(a) Original problem.

(b) Finite-element idealization.
(Only a quarter of the plate is idealized due to symmetry).

Figure 1.16 Modeling of a plate with a hole.

1.7.11 Finite-Element Method

In many engineering systems, the governing equations will be in the form of partial differential equations, and the solution domain will not be regular. These problems can be solved conveniently by replacing the solution domain by several regular subdomains or geometric figures, known as finite elements, and assuming a simple solution in each finite element. When equilibrium and compatibility conditions are enforced, the process leads to a system of algebraic (or a system of ordinary differential) equations that can be solved easily. For example, the stresses induced in a plate with a hole (Fig. 1.16(a)) can be found by modeling the plate by a number of triangular elements as shown in Fig. 1.16(b).

1.8 Software for Numerical Analysis

Several commercial, as well as public-domain, software packages are available for the solution of numerical analysis problems. Although most of the software was originally written for mainframe computers, many of the programs have been modified for use on all types of computers in recent years. Most of the numerical analysis software was developed in Fortran; however, some software is available in C and Pascal. The International Mathematical and Scientific Library (IMSL) consists of more than 700 subroutines, originally written in Fortran, in the areas of general applied mathematics, statistics, and special functions. Most of the programs of IMSL are available in both single- and double-precision versions and can be used on a variety of computers ranging from personal computers to supercomputers. Currently, IMSL programs are available in C also. LINPACK and EISPACK are public-domain Fortran packages available from Argonne National Laboratory. EISPACK is the first large-scale public-domain package made available for the solution of algebraic eigenvalue problems. LINPACK can be used for the solution of systems of linear equations and least square problems.

Several software packages have been developed for the symbolic solution of mathematical problems. Although Macsyma, Derive, Mathematica, and MAPLE are popular for the symbolic solution, only MAPLE is considered in this book for demonstrating the symbolic or numerical solution of different types of practical engineering problems. The basic concepts of MAPLE are summarized in Appendix C. For the interactive solution of problems representing different types of engineering and scientific applications, the software MATLAB can be used very conveniently. MATLAB is useful, especially for problems involving vector and matrix manipulation. The basic ideas and procedures of MATLAB are indicated in Appendix D. MATHCAD is another software that can be used for the solution of mathematical problems, both numerically and symbolically. It also provides two- and three-dimensional plots. A brief outline of the concepts of MATHCAD are given in Appendix E.

1.9 Use of Software Packages

To illustrate the use of software packages MATLAB, MAPLE, and MATHCAD through a simple example, the multiplication of the following matrices is considered:

$$A = \begin{bmatrix} 2 & 3 & 4 \\ 1 & -5 & 6 \end{bmatrix}, \quad B = \begin{bmatrix} 8 & 0 \\ 2 & 7 \\ -1 & 4 \end{bmatrix}.$$

The result is

$$C = AB = \begin{bmatrix} 18 & 37 \\ -8 & -11 \end{bmatrix}.$$

1.9.1 MATLAB

When the ENTER key is pressed, the prompt symbol "≫" is displayed to indicate that MATLAB is ready to begin. The method of entering matrices, the display of matrices, the implementation of matrix multiplication, and the display of the result are shown in the following example:

▶Example 1.12

```
>> A=[2. 3. 4.; 1. -5. 6.]

   A =
          2    3    4
          1   -5    6

>> B = [8. 0.; 2. 7.; -1. 4.]

   B =
          8   0
          2   7
         -1   4

>> C = A*B

   C =
         18    37
         -8   -11

>>
```

1.9.2 MAPLE

In this case, the prompt symbol is ">" and the linear algebra module, *linalg*, is to be initialized. In MAPLE, a matrix can be created in several ways. Two methods are indicated in Example 1.13. Note that the elements are separated by a comma. The operations involved in carrying out the matrix multiplication and the results are also indicated in Example 1.13.

▶Example 1.13

```
> A:=linalg [matrix] ([[2., 3., 4.], [1., -5., 6.]]);
```

$$A := \begin{bmatrix} 2. & 3. & 4. \\ 1. & -5. & 6. \end{bmatrix}$$

```
> B:=linalg [matrix] (3, 2, [8., 0., 2., 7., -1., 4.]);
```

$$B := \begin{bmatrix} 8. & 0 \\ 2. & 7. \\ -1. & 4. \end{bmatrix}$$

```
> C:=linalg [multiply] (A, B);
>
```

$$C := \begin{bmatrix} 18. & 37. \\ -8. & -11. \end{bmatrix}$$

1.9.3 MATHCAD

In MATHCAD, the matrices A and B are defined (using the procedure outlined in Appendix E), and the matrix C is defined as C := A*B. When "C =" is entered, the result will be displayed as indicated in the following example:

▶Example 1.14

$$\mathrm{A} := \begin{pmatrix} 2 & 3 & 4 \\ 1 & -5 & 6 \end{pmatrix}$$

$$\mathrm{B} := \begin{pmatrix} 8 & 0 \\ 2 & 7 \\ -1 & 4 \end{pmatrix}$$

$$\mathrm{C} := \mathrm{A} \cdot \mathrm{B}$$

$$\mathrm{C} = \begin{pmatrix} 18 & 37 \\ -8 & -11 \end{pmatrix}$$

1.10 Computer Programs

Fortran and C programs are given for the multiplication of the following matrices:

$$A = \begin{bmatrix} 2 & 3 & 4 \\ 1 & -5 & 6 \end{bmatrix}, \quad B = \begin{bmatrix} 8 & 0 \\ 2 & 7 \\ -1 & 4 \end{bmatrix}.$$

1.10.1 Fortran Program

Program 1.1 Matrix multiplication:

Input data required:

Sizes of matrices $A(L, M)$ and $B(M, N)$
Matrices A and B

Listing and output of program:

```
      PROGRAM MATRIX_MULT
      ! MULTIPLICATION OF MATRICES A AND B
      INTEGER, PARAMETER :: L=2, M=3, N=2
      REAL, DIMENSION (2, 3) :: A = (/2., 1., 3., -5., 4., 6./)
      REAL, DIMENSION (3, 2) :: B = (/8., 2., -1., 0., 7., 4./)
      REAL, DIMENSION (2, 2) :: C
      PRINT 10
10    FORMAT (/,'MULTIPLICATION OF MATRICES: C=A*B',//,
     2'MATRIX A', /)
      DO I=1, L
      PRINT 30, (A(I,J), J=1, M)
30    FORMAT (6E14.6)
      END DO
      PRINT 40
40    FORMAT (/,'MATRIX B', /)
      DO I=1, M
      PRINT 30, (B(I,J), J=1,N)
      END DO
      C = MATMUL (A, B)
      PRINT 60
60    FORMAT (/,'MATRIX C', /)
      DO I=1,L
      PRINT 30, (C(I,J), J=1,N)
      END DO
      STOP
      END PROGRAM
```

Output:

```
MULTIPLICATION OF MATRICES: C=A*B

MATRIX A

  0.200000E+01   0.300000E+01   0.400000E+01
  0.100000E+01  -0.500000E+01   0.600000E+01

MATRIX B

  0.800000E+01   0.000000E+00
  0.200000E+01   0.700000E+01
 -0.100000E+01   0.400000E+01
```

```
MATRIX C

  0.180000E+02     0.370000E+02
 -0.800000E+01    -0.110000E+02
```

1.10.2 C Program

Program 1.2 Matrix multiplication:

Input data required:

Same as in the case of Program 1.1

Program listing and output:

```
//Multiplication of matrices A and B

#include <stdio.h>
#include <stdlib.h>
#include <math.h>

void matmul(a, b, c, l, m, n)
double a[][3], b[][2], c[][2];
int l, m, n;
{
    int i, j, k;
    for(i=0; i<l; i++)
    {
        for(j=0; j<n; j++)
        {
            c[i][j]=0.;
            for(k=0; k<m; k++)
            {
                c[i][j]=c[i][j] + a[i][k]*b[k][j];
            }
        }
    }
return;
}
int main()
{
    int i, j, l=2, m=3, n=2;
    double a[2][3] = {{2., 3., 4.,}, {1., -5., 6.}},
           b[3][2] = {{8., 0.}, {2., 7.}, {-1., 4.}},
           c[2][2];
    void matmul();
```

```
        matmul(a, b, c, l, m, n);

        printf(``Multiplication of matrices: C=A*B\n'');
        printf(``\nMatrix A\n'');

        for(i=0; i<l; i++)
        {
                for(j=0; j<m; j++)
                        printf(``%14.5e'',  a[i][j]);
                printf(``\n'');
        }
        printf(``\nMatrix B\n'');
        for (i=0; i<m; i++)
        {
                for(j=0; j<n; j++)
                        printf(``%14.5e'', b[i][j]);
                printf(``\n'');
        }
        printf(``\nMatrix C\n'');
        for(i=0; i<l; i++)
        {
                for(j=0; j<n; j++)
                        printf(``%14.5e'', c[i][j]);
                printf(``\n'');
        }
}
```

output:

```
Multiplication of matrices:  C=A*B

Matrix A
   2.00000e+00    3.00000e+00    4.00000e+00
   1.00000e+00   -5.00000e+00    6.00000e+00

Matrix B
   8.00000e+00    0.00000e+00
   2.00000e+00    7.00000e+00
  -1.00000e+00    4.00000e+00

Matrix C
   1.80000e+01    3.70000e+01
  -8.00000e+00   -1.10000e+01
```

REFERENCES AND BIBLIOGRAPHY

1.1. B. W. Char et al, *First Leaves: A Tutorial Introduction to MAPLE V*, Springer-Verlag, New York, 1992.

1.2. B. W. Char et al, *MAPLE V—Library Reference Manual*, Springer-Verlag, New York, 1991.

1.3. J. Penny and G. Lindfield, *Numerical Methods Using MATLAB*, Ellis Horwood, New York, 1995.

1.4. C. Tocci and S. Adams, *Applied MAPLE for Engineers and Scientists*, Artech House, Boston, 1996.

1.5. J. S. Robertson, *Engineering Mathematics with MAPLE*, McGraw-Hill, New York, 1996.

1.6. *User's Manual—IMSL MATH/LIBRARY*, Volume 1, IMSL, Inc., Houston, TX, December 1989.

1.7. A. Cavallo, R. Setola, and F. Vasca, *Using MATLAB, Simulink and Control System Toolbox*, Prentice Hall, London, 1996.

1.8. E. C. Ackermann, *The Essentials of C Programming Language*, Research and Education Association, Piscataway, NJ, 1994.

1.9. M. L. James, G. M. Smith, M. Gerald, and J. C. Wolford, *Applied Numerical Methods for Digital Computation*, HarperCollins, New York, 1993.

1.10. R. Johnsonbaugh and M. Kalin, *C for Scientists and Engineers*, Prentice Hall, Upper Saddle River, NJ, 1997.

1.11. S. J. Chapman, *Introduction to Fortran 90/95*, WCB/McGraw-Hill, Boston, 1998.

1.12. *MATHCAD User's Guide: MATHCAD 2000 Professional and MATHCAD 2000 Standard*, MathSoft, Inc., Cambridge, MA, 1999.

1.13. J. R. Rice, *Numerical Methods, Software, and Analysis*, Second Edition, Academic Press, San Diego, CA, 1993.

REVIEW QUESTIONS

The following questions along with corresponding answers are available in an interactive format at the Companion Web site at http://www.prenhall.com/rao.

1.1. Indicate whether the following statement is **true or false:**
1. The programs written in a low level language are machine independent.
2. A binary number consists of the digits 0, 1, and 2.
3. A variable can be assigned different values during program execution.
4. A variable name can be an integer.
5. A variable name in Fortran can consist of any number of alphabetic or numeric characters.
6. Fortran IV was developed in 1966.

1.2. **Fill in the blank space** by selecting the proper word:
1. The system software is written in a ______ language. (low-level, high-level)
2. User software is developed using a ______ language. (low-level, high-level)
3. In a computer, integers are stored as signed ______ numbers. (binary, decimal)
4. Data are represented as constants or variables in a ______ language such as Fortran. (low-level, high-level)
5. In Fortran, "GO TO" denotes a ______ statement. (control, input)
6. The difference between the true value and the approximate value is called the ______. (error, absolute error)
7. IMSL programs are written in the form of ______. (subroutines, functions)
8. In MAPLE, all expressions must end with a ______. (colon, semicolon)
9. In MATLAB, a script file is also known as a(n) ______. (m-file, matlab file)
10. Basic was developed in ______. (1964, 1980)

1.3. Give a **brief answer** to each of the following questions:
1. What is the first programming language standardized by the ANSI?
2. Give a number that requires an infinite number of digits to represent its fractional part.

(*continued, page 39*)

3. Identify the fraction whose binary representation is given by 0.1001 1001 1001 . . ..
4. How is the complex number, $-2.41 + 3.25i$, represented in Fortran?
5. If x and $\tilde{x}$ denote the true and approximate values of a quantity, respectively, define the relative error.
6. What is error analysis?
7. What is the round-off number corresponding to 9.46932 that has four significant digits?
8. Name some of the dialects of Basic.

1.4. What does each of the following **acronyms stand for?**
1. CPU
2. DOS
3. CAD
4. ANSI
5. ASCII
6. IMSL
7. MATLAB
8. COBOL
9. BASIC
10. FORTRAN

1.5. Identify the **most appropriate answer** out of the choices given:
1. The physical component of a computer is called (a) hardware, (b) software, or (c) memory unit.
2. User programs are known as (a) source codes, (b) system software, or (c) machine program.
3. The programming language developed at Dartmouth College is (a) Fortran, (b) C, or (c) Basic.
4. Fortran was developed at (a) IBM, (b) Dartmouth College, or (c) Bell Laboratories.
5. Fortran 90 was developed in (a) 1977, (b) 1990, or (c) 1997.
6. Fixed-point numbers are the same as (a) integers, (b) real numbers, or (c) complex numbers.
7. Floating-point numbers are the same as (a) integers, (b) real numbers, or (c) complex numbers.

(*continued, page 40*)

1.6. **Match the following** programming languages with their primary uses:

1. Fortran	(a) used to generate codes for computers
2. Cobol	(b) used to develop good programs
3. Basic	(c) used in engineering and science
4. C	(d) used on personal (micro) computers
5. Pascal	(e) used to standardize military software
6. Ada	(f) used in business applications

1.7. **Match the following** programming languages with the names of their developers:

1. Ada	(a) Niklaus Wirth
2. Fortran	(b) John Kemeny and Thomas Kurtz
3. Basic	(c) Dennis Ritchie
4. C	(d) Jean Ichbian
5. Pascal	(e) John Backus

1.8. **Match the following** errors with their sources:

1. Round-off error	(a) error in the output due to errors in input data
2. Input error	(b) error due to storage of finite number of digits
3. Procedural error	(c) error during arithmetic operations
4. Propagation error	(d) error due to the use of approximate expression in place of an exact expression
5. Computational error	(e) error in data transfer
6. Truncation error	(f) error due to the use of approximate model

PROBLEMS

Section 1.4

1.1. Find the binary number that approximates e to within 10^{-2}.

1.2. Determine the next larger integer after the following number in binary system: (a) 111, (b) 1111, and (c) 11111.

1.3. Convert the following decimal numbers to binary form: (a) 0.8 (b) 0.7.

Section 1.6

1.4. How many significant digits are there in each of the following numbers?
(a) 702.10, (b) 0.7021×10^3, (c) 0.70210×10^5, (d) 0.000913, and (e) 0.913×10^{-3}.

1.5. Round off each of the following numbers to four significant digits:
(a) 92.4552, (b) 9245.71, (c) 0.924552×10^2,
(d) 0.00924552×10^4, and (e) 924552.0×10^{-4}.

1.6. Perform each of the following arithmetic operations using (a) exact computations, (b) four-digit approximation (chopping), and (c) four-digit approximation (rounding off). Determine the absolute and relative errors involved in parts (b) and (c):
(i) $\frac{1}{3} + \frac{1}{7}$, (ii) $\frac{1}{3} \times \frac{1}{7}$, (iii) $\frac{1}{3} - \frac{1}{7}$, and (iv) $\frac{1}{3}/\frac{1}{7}$.

1.7. Consider the decimal numbers $x = 0.5792 \times 10^4$, $y = 0.1337 \times 10^{-3}$, and $z = 0.4495 \times 10^2$ with a four-digit accuracy. Perform the following operations using a four-digit accuracy (chopping) and determine the error involved in each case:
(a) $x + y + z$, (b) $x - y + z$, (c) $(x/y) + z$, (d) $(xy)/z$, and (e) $x/(yz)$.

1.8. Compute the sum $x = \sum_{i=1}^{10} \frac{1}{i}$ first by using $1 + \frac{1}{2} + \frac{1}{3} + \cdots + \frac{1}{10}$ and then by using $\frac{1}{10} + \frac{1}{9} + \frac{1}{8} + \cdots + 1$ with three-digit arithmetic (chopping). Determine which method is more accurate and why?

1.9. Consider the evaluation of the expression $X = (A - B) * C$, where $A = 0.2345$, $B = 0.2344$, and $C = 10^5$. If an approximate value of 0.2346 is used for A, determine the absolute and relative errors in A and X.

1.10. Evaluate the value of $f = x^4 - 2x^3 + 6x^2 - 4x - 1.42$ at $x = 1.69$ using three-digit arithmetic, and determine the absolute and relative errors involved.

1.11. The numbers 0.6000×10^1, 0.6000×10^4, and 0.6000×10^{-3} are represented approximately as 0.6020×10^1, 0.6020×10^4, and 0.6020×10^{-3}, respectively. Determine the absolute error, relative error and the number of significant digits of approximation in each case. Also indicate whether absolute error or relative error is more meaningful of the error involved.

1.12. The value of e^x can be determined as

$$e^x = 1 + x + \frac{x^2}{2!} + \frac{x^3}{3!} + \frac{x^4}{4!} + \cdots. \tag{a}$$

By considering a different number of terms ranging from 1 to 10 on the right-hand side of Eq. (a), determine the value of e. Compare the result with the value 2.71828183, and find the absolute and relative error in each case.

1.13. The Maclaurin expansion of $\sin x$ is given by

$$\sin x = x - \frac{x^3}{3!} + \frac{x^5}{5!} - \frac{x^7}{7!} + \frac{x^9}{9!} - \cdots.$$

By considering one, two, three, four, and five terms on the right-hand side, estimate the value of $\sin x$ at $x = \frac{\pi}{6}$, and determine the error in each case. Compare this with the true value of $\sin \frac{\pi}{6}$, which is 0.5.

1.14. The Maclaurin expansion of $\cos x$ is given by

$$\cos x = 1 - \frac{x^2}{2!} + \frac{x^4}{4!} - \frac{x^6}{6!} + \frac{x^8}{8!} - \cdots.$$

By considering one, two, three, four, and five terms on the right-hand side, estimate the value of $\cos x$ at $x = \frac{\pi}{3}$, and determine the error in each case. Compare this with the true value of $\cos \frac{\pi}{3}$, which is 0.5.

Section 1.9

1.15. Expand the expression $(x + 2y)^3$ using MAPLE.

1.16. Find the derivative of the function $f = \frac{x^2}{x^3+1}$ using MAPLE and MATHCAD.

1.17. Carry out the following operations using MATLAB and MAPLE.

(a) $[A] + [B]$, (b) $[A] - [B]$, and (c) $[A]\,[B]$,

where

$$[A] = \begin{bmatrix} 1 & 0 & 2 \\ 3 & 4 & 6 \\ 5 & 2 & 7 \end{bmatrix}, \text{ and } [B] = \begin{bmatrix} 6 & 1 & 8 \\ 7 & 5 & 3 \\ 2 & 9 & 4 \end{bmatrix}.$$

Section 1.10

1.18. Compute $[C] = [A][B]$ for the matrices $[A]$ and $[B]$ defined in Problem 1.17 using the Fortran program of Section 1.10.1.

1.19. Compute $[C] = [A][B]$ for the matrices $[A]$ and $[B]$ defined in Problem 1.17 using the C program of Section 1.10.2.

1.20. Write a Fortran program to compute
(a) $[A]+[B]$, (b) $[A]-[B]$, and (c)$[A][B]$,
where $[A]=[a_{ij}]$, $[B]=[b_{ij}]$,
$a_{ij}=\frac{1}{i+j-1}$, $i,j=1,2,\ldots,5$,
and $b_{ij}=\max(i,j)$, $i,j=1,2,\ldots,5$.

1.21. Write a C program to compute the following:
(a) $[A]+[B]$, (b) $[A]-[B]$ (c) $[A]\,[B]$,
where $[A]$ and $[B]$ are as defined in Problem 1.20.

General

1.22. Represent the following binary numbers in decimal form: (a) 10110, (b) 110011.

1.23. Determine the largest interval in which the approximate value ($\tilde{x}$) of a number x must lie in order to limit the relative error to 10^{-4} for the following cases:
(a) $x=\sqrt{3}$, (b) $x=\pi$, (c) $x=\sqrt{5}$, and (d) $x=10$.

1.24. Consider the Fibonacci numbers, F_i, defined as $F_0=F_1=1$ and $F_{i+2}=F_{i+1}+F_i$, $i=1,2,\ldots$ These numbers are used in many applications including the minimization of a function. It is known that the ratio $x_n=\frac{F_{n+1}}{F_n}$ converges to $\frac{1+\sqrt{5}}{2}$ as $n\to\infty$. This is known as the golden ratio. Determine the relative error involved in approximating x_∞ by x_n for $n=1,2,\ldots,10$.

1.25. The Taylor's series expansion of $f(x)$ about x_0 is given by

$$f(x)=f(x_0)+\left.\frac{df}{dx}\right|_{x_0}(x-x_0)+\frac{1}{2!}\left.\frac{d^2f}{dx^2}\right|_{x_0}(x-x_0)^2+\cdots+\frac{1}{n!}\left.\frac{d^nf}{dx^n}\right|_{x_0}(x-x_0)^n+R_n,$$

where R_n is called the remainder term, given by,

$$R_n=\frac{1}{(n+1)!}\frac{d^{n+1}f}{dx^{n+1}}(\xi),\quad x_0<\xi<x.$$

Expand the function $f(x)=0.1x^4+0.2x^3+0.3x^2-0.4x-0.5$ in Taylor's series about $x=0$ and use the resulting expression to estimate the value of f at $x=1$ by retaining one, two, three, and four terms of the expansion. Determine the absolute and relative errors involved in each case.

1.26. Suggest a method of computing the value of the following expression without losing accuracy in the interval indicated:
(a) $f(x)=\cos^2 x-\sin^2 x,\quad 0\le x\le\frac{\pi}{2}$,
(b) $f(x)=\sqrt{2x^2+1}-1,\quad -1\le x\le 1$,
(c) $f(x)=\sqrt{x+4}-\sqrt{x+3},\quad 10^5\le x\le 10^6$.

1.27. Use Taylor's series expansion to approximate the following functions as (i) linear functions and (ii) quadratic functions:

(a) $$\frac{1-x}{1+y};$$

(b) $$\sqrt{1-4x+3y}.$$

Estimate the error and the relative error involved in each case.

1.28. Write a Fortran computer program to cause an overflow and an underflow of numbers.

1.29. Find the roots of the quadratic equation $x^2 - 90x + 1 = 0$ using a four-digit arithmetic and compute the errors involved.

1.30. Consider the function $f(x) = \sqrt{1-x^3}$. If an approximate value $\bar{x} = 0.998$ is used instead of the true value $x = 0.999$, determine the errors involved in the computation of the function $f(x)$.

1.31. The value of $f(x) = \log_e \frac{x+1}{x-1}$ can be computed using the series expansion:

$$\log_e(x+1) - \log_e(x-1) = 2\left[\frac{1}{x} + \frac{1}{3x^3} + \frac{1}{5x^5} + \cdots\right].$$

Use $x = 1.1$ and compare the values of $f(x)$ obtained with a different number of terms in the series.

1.32. (a) The equivalent spring constant, k_{eq}, of a system of springs connected in series is given by

$$\frac{1}{k_{eq}} = \frac{1}{k_1} + \frac{1}{k_2} + \cdots,$$

where $k_1, k_2, \ldots$, are the spring constants of the individual springs. Find the value of k_{eq} for a system of 10 springs in series with $k_i = 10^i, i = 1, 2, \ldots, 10$. Discuss the accuracy of the computed value of k_{eq}.

(b) The girder of an overhead traveling crane and the wire ropes 1 and 2 can be considered as springs in series (Fig. 1.17) with $k_g = 7 \times 10^6$ lb/in, $k_1 = 3 \times 10^4$ lb/in and $k_2 = 9 \times 10^2$ lb/in. Determine the value of the equivalent spring constant, k_{eq}, using four-digit arithmetic.

1.33. A beam of length L_b, attached to a wall by a pivot at one end, carrys a load W as shown in Fig. 1.18. One end of a cable is connected to the beam at a distance of x from the wall, and the other end is attached to the wall. The tension T in the cable is given by

$$T = \frac{W\,L_c\,L_b}{x\sqrt{L_b^2 - x^2}},$$

where $L_c = 15$ ft, $L_b = 20$ ft, and $W = 1000$ lb.

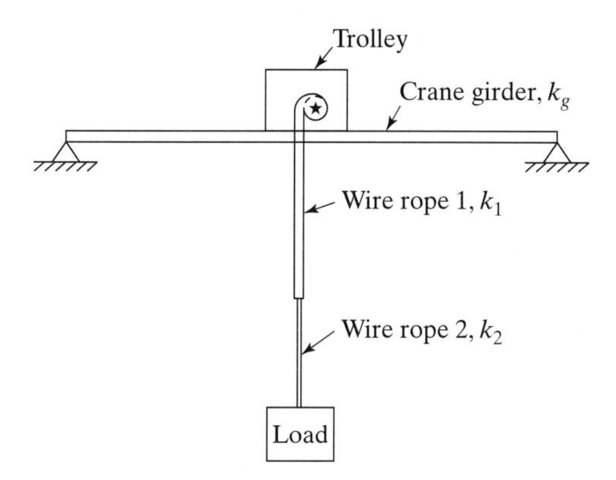

Figure 1.17 Overhead traveling crane.

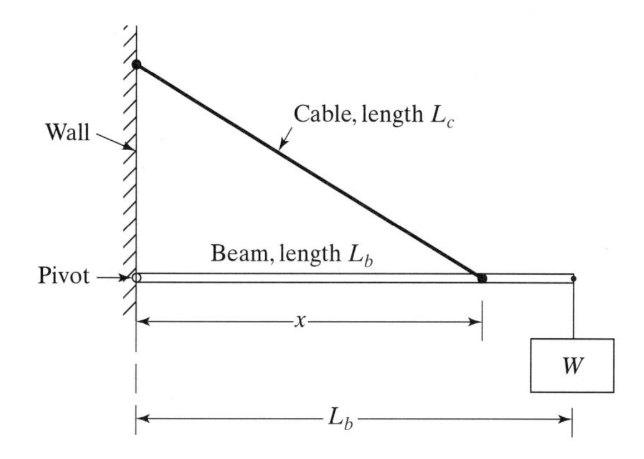

Figure 1.18 Beam under load.

(a) Determine the distance x at which the cable is to be attached to the beam for minimizing the tension in the cable.

(b) Determine the tension in the cable at 0.1-ft intervals from $x = 1$ to 12 ft, and find the value of x that corresponds to minimum tension.

(c) Discuss the accuracy of the result found in part (b).

1.34. When several data points (x_i, y_i), $i = 1, 2, \ldots, n$ are observed in an experiment, a least squares approach can be used to fit a straight line through the data points (Fig. 1.19):

$$y = a\,x + b, \tag{E1}$$

where

$$a = \frac{\left(\sum x_i\, y_i\right) - \bar{y}\left(\sum x_i\right)}{\left(\sum x_i^2\right) - \bar{x}\left(\sum x_i\right)}, \tag{E2}$$

and

$$b = \bar{y} - a\,\bar{x}, \tag{E3}$$

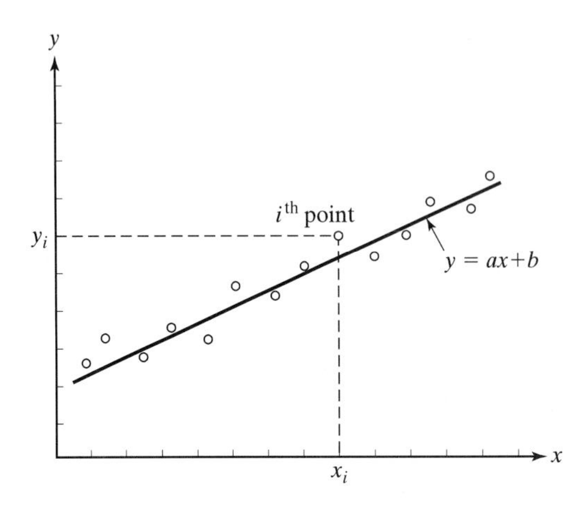

Figure 1.19 Least squares fit.

where the summations extend from $i = 1$ to n, and $\bar{x}(\bar{y})$ is the average value of $x(y)$. Using Eqs. (E1) through (E3), determine the straight line corresponding to the following data:

i	1	2	3	4	5	6	7	8	9	10
x_i	1.1	1.4	2.6	3.3	3.6	4.9	5.3	6.0	6.6	7.5
y_i	2.4	3.8	3.2	4.1	5.2	5.0	6.4	7.7	7.1	7.4

Suggest a method of estimating the accuracy of the straight-line fit.

PROJECTS

1.1. (a) Develop a procedure for converting a decimal number X to its equivalent representation Y in base B when $2 \leq B \leq 9$.

(b) Develop a procedure for converting a number Y given in base B to its equivalent decimal representation.

(c) Write a computer program to implement the procedures of parts (a) and (b). Assume that the given number has up to five digits in the integral part and up to five digits in the fractional part.

1.2. The circumference S of a circle of radius R can be approximated by the perimeters of inscribed and circumscribed n-sided polygons as shown in Fig. 1.20 with $S^{(l)} = n\,r = 2\,n\,R \tan \frac{\pi}{n}$, $S^{(u)} = n\,s = 2\,n\,R\,\sin \frac{\pi}{n}$,

(a) Prove $S^{(l)} \leq S \leq S^{(u)}$,

(b) Prove $\lim_{n\to\infty} S^{(l)} = S$, and $\lim_{n\to\infty} S^{(u)} = S$.

(c) Write a computer program to evaluate the bounds on the circumference of a circle of radius 1 by considering values of n ranging from 3 to 30.

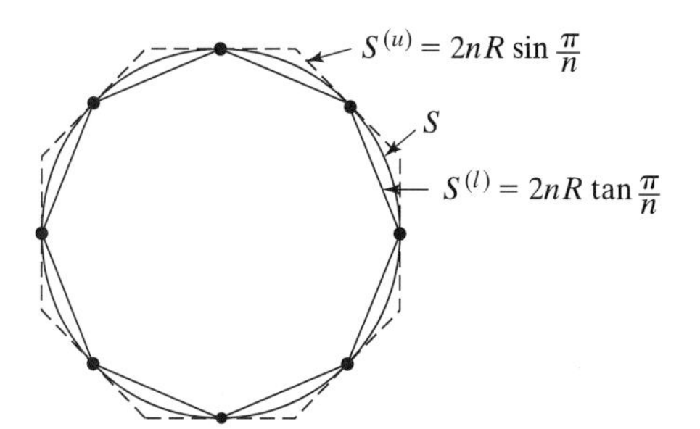

Figure 1.20 Approximation of circumference of a circle.

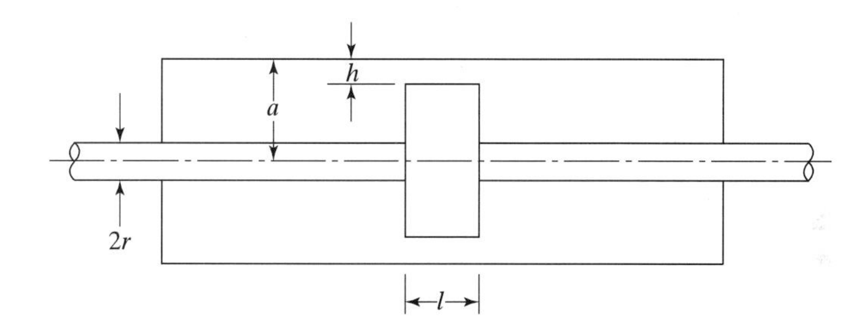

Figure 1.21 Dashpot.

1.3. The damping constant (c) of the dashpot shown in Fig. 1.21 is given by

$$c = \frac{6\pi\ \mu\ l}{h^3} \left\{ \left(a - \frac{h}{2} \right)^2 - r^2 \right\} \left(\frac{a^2 - r^2}{a - \frac{h}{2}} - h \right),$$

where μ is the viscosity of the fluid. Suggest a procedure for finding the influence of small errors in a, h, r, and l on c from the reference values, $\mu = 0.3445$ Pa-s, $l = 10$ cm, $h = 0.1$ cm, $a = 2$ cm, and $r = 0.5$ cm.

Use the procedure to predict the value of c under the following conditions:

(a) $l = 9.999$ cm, $h = 0.009$ cm, $a = 1.999$ cm, and $r = 0.499$ cm.

(b) $l = 10.001$ cm, $h = 0.101$ cm, $a = 2.001$ cm, and $r = 0.501$ cm.

(c) $l = 9.999$ cm, $h = 0.101$ cm, $a = 2.001$ cm, and $r = 0.499$ cm.

2

Solution of Nonlinear Equations

2.1 Introduction

Many engineering analyses require the determination of the value(s) of the variable x that satisfies a nonlinear equation:

$$f(x) = 0. \tag{2.1}$$

The values of x that satisfy Eq. (2.1) are known as the roots of Eq. (2.1); the number of roots may be finite or infinite and may be real or complex depending on the nature of the equation and the physical problem. The function $f(x)$ may or may not be available in explicit form. If it is available in explicit form, it may be in the form of a general nonlinear equation, or a polynomial or a transcendental equation:

$$x^4 - 80x + 120 = 0. \text{(polynomial equation)} \tag{2.2}$$

and

$$\tan x - \tanh x = 0. \text{(transcendental equation)}. \tag{2.3}$$

CHAPTER CONTENTS

The roots of Eq. (2.1) are also known as the zeros of the function $f(x)$. Thus, a function is said to have zeros, while an equation is said to have roots. A transcendental function is one whose value cannot be determined for any specified value of the argument by a finite number of additions, subtractions, multiplications, or divisions. Exponential, logarithmic, trigonometric, and hyperbolic functions are examples of transcendental functions. Any equation containing transcendental functions is called a transcendental equation. If the function is known only implicitly, we do not have an explicit mathematical expression available for $f(x)$, but a rule will be known to evaluate the value of $f(x)$ for any specified value of the argument. The methods of finding the approximate values of the roots of Eq. (2.1) are considered in this chapter.

2.1.1 Polynomials in x

As a special case of the equation $f(x) = 0$, we consider a polynomial equation

$$f(x) = a_n x^n + a_{n-1} x^{n-1} + \cdots + a_2 x^2 + a_1 x + a_0 = 0, \tag{2.4}$$

where n denotes the degree of the polynomial and $a_0, a_1, a_2, \ldots, a_n$ are the coefficients of the polynomial that are real (complex, in some cases). Equation (2.4), in general, will have n roots, some or all of which may be complex. The roots may be independent, or some of them may be repeated. If $x_1, x_2, \ldots, x_n$ denote the roots of Eq. (2.4), they are related to the coefficients of the polynomial as [2.8, 2.16]:

$$\sum_{i=1}^{n} x_i = -\frac{a_{n-1}}{a_n},$$

$$\sum_{i=1}^{n} \sum_{j=1, j \neq i}^{n} x_i x_j = \frac{a_{n-2}}{a_n},$$

$$\sum_{i=1}^{n} \sum_{j=1, j \neq i}^{n} \sum_{k=1, k \neq j}^{n} x_i x_j x_k = -\frac{a_{n-3}}{a_n},$$

$$\vdots,$$

$$x_1 x_2 \cdots x_{n-1} x_n = (-1)^n \frac{a_0}{a_n}. \tag{2.5}$$

Equations (2.5) are known as Newton's relations.

2.1.2 Graphical Interpretation

The real roots of an equation can be interpreted graphically as shown in Fig. 2.1. (Graphical interpretation is not possible if the roots are complex.) The roots occur where the graph of $f(x)$ crosses or touches the x axis. If the graph of $f(x)$ touches or coincides with the x axis at x_1, it indicates that the equation $f(x) = 0$ has a multiple root at x_1. (See Fig. 2.2.)

In some cases, the equation $f(x) = 0$ can be split into two separate parts. In such a case, the two parts can be represented graphically, and the point of

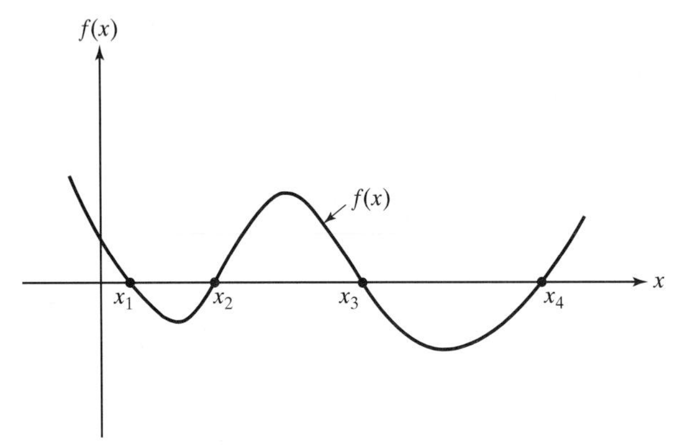

Figure 2.1 x_1, x_2, x_3, x_4 - roots of the equation.

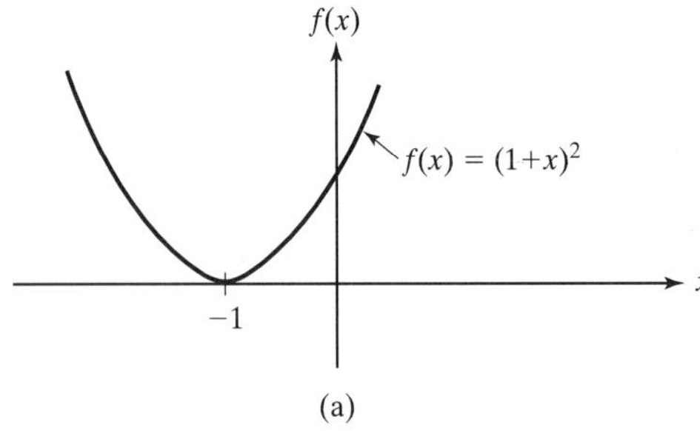

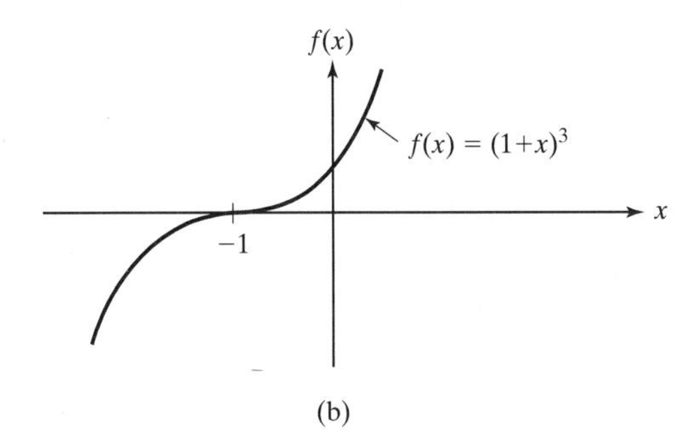

Figure 2.2 Multiple roots of $f(x) = 0$.

intersection of the two parts denotes the root of the equation. As an example, consider the equation

$$f(x) = x^3 - 2x + 1 = 0,$$

which can be rewritten as

$$x^3 = 2x - 1. \tag{2.6}$$

When the graphs of the functions $f_1 = x^3$ and $f_2 = 2x - 1$ are drawn as shown in Fig. 2.3, the points of intersection, x_1 and x_2, indicate the roots of the equation, $f(x) = 0$.

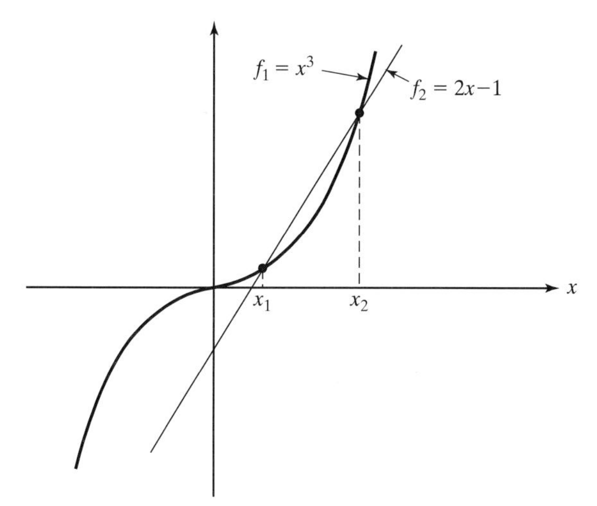

Figure 2.3 Point of intersection of two graphs.

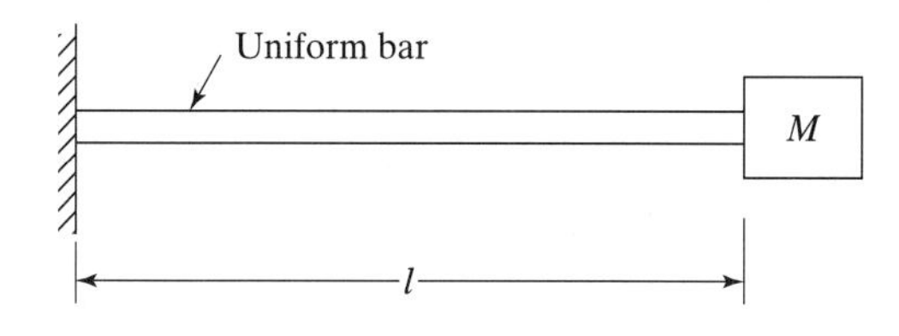

Figure 2.4 Bar carrying an end mass.

▶Example 2.1

The natural frequencies of axial vibration of a bar, fixed at one end and carrying a mass M at the other end (see Fig. 2.4), are given by the frequency equation [2.1, 2.2]

$$\cot\left(\frac{\omega l}{c}\right) = \frac{M}{m}\frac{\omega l}{c}, \tag{E1}$$

where ω is the natural frequency, l is the length, $c = \sqrt{\frac{E}{\rho}}$, E is Young's modulus, ρ is the density, $m = \rho A l$ is the mass of the bar, and A is the cross sectional area of the bar. Find the first three natural frequencies of the bar using a graphical procedure for $\frac{M}{m} = 0.1, 0.5,$ and 1.0.

Solution

The frequency equation, Eq. (E1), has the parameters $\frac{M}{m}$ and $\frac{\omega l}{c}$. The left-hand side of Eq. (E1), $\cot\left(\frac{\omega l}{c}\right)$, can be plotted as a function of $\left(\frac{\omega l}{c}\right)$, and the relationship is shown by the solid lines in Fig. 2.5. The right-hand side of Eq. (E1) represents a straight line in $\left(\frac{\omega l}{c}\right)$ with the slope equal to $\frac{M}{m}$. It is shown by dotted lines for three different values of $\frac{M}{m}$. The intersection points of the solid and dotted lines represent

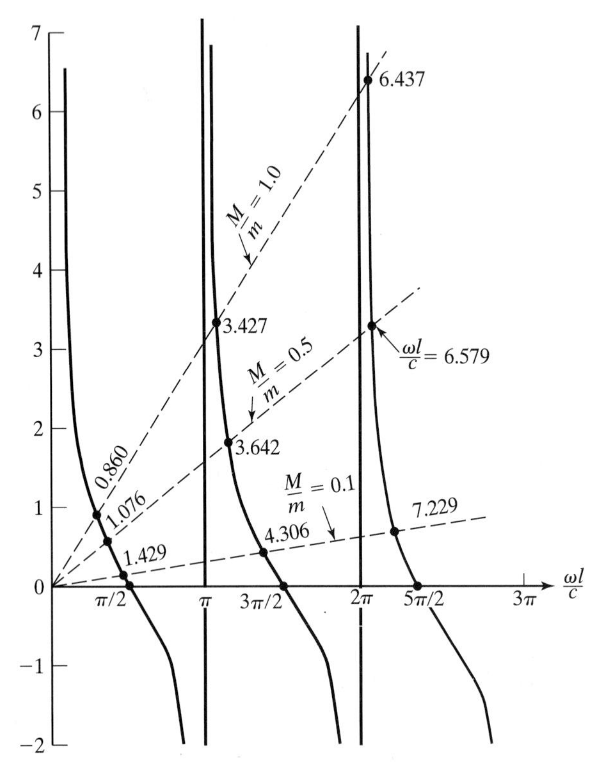

Figure 2.5 Solution of frequency equation.

the roots of the frequency equation. The following results can be obtained from Fig. 2.5.

Frequency parameter	Value of $\frac{M}{m}$ 0.1	0.5	1.0
$\frac{\omega_1 l}{c}$	1.429	1.076	0.860
$\frac{\omega_2 l}{c}$	4.306	3.642	3.427
$\frac{\omega_3 l}{c}$	7.229	6.579	6.437

Note that although the graphical method can be used to find the roots of any equation $f(x) = 0$, it has the following limitations:

1. It requires extensive computations to plot the graph of $f(x)$.
2. The roots found from the graphical method will be less accurate.

However, in some cases, the roots found from the graphical method can be used as good initial approximations for some of the iterative procedures presented in subsequent sections. ◀

2.2 Engineering Applications

The solutions of many engineering analysis problems require the determination of the roots of nonlinear equations. The following examples illustrate representative applications.

▶Example 2.2

Water is discharged from a reservoir through a long pipe as shown in Fig. 2.6. By neglecting the change in the level of the reservoir, the transient velocity of the water flowing from the pipe, $v(t)$, can be expressed as [2.4]:

$$\frac{v(t)}{\sqrt{2gh}} = \tanh\left(\frac{t}{2L}\sqrt{2gh}\right), \tag{E1}$$

where h is the height of the fluid in the reservoir, L is the length of the pipe, g is the acceleration due to gravity, and t is the time elapsed from the beginning of the flow. Find the value of h necessary for achieving a velocity of $v = 5$ m/sec at time $t = 3$ sec when $L = 5$ m and $g = 9.81$ m/sec^2.

Solution

Equation (E1) is rewritten as $f_1 = f_2$ with $f_1 = \frac{v(t)}{\sqrt{2gh}} = \frac{5}{\sqrt{2(9.81)h}} = \frac{1.1288}{\sqrt{h}}$ and $f_2 = \tanh\left(\frac{3}{2(5)}\sqrt{2(9.81)h}\right) = \tanh(1.3288\sqrt{h})$. The functions f_1 and f_2 are plotted by varying h as shown in Fig. 2.7. It can be seen that the two graphs intersect at $h = 1.45$ m, thereby indicating that Eq. (E1) is satisfied when $h = 1.45$ m. ◀

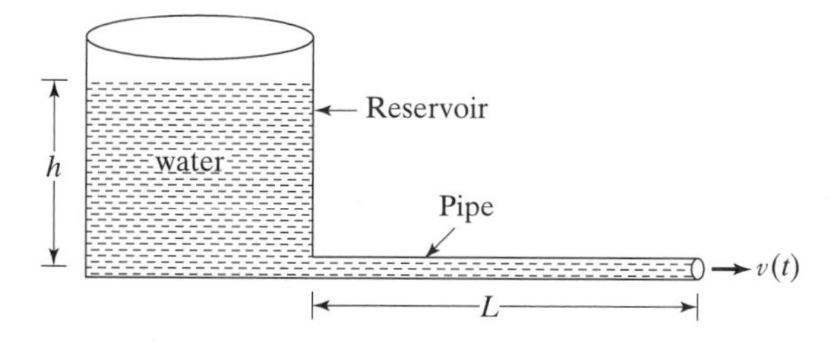

Figure 2.6 Discharge of water from reservoir.

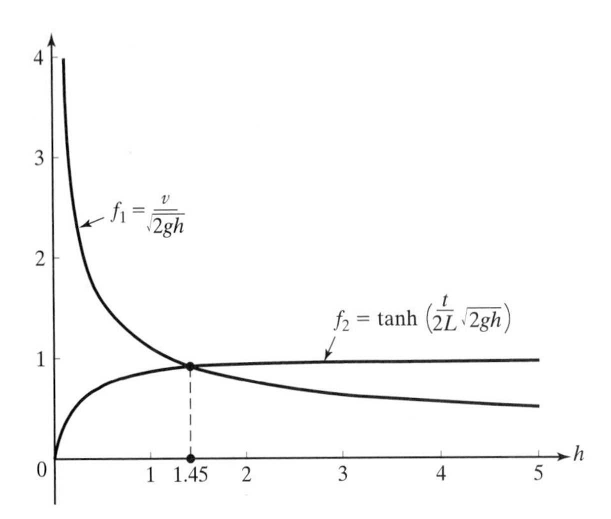

Figure 2.7 Solution of Eq. (E1).

▶Example 2.3

The length of a belt in an open-belt drive, L, is given by [2.5]:

$$L = \sqrt{4c^2 - (D - d)^2} + \frac{1}{2}\{D\theta_D + d\theta_d\}, \tag{E1}$$

where

$$\theta_D = \pi + 2\sin^{-1}\left(\frac{D - d}{2c}\right) \tag{E2}$$

and

$$\theta_d = \pi - 2\sin^{-1}\left(\frac{D - d}{2c}\right). \tag{E3}$$

c is the center distance, D is the diameter of the larger pulley, d is the diameter of the smaller pulley, θ_D is the angle of contact of the belt with the larger pulley, and θ_d is the angle of contact of the belt with the smaller pulley. (See Fig. 2.8.) If a belt

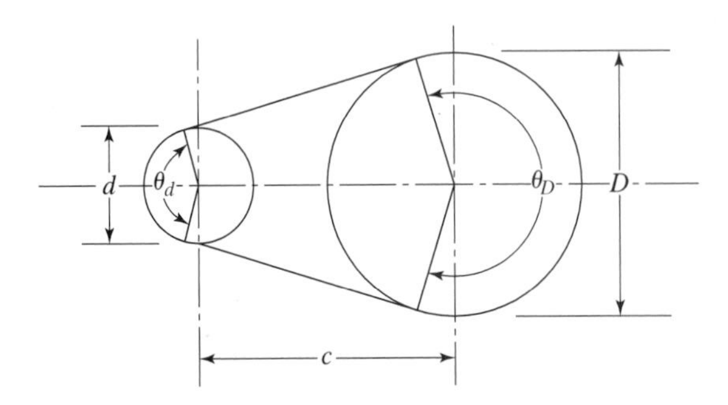

Figure 2.8 Open belt drive.

having a length 11 m is used to connect two pulleys with diameters 0.4 m and 0.2 m, determine the center distance between the pulleys.

Solution

For the given data, the angles of contact of the belt with the two pulleys can be expressed, from Eqs. (E2) and (E3) as

$$\theta_D = \pi + 2\sin^{-1}\left(\frac{0.4-0.2}{2c}\right) = 3.1416 + 2\sin^{-1}\left(\frac{0.1}{c}\right) \tag{E4}$$

and

$$\theta_d = \pi - 2\sin^{-1}\left(\frac{0.4-0.2}{2c}\right) = 3.1416 - 2\sin^{-1}\left(\frac{0.1}{c}\right). \tag{E5}$$

Equation (E1) can be rewritten as

$$f(c) = \sqrt{4c^2 - (D-d)^2} + \frac{1}{2}\{D\theta_D + d\theta_d\} - L = 0. \tag{E6}$$

Using the given data and Eqs. (E4) and (E5), Eq. (E6) can be expressed as

$$f(c) = \sqrt{4c^2 - (0.4-0.2)^2} + 0.5\left[0.4\left\{3.1416 + 2\sin^{-1}\left(\frac{0.1}{c}\right)\right\}\right.$$
$$\left.+0.2\left\{3.1416 - 2\sin^{-1}\left(\frac{0.1}{c}\right)\right\}\right] - 11 = 0$$

or

$$f(c) = \sqrt{4c^2 - 0.04} + 0.24\sin^{-1}\left(\frac{0.1}{c}\right) - 10.055752 = 0 \tag{E7}$$

The function $f(c)$ is shown plotted in Fig. 2.9. It can be seen that $f(c) = 0$ at the center distance, $c = 5.05$ m. ◀

▶Example 2.4

An electrical circuit consists of a resistor with resistance 14.85 ohms, an inductor with inductance 7 henries, and a capacitor with capacitance $\frac{1}{42}$ farad in parallel as shown in Fig. 2.10(a). If the initial current through the inductance, $i(0)$, and the initial voltage across the capacitor, $e(0)$, are specified as 10 amperes and 0 volts, respectively, the response of the circuit is given by [2.7]:

$$e(t) = 210e^{-1.414t}\sin 2t. \tag{E1}$$

Determine the time at which the response of the circuit will be zero.

Solution

The response of the circuit, $e(t)$, can be plotted as a function of time, t, as shown in Fig. 2.10(b). It can be seen that the response will be zero at $t_1 = 1.6$ sec, $t_2 = 3.1$ sec, etc. ◀

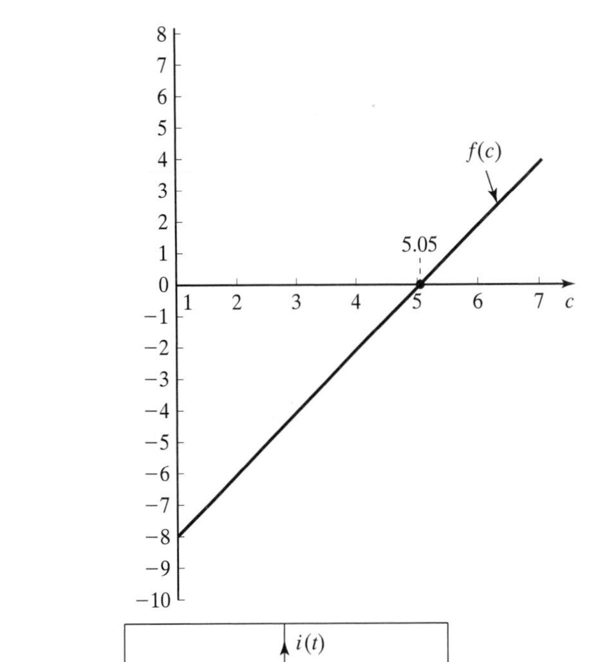

Figure 2.9 Graph of Eq. (E7).

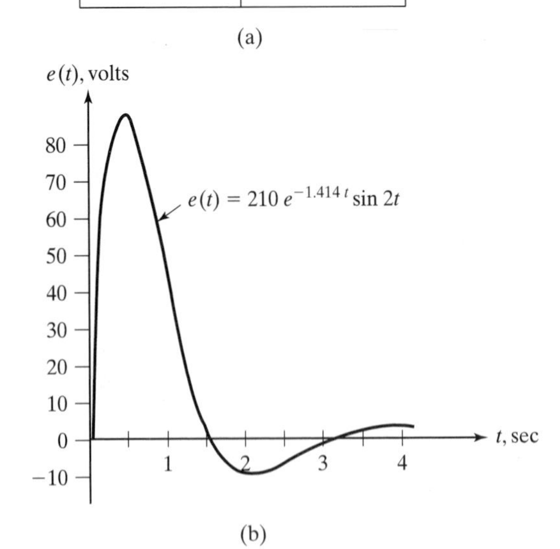

Figure 2.10 Electrical circuit and its response.

▶Example 2.5

The shear stress induced along the z axis when two spheres are in contact with each other, while carrying a load F, is given by

$$h(\lambda) = \frac{0.75}{1+\lambda^2} + 0.65\lambda \tan^{-1}\left(\frac{1}{\lambda}\right) - 0.65, \tag{E1}$$

where

$$h(\lambda) = \frac{\tau_{zx}}{p_{\max}} \text{ and } \lambda = \frac{z}{a},$$

in which τ_{zx} is the shear force,

$$p_{\max} = \frac{3F}{2\pi a^2} \tag{E2}$$

is the maximum pressure developed at the center of the contact area, and the radius of the contact area (see Fig. 2.11) is

$$a = \left\{ 0.34125F \frac{\left(\frac{1}{E_1} + \frac{1}{E_2}\right)}{\left(\frac{1}{d_1} + \frac{1}{d_2}\right)} \right\}^{\frac{1}{3}}, \tag{E3}$$

where E_1 and E_2 are Young's moduli of the two spheres, and d_1 and d_2 are diameters of the two spheres. Poisson's ratios of the spheres was assumed to be 0.3 in deriving Eqs. (E1) and (E3). In many practical applications such as ball bearings, when the contact load F is large, a crack originates at the point of maximum shear stress and propagates to the surface leading to a fatigue failure. Thus it becomes necessary to

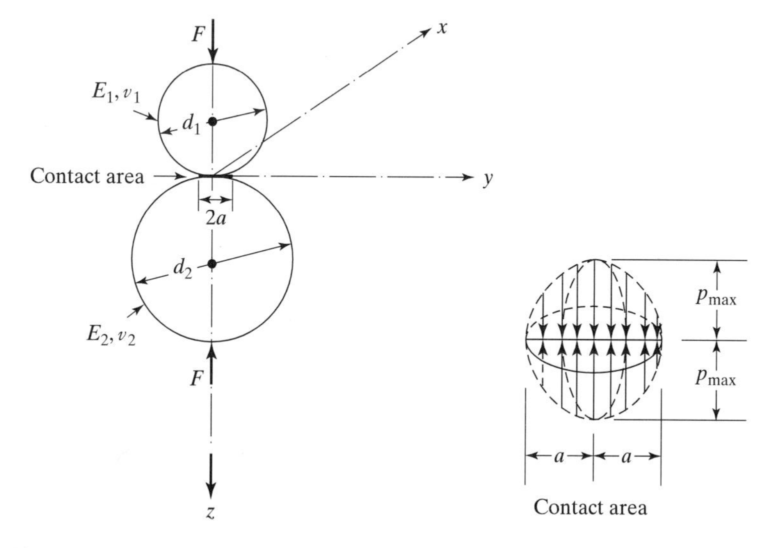

Figure 2.11 Contact stress between spheres.

find the point at which the shear stress attains a maximum value in order to locate the origin of the crack. Determine the value of λ at which the shear stress, given by Eq. (E1), attains its maximum value.

Solution

To find the value of λ at which the shear stress attains its maximum value, the condition $\frac{dh}{d\lambda} = 0$ is used. This yields the equation

$$\frac{dh(\lambda)}{d\lambda} = f(\lambda) = \frac{1.5\lambda}{(1+\lambda^2)^2} - 0.65\tan^{-1}\left(\frac{1}{\lambda}\right) + \frac{0.65\lambda}{(1+\lambda^2)} = 0. \qquad \text{(E4)}$$

This equation is plotted in Fig. 2.12. It can be observed that the root of Eq. (E4) lies at $\lambda \approx 0.48$, which implies that the shear stress will attain its maximum value at $z \approx 0.48\,a$. ◀

2.3 Incremental Search Method

In the incremental search method, the value of x is incremented from an initial value x_1, successively until a change in the sign of the function $f(x)$ is observed. The idea

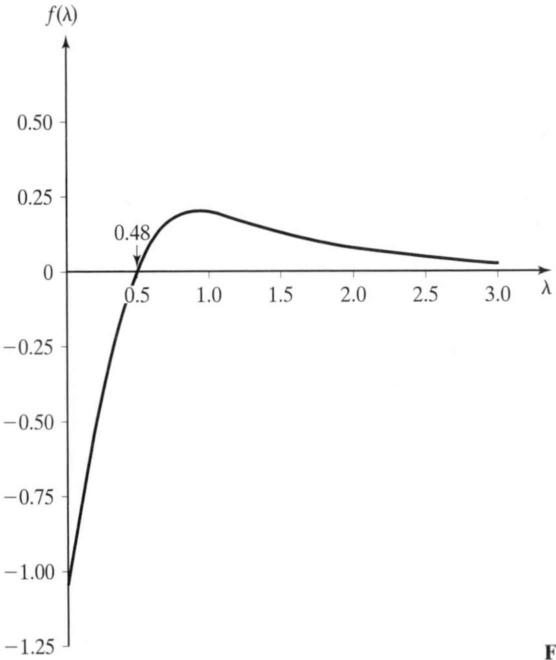

Figure 2.12 Graph of Eq. (E4).

is that $f(x)$ changes sign between x_i and x_{i+1}, if it has a root in the interval $[x_i, x_{i+1}]$ as shown in Fig. 2.13. This implies that

$$f(x_i)f(x_{i+1}) < 0, \tag{2.7}$$

whenever a root is crossed. Here, $x_{i+1} = x_i + \Delta x$, where $\Delta x \equiv \Delta x^{(1)}$ denotes the initial step size. Once the location of the root between x_i and x_{i+1} is observed, the process can be repeated with a smaller step size $\Delta x^{(2)}$ from the new starting point, x_i. By successive application of this procedure, we can make the interval $\Delta x^{(i)}$, in which the root lies, as small as possible to satisfy the specified convergence requirement.

Notes

1. The incremental search method can be used to find only the real roots of a function $f(x)$. The starting value for the incremental search method, x_1, is usually taken as the smallest value in the range of interest. After finding one root as $x = x_1^*$, additional roots can be found by using x_1^* as the new starting point, x_1, and the original $\Delta x^{(1)}$ as the new step length. The procedure is applied until another change in sign is observed that indicates the crossing of the second root, $x = x_2^*$.
2. If the graph of $f(x)$ touches the x axis tangentially, the function $f(x)$ does not undergo a sign change. Hence, the incremental search method cannot find the root.
3. The method may not be able to distinguish between a root and a singular point. (See Fig. 2.14.)
4. An accelerated step size can be used to bracket the root quickly. In this case, the step size is increased (for example, doubled) at each stage.

▶Example 2.6

Find the root of the equation

$$f(x) = \frac{1.5x}{(1+x^2)^2} - 0.65\tan^{-1}\left(\frac{1}{x}\right) + \frac{0.65x}{1+x^2} = 0, \tag{E1}$$

using the incremental search method with $x_1 = 0.0$ and $\Delta x^{(1)} = 0.1$.

Solution

Since the initial value is $x_1 = 0.0$ and the initial step size is $\Delta x^{(1)} = 0.1$, the function is evaluated at 0.0, 0.1, 0.2, ... to obtain the following results:

$$x_1 = 0.0, \quad f_1 = f(x_1) = -1.021017;$$

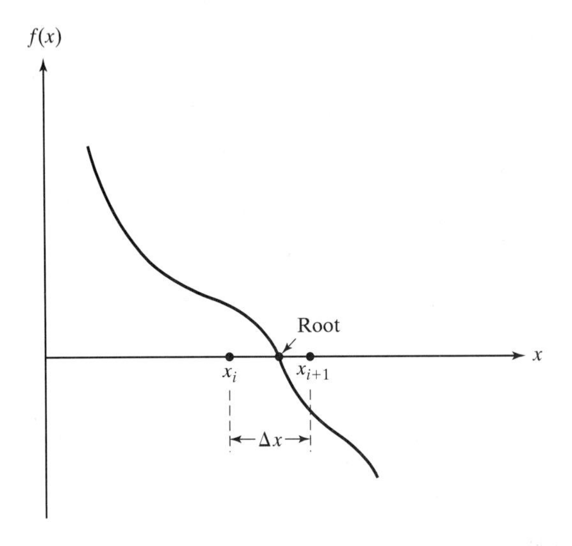

Figure 2.13 Change in sign of $f(x)$ at a root.

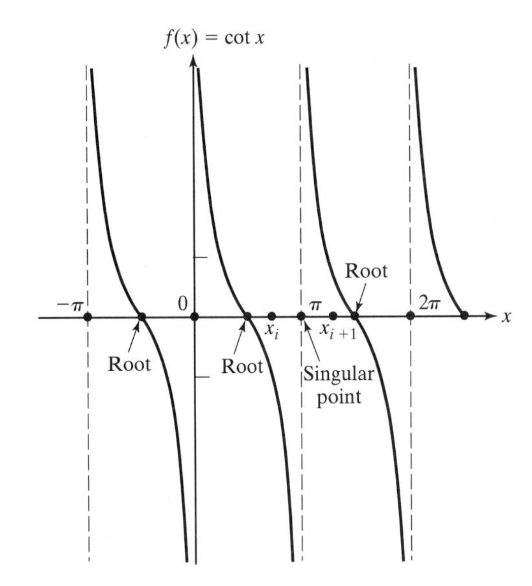

Figure 2.14 Roots and singular point.

$$x_2 = 0.1, \quad f_2 = f(x_2) = -0.744832;$$

$$x_3 = 0.2, \quad f_3 = f(x_3) = -0.490343;$$

$$x_4 = 0.3, \quad f_4 = f(x_4) = -0.273916;$$

$$x_5 = 0.4, \quad f_5 = f(x_5) = -0.103652;$$

$$x_6 = 0.5, \quad f_5 = f(x_5) = +0.020354.$$

Since the sign of the function f changes between x_5 and $x_6(f_5 f_6 < 0)$, the step size is reduced to $\Delta x^{(2)} = 0.01$, and the procedure is repeated from $x_5 = 0.40$. This yields the following results:

$$x_5 = 0.40, \quad f_5 = f(x_5) = -0.103652;$$

$$x_6 = 0.41, \quad f_6 = f(x_6) = -0.089227;$$

$$\vdots;$$

$$x_{13} = 0.48, \quad f_{13} = f(x_{13}) = -0.000956;$$

$$x_{14} = 0.49, \quad f_{14} = f(x_{14}) = +0.009907.$$

Since a change in sign of f is observed between x_{13} and $x_{14}(f_{13} f_{14} < 0)$, the procedure can be continued from x_{13} with the reduced step size, $\Delta x^{(3)} = \frac{\Delta x^{(2)}}{10} = 0.001$. However, the procedure is terminated at this stage, and hence, the root of Eq. (E1) can be taken as $x_{13} = 0.48$ or $x_{14} = 0.49$, or the midpoint of x_{13} and x_{14} (0.485). ◀

2.4 Bisection Method

In order to find the roots of the equation $f(x) = 0$ using the bisection method, the function $f(x)$ is first evaluated at equally spaced intervals of x until two successive function values are found with opposite signs. Let $a = x_k$ and $b = x_{k+1}$ be the values of x at which the function values $f(a)$ and $f(b)$ have opposite signs. This implies that the function $f(x)$ has a root between $a = x_k$ and $b = x_{k+1}$ (see Fig. 2.15). The interval (x_k, x_{k+1}), in which the root is expected to lie, is called the interval of uncertainty. The midpoint of the current interval of uncertainty (a, b) is computed as

$$x_{\text{mid}} = \frac{a+b}{2}, \tag{2.8}$$

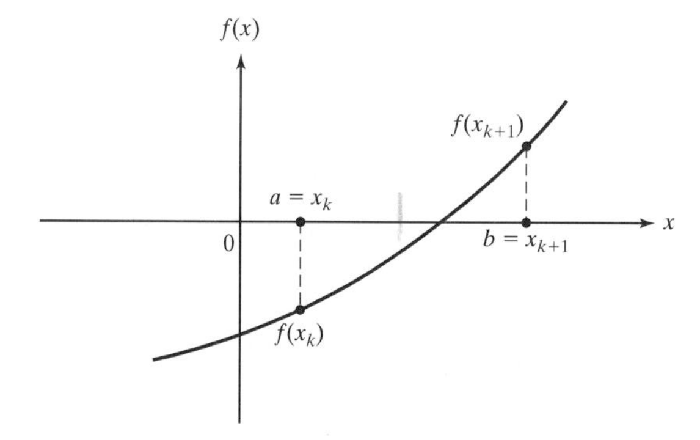

Figure 2.15 Opposite signs of function at a and b.

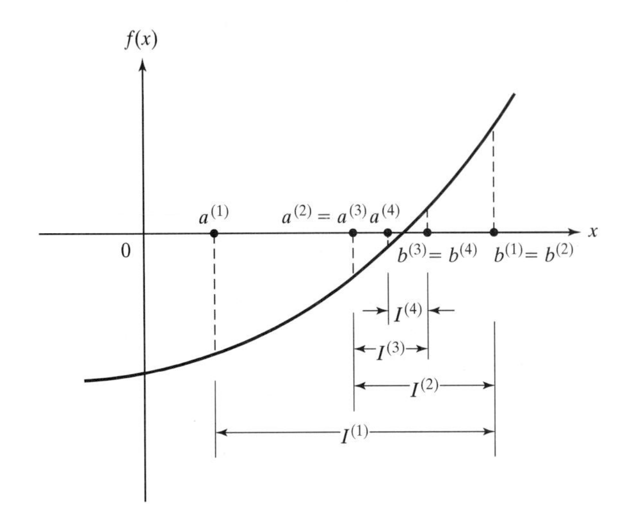

Figure 2.16 Bisection method.

and the function value $f(x_{\text{mid}})$ is determined. If $f(x_{\text{mid}}) = 0, x_{\text{mid}}$ will be a root of $f(x) = 0$. If $f(x_{\text{mid}}) \neq 0$, then the sign of $f(x_{\text{mid}})$ will coincide with that of $f(a)$ or $f(b)$. If the signs of $f(x_{\text{mid}})$ and $f(a)$ coincide, then a is replaced by x_{mid}. Otherwise (that is, if the signs of $f(x_{\text{mid}})$ and $f(b)$ coincide), b is replaced by x_{mid}. Thus the interval of uncertainty is reduced to half of its original value. Again the midpoint of the current interval of uncertainty is computed using Eq. (2.8), and the procedure is repeated until a specified convergence criterion is satisfied. The reduction of the intervals of uncertainty (i.e., the progress of the iterative process) is shown in Fig. 2.16. The following convergence criterion can be used to stop the iterative procedure:

$$|f(x_{\text{mid}})| \leq \varepsilon. \tag{2.9}$$

Here ε is a specified small number.

Assuming that the values of a and b, at which $f(a)$ and $f(b)$ have opposite signs, are known, the iterative procedure used to find the roots of $f(x) = 0$ can be summarized as follows:

1. Set $a^{(1)} = a, b^{(1)} = b$, and $i = 0$.
2. Set iteration number $i = i + 1$.
3. Find $x_{\text{mid}} = \frac{a^{(i)}+b^{(i)}}{2}$.
4. If x_{mid} satisfies the convergence criterion

   $$|f(x_{\text{mid}})| \leq \varepsilon,$$

   take the desired root as $x_{\text{root}} = x_{\text{mid}}$ and stop the procedure. Otherwise, go to step 5.

5. If $f(x_{\text{mid}}) \cdot f(a^{(i)}) > 0$, both $f(x_{\text{mid}})$ and $f(a^{(i)})$ will have the same sign, hence, set $a^{(i+1)} = x_{\text{mid}}$ and $b^{(i+1)} = b^{(i)}$, and go to step 2.
6. If $f(x_{\text{mid}}) \cdot f(a^{(i)}) < 0$, $f(x_{\text{mid}})$ and $f(a^{(i)})$ will have opposite signs, hence, set $b^{(i+1)} = x_{\text{mid}}$ and $a^{(i+1)} = a^{(i)}$, and go to step 2.

Notes

1. The bisection method works when the initial interval of uncertainty (a, b) contains an odd number of roots. The method will not work if the interval (a, b) contains a double root (as in Fig. 2.2(a)), since $f(a)$ and $f(b)$ will have the same sign.
2. The method will not be able to distinguish between a singularity and a root. (See Fig. 2.14.)

▶Example 2.7

Find the root of the equation

$$f(x) = \frac{1.5x}{(1+x^2)^2} - 0.65\tan^{-1}\left(\frac{1}{x}\right) + \frac{0.65x}{1+x^2} = 0 \tag{E1}$$

using the bisection method with $a = 0.0$, $b = 2.0$, and $\varepsilon = 0.005$.

Solution

The function values at a and b can be found as $f(a) = f(0.0) = -1.021017$ and $f(b) = f(2.0) = +0.078629$. Using $a^{(1)} = a = 0.0$ and $b^{(1)} = b = 2.0$, we set the iteration number as $i = 1$ and find that

$$x_{\text{mid}} = \frac{a^{(1)} + b^{(1)}}{2} = 1.0 \text{ with } f(x_{\text{mid}}) = f(1.0) = +0.189491.$$

Since $|f(x_{\text{mid}})| > \varepsilon = 0.005$, and $f(a^{(1)})$ and $f(x_{\text{mid}})$ have opposite signs, we choose $a^{(2)} = a^{(1)} = 0.0$ and $b^{(2)} = x_{\text{mid}} = 1.0$.

In the next iteration ($i = 2$), we use $x_{\text{mid}} = \frac{a^{(2)}+b^{(2)}}{2} = 0.5$ and find the corresponding function value as $f(x_{\text{mid}}) = +0.020353$. Since $|f(x_{\text{mid}})| > \varepsilon = 0.005$, and $f(a^{(2)})$ and $f(x_{\text{mid}})$ have opposite signs, we select $a^{(3)} = a^{(2)} = 0.0$ and $b^{(3)} = x_{\text{mid}} = 0.5$.

In the third iteration ($i = 3$), we use $x_{\text{mid}} = \frac{a^{(3)}+b^{(3)}}{2} = 0.25$ and find $f(x_{\text{mid}}) = -0.376661$. Since $|f(x_{\text{mid}})| > \varepsilon = 0.005$, the process is repeated with $i = 4$. The results are summarized as follows:

Iteration	$a^{(i)}$	$b^{(i)}$	$x_{\text{mid}} = \frac{a^{(i)} + b^{(i)}}{2}$	$f(x_{\text{mid}})$	Is $\lvert f(x_{\text{mid}})\rvert < \varepsilon$?
1	0.0	2.0	1.0	0.189491	no
2	0.0	1.0	0.5	0.020353	no
3	0.0	0.5	0.25	−0.376661	no
4	0.25	0.50	0.375	−0.141767	no
5	0.375	0.500	0.4375	−0.051935	no
6	0.4375	0.5000	0.46875	−0.013684	no
7	0.46875	0.50000	0.484375	0.003848	yes

It can be noted that the solution satisfies the convergence criterion, and hence, the root of Eq. (E1) can be taken as $x \approx 0.484375$. ◀

2.5 Newton–Raphson Method

The Newton–Raphson method, or simply the Newton's method, is a well-known and most powerful method used for finding the root of the equation $f(x) = 0$. The Newton's method can be derived by considering the Taylor's series expansion of the function $f(x)$ about an arbitrary point x_1 as

$$f(x) = f(x_1) + (x - x_1)f'(x_1) + \frac{1}{2!}(x - x_1)^2 f''(x_1) + \cdots, \tag{2.10}$$

where the function, f, and its derivatives, $f', f'', \ldots$ on the right-hand side of Eq. (2.10) are evaluated at x_1. By considering only the first two terms in the expansion, we have

$$f(x) \approx f(x_1) + (x - x_1)f'(x_1). \tag{2.11}$$

In order to find the root of $f(x) = 0$, we set $f(x)$ equal to zero in Eq. (2.11) to obtain

$$f(x_1) + (x - x_1)f'(x_1) = 0. \tag{2.12}$$

Since the higher order derivative terms were neglected in the approximation of $f(x)$ in Eq. (2.11), the solution of Eq. (2.12) yields the next approximation to the root (instead of the exact root) as

$$x = x_2 = x_1 - \frac{f(x_1)}{f'(x_1)}, \tag{2.13}$$

where x_2 denotes an improved approximation to the root. To further improve the root, we use x_2 in place of x_1 on the right-hand side of Eq. (2.13) to obtain x_3. This

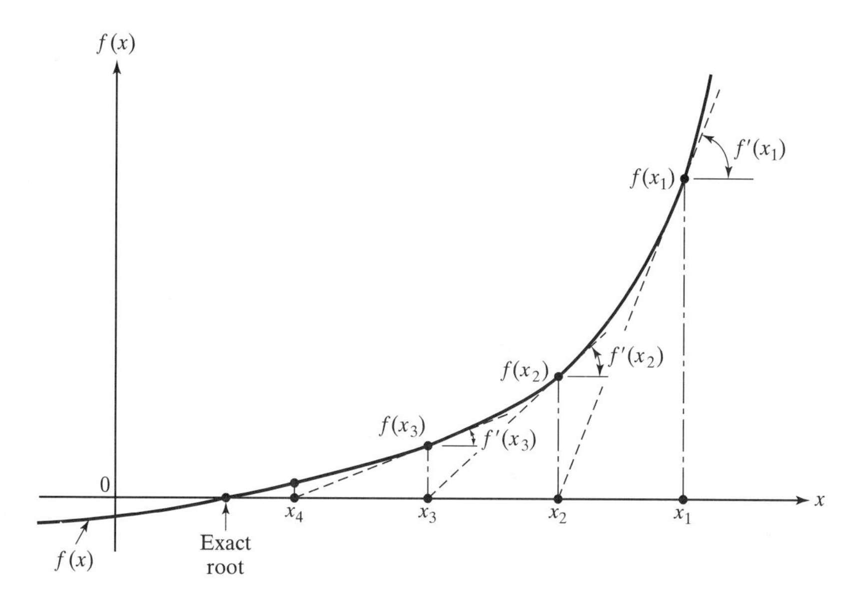

Figure 2.17 Newton's method.

iterative procedure can be generalized as

$$x_{i+1} = x_i - \frac{f(x_i)}{f'(x_i)} : i = 1, 2, \cdots. \tag{2.14}$$

The procedure is shown graphically in Fig. 2.17 assuming a real root for the equation $f(x) = 0$. If x_1 is the initial guess for the root of $f(x) = 0$, the point of intersection of the tangent to the curve at x_1 with the x axis gives the next approximation to the root, x_2. The convergence of the procedure to the exact root can also be seen in Fig. 2.17.

▶Example 2.8

Find the root of the equation

$$f(x) = \frac{1.5x}{(1+x^2)^2} - 0.65\tan^{-1}\left(\frac{1}{x}\right) + \frac{0.65x}{1+x^2} = 0 \tag{E1}$$

using Newton–Raphson method with the starting point $x_1 = 0.0$ and the convergence criterion $\varepsilon = 10^{-5}$.

Solution

The derivative of $f(x)$ can be found as

$$\frac{df(x)}{dx} = \frac{1.5}{(1+x^2)^2} - \frac{6x^2}{(1+x^2)^3} + \frac{0.65}{(1+x^2)} + \frac{0.65}{(1+x^2)} - \frac{1.3x^2}{(1+x^2)^2},$$

which can be simplified as

$$f'(x) = \frac{df(x)}{dx} = \frac{2.8 - 3.2x^2}{(1 + x^2)^3}. \tag{E2}$$

The iterative process of Newton–Raphson method is given by Eq. (2.14):

$$x_{i+1} = x_i - \frac{f(x_i)}{f'(x_i)}. \tag{E3}$$

At the starting point $x_1 = 0.0$, we have

$$x_1 = 0.0,\ f(x_1) = -1.021017, \text{ and } f'(x_1) = 2.800000,$$

and hence,

$$x_2 = x_1 - \frac{f(x_1)}{f'(x_1)} = 0.0 - \frac{(-1.021017)}{2.800000} = 0.364649.$$

Since $f(x_2) = -0.158413$ and $f'(x_2) = 1.632745$, Eq. (E3) gives

$$x_3 = x_2 - \frac{f(x_2)}{f'(x_2)} = 0.364649 - \frac{(-0.158413)}{1.632745} = 0.461671.$$

The values of f and f' at x_3 are determined as $f(x_3) = -0.021972$ and $f'(x_3) = 1.186268$, and Eq. (E3) yields

$$x_4 = x_3 - \frac{f(x_3)}{f'(x_3)} = 0.461671 - \frac{(-0.021972)}{1.186268} = 0.480193.$$

At x_4, the value of f and f' are given by

$$f(x_4) = -0.000742 \text{ and } f'(x_4) = 1.106571.$$

Equation (E3) yields the next approximation as

$$x_5 = x_4 - \frac{f(x_4)}{f'(x_4)} = 0.480193 - \frac{(-0.000742)}{1.106571} = 0.480864.$$

Since the value of $f(x_5) = f(0.480864) = -0.000001$ satisfies the convergence criterion $|f(x_5)| = 0.000001 < 10^{-5}$, x_5 is taken as the root of Eq. (E1). ◀

Notes

1. Newton's method requires the derivative of the function, $f' = \frac{df}{dx}$. In many cases, such as polynomial equations, the derivative of $f(x)$ can be obtained easily. However, in some problems, such as those involving transcendental functions, the differentiation of the function $f(x)$ may be quite complicated. In some cases, the function $f(x)$ may not be available in explicit form. In such cases, the derivative of $f(x)$ can be obtained numerically by using a finite difference approximation. (See Chapter 7.)
2. Newton's method converges very fast in most cases. However, it may not converge, if the initial guess (x_1) is very far from the exact root. Also, the method may not converge, if the value of the derivative is close to zero or varies substantially near the root. A few instances in which Newton's method

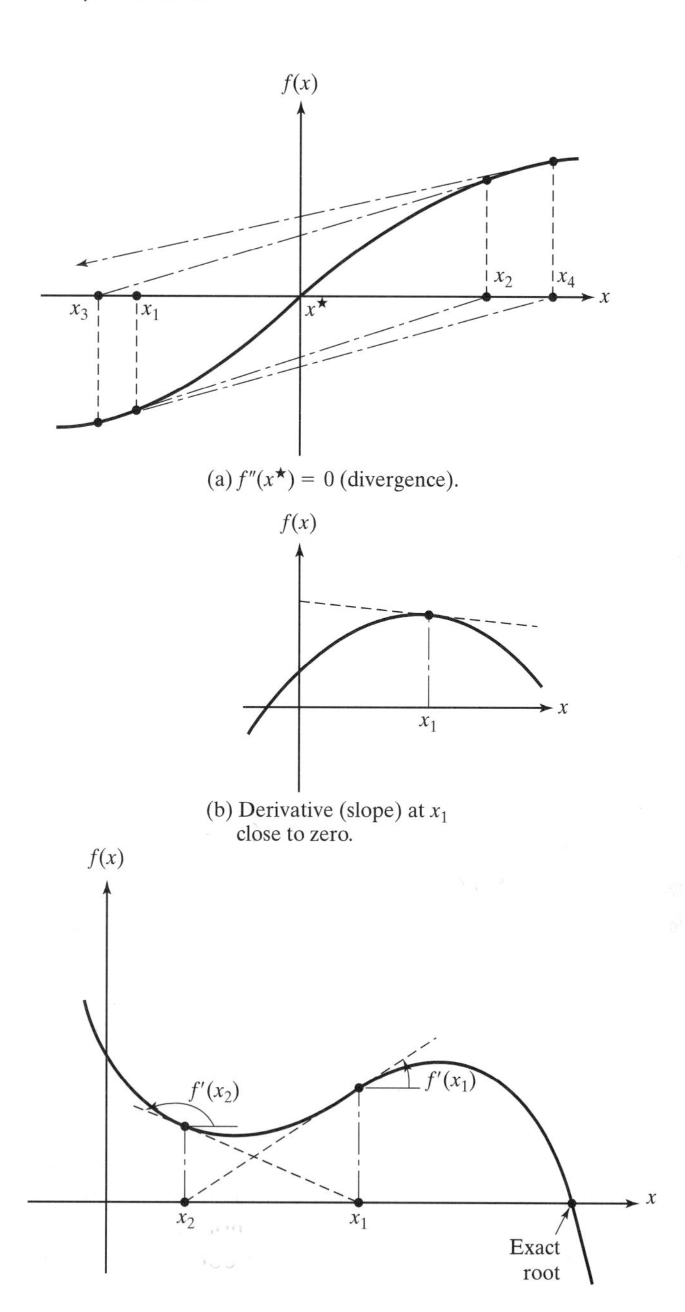

Figure 2.18 Non-convergence of Newton's method.

does not converge are illustrated in Fig. 2.18. These difficulties can usually be corrected by starting from a new initial guess point. The following convergence criteria can be used for stopping the iterative procedure of Eq. (2.14):

$$|x_i - x_{i-1}| \leq \varepsilon, \tag{2.15}$$

$$\left|\frac{x_i - x_{i-1}}{x_i}\right| \leq \varepsilon; \quad x_i \neq 0, \tag{2.16}$$

and

$$|f(x_i)| \leq \varepsilon. \tag{2.17}$$

3. Newton's method can also be used for finding complex roots. In this case, we need to perform the computations using complex arithmetic with a complex number as the initial guess for the root.

▶Example 2.9

Find the root of the equation $f(x) = \tan^{-1} x = 0$ using the starting points $x_1 = 1.6$ and $x_1 = 0.2$.

Solution

The sequence of approximations generated by the Newton–Raphson method from the starting point $x_1 = 1.6$ is given by $x_1 = 1.6, x_2 = -2.0034213, x_3 = 3.5509133, x_4 = -14.090254, x_5 = 285.20148, x_6 = -127199.55, x_7 = 2.5414801 \times 10^{10}, \ldots$, which can be seen to be diverging. On the other hand, when the starting point $x_1 = 0.2$ (which is closer to the root $x^* = 0.0$) is used, the Newton–Raphson method converged as follows: $x_1 = 0.2, x_2 = -0.00529137, x_3 = 0.98720193 \times 10^{-7}$, and $x_4 = 0.0$.

This example indicates that, when $f''(x^*) = 0$, the Newton–Raphson method might diverge, unless the starting point is sufficiently close to the root. ◀

▶Example 2.10

Find the root of the equation $f(x) = 5x^3 - 39x^2 + 22x - 60 = 0$ using the starting points $x_1 = 0.95$ and $x_1 = 0.0$.

Solution

The Newton–Raphson method produced the following sequence of approximations from the first starting point: $x_1 = 0.95$, $x_2 = 1.5569093$, $x_3 = -1.0405447$, $x_4 = 0.03627062$, $x_5 = 0.75702357$, $x_6 = 1.3309088$, $x_7 = 3.0455060$, $x_8 = 1.1724654$, $x_9 = 2.0151870, \ldots$ It can be seen that the method never converged. However, the use of the second starting point produced the root, $x^* = 5.0$. The convergence can be seen from the results: $x_1 = 0.0$, $x_2 = 0.73170733$, $x_3 = 1.3058259$, $x_4 = 2.75444658$, $x_5 = 1.4133221$, $x_6 = 5.9946117$, $x_7 = 5.2962012$, $x_8 = 5.0381327$, $x_9 = 5.0007577$, and $x_{10} = 5.0000005$. ◀

▶Example 2.11

Find the root of the equation $f(x) = x^3 - x - 3 = 0$ using the starting points $x_1 = 0.0$ and $x_1 = 0.5$.

Solution

The following sequence of approximations were generated by the Newton–Raphson method from $x_1 = 0.0$: $x_1 = 0.0$, $x_2 = -3.0$, $x_3 = -1.9615384$, $x_4 = -1.1471759$, $x_5 = -0.00657928$, $x_6 = -3.0003891$, $x_7 = -1.9618182$, $x_8 = -1.1474303$, $x_9 = -0.00725639$, $x_{10} = -3.0004730, \ldots$ It can be seen that the method has not converged; rather the method resulted in a cyclic process with $x_{i+4} \approx x_i (i = 1, 2, 3, \ldots)$. In such a case, a different starting point is likely to result in the convergence of the process. The use of $x_1 = 0.5$ led to the following sequence of approximations: $x_1 = 0.5$, $x_2 = -13.0$, $x_3 = -8.6778660$, $x_4 = -5.7976274$, $x_5 = -3.8737497$, $x_6 = -2.5730152$, $x_7 = -1.6472325$, $x_8 = -0.83179486$, $x_9 = 1.7189554$, $x_{10} = 1.6731508$, $x_{11} = 1.6717013$, $x_{12} = 1.6716999$, and $x_{13} = 1.6716999$. This indicates the root as $x^* = 1.6716999$. ◀

▶Example 2.12

Find the root of the equation $f(x) = \sin 2x = 0$ using the starting point $x_1 = 0.75$.

Solution

The roots of $f(x)$ nearest to the starting point x_1 are 0.0 and $\frac{\pi}{2} \approx 1.5708$. However, the Newton–Raphson method converged to a root far from x_1 as follows: $x_1 = 0.75$, $x_2 = -6.3007$, $x_3 = -6.2832$, and $x_4 = -6.2832 \approx -2\pi$. This example illustrates that when $f'(x_1)$ is small (that is, when the tangent to the curve $f(x)$ is nearly horizontal), the Newton–Raphson method might converge to a root farther than the one closer to the starting point. Thus, it is a good practice to verify that the root given by the Newton–Raphson method is the desired one. ◀

▶Example 2.13

Find the root of the equation $f(x) = x^2 + x + 5 = 0$ using the starting points $x_1 = 1.0$ and $x_1 = 1 + i$.

Solution

When the first starting point was used, the Newton–Raphson method gave $x_1 = 1.0$, $x_2 = -1.3333$, $x_3 = 1.9333$, $x_4 = -0.2594$, $x_5 = -10.2492$, $x_6 = -5.1310$, $x_7 =$

$-2.3026, \ldots$ with no convergence. The exact roots of the equation are given by $-0.5 \pm 2.1794i$. By using the complex number as a starting point ($x_1 = 1 + i$), the Newton–Raphson method converged to the correct root, $x^* = -0.5 + 2.179449i$, as follows: $x_1 = 1 + i$, $x_2 = -0.8461539 + 1.230769i$, $x_3 = -0.1701357 + 2.403620i$, $x_4 = -0.4681635 + 2.171638i$, $x_5 = -0.5001113 + 2.179229i$, and $x_6 = -0.5 + 2.179449i$. ◀

2.6 Secant Method

The secant method is similar to the Newton's method, but is different in that the derivative f' is approximated by using two consecutive iterative values of f. By using an approximation for the derivative $f'(x_i)$ as

$$f'(x_i) \approx \frac{f(x_i) - f(x_{i-1})}{x_i - x_{i-1}}, \tag{2.18}$$

Eq. (2.14) can be rewritten as

$$x_{i+1} = x_i - \frac{f(x_i)}{f'(x_i)} = x_i - f(x_i)\left[\frac{x_i - x_{i-1}}{f(x_i) - f(x_{i-1})}\right], i = 2, 3, 4, \ldots. \tag{2.19}$$

Thus, Eq. (2.19) represents the general expression for the iterative process of the secant method. Notice that two initial guesses x_1 and x_2 are required to start the iterative process according to Eq. (2.19). The graphical illustration of the secant method and its convergence is shown in Fig. 2.19. It can be noted that geometrically

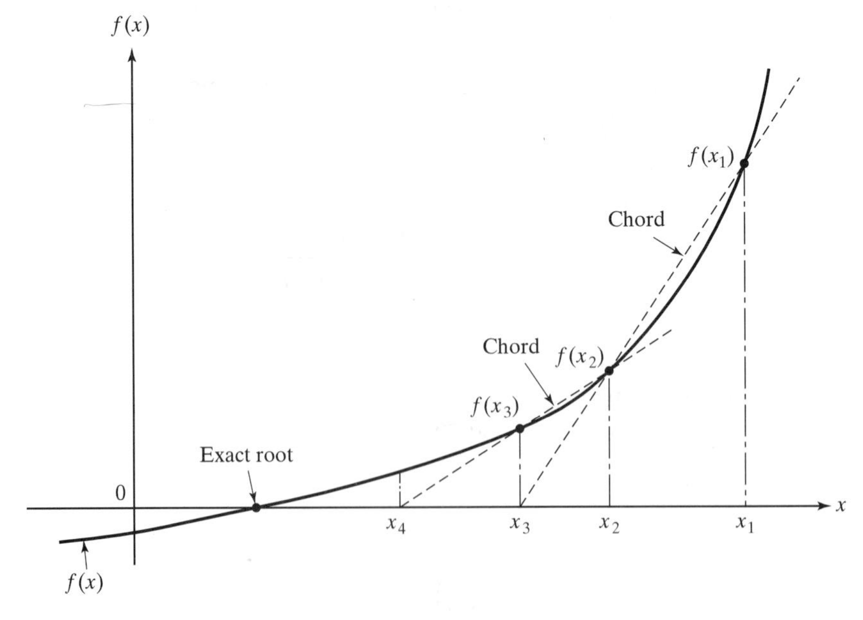

Figure 2.19 Convergence of secant method.

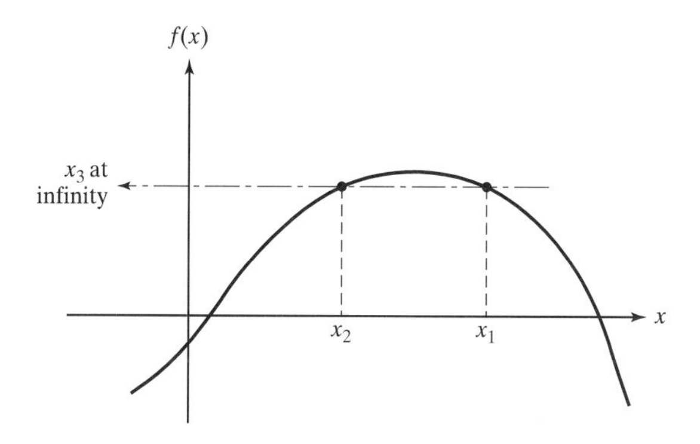

Figure 2.20 Nonconvergence of secant method.

the intersection of the chord joining the points $(x_i, f(x_i))$ and $(x_{i-1}, f(x_{i-1}))$ with the x axis is taken as the next approximation in the secant method.

The iterative process can be stopped whenever any or all of the convergence criteria, given by Eqs. (2.15) to (2.17), are satisfied. The process may not converge, if $f(x_1) \approx f(x_2)$ as shown in Fig. 2.20. In such a case, the next approximation (x_3) will be near infinity. Another difficulty might arise when two consecutive values of x_{i-1} and x_i become very close, thereby causing $f(x_{i-1})$ and $f(x_i)$ also to be very close, and ultimately resulting in a significant round-off error in Eq. (2.19). This difficulty can be avoided by fixing or freezing the values of x_{i-1} and $f(x_{i-1})$ whenever $|f(x_{i+1})|$ becomes smaller than a prescribed number, ε. This means that Eq. (2.19) is modified as

$$x_{i+k} = x_{i+k-1} - f(x_{i+k-1})\left[\frac{x_{i+k-1} - x_{i-1}}{f(x_{i+k-1}) - f(x_{i-1})}\right]; k = 2, 3, 4, \ldots, \tag{2.20}$$

if $|f(x_{i+k})| \le \varepsilon$.

The following iterative process can be used to implement the secant method:

1. Start with two initial approximations x_1 and x_2 for the root of $f(x) = 0$ and a small number ε to test the convergence of the process. Set $i = 2$.
2. Find the new approximation, x_{i+1}, as
$$x_{i+1} = x_i - \frac{f(x_i)(x_i - x_{i-1})}{f(x_i) - f(x_{i-1})}.$$
3. Verify the convergence of the process. If
$$|f(x_{i+1})| \le \varepsilon,$$
stop the process by taking x_{i+1} as the root. Otherwise, update the iteration number as $i = i + 1$ and go to step 2.

▶Example 2.14

Find the root of the equation

$$f(x) = \frac{1.5x}{(1 + x^2)^2} - 0.65 \tan^{-1}\left(\frac{1}{x}\right) + \frac{0.65x}{(1 + x^2)} = 0$$

using 0.0 and 0.5 as the initial approximations with $\varepsilon = 10^{-5}$.

Solution

At the initial approximate points, the function values are given by

$$f(x_1) = f(0.0) = -1.02101707 \text{and} f(x_2) = f(0.5) = 0.02035314.$$

Since $|f(0.0)| > \varepsilon$ and $|f(0.5)| > \varepsilon$, we proceed to find the next approximation using Eq. (2.19):

$$x_3 = x_2 - \frac{f(x_2)(x_2 - x_1)}{f(x_2) - f(x_1)}$$

$$= 0.5 - \frac{(0.02035314)(0.5 - 0.0)}{0.02035314 - (-1.02101707)} = 0.49022764.$$

The value of $f(x_3)$ is found to be

$$|f(x_3)| = |f(0.49022764)| = |0.01015013| > \varepsilon,$$

and hence, we compute the next approximate value as

$$x_4 = x_3 - \frac{f(x_3)(x_3 - x_2)}{f(x_3) - f(x_2)}$$

$$= 0.49022764 - \frac{(0.01015013)(0.49022764 - 0.5)}{0.01015013 - 0.02035314} = 0.48540229.$$

The value of $f(x_4)$ is found as

$$|f(x_4)| = |f(0.48540229)| = |0.00496507| > \varepsilon.$$

Hence, we proceed to the next iteration. The results of the iterative process are as follows:

Value of i	Point (x_i)	$f(x_i)$	Is $\lvert f(x_i)\rvert \le \varepsilon$?	x_{i+1} given by Eq. (2.19)
1	0.0	−1.02101707	no	—
2	0.5	0.02035314	no	0.49022764
3	0.49022764	0.01015013	no	0.48540229
4	0.48540229	0.00496507	no	0.48305339
5	0.48305339	0.00240600	no	0.48191774
6	0.48191774	0.00116086	no	0.48137063
7	0.48137063	0.0055802	no	0.48110777
8	0.48110777	0.00026876	no	0.48098111
9	0.48098111	0.00012833	no	0.48092049
10	0.48092049	0.00006276	no	0.48089087
11	0.48089087	0.00002861	no	0.48087740
12	0.48087740	0.00001460	no	0.48087072
13	0.48087072	0.00000721	yes	—

◀

2.7 Regula Falsi Method

The regula falsi method (also known as the method of false position or the method of linear interpolation) is based on linear interpolation. The method is similar to the bisection method, in that the initial interval (a, b) is generated so as to bracket a root of the equation $f(x) = 0$. However, instead of continuously bisecting the interval, the linear interpolation of $f(x)$ passing through the two end points of the current interval of uncertainty is used to find a new approximation for the root. Thus, if $x_1 = a$ and $x_2 = b$ denote two points on either side of the exact root, a linear equation (straight line) is assumed through the points $[x_1, f(x_1)]$ and $[x_2, f(x_2)]$ as

$$y(x) = px + q, \tag{2.21}$$

where $y = f(x_1)$ at $x = x_1$ and $y = f(x_2)$ at $x = x_2$. This yields

$$y(x) = \left[\frac{f(x_2) - f(x_1)}{x_2 - x_1}\right] x + \left\{\frac{x_2 f(x_1) - x_1 f(x_2)}{x_2 - x_1}\right\}. \tag{2.22}$$

The point of intersection of this straight line, $y(x)$, with the x axis, x_3, can be found by setting $y(x)$, given by Eq. (2.22), to zero:

$$y(x_3) = \left[\frac{f(x_2) - f(x_1)}{(x_2 - x_1)}\right] x_3 + \left\{\frac{x_2 f(x_1) - x_1 f(x_2)}{x_2 - x_1}\right\} = 0. \tag{2.23}$$

This gives the new approximation to the root (x_3) as

$$x_3 = x_2 - \frac{f(x_2)(x_2 - x_1)}{f(x_2) - f(x_1)}. \tag{2.24}$$

It can be seen that x_3 will be closer to the exact root, x^*, than either x_1 or x_2. Equation (2.24) can be generalized as

$$x_{i+1} = x_i - \frac{f(x_i)(x_i - x_{i-1})}{f(x_i) - f(x_{i-1})}. \tag{2.25}$$

For the next iteration, x_{i+1} replaces x_i if $f(x_{i+1})f(x_i) > 0$ or x_{i-1} if $f(x_{i+1})f(x_{i-1}) > 0$. The convergence of the process is shown in Fig. 2.21. The iterative process can be stopped whenever the convergence criteria

$$|f(x_i)| \leq \varepsilon \tag{2.26}$$

and

$$|x_i - x_{i-1}| \leq \varepsilon \tag{2.27}$$

are satisfied. In these equations, ε is a small number on the order of 10^{-3} to 10^{-6}.

Notes

1. In some cases, the convergence of the regula falsi method may be very slow as indicated in Fig. 2.21. Here one of the end points of interpolation (x_1) does not change from the original end point (such a point is called a stagnation point). Assuming that x_1 is the stagnation point and x_i is the current approximation

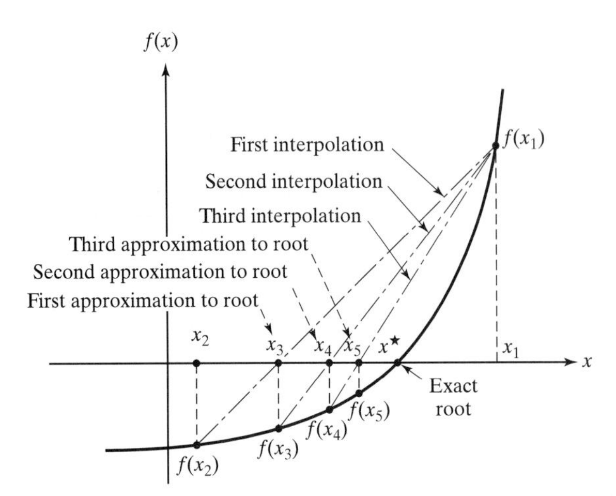

Figure 2.21 Slow convergence of regula falsi method.

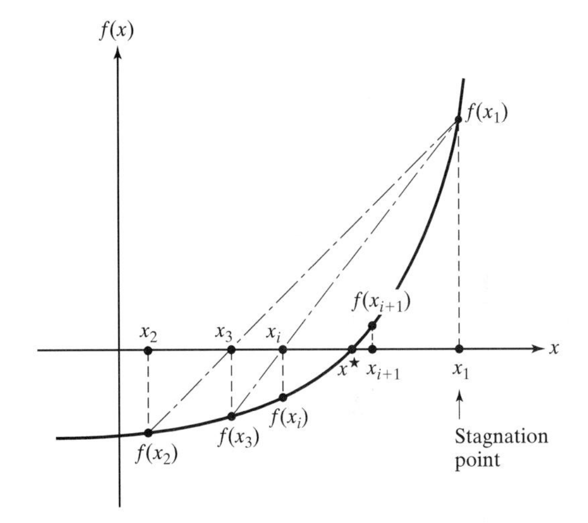

Figure 2.22 Improvement of stagnation point. (x_{i+1} is determined as middle point of x_1 and x_i, $x_{i+1} = \frac{x_1+x_i}{2}$.)

to the root, the convergence of the process can be improved by taking the next value, x_{i+1}, as $\left(\frac{x_1+x_i}{2}\right)$. This procedure is illustrated in Fig. 2.22.

2. Equation (2.25) can be seen to be identical to Eq. (2.19) of the secant method. However, the regula falsi method converges in cases where the secant method might diverge. Since the estimates x_i and x_{i+1} in the regula falsi method are selected such that the values $f(x_i)$ and $f(x_{i+1})$ have opposite signs, the method converges all the time (Fig. 2.23(a)). In the secant method, on the other hand, the current estimates, x_i and x_{i+1}, may not bracket the minimum (as shown in Fig. 2.23(b)). Hence, the method may diverge in some cases.

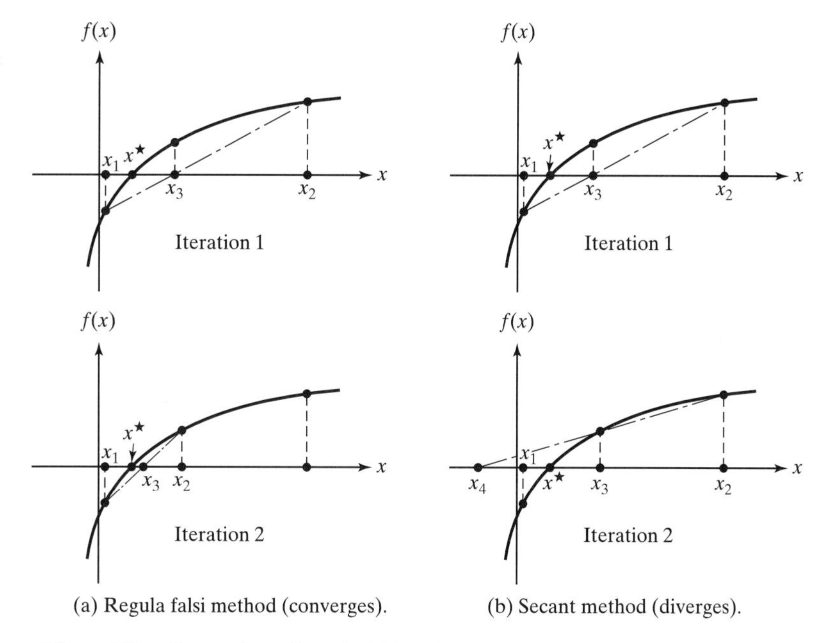

(a) Regula falsi method (converges). (b) Secant method (diverges).

Figure 2.23 Comparison of regula falsi and secant methods.

2.8 Fixed Point Iteration or Successive Substitution Method

In this method, the equation $f(x) = 0$ is rewritten in the form

$$x = g(x), \tag{2.28}$$

and an iterative procedure is adopted using the relation

$$x_{i+1} = g(x_i); i = 1, 2, 3, \cdots, \tag{2.29}$$

where a new approximation to the root, x_{i+1}, is found using the previous approximation, x_i (x_1 denotes the initial guess). The iterative process can be stopped whenever the convergence criterion

$$|x_{i+1} - g(x_{i+1})| \leq \varepsilon \tag{2.30}$$

is satisfied, where ε is a small number on the order of 10^{-3} to 10^{-6}. The method is very simple; however, it may not always converge with an arbitrarily chosen form of the function $g(x)$. The condition to be satisfied for convergence to the correct root is given by [2.14]

$$|g'(x)| < 1 \tag{2.31}$$

in the neighborhood of the correct root. It has been observed that convergence of the process is asymptotic if $0 < g'(x) < 1$ and oscillatory if $-1 < g'(x) < 0$.

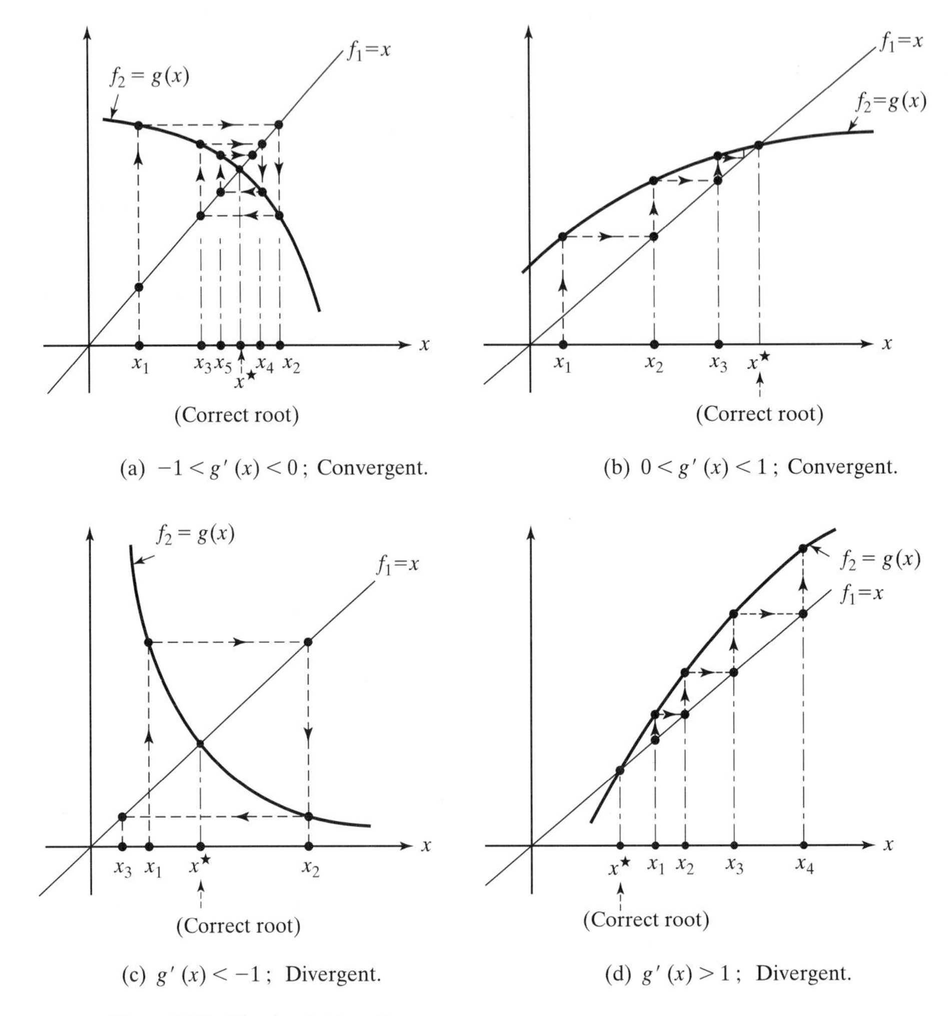

Figure 2.24 Fixed point iteration.

In addition, the convergence will be faster as $|g'(x)| \to 0$ in the neighborhood of the correct root. The convergence of the process in different cases is illustrated in Fig. 2.24.

▶Example 2.15

Find the root of the equation

$$f(x) = \frac{1.5x}{(1+x^2)^2} - 0.65\tan^{-1}\left(\frac{1}{x}\right) + \frac{0.65x}{1+x^2} = 0 \tag{E1}$$

using the fixed point iteration with $x_1 = 0.0$ and $\varepsilon = 10^{-5}$.

Solution

Equation (E1) can be rearranged as

$$x = g(x),$$

where

$$g(x) = \frac{13}{30}(1 + x^2)^2 \tan^{-1}\left(\frac{1}{x}\right) - \frac{13}{30}x(1 + x^2).$$

Thus, the iterative process can be expressed as

$$x_{i+1} = g(x_i).$$

At the starting point, $x_1 = 0.0$, the value of g can be found as $g(0.0) = 0.606535$. Since $x_1 \neq g(x_1)$, we set $x_2 = g(x_1) = 0.606535$ and find $g(x_2) = 0.472035$. Again $x_2 \neq g(x_2)$, and hence we proceed to the next iteration with $x_3 = 0.472035$. The results of the iterative process are as follows:

Iteration number (i)	x_i	$g(x_i)$	Is $\lvert x_i - g(x_i)\rvert \leq \varepsilon$?
1	0.100000	0.606535	no
2	0.606535	0.472035	no
3	0.472035	0.481914	no
4	0.481914	0.480744	no
5	0.480744	0.480878	no
6	0.480878	0.480863	no
7	0.480863	0.480864	yes

It can be seen that the specified convergence criterion is satisfied at the end of seventh iteration; as such, the root of Eq. (E1) is taken as $x_7 = 0.480863$. ◀

2.9 Determination of Multiple Roots

A function $f(x)$ is said to have a root x^* of multiplicity $p > 1$ if

$$f(x) = (x - x^*)^p g(x), \tag{2.32}$$

where $g(x)$ is continuous at $x = x^*$ with $g(x^*) \neq 0$. If p is an integer, Eq. (2.32) gives

$$f(x^*) = f'(x^*) = \cdots = f^{(p-1)}(x^*) = 0;\ f^{(p)}(x^*) \neq 0. \tag{2.33}$$

The function $f(x)$ does not change sign at x^*, if x^* is an even multiple root (p = even number). Since the function value $f(x)$ and its derivative $f'(x)$ are zero at a multiple root x^*, the Newton–Raphson and secant methods, which contain derivatives in the denominator, cannot be applied conveniently to find the root x^*.

It was shown [2.18] that the Newton–Raphson and secant methods converge only linearly for multiple roots. However, a slightly modified formula, given by Eq. (2.34), has been shown to lead to quadratic convergence,

$$x_{i+1} = x_i - p\frac{f(x_i)}{f'(x_i)}, \tag{2.34}$$

where p denotes the multiplicity of the root ($p = 2$ for double root, $p = 3$ for triple root, etc.). Equation (2.34) is not very practical since its use requires a prior knowledge of the multiplicity of the root.

Another approach has been suggested [2.18] for modifying Newton–Raphson method to find multiple roots of a function. In this approach, we consider the function

$$g(x) = \frac{f(x)}{f'(x)}. \tag{2.35}$$

The function $g(x)$ can be shown to have all its roots at the same locations as those of $f(x)$. Differentiation of $g(x)$ gives

$$g'(x) = \frac{f'(x)f'(x) - f(x)f''(x)}{\{f'(x)\}^2}. \tag{2.36}$$

The use of $g(x)$ in place of $f(x)$ in Eq. (2.14) leads to the desired formula:

$$x_{i+1} = x_i - \frac{g(x_i)}{g'(x_i)}. \tag{2.37}$$

▶Example 2.16

Find the roots of the equation

$$f(x) = (x-2)^2(x-5) = x^3 - 9x^2 + 24x - 20 = 0 \tag{E1}$$

starting from $x_1 = 0.5$ using Eqs. (2.14), (2.34) with $p = 2$, and (2.37).

Solution

Noting that

$$f'(x) = 3x^2 - 18x + 24 \tag{E2}$$

and

$$f''(x) = 6x - 18, \tag{E3}$$

Eqs. (2.14), (2.34), and (2.37) can be expressed as

$$x_{i+1} = x_i - \left(\frac{x_i^3 - 9x_i^2 + 24x_i - 20}{3x_i^2 - 18x_i + 24}\right), \tag{E4}$$

$$x_{i+1} = x_i - 2\left(\frac{x_i^3 - 9x_i^2 + 24x_i - 20}{3x_i^2 - 18x_i + 24}\right), \tag{E5}$$

and

$$x_{i+1} = x_i - \left\{ \frac{(x_i^3 - 9x_i^2 + 24x_i - 20)(3x_i^2 - 18x_i + 24)}{(3x_i^2 - 18x_i + 24)^2 - (x_i^3 - 9x_i^2 + 24x_i - 20)(6x_i - 18)} \right\}$$

$$= x_i - \left(\frac{3x_i^5 - 45x_i^4 + 258x_i^3 - 708x_i^2 + 936x_i - 480}{3x_i^4 - 36x_i^3 + 162x_i^2 - 312x_i + 216} \right). \qquad \text{(E6)}$$

Using $x_1 = 0.5$, the sequence of approximations produced by Eqs. (E4), (E5), and (E6) are shown in the following table.

Iteration i	Eq. (2.14) or (E4)	Eq. (2.34) or (E5)	Eq. (2.37) or (E6)
1	0.5	0.5	0.5
2	1.1428571	1.7857143	2.1578946
3	1.5285714	1.9930871	2.0044441
4	1.7492974	1.9998698	1.5044441
5	1.8699937	2.0047598	2.0298657
6	1.9336742	—	1.9989179
7	1.9664814		—
8	1.9831409		
9	1.9915178		
10	1.9958467		
11	1.9979852		
12	1.9990886		
13	1.9994373		
14	2.0000024		

It can be seen that the modified Newton–Raphson methods, Eqs. (2.34) and (2.37), are more efficient in finding the multiple root compared with the standard Newton–Raphson method, Eq. (2.14). ◀

2.10 Bairstow's Method

Bairstow's method is an iterative method in which a quadratic factor of the polynomial is found in each stage. When the procedure is applied repeatedly to the deflated polynomials, all the quadratic factors of the polynomial can be determined. Consider an nth degree polynomial equation,

$$f(x) = a_n x^n + a_{n-1} x^{n-1} + a_{n-2} x^{n-2} + \cdots + a_2 x^2 + a_1 x + a_0. \tag{2.38}$$

By factoring an arbitrary quadratic term, $f(x)$ can be expressed as

$$f(x) = (x^2 + \alpha x + \beta)(b_n x_{n-2} + b_{n-1} x^{n-3} + \cdots + b_4 x^2 + b_3 x + b_2) + R(x), \tag{2.39}$$

where $R(x)$ is the remainder. The remainder term can be found to be of the form

$$R(x) = b_1 x + b_0. \tag{2.40}$$

By multiplying the quadratic term and the (n-2)nd degree polynomial in Eq. (2.39) and equating the coefficients of like powers of x, we obtain the following system of equations:

$$\begin{aligned} a_n &= b_n, \\ a_{n-1} &= b_{n-1} + \alpha b_n, \\ a_{n-2} &= b_{n-2} + \alpha b_{n-1} + \beta b_n, \\ &\vdots, \\ a_2 &= b_2 + \alpha b_3 + \beta b_4, \\ a_1 &= b_1 + \alpha b_2 + \beta b_3, \end{aligned}$$

and

$$a_0 = b_0 + \beta b_2. \tag{2.41}$$

Equations (2.41) can be solved to find the coefficients $b_n, b_{n-1}, \ldots, b_1, b_0$ as

$$\begin{aligned} b_n &= a_n, \\ b_{n-1} &= a_{n-1} - \alpha b_n, \\ b_{n-2} &= a_{n-2} - \alpha b_{n-1} - \beta b_n, \\ &\vdots, \\ b_2 &= a_2 - \alpha b_3 - \beta b_4, \\ b_1 &= a_1 - \alpha b_2 - \beta b_3, \end{aligned}$$

and

$$b_0 = a_0 - \beta b_2. \tag{2.42}$$

If we require $(x^2 + \alpha x + \beta)$ to be an exact quadratic factor of $f(x)$, then the remainder, $R(x)$, given by Eq. (2.40) should be zero. This requires that

$$b_0 = 0; b_1 = 0. \tag{2.43}$$

It can be seen from Eq. (2.42) that b_0 and b_1 are functions of α and β. Thus, if $\bar{\alpha}$ and $\bar{\beta}$ denote the correct values of α and β, Eq. (2.43) gives

$$b_0(\bar{\alpha}, \bar{\beta}) = 0; \; b_1(\bar{\alpha}, \bar{\beta}) = 0. \tag{2.44}$$

If α and β denote initial approximations of $\bar{\alpha}$ and $\bar{\beta}$, we seek their improved values as $\alpha + d\alpha$ and $\beta + d\beta$, which drive Eqs. (2.44) closer to zero. By expanding b_0 and b_1 in Taylor's series about α and β, we have

$$b_0(\alpha + d\alpha, \beta + d\beta) = b_0(\alpha, \beta) + \Delta\alpha\left(\frac{\partial b_0}{\partial \alpha}\right) + \Delta\beta\left(\frac{\partial b_0}{\partial \beta}\right) \tag{2.45}$$

and

$$b_1(\alpha + d\alpha, \beta + d\beta) = b_1(\alpha, \beta) + \Delta\alpha\left(\frac{\partial b_1}{\partial \alpha}\right) + \Delta\beta\left(\frac{\partial b_1}{\partial \beta}\right), \tag{2.46}$$

where the partial derivatives are evaluated at α and β, and $\Delta\alpha$ and $\Delta\beta$ are given by

$$\Delta\alpha = \bar{\alpha} - \alpha$$

and

$$\Delta\beta = \bar{\beta} - \beta. \tag{2.47}$$

By setting Eqs. (2.45) and (2.46) to zero (assuming that $\alpha + d\alpha \approx \bar{\alpha}$ and $\beta + d\beta \approx \bar{\beta}$), we obtain

$$\Delta\alpha\left(\frac{\partial b_0}{\partial \alpha}\right) + \Delta\beta\left(\frac{\partial b_0}{\partial \beta}\right) = -b_0(\alpha, \beta)$$

and

$$\Delta\alpha\left(\frac{\partial b_1}{\partial \alpha}\right) + \Delta\beta\left(\frac{\partial b_1}{\partial \beta}\right) = -b_1(\alpha, \beta). \tag{2.48}$$

It can be seen that the evaluation of the partial derivatives of b_0 and b_1 in Eqs. (2.48) requires a recursive calculation of the partial derivatives of all the $b_i (i = n, n-1, \ldots, 1, 0)$ in Eq. (2.42):

$$\frac{\partial b_n}{\partial \alpha} = 0;$$

$$\frac{\partial b_n}{\partial \beta} = 0;$$

$$\frac{\partial b_{n-1}}{\partial \alpha} = -b_n - \alpha\frac{\partial b_n}{\partial \alpha} = -b_n;$$

$$\frac{\partial b_{n-1}}{\partial \beta} = -\alpha\frac{\partial b_n}{\partial \beta} = 0;$$

$$\frac{\partial b_{n-2}}{\partial \alpha} = -b_{n-1} - \alpha\frac{\partial b_{n-1}}{\partial \alpha} - \beta\frac{\partial b_n}{\partial \alpha} = -b_{n-1} + \alpha b_n;$$

$$\frac{\partial b_{n-2}}{\partial \beta} = -\alpha\frac{\partial b_{n-1}}{\partial \beta} - b_n - \beta\frac{\partial b_n}{\partial \beta} = -b_n;$$

$$\vdots$$

$$\frac{\partial b_2}{\partial \alpha} = -b_3 - \alpha\frac{\partial b_3}{\partial \alpha} - \beta\frac{\partial b_4}{\partial \alpha};$$

$$\frac{\partial b_2}{\partial \beta} = -\alpha\frac{\partial b_3}{\partial \beta} - b_4 - \beta\frac{\partial b_4}{\partial \beta};$$

$$\frac{\partial b_1}{\partial \alpha} = -b_2 - \alpha\frac{\partial b_2}{\partial \alpha} - \beta\frac{\partial b_3}{\partial \alpha};$$
$$\frac{\partial b_1}{\partial \beta} = -\alpha\frac{\partial b_2}{\partial \beta} - b_3 - \beta\frac{\partial b_3}{\partial \beta};$$
$$\frac{\partial b_0}{\partial \alpha} = -\beta\frac{\partial b_2}{\partial \alpha};$$
$$\frac{\partial b_0}{\partial \beta} = -b_2 - \beta\frac{\partial b_2}{\partial \beta}. \tag{2.49}$$

Here all the partial derivatives are evaluated at (α, β). Once the partial derivatives of b_0 and b_1 with respect to α and β are evaluated, Eqs. (2.48) can be solved for $\Delta\alpha$ and $\Delta\beta$ as

$$\Delta\alpha = \frac{b_1\left(\frac{\partial b_0}{\partial \beta}\right) - b_0\left(\frac{\partial b_1}{\partial \beta}\right)}{\left(\frac{\partial b_0}{\partial \alpha}\frac{\partial b_1}{\partial \beta} - \frac{\partial b_1}{\partial \alpha}\frac{\partial b_0}{\partial \beta}\right)} \tag{2.50}$$

and

$$\Delta\beta = \frac{b_0\left(\frac{\partial b_1}{\partial \alpha}\right) - b_1\left(\frac{\partial b_0}{\partial \alpha}\right)}{\left(\frac{\partial b_1}{\partial \beta}\frac{\partial b_0}{\partial \alpha} - \frac{\partial b_0}{\partial \beta}\frac{\partial b_1}{\partial \alpha}\right)}, \tag{2.51}$$

Once $\Delta\alpha$ and $\Delta\beta$ are known, new approximations to the correct values of $\bar{\alpha}$ and $\bar{\beta}$ are computed as $\alpha + \Delta\alpha$ and $\beta + \Delta\beta$, respectively.

The quadratic factor $(x^2+\alpha x+\beta)$ of the polynomial $f(x)$ and the corresponding roots are computed iteratively as follows:

1. Choose initial approximations α and β to $\bar{\alpha}$ and $\bar{\beta}$, respectively.
2. Compute the coefficients b_1 and b_0 using Eqs. (2.42).
3. Compute the partial derivatives of b_1 and b_0 with respect to α and β using Eqs. (2.49).
4. Solve Eqs. (2.50) and (2.51) for $\Delta\alpha$ and $\Delta\beta$.
5. Compute the new approximations for $\bar{\alpha}$ and $\bar{\beta}$ as $\alpha + \Delta\alpha$ and $\beta + \Delta\beta$.
6. By using the new values of α and β computed in step 5, repeat steps 2 through 5 until convergence occurs (i.e., until the values of $\Delta\alpha$ and $\Delta\beta$ are found to be very small).
7. Using the converged values of α and β for $\bar{\alpha}$ and $\bar{\beta}$, the roots of $f(x)$ from the quadratic factor are computed as

$$x^2 + \bar{\alpha}x + \bar{\beta} = 0; x = \frac{-\bar{\alpha} \pm \sqrt{(\bar{\alpha})^2 - 4\bar{\beta}}}{2}. \tag{2.52}$$

Next the reduced polynomial of degree $(n-2)$ is considered to obtain another quadratic factor. The procedure is continued until all the roots of $f(x) = 0$ are found.

The Bairstow's method can be used to determine the real, complex, or repeated roots of a polynomial equation. However, often it is difficult to select good initial approximations of α and β. Another disadvantage with the method is that the roots found may not be very accurate.

▶Example 2.17

Find the roots of the polynomial equation

$$f(x) = x^4 - 10x^3 + 35x^2 - 50x + 24 = 0$$

using Bairstow's method. Assume the initial approximations of $\bar{\alpha}$ and $\bar{\beta}$ as $\alpha = -2.0$ and $\beta = 1.0$.

Solution

The coefficients of the polynomial equation are given by

$$a_4 = 1.0, a_3 = -10.0, a_2 = 35.0, a_1 = -50, \text{ and } a_0 = 24.0.$$

The coefficients $b_0, b_1, \cdots$ can be computed from Eqs. (2.42) as

$$b_4 = 1.0, b_3 = -8.0, b_2 = 18.0, b_1 = -6.0, \text{ and } b_0 = 6.0.$$

Equations (2.49) can be used to compute the partial derivatives of b_i with respect to α and β as

$$\frac{\partial b_i}{\partial \alpha}, i = 0, 1, 2, 3, 4 : -6.0, -5.0, 6.0, -1.0, 1.0$$

and

$$\frac{\partial b_i}{\partial \beta}, i = 0, 1, 2, 3, 4 : -17.0, 6.0, -1.0, 0.0, 0.0.$$

The values of $\Delta\alpha$ and $\Delta\beta$ can be determined from Eqs. (2.50) and (2.51) as

$$\Delta\alpha = -0.54545456 \text{ and } \Delta\beta = 0.54545456.$$

The new approximations to $\bar{\alpha}$ and $\bar{\beta}$ are given by

$$\alpha = -2.54545456, \text{ and } \beta = 1.54545456.$$

Since $\Delta\alpha$ and $\Delta\beta$ are not small, we proceed to the next iteration with the new values of $\alpha = -2.54545456$ and $\beta = 1.54545456$. The values of α and β found in different iterations are indicated in the following table:

Iteration (i)	α	β
—	−2.0	1.0
1	−2.545455	1.545455
2	−2.848951	1.848951
3	−2.974668	1.974668
4	−2.999086	1.999086
5	−2.999984	1.000086
6	−3.000010	2.000008

From the converged values of $\alpha = -3.000010$ and $\beta = 2.000008$, the roots of the quadratic factor can be obtained from Eq. (2.52) as $x_1 = 1.000002$ and $x_2 = 1.999982$. The coefficients of the remaining polynomial are found to be

$$a_2 = 1.0, a_1 = -7.000016, \text{ and } a_0 = 12.000075.$$

Since the order of this remaining polynomial is 2, its roots can be determined readily as $x_3 = 3.000028$ and $x_4 = 3.999900$. The roots found can be compared with the exact values: $x_1 = 1, x_2 = 2, x_3 = 3$, and $x_4 = 4$. ◀

2.11 Muller's Method

This method is based on quadratic interpolation and can find the roots of any equation $f(x) = 0$. Let p_1, p_2, and p_3 denote three approximations to the root with the corresponding values of f given by $f_1 = f(p_1)$, $f_2 = f(p_2)$ and $f_3 = f(p_3)$, respectively. Then a quadratic equation is assumed to pass through the three points. The proper root of the quadratic equation is then used to define another quadratic equation for the next approximation (or iteration). It can be seen that Muller's method can be considered a generalization of the regula falsi method where a quadratic equation is used instead of a linear equation for interpolation. Since a higher degree equation is used for the interpolation, Muller's method converges faster than the secant method. The method can be used to find both real and complex roots of a given equation.

To define the approximation quadratic, a new variable t is defined for convenience as

$$t = x - p_3, \tag{2.53}$$

so that

$$t_1 = p_1 - p_3,$$

$$t_2 = p_2 - p_3, \tag{2.54}$$

and

$$t_3 = p_3 - p_3 = 0.$$

The quadratic equation is assumed as

$$h(t) = at^2 + bt + c. \tag{2.55}$$

Since $h(t)$ passes through the three points, (t_1, f_1), (t_2, f_2), and (t_3, f_3), we have

$$f_1 = at_1^2 + bt_1 + c, \tag{2.56}$$

$$f_2 = at_2^2 + bt_2 + c, \tag{2.57}$$

and

$$f_3 = at_3^2 + bt_3 + c. \tag{2.58}$$

Equations (2.56) to (2.58) can be solved to obtain

$$c = f_3, \tag{2.59}$$

$$b = \frac{t_2^2(f_1 - f_3) - t_1^2(f_2 - f_3)}{(t_1 t_2^2 - t_2 t_1^2)}, \tag{2.60}$$

and

$$a = \frac{t_2(f_1 - f_3) - t_1(f_2 - f_3)}{(t_2 t_1^2 - t_1 t_2^2)}. \tag{2.61}$$

By setting $h(t) = 0$, the roots of Eq. (2.55) can be found as

$$t = \frac{-b \pm \sqrt{b^2 - 4ac}}{2a}. \tag{2.62}$$

The numerator and denominator of Eq. (2.62) are multiplied by $b \pm \sqrt{b^2 - 4ac}$, and Eq. (2.62) is rewritten as

$$t = \frac{-2c}{b \pm \sqrt{b^2 - 4ac}}. \tag{2.63}$$

The roots are expressed in the form of Eq. (2.63) to avoid loss of significant digits in subtraction. The root having the smallest absolute value is selected in Eq. (2.63) for stability. The root can be expressed in terms of x as

$$t = t_4 = x - p_3$$

or

$$x = p_4 = p_3 + t_4 = p_3 - \frac{2c}{b \pm \sqrt{b^2 - 4ac}}. \tag{2.64}$$

For the next approximation, the two values out of p_1, p_2, and p_3 that are closest to p_4 are chosen as p_1 and p_2, and p_4 is set equal to p_3. In Fig. 2.25, for example, p_2, p_3, and p_4 are chosen as p_1, p_2, and p_3, respectively, for the next approximation (that is, p_1, which lies farthest to p_4 is discarded). Note that only one function evaluation is needed (at the new p_3 or the previous p_4) in each new iteration.

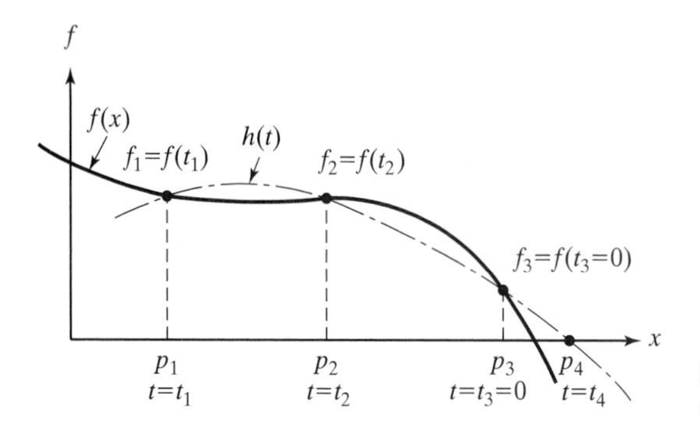

Figure 2.25 Selection of points for quadratic interpolation.

The iterative process of Muller's method can be described as follows:

1. Start with three approximations p_1, p_2, and p_3 to the root of the equation $f(x) = 0$ and a small number ε to test the convergence of the method. Choose $p_1 = -1$, $p_2 = 0$, and $p_3 = 1$, if no better approximations are available. Set $i = 3$ ($i = 3$ indicates that the next approximation to the root will be $p_{i+1} = p_4$).
2. Define $t_{i-2} = p_{i-2} - p_i$, $t_{i-1} = p_{i-1} - p_i$, and $t_i = p_i - p_i = 0$.
3. Find the function values

$$f_{i-2} = f(p_{i-2}),\ f_{i-1} = f(p_{i-1}),\ \text{and}\ f_i = f(p_i).$$

4. Find the coefficients of the interpolating quadratic function (using relations similar to Eqs. (2.59) to (2.61)) as

$$a = \frac{t_{i-1}(f_{i-2} - f_i) - t_{i-2}(f_{i-1} - f_i)}{t_{i-1}t_{i-2}^2 - t_{i-2}t_{i-1}^2}, \tag{2.65}$$

$$b = \frac{t_{i-1}^2(f_{i-2} - f_i) - t_{i-2}^2(f_{i-1} - f_i)}{t_{i-2}t_{i-1}^2 - t_{i-1}t_{i-2}^2}, \tag{2.66}$$

and

$$c = f_i. \tag{2.67}$$

5. Compute the root of the interpolating quadratic equation as

$$t = \frac{-2c}{b \pm \sqrt{b^2 - 4ac}} \tag{2.68}$$

by choosing the sign in the denominator that gives the smaller absolute value of t.

6. Find the new approximation to the root as

$$p_{i+1} = p_i - \frac{2c}{b \pm \sqrt{b^2 - 4ac}}. \tag{2.69}$$

7. Determine $f_{i+1} = f(p_{i+1})$ and test the convergence of the method:

$$|f_{i+1}| \le \varepsilon. \tag{2.70}$$

If convergence is achieved, stop the process by taking p_{i+1} as the root of the equation $f(x) = 0$. Otherwise, set the two values out of p_i, p_{i-1}, and p_{i-2} that are closest to p_{i+1} as p_i and p_{i-1}, update the iteration number as $i = i + 1$, and go to step 2.

▶Example 2.18

Find the root of the equation

$$f(x) = x^3 - 26x^2 + 175x - 150 = 0$$

with the initial approximations $p_1 = -2$, $p_2 = 0$, and $p_3 = 3$ and the convergence parameter $\varepsilon = 10^{-4}$ using Muller's method.

Solution

The values of $f(x)$ corresponding to p_1, p_2, and p_3 are given by

$$f_1 = f(-2.0) = -352.0,\ f_2 = f(0.0) = -150.0,\ \text{and}\ f_3 = f(3.0) = 168.0.$$

The values of t_1, t_2, and t_3 can be determined as

$$t_1 = p_1 - p_3 = -1.0 - 3.0 = -4.0,\ t_2 = p_2 - p_3 = 0.0 - 3.0 = -3.0,$$

$$\text{and}\ t_3 = p_3 - p_3 = 3.0 - 3.0 = 0.0.$$

The coefficients of the approximating quadratic can be computed, using Eqs. (2.59) to (2.61), as

$$a = -24.0,\ b = 34.0,\ \text{and}\ c = 168.0.$$

The value of t, given by Eq. (2.63), can be found as $t = -2.03059673$, which corresponds to $p_4 = p_3 + t = 3.0 - 2.03059673 = 0.06940327$ and $f_4 = f(p_4) = -3.87673950$. Since $|f_4| > \varepsilon$, we proceed to the next iteration with the new $p_1 = 0.0$, $p_2 = 3.0$, and $p_3 = 0.96940327$.

The corresponding values of the function $f(x)$ are given by

$$f_1 = f(0.0) = -150.0,\ f_2 = f(3.0) = 168.0,$$

$$\text{and}\ f_3 = f(0.96940327) = -3.87673950.$$

The values of t_1, t_2, and t_3 can be determined as

$$t_1 = p_1 - p_3 = 0.0 - 0.96940327 = -0.96940327,$$

$$t_2 = p_2 - p_3 = 3.0 - 0.96940327 = 2.03059673,$$

and

$$t_3 = p_3 - p_3 = 0.0.$$

The coefficients of the approximating quadratic can be found from Eqs. (2.59) to (2.61) as $a = -22.03062439$, $b = 129.37876892$, and $c = -3.87673950$. The value of t can be found, using Eq. (2.63), as $t = 0.03011874$, which yields $p_4 = p_3 + t = 0.99952203$ and $f_4 = f(p_4) = -0.06024170$. Since $|f_4| > \varepsilon$, we continue with the iterative process.

For the next iteration, we take $p_1 = 3.0, p_2 = 0.96940327$, and $p_3 = 0.99952203$. These correspond to $f_1 = f(p_1) = 168.0, f_2 = f(p_2) = -3.87673950, f_3 = f(p_3) = -0.06024170, t_1 = p_1 - p_3 = 2.00047779, t_2 = p_2 - p_3 = -0.03011876$, and $t_3 = p_3 - p_3 = 0.0$. The coefficients of the approximating quadratic are given by Eqs. (2.59) to (2.61) as $a = -21.03071594, b = 126.08151245$, and $c = -0.06024170$. These result in a value of $t = 0.00047784$, which corresponds to $p_4 = p_3 + t = 0.99999988$ and $f_4 = f(p_4) = -0.00001526$. Since $|f_4| < \varepsilon = 10^{-5}$, we stop the iterative process and take the root as $x = 0.99999988$. This value can be compared with the exact value of 1; the other two roots being $x = 10$ and $x = 15$. ◀

2.12 Newton–Raphson Method for Simultaneous Nonlinear Equations

Many engineering analyses require the solution of a set of simultaneous nonlinear algebraic equations. The Newton–Raphson method can be extended to solve a system of nonlinear equations. Consider the following system of nonlinear equations:

$$
\begin{aligned}
f_1(x_1, x_2, \ldots, x_n) &= 0; \\
f_2(x_1, x_2, \ldots, x_n) &= 0; \\
&\vdots \\
f_n(x_1, x_2, \ldots, x_n) &= 0.
\end{aligned}
\tag{2.71}
$$

If the functions $f_1, f_2, \ldots, f_n$ are expanded in Taylor's series about an arbitrary (guess or approximate) solution $(x_1, x_2, \ldots, x_n)$ and if only the linear terms are retained, we can obtain the values of the functions at the exact solution $(\bar{x}_1, \bar{x}_2, \ldots, \bar{x}_n)$ as

$$
\begin{aligned}
f_1(\bar{x}_1, \bar{x}_2, \ldots, \bar{x}_n) &\approx f_1(x_1, x_2, \ldots, x_n) + \left(\frac{\partial f_1}{\partial x_1}\right)(\bar{x}_1 - x_1) + \left(\frac{\partial f_1}{\partial x_2}\right)(\bar{x}_2 - x_2) + \\
&\cdots + \left(\frac{\partial f_1}{\partial x_n}\right)(\bar{x}_n - x_n);
\end{aligned}
\tag{2.72}
$$

$$\vdots$$

$$
\begin{aligned}
f_n(\bar{x}_1, \bar{x}_2, \ldots, \bar{x}_n) &\approx f_n(x_1, x_2, \ldots, x_n) + \left(\frac{\partial f_n}{\partial x_1}\right)(\bar{x}_1 - x_1) + \left(\frac{\partial f_n}{\partial x_2}\right)(\bar{x}_2 - x_2) + \\
&\cdots + \left(\frac{\partial f_n}{\partial x_n}\right)(\bar{x}_n - x_n),
\end{aligned}
\tag{2.73}
$$

where the partial derivatives are evaluated at $(x_1, x_2, \ldots, x_n)$. Equations (2.72) to (2.73) provide corrections to the assumed solution $(x_1, x_2, \ldots, x_n)$. By setting Eqs. (2.72) to (2.73) to zero and denoting $\Delta x_i = \bar{x}_i - x_i (i = 1, 2, \ldots, n)$ as the corrections to the variables $x_i (i = 1, 2, \ldots, n)$, we can rewrite Eqs. (2.72) to (2.73) as

$$\begin{bmatrix} \frac{\partial f_1}{\partial x_1} & \frac{\partial f_1}{\partial x_2} & & \frac{\partial f_1}{\partial x_n} \\ \frac{\partial f_2}{\partial x_1} & \frac{\partial f_2}{\partial x_2} & & \frac{\partial f_2}{\partial x_n} \\ \cdot & \cdot & & \cdot \\ & & \cdots & \\ \cdot & \cdot & & \cdot \\ \cdot & \cdot & & \cdot \\ \frac{\partial f_n}{\partial x_1} & \frac{\partial f_n}{\partial x_2} & & \frac{\partial f_n}{\partial x_n} \end{bmatrix} \begin{Bmatrix} \Delta x_1 \\ \Delta x_2 \\ \cdot \\ \cdot \\ \cdot \\ \Delta x_n \end{Bmatrix} = \begin{Bmatrix} -f_1 \\ -f_2 \\ \cdot \\ \cdot \\ \cdot \\ -f_n \end{Bmatrix}, \tag{2.74}$$

where the partial derivatives $\frac{\partial f_i}{\partial x_j}$ and the functions f_i are evaluated at the current approximation $(x_1, x_2, \ldots, x_n)$. The matrix of first partial derivatives of f_i, on the left-hand side of Eq. (2.74), is known as the Jacobian of f_i. Equations (2.74) denote a system of linear equations in the unknowns $\Delta x_1, \Delta x_2, \ldots, \Delta x_n$. By solving Eqs. (2.74), we find $\Delta x_i (i = 1, 2, \ldots, n)$ and determine the next approximate solution as

$$\begin{aligned} x_1^{(i+1)} &= x_1^{(i)} + \Delta x_1^{(i)}; \\ x_2^{(i+1)} &= x_2^{(i)} + \Delta x_2^{(i)}; \\ &\vdots \\ x_n^{(i+1)} &= x_n^{(i)} + \Delta x_n^{(i)}, \end{aligned} \tag{2.75}$$

where the superscripts i and $i+1$ denote the current and next iteration, respectively. The iterative process can be stated as follows:

1. Start with an initial guess solution $(x_1^{(1)}, x_2^{(1)}, \ldots, x_n^{(1)})$ and a small number ε to test the convergence of the process. Set the iteration number, $i = 1$.
2. Evaluate the function values:
$$f_j(x_1^{(i)}, x_2^{(i)}, \ldots, x_n^{(i)}); j = 1, 2, \ldots, n. \tag{2.76}$$
3. Find the partial derivatives of the functions f_j as
$$\frac{\partial f_j(x_1^{(i)}, x_2^{(i)}, \ldots, x_n^{(i)})}{\partial x_k}; j = 1, 2, \ldots, n; k = 1, 2, \ldots, n. \tag{2.77}$$
4. Solve the Eqs. (2.74) and find $\Delta x_j^{(i)}; j = 1, 2, \ldots, n$.

5. Find the new solution as

$$x_j^{(i+1)} = x_j^{(i)} + \Delta x_j^{(i)}; j = 1, 2, \ldots, n. \tag{2.78}$$

6. Evaluate the function values

$$f_j(x_1^{(i+1)}, x_2^{(i+1)}, \ldots, x_n^{(i+1)}); j = 1, 2, \ldots, n, \tag{2.79}$$

and test for the convergence of the process:

$$|f_j(x_1^{(i+1)}, x_2^{(i+1)}, \ldots, x_n^{(i+1)})| \le \varepsilon; j = 1, 2, \ldots, n. \tag{2.80}$$

If the convergence criteria of Eq. (2.80) are satisfied, stop the process by taking the solution as $(x_1^{(i+1)}, x_2^{(i+1)}, \ldots, x_n^{(i+1)})$. Otherwise, update the iteration number as $i = i + 1$ and go to step 3.

▶Example 2.19

Find the solution of the equations

$$f(x_1, x_2) = x_1^2 + x_2^2 - 8x_1 - 4x_2 + 11 = 0 \tag{E1}$$

and

$$f_2(x_1, x_2) = x_1^2 + x_2^2 - 20x_1 + 75 = 0 \tag{E2}$$

by taking the starting point as $(x_1 = 2, x_2 = 4)$ and the value of ε as 10^{-5}.

Solution

The partial derivatives of f_1 and f_2 can be found as

$$\frac{\partial f_1}{\partial x_1} = 2x_1 - 8, \quad \frac{\partial f_1}{\partial x_2} = 2x_2 - 4$$

and

$$\frac{\partial f_2}{\partial x_1} = 2x_1 - 20, \quad \frac{\partial f_2}{\partial x_2} = 2x_2.$$

At the starting point $(x_1^{(1)} = 2, x_2^{(1)} = 4)$, the function values are given by

$$f_1(2, 4) = 2^2 + 4^2 - 8(2) - 4(4) + 11 = -1$$

and

$$f_2(2, 4) = 2^2 + 4^2 - 20(2) + 75 = 55.$$

Since $|f_1| > \varepsilon$ and $|f_2| > \varepsilon$, we compute the Jacobian as

$$[J]|_{(x_1^{(1)}, x_2^{(1)})} = \begin{bmatrix} 2x_1 - 8 & 2x_2 - 4 \\ 2x_1 - 20 & 2x_2 \end{bmatrix}\Bigg|_{(2,4)} = \begin{bmatrix} -4 & 4 \\ -16 & 8 \end{bmatrix}$$

and solve the equations

$$\begin{bmatrix} -4 & 4 \\ -16 & 8 \end{bmatrix} \begin{Bmatrix} \Delta x_1^{(1)} \\ \Delta x_2^{(1)} \end{Bmatrix} = \begin{Bmatrix} -f_1 \\ -f_2 \end{Bmatrix} = \begin{Bmatrix} 1 \\ -55 \end{Bmatrix}$$

to obtain

$$\Delta x_1^{(1)} = 7.125 \text{ and } \Delta x_2^{(1)} = 7.375.$$

This yields the new point as

$$x_1^{(2)} = x_1^{(1)} + \Delta x_1^{(1)} = 2 + 7.125 = 9.125$$

and

$$x_2^{(2)} = x_2^{(1)} + \Delta x_2^{(1)} = 4 + 7.375 = 11.375.$$

The values of the functions f_1 and f_2 at $(x_1^{(2)} = 9.125, x_2^{(2)} = 11.375)$ are given by $f_1 = 105.15625$ and $f_2 = 105.15625$. Since $|f_1| > \varepsilon$ and $|f_2| > \varepsilon$, the Jacobian is computed as

$$[J]|_{(x_1^{(2)}, x_2^{(2)})} = \begin{bmatrix} 10.25 & 18.75 \\ -1.75 & 22.75 \end{bmatrix}.$$

The solution of the equations

$$\begin{bmatrix} 10.25 & 18.75 \\ -1.75 & 22.75 \end{bmatrix} \begin{Bmatrix} \Delta x_1^{(2)} \\ \Delta x_2^{(2)} \end{Bmatrix} = \begin{Bmatrix} -f_1 \\ -f_2 \end{Bmatrix} = \begin{Bmatrix} -105.15625 \\ -105.15625 \end{Bmatrix}$$

is given by

$$\Delta x_1^{(2)} = -1.581295, \Delta x_2^{(2)} = -4.743892.$$

The new approximate solution can be found as

$$x_1^{(3)} = x_1^{(2)} + \Delta x_1^{(2)} = 9.125 - 1.581295 = 7.543705$$

and

$$x_2^{(3)} = x_2^{(2)} + \Delta x_2^{(2)} = 11.375 - 4.743892 = 6.631108.$$

Since

$$|f_1(x_1^{(3)}, x_2^{(3)})| = |5.141006| = 5.141006 > \varepsilon$$

and

$$|f_2(x_1^{(3)}, x_2^{(3)})| = |5.141037| = 5.141037 > \varepsilon,$$

we proceed to the next iteration. The results of the iterative process are given in the following table (note that the iterative process converged in eight iterations): ◀

Iteration (i)	$x_1^{(i)}$	$x_2^{(i)}$	$f_1(x_1^{(i)}, x_2^{(i)})$	$f_2(x_1^{(i)}, x_2^{(i)})$	$[J]\|_{(x_1^{(i)}, x_2^{(i)})}$	Δx_1	Δx_2
1	2.0	4.0	−1.0	55.0	$\begin{bmatrix} -4 & 4 \\ -16 & 8 \end{bmatrix}$	7.125	7.375
2	9.125	11.375	105.15625	105.15625	$\begin{bmatrix} 10.25 & 18.75 \\ -1.75 & 22.75 \end{bmatrix}$	−1.581295	−4.743892
3	7.543705	6.631108	25.00502	25.00499	$\begin{bmatrix} 7.087410 & 9.262217 \\ -4.912590 & 13.262217 \end{bmatrix}$	−0.717011	−2.151025
4	6.826694	4.480083	5.141006	5.141037	$\begin{bmatrix} 5.653387 & 4.960167 \\ -6.346613 & 8.960167 \end{bmatrix}$	−0.250366	−0.751102
5	6.576328	3.728981	0.626846	0.626831	$\begin{bmatrix} 5.152657 & 3.457962 \\ -6.847343 & 7.457962 \end{bmatrix}$	−0.040374	−0.121117
6	6.535954	3.607864	0.016280	0.016296	$\begin{bmatrix} 5.071909 & 3.215729 \\ -6.928091 & 7.215729 \end{bmatrix}$	−0.001105	−0.003319
7	6.534849	3.604545	0.000019	0.000015	$\begin{bmatrix} 5.069698 & 3.209089 \\ -6.930302 & 7.209089 \end{bmatrix}$	−0.000002	−0.000004
8	6.543847	3.604541	0.000004	0.000000	—	—	—

2.13 Unconstrained Minimization

It can be seen that the root of the equation $f(x) = 0$ is same as the minimum of the function

$$F(x) = f^2(x). \tag{2.81}$$

Similarly the solution of the simultaneous linear or nonlinear equations

$$\begin{aligned} f_1(x_1, x_2, \ldots, x_n) &= 0; \\ f_2(x_1, x_2, \ldots, x_n) &= 0; \\ &\vdots; \\ f_n(x_1, x_2, \ldots, x_n) &= 0 \end{aligned} \tag{2.82}$$

is same as the unconstrained minimum of the function

$$\begin{aligned} F(x_1, x_2, \ldots, x_n) &= f_1^2(x_1, x_2, \ldots, x_n) \\ &+ f_2^2(x_1, x_2, \ldots, x_n) + \cdots + f_n^2(x_1, x_2, \ldots, x_n). \end{aligned} \tag{2.83}$$

The necessary and sufficient conditions for the minimum of $F(x)$, given by Eq. (2.81), are

$$\frac{dF(x^*)}{dx} = 0 \tag{2.84}$$

and

$$\frac{d^2F(x^*)}{dx^2} > 0, \tag{2.85}$$

where x^* is the minimum of $F(x)$ or the root of the equation $f(x) = 0$. Similarly, the necessary and sufficient conditions for the minimum of $F(x_1, x_2, \ldots, x_n)$, given by Eq. (2.82), are

$$\frac{\partial F(x_1^*, x_2^*, \ldots, x_n^*)}{\partial x_i} = 0; i = 1, 2, .., n \tag{2.86}$$

and

$$[H(x_1^*, x_2^*, \ldots, x_n^*)] = \left[\frac{\partial^2 F(x_1^*, x_2^*, \ldots, x_n^*)}{\partial x_i \partial x_j}\right] = \text{positive definite.} \tag{2.87}$$

Although Eqs. (2.84) and (2.85) or (2.86) and (2.87) can be solved to find the minimum of F, it is more convenient to find the minimum of $F(x)$ or $F(x_1, x_2, \ldots, x_n)$ through numerical methods. The numerical methods of finding the solution of simultaneous equations through the minimization of the function $F(x)$ or $F(X_1, x_2, \ldots, x_n)$ are presented in Chapter 12.

2.14 Convergence of Methods

2.14.1 Bisection Method

The interval at the end of ith bisection, h_{i+1}, can be expressed as

$$h_{i+1} = b^{(i+1)} - a^{(i+1)} = \frac{1}{2}h_i = \frac{1}{2^2}h_{i-1} = \frac{1}{2^{i+1}}h_0 = \frac{1}{2^{i+1}}(b^{(1)} - a^{(1)}). \tag{2.88}$$

Since the root of $f(x) = 0$ can lie anywhere in the interval h_{i+1}, the error, after ith bisection e_i, can be expressed as

$$|e_i| \le h_{i+1} = \frac{1}{2^{i+1}}h_0. \tag{2.89}$$

The bisection method is independent of the nature of the function $f(x)$ and always converges at the same rate. The inequality given by Eq. (2.89) is called an a priori error bound. By taking logarithms on both sides of the inequality (2.89), we obtain

$$\log_{10}|e_i| \le \log_{10} h_0 - (i + 1)\log_{10} 2, \tag{2.90}$$

which can be rearranged to obtain

$$i \le \left\{\frac{\log_{10} h_0 - \log_{10}|e_i|}{\log_{10} 2}\right\} - 1. \tag{2.91}$$

Thus, the inequality (2.90) gives an a priori bound on the number of bisections or iterations required to achieve a specified accuracy, $|e_i|$.

▶Example 2.20

Determine the number of iterations (bisections) required in a problem with $h_0 = 0.05$, if an accuracy of 0.0001 is desired.

Solution

In this example, $h_0 = 0.5$. Since the accuracy desired is 0.0001, we have $\varepsilon = 0.00005$. Hence, Eq. (2.90) gives

$$
\begin{aligned}
i &\leq \left\{\frac{\log_{10} 0.5 - \log_{10} 0.00005}{\log_{10} 2}\right\} - 1 \\
&\leq \left\{\frac{\log_{10} 10{,}000}{\log_{10} 2}\right\} - 1 \\
&\leq 12.28771238.
\end{aligned}
$$

By rounding the number of bisections to an integral number, we find that at most 13 bisections are required to achieve the specified accuracy. ◀

2.14.2 Newton's Method

The Taylor's series expansion can be used to find the value of $f(x)$ at x_{i+1} as

$$f(x_{i+1}) = f(x_i) + (x_{i+1} - x_i)f'(x_i) + \frac{1}{2}(x_{i+1} - x_i)^2 f''(x_i); x_i \leq \xi \leq x_{i+1}. \tag{2.92}$$

If $x_{i+1} = x^*$ is the root of $f(x) = 0$, Eq. (2.92) gives

$$f(x^*) = 0 = f(x_i) + (x^* - x_i)f'(x_i) + \frac{1}{2}(x^* - x_i)^2 f''(\xi). \tag{2.93}$$

Newton's iterative formula, Eq. (2.14), can be rewritten as

$$f(x_i) + (x_{i+1} - x_i)f'(x_i) = 0. \tag{2.94}$$

Subtraction of Eq. (2.94) from (2.93) yields

$$(x^* - x_{i+1})f'(x_i) + \frac{1}{2}(x^* - x_i)^2 f''(\xi) = 0. \tag{2.95}$$

Denoting the error in iteration i, e_i, as

$$e_i = x^* - x_i. \tag{2.96}$$

Eq. (2.95) can be expressed as

$$e_{i+1} f'(x_i) + \frac{1}{2} e_i^2 f''(\xi) = 0. \tag{2.97}$$

As the method converges, both x_i and ξ can be replaced by the actual root x^* so that Eq. (2.96) can be rewritten as

$$e_{i+1} = -\frac{f''(x^*)}{2f'(x^*)} e_i^2. \tag{2.98}$$

Equation (2.98) indicates that the error in an iteration is proportional to the square of the error in the previous iteration. This means that the number of correct decimal places in the numerical solution nearly doubles with each iteration. This type of convergence is termed *quadratic convergence*, and hence Newton's method is considered to be a *quadratically convergent method*.

▶Example 2.21

Show that the Newton–Raphson method is quadratically convergent using the results of Example 2.8.

Solution

The results of Example 2.8 can be summarized as follows:

i	x_i	Error, $\bar{e}_i = x^* - x_i$	Percent error
1	0.0	0.480864	100.0
2	0.364649	0.116215	24.17
3	0.461671	0.019193	3.99
4	0.480193	0.000671	0.14
5	0.480864	≈ 0.0	≈ 0.00

Equation (2.98) gives the relation between errors in two consecutive iterations as

$$e_{i+1} = -\frac{f''(x^*)}{2f'(x^*)} e_i^2. \tag{E1}$$

The first and second derivatives of $f(x)$ considered in Example 2.8 are given by

$$f'(x) = \frac{2.8 - 3.2x^2}{(1 + x^2)^3} \tag{E2}$$

and

$$f''(x) = \frac{-23.2x + 12.8x^2}{(1 + x^2)^4}. \tag{E3}$$

Since $x_1 = 0.0$, $f'(x_1) = 2.8$, and $f''(x_1) = 0.0$, Eq. (E1) gives $e_2 = (0)\ e_1^2 = 0$. Hence we start the computations with x_2. At $x_2 = 0.364649$, $f'(x_2) = 1.6327441$ and $f''(x_2) = -4.7577443$, so that $e_3 = 1.4569779\, e_2^2$. Using $e_2 = 0.116215$, we obtain $e_3 = 0.0196778$, which can be seen to be close to the true error, $\bar{e}_3 = 0.019193$. For the next iteration, $x_3 = 0.461671$, $f'(x_3) = 1.186268917$, $f''(x_3) = -4.363599799$, and Eq. (E1) gives, with $e_3 = 0.019193$, $e_4 = 1.839211892\, e_3^2 = 0.000677513$, which compares very well with the true error of $\bar{e}_4 = 0.000671$. For the next iteration

$x_4 = 0.480193$, $f'(x_4) = 1.106573381$, $f''(x_4) = -4.239961723$, and Eq. (E1) gives, with $e_4 = 0.000671$, $e_5 = 1.915806848\, e_4^2 = 0.000000863$, which can be considered to be approximately equal to zero. Thus, the results prove the quadratic convergence of the Newton–Raphson method. ◀

2.14.3 Fixed Point Iteration Method

The iterative procedure gives

$$x_{i+1} = g(x_i). \tag{2.99}$$

If x^* denotes the true solution, then

$$x^* = g(x^*). \tag{2.100}$$

Subtraction of Eq. (2.99) from (2.100) gives

$$x^* - x_{i+1} = g(x^*) - g(x_i). \tag{2.101}$$

Dividing both sides of Eq. (2.101) by $x^* - x_i$ yields

$$\frac{x^* - x_{i+1}}{x^* - x_i} = \frac{g(x^*) - g(x_i)}{x^* - x_i}. \tag{2.102}$$

From the mean value theorem, the right-hand side of Eq. (2.102) can be rewritten as

$$\frac{g(x^*) - g(x_i)}{x^* - x_i} = g'(\xi); x^* \le \xi \le x_i, \tag{2.103}$$

where $g'(\xi)$ denotes the derivative of $g(x)$ at $x = \xi$, which lies somewhere between x and x_i. Equations (2.102) and (2.103) yield

$$e_{i+1} = g'(\xi)e_i, \tag{2.104}$$

where e_i and e_{i+1} denote the errors in iterations i and $i + 1$, respectively. Note that the method converges only if $g'(\xi) < 1$, since in this case, the error in iteration $i + 1$ will be smaller than the error in iteration i. If $g'(\xi) < 1$, the error decreases with successive iteration. Hence the method is said to exhibit *linear convergence*.

2.14.4 Secant Method

The error in the secant method can be shown to follow the relation [2.17]

$$|e_{i+1}| = c|e_i|^p, \tag{2.105}$$

where

$$c = \left| \frac{f''(x^*)}{2f'(x^*)} \right| \tag{2.106}$$

and $p = 1.618$. Thus, the secant method is considered to have order of convergence $p = 1.618$.

2.15 Choice of the Method

The graphical method can be used to find the approximate, real roots of nonlinear equations. As such, the method can be used to find good starting points for use with other more accurate and efficient methods. The incremental search method can be used to find the real roots of any function. This method cannot find complex roots. The method fails to find the root when the function $f(x)$ does not undergo a sign change at the root. Although the method is not efficient, it can be used to find an approximate value of the root for use as a starting point for other more efficient methods. The bisection method, although slow, always finds the root. If $f(x)$ can be evaluated quickly, the bisection method is recommended. The accuracy improves with each iteration in this method. The bisection method cannot be used to find the complex roots of an equation.

If the function $f(x)$ is continuous with continuous derivatives, the Newton–Raphson method can be used to find the root efficiently. However, the method may not converge if the value of the derivative of f is close to zero. The method may also not converge if the initial guess is not good. The secant method is similar to the Newton–Raphson method where the derivative is approximated by the chord. This method is preferred over the Newton–Raphson method, when the evaluation of $f'(x)$ is difficult. The regula falsi method uses linear interpolation of the function in finding the root. The Newton–Raphson, secant, and regula falsi methods require smaller number of iterations to find a root. Although the secant method gives better results than the regula falsi method, it may not converge in some cases. The Newton–Raphson, secant, and regula falsi methods can find complex roots provided the initial approximations are given as complex numbers. Note that a complex root cannot be found by giving a real number as the initial guess. The secant and regula falsi methods require two initial guess values of the root, while the Newton–Raphson method requires only one initial guess value of the root.

The fixed point iteration is simple and efficient, if the equation $f(x) = 0$ can be rewritten as $x = g(x)$. This method may not converge if $|g'(x)| > 1$. Although Muller's method requires three initial approximations for the root, it is particularly efficient in finding the complex roots of equations. Since the method is based on a quadratic approximation of $f(x)$, it converges faster than the secant method.

For a polynomial, if all the roots are required, Bairstow's method is recommended. After finding a quadratic factor, Bairstow's method is to be applied to the deflated polynomial. In this method, if the locations of some of the roots are known, these roots must be found first.

The Newton–Raphson method is very efficient in solving systems of nonlinear equations. The method may not converge if the initial approximation is not good. Since the method requires the first derivatives of the functions, we need to use either the exact expressions or the finite difference equivalents for the derivatives of the functions involved.

2.16 Use of Software Packages

All the software packages have built-in programs (functions) for finding roots of nonlinear equations, roots of polynomial equations, and solutions of simultaneous nonlinear equations. Usually, all the roots of a polynomial equation are determined by the program without requiring initial guess values for the roots. However, most programs used for finding the roots of nonlinear equations require input in the form of lower and upper bounds (range) or an initial guess for the root. In addition, some symbolic programs can find the solution in the form of a mathematical expression.

2.16.1 MATLAB

To find the roots of a polynomial equation use the function *roots(p)* where p is a row vector of coefficients of the polynomial in descending order of the power of the variable. The roots of the polynomial are returned in the form of a column vector. To find the roots of a nonlinear equation, the function *fzero(y, x1)* is to be used. Here y defines the nonlinear function and *x1* indicates the initial estimate (starting value) of the root. Several illustrative examples follow:

▶Example 2.22

The roots of $f(x) = x^{12} - 2 = 0$ are as follows

```
>> roots ([1 0 0 0 0 0 0 0 0 0 0 0 -2])
ans =
  -1.0595
  -0.9175 + 0.5297i
  -0.9175 - 0.5297i
  -0.5297 + 0.9175i
  -0.5297 - 0.9175i
   0.0000 + 1.0595i
   0.0000 - 1.0595i
   0.5297 + 0.9175i
   0.5297 - 0.9175i
   1.0595
   0.9175 + 0.5297i
   0.9175 - 0.5297i
```

▶Example 2.23

The roots of $f(x) = \tan^{-1} x = 0$.
Note: To find the roots of a general nonlinear equation, the lower bound, increment for search and upper bound on x are to be given before using the function *fzero*. In addition, the value of x in the vicinity of the desired root is to be given as an argument of *fzero*.

```
>> x=-10:0.01:10;
>> f='atan(x)';
>> root=fzero(f,1.6)
Zero found in the interval: [-0.448, 3.0482].

root =

   1.5040e-20

>> root=fzero(f,0.2)
Zero found in the interval: [-0.056, 0.38102].

root =

  -5.9260e-24

>>
```

▶Example 2.24

The roots of $f(x) = \sin 2x = 0$ are

```
>> x=-5:0.01:5;
>> y4='sin(2*x)';
>> x4=fzero(y4,0.75)
x4 =
  9.2059e-023
>> x4=fzero(y4,2.75)
x4 =
    3.1416
>> x4=fzero(y4,-2.75)
x4 =
   -3.1416
```

▶Example 2.25

The roots of

$$f(x) = \frac{1.5x}{(1+x^2)^2} - 0.65\tan^{-1}\frac{1}{x} + \frac{0.65x}{(1+x^2)} = 0:$$

```
>> x=-2:0.01:2;
>> y2='1.5*x./(1+x.^2).^2-0.65*(atan(1./x))+0.65*x./(1+x.^2)';
>> fplot(y2,[-2 2])
>> xlabel('x'),ylabel('y=f(x)'),
>> title('y=1.5*x/(1+x^2)^2-0.65* (atan(1/x))+0.65*x/(1+x^2)')
>> x2=fzero(y2,-0.1)
```

```
x2 =

  2.4831e-016

>> x2=fzero(y2,0.1)

x2 =

 -2.4831e-016

>> x2=fzero(y2,0.8)

x2 =

     0.4809

>> x2=fzero(y2,-0.8)

x2 =

     -0.4809
```

▶Example 2.26

Solve the following simultaneous nonlinear equations:

$$f(x, y) = x^2 + y^2 - 8x - 4y + 11 = 0$$

and

$$g(x, y) = x^2 + y^2 - 20x + 75 = 0.$$

(a) By minimizing $F = f^2(x, y) + g^2(x, y)$ with initial estimates of $(x = 2, y = 4)$, and $(x = 1, y = 1)$.

Note: The function F is defined as $f3$ in an m-file.

Listing of *f3.m*:

```
function f=f3(x)

f=(x(1)^2+x(2)^2-8*x(1)-4*x(2)+11)^2+(x(1)^2+x(2)^2-20*x(1)
+75)^2;

>> [x3]=fmins('f3',[2,4]);
>> x3
```

```
x3 =

     5.0652   -0.8045

>> [x3]=fmins('f3',[1,1]);
>> x3

x3 =

     5.0652   -0.8045

>>
```

(b) By using the function *solve*. This gives all the solutions of the equations.

Note: A blank space is to be left on both sides of the equality sign in defining the equations.

```
>> [x,y] = solve('x^2+y^2-8*x-4*y+11=0','x^2+y^2-20*x+75 = 0')

x =

[ 29/5+3/10*6^(1/2)]
[ 29/5-3/10*6^(1/2)]

y =

[ 7/5+9/10*6^(1/2)]
[ 7/5-9/10*6^(1/2)]

>>
```

2.16.2 MAPLE

The *solve* command finds all the possible roots of a nonlinear equation, including polynomial equations. This command provides symbolic solutions of equations, if they exist. The command *fsolve* uses Newton–Raphson method to find the roots of a general nonlinear equation. For a nonlinear equation, it gives one root, while the command provides all the real roots for a polynomial equation. If all the roots of a polynomial equation, including real and complex ones are required, the word *complex* is to be included in the *fsolve* command. The following examples illustrate the procedures.:

▶Example 2.27

Roots of $f(x) = x^{12} - 2 = 0$:

```
> eqn1:=x^12-2;
```

$$eqn1 := x^{12} - 2$$

```
> fsolve(eqn1 =0, x, complex);
```

$\{x = -1.059463094\}, \{x = -.9175219541 - .5297315472\, I\},$
$\{x = -.9175219541 + .5297315472\, I\}, \{x = -.5297315472 - .9175219541\, I\},$
$\{x = .5297315472 + .9175219541\, I\}, \{x = -1.059463094\, I\}, \{x = 1.059463094\, I\},$
$\{x = .5297315472 - .9175219541\, I\}, \{x = .5297315472 + .9175219541\, I\},$
$\{x = .9175219541 - .5297315472\, I\}, \{x = .9175219541 + .5297315472\, I\},$
$\{x = 1.059463094\}$

```
> evalf(solve(eqn1=0, x));
```

$\{x = 1.059463094\}, \{x = -1.059463094\}, \{x = 1.059463094\, I\}, \{x = -1.059463094\, I\},$
$\{x = .9175219535 + .5297315470\, I\}, \{x = -.9175219535 - .5297315470\, I\},$
$\{x = .5297315475 - .9175219545\, I\}, \{x = -.5297315475 + .9175219545\, I\},$
$\{x = .9175219535 - .5297315470\, I\}, \{x = -.9175219535 + .5297315470\, I\},$
$\{x = .5297315475 + .9175219545\, I\}, \{x = -.5297315475 - .9175219545\, I\}$

▶Example 2.28

Roots of $f(x) = \tan^{-1} x = 0$:

```
> eqn:=arctan(x);
```

$$eqn := arctan(x)$$

```
> fsolve(eqn=0, x);
```

$$\{x = 0\}$$

▶Example 2.29

Roots of

$$f(x) = \frac{1.5x}{(1+x^2)^2} - 0.65 \tan^{-1}\left(\frac{1}{x}\right) + \frac{0.65x}{1+x^2} = 0:$$

(a) By giving different ranges for the root:

```
> eq:=1.5*x/(1.0+x^2.0)^2.0-0.65*arctan(1.0/x)+0.65*x/(1+x^2)
  =0:
> fsolve(eq, x, -1.0..1.0);
                                   -.4808644853
```

If the range does not contain a root, the program returns just the equation and the range given:

```
> eq:=1.5*x/(1.0+x^2.0)^2.0-0.65*arctan(1.0/x)+0.65*x/(1+x^2)
  =0:
> fsolve(eq, x, -0.48..0.48);

                  x                                            x
fsolve(1.5 ------------ - .65 arctan(1.01/x)+ .65 ------ = 0.
                2.0 2.0                                        2
           (1.0 + x    )                                   1 + x

       x, -.48 .. .48)
```

```
>
> f:=1.5*x/(1+x^2)^2-0.65*(arctan(1/x))+0.65*x/(1+x^2):
> fsolve(f,x,0..1);
```

$$.4808644853$$

```
>
```

(b) Without giving a range for the root:

```
> eqn2:=1.5*x/(1+x^2)^2-0.65*(arctan(1/x))+0.65*x/(1+x^2);
```

$$eqn2 := 1.5\frac{x}{(1+x^2)^2} - .65\arctan\left(\frac{1}{x}\right) + .65\frac{x}{1+x^2}$$

```
> fsolve(eqn2=0,x);
```

$$\{x = -.4808644853\}$$

▶Example 2.30

Roots of $f(x) = x^4 - 12x^3 + 49x^2 - 78x + 40 = 0$:

```
> eqn:=x^4.0-12.0*x^3.0+49.0*x^2.0-78.0*x+40.0=0;
                  4.0           3.0           2.0
        eqn := x     - 12.0 x      + 49.0 x      - 78.0 x + 40.0 = 0
> solve(eqn, {x});
                 {x = 5.}, {x = 1.}, {x = 2.}, {x = 4.}

> poly :=x^4-12.*x^3+49.*x^2-78.*x+40.:fsolve(poly,x,maxsols=4 );
                 1., 2.000000000, 4.000000000, 5.000000000
  solve(x^4-12.*x^3+49.*x^2-78.*x+40.,x);
                              5., 1., 2., 4.
```

```
roots(x^4-12*x^3+49*x^2-78*x+40);
                   [[1, 1], [2, 1], [4, 1], [5, 1]]
```

▶Example 2.31

Solution of simultaneous nonlinear equations:

$$f(x, y) = x^2 + y^2 - 8x - 4y + 11 = 0$$

and

$$g(x, y) = x^2 + y^2 - 20x + 75 = 0$$

(a) By giving ranges for x and y:

A different range gives a different answer:

```
> f:=x^2+y^2-8*x-4*y+11=0:
> g:=x^2+y^2-20*x+75=0:
> fsolve(f,g,x,y,x=0..100,y=-29.8..100);
                {x = 6.534846923, y = 3.604540769}
> fsolve(f,g,x,y,x=0..100,y=-29.9..100);
                {x = 5.065153077, y = -.8045407685}
> fsolve(f,g,x,y);
                {x = 5.065153077, y = -.8045407685}
>
```

(b) Without giving ranges for variables:

```
> fsolve(x^2+y^2-8*x-4*y+11=0,x^2+y^2-20*x+75=0);
                {x = 5.065153077, y = -.8045407685}
>
```

2.16.3 MATHCAD

To find the roots of polynomial equations, one can use the *polyroots* (v) command. Here v denotes the vector of coefficients of the polynomial starting with the constant term. All the roots of the equation are returned as a vector. To find the root of a nonlinear equation, the *root(f(x),x)* command can be used. Here $f(x)$ is the nonlinear function and x is the variable. By giving a guess value of the root, the program returns the root closest to the guess value. To find the solution of several nonlinear equations, we need to use the appropriate solve blocks. The procedures are illustrated with the following examples.

▶Example 2.32

Roots of $f(x) = x^{12} - 2 = 0$ (V1 denotes the vector of polynomial coefficients):

$x^{12} - 2$

$$V1 := \begin{pmatrix} -2 \\ 0 \\ 0 \\ 0 \\ 0 \\ 0 \\ 0 \\ 0 \\ 0 \\ 0 \\ 0 \\ 0 \\ 1 \end{pmatrix}$$

polyroots(V1) =

	0
0	−1.059
1	−0.918 + 0.53i
2	−0.918 − 0.53i
3	−0.53 − 0.918i
4	−0.53 + 0.918i
5	$-3.841 \cdot 10^{-9} + 1.059i$
6	−1.059i
7	0.53 − 0.918i
8	0.53 + 0.918i
9	0.918 + 0.53i
10	0.918 − 0.53i
11	1.059

▶Example 2.33

The roots of $f(x) = \tan^{-1} x = 0$: with different initial guess values:

$x := 1.6$

$\text{root}(\text{atan}(x),x) = -5.335 \times 10^{-5}$

$x := 0.2$

$\text{root}(\text{atan}(x),x) = 6.804 \times 10^{-5}$

▶Example 2.34

Roots of $f(x) = \sin 2x = 0$: with different initial guess values:

$x := 0.75$

$\text{root}(\sin(2\cdot x),x) = -6.283$

$x := 0.0$

$\text{root}(\sin(2\cdot x),x) = 0$

▶Example 2.35

Roots of $f(x) = \frac{1.5x}{(1+x^2)^2} - 0.65\tan^{-1} x + \frac{0.65x}{1+x^2} = 0$ using different initial guess values:

$x := -1.2$

$$\text{root}\left[\left[\frac{(1.5 \cdot x)}{[(1+x^2)^2]}\right] - 0.65 \cdot \text{atan}\left(\frac{1}{x}\right) + \left[\frac{(0.65 \cdot x)}{(1+x^2)}\right], x\right] = -11.222$$

$x := 0.8$

$$\text{root}\left[\left[\frac{(1.5 \cdot x)}{[(1+x^2)^2]}\right] - 0.65 \cdot \text{atan}\left(\frac{1}{x}\right) + \left[\frac{(0.65 \cdot x)}{(1+x^2)}\right], x\right] = 12.001$$

$x := -0.8$

$$\text{root}\left[\left[\frac{(1.5 \cdot x)}{[(1+x^2)^2]}\right] - 0.65 \cdot \text{atan}\left(\frac{1}{x}\right) + \left[\frac{(0.65 \cdot x)}{(1+x^2)}\right], x\right] = -12.001$$

▶Example 2.36

Solution of simultaneous nonlinear equations:

$$f(x, y) = x^2 + y^2 - 8x - 4y + 11 = 0$$

and

$$g(x, y) = x^2 + y^2 - 20x + 75 = 0.$$

$x := 2$ $y := 4$

Given

$$x^2 + y^2 - 8 \cdot x - 4 \cdot y + 11 = 0$$

$$x^2 + y^2 - 20 \cdot x + 75 = 0$$

$$\begin{pmatrix} xval \\ yval \end{pmatrix} := Find(x,y)$$

$$xval = 6.535$$

$$yval = 3.605$$

2.17 Computer Programs

2.17.1 Fortran Programs

Program 2.1 Secant method

The input data required are as follows:

X1,X2 = initial (guess) values of the root
MITER = maximum number of iterations permitted (usual value: 100)
EPS = convergence criterion (usual value: 0.00001)

Subprogram to evaluate the function $f(x)$ in the following form:

```
FUNCTION F(X)
F=...
RETURN
END
```

Illustration (Example 2.5)

The listing and output of Program 2.1 are as follows:

```
      PROGRAM FOR SECANT METHOD
      DATA X1,X2,MITER,EPS/0.1,0.5,100,0.00001/
      I=0
      PRINT 10
10    FORMAT ('SOLUTION BY SECANT METHOD',/)
      Y=X1
      YY=X2
      Z=F(Y)
      ZZ=F(YY)
      DO WHILE(1)
         I=I+1
```

```
        IF (I .GT. MITER) Then
        PRINT 90
90      FORMAT (/, 'NO CONVERGENCE IN MITER ITERATIONS')
        STOP
        END IF
        X3=YY-ZZ*(YY-Y)/(ZZ-Z)
        F3=F(X3)
        PRINT 30, I, Y, Z, YY, ZZ
30      FORMAT('ITER=',I3,'x1=',E12.6,'F(x1)=',E12.6,/,10x,  &
    &   ' x2 = ', E12.6,' F (x2) = ', E12.6)
        PRINT 40, X3, F3
40      FORMAT (9X, 'NEWX = ', E12.6, 1X, 'NEWF = ', E12.6,/)
        IF (ABS (F3) .LE. EPS) THEN
        PRINT 60, X3, F3
60      FORMAT ('CONVERGED SOLUTION: X = ', E15.8,' F = ', E15.8)
     STOP
     ELSE
        Y=YY
        YY=X3
        Z=ZZ
        ZZ=F3
        END IF
     END DO
     END

     FUNCTION F(X)
     F=1.5*X/((1.0+X*X)**2)-0.65*ATAN(1.0/X)+0.65*X/(X*X+1.0)
     RETURN
     END
```

```
SOLUTION BY SECANT METHOD

ITER =     1 X1 = 0.100000E+00 F(X1) = -.744832E+00
             X2 = 0.500000E+00 F(X2) = 0.203533E-01
           NEWX = 0.489360E+00 NEWF  = 0.922492E-02

ITER =     2 X1 = 0.500000E+00 F(X1) = 0.203533E-01
             X2 = 0.489360E+00 F(X2) = 0.922492E-02
           NEWX = 0.480541E+00 NEWF  = -.357777E-03

ITER =     3 X1 = 0.489360E+00 F(X1) = 0.922492E-02
             X2 = 0.480541E+00 F(X2) = -.357777E-03
           NEWX = 0.480870E+00 NEWF  = 0.587106E-05
```

```
CONVERGED SOLUTION: X = 0.48086980E+00 F = 0.58710575E-05
```

Program 2.2 Bairstow's method

The input data required are

N = degree of the polynomial
NP1 = N + 1
NM1 = N − 1
ALP,BET = initial approximations to $\bar{\alpha}$ and $\bar{\beta}$, respectively
MITER = maximum number of iterations permitted (usual value: 100)
EPS = convergence criterion (usual value: 0.0001)
A(I),I = $1, 2, \ldots,$ NP1 = coefficients of the polynomial (A(1): constant, A(NP1): coefficient of x**N)

Illustrative Example

Find the roots of the equation

$$f(x) = x^4 - 8x^3 + 18x^2 - 6x + 6 = 0.$$

Using N = 4, ALP = −2.0, BET = 1.0, MITER = 100, EPS = 10^{-4}, A(1) = 6, A(2) = −6, A(3) = 18, A(4) = −8, and A(5) = 1, the output of Program 2.2 follows:

```
ITERATION NO:      1 B(I):
    6.00000000   -6.00000000  18.00000000  -8.00000000  1.00000000
ITERATION NO:      2 B(I):
    1.62283897   -1.62283993  14.47933960  -7.45454550  1.00000000
ITERATION NO:      3 B(I):
    0.37398720   -0.37398624  12.77804947  -7.15104675  1.00000000
ITERATION NO:      4 B(I):
    0.05259323   -0.05259418  12.12726974  -7.02532578  1.00000000
ITERATION NO:      5 B(I):
    0.00181770   -0.00181770  12.00453663  -7.00090694  1.00000000
ITERATION NO:      6 B(I):
0.00000572       -0.00000477  12.00000286  -7.00000095  1.00000000

QUADRATIC EQUATION: SOLUTION IS      2.000000       1.000000

COEFFICIENTS OF REMAINING POLYNOMIAL
  12.000003    -7.000001    1.000000
QUADRATIC EQUATION: SOLUTION IS      4.000001       3.000000
```

Program 2.3 Newton–Raphson method for solving a system of nonlinear equations:

The following input data are required

N = number of equations
NP1 = N + 1
X(1), X(2), ..., X(N) = initial (guess) values for the N unknown variables
DX(1), DX(2), ..., DX(N) = increments of X(1),X(2),...,X(N) for use in finite difference derivatives (usual value: 0.01 for each)
MITER = maximum number of iterations permitted (usual value: 50)
EPS = convergence criterion (usual value: 0.00001)

Subprogram to evaluate the functions $f_i(\vec{X})$ in the following form:

```
SUBROUTINE FUN(X,F,N)
DIMENSION X(N),F(N)
F(1)=...
F(2)=...
.
.
.
F(N)=...
RETURN
END
```

Illustration (Example 2.19)

Using N = 2, X(1) = 2, X(2) = 4, DX(1) = DX(2) = 0.01, MITER = 50, and EPS = 10^{-5}, the output of Program 2.3 is as follows:

```
ITER =  1 VALUES OF X (I):   0.20000000E+01    0.15000000E+02
          VALUES OF F (I):   0.40000000E+01   -0.43890562E+01
          NEW X (I):         0.12145693E+01    0.25270457E+01
          NEW F (I):         0.28611388E+01   -0.18417971E+01
ITER =  2 VALUES OF X (I):   0.12145693E+01    0.28611388E+01
          VALUES OF F (I):   0.25270457E+01   -0.18417971E+01
          NEW X (I):         0.59310120E+00    0.22624547E+01
          NEW F (I):         0.47047043E+00   -0.54713690E+00
ITER =  3 VALUES OF X (I):   0.59310120E+00    0.47047043E+00
          VALUES OF F (I):   0.22624547E+01   -0.54713690E+00
          NEW X (I):         0.27988195E+00    0.22411087E+01
          NEW F (I):         0.10090208E+00   -0.81864953E-01
ITER =  4 VALUES OF X (I):   0.27988195E+00    0.10090208E+00
          VALUES OF F (I):   0.22411087E+01   -0.81864953E-01
          NEW X (I):         0.20796204E+00    0.22276907E+01
          NEW F (I):         0.58541298E-02   -0.34757853E-02
ITER =  5 VALUES OF X (I):   0.20796204E+00    0.58541298E-02
          VALUES OF F (I):   0.22276907E+01   -0.34757853E-02
          NEW X (I):         0.20435423E+00    0.22267201E+01
```

```
            NEW F (I):           0.42915344E-04    -0.12516975E-04
ITER =   6 VALUES OF X (I):      0.20435423E+00     0.42915344E-04
            VALUES OF F (I):     0.22267201E+01    -0.12516975E-04
            NEW X (I):           0.20433748E+00     0.22267120E+01
            NEW F (I):           0.00000000E+00    -0.11920929E-06

SOLUTION CONVERGED IN     6    ITERATIONS
SOLUTION IS: X(I) ARE:  0.20433748E+00     0.22267120E+01
             F(I) ARE:  0.00000000E+00    -0.11920929E-06
```

2.17.2 C Programs

Program 2.4 Newton–Raphson method
The following input data is to be given interactively by the user:

X1 = initial (guess) value of the root
DX = increment of x (step size) to be used in finite difference formula (usual value: 0.01)
MITER = maximum number of iterations permitted (usual value: 100)
EPS = convergence criterion (usual value: 0.00001)
Subprogram to evaluate the function $f(x)$.
Illustration (Example 2.8)
The following is a program listing and output:

```
/* Newton-Raphson Method - C Program */
#include <stdio.h>
#include <stdlib.h>
#include <math.h>

/* X1, X2    - current and next x value
   F1, F2    - current and next f value
   DX        - percentage increment of x
   I, MITER - number of iteration and maximum number
   FD1       - current derivative of f value */
void func1 (X, F)
float X, *F;
{
      *F=1.5*X/pow((1.0+X*X),2)-0.65*atan(1.0/X)+0.65*X/(X*X+1.0);
      return;
}
void func2 (X, DX, FD)
float X, DX, *FD;
{
        float FR, FP, Y;
        void func1 ();
        func1 (X, &FR);
        Y=(1.0+DX)*X;
        if (fabs (X) > = 0.001)
```

```
        {
                func1 (Y, &FP);
                *FD=(FP-FR)/(DX*X);
        }
        else
        {
                Y=DX;
                func1 (Y, &FP);
                *FD=(FP-FR)/DX;
        }
        return;
}
main ()
{
        int I, MITER;
        float X1, X2, F1, F2, DX, EPS, FD1;
        void func1 (), func2 ();
        printf (''Initial data for Newton-Raphson Method\n'');
        printf (''\nMaximum number of Iterations?\n'');
        scanf (''%d'', &MITER);
        printf (''Tolerance of result?\n'');
        scanf (''%f'', &EPS);
        printf (''Initial guess value of X?\n'');
        scanf (''%f'', &X1);
        printf (''Increamental percentage of X?\n'');
        scanf (''%f'', &DX);
        F2=1.0E10;
        func1 (X1, &F1);
        printf (''\nIte No.   X(I)    F(I)    FD(I)   X(I + 1)   F(
I+1)\n'');
        for (I = 1; (fabs (F2) >EPS) &&(I}<=MITER); I++)
        {
           func2 (X1, DX, &FD1);
           X2 = X1 - F1/FD1;
           func1 (X2, &F2);
           printf (''%d %12.6f %12.6f %12.6f %12.6f %12.6f\n'', I,
X1, F1, FD1, X2, F2);
           X1=X2;
           F1=F2;
        }
        if  (I>MITER)
                printf(''\nSolution not converged in MITER iterations'');
        else
                printf(''\nSolution converged in \%d iterations'', I - 1);
                printf(''\nSolution is: X = \%12.6F F = \%12.6F$'', X1, F1);
}
```

```
Initial data for Newton-Raphson Method

Maximum number of Iterations?
20
Tolerance of result?
0.0001
```

```
Initial guess value of X?
0.5
Incremental percentage of X?
0.1

Ite No.    X(I)        F(I)        FD(I)       X(I+1)       F(I+1)
1        0.500000    0.290353    4.149514    0.430027     0.161431
2        0.430027    0.161431    3.798167    0.387525     0.047333
3        0.387525    0.047333    3.609621    0.374412     0.012016
4        0.374412    0.012016    3.555357    0.371032     0.002904
5        0.371032    0.002904    3.541673    0.370212     0.000693
6        0.370212    0.000693    3.538372    0.370016     0.000165
7        0.370016    0.000165    3.537585    0.369970     0.000039

Solution converged in 7 iterations
Solution is: X=   0.369970 F =   0.000039
```

Program 2.5 Fixed–point iteration method
The following input data are required:

X1 = initial (guess) value of the root
MITER = maximum number of iterations permitted (usual value: 100)
EPS = convergence criterion (usual value: 0.00001)

Subprogram to evaluate $g(x)$ and $g'(x)$ where $f(x) = x - g(x) = 0$.

Illustration (Example 2.15)

The output of Program 2.5 is follows:

```
SOLUTION BY FIXED POINT ITERATION METHOD

I     X(I)           F(I)           GD(I)
1   0.10000000     0.60653551     0.16134542
2   0.60653551     0.47203515     0.43700819
3   0.47203515     0.48191504     0.43697774
4   0.48191504     0.48074364     0.43867080
5   0.48074364     0.48087844     0.43848403
6   0.48087844     0.48086287     0.43850571

Solution converged in 6 iterations
Solution: X = 0.48087844 F = 0.48086287 GPRIME = 0.43850571
```

REFERENCES AND BIBLIOGRAPHY

2.1. S. S. Rao, *Mechanical Vibrations*, 3rd edition, Addison-Wesley, Reading, MA, 1995.

2.2. E. Volterra and E. C. Zachmanoglou, *Dynamics of Vibrations*, Charles E. Merrill Books, Columbus, OH, 1965.

2.3. J. R. Howell and R. O. Buckius, *Fundamentals of Engineering Thermodynamics*, 2d edition, McGraw-Hill, New York, 1992.

2.4. R. W. Fox and A. T. McDonald, *Introduction to Fluid Mechanics*, 4th edition, John Wiley, New York, 1992.

2.5. J. E. Shigley and C. R. Mischke, *Mechanical Engineering Design*, 5th edition, McGraw-Hill, New York, 1989.

2.6. R. L. Norton, *Design of Machinery*, McGraw-Hill, New York, 1992.

2.7. W. H. Hayt and J. E. Kemmerly, *Engineering Circuit Analysis*, McGraw-Hill, New York, 1986.

2.8. A. Constantinides, *Applied Numerical Methods with Personal Computers*, McGraw-Hill, New York, 1987.

2.9. B. Carnahan, H. A. Luther, and J. O. Wilkes, *Applied Numerical Methods*, John Wiley, New York, 1969.

2.10. L. Bairstow, "Investigations Relating to the Stability of the Aeroplane," *Reports and Memoranda No. 154*, Advisory Committee of Aeronautics, 1914.

2.11. F. L. Hitchcock, "An Improvement on the G. C. D. Method for Complex Roots," *J. Math and Phys*, Vol. 23, pp. 69–74, 1944.

2.12. D. Muller, "A Method for Solving Algebraic Equations Using an Automatic Computer," *Math Tables Aids Comput*, Vol. 10, pp. 208–215, 1956.

2.13. W. L. Frank, "Finding Zeros of Arbitrary Functions," *J. Assoc Comput Math*, Vol. 5, pp. 154–165, 1958.

2.14. S. Nakamura, *Applied Numerical Methods with Software*, Prentice Hall, Englewood Cliffs, NJ, 1991.

2.15. F. M. White, *Heat and Mass Transfer*, Addison-Wesley, Reading, MA, 1988.

2.16. M. G. Salvadori, *Numerical Methods in Engineering*, Prentice-Hall, Englewood Cliffs, NJ, 1952.

2.17. K. E. Atkinson, *An Introduction to Numerical Analysis*, 2d edition, John Wiley, New York, 1989.

2.18. A. Ralston and P. Rabinowitz, *A First Course in Numerical Analysis*, 2d edition, McGraw-Hill, New York, 1978.

2.19. J. H. Wilkinson, "The Evaluation of Zeros of Ill-Conditioned Polynomials," *Numer. Math.*, Vol. 1, pp. 150–180, 1959.

REVIEW QUESTIONS

The following questions along with corresponding answers are available in an interactive format at the Companion Web site at http://www.prenhall.com/rao.

2.1. **Give short answers:**

1. What is the difference between a polynomial equation and a transcendental equation?
2. How many roots can an nth degree polynomial equation have?
3. How many roots can a transcendental equation have?
4. Can you use the incremental search method to find the complex roots of an equation?
5. Can you use the Newton–Raphson method to find the roots of a nondifferentiable function?
6. What is the difference between the secant and the regula falsi methods?
7. What is the basic principle used in Bairstow's method?
8. What is the similarity between the secant and Muller's methods?
9. What is the role of Taylor's series expansion in the development of Newton–Raphson method?
10. What is the limitation of the fixed point iteration method in finding the roots of an equation?

2.2. Answer **true or false** to each of the following:

1. A polynomial with real coefficients can have all complex roots.
2. A polynomial with complex coefficients can have all real roots.
3. The root of the equation $f(x) = x^3 - \sin x = 0$ can be found by finding the intersection point of the two graphs $f_1(x) = x^3$ and $f_2 = \sin x$.
4. The graphical method of finding the roots of an equation is computationally simple.
5. If $f(x_i)f(x_{i+1}) < 0$, the root of $f(x)$ is guaranteed to lie between x_i and x_{i+1}.
6. The incremental method can be used to find both real and complex roots of $f(x)$.
7. The incremental method can find the root x_1 even when $f(x_1)$ touches the x axis tangentially.
8. A complex root can be determined even by giving a real value as the initial guess in Newton's method.
9. The regula falsi method is the same as the method of linear interpolation.

(*continued, page 116*)

10. The fixed point iteration method and the method of successive substitutions are the same.
11. The Newton–Raphson method requires the inversion of the $n \times n$ Jacobian matrix.
12. If the initial guesses are real numbers, Muller's method cannot find the complex roots.

2.3. **Define the following:**
1. Jacobian of $f_i(x_1, x_2, \ldots, x_n)$.
2. stagnation point
3. interval of uncertainty
4. deflated polynomial
5. bisection method

2.4. **Fill in the blanks** with the proper words:
1. If the graph of $f(x)$ touches the x axis at x_1, then x_1 denotes a ______ root. (single/multiple)
2. The ______ method requires two initial guesses to find a root. (Newton's/secant)
3. The secant method is similar to the ______ method. (bisection/Newton's)
4. The regula falsi method is similar to the ______ method in bracketing a root. (bisection/secant).
5. Muller's method can be considered to be more general than the ______ method. (regula falsi/Newton's)

2.5. **Match the following:**

1. $\alpha \cosh \dfrac{100}{\alpha} - \alpha - 1 = 0$	(a) Polynomial equation
2. $2\alpha^5 - 3\alpha^3 + 5\alpha^2 + 4 = 0$	(b) General nonlinear equation
3. $4^{x^2} + 20 \ln x - x - 1 = 0$	(c) Transcendental equation

2.6. **Match the following:**

1. Fixed point iteration method	(a) requires no derivatives of $f(x)$
2. Newton's method	(b) requires the equation to be written in the form $x = g(x)$ to find the root
3. Bisection method	(c) requires the first derivative of $f(x)$
4. Muller's method	(d) requires two initial guess values for the root
5. Secant method	(e) requires three initial guess values for the root

(*continued, page 117*)

2.7. **Match the following:**

1. Regula falsi method	(a) successive substitution method
2. Fixed point iteration method	(b) accelerated step size can be used
3. Muller's method	(c) quadratic factors are found
4. Bairstow's method	(d) linear interpolation method
5. Newton–Raphson method	(e) quadratic interpolation method
6. Incremental search method	(f) intersection of chord and x axis is sought
7. Secant method	(g) based on Taylor's series expansion of $f(x)$

2.8. Select the most appropriate answer out of the multiple choices given:

1. The values of x and the corresponding values of the function, $f(x) = x^3 - 4x + 2$, in the bisection method are $x = 0.0$, $f(0) = 2.0$; $x = 1.0$, $f(1.0) = -1.0$; $x = 0.5$, $f(0.5) = 0.125$; and $x = 0.75$, $f(0.75) = -0.578125$.

   The next value of x is (a) 0.625 (b) 0.875 (c) 1.250.
2. In the incremental search method using an accelerated step size, the function values are evaluated at $x = 0.00, 0.05, 0.15,$ and 0.35. The next function value is to be evaluated at (a) 0.55 (b) 0.75 (c) 0.45.
3. The forward finite difference approximation of the derivative of $f(x) = x^3-4x+2$ at $x = 1$ with a step size (h) of 0.1 is given by

   (a) −0.99 (b) −1.29 (c) −0.69.
4. The backward finite difference approximation of the derivative of $f(x) = x^3 - 4x + 2$ at $x = 1$ with a step size (h) of 0.1 is given by

   (a) −0.99 (b) −1.29 (c) −0.69.
5. The central finite difference approximation of the derivative of $f(x) = x^3-4x+2$ at $x = 1$ with a step size (h) of 0.1 is given by

   (a) −0.99 (b) −1.29 (c) −0.69.
6. Newton's iteration, $x_{i+1} = x_i - \frac{f(x_i)}{f'(x_i)}$, applied to the function $f(x) = x^3 - 4x + 2$ with $x_1 = 1.0$ gives x_2 as

   (a) 0.0 (b) 2.0 (c) 1.0.
7. The secant method, $x_{i+1} = x_i - f(x_i)\frac{x_i - x_{i-1}}{f(x_i)-f(x_{i-1})}$, applied to the function $f(x) = x^3 - 4x + 2$ with $x_1 = 0.0$ and $x_2 = 1.0$ gives x_3 as

   (a) 0.6667 (b) 0.3333 (c) 1.3333.

(*continued, page 118*)

8. The regula falsi method, $x_{i+1} = x_i - f(x_i)\frac{x_i - x_{i-1}}{f(x_i) - f(x_{i-1})}$, applied to the function $f(x) = e^x - \sin x$ with $x_1 = 0.0$ and $x_2 = 1.0$ gives x_3 as
   (a) 3.1405 (b) −2.1405 (c) −1.1405.
9. The fixed point iteration, $x_{i+1} = g(x_i)$, applied to the function $f(x) = x^3 - 4x + 2$ or $x = (x^3 + 2)/4$ with $x_1 = 0.0$ gives x_2 as
   (a) 0.5 (b) 2.0 (c) 1.0.
10. Newton's method, $x_{i+1} = x_i - \frac{f(x_i)}{f'(x_i)}$, applied to the function $f(x) = x^2 - 3x + 3$ with $x_1 = i$ gives x_2 as
    (a) $2.0 - 3.0i$ (b) $0.9231 - 0.3846i$ (c) $0.9231 + 0.6154i$.
11. The Jacobian of the functions $f_1 = x_1^2 - x_2 + 2$ and $f_2 = x_1^3 - x_2^2$ at $x_1 = 0$ and $x_2 = 1$ is given by
    (a) $\begin{bmatrix} 0 & -1 \\ 0 & -2 \end{bmatrix}$ (b) $\begin{bmatrix} 2 & -1 \\ 3 & 0 \end{bmatrix}$ (c) $\begin{bmatrix} 0 & -1 \\ 0 & 0 \end{bmatrix}$.
12. The Newton–Raphson method, applied to the functions $f_1 = x_1 - x_2$ and $f_2 = x_1^2 + x_2^2 - 4$ with $x_1^{(1)} = 0$ and $x_2^{(1)} = 1$ gives $x_1^{(2)}$ and $x_2^{(2)}$ as
    (a) 2.5, 1.5 (b) 2.5, 2.5 (c) −2.5, −0.5.
13. If $x_1 = 1$ is a double root of the polynomial $f(x) = x^4 - 7x^3 + 17x^2 - 17x + 6$, the function $f_1(x)$ that gives $f(x) = f_1(x)(x-1)^2$ is
    (a) $x^2 - 2x + 1$ (b) $x^2 - 5x + 6$ (c) $x^2 - 5x - 6$.

PROBLEMS

Section 2.1

2.1. For a curved beam subjected to bending, such as a crane hook lifting a load, the location of the neutral axis (r_n) is given by (see Fig. 2.26)

$$4r_n(2R - \sqrt{4R^2 - d^2}) = d^2,$$

where R is the radius of the centroidal axis and d is the diameter of the cross section (assumed to be circular) of the curved beam. Use a graphical procedure to find the value of d for which $r_n = 4d$ when $R = 10$ in.

2.2. When two spheres in contact, such as ball bearings, transmit a load P, the normal stresses developed at a point along the z axis are given by (see Fig. 2.11)

$$\frac{\sigma_x}{p_{\max}} = \frac{\sigma_y}{p_{\max}} = -\left[1 - \left(\frac{z}{a}\right)\tan^{-1}\frac{1}{\left(\frac{z}{a}\right)}\right](1+\mu) + \frac{1}{2\left(1 + (\frac{z}{a})^2\right)},$$

where σ_x and σ_y are the normal stresses in the x and y directions, respectively, $p_{\max}$ is the maximum pressure at the contact point, and a is the radius of the deformed flat circular surface around the contact point. Using a graphical procedure, find the value of $(\frac{z}{a})$ at which $\frac{\sigma_x}{p_{\max}}$ attains a value of -0.5. Assume the value of Poisson's ratio (μ) as 0.3.

2.3. The design of a mechanical component requires the maximum principal stress to be less than the material strength. For a component subjected to arbitrary loads, the principal stresses are given by the solution of the cubic equation

$$\sigma^3 - I_1\sigma^2 + I_2\sigma - I_3 = 0 \qquad \text{(E1)}$$

where

$$I_1 = \sigma_x + \sigma_y + \sigma_z, \qquad \text{(E2)}$$

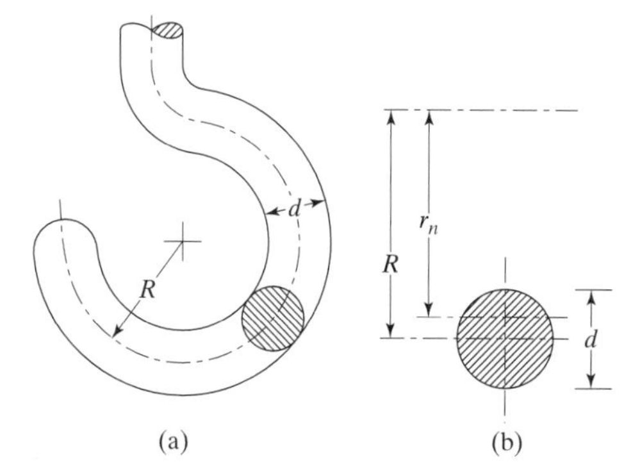

Figure 2.26 Crane hook.

$$I_2 = \sigma_x\sigma_y + \sigma_x\sigma_z + \sigma_y\sigma_z - \tau_{xy}^2 - \tau_{yz}^2 - \tau_{zx}^2, \tag{E3}$$

and

$$I_3 = \sigma_x\sigma_y\sigma_z + 2\tau_{xy}\tau_{yz}\tau_{zx} - \sigma_x\tau_{yz}^2 - \sigma_y\tau_{zx}^2 - \sigma_z\tau_{xy}^2, \tag{E4}$$

with $\sigma_x, \sigma_y, \sigma_z$ denoting the normal stresses acting along the *x, y, z* directions, and $\tau_{xy}, \tau_{yz}, \tau_{zx}$ indicating the shear stresses acting in the *xy, yz, zx* planes, respectively, as shown in Fig. 2.27. Find the principal stresses in a machine component for the following data using a graphical procedure: $\sigma_x = 3$ ksi, $\sigma_y = 0$, $\sigma_z = 0$, $\tau_{xy} = 1$ ksi, $\tau_{yz} = 2$ ksi, and $\tau_{zx} = 1$ ksi.

Section 2.2

In Problems 2.4 to 2.21, find an approximate solution using a graphical procedure.

2.4. The stress induced in a column subjected to an eccentric load (as shown in Fig. 2.28) is given by the secant formula [2.5]

$$\frac{P}{A} = \frac{S_{yc}}{1 + \left(\frac{ec}{k^2}\right)\sec\left(\frac{l}{2k}\sqrt{\frac{P}{AE}}\right)},$$

where P is the axial load applied, A is the cross-sectional area of the column, S_{yc} is the yield stress of the material in compression, e is the eccentricity of the load, l is the length of the column, E is Young's modulus, c is the distance of the outermost fiber from the neutral axis of the column, and k is the radius of gyration of the cross section of the column. Find the value of $\frac{P}{A}$ for the following data: $S_{yc} = 40,000$ psi, $\frac{ec}{k^2} = 0.2$, $\frac{l}{k} = 50$, and $E = 30 \times 10^6$ psi.

2.5. The specific volume of a gas (v) can be expressed as a function of its pressure p and temperature T, using van der Waals equation, as [2.3]

$$p = \frac{KT}{v - b} - \frac{a}{v^2}, \tag{E1}$$

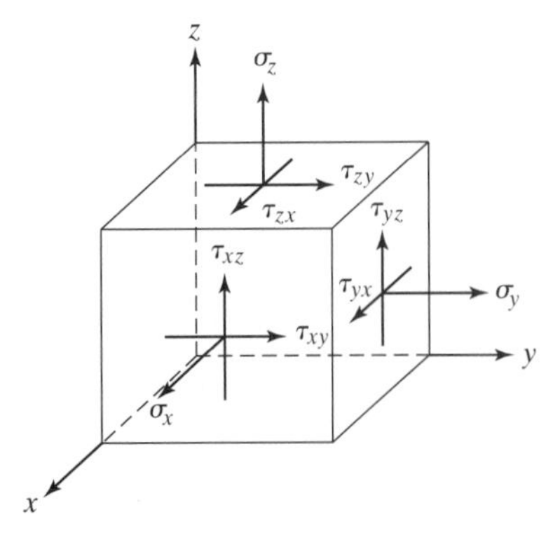

Figure 2.27 Stresses acting on an element.

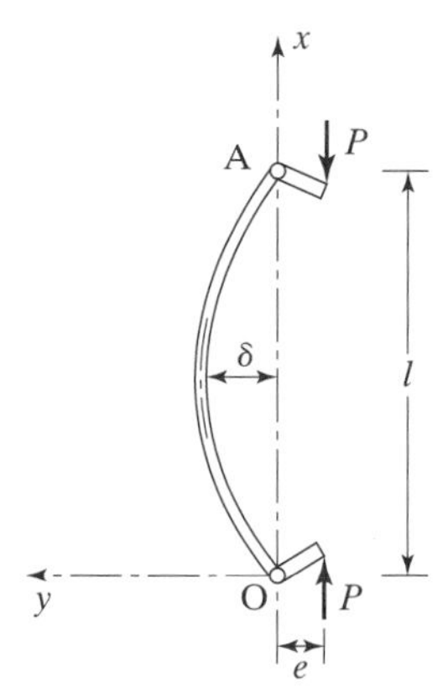

Figure 2.28 Column subjected to eccentric load.

where K is the gas constant, and a and b are constants. Find the specific volume of methane gas at a pressure of 100 atm and temperature 0° F with $a = 4787.4$ $\text{ft}^4 - \text{lbf/lbm}^2$, $b = 0.0496\text{ft}^3/\text{ lbm}$, and $K = 96.35$ ft − lbf/lbm −° R.

2.6. If two plates are fastened by a bolted joint, the stiffness of the fastened members or plates (k_m) is given by

$$k_m = \frac{0.577\pi Ed}{2\ln\left(5\left\{\frac{0.577l + 0.5d}{0.577l + 2.5d}\right\}\right)},$$

where E is Young's modulus of the fastened members, d is the diameter of the bolt, and l is the thickness of the fastened members. Find the value of d that corresponds to a value of $k_m = 7.5 \times 10^6$ lb/in when $E = 12 \times 10^6$ psi and $l = 1.5$ in.

2.7. The normal stress induced at the inner fiber of a torsional helical spring is given by

$$\sigma_i = \left\{\frac{4C^2 - C - 1}{4C(C-1)}\right\}\frac{Mc}{I},$$

where $I = \frac{\pi d^4}{64}$, $c = \frac{d}{2}$, $C = \frac{D}{d}$, M is the bending moment, D is the mean coil diameter, and d is the wire diameter. Find the value of C that corresponds to a stress of $\sigma_i = 55 \times 10^3$ psi, when $M = 5$ lb-in and $d = 0.1$ in.

2.8. The angular position of the output link (θ_4) of a four-bar linkage corresponding to any specified angular position of the input link (θ_2) can be computed using the Freudenstein's equation [2.6]

$$k_1 \cos\theta_4 - k_2 \cos\theta_2 + k_3 = \cos(\theta_2 - \theta_4), \tag{E1}$$

where

$$k_1 = \frac{d}{a},\ k_2 = \frac{d}{c},\ \text{and } k_3 = \frac{a^2 - b^2 + c^2 + d^2}{2ac}. \tag{E2}$$

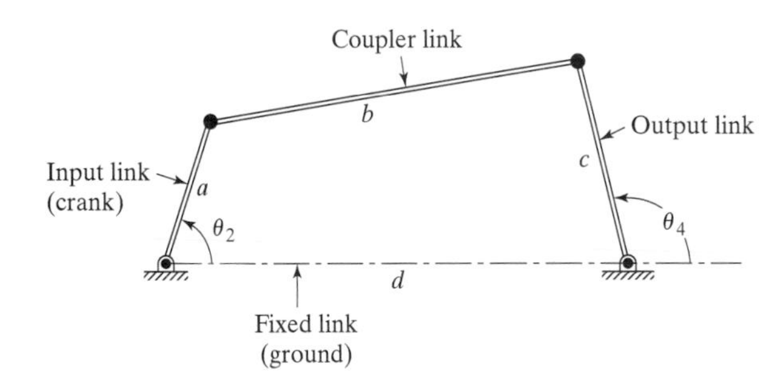

Figure 2.29 Four-bar linkage.

a, b, c and d are the link lengths shown in Fig. 2.29. Find the value of θ_4 for the following data: $\theta_2 = 30°$, $a = 1$, $b = 2$, $c = 4$, and $d = 5$.

2.9. The natural frequencies of radial vibration of an elastic sphere (ω) are given by [2.2]

$$\frac{\tan(ka)}{ka} = \frac{4G}{4G - (\lambda + 2G)(ka)^2}, \tag{E1}$$

where

$$\lambda = \frac{vE}{(1+\nu)(1-2\nu)} = \text{Lame's constant}$$

and $G = \frac{E}{2(1+\nu)}$ = shear modulus, a = radius of the sphere, ρ = density, and

$$k^2 = \left(\frac{\omega^2 \rho}{\lambda + 2G}\right).$$

When v is equal to 0.25, which makes λ equal to G, Eq. (E1) reduces to

$$\frac{\tan ka}{ka} = \frac{1}{1 - \frac{3}{4}k^2a^2}.$$

Find the first natural frequency of vibration (value of ω) of a sphere with the following data. $a = 10$ in, $\nu = 0.25$, $\rho g = 0.28$ lb/in^3, and $E = 30 \times 10^6$ psi.

2.10. The variation of temperature with time along the thickness of a slab, which is suddenly immersed in a hot fluid, requires the solution of the transcendental equation

$$\beta \tan \beta = \frac{h_0 l}{k}, \tag{E1}$$

where h_0 is the convection coefficient, k is the thermal conductivity, and $2\,l$ is the thickness of the slab. (See Fig. 2.30.) Determine the smallest root (value of β) of Eq. (E1) for a slab with $\frac{h_0 l}{k} = 0$.

2.11. An open box is to be constructed from a rectangular piece of cardboard of size $30'' \times 50''$. If four square pieces are to be cut from the corners of the cardboard

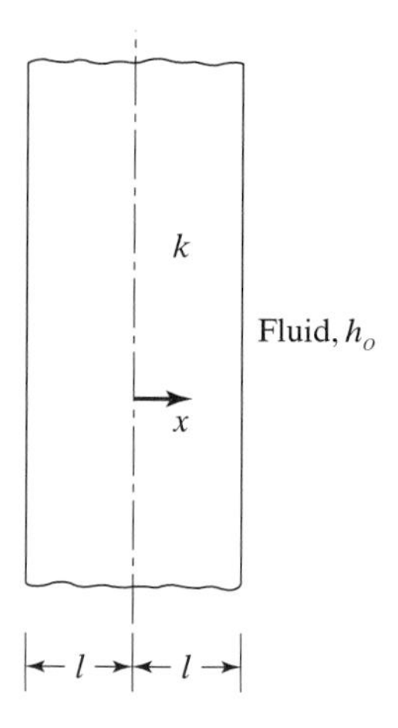

Figure 2.30 A slab of thickness $2l$ (infinite size in yz-plane).

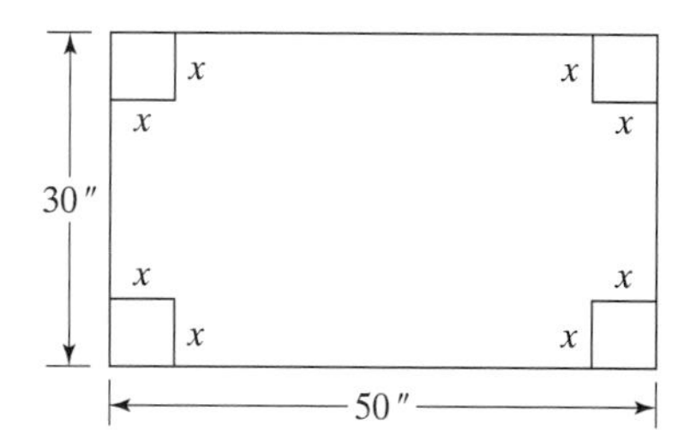

Figure 2.31 Open box from rectangular cardboard.

to obtain a height of x'' (see Fig. 2.31), determine the value of x that gives a box of volume 3000 in^3.

2.12. The natural frequencies of a 3 degree of freedom spring–mass system, shown in Fig. 2.32, are given by the equation

$$\alpha^3 - 5\alpha^2 + 6\alpha - 1 = 0, \tag{E1}$$

where $\alpha = \frac{m\omega^2}{k}$, m is the mass, ω is the natural frequency, and k is the spring stiffness. Find the roots of Eq. (E1).

2.13. The determination of the temperature variation with time in the radial direction of a sphere of radius R and thermal conductivity k, when suddenly immersed in a hot fluid with a heat transfer coefficient h_0, requires the solution of the transcendental equation

$$1 - \beta \cot \beta = \frac{h_0 R}{k}. \tag{E1}$$

Determine the first root of Eq. (E1) for a sphere with $\frac{h_0 R}{k} = 1.0$.

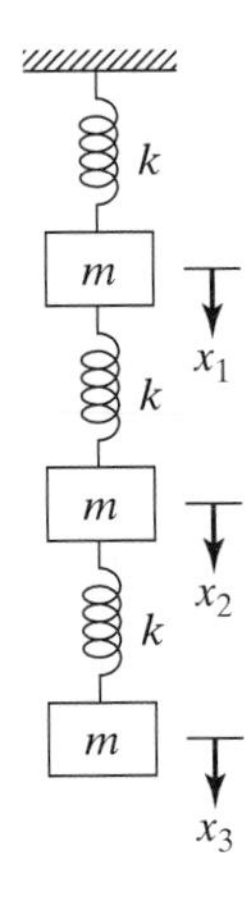

Figure 2.32 Spring-mass system.

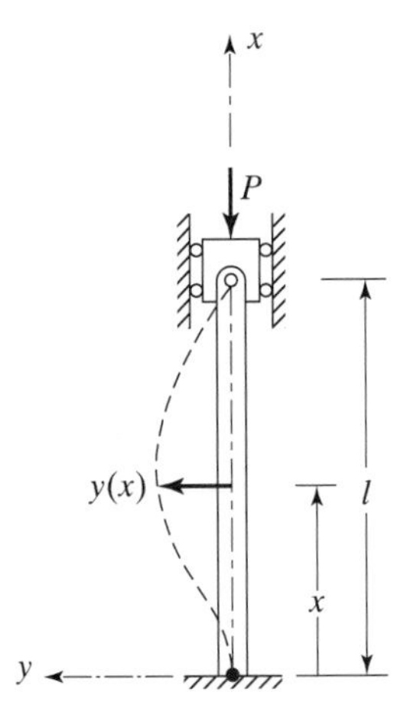

Figure 2.33 Transverse deflection of column.

2.14. The transverse deflection of a long column, fixed at one end and pinned at the other end (Fig. 2.33), under axial load is given by

$$y(x) = 0.77 y_0 \left[1 - \cos \frac{ax}{l} + \frac{1}{a} \sin \frac{ax}{l} - \frac{x}{l} \right],$$

where $a = 4.49$, l is the length of the column, and y_0 is a constant. Find the value of $\left(\frac{x}{l}\right)$ at which $y(x) = 0.5\, y_0$.

2.15. The velocity of a nonNewtonian fluid flowing in a circular tube (u) can be expressed as

$$\frac{u}{u_{\text{mean}}} = \left(\frac{3n+1}{n+1} \right) \left[1 - \left(\frac{r}{R} \right)^{\frac{n+1}{n}} \right],$$

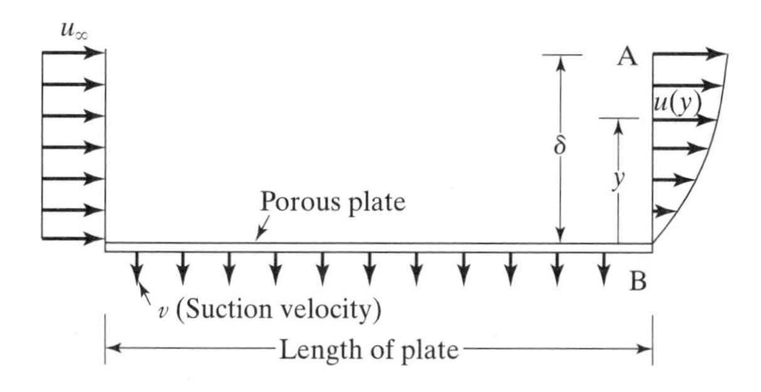

Figure 2.34 Flow past a porous flat plate.

where u_{mean} is the mean velocity of the fluid, r is radial distance from the center of the tube, R is the radius of the tube, and n is a constant whose value depends on the fluid (for example, $n = 1$ for a Newtonian fluid, $n = 3$ for a dilatant fluid, and $n = \frac{1}{3}$ for a pseudoplastic fluid). Determine the value of n of a fluid for which $\frac{u}{u_{\text{mean}}} = 0.9$ at $\frac{r}{R} = 0.8$.

2.16. When water flows steadily past a porous flat plate as shown in Fig. 2.34, the variation of velocity (u) with the distance y at section AB is given by [2.4]

$$\frac{u}{u_\infty} = 3\left(\frac{y}{\delta}\right) - 2\left(\frac{y}{\delta}\right)^{1.5}.$$

Find the value of $\frac{y}{\delta}$ at which $\frac{u}{u_\infty}$ attains a value of 0.5.

2.17. The discharge coefficient (C) of an orifice (shown in Fig. 2.35) is defined as

$$C = \frac{\text{actual mass flow rate}}{\text{theoretical mass flow rate}}$$

and is given by [2.4]

$$C = 0.5959 + 0.0312\beta^{2.1} - 0.184\beta^8 + \frac{91.71\beta^{2.5}}{\text{Re}^{0.75}}, \tag{E1}$$

where β denotes the ratio of diameters,

$$\beta = \frac{d}{D},$$

and Re indicates Reynold's number. Find the value of β that yields a discharge coefficient of $C = 0.5959$ at a Reynold's number of $\text{Re} = 10^4$.

2.18. The ratio of stagnation pressure (p_0) to static pressure (p) in a compressible fluid flow can be expressed as [2.4]:

$$\frac{p_0}{p} = 1 + \frac{k}{2}M^2\left\{1 + \frac{1}{4}M^2 + \left(\frac{2-k}{24}\right)M^4\right\},$$

where k is the ratio of specific heats of the fluid and M is the Mach number.

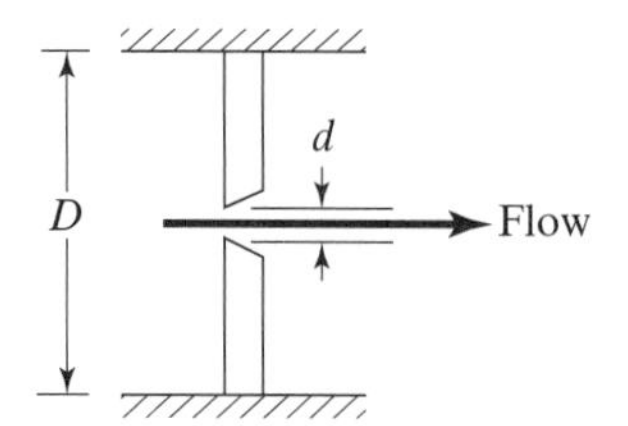

Figure 2.35 Flow through orifice.

Find the value of the Mach number at which the ratio of pressures, $\left(\frac{p_0}{p}\right)$, will be equal to 1.2 when the fluid is air for which $k = 1.4$.

2.19. A set of three mutually perpendicular axes in a body for which the products of inertia about these axes are zero are known as the principal axes of inertia of the body. They are given by the solution of the following determinantal equation:

$$\begin{vmatrix} I_{xx} - I & I_{xy} & I_{xz} \\ I_{yx} & I_{yy} - I & I_{yz} \\ I_{zx} & I_{zy} & I_{zz} - I \end{vmatrix} = 0.$$

Find the principal moments of inertia of a body for which $I_{xx} = 12$, $I_{yy} = 16$, $I_{zz} = 16$, $I_{xy} = 6$, $I_{yz} = 2$, and $I_{xz} = -6$.
Hint: $I_{ij} = I_{ji}$ for any body.

2.20. The vertical velocity of a motorcycle due to a road bump is given by (see Fig. 2.36)

$$v(t) = Xe^{-\zeta\omega_n t}(-\zeta\omega_n \sin\omega_d t + \omega_d \cos\omega_d t),$$

where X is the maximum displacement of the motor cycle, ζ is the damping constant of the suspension, ω_n is the undamped natural frequency of the system, $\omega_d = \omega_n\sqrt{1-\zeta^2}$ is the damped natural frequency of the system, and t is time. Determine the time, t, at which the velocity of the motorcycle attains a value of 1 m/s for the following data: $X = 0.4550$ m, $\zeta = 0.4037$, and $\omega_n = 3.4338$ rad/s.

2.21. The schematic diagram of a large cannon is shown in Fig. 2.37 [2.1]. When the gun is fired, high-pressure gases accelerate the projectile inside the barrel to a high velocity. The reaction force pushes the gun barrel in the opposite direction of the projectile. Since it is desirable to bring the gun barrel to rest in the shortest time without oscillation, it is made to translate backward against a critically damped spring–damper system called the recoil mechanism. In a particular case, the recoil distance (x) at time t is given by

$$x = vte^{-\omega_n t},$$

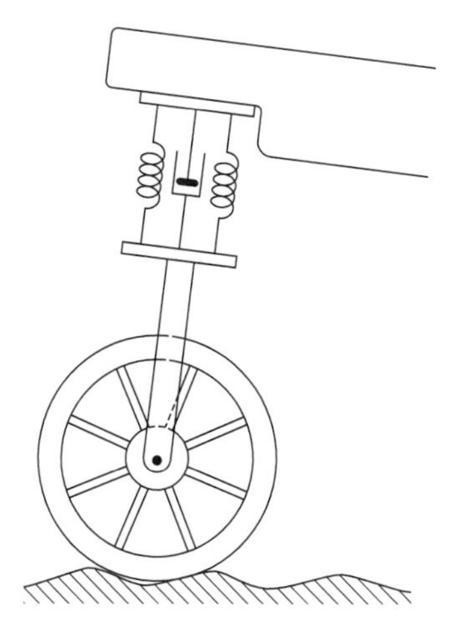

Figure 2.36 Motor cycle over a road bump.

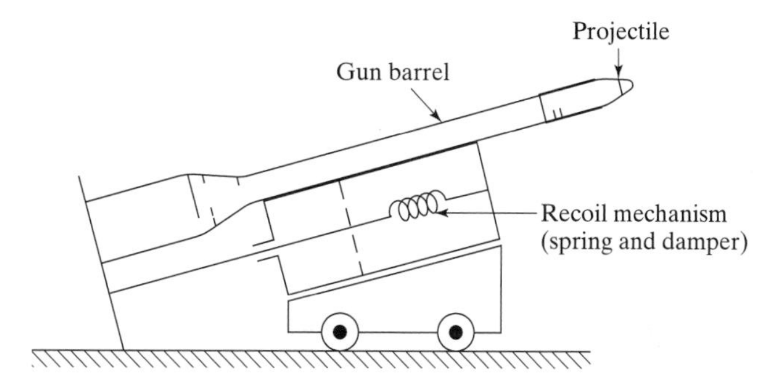

Figure 2.37 Cannon with a projectile.

where v is the initial recoil velocity, ω_n is the natural frequency of the recoil mechanism, and t is the time. Find the time corresponding to a recoil distance of 0.1 m when $v = 4.8626$ m/s and $\omega_n = 4.4721$ rad/s.

Section 2.3

2.22. Find the root of the equation (see Example 2.2)

$$f(x) = \frac{1.1288}{\sqrt{x}} - \tanh(1.3288\sqrt{x}) = 0$$

using incremental search method with $x_1 = 1.0$ and $\Delta x^{(1)} = 0.1$.

2.23. Find the root of the equation (see Problem 2.3)

$$f(x) = x^3 - 3x^2 - 6x + 8 = 0$$

using incremental search method with $x_1 = 0.0$ and $\Delta x^{(1)} = 0.5$.

2.24. Find the interval $a \leq x \leq b$ such that $f(a)$ and $f(b)$ have opposite signs for the function

$$f(x) = e^{2x} - 3x - 8$$

using incremental search procedure.

2.25. Find the solution of Problem 2.15 using incremental search method with $n_1 = 0.0$ and $\Delta n^{(1)} = 0.1$.

Section 2.4

2.26. Find the root of the equation (see Example 2.3)

$$f(x) = \sqrt{4x^2 - 0.04} + 0.24 \sin^{-1}\left(\frac{0.1}{x}\right) - 10.055752 = 0$$

using bisection method with $a = 2$, $b = 10$, and $\varepsilon = 0.001$.

2.27. Find the solution of Problem 2.1 using bisection method with $a = 0.0$, $b = 5.0$, and $\varepsilon = 0.001$.

2.28. Find the solution of Problem 2.7 using bisection method with the approximations 5.0 and 12.0 for C, and $\varepsilon = 0.001$.

Section 2.5

2.29. Find the roots of the equation

$$f(x) = x^3 - 0.2589x^2 + 0.02262x - 0.001122 = 0$$

using Newton–Raphson method with $x_1 = 0.0$ and $\varepsilon = 0.001$.

2.30. Find the solution of Problem 2.2 using Newton–Raphson method with the starting point $\left(\frac{z}{a}\right)_1 = 0.1$ and $\varepsilon = 0.001$.

2.31. Find the solution of Problem 2.8 using Newton–Raphson method with the starting value of θ_4 as 90° and $\varepsilon = 0.001$.

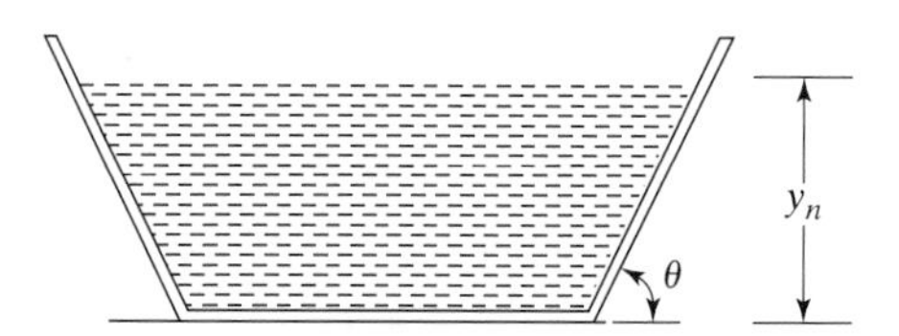

Figure 2.38 Flow through trapezoidal channel.

2.32. Find the solution of Problem 2.5 using Newton–Raphson method with the starting value of $v = 0.0$ and $\varepsilon = 0.001$.

Section 2.6

2.33. Find the root of the equation $f(x) = 8x^3 + x^2 + 8x - 3 = 0$ using secant method with 0.0 and 0.6 as the initial approximations for x and $\varepsilon = 0.001$.

2.34. Find the solution of Problem 2.6 using secant method with 0.0 and 1.0 as the initial approximations for d and $\varepsilon = 0.001$.

2.35. When water flows through a trapezoidal channel section (Fig. 2.38), the cross-sectional area of the fluid (A) and the wetted perimeter of the fluid (P) are related as

$$P = \frac{A}{y_n} - y_n \cot\theta + \frac{2y_n}{\sin\theta}.$$

Find the value of θ for which $P = 3$ m, when $A = 1$ m^2 and $y_n = 0.5$ m with 10° and 60° as the initial approximations for θ and $\varepsilon = 0.001$.

Section 2.7

2.36. Find the solution of the equation $f(\lambda) = 0.8 - \tan\lambda\left(\frac{1-0.08\tan\lambda}{0.08+\tan\lambda}\right) = 0$ using regula falsi method with 10° and 30° as the initial approximations for λ and $\varepsilon = 0.001$.

2.37. Find the solution of Problem 2.16 using regula falsi method with the initial approximations 0.0 and 1.0 for $\frac{y}{\delta}$ and $\varepsilon = 0.001$.

2.38. Find the root of the equation $f(x) = \ln x$ with the initial approximations $x_1 = 0.6$ and $x_2 = 6.0$ using

(a) secant method and

(b) regula falsi method.

Discuss your observations.

Section 2.8

2.39. Find the root of the equation

$$f(x) = x - \frac{1}{x^2 + 9} = 0$$

using fixed point iteration with $x_1 = 3.0$ and $\varepsilon = 0.001$.

2.40. Find the root of the equation

$$f(x) = x^4 - 31x^3 + 305x^2 - 1025x + 750 = 0$$

using fixed point iteration with $x_1 = 20.0$ and $\varepsilon = 0.001$.

2.41. A person plans to deposit $ 100 a month into a bank account for 10 years. Find the interest rate needed to accumulate a total amount of $ 20,000, including interest at the end of 10 years. Use fixed-point iteration by taking the initial value of the interest rate as 0.05 and $\varepsilon = 0.001$.

Hint: Formula is $S = R\left\{\frac{(1+i)^n - 1}{i}\right\}$, where S = future sum, R = uniform payments, n = number of payments, and i = interest rate.

2.42. The fixed point iteration method can be extended to find the solution of several nonlinear equations. If $f_1(x, y) = 0$ and $f_2(x, y) = 0$ denote the equations to be solved, they are rewritten in the form:

$$f_1(x, y) = 0 \longrightarrow x = g_1(x, y)$$

and

$$f_2(x, y) = 0 \longrightarrow y = g_2(x, y).$$

If (x_1, y_1) denotes the initial approximation to the solution, then the fixed-point iteration can be used to find the next approximation as

$$x_{i+1} = g_1(x_i, y_i)$$

and

$$y_{i+1} = g_2(x_i, y_i).$$

Using this technique, solve the following system of equations:

$$y^2 = 8x$$

and

$$(x - 1)^2 + (y - 2)^2 - 4 = 0.$$

Section 2.9

2.43. Find the root of multiplicity two of the following equation using the starting point, $x_1 = 1.0$:

$$f(x) = x^4 - 27x^3 + 265x^2 - 1125x + 1750 = 0.$$

2.44. Find the root of multiplicity three of the following equation using the starting point, $x_1 = 2.0$:

$$f(x) = x^4 - 8x^3 + 18x^2 - 16x + 5 = 0.$$

2.45. Find the root of muliplicity four of the following equation using $x_1 = 0.0$:

$$f(x) = x^4 - 8x^3 + 24x^2 - 32x + 16 = 0.$$

Section 2.10

2.46. Find the roots of the equation

$$x^4 - 32x^3 + 244x^2 - 20x - 1200 = 0$$

using Bairstow's method. Assume the initial approximations of $\bar{\alpha}$ and $\bar{\beta}$ as -1 and 1, respectively.

2.47. Find the roots of the equation

$$x^3 - 7x^2 + 12x - 10 = 0$$

using Bairstow's method. Assume the initial values of $\bar{\alpha}$ and $\bar{\beta}$ as -1 and 1, respectively.

Section 2.11

2.48. Solve Problem 2.17 using Muller's method with the starting values $\beta_1 = 0.0$, $\beta_2 = 0.2$, and $\beta_3 = 0.8$.

2.49. Use Muller's method to find a root of the equation

$$f(x) = e^x - x^2 = 0$$

using the initial approximations $x_1 = -1$, $x_2 = 0$, and $x_3 = 1$.

2.50. Use Muller's method to find a root of the equation

$$f(x) = x^4 - 31x^3 + 305x^2 - 1025x + 750 = 0$$

using the starting points $x_1 = 4$, $x_2 = 6$, and $x_3 = 12$.

Section 2.12

2.51. Find the Jacobian matrix for the following functions at the points (i) $x = 1$, $y = 1$, and (ii) $x = 0$, $y = 0$:

$$f_1(x, y) = 10x^3 - y \text{ and } f_2(x, y) = xy - 1.$$

2.52. Find the solution of the following equations using Newton–Raphson method with the starting point (0.5, 0.5):

$$10x^3 - y = 0 \text{ and } xy - 1 = 0.$$

2.53. Find the solution of the following equations using Newton–Raphson method with the starting point (1.0, 1.0):

$$\cosh x - y = 0 \text{ and } xy - 1 = 0.$$

2.54. Find the solution of the following equations using Newton–Raphson method with the starting point (0.1, 0.1):

$$10x^2 + y^2 - 10 = 0 \text{ and } x^2 + y^2 - 4x - 2y + 1 = 0.$$

Section 2.13

2.55. Consider the simultaneous equations:

$$f_1(x, y) = x^2 - y = 0 \tag{E1}$$

and

$$f_2(x, y) = y^2 - x = 0. \tag{E2}$$

Construct a function $F(x, y)$ whose minimum will yield the solution of Eqs. (E1) and (E2). Derive the necessary and sufficient conditions corresponding to the minimum of $F(x, y)$.

Section 2.14

2.56. Prove that the results of Example 2.7 satisfy the convergence relationship given by Eq. (2.91).

2.57. Show that the secant method converges with an order of convergence 1.618 using the results of Example 2.14.

2.58. Show that the fixed point iteration converges linearly using the results of Example 2.15.

Section 2.16

2.59. Find the solution of Problem 2.36 using MAPLE.

2.60. Solve Problem 2.18 using MAPLE.

2.61. Find the solution of the equations $f_1(x, y) = 2e^x + xy - 2 = 0$ and $f_2(x, y) = \sin(x\,y) + 2x + y - 1 = 0$ using MAPLE.

2.62. Solve Problem 2.36 using MATLAB.

2.63. Solve Problem 2.18 using MATLAB.

2.64. Find the solution of the equations given in Problem 2.61 using MATLAB.

2.65. Solve Problem 2.36 using MATHCAD.

2.66. Solve Problem 2.18 using MATHCAD.

2.67. Find the solution of the equations given in Problem 2.61 using MATHCAD.

Section 2.17

2.68. Find the solution of Problem 2.8 using PROGRAM 2.1 (Secant method) with $X1 = 0.0$, $X2 = 5.0$, MITER = 100, and EPS = 0.001.

2.69. Solve Problem 2.18 using PROGRAM 2.1 (Sécant method) with $X1 = 0.1$, $X2 = 2.0$, MITER = 100, and EPS = 0.0001.

2.70. Solve Problem 2.40 using PROGRAM 2.2 (Bairstow's method). Assume the initial values of $\bar{\alpha}$ and $\bar{\beta}$ as 0.0 and 2.0 respectively.

2.71. Solve Problem 2.44 using PROGRAM 2.2 (Bairstow's method) with $\bar{\alpha} = 1.0$ and $\bar{\beta} = 2.0$.

2.72. Using PROGRAM 2.3 (Newton–Raphson method), find the solution of the following equations:

$$\cosh x - y = 0 \text{ and } xy - 1 = 0.$$

Assume the initial values of x and y as 1.0 and 1.0, respectively.

2.73. Solve Problem 2.61 using PROGRAM 2.3.

2.74. Find the solution of Problem 2.19 using PROGRAM 2.4 (Newton–Raphson method) with X1 = 0.1, DX = 0.01, MITER = 100, and EPS = 0.001.

2.75. Find the solution of Problem 2.26 using PROGRAM 2.4 (Newton–Raphson method) with X1 = 0.1, DX = 0.01, MITER = 100, and EPS = 0.001.

2.76. Solve Problem 2.14 using PROGRAM 2.5 (Fixed point iteration) with X1 = 0.1, MITER = 100, and EPS = 0.0001.

2.77. Solve Problem 2.22 using PROGRAM 2.5 (Fixed point iteration) with X1 = 0.01, MITER = 100, and EPS = 0.001.

General

2.78. Write a computer program to implement the incremental search method for finding the root of the equation, $f(x) = 0$, using x_1 as the starting point and $\Delta x^{(1)}$ as the initial step length. Use this program to find the root of the equation

$$f(x) = \frac{1.5x}{(1+x^2)^2} - 0.65\tan^{-1}\left(\frac{1}{x}\right) + \frac{0.65x}{1+x^2} = 0$$

using $x_1 = 0.0$ and $\Delta x^{(1)} = 0.1$.

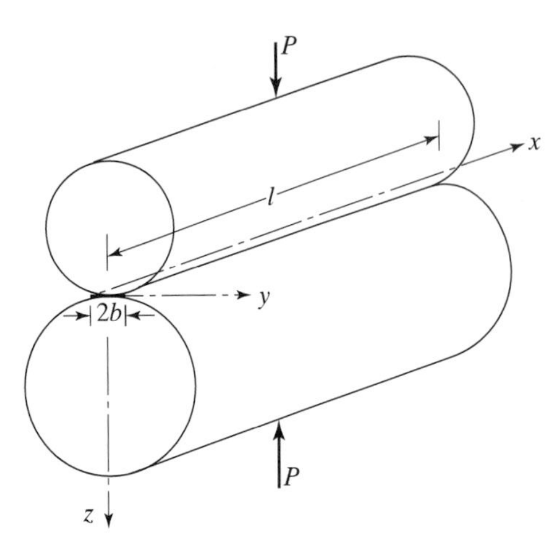

Figure 2.39 Contact stress between cylinders.

2.79. Write a computer program to implement the regula falsi method for finding the root of the equation, $f(x) = 0$, using x_1 and x_2 as the starting points. Using this program, find the root of the function given in Problem 2.78 with $x_1 = 0.0$ and $x_2 = 1.0$.

2.80. Write a computer program to find the root of the equation, $f(x) = 0$, using Muller's method. Using this program, find the roots of the polynomial equation

$$f(x) = x^4 - 12x^3 + 49x^2 - 78x + 40 = 0.$$

2.81. When two cylinders in contact, such as roller bearings, transmit a load P as shown in Fig. 2.39, the normal stress developed at a point along the z-axis is given by

$$\frac{\sigma_y}{p_{\max}} = -\left(2 - \frac{1}{1 + \left(\frac{z}{b}\right)^2}\right)\sqrt{1 + \left(\frac{z}{b}\right)^2} + 2\left(\frac{z}{b}\right),$$

where $p_{\max}$ is the maximum pressure along the contact line and $2b$ is the width of the deformed flat rectangular surface around the contact line. Find the value of $\left(\frac{z}{b}\right)$ at which $\frac{\sigma_y}{p_{\max}} = -0.5$ using a graphical procedure.

2.82. The efficiency of a square-threaded screw (η) is given by

$$\eta = \tan\lambda \left(\frac{1 - \mu \tan\lambda}{\tan\lambda + \mu}\right),$$

where μ is the coefficient of friction and λ is the lead angle of the threads. Find the lead angle λ which gives an efficiency of 0.8 when $\mu = 0.08$.

2.83. The natural frequencies of vibration of a circular plate with a clamped edge (λ) are given by [2.2]

$$\tan\left(\lambda a - \frac{\pi}{4}(n+1)\right) = \frac{1 - \dfrac{Q}{(8\lambda a)^2} + \dfrac{R}{(8\lambda a)^3} - \dfrac{S}{(8\lambda a)^4}}{1 - \dfrac{P}{(8\lambda a)} + \dfrac{Q}{(8\lambda a)^2} - \dfrac{S}{(8\lambda a)^4}},$$

with

$$P = 2(4n^2 - 1),$$
$$Q = (4n^2 - 1)(4n^2 - 9),$$
$$R = \frac{2}{3}(4n^2 - 1)(4n^2 - 9)(4n^2 - 13),$$

and

$$S = \frac{1}{6}(4n^2 - 1)(4n^2 - 9)(4n^2 - 25),$$

where ν is the Poisson's ratio, a is the radius of the plate, and n is the number of nodal lines that define the mode shape. Find the natural frequency (value of λa) corresponding to $n = 1$ for a plate with $\nu = 0.3$.

2.84. The thermal efficiency of a uniform fin (η), considering convection through the tip, is given by [2.15]

$$\eta = \left(\frac{\sinh\dfrac{l}{\lambda} + \alpha\cosh\dfrac{l}{\lambda}}{\cosh\dfrac{l}{\lambda} + \alpha\sinh\dfrac{l}{\lambda}}\right)\left(\frac{\lambda}{l + \dfrac{A}{P}}\right),$$

with

$$\lambda = \left(\frac{kA}{h_\infty P}\right)^{\frac{1}{2}}$$

and

$$\alpha = \left(\frac{h_\infty A}{kP}\right)^{\frac{1}{2}},$$

where l = length, A = cross-sectional area, P = perimeter of the cross section, k = thermal conductivity, and h_∞ = heat transfer coefficient of the fin (Fig. 2.40). If the fin is made of aluminum with a square cross section with $l = 0.1\text{m}$, $k = 240\text{W}/(\text{m} -^\circ \text{C})$, and $h_\infty = 9\text{W}/(\text{m}^2 -^\circ \text{C})$, determine the necessary cross-sectional dimensions of the fin to achieve an efficiency of 0.95.

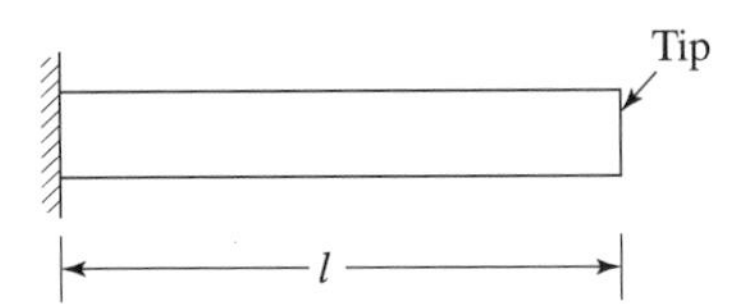

Figure 2.40 Uniform fin.

2.85. The equation governing the natural frequencies of a free-free beam with continuous material distribution is given by [2.1]

$$\cos \beta_n l \cosh \beta_n l = 1,$$

where the nth natural frequency, ω_n, is given by

$$\omega_n = (\beta_n l)^2 \left(\frac{EI}{\rho A l^4} \right)^{\frac{1}{2}} \quad \text{and } n = 1, 2, \ldots,$$

where l is the length, E is Young's modulus, I is the area moment of inertia of the cross section, ρ is the density, and A is the area of cross section of the beam. Find the fundamental (smallest) natural frequency of vibration of the beam, $\beta_1 l$.

2.86. The pressure-volume (p-v) diagram of an ideal air-standard dual-cycle heat engine is shown in Fig. 2.41. In this engine, the piston starts compressing the air isentropically at state 1 and continues to state 2. Fuel is injected into the air at state 2. Because of the high compression ratio used, the temperature of the compressed air will be sufficiently high to ignite the fuel without any ignition source. Thus, combustion occurs during the constant volume process of fuel injection and continues at constant pressure while the piston moves to reach state 4. The combustion process ends at state 4, and expansion starts at point 4 and continues to point 5 isentropically. At state 5, the exhaust valve opens and the pressure drops to the initial pressure at state 1. The cycle then repeats. The efficiency of the engine (η) can be expressed as [2.3]

$$\eta = 1 - \frac{\left[\dfrac{1}{r_1^{k-1}} \right] \left[\dfrac{p_3}{p_2} \left(\dfrac{r_1}{r_2} \right)^k - 1 \right]}{\dfrac{p_3}{p_2} - 1 + k \left(\dfrac{p_3}{p_2} \right) \left\{ \left(\dfrac{r_1}{r_2} \right) - 1 \right\}},$$

where r_1 is the volume compression ratio,

$$r_1 = \frac{v_1}{v_2};$$

r_2 is the volume expansion ratio

$$r_2 = \frac{v_5}{v_4};$$

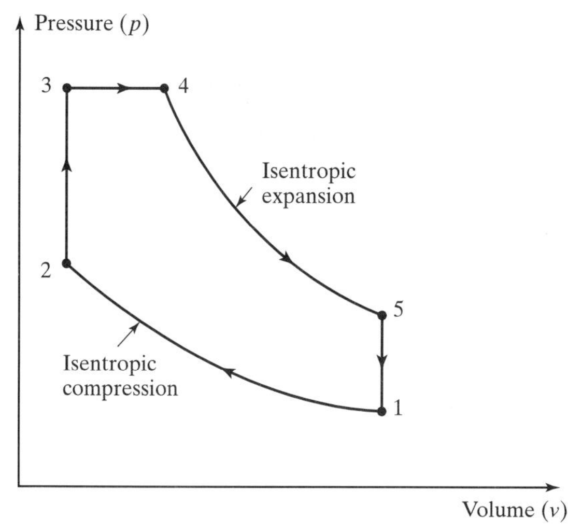

Figure 2.41 $p - v$ diagram of the dual cycle.

p_2 and p_3 are the pressures at states 2 and 3, respectively, and k is the ratio of specific heats of air. If the efficiency of a dual-cycle heat engine is known to be 0.651, find the value of r_1 when $k = 1.4$, $r_2 = 11.82$, and $\frac{p_3}{p_2} = 1.3657$.

2.87. The speed of wave propagation of a surface wave in a fluid flowing in a channel is given by [2.4].

$$v = \sqrt{\frac{g\lambda}{2\pi} \tanh\left(\frac{2\pi y}{\lambda}\right)},$$

where v is the speed of the wave, λ is the length of the wave, y is the depth of the fluid and g is the acceleration due to gravity. Find the value of $\frac{\lambda}{y}$ for which $\frac{v}{\sqrt{gy}} = 0.99$ assuming the fluid to be water.

2.88. Consider the simultaneous equations

$$f_1(x, y) = y^2 - 2y - 8x - 23 = 0 \tag{E1}$$

and

$$f_2(x, y) = y^2 + 4x^2 + 4y - 8x - 8 = 0 \tag{E2}$$

(a) Plot the functions $f_1(x, y) = 0$ and $f_2(x, y) = 0$ and find an approximate solution of the problem. Using the approximate solution as a starting point, find the solution of Eqs. (E1) and (E2) by applying the Newton–Raphson method.

(b) Construct a function $F(x, y)$ whose minimum will yield the solution of Eqs. (E1) and (E2). Derive the necessary and sufficient conditions corresponding to the minimum of $F(x, y)$.

PROJECTS

2.1. The thermal efficiency of a uniform fin (η), with an insulated tip, is given by [2.15]

$$\eta = \frac{\tanh \frac{l}{\lambda}}{\frac{l}{\lambda}}$$

with

$$\lambda = \left(\frac{kA}{h_\infty P} \right)^{\frac{1}{2}},$$

where l = length, A = cross-sectional area, P = perimeter of the cross section, k = thermal conductivity, and h_∞ = heat transfer coefficient of the fin. If the fin is made of aluminum with a square cross section with $l = 0.1$ m, $k = 240$ W/(m-°C), and $h_\infty = 9$ W/(m^2-°C), determine the necessary cross-sectional dimensions of the fin to achieve an efficiency of 0.95 using the following methods:

(a) Graphical method
(b) Bisection method
(c) Newton–Raphson method
(d) Regula Falsi method
(e) Muller's method

2.2. The natural frequencies of a beam, fixed at one end (x = 0), and carrying a mass M at the other end ($x = l$), are given by the frequency equation

$$1 + \frac{1}{\cos \beta_n l \cosh \beta_n l} - \alpha \beta_n l (\tan \beta_n l - \tanh \beta_n l) = 0,$$

where

$$\alpha = \frac{\text{mass attached at the end of the beam}}{\text{mass of the beam}} = \frac{M}{\rho A l}$$

is the mass ratio,

$$\beta_n = \frac{\sqrt{\omega_\text{n}}}{l} \left\{ \frac{\rho A l^4}{EI} \right\}^{\frac{1}{4}},$$

l is the length, E is the Young's modulus I is the area moment of inertia of the cross section, ρ is the density and A is the cross-sectional area of the beam.

(a) Determine the fundamental frequency of vibration of the beam, $\beta_1 l$, for $\alpha = 0.0, 1.0, 10.0, 100.0$, and 1000.0.
(b) Plot a graph to show the variation of $\beta_1 l$ with α.

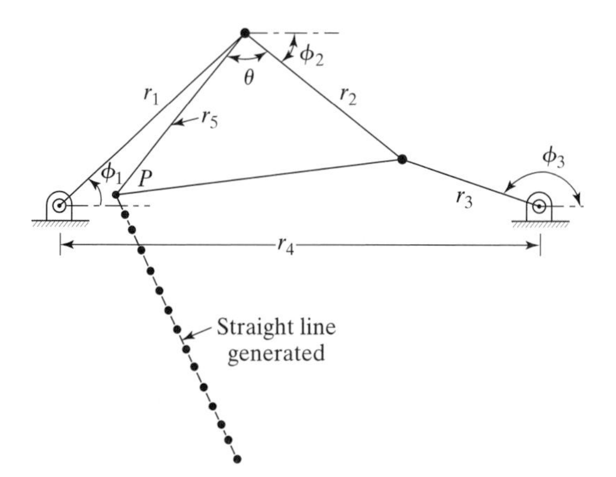

Figure 2.42 Four - bar straight - line mechanism.

(c) Determine how each of the parameters of the beam need to be changed to maximize the fundamental frequency of vibration of the beam.

2.3. A four-bar straight-line mechanism is shown in Fig. 2.42. For specified values of the input link length r_1 and the input angle ϕ_1, the coupler angle ϕ_2 and the follower angle ϕ_3 can be computed by solving the following vector loop equations:

$$r_1 \cos \phi_1 + r_2 \cos \phi_2 - r_3 \cos \phi_3 - r_4 = 0$$

and

$$r_1 \sin \phi_1 + r_2 \sin \phi_2 - r_3 \sin \phi_3 = 0. \tag{E1}$$

The x and y components of the position of the coupler point P can be determined as

$$x_P = r_1 \cos \phi_1 + r_5 \cos(\phi_2 + \theta)$$

and

$$y_P = r_1 \sin \phi_1 + r_5 \sin(\phi_2 + \theta). \tag{E2}$$

Solve Eqs. (E1) using Newton–Raphson method, find the (x, y) coordinates of P using Eqs. (E2), and plot the position of P for the following data:

$$r_1 = 0.963, r_2 = 0.764, r_3 = 0.528, r_4 = 1.815,$$

$$r_5 = 0.778, \theta = -89.65^\circ, \text{ and } 0 \leq \phi_1 \leq 90^\circ.$$

2.4. Discuss the influence of small errors in the coefficients a_i of the polynomial equation

$$f(x) = a_n x^n + a_{n-1} x^{n-1} + \cdots + a_1 x + a_0 = 0.$$

on the accuracy of the computed roots of the equation.

Note: If the new coefficients are given by $b_i = a_i + \varepsilon_i, i = 0, 1, 2, \ldots n$, determine the new roots $y_i = x_i + \delta_i, i = 1, 2, \ldots, n$, where ε_i and δ_i are small numbers and x_i are the roots of the original polynomial.

(a) Find the roots of the equation

$$f(x) = a_2 x^2 + a_1 x + a_0 = 0$$

using $a_2 = 1$, $a_1 = 5$, $a_0 = 6.24$, $\varepsilon_2 = 0.01$, $\varepsilon_1 = -0.03$, and $\varepsilon_0 = 0.02$.

(b) Find the roots of the equation [2.19]

$$f(x) = \prod_{i=1}^{20} (x - i) + 2^{-23} x^{19} = 0$$

and discuss the results.

3

Solution of Simultaneous Linear Algebraic Equations

CHAPTER CONTENTS

3.1 Introduction

Many practical applications of engineering and science, quantitative problems of business and economics, and mathematical models of social sciences lead to a system of linear algebraic equations. For example, an accurate finite-element analysis of aerodynamic loads acting on an airplane in a specific flight condition involves thousands of simultaneous linear equations. A set of simultaneous linear algebraic equations can be expressed in a general form as

$$\begin{aligned} a_{11}x_1 + a_{12}x_2 + \cdots + a_{1n}x_n &= b_1, \\ a_{21}x_1 + a_{22}x_2 + \cdots + a_{2n}x_n &= b_2, \\ &\vdots \\ \text{and} \quad a_{n1}x_1 + a_{n2}x_2 + \cdots + a_{nn}x_n &= b_n, \end{aligned} \tag{3.1}$$

where the coefficients $a_{ij}(i = 1, 2, \ldots, n; j = 1, 2, \ldots, n)$ and the constants b_i $(i = 1, 2, \ldots, n)$ are known, $x_i(i = 1, 2, \ldots, n)$ are the unknowns, and n is the number of equations. Equation (3.1) can be expressed in matrix form as†

$$[A]\vec{x} = \vec{b}, \tag{3.2}$$

where the $n \times n$ coefficient matrix $[A]$, the n-component constant vector $\vec{b}$ and the n-component vector of unknowns $\vec{x}$ are given by

$$[A] = \begin{bmatrix} a_{11} & a_{12} & \cdot & \cdot & \cdot & a_{1n} \\ a_{21} & a_{22} & \cdot & \cdot & \cdot & a_{2n} \\ \cdot & \cdot & \cdot & \cdot & \cdot & \cdot \\ \cdot & \cdot & \cdot & \cdot & \cdot & \cdot \\ \cdot & \cdot & \cdot & \cdot & \cdot & \cdot \\ a_{n1} & a_{n2} & \cdot & \cdot & \cdot & a_{nn} \end{bmatrix}, \quad \vec{b} = \begin{Bmatrix} b_1 \\ b_2 \\ \cdot \\ \cdot \\ \cdot \\ b_n \end{Bmatrix}, \quad \vec{x} = \begin{Bmatrix} x_1 \\ x_2 \\ \cdot \\ \cdot \\ \cdot \\ x_n \end{Bmatrix}.$$

Several methods can be used for solving systems of linear algebraic equations, Eq. (3.1). These methods can be divided into two types: direct and iterative. Direct methods are those that in the absence of round-off and other errors, will yield the exact solution in a finite number of elementary arithmetic operations. In practice, because a computer works with a finite word length, sometimes the direct methods do not give good solutions. Indeed the errors arising from round-off and truncation may lead to extremely poor or even useless results.

† A brief review of matrix algebra is given in Appendix F.

The fundamental method used for direct solutions is Gaussian elimination, but even within this class there are a variety of methods that vary in computational efficiency and accuracy. Iterative methods are those which start with an initial approximation and which by applying a suitably chosen algorithm, lead to successively better approximations. When the process converges, we can expect to get a good approximate solution. The accuracy and the rate of convergence of iterative methods vary with the algorithm chosen. The main advantages of iterative methods are the simplicity and uniformity of the operations to be performed, which make them well suited for use on digital computers and their insensitivity to the growth of round-off errors. Some of the commonly used direct and iterative methods are as follows:

Direct (elimination) methods	Indirect (iterative) methods
(1) Gauss elimination method	1. Jacobi method
(2) Gauss–Jordan method	2. Gauss–Seidel method
(3) LU decomposition method	3. Relaxation (Southwell) method

3.2 Engineering Applications

As stated earlier, simultaneous linear algebraic equations arise in a variety of practical applications. In engineering, for example, the force analysis of statically determinate structures, flow of current in electrical circuits, force analysis of mechanisms, finite-element analysis of heat flow/stress analysis problems, flow of fluids in pipe networks, and least-squares curve-fitting calculations require the solution of a set of simultaneous linear algebraic equations using the methods described in this chapter. The examples that follow illustrate some of these applications.

▶Example 3.1

A scaffolding system, consisting of three rigid bars and six wire ropes, is used to support the loads P_1, P_2, and P_3 as shown in Fig. 3.1(a). Find the tensions developed in the ropes A, B, C, D, E, and F where $P_1 = 2000$ lb, $P_2 = 1000$ lb, and $P_3 = 500$ lb.

Solution

The free-body diagrams of the three bars are shown in Fig. 3.1(b), where P_i denote the applied loads and T_i represent the tensions in the ropes ($i = A, B, \ldots, F$). The equations of equilibrium for vertical forces and moments (about the left support point) for each of the three rigid bars yield the following equations.

For bar 3,

$$T_E + T_F = P_3 \tag{E1}$$

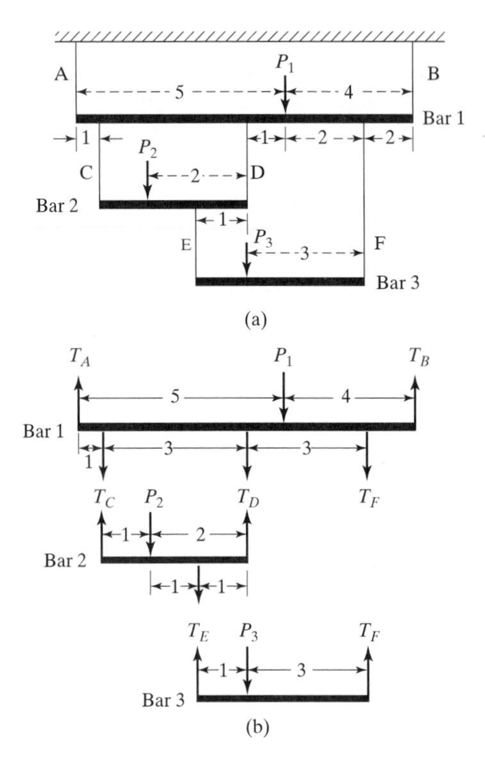

Figure 3.1 Scaffolding system.

$$P_3(1) - T_F(4) = 0. \tag{E2}$$

For bar 2,

$$T_C + T_D - T_E = P_2, \tag{E3}$$

$$P_2(1) + T_E(2) - T_D(3) = 0. \tag{E4}$$

For bar 1,

$$T_A + T_B - T_C - T_D - T_F = P_1, \tag{E5}$$

$$P_1(5) - T_B(9) + T_C(1) + T_D(4) + T_F(7) = 0. \tag{E6}$$

Equations (E1) through (E6) can be expressed in matrix form as

$$\begin{bmatrix} 0 & 0 & 0 & 0 & 1 & 1 \\ 0 & 0 & 0 & 0 & 0 & -4 \\ 0 & 0 & 1 & 1 & -1 & 0 \\ 0 & 0 & 0 & -3 & 2 & 0 \\ 1 & 1 & -1 & -1 & 0 & -1 \\ 0 & -9 & 1 & 4 & 0 & 7 \end{bmatrix} \begin{Bmatrix} T_A \\ T_B \\ T_C \\ T_D \\ T_E \\ T_F \end{Bmatrix} = \begin{Bmatrix} P_3 \\ -P_3 \\ P_2 \\ -P_2 \\ P_1 \\ -5P_1 \end{Bmatrix}. \tag{E7}$$

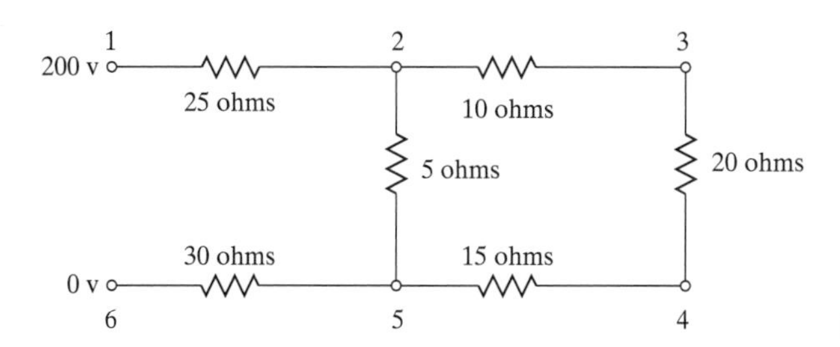

Figure 3.2 Electrical network.

When $P_1 = 2000$ lb, $P_2 = 1000$ lb, and $P_3 = 500$ lb, the solution of Eq. (E7) can be found as $T_A = 1944.45$ lb, $T_B = 1555.55$ lb, $T_C = 791.67$ lb, $T_D = 583.33$ lb, $T_E = 375.00$ lb, and $T_F = 125.00$ lb. ◀

▶Example 3.2

An electrical network consists of six resistors as shown in Fig. 3.2. If the voltages at nodes 1 and 6 are specified as 200 and 0 volts, respectively, determine the voltages at the nodes 2, 3, 4, and 5.

Solution

Let v_i denote the voltage at node i and I_{ij} represent the current flowing from node i to node j. Then the application of Kirchhoff's law, which states that the sum of the currents arriving at any node must be equal to zero, to various nodes leads to the following equations:

Node 2:

$$I_{12} + I_{32} + I_{52} = \frac{v_1 - v_2}{25} + \frac{v_3 - v_2}{10} + \frac{v_5 - v_2}{5} = 0$$

or

$$-17\, v_2 + 5\, v_3 + 10\, v_5 = -400. \tag{E1}$$

Node 3:

$$I_{23} + I_{43} = \frac{v_2 - v_3}{10} + \frac{v_4 - v_3}{20} = 0$$

or

$$2v_2 - 3v_3 + v_4 = 0. \tag{E2}$$

Node 4:

$$I_{34} + I_{54} = \frac{v_3 - v_4}{20} + \frac{v_5 - v_4}{15} = 0$$

or

$$3\, v_3 - 7\, v_4 + 4\, v_5 = 0. \tag{E3}$$

Node 5:

$$I_{65} + I_{25} + I_{45} = \frac{v_6 - v_5}{30} + \frac{v_2 - v_5}{5} + \frac{v_4 - v_5}{15} = 0$$

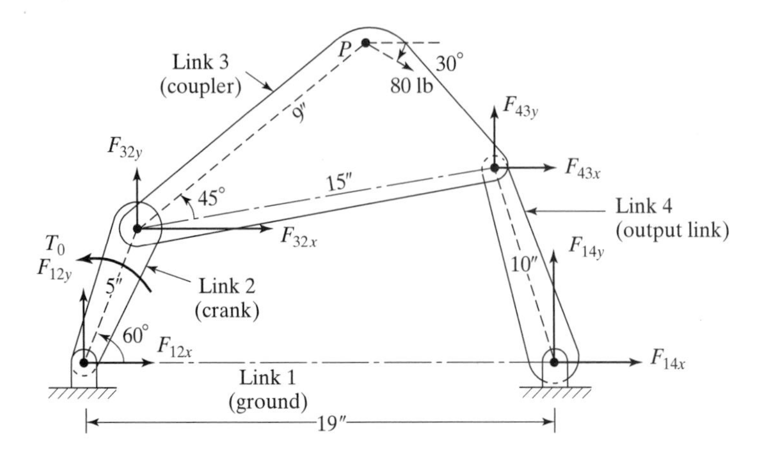

Figure 3.3 Four-bar mechanism.

or

$$6v_2 + 2v_4 - 9v_5 = 0. \tag{E4}$$

Equations (E1) through (E4) can be expressed in matrix form as

$$\begin{bmatrix} -17 & 5 & 0 & 10 \\ 2 & -3 & 1 & 0 \\ 0 & 3 & -7 & 4 \\ 6 & 0 & 2 & -9 \end{bmatrix} \begin{Bmatrix} v_2 \\ v_3 \\ v_4 \\ v_5 \end{Bmatrix} = \begin{Bmatrix} -400 \\ 0 \\ 0 \\ 0 \end{Bmatrix}. \tag{E5}$$

The nodal voltages can be found by solving Eq. (E5) as $v_2 = 115.9663$ volts, $v_3 = 112.6049$ volts, $v_4 = 105.8823$ volts, and $v_5 = 100.8403$ volts. ◀

▶Example 3.3

The crank of the four-bar mechanism shown in Fig. 3.3 has an angular velocity of 25 rad/sec and angular acceleration of - 40 rad/sec^2. The crank, coupler, and output links weigh 1.5, 7.7, and 5.8 lb, respectively. A torque T_0 is applied to the crank while maintaining the stated motion, and an external force of 80 lb acts at point P on the coupler as indicated in Fig. 3.3. Find the joint forces and the external torque acting on the crank.

Solution

The application of dynamic equilibrium equations to the free bodies of the crank, coupler, and the output links yields the following [3.7]:

$$\begin{bmatrix} 1 & 0 & 1 & 0 & 0 & 0 & 0 & 0 & 0 \\ 0 & 1 & 0 & 1 & 0 & 0 & 0 & 0 & 0 \\ 3 & 0 & -1.333 & 2.5 & 0 & 0 & 0 & 0 & 1 \\ 0 & 0 & -1 & 0 & 1 & 0 & 0 & 0 & 0 \\ 0 & 0 & 0 & -1 & 0 & 1 & 0 & 0 & 0 \\ 0 & 0 & -8.217 & 3.672 & 2.858 & 10.332 & 0 & 0 & 0 \\ 0 & 0 & 0 & 0 & -1 & 0 & 1 & 0 & 0 \\ 0 & 0 & 0 & 0 & 0 & -1 & 0 & 1 & 0 \\ 0 & 0 & 0 & 0 & 4.843 & 1.244 & 4.843 & 1.244 & 0 \end{bmatrix} \begin{Bmatrix} F_{12_x} \\ F_{12_y} \\ F_{32_x} \\ F_{32_y} \\ F_{43_x} \\ F_{43_y} \\ F_{14_x} \\ F_{14_y} \\ T_0 \end{Bmatrix}$$

$$= \begin{Bmatrix} 0.480 \\ -7.500 \\ -16.000 \\ -122.927 \\ -10.909 \\ 298.003 \\ -18.895 \\ -9.728 \\ 101.032 \end{Bmatrix}. \tag{E1}$$

The solution of Eq. (E1) gives: $F_{12_x} = -120.51$ lb, $F_{12_y} = -108.21$ lb, $F_{32_x} = 120.99$ lb, $F_{32_y} = 100.71$ lb, $F_{43_x} = -1.94$ lb, $F_{43_y} = 89.81$ lb, $F_{14_x} = -20.84$ lb, $F_{14_y} = 80.08$ lb, and $T_0 = 255.04$ lb-in. ◀

▶Example 3.4

Two sides of a square plate of uniform thickness are maintained at 200°C, and the other two sides are exposed to an atmospheric temperature of 0°C. (See Fig. 3.4(a).) Derive the equations for finding the temperature distribution in the plate.

Solution

The steady-state temperature distribution in a flat plate (or any two-dimensional body) is governed by the Laplace Equation

$$\frac{\partial^2 T}{\partial x^2} + \frac{\partial^2 T}{\partial y^2} = 0, \tag{E1}$$

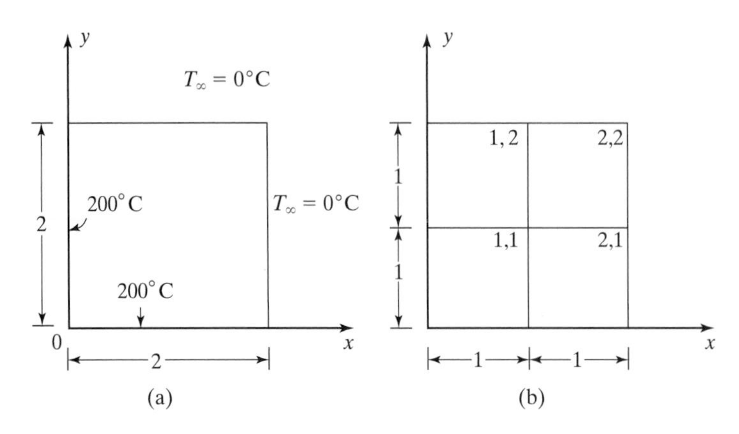

Figure 3.4 Temperature distribution in square plate.

where $T(x, y)$ is the temperature, and x and y are the coordinate directions. An approximate solution of Eq. (E1) can be found by using a finite difference scheme. In this method, the body is divided into a two-dimensional grid as shown in Fig. 3.5 and the partial derivatives of Eq. (E1) are replaced at the grid point (i, j) by their finite difference equivalents as

$$\left.\frac{\partial^2 T}{\partial x^2}\right|_{(i,j)} = \frac{T_{i+1,j} - 2T_{i,j} + T_{i-1,j}}{(\Delta x)^2}, \tag{E2}$$

$$\left.\frac{\partial^2 T}{\partial y^2}\right|_{(i,j)} = \frac{T_{i,j+1} - 2T_{i,j} + T_{i,j-1}}{(\Delta y)^2}. \tag{E3}$$

(See Section 7.5.) For a uniform grid with $\Delta x = \Delta y$, the use of Eqs. (E2) and (E3) in (E1) yields

$$T_{i+1,j} + T_{i-1,j} + T_{i,j+1} + T_{i,j-1} - 4T_{i,j} = 0. \tag{E4}$$

The application of Eq. (E4) to the node points (2,1), (1,1), (2,2), and (1,2) shown in Fig. 3.4(b) yields the following equations:

$$\begin{bmatrix} -4 & 1 & 1 & 0 \\ 1 & -4 & 0 & 1 \\ 1 & 0 & -4 & 1 \\ 0 & 1 & 1 & -4 \end{bmatrix} \begin{Bmatrix} T_{2,1} \\ T_{1,1} \\ T_{2,2} \\ T_{1,2} \end{Bmatrix} = \begin{Bmatrix} -200 \\ -400 \\ 0 \\ -200 \end{Bmatrix}. \tag{E5}$$

The solution of Eq. (E5) can be found as $T_{2,1} = 100°C$, $T_{1,1} = 150.0°C$, $T_{2,2} = 50.0°C$, and $T_{1,2} = 100.0°C$. ◀

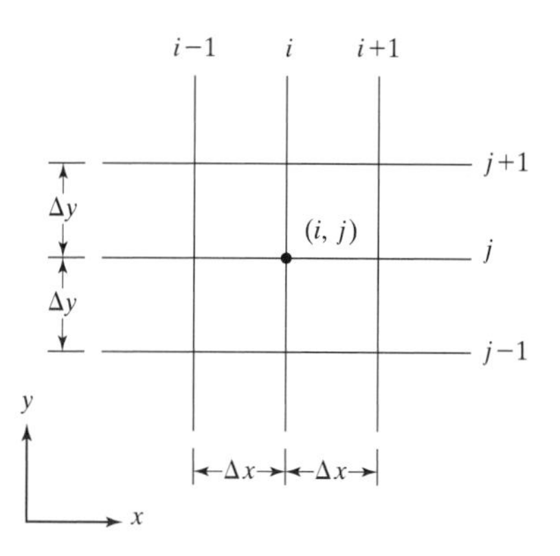

Figure 3.5 A two-dimensional grid.

3.3 Vector and Matrix Norms

3.3.1 Vector Norm

A norm or length is a measure of the size of a vector or matrix. The Euclidean norm or length of a vector

$$\vec{x} = \begin{Bmatrix} x_1 \\ x_2 \\ \cdot \\ \cdot \\ \cdot \\ x_n \end{Bmatrix}$$

is defined as

$$\|\vec{x}\| = (x_1^2 + x_2^2 + \cdots + x_n^2)^{\frac{1}{2}}. \tag{3.3}$$

The norm of a vector $\vec{x}$, $\|\vec{x}\|$, has the following properties:

(1) $\|\vec{x}\| \geq 0$ for any $\vec{x}$ and $\|\vec{x}\| = 0$ if and only if $\vec{x} = \vec{0}$.

(2) $\|k\vec{x}\| = |k|\|\vec{x}\|$ for any real or complex number k.

(3) $\|\vec{x} + \vec{y}\| \leq \|\vec{x}\| + \|\vec{y}\|$ for any two vectors $\vec{x}$ and $\vec{y}$ of the same order.

In general, the L_p- norm of a vector $\vec{x}$ is defined as

$$L_p = \left\{ \sum_{i=1}^{n} |x_i|^p \right\}^{\frac{1}{p}}. \tag{3.4}$$

It can be seen that L_2 corresponds to the Euclidean norm of $\vec{x}$ and L_1 corresponds to the sum of the absolute values of the components of $\vec{x}$. If p is increased to infinity, the value of L_p tends toward the absolute value of the largest component of $\vec{x}$:

$$L_\infty = \max_i |x_i|. \tag{3.5}$$

Thus, the L_∞ norm is simply the absolute value of the largest component of the vector $\vec{x}$.

▶Example 3.5

Find the norms L_1, L_2, and L_∞ of the vector $\vec{x} = \begin{Bmatrix} 2 \\ -5 \\ 3 \end{Bmatrix}$.

Solution

The application of Eq. (3.4) gives

$$L_1 = \sum_{i=1}^{3} |x_i| = 2 + 5 + 3 = 10,$$

$$L_2 = \left(\sum_{i=1}^{3} x_i^2 \right)^{\frac{1}{2}} = \sqrt{(2)^2 + (-5)^2 + (3)^2} = \sqrt{38} = 6.1644,$$

$$L_\infty = \max |x_i| = 5.$$ ◀

3.3.2 Matrix Norm

The matrix norm is defined in terms of its effect on a vector. For example, a vector $\vec{x}$ premultiplied by a square matrix $[A]$ yields another vector $\vec{y}$ that differs from $\vec{x}$. When all combinations of $\vec{x}$ and $\vec{y}$ are considered, the largest ratio of $\frac{\|\vec{y}\|}{\|\vec{x}\|}$ is defined as the norm of the matrix $[A]$:

$$\|[A]\| = \max_{\vec{x}=1} \frac{\|\vec{y}\|}{\|\vec{x}\|} = \max_{\vec{x}=1} \frac{\|[A]\vec{x}\|}{\|\vec{x}\|}.$$

The norm of a matrix is useful for defining the condition number of a matrix which, in turn, can be used to quantify the degree of ill conditioning of a set of linear equations. (See Section 3.6.1.) The norm of a matrix has the following properties:

(1) For two square matrices $[A]$ and $[B]$,

$$\|[A][B]\| \le \|[A]\| \, \|[B]\|. \tag{3.6}$$

(2) For any matrix $[A]$ and vector $\vec{x}$,

$$\|[A]\vec{x}\| \le \|[A]\| \, \|\vec{x}\|. \tag{3.7}$$

The norm of a matrix $[A]$ can also be defined similar to that of a vector. The L_1 and L_∞ norms of a square matrix $[A]$ are defined as

$$\|[A]\|_1 = \max_{1\le j\le n} \sum_{i=1}^{n} |a_{ij}| = \text{maximum column sum}, \tag{3.8}$$

$$\|[A]\|_\infty = \max_{1\le j\le n} \sum_{j=1}^{n} |a_{ij}| = \text{maximum row sum}. \tag{3.9}$$

The L_2 norm, also known as the spectral radius or spectral norm, of a square matrix $[A]$ is defined as

$$\|[A]\|_2 = \max_{1\le i\le n} \lambda_i, \tag{3.10}$$

where λ_i is the ith eigenvalue of the matrix $[A]$. (See Chapter 4 for methods of computing the eigenvalues of a matrix.)

For an $m \times n$ matrix $[A]$, the Euclidean norm, L_e, also known as the Frobenius norm, is defined as

$$\|[A]\|_e = \left(\sum_{i=1}^{m} \sum_{j=1}^{n} a_{ij}^2 \right)^{\frac{1}{2}}. \tag{3.11}$$

It is to be noted that $\|[A]\|_2 \ne \|[A]\|_e$ for a matrix; however, the L_2 norm is same as L_e for a vector.

▶Example 3.6

Find the L_1, L_2, L_∞, and L_e norms of the matrix $[A] = \begin{bmatrix} 4 & 3 \\ 7 & 6 \end{bmatrix}$.

Solution

The application of Eqs. (3.8) through (3.11) yields

$$\|[A]\|_1 = \max\{4+7,\ 3+6\} = 11.$$

The eigenvalues of the matrix $[A]$ are defined by (see Chapter 4)

$$|[A] - \lambda[I]| = \begin{vmatrix} 4-\lambda & 3 \\ 7 & 6-\lambda \end{vmatrix} = (4-\lambda)(6-\lambda) - 21$$

$$= \lambda^2 - 10\lambda + 3 = 0,$$

which gives the solution: $\lambda_1 = 0.3096$, $\lambda_2 = 9.6904$. This yields

$$\|[A]\|_2 = 9.6904,$$

$$\|[A]\|_\infty = \max\{4+3, 7+6\} = 13,$$

$$\|[A]\|_e = \left\{(4)^2 + (7)^2 + (3)^2 + (6)^2\right\}^{\frac{1}{2}} = \sqrt{110} = 10.4881.$$ ◀

3.4 Basic Concepts of Solution

Homogeneous equations

If all the constants b_i are zero (i.e. $\vec{b} = \vec{0}$), Eq. (3.1) is called a set of homogeneous equations. In this case, a simple (trivial) solution is given by $\vec{x} = \vec{0}$. A nontrivial solution exists only if the determinant of the coefficient matrix is zero:

$$|[A]| = 0.$$

Nonhomogeneous equations

If at least one b_i is nonzero, Eq. (3.1) is called a set of nonhomogeneous equations.

Augmented matrix of [A]

The augmented matrix $[A']$ is defined as the $n \times (n+1)$ matrix obtained by adding the constant vector $\vec{b}$ as the $(n+1)$st column to the columns of the matrix $[A]$:

$$[A'] = \begin{bmatrix} a_{11} & a_{12} & \cdot & \cdot & \cdot & a_{1n} & b_1 \\ a_{21} & a_{22} & \cdot & \cdot & \cdot & a_{2n} & b_2 \\ \cdot & \cdot & \cdot & \cdot & \cdot & \cdot & \cdot \\ \cdot & \cdot & \cdot & \cdot & \cdot & \cdot & \cdot \\ \cdot & \cdot & \cdot & \cdot & \cdot & \cdot & \cdot \\ a_{n1} & a_{n2} & \cdot & \cdot & \cdot & a_{nn} & b_n \end{bmatrix}.$$

Existence of solution to Eq. (3.1)

The system of equations (3.1) has a solution if and only if the rank of the augmented matrix $[A']$ is equal to the rank of the coefficient matrix $[A]$. Further, the solution of Eq. (3.1) is unique if the rank (r) is equal to n. If the rank r is less than n, then no unique solution exists, implying that the equations are either inconsistent or one or more equations are redundant.

3.5 Linearly Independent Equations

If no equation can be expressed as a linear combination of other equations in the system, the system of equations is said to be linearly independent. For example, consider the following system of equations:

$$\begin{aligned} x_1 - x_2 + x_3 &= 3; \\ 2x_1 + x_2 - x_3 &= 0; \\ 3x_1 + 2x_2 + 2x_3 &= 15. \end{aligned} \tag{3.12}$$

These equations have a unique solution: $x_1 = 1, x_2 = 2$, and $x_3 = 4$. Next, consider the following system of equations:

$$\begin{aligned} x_1 - x_2 + x_3 &= 3; \\ 2x_1 + x_2 - x_3 &= 0; \\ 8x_1 + x_2 - x_3 &= 6. \end{aligned} \tag{3.13}$$

These equations are not independent; the third equation can be obtained by adding twice the first equation and thrice the second equation. Thus, Eq. (3.13) does not have a unique solution. The determinant of the coefficient matrix of the variables in Eq. (3.13) can be seen to be zero; That is,

$$\left\| \begin{bmatrix} 1 & -1 & 1 \\ 2 & 1 & -1 \\ 8 & 1 & -1 \end{bmatrix} \right\| = 0.$$

Note: If any equation is a linear combination of others, the coefficient matrix will be singular and the solution cannot be found. Even when the equations are close to being linearly dependent, they might become linearly dependent during numerical computations, due to round-off errors, and an accurate solution cannot be found.

3.6 Ill Conditioned Equations

A system of equations is said to be ill conditioned if small changes in the coefficients lead to very large variations in the solution. For example, consider the following set of equations:

$$\begin{aligned} x_1 - x_2 &= 5; \\ kx_1 - x_2 &= 4, \end{aligned} \tag{3.14}$$

where the coefficient k will be specified later. The determinant of the coefficient matrix in Eq. (3.14) is given by

$$\begin{vmatrix} 1 & -1 \\ k & -1 \end{vmatrix} = (k - 1), \tag{3.15}$$

and the solution of Eq. (3.14) can be expressed as

$$x_1 = \frac{1}{1-k}, \quad x_2 = \frac{5k-4}{1-k}. \tag{3.16}$$

The values of x_1 and x_2 when k varies by a small amount around 1.0 are shown in Table 3.1.

It can be seen that the solution varies by a large amount when the coefficient k is changed even by a small amount. This indicates that the set of equations (3.14) are ill conditioned when k is nearly unity. It is to be noted that although ill conditioning in Eq. (3.14) is introduced through the coefficient matrix, the constant vector, $\vec{b}$, can also cause ill conditioning of equations.

Table 3.1 Solution of Eq. (3.14).

Value of k	Solution		Determinant of the coefficient matrix, $(k-1)$
	$x_1 = \dfrac{1}{(1-k)}$	$x_2 = \dfrac{(5k-4)}{(1-k)}$	
1.0000	No solution		0.0000
0.9997	3333.3333	3328.3333	−0.0003
0.9998	5000.0000	4995.0000	−0.0002
0.9999	10000.0000	9995.0000	−0.0001
1.0001	−10000.0000	−10005.0000	+0.0001
1.0002	−5000.0000	−5005.0000	+0.0002
1.0003	−3333.3333	−3338.3333	+0.0003

3.6.1 Quantification of the Degree of Ill Conditioning

Several methods are available to measure the degree of ill conditioning of a set of equations. Two methods are indicated next.

Method 1

The condition number of the coefficient matrix, cond (A), is often used as a measure of the degree of ill conditioning of a set of equations. It is defined as

$$\text{cond}\,(A) = \|A\|\|A^{-1}\|, \tag{3.17}$$

where $\|A\|$ is the norm of the matrix $[A]$. Since the accurate computation of the inverse of the matrix $[A]$ is needed, the evaluation of Eq. (3.17) is usually expensive. In addition, the inversion of the matrix $[A]$ of an ill conditioned system of equations may result in an inaccurate solution. Different norms such as $\|[A]\|_1$ and $\|[A]\|_\infty$ can be used in Eq. (3.17). The condition number of an identity matrix is one, and the condition number of any other matrix will always be greater than one. Thus, a smaller condition number indicates that the equations are well conditioned, and a larger condition number denotes that the equations are poorly conditioned, or ill conditioned. In general, if the condition number, cond (A), is equal to 10^s, then s decimal digits may be lost during the computational process.

▶Example 3.7

Find the condition number of the set of equations (3.14).

Solution

The coefficient matrix of Eq. (3.14) is given by

$$[A] = \begin{bmatrix} 1 & -1 \\ k & -1 \end{bmatrix}$$

and its inverse by

$$A^{-1} = \frac{1}{(k-1)} \begin{bmatrix} -1 & 1 \\ -k & 1 \end{bmatrix}.$$

For $k > 1$, the ∞-norms of A and $[A^{-1}]$ are given by

$$\|[A]\| = (k+1) \text{and} \|[A^{-1}]\| = \frac{k+1}{k-1},$$

and hence the condition number of the system of equations (3.14) is given by

$$\text{cond}\ (A) = \|[A]\|\|[A^{-1}]\| = (k+1)\left(\frac{k+1}{k-1}\right).$$

This indicates that

$$\text{cond}\ (A) = (2.0001)(20001) = 40004.0001 \text{ if } k = 1.0001$$

and

$$\text{cond}\ (A) = (2.0002)(10001) = 20004.0002 \text{ if } k = 1.0002.$$

Similarly, for $k < 1$, the ∞-norms of $[A]$ and $[A^{-1}]$ and cond (A) are given by

$$\|A\| = 2,\ \|[A^{-1}]\| = \left(\frac{2}{1-k}\right),\ \text{cond}\ (A) = \|[A]\|\|[A^{-1}]\| = \left(\frac{4}{1-k}\right).$$

This shows that

$$\text{cond}\ (A) = 40000.0 \text{ if } k = 0.9999 \text{ and cond}\ (A) = 20000.0 \text{ if } k = 0.9998.$$ ◀

Method 2

The determinant of the coefficient matrix can also be used as a measure of the degree of ill conditioning of a set of equations. A normalized form of this measure, denoted as $D(A)$, can be defined as

$$D(A) = \frac{|\det(A)|}{A_1 A_2 \cdots A_n}, \tag{3.18}$$

where $[A]$ is the coefficient matrix and the A_i are defined as

$$A_i = \left(a_{i1}^2 + a_{i2}^2 + \cdots + a_{in}^2\right)^{\frac{1}{2}}; \quad i = 1, 2, \ldots, n. \tag{3.19}$$

It can be seen that $D(A) = 0$ if the matrix $[A]$ is singular and $D(A) = 1$ if the matrix $[A]$ is diagonal. Thus, the closer the value of $D(A)$ is to unity, the better the condition of the equations.

▶Example 3.8

Determine the value of $D(A)$ for the system of equations (3.14).

Solution

The coefficient matrix of Eq. (3.14) is given by

$$[A] = \begin{bmatrix} 1 & -1 \\ k & -1 \end{bmatrix}$$

and its determinant by det $(A) = (k-1)$. The values of A_1 and A_2 can be computed as

$$A_1 = \left(1^2 + (-1)^2\right)^{\frac{1}{2}} = \sqrt{2} \text{ and } A_2 = \left(k^2 + (-1)^2\right)^{\frac{1}{2}} = \sqrt{1+k^2}.$$

Thus, the value of $D(A)$ is given by

$$D(A) = \frac{|\det(A)|}{A_1 A_2} = \frac{|(k-1)|}{\sqrt{2(1+k^2)}}.$$

This shows that $D(A)$ is equal to 0 for $k = 1$, 5.00025×10^{-5} for $k = 0.9999$, 4.99975×10^{-5} for $k = 1.0001$, 1.00010×10^{-4} for $k = 0.9998$, and 0.99990×10^{-4} for $k = 1.0002$. ◀

Notes

(a) The condition number of the coefficient matrix gives an idea of the degree of ill conditioning present in a system of equations. However, it requires the evaluation of the inverse of the coefficient matrix, which is computationally expensive. Some simple tests can be used to identify the presence of ill conditioning in a set of equations as indicated next.

(b) If a set of equations is ill conditioned, the resulting solution will be inaccurate. In the absence of a measure of ill conditioning, such as the condition number of the coefficient matrix, it is not easy to examine the solution and determine whether it is in error. However, often, we may be able to identify a set of ill conditioned equations by looking at the following symptoms:

(1) The diagonal elements will be smaller than the off-diagonal elements in the coefficient matrix.

(2) A small change in the elements of the coefficient matrix will result in significantly larger changes in the solution.

(3) A small change in the right-hand-side vector results in a significantly larger changes in the solution vector.

(4) If $[A]$ denotes the coefficient matrix, the product $[A][A]^{-1}$ or $[A]^{-1}[A]$ will be significantly different from $[I]$.

(5) If $[A]$ denotes the coefficient matrix, the product, det $([A])$ det $([A]^{-1})$, will be significantly different from unity.

3.7 Graphical Interpretation of the Solution

Any linear equation in n unknowns (x_i) can be considered as a hyperplane in an n-dimensional Cartesian space in which each coordinate axis represents an unknown. Thus, a set of n equations denotes n hyperplanes in the n-dimensional space. Seeking a solution of the set of equations amounts to finding a common point of intersection of all the hyperplanes. These aspects can be illustrated graphically by considering sets of equations involving two variables. For this, we first consider the following set of equations:

$$3x_1 + x_2 = 5, \tag{3.20}$$

and

$$x_1 - 2x_2 = -3. \tag{3.21}$$

These equations denote two lines (special case of hyperplanes) in the two-dimensional space in which x_1 and x_2 represent the coordinate axes. (See Fig. 3.6.) Equations (3.20) and (3.21) can be seen to be linearly independent (denoting two different intersecting straight lines) and have a unique solution at the point of intersection $(x_1 = 1, x_2 = 2)$. Next we consider the following set of equations:

$$3x_1 + x_2 = 5, \tag{3.22}$$

and

$$6x_1 + 2x_2 = 10. \tag{3.23}$$

Since the second equation is equal to twice the first equation, Eqs. (3.22) and (3.23) are not linearly independent. They both represent the straight line shown in Fig. 3.7. Since any point on the straight line satisfies Eqs. (3.22) and (3.23), there will be

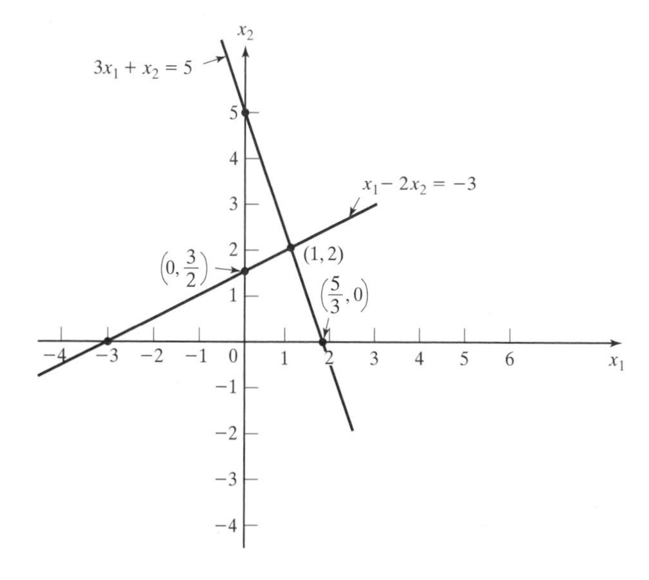

Figure 3.6 Graphs of linearly independent equations.

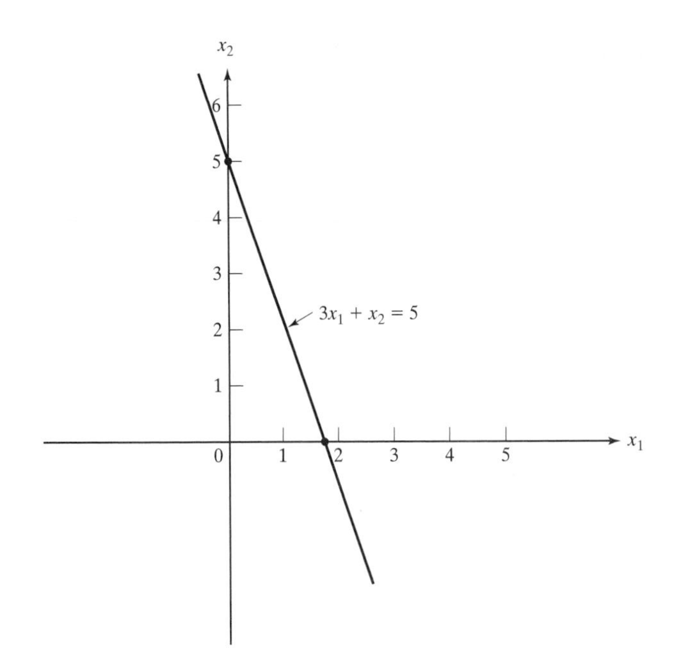

Figure 3.7 Graph of linearly dependent equations.

an infinite number of solutions to Eqs. (3.22) and (3.23). Thus, the solution of Eqs. (3.22) and (3.23) is not unique. In some cases, the set of equations may not be consistent and hence may not have a solution. For example, the set of equations

$$3x_1 + x_2 = 5, \tag{3.24}$$

and

$$6x_1 + 2x_2 = 15 \tag{3.25}$$

denotes two parallel straight lines as indicated graphically in Fig. 3.8. Since these equations do not have a solution, they are said to be inconsistent. Another example of a set of equations that do not have a solution is given by the following equations:

$$3x_1 + x_2 = 5, \tag{3.26}$$

$$x_1 - 2x_2 = -3, \tag{3.27}$$

and

$$x_1 - x_2 = 1. \tag{3.28}$$

These equations also do not have a solution (no common intersection point in Fig. 3.9); they are said to be an overdetermined system of equations.

Note: The graphical method can be used to find the solution only when the number of equations is two. Further, the solution time will be more compared with numerical procedures, and the resulting solution will not be very accurate.

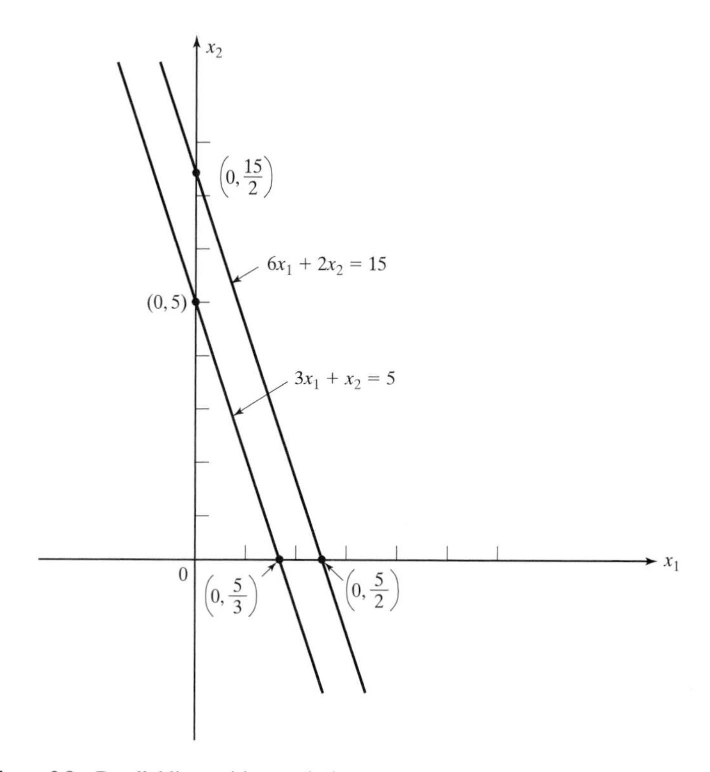

Figure 3.8 Parallel lines with no solution.

3.8 Solution Using Cramer's Rule

The system of equations (3.1) can be solved by using Cramer's rule, which gives x_i as

$$x_i = \frac{|[A_i]|}{|[A]|}; i = 1, 2, \cdots, n, \tag{3.29}$$

where $[A_i]$ is the matrix obtained by replacing the ith column of matrix $[A]$ with the constant vector $\vec{b}$:

$$A_i = \begin{bmatrix} a_{11} & a_{12} & \cdot & \cdot & \cdot & a_{1,i-1} & b_1 & a_{1,i+1} & \cdot & \cdot & \cdot & a_{1n} \\ a_{21} & a_{22} & \cdot & \cdot & \cdot & a_{2,i-1} & b_2 & a_{2,i+1} & \cdot & \cdot & \cdot & a_{2n} \\ \cdot & \cdot & \cdot & \cdot & \cdot & \cdot & \cdot & \cdot & \cdot & \cdot & \cdot & \cdot \\ \cdot & \cdot & \cdot & \cdot & \cdot & \cdot & \cdot & \cdot & \cdot & \cdot & \cdot & \cdot \\ \cdot & \cdot & \cdot & \cdot & \cdot & \cdot & \cdot & \cdot & \cdot & \cdot & \cdot & \cdot \\ a_{n1} & a_{n2} & \cdot & \cdot & \cdot & a_{n,i-1} & b_n & a_{n,i+1} & \cdot & \cdot & \cdot & a_{nn} \end{bmatrix}. \tag{3.30}$$

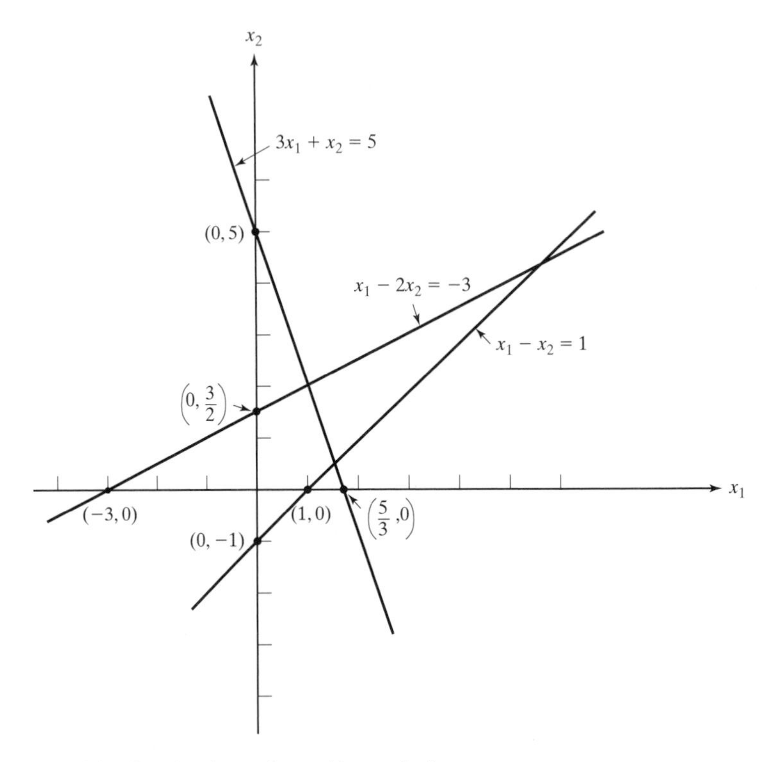

Figure 3.9 Graphs of equations with no solution.

Note: The solution given by Eq. (3.29) assumes that the coefficient matrix A is nonsingular. The determination of all x_i in Eq. (3.29) requires the evaluation of $n + 1$ determinants of matrices of order $n \times n$. If the determinants are evaluated in terms of their minors, the computation of each determinant requires $(n - 1)$ $n\,!$ arithmetic operations [3.8, 3.15]. Thus, the solution of a set of linear equations using Eq. (3.29) requires a total of $(n - 1)(n + 1)!$ arithmetic operations. This represents a very large number of operations, and hence Cramer's rule becomes impractical for all but small values of n. For example, even to solve eight equations, the number of arithmetic operations required will be $7(9!) = 2,540,160$, which is quite large. In addition to the large computational effort involved, the accuracy of the solution will be severely affected by the round-off errors. Hence, Cramer's rule is not recommended for solving more than three or four simultaneous equations.

▶Example 3.9

Find the solution of the following system of equations using Cramer's rule:

$$x_1 - x_2 + x_3 = 3;$$

$$2x_1 + x_2 - x_3 = 0;$$

$$3x_1 + 2x_2 + 2x_3 = 15.$$

Solution

The given system of equations can be expressed in matrix form as $[A]\vec{x} = \vec{b}$, where

$$[A] = \begin{bmatrix} 1 & -1 & 1 \\ 2 & 1 & -1 \\ 3 & 2 & 2 \end{bmatrix}, \vec{x} = \begin{Bmatrix} x_1 \\ x_2 \\ x_3 \end{Bmatrix}, \text{ and } \vec{b} = \begin{Bmatrix} 3 \\ 0 \\ 15 \end{Bmatrix}.$$

The matrices $[A_1]$, $[A_2]$, and $[A_3]$ can be defined by replacing columns 1, 2, and 3 of $[A]$ by the vector $\vec{b}$, respectively, as

$$[A_1] = \begin{bmatrix} 3 & -1 & 1 \\ 0 & 1 & -1 \\ 15 & 2 & 2 \end{bmatrix};$$

$$[A_2] = \begin{bmatrix} 1 & 3 & 1 \\ 2 & 0 & -1 \\ 3 & 15 & 2 \end{bmatrix};$$

$$[A_3] = \begin{bmatrix} 1 & -1 & 3 \\ 2 & 1 & 0 \\ 3 & 2 & 15 \end{bmatrix}.$$

The determinants of the various matrices can be computed as

$$|[A]| = 1\begin{vmatrix} 1 & -1 \\ 2 & 2 \end{vmatrix} + 1\begin{vmatrix} 2 & -1 \\ 3 & 2 \end{vmatrix} + 1\begin{vmatrix} 2 & 1 \\ 3 & 2 \end{vmatrix} = 1(2+2) + 1(4+3) + 1(4-3) = 12;$$

$$|[A_1]| = 3\begin{vmatrix} 1 & -1 \\ 2 & 2 \end{vmatrix} + 1\begin{vmatrix} 0 & -1 \\ 15 & 2 \end{vmatrix} + 1\begin{vmatrix} 0 & 1 \\ 15 & 2 \end{vmatrix} = 3(2+2) + 1(0+15) + 1(0-15) = 12;$$

$$|[A_2]| = 1\begin{vmatrix} 0 & -1 \\ 15 & 2 \end{vmatrix} - 3\begin{vmatrix} 2 & -1 \\ 3 & 2 \end{vmatrix} + 1\begin{vmatrix} 2 & 0 \\ 3 & 15 \end{vmatrix} = 1(0+15) - 3(4+3) + 1(30-0) = 24;$$

$$|[A_3]| = 1\begin{vmatrix} 1 & 0 \\ 2 & 15 \end{vmatrix} + 1\begin{vmatrix} 2 & 0 \\ 3 & 15 \end{vmatrix} + 3\begin{vmatrix} 2 & 1 \\ 3 & 2 \end{vmatrix} = 1(15-0) + 1(30-0) + 3(4-3) = 48.$$

The application of Eq. (3.29) gives

$$x_1 = \frac{|[A_1]|}{|[A]|} = \frac{12}{12} = 1;$$

$$x_2 = \frac{|[A_2]|}{|[A]|} = \frac{24}{12} = 2;$$

$$x_3 = \frac{|[A_3]|}{|[A]|} = \frac{48}{12} = 4.$$

◀

3.9 Gauss Elimination Method

The most frequently used direct method for the solution of simultaneous algebraic linear equations is the Gaussian elimination method. This method is based on the principle of reducing a set of n equations in n unknowns to an equivalent triangular form. If the original equations are given by Eqs. (3.1), the triangular form of equations appears as follows:

$$\begin{aligned} a_{11}x_1 + a_{12}x_2 + a_{13}x_3 + \cdots + a_{1n}x_n &= b_1; \\ a'_{22}x_2 + a'_{23}x_3 + \cdots + a'_{2n}x_n &= b'_2; \\ a'_{33}x_3 + \cdots + a'_{3n}x_n &= b'_3; \\ &\vdots \\ a'_{n-1,n-1}x_{n-1} + a'_{n-1,n}x_n &= b'_{n-1}; \\ a'_{nn}x_n &= b'_n. \end{aligned} \tag{3.31}$$

The reduction of Eq. (3.1) to the triangular form of Eq. (3.31), known as *forward elimination*, is made such that the solution given by Eq. (3.31) is same as that of Eq. (3.1). The solution of Eq. (3.31) can be determined in a simple manner, using a process known as *back substitution*. Since the last equation of system (3.31) contains only one unknown, x_n, it is solved first. The remaining unknowns are found by using back substitution starting with the variable x_n and proceeding backwards until the variable x_1 is determined.

3.9.1 Basic Approach

The operations used in reducing the equations to a triangular form are known as elementary operations which can be stated as follows:

(1) Any equation can be multiplied (or divided) by a nonzero scalar.

(2) Any equation can be added to (or subtracted from) another equation.

(3) The positions of any two equations in the set can be interchanged.

The method of reducing a set of equations to a triangular form is illustrated with the following example:

▶Example 3.10

Reduce the following set of equations to a triangular form:

$$2x_1 - x_2 + x_3 = 4; \tag{E1}$$

$$4x_1 + 3x_2 - x_3 = 6; \tag{E2}$$

and

$$3x_1 + 2x_2 + 2x_3 = 15. \tag{E3}$$

Solution

First, we eliminate x_1 from Eq. (E2). For this, we multiply Eq. (E1) by -2 and add it to Eq. (E2). Next, we eliminate x_1 from Eq. (E3). For this, we multiply Eq. (E1) by $(-3/2)$ and add it to Eq. (E3). These operations yield:

$$2x_1 - x_2 + x_3 = 4, \tag{E4}$$

$$5x_2 - 3x_3 = -2, \tag{E5}$$

and

$$\frac{7}{2}x_2 + \frac{1}{2}x_3 = 9. \tag{E6}$$

Equation (E1), which was used to eliminate x_1 from the remaining equations, is called the pivot equation and the coefficient of x_1 in Eq. (E1), namely, $a_{11} = 2$, is called the pivot element or pivot coefficient.

In the next step, we eliminate x_2 from Eq. (E6). For this, we multiply Eq. (E5) by $-7/10$ and add it to Eq. (E6) to obtain the new system of equations:

$$2x_1 - x_2 + x_3 = 4, \tag{E7}$$

$$5x_2 - 3x_3 = -2, \tag{E8}$$

and

$$\frac{13}{5}x_3 = \frac{52}{5}. \tag{E9}$$

These equations can be seen to be in triangular form. They can be solved using back substitution starting with Eq. (E9), which gives $x_3 = 4$. Next, Eq. (E8) is solved to find $x_2 = 2$. Finally, Eq. (E7) is solved for the last unknown to obtain $x_1 = 1$. Thus, the solution of Eqs. (E1)–(E3) is given by

$$\vec{x} = \begin{Bmatrix} x_1 \\ x_2 \\ x_3 \end{Bmatrix} = \begin{Bmatrix} 1 \\ 2 \\ 4 \end{Bmatrix}. \tag{E10}$$

◀

3.9.2 Generalization

To use the Gauss elimination method to solve the equations $[A]\vec{x} = \vec{b}$, we consider the augmented matrix, $[C]$, of order $n \times (n+1)$:

$$[C] = \begin{bmatrix} a_{11} & a_{12} & \cdot & \cdot & \cdot & a_{1n} & a_{1,n+1} \\ a_{21} & a_{22} & \cdot & \cdot & \cdot & a_{2n} & a_{2,n+1} \\ \cdot & \cdot & \cdot & \cdot & \cdot & \cdot & \cdot \\ \cdot & \cdot & \cdot & \cdot & \cdot & \cdot & \cdot \\ \cdot & \cdot & \cdot & \cdot & \cdot & \cdot & \cdot \\ a_{n1} & a_{n2} & \cdot & \cdot & \cdot & a_{nn} & a_{n,n+1} \end{bmatrix},$$

where the first n columns of $[C]$ are same as the columns of $[A]$ and the last $(n+1)$st column is same as the constant vector $\vec{b}$:

$$\begin{Bmatrix} a_{1,n+1} \\ a_{2,n+1} \\ \cdot \\ \cdot \\ \cdot \\ a_{n,n+1} \end{Bmatrix} = \begin{Bmatrix} b_1 \\ b_2 \\ \cdot \\ \cdot \\ \cdot \\ b_n \end{Bmatrix}.$$

By using $(n-1)$ elimination steps (through suitable elementary operations), the matrix $[C]$ is reduced to an upper triangular augmented matrix $[C^{(n-1)}]$ of the form

$$[C^{(n-1)}] = \begin{bmatrix} a_{11}^{(0)} & a_{12}^{(0)} & a_{13}^{(0)} & \cdot & \cdot & \cdot & a_{1,n-1}^{(0)} & a_{1n}^{(0)} & a_{1,n+1}^{(0)} \\ 0 & a_{22}^{(1)} & a_{23}^{(1)} & \cdot & \cdot & \cdot & a_{2,n-1}^{(1)} & a_{2n}^{(1)} & a_{2,n+1}^{(1)} \\ \cdot & \cdot & \cdot & \cdot & \cdot & \cdot & \cdot & \cdot & \cdot \\ \cdot & \cdot & \cdot & \cdot & \cdot & \cdot & \cdot & \cdot & \cdot \\ \cdot & \cdot & \cdot & \cdot & \cdot & \cdot & \cdot & \cdot & \cdot \\ 0 & 0 & 0 & \cdot & \cdot & \cdot & 0 & a_{nn}^{(n-1)} & a_{n,n+1}^{(n-1)} \end{bmatrix}, \tag{3.32}$$

where the superscript indicates the elimination step in which the particular coefficient is obtained. Initially, the elements of $[C^{(0)}] = [C]$ are given by

$$a_{ij}^{(0)} = a_{ij}; i = 1, 2, \cdots, n; j = 1, 2, \ldots, n,$$

$$a_{i,n+1}^{(0)} = b_i; i = 1, 2, \ldots, n. \tag{3.33}$$

The first elimination step is then applied to obtain the elements $a_{ij}^{(1)}$ as

$$a_{ij}^{(1)} = a_{ij}^{(0)} - \frac{a_{i1}^{(0)}}{a_{11}^{(0)}} a_{1j}^{(0)}; i = 2, 3, \ldots, n; j = 1, 2, \ldots, n+1. \tag{3.34}$$

Equation (3.34) implies that the first row has been used as the pivot row with a_{11} as the pivot element. It can be seen that the first elements in rows $2, 3, \ldots, n$ become zero at the end of the first elimination step (i.e., $a_{i1}^{(1)} = 0$, for $i = 2, 3, \ldots, n$). The elimination procedure is continued by using rows $1, 2, \ldots, n-1$ in sequence as pivot rows with $a_{ii}^{(k)}$ as the pivot element in the kth elimination step. The general formula for the elements $a_{ij}^{(k)}$ at the end of kth elimination step can be expressed as

$$a_{ij}^{(k)} = a_{ij}^{(k-1)} - \frac{a_{ik}^{(k-1)}}{a_{kk}^{(k-1)}} a_{kj}^{(k-1)}; i = k+1, k+2, \ldots, n, \tag{3.35}$$

$$j = k, k+1, \ldots, n+1; k = 1, 2, \ldots, n-1.$$

Note that the pivot element $a_{kk}^{(k-1)}$ appearing in the denominator of Eq. (3.35) cannot be zero. Since it is always the diagonal element, it should be possible to rearrange the equations in each step to achieve diagonal dominance of the coefficient matrix; that is, the row containing the largest pivot element can be chosen as the pivot row. This procedure of rearranging the equations is called *partial pivoting*, and it serves two purposes: It avoids the possibility of division by zero and also improves the accuracy of the resulting solution. Sometimes, the columns are also searched in addition to the rows to locate the maximum pivot element. In such a case, the procedure is called *complete pivoting*. If the pivoting procedure cannot locate a nonzero element for pivoting, the matrix must be singular, and, in that case, the system of equations will not have a solution.

At the end of $(n-1)$ elimination steps, the system of equations will be reduced to the triangular form, Eq. (3.32). By using back substitution, we compute the values of the unknowns x_i as

$$x_n = \frac{a_{n,n+1}}{a_{nn}}, \tag{3.36}$$

$$x_i = \frac{a_{i,n+1} - \sum_{j=i+1}^{n} a_{ij} x_j}{a_{ii}}; i = n-1, n-2, \ldots, 1, \tag{3.37}$$

where the superscripts of a_{ij} are dropped for simplicity of notation. Note that there is no need to use superscripts in a computer program, since the old coefficients can be replaced by the new ones during the elimination process.

Notes

(a) It can be shown (Section 3.19) that the Gauss elimination method requires $n^3/3$ multiplications to find the solution $x_1, x_2, \ldots, x_n$, whereas the multiplication of two $n \times n$ matrices requires n^3 multiplications.

(b) The determinant of the coefficient matrix can also be computed during Gauss elimination. At the end of the forward elimination process, the original matrix will be transformed into an upper triangular form. The determinant can be found by computing the product of all the numbers along the diagonal and multiplying the result by -1 or $+1$, depending on whether the number of pivot operations performed is odd or even, respectively.

(c) Several techniques can be used to improve the solution found by Gauss elimination method. These include (i) using more significant figures during the computations, (ii) using partial or complete pivoting during the computations, (iii) using scaling during the calculations, and (iv) using an error correction method after finding the solution.

▶Example 3.11

Find the solution of the following set of equations using the Gauss elimination method:

$$2x_1 - x_2 + x_3 = 4, \tag{E1}$$

$$4x_1 + 3x_2 - x_3 = 6, \tag{E2}$$

and

$$3x_1 + 2x_2 + 2x_3 = 15. \tag{E3}$$

Solution

The augmented matrix $[C]$ corresponding to Eqs. (E1) through (E3) can be identified as

$$[C] = \begin{bmatrix} 2 & -1 & 1 & 4 \\ 4 & 3 & -1 & 6 \\ 3 & 2 & 2 & 15 \end{bmatrix}. \tag{E4}$$

The application of Eq. (3.35) for $k = 1$ yields the augmented matrix

$$[C] = \begin{bmatrix} 2 & -1 & 1 & 4 \\ 0 & 5 & -3 & -2 \\ 0 & \frac{7}{2} & \frac{1}{2} & 9 \end{bmatrix}. \tag{E5}$$

The use of Eq. (3.35) for $k = 2$ gives the reduced augmented matrix

$$[C] = \begin{bmatrix} 2 & -1 & 1 & 4 \\ 0 & 5 & -3 & -2 \\ 0 & 0 & \frac{13}{5} & \frac{52}{5} \end{bmatrix}. \tag{E6}$$

Using back substitution, we get from Eq. (E6) the solution as

$$x_3 = \frac{52}{13} = 4; 5x_2 = -2 + 3x_3 = -2 + 12 = 10 \text{ or } x_2 = 2;$$

$$2x_1 = 4 + x_2 - x_3 = 4 + 2 - 4 = 2 \text{ or } x_1 = 1.$$

Thus, the solution vector $\vec{x}$ is given by

$$\vec{x} = \begin{Bmatrix} x_1 \\ x_2 \\ x_3 \end{Bmatrix} = \begin{Bmatrix} 1 \\ 2 \\ 4 \end{Bmatrix}.$$

◀

Note: The solution (using hand calculations) in Example 3.11 was exact, since the number of equations was small and no round-off errors were involved (exact fractions were used) in the computations. If the problem is solved on a computer, fractions will be stored as decimal numbers using a limited number of significant digits. The round-off error will be introduced, theoretically, in each arithmetic operation that leads to a final solution that may not be exact. The error due to round-off will increase rapidly with an increase in the number of equations.

3.9.3 Error Correction

It is possible to improve the numerical solution of a set of equations, $[A]\vec{x} = \vec{b}$. Let $\vec{x}_1$ be the solution obtained by solving these equations using a numerical method, such as the Gauss elimination method. If $\vec{x}_1$ is not exact, we can compute a vector $\vec{b}_1$ as $\vec{b}_1 = [A]\vec{x}_1$. If $\vec{x}_1$ is exact, $\vec{b}_1 = \vec{b}$. If $\vec{x}_1$ is not exact, then $\vec{b}_1$ differs from $\vec{b}$ slightly. In this case, we can subtract the equations $[A]\vec{x}_1 = \vec{b}_1$ from $[A]\vec{x} = \vec{b}$ to obtain

$$[A](\vec{x} - \vec{x}_1) = \vec{b} - \vec{b}_1, \tag{3.38}$$

which can be rewritten as

$$[A]\vec{e} = \vec{b}_2, \tag{3.39}$$

where $\vec{e} = \vec{x} - \vec{x}_1$ is the error vector and $\vec{b}_2 = \vec{b} - \vec{b}_1$, Equations (3.39) can now be solved to find $\vec{e}$. Once $\vec{e}$ is known, an improved or more accurate solution can be determined as $\vec{x} = \vec{x}_1 + \vec{e}$. Since Eq. (3.39) differs from the original equations $[A]\vec{x} = \vec{b}$, only on the right-hand side, Gauss elimination can be used to solve Eq. (3.39) efficiently. Since the coefficient matrix is the same, by retaining multiplication factors $\left(\frac{a_{ik}^{(k-1)}}{a_{kk}^{(k-1)}}\right)$ used in the solution of the original equations, Eq. (3.35) can be used by considering $\vec{b}_2$ as the new $(n+1)$st column to find the solution.

3.9.4 Scaling

Scaling is the process of adjusting the coefficients of a set of equations so that they will all have the same order of magnitude. Usually, scaling will be required if the system of equations represent relationships between parameters that are measured in widely different units (such as micrometer and meter, and Pascal and GigaPascal). The following example illustrates the procedure of scaling.

▶Example 3.12

Find the solution of the following equations using Gauss elimination:

$$\begin{bmatrix} 6 & 3 & 150 \\ -2 & 5 & 200 \\ 1 & 2 & -1 \end{bmatrix} \begin{Bmatrix} x_1 \\ x_2 \\ x_3 \end{Bmatrix} = \begin{Bmatrix} 462 \\ 608 \\ 2 \end{Bmatrix}. \tag{E1}$$

Solution

For comparison, the problem is first solved without using scaling and then using scaling. By retaining only three significant figures to indicate the round-off error, we can reduce Eq. (E1) by using the forward elimination process to

$$\begin{bmatrix} 1 & 0.5 & 25 \\ 0 & 1 & 41.6 \\ 0 & 0 & -101 \end{bmatrix} \begin{Bmatrix} x_1 \\ x_2 \\ x_3 \end{Bmatrix} = \begin{Bmatrix} 77 \\ 127 \\ -306 \end{Bmatrix}. \tag{E2}$$

The solution of Eq. (E2) is $x_3 = 3.02$, $x_2 = 1.36$, and $x_1 = 0.82$, which can be compared to the exact solution, $x_3 = 3$, $x_2 = 2$, and $x_1 = 1$.

Next we use the scaling process and divide each row of Eq. (E1) by the magnitude of the largest coefficient appearing in that equation. This gives

$$\begin{bmatrix} 0.04 & 0.02 & 1.00 \\ -0.01 & 0.025 & 1.00 \\ 0.50 & 1.00 & -0.50 \end{bmatrix} \begin{Bmatrix} x_1 \\ x_2 \\ x_3 \end{Bmatrix} = \begin{Bmatrix} 3.08 \\ 3.04 \\ 1.00 \end{Bmatrix}. \tag{E3}$$

Again, by retaining only three significant digits, the forward elimination process is used to obtain

$$\begin{bmatrix} 1 & 0.5 & 25.0 \\ 0 & 1.0 & 41.6 \\ 0 & 0.0 & -44.2 \end{bmatrix} \begin{Bmatrix} x_1 \\ x_2 \\ x_3 \end{Bmatrix} = \begin{Bmatrix} 77.0 \\ 127.0 \\ -132.7 \end{Bmatrix}. \tag{E4}$$

The solution of Eq. (E4) gives $x_3 = 3.00$, $x_2 = 2.20$, and $x_1 = 0.90$, which can be seen to be a better solution than the one obtained without using scaling. ◀

3.10 Gauss–Jordan Elimination Procedure

The Gauss–Jordan elimination method is an extension of the Gauss elimination method. It uses the same elementary operations as Gauss elimination method. However, while the Gauss elimination method reduces all the off-diagonal elements below the diagonal to zero, the Gauss–Jordan elimination method reduces all the off-diagonal elements both below and above the diagonal to zero. Thus, in the Gauss–Jordan method, the solution can be determined without any need for back substitution. The procedure is illustrated with the help of the following example:

▶Example 3.13

Find the solution of the following set of equations using the Gauss–Jordan elimination method:

$$2x_1 - x_2 + x_3 = 4, \tag{E1}$$

$$4x_1 + 3x_2 - x_3 = 6, \tag{E2}$$

and

$$3x_1 + 2x_2 + 2x_3 = 15. \tag{E3}$$

Solution

The augmented matrix $[C]$ corresponding to Eqs. (E1) through (E3) can be written as

$$[C] = \begin{bmatrix} 2 & -1 & 1 & 4 \\ 4 & 3 & -1 & 6 \\ 3 & 2 & 2 & 15 \end{bmatrix}. \tag{E4}$$

First, we normalize row 1 by dividing it by 2 (to reduce the coefficient a_{11} to 1):

$$\begin{bmatrix} 1 & -\frac{1}{2} & \frac{1}{2} & 2 \\ 4 & 3 & -1 & 6 \\ 3 & 2 & 2 & 15 \end{bmatrix}. \tag{E5}$$

To reduce the coefficient 4 (a_{21}) to zero, we multiply the first row by –4 and add it to the second row:

$$\begin{bmatrix} 1 & -\frac{1}{2} & \frac{1}{2} & 2 \\ 0 & 5 & -3 & -2 \\ 3 & 2 & 2 & 15 \end{bmatrix}. \tag{E6}$$

To reduce the coefficient 3 (a_{31}) to zero, we multiply the first row by -3 and add it to the third row:

$$\begin{bmatrix} 1 & -\frac{1}{2} & \frac{1}{2} & 2 \\ 0 & 5 & -3 & -2 \\ 0 & \frac{7}{2} & \frac{1}{2} & 9 \end{bmatrix}. \tag{E7}$$

Next we normalize row 2 by dividing it by 5 (to reduce the new coefficient a_{22} to 1):

$$\begin{bmatrix} 1 & -\frac{1}{2} & \frac{1}{2} & 2 \\ 0 & 1 & -\frac{3}{5} & -\frac{2}{5} \\ 0 & \frac{7}{2} & \frac{1}{2} & 9 \end{bmatrix}. \tag{E8}$$

To reduce the coefficient $-\frac{1}{2}$ (new a_{12}) to zero, we multiply the second row by $\frac{1}{2}$ and add it to the first row:

$$\begin{bmatrix} 1 & 0 & \frac{1}{5} & \frac{9}{5} \\ 0 & 1 & -\frac{3}{5} & -\frac{2}{5} \\ 0 & \frac{7}{2} & \frac{1}{2} & 9 \end{bmatrix}. \tag{E9}$$

To reduce the coefficient $\frac{7}{2}$ (new a_{32}) to zero, we multiply the second row by $-\frac{7}{2}$ and add it to the third row:

$$\begin{bmatrix} 1 & 0 & \frac{1}{5} & \frac{9}{5} \\ 0 & 1 & -\frac{3}{5} & -\frac{2}{5} \\ 0 & 0 & \frac{13}{5} & \frac{52}{5} \end{bmatrix}. \tag{E10}$$

Next we normalize row 3 by dividing it by $\frac{13}{5}$ (to reduce the new coefficient a_{33} to 1):

$$\begin{bmatrix} 1 & 0 & \frac{1}{5} & \frac{9}{5} \\ 0 & 1 & -\frac{3}{5} & -\frac{2}{5} \\ 0 & 0 & 1 & 4 \end{bmatrix}. \tag{E11}$$

To reduce the coefficient $\frac{1}{5}$ (new a_{13}) to zero, we multiply the third row by $-\frac{1}{5}$ and add it to the first row:

$$\begin{bmatrix} 1 & 0 & 0 & 1 \\ 0 & 1 & -\frac{3}{5} & -\frac{2}{5} \\ 0 & 0 & 1 & 4 \end{bmatrix}. \tag{E12}$$

Finally, to reduce the coefficient $-\frac{3}{5}$ (new a_{23}) to zero, we multiply the third row by $\frac{3}{5}$ and add it to the second row:

$$\begin{bmatrix} 1 & 0 & 0 & 1 \\ 0 & 1 & 0 & 2 \\ 0 & 0 & 1 & 4 \end{bmatrix}. \tag{E13}$$

This matrix can be considered as the augmented matrix corresponding to the system of equations

$$[I]\vec{x} = \vec{\tilde{b}}, \tag{E14}$$

where the current vector $\vec{\tilde{b}}$ can be identified as the solution of the original equations:

$$\vec{x} = \vec{\tilde{b}} = \begin{Bmatrix} 1 \\ 2 \\ 4 \end{Bmatrix}. \tag{E15}$$

As can be seen, there is no need to use back substitution to determine the solution $\vec{x}$. ◀

3.10.1 Generalization

The Gauss–Jordan elimination method of solving the equations $[A]\vec{x} = \vec{b}$ first involves the construction of the augmented matrix $[C]$ as in in the case of Gauss elimination method:

$$[C] = \begin{bmatrix} a_{11} & a_{12} & \cdot & \cdot & \cdot & a_{1n} & a_{1,n+1} \\ a_{21} & a_{22} & \cdot & \cdot & \cdot & a_{2n} & a_{2,n+1} \\ \cdot & \cdot & \cdot & \cdot & \cdot & \cdot & \cdot \\ \cdot & \cdot & \cdot & \cdot & \cdot & \cdot & \cdot \\ \cdot & \cdot & \cdot & \cdot & \cdot & \cdot & \cdot \\ a_{n1} & a_{n2} & \cdot & \cdot & \cdot & a_{nn} & a_{n,n+1} \end{bmatrix},$$

where $a_{i,n+1} = b_i (i = 1, 2, \ldots, n)$. Then the normalization and reduction procedures are used as indicated in Example 3.13. The general procedure can be summarized by the following steps:

(1) Initialization (to define the augmented matrix using the coefficient matrix and the constant vector):

$$a_{ij}^{(0)} = a_{ij}; i = 1, 2, \ldots, n; j = 1, 2, \ldots, n,$$

$$a_{ij}^{(0)} = b_i; i = 1, 2, \ldots, n; j = n + 1. \tag{3.40}$$

(2) Normalization (to reduce the diagonal coefficients of the augmented matrix to unity by dividing by its pivot element):

$$a_{kj}^{(k)} = \frac{a_{kj}^{(k-1)}}{a_{kk}^{(k-1)}}; j = k, k + 1, \ldots, n, n + 1. \tag{3.41}$$

(3) Reduction (to reduce all off-diagonal coefficients in each row and column to zero by using elementary operations):

$$a_{ij}^{(k)} = a_{ij}^{(k-1)} - a_{ik}^{(k-1)} a_{kj}^{(k)}; j = k, k + 1, \ldots, n, n + 1;$$
$$i = 1, 2, \ldots, n (i \neq k). \tag{3.42}$$

Note that Equations (3.41) and (3.42) are to be used with $k = 1, 2, \ldots, n$.

3.11 *LU* Decomposition Method

A modification of the elimination method, known as the (LU) decomposition method, is frequently used for solving a system of linear algebraic equations. The method is based on the fact that any square matrix $[A]$ can be written as a product of two matrices as

$$[A] = [L][U] \tag{3.43}$$

where $[L]$ is a lower triangular matrix and $[U]$ is an upper triangular matrix given by

$$[L] = \begin{bmatrix} l_{11} & 0 & 0 & \cdot & \cdot & \cdot & 0 \\ l_{21} & l_{22} & 0 & \cdot & \cdot & \cdot & 0 \\ l_{31} & l_{32} & l_{33} & \cdot & \cdot & \cdot & 0 \\ \cdot & \cdot & \cdot & \cdot & \cdot & \cdot & \cdot \\ \cdot & \cdot & \cdot & \cdot & \cdot & \cdot & \cdot \\ \cdot & \cdot & \cdot & \cdot & \cdot & \cdot & \cdot \\ l_{n1} & l_{n2} & l_{n3} & \cdot & \cdot & \cdot & l_{nn} \end{bmatrix}, \tag{3.44}$$

$$[U] = \begin{bmatrix} u_{11} & u_{12} & u_{13} & \cdot & \cdot & \cdot & u_{1n} \\ 0 & u_{22} & u_{23} & \cdot & \cdot & \cdot & u_{2n} \\ 0 & 0 & u_{33} & \cdot & \cdot & \cdot & u_{3n} \\ \cdot & \cdot & \cdot & \cdot & \cdot & \cdot & \cdot \\ \cdot & \cdot & \cdot & \cdot & \cdot & \cdot & \cdot \\ \cdot & \cdot & \cdot & \cdot & \cdot & \cdot & \cdot \\ 0 & 0 & 0 & \cdot & \cdot & \cdot & u_{nn} \end{bmatrix}. \quad (3.45)$$

It can be seen that there are a total of (n^2+n) entries (unknowns) in $[L]$ and $[U]$, while there are only n^2 entries (known values) in $[A]$. Hence, n additional conditions are to be imposed in order to determine the matrices $[L]$ and $[U]$ uniquely. Several methods have been proposed to determine the matrices $[L]$ and $[U]$ uniquely. In the Crout's method, all the diagonal elements of $[U]$ are chose as one; that is, $u_{ii} = 1$ for $i = 1, 2, \ldots, n$. In the Doolittle's method, all the diagonal elements of $[L]$ are selected as one; that is, $l_{ii} = 1$ for $i = 1, 2, \ldots, n$. In the Choleski's method, the diagonal elements of $[U]$ and $[L]$ are assumed to be same; that is, $u_{ii} = l_{ii}$ for $i = 1, 2, \ldots, n$.

3.11.1 Crout's Method

To illustrate Crout's method of decomposition, we consider the LU decomposition of a 3×3 matrix as

$$\begin{bmatrix} a_{11} & a_{12} & a_{13} \\ a_{21} & a_{22} & a_{23} \\ a_{31} & a_{32} & a_{33} \end{bmatrix} = \begin{bmatrix} l_{11} & 0 & 0 \\ l_{21} & l_{22} & 0 \\ l_{31} & l_{32} & l_{33} \end{bmatrix} \begin{bmatrix} 1 & u_{12} & u_{13} \\ 0 & 1 & u_{23} \\ 0 & 0 & 1 \end{bmatrix}. \quad (3.46)$$

By carrying out the matrix multiplication on the right-hand side of Eq. (3.46), we obtain

$$\begin{bmatrix} a_{11} & a_{12} & a_{13} \\ a_{21} & a_{22} & a_{23} \\ a_{31} & a_{32} & a_{33} \end{bmatrix} = \begin{bmatrix} l_{11} & (l_{11}u_{12}) & (l_{11}u_{13}) \\ l_{21} & (l_{21}u_{12} + l_{22}) & (l_{21}u_{13} + l_{22}u_{23}) \\ l_{31} & (l_{31}u_{12} + l_{32}) & (l_{31}u_{13} + l_{32}u_{23} + l_{33}) \end{bmatrix}. \quad (3.47)$$

By equating the corresponding elements of the two matrices of Eq. (3.47), we can find the elements of the matrices $[L]$ and $[U]$ as follows:

$$l_{11} = a_{11}, l_{21} = a_{21}, l_{31} = a_{31},$$

$$l_{11}u_{12} = a_{12}, \text{ or } u_{12} = \frac{a_{12}}{l_{11}} = \frac{a_{12}}{a_{11}},$$

$$l_{21}u_{12} + l_{22} = a_{22}, \text{ or } l_{22} = a_{22} - l_{21}u_{12},$$

$$l_{31}u_{12} + l_{32} = a_{32}, \text{ or } l_{32} = a_{32} - l_{31}u_{12},$$

$$l_{11}u_{13} = a_{13}, \text{ or } u_{13} = \frac{a_{13}}{l_{11}} = \frac{a_{13}}{a_{11}},$$

$$l_{21}u_{13} + l_{22}u_{23} = a_{23}, \text{ or } u_{23} = \frac{a_{23} - l_{21}u_{13}}{l_{22}},$$

and

$$l_{31}u_{13} + l_{32}u_{23} + l_{33} = a_{33} \text{ or } l_{33} = a_{33} - l_{31}u_{13} - l_{32}u_{23}. \tag{3.48}$$

Decomposition of a General $n \times n$ Matrix

If $[A]$ is an $n \times n$ matrix given by

$$[A] = \begin{bmatrix} a_{11} & a_{12} & \cdot & \cdot & \cdot & a_{1n} \\ a_{21} & a_{22} & \cdot & \cdot & \cdot & a_{2n} \\ \cdot & \cdot & \cdot & \cdot & \cdot & \cdot \\ \cdot & \cdot & \cdot & \cdot & \cdot & \cdot \\ \cdot & \cdot & \cdot & \cdot & \cdot & \cdot \\ a_{n1} & a_{n2} & \cdot & \cdot & \cdot & a_{nn} \end{bmatrix}, \tag{3.49}$$

the $[L]$ and $[U]$ matrices can be expressed as indicated in Eq. (3.44) and (3.45) with $u_{ii} = 1, i = 1, 2, \ldots, n$. The elements of $[L]$ and $[U]$ can be determined as follows:

$$l_{ij} = \left\{ a_{ij} - \sum_{k=1}^{j-1} l_{ik}u_{kj} \right\}; i \geq j; i = 1, 2, \ldots, n, \tag{3.50}$$

$$u_{ij} = \left\{ \frac{a_{ij} - \sum_{k=1}^{i-1} l_{ik}u_{kj}}{l_{ii}} \right\}; i < j; j = 2, 3, \ldots, n, \tag{3.51}$$

and

$$u_{ii} = 1; i = 1, 2, \ldots, n. \tag{3.52}$$

For the relevant indices i and j, it is convenient to compute the elements in the order $l_{i1}, u_{1j}; l_{i2}, u_{2j}; l_{i3}, u_{3j}; \ldots; l_{i,n-1}, u_{n-1,j}; l_{nn}$. The application of Eqs. (3.50) through (3.52) is illustrated through the following example:

▶Example 3.14

Find the LU decomposition of the following matrix using Crout's method:

$$[A] = \begin{bmatrix} a_{11} & a_{12} & a_{13} \\ a_{21} & a_{22} & a_{23} \\ a_{31} & a_{32} & a_{33} \end{bmatrix} = \begin{bmatrix} 2 & -1 & 1 \\ 4 & 3 & -1 \\ 3 & 2 & 2 \end{bmatrix}.$$

Solution

Applying Eqs. (3.50) through (3.52), we obtain

$$l_{11} = a_{11} = 2, l_{21} = a_{21} = 4, l_{31} = a_{31} = 3;$$

$$u_{11} = \frac{a_{11}}{l_{11}} = \frac{2}{2} = 1, u_{12} = \frac{a_{12}}{l_{11}} = -\frac{1}{2}, \text{ and } u_{13} = \frac{a_{13}}{l_{11}} = \frac{1}{2};$$

$$l_{22} = a_{22} - l_{21}u_{12} = 3 - (4)\left(\frac{-1}{2}\right) = 5, \; l_{32} = a_{32} - l_{31}u_{12} = 2 - (3)\left(\frac{-1}{2}\right) = \frac{7}{2};$$

$$u_{22} = 1;$$

$$u_{23} = \frac{a_{23} - l_{21}u_{13}}{l_{22}} = \frac{-1 - (4)\left(\frac{1}{2}\right)}{5} = -\frac{3}{5},$$

$$l_{33} = a_{33} - l_{31}u_{13} - l_{32}u_{23} = 2 - (3)\left(\frac{1}{2}\right) - \left(\frac{7}{2}\right)\left(\frac{-3}{5}\right) = 2 - \frac{3}{2} + \frac{21}{10} = \frac{13}{5}.$$

Thus, the matrices $[L]$ and $[U]$ are given by

$$[L] = \begin{bmatrix} 2 & 0 & 0 \\ 4 & 5 & 0 \\ 3 & \frac{7}{2} & \frac{13}{5} \end{bmatrix},$$

$$\begin{matrix} \uparrow & \uparrow & \uparrow \\ ① & ③ & ⑤ \end{matrix}$$

and

$$[U] = \begin{bmatrix} 1 & \frac{-1}{2} & \frac{1}{2} \\ 0 & 1 & \frac{-3}{5} \\ 0 & 0 & 1 \end{bmatrix} \begin{matrix} \leftarrow ② \\ \leftarrow ④ \\ \leftarrow ⑥ \end{matrix}.$$

Note that the encircled numbers indicate the order in which the columns and rows of the matrices $[L]$ and $[U]$ are generated. ◀

Solution of equations

The system of linear equations to be solved can be expressed, using LU decomposition of the matrix $[A]$ as

$$[A]\vec{x} = [L][U]\vec{x} = \vec{b}. \tag{3.53}$$

To find the solution $\vec{x}$, we define a vector $\vec{z}$ as

$$\vec{z} = [U]\vec{x} \tag{3.54}$$

and express Eq. (3.53) as

$$[L]\vec{z} = \vec{b}, \tag{3.55}$$

or, in expanded form,

$$\begin{array}{|ll|}\hline l_{11}z_1 & = b_1, \\ l_{21}z_1 + l_{22}z_2 & = b_2, \\ l_{31}z_1 + l_{32}z_2 + l_{33}z_3 & = b_3, \\ \vdots & \\ l_{n1}z_1 + l_{n2}z_2 + l_{n3}z_3 + \cdots + l_{nn}z_n & = b_n. \\ \hline \end{array} \tag{3.56}$$

The first of these equations can be solved for z_1, after which the second equation can be solved for z_2, the third equation for z_3, and so on. Thus, the values of $z_1, z_2, z_3, \ldots, z_n$ can be determined in succession. Once z_i are known, the values of x_i can be found by rewriting Eq. (3.54) in scalar form as

$$\begin{array}{|ll|}\hline x_1 + u_{12}x_2 + u_{13}x_3 + \cdots + u_{1n}x_n & = z_1, \\ x_2 + u_{23}x_3 + \cdots + u_{2n}x_n & = z_2, \\ x_3 + \cdots + u_{3n}x_n & = z_3, \\ \vdots & \\ x_{n-1} + u_{n-1,n}x_n & = z_{n-1}, \\ x_n & = z_n. \\ \hline \end{array} \tag{3.57}$$

These equations can be solved using back substitution (as in the case of Gauss elimination process) for $x_n, x_{n-1}, \ldots, x_2, x_1$ in that order. The general equations for z_i (from Eqs. (3.56)) and x_i (from Eqs. (3.57)) can be expressed as

$$z_1 = \frac{b_1}{l_{11}},$$

$$z_i = \frac{b_i - \sum_{k=l}^{i-1} l_{ik}z_k}{l_{ii}}; i = 2, 3, \ldots, n, \tag{3.58}$$

and

$$x_n = z_n,$$

$$x_i = z_i - \sum_{k=i+1}^{n} u_{ik}x_k; i = n-1, n-2, \ldots, 2, 1. \tag{3.59}$$

▶Example 3.15

Solve the following equations using the LU decomposition procedure:

$$2x_1 - x_2 + x_3 = 4;$$

$$4x_1 + 3x_2 - x_3 = 6;$$

$$3x_1 + 2x_2 + 2x_3 = 15.$$

Solution

Using the result of Example 3.14, we can express the given system of equations

$$[A]\vec{x} \equiv [L][U]\vec{x} \equiv \begin{bmatrix} 2 & 0 & 0 \\ 4 & 5 & 0 \\ 3 & \frac{7}{2} & \frac{13}{5} \end{bmatrix} \begin{bmatrix} 1 & -\frac{1}{2} & \frac{1}{2} \\ 0 & 1 & -\frac{3}{5} \\ 0 & 0 & 1 \end{bmatrix} \vec{x} = \begin{Bmatrix} 4 \\ 6 \\ 15 \end{Bmatrix}. \tag{E1}$$

Using Eq. (3.58), we can obtain the elements of the vector $\vec{z} = [U]\vec{x}$ as

$$z_1 = \frac{b_1}{l_{11}} = \frac{4}{2} = 2;$$

$$z_2 = \frac{b_2 - l_{21}z_1}{l_{22}} = \frac{6 - (4)(2)}{5} = -\frac{2}{5};$$

$$z_3 = \frac{b_3 - l_{31}z_1 - l_{32}z_2}{l_{33}} = \frac{15 - (3)(2) - \left(\frac{7}{2}\right)\left(-\frac{2}{5}\right)}{\frac{13}{5}} = 4.$$

Next the application of Eq. (3.59) yields the desired solution

$$x_3 = z_3 = 4;$$

$$x_2 = z_2 - u_{23}x_3 = -\frac{2}{5} - \left(-\frac{3}{5}\right)(4) = 2;$$

$$x_1 = z_1 - u_{12}x_2 - u_{13}x_3 = 2 - \left(-\frac{1}{2}\right)(2) - \left(\frac{1}{2}\right)(4) = 1.$$

◀

Notes

(1) The *LU* decomposition method is more efficient than the Gauss elimination method when the solution of a set of simultaneous linear equations is to be found with the same coefficient matrix, but with different right-hand-side vectors, as in

$$[A]\vec{x}_i = \vec{b}_i (i = 1, 2, \ldots, p). \tag{3.60}$$

In this case, the matrix $[A]$ needs to be decomposed only once; the solution vectors $\vec{x}_i$ can be determined by applying Eqs. (3.58) and (3.59) separately for each $\vec{b}_i$.

(2) The upper triangular matrix $[U]$ in Doolittle method (see Problem 3.68) is same as the upper triangular matrix found in the Gauss elimination method.

Except for the diagonal elements l_{ij} that have a value of one, the elements l_{ij} of $[L]$ are same as the multipliers $\frac{a_{ik}^{(k-1)}}{a_{kk}^{(k-1)}}$ used in Eq. (3.35) of the Gauss elimination process.

(3) In practice, the matrices $[L]$ and $[U]$ need not be stored separately. By omitting the zeroes in $[L]$ and $[U]$ and the ones in the diagonal of $[U]$ we can store the other essential elements of $[L]$ and $[U]$ in a single $n \times n$ matrix as

$$\begin{bmatrix} l_{11} & u_{12} & u_{13} & \cdot & \cdot & \cdot & u_{1n} \\ l_{21} & l_{22} & u_{23} & \cdot & \cdot & \cdot & u_{2n} \\ l_{31} & l_{32} & l_{33} & \cdot & \cdot & \cdot & u_{3n} \\ \cdot & \cdot & \cdot & \cdot & \cdot & \cdot & \cdot \\ \cdot & \cdot & \cdot & \cdot & \cdot & \cdot & \cdot \\ \cdot & \cdot & \cdot & \cdot & \cdot & \cdot & \cdot \\ l_{n1} & l_{n2} & l_{n3} & \cdot & \cdot & \cdot & l_{nn} \end{bmatrix}. \tag{3.61}$$

Further, an examination of the formulas used for generating the elements l_{ij} and u_{ij} shows that once an element of the matrix $[A]$, a_{ij}, is used, it is not needed in subsequent computations. Hence, the elements of the matrix of Eq. (3.61) can be stored in place of the elements of the original matrix, $[A]$.

3.11.2 Choleski's Method for Symmetric Matrices

In many engineering applications, such as the finite-element analysis, the matrices involved will be symmetric, banded, and positive definite. The Choleski's method of decomposition is particularly useful for symmetric matrices. In Choleski's method, a symmetric positive definite matrix $[A]$ is decomposed uniquely as

$$[A] = [U]^T[U], \tag{3.62}$$

where

$$[U] = \begin{bmatrix} u_{11} & u_{12} & u_{13} & \cdot & \cdot & \cdot & u_{1n} \\ 0 & u_{22} & u_{23} & \cdot & \cdot & \cdot & u_{2n} \\ 0 & 0 & u_{33} & \cdot & \cdot & \cdot & u_{3n} \\ \cdot & \cdot & \cdot & \cdot & \cdot & \cdot & \cdot \\ \cdot & \cdot & \cdot & \cdot & \cdot & \cdot & \cdot \\ \cdot & \cdot & \cdot & \cdot & \cdot & \cdot & \cdot \\ 0 & 0 & 0 & \cdot & \cdot & \cdot & u_{nn} \end{bmatrix} \tag{3.63}$$

is an upper triangular matrix including the major diagonal. The elements of $[U] = [u_{ij}]$ are given by

$$
\begin{aligned}
u_{11} &= (a_{11})^{\frac{1}{2}};\\
u_{1j} &= \frac{a_{1j}}{u_{11}}; j = 2, 3, \ldots, n;\\
u_{ii} &= \left(a_{ii} - \sum_{k=1}^{i=1} u_{ki}^2\right)^{\frac{1}{2}}; i = 2, 3, \ldots, n; \qquad (3.64)\\
u_{ij} &= \frac{1}{u_{ii}}\left(a_{ij} - \sum_{k=1}^{i-1} u_{ki}u_{kj}\right); i = 2, 3, \ldots, n; j = i+1, i+2, \ldots, n;\\
u_{ij} &= 0; i > j.
\end{aligned}
$$

(See Problem 3.50) The matrix $[A]$ can also be decomposed as $[A] = [L][L]^T$, where $[L]$ represents a lower triangular matrix. The elements of $[L]$ can be found in Ref. [3.6]. (See also Problem 3.51.)

▶Example 3.16

Decompose the following matrix into the form $[A] = [U]^T[U]$:

$$
[A] = \begin{bmatrix} 4 & -1 & 1 \\ -1 & 6 & -4 \\ 1 & -4 & 5 \end{bmatrix}.
$$

Solution

Using the relations of Eq. (3.64), we obtain

$$
\begin{aligned}
u_{11} &= (a_{11})^{\frac{1}{2}} = \sqrt{4} = 2;\\
u_{12} &= \frac{a_{12}}{u_{11}} = \frac{-1}{2} = -0.5;\\
u_{13} &= \frac{a_{13}}{u_{11}} = \frac{1}{2} = 0.5;\\
u_{22} &= \left(a_{22} - u_{12}^2\right)^{\frac{1}{2}} = \left(6 - (-0.5)^2\right)^{\frac{1}{2}} = \sqrt{5.75} = 2.3979;\\
u_{23} &= \frac{1}{u_{22}}[a_{23} - u_{12}u_{13}] = \frac{1}{2.3979}(-4 + 0.5 \times 0.5) = -1.5639;\\
u_{33} &= \left(a_{33} - u_{13}^2 - u_{23}^2\right)^{\frac{1}{2}} = \left(5 - (0.5)^2 - (-1.5639)^2\right)^{\frac{1}{2}} = \sqrt{2.3042} = 1.5180.
\end{aligned}
$$

This yields the desired form:

$$\begin{bmatrix} 4 & -1 & 1 \\ -1 & 6 & -4 \\ 1 & -4 & 5 \end{bmatrix} = \begin{bmatrix} 2 & 0 & 0 \\ -0.5 & 2.3979 & 0 \\ 0.5 & -1.5639 & 1.5180 \end{bmatrix} \begin{bmatrix} 2 & -0.5 & 0.5 \\ 0 & 2.3979 & -1.5639 \\ 0 & 0 & 1.5180 \end{bmatrix}.$$

3.11.3 Inverse of a Symmetric Matrix

If the inverse of a symmetric matrix $[A]$ is needed, we first decompose it as $[A] = [U]^T[U]$ using Eq. (3.64) and then find $[A]^{-1}$ as

$$[A]^{-1} = \left([U]^T[U]\right)^{-1} = [U]^{-1}\left([U]^T\right)^{-1}. \tag{3.65}$$

The elements, λ_{ij}, of $[U]^{-1}$ can be determined from $[U][U]^{-1} = [I]$, which leads to

$$\begin{aligned} \lambda_{ii} &= \frac{1}{u_{ii}}; \\ \lambda_{ij} &= \frac{-\left(\sum_{k=i+1}^{j} u_{ik}\lambda_{kj}\right)}{u_{ii}}; i < j; \\ \lambda_{ij} &= 0; i > j. \end{aligned} \tag{3.66}$$

Hence, the inverse of $[U]$ is also an upper triangular matrix. The inverse of $[U]^T$ can be obtained from the relation

$$\left([U]^T\right)^{-1} = \left([U]^{-1}\right)^T. \tag{3.67}$$

Finally, the inverse of the symmetric matrix $[A]$ can be calculated as

$$[A]^{-1} = [U]^{-1}\left([U]^{-1}\right)^T. \tag{3.68}$$

▶Example 3.17

Find the inverse of the matrix

$$[A] = \begin{bmatrix} 4 & -1 & 1 \\ -1 & 6 & -4 \\ 1 & -4 & 5 \end{bmatrix}$$

using the relation

$$[A]^{-1} = [U]^{-1}\left([U]^{-1}\right)^T.$$

Solution

From Example 3.16, the matrix $[U]$ is given by

$$[U] = \begin{bmatrix} 2 & -0.5 & 0.5 \\ 0 & 2.3979 & -1.5639 \\ 0 & 0 & 1.5180 \end{bmatrix}.$$

The elements of the matrix $[U]^{-1} = [\lambda_{ij}]$ are given by Eq. (3.66) as

$$\begin{aligned}
\lambda_{11} &= \frac{1}{u_{11}} = \frac{1}{2} = 0.5; \\
\lambda_{22} &= \frac{1}{u_{22}} = \frac{1}{2.3979} = 0.4170; \\
\lambda_{33} &= \frac{1}{u_{33}} = \frac{1}{1.5180} = 0.6588; \\
\lambda_{12} &= -\frac{1}{u_{11}}(u_{12}\lambda_{22}) = -\frac{(-0.5)(0.4170)}{2} = 0.1042; \\
\lambda_{23} &= -\frac{1}{u_{22}}(u_{23}\lambda_{33}) = -\frac{1}{2.3979}(-1.5639 \times 0.6588) = 0.4297; \\
\lambda_{13} &= -\frac{1}{u_{11}}(u_{12}\lambda_{23} + u_{13}\lambda_{33}) \\
&= -\frac{1}{2}(-0.5 \times 0.4297 + 0.5 \times 0.6588) = -0.0573; \\
\lambda_{21} &= \lambda_{31} = \lambda_{32} = 0.
\end{aligned}$$

Thus, the inverse of $[A]$ is given by

$$\begin{aligned}
[A]^{-1} &= \begin{bmatrix} 0.5 & 0.1042 & -0.0573 \\ 0 & 0.4170 & 0.4297 \\ 0 & 0 & 0.6588 \end{bmatrix} \begin{bmatrix} 0.5 & 0 & 0 \\ 0.1042 & 0.4170 & 0 \\ -0.0573 & 0.4297 & 0.6588 \end{bmatrix} \\
&= \begin{bmatrix} 0.2641 & 0.0188 & -0.0377 \\ 0.0188 & 0.3585 & 0.2831 \\ -0.0377 & 0.2831 & 0.4340 \end{bmatrix}.
\end{aligned}$$

◀

3.12 Jacobi Iteration Method

Consider the following system of linear equations,

$$\begin{aligned} a_{11}x_1 + a_{12}x_2 + \cdots + a_{1n}x_n &= b_1; \\ a_{21}x_1 + a_{22}x_2 + \cdots + a_{2n}x_n &= b_2; \\ &\vdots \\ a_{n1}x_1 + a_{n2}x_2 + \cdots + a_{nn}x_n &= b_n. \end{aligned} \tag{3.69}$$

These equations can be rewritten, by solving the ith equation for the unknown $x_i (i = 1, 2, \ldots, n)$, as follows:

$$\begin{aligned} x_1 &= \frac{1}{a_{11}}[b_1 - a_{12}x_2 - a_{13}x_3 - \cdots - a_{1n}x_n]; \\ x_2 &= \frac{1}{a_{22}}[b_2 - a_{21}x_1 - a_{23}x_3 - \cdots - a_{2n}x_n]; \\ &\vdots \\ x_n &= \frac{1}{a_{nn}}[b_n - a_{n1}x_1 - a_{n2}x_2 - \cdots - a_{n,n-1}x_{n-1}]. \end{aligned} \tag{3.70}$$

Equation (3.70) can be expressed in a compact form as

$$x_i = \frac{1}{a_{ii}}\left[b_i - \sum_{j=1, j\neq i}^{n} a_{ij}x_j\right]; \quad i = 1, 2, \ldots, n. \tag{3.71}$$

In the Jacobi iteration method, we start with a set of initial (approximate) values $x_1^{(1)}, x_2^{(1)}, \ldots, x_n^{(1)}$ for $x_1, x_2, \ldots, x_n$, respectively. If no better initial estimate is available, we can assume each component to be zero [i.e., $x_i^{(1)} = 0 (i = 1, 2, \ldots, n)$]. The assumed solution $x_i^{(1)}$ is substituted into the right-hand side of Eq. (3.70) or (3.71) to generate a new solution as

$$x_i^{(2)} = \frac{1}{a_{ii}}\left[b_i - \sum_{j=1, j\neq i}^{n} a_{ij}x_j^{(1)}\right]; \quad i = 1, 2, \ldots, n. \tag{3.72}$$

These values, $x_i^{(2)}$, obtained after the first iteration, are substituted into the right-hand side of Eq. (3.70) or (3.71) to generate the next set of values, $x_i^{(3)}$. This iterative process can be expressed as

$$x_i^{(k+1)} = \frac{1}{a_{ii}}\left[b_i - \sum_{j=1, j\neq i}^{n} a_{ij}x_j^{(k)}\right]; \quad i = 1, 2, \ldots, n; k = 1, 2, \ldots. \tag{3.73}$$

The iterative process is continued until the values of $x_i^{(k)}$ determined in two successive iterations are sufficiently close to one another. Thus, the convergence criterion for stopping the iterative process can be assumed as

$$\left| x_i^{(k+1)} - x_i^{(k)} \right| \leq \varepsilon; \quad i = 1, 2, \ldots, n, \tag{3.74}$$

or

$$\left| \frac{x_i^{(k+1)} - x_i^{(k)}}{x_i^{(k)}} \right| \leq \varepsilon; \quad i = 1, 2, \ldots, n, \tag{3.75}$$

where ε is a small number. It has been found [3.11, 3.12] that a sufficient condition for the convergence of Jacobi method is

$$|a_{ii}| > \sum_{j=1, j \neq i}^{n} |a_{ij}|. \tag{3.76}$$

Thus, the values $x_i^{(k)}$ will converge to the correct solution irrespective of the initial values $x_i^{(1)}$ used when the condition of Eq. (3.76) is satisfied. Note that the condition of Eq. (3.76) implies that the equations are diagonally dominant; that is, the coefficient on the diagonal in any row is larger in absolute value than the sum of the absolute values of the other coefficients in the same row.

▶Example 3.18

Determine whether the following system of equations is diagonally dominant:

$$-5x_1 - x_2 + 2x_3 = 1; \tag{E1}$$

$$2x_1 + x_2 + 7x_3 = 32; \tag{E2}$$

$$2x_1 + 6x_2 - 3x_3 = 2. \tag{E3}$$

Solution

It can be seen that the diagonal coefficients in Eqs. (E2) and (E3) are not large compared with the other coefficients of the same equation. However, interchanging Eqs. (E2) and (E3) yields the following system:

$$-5x_1 - x_2 + 2x_3 = 1; \tag{E4}$$

$$2x_1 + 6x_2 - 3x_3 = 2; \tag{E5}$$

$$2x_1 + x_2 + 7x_3 = 32. \tag{E6}$$

This system can be observed to be diagonally dominant, because

$$|-5| > |-1| + |2|;$$

$$|6| > |2| + |-3|;$$

$$|7| > |2| + |1|.$$

◀

▶Example 3.19

Solve the following system of equations using Jacobi iteration method:

$$-5x_1 - x_2 + 2x_3 = 1; \quad \text{(E1)}$$

$$2x_1 + 6x_2 - 3x_3 = 2; \quad \text{(E2)}$$

$$2x_1 + x_2 + 7x_3 = 32. \quad \text{(E3)}$$

Solution

Equations (E1), (E2), and (E3) can be expressed, by solving for x_1, x_2, and x_3, respectively, as

$$x_1 = -\frac{1}{5}(1 + x_2 - 2x_3); \quad \text{(E4)}$$

$$x_2 = \frac{1}{6}(2 - 2x_1 + 3x_3); \quad \text{(E5)}$$

$$x_3 = \frac{1}{7}(32 - 2x_1 - x_2). \quad \text{(E6)}$$

By choosing the initial approximation for the variables as $x_i^{(1)} = 0, i = 1, 2$, and 3 and, substituting them into the right-hand side of Eqs. (E4) through (E6), we generate the new approximation as

$$x_1^{(2)} = -\frac{1}{5}(1 + 0 - 2(0)) = -\frac{1}{5} = -0.2000;$$

$$x_2^{(2)} = \frac{1}{6}(2 - 2(0) + 3(0)) = \frac{2}{6} = 0.3333;$$

$$x_3^{(2)} = \frac{1}{7}(32 - 2(0) - 0) = \frac{32}{7} = 4.5714.$$

These values are again substituted into the right-hand sides of Eqs. (E4) through (E6) to generate a second approximation:

$$x_1^{(3)} = -\frac{1}{5}(1 + 0.3333 - 2(4.5714)) = 1.5619;$$

$$x_2^{(3)} = \frac{1}{6}(2 - 2(-0.2000) + 3(4.5714)) = 2.6857;$$

$$x_3^{(3)} = \frac{1}{7}(32 - 2(-0.2000) - 0.3333) = 4.5810.$$

When the process is repeated, the results shown in Table 3.2 are obtained.

The process is stopped by using the convergence criterion

$$\|\vec{x}^{(k+1)} - \vec{x}^{(k)}\| \leq 10^{-6}.$$

Table 3.2 Results of Jacobi iteration process.

Iteration number	Variables x_1	x_2	x_3
1	0.0000	0.0000	0.0000
2	−0.2000	0.3333	4.5714
3	1.5619	2.6857	4.5810
4	1.0952	2.1032	3.7415
5	0.8760	1.8390	3.9580
6	1.0154	2.0204	4.0584
7	1.0193	2.0241	3.9927
8	0.9923	1.9899	3.9910
9	0.9984	1.9981	4.0037
10	1.0018	2.0023	4.0007
11	0.9998	1.9997	3.9991

◀

3.13 Gauss–Seidel Iteration Method

It can be seen that, in Jacobi iteration method, *all* the variables $x_i^{(k+1)}$ are computed using the values of the previous iteration $x_i^{(k)}$. This implies that both the present, as well as the previous set of values, are to be stored. The storage requirement and the rate of convergence can be improved using the Gauss–Seidel iteration method in which the values of $x_1^{(k+1)}, x_2^{(k+1)}, \ldots, x_i^{(k+1)}$ computed in the current iteration, along with the values of $x_{i+2}^{(k)}, x_{i+3}^{(k)}, \ldots, x_n^{(k)}$, are used in finding the value $x_{i+1}^{(k+1)}$. This implies that always the most recent approximations to the variables are used during the computations. The iterative process can be expressed as follows:

$$x_i^{(k+1)} = \frac{1}{a_{ii}}\left[b_i - \sum_{j=1}^{i-1} a_{ij}x_j^{(k+1)} - \sum_{j=i+1}^{n} a_{ij}x_i^{(k)}\right]; \; i = 1, 2, \ldots, n; \; k = 1, 2, \ldots. \tag{3.77}$$

Notes

(1) The method requires an initial guess for the values of the unknowns $x_2, x_3, \ldots, x_n$ (x_1 is not required) as $x_2^{(1)}, x_3^{(1)}, \ldots, x_n^{(1)}$.

(2) The iteration process is to be continued until all the newly computed values of $x_i^{(k+1)}$ converge to within a convergence criterion ε of their previous values $x_i^{(k)}$ as indicated in Eqs. (3.74) and (3.75).

(3) The Gauss–Seidel method also converges to the correct solution, irrespective of the initial estimate, if the system of equations is diagonally dominant. Note that the diagonal dominance is only a sufficient condition; that is, if the

condition holds, the iterative process always converges. But, in many cases, the solution converges, even if the system is weakly diagonally dominant. If the initial guess is close to the correct solution, the iteration method converges fast.

(4) The Gauss–Seidel method is simple to program and implement. An accurate solution can be expected even when the number of equations is very large, provided that the system of equations is diagonally dominant.

▶Example 3.20

Find the solution of the following equations using the Gauss–Seidel iteration method:

$$-5x_1 - x_2 + 2x_3 = 1; \tag{E1}$$

$$2x_1 + 6x_2 - 3x_3 = 2; \tag{E2}$$

$$2x_1 + x_2 + 7x_3 = 32. \tag{E3}$$

Solution

Equations (E1), (E2), and (E3) can be expressed by solving for x_1, x_2, and x_3, respectively, as

$$x_1 = -\frac{1}{5}[1 + x_2 - 2x_3]; \tag{E4}$$

$$x_2 = \frac{1}{6}[2 - 2x_1 + 3x_3]; \tag{E5}$$

$$x_3 = \frac{1}{7}[32 - 2x_1 - x_2]. \tag{E6}$$

By choosing the initial guess as $x_2^{(1)} = 0$ and $x_3^{(1)} = 0$ and substituting them into the right-hand side of Eq. (E4), we obtain

$$x_1^{(2)} = -\frac{1}{5}[1 + 0 - 2(0)] = -0.2000.$$

By substituting $x_1^{(2)} = -0.2000, x_2^{(1)} = 0$ and $x_3^{(1)} = 0$ into the right-hand side of Eq. (E5), we obtain

$$x_2^{(2)} = \frac{1}{6}[2 - 2(-0.2000) + 3(0)] = 0.4000.$$

Next, by substituting $x_1^{(2)} = -0.2000, x_2^{(2)} = 0.4000$, and $x_3^{(1)} = 0$ into the right-hand side of Eq. (E6), we obtain

$$x_3^{(2)} = \frac{1}{7}[32 - 2(-0.2000) - (0.4000)] = 4.5714.$$

These values are again used to generate $x_i^{(3)}, i = 1, 2, 3$ using the general iteration, Eq. (3.77), to obtain the results indicated in Table 3.3.

Table 3.3 Convergence of Gauss–Seidel iteration procedure.

Iteration number	Variables x_1	x_2	x_3
1	0.0000	0.0000	0.0000
2	−0.2000	0.4000	4.5714
3	1.5486	2.1029	3.8286
4	0.9109	1.9440	4.0335
5	1.0246	2.0085	3.9918
6	0.9950	1.9975	4.0018
7	1.0012	2.0005	3.9996
8	0.9997	1.9999	4.0001

The process is stopped by using the convergence criterion

$$\|\vec{x}^{(k+1)} - \vec{x}^{(k)}\| \leq 10^{-6}.$$

Note that the Gauss–Seidel iteration method required 8 iterations, while the Jacobi iteration method required 11 iterations to satisfy the same convergence criterion. ◀

3.14 Relaxation Methods

The convergence characteristics of the Gauss–Seidel method can be improved by using a modification suggested by Southwell. The procedure is known as relaxation method. In the Gauss–Seidel method, the order in which the equations are used in finding the solution is important. In any iteration, it is desirable to improve the variable x_i that has the largest error. Since the variable x_i does not appear on the right-hand side, its own error will not influence its value in the next iteration. Thus, by selecting a proper equation for x_i, we introduce smaller errors in the values of x_i in the next iteration. The *relaxation method* is a procedure that permits the selection of the best equation for x_i to achieve a faster convergence. The iterative process in the relaxation method can be stated as [3.13]

$$x_i^{(k+1)} = \omega \left\{ \frac{1}{a_{ii}} \left[b_i - \sum_{j=1}^{i-1} a_{ij} x_j^{(k+1)} - \sum_{j=i+1}^{n} a_{ij} x_j^{(k)} \right] \right\} + (1-\omega) x_i^{(k)};$$

$$i = 1, 2, \ldots, n; k = 1, 2, \ldots, \tag{3.78}$$

where ω is a constant in the range $0 < \omega < 2$. It can be seen that Eq. (3.78) reduces to the Gauss–seidel iteration method, Eq. (3.77), for $\omega = 1$. Equation (3.78) can be rewritten as

$$x_i^{(k+1)} = x_i^{(k)} + \frac{\omega}{a_{ii}} \left[b_i - \sum_{j=1}^{i-1} a_{ij} x_j^{(k+1)} - \sum_{j=i}^{n} a_{ij} x_j^{(k)} \right];$$

$$i = 1, 2, \ldots, n; \; k = 1, 2, \ldots. \tag{3.79}$$

It has been shown that the process diverges for $\omega > 2$ [3.13]. The method is known as successive underrelaxation for $0 < \omega < 1$ and successive overrelaxation for $1 < \omega < 2$. The successive overrelaxation is commonly used for the solution of linear equations. However, the optimum value of the relaxation factor, ω, which yields the fastest convergence, is not known; it is usually determined by a trial-and-error process. The following example is considered to illustrate the successive overrelaxation method.

▶Example 3.21

Find the solution of the following system of equations using the relaxation method:

$$-5x_1 - x_2 + 2x_3 = 1, \tag{E1}$$

$$2x_1 + 6x_2 - 3x_3 = 2, \tag{E2}$$

$$2x_1 + x_2 + 7x_3 = 32. \tag{E3}$$

Also find the optimum value of the relaxation factor, ω.

Solution

By using the initial approximation $x_i^{(1)} = 0; i = 1, 2,$ and 3, the successive overrelaxation procedure (with a specified value of ω) given by Eq. (3.79), is used to find a solution that satisfies the convergence criterion:

$$\|\vec{x}^{(k+1)} - \vec{x}^{(k)}\| \leq 10^{-6}.$$

The equations are solved for a series of values of ω ranging from 0.1 to 1.4 in increments of 0.1. The convergence rates are shown in Table 3.4. If the process does not converge in 51 iterations, the solution is taken as the one found in the 51st iteration. It can be seen from Table 3.4 that the process has not converged for ω equal to 0.1 and 1.4. The process was found to diverge for $\omega > 1.4$ in this example. ◀

3.15 Simultaneous Linear Equations with Complex Coefficients and Constants

Many engineering applications lead to systems of linear equations involving complex coefficients, a_{ij}, and right-hand side constants, b_i. In such cases, the solution vector

Table 3.4 Convergence of the relaxation method for different values of ω.

Value of ω	Number of iterations required for convergence	Converged Solution Obtained x_1	x_2	x_3
0.1	50	0.9996	1.9832	4.0051
0.2	31	1.0025	1.9956	4.0014
0.3	22	1.0022	1.9973	4.0004
0.4	17	1.0018	1.9979	4.0000
0.5	14	1.0013	1.9985	3.9998
0.6	12	1.0008	1.9990	3.9998
0.7	10	1.0007	1.9991	3.9998
0.8	8	1.0009	1.9988	3.9998
0.9	7	1.0005	1.9994	3.9999
1.0	8	0.9997	1.9999	4.0001
1.1	12	0.9997	1.9999	4.0001
1.2	18	0.9996	1.9998	4.0003
1.3	33	1.0004	2.0001	3.9997
1.4	51	1.0445	2.0112	3.9644
1.5	solution diverged			

$\vec{x}$ will also be complex. Let the system of equations be expressed as

$$[A]\vec{x} = \vec{b}, \tag{3.80}$$

where the complex coefficient matrix, $[A]$, and the complex vectors $\vec{x}$ and $\vec{b}$ can be written as combinations of real and imaginary parts as

$$[A] = [C] + i[D]; \tag{3.81}$$

$$\vec{x} = \vec{y} + i\vec{z}; \tag{3.82}$$

$$\vec{b} = \vec{e} + i\vec{f}. \tag{3.83}$$

where $i = \sqrt{-1}$ and the $n \times n$ matrices $[C]$ and $[D]$, and the n-component vectors $\vec{y}, \vec{z}, \vec{e}$, and $\vec{f}$ are real. Substituting Eqs. (3.81) through (3.83) into Eq. (3.80) leads to

$$([C] + i[D])(\vec{y} + i\vec{z}) = ([C]\vec{y} - [D]\vec{z}) + i([D]\vec{y} + [C]\vec{z}) = \vec{e} + i\vec{f}. \tag{3.84}$$

Equating the real and imaginary parts in Eq. (3.84) yields

$$[C]\vec{y} - [D]\vec{z} = \vec{e}; \tag{3.85}$$

$$[D]\vec{y} + [C]\vec{z} = \vec{f}. \tag{3.86}$$

Equations (3.85) and (3.86) can be combined into a single matrix equation as

$$\begin{bmatrix} [C] & -[D] \\ [D] & [C] \end{bmatrix} \begin{Bmatrix} \vec{y} \\ \vec{z} \end{Bmatrix} = \begin{Bmatrix} \vec{e} \\ \vec{f} \end{Bmatrix}. \tag{3.87}$$

Equation (3.87) represents a system of $2n$ real linear algebraic equations that can be solved by any of the methods discussed earlier.

▶Example 3.22

Find the solution of the following set of complex equations:

$$\begin{bmatrix} 2+i & 1-4i \\ 4+2i & 5+3i \end{bmatrix} \begin{Bmatrix} x_1 \\ x_2 \end{Bmatrix} = \begin{Bmatrix} 3+2i \\ 2-2i \end{Bmatrix}. \tag{E1}$$

Solution

The application of Eqs. (3.81) through (3.83) to Eq. (E1) leads to

$$[A] = \begin{bmatrix} 2+i & 1-4i \\ 4+2i & 5+3i \end{bmatrix},$$

$$[C] = \begin{bmatrix} 2 & 1 \\ 4 & 5 \end{bmatrix},$$

$$[D] = \begin{bmatrix} 1 & -4 \\ 2 & 3 \end{bmatrix},$$

$$\vec{e} = \begin{Bmatrix} 3 \\ 2 \end{Bmatrix},$$

and

$$\vec{f} = \begin{Bmatrix} 2 \\ -2 \end{Bmatrix}.$$

The application of Eq. (3.87) yields the set of real equations

$$\begin{bmatrix} 2 & 1 & -1 & 4 \\ 4 & 5 & -2 & -3 \\ 1 & -4 & 2 & 1 \\ 2 & 3 & 4 & 5 \end{bmatrix} \begin{Bmatrix} y_1 \\ y_2 \\ z_1 \\ z_2 \end{Bmatrix} = \begin{Bmatrix} 3 \\ 2 \\ 2 \\ -2 \end{Bmatrix}. \tag{E2}$$

Using the Gauss elimination method, we can find the solution of Eq. (E2): $y_1 = 1.0$, $y_2 = -0.6$, $z_1 = -0.8$, and $z_2 = 0.2$. ◀

3.16 Matrix Inversion

As indicated in Appendix F, the inverse of a square matrix $[A]$ is defined by the relation

$$[A][A]^{-1} = [A]^{-1}[A] = [I], \tag{3.88}$$

where $[I]$ is the identity matrix. The inverse of a matrix $[A]$ may not exist always; $[A]^{-1}$ will not exist if the determinant of $[A]$ is zero. The inverse matrix of $[A]$ can be determined as

$$[A]^{-1} = \frac{\text{adjoint}[A]}{\det[A]}, \tag{3.89}$$

where det $[A]$ is the determinant of the matrix $[A]$ (assumed to be nonzero) and adjoint $[A]$ is the adjacent matrix of $[A] = [a_{ij}]$. It can be seen from Appendix F, the computation of the adjoint matrix of $[A]$ requires the evaluation of n^2 determinants of order $n - 1$. Thus, the computation of the inverse matrix using Eq. (3.89) is not very efficient. The Gauss elimination method is more efficient and hence is used commonly to find the inverse of a matrix. The inverse matrix can also be used to find the solution of the equations $[A]\vec{x} = \vec{b}$, where

$$\vec{x} = [A]^{-1}\vec{b}. \tag{3.90}$$

Note that the computation of the solution using Eq. (3.90) involves $\approx \frac{8n^3}{3}$ arithmetic operations in computing $[A]^{-1}$ followed by the multiplication of $[A]^{-1}$ and $\vec{b}$. On the other hand, the solution using Gauss elimination requires $\approx \frac{2n^3}{3}$ arithmetic operations. Thus, Eq. (3.90) requires four times the computational time than the Gauss elimination procedure.

3.16.1 Inverse of a Matrix by Gauss–Jordan Elimination

To find the inverse of a $n \times n$ square matrix $[A]$, we augment the matrix $[A]$ with an identity matrix of the same order $(n \times n)$. Then the elementary row operations are used to reduce the original matrix $[A]$ to the identity matrix. By then, the identity matrix will be located in the left half of the augmented matrix, and the right half contains the inverse of the matrix $[A]$. It is to be noted that the procedure is equivalent to solving a set of linear equations with n right-hand-side vectors; the ith right-hand-side vector being a n-component vector with unity in the ith position and zeros everywhere else. The procedure is illustrated through the following example:

▶Example 3.23

Find the inverse of the matrix

$$[A] = \begin{bmatrix} 2 & -1 & 1 \\ 4 & 3 & -1 \\ 3 & 2 & 2 \end{bmatrix} \tag{E1}$$

using the Gauss–Jordan elimination method.

Solution

By augmenting the 3 × 3 matrix $[A]$, with the identity matrix of order 3 × 3, we obtain the 3 × 6 matrix, denoted as

$$[C] = [[A]|[I]] = \begin{bmatrix} 2 & -1 & 1 & 1 & 0 & 0 \\ 4 & 3 & -1 & 0 & 1 & 0 \\ 3 & 2 & 2 & 0 & 0 & 1 \end{bmatrix}. \tag{E2}$$

First, we divide row 1 by 2 to reduce the coefficient a_{11} to unity:

$$\begin{bmatrix} 1 & -\frac{1}{2} & \frac{1}{2} & \frac{1}{2} & 0 & 0 \\ 4 & 3 & -1 & 0 & 1 & 0 \\ 3 & 2 & 2 & 0 & 0 & 1 \end{bmatrix}. \tag{E3}$$

Adding −4 times row 1 to row 2 and −3 times row 1 to row 3 leads to

$$\begin{bmatrix} 1 & -\frac{1}{2} & \frac{1}{2} & \frac{1}{2} & 0 & 0 \\ 0 & 5 & -3 & -2 & 1 & 0 \\ 0 & \frac{7}{2} & \frac{1}{2} & -\frac{3}{2} & 0 & 1 \end{bmatrix}. \tag{E4}$$

Next, we divide row 2 by 5 to reduce the modified a_{22} to unity:

$$\begin{bmatrix} 1 & -\frac{1}{2} & \frac{1}{2} & \frac{1}{2} & 0 & 0 \\ 0 & 1 & -\frac{3}{5} & -\frac{2}{5} & \frac{1}{5} & 0 \\ 0 & \frac{7}{2} & \frac{1}{2} & -\frac{3}{2} & 0 & 1 \end{bmatrix}. \tag{E5}$$

Adding $\frac{1}{2}$ times row 2 to row 1 and $-\frac{7}{2}$ times row 2 to row 3 yields

$$\begin{bmatrix} 1 & 0 & \frac{1}{5} & \frac{3}{10} & \frac{1}{10} & 0 \\ 0 & 1 & -\frac{3}{5} & -\frac{2}{5} & \frac{1}{5} & 0 \\ 0 & 0 & \frac{13}{5} & -\frac{1}{10} & -\frac{7}{10} & 1 \end{bmatrix}. \tag{E6}$$

Then, we multiply row 3 by $\frac{5}{13}$ to reduce the modified a_{33} to unity:

$$\begin{bmatrix} 1 & 0 & \frac{1}{5} & \frac{3}{10} & \frac{1}{10} & 0 \\ 0 & 1 & -\frac{3}{5} & -\frac{2}{5} & \frac{1}{5} & 0 \\ 0 & 0 & 1 & -\frac{1}{26} & -\frac{7}{26} & \frac{5}{13} \end{bmatrix}. \tag{E7}$$

Finally, the adding $-\frac{1}{5}$ times row 3 to row 1 and $\frac{3}{5}$ times row 3 to row 2 gives

$$\begin{bmatrix} 1 & 0 & 0 & \frac{4}{13} & \frac{2}{13} & -\frac{1}{13} \\ 0 & 1 & 0 & -\frac{11}{26} & \frac{1}{26} & \frac{3}{13} \\ 0 & 0 & 1 & -\frac{1}{26} & -\frac{7}{26} & \frac{5}{13} \end{bmatrix}. \tag{E8}$$

Since the left half of this matrix is reduced to the identity matrix, the right half gives the required inverse,

$$[A]^{-1} = \frac{1}{26}\begin{bmatrix} 8 & 4 & -2 \\ -11 & 1 & 6 \\ -1 & -7 & 10 \end{bmatrix}. \tag{E9}$$

This result can be verified by multiplying with $[A]$, which yields

$$[A]^{-1}[A] = \frac{1}{26}\begin{bmatrix} 8 & 4 & -2 \\ -11 & 1 & 6 \\ -1 & -7 & 10 \end{bmatrix}\begin{bmatrix} 2 & -1 & 1 \\ 4 & 3 & -1 \\ 3 & 2 & 2 \end{bmatrix}$$

$$= \frac{1}{26}\begin{bmatrix} 26 & 0 & 0 \\ 0 & 26 & 0 \\ 0 & 0 & 26 \end{bmatrix} = \begin{bmatrix} 1 & 0 & 0 \\ 0 & 1 & 0 \\ 0 & 0 & 1 \end{bmatrix}. \tag{E10}$$

◀

3.17 Equations with Special Form of Coefficient Matrix

In many engineering applications such as the finite-element analysis of stress, fluid flow, or heat transfer problems, the equilibrium or steady state equations will be of the form

$$[A]\vec{x} = \vec{b}, \tag{3.91}$$

where the coefficient matrix $[A]$ is symmetric i.e., $[A]^T = [A]$. In addition, the matrix will be banded. (That is, the nonzero elements are located in a band about the main diagonal as shown in Fig. 3.10.) A particular case of a banded matrix in which all elements except those on the leading (main) diagonal and the two diagonals adjacent to it are zero is known as a tridiagonal or Jacobi matrix. (See Fig. 3.11.) A banded matrix can be considered to be an example of a sparse matrix in which a large proportion of the elements are zero. The banded matrices have significant advantages in that the zero elements outside the band neither have to be stored nor have to be operated upon in the computer during the solution procedure. Several algorithms have been developed to solve systems of linear equations $[A]\vec{x} = \vec{b}$ involving symmetric, banded, tridiagonal, or sparse coefficient matrix, $[A]$. (See Refs. [3.3] and [3.18].) Special procedures have also been developed to reduce the storage requirements, as well as the number of calculations in solving equations with sparse matrices. The procedure used in reducing the storage requirements can be illustrated with the help of the sparse matrix shown in Fig. 3.12. In this symmetric matrix, only the necessary elements of the upper triangular matrix are stored. One possibility is to store the elements in the following row vector:

$$\begin{matrix} \vec{c} \\ \text{element number} \end{matrix} = \left\{ \begin{matrix} 8 & 2 & 1 & 4 & 3 & 7 & 5 & 0 & 2 & 0 \\ 1 & 2 & 3 & 4 & 5 & 6 & 7 & 8 & 9 & 10 \end{matrix} \right.$$

$$\left. \begin{matrix} 4 & 9 & 5 & 0 & 2 & 3 & 3 & 2 & 7 & 1 & 6 & 2 & 2 \\ 11 & 12 & 13 & 14 & 15 & 16 & 17 & 18 & 19 & 20 & 21 & 22 & 23 \end{matrix} \right\} \tag{3.92}$$

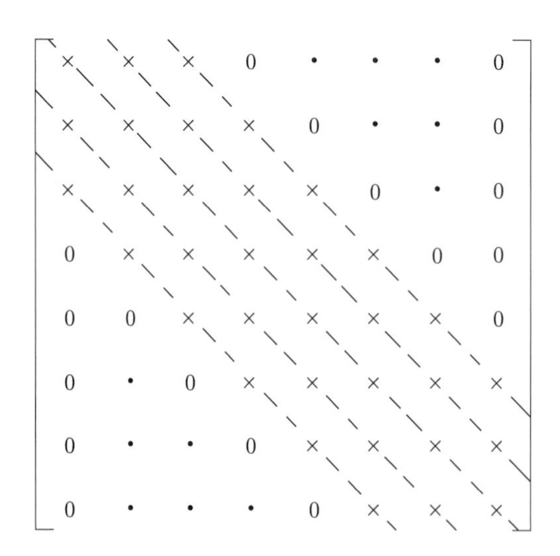

Figure 3.10 A banded matrix.

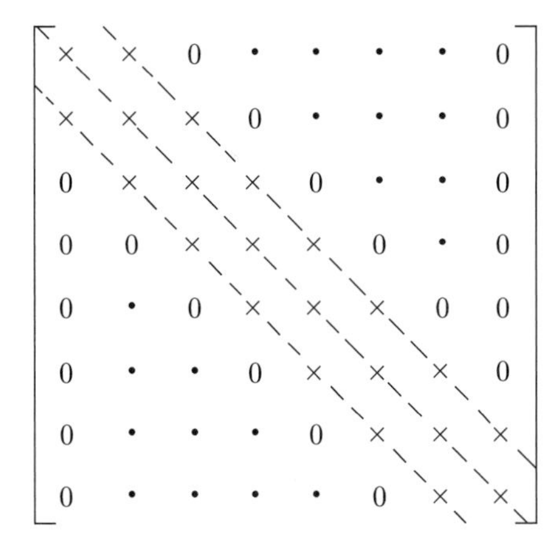

Figure 3.11 A tridiagonal or Jacobi matrix.

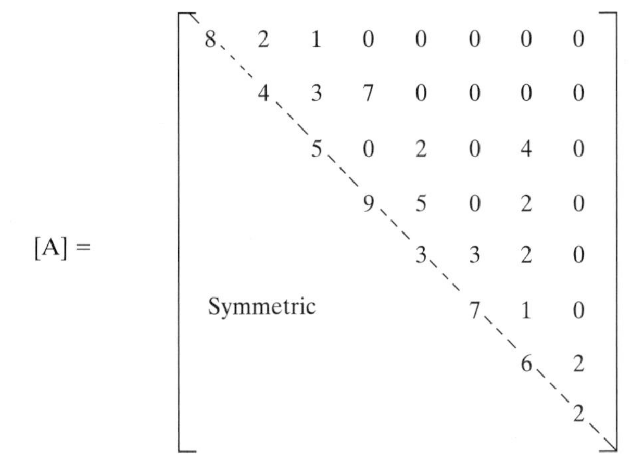

Figure 3.12 A sparse matrix.

Since different number of elements are stored from different rows in the matrix, the locations of the diagonal elements of $[A]$ in the row vector $\vec{c}$ are defined through another row vector $\vec{d}$ as follows:

$$\vec{d} = \{1\ 4\ 7\ 12\ 16\ 19\ 21\ 23\}. \tag{3.93}$$

This vector indicates that the $1^{\text{st}}, 4^{\text{th}}, 7^{\text{th}}, \ldots, 23^{\text{rd}}$ elements of the vector $\vec{c}$ are same as the diagonal elements of the $1^{\text{st}}, 2^{\text{nd}}, 3^{\text{rd}}, \ldots, 8^{\text{th}}$ rows of the sparse coefficient matrix $[A]$.

3.17.1 *LU* Decomposition of a Tridiagonal Matrix

Let the matrix $[A]$ be tridiagonal as

$$[A] = \begin{bmatrix} a_{11} & a_{12} & 0 & 0 & \cdot & \cdot & \cdot & 0 & 0 & 0 \\ a_{21} & a_{22} & a_{23} & 0 & \cdot & \cdot & \cdot & 0 & 0 & 0 \\ 0 & a_{32} & a_{33} & a_{34} & \cdot & \cdot & \cdot & 0 & 0 & 0 \\ \cdot & \cdot & \cdot & \cdot & \cdot & \cdot & \cdot & \cdot & \cdot & \cdot \\ \cdot & \cdot & \cdot & \cdot & \cdot & \cdot & \cdot & \cdot & \cdot & \cdot \\ \cdot & \cdot & \cdot & \cdot & \cdot & \cdot & \cdot & \cdot & \cdot & \cdot \\ 0 & 0 & 0 & 0 & \cdot & \cdot & \cdot & a_{n-1,n-2} & a_{n-1,n-1} & a_{n-1,n} \\ 0 & 0 & 0 & 0 & \cdot & \cdot & \cdot & 0 & a_{n,n-1} & a_{nn} \end{bmatrix}. \tag{3.94}$$

The LU decomposition of the matrix $[A]$ leads to $[L]$ and $[U]$ that are not only triangular, but also of band type. Crout's method, for example, gives

$$[L] = \begin{bmatrix} l_{11} & 0 & 0 & \cdot & \cdot & \cdot & 0 & 0 & 0 \\ l_{21} & l_{22} & 0 & \cdot & \cdot & \cdot & 0 & 0 & 0 \\ 0 & l_{32} & l_{33} & \cdot & \cdot & \cdot & 0 & 0 & 0 \\ \cdot & \cdot & \cdot & \cdot & \cdot & \cdot & \cdot & \cdot & \cdot \\ \cdot & \cdot & \cdot & \cdot & \cdot & \cdot & \cdot & \cdot & \cdot \\ \cdot & \cdot & \cdot & \cdot & \cdot & \cdot & \cdot & \cdot & \cdot \\ 0 & 0 & 0 & \cdot & \cdot & \cdot & l_{n-1,n-2} & l_{n-1,n-1} & 0 \\ 0 & 0 & 0 & \cdot & \cdot & \cdot & 0 & l_{n,n-1} & l_{nn} \end{bmatrix}, \tag{3.95}$$

and

$$[U] = \begin{bmatrix} 1 & u_{12} & 0 & \cdot & \cdot & \cdot & 0 & 0 \\ 0 & 1 & u_{23} & \cdot & \cdot & \cdot & 0 & 0 \\ 0 & 0 & 1 & \cdot & \cdot & \cdot & 0 & 0 \\ \cdot & \cdot & \cdot & \cdot & \cdot & \cdot & \cdot & \cdot \\ \cdot & \cdot & \cdot & \cdot & \cdot & \cdot & \cdot & \cdot \\ \cdot & \cdot & \cdot & \cdot & \cdot & \cdot & \cdot & \cdot \\ 0 & 0 & 0 & \cdot & \cdot & \cdot & 1 & u_{n-1,n} \\ 0 & 0 & 0 & \cdot & \cdot & \cdot & 0 & 1 \end{bmatrix}, \tag{3.96}$$

where the nonzero elements lie along the main diagonal and along the first subdiagonal in $[L]$ and the ones lie along the main diagonal and nonzero elements along the first superdiagonal in $[U]$. It can be seen that the matrix $[A]$ has only $(3n-2)$ nonzero elements, while $[L]$ has $(2n-1)$ unknown entries and $[U]$ has $(n-1)$ unknown entries. Thus, multiplying $[L]$ and $[U]$ and equating the result to $[A]$ gives $(3n-2)$ conditions to be satisfied in $(2n-1)+(n-1)=(3n-2)$ unknowns. Noting that

$$[L][U] = \begin{bmatrix} (l_{11}) & (l_{11}u_{12}) & 0 & \cdots & 0 & 0 \\ (l_{21}) & (u_{12}l_{21}+l_{22}) & (u_{23}l_{22}) & \cdots & 0 & 0 \\ 0 & (l_{32}) & (u_{23}l_{32}+l_{33}) & \cdots & 0 & 0 \\ \cdot & \cdot & \cdot & \cdots & \cdot & \cdot \\ \cdot & \cdot & \cdot & \cdots & \cdot & \cdot \\ \cdot & \cdot & \cdot & \cdots & \cdot & \cdot \\ 0 & 0 & 0 & \cdots & (l_{n,n-1}) & (u_{n-1,n}\, l_{n,n-1}+l_{nn}) \end{bmatrix} \tag{3.97}$$

and equating the nonzero elements of $[L][U]$ with those of $[A]$ yields

$$a_{11} = l_{11}; \tag{3.98}$$

$$a_{i,i-1} = l_{i,i-1}; i = 2, 3, \ldots, n; \tag{3.99}$$

$$a_{ii} = l_{i,i-1}u_{i-1,i} + l_{ii}; i = 2, 3, \ldots, n; \tag{3.100}$$

$$a_{i,i+1} = l_{ii}u_{i,i+1}; i = 1, 2, \ldots, n-1. \tag{3.101}$$

These elements are computed by first finding the nonzero off-diagonal terms of $[L](l_{i,i-1})$ using Eq. (3.99), and then finding the remaining elements of $[U]$ and $[L](u_{i,i+1}$ and $l_{ii})$ using Eqs. (3.101) and (3.100) alternatively.

3.17.2 Solution of a Tridiagonal System of Equations

To solve a tridiagonal system of equations $[A]\vec{x} = \vec{b}$ using LU decomposition, we write

$$[A]\vec{x} = [L][U]\vec{x} = \vec{b}. \tag{3.102}$$

By defining a vector $\vec{z}$ as

$$\vec{z} = [U]\vec{x}, \tag{3.103}$$

so that Eq. (3.102) becomes

$$[L]\vec{z} = \vec{b}, \tag{3.104}$$

we can determine the values of z_i (components of the vector $\vec{z}$) by using the following computational sequence (assuming that $l_{ii} \neq 0$):

Step 1

$$l_{11} = a_{11};$$
$$u_{12} = \frac{a_{12}}{l_{11}}.$$

Step 2 For $i = 2, 3, \ldots, n-1$, compute the following:

$$l_{i,i-1} = a_{i,i-1};$$
$$l_{ii} = a_{ii} - l_{i,i-1}u_{i-1,i};$$
$$u_{i,i+1} = \frac{a_{i,i+1}}{l_{ii}}.$$

Step 3

$$l_{n,n-1} = a_{n,n-1};$$
$$l_{nn} = a_{nn} - l_{n,n-1}u_{n-1,n}.$$

Step 4

$$z_1 = \frac{b_1}{l_{11}};$$
$$z_i = \frac{1}{l_{ii}}(b_i - l_{i,i-1}z_{i-1}); i = 2, 3, \ldots, n.$$

Once $\vec{z}$ is determined, Eq. (3.103) can be solved for $\vec{x}$ using back substitution. This yields

$$x_n = z_n;$$
$$x_i = z_i - u_{i,i+1}x_{i+1}; i = n-1, n-2, \ldots, 1.$$

3.17.3 Determinant of a Tridiagonal Matrix

Once $[A]$ is decomposed as $[L][U]$, the determinant of $[A]$ can be determined as

$$|[A]| = |[L]||[U]|. \tag{3.105}$$

Since the determinant of a triangular matrix is given by the product of the diagonal elements (see Problem 3.17), we have

$$|[L]| = l_{11}l_{22}\cdots l_{nn}, \tag{3.106}$$
$$|[U]| = 1, \tag{3.107}$$

and hence,

$$|[A]| = l_{11} l_{22} \cdots l_{nn}. \tag{3.108}$$

This indicates that the matrix $[A]$ will be singular if any $l_{ii} = 0$.

▶Example 3.24

Find the solution of the following equations using LU decomposition:

$$\begin{bmatrix} 2 & -1 & 0 & 0 \\ -1 & 2 & -1 & 0 \\ 0 & -1 & 2 & -1 \\ 0 & 0 & -1 & 2 \end{bmatrix} \begin{Bmatrix} x_1 \\ x_2 \\ x_3 \\ x_4 \end{Bmatrix} = \begin{Bmatrix} 1 \\ 0 \\ 0 \\ 0 \end{Bmatrix}.$$

Also find the determinant of the matrix $[A]$.

Solution

Noting that $a_{ii} = 2 (i = 1, 2, 3, 4)$, $a_{i,i+1} = -1 (i = 1, 2, 3)$, and $a_{i,i-1} = -1 (i = 2, 3, 4)$, we find that the application of the equations of Steps 1–3 of Section 3.17.2 yields the following:

$$l_{11} = 2;$$

$$u_{11} = 1;$$

$$u_{12} = \frac{a_{12}}{l_{11}} = -\frac{1}{2};$$

$$l_{21} = a_{21} = -1;$$

$$l_{22} = a_{22} - l_{21} u_{12} = 2 - (-1) \left(-\frac{1}{2}\right) = \frac{3}{2};$$

$$u_{23} = \frac{a_{23}}{l_{22}} = \frac{-1}{(\frac{3}{2})} = -\frac{2}{3};$$

$$l_{32} = a_{32} = -1;$$

$$l_{33} = a_{33} - l_{32} u_{23} = 2 - (-1) \left(-\frac{2}{3}\right) = \frac{4}{3};$$

$$u_{34} = \frac{a_{34}}{l_{33}} = -\frac{3}{4};$$

$$l_{43} = a_{43} = -1;$$

$$l_{44} = a_{44} - l_{43} u_{34} = 2 - (-1) \left(-\frac{3}{4}\right) = \frac{5}{4}.$$

This gives $[A] = [L][U]$:

$$\begin{bmatrix} 2 & -1 & 0 & 0 \\ -1 & 2 & -1 & 0 \\ 0 & -1 & 2 & -1 \\ 0 & 0 & -1 & 2 \end{bmatrix} = \begin{bmatrix} 2 & 0 & 0 & 0 \\ -1 & \frac{3}{2} & 0 & 0 \\ 0 & -1 & \frac{4}{3} & 0 \\ 0 & 0 & -1 & \frac{5}{4} \end{bmatrix} \begin{bmatrix} 1 & -\frac{1}{2} & 0 & 0 \\ 0 & 1 & -\frac{2}{3} & 0 \\ 0 & 0 & 1 & -\frac{3}{4} \\ 0 & 0 & 0 & 1 \end{bmatrix}.$$

The use of the equations of Step 4 of Section 3.17.2 leads to the following:

$$z_1 = \frac{b_1}{l_{11}} = \frac{1}{2};$$

$$z_2 = \frac{1}{l_{22}}(b_2 - l_{21}z_1) = \frac{2}{3}\left(0 - (-1)\left(\frac{1}{2}\right)\right) = \frac{1}{3};$$

$$z_3 = \frac{1}{l_{33}}(b_3 - l_{32}z_2) = \frac{3}{4}\left(0 - (-1)\left(\frac{1}{3}\right)\right) = \frac{1}{4};$$

$$z_4 = \frac{1}{l_{44}}(b_4 - l_{43}z_3) = \frac{4}{5}\left(0 - (-1)\left(\frac{1}{4}\right)\right) = \frac{1}{5}.$$

The components of $\vec{x}$ can be found from the equations of Section 3.17.2:

$$x_4 = z_4 = \frac{1}{5};$$

$$x_3 = z_3 - u_{34}x_4 = \frac{1}{4} - \left(-\frac{3}{4}\right)\left(\frac{1}{5}\right) = \frac{2}{5};$$

$$x_2 = z_2 - u_{23}x_3 = \frac{1}{3} - \left(-\frac{2}{3}\right)\left(\frac{2}{5}\right) = \frac{3}{5};$$

$$x_1 = z_1 - u_{12}x_2 = \frac{1}{2} - \left(-\frac{1}{2}\right)\left(\frac{3}{5}\right) = \frac{4}{5}.$$

Finally, the determinant of the matrix $[A]$ can be found from Eq. (3.108) as

$$|[A]| = l_{11}l_{22}l_{33}l_{44} = (2)\left(\frac{3}{2}\right)\left(\frac{4}{3}\right)\left(\frac{5}{4}\right) = 5.$$

◀

3.18 Overdetermined, Underdetermined, and Homogeneous Equations

In the previous sections, we saw that a unique solution can be found for the system of linear equations $[A]\vec{x} = \vec{b}$ if $[A]$ is a square matrix and the determinant of

$[A]$ is nonzero. In some situations, the coefficient matrix may be rectangular. If the number of equations (m) is smaller than the number of unknowns (n), the matrix $[A]$ will be order $m \times n$, and the equations are said to be *underdetermined system of equations*. On the other hand, if the number of equations (m) is larger than the number of unknowns (n), the matrix $[A]$ will be of order $m \times n$, and the equations are said to be *overdetermined system of equations*. If the constant vector $\vec{b} = \vec{0}$ (i.e., $[A]\vec{x} = \vec{0}$ with a square matrix $[A]$), the equations are said to be *homogeneous system of equations*. In all these cases, the methods described in the previous sections cannot be applied directly. The equations need to be modified suitably before the solution can be found as indicated in the following sections.

3.18.1 Overdetermined System of Equations

As previously stated, an overdetermined system has more linearly independent equations than unknowns ($m > n$). The following equations denote an overdetermined system with $m = 4$ and $n = 2$:

$$3x_1 + x_2 = 5;$$

$$x_1 - 2x_2 = -3;$$

$$x_1 - x_2 = 1;$$

$$2x_1 + 3x_2 = 6.$$

This system has no solution, since the equations do not intersect at a common point when represented graphically. (See Fig. 3.13.)

In general, an overdetermined system can be expressed as

$$\underset{m \times n}{[A]}\ \underset{n \times 1}{\vec{x}} = \underset{m \times 1}{\vec{b}}\ ; m > n. \tag{3.109}$$

Since no solution ($\vec{x}$) satisfies all the equations exactly, we define a residual or error vector as

$$\vec{e} = [A]\vec{x} - \vec{b}. \tag{3.110}$$

An approximate solution of Eq. (3.109) can be determined by minimizing the error vector using a suitable criterion. A commonly used procedure involves minimizing the sum of squares of the components of the error vector (known as the least-squares approach). The sum of squares of the components of $\vec{e}$ can be expressed as

$$\begin{aligned} f = \vec{e}^T\vec{e} &= \left(\vec{x}^T[A]^T - \vec{b}^T\right)\left([A]\vec{x} - \vec{b}\right) \\ &= \vec{x}^T[A]^T[A]\vec{x} - \vec{b}^T[A]\vec{x} - \vec{x}^T[A]^T\vec{b} + \vec{b}^T\vec{b}. \end{aligned} \tag{3.111}$$

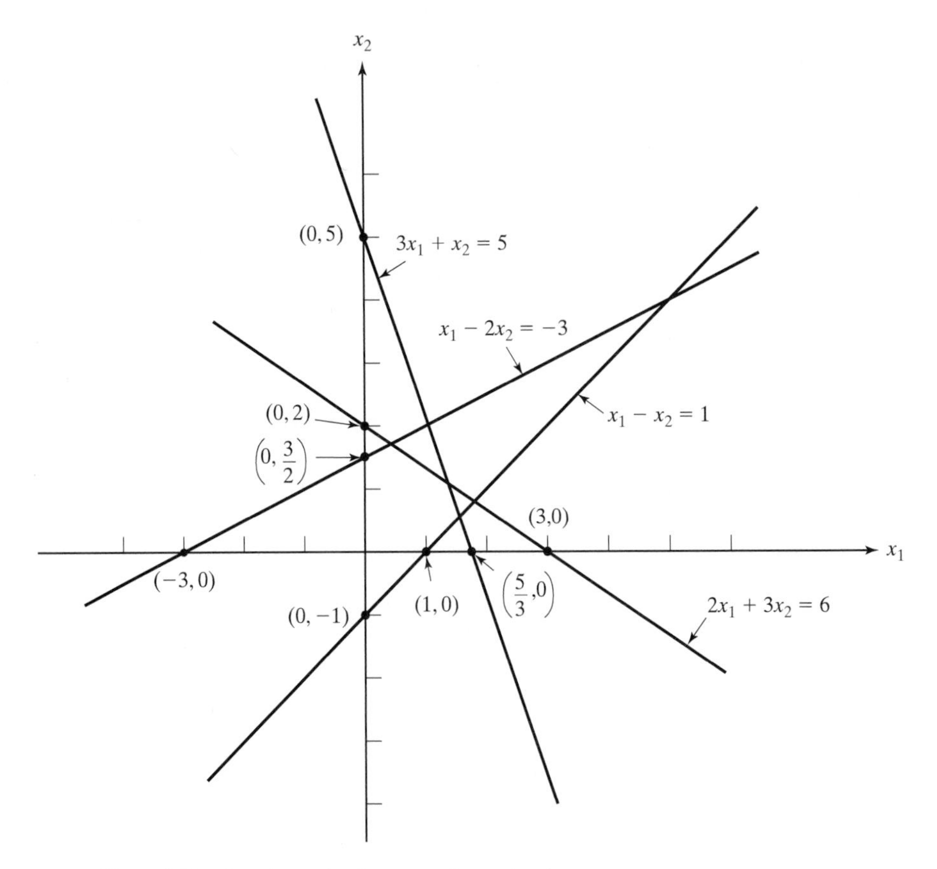

Figure 3.13 Overdetermined system of equations.

For minimizing f, we set the partial derivatives of f with respect to each of the components of $\vec{x}$ equal to zero. This leads to the equations

$$\underset{n \times m}{[A]^T} \underset{m \times n}{[A]} \underset{n \times 1}{\vec{x}} = \underset{n \times m}{[A]^T} \underset{m \times 1}{\vec{b}},$$

or

$$\underset{n \times n}{[C]} \underset{n \times 1}{\vec{x}} = \underset{n \times 1}{\vec{d}}. \tag{3.112}$$

$[C] = [A]^T[A]$ and $\vec{d} = [A]^T\vec{b}$. (See Problem 3.74.) Equation (3.112) denotes a system of n equations in n unknowns; hence, it can be solved using any of the methods described earlier. As already indicated, the resulting solution $\vec{x}$ will only be an approximate solution to the original overdetermined system of equations (Eq. (3.109)).

3.18.2 Underdetermined System of Equations

There will be an infinite number of solutions to an underdetermined system of equations. For example, consider an equation in two unknowns ($m = 1, n = 2$):

$$3x_1 + x_2 = 5. \tag{3.113}$$

This equation is shown graphically in Fig. 3.7. Since any point on the straight line of Fig. 3.7 satisfies Eq. (3.113), there will be an infinite number of solutions. Equation (3.113) can be solved for x_1 as

$$x_1 = \frac{1}{3}(5 - x_2). \tag{3.114}$$

In Eq. (3.114), the variable on the right-hand side (x_2) is called a free, or independent, or nonbasic, variable, and the variable on the left-hand side is called a basic variable. For any arbitrary value of x_2, the value of x_1 can be determined using Eq. (3.114). In general, if we have m linearly independent equations in n unknowns ($m < n$), we can select any m of the n variables as basic variables and express them in terms of the remaining $(n - m)$ nonbasic, or free, variables. The main requirement in selecting the basic variables is that the coefficient matrix corresponding to these variables must not be singular. Thus, the choice of the m basic variables may not be apparent at a glance. However, the Gauss or Gauss–Jordan elimination method can be used (the Gauss–Jordan method is more advantageous) to identify the set of basic variables as described in the following steps:

Step 1 Let the original m equations in n unknowns be expressed as follows: ($m \leq n$)

$$\begin{aligned}
a_{11}x_1 + a_{12}x_2 + \cdots + a_{1n}x_n &= b_1; \\
a_{21}x_1 + a_{22}x_2 + \cdots + a_{2n}x_n &= b_2; \\
&\vdots \\
a_{m1}x_1 + a_{m2}x_2 + \cdots + a_{mn}x_n &= b_m.
\end{aligned} \tag{3.115}$$

Step 2 Construct the $m \times (n + 1)$ augmented matrix

$$[C] = \begin{bmatrix}
a_{11} & a_{12} & \cdot & \cdot & \cdot & a_{1n} & a_{1,n+1} \\
a_{21} & a_{22} & \cdot & \cdot & \cdot & a_{2n} & a_{2,n+1} \\
\cdot & \cdot & \cdot & \cdot & \cdot & \cdot & \cdot \\
\cdot & \cdot & \cdot & \cdot & \cdot & \cdot & \cdot \\
\cdot & \cdot & \cdot & \cdot & \cdot & \cdot & \cdot \\
a_{m1} & a_{m2} & \cdot & \cdot & \cdot & a_{mn} & a_{m,n+1}
\end{bmatrix}, \tag{3.116}$$

where $a_{i,n+1} = b_i (i = 1, 2, \ldots, m)$.

Step 3 Apply the Gauss–Jordan elimination procedure, and reduce the matrix $[C]$ to the following form:

$$[C] = \begin{bmatrix} 1 & 0 & 0 & \cdot & \cdot & \cdot & 0 & a'_{1,m+1} & \cdot & \cdot & a'_{1n} & a'_{1,n+1} \\ 0 & 1 & 0 & \cdot & \cdot & \cdot & 0 & a'_{2,m+1} & \cdot & \cdot & a'_{2n} & a'_{2,n+1} \\ 0 & 0 & 1 & \cdot & \cdot & \cdot & 0 & a'_{3,m+1} & \cdot & \cdot & a'_{3n} & a'_{3,n+1} \\ \cdot & \cdot & \cdot & \cdot & \cdot & \cdot & \cdot & \cdot & \cdot & \cdot & \cdot & \cdot \\ \cdot & \cdot & \cdot & \cdot & \cdot & \cdot & \cdot & \cdot & \cdot & \cdot & \cdot & \cdot \\ \cdot & \cdot & \cdot & \cdot & \cdot & \cdot & \cdot & \cdot & \cdot & \cdot & \cdot & \cdot \\ 0 & 0 & 0 & \cdot & \cdot & \cdot & 1 & a'_{m,m+1} & \cdot & \cdot & a'_{mn} & a'_{m,n+1} \end{bmatrix}. \tag{3.117}$$

Step 4 Equation (3.117) assumes that $x_1, x_2, \ldots, x_m$ are the basic variables and that $x_{m+1}, x_{m+2}, \ldots, x_n$ are the free (nonbasic) variables, and hence, the solution can be expressed as

$$\begin{aligned} x_1 &= a'_{1,n+1} - a'_{1,m+1}x_{m+1} - \cdots - a'_{1n}x_n; \\ x_2 &= a'_{2,n+1} - a'_{2,m+1}x_{m+1} - \cdots - a'_{2n}x_n; \\ &\vdots \\ x_m &= a'_{m,n+1} - a'_{m,m+1}x_{m+1} - \cdots - a'_{mn}x_n. \end{aligned} \tag{3.118}$$

Step 5 Any set of values can be chosen for the free variables $x_{m+1}, x_{m+2}, \ldots, x_n$, and the corresponding values of the basic variables $x_1, x_2, \ldots, x_m$ can be determined. In particular, if $x_{m+1} = x_{m+2} = \cdots = x_n = 0$ is chosen, we obtain

$$x_i = a'_{i,n+1}; \quad i = 1, 2, \ldots, m. \tag{3.119}$$

The procedure is illustrated in the following example.

▶Example 3.25

Determine the solution of the following system of equations ($m = 3, n = 4$):

$$3x_1 + 2x_2 - 2x_3 - 7x_4 = -1; \tag{E1}$$

$$x_1 - x_2 + x_3 + x_4 = 3; \tag{E2}$$

$$2x_1 + x_2 + 4x_3 + 2x_4 = 20. \tag{E3}$$

Solution

The augmented matrix can be written as

$$\begin{bmatrix} 3 & 2 & -2 & -7 & -1 \\ 1 & -1 & 1 & 1 & 3 \\ 2 & 1 & 4 & 2 & 20 \end{bmatrix}.$$

The first row is normalized (a_{11} reduced to unity) by dividing it by 3:

$$\begin{bmatrix} 1 & \frac{2}{3} & -\frac{2}{3} & -\frac{7}{3} & -\frac{1}{3} \\ 1 & -1 & 1 & 1 & 3 \\ 2 & 1 & 4 & 2 & 20 \end{bmatrix}.$$

To reduce the coefficient 1 (a_{21}) to zero, we multiply the 1st row by -1 and add it to the 2nd row. Similarly, the coefficient 2 (a_{31}) is reduced to zero by multiplying the 1st row by -2 and adding it to the 3rd row. This results in the matrix

$$\begin{bmatrix} 1 & \frac{2}{3} & -\frac{2}{3} & -\frac{7}{3} & -\frac{1}{3} \\ 0 & -\frac{5}{3} & \frac{5}{3} & \frac{10}{3} & \frac{10}{3} \\ 0 & -\frac{1}{3} & \frac{16}{3} & \frac{20}{3} & \frac{62}{3} \end{bmatrix}.$$

The second row is normalized (current a_{22} is reduced to unity) by multiplying it by $-\frac{3}{5}$:

$$\begin{bmatrix} 1 & \frac{2}{3} & -\frac{2}{3} & -\frac{7}{3} & -\frac{1}{3} \\ 0 & 1 & -1 & -2 & -2 \\ 0 & -\frac{1}{3} & \frac{16}{3} & \frac{20}{3} & \frac{62}{3} \end{bmatrix}.$$

The coefficient $\frac{2}{3}$ (current a_{12}) is reduced to zero by multiplying the 2nd row by $-\frac{2}{3}$ and adding it to the 1st row. Similarly, the coefficient $-\frac{1}{3}$ (current a_{32}) is reduced to zero by multiplying the 2nd row by $\frac{1}{3}$ and adding it to the 3rd row. These operations give the following matrix:

$$\begin{bmatrix} 1 & 0 & 0 & -1 & 1 \\ 0 & 1 & -1 & -2 & -2 \\ 0 & 0 & 5 & 6 & 20 \end{bmatrix}.$$

The third row is normalized (current a_{33} is reduced to unity) by dividing it by 5:

$$\begin{bmatrix} 1 & 0 & 0 & -1 & 1 \\ 0 & 1 & -1 & -2 & -2 \\ 0 & 0 & 1 & \frac{6}{5} & 4 \end{bmatrix}.$$

Finally, the coefficient -1 (current a_{23}) is reduced to zero by adding 3rd row to the 2nd row:

$$\begin{bmatrix} 1 & 0 & 0 & -1 & 1 \\ 0 & 1 & 0 & -\frac{4}{5} & 2 \\ 0 & 0 & 1 & \frac{6}{5} & 4 \end{bmatrix}.$$

From this matrix, the basic variables (x_1, x_2, and x_3) can be expressed in terms of the free variable (x_4) as follows:

$$x_1 = 1 + x_4;$$
$$x_2 = 2 + \frac{4}{5}x_4;$$
$$x_3 = 4 - \frac{6}{5}x_4.$$

If x_4 is chosen as zero, the solution of the underdetermined system can be expressed as

$$x_1 = 1, x_2 = 2, x_3 = 4, \text{ and } x_4 = 0.$$ ◀

3.18.3 Homogeneous System of Equations

When the constant vector is zero, the equations

$$[A]\vec{x} = \vec{0} \tag{3.120}$$

are known as homogeneous system of equations. These equations are satisfied by the trivial solution $\vec{x} = \vec{0}$. It can be proved that these equations will have a nontrivial (or nonzero) solution, if and only if the matrix $[A]$ is singular i.e., if the rank (r) of the $n \times n$ matrix $[A]$ is less than n. In such a case, Eq. (3.120) will contain r independent equations ($r < n$). Hence, any set of r variables can be considered as the basic variables and can be expressed in terms of the remaining $(n - r)$ free, or nonbasic, variables. When nonzero values are chosen for the free variables, the system of r independent equations can be transformed to a nonhomogeneous form. Then the Gauss–Jordan (or Gauss) elimination method can be applied as in the previous section to determine the solution. Note that the initial augmented matrix will be of the form

$$\begin{bmatrix} a_{11} & a_{12} & \cdot & \cdot & \cdot & a_{1r} & a_{1,r+1} & \cdot & \cdot & \cdot & a_{1n} & a_{1,n+1} = 0 \\ a_{21} & a_{22} & \cdot & \cdot & \cdot & a_{2r} & a_{2,r+1} & \cdot & \cdot & \cdot & a_{2n} & a_{2,n+1} = 0 \\ \cdot & \cdot & \cdot & \cdot & \cdot & \cdot & \cdot & \cdot & \cdot & \cdot & \cdot & \cdot \\ \cdot & \cdot & \cdot & \cdot & \cdot & \cdot & \cdot & \cdot & \cdot & \cdot & \cdot & \cdot \\ \cdot & \cdot & \cdot & \cdot & \cdot & \cdot & \cdot & \cdot & \cdot & \cdot & \cdot & \cdot \\ a_{n1} & a_{n2} & \cdot & \cdot & \cdot & a_{nr} & a_{n,r+1} & \cdot & \cdot & \cdot & a_{nn} & a_{n,n+1} = 0 \end{bmatrix}, \tag{3.121}$$

where the last column can be seen to be zero (since the constant vector is zero for homogeneous equations). The final modified augmented matrix will have the following form:

$$\begin{bmatrix} 1 & 0 & 0 & \cdot & \cdot & \cdot & 0 & a'_{1,r+1} & \cdot & \cdot & a'_{1n} & a'_{1,n+1} = 0 \\ 0 & 1 & 0 & \cdot & \cdot & \cdot & 0 & a'_{2,r+1} & \cdot & \cdot & a'_{2n} & a'_{2,n+1} = 0 \\ \cdot & \cdot & \cdot & \cdot & \cdot & \cdot & \cdot & \cdot & \cdot & \cdot & \cdot & \cdot \\ 0 & 0 & 0 & \cdot & \cdot & \cdot & 1 & a'_{r,r+1} & \cdot & \cdot & a'_{rn} & a'_{r,n+1} = 0 \\ 0 & 0 & 0 & \cdot & \cdot & \cdot & 0 & 0 & \cdot & \cdot & 0 & 0 \\ \cdot & \cdot & \cdot & \cdot & \cdot & \cdot & \cdot & \cdot & \cdot & \cdot & \cdot & \cdot \\ 0 & 0 & 0 & \cdot & \cdot & \cdot & 0 & 0 & \cdot & \cdot & 0 & 0 \end{bmatrix}. \tag{3.122}$$

In this matrix, the null rows represent the linearly dependent equations that have been eliminated. From this matrix, the final solution of the original homogeneous system of equations can be expressed as

$$\begin{aligned} x_1 &= -a'_{1,r+1}x_{r+1} - \cdots - a'_{1,n}x_n; \\ x_2 &= -a'_{2,r+1}x_{r+1} - \cdots - a'_{2,n}x_n; \\ x_r &= -a'_{r,r+1}x_{r+1} - \cdots - a'_{r,n}x_n. \end{aligned} \tag{3.123}$$

▶Example 3.26

Find the solution of the following equations:

$$\begin{aligned} 3x_1 - 2x_2 + 2x_3 &= 0; \\ x_1 + x_2 + 4x_3 &= 0; \\ 4x_1 - x_2 + 6x_3 &= 0. \end{aligned}$$

Solution

The determinant of the coefficient matrix $[A]$ is given by

$$|[A]| = \left| \begin{bmatrix} 3 & -2 & 2 \\ 1 & 1 & 4 \\ 4 & -1 & 6 \end{bmatrix} \right|$$

$$= 3 \begin{vmatrix} 1 & 4 \\ -1 & 6 \end{vmatrix} - (-2) \begin{vmatrix} 1 & 4 \\ 4 & 6 \end{vmatrix} + 2 \begin{vmatrix} 1 & 1 \\ 4 & -1 \end{vmatrix}$$

$$= 3(6 + 4) + 2(6 - 16) + 2(-1 - 4) = 0,$$

and hence, a nontrivial solution can be sought to the system of equations. The augmented matrix is given by

$$\begin{bmatrix} 3 & -2 & 2 & 0 \\ 1 & 1 & 4 & 0 \\ 4 & -1 & 6 & 0 \end{bmatrix}.$$

The first row is normalized (a_{11} reduced to 1) by dividing it by 3:

$$\begin{bmatrix} 1 & -\frac{2}{3} & \frac{2}{3} & 0 \\ 1 & 1 & 4 & 0 \\ 4 & -1 & 6 & 0 \end{bmatrix}.$$

The coefficients 1 (a_{21}) and 4 (a_{31}) are reduced to zero by adding -1 times and -4 times the first equation to equations 2 and 3, respectively, to obtain

$$\begin{bmatrix} 1 & -\frac{2}{3} & \frac{2}{3} & 0 \\ 0 & \frac{5}{3} & \frac{10}{3} & 0 \\ 0 & \frac{5}{3} & \frac{10}{3} & 0 \end{bmatrix}.$$

The second row is normalized (current a_{22} is reduced to 1) by multiplying the 2nd equation by $\frac{3}{5}$:

$$\begin{bmatrix} 1 & -\frac{2}{3} & \frac{2}{3} & 0 \\ 0 & 1 & 2 & 0 \\ 0 & \frac{5}{3} & \frac{10}{3} & 0 \end{bmatrix}.$$

Next, the coefficients $-\frac{2}{3}$ (current a_{12}) and $\frac{5}{3}$ (current a_{32}) are reduced to zero by adding $\frac{2}{3}$ times and $-\frac{5}{3}$ times the 2nd equation to 1st and 3rd equations, respectively, to find

$$\begin{bmatrix} 1 & 0 & 2 & 0 \\ 0 & 1 & 2 & 0 \\ 0 & 0 & 0 & 0 \end{bmatrix}.$$

From this, the solution of the homogeneous equations can be determined as

$$x_1 = -2x_3;$$

$$x_2 = -2x_3.$$

This indicates that there are two independent equations ($r = 2$) in which the basic variables are x_1 and x_2 and the free or nonbasic variable is x_3. ◀

3.19 Comparative Efficiencies of Various Methods

The computational effort involved in the solution of a set of n linear equations, $[A]\vec{x} = \vec{b}$, varies with the method. The number of arithmetic operations required by a particular method reflects the efficiency of that method. Note that in addition to the time taken for the arithmetic operations, an algorithm requires additional time for tasks such as storing the values, incrementing of variables in loops, and accessing the elements of a matrix or vector. The number of arithmetic operations can be determined by examining the details of computations in various loops of the algorithm as indicated next.

3.19.1 Arithmetic Operations Count for Gauss Elimination Method

As seen in Section 3.9, the Gaussian elimination method basically involves two stages: forward elimination to reduce the matrix to a triangular form and backward substitution to determine the solution. During the forward elimination process, in the first equation, each of $a_{12}, a_{13}, \ldots, a_{1n}$ and $a_{1,n+1} = b_1$ is to be multiplied by the quantity $\frac{a_{21}}{a_{11}}$ and the result is to be subtracted from $a_{21}, a_{22}, \ldots, a_{2,n+1}$, respectively. This amounts to having n multiplications and n subtractions (in addition to one division in finding the value of $\frac{a_{21}}{a_{11}}$). Thus, a total of $(n + 1)$ multiplications and divisions and n subtractions are involved. This is repeated for the 2nd, 3rd, ..., nth equation (a total of $n-1$ times) so that the number of operations involved in reducing the $(n-1)$ coefficients in the first column is $(n+1)(n-1)$ multiplications and divisions and $n(n-1)$ subtractions. Noting that there are $(n-1)$ coefficients to the right of the pivot element (including $a_{2,n+1} = b_2$), and $(n-2)$ equations for the second equation, the number of operations involved in reducing the $(n-2)$ coefficients in the second column is $n(n-2)$ multiplications and divisions and $(n-1)(n-2)$ subtractions.

This gives the total number of multiplications and divisions involved in reducing the equations to triangular form as:

$$
\begin{aligned}
N_{m1} &= (n+1)(n-1) + n(n-2) + (n-1)(n-3) + \cdots + (3)(1) \\
&= \sum_{i=1}^{n-1} (n-i+2)(n-i) \\
&= \sum_{i=1}^{n-1} (n^2 - 2ni + i^2 + 2n - 2i) \\
&= (2n+n^2)\sum_{i=1}^{n-1} 1 - (2n+n)\sum_{i=1}^{n-1} i + \sum_{i=1}^{n-1} i^2.
\end{aligned}
\tag{3.124}
$$

Since

$$
\sum_{i=1}^{p} 1 = p, \sum_{i=1}^{p} i = \frac{p(p+1)}{2} \text{ and, } \sum_{i=1}^{p} i^2 = \frac{p(p+1)(2p+1)}{6}, \tag{3.125}
$$

Eq. (3.124) can be expressed as

$$
N_{m1} = (n^2+2n)(n-1) - (2n+2)\frac{(n-1)n}{2} + \frac{(n-1)n(2n-1)}{6} = \frac{2n^3+3n^2-5n}{6}. \tag{3.126}
$$

Similarly, the total number of additions and subtractions involved in the forward elimination process is

$$
\begin{aligned}
N_{a1} &= n(n-1) + (n-1)(n-2) + \cdots + (2)(1) \\
&= \sum_{i=1}^{n-1} (n-i)(n-i+1) = (n^2+n)\sum_{i=1}^{n-1} 1 - (2n+1)\sum_{i=1}^{n-1} i + \sum_{i=1}^{n-1} i^2 \\
&= (n^2+n)(n-1) - (2n+1)\frac{(n-1)(n)}{2} + \frac{(n-1)(n)(2n-1)}{6} \\
&= \frac{(n^3-n)}{3}.
\end{aligned}
\tag{3.127}
$$

During the backward substitution process, first x_n is found as

$$
x_n = \frac{a'_{n,n+1}}{a'_{n,n}} = \frac{b'_n}{a'_{n,n}} \text{ (one division and zero subtraction involved).} \tag{3.128}
$$

Then x_{n-1} is found as

$$
x_{n-1} = \frac{b'_{n-1} - a'_{n-1,n} x_n}{a'_{n-1,n-1}}
$$

$$
\text{(one division, one multiplication, and one subtraction involved).} \tag{3.129}
$$

As we find $x_{n-2}, x_{n-3}, \ldots, x_1$, at each step, we need to consider one more term in the numerator involving a multiplication and a subtraction. The total number of multiplications and divisions is

$$N_{m2} = 1 + 2 + 3 + \cdots + n = \frac{n^2 + n}{2} \tag{3.130}$$

and the total number of additions and subtractions is

$$N_{a2} = 0 + 1 + 2 + \cdots + (n - 1) = \frac{n^2 - n}{2}. \tag{3.131}$$

Thus, the total number of arithmetic operations involved in Gaussian elimination method are given by
Number of multiplications and divisions (N_m):

$$N_m = N_{m1} + N_{m2} = \frac{2n^3 + 3n^2 - 5n}{6} + \frac{n^2 + n}{2} = \frac{1}{3}(n^3 + 3n^2 - n). \tag{3.132}$$

Number of additions and subtractions (N_a):

$$N_a = N_{a1} + N_{a2} = \frac{n^3 - n}{3} + \frac{n^2 - n}{2} = \frac{1}{6}(2n^3 + 3n^2 - 5n). \tag{3.133}$$

3.19.2 Arithmetic Operations Count for Other Methods

It can be shown that Crout's method requires a total of

$$N = (n - 1)(n + 1)! \tag{3.134}$$

arithmetic operations for solving the equations. The Gauss–Jordan elimination method, on the other hand, requires (see Problem 3.77)

$$N_1 = \frac{1}{2}n^3 + n^2 - \frac{1}{2}n \tag{3.135}$$

multiplications and divisions and

$$N_2 = \frac{1}{2}n^3 - \frac{1}{2}n \tag{3.136}$$

additions and subtractions. If Choleski method is used, the number of arithmetic operations required to solve the equations $[A]\vec{x} = [U][U]^T\vec{x} = \vec{b}$ are given by

$$N_1 = n \text{ square roots}, \tag{3.137}$$

$$N_2 = \frac{1}{6}n^3 + \frac{3}{2}n^2 + \frac{1}{3}n \text{ multiplications and divisions}, \tag{3.138}$$

and

$$N_3 = \frac{1}{6}n^3 + n^2 - \frac{7}{6}n \text{ additions and subtractions.} \tag{3.139}$$

(See Problem 3.78). Similarly, if the augmented matrix, $[C] = [[A]|[I]]$, is reduced to the form $[C'] = [[I]|[A]^{-1}]$ using a series of elementary operations, the number of arithmetic operations involved (in finding the inverse of $[A]$) can be found as follows:

In Gauss elimination method,

$$N_1 = \frac{4}{3}n^3 - \frac{1}{3}n \text{ multiplications and divisions} \tag{3.140}$$

and

$$N_2 = \frac{4}{3}n^3 - \frac{3}{2}n^2 + \frac{1}{6}n \text{ additions and subtractions.} \tag{3.141}$$

In the Gauss–Jordan elimination method,

$$N_1 = \frac{3}{2}n^3 - \frac{1}{2}n \text{ multiplications and divisions,} \tag{3.142}$$

and

$$N_2 = \frac{3}{2}n^3 - 2n^2 + \frac{1}{2}n \text{ additions and subtractions.} \tag{3.143}$$

The number of arithmetic operations required by different methods in solving a system of linear equations for few specific values of n are shown in Table 3.5. The computational effort involved in the Gauss elimination and Gauss–Jordan elimination methods compared with that of Choleski decomposition method is shown in Fig. 3.14 for different sizes ($n \times n$) of the coefficient matrix.

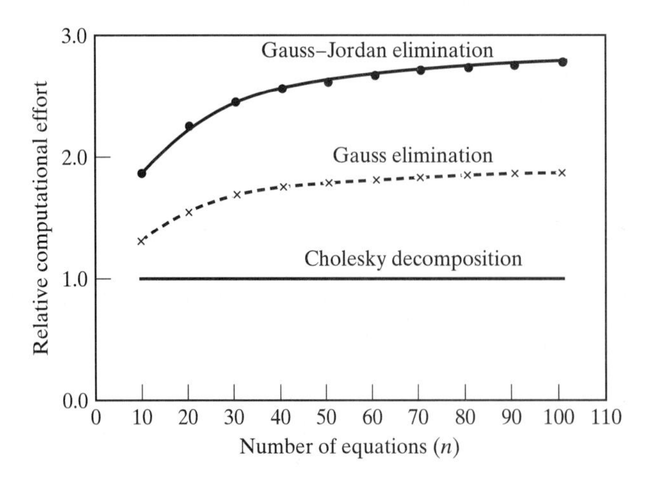

Figure 3.14 Comparison of computational efforts.

Table 3.5 Number of arithmetic operations required in solving a system of linear equations.

Type of operations	Value of n	Gauss elimination method	Gauss–Jordan elimination method	Choleski decomposition method
Multiplications/ divisions	6	106	141	92
	30	9890	14385	5860
	60	75580	111570	41420
	100	343300	509950	181699
Additions/ subtractions	6	85	105	65
	30	9425	13485	5365
	60	73750	107970	39530
	100	338250	499950	176550
Square roots	6	—	—	6
	30	—	—	30
	60	—	—	60
	100	—	—	100

3.19.3 Observations and Recommendations

The following observations and recommendations can be made in connection with the solution of a set of linear algebraic equations:

(1) The round-off error, as well as the time required to solve the equations, depends on the total number of arithmetic operations performed. In general, the time required to perform a multiplication or division operation is approximately same. On the other hand, the time required to perform an addition or subtraction is considerably small compared with that required for a multiplication or division. Hence, a good estimate of the time required to solve a system of equations can be obtained by considering only the multiplication and division operations involved in the process. Thus, for the Gauss elimination method, the total computational time can be seen to be proportional to $\frac{n^3}{3}$.

(2) In general, if the number of equations is very large, the round-off errors might accumulate to such an extent that the resulting solution might not be accurate. This means that although the solution found is algorithmically correct, it will be wrong, since it won't satisfy the equations when substituted back.

(3) If the number of equations is less than approximately 50, an accurate solution can be found by using even single-precision arithmetic. If the value of n lies between 50 and few hundred, it is better to use double-precision arithmetic to obtain a reasonably accurate solution. However, if n is on the order of thousands, numerical errors will result even with double-precision arithmetic. If the matrices have special structure, such as symmetric, banded, tridiagonal, or sparse structure, special techniques can be used to obtain accurate solutions efficiently even for large values of n.

(4) The Gauss elimination procedure can be applied, theoretically, to any number of equations. However, the accuracy of the solution suffers from the round-off errors when $n > 50$.

(5) The Gauss–Jordan elimination method is also a general procedure that can be used not only to find the solution, but also to find the inverse of the coefficient matrix. As with the Gauss elimination method, this method also is expected to yield accurate solutions only for $n \leq 50$. For larger n, the accuracy of the solution suffers from the round-off errors.

3.20 Choice of the Method

(1) If the equations are to be solved for different right-hand-side vectors, a direct method is preferred. In this case, we find the inverse of the coefficient matrix once and determine the solution for different cases by matrix multiplication.

(2) The Gauss–Seidel method will give accurate solution, even when the number of equations is several thousands (if the system is diagonally dominant). It is usually twice as fast as the Jacobi method.

(3) The determination of the optimal relaxation factor, especially for large systems of equations, is difficult. However, if the optimal value of the relaxation factor is known, then the successive overrelaxation method works many times faster than the Gauss–Seidel method.

(4) If the equations are tridiagonal, the direct method of solution, which requires only $5n - 4$ operations, is to be used.

(5) For large systems of equations, we should compute the condition number of the coefficient matrix and reformulate the problem, if possible, to avoid ill conditioning of the equations.

3.21 Use of Software Packages

3.21.1 MATLAB

▶Example 3.27

Find the solution of the following system of equations:

$$\begin{bmatrix} -4 & 1 & 1 & 0 \\ 1 & -4 & 0 & 1 \\ 1 & 0 & -4 & 1 \\ 0 & 1 & 1 & -4 \end{bmatrix} \begin{Bmatrix} x_1 \\ x_2 \\ x_3 \\ x_4 \end{Bmatrix} = \begin{Bmatrix} -200 \\ -400 \\ 0 \\ -200 \end{Bmatrix}. \tag{E1}$$

Solution

```
>> A2 = [-4  1  1 0;  1  -4  0  1;  1  0  -4  1;  0  1  1  -4]
A2 =
    -4   1   1   0
     1  -4   0   1
     1   0  -4   1
     0   1   1  -4
>> b2 = [-200;  -400;  0; -200]
b2 =
  -200
  -400
     0
  -200
>> C2 = inv(A2)
C2 =
  -0.2917  -0.0833  -0.0833  -0.0417
  -0.0833  -0.2917  -0.0417  -0.0833
  -0.0833  -0.0417  -0.2917  -0.0833
  -0.0417  -0.0833  -0.0833  -0.2917
>> Sol2 = C2*b2
Sol2 =
  100.0000
  150.0000
   50.0000
  100.0000
```

◀

▶Example 3.28

Find the solution of Eq. (E1) of Example 3.27 using LU decomposition.

Solution

First, the inverse of the coefficient matrix $[A]$ is determined as $[C] = [A]^{-1}$ and the solution vector is defined as $\vec{x} = [C]\vec{b}$. Next the LU-decomposition of the matrix $[A]$ is found by using the command

$$>> [L, U, P] = lu(A),$$

where L, U, and P represent the lower triangular, upper triangular, and the permutation matrix, respectively, with $[L][U] = [P][A]$. Finally, the MATLAB operator \ is used to find the solution of $[A]\vec{x} = \vec{b}$ by using the LU-decomposition technique.

```
>> A = [-4  1  1  0;  1  -4  0  1;  1  0  -4  1;  0  1  1  -4]
A =
```

```
    -4   1   1   0
     1  -4   0   1
     1   0  -4   1
     0   1   1  -4
>> b = [-200;  -400;  0;  -200]
b =
    -200
    -400
       0
    -200
>> C = inv(A)
C =
    -0.2917  -0.0833  -0.0833  -0.0417
    -0.0833  -0.2917  -0.0417  -0.0833
    -0.0833  -0.0417  -0.2917  -0.0833
    -0.0417  -0.0833  -0.0833  -0.2917
>> x = C*b
x =
    100.0000
    150.0000
     50.0000
    100.0000
>> [L, U, P] = lu (A)
L =
    1.0000         0         0         0
   -0.2500    1.0000         0         0
   -0.2500   -0.0667    1.0000         0
         0   -0.2667   -0.2857    1.0000
U =
   -4.0000    1.0000    1.0000         0
         0   -3.7500    0.2500    1.0000
         0         0   -3.7333    1.0667
         0         0         0   -3.4286
P =
       1     0     0     0
       0     1     0     0
       0     0     1     0
       0     0     0     1
>> x=A \ b
x =
      100.0000
      150.0000
       50.0000
      100.0000
>>
```

◀

▶Example 3.29

Use MATLAB to find the inverse, $[B] = [A]^{-1}$, of the matrix $[A]$ given by

$$[A] = \begin{bmatrix} 1 & \frac{1}{2} & \frac{1}{3} & & \frac{1}{n} \\ \frac{1}{2} & \frac{1}{3} & \frac{1}{4} & \cdots & \frac{1}{n+1} \\ \frac{1}{3} & \frac{1}{4} & \frac{1}{5} & \cdots & \frac{1}{n+2} \\ \cdot & \cdot & \cdot & \cdots & \cdot \\ \cdot & \cdot & \cdot & \cdots & \cdot \\ \cdot & \cdot & \cdot & \cdots & \cdot \\ \frac{1}{n} & \frac{1}{n+1} & \frac{1}{n+2} & & \frac{1}{2n-1} \end{bmatrix}, n = 50. \tag{E1}$$

Find $[C] = [A][B]$ and compute the measure of error E:

$$E = \sum_{i=1}^{n} \sum_{j=1}^{n} |c_{ij}|, \quad n = 50.$$

(*Note*: $E = 50$ if the solution is correct with $[C] = [I]$).

Solution

The matrix $[A]$ is defined in an m.file labeled *Rao26.m*.

Listing of *Rao26.m*:

```
function [A, AA, B, E] = Rao26 (n)

A=zeros (n,n);
AA=zeros (n,n);

for j=1:n
   for i=1:n
     A (i,j) = 1 / (i + j + 1);
   end
end

AA=inv(A);
B=A*AA;

E=0;
for j=1:n
```

```
   for i=1:n
      E=E+abs (B (i,j));
   end
end

>> n=50;
>> [A, AA, B, E] = Rao26(n);
>> E

E =

   2.350100203137781e+011
```

(E is very large, indicating that the computed inverse is in error. The inverse matrix, AA, is not printed to save space). ◀

3.21.2 MAPLE

▶Example 3.30

Find the inverse of the following matrix in symbolic form using MAPLE:

$$[A] = \begin{bmatrix} a_{11} & a_{12} \\ a_{21} & a_{22} \end{bmatrix}. \tag{E1}$$

Solution

```
> with (linalg):
Warning: new definition for   norm
Warning: new definition for   trace
> A := array( [[a11,a12], [a21,a22]] );

                                   [ a11 a12 ]
                              A := [         ]
                                   [ a21 a22 ]
> B := inverse ( A );

                  [          a22                       a12          ]
                  [ - -------------------   ------------------- ]
                  [   - a11 a22 + a12 a21   - a11 a22 + a12 a21 ]
             B := [                                             ]
                  [          a21                       a11          ]
                  [   -------------------   - ------------------- ]
                  [   - a11 a22 + a12 a21   - a11 a22 + a12 a21 ]
```

> **Note:** Before solving problems of linear algebra, the package *linalg* is to be loaded using the command
>
> > with (linalg);
>
> This results in the display of all the commands contained in the *linalg* package. The display can be suppressed by using colon in place of semi-colon as
>
> > with (linalg):
>
> We can then use the commands needed for the solution of the problem as indicated above.

◀

▶Example 3.31

Find the inverse of the coefficient matrix and the solution of the equations given by Eq. (E1) of Example 3.27.

Solution

```
> with (linalg):
Warning: new definition for  norm
Warning: new definition for  trace
> A := array( [[-4, 1, 1, 0], [1, -4, 0, 1], [1, 0, -4, 1],
[0, 1, 1, -4]] );
                              [ -4   1   1   0 ]
                              [                ]
                              [  1  -4   0   1 ]
                         A := [                ]
                              [  1   0  -4   1 ]
                              [                ]
                              [  0   1   1  -4 ]

> b := array( [-200, -400, 0, -200] );

                    b := [ -200, -400, 0, -200 ]
> linsolve( A,b );
                       [ 100, 150, 50, 100 ]

> C := inverse( A );
                   [ -7/24  -1/12  -1/12  -1/24 ]
                   [                            ]
                   [ -1/12  -7/24  -1/24  -1/12 ]
              C := [                            ]
                   [ -1/12  -1/24  -7/24  -1/12 ]
                   [                            ]
                   [ -1/24  -1/12  -1/12  -7/24 ]
```

◀

▶Example 3.32

Find the solution of Eq. (E1) of Example 3.27 using *LU* decomposition.

Solution

```
> A4 := array( [[-4, 1, 1, 0], [1, -4, 0, 1], [1, 0, -4, 1],
  [0, 1, 1, -4]] );
```

$$A4 := \begin{bmatrix} -4 & 1 & 1 & 0 \\ 1 & -4 & 0 & 1 \\ 1 & 0 & -4 & 1 \\ 0 & 1 & 1 & -4 \end{bmatrix}$$

```
> b4 := array( [-200, -400, 0, -200]);
```

$$b4 := [-200, -400, 0, -200]$$

```
> Sol4 := linsolve (A4, b4);
```

$$Sol4 := [100, 150, 50, 100]$$

```
> C4 := inverse (A4);
```

$$C4 := \begin{bmatrix} \frac{-7}{24} & \frac{-1}{12} & \frac{-1}{12} & \frac{-1}{24} \\ \frac{-1}{12} & \frac{-7}{24} & \frac{-1}{24} & \frac{-1}{12} \\ \frac{-1}{12} & \frac{-1}{24} & \frac{-7}{24} & \frac{-1}{12} \\ \frac{-1}{24} & \frac{-1}{12} & \frac{-1}{12} & \frac{-7}{24} \end{bmatrix}$$

```
> LUdecomp (A4, L = 'l', U = 'u',P= 'p');
```

$$\begin{bmatrix} -4 & 1 & 1 & 0 \\ 0 & \frac{-15}{4} & \frac{1}{4} & 1 \\ 0 & 0 & \frac{-56}{15} & \frac{16}{15} \\ 0 & 0 & 0 & \frac{-24}{7} \end{bmatrix}$$

```
> evalm (l);
```

$$\begin{bmatrix} 1 & 0 & 0 & 0 \\ \frac{-1}{4} & 1 & 0 & 0 \\ \frac{-1}{4} & \frac{-1}{15} & 1 & 0 \\ 0 & \frac{-4}{15} & \frac{-2}{7} & 1 \end{bmatrix}$$

```
> evalm (p);
```

$$\begin{bmatrix} 1 & 0 & 0 & 0 \\ 0 & 1 & 0 & 0 \\ 0 & 0 & 1 & 0 \\ 0 & 0 & 0 & 1 \end{bmatrix}$$

```
>
```

◀

3.21.3 MATHCAD

▶Example 3.33

Find the inverse of the matrix $[A]$ given by Eq. (E1) of Example 3.30.

Solution

$$\begin{pmatrix} a_{11} & a_{12} \\ a_{21} & a_{22} \end{pmatrix}^{-1} \rightarrow \begin{bmatrix} \frac{a_{22}}{(a_{11} \cdot a_{22} - a_{12} \cdot a_{21})} & \frac{-a_{12}}{(a_{11} \cdot a_{22} - a_{12} \cdot a_{21})} \\ \frac{-a_{21}}{(a_{11} \cdot a_{22} - a_{12} \cdot a_{21})} & \frac{a_{11}}{(a_{11} \cdot a_{22} - a_{12} \cdot a_{21})} \end{bmatrix}$$ ◀

▶Example 3.34

Find the solution of the system of Eq. (E1) of Example 3.27 using LU decomposition.

Solution

$$A_4 := \begin{pmatrix} -4 & 1 & 1 & 0 \\ 1 & -4 & 0 & 1 \\ 1 & 0 & -4 & 1 \\ 0 & 1 & 1 & -4 \end{pmatrix}$$

$$b_4 := \begin{pmatrix} -200 \\ -400 \\ 0 \\ -200 \end{pmatrix}$$

$$(A_4)^{-1} = \begin{pmatrix} -0.292 & -0.083 & -0.083 & -0.042 \\ -0.083 & -0.292 & -0.042 & -0.083 \\ -0.083 & -0.042 & -0.292 & -0.083 \\ -0.042 & -0.083 & -0.083 & -0.292 \end{pmatrix}$$

$$\text{lsolve}\,(A_4, b_4) = \begin{pmatrix} 100 \\ 150 \\ 50 \\ 100 \end{pmatrix}$$

lu(A₄) =

	0	1	2	3	4	5	6	7	8
0	1	0	0	0	1	0	0	0	–4
1	0	1	0	0	–0.25	1	0	0	0
2	0	0	1	0	–0.25	–0.067	1	0	0
3	0	0	0	1	0	–0.267	–0.286	1	0

◀

▶Example 3.35

Solve Example 3.29 using MATHCAD.

Solution

```
n:=50
i:=0..n–1     j:=0..n–1
```

A =

	0	1	2	3	4	5	6	7	8	9
0	1	0.5	0.333	0.25	0.2	0.167	0.143	0.125	0.111	0.1
1	0.5	0.333	0.25	0.2	0.167	0.143	0.125	0.111	0.1	0.091
2	0.333	0.25	0.2	0.167	0.143	0.125	0.111	0.1	0.091	0.083
3	0.25	0.2	0.167	0.143	0.125	0.111	0.1	0.091	0.083	0.077
4	0.2	0.167	0.143	0.125	0.111	0.1	0.091	0.083	0.077	0.071
5	0.167	0.143	0.125	0.111	0.1	0.091	0.083	0.077	0.071	0.067
6	0.143	0.125	0.111	0.1	0.091	0.083	0.077	0.071	0.067	0.063
7	0.125	0.111	0.1	0.091	0.083	0.077	0.071	0.067	0.063	0.059
8	0.111	0.1	0.091	0.083	0.077	0.071	0.067	0.063	0.059	0.056
9	0.1	0.091	0.083	0.077	0.071	0.067	0.063	0.059	0.056	0.053
10	0.091	0.083	0.077	0.071	0.067	0.063	0.059	0.056	0.053	0.05
11	0.083	0.077	0.071	0.067	0.063	0.059	0.056	0.053	0.05	0.048
12	0.077	0.071	0.067	0.063	0.059	0.056	0.053	0.05	0.048	0.045
13	0.071	0.067	0.063	0.059	0.056	0.053	0.05	0.048	0.045	0.043
14	0.067	0.063	0.059	0.056	0.053	0.05	0.048	0.045	0.043	0.042
15	0.063	0.059	0.056	0.053	0.05	0.048	0.045	0.043	0.042	0.04

(Only 16 rows and 10 columns of $[A]$ are shown.)

$A^{-1} =$

	0	1	2	3	4
0	162.206	$-1.298.10^{4}$	$3.327.10^{5}$	$-3.961.10^{6}$	$2.522.10^{7}$
1	$-1.306.10^{4}$	$1.396.10^{6}$	$-4.027.10^{7}$	$5.116.10^{8}$	$-3.395.10^{9}$
2	$3.395.10^{5}$	$-4.09.10^{7}$	$1.259.10^{9}$	$-1.666.10^{10}$	$1.138.10^{11}$
3	$-4.163.10^{6}$	$5.356.10^{8}$	$-1.718.10^{10}$	$2.336.10^{11}$	$-1.627.10^{12}$
4	$2.79.10^{7}$	$-3.741.10^{9}$	$1.233.10^{11}$	$-1.707.10^{12}$	$1.207.10^{13}$
5	$-1.08.10^{8}$	$1.489.10^{10}$	$-4.988.10^{11}$	$6.98.10^{12}$	$-4.98.10^{13}$
6	$2.377.10^{8}$	$-3.32.10^{10}$	$1.118.10^{12}$	$-1.561.10^{13}$	$1.115.10^{14}$
7	$-2.506.10^{8}$	$3.432.10^{10}$	$-1.113.10^{12}$	$1.478.10^{13}$	$-1.018.10^{14}$
8	$-1.131.10^{7}$	$6.344.10^{9}$	$-4.176.10^{11}$	$9.178.10^{12}$	$-8.504.10^{13}$
9	$3.122.10^{8}$	$-5.44.10^{10}$	$2.265.10^{12}$	$-3.881.10^{13}$	$3.293.10^{14}$
10	$-3.561.10^{8}$	$6.356.10^{10}$	$-2.73.10^{12}$	$4.894.10^{13}$	$-4.486.10^{14}$
11	$2.081.10^{8}$	$-4.198.10^{10}$	$2.024.10^{12}$	$-4.088.10^{13}$	$4.278.10^{14}$
12	$3.12.10^{8}$	$-3.798.10^{10}$	$9.783.10^{11}$	$-6.085.10^{12}$	$-6.294.10^{13}$
13	$-8.868.10^{8}$	$1.297.10^{11}$	$-4.56.10^{12}$	$6.505.10^{13}$	$-4.335.10^{14}$
14	$4.348.10^{8}$	$-6.504.10^{10}$	$2.319.10^{12}$	$-3.347.10^{13}$	$2.302.10^{14}$
15	$-4.38.10^{7}$	$7.577.10^{9}$	$-3.209.10^{11}$	$5.942.10^{12}$	$-6.3.10^{13}$

(Only 16 rows and 5 columns of $[A]^{-1}$ are shown.)

$$B := A \cdot A^{-1}$$

$$E := \left| \begin{array}{l} E \leftarrow 0 \\ \text{for } i \in 0..n-1 \\ \quad \text{for } j \in 0..n-1 \\ \quad \quad E \leftarrow E + |B_{i,j}| \end{array} \right.$$

$$E = 1.913 \times 10^{3}$$

(The value of $E = 1913$ is large compared with the ideal value of 50. This shows that the computed inverse is in error. However, the error is much less compared with that of MATLAB). ◀

3.22 Computer Programs

3.22.1 Fortran Programs

Program 3.1 Gauss–Jordan elimination method:

Input data required: Indicated at the beginning of program listing.

Illustration (Example 3.10):

Output of the program:

```
SOLUTION BY GAUSS-JORDAN ELIMINATION METHOD

COEFFICIENT MATRIX:

 0.20000000E+01   -0.10000000E+01   0.10000000E+01
 0.40000000E+01    0.30000000E+01  -0.10000000E+01
 0.30000000E+01    0.20000000E+01   0.20000000E+01

RIGHT HAND SIDE VECTOR

 0.40000000E+01    0.60000000E+01   0.15000000E+02

SOLUTION VECTOR

 0.10000001E+01    0.19999998E+01   0.39999995E+01
```

Program 3.2 Solution of tridiagonal system of equations:

Input data required:

N = number of equations to be solved
$D1(N)$ = array defining the left diagonal elements; first element of $D1$ must be zero
$D2(N)$ = array defining the major diagonal elements
$D3(N)$ = array defining the right diagonal elements; last element of $D3$ must be zero
$B(N)$ = array defining the right hand side vector

Illustration (Example 3.24). The output of Program 3.2 as follows:

```
SOLUTION OF TRIDIAGONAL SYSTEM OF EQUATIONS

COEFFICIENT MATRIX (THREE MAJOR DIAGONALS):

 0.00000000E+00  -0.10000000E+01  -0.10000000E+01  -0.10000000E+01
 0.20000000E+01   0.20000000E+01   0.20000000E+01   0.20000000E+01
-0.10000000E+01  -0.10000000E+01  -0.10000000E+01   0.00000000E+00

RIGHT HAND SIDE VECTOR

0.10000000E+01   0.00000000E+00   0.00000000E+00   0.00000000E+00

SOLUTION VECTOR

0.80000019E+00   0.60000008E+00   0.40000010E+00   0.20000002E+00
```

3.22.2 C Programs

Program 3.3 Solution using LU *decomposition:*

Input data required: Indicated at the beginning of program listing.

Illustration (Example 3.15).

Program output:

```
Solution by LU Decomposition

Coefficient Matrix:
    2.000000e+00   -1.000000e+00    1.000000e+00
    4.000000e+00    3.000000e+00   -1.000000e+00
    3.000000e+00    2.000000e+00    2.000000e+00

Right hand side vector:
    4.000000e+00    6.000000e+00    1.500000e+01

Lower triangular matrix:
    2.000000e+00    0.000000e+00    0.000000e+00
    4.000000e+00    5.000000e+00    0.000000e+00
    3.000000e+00    3.500000e+00    2.600000e+00

Upper triangular matrix:
    1.000000e+00   -5.000000e-01    5.000000e-01
    0.000000e+00    1.000000e+00   -6.000000e-01
    0.000000e+00    0.000000e+00    1.000000e+00

Solution vector:
    1.000000e+00    2.000000e+00    4.000000e+00
```

Program 3.4 Relaxation method:

Input data required:

n = number of equations to be solved
maxiter = maximum number of iterations permitted
$a[n][n]$ = coefficient matrix of given equations
$b[n]$ = right hand side vector
$bb[n]$ = initial guess for solution vector (can be taken as zero vector in most cases)
$x[n]$ = dummy vector (values are not needed)
eps = convergence criterion, usually set equal to 1.0e-06
omeg = value of omega ($1 <$ omega < 2)

```
Solution by Relaxation Method

omeg = 1.100000e+00

Coefficient matrix:
  -5.000000e+00    -1.000000e+00     2.000000e+00
   2.000000e+00     6.000000e+00    -3.000000e+00
   2.000000e+00     1.000000e+00     7.000000e+00

Right hand side vector:

   1.000000e+00     2.000000e+00     3.200000e+01
   0.000000e+00     0.000000e+00     0.000000e+00
  -2.200000e-01     4.473333e-01     5.027419e+00
   1.915651e+00     2.384608e+00     3.549044e+00
   6.254002e-01     1.850866e+00     4.186262e+00
   1.152225e+00     2.061542e+00     3.923861e+00
   9.377371e-01     1.974799e+00     4.031142e+00
   1.025473e+00     2.010308e+00     3.987260e+00
   9.895793e-01     1.995783e+00     4.005212e+00

Solution vector:
   1.004263e+00     2.001725e+00     3.997868e+00

Iterations required to get convergence = 8
```

REFERENCES AND BIBLIOGRAPHY

3.1. S. H. Crandall, *Engineering Analysis: A Survey of Numerical Procedures*, McGraw-Hill, New York, 1956.

3.2. R. L. Burden and J. D. Faires, *Numerical Analysis*, 5th edition, Prindle, Weber & Schmidt, Boston, 1993.

3.3. G. R. Lindfield and J. E. T. Penny, *Microcomputers in Numerical Analysis*, Ellis Horwood, Chichester, 1989.

3.4. J. R. Rice, *Numerical Methods, Software, and Analysis: IMSL Reference Edition*, McGraw-Hill, New York, 1983.

3.5. C. F. Gerald and P. O. Wheatley, *Applied Numerical Analysis*, 3d edition, Addison-Wesley, Reading, MA, 1984.

3.6. S. S. Rao, *The Finite Element Method in Engineering*, Pergamon Press, Oxford, 1989.

3.7. R. L. Norton, *Design of Machinery*, McGraw-Hill, New York, 1992.

3.8. M. L. James, G. M. Smith and J. C. Wolford, *Applied Numerical Methods for Digital Computation*, 4th edition, Harper Collins, New York, 1993.

3.9. K. Atkinson, *An Introduction to Numerical Analysis*, John Wiley, New York, 1978.

3.10. B. Carnahan, H. A. Luther, and J. O. Wilkes, *Applied Numerical Methods*, John Wiley, New York, 1969.

3.11. G. W. Stewart, *Introduction to Matrix Computations*, Academic Press, New York, 1973.

3.12. R. S. Varga, *Matrix Iterative Analysis*, Prentice-Hall, Englewood Cliffs, New Jersy, 1962.

3.13. R. V. Southwell, *Relaxation Methods in Engineering Science: A Treatise on Approximate Computation*, Oxford University Press, Oxford, 1951.

3.14. R. L. Sack, *Matrix Structural Analysis*, PWS-KENT Publishing Co., Boston, 1989.

3.15. J. H. Mathews, *Numerical Methods for Mathematics, Science, and Engineering*, 2d edition, Prentice Hall, Englewood Cliffs, NJ, 1992.

3.16. S. C. Chapra and R. P. Canale, *Numerical Methods for Engineers with Personal Computer Applications*, McGraw-Hill, New York, 1985.

3.17. A. W. Al-Khafaji and J. R. Tooley, *Numerical Methods in Engineering Practice*, Holt, Rinehart and Winston, Inc., New York, 1986.

3.18. R. P. Tewarson, *Sparse Matrices*, Academic Press, New York, 1973.

3.19. L. Fox, *An Introduction to Numerical Linear Algebra*, Oxford University Press, New York, 1965.

3.20. R. T. Gregory and D. L. Karney, *A Collection of Matrices for Testing Computational Algorithms*, Wiley, New York, 1969.

3.21. W. Cheney and D. Kincaid, *Numerical Mathematics and Computing*, 3d edition, Brooks/Cole, Pacific Grove, CA, 1994.

3.22. N. S. Asaithambi, *Numerical Analysis. Theory and Practice*, Saunders College Publishing, Orlando, FL, 1995.

3.23. A. Biran and M. Breiner, MATLAB *for Engineers*, Addison-Wesley, Workingham, UK, 1995.

3.24. D. M. Etter, *Engineering Problem Solving with* MATLAB, Prentice Hall, Englewood Cliffs, NJ, 1993.

3.25. B. W. Char, K. O. Geddes, G. H. Gonnet, B. L. Leong, M. B. Monagan, and S. M. Watt, *First Leaves: A Tutorial Introduction to* MAPLE*V*, Springer-Verlag, New York, 1992.

3.26. B. W. Char, K. O. Geddes, G. H. Gonnet, B. L. Leong, M. B. Monagan, and S. M. Watt, MAPLE*V Library Reference Manual*, Springer-Verlag, New York, 1991.

REVIEW QUESTIONS

The following questions along with corresponding answers are available in an interactive format at the Companion Web site at http://www.prenhall.com/rao.

3.1. Give **brief answers** to the following:

1. What is the difference between, the Gauss and Gauss–Jordan elimination methods?
2. State three engineering analyses that require the solution of a set of linear algebraic equations.
3. State the condition for the existence of a nontrivial solution to a system of homogeneous linear equations.
4. What is Cramer's rule?
5. How do you find whether a given set of linear equations is linearly independent?
6. What is the difference between direct and indirect methods of solving a system of linear algebraic equations?
7. What is meant by ill conditioning of a set of linear equations?
8. How do you quantify the degree of ill conditioning of a set of linear equations?
9. What is the difference between partial pivoting and complete pivoting in the elimination methods?
10. What is LU decomposition and in what way it is superior to the Gauss elimination method?
11. What is a diagonally dominant matrix?
12. What is the difference between successive under- and overrelaxation methods?
13. Indicate a procedure for solving a system of complex simultaneous linear equations.
14. How is the inverse of a square matrix defined?
15. How do you graphically interpret a system of linear equations?
16. Define the following matrices: sparse matrix, banded matrix, and tridiagonal matrix.
17. What is the difference between overdetermined and underdetermined systems of linear equations?
18. How is the determinant of a matrix defined? What is the value of the determinant of a triangular matrix?
19. What is Jacobi method of iteration?

(*continued, page 229*)

20. What type of convergence criteria can be used in the iterative methods of solving systems of linear equations?

3.2. Indicate whether the following statements are **true or false**:

1. The system of equations $[A]\vec{x} = \vec{0}$ will have the unique solution $\vec{x} = \vec{0}$ only if $[A]$ is nonsingular.
2. The system of equations $[A]\vec{x} = \vec{b}$ will always have a unique solution.
3. The Gauss elimination method cannot be used if any diagonal element of $[A]$ is zero.
4. Ill conditioning of equations occurs when the determinant of $[A]$ is close to zero.
5. The Jacobi iteration method converges if the matrix $[A]$ is diagonally dominant.
6. The Jacobi iteration method converges if the matrix $[A]$ is nonsingular.
7. The Jacobi or Gauss–Seidel iteration method will not converge if the matrix $[A]$ is not diagonally dominant.
8. The relaxation method can be used to improve the convergence characteristics of Gauss–Seidel method of iteration.
9. The equations $x_1 - 5x_2 = 3$ and $-2x_1 + 10x_2 = -6$ are linearly independent.
10. Back substitution is not required in the Gauss–Jordan method.
11. Back substitution is not required in the Gauss elimination method.
12. A symmetric matrix $[A]$ can always be decomposed as $[A] = [L][U]$ with $[L] = [U]^T$.
13. $([A][B])^T = [A]^T[B]^T$
14. $([A][B])^{-1} = [B]^{-1}[A]^{-1}$
15. The system of equations $[A]\vec{x} = \vec{0}$ cannot be solved for a unique solution $\vec{x}$.
16. The equations $x_1 - 2x_2 = 4$ and $kx_1 - x_2 = 3$ are ill conditioned, around $k = 0.5$.

3.3. Indicate whether the following statements are **true or false** for a square matrix:

1. The inverse of the matrix $[A]$ exists only if $|[A]| \neq 0$.
2. If all the entries of a column or row are zero in the matrix $[A]$, then det $[A] = 0$.
3. If two columns are the same in $[A]$, then det $[A] = 0$.
4. If two rows are the same in $[A]$, then det $[A] = 0$.
5. If all the elements of $[A]$ are multiplied by a constant a, then the determinant of the resulting matrix is equal to a times det $[A]$.
6. If $[A]$ and $[B]$ are both $n \times n$ matrices, then det $([A][B])$ = det $[A] \cdot$ det $[B]$.
7. If both $[A]$ and $[B]$ are $n \times n$ matrices, then $[A][B] = [B][A]$.
8. When two rows are interchanged in $[A]$, then the determinant of the new matrix is same as det $[A]$.
9. The determinant of $[A]$ can be defined in $2n$ different ways.

(*continued, page 230*)

10. The minor of an element of a matrix of size $n \times n$ is a matrix of size $n-1 \times n-1$.
11. The determinant of an upper triangular matrix $[A]$ with $a_{ii} = 1 (i = 1, 2, \ldots, n)$ is equal to unity.
12. The identity matrix is a matrix whose elements are all equal to unity.

3.4. State whether the following statements are **true or false**:
1. An overdetermined system has more unknowns than the equations.
2. A matrix without an inverse is called a singular matrix.
3. The matrix $[B] = [A][A]^T$ is a square matrix for any order of the matrix $[A]$.
4. An $n \times n$ matrix $[A]$ is considered to be diagonally dominant only when its diagonal elements are greater than every offdiagonal element.
5. The Euclidean norm and the L_2 norm are same for a vector.
6. The Euclidean and L_2 norms are the same for a square matrix.
7. A tridiagonal matrix is a banded matrix.

3.5. Choose the **appropriate answer**:
1. In a system of n simultaneous equations in n unknowns, $[A]\vec{x} = \vec{b}$, the matrix $[A]$ is a ______ matrix. (square, rectangular).
2. Gaussian elimination is a ______ method of solution. (direct, indirect).
3. Indirect methods of solution ______ require an initial starting solution. (do, do not).
4. The Jacobi method is a ______ method of solution. (direct, indirect).
5. The identity matrix is a ______ matrix. (diagonal, full).

3.6. **Match the following:**

(1)	Symmetric matrix	(a)	Consists of cofactors as elements
(2)	Diagonal matrix	(b)	Rows and columns are interchanged
(3)	Transpose of a matrix	(c)	Off-diagonal elements are symmetric about the diagonal
(4)	Trace of a matrix	(d)	Defined in terms of minors and cofactors
(5)	Determinant of a matrix	(e)	Sum of the elements in the main diagonal
(6)	Adjoint matrix	(f)	All elements are zero except the diagonal ones

3.7. **Define** the following terms:

Pivot element, tridiagonal matrix, elementary operations, augmented matrix, lower triangular matrix, upper triangular matrix.

(*continued, page 231*)

3.8. Select the most **appropriate answer**:

1. If $[A] = \begin{bmatrix} 1 & 2 \\ 3 & 4 \end{bmatrix}$, then $[B] = \frac{1}{2}\begin{bmatrix} -4 & 2 \\ 3 & -1 \end{bmatrix}$ is

   (a) inverse of $[A]$, (b) adjoint of $[A]$, (c) transpose of $[A]$.

2. If $[A] = \begin{bmatrix} 4 & 3 \\ 2 & 1 \end{bmatrix}$ and $[B] = \begin{bmatrix} 2 & 1 \\ 3 & 4 \end{bmatrix}$, then $[C] = \begin{bmatrix} 17 & 16 \\ 7 & 6 \end{bmatrix}$ denotes

   (a) product of $[A]$ and $[B]$, (b) sum of $[A]$ and $[B]$, (c) difference of $[A]$ and $[B]$.

3. The L_1 norm of the matrix $[A] = \begin{bmatrix} 1 & 8 \\ 2 & 4 \end{bmatrix}$ is

   (a) 3, (b) 12, (c) 9, (d) 6.

4. The L_2 norm of the matrix $[A] = \begin{bmatrix} 1 & 8 \\ 2 & 4 \end{bmatrix}$ is

   (a) 3, (b) 9, (c) 6.772.

5. The L_∞ norm of the matrix $[A] = \begin{bmatrix} 1 & 8 \\ 2 & 4 \end{bmatrix}$ is

   (a) 9, (b) 6, (c) 3, (d) 12.

6. The L_e norm of the matrix $[A] = \begin{bmatrix} 1 & 8 \\ 2 & 4 \end{bmatrix}$ is

   (a) $\sqrt{85}$, (b) $\sqrt{65}$, (c) $\sqrt{20}$, (d) $\sqrt{80}$.

7. The rank of the matrix $[A] = \begin{bmatrix} 2 & 0 & 4 \\ 3 & 5 & 1 \\ 0 & 6 & 0 \end{bmatrix}$ is

   (a) 3, (b) 2, (c) 1.

(*continued, page 232*)

8. The condition number of $[A] = \begin{bmatrix} 1 & 2 \\ 3 & 4 \end{bmatrix}$, based on ∞- norms of the matrices $[A]$ and $[A]^{-1}$ is
   (a) 7, (b) 3, (c) 21.
9. Consider the system of equations $x_1 + 2x_2 = 1$ and $3x_1 + 4x_2 = 1$. The value of $\|[A_1]\|$ to be used in Cramer's rule to find $x_1 = \frac{\|[A_1]\|}{\|[A]\|}$ is
   (a) −2, (b) 2, (c) 1.
10. For the system of equations given in part 9 of this question, the value of $\|[A_2]\|$ to be used in Cramer's rule to find $x_2 = \frac{\|[A_2]\|}{\|[A]\|}$ is
    (a) −2, (b) 2, (c) 1.
11. The L_1 norm of the vector $\vec{x} = \begin{Bmatrix} 1 \\ -2 \\ 3 \end{Bmatrix}$ is
    (a) 6, (b) $\sqrt{14}$, (c) 3, (d) 2.
12. The L_2 norm of the vector $\vec{x}$ given in part 11 of this question is
    (a) 6, (b) $\sqrt{14}$, (c) 3, (d) 2.
13. The L_∞ norm of the vector $\vec{x}$ given in part 11 of this question is
    (a) 6, (b) $\sqrt{14}$, (c) 3, (d) 2.
14. If $\|\vec{x}\| = 3$ and $\|\vec{y}\| = 5$ and both $\vec{x}$ and $\vec{y}$ have the same order, then $\|\vec{x} + \vec{y}\|$ will be
    (a) $= 8$, (b) ≤ 8, (c) ≥ 8.

3.9. **Match the following:**

1.	Doolittle method	(a)	Diagonal elements of $[U]$ and $[L]$ are same.
2.	Choleski method	(b)	Each diagonal element of $[L]$ is 1.
3.	Crout method	(c)	Each diagonal element of $[U]$ is 1.

PROBLEMS

Section 3.2

3.1. Derive the equations necessary for finding the currents flowing in the three loops of the electrical network shown in Fig. 3.15.

3.2. A direct current network is shown in Fig. 3.16 in which the currents flowing in various loops are indicated as $I_i, i = 1, 2, 3, 4$. The application of Kirchhoff's voltage law, which states that the net voltage drop in any closed loop is zero, to various loops yields the following equations:

$$V_0 - I_1 - 2(I_1 - I_2) - 9(I_1 - I_4) = 0;$$

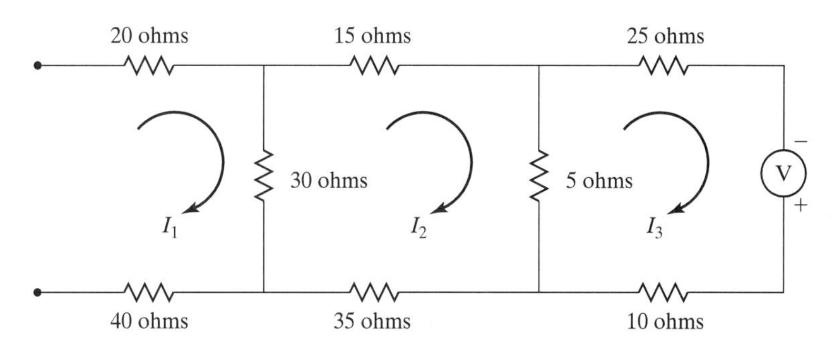

Figure 3.15 Electrical network.

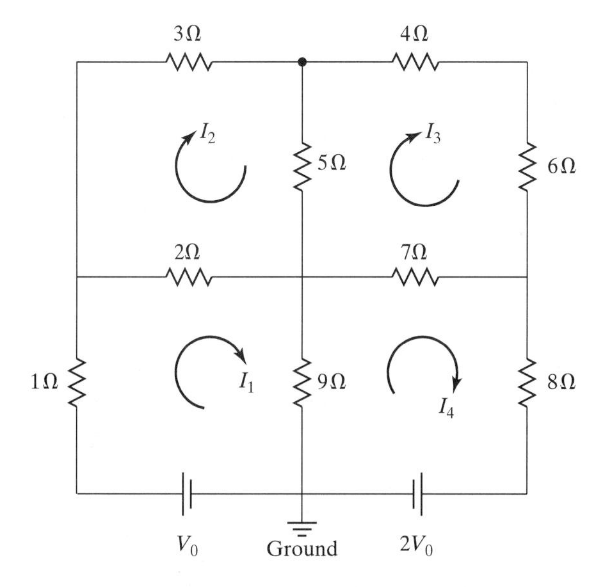

Figure 3.16 Direct current network.

$$-3I_2 - 5(I_2 - I_3) - 2(I_2 - I_1) = 0;$$

$$-4I_3 - 6I_3 - 7(I_3 - I_4) - 5(I_3 - I_2) = 0;$$

$$-2V_0 - 7(I_4 - I_3) - 8I_4 - 9(I_4 - I_1) = 0.$$

Determine the currents flowing in the various loops for $V_0 = 100$ volts using Gauss–Jordan elimination method.

3.3. Coordinate transformation from (X, Y, Z)-system to (x, y, z)-system can be achieved using the Euler rotations ϕ, θ and ψ:

$$\begin{Bmatrix} x \\ y \\ z \end{Bmatrix} = \begin{bmatrix} 1 & 0 & 0 \\ 0 & \cos\phi & \sin\phi \\ 0 & -\sin\phi & \cos\phi \end{bmatrix} \begin{bmatrix} \cos\theta & \sin\theta & 0 \\ -\sin\theta & \cos\theta & 0 \\ 0 & 0 & 1 \end{bmatrix}$$

$$\times \begin{bmatrix} 1 & 0 & 0 \\ 0 & \cos\psi & \sin\psi \\ 0 & -\sin\psi & \cos\psi \end{bmatrix} \begin{Bmatrix} X \\ Y \\ Z \end{Bmatrix}.$$

Using this relation, find the values of X, Y, and Z when $x = 1, y = 2$, and $z = 3$ with $\phi = 30°, \theta = 20°$, and $\psi = 10°$.

3.4. Three carts, interconnected by springs, are subjected to the loads P_1, P_2, and P_3 as shown in Fig. 3.17. The displacements of the carts are governed by the equilibrium equations

$$P_1 - k_1u_1 + k_5(u_3 - u_1) + k_4(u_2 - u_1) = 0;$$

$$P_2 - k_2u_2 - k_4(u_2 - u_1) + k_6(u_3 - u_2) = 0;$$

$$P_3 - k_7u_3 - k_8u_3 - k_3u_3 - k_6(u_3 - u_2) - k_5(u_3 - u_1) = 0.$$

Find the displacements of the carts for the following data using the Gauss–Jordan elimination method:

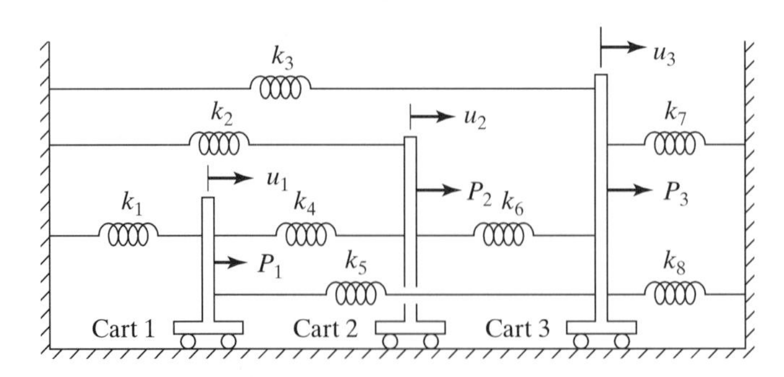

Figure 3.17 Carts interconnected by springs.

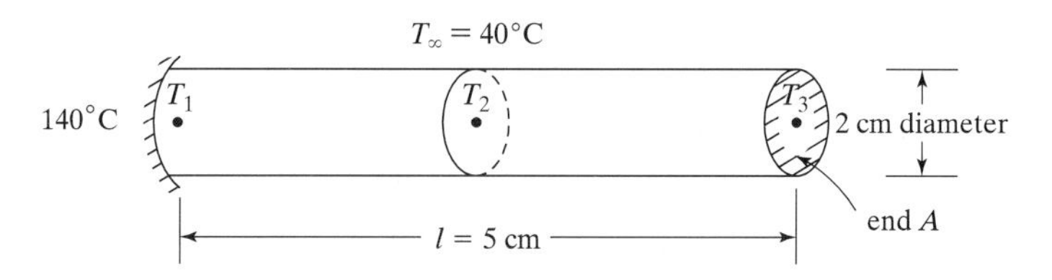

Figure 3.18 Uniform circular fin.

$k_1 = 5000$ N/m, $k_2 = 1500$ N/m, $k_3 = 2000$ N/m, $k_4 = 1000$ N/m, $k_5 = 2500$ N/m, $k_6 = 500$ N/m, $k_7 = 3000$ N/m, $k_8 = 3500$ N/m, $P_1 = 1000$ N, $P_2 = 2000$ N, $P_3 = 3000$ N.

3.5. A fin, with a uniform circular section (shown in Fig. 3.18), has a root temperature of 140°C and an ambient temperature of 40°C. It has a thermal conductivity of $k = 70 \frac{\text{watts}}{\text{cm}-^\circ\text{K}}$ and a heat-transfer coefficient of $h = 5 \frac{\text{watts}}{\text{cm}^2-^\circ\text{K}}$. When the convection loss from the end A is also considered, the nodal temperatures T_1, T_2, and T_3 are governed by the equation

$$\begin{bmatrix} 1 & 0 & 0 \\ 0 & 72.6668 & -23.8333 \\ 0 & -23.8333 & 41.3334 \end{bmatrix} \begin{Bmatrix} T_1 \\ T_2 \\ T_3 \end{Bmatrix} = \begin{Bmatrix} 140 \\ 4336 \\ 700 \end{Bmatrix}.$$

Determine the values of the nodal temperatures using the Gauss–Jordan elimination method.

3.6. A steel cylinder, of weight $W = 500$ lb and diameter 1 ft, is placed in a steel V-block as shown in Fig. 3.19. When a moment M is applied about the axis of the cylinder, the cylinder starts to rotate. The force equilibrium equations along the x- and y-directions and the moment equilibrium equation about the axis of the cylinder are given by

$$(R_1 + \mu R_1 - R_2 + \mu R_2) \sin 45^\circ = 0,$$

$$(R_1 - \mu R_1 + R_2 + \mu R_2) \cos 45^\circ = \text{W},$$

and

$$6(\mu R_1 + \mu R_2) - M = 0,$$

where R_1 and R_2 are the normal reactions at the points of contact 1 and 2, respectively, and μ is the coefficient of friction. Find the magnitude of the moment (M) and the reactions at the points of contact between the cylinder and the V-block (R_1 and R_2) using the Gauss elimination method. Assume that the coefficient of friction between the cylinder and the V-block is $\mu = 0.25$.

3.7. A truss structure is composed of five members and is subjected to two loads as shown in Fig. 3.20. The cross-sectional area is 1 in^2 for members 1, 2,

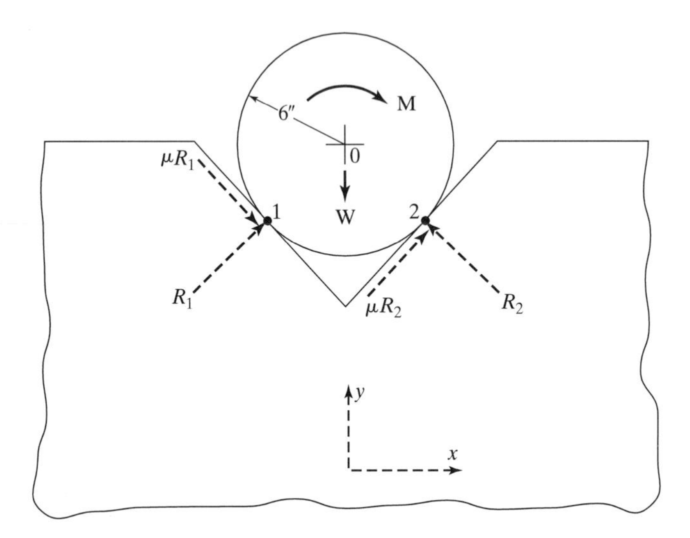

Figure 3.19 Cylinder in a V-block.

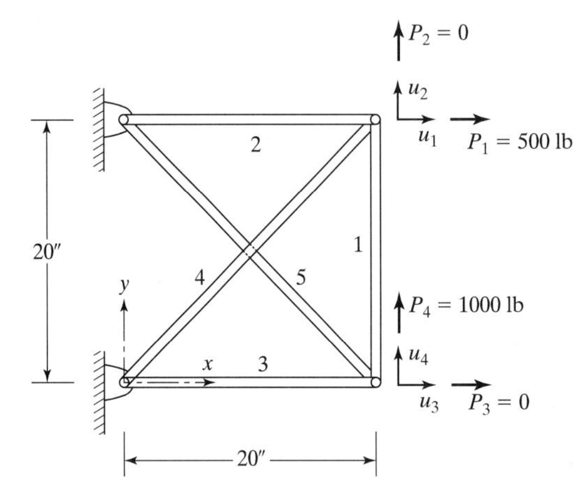

Figure 3.20 Five-member truss.

and 3, and $\frac{1}{\sqrt{2}}$ in^2 for members 4 and 5. Young's modulus of the material is 30×10^6 psi. The displacements of the two unsupported nodes are given by the following equilibrium equations:

$$0.375 \times 10^6 \begin{bmatrix} 5 & 1 & 0 & 0 \\ 1 & 5 & 0 & -4 \\ 0 & 0 & 5 & -1 \\ 0 & -4 & -1 & 5 \end{bmatrix} \begin{Bmatrix} u_1 \\ u_2 \\ u_3 \\ u_4 \end{Bmatrix} = \begin{Bmatrix} 500 \\ 0 \\ 0 \\ 1000 \end{Bmatrix}.$$

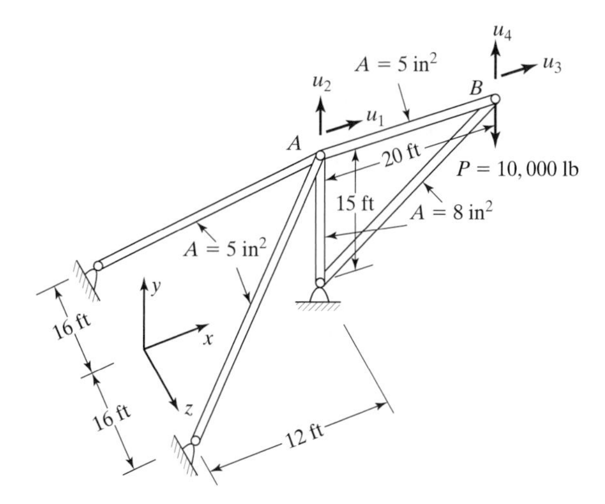

Figure 3.21 Three-dimensional truss.

Determine the nodal displacements, u_i, $i = 1, 2, 3, 4$, using the Gauss–Jordan method.

3.8. The displacements of the unsupported nodes A and B of the three-dimensional truss shown in Fig. 3.21 can be determined by solving the following equilibrium equations:

$$10^6 \begin{bmatrix} 0.8554 & 0.2880 & -0.6250 & 0.0000 \\ 0.2880 & 1.6934 & 0.0000 & 0.0000 \\ -0.6250 & 0.0000 & 1.1370 & 0.3840 \\ 0.0000 & 0.0000 & 0.3840 & 0.2880 \end{bmatrix} \begin{Bmatrix} u_1 \\ u_2 \\ u_3 \\ u_4 \end{Bmatrix}$$

$$= \begin{Bmatrix} 0 \\ 0 \\ 0 \\ -10000 \end{Bmatrix}.$$

Find the solution of these equations using the Gauss–Jordan elimination method.

3.9. A rigid frame consists of two vertical members, each with a length 288 in, cross-sectional area 20 in^2, and area moment of inertia 1200 in^4 and a horizontal member with a length 480 in, cross-sectional area 30 in^2, and area moment of inertia 1200 in^4. (See Fig. 3.22.) All the members have a Young's modulus of 30×10^6 psi. The equilibrium equations of the frame can be derived as [3.6, 3.14]:

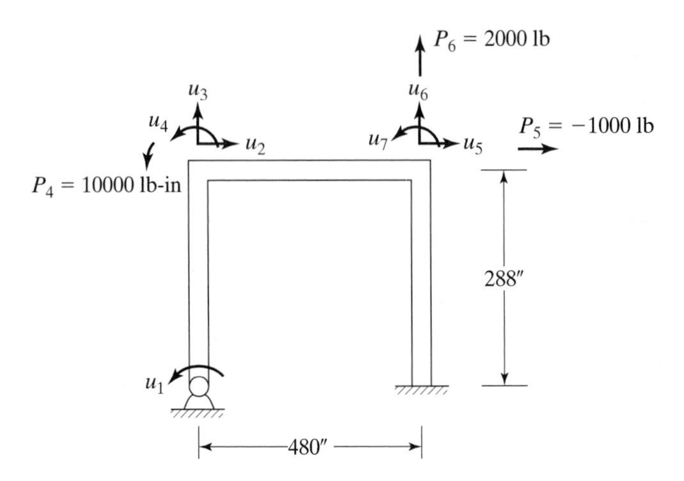

Figure 3.22 Rigid frame.

$$
10 \times 10^6 \begin{bmatrix} 4833.3 & 50.3 & 0.0 & 2416.7 & 0.0 & 0.0 & 0.0 \\ 50.3 & 36.9 & 0.0 & 50.3 & -36.3 & 0.0 & 0.0 \\ 0.0 & 0.0 & 40.6 & 36.3 & 0.0 & -0.3 & 36.3 \\ 2416.7 & 50.3 & 36.3 & 10633.0 & 0.0 & -36.3 & 2900.0 \\ 0.0 & -36.3 & 0.0 & 0.0 & 36.9 & 0.0 & 50.3 \\ 0.0 & 0.0 & -0.3 & -36.3 & 0.0 & 40.6 & -36.3 \\ 0.0 & 0.0 & 36.3 & 2900.0 & 50.3 & -36.3 & 10633.0 \end{bmatrix}
$$

$$
\times \begin{Bmatrix} u_1 \\ u_2 \\ u_3 \\ u_4 \\ u_5 \\ u_6 \\ u_7 \end{Bmatrix} = \begin{Bmatrix} 0 \\ 0 \\ 0 \\ 10000 \\ -1000 \\ 2000 \\ 0 \end{Bmatrix}.
$$

Determine the displacement components of the frame under the given loading using the computer program 3.1.

3.10. An incompressible fluid is pumped into a network of pipes at a flow rate Q_{in} and pressure $p_{\text{in}} = 200$ psi by a pump as shown in Fig. 3.23. For an

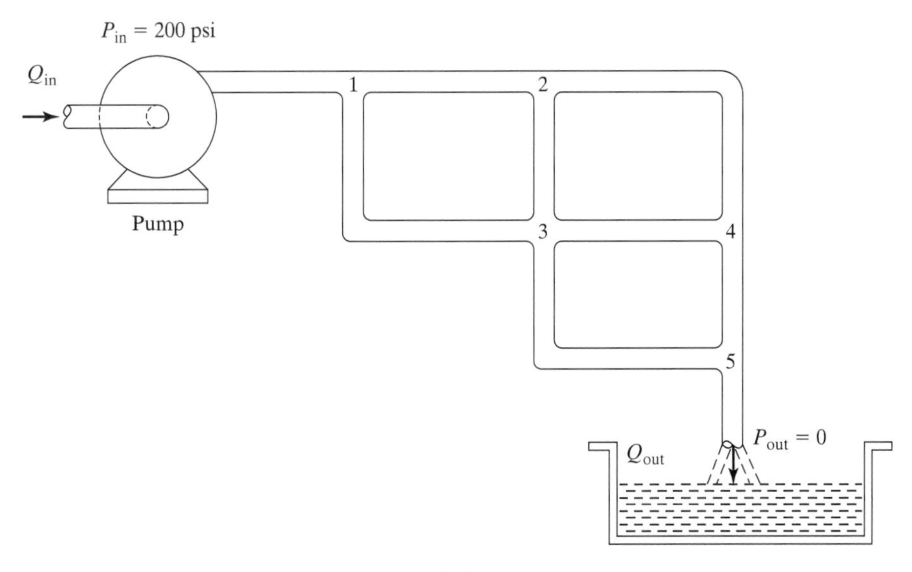

Figure 3.23 Flow through a network of pipes.

incompressible fluid, the pressure drop in any branch of the pipe (Δp) is proportional to the square of the flow rate (q), so that

$$\Delta p = \alpha q^2, \quad \text{or} \quad q = c\sqrt{\Delta p},$$

where c is a constant, known as the resistance coefficient. For continuity of fluid flow, the inflow rate must be equal to the outflow rate at any junction or node of the pipe network. This leads to the following equations:

Node 1:

$$c_{i1}\sqrt{p_{\text{in}} - p_1} = c_{12}\sqrt{p_1 - p_2} + c_{13}\sqrt{p_1 - p_3}; \tag{E1}$$

Node 2:

$$c_{12}\sqrt{p_1 - p_2} = c_{23}\sqrt{p_2 - p_3} + c_{24}\sqrt{p_2 - p_4}; \tag{E2}$$

Node 3:

$$c_{23}\sqrt{p_2 - p_3} + c_{13}\sqrt{p_1 - p_3} = c_{34}\sqrt{p_3 - p_4} + c_{35}\sqrt{p_3 - p_5}; \tag{E3}$$

Node 4:

$$c_{24}\sqrt{p_2 - p_4} + c_{34}\sqrt{p_3 - p_4} = c_{45}\sqrt{p_4 - p_5}; \tag{E4}$$

Node 5:

$$c_{35}\sqrt{p_3 - p_5} + c_{45}\sqrt{p_4 - p_5} = c_{50}\sqrt{p_5 - p_{\text{out}}}. \tag{E5}$$

With known values of the resistance coefficients c_{ij} (in terms of the diameter, length, material, roughness of the pipe segment and the nature of the fluid), suggest a method of solving Eqs. (E1)–(E5).

Section 3.3

3.11. Consider the matrices $[A]$ and $[B]$ given by

$$[A] = \begin{bmatrix} -2 & 8 & 3 \\ 2 & -4 & -6 \\ 5 & 0 & 7 \end{bmatrix};$$

$$[B] = \begin{bmatrix} -3 & 7 & 3 \\ 4 & -6 & 8 \\ 9 & 2 & -5 \end{bmatrix}.$$

Determine the following:
(a) $[A] + [B]$.
(b) $[A] - [B]$.
(c) $2[A] - 3[B]$.
(d) $[A]^T - 2[B]^T$.

3.12. Determine the matrix products (a) $[A][B]$ and (b) $[B][A]$ for the matrices $[A]$ and $[B]$ given in Problem 3.11.

3.13. Consider the following matrices:

$$[A] = \begin{bmatrix} 2 & -3 & 4 \\ 3 & -1 & 5 \end{bmatrix}; \quad [B] = \begin{bmatrix} 4 & 0 \\ -1 & 6 \\ 3 & -2 \end{bmatrix}.$$

Find the values of (a) $[A]\,[B]$ and (b) $[B]\,[A]$.

3.14. Find the determinants of the following matrices:

$$[A] = \begin{bmatrix} -2 & -6 \\ 5 & 1 \end{bmatrix}; \quad [B] = \begin{bmatrix} 3 & 0 & 5 \\ 0 & 5 & -4 \\ 4 & -5 & 1 \end{bmatrix}.$$

3.15. Show that $|[A]| = |[A]^T|$ by considering the following matrix:

$$[A] = \begin{bmatrix} 1 & 2 & 3 \\ 4 & 5 & 6 \\ 7 & 8 & 0 \end{bmatrix}.$$

3.16. Find the determinant of the following triangular matrix:

(a)
$$[A] = \begin{bmatrix} 1 & 2 & 3 \\ 0 & 4 & 5 \\ 0 & 0 & 6 \end{bmatrix}.$$

(b)
$$[B] = \begin{bmatrix} 1 & 0 & 0 \\ 4 & 2 & 0 \\ 6 & 5 & 3 \end{bmatrix}.$$

(c)
$$[C] = \begin{bmatrix} -1 & 4 & 2 & 7 \\ 0 & 3 & 5 & -3 \\ 0 & 0 & 2 & 6 \\ 0 & 0 & 0 & 8 \end{bmatrix}.$$

3.17. Prove that the determinant of a lower or upper triangular matrix is given by the product of its diagonal elements.

3.18. The area of a triangle, A, with vertices located at (x_i, y_i), $i = 1, 2$, and 3, is given by [3.6] as

$$A = \frac{1}{2} \begin{vmatrix} x_1 & y_1 & 1 \\ x_2 & y_2 & 1 \\ x_3 & y_3 & 1 \end{vmatrix}.$$

Using this relation, find the area of the triangle whose vertices are located at (5,4), (8,6), and (4,8).

3.19. Determine the rank of the following matrix:

(a)
$$[A] = \begin{bmatrix} 1 & 2 & 3 \\ 4 & 0 & 5 \\ 6 & 7 & 8 \end{bmatrix}.$$

(b)
$$[B] = \begin{bmatrix} 9 & 4 & 2 & -1 \\ 3 & 6 & 5 & 8 \end{bmatrix}.$$

(c)
$$[C] = \begin{bmatrix} 2 & 4 & 6 \\ 1 & 3 & 7 \\ 3 & 7 & 13 \\ 1 & 1 & -1 \end{bmatrix}.$$

3.20. Find the adjoint matrix of
$$[A] = \begin{bmatrix} 4 & -1 & 2 \\ 4 & -8 & 2 \\ -2 & 1 & 6 \end{bmatrix}.$$

3.21. Find the inverse of the matrix
$$[A] = \begin{bmatrix} 4 & -1 & 2 \\ 4 & -8 & 2 \\ -2 & 1 & 6 \end{bmatrix}.$$
using Eq. (15) of Appendix F.

3.22. Find the L_1, L_2, and L_∞ norms of the following vectors:
$$\text{(a) } \vec{x} = \begin{Bmatrix} 1 \\ 2 \\ 3 \end{Bmatrix}, \quad \text{(b) } \vec{x} = \begin{Bmatrix} -5 \\ 4 \\ -2 \end{Bmatrix}, \quad \text{(c) } \vec{x} = \begin{Bmatrix} 7 \\ -7 \\ 7 \end{Bmatrix}.$$

3.23. Find the L_1, L_2, L_∞, and L_e of the matrix
$$[A] = \begin{bmatrix} 1 & 2 & 3 \\ 4 & 5 & 6 \\ 7 & 8 & 9 \end{bmatrix}.$$

Section 3.4

In Problems 3.24 and 3.25, determine whether the given system of equations has a solution.

3.24.
$$\begin{aligned} x_1 + 2x_2 - x_3 &= 5; \\ 2x_1 - x_2 + 2x_3 &= 0; \\ x_1 + 3x_2 - x_3 &= 7. \end{aligned}$$

3.25.

$$5x_1 - 4x_2 + 2x_3 = 1;$$
$$-2x_1 + 3x_2 + x_3 = 10;$$
$$6x_1 + x_2 - x_3 = -10.$$

Section 3.5

In Problems 3.26 and 3.27, determine whether the given set of equations are independent.

3.26.

$$5x_1 - 4x_2 + 2x_3 = 11;$$
$$-2x_1 + 3x_2 + x_3 = 1;$$
$$16x_1 - 17x_2 + x_3 = 19.$$

3.27.

$$4x_1 - 2x_2 + x_3 = 11;$$
$$x_1 + 5x_2 - 3x_3 = -6;$$
$$3x_1 - 7x_2 + 4x_3 = 17.$$

Section 3.6

3.28. Find the solution of the following equations for values of k around 1.25:

$$4x_1 - x_2 = 3;$$
$$5x_1 - kx_2 = 0.$$

3.29. Find (a) the condition number and (b) the value of $D(A)$ for the system of equations given in Problem 3.28.

3.30. Consider the system of equations:

$$x_1 + 3x_2 = 8.0;$$
$$1.0001x_1 + 3x_2 = 8.0002.$$

Find (a) the condition number and (b) the value of $D(A)$ of these equations.

3.31. Consider the following systems of equations:

(a)
$$\begin{bmatrix} 1 & \frac{1}{2} & \frac{1}{3} & \frac{1}{4} \\ \frac{1}{2} & \frac{1}{3} & \frac{1}{4} & \frac{1}{5} \\ \frac{1}{3} & \frac{1}{4} & \frac{1}{5} & \frac{1}{6} \\ \frac{1}{4} & \frac{1}{5} & \frac{1}{6} & \frac{1}{7} \end{bmatrix} \begin{Bmatrix} x_1 \\ x_2 \\ x_3 \\ x_4 \end{Bmatrix} = \begin{Bmatrix} 1 \\ 0 \\ 0 \\ 0 \end{Bmatrix};$$

(b)
$$\begin{bmatrix} 1.00 & 0.50 & 0.33 & 0.25 \\ 0.50 & 0.33 & 0.25 & 0.20 \\ 0.33 & 0.25 & 0.20 & 0.16 \\ 0.25 & 0.20 & 0.16 & 0.14 \end{bmatrix} \begin{Bmatrix} x_1 \\ x_2 \\ x_3 \\ x_4 \end{Bmatrix} = \begin{Bmatrix} 1 \\ 0 \\ 0 \\ 0 \end{Bmatrix}.$$

Find the solution vector in each case, compare the results, and determine whether the coefficient matrix given in part (a) is ill conditioned.

3.32. Find the condition number and the value of $D(A)$ for the following set of equations:

$$x_1 + x_2 = 1;$$
$$1.001x_1 + x_2 = 2.$$

3.33. Consider the following equations:

$$8x_1 + 8x_2 + 7x_3 = 23;$$
$$8x_1 + 7x_2 + 6x_3 = 21;$$
$$7x_1 + 6x_2 + 5x_3 = 18.$$

(a) Find the solution of these equations to within 0.01.
(b) Find the exact solution of these equations.
(c) Show that these equations are ill conditioned.

3.34. Solve the following systems of equations using a three-digit arithmetic and compare the resulting solutions:

(a)
$$x_1 + x_2 = 1;$$
$$1.001x_1 + x_2 = 2;$$

(b)
$$x_1 + x_2 = 1;$$
$$0.999x_1 + x_2 = 2.$$

Section 3.7

3.35. Plot the following equations and give a graphical interpretation of the ill conditioning of the equations:

$$x_1 + 3x_2 = 8.0;$$
$$1.0001x_1 + 3x_2 = 8.0002.$$

3.36. Solve the following equations graphically:

$$8x_1 + 9x_2 = -1;$$
$$5x_1 - x_2 = 6.$$

Section 3.8

3.37. Find the solution of the following equations using Cramer's rule:

(a)
$$4x_1 + 5x_2 = 7;$$
$$-2x_1 + 3x_2 = 13.$$

(b)
$$8x_1 + 9x_2 = -1;$$
$$5x_1 - x_2 = 6.$$

(c)
$$7x_1 + 3x_2 = 13;$$
$$2x_1 - x_2 = 0.$$

Find the solution of the given equations using Cramer's rule in Problems 3.38 and 3.39.

3.38.
$$5x_1 - 4x_2 + 2x_3 = 11;$$
$$-2x_1 + 3x_2 + x_3 = 1;$$
$$6x_1 + x_2 - x_3 = 3.$$

3.39.
$$4x_1 - 2x_2 + x_3 = 11;$$
$$x_1 + 5x_2 - 3x_3 = -6;$$
$$3x_1 + 3x_2 + 2x_3 = 5.$$

Section 3.9

3.40. Reduce the following system of equations into an equivalent upper triangular form using elementary operations:

$$x_1 + 3x_2 + x_3 + 5x_4 = 4;$$
$$2x_1 + x_2 + 3x_4 = 5;$$
$$4x_1 + 2x_2 + 2x_3 + x_4 = 11;$$
$$-3x_1 + x_2 + 3x_3 + 2x_4 = 3.$$

3.41. Find the solution of the equations given in Problem 3.40 using the Gauss elimination method.

3.42. Solve the following system of equations using the Gauss elimination method:

$$x_1 + x_2 + x_3 = 6;$$
$$3x_1 + 2x_2 + x_3 = 10;$$
$$-2x_1 + 3x_2 - 2x_3 = -2.$$

Section 3.10

3.43. Find the solution of the equations given in Problem 3.40 using the Gauss–Jordan elimination method.

3.44. Find the solution of the equations given in Problem 3.42 using the Gauss–Jordan elimination method.

Section 3.11

In Problems 3.45 and 3.46, express the given system of equations in the form $[L][U]\vec{x} = \vec{b}$ using Crout's method, and solve the resulting equations:

3.45.

$$\begin{bmatrix} 4 & -1 & 1 \\ 8 & 3 & -1 \\ 3 & 1 & 1 \end{bmatrix} \begin{Bmatrix} x_1 \\ x_2 \\ x_3 \end{Bmatrix} = \begin{Bmatrix} 6 \\ 10 \\ 9 \end{Bmatrix}.$$

3.46.

$$\begin{bmatrix} 1 & 2 & 0 & -1 \\ 2 & 3 & -1 & 0 \\ 0 & 4 & 2 & -5 \\ 5 & 5 & 2 & -4 \end{bmatrix} \begin{Bmatrix} x_1 \\ x_2 \\ x_3 \\ x_4 \end{Bmatrix} = \begin{Bmatrix} 1 \\ -2 \\ 3 \\ 6 \end{Bmatrix}.$$

3.47. Express the following equations in the form $[L][U]\vec{x} = \vec{b}$ using the Choleski method, and find the solution of the resulting equations.

$$\begin{bmatrix} 16 & -4 & 4 \\ -4 & 17 & 11 \\ 4 & 11 & 14 \end{bmatrix} \begin{Bmatrix} x_1 \\ x_2 \\ x_3 \end{Bmatrix} = \begin{Bmatrix} -4 \\ -3 \\ -16 \end{Bmatrix}.$$

In Problems 3.48 and 3.49, find the inverse of the given matrix using the relation $[A]^{-1} = [U]^{-1}([U]^{-1})^T$.

3.48.

$$[A] = \begin{bmatrix} 5 & -1 & 1 \\ -1 & 6 & -4 \\ 1 & -4 & 3 \end{bmatrix}.$$

3.49.

$$[A] = \begin{bmatrix} 2 & -1 & 0 & 0 \\ -1 & 2 & -1 & 0 \\ 0 & -1 & 2 & -1 \\ 0 & 0 & -1 & 2 \end{bmatrix}.$$

3.50. Derive Eq. (3.64).

3.51. Derive the equations necessary for decomposing a symmetric matrix $[A]$ as $[A] = [L][L]^T$ using Choleski's method.

3.52 - 3.53

In the Doolittle method, a general matrix $[A] = [a_{ij}]$ is decomposed as $[A] = [L][U]$ with the elements of $[L] = [l_{ij}]$ and $[U] = [u_{ij}]$ given by

$$l_{ii} = 1, i = 1, 2, \ldots, n$$

For $k = 1, 2, \ldots, n$:

$$\text{Diagonal element of } [U]: \; u_{kk} = \frac{1}{l_{kk}}\left(a_{kk} - \sum_{m=1}^{k-1} l_{km}u_{mk}\right)$$

$$k^{\text{th}} \text{ column of } [L]: \; l_{ik} = \frac{1}{u_{kk}}\left(a_{ik} - \sum_{m=1}^{k-1} l_{im}u_{mk}\right), k \le i \le n$$

$$k^{\text{th}} \text{ row of } [U]: \; u_{kj} = \frac{1}{l_{kk}}\left(a_{kj} - \sum_{m=1}^{k-1} l_{km}u_{mj}\right), k \le j \le n.$$

Using this method, express the system of equations given in Problems 3.45 and 3.46 in the form $[L][U]\vec{x} = \vec{b}$.

Section 3.12

In problems 3.54 and 3.42, determine whether the given matrix is diagonally dominant.

3.54.

$$[A] = \begin{bmatrix} 4 & -2 & 1 \\ 1 & 5 & -3 \\ 2 & 2 & 5 \end{bmatrix}.$$

3.55.

$$[A] = \begin{bmatrix} 3 & -1 & 0 & -1 \\ -1 & 3 & -1 & 0 \\ 0 & -1 & 4 & -1 \\ -1 & 0 & -1 & 4 \end{bmatrix}.$$

In Problems 3.56 and 3.57, solve the given system of equations using the Jacobi iteration method with the initial approximation, $\vec{x}^{(1)} = \vec{0}$.

3.56.

$$\begin{aligned} 4x_1 - 2x_2 + x_3 &= 11; \\ x_1 + 5x_2 - 3x_3 &= -6; \\ 2x_1 + 2x_2 + 5x_3 &= 7. \end{aligned}$$

3.57.

$$3x_1 - x_2 - x_4 = -3;$$
$$-x_1 + 3x_2 - x_3 = 2;$$
$$-x_2 + 4x_3 - x_4 = 6;$$
$$-x_1 - x_3 + 4x_4 = 12.$$

Section 3.13

3.58. Solve Problem 3.56 using the Gauss–Seidel iteration method.

3.59. Solve Problem 3.57 using the Gauss–Seidel iteration method.

Section 3.14

3.60. Solve Problem 3.56 using the relaxation method. Also, find the optimum value of the relaxation factor, ω.

3.61. Solve Problem 3.57 using the relaxation method. Also, find the optimum value of the relaxation factor, ω.

Section 3.15

3.62. Solve the following system of complex equations:

$$(5 - 5i)x_1 - (2 - 2i)x_2 = 1;$$
$$-(2 - 2i)x_1 + (3 + 2i)x_2 = 0.$$

3.63. Figure 3.24 shows an alternating current electrical circuit consisting of resistors, inductors, and capacitors. Such circuits can be analyzed using complex arithmetic, since the sinusoidal variation of the voltage/current can be represented as a complex variable. Kirchhoff's laws can be applied to these circuits

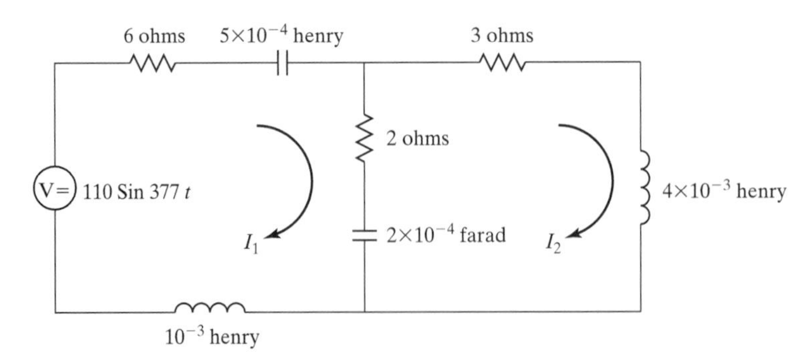

Figure 3.24 Electrical circuit.

by taking the voltage drop across a resistance of R ohms, an inductance of L henries and a capacitance of C farads as IR, $i\omega L$, and $-\frac{i}{\omega C}$, respectively, where $i = \sqrt{-1}$, I is the alternating current, and ω is the frequency of the current. Using Kirchhoff's law, the sum of the voltages around each of the loops is set equal to zero to obtain the equations:

$$V - 6I_1 - \left(-\frac{i}{5 \times 10^{-4}\omega}\right) I_1 - 2(I_1 - I_2)$$

$$- \left(-\frac{i}{2 \times 10^{-4}\omega}\right)(I_1 - I_2) - (10^{-3} i\omega) I_1 = 0; \quad \text{(E1)}$$

$$-2(I_1 - I_2) - \left(-\frac{i}{2 \times 10^{-4}\omega}\right)(I_1 - I_2) - 3I_2 - (4 \times 10^{-3} i\omega) I_2 = 0. \quad \text{(E2)}$$

Using $\omega = 377$ and denoting I_1 and I_2 as $x_1 \sin \omega t$ and $x_2 \sin \omega t$, respectively, Eqs. (E1) and (E2) can be rewritten as

$$(8 - 18.1906i)x_1 - (2 - 13.2626i)x_2 = 100; \quad \text{(E3)}$$

$$(2 - 13.2626i)x_1 + (1 + 14.7706i)x_2 = 0. \quad \text{(E4)}$$

Find the solution of Eqs. (E3) and (E4).

Section 3.16

3.64. Find the inverse of the matrix given in Problem 3.54 using Gauss–Jordan elimination method.

3.65. Find the inverse of the matrix given in Problem 3.49 using Gauss–Jordan elimination method.

Section 3.17

3.66. Solve the following tridiagonal system of equations using LU decomposition:

$$5x_1 + 4x_2 = 13;$$
$$4x_1 - 3x_2 + 7x_3 = 19;$$
$$x_2 - 6x_3 + 4x_4 = 0;$$
$$12x_3 + 2x_4 = 44.$$

3.67. Solve the following tridiagonal system of equations using LU decomposition:

$$6x_1 - 4x_2 = 4;$$
$$2x_1 + 8x_2 - 3x_3 = 5;$$
$$x_2 + 6x_3 - 2x_4 = 2;$$
$$3x_3 + 16x_4 = 5.$$

3.68. The tridiagonal matrix

$$[A] = \begin{bmatrix} a_{11} & a_{12} & 0 & \cdot & \cdot & \cdot & 0 & 0 & 0 \\ a_{21} & a_{22} & a_{23} & \cdot & \cdot & \cdot & 0 & 0 & 0 \\ 0 & a_{32} & a_{33} & \cdot & \cdot & \cdot & 0 & 0 & 0 \\ \cdot & \cdot & \cdot & \cdot & \cdot & \cdot & \cdot & \cdot & \cdot \\ \cdot & \cdot & \cdot & \cdot & \cdot & \cdot & \cdot & \cdot & \cdot \\ \cdot & \cdot & \cdot & \cdot & \cdot & \cdot & \cdot & \cdot & \cdot \\ 0 & 0 & 0 & \cdot & \cdot & \cdot & a_{n-1,n-2} & a_{n-1,n-1} & a_{n-1,n} \\ 0 & 0 & 0 & \cdot & \cdot & \cdot & 0 & a_{n,n-1} & a_{nn} \end{bmatrix} \tag{E1}$$

can be decomposed, using the Doolittle method, as

$$[A] = [L][U], \tag{E2}$$

where

$$[L] = \begin{bmatrix} 1 & \cdot & \cdot & \cdot & \cdot & \cdot & \cdot & \cdot & \cdot \\ l_{21} & 1 & \cdot & \cdot & \cdot & \cdot & \cdot & \cdot & \cdot \\ 0 & l_{32} & 1 & \cdot & \cdot & \cdot & \cdot & \cdot & \cdot \\ 0 & 0 & l_{43} & 1 & \cdot & \cdot & \cdot & \cdot & \cdot \\ \cdot & \cdot & \cdot & \cdot & \cdot & \cdot & \cdot & \cdot & \cdot \\ \cdot & \cdot & \cdot & \cdot & \cdot & \cdot & \cdot & \cdot & \cdot \\ \cdot & \cdot & \cdot & \cdot & \cdot & \cdot & \cdot & \cdot & \cdot \\ 0 & 0 & 0 & 0 & \cdot & \cdot & \cdot & l_{n,n-1} & 1 \end{bmatrix} \tag{E3}$$

and

$$[U] = \begin{bmatrix} u_{11} & u_{12} & 0 & 0 & \cdot & \cdot & \cdot & 0 & 0 \\ \cdot & u_{22} & u_{23} & 0 & \cdot & \cdot & \cdot & 0 & 0 \\ \cdot & \cdot & u_{33} & u_{34} & \cdot & \cdot & \cdot & 0 & 0 \\ \cdot & \cdot & \cdot & \cdot & \cdot & \cdot & \cdot & \cdot & \cdot \\ \cdot & \cdot & \cdot & \cdot & \cdot & \cdot & \cdot & \cdot & \cdot \\ \cdot & \cdot & \cdot & \cdot & \cdot & \cdot & \cdot & \cdot & \cdot \\ \cdot & \cdot & \cdot & \cdot & \cdot & \cdot & \cdot & u_{n-1,n-1} & u_{n-1,n} \\ \cdot & \cdot & \cdot & \cdot & \cdot & \cdot & \cdot & \cdot & u_{nn} \end{bmatrix}. \tag{E4}$$

The Thomas algorithm uses Eq. (E2) and finds the elements l_{ij} and u_{ij} as

$$u_{11} = a_{11};$$

$$l_{i,i-1} = \frac{a_{i,i-1}}{u_{i-1,i-1}};$$

$$u_{ii} = a_{ii} - l_{i,i-1}a_{i-1,i}, i = 2, 3, \ldots, n.$$

The solution of equations, $[A]\vec{x} = \vec{b}$, is given by

$$x_n = \frac{\beta_n}{\alpha_n};$$

$$x_i = \frac{(\beta_i - a_{i,i+1}x_{i+1})}{\alpha_i}; \quad i = n-1, n-2, \ldots, 1,$$

where

$$\beta_1 = b_1;$$

$$\beta_i = b_i - l_{i,i-1}\beta_{i-1}; \; i = 2, 3, \ldots, n;$$

$$\alpha_i = u_{ii}; \; i = 1, 2, \ldots, n.$$

Solve Problem 3.67 using Thomas algorithm.

3.69. Find the solution of the equations given in Example 3.24 using Thomas' algorithm.

Section 3.18

3.70. Decide whether the following system of equations is overdetermined:

$$4x_1 + 3x_2 = 1;$$

$$-3x_1 + 4x_2 = 1;$$

$$8x_1 + 8x_2 = 1.$$

3.71. Solve the system of equations given in Problem 3.70 using a least-squares approach.

3.72. Determine the solution of the following equations using the Gauss–Jordan elimination technique by treating x_4 as the free variable:

$$4x_1 + 3x_2 + 2x_3 + x_4 = 16;$$

$$3x_1 + 3x_2 + 2x_3 + x_4 = 15;$$

$$2x_1 + 2x_2 + x_3 + x_4 = 9.$$

3.73. Find the solution of the following homogeneous equations:

$$x_1 + 2x_2 + 3x_3 = 0;$$
$$4x_1 + 5x_2 + 6x_3 = 0;$$
$$7x_1 + 8x_2 + 9x_3 = 0.$$

3.74. Derive Eq. (3.112) from Eq. (3.111).

3.75. Find the solution of the equations

$$3x_1 + x_2 = 5;$$
$$x_1 - 2x_2 = -3;$$
$$x_1 - x_2 = 1;$$
$$2x_1 + 3x_2 = 6.$$

using the least-squares approach.

Section 3.19

3.76. Estimate the number of arithmetic operations involved in finding the inverse of a matrix of size $n \times n$ using:
(a) Gauss elimination method, and (b) Gauss–Jordan elimination method for values of $n = 3, 10, 50$, and 200.

3.77. Prove Eqs. (3.135) and (3.136) for the Gauss–Jordan elimination method.

3.78. Prove Eqs. (3.137) through (3.139) for the Choleski method.

Section 3.21

3.79. Find the solution of Eq. (E7) of Example 3.1 using MATLAB.

3.80. Find the inverse of the coefficient matrix and the solution of Eq. (E5) in Example 3.2 using MATLAB.

3.81. Find the inverse of the matrix $[A]$ given in Problem 3.21 using MATLAB.

3.82. Solve the following system of equations using MATLAB:

$$0.8765x_1 + 0.9903x_2 + 0.7456x_3 + 1.0001x_4 = 2.6125;$$
$$0.9876x_1 + 1.0234x_2 + 0.3333x_3 + 0.7654x_4 = 3.1097;$$
$$0.7799x_1 + 0.5631x_2 + 0.7125x_3 + 0.8973x_4 = 2.9528;$$
$$0.5793x_1 + 0.0004x_2 + 0.9371x_3 + 0.8214x_4 = 2.3382.$$

3.83. Use MATLAB to find the inverse, $[B] = [A]^{-1}$, of the matrix $[A]$ given by

$$[A] = \begin{bmatrix} \left(\frac{n+2}{2n+2}\right) & -\frac{1}{2} & 0 & 0 & \cdots & 0 & 0 & \left(\frac{1}{2n+2}\right) \\ -\frac{1}{2} & 1 & -\frac{1}{2} & 0 & \cdots & 0 & 0 & 0 \\ 0 & -\frac{1}{2} & 1 & -\frac{1}{2} & \cdots & 0 & 0 & 0 \\ \cdot & \cdot & \cdot & \cdot & \cdots & \cdot & \cdot & \cdot \\ \cdot & \cdot & \cdot & \cdot & \cdots & \cdot & \cdot & \cdot \\ \cdot & \cdot & \cdot & \cdot & \cdots & \cdot & \cdot & \cdot \\ 0 & 0 & 0 & 0 & \cdots & -\frac{1}{2} & 1 & -\frac{1}{2} \\ \left(\frac{1}{2n+2}\right) & 0 & 0 & 0 & \cdots & 0 & -\frac{1}{2} & \left(\frac{n+2}{2n+2}\right) \end{bmatrix},$$

with $n = 50$. Find $[C] = [A][B]$ and compute the measure of error, E:

$$E = \sum_{i=1}^{n} \sum_{j=1}^{n} |c_{ij}|, \; n = 50.$$

(*Note*: $E = 50$ if the solution is correct with $[C] = [I]$.)

3.84. Find the solution of Eq. (E1) of Example 3.3 using MAPLE.

3.85. Find the inverse of the coefficient matrix and the solution of the equations using MAPLE

$$[A]\vec{x} = \vec{b},$$

where

$$[A] = \begin{bmatrix} 1 & -2 & 3 & 1 \\ -2 & 1 & -2 & -1 \\ 3 & -2 & 1 & 5 \\ 1 & -1 & 5 & 3 \end{bmatrix}$$

and

$$\vec{b} = \begin{Bmatrix} 30 \\ -40 \\ 70 \\ 80 \end{Bmatrix}.$$

3.86. Find the inverse of the matrix $[A]$ given in Problem 3.23 using MAPLE.

3.87. Find the inverse of the following matrix in symbolic form using MAPLE:

$$[A] = \begin{bmatrix} a_{11} & a_{12} & a_{13} \\ a_{21} & a_{22} & a_{23} \\ a_{31} & a_{32} & a_{33} \end{bmatrix}.$$

3.88. Find the solution of the following equations in terms of k using MAPLE:

$$x_1 + x_2 + x_3 = 4;$$
$$3x_1 + 2x_2 + kx_3 = 6;$$
$$-2x_1 + 3x_2 - 2x_3 = -8.$$

3.89. Find the solution of the equations given in Problem 3.82 using MAPLE.

3.90. Find the solution of the equations given in Problem 3.82 using MATHCAD.

3.91. Solve Problem 3.88 using MATHCAD.

3.92. Solve Problem 3.83 using MATHCAD.

Section 3.22

3.93. Write a Fortran program for finding the inverse of a square matrix using the Gauss elimination method. Using this program, find the inverse of the following matrix with $n = 20$:

$$[A] = \begin{bmatrix} n & n-1 & n-2 & \cdot & \cdot & \cdot & 2 \\ n-1 & n-1 & n-2 & \cdot & \cdot & \cdot & 2 \\ n-2 & n-2 & n-2 & \cdot & \cdot & \cdot & 2 \\ \cdot & \cdot & \cdot & \cdot & \cdot & \cdot & \cdot \\ \cdot & \cdot & \cdot & \cdot & \cdot & \cdot & \cdot \\ \cdot & \cdot & \cdot & \cdot & \cdot & \cdot & \cdot \\ 2 & 2 & 2 & \cdot & \cdot & \cdot & 2 \\ 1 & 1 & 1 & \cdot & \cdot & \cdot & 1 \end{bmatrix}.$$

Determine the accuracy of the result by comparing the value of $[A][A]^{-1}$ with $[I]$.

3.94. Write a computer program to solve an overdetermined system of n equations in m unknowns $(n > m)$ using least-squares approach.

3.95. Write a computer program to solve a system of complex linear equations.

3.96. Write a computer program to solve a system of linear equations using the Jacobi iteration method.

3.97. Write a computer program to (a) find the inverse of a square matrix using the Gauss–Jordan elimination method and (b) use the inverse in finding the solution of a set of linear equations.

3.98. Solve Problem 3.42 using computer program 3.1.

3.99. Solve Problem 3.115 using computer program 3.1.

3.100. Solve Problem 3.67 using computer program 3.2.

3.101. Solve Problem 3.66 using computer program 3.2.

3.102. Solve Problem 3.46 using computer program 3.3.

3.103. Solve Problem 3.57 using computer program 3.3.

3.104. Solve Problem 3.116 using computer program 3.4.

3.105. Solve Problem 3.46 using computer program 3.4.

3.106. Solve a system of n linear equations

$$[A]\vec{x} = \vec{b}, \tag{E1}$$

where n is very large, Eq. (E1) can be partitioned as

$$\begin{bmatrix} \underset{p \times p}{[A_{11}]} & \underset{p \times q}{[A_{12}]} \\ \cdot & \cdot \\ \underset{q \times p}{[A_{21}]} & \underset{q \times q}{[A_{22}]} \end{bmatrix} \begin{Bmatrix} \underset{p \times 1}{\vec{x}_1} \\ \cdot \\ \underset{q \times 1}{\vec{x}_2} \end{Bmatrix} = \begin{Bmatrix} \underset{p \times 1}{\vec{b}_1} \\ \cdot \\ \underset{q \times 1}{\vec{b}_2} \end{Bmatrix}, \tag{E2}$$

where the orders of the submatrices and subvectors are also indicated. Equation (E2) can be rewritten as two separate sets of matrix equations:

$$[A_{11}]\vec{x}_1 + [A_{12}]\vec{x}_2 = \vec{b}_1; \tag{E3}$$

$$[A_{21}]\vec{x}_1 + [A_{22}]\vec{x}_2 = \vec{b}_2. \tag{E4}$$

Equation (E4) can be used to express $\vec{x}_2$

$$\vec{x}_2 = [A_{22}]^{-1}(\vec{b}_2 - [A_{21}]\vec{x}_1). \tag{E5}$$

Substituting of Eq. (E5) into Eq. (E3) leads to the solution

$$\vec{x}_1 = [[A_{11}] - [A_{12}][A_{22}]^{-1}[A_{21}]]^{-1}(\vec{b}_1 - [A_{12}][A_{22}]^{-1}\vec{b}_2). \tag{E6}$$

Equations (E6) and (E5) give the desired solution

$$\vec{x} = \begin{Bmatrix} \vec{x}_1 \\ \vec{x}_2 \end{Bmatrix}. \tag{E7}$$

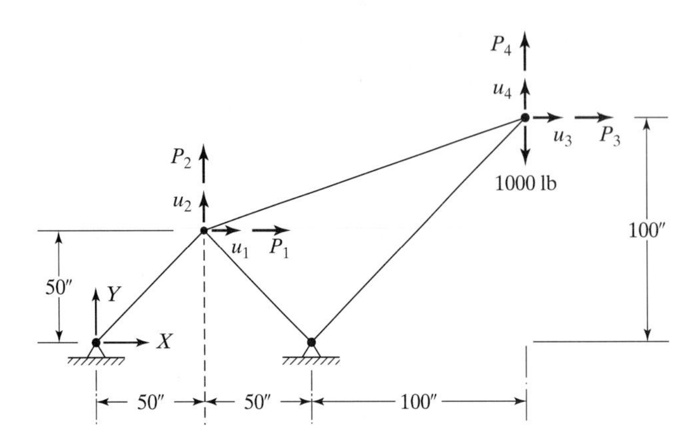

Figure 3.25 Crane.

This procedure is more economical, since it requires the inverses of matrices of smaller orders.

(a) Write a computer program to implement the block matrix method.

(b) Using the program of part (a), solve the equations (E1) of Example 3.3 with $p = 4$ and $q = 5$.

General

3.107. The nodal displacements of the crane shown in Fig. 3.25 can be found by solving the equilibrium equations:

$$\begin{bmatrix} 170.4105 & 37.9473 & -113.8420 & -37.9473 \\ 37.9473 & 69.2176 & -37.9473 & -12.6491 \\ -113.8420 & -37.9473 & 120.9131 & 45.0184 \\ -37.9473 & -12.6491 & 45.0184 & 19.7202 \end{bmatrix} \begin{Bmatrix} u_1 \\ u_2 \\ u_3 \\ u_4 \end{Bmatrix} = \begin{Bmatrix} P_1 \\ P_2 \\ P_3 \\ P_4 \end{Bmatrix},$$

where $u_i, i = 1, 2, 3, 4$, are the components of nodal displacement (inch) and P_i is the load applied (lb) along the direction $u_1, i = 1, 2, 3, 4$. Find the nodal displacements when the load lifted is given by 1000 lb (i.e., $P_1 = P_2 = P_3 = 0, P_4 = -1000$).

3.108. A rigid bar, of negligible weight, is supported on three springs and is subjected to a load of 1000 N as shown in Fig. 3.26(a). The determination of the deflections of the springs involves solving the vertical force and moment equilibrium equations, as well as the compatibility of displacements of the springs (Fig. 3.26(b)), which can be written as

$$k_A\delta_A + k_B\delta_B + k_C\delta_C = 1000,$$

$$2k_C\delta_C + k_B\delta_B = 400,$$

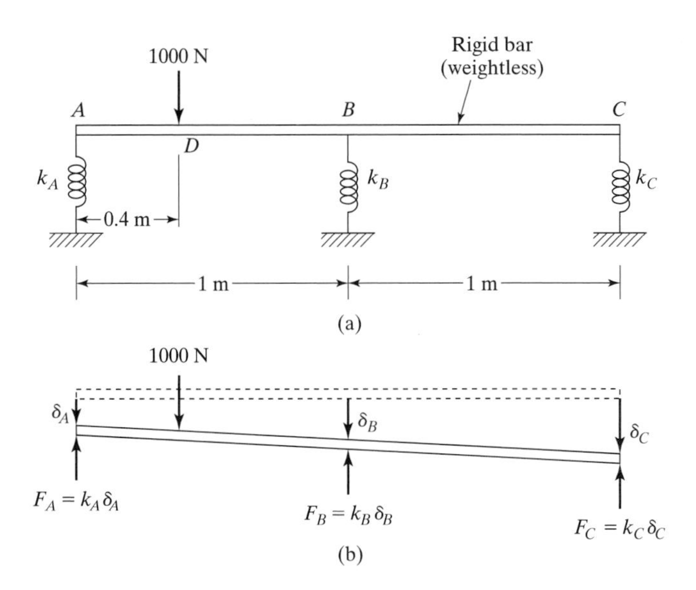

Figure 3.26 Rigid bar on three springs.

and

$$\delta_A - 2\delta_B + \delta_C = 0,$$

where k_A, k_B, and k_C are the stiffnesses, and δ_A, δ_B, and δ_C are the deflections of the springs A, B, and C, respectively. Find the values of δ_A, δ_B and δ_C when $k_A = 1000$ lb/in, $k_B = 500$ lb/in, and $k_C = 2000$ lb/in.

3.109. The volume of a tetrahedron, V, with vertices located at (x_i, y_i, z_i), $i = 1, 2, 3$, is given by [3.6]:

$$V = \frac{1}{6} \begin{vmatrix} x_1 & y_1 & z_1 & 1 \\ x_2 & y_2 & z_2 & 1 \\ x_3 & y_3 & z_3 & 1 \\ x_4 & y_4 & z_4 & 1 \end{vmatrix}.$$

Using this relation, find the volume of the tetrahedron whose vertices are located at (2,4,2), (0,0,0), (4,0,0), and (2,0,6).

3.110. Find the inverse of the matrix

$$[A] = \begin{bmatrix} 2 & 1 & 1 \\ 3 & 3 & 5 \\ 1 & 0 & 1 \end{bmatrix}$$

using Eq. (15) of Appendix F.

3.111. Find the L_1, L_2, L_∞, and L_e of the following matrix:

$$[A] = \begin{bmatrix} 2 & 1 & 6 \\ -1 & 6 & -4 \\ 3 & -4 & 2 \end{bmatrix}.$$

3.112. Determine whether the following system of equations are independent:

$$5x_1 - x_2 + 3x_3 = 12;$$
$$3x_1 + 2x_2 + x_3 = 10;$$
$$-2x_1 + 3x_2 - 2x_3 = -2.$$

3.113. In certain applications, some of the coefficients of the equations may be larger than others by an order of magnitude. Such situations occur when different units are used in formulating the mathematical model. The resulting coefficient matrix might exhibit ill conditioning. Such problems can be handled by using a proper scaling of the variables. Use a suitable scaling of variables and solve the following system of equations:

$$3x_1 - 6x_2 + 5000x_3 = -5;$$
$$8x_1 + 0x_2 - 5000x_3 = -2;$$
$$5x_1 - 8x_2 + 6000x_3 = -7.$$

3.114. Use Cramer's rule to solve the following equations:

$$x_1 + x_2 + x_3 = 6;$$
$$3x_1 + 2x_2 + x_3 = 10;$$
$$-2x_1 + 3x_2 - 2x_3 = -2.$$

3.115. Solve the following equations using a suitable method:

$$5x_1 - 4x_2 + 2x_3 = 1;$$
$$-2x_1 + 3x_2 + x_3 = 10;$$
$$6x_1 + x_2 - x_3 = -10.$$

3.116. Solve the following equations using $[L][U]$ decomposition:

$$\begin{bmatrix} 1 & -1 & 0 & 0 & 0 \\ -1 & 2 & -1 & 0 & 0 \\ 0 & -1 & 2 & -1 & 0 \\ 0 & 0 & -1 & 2 & -1 \\ 0 & 0 & 0 & -1 & 2 \end{bmatrix} \begin{Bmatrix} x_1 \\ x_2 \\ x_3 \\ x_4 \\ x_5 \end{Bmatrix} = \begin{Bmatrix} 1 \\ 0 \\ 0 \\ 0 \\ 0 \end{Bmatrix}.$$

3.117. Find the solution of the following system of homogeneous equations:

$$\begin{bmatrix} 1 & 4 & 13 & 16 \\ 15 & 14 & 3 & 2 \\ 12 & 9 & 8 & 5 \\ 6 & 7 & 10 & 11 \end{bmatrix} \begin{Bmatrix} x_1 \\ x_2 \\ x_3 \\ x_4 \end{Bmatrix} = \begin{Bmatrix} 0 \\ 0 \\ 0 \\ 0 \end{Bmatrix}.$$

3.118. The temperatures measured at various points inside a heated wall are as follows:

Distance from heated surface, x (of wall thickness), percentage	0	25	50	75	100
Temperature, T (°C)	400	300	150	80	20

If the temperature variation is to be approximated by the relation, $T(x) = a + bx$, determine the constants a and b using a least-squares technique.

3.119. A bearing manufacturer produces three types of bearings, B_1, B_2, and B_3, using . vo machines A_1 and A_2. Each unit of bearing B_1, B_2, and B_3 require processing times of 10, 6, and 12 minutes on machine A_1 and 8, 4, and 4 minutes on machine A_2, respectively. If the times available on machines A_1 and A_2 per day are 1200 and 1000 minutes, respectively, determine the number of bearings B_1, B_2, and B_3 that can be produced per day.

3.120. The displacements x_1, x_2, and x_3 (mode shapes) of the three masses of a vibrating system are governed by the equation

$$\lambda \begin{bmatrix} 1 & 0 & 0 \\ 0 & 1 & 0 \\ 0 & 0 & 1 \end{bmatrix} \begin{Bmatrix} x_1 \\ x_2 \\ x_3 \end{Bmatrix} = \begin{bmatrix} 1 & 1 & 1 \\ 1 & 2 & 2 \\ 1 & 2 & 3 \end{bmatrix} \begin{Bmatrix} x_1 \\ x_2 \\ x_3 \end{Bmatrix},$$

where λ is the natural frequency of vibration. Determine the mode shapes corresponding to the natural frequencies $\lambda_1 = 5.0489, \lambda_2 = 0.6430$, and $\lambda_3 = 0.3078$.

3.121. Prove the equivalence of the following relations:

(a) $\sum_{i=1}^{p} i = \frac{1}{2} p(p+1)$.

(b) $\sum_{i=1}^{p} i^2 = \frac{1}{6} p(p+1)(2p+1)$.

3.122. Determine the quadratic equation $y = a + bx + cx^2$ that passes through the points (1,1), (2,3), and (3,5).

PROJECTS

3.1. (a) The method of least squares can be used to fit a curve through n data points, $(x_i, y_i), i = 1, 2, \ldots, n$. If a quadratic equation (or parabola) is used for curve fitting, $y(x)$ is taken as

$$y(x) = a + bx + cx^2, \tag{E1}$$

and the error between the theoretical curve and the data points is used to construct the function

$$E = \sum_{i=1}^{n} (y_i - a - bx_i - cx_i^2)^2. \tag{E2}$$

The function E is minimized with respect to a, b, and c to find the best-fitting parabola. For this, the conditions

$$\frac{\partial E}{\partial a} = 0, \tag{E3}$$

$$\frac{\partial E}{\partial b} = 0, \tag{E4}$$

and

$$\frac{\partial E}{\partial c} = 0 \tag{E5}$$

are used to obtain the following linear equations in a, b, and c:

$$(n)a + \left(\sum_{i=1}^{n} x_i\right) b + \left(\sum_{i=1}^{n} x_i^2\right) c = \sum_{i=1}^{n} y_i; \tag{E6}$$

$$\left(\sum_{i=1}^{n} x_i\right) a + \left(\sum_{i=1}^{n} x_i^2\right) b + \left(\sum_{i=1}^{n} x_i^3\right) c = \sum_{i=1}^{n} x_i y_i; \tag{E7}$$

$$\left(\sum_{i=1}^{n} x_i^2\right) a + \left(\sum_{i=1}^{n} x_i^3\right) b + \left(\sum_{i=1}^{n} x_i^4\right) c = \sum_{i=1}^{n} x_i^2 y_i. \tag{E8}$$

Write a computer program to fit a quadratic equation through n data points.

(b) Use the computer program of part (a) to express the variation of temperature along the thickness of a wall for the following data:

i (point number)	1	2	3	4	5	6
x_i (distance, inches)	0	1	2	3	4	5
y_i (temperature°C)	2.5	9.2	13.3	26.7	31.8	50.4

(c) In general, it is possible to fit a polynomial of order m through n data points $(x_i, y_i), i = 1, 2, \ldots, n (m \le n - 1)$. Develop a computer program to determine polynomials of orders 1, 2, 3, 4, and 5 using 6 data points. Use the resulting program to express the variation of temperature along the thickness of a wall using the data points of part (a). Determine the best polynomial for the given data.

3.2. (a) Systems of equations of the form

$$[A]\vec{x} = \vec{b}, \tag{E1}$$

with symmetric and banded matrices $[A]$ are encountered in several engineering applications involving finite element analysis. Write a computer program to solve equations (E1) using LU decomposition by storing only the elements of the diagonals having nonzero elements.

(b) Generalize the computer program of part (a) to solve the equations

$$[A]\vec{x}_i = \vec{b}_i; i = 1, 2, \ldots, k \tag{E2}$$

using minimal amount of computations.

(c) Use the program of part (b) to solve equations (E2) with $k = 4$ and

$$[A] = \begin{bmatrix} 5 & -5 & 0 & 0 & 0 \\ -5 & 10 & -5 & 0 & 0 \\ 0 & -5 & 10 & -5 & 0 \\ 0 & 0 & -5 & 10 & -5 \\ 0 & 0 & 0 & -5 & 10 \end{bmatrix},$$

$$\vec{b}_1 = \begin{Bmatrix} 0 \\ 1 \\ 0 \\ 0 \\ 0 \end{Bmatrix}, \quad \vec{b}_2 = \begin{Bmatrix} 0 \\ 0 \\ 1 \\ 0 \\ 0 \end{Bmatrix}, \quad \vec{b}_3 = \begin{Bmatrix} 0 \\ 0 \\ 0 \\ 1 \\ 0 \end{Bmatrix}, \quad \text{and} \quad \vec{b}_4 = \begin{Bmatrix} 1 \\ 1 \\ 1 \\ 1 \\ 1 \end{Bmatrix}.$$

3.3. (a) The equilibrium equations of the truss shown in Fig. 3.27 can be expressed as

$$[A]\vec{x} = \vec{b}, \tag{E1}$$

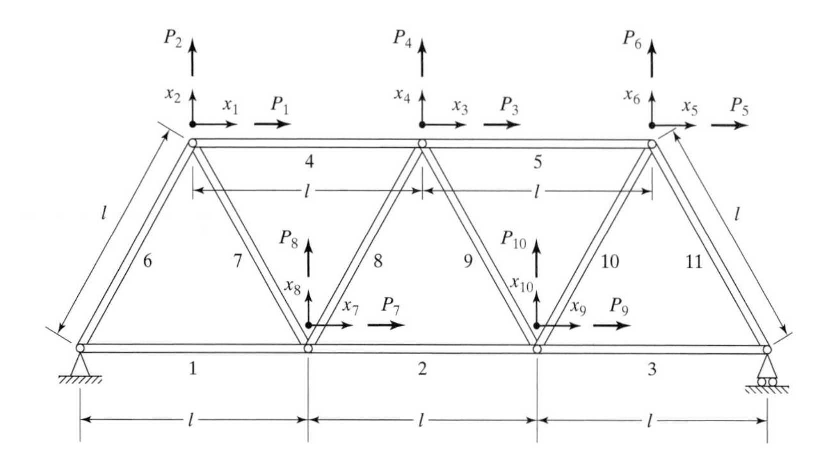

Figure 3.27 Planar truss.

where

$$A = \left[\begin{matrix} (4A_4+A_6+A_7) & \sqrt{3}(A_6-A_7) & -4A_4 & 0 & 0 \\ \sqrt{3}(A_6-A_7) & 3(A_6+A_7) & 0 & 0 & 0 \\ -4A_4 & 0 & (4A_4+4A_5+A_8+A_9) & \sqrt{3}(A_8-A_9) & -4A_5 \\ 0 & 0 & \sqrt{3}(A_8-A_9) & 3(A_8+A_9) & 0 \\ 0 & 0 & -4A_5 & 0 & (4A_5+A_{10}+A_{11}) \\ 0 & 0 & 0 & 0 & \sqrt{3}(A_{10}-A_{11}) \\ -A_7 & \sqrt{3}A_7 & -A_8 & -\sqrt{3}A_8 & 0 \\ \sqrt{3}A_7 & -3A_7 & -\sqrt{3}A_8 & -3A_8 & 0 \\ 0 & 0 & -A_9 & \sqrt{3}A_9 & -A_{10} \\ 0 & 0 & \sqrt{3}A_9 & -\sqrt{3}A_9 & -\sqrt{3}A_{10} \end{matrix}\right.$$

$$\left.\begin{matrix} 0 & -A_7 & \sqrt{3}A_7 & 0 & 0 \\ 0 & \sqrt{3}A_7 & -3A_7 & 0 & 0 \\ 0 & -A_8 & -\sqrt{3}A_8 & -A_9 & \sqrt{3}A_9 \\ 0 & -\sqrt{3}A_8 & -3A_8 & \sqrt{3}A_9 & -3A_9 \\ \sqrt{3}(A_{10}-A_{11}) & 0 & 0 & -A_{10} & -\sqrt{3}A_{10} \\ 3(A_{10}+A_{11}) & 0 & 0 & -\sqrt{3}A_{10} & -3A_{10} \\ 0 & (4A_1+4A_2+A_7+A_8) & -\sqrt{3}(A_7-A_8) & -4A_2 & 0 \\ 0 & -\sqrt{3}(A_7-A_8) & 3(A_7+A_8) & 0 & 0 \\ -\sqrt{3}A_{10} & -4A_2 & 0 & (4A_2+4A_3+A_9+A_{10}) & -\sqrt{3}(A_9-A_{10}) \\ -3A_{10} & 0 & 0 & -\sqrt{3}(A_9-A_{10}) & 3(A_9+A_{10}) \end{matrix}\right]$$

and

$$\vec{x} = \begin{Bmatrix} x_1 \\ x_2 \\ \cdot \\ \cdot \\ \cdot \\ x_{10} \end{Bmatrix}, \quad \vec{b} = \frac{4l}{E} \begin{Bmatrix} P_1 \\ P_2 \\ \cdot \\ \cdot \\ \cdot \\ P_{10} \end{Bmatrix},$$

in which E is Young's modulus, l is the length, A_i is the area of cross section of member i, P_j is the jth load component, and x_j is the jth displacement component. Assuming that Eq. (E1) is solved (by finding $[A]^{-1}$ explicitly) at a base design $\vec{A}_0 = \begin{Bmatrix} A_1 \\ A_2 \\ \cdot \\ \cdot \\ \cdot \\ A_{10} \end{Bmatrix}$, develop a procedure for finding the approximate values of the displacements at a perturbed design $\vec{A}_0 + \Delta\vec{A}_i$, where $\Delta\vec{A}_i = \begin{Bmatrix} 0 \\ \cdot \\ \cdot \\ \cdot \\ \delta A_i \\ 0 \\ \cdot \\ \cdot \\ \cdot \\ 0 \end{Bmatrix}$ without solving Eq. (E1) directly.

(b) Using the procedure developed in part (a), find the partial derivatives $\frac{\partial x_j}{\partial A_i}$, $i = 1, 2, \ldots, 11$, $j = 1, 2, \ldots, 10$ for the following base design:

$$A_i = 1 \text{ in}^2, i = 1, 2, \ldots, 5;$$

$$A_i = 2 \text{ in}^2, i = 6, 7, \ldots, 11;$$

$$E = 30 \times 10^6 \text{ psi};$$

$$l = 30 \text{ in};$$

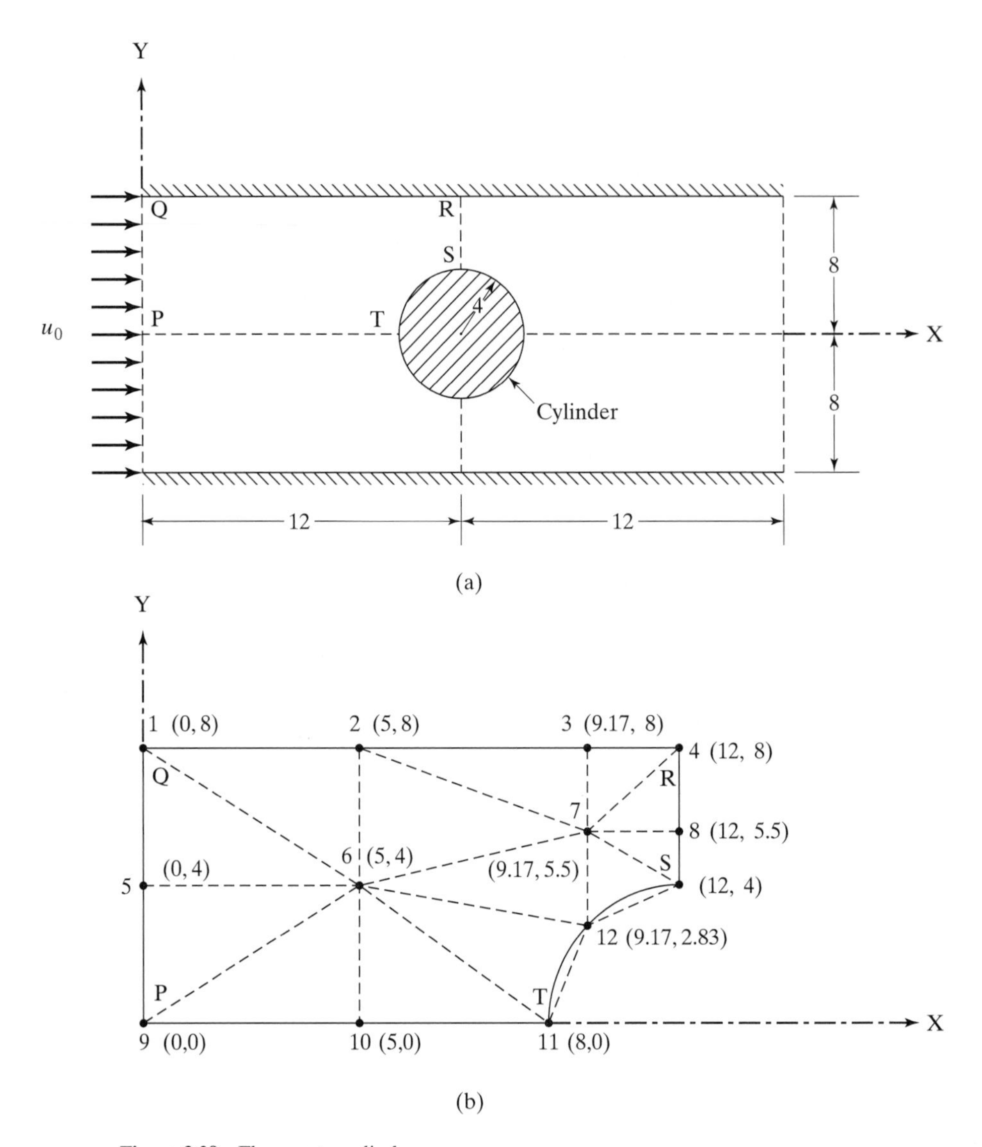

Figure 3.28 Flow past a cylinder.

$$P_i = -1000 \text{ lb}, i = 2, 4, 6, 8, 10;$$

$$P_i = 0, i = 1, 3, 5, 7, 9.$$

3.4. The equations governing the ideal fluid flow past a cylinder, based on the finite-element idealization shown in Fig. 3.28(b), are given by

$$[A]\vec{x} = \vec{b},$$

where

$$
[A] = \left[\begin{array}{cccccc}
1.0250 & -0.4000 & 0 & 0 & -0.6250 & 0 \\
-0.4000 & 1.9135 & -0.2998 & 0 & 0 & -1.0338 \\
0 & -0.2998 & 2.1415 & -0.4417 & 0 & 0 \\
0 & 0 & -0.4417 & 1.0077 & 0 & 0 \\
-0.6250 & 0 & 0 & 0 & 2.0500 & -0.8000 \\
0 & -1.0338 & 0 & 0 & -0.8000 & 3.8097 \\
0 & -0.1799 & -1.4000 & 0 & 0 & -0.4401 \\
0 & 0 & 0 & -0.5660 & 0 & 0 \\
0 & 0 & 0 & 0 & -0.6250 & 0 \\
0 & 0 & 0 & 0 & 0 & -1.0000 \\
0 & 0 & 0 & 0 & 0 & -0.0595 \\
0 & 0 & 0 & 0 & 0 & -0.4764 \\
0 & 0 & 0 & 0 & 0 & 0
\end{array}\right.
$$

$$
\left.\begin{array}{ccccccc}
0 & 0 & 0 & 0 & 0 & 0 & 0 \\
-0.1799 & 0 & 0 & 0 & 0 & 0 & 0 \\
-1.4000 & 0 & 0 & 0 & 0 & 0 & 0 \\
0 & -0.5660 & 0 & 0 & 0 & 0 & 0 \\
0 & 0 & -0.6250 & 0 & 0 & 0 & 0 \\
-0.4401 & 0 & 0 & -1.0000 & -0.0595 & -0.4764 & 0 \\
4.0492 & -0.7067 & 0 & 0 & 0 & -1.1159 & -0.2067 \\
-0.7067 & 2.2161 & 0 & 0 & 0 & 0 & -0.9433 \\
0 & 0 & 1.0250 & -0.4000 & 0 & 0 & 0 \\
0 & 0 & -0.4000 & 2.0667 & -0.6667 & 0 & 0 \\
0 & 0 & 0 & -0.6667 & 1.3788 & -0.6526 & 0 \\
-1.1159 & 0 & 0 & 0 & -0.6526 & 2.5098 & -0.2650 \\
-0.2067 & -0.9433 & 0 & 0 & 0 & -0.2650 & 1.4150
\end{array}\right],
$$

$$\vec{x} = \begin{Bmatrix} x_1 \\ x_2 \\ \cdot \\ \cdot \\ \cdot \\ x_{13} \end{Bmatrix}, \quad \text{and} \quad \vec{b} = u_0 \begin{Bmatrix} 2 \\ 0 \\ 0 \\ 0 \\ 4 \\ 0 \\ 0 \\ 0 \\ 2 \\ 0 \\ 0 \\ 0 \\ 0 \end{Bmatrix},$$

in which x_i is the potential at node i, and u_0 is the inflow velocity. The velocity of the fluid between nodes i and j can be computed as

$$u_{ij} = \frac{\partial x}{\partial X} \approx \frac{x_j - x_i}{X_j - X_i}.$$

Determine the velocity distribution of the fluid in the region shown in Fig. 3.28(a) for the following cases:

(a) $u_0 = 1$; (b) $u_0 = 5$; (c) $u_0 = 100$.

3.5. Two sides of a square plate, with uniform heat generation, are insulated as shown in Fig. 3.29. The heat conduction analysis of the plate, using the finite-element grid shown by the dashed lines, leads to the equations:

$$[A]\vec{x} = \vec{b},$$

where

$$[A] = \begin{bmatrix} 2 & -1 & 0 & -1 & 0 & 0 & 0 & 0 & 0 \\ -1 & 4 & 0 & 0 & -2 & 0 & 0 & 0 & 0 \\ 0 & 0 & 1 & 0 & 0 & 0 & 0 & 0 & 0 \\ -1 & 0 & 0 & 4 & -2 & 0 & 0 & 0 & 0 \\ 0 & -2 & 0 & -2 & 8 & 0 & 0 & 0 & 0 \\ 0 & 0 & 0 & 0 & 0 & 1 & 0 & 0 & 0 \\ 0 & 0 & 0 & 0 & 0 & 0 & 1 & 0 & 0 \\ 0 & 0 & 0 & 0 & 0 & 0 & 0 & 1 & 0 \\ 0 & 0 & 0 & 0 & 0 & 0 & 0 & 0 & 1 \end{bmatrix},$$

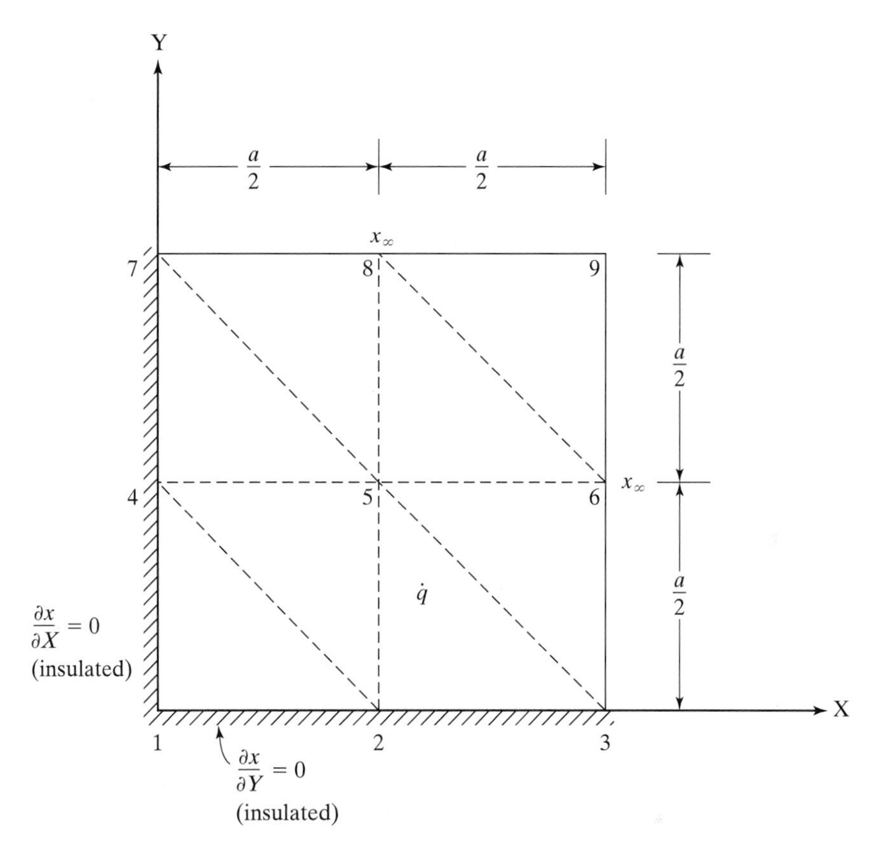

Figure 3.29 Square plate with uniform heat generation.

and

$$\vec{x} = \begin{Bmatrix} x_1 \\ x_2 \\ \cdot \\ \cdot \\ \cdot \\ x_9 \end{Bmatrix}, \quad \vec{b} = \frac{\dot{q}a^2}{12k} \begin{Bmatrix} 1 \\ 3 \\ 0 \\ 3 \\ 6 \\ 0 \\ 0 \\ 0 \\ 0 \end{Bmatrix} + x_\infty \begin{Bmatrix} 0 \\ 1 \\ 1 \\ 1 \\ 4 \\ 1 \\ 1 \\ 1 \\ 1 \end{Bmatrix},$$

in which x_i is the temperature at node i, $\dot{q}$ is the rate of heat generation, a is the side of the plate, and k is the thermal conductivity. Write a computer program to determine the temperature and temperature gradient in the plate for varying values of the parameters as indicated:

$a = 5 - 25$ cm, $k = 20 - 100$ watts/cm $-^\circ$ K, $x_\infty = 20 - 100^\circ$C, and $\dot{q} = 0 - 200$ watts/cm^3.

3.6. For any square matrix $[A]$, the first, second, ..., nth columns of $[A]^{-1}$ are nothing, but the solutions $\vec{X}_1, \vec{X}_2, \ldots, \vec{X}_n$ corresponding to the right-hand-side vectors $\vec{b}_1, \vec{b}_2, \ldots, \vec{b}_n$, respectively, where

$$\vec{b}_1 = \begin{Bmatrix} 1 \\ 0 \\ 0 \\ \cdot \\ \cdot \\ \cdot \\ 0 \end{Bmatrix}, \quad \vec{b}_2 = \begin{Bmatrix} 0 \\ 1 \\ 0 \\ \cdot \\ \cdot \\ \cdot \\ 0 \end{Bmatrix}, \ldots, \text{ and } \vec{b}_n = \begin{Bmatrix} 0 \\ 0 \\ 0 \\ \cdot \\ \cdot \\ \cdot \\ 1 \end{Bmatrix}.$$

Using this principle, write a computer program to find the inverse of a square matrix. Using this program, find the inverse of the matrix $[A]$ given in Problem 3.83 with $n = 100$. Determine the accuracy of the result by comparing the value of $[A][A]^{-1}$ with $[I]$.

3.7. The Hilbert matrix, $[A] = [a_{ij}]$, is defined as

$$a_{ij} = \frac{1}{i + j - 1}; \quad i, j = 1, 2, \ldots, n, \tag{E1}$$

and its inverse is given by $[B] = [A]^{-1} = [b_{ij}]$ with [3.20]

$$b_{ij} = \frac{(-1)^{i+j}(n + i - 1)!(n + j - 1)!}{(i + j - 1)[(i - 1)!(j - 1)!]^2(n - i)!(n - j)!}; \; i, j = 1, 2, ..., n. \tag{E2}$$

(a) Generate the matrix $[A]$ for $n = 1, 2, \ldots, 8$ and, in each case, find its inverse using the Gauss elimination method with single precision arithmetic (using any computer program).

(b) Generate the inverse of $[A]$ in each case using the Gauss elimination method with double-precision arithmetic (using any computer program).

(c) Examine, in each case, the accuracy of the results obtained in (a) and (b) by comparing them with the exact inverse given by Eq. (E2).

(d) What conclusions can be drawn from this study?

4

Solution of Matrix Eigenvalue Problem

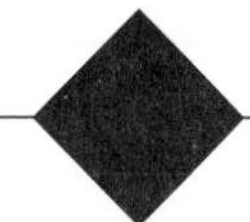

4.1 Introduction

CHAPTER CONTENTS

Definition

The standard eigenvalue problem is defined by the homogeneous equations

$$
\begin{aligned}
&(a_{11} - \lambda)x_1 + a_{12}x_2 + a_{13}x_3 + \cdots \\
&\quad + a_{1n}x_n = 0; \\
&a_{21}x_1 + (a_{22} - \lambda)x_2 + a_{23}x_3 + \cdots \\
&\quad + a_{2n}x_n = 0; \\
&\qquad \vdots \\
&a_{n1}x_1 + a_{n2}x_2 + a_{n3}x_3 + \cdots \\
&\quad + (a_{nn} - \lambda)x_n = 0. \qquad (4.1)
\end{aligned}
$$

This set of equations can be expressed in matrix form as

$$[A]\vec{X} = \lambda\vec{X}, \qquad (4.2)$$

where

$$[A] = \begin{bmatrix} a_{11} & a_{12} & \cdot & \cdot & \cdot & a_{1n} \\ a_{21} & a_{22} & \cdot & \cdot & \cdot & a_{2n} \\ \cdot & \cdot & \cdot & \cdot & \cdot & \cdot \\ \cdot & \cdot & \cdot & \cdot & \cdot & \cdot \\ \cdot & \cdot & \cdot & \cdot & \cdot & \cdot \\ a_{n1} & a_{n2} & \cdot & \cdot & \cdot & a_{nn} \end{bmatrix} \tag{4.3}$$

is a known square matrix of order n,

$$\vec{X} = \begin{Bmatrix} x_1 \\ x_2 \\ \cdot \\ \cdot \\ \cdot \\ x_n \end{Bmatrix} \tag{4.4}$$

is an unknown n-component vector, called the eigenvector, and λ is an unknown scalar, called the eigenvalue.

Characteristic Equation

Equation (4.2) can be rewritten as

$$[[A] - \lambda[I]]\vec{X} = \vec{0}, \tag{4.5}$$

where $[I]$ is an identity matrix of order n. It can be seen that Eq. (4.5) represents a system of n homogeneous linear equations in $n + 1$ unknowns, namely, the n components of $\vec{X}$ and λ. As stated in Section 3.18.3, for a nontrivial solution, the determinant of the coefficient matrix of $\vec{X}$ in Eq. (4.5) must be zero; that is,

$$\big|[A] - \lambda[I]\big| = 0, \tag{4.6}$$

or

$$\begin{vmatrix} (a_{11}-\lambda) & a_{12} & \cdots & a_{1n} \\ a_{21} & (a_{22}-\lambda) & \cdots & a_{2n} \\ \cdot & \cdot & \cdots & \cdot \\ \cdot & \cdot & \cdots & \cdot \\ \cdot & \cdot & \cdots & \cdot \\ a_{n1} & a_{n2} & \cdots & (a_{nn}-\lambda) \end{vmatrix} = 0. \tag{4.7}$$

Equation (4.7), when expanded, gives an nth order polynomial in λ, called the characteristic equation. Assuming that $\lambda_1, \lambda_2, \ldots, \lambda_n$ are the roots of this polynomial, we find that there are n solutions (eigenvalues, $\lambda_i, i = 1, 2, \ldots, n$) to Eq. (4.2). Corresponding to each distinct eigenvalue λ_i, a nontrivial solution of the system of linear equations (4.2) can be determined as

$$[A]\vec{X}^{(i)} = \lambda_i \vec{X}^{(i)}, \tag{4.8}$$

where

$$\vec{X}^{(i)} = \begin{Bmatrix} x_1^{(i)} \\ x_2^{(i)} \\ \cdot \\ \cdot \\ \cdot \\ x_n^{(i)} \end{Bmatrix} \tag{4.9}$$

is called the eigenvector corresponding to the eigenvalue λ_i.

Right and Left Eigenvectors

The eigenvector defined by Eqs. (4.2), (4.5), and (4.8) is more accurately called the right eigenvector. The term eigenvector is exclusively used in the literature to refer to the right eigenvector. If we consider the eigenvalue problem as

$$[A]^T \vec{Y} = \lambda \vec{Y}, \tag{4.10}$$

where $[A]^T$ is the transpose of the matrix $[A]$, $\vec{Y}$ will be the eigenvector of the matrix $[A]^T$ and λ will be the eigenvalue. The characteristic equation corresponding to Eq. (4.10) is given by

$$|[A]^T - \lambda[I]| = 0. \tag{4.11}$$

Since

$$|[A]^T - \lambda[I]| = |[A] - \lambda[I]|, \tag{4.12}$$

the eigenvalues of $[A]^T$ will be identical to those of $[A]$. However, the eigenvectors of $[A]^T$, $\vec{Y}$, will be different (unless $[A]$ is a symmetric matrix) because they are given by the solution of

$$[A]^T \vec{Y}^{(i)} = \lambda_i \vec{Y}^{(i)}, \tag{4.13}$$

or

$$\vec{Y}^{(i)^T}[A] = \lambda_i \vec{Y}^{(i)^T} \tag{4.14}$$

The eigenvectors of $[A]^T$, $\vec{Y}^{(i)}$, are known as the left eigenvectors of $[A]$. If $\vec{X}^{(i)}$ and $\vec{Y}^{(j)}$ denote the right and left eigenvectors corresponding to distinct eigenvalues λ_i and λ_j, respectively, it can be shown that $\vec{X}^{(i)}$ and $\vec{Y}^{(j)}$ are orthogonal. (See Problem 4.1.)

▶Example 4.1

Derive the characteristic equations corresponding to the problems

$$[A]\vec{X} = \lambda[I]\vec{X} \tag{a}$$

and

$$[A]^T \vec{Y} = \mu[I]\vec{Y}, \tag{b}$$

where

$$[A] = \begin{bmatrix} 4 & 1 \\ 3 & 2 \end{bmatrix}. \tag{c}$$

Solution

The characteristic equation corresponding to Eq. (a) is given by

$$|[A]-\lambda[I]| = \left| \begin{bmatrix} 4 & 1 \\ 3 & 2 \end{bmatrix} - \lambda \begin{bmatrix} 1 & 0 \\ 0 & 1 \end{bmatrix} \right| = \begin{vmatrix} 4-\lambda & 1 \\ 3 & 2-\lambda \end{vmatrix} = \lambda^2 - 6\lambda + 5 = 0. \tag{d}$$

Similarly, the characteristic equation corresponding to Eq. (b) is given by

$$|[A]^T - \mu[I]| = \left| \begin{bmatrix} 4 & 3 \\ 1 & 2 \end{bmatrix} - \mu \begin{bmatrix} 1 & 0 \\ 0 & 1 \end{bmatrix} \right| = \begin{vmatrix} 4-\mu & 3 \\ 1 & 2-\mu \end{vmatrix} = \mu^2 - 6\mu + 5 = 0. \tag{e}$$

It can be seen that the polynomials in Eqs. (d) and (e) are identical. ◀

General Eigenvalue Problem

The eigenvalue problem stated in Eq. (4.2) is known as the standard eigenvalue problem. However, many physical systems give rise to eigenvalue problems, which are expressed in a more general form as

$$[A]\vec{X} = \lambda[B]\vec{X}, \tag{4.15}$$

where $[A]$ and $[B]$ are symmetric matrices of order n, $\vec{X}$ is an n-component eigenvector, and λ is an eigenvalue. Equation (4.15) is known as the general

Table 4.1 Characteristics of eigenvalue problems.

Characteristic of the matrix, $[A]$	Nature of the eigenvalues
Singular matrix, $\|[A]\| = 0$	At least one λ_i is zero
Nonsingular matrix, $\|[A]\| \neq 0$	No zero eigenvalue
Symmetric matrix $[A]^T = [A]$	All eigenvalues are real
Zero matrix, $[A] = [0]$	All eigenvalues are zero
Identity matrix, $[A] = [I]$	All eigenvalues are equal to one
Diagonal matrix, $[A]$	Eigenvalues are equal to the diagonal elements of $[A]$
Inverse of $[A]$, $[A]^{-1}$	Eigenvalues of $[A]^{-1}$ are equal to the reciprocals of those of $[A]$
Transformed matrix, $[B] = [P]^T[A][P]$ ($[P]$ is an orthogonal matrix)	Eigenvalues of $[B]$ are same as those of $[A]$
Hermitian matrix	All eigenvalues are real

eigenvalue problem. A general eigenvalue problem can be reduced to the standard form, as indicated in Section 4.3.

Characteristics of Eigenvalue Problems

The characteristics of some of the eigenvalue problems are summarized in Table 4.1.

4.2 Engineering Applications

▶Example 4.2

A forging hammer of mass m_1 is mounted on a concrete foundation block of mass m_2. The stiffnesses of the springs underneath the forging hammer and the foundation block are given by k_2 and k_1, respectively (Fig. 4.1). The system undergoes simple harmonic motion at one of its natural frequencies (ω), which are given by:

$$\begin{bmatrix} k_1 + k_2 & -k_2 \\ -k_2 & k_2 \end{bmatrix} \begin{Bmatrix} x_1 \\ x_2 \end{Bmatrix} = \omega^2 \begin{bmatrix} m_1 & 0 \\ 0 & m_2 \end{bmatrix} \begin{Bmatrix} x_1 \\ x_2 \end{Bmatrix}, \tag{a}$$

where ω^2 is the eigenvalue and $\vec{X} = \begin{Bmatrix} x_1 \\ x_2 \end{Bmatrix}$ is the eigenvector or mode shape (displacement pattern) of the system. Determine the natural frequencies and mode shapes of the system for the following data:

$$m_1 = 20,000 \text{ kg}, \; m_2 = 5,000 \text{ kg}, \; k_1 = 10^7 \text{ N/m}, \; \text{and } k_2 = 5 \times 10^6 \text{ N/m}.$$

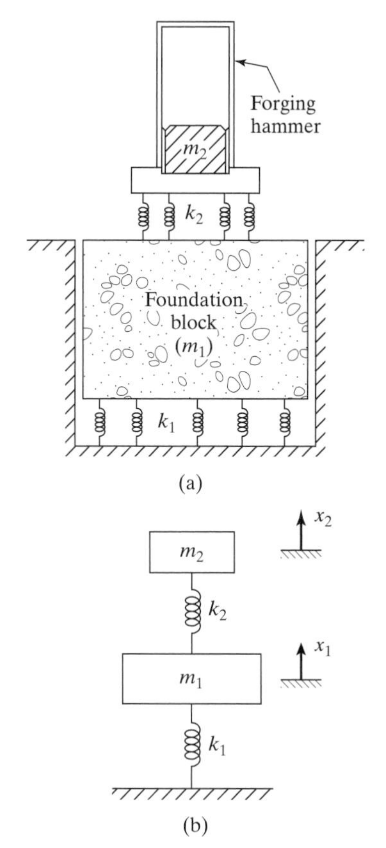

Figure 4.1 Forging hammer.

Solution

The natural frequencies are given by the solution of Eq. (a):

$$\begin{bmatrix} 15 \times 10^6 & -5 \times 10^6 \\ -5 \times 10^6 & 5 \times 10^6 \end{bmatrix} \begin{Bmatrix} x_1 \\ x_2 \end{Bmatrix} = \omega^2 \begin{bmatrix} 20,000 & 0 \\ 0 & 5000 \end{bmatrix} \begin{Bmatrix} x_1 \\ x_2 \end{Bmatrix}. \tag{b}$$

The solution of these equations yields

$$\omega_1 = 18.9634 \text{ rad/sec}, \ \vec{X}^{(1)} = \begin{Bmatrix} x_1 \\ x_2 \end{Bmatrix}^{(1)} = \begin{Bmatrix} 1.0 \\ 1.5615 \end{Bmatrix}$$

and

$$\omega_2 = 37.2879 \text{ rad/sec}, \ \vec{X}^{(2)} = \begin{Bmatrix} x_1 \\ x_2 \end{Bmatrix}^{(2)} = \begin{Bmatrix} 1.0 \\ -2.5615 \end{Bmatrix}.$$

(See Problem 4.32.) ◀

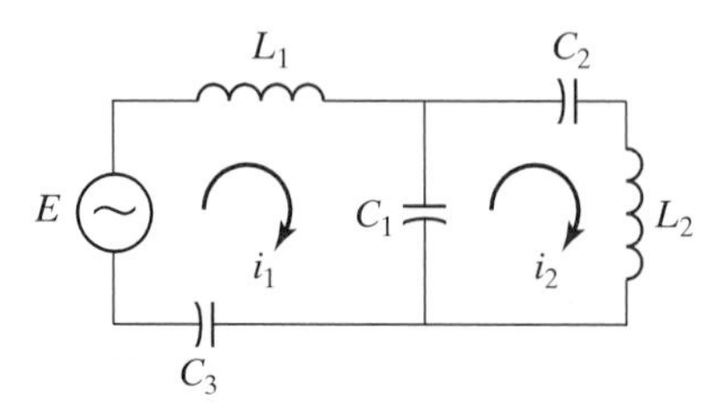

Figure 4.2 Electrical circuit.

▶Example 4.3

Consider the electrical circuit shown in Fig. 4.2 where L_i denotes inductances, C_i represent the capacitances, i_i indicates the currents, and E is the applied voltage. The voltage drops across an inductance L and a capacitance C are given, respectively, by

$$v_L = L\frac{di}{dt}, \quad v_C = \frac{1}{C}\int i\,dt.$$

By considering the voltage drop across the loops of Fig. 4.2 by using Kirchhoff's law, derive the corresponding eigenvalue problem.

Solution

The application of Kirchhoff's law gives the following:

$$\text{Loop 1:}\ E - L_1\frac{di_1}{dt} - \frac{1}{C_1}\int (i_1 - i_2)dt - \frac{1}{C_3}\int i_1\,dt = 0; \qquad \text{(a)}$$

$$\text{Loop 2:}\ \frac{1}{C_1}\int (i_1 - i_2)dt - \frac{1}{C_2}\int i_2 dt - L_2\frac{di_2}{dt} = 0. \qquad \text{(b)}$$

Differentiating Eqs. (a) and (b) with respect to time and rearranging the terms yields

$$L_1\frac{d^2i_1}{dt^2} + i_1\left(\frac{1}{C_1} + \frac{1}{C_3}\right) - i_2\left(\frac{1}{C_1}\right) = \frac{dE}{dt}, \qquad \text{(c)}$$

and

$$L_2\frac{d^2i_2}{dt^2} - i_1\left(\frac{1}{C_1}\right) + i_2\left(\frac{1}{C_1} + \frac{1}{C_2}\right) = 0. \qquad \text{(d)}$$

When $\frac{dE}{dt} = 0$, Eqs. (c) and (d) can be stated in matrix form as

$$\begin{bmatrix} L_1 & 0 \\ 0 & L_2 \end{bmatrix}\begin{Bmatrix} \dfrac{d^2i_1}{dt^2} \\ \dfrac{d^2i_2}{dt^2} \end{Bmatrix} + \begin{bmatrix} \dfrac{1}{C_1} + \dfrac{1}{C_3} & -\dfrac{1}{C_1} \\ -\dfrac{1}{C_1} & \dfrac{1}{C_1} + \dfrac{1}{C_2} \end{bmatrix}\begin{Bmatrix} i_1 \\ i_2 \end{Bmatrix} = \begin{Bmatrix} 0 \\ 0 \end{Bmatrix}. \qquad \text{(e)}$$

By assuming a sinusoidal variation of current in each loop as

$$i_j = I_j \sin(\omega t + \phi), \quad j = 1, 2, \qquad \text{(f)}$$

where I_1 and I_2 are the amplitudes of currents i_1 and i_2, respectively, ω is the natural frequency of vibration, ϕ is the phase angle, and t is time, Eq. (e) can be expressed as an eigenvalue problem:

$$\omega^2 \begin{bmatrix} L_1 & 0 \\ 0 & L_2 \end{bmatrix} \begin{Bmatrix} I_1 \\ I_2 \end{Bmatrix} = \begin{bmatrix} \frac{1}{C_1} + \frac{1}{C_3} & -\frac{1}{C_1} \\ -\frac{1}{C_1} & \frac{1}{C_1} + \frac{1}{C_2} \end{bmatrix} \begin{Bmatrix} I_1 \\ I_2 \end{Bmatrix}. \quad \text{(g)}$$

◀

▶Example 4.4

The pin-ended column shown in Fig. 4.3(a) is subjected to a compressive (axial) force, P. If the column is perturbed slightly as shown in Fig. 4.3(b), as might occur from a slight vibration of the supports, it may not return to the horizontal position even after the removal of the disturbance; instead, the deflection might grow if the load is sufficiently large. Such a load is called the buckling load. The equation governing the transverse deflection of the column is given by

$$\frac{d^2y}{dx^2} + \frac{P}{EI}y = 0, \quad \text{(a)}$$

where E is Young's modulus and I is the moment of inertia of the cross section of the column about the z-axis. Determine the solution of Eq. (a) and the buckling load of the column.

Solution

The solution of Eq. (a) is given by (can be verified by substitution)

$$y(x) = c_1 \sin \lambda x + c_2 \cos \lambda x, \quad \text{(b)}$$

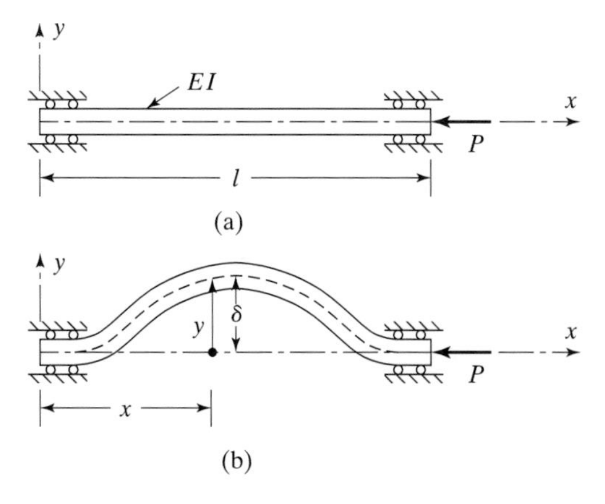

Figure 4.3 Pin-ended column.

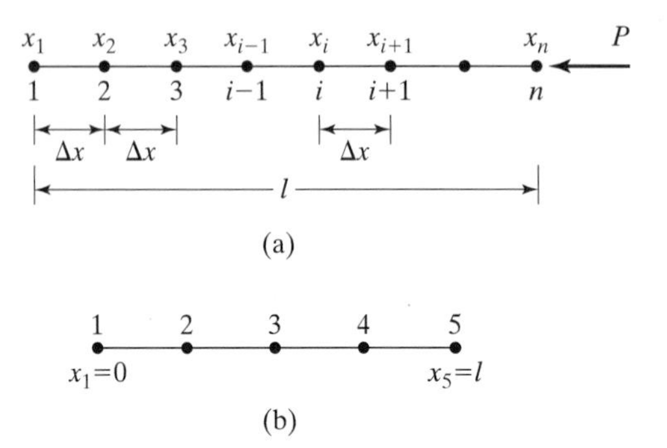

Figure 4.4 Discretization of column.

where

$$\lambda = \sqrt{\frac{P}{EI}} \tag{c}$$

and c_1 and c_2 are constants to be determined from the boundary conditions

$$y(x = 0) = 0 \text{ and } y(x = l) = 0, \tag{d}$$

which yield

$$c_2 = 0; \tag{e}$$

$$c_1 \sin \lambda l = 0. \tag{f}$$

For a nontrivial solution, Eq. (f) gives

$$\lambda l = n\pi, \quad n = 1, 2, \ldots, \tag{g}$$

or

$$P = P_{\text{cri}} = \left(\frac{n\pi}{l}\right)^2 EI, \quad n = 1, 2, \ldots. \tag{h}$$

where P_{cri} denotes the critical load. The smallest value of P satisfying Eq. (h), known as Euler's buckling load, is given by

$$P_{\text{cri}} = \frac{\pi^2 EI}{l^2}, \tag{i}$$

and the corresponding deflection shape is given by

$$y(x) = c_1 \sin \frac{\pi x}{l}. \tag{j}$$

Note: This problem represents an eigenvalue problem arising from the solution of a differential equation (not an algebraic eigenvalue problem). The solutions given by Eq. (h) are known as the eigenvalues of the problem. The problem can be converted into an algebraic eigenvalue problem by approximating the derivative in Eq. (a) by a finite difference formula as illustrated in the following example. ◀

▶Example 4.5

Divide the column of Fig. 4.3 into several segments as shown in Fig. 4.4 and evaluate Eq. (a) of Example 4.4 at each discrete station using the central difference formula

$$\frac{d^2y}{dx^2} \approx \frac{y_{i+1} - 2y_i + y_{i-1}}{(\Delta x)^2}, \tag{a}$$

where $y_i = y(x = x_i), i = 1, 2, \ldots, n$, and $\Delta x = \frac{1}{n-1}$. Derive the associated eigenvalue problem for $n = 5$.

Solution

The evaluation of Eq. (a) of Example 4.4 at station i yields

$$\frac{y_{i+1} - 2y_i + y_{i-1}}{(\Delta x)^2} + \lambda y_i = 0 \text{ or } y_{i-1} - (2 - \lambda(\Delta x)^2)y_i + y_{i+1} = 0, \tag{b}$$

where $\lambda = (P/EI)$. Using $n = 5$ and $\Delta x = \frac{1}{4}$, and applying Eq. (b) at stations $i = 2, 3$, and 4 yields

$$y_1 - \left(2 - \frac{\lambda l^2}{16}\right) y_2 + y_3 = 0,$$

$$y_2 - \left(2 - \frac{\lambda l^2}{16}\right) y_3 + y_4 = 0, \tag{c}$$

and

$$y_3 - \left(2 - \frac{\lambda l^2}{16}\right) y_4 + y_5 = 0.$$

Using the boundary conditions $y_1 = y_5 = 0$, Eqs. (c) can be stated in matrix form as

$$\begin{bmatrix} \left(2 - \frac{\lambda l^2}{16}\right) & 1 & 0 \\ 1 & \left(2 - \frac{\lambda l^2}{16}\right) & 1 \\ 0 & 1 & \left(2 - \frac{\lambda l^2}{16}\right) \end{bmatrix} \begin{Bmatrix} y_2 \\ y_3 \\ y_4 \end{Bmatrix} = \begin{Bmatrix} 0 \\ 0 \\ 0 \end{Bmatrix}. \tag{d}$$

Equation (d) represents an algebraic eigenvalue problem whose solution gives different values of the critical load. ◀

4.3 Conversion of General Eigenvalue Problem to Standard Form

The following procedures can be used to convert a general eigenvalue problem to an equivalent standard eigenvalue problem.

4.3.1 When the Matrix $[B]$ is Diagonal

When $[B]$ is diagonal with diagonal elements equal to $b_{ii} > 0$, it can be expressed as

$$[B] = [C]^T[C] = [C][C], \tag{4.16}$$

where $[C]$ is also a diagonal matrix. The diagonal elements of the matrix $[C]$, c_{ii}, can be expressed as

$$c_{ii} = \sqrt{b_{ii}}. \tag{4.17}$$

Substituting Eq. (4.16) into Eq. (4.15) yields

$$[A]\vec{X} = \lambda[C]\,[C]\vec{X}, \tag{4.18}$$

or

$$[A]\,[C]^{-1}\,[C]\vec{X} = \lambda[C]\,[C]\vec{X}. \tag{4.19}$$

Premultiplying Eq. (4.19) by $[C]^{-1}$, we obtain

$$[C]^{-1}[A]\,[C]^{-1}\,[C]\vec{X} = \lambda[C]\vec{X}, \tag{4.20}$$

or

$$[P]\vec{Y} = \lambda\vec{Y}, \tag{4.21}$$

where

$$[P] = [C]^{-1}[A]\,[C]^{-1} \tag{4.22}$$

and

$$\vec{Y} = [C]\vec{X}. \tag{4.23}$$

Equation (4.21) represents a standard eigenvalue problem and its solution gives the eigenvalues of the original problem, Eq. (4.15). The eigenvectors of the original problem, Eq. (4.15), however, are given by

$$\vec{X} = [C]^{-1}\vec{Y}. \tag{4.24}$$

▶Example 4.6

Convert the eigenvalue problem

$$\begin{bmatrix} 15000 & -5000 \\ -5000 & 5000 \end{bmatrix} \begin{Bmatrix} x_1 \\ x_2 \end{Bmatrix} = \lambda \begin{bmatrix} 20 & 0 \\ 0 & 5 \end{bmatrix} \begin{Bmatrix} x_1 \\ x_2 \end{Bmatrix} \tag{a}$$

to a standard eigenvalue problem. (See Eq. (b) of Example 4.3.)

Solution

The eigenvalue problem of Eq. (a) can be identified to be of the form of Eq. (4.15) with

$$[A] = \begin{bmatrix} 15000 & -5000 \\ -5000 & 5000 \end{bmatrix}, \tag{b}$$

and

$$[B] = \begin{bmatrix} 20 & 0 \\ 0 & 5 \end{bmatrix}. \tag{c}$$

Since the matrix $[B]$ is diagonal with $b_{ii} > 0 (i = 1, 2)$, it can be expressed as

$$[B] = [C]\,[C], \tag{d}$$

where

$$[C] = \begin{bmatrix} \sqrt{20} & 0 \\ 0 & \sqrt{5} \end{bmatrix} = \begin{bmatrix} 4.4721 & 0 \\ 0 & 2.2361 \end{bmatrix}. \tag{e}$$

Equation (e) gives

$$[C]^{-1} = \begin{bmatrix} \dfrac{1}{4.4721} & 0 \\ 0 & \dfrac{1}{2.2361} \end{bmatrix} = \begin{bmatrix} 0.2236 & 0 \\ 0 & 0.4472 \end{bmatrix}. \tag{f}$$

The matrix $[P]$ of Eq. (4.22) is given by

$$[P] = [C]^{-1}\,[A]\,[C]^{-1} = \begin{bmatrix} 0.2236 & 0 \\ 0 & 0.4472 \end{bmatrix} \begin{bmatrix} 15000 & -5000 \\ -5000 & 5000 \end{bmatrix} \begin{bmatrix} 0.2236 & 0 \\ 0 & 0.4472 \end{bmatrix}$$

$$= \begin{bmatrix} 749.9544 & -499.9696 \\ -499.9696 & 999.9392 \end{bmatrix}. \tag{g}$$

Thus, the equivalent standard eigenvalue problem can be stated as

$$[P]\vec{Y} = \lambda \vec{Y}, \tag{h}$$

where

$$\vec{Y} = [C]\vec{X} = \begin{bmatrix} 4.4721 & 0 \\ 0 & 2.2361 \end{bmatrix} \begin{Bmatrix} x_1 \\ x_2 \end{Bmatrix} = \begin{Bmatrix} 4.4721 & x_1 \\ 2.2361 & x_2 \end{Bmatrix}. \tag{i}$$

◀

4.3.2 When the Matrix $[B]$ is Symmetric and Positive Definite

If $[B]$ is symmetric and positive definite, the general eigenvalue problem of Eq. (4.15) can be reduced to the standard form by multiplying throughout by $[B]^{-1}$—that is,

$$[B]^{-1}[A]\vec{X} = \lambda[B]^{-1}[B]\vec{X},$$

or

$$[D]\vec{X} = \lambda\vec{X}, \tag{4.25}$$

where

$$[D] = [B]^{-1}[A]. \tag{4.26}$$

This procedure is not recommended because of the following reasons:

1. The matrix $[D]$, given by Eq. (4.26), will be nonsymmetric, although $[A]$ and $[B]$ are symmetric. It is desirable to maintain $[D]$ as a symmetric matrix to simplify the solution procedure.
2. The inversion of the matrix $[B]$ is computationally expensive, especially when the order of $[B]$ is large, compared with the procedure described next.

The common way to reduce the general eigenvalue problem to the special form is to decompose the matrix $[B]$ by using Choleski decomposition as

$$[B] = [U]^T[U], \tag{4.27}$$

where $[U]$ is an upper triangular matrix. By using Eq. (4.27), the problem of Eq. (4.15) can be stated as

$$[A]\vec{X} = \lambda[U]^T[U]\vec{X}. \tag{4.28}$$

Premultiplying this equation by $([U]^T)^{-1}$, we obtain

$$([U]^T)^{-1}[A]\vec{X} = \lambda([U]^T)^{-1}[U]^T[U]\vec{X} = \lambda[U]\vec{X}. \tag{4.29}$$

By defining a new vector $\vec{Y}$ as

$$\vec{Y} = [U]\vec{X}, \tag{4.30}$$

Eq. (4.29) can be expressed as a standard eigenvalue problem

$$[D]\vec{Y} = \lambda\vec{Y}, \tag{4.31}$$

where

$$[D] = ([U]^T)^{-1}[A][U]^{-1}. \tag{4.32}$$

Thus, to derive the matrix $[D]$, we first decompose the symmetric and positive definite matrix $[B]$ as indicated in Eq. (4.27), find $[U]^{-1}$ and $([U]^T)^{-1} = ([U]^{-1})^T$, and then carry out the matrix multiplications as indicated in Eq. (4.32). The solution of the standard eigenvalue problem of Eq. (4.31) yields the eigenvalues λ_i and the corresponding eigenvectors $\vec{Y}^{(i)}$. Once $\vec{Y}^{(i)}$ are known, the eigenvectors of the original problem can be determined, using Eq. (4.30), as

$$\vec{X}^{(i)} = [U]^{-1}\vec{Y}^{(i)}. \tag{4.33}$$

Note that the determination of the upper triangular matrix $[U]$ and its inverse $[U]^{-1}$ are relatively simple as indicated in Section 3.11.

▶Example 4.7

Convert the following general eigenvalue problem into the standard form using Eq. (4.25) and (4.26):

$$\lambda \begin{bmatrix} 4 & -1 & 1 \\ -1 & 6 & -4 \\ 1 & -4 & 5 \end{bmatrix} \begin{Bmatrix} x_1 \\ x_2 \\ x_3 \end{Bmatrix} = \begin{bmatrix} 300 & -200 & 0 \\ -200 & 500 & -300 \\ 0 & -300 & 300 \end{bmatrix} \begin{Bmatrix} x_1 \\ x_2 \\ x_3 \end{Bmatrix}. \tag{a}$$

Solution

By comparing Eq. (a) with Eq. (4.15), we can identify the matrices $[A]$ and $[B]$ as

$$[A] = \begin{bmatrix} 300 & -200 & 0 \\ -200 & 500 & -300 \\ 0 & -300 & 300 \end{bmatrix}, \tag{b}$$

and

$$[B] = \begin{bmatrix} 4 & -1 & 1 \\ -1 & 6 & -4 \\ 1 & -4 & 5 \end{bmatrix}. \tag{c}$$

The inverse of $[B]$ is given by (see Example 3.17)

$$[B]^{-1} = \begin{bmatrix} 0.2641 & 0.0188 & -0.0377 \\ 0.0188 & 0.3585 & 0.2831 \\ -0.0377 & 0.2831 & 0.4340 \end{bmatrix}, \tag{d}$$

and the matrix $[D]$ can be determined as

$$[D] = [B]^{-1}[A] = \begin{bmatrix} 75.47 & -32.11 & -16.95 \\ -66.06 & 90.56 & -22.62 \\ -67.93 & 18.89 & 45.27 \end{bmatrix}. \tag{e}$$

Thus, the standard eigenvalue problem can be stated as (Eq. (4.25)):

$$\begin{bmatrix} 75.47 & -32.11 & -16.95 \\ -66.06 & 90.56 & -22.62 \\ -67.93 & 18.89 & 45.27 \end{bmatrix} \begin{Bmatrix} x_1 \\ x_2 \\ x_3 \end{Bmatrix} = \lambda \begin{Bmatrix} x_1 \\ x_2 \\ x_3 \end{Bmatrix}. \tag{f}$$

Note that the matrix $[D]$ is nonsymmetric, although $[A]$ and $[B]$ are symmetric. ◀

▶Example 4.8

Convert the general eigenvalue problem given in Eq. (a) of Example 4.7 into the standard form using Eqs. (4.31) and (4.32).

Solution

The matrix $[B]$, given by Eq. (c) of Example 4.7 can be decomposed using Choleski decomposition as

$$[A] = [U]^T[U], \tag{a}$$

where the upper triangular matrix $[U]$ is given by

$$[U] = \begin{bmatrix} 2 & -0.5 & 0.5 \\ 0 & 2.3979 & -1.5639 \\ 0 & 0 & 1.5180 \end{bmatrix}. \tag{b}$$

(See Example 3.16.) The inverse of $[U]$ can be found as

$$[U]^{-1} = \begin{bmatrix} 0.5 & 0.1042 & -0.0573 \\ 0 & 0.4170 & 0.4297 \\ 0 & 0 & 0.6588 \end{bmatrix}. \tag{c}$$

(See Example 3.17.) Noting that $[[U]^T]^{-1} = [[U]^{-1}]^T$, we obtain the matrix

$$[D] = ([U]^T)^{-1}[A][U]^{-1} = \begin{bmatrix} 75.0000 & -26.0700 & -51.5650 \\ -26.0700 & 72.8212 & 1.2093 \\ -51.5650 & 1.2093 & 63.5082 \end{bmatrix}. \tag{d}$$

Thus, the standard eigenvalue problem corresponding to Eq. (a) of Example 4.7 can be stated as

$$\begin{bmatrix} 75.0000 & -26.0700 & -51.5650 \\ -26.0700 & 72.8212 & 1.2093 \\ -51.5650 & 1.2093 & 63.5082 \end{bmatrix} \begin{Bmatrix} y_1 \\ y_2 \\ y_3 \end{Bmatrix} = \lambda \begin{Bmatrix} y_1 \\ y_2 \\ y_3 \end{Bmatrix}, \tag{e}$$

where

$$\begin{Bmatrix} y_1 \\ y_2 \\ y_3 \end{Bmatrix} = [U]\vec{X} = \begin{Bmatrix} 2\,x_1 - 0.5\,x_2 + 0.5\,x_3 \\ 2.3979\,x_2 - 1.5639\,x_3 \\ 1.5180\,x_3 \end{Bmatrix}. \tag{f}$$

Note that the matrix $[D]$ is symmetric in this case. ◀

4.4 Methods of Solving Eigenvalue Problems

Several methods are available for solving an eigenvalue problem. The methods can be classified broadly into three categories:

1. Transformation methods
2. Iterative methods
3. Determinant search methods

The transformation methods find *all* the eigenvalues and eigenvectors simultaneously. In general, these methods are useful when the order of the matrices involved is small and when the matrices are fully populated. These methods are not suitable when the order of the matrices is large and only a small number of eigenvalues are desired. Most of the transformation methods are applicable only for the solution of standard eigenvalue problems. The basic idea used in the transformation methods is that under certain transformations, such as a similarity transformation, the eigenvalues of a matrix remain unchanged. Some of the commonly used transformation methods include the Jacobi diagonalization, Givens tridiagonalization, Householder's transformation, LR transformation, and the QR transformation methods. The Jacobh diagonalization method can find the eigenvalues and eigenvectors of symmetric matrices. The Givens and Householder's methods first reduce a symmetric matrix to a tridiagonal matrix before finding the eigenvalues and eigenvectors.

The iterative methods are useful for finding a small number of eigenvalues of large matrices that are sparse or banded. Some of the commonly used iterative methods include the power method (or direct vector iteration method), the vector iteration method with a shift, the subspace iteration method, and the Lanczos method.

The determinant search methods use an iterative procedure to locate as many zeros of the characteristic polynomial (or determinantal equation) as desired. Any of the iterative methods of solving polynomial equations described in Chapter 2, such as the secant method, can be used for finding the zeros. The Sturm sequence property can be used to avoid jumping over two or more roots.

4.5 Solution of the Characteristic Polynomial Equation

The eigenvalues of the problem

$$[A]\vec{X} = \lambda[I]\vec{X} = \lambda\vec{X} \tag{4.34}$$

are given by the roots of the determinantal equation

$$\big|[A] - \lambda[I]\big| = 0, \tag{4.35}$$

or

$$\begin{vmatrix} (a_{11}-\lambda) & a_{12} & \cdots & a_{1n} \\ a_{21} & (a_{22}-\lambda) & \cdots & a_{2n} \\ \cdot & \cdot & \cdots & \cdot \\ \cdot & \cdot & \cdots & \cdot \\ \cdot & \cdot & \cdots & \cdot \\ a_{n1} & a_{n2} & \cdots & (a_{nn}-\lambda) \end{vmatrix} = 0. \tag{4.36}$$

Equation (4.36), when expanded, gives an nth order polynomial in λ, called the characteristic polynomial,

$$(-1)^n[\lambda^n - p_1\lambda^{n-1} - p_2\lambda^{n-2} - \cdots - p_{n-1}\lambda - p_n] = 0, \tag{4.37}$$

where

$$p_1 = a_{11} + a_{22} + \cdots + a_{nn} \tag{4.38}$$

and other coefficients of the polynomial $p_i (i = 2, 3, \ldots, n)$ are given by the sum, with sign $(-1)^{i-1}$, of all the principal minors of order i of the matrix $[A]$. The solution of Eq. (4.37) gives, in general, n roots, $\lambda_1, \lambda_2, \ldots, \lambda_n$. Each λ_i has a corresponding eigenvector $\vec{X}^{(i)}$ that satisfies Eq. (4.34):

$$[A]\vec{X}^{(i)} = \lambda_i[I]\vec{X}^{(i)}. \tag{4.39}$$

The n components of the eigenvector $\vec{X}^{(i)}$ can be found by solving the n homogeneous equations represented by Eq. (4.39). For eigenvalue problems involving larger order matrices $[A]$, it is tedius to derive the characteristic polynomial, find the roots of the polynomial (λ_i), and solve systems of homogeneous equations for $\vec{X}^{(i)}$.

▶Example 4.9

By expanding the determinantal equation (4.36), generate the characteristic equation corresponding to the following eigenvalue problem:

$$\begin{bmatrix} 12 & 6 & -6 \\ 6 & 16 & 2 \\ -6 & 2 & 16 \end{bmatrix} \begin{Bmatrix} x_1 \\ x_2 \\ x_3 \end{Bmatrix} = \lambda \begin{Bmatrix} x_1 \\ x_2 \\ x_3 \end{Bmatrix}. \tag{a}$$

Solution

The determinantal equation corresponding to the eigenvalue problem of Eq. (a) is given by

$$\begin{vmatrix} (12-\lambda) & 6 & -6 \\ 6 & (16-\lambda) & 2 \\ -6 & 2 & (16-\lambda) \end{vmatrix} = 0. \tag{b}$$

The determinant of Eq. (b) can be expanded as

$$(12-\lambda)\begin{vmatrix} (16-\lambda) & 2 \\ 2 & (16-\lambda) \end{vmatrix} - 6\begin{vmatrix} 6 & 2 \\ -6 & (16-\lambda) \end{vmatrix} - 6\begin{vmatrix} 6 & (16-\lambda) \\ -6 & 2 \end{vmatrix} = 0, \tag{c}$$

which can be expressed as

$$(12-\lambda)\{(16-\lambda)^2-(2)^2\}-6\{(6)(16-\lambda)-(2)(-6)\}-6\{(6)(2)-(-6)(16-\lambda)\} = 0. \tag{d}$$

The simplification of Eq. (d) gives the desired characteristic equation:

$$-\lambda^3 + 44\lambda^2 - 564\lambda + 1728 = 0. \tag{e}$$

◀

4.5.1 Faddeev–Leverrier Method

The polynomial coefficients $p_1, p_2, \ldots, p_n$ indicated in Eq. (4.37) can be generated more conveniently using the Faddeev–Leverrier method [4.1]. In this method, a sequence of matrices $[P_i]$ are defined and the polynomial coefficients p_i are generated as follows:

$$[P_1] = [A],\ p_1 = \text{trace}\ [P_1]; \tag{4.40}$$

$$[P_2] = [A]([P_1] - p_1[I]),\ p_2 = \frac{1}{2}\ \text{trace}\ [P_2]; \tag{4.41}$$

$$[P_3] = [A]([P_2] - p_2[I]),\ p_3 = \frac{1}{3}\ \text{trace}\ [P_3]; \tag{4.42}$$

$$\vdots$$

$$[P_i] = [A]([P_{i-1}] - p_{i-1}[I]),\ p_i = \frac{1}{i}\ \text{trace}\ [P_i]; \tag{4.43}$$

$$\vdots$$

$$[P_n] = [A]([P_{n-1}] - p_{n-1}[I]),\ p_n = \frac{1}{n}\ \text{trace}\ [P_n]. \tag{4.44}$$

The method is illustrated with the following example.

▶Example 4.10

By using the Faddeev–Leverrier method, generate the characteristic polynomial equation corresponding to the following eigenvalue problem:

$$\begin{bmatrix} 12 & 6 & -6 \\ 6 & 16 & 2 \\ -6 & 2 & 16 \end{bmatrix} \begin{Bmatrix} x_1 \\ x_2 \\ x_3 \end{Bmatrix} = \lambda \begin{Bmatrix} x_1 \\ x_2 \\ x_3 \end{Bmatrix}.$$

Solution

Using Eqs. (4.40) through (4.44), we obtain

$$[P_1] = [A] = \begin{bmatrix} 12 & 6 & -6 \\ 6 & 16 & 2 \\ -6 & 2 & 16 \end{bmatrix},\ p_1 = \text{trace}\ [P_1] = 12 + 16 + 16 = 44;$$

$$[P_2] = [A]([P_1] - p_1[I]) = \begin{bmatrix} 12 & 6 & -6 \\ 6 & 16 & 2 \\ -6 & 2 & 16 \end{bmatrix} \left(\begin{bmatrix} 12 & 6 & -6 \\ 6 & 16 & 2 \\ -6 & 2 & 16 \end{bmatrix} - 44 \begin{bmatrix} 1 & 0 & 0 \\ 0 & 1 & 0 \\ 0 & 0 & 1 \end{bmatrix} \right)$$

$$= \begin{bmatrix} -312 & -108 & 108 \\ -108 & -408 & -60 \\ 108 & -60 & -408 \end{bmatrix},\ p_2 = \frac{1}{2}\ \text{trace}\ [P_2] = \frac{1}{2}(-312 - 408 - 408) = -564;$$

$$[P_3] = [A]([P_2] - p_2[I])$$

$$= \begin{bmatrix} 12 & 6 & -6 \\ 6 & 16 & 2 \\ -6 & 2 & 16 \end{bmatrix} \left(\begin{bmatrix} -312 & -108 & 108 \\ -108 & -408 & -60 \\ 108 & -60 & -408 \end{bmatrix} + 564 \begin{bmatrix} 1 & 0 & 0 \\ 0 & 1 & 0 \\ 0 & 0 & 1 \end{bmatrix} \right)$$

$$= \begin{bmatrix} 1728 & 0 & 0 \\ 0 & 1728 & 0 \\ 0 & 0 & 1728 \end{bmatrix}, \ p_3 = \frac{1}{3}\ \text{trace}\ [P_3] = \frac{1}{3}(1728 + 1728 + 1728) = 1728.$$

Thus, the characteristic polynomial equation is given by

$$(-1)^3[\lambda^3 - 44\,\lambda^2 + 564\,\lambda - 1728] = 0,$$

or

$$-\lambda^3 + 44\,\lambda^2 - 564\,\lambda + 1728 = 0.$$

◀

▶Example 4.11

Find the eigenvalues of the problem considered in Example 4.9 by finding the roots of the polynomial equation. Also, determine the eigenvectors of the problem.

Solution

The characteristic polynomial equation is given by

$$\lambda^3 - 44\,\lambda^2 + 564\,\lambda - 1728 = 0. \tag{a}$$

The roots of this cubic equation can be found as

$$\lambda_1 = 4.4560, \lambda_2 = 18.0000, \ \text{and}\ \lambda_3 = 21.5440. \tag{b}$$

(The methods of solving nonlinear equations considered in Chapter 2 can be used to find these roots).

First Eigenvector

The eigenvector corresponding to $\lambda_1 = 4.4560$ is defined by

$$\begin{bmatrix} 12 & 6 & -6 \\ 6 & 16 & 2 \\ -6 & 2 & 16 \end{bmatrix} \begin{Bmatrix} x_1 \\ x_2 \\ x_3 \end{Bmatrix} = 4.456 \begin{Bmatrix} x_1 \\ x_2 \\ x_3 \end{Bmatrix},$$

or

$$\begin{bmatrix} 7.544 & 6 & -6 \\ 6 & 11.544 & 2 \\ -6 & 2 & 11.544 \end{bmatrix} \begin{Bmatrix} x_1 \\ x_2 \\ x_3 \end{Bmatrix} = \begin{Bmatrix} 0 \\ 0 \\ 0 \end{Bmatrix}. \tag{c}$$

The first two formulas of Eq. (c) can be expressed as

$$7.544\,x_1 + 6\,x_2 - 6\,x_3 = 0 \tag{d}$$

and

$$6x_1 + 11.544x_2 + 2x_3 = 0. \tag{e}$$

Equations (d) and (e) can be solved to express x_2 and x_3 in terms of x_1;

$$x_2 = -0.6287x_1; \quad x_3 = 0.6288x_1. \tag{f}$$

Thus, the first eigenvector is given by

$$\vec{X}^{(1)} = \begin{Bmatrix} x_1 \\ x_2 \\ x_3 \end{Bmatrix}^{(1)} = x_1^{(1)} \begin{Bmatrix} 1.0 \\ -0.6287 \\ 0.6288 \end{Bmatrix}. \tag{g}$$

Second eigenvector

The eigenvector corresponding to $\lambda_2 = 18.0$ is given by

$$\begin{bmatrix} 12 & 6 & -6 \\ 6 & 16 & 2 \\ -6 & 2 & 16 \end{bmatrix} \begin{Bmatrix} x_1 \\ x_2 \\ x_3 \end{Bmatrix} = 18 \begin{Bmatrix} x_1 \\ x_2 \\ x_3 \end{Bmatrix},$$

or

$$\begin{bmatrix} -6 & 6 & -6 \\ 6 & -2 & 2 \\ -6 & 2 & -2 \end{bmatrix} \begin{Bmatrix} x_1 \\ x_2 \\ x_3 \end{Bmatrix} = \begin{Bmatrix} 0 \\ 0 \\ 0 \end{Bmatrix}. \tag{h}$$

The first two formulas of Eq. (h) can be written as

$$-6x_1 + 6x_2 - 6x_3 = 0 \tag{i}$$

and

$$6x_1 - 2x_2 + 2x_3 = 0. \tag{j}$$

Equations (i) and (j) can be solved to obtain

$$x_1 = 0, \text{ and } x_2 = x_3. \tag{k}$$

Thus, the second eigenvector is given by

$$\vec{X}^{(2)} = \begin{Bmatrix} x_1 \\ x_2 \\ x_3 \end{Bmatrix}^{(2)} = x_2^{(2)} \begin{Bmatrix} 0 \\ 1 \\ 1 \end{Bmatrix}. \tag{l}$$

Third eigenvector

The eigenvector corresponding to $\lambda_3 = 21.544$ is given by

$$\begin{bmatrix} 12 & 6 & -6 \\ 6 & 16 & 2 \\ -6 & 2 & 16 \end{bmatrix} \begin{Bmatrix} x_1 \\ x_2 \\ x_3 \end{Bmatrix} = 21.544 \begin{Bmatrix} x_1 \\ x_2 \\ x_3 \end{Bmatrix},$$

or

$$\begin{bmatrix} -9.544 & 6 & -6 \\ 6 & -5.544 & 2 \\ -6 & 2 & -5.544 \end{bmatrix} \begin{Bmatrix} x_1 \\ x_2 \\ x_3 \end{Bmatrix} = \begin{Bmatrix} 0 \\ 0 \\ 0 \end{Bmatrix}. \tag{m}$$

The first two formulas of Eq. (m) can be written as

$$6\,x_1 - 5.544\,x_2 + 2\,x_3 = 0 \tag{n}$$

and

$$-6\,x_1 + 2\,x_2 - 5.544\,x_3 = 0. \tag{o}$$

Equations (n) and (o) can be solved to obtain

$$x_1 = 1.2573\,x_3, \quad \text{and} \quad x_2 = -x_3. \tag{p}$$

Thus, the third eigenvector is given by

$$\vec{X}^{(3)} = \begin{Bmatrix} x_1 \\ x_2 \\ x_3 \end{Bmatrix}^{(3)} = x_3^{(3)} \begin{Bmatrix} 1.2573 \\ -1.0 \\ 1.0 \end{Bmatrix}. \tag{q}$$

Note that, in Eqs. (g), (l), and (q), the components x_1, x_2 and x_3 were chosen as reference values for simplicity. ◀

4.6 Jacobi Method

4.6.1 Transformation of Coordinates

The Jacobi method uses the concept of coordinate transformation. For this, consider two coordinate systems (x_1, x_2) and $(\tilde{x}_1, \tilde{x}_2)$ as indicated in Fig. 4.5. The coordinates of a typical point Q in the two-dimensional space can be expressed in either coordinate system. The two sets of coordinates of Q are related as follows:

$$x_1 = \tilde{x}_1 \cos\theta - \tilde{x}_2 \sin\theta \tag{4.45}$$

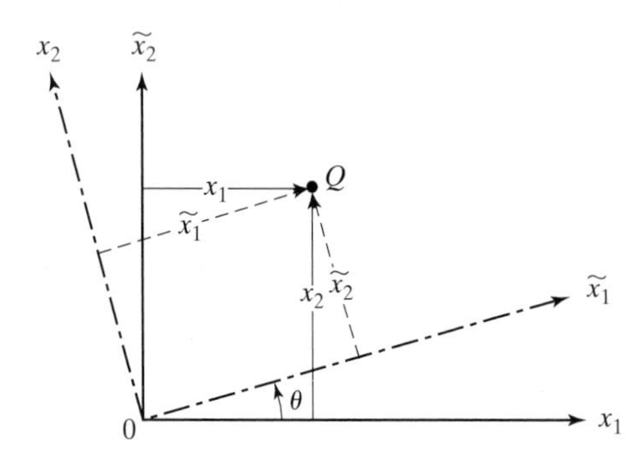

Figure 4.5 Transformation of coordinates.

and

$$x_2 = \tilde{x}_1 \sin\theta + \tilde{x}_2 \cos\theta. \tag{4.46}$$

These equations can be expressed in matrix form as

$$\vec{x} = [P]\vec{\tilde{x}}, \tag{4.47}$$

where

$$\vec{x} = \begin{Bmatrix} x_1 \\ x_2 \end{Bmatrix}, \tag{4.48}$$

$$\vec{\tilde{x}} = \begin{Bmatrix} \tilde{x}_1 \\ \tilde{x}_2 \end{Bmatrix}, \tag{4.49}$$

and

$$[P] = \begin{bmatrix} \cos\theta & -\sin\theta \\ \sin\theta & \cos\theta \end{bmatrix}. \tag{4.50}$$

The coordinate transformation matrix, $[P]$, can be verified to be an orthogonal matrix

$$[P]^T[P] = [I]. \tag{4.51}$$

(See Problem 4.33.) An important property of any orthogonal matrix, such as $[P]$, is that the eigenvalues of the matrix, $[P][A][P]^T$, will be same as those of the matrix $[A]$. To prove this, consider the matrix $[[A] - \lambda[I]]$. By premultiplying this matrix by $[P]$ and postmultiplying by $[P]^T$, we obtain

$$[[P]\,[A][P]^T - \lambda[P]\,[I][P]^T] = [[P]\,[A][P]^T - \lambda[I]] \tag{4.52}$$

since $[P]\,[P]^T = [I]$. This shows that the eigenvalues of $[P]\,[A][P]^T$ will be identical to those of $[A]$.

4.6.2 Similarity Transformation

A matrix $[B]$ is said to be *similar* to an arbitrary matrix $[A]$ if and only if there exists a matrix $[P]$ such that

$$[B] = [P]^{-1}[A][P]. \tag{4.53}$$

It can be shown that if the eigenvalues of an arbitrary matrix $[A]$ are all distinct and equal to $\lambda_1, \lambda_2, \ldots, \lambda_n$, then a similarity transformation $[P]^{-1}[A][P]$ exists such that

$$[P]^{-1}[A][P] = [D] = \begin{bmatrix} \lambda_1 & 0 & \cdots & 0 \\ 0 & \lambda_2 & \cdots & 0 \\ \cdot & \cdot & \cdots & \cdot \\ \cdot & \cdot & \cdots & \cdot \\ \cdot & \cdot & \cdots & \cdot \\ 0 & 0 & \cdots & \lambda_n \end{bmatrix}, \tag{4.54}$$

where $[D]$ is a diagonal matrix with its diagonal elements denoting the eigenvalues of $[A]$. If the eigenvectors of $[A]$ are denoted as $\vec{X}^{(1)}, \vec{X}^{(2)}, \ldots, \vec{X}^{(n)}$ such that

$$[A]\vec{X}^{(i)} = \lambda_i \vec{X}^{(i)}; \quad i = 1, 2, \ldots, n, \tag{4.55}$$

the matrix $[P]$ can be formulated as

$$[P] = \left[\vec{X}^{(1)} \vec{X}^{(2)} \cdots \vec{X}^{(n)}\right], \tag{4.56}$$

where the eigenvectors define the columns of the matrix $[P]$. Since the eigenvalues λ_i are distinct, the eigenvectors and hence the matrix $[P]$ exists (these may be complex). Equation (4.55) can be represented, in view of Eq. (4.56), as

$$[A]\,[P] = [P]\,[D]. \tag{4.57}$$

By premultiplying Eq. (4.57) by $[P]^{-1}$, we obtain the desired result

$$[P]^{-1}[A]\,[P] = [P]^{-1}[P]\,[D] = [D]. \tag{4.58}$$

Here the rows of the matrix $[P]^{-1}$ denote the left eigenvectors of $[A]$. In deriving Eq. (4.58), the matrix $[A]$ was assumed to be arbitrary. However, if the matrix $[A]$ is assumed to be symmetric, it can be proved that an orthogonal matrix $[Q]$ exists such that

$$[Q]^T[A]\,[Q] = [D], \tag{4.59}$$

where $[D]$ is a diagonal matrix whose diagonal elements denote the eigenvalues of $[A]$. Notice that when $[A]$ is symmetric, the similarity transformation of Eq. (4.54) is replaced by an orthogonal one, and also there is no requirement that the eigenvalues be distinct in order for Eq. (4.59) to be valid.

4.6.3 Jacobi Diagonalization Procedure

As stated earlier, the Jacobi diagonalization method can be used to find the eigenvalues and eigenvectors of symmetric matrices. It is based on the principle that an orthogonal matrix $[Q]$ can always be found such that Eq. (4.59) holds with $[D]$ indicating a diagonal matrix whose diagonal elements are the eigenvalues of $[A]$. The matrix $[Q]$ is found by generating a sequence of orthogonal matrices $[R^{(i)}]$, $i = 1, 2, \ldots$ such that

$$\lim_{k\to\infty} [R^{(1)}][R^{(2)}]\cdots[R^{(k)}] = [Q]. \tag{4.60}$$

Although an infinite sequence of $[R^{(i)}]$ is needed for theoretical convergence to $[Q]$, in practice, the process can be stopped whenever a suitable convergence criterion is satisfied. The orthogonal matrices $[R^{(i)}]$ are used to modify the matrix $[A]$ and generate a sequence of new matrices $[A^{(i)}]$ as

$$[A^{(1)}] = [A]; \tag{4.61}$$

$$[A^{(2)}] = [R^{(1)}]^T[A^{(1)}][R^{(1)}]; \tag{4.62}$$

$$[A^{(3)}] = [R^{(2)}]^T[A^{(2)}][R^{(2)}] \equiv [R^{(2)}]^T[R^{(1)}]^T[A^{(1)}][R^{(1)}][R^{(2)}]; \tag{4.63}$$

$$\vdots$$

$$[A^{(k)}] = [R^{(k-1)}]^T[A^{(k-1)}][R^{(k-1)}] = [R^{(k-1)}]^T[R^{(k-2)}]^T\cdots$$
$$[R^{(2)}]^T[R^{(1)}]^T[A][R^{(1)}][R^{(2)}]\cdots[R^{(k-2)}][R^{(k-1)}]. \tag{4.64}$$

Let the elements of the matrix $[A^{(k)}]$ be denoted as a_{ij}. The orthogonal matrix $[R^{(k)}]$ is taken as a rotation matrix and is constructed in such a way that the largest off-diagonal coefficient (a_{st}, for example) in $[A^{(k)}]$ is reduced to zero. The elements R_{ij} of a simple orthogonal matrix $[R^{(k)}]$ are given by

$$R_{ii} = 1; \text{ for } i = 1, 2, \ldots, n \text{ and } i \neq s, t;$$

$$R_{ij} = 0; \text{ for all } i, j = 1, 2, \ldots, n, \text{ except the following:}$$

$$R_{ss} = \cos\theta,$$

$$R_{tt} = \cos\theta,$$

$$R_{st} = -\sin\theta,$$

and

$$R_{ts} = \sin\theta, \tag{4.65}$$

so that

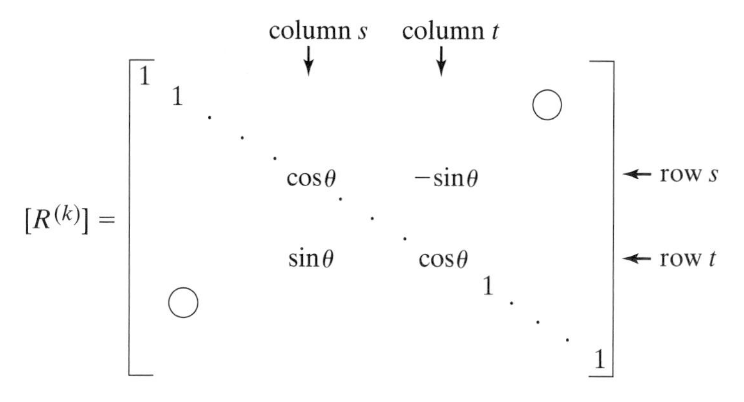

(4.66)

Denoting the elements of the matrix $[A^{(k+1)}]$ as $\bar{a}_{ij}$, the product $[R^{(k)}]^T[A^{(k)}][R^{(k)}] = [A^{(k+1)}]$ leads to the scalar equations

$$\bar{a}_{ss} = a_{ss}\cos^2\theta + 2a_{st}\sin\theta\cos\theta + a_{tt}\sin^2\theta; \tag{4.67}$$

$$\bar{a}_{tt} = a_{ss}\sin^2\theta - 2a_{st}\sin\theta\cos\theta + a_{tt}\cos^2\theta; \tag{4.68}$$

$$\bar{a}_{st} = a_{st}\cos 2\theta - \frac{1}{2}(a_{ss} - a_{tt})\sin 2\theta; \tag{4.69}$$

$$\bar{a}_{is} = a_{is}\cos\theta + a_{it}\sin\theta; \quad i \neq s, t; \tag{4.70}$$

$$\bar{a}_{it} = -a_{is}\sin\theta + a_{it}\cos\theta; \quad i \neq s, t; \tag{4.71}$$

$$\bar{a}_{si} = a_{si}\cos\theta + a_{ti}\sin\theta; \quad i \neq s, t; \tag{4.72}$$

$$\bar{a}_{ti} = -a_{si}\sin\theta + a_{ti}\cos\theta; \quad i \neq s, t; \tag{4.73}$$

$$\bar{a}_{ij} = a_{ij}; \quad i, j \neq s, t. \tag{4.74}$$

Equations (4.67) through (4.74) indicate that only the elements in the s and t row and column of $[A^{(k+1)}]$ differ from those in the s and t row and column of $[A^{(k)}]$. In order to make the element $\bar{a}_{st}$ equal to zero, we must select

$$\tan 2\theta = \frac{2a_{st}}{a_{ss} - a_{tt}} \tag{4.75}$$

Using Eq. (4.75), the values of $\cos\theta$ and $\sin\theta$ for use in Eq. (4.66) are determined as follows:

$$\sin 2\theta = \frac{2a_{st}}{d}, \cos 2\theta = \frac{a_{ss} - a_{tt}}{d}, \tag{4.76}$$

where

$$d = \sqrt{(a_{ss} - a_{tt})^2 + 4a_{st}^2}. \tag{4.77}$$

Since

$$\tan\theta = \frac{\sin 2\theta}{1+\cos 2\theta}, \tag{4.78}$$

we have

$$\sin\theta = \frac{|\sin 2\theta|}{\tilde{d}}, \ \cos\theta = \pm\frac{1+\cos 2\theta}{\tilde{d}}, \tag{4.79}$$

where

$$\tilde{d} = \sqrt{2(1+\cos 2\theta)} \tag{4.80}$$

and the sign of $\cos\theta$ is the same as the sign of a_{st}. As indicated by Eq. (4.60), theoretically, there are an infinite number of θ values corresponding to the infinite $[A^{(k)}]$ matrices. However, as θ approaches zero, the $[R^{(k)}]$ matrices tend to be same as the identity matrix, thereby, indicating that no further transformations are required. Once convergence is achieved, the eigenvalues of $[A]$ are given by the diagonal elements of the matrix $[D^{(k)}]$

$$[D^{(k)}] = [R^{(k)}]^T[R^{((k-1)}]^T\cdots[R^{(2)}]^T[R^{(1)}]^T[A][R^{(1)}][R^{(2)}]\cdots[R^{(k-1)}][R^{(k)}] \tag{4.81}$$

and the corresponding eigenvectors by the columns of the matrix

$$[Q^{(k)}] = [R^{(1)}][R^{(2)}]\cdots[R^{(k-1)}][R^{(k)}]. \tag{4.82}$$

4.6.4 Algorithm

1. To find the eigenvalues and eigenvectors of the matrix $[A]$, set $[A^{(1)}] = [a_{ij}] \equiv [A]$, $[Q^{(1)}] = [I]$, and $k = 1$.
2. Find integers s and t such that

$$a_{st} = \max_{\substack{i,j \\ i\neq j}}(a_{ij}).$$

3. Find d using Eq. (4.77).
4. Determine the values of $\sin 2\theta$ and $\cos 2\theta$ using Eq. (4.76).
5. Find $\tilde{d}$ from Eq. (4.80).
6. Find the values of $\sin\theta$ and $\cos\theta$ using Eq. (4.79) with the sign of $\cos\theta$ taken to be same as that of a_{st}.
7. Formulate the matrix $[R^{(k)}]$ using Eq. (4.66).

8. Determine $[A^{(k+1)}] = [R^{(k)}]^T[A^{(k)}][R^{(k)}] \equiv [a_{ij}]$ and $[Q^{(k+1)}] = [Q^{(k)}][R^{(k)}]$.

9. If $\max\limits_{\substack{i,j \\ i \neq j}} |a_{ij}| > \varepsilon$, set $k = k + 1$ and go to step 2. The value of ε can be selected as a small quantity, such as 10^{-6}.

10. If $\max\limits_{\substack{i,j \\ i \neq j}} |a_{ij}| \leq \varepsilon$, the eigenvalues are given by $\lambda_i = a_{ii}, i = 1, 2, \ldots, n$ and the eigenvectors, $\vec{X}^{(i)}$, by $\left[\vec{X}^{(1)}\vec{X}^{(2)}\ldots\vec{X}^{(n)}\right] \equiv [Q^{(k+1)}]$.

▶Example 4.12

Find the eigenvalues and eigenvectors of the matrix

$$[A] = \begin{bmatrix} 12 & 6 & -6 \\ 6 & 16 & 2 \\ -6 & 2 & 16 \end{bmatrix} \tag{a}$$

by using the Jacobi method.

Solution

The iteration number k, the values of $s, t, \cos\theta, \sin\theta$, and the matrices $[A^{(k)}] = [D^{(k)}]$, and $[Q^{(k)}]$ in each iteration are shown in Table 4.2. The eigenvalues of the matrix $[A]$ are given by the diagonal elements of $[A^{(6)}]$ and the eigenvectors by the columns of the matrix $[Q^{(6)}]$:

$$\lambda_1 = 4.45599,\ \lambda_2 = 18.0,\ \text{and}\ \lambda_3 = 21.544; \tag{b}$$

$$\vec{X}^{(1)} = \begin{Bmatrix} 0.747374 \\ -0.469805 \\ 0.469804 \end{Bmatrix},\ \vec{X}^{(2)} = \begin{Bmatrix} 0.0 \\ 0.707108 \\ 0.707106 \end{Bmatrix},\ \vec{X}^{(3)} = \begin{Bmatrix} 0.664404 \\ 0.528472 \\ -0.528474 \end{Bmatrix}. \tag{c}$$

◀

4.7 Given's Method

In the Jacobi method, the solution of the eigenvalue problem associated with a real symmetric matrix $[A]$ is obtained by diagonalizing the matrix. The diagonalization is

Table 4.2

Iter (k)	s	t	$\cos\theta$	$\sin\theta$	$[A^{(k)}]$	$[Q^{(k)}]$
1	1	2	0.8112	0.5847	$\begin{bmatrix} 20.3245 & 0.0 & -1.8858 \\ 0.0 & 7.6754 & 6.0369 \\ -1.8858 & 6.0369 & 16.0 \end{bmatrix}$	$\begin{bmatrix} 0.5847 & 0.8112 & 0.0 \\ -0.8112 & 0.5847 & 0.0 \\ 0.0 & 0.0 & 1.0 \end{bmatrix}$
2	2	3	0.8853	0.4650	$\begin{bmatrix} 20.3246 & -1.6695 & -0.8768 \\ -1.6695 & 19.1704 & 0.0 \\ -0.8768 & 0.0 & 4.5050 \end{bmatrix}$	$\begin{bmatrix} 0.5847 & 0.8112 & 0.0 \\ -0.3772 & 0.2719 & 0.8853 \\ 0.7182 & -0.5177 & 0.4650 \end{bmatrix}$
3	2	1	0.8145	-0.5802	$\begin{bmatrix} 17.9810 & 0.0 & 0.5087 \\ 0.0 & 21.5139 & -0.7141 \\ 0.5087 & -0.7141 & 4.5050 \end{bmatrix}$	$\begin{bmatrix} -0.0321 & -0.6921 & -0.7211 \\ 0.6951 & 0.5030 & -0.5137 \\ 0.7182 & -0.5177 & 0.4650 \end{bmatrix}$
4	3	2	0.9991	-0.0419	$\begin{bmatrix} 17.9810 & -0.5083 & -0.0213 \\ -0.5083 & 4.4751 & 0.0 \\ -0.0213 & 0.0 & 21.5441 \end{bmatrix}$	$\begin{bmatrix} -0.0321 & -0.6921 & -0.7211 \\ -0.7467 & 0.4961 & -0.4430 \\ 0.6644 & 0.5242 & -0.5327 \end{bmatrix}$
5	1	2	0.3756	-0.9993	$\begin{bmatrix} 18.0001 & 0.0 & 0.0213 \\ 0.0 & 4.4560 & 0.0008 \\ 0.0213 & 0.0008 & 21.5441 \end{bmatrix}$	$\begin{bmatrix} 0.0040 & 0.7103 & 0.7039 \\ 0.7474 & -0.4698 & 0.4698 \\ 0.6644 & 0.5242 & -0.5327 \end{bmatrix}$
6	1	3	0.9998	0.0060	$\begin{bmatrix} 21.5380 & 0.0008 & 0.0 \\ 0.0008 & 4.4560 & 0.0 \\ 0.0 & 0.0 & 17.9948 \end{bmatrix}$	$\begin{bmatrix} 0.6643 & 0.5284 & -0.5284 \\ 0.7474 & -0.4698 & 0.4698 \\ 0.0 & -0.7070 & -0.7070 \end{bmatrix}$

done iteratively by making use of rotation matrices. Given's method is also based on the concept of rotation matrices. However, Given's method does not give the actual solution of the problem, but only reduces the symmetric matrix $[A]$ to a tridiagonal form. Another difference is that although both the methods are iterative in nature, the Jacobi method might require a very large number of iterations to reduce the matrix $[A]$ to a diagonal form, while Given's method requires only $(n-1)(n-2)/2$ rotations to reduce the matrix $[A]$ to a tridiagonal form. In fact, the Jacobi method is called an infinite iterative method. The reason is that in the Jacobi method, an element, a_{st}, which is reduced to zero in a particular iteration, may again assume a nonzero value in the next iteration. On the other hand, in Given's method, all the elements reduced to zero in the previous iterations remain as zero in subsequent iterations. In Given's method, a rotation in the plane (s, t) is used to reduce the elements in positions $(s-1, t)$ and $(s, t-1)$ to zero in the matrix $[A^{(k)}]$. It can be verified that, once a pair of elements are reduced to zero, they remain zero and hence do not need to be operated again. Let the elements of the rotation (orthogonal) matrix $[R^{(k)}]$ be given by Eq. (4.65):

$$R_{ii} = 1; \quad i = 1, 2, \ldots, n; \quad i \neq s, t;$$

$$R_{ij} = 0 \text{ for all } i, j = 1, 2, \ldots, n, \text{ except}$$

$$R_{ss} = R_{tt} = \cos\theta$$

and

$$R_{st} = -R_{ts} = -\sin\theta. \tag{4.83}$$

The element $\bar{a}_{it}$ in the matrix $[A^{(k+1)}] = [R^{(k)}]^T[A^{(k)}][R^{(k)}]$ is given by Eq. (4.71):

$$\bar{a}_{it} = -a_{is}\sin\theta + a_{it}\cos\theta; \quad i \neq s, t. \tag{4.84}$$

This can be reduced to zero if θ is selected such that

$$\sin\theta = \alpha\, a_{it},$$

and

$$\cos\theta = \alpha\, a_{is}, \tag{4.85}$$

where

$$\alpha = \frac{1}{\sqrt{(a_{is}^2 + a_{it}^2)}}. \tag{4.86}$$

It can be seen that the computation of the elements of the rotation matrix is much simpler in Given's method. This procedure is to be carried until all the elements in a given row, except the tridiagonal elements, are reduced to zero, which automatically causes the corresponding elements of the associated column also reduce to zero. It can be seen that $(n-2)$ rotations are needed to reduce the original matrix $[A] = [A^{(1)}]$ to $[A^{(2)}]$ with all the elements in the first row and first column, except the tridiagonal elements, equal to zero. Similarly, $(n-3)$ rotations are needed to reduce $[A^{(2)}]$ to $[A^{(3)}]$ in which all the elements in the second row and second column, except the tridiagonal elements, are zero. Thus, the total number of rotations (M) required to reduce the original matrix $[A]$ to the tridiagonal form is equal to

$$M = \frac{1}{2}(n-1)(n-2). \tag{4.87}$$

The final tridiagonal matrix $[A^{(M)}]$ can be represented as

$$[A^{(M)}] = [R^{(M)}]^T[A^{(M-1)}]\,[R^{(M)}]$$

$$= \begin{bmatrix} \alpha_1 & \beta_2 & 0 & \cdot & \cdot & \cdot & 0 & 0 \\ \beta_2 & \alpha_2 & \beta_3 & \cdot & \cdot & \cdot & 0 & 0 \\ 0 & \beta_3 & \alpha_3 & \cdot & \cdot & \cdot & 0 & 0 \\ \cdot & \cdot & \cdot & \cdot & \cdot & \cdot & \cdot & \cdot \\ \cdot & \cdot & \cdot & \cdot & \cdot & \cdot & \cdot & \cdot \\ \cdot & \cdot & \cdot & \cdot & \cdot & \cdot & \cdot & \cdot \\ 0 & 0 & 0 & \cdot & \cdot & \cdot & \alpha_{n-1} & \beta_n \\ 0 & 0 & 0 & \cdot & \cdot & \cdot & \beta_n & \alpha_n \end{bmatrix}. \tag{4.88}$$

Since the final matrix $[A^{(M)}]$ is generated from the matrix $[A]$ by using similarity transformations, the eigenvalues of $[A^{(M)}]$ will be same as those of $[A]$. However, the eigenvectors of $[A^{(M)}]$ will be different from those of $[A]$. To find the relation between the eigenvectors of the matrices $[A^{(M)}]$ and $[A]$, consider the two eigenvalue problems

$$[A]\vec{X} = \lambda\vec{X} \tag{4.89}$$

and

$$[A^{(M)}]\vec{Y} = \lambda\vec{Y}, \tag{4.90}$$

where $\vec{X}$ and $\vec{Y}$ denote the eigenvectors of the matrices $[A]$ and $[A^{(M)}]$, respectively. The matrix $[A^{(M)}]$ can be represented as

$$\begin{aligned}[A^{(M)}] &= [R^{(M)}]^T[A^{(M-1)}]\,[R^{(M)}] \\ &= [R^{(M)}]^T[R^{(M-1)}]^T\cdots[R^{(2)}]^T[R^{(1)}]^T[A]\,[R^{(1)}]\,[R^{(2)}]\cdots[R^{(M-1)}]\,[R^{(M)}] \\ &\equiv [R]^T[A]\,[R],\end{aligned} \tag{4.91}$$

where

$$[R] = [R^{(1)}]\,[R^{(2)}]\cdots[R^{(M-1)}]\,[R^{(M)}]. \tag{4.92}$$

The matrix $[R]$ is an orthogonal matrix and, unlike in the Jacobi method, does not represent the matrix of eigenvectors since the matrix $[A^{(M)}]$ is a tridiagonal matrix (but not a diagonal matrix). Using Eq. (4.91), we can write Eq. (4.90) as

$$[R]^T[A]\,[R]\vec{Y} = \lambda\vec{Y}. \tag{4.93}$$

Premultiplying Eq. (4.93) by $[R]$ and noting that $[R]$ is an orthogonal matrix with

$$[R][R]^T = [I], \tag{4.94}$$

we obtain

$$[A]\,[R]\vec{Y} = \lambda[R]\vec{Y}. \tag{4.95}$$

By comparing Eq. (4.89) and (4.95), we find that the eigenvectors of $[A]$ and $[A^{(M)}]$ are related as

$$\vec{X} = [R]\vec{Y}. \tag{4.96}$$

We shall consider the eigenvalue solution of a tridiagonal matrix in Section 4.9.

▶Example 4.13

Reduce the matrix

$$[A] = \begin{bmatrix} 12 & 6 & -6 \\ 6 & 16 & 2 \\ -6 & 2 & 16 \end{bmatrix}$$

to tridiagonal form by using Given's method.

Solution

For this matrix, only one rotation is required. Since the elements in positions $(p-1, q) = (1, 3)$ and $(q, p-1) = (3, 1)$ are to be reduced to zero, we perform a rotation in the $(p, q) = (2, 3)$ plane. Assuming that $[A^{(1)}] = [A]$, we have

$$a_{p-1,p}^{(1)} = a_{1,2}^{(1)} = 6 \text{ and } a_{p-1,q}^{(1)} = a_{1,3}^{(1)} = -6.$$

Hence,

$$\alpha = \frac{1}{\sqrt{(6)^2 + (-6)^2}} = \frac{1}{6\sqrt{2}}, \cos\alpha = \frac{6}{6\sqrt{2}} = \frac{1}{\sqrt{2}}, \text{ and } \sin\alpha = \frac{-6}{6\sqrt{2}} = -\frac{1}{\sqrt{2}},$$

and the transformation matrix is given by

$$[R^{(1)}] = \begin{bmatrix} 1 & 0 & 0 \\ 0 & \cos\alpha & -\sin\alpha \\ 0 & \sin\alpha & \cos\alpha \end{bmatrix} = \frac{1}{\sqrt{2}} \begin{bmatrix} \sqrt{2} & 0 & 0 \\ 0 & 1 & 1 \\ 0 & -1 & 1 \end{bmatrix}.$$

This gives the desired tridiagonal matrix:

$$[A^{(2)}] = [R^{(1)}]^T [A^{(1)}] [R^{(1)}] = \begin{bmatrix} 12 & 6\sqrt{2} & 0 \\ 6\sqrt{2} & 14 & 0 \\ 0 & 0 & 18 \end{bmatrix}.$$

◀

▶Example 4.14

Verify that the eigenvalues of $[A^{(2)}]$ in Example 4.13 are identical to those of $[A]$ in Example 4.12.

Solution

The eigenvalues of the matrix $[A^{(2)}]$ are given by

$$\begin{vmatrix} (12-\lambda) & 6\sqrt{2} & 0 \\ 6\sqrt{2} & (14-\lambda) & 0 \\ 0 & 0 & (18-\lambda) \end{vmatrix} = 0.$$

This yields the characteristic polynomial,

$$-\lambda^3 + 44\lambda^2 - 564\lambda + 1728 = 0,$$

which can be seen to be identical to the characteristic polynomial equation corresponding to the matrix $[A]$. (See Example 4.10.) ◀

4.8 Householder's Method

This method is a variation of Given's method and reduces a given symmetric matrix $[A]$ into a tridiagonal form having the same eigenvalues. Whereas Given's method uses a series of orthogonal matrices, each of which is a plane rotation matrix, Householder's method uses a series of orthogonal matrices that do not represent plane rotations. Householder's method is considered to be the fastest and most accurate of all methods. It is simpler and requires less computer storage than

Given's method [4.2]. This method reduces a whole row and column (except for the tridiagonal elements) to zero at a time, without affecting previous rows and columns. The method requires only $(n-2)$ transformations, although each involves more computations than Given's method. The transformations are given by

$$[A_1] = [A], \tag{4.97}$$

$$[A_2] = [P_1]^T[A_1][P_1], \tag{4.98}$$

$$\vdots$$

$$[A_k] = [P_k]^T[A_{k-1}][P_k], \tag{4.99}$$

$$\vdots$$

and

$$[A_{n-1}] = [P_{n-2}]^T[A_{n-2}][P_{n-1}], \tag{4.100}$$

where n is the order of the matrix $[A]$ and $[P_k]$ denotes the kth transformation matrix, which is constructed as

$$[P_k] = [I] - 2\vec{v}_k\,\vec{v}_k^T \tag{4.101}$$

with

$$\vec{v}_k^T\,\vec{v}_k = 1 \tag{4.102}$$

and $\vec{v}_k$ is a n-component vector whose first $k-1$ components are zero:

$$\vec{v}_k = \begin{Bmatrix} 0 \\ 0 \\ \cdot \\ \cdot \\ \cdot \\ 0 \\ v_k^{(k)} \\ v_k^{(k+1)} \\ \cdot \\ \cdot \\ \cdot \\ v_k^{(n)} \end{Bmatrix}. \tag{4.103}$$

The matrix $[P_k]$ is symmetric and orthogonal, since

$$[P_k]^T[P_k] = ([I] - 2\vec{v}_k\vec{v}_k^T)^T([I] - 2\vec{v}_k\vec{v}_k^T) = [I] - 4\vec{v}_k\,\vec{v}_k^T + 4\vec{v}_k\,\vec{v}_k^T = [I]. \tag{4.104}$$

Let the symmetric matrix $[A_{k-1}] \equiv [\alpha_{ij}]$ have zeros in its first $(k-2)$ rows and columns, except for the tridiagonal elements:

$$[A_{k-1}] = \begin{bmatrix} \alpha_{11} & \alpha_{12} & 0 & \cdot & \cdot & \cdot & & 0 \\ \alpha_{12} & \alpha_{22} & \alpha_{23} & \cdot & \cdot & \cdot & & \\ 0 & \alpha_{23} & \alpha_{33} & \cdot & \cdot & \cdot & & \\ \cdot & & \ddots & & & & & \\ \cdot & & & \alpha_{k-2,k-2} & \alpha_{k-2,k-1} & 0 & & 0 \\ \cdot & & & \alpha_{k-2,k-1} & \alpha_{k-1,k-1} & & & \alpha_{k-1,n} \\ & & & 0 & & & & \\ & & & \vdots & & \ddots & & \\ 0 & & & 0 & \alpha_{k-1,n} & & & \alpha_{nn} \end{bmatrix} \leftarrow \text{row } k-1$$

$$\uparrow \text{ column } k-1 \tag{4.105}$$

The matrix $[P_k]$, given by Eq. (4.101), will have the following form:

$$[P_k] = [I] - 2\vec{v}_k \vec{v}_k^T = \begin{bmatrix} 1 & 0 & & & & & \\ 0 & \cdot & & & & & \\ \cdot & & \cdot & & & & \\ \cdot & & & \cdot & & & \\ \cdot & & & 1 & 0 & \cdot\ \cdot\ \cdot & 0 \\ & & & 0 & \{1-2(v_k^{(k)})^2\} & & \{-2\, v_k^{(n)} v_k^{(k)}\} \\ & & & \vdots & \vdots & \ddots & \vdots \\ 0 & & & 0 & \{-2\, v_k^{(n)}\, v_k^{(k)}\} & & \{1-2(v_k^{(n)})^2\} \end{bmatrix} \leftarrow \text{row } k$$

$$\uparrow \text{ column } k \tag{4.106}$$

It can be verified that the matrix $[A_k]$ will have zeros in the positions indicated as zero for $[A_{k-1}]$ in Eq. (4.105). The aim of the transformation is to choose the $(n-k+1)$ constants $v_k^{(k)}, v_k^{(k+1)}, \ldots, v_k^{(n)}$ so as to satisfy the normalization condition, Eq. (4.102), and also to make the $(n-k)$ off-tridiagonal elements of the $(k-1)$th row (or column) of $[A_k]$ equal to zero. The following choice satisfies these requirements:

$$(v_k^{(k)})^2 = \frac{1}{2}\left[1 \pm \frac{\alpha_{k-1,k}}{\sqrt{T}}\right]; \tag{4.107}$$

$$v_k^{(i)} = \pm \frac{\alpha_{k-1,i}}{2 v_k^{(k)} \sqrt{T}}; \quad i = k+1, k+2, \ldots, n, \tag{4.108}$$

where

$$T = \sum_{i=k}^{n} \alpha_{k-1,i}^2. \quad (4.109)$$

The sign preceding the quantity $\sqrt{T}$ in Eqs. (4.107) and (4.108) is taken to be same as that of the coefficient $\alpha_{k-1,k}$.

The application of Eqs. (4.97) to Eq. (4.100) finally yields the desired tridiagonal matrix, $[A_{k-1}]$.

Algorithm

1. Let $[A_1] \equiv [A]$ and $[A_1] = [a_{ij}^{(1)}]$. Set $k = 1$.
2. Compute $R = \sum_{i=k+1}^{n} (a_{ik}^{(k)})^2$.
3. If $R = 0$, set $[A_{k+1}] = [A_k]$, $[A_{k+1}] = [a_{ij}^{(k+1)}]$, $k = k + 1$, and go to step 2.

   If $R \neq 0$, set $S = \frac{a_{k+1,k}^{(k)}}{|a_{k+1,k}^{(k)}|} \sqrt{R}$.
4. Compute $Q = \frac{1}{2}(S^2 + S a_{k+1,k}^{(k)})$.
5. Find the vector $\vec{u} = \{u_1, u_2 \ldots, u_n\}^T$ with

$$u_i = \left\{ \begin{array}{l} 0 \text{ if } i = 1, 2, \ldots, k \\ a_{k+1,k}^{(k)} + S \text{ if } i = k + 1 \\ a_{ik}^{(k)} \text{ if } i = k + 2, k + 3, \ldots, n \end{array} \right\}.$$

6. Determine the vector $\vec{v} = \{v_1, v_2 \ldots, v_n\}^T$ with

$$v_i = \frac{1}{2Q} \sum_{j=k+1}^{n} a_{ij}^{(k)} u_j; \quad i = 1, 2, \ldots, n.$$

7. Calculate the vector $\vec{w} = \{w_1, w_2 \ldots, w_n\}^T$ with

$$w_i = v_i - \frac{u_i}{4Q} \left(\sum_{j=k+1}^{n} u_j v_j \right), i = 1, 2, \ldots, n.$$

8. Determine the matrix $[A_{k+1}] = [a_{ij}^{(k+1)}]$ as

$$a_{ij}^{(k+1)} = a_{ij}^{(k)} - u_i w_j - w_i u_j, \quad i + 1, 2, \ldots, n, \ j = 1, 2, \ldots, n.$$

9. Set $k = k + 1$. If $k < n - 1$, go to step 2. If $k = n - 1$, the matrix $[A_{k+1}] \equiv [A_n]$ is the desired tridiagonal matrix and hence stop the procedure.

▶Example 4.15

Reduce the matrix

$$[A] = \begin{bmatrix} 4 & -2 & 6 & 4 \\ -2 & 2 & -1 & 3 \\ 6 & -1 & 22 & 13 \\ 4 & 3 & 13 & 46 \end{bmatrix}$$

to tridiagonal form using Householder's method.

Solution

1. Set $[A_1] \equiv [a_{ij}^{(1)}] \equiv [A]$ and $k = 1$.
2.

$$R = \sum_{i=2}^{4} (a_{i1}^{(1)})^2 = (-2)^2 + (6)^2 + (4)^2 = 56.$$

3. Since $R \neq 0$, we set

$$S = \frac{a_{21}^{(1)}}{\left|a_{21}^{(1)}\right|}\sqrt{R} = \frac{-2}{|-2|}\sqrt{56} = -7.4833.$$

4.

$$Q = \frac{1}{2}(S^2 + Sa_{21}) = \frac{1}{2}(56 - 7.4833(-2)) = 35.4833.$$

5.

$$\vec{u} = \begin{Bmatrix} u_1 \\ u_2 \\ u_3 \\ u_4 \end{Bmatrix} = \begin{Bmatrix} 0 \\ -9.4833 \\ 6 \\ 4 \end{Bmatrix}.$$

6.

$$\begin{aligned} v_1 &= \frac{1}{2Q}\left(\sum_{j=2}^{4} a_{1j}^{(1)} u_j\right) \\ &= \frac{1}{2(35.4833)}((-2)(-9.4833) + (6)(6) + (4)(4)) = 1.0; \\ v_2 &= \frac{1}{2Q}\left(\sum_{j=2}^{4} a_{2j}^{(1)} u_j\right) \end{aligned}$$

$$= \frac{1}{2(35.4833)}((2)(-9.4833) + (-1)(6) + (3)(4)) = -0.1827;$$

$$v_3 = \frac{1}{2Q}\left(\sum_{j=2}^{4} a_{3j}^{(1)} u_j\right)$$

$$= \frac{1}{2(35.4833)}((-1)(-9.4833) + (22)(6) + (13)(4)) = 2.7264;$$

$$v_4 = \frac{1}{2Q}\left(\sum_{j=2}^{4} a_{4j}^{(1)} u_j\right)$$

$$= \frac{1}{2(35.4833)}((3)(-9.4833) + (13)(6) + (46)(4)) = 3.2910.$$

7.

$$w_1 = v_1 - \frac{u_1}{4Q}\left(\sum_{j=2}^{4} u_j v_j\right) = 1.0 - 0 = 1.0;$$

$$w_2 = v_2 - \frac{u_2}{4Q}\left(\sum_{j=2}^{4} u_j v_j\right)$$

$$= -0.1827 - \frac{(-9.4833)}{4(35.4833)}((-9.4833)(-0.1827) + (6)(2.7264) + (4)(3.2910))$$

$$= -0.1827 + 0.06681\,(31.2550) = 1.9054;$$

$$w_3 = v_3 - \frac{u_3}{4Q}\left(\sum_{j=2}^{4} u_j v_j\right)$$

$$= 2.7264 - \frac{6}{4(35.4833)}(31.2550) = 1.4051;$$

$$w_4 = v_4 - \frac{u_4}{4Q}\left(\sum_{j=2}^{4} u_j v_j\right)$$

$$= 3.2910 - \frac{4}{4(35.4833)}(31.2550) = 2.4102.$$

8.

$$a_{11}^{(2)} = a_{11}^{(1)} - u_1 w_1 - w_1 u_1 = 4 - (0)(1.0) - (1.0)(0) = 4;$$

$$a_{12}^{(2)} = a_{12}^{(1)} - u_1 w_2 - w_1 u_2 = -2 - (0)(1.9054) - (1.0)(-9.4833) = 7.4833;$$

$$a_{13}^{(2)} = a_{13}^{(1)} - u_1 w_3 - w_1 u_3 = 6 - (0)(1.4051) - (-1.0)(6) = 0;$$

$$a_{14}^{(2)} = a_{14}^{(1)} - u_1 w_4 - w_1 u_4 = 4 - (0)(2.4102) - (-1.0)(4) = 0;$$

$$a_{21}^{(2)} = a_{21}^{(1)} - u_2 w_1 - w_2 u_1 = -2 - (-9.4833)(1.0) - (1.9054)(0) = 7.4833;$$

$$a_{22}^{(2)} = a_{22}^{(1)} - u_2 w_2 - w_2 u_2 = 2 - (2)(-9.4833)(1.9054) = 38.1389;$$

$$a_{23}^{(2)} = a_{23}^{(1)} - u_2 w_3 - w_2 u_3 = -1 - (-9.4833)(1.4051) - (1.9054)(6) = 0.8926;$$

$$a_{24}^{(2)} = a_{24}^{(1)} - u_2 w_4 - w_2 u_4 = 3 - (-9.4833)(2.4102) - (1.9054)(4) = 18.2350;$$

$$a_{31}^{(2)} = a_{31}^{(1)} - u_1 w_3 - w_1 u_3 = 6 - (0)(1.4051) - (1.0)(6) = 0;$$

$$a_{32}^{(2)} = a_{32}^{(1)} - u_3 w_2 - w_3 u_2 = -1 - (6)(1.9054) - (1.4051)(-9.4833) = 0.8926;$$

$$a_{33}^{(2)} = a_{33}^{(1)} - 2u_3 w_3 = 22 - 2(6)(1.4051) = 5.1388;$$

$$a_{34}^{(2)} = a_{34}^{(1)} - u_3 w_4 - w_3 u_4 = 13 - (6)(2.4102) - (1.4051)(4) = -7.0816;$$

$$a_{41}^{(2)} = a_{41}^{(1)} - u_4 w_1 - w_4 u_1 = 4 - (4)(1.0) - (2.4102)(0) = 0;$$

$$a_{42}^{(2)} = a_{42}^{(1)} - u_4 w_2 - w_4 u_2 = 3 - (4)(1.9054) - (2.4102)(-9.4833) = 18.2350;$$

$$a_{43}^{(2)} = a_{43}^{(1)} - u_4 w_3 - w_4 u_3 = 13 - (4)(1.4051) - (2.4102)(6) = -7.0816;$$

$$a_{44}^{(2)} = a_{44}^{(1)} - 2u_4 w_4 = 46 - 2(4)(2.4102) = 26.7184.$$

Thus,

$$[A_2] = [a_{ij}^{(2)}] = \begin{bmatrix} 4 & 7.4833 & 0 & 0 \\ 7.4833 & 38.1389 & 0.8926 & 18.2350 \\ 0 & 0.8926 & 5.1388 & -7.0816 \\ 0 & 18.2350 & -7.0816 & 26.7184 \end{bmatrix}.$$

9. $k = 2$. since $k < 3$, we go to step 2.

2.

$$R = \sum_{i=3}^{4} \left(a_{i2}^{(2)}\right) = \left(a_{32}^{(2)}\right)^2 + \left(a_{42}^{(2)}\right)^2 = (0.8926)^2 + (18.2350)^2 = 333.3119.$$

3. Since $R \neq 0$, we set

$$S = \frac{a_{32}^{(2)}}{\left|a_{32}^{(2)}\right|}\sqrt{R} = \frac{0.8926}{|0.8926|}\left(\sqrt{333.3119}\right) = 18.2568.$$

4.

$$Q = \frac{1}{2}(333.3119 + 18.2568 \times 0.8926) = 174.8040.$$

5.

$$\begin{aligned}
u_1 &= 0;\\
u_2 &= 0;\\
u_3 &= a_{32}^{(2)} + S = 0.8926 + 18.2568 = 19.1494;\\
u_4 &= a_{24}^{(2)} = 18.2350.
\end{aligned}$$

6.

$$\begin{aligned}
v_1 &= \frac{1}{2Q}\left(a_{13}^{(2)}u_3 + a_{14}^{(2)}u_4\right) = \frac{1}{2(174.8040)}(0 + 0) = 0;\\
v_2 &= \frac{1}{2Q}\left(a_{23}^{(2)}u_3 + a_{24}^{(2)}u_4\right)\\
&= \frac{1}{2(174.8040)}(0.8926 \times 19.1494 + 18.2350 \times 18.2350) = 1.0;\\
v_3 &= \frac{1}{2Q}\left(a_{33}^{(2)}u_3 + a_{34}^{(2)}u_4\right)\\
&= \frac{1}{2(174.8040)}(5.1388 \times 19.1494 - 7.0816 \times 18.2350) = -0.08789;\\
v_4 &= \frac{1}{2Q}\left(a_{43}^{(2)}u_3 + a_{44}^{(2)}u_4\right)\\
&= \frac{1}{2(174.8040)}(-7.0816 \times 19.1494 + 26.7184 \times 18.2350) = 1.0057.
\end{aligned}$$

7.

$$\begin{aligned}
w_1 &= v_1 - \frac{u_1}{4Q}(u_3v_3 + u_4v_4) = 0 - 0 = 0;\\
w_2 &= v_2 - \frac{u_2}{4Q}(u_3v_3 + u_4v_4) = 1.0 - 0 = 1.0;\\
w_3 &= v_3 - \frac{u_3}{4Q}(u_3v_3 + u_4v_4)\\
&= -0.08789 - \frac{19.1494}{4(174.8040)}\\
&\quad \times (19.1494 \times (-0.08789) + 18.2350 \times 1.0057)
\end{aligned}$$

$$= -0.5440;$$

$$w_4 = v_4 - \frac{u_4}{4Q}(u_3 v_3 + u_4 v_4)$$

$$= 1.0057 - \frac{18.2350}{4(174.8040)}(19.1494 \times (-0.08789) + 18.2350 \times 1.0057)$$

$$= 0.5713.$$

8.

$$a_{11}^{(3)} = a_{11}^{(2)} - 2u_1 w_1 = 4 - 2(0)(0) = 4;$$

$$a_{12}^{(3)} = a_{12}^{(2)} - u_1 w_2 - w_1 u_2 = 7.4833 - 0 - 0 = 7.4833;$$

$$a_{13}^{(3)} = a_{13}^{(2)} - u_1 w_3 - w_1 u_3 = 0 - 0 - 0 = 0;$$

$$a_{14}^{(3)} = a_{14}^{(2)} - u_1 w_4 - w_1 u_4 = 0 - 0 - 0 = 0;$$

$$a_{21}^{(3)} = a_{21}^{(2)} - u_2 w_1 - w_2 u_1 = 7.4833 - 0 - 0 = 7.4833;$$

$$a_{22}^{(3)} = a_{22}^{(2)} - 2u_2 w_2 = 38.1389 - 0 = 38.1389;$$

$$a_{23}^{(3)} = a_{23}^{(2)} - u_2 w_3 - w_2 u_3$$

$$= 0.8926 - 0 - (1.0)(19.1494) = -18.2568;$$

$$a_{24}^{(3)} = a_{24}^{(2)} - u_2 w_4 - w_2 u_4$$

$$= 18.2350 - 0 - (1.0)(18.2350) = 0.0;$$

$$a_{31}^{(3)} = a_{31}^{(2)} - u_3 w_1 - w_3 u_1 = 0 - 0 - 0 = 0;$$

$$a_{32}^{(3)} = a_{32}^{(2)} - u_3 w_2 - w_3 u_2 = 0.8926 - 19.1494(1.0) - 0 = -18.2568;$$

$$a_{33}^{(3)} = a_{33}^{(2)} - 2u_3 w_3 = 5.1388 - 2(19.1494)(-0.5440) = 25.9733;$$

$$a_{34}^{(3)} = a_{34}^{(2)} - u_3 w_4 - w_3 u_4$$

$$= -7.0816 - 19.1494(0.5713) - (-0.5440)(18.2350) = -8.1018;$$

$$a_{41}^{(3)} = a_{41}^{(2)} - u_4 w_1 - w_4 u_1 = 0 - 0 - 0 = 0;$$

$$a_{42}^{(3)} = a_{42}^{(2)} - u_4 w_2 - w_4 u_2$$

$$= 18.2350 - (18.2350)(1.0) - 0 = 0.0;$$

$$a_{43}^{(3)} = a_{43}^{(2)} - u_4 w_3 - w_4 u_3$$

$$= -7.0816 - (18.2350)(-0.5440) - (0.5713)(19.1494) = -8.1018;$$

$$a_{44}^{(3)} = a_{44}^{(2)} - 2u_4 w_4 = 26.7184 - 2(18.2350)(0.5713) = 5.8831.$$

Thus,

$$[A_3] = [a_{ij}^{(3)}] = \begin{bmatrix} 4 & 7.4833 & 0 & 0 \\ 7.4833 & 38.1389 & -18.2568 & 0 \\ 0 & -18.2568 & 25.9733 & -8.1018 \\ 0 & 0 & -8.1018 & 5.8831 \end{bmatrix}.$$

9. Set $k = 3$. Since $k = n - 1 = 3$, $[A_3]$ is the desired tridiagonal matrix and the procedure is complete.

4.9 Eigenvalues of a Tridiagonal Matrix

The tridiagonal matrix obtained in Given's or Householder's method can be represented as

$$[B] = \begin{bmatrix} \alpha_1 & \beta_2 & & & & & \bigcirc \\ \beta_2 & \alpha_2 & \beta_3 & & & & \\ & \beta_3 & \alpha_3 & \beta_4 & & & \\ & & \beta_4 & \alpha_4 & \beta_5 & & \\ & & & \ddots & \ddots & \ddots & \\ & & & & & & \beta_n \\ \bigcirc & & & & & \beta_n & \alpha_n \end{bmatrix}. \quad (4.110)$$

The eigenvalues of $[B]$ are same as those of the original symmetric matrix $[A]$. The eigenvalues of the matrix $[B]$ given by Eq. (4.110), can be determined from the following equation:

$$|[B] - \lambda\,[I]| = \begin{bmatrix} \alpha_1-\lambda & \beta_2 & & & \\ \beta_2 & \alpha_2-\lambda & \beta_3 & & \\ & \beta_3 & \ddots & \ddots & \\ & & \ddots & \ddots & \beta_n \\ & & & \beta_n & \alpha_n-\lambda \end{bmatrix}. \quad (4.111)$$

The values of the leading principal minors of the matrix, $[B] - \lambda[I]$, are given by

$$f_0(\lambda) = 1; \quad (4.112)$$

$$f_1(\lambda) = \alpha_1 - \lambda; \quad (4.113)$$

$$f_2(\lambda) = (\alpha_2 - \lambda) f_1(\lambda) - \beta_2^2 f_0(\lambda); \tag{4.114}$$

$$\vdots$$

$$f_r(\lambda) = (\alpha_r - \lambda) f_{r-1}(\lambda) - \beta_r^2 f_{r-2}(\lambda); \quad r = 2, 3, \ldots, n. \tag{4.115}$$

The nth degree function gives the characteristic equation

$$f_n(\lambda) = 0. \tag{4.116}$$

The roots of the nth degree polynomial equation, Eq. (4.116), gives the eigenvalues of the tridiagonal matrix $[B]$, which are same as those of the original matrix, $[A]$.

▶Example 4.16

Generate the functions $f_i(\lambda), i = 0, 1, 2, 3,$ and 4 for the tridiagonal matrix

$$[B] = \begin{bmatrix} 3 & -2 & 0 & 0 \\ -2 & 5 & -3 & 0 \\ 0 & -3 & 7 & -4 \\ 0 & 0 & -4 & 9 \end{bmatrix}. \tag{a}$$

Solution

A comparison of Eq. (a) with Eq. (4.110) indicates that

$$\alpha_1 = 3, \alpha_2 = 5, \alpha_3 = 7, \alpha_4 = 9, \beta_2 = -2, \beta_3 = -3, \beta_4 = -4.$$

Equations (4.112) to (4.115) yield the desired functions

$$f_0(\lambda) = 1; \tag{b}$$

$$f_1(\lambda) = \alpha_1 - \lambda = 3 - \lambda; \tag{c}$$

$$f_2(\lambda) = (\alpha_2 - \lambda) f_1(\lambda) - \beta_2^2 f_0(\lambda) = (5 - \lambda)(3 - \lambda) - (-2)^2(1) = \lambda^2 - 8\lambda + 11; \tag{d}$$

$$\begin{aligned} f_3(\lambda) &= (\alpha_3 - \lambda) f_2(\lambda) - \beta_3^2 f_1(\lambda) \\ &= (7 - \lambda)(\lambda^2 - 8\lambda + 11) - (-3)^2(3 - \lambda) = -\lambda^3 + 15\lambda^2 - 58\lambda + 50; \end{aligned} \tag{e}$$

$$\begin{aligned} f_4(\lambda) &= (\alpha_4 - \lambda) f_3(\lambda) - \beta_4^2 f_2(\lambda) \\ &= (9 - \lambda)(-\lambda^3 + 15\lambda^2 - 58\lambda + 50) - (-4)^2(\lambda^2 - 8\lambda + 11) \\ &= \lambda^4 - 24\lambda^3 + 177\lambda^2 - 444\lambda + 274. \end{aligned} \tag{f}$$

The roots of the polynomial equation, Eq. (4.116), can be determined by using the methods we studied in Chapter 2. Sometimes, it is more convenient to use the properties of the Sturm sequences to locate the roots of Eq. (4.116). We assume that no β_i in Eq. (4.110) is zero. If any $\beta_i = 0$, the determinant of Eq. (4.111) can be written as the product of two determinants of smaller tridiagonal matrices and the procedure outlined next can be applied to each. ◀

4.9.1 Sturm Sequences

A sequence of polynomials $\{f_i(x)\} = \{f_1(x), f_2(x), \ldots, f_n(x)\}$ is called a *Sturm sequence* on the interval (a, b), where either a or b may be infinite, if (i) $f_n(x)$ does not vanish in (a, b) and (ii) at any zero of $f_k(x)$, $k = 2, 3, \ldots, n-1$, the two adjacent functions are nonzero and have opposite signs so that $f_{k-1}(x) f_{k+1}(x) < 0$.

It can be shown that the sequence of polynomials given by Eqs. (4.112) to (4.115) is a Sturm sequence and for any specific value of λ, such as $\lambda = p$, the number of changes of sign in the sequence is equal to the number of roots of $f_n(\lambda)$ smaller than p. This property can be used to find either all the roots $(\lambda_1, \lambda_2, \ldots, \lambda_n)$ or any particular root (λ_k). For example, if the value of $\lambda_k (\lambda_1 > \lambda_2 > \cdots > \lambda_k > \cdots > \lambda_n)$ is to be determined, we start with two trial values a and b such that $a > \lambda_1$ and $b < \lambda_n$. Since the number of sign changes in the sequence is n for a and zero for b, we try $p = \frac{1}{2}(a + b)$ and bisect the appropriate interval containing λ_k and repeat the process. The selection of the trial values a and b can be based on a theorem that states that the root of largest modulus is smaller than any norm of the matrix, $[A]$, corresponding to the sequence of polynomials [4.5]. This gives

$$-b = a = \max_r (|\alpha_r| + |\beta_r| + |\beta_{r+1}|). \tag{4.117}$$

If the tridiagonal matrix $[B]$ is known to be positive definite so that the roots are positive, we can take $b = 0$. The following example illustrates the method of finding the roots of a polynomial equation using the property of Sturm sequence.

▶Example 4.17

Use the property of Sturm sequence to locate all the real roots of the characteristic equation of Example 4.16:

$$f_4(\lambda) = \lambda^4 - 24\lambda^3 + 177\lambda^2 - 444\lambda + 274.$$

Solution

The Sturm sequence corresponding to the characteristic equation is given by Eqs. (b) to (f) of Example 4.16. Considering the tridiagonal matrix $[B]$ of Example 4.16, its row norm can be found as 14. Thus, the roots are expected to lie between -14 and 14. By giving some representative values to λ in the range $(-14, 14)$, the signs of Eqs. (b) to (f) of Example 4.16 are determined. The results are shown in Table 4.3. It can be seen that there is one sign change in each of the intervals (0, 1), (1, 5), (5, 10), and (10, 14). This indicates that there is one root in each of these intervals. By choosing more points (values of λ) within each of these intervals, the roots can be more precisely located. It can be seen that this method becomes uneconomical if accurate values of the roots are to be found. However, the Sturm sequence method can be used to find the approximate locations of the roots that can be used to provide

Table 4.3

Value of p	$f_0(p)$	$f_1(p)$	$f_2(p)$	$f_3(p)$	$f_4(p)$	Comments
−14.00	0.10E + 01	0.17E + 02	0.32E + 03	0.65E + 04	0.15E + 06	No change of sign, no root < −14
−10.00	0.10E + 01	0.13E + 02	0.19E + 03	0.31E + 04	0.56E + 05	
−5.00	0.10E + 01	0.80E + 01	0.76E + 02	0.84E + 03	0.11E + 05	
−1.00	0.10E + 01	0.40E + 01	0.20E + 02	0.12E + 03	0.92E + 03	
0.00	0.10E + 01	0.30E + 01	0.11E + 02	0.50E + 02	0.27E + 03	No change in sign, no root < 0
1.00	0.10E + 01	0.20E + 01	0.40E + 01	0.60E + 01	−0.16E + 02	One change in sign, one root < 1
5.00	0.10E + 01	−0.20E + 01	−0.40E + 01	0.10E + 02	0.10E + 03	Two changes in sign, two roots < 5
10.00	0.10E + 01	−0.70E + 01	0.31E + 02	−0.30E + 02	−0.47E + 03	Three changes in sign, three roots < 10
14.00	0.10E + 01	−0.11E + 02	0.95E + 02	−0.57E + 03	0.13E + 04	Four changes in sign, four roots < 14

good starting points for more accurate root finding methods discussed in Chapter 2. In the present case, the roots of the characteristic equation are given by

$$\lambda_1 = 0.90516,\ \lambda_2 = 3.38984,\ \lambda_3 = 7.06446, \text{ and } \lambda_4 = 12.6405.$$

◀

4.10 Eigenvectors of a Tridiagonal Matrix

The eigenvalue problem associated with a tridiagonal matrix $[B]$, given by Eq. (4.110), can be stated the following way:

$$[[B] - \lambda[I]]\vec{X} = \vec{0}. \tag{4.118}$$

If an eigenvalue is known to be λ_i, the corresponding eigenvector $\vec{X}^{(i)}$ can be determined by first writing Eq. (4.118) in explicit form as

$$(\alpha_1 - \lambda_i)x_1 + \beta_2 x_2 = 0; \tag{4.119}$$

$$\beta_2 x_1 + (\alpha_2 - \lambda_i)x_2 + \beta_3 x_3 = 0; \tag{4.120}$$

$$\beta_4 x_2 + (\alpha_3 - \lambda_i)x_3 + \beta_5 x_4 = 0; \tag{4.121}$$

$$\vdots$$

$$\beta_n x_{n-1} + (\alpha_n - \lambda_i)x_n = 0, \tag{4.122}$$

where $x_1, x_2, \ldots, x_n$ are the components of $\vec{X}^{(i)}$. By setting the first component of the eigenvector equal to one ($x_1 = 1$) arbitrarily, Eq. (4.119) can be solved to find

$$x_2 = -\frac{(\alpha_1 - \lambda_i)}{\beta_2} = -\frac{f_1(\lambda_i)}{\beta_2}, \tag{4.123}$$

where $f_1(\lambda))$ is given by Eq. (4.113). Equation (4.120) gives

$$x_3 = \frac{1}{\beta_3}[-\beta_2 x_1 - (\alpha_2 - \lambda_i)x_2] = \frac{1}{\beta_3}\left[-\beta_2 + (\alpha_2 - \lambda_i)\frac{f_1(\lambda_i}{\beta_2}\right]$$
$$= \frac{1}{\beta_2\beta_3}[(\alpha_2 - \lambda_i)f_1(\lambda_i) - \beta_2^2]. \tag{4.124}$$

This process can be continued until all the $x_j (j = 1, 2, \ldots, n)$ are found from the first $(n-1)$ equations of (4.119) to (4.122). The components of $\vec{X}^{(i)}$ can be expressed as

$$x_1 = 1 \text{ and } x_j = \frac{(-1)^{j-1} f_{j-1}(\lambda_i)}{\beta_2\beta_3\ldots\beta_j}; \quad j = 2, 3, \ldots, n. \tag{4.125}$$

The solution given by Eq. (4.125) can be verified to satisfy the last equation of Eqs. (4.119) to (4.122), namely Eq. (4.122).

▶Example 4.18

Find the eigenvectors of the matrix

$$[A] = \begin{bmatrix} 3 & -2 & 0 & 0 \\ -2 & 5 & -3 & 0 \\ 0 & -3 & 7 & -4 \\ 0 & 0 & -4 & 9 \end{bmatrix},$$

corresponding to the eigenvalues $\lambda_1 = 0.90516$, $\lambda_2 = 3.38984$, $\lambda_3 = 7.06446$, and $\lambda_4 = 12.6405$ (given in Example 4.17).

Solution

Noting that $\alpha_1 = 3$, $\alpha_2 = 5$, $\alpha_3 = 7$, $\alpha_4 = 9$, $\beta_2 = -2$, $\beta_3 = -3$, and $\beta_4 = -4$, Eqs. (4.123) and (4.125) yield the eigenvectors

$$\vec{X}^{(1)} = \begin{Bmatrix} 1.0 \\ 1.0474 \\ 0.7630 \\ 0.3770 \end{Bmatrix}, \vec{X}^{(2)} = \begin{Bmatrix} 1.0 \\ -0.1949 \\ -0.7713 \\ -0.5499 \end{Bmatrix}, \vec{X}^{(3)} = \begin{Bmatrix} 1.0 \\ -2.0322 \\ 0.7318 \\ 1.5124 \end{Bmatrix}, \vec{X}^{(4)} = \begin{Bmatrix} 1.0 \\ -4.8203 \\ 11.6097 \\ -12.7559 \end{Bmatrix}.$$

◀

4.11 Power Method

Usually the n eigenvalues of a matrix $[A]$ are numbered such that the largest or the dominant eigenvalue is denoted as λ_1 and the smallest as λ_n, that is,

$$|\lambda_1| \geq |\lambda_2| \geq \ldots |\lambda_n|. \tag{4.126}$$

The power method is an iterative technique that can be used to find the largest eigenvalue and the corresponding eigenvector. This method involves choosing an arbitrary trial vector $\vec{X}_1$ and refining it by using the properties of the eigenvalue problem. Consider the standard eigenvalue problem

$$[A]\vec{X} = \lambda\vec{X}. \tag{4.127}$$

If the n eigenvectors, $\vec{X}^{(i)}, i = 1, 2, \ldots, n$, of $[A]$ are independent, they can be used as basis vectors to express any n-component trial vector

$$\vec{X}_1 = c_1\vec{X}^{(1)} + c_2\vec{X}^{(2)} + \cdots + c_n\vec{X}^{(n)}, \tag{4.128}$$

where $c_1, c_2, \ldots, c_n$ are unknown constants and $\vec{X}^{(1)}, \vec{X}^{(2)}, \ldots, \vec{X}^{(n)}$ are unknown eigenvectors of the matrix $[A]$. By premultiplying the initial trial vector $\vec{X}_1$ by $[A]$, we obtain a second trial vector

$$\vec{X}_2 = [A]\vec{X}_1. \tag{4.129}$$

By using Eq. (4.128), Eq. (4.129) can be expressed as

$$\vec{X}_2 = [A]\vec{X}_1 = [A]\sum_{i=1}^{n} c_i\vec{X}^{(i)} = \sum_{i=1}^{n} c_i[A]\vec{X}^{(i)} = \sum_{i=1}^{n} x_i\lambda_i\vec{X}^{(i)}, \tag{4.130}$$

since

$$[A]\vec{X}^{(i)} = \lambda_i\vec{X}^{(i)}, \tag{4.131}$$

from Eq. (4.127). The process of generating a new trial vector by premultiplying the current vector by $[A]$ is continued as

$$\vec{X}_3 = [A]\vec{X}_2; \tag{4.132}$$

$$\vdots$$

$$\vec{X}_{r+1} = [A]\vec{X}_r. \tag{4.133}$$

Using Eq. (4.130) in Eq. (4.132), we obtain

$$\vec{X}_3 = [A]\left(\sum_{i=1}^{n} c_i\lambda_i\vec{X}^{(i)}\right) = \sum_{i=1}^{n} c_i\lambda_i\left([A]\vec{X}^{(i)}\right) = \sum_{i=1}^{n} c_i\lambda_i^2\vec{X}^{(i)}. \tag{4.134}$$

In general, $\vec{X}_{(r+1)}$ can be expressed as follows:

$$\vec{X}_{(r+1)} = \sum_{i=1}^{n} c_i\lambda_i^r\vec{X}^{(i)} = c_1\lambda_1^r\vec{X}^{(1)} + c_2\lambda_2^r\vec{X}^{(2)} + \cdots + c_n\lambda_n^r\vec{X}^{(n)}. \tag{4.135}$$

Equation (4.135) can be rewritten as

$$\vec{X}_{r+1} = \lambda_1^r \left[c_1 \vec{X}^{(1)} + \left(\frac{\lambda_2}{\lambda_1}\right)^r c_2 \vec{X}^{(2)} + \left(\frac{\lambda_3}{\lambda_1}\right)^r c_3 \vec{X}^{(3)} + \cdots + \left(\frac{\lambda_n}{\lambda_1}\right)^r c_n \vec{X}^{(n)} \right]. \tag{4.136}$$

In view of Eq. (4.126), the ratios $\frac{\lambda_2}{\lambda_1}, \frac{\lambda_3}{\lambda_1}, \ldots,$ and $\frac{\lambda_n}{\lambda_1}$ are less than one. For large values of r, the values of $\left(\frac{\lambda_2}{\lambda_1}\right)^r, \left(\frac{\lambda_3}{\lambda_1}\right)^r, \ldots,$ and $\left(\frac{\lambda_n}{\lambda_1}\right)^r$ approach zero. Thus,

$$\vec{X}_{r+1} = \lambda_1^r c_1 \vec{X}^{(1)}. \tag{4.137}$$

This indicates that the $(r+1)$th trial vector becomes identical to the first eigenvector to within a multiplicative constant. Since

$$\vec{X}_r = \lambda_1^{r-1} c_1 \vec{X}^{(1)}, \tag{4.138}$$

the first eigenvalue λ_1 can be found by taking the ratio of any two corresponding components in the vectors $\vec{X}_r$ and $\vec{X}_{r+1}$, that is,

$$\lambda_1 \approx \frac{x_{i,r+1}}{x_{i,r}}, \tag{4.139}$$

where $x_{i,r}$ and $x_{i,r+1}$ are the ith components of the vectors $\vec{X}_r$ and $\vec{X}_{r+1}$, respectively. Alternatively, λ_1 can be found as

$$\lambda_1 \approx \left(\frac{\vec{X}_{r+1}^T \vec{X}_{r+1}}{\vec{X}_{r+1}^T \vec{X}_r} \right). \tag{4.140}$$

Although it is theoretically necessary to have $r \to \infty$ for the convergence of the method, in practice only a finite number of iterations are sufficient to find a good estimate of λ_1. The actual number of iterations (r) necessary to find λ_1 to within a desired degree of accuracy depends on the closeness of the initial trial vector $\vec{X}_1$ to the dominant eigenvector, $\vec{X}^{(1)}$, and, also, on how well λ_1 and λ_2 are separated. Fewer iterations will be required if λ_2 is very small compared with λ_1. It can be noted that the power method has a distinct advantage in which any errors made during the computational process will not yield incorrect results. For example, if an error is made in multiplying $[A]$ and $\vec{X}_i$, the result will not be the desired trial vector $\vec{X}_{i+1}$. But the wrong vector can be considered as a new trial vector. This may delay the convergence, but will not produce wrong results. In general, we can take any set of n numbers for the first trial vector, $\vec{X}_1$, and still achieve convergence to the dominant eigenvector. Only in the unusual case in which the trial vector $\vec{X}_1$ is exactly proportional to one of the other eigenvectors $\vec{X}^{(i)} (i \neq 1)$, the method fails to converge to $\vec{X}^{(1)}$. In such a case, the premultiplication of $\vec{X}^{(i)} (i \neq 1)$ by $[A]$ results in a vector proportional to $\vec{X}^{(i)}$ itself.

Algorithm

1. To find the dominant or the largest eigenvalue of the given matrix $[A]$, start with a nonzero trial vector $\vec{X}_1$ such that the largest component of $\vec{X}_1$, labelled as c_1, is equal to one. Set $k = 1$.

2. Find $\vec{X}_{k+1} = [A]\vec{X}_k$.

3. If $\left|\frac{x_{i,k+1} - x_{i,k}}{x_{i,k}}\right| < \varepsilon$ for $i = 1, 2, \ldots, n$, convergence is achieved and hence go to step 4. Otherwise, find

$$c_{k+1} = \max_{i=1,n} |x_{i,k+1}|$$

   and normalize $\vec{X}_{k+1}$ by dividing each of its components by c_{k+1} as $\frac{x_{i,k+1}}{c_{k+1}}$.

   Set $k = k + 1$ and go to step 2.

4. Find the converged value of the largest eigenvalue as

$$\lambda_1 = \frac{x_{i,k+1}}{x_{i,k}}$$

   and stop the process.

▶Example 4.19

Find the largest eigenvalue and the corresponding eigenvector of the matrix

$$[A] = \begin{bmatrix} 12 & 6 & -6 \\ 6 & 16 & 2 \\ -6 & 2 & 16 \end{bmatrix}.$$

Solution

By using the initial trial vector, $\vec{X}_1 = \{1.0 \quad 0.5 \quad -0.5\}^T$, the second trial vector can be obtained as

$$\vec{X}_2 = [A]\vec{X}_1 = \begin{Bmatrix} 18.0 \\ 13.0 \\ -13.0 \end{Bmatrix}.$$

The estimate of λ_1 is given by the ratios $\left(\frac{x_{i,2}}{x_{i,1}}\right) = \frac{18.0}{1.0} = 18.0$ for $i = 1$, $\frac{13.0}{0.5} = 26.0$ fot $i = 2$, and $\frac{-13.0}{-0.5} = 26.0$ for $i = 3$. Since these values have not converged to a single value, the iterative process is continued until a converged value of $\lambda_1 = 21.5439$ with

Table 4.4

Iteration (k)	$\vec{X}_k$			$\vec{X}_{k+1} = [A]\vec{X}_k$			$\frac{x_{1,k+1}}{x_{1,k}}$	$\frac{x_{2,k+1}}{x_{2,k}}$	$\frac{x_{3,k+1}}{x_{3,k}}$
1	1.000000	0.500000	−0.500000	18.000000	13.000000	−13.000000	18.000000	26.000000	26.000000
2	1.000000	0.722222	−0.722222	20.666664	16.111111	−16.111111	20.666664	22.307692	22.307692
3	1.000000	0.779570	−0.779570	21.354839	16.913980	−16.913979	21.354839	21.696552	21.696548
4	1.000000	0.792044	−0.792044	21.504532	17.088621	−17.088619	21.504532	21.575333	21.575333
5	1.000000	0.794652	−0.794652	21.535826	17.125130	−17.125128	21.535826	21.550474	21.550474
6	1.000000	0.795193	−0.795193	21.542313	17.132696	−17.132694	21.542313	21.545341	21.545341
7	1.000000	0.795304	−0.795304	21.543653	17.134262	−17.134260	21.543653	21.544281	21.544283
8	1.000000	0.795328	−0.795327	21.543930	17.134588	−17.134584	21.543930	21.544064	21.544062

$\vec{X}^{(1)} = \{1.0000 \quad 0.7953 \quad -0.7953\}^T$ is obtained. The details of the iterative process are shown in Table 4.4. ◀

4.11.1 Inverse Power Method

In many practical situations, the computation of the smallest eigenvalue (λ_n) will be of interest. In such a case, the inverse power method can be used. By premultiplying Eq. (4.2) by $[A]^{-1}$, we obtain

$$[A]^{-1}[A]\vec{X} = \lambda[A]^{-1}\vec{X}$$

or

$$[\tilde{A}]\vec{X} = \tilde{\lambda}\vec{X}, \tag{4.141}$$

where $[\tilde{A}] = [A]^{-1}$ and $\tilde{\lambda} = \frac{1}{\lambda}$. According to Eq. (4.126),

$$|\lambda_n| \leq |\lambda_{n-1}| \leq \cdots \leq |\lambda_2| \leq |\lambda_1|. \tag{4.142}$$

By defining

$$\tilde{\lambda}_i = \frac{1}{\lambda_i}; \quad i = 1, 2, \ldots, n. \tag{4.143}$$

Eq. (4.142) can be restated as

$$|\tilde{\lambda}_n| \geq |\tilde{\lambda}_{n-1}| \geq \cdots \geq |\tilde{\lambda}_2| \geq |\tilde{\lambda}_1|. \tag{4.144}$$

It can be seen that all the relations, such as Eq. (4.141) and (4.144), are similar to those used in the power method provided that quantities with tildes are used. As in the power method, we start with any arbitrary trial vector $\vec{X}_1$ and premultiply it by $[\tilde{A}]$ to generate a second trial vector

$$\vec{X}_2 = [\tilde{A}]\vec{X}_1. \tag{4.145}$$

Since the arbitrary vector $\vec{X}_1$ can be represented as a linear combination of the n eigenvectors of the matrix $[\tilde{A}]$, we have

$$\vec{X}_1 = \sum_{i=1}^{n} c_i \vec{X}^{(i)}, \tag{4.146}$$

where c_i are the unknown constants and $\vec{X}^{(i)}$ are the unknown eigenvectors. Equations (4.145) and (4.146) yield

$$\vec{X}_2 = \sum_{i=1}^{n} c_i \left([\tilde{A}]\vec{X}^{(i)}\right) = \sum_{i=1}^{n} c_i \tilde{\lambda}_i \vec{X}^{(i)}. \tag{4.147}$$

Continuation of the procedure gives, after r iterations,

$$\begin{aligned}\vec{X}_{r+1} = [\tilde{A}]\vec{X}_r &= \sum_{i=1}^{n} c_i \tilde{\lambda}_i^r \vec{X}^{(i)} = c_1 \tilde{\lambda}_1^r \vec{X}^{(1)} + c_2 \tilde{\lambda}_2^r \vec{X}^{(2)} + \cdots + c_n \tilde{\lambda}_n^r \vec{X}^{(n)} \\ &= \tilde{\lambda}_n^r \left[c_1 \left(\frac{\tilde{\lambda}_1}{\tilde{\lambda}_n}\right)^r \vec{X}^{(1)} + c_2 \left(\frac{\tilde{\lambda}_2}{\tilde{\lambda}_n}\right)^r + \cdots \right. \\ &\left. + c_{n-1} \left(\frac{\tilde{\lambda}_{n-1}}{\tilde{\lambda}_n}\right)^r \vec{X}^{(n-1)} + c_n \vec{X}^{(n)} \right].\end{aligned} \tag{4.148}$$

In view of Eq. (4.144),

$$\left(\frac{\tilde{\lambda}_i}{\tilde{\lambda}_n}\right)^r \to 0; \quad i = 1, 2, \ldots, n-1 \tag{4.149}$$

as $r \to \infty$, and Eq. (4.148) gives

$$\vec{X}_{r+1} = c_n \tilde{\lambda}_n^r \vec{X}^{(n)}. \tag{4.150}$$

As before, $\tilde{\lambda}_n$ is determined as the ratio of the ith components of the vectors $\vec{X}_{r+1}$ and $\vec{X}_r$:

$$\tilde{\lambda}_n = \frac{x_{i,r+1}}{x_{i,r}}. \tag{4.151}$$

Since $\tilde{\lambda}_n = \frac{1}{\lambda_n}$, Eq. (4.151) gives the desired smallest eigenvalue

$$\lambda_n = \frac{x_{i,r}}{x_{i,r+1}}, \tag{4.152}$$

and the corresponding eigenvector, $\vec{X}^{(n)}$, is given by

$$\vec{X}^{(n)} = \vec{X}_{r+1} = [\tilde{A}]\vec{X}_r. \tag{4.153}$$

▶Example 4.20

Find the smallest eigenvalue and the corresponding eigenvector of the matrix

$$[A] = \begin{bmatrix} 12 & 6 & -6 \\ 6 & 16 & 2 \\ -6 & 2 & 16 \end{bmatrix}$$

using the inverse power method.

Table 4.5

Iteration (k)	$\vec{X}_k$			$\vec{X}_{k+1} = [A]\vec{X}_k$			$\frac{x_{1,k+1}}{x_{1,k}}$	$\frac{x_{2,k+1}}{x_{2,k}}$	$\frac{x_{3,k+1}}{x_{3,k}}$
1	1.000000	−0.500000	0.500000	0.208333	−0.125000	0.125000	0.208333	0.250000	0.250000
2	1.000000	−0.600000	0.600000	0.220833	−0.137500	0.137500	0.220833	0.229167	0.229167
3	1.000000	−0.622642	0.622642	0.223664	−0.140330	0.140330	0.223664	0.225379	0.225379
4	1.000000	−0.627417	0.627417	0.224260	−0.140927	0.140927	0.224260	0.224615	0.224615
5	1.000000	−0.628408	0.628408	0.224384	−0.141051	0.141051	0.224384	0.224458	0.224458
6	1.000000	−0.628613	0.628613	0.224410	−0.141077	0.141077	0.224410	0.224425	0.224425
7	1.000000	−0.628656	0.628656	0.224415	−0.141082	0.141082	0.224415	0.224418	0.224418
8	1.000000	−0.628665	0.628665	0.224416	−0.141083	0.141083	0.224416	0.224417	0.224417

Solution

The inverse of $[A]$ is given by

$$[\tilde{A}] = \begin{bmatrix} 0.14583333 & -0.0625 & 0.0625 \\ -0.0625 & 0.090277784 & -0.034722228 \\ 0.0625 & -0.034722224 & 0.090277784 \end{bmatrix}.$$

By using the initial trial vector, $\vec{X}_1 = \{1.0 - 0.5\ 0.5\}^T$, the second trial vector can be obtained as follows:

$$\vec{X}_2 = [\tilde{A}]\vec{X}_1 = \begin{Bmatrix} 0.208333 \\ -0.125000 \\ 0.125000 \end{Bmatrix}.$$

The estimate of $\tilde{\lambda}_n$ is given by the ratios $\left(\frac{x_{i,2}}{x_{i,1}}\right) = \frac{0.208333}{1.0} = 0.208333$ for $i = 1$, $\frac{-0.125000}{-0.5} = 0.25$ for $i = 2$, and $\frac{0.125000}{0.5} = 0.25$ for $i = 3$. Since these values have not converged to a single value, the iterative process is continued until a converged value of $\tilde{\lambda}_n = 0.22441642$ with $\vec{X}^{(n)} = \{\ 1.0 \quad -0.628667 \quad 0.628667\ \}^T$ is obtained. The details of the iterative process are shown in Table 4.5. ◀

4.11.2 Computation of Intermediate Eigenvalues

Once the largest eigenvalue (λ_1) and the corresponding eigenvector $\vec{X}^{(1)}$ are determined, it is possible to find the next largest eigenvalue using a modified iterative process. Before we proceed, it is to be remembered that any arbitrary trial vector premultiplied by $[A]$ would lead again to the largest eigenvalue. Therefore, it is necessary to remove the largest eigenvalue from the characteristic equation corresponding to $[A]$. Thus, the matrix $[A_1] = [A]$ is to be transformed to $[A_2]$, which possesses only the remaining eigenvalues. In general, after finding the eigenvalues $\lambda_1 > \lambda_2 > \cdots > \lambda_i$ and the corresponding eigenvectors $\vec{X}^{(1)}, \vec{X}^{(2)}, \ldots, \vec{X}^{(i)}$, the matrix $[A]$ is to be transformed to $[A_{i+1}]$ to find the remaining eigenvalues $\lambda_{i+1}, \lambda_{i+2}, \ldots, \lambda_n$ and the corresponding eigenvectors $\vec{X}^{(i+1)}, \vec{X}^{(i+2)}, \ldots, \vec{X}^{(n)}$. The procedure is called the *deflation* method. We consider a particular method, known as *Hotelling deflation*, which is applicable only to real symmetric matrices, in this section. According to this procedure, the so-called deflated matrix $[A_i]$ to be used in finding the eigenvector $\vec{X}^{(i)}$ is determined by using the relation:

$$[A_{i+1}] = [A_i] - \lambda_i \vec{X}^{(i)} \vec{X}^{(i)^T}; \quad i = 1, 2, \ldots, n-1, \tag{4.154}$$

where $[A_1] = [A]$ and the eigenvector $\vec{X}^{(i)}$ has been normalized as

$$\vec{X}^{(i)^T} \vec{X}^{(i)} = 1; \quad i = 1, 2, \ldots, n-1. \tag{4.155}$$

For example, when Eq. (4.154) is considered with $i = 1$,

$$[A_2] = [A_1] - \lambda_1 \vec{X}^{(1)} \vec{X}^{(1)^T}. \tag{4.156}$$

If the dominant or the first eigenvector $\vec{X}^{(1)}$ is premultiplied by $[A_2]$, we obtain

$$[A_2]\vec{X}^{(1)} = [A_1]\vec{X}^{(1)} - \lambda_1 \vec{X}^{(1)} \left(\vec{X}^{(1)^T} \vec{X}^{(1)}\right) = [A]\vec{X}^{(1)} - \lambda_1 \vec{X}^{(1)} = \vec{0} \tag{4.157}$$

in view of Eqs. (4.155) and (4.127). When an arbitrary trial vector $\vec{X}_1$ is used in conjunction with $[A_2]$, the iterative process yields the second trial vector

$$\vec{X}_2 = [A_2]\vec{X}_1. \tag{4.158}$$

Noting that $\vec{X}_1$ can be expressed by Eq. (4.128), we have

$$\begin{aligned} \vec{X}_2 &= [A_2]\left(\sum_{i=1}^{n} c_i \vec{X}^{(i)}\right) = \left[[A_1] - \lambda_1 \vec{X}^{(1)} \vec{X}^{(1)^T}\right] \sum_{i=1}^{n} c_i \vec{X}^{(i)} \\ &= \sum_{i=1}^{n} c_i [A_1]\vec{X}^{(i)} - \lambda_1 \vec{X}^{(1)} \left(\sum_{i=1}^{n} c_i \vec{X}^{(1)^T} \vec{X}^{(i)}\right). \end{aligned} \tag{4.159}$$

Since the eigenvectors are orthogonal, the quantity in parenthesis will be equal to c_1. This gives

$$\vec{X}_2 = \sum_{i=1}^{n} c_i \lambda_i \vec{X}^{(i)} - c_1 \lambda_1 \vec{X}^{(1)} = \sum_{i=2}^{n} c_i \lambda_i \vec{X}^{(i)}, \tag{4.160}$$

which can be seen to be entirely free of the first eigenvector $\vec{X}^{(1)}$. In general, the deflated matrix $[A_{i+1}]$ will be free of all the previous eigenvalues $\lambda_1, \lambda_2, \ldots, \lambda_i$ and hence $[A_{i+1}]$ can be used in the iterative process with any arbitrary initial trial vector $\vec{X}_1$ to seek the eigenvector $\vec{X}^{(i+1)}$. The main drawback of the deflation procedure is that the deflated matrices $[A_{i+1}]$ are based on the previously computed eigenvalues and eigenvectors, and hence, any error made in the computation of any eigenvalue/eigenvector adversely influences the accuracy of the subsequent eigenvalues and eigenvectors. This is the reason why the Hotelling method is generally used only for small eigenvalue problems. For large problems, more accurate and efficient methods are available [4.6].

▶Example 4.21

Find the second largest eigenvalue and the corresponding eigenvector of the matrix

$$[A] = \begin{bmatrix} 12 & 6 & -6 \\ 6 & 16 & 2 \\ -6 & 2 & 16 \end{bmatrix}$$

if the largest eigenvalue and the corresponding eigenvector are known to be

$$\lambda_1 = 21.543930, \quad \vec{X}^{(1)} = \begin{Bmatrix} 1.0 \\ 0.7953326 \\ -0.7953324 \end{Bmatrix}.$$

Solution

By assuming the eigenvector $\vec{X}^{(1)}$ to be $k \begin{Bmatrix} 1.0 \\ 0.7953326 \\ -0.7953326 \end{Bmatrix}$, where k is a constant, it is normalized as $\vec{X}^{(1)^T}\vec{X}^{(1)} = 1$, which gives $k = 0.66443967$. Using $\vec{X}^{(1)} = \begin{Bmatrix} 0.66443967 \\ 0.52845053 \\ -0.52845053 \end{Bmatrix}$ with $[A_1] = [A]$, the deflated matrix, $[A_2]$, can be found as

$$[A_2] = [A_1] - \lambda_1 \vec{X}^{(1)} \vec{X}^{(1)^T}$$

$$= \begin{bmatrix} 2.4887819 & -1.5645809 & 1.5645790 \\ -1.5645809 & 9.9836426 & 8.0163555 \\ 1.5645790 & 8.0163555 & 9.9836445 \end{bmatrix}.$$

Using the initial trial vector $\vec{X}_1 = \{\ 1.0 \quad 1.0 \quad 1.0\ \}^T$, the iterative process is

$$\vec{X}_{k+1} = [A_2]\vec{X}_k,\ k = 1, 2, \ldots.$$

When convergence is achieved, the converged value of $\left(\frac{x_{i,k+1}}{x_{i,k}}\right)$ gives the eigenvalue λ_2. The details of the iterative process are shown in Table 4.6. ◀

4.11.3 Deflation Procedure for General Matrices

The Hotelling deflation procedure is applicable to symmetric matrices only. In this section, we consider another deflation procedure, which is applicable to both

Table 4.6

Iteration (k)	$\vec{X}_k$			$\vec{X}_{k+1} = [A_2]\vec{X}_k$			$\frac{x_{1,k+1}}{x_{1,k}}$	$\frac{x_{2,k+1}}{x_{2,k}}$	$\frac{x_{3,k+1}}{x_{3,k}}$
1	1.000000	1.000000	1.000000	2.488780	16.435417	19.564579	2.488780	16.435417	19.564579
2	0.127208	0.840060	1.000000	0.566832	16.204185	16.916889	4.455927	19.289324	16.916889
3	0.033507	0.957870	1.000000	0.149305	17.526966	17.714697	4.455944	18.297850	17.714697
4	0.008428	0.989403	1.000000	0.037555	17.881010	17.928234	4.455810	18.072533	17.928234
5	0.002095	0.997366	1.000000	0.009333	17.970423	17.982162	4.455297	18.017883	17.982162
6	0.000519	0.999347	1.000000	0.002311	17.992668	17.995579	4.453048	18.004423	17.995579
7	0.000128	0.999838	1.000000	0.000571	17.998182	17.998905	4.444361	18.001095	17.998905
8	0.000032	0.999960	1.000000	0.000140	17.999546	17.999727	4.409506	18.000269	17.999727
9	0.000008	0.999990	1.000000	0.000033	17.999886	17.999931	4.265920	18.000067	17.999931

symmetric and nonsymmetric matrices. We assume that the dominant eigenvalue λ_1 and the corresponding eigenvector $\vec{X}^{(1)}$ have been found, and that the eigenvector $\vec{X}^{(1)}$ is scaled by making the largest element (say, the pth element) of $\vec{X}^{(1)}$ equal to one. Then the deflated matrix $[A_2]$ is constructed as

$$[A_2] = [A_1] - \vec{X}^{(1)}\vec{r}_p^{\,T}, \tag{4.161}$$

where $[A_1] = [A]$, and $\vec{r}_p^{\,T}$ is the pth row of the matrix $[A]$. Since any arbitrary trial vector $\vec{X}_1$ can be expressed, as indicated in Eq. (4.128), as follows:

$$\vec{X}_1 = \sum_{i=1}^{n} c_i \vec{X}^{(i)}. \tag{4.162}$$

It is assumed that, although, all the eigenvectors $\vec{X}^{(i)}$ in Eq. (4.162) are not known, they are considered to be normalized so that their pth element is equal to one. Note that this normalization is different from the one used in Eq. (4.155). By premultiplying Eq. (4.162) by $[A_2]$ of Eq. (4.161), we obtain

$$[A_2]\vec{X}_1 = \sum_{i=1}^{n} c_i [A_1]\vec{X}^{(i)} - X^{(1)} \sum_{i=1}^{n} c_i \left(\vec{r}_p^{\,T}\vec{X}^{(i)}\right). \tag{4.163}$$

Note that the pth row of $[A]$ multiplied by $\vec{X}^{(i)}$ is equal to λ_i times the pth element of $\vec{X}^{(i)}$, which is equal to one. This gives

$$\vec{r}_p^{\,T}\vec{X}^{(i)} = \lambda_i. \tag{4.164}$$

Thus,

$$[A_2]\vec{X}_1 = \sum_{i=1}^{n} c_i\lambda_i \vec{X}^{(i)} - \vec{X}^{(1)} \sum_{i=1}^{n} c_i\lambda_i = \sum_{i=2}^{n} c_i\lambda_i \left(\vec{X}^{(i)} - \vec{X}^{(1)}\right). \tag{4.165}$$

Equation (4.165) indicates that $[A_2]\vec{X}_1$ is always deficient in $\vec{X}^{(1)}$, and hence, the iterative process converges to λ_2 and $\vec{X}^{(2)}$. The procedure can be repeated to find all the intermediate eigenvalues and eigenvectors. It is to be noted that the application of the iterative process

$$\vec{X}_{k+1} = [A_2]\vec{X}_k \tag{4.166}$$

converges to the vector $\vec{X}_{k+1} = (\vec{X}^{(2)} - \vec{X}^{(1)})$ instead of $\vec{X}^{(2)}$. Then $\vec{X}^{(2)}$ can be found easily since $\vec{X}^{(1)}$ is already known.

4.12 Choice of Method

The method of determining the eigenvalues by solving the characteristic polynomial can be used for matrices of small order (usually less than 5 or 6). It becomes increasingly difficult to find eigenvalues very accurately with large size matrices. The power method or the inverse power method is most efficient when only the largest or the smallest eigenvalue and the corresponding eigenvector are to be found. The power method is also efficient when only a few of the smallest or largest eigenvalues and the corresponding eigenvectors are required. An advantage of the power method is that the eigenvectors are obtained simultaneously with the eigenvalues.

If all the eigenvalues and eigenvectors of a matrix are required, the Jacobi method is to be preferred. In this method, all the eigenvalues and the eigenvectors are determined simultaneously. A disadvantage of the Jacobi method is that the elements that are annihilated or reduced to zero by a particular rotation may not necessarily remain zero during subsequent rotations. Hence, it might take a large number of rotations (iterations) in finding the solution of an eigenvalue problem. This method is not efficient if the size of the matrix is large.

Given's and Householder methods find one eigenvalue at a time and hence are useful for finding only a few of the eigenvalues and eigenvectors. Unlike Jacobi method, Given's method preserves the zeros of the off-diagonal locations once they are created. It requires $\frac{1}{2}(n-1)(n-2)$ rotations, while the Householder method requires only $(n-2)$ transformations to reduce a full matrix of order n to a tridiagonal form. But each transformation in the Householder method is more complex than the rotations used in Given's method. However, the Householder method is considered to be the fastest and most accurate of all the methods. Hence, it is the preferred method for large order matrices.

4.13 Use of Software Packages

4.13.1 MATLAB

▶Example 4.22

Find the eigenvalues and eigenvectors of the following matrix:

$$[A] = \begin{bmatrix} 2 & 1 & 3 & 4 \\ 1 & -3 & 1 & 5 \\ 3 & 1 & 6 & -2 \\ 4 & 5 & -2 & -1 \end{bmatrix}. \tag{E1}$$

Solution

```
>> A=[2 1 3 4; 1 -3 1 5; 3 1 6 -2; 4 5 -2  -1]

A =

     2     1     3     4
     1    -3     1     5
     3     1     6    -2
     4     5    -2    -1

>> b=eig(A)

b =

    7.9329
    5.6689
   -1.5732
   -8.0286

>> [V, d] = eig(A)

V =

    0.5601    0.3787     0.6880    0.2635
    0.2116    0.3624    -0.6241    0.6590
    0.7767   -0.5379    -0.2598   -0.1996
    0.1954    0.6602    -0.2638   -0.6756

d =

    7.9329         0          0         0
         0    5.6689          0         0
         0         0    -1.5732         0
         0         0          0   -8.0286

>>
```

Note: The use of the command "`b = eig(A)`" gives the eigenvalues of the matrix $[A]$. The use of the command "`[V,d] = eig(A)`" gives the eigenvalues as diagonal elements of the matrix $[D]$ and the eigenvectors as corresponding columns of the matrix $[V]$.

◀

▶Example 4.23

Find the eigenvalues and eigenvectors of the following nonsymmetric matrix:

$$[A] = \begin{bmatrix} 8 & -1 & -5 \\ -4 & 4 & -2 \\ 18 & -5 & -7 \end{bmatrix} \quad \text{(E1)}$$

Solution

```
>> A=[8 -1 -5; -4 4 -2; 18 -5 -7]

A =

     8    -1    -5
    -4     4    -2
    18    -5    -7

>> b=eig(A)

b =

   2.0000 + 4.0000i
   2.0000 - 4.0000i
   1.0000

>> [V, D]=eig(A)

V =

   0.0484 + 0.4446i   0.0484 - 0.4446i   0.4082
  -0.3962 + 0.4930i  -0.3962 - 0.4930i   0.8165
   0.4930 + 0.3962i   0.4930 - 0.3962i   0.4082

D =

   2.0000 + 4.0000i        0                  0
        0             2.0000 - 4.0000i        0
        0                  0             1.0000

>>
```

Note: Eigenvalues and eigenvectors are complex.

◀

▶Example 4.24

Determine the eigenvalues and eigenvectors of the following matrix with $n = 20$:

$$[A] = [a_{ij}] = \left[\frac{1}{i+j-1}\right]; \quad i, j = 1, 2, \ldots, n. \tag{E1}$$

Solution

The matrix $[A]$ is defined in an m.file labeled *Rao41.m*, whose listing is as follows:

```
function [A] = Rao41(n)
A=zeros(n,n);
for i=1:n
   for j=1: n
      A(i,j)=1/(i+j-1);
   end
end
```

```
Operations with matrix [A]:

>> n=20;
>> [A]=Rao41(n);
>> b=eig(A)


b =

   -0.0000
    0.0000
   -0.0000
    0.0000
    0.0000
   -0.0000
    0.0000
    0.0000
    0.0000
    0.0000
```

```
    0.0000
    0.0000
    0.0000
    0.0000
    0.0001
    0.0009
    0.0090
    0.0756
    0.4870
    1.9071

>> [V,D]=eig(A)

V =

    0.6412
    0.4049
    0.3089
    0.2538
    0.2171
    0.1906
    0.1704
    0.1544
    0.1413
    0.1304      Only column 20 (eigenvalue 20) is shown here.
    0.1212
    0.1132
    0.1063
    0.1002
    0.0948
    0.0900
    0.0857
    0.0818
    0.0782
    0.0749

D =

         0
         0
```

```
          0
          0
          0
          0
          0
          0
          0
          0      Only column 20 (containing eigenvalue 20) is
          0      shown here.
          0
          0
          0
          0
          0
          0
          0
          0
     1.9071
```

◀

4.13.2 MAPLE

▶Example 4.25

Find the eigenvalues and eigenvectors of the matrix:

$$[A] = \begin{bmatrix} \mathrm{A} & \mathrm{B} \\ \mathrm{C} & \mathrm{D} \end{bmatrix} \tag{E1}$$

Solution

```
> A7:=array([[A,B],[C,D]]);
```

$$A7 := \begin{bmatrix} A & B \\ C & D \end{bmatrix}$$

```
> eigenvalues(A7);
```

$$\frac{1}{2}A + \frac{1}{2}\mathrm{D} + \frac{1}{2}\sqrt{A^2 - 2A\mathrm{D} + \mathrm{D}^2 + 4BC},\ \frac{1}{2}A + \frac{1}{2}\mathrm{D} - \frac{1}{2}\sqrt{A^2 - 2A\mathrm{D} + \mathrm{D}^2 + 4BC}$$

```
> eigenvectors(A7);
```

$$\left[\frac{1}{2}A+\frac{1}{2}\mathrm{D}+\frac{1}{2}\sqrt{A^2-2A\mathrm{D}+\mathrm{D}^2+4BC},1,\left\{\left[-\frac{-\frac{1}{2}A+\frac{1}{2}\mathrm{D}-\frac{1}{2}\sqrt{A^2-2A\mathrm{D}+\mathrm{D}^2+4BC}}{C},1\right]\right\}\right],\left[\frac{1}{2}A+\frac{1}{2}\mathrm{D}-\frac{1}{2}\sqrt{A^2-2A\mathrm{D}+\mathrm{D}^2+4BC},1,\left\{\left[-\frac{-\frac{1}{2}A+\frac{1}{2}\mathrm{D}+\frac{1}{2}\sqrt{A^2-2A\mathrm{D}+\mathrm{D}^2+4BC}}{C},1\right]\right\}\right]$$

```
>
```

◀

▶Example 4.26

Find the eigenvalues and eigenvectors of the matrix $[A]$ given by Eq. (E1) of Example 4.24.

Solution

```
> with(linalg):
Warning, new definition for norm
Warning, new definition for trace
> n:=20;
```

$$n := 20$$

```
> f:=(i,j)->(i+j-1)^(-1):
> A1:=matrix(n,n,f);
> V1:=evalf(Eigenvals(A1,VC1));
```

$V1 := [-.3769593742\ 10^{-9}, -.8031729369\ 10^{-10}, -.1584963158\ 10^{-10}, -.9808808663\ 10^{-11}, -.4135054336\ 10^{-11}, .3468170793\ 10^{-11}, .7817189510\ 10^{-11}, .1679696767\ 10^{-10}, .5406426261\ 10^{-10}, .1768945321\ 10^{-9}, .5711599357\ 10^{-9}, .1412931042\ 10^{-7}, .2827407338\ 10^{-6}, .4830466825\ 10^{-5}, .00007033434244, .0008676710081, .008961128618, .07559582088, .4870384065, 1.907134722]$

```
> print(transpose(VC1));>
```

(Only the 20th eigenvector is shown here)

[.6411980287, .4049415181, .3088686324, .2537729671, .2171356341, .1906463803, .1704291922, .1544011793, .1413304807, .1304361525, .1211962782, .1132472619, .1063271572, .1002419493, .09484455193, .09002122523, .08568249749, .08175692982, .07818673127, .07492459622]

```
>
```

◀

▶Example 4.27

Find the eigenvalues and eigenvectors of the matrix $[A]$ defined by Eq. (E1) in Example 4.23.

Solution

```
>
>
> A4:=array([[8,-1,-5],[-4,4,-2],[18,-5,-7]]);
```

$$A4 := \begin{bmatrix} 8 & -1 & -5 \\ -4 & 4 & -2 \\ 18 & -5 & -7 \end{bmatrix}$$

```
> V4:=evalf(Eigenvals(A4));
```

$$V4 := [2.000000014 + 4.000000001\, I, 2.000000014 - 4.000000001\, I, 1.000000001]$$

```
> vc4:=eigenvectors(A4);
```

$$vc4 := \left[2+4I, 1, \left\{\left[\frac{1}{2}+\frac{1}{2}I, I, 1\right]\right\}\right], \left[2-4I, 1, \left\{\left[\frac{1}{2}-\frac{1}{2}I, -I, 1\right]\right\}\right], [1, 1, \{[1, 2, 1]\}]$$

```
>
```

◀

4.13.3 MATHCAD

▶Example 4.28

Find the eigenvalues and eigenvectors of the matrix $[A]$ given by Eq. (E1) of Example 4.24.

Solution

```
n:=10
i:=0..9
j:=0..9
```

$$A_{i,j} := \frac{1}{i+j+1}$$

A =

	0	1	2	3	4	5	6	7	8	9
0	1	0.5	0.333	0.25	0.2	0.167	0.143	0.125	0.111	0.1
1	0.5	0.333	0.25	0.2	0.167	0.143	0.125	0.111	0.1	0.091
2	0.333	0.25	0.2	0.167	0.143	0.125	0.111	0.1	0.091	0.083
3	0.25	0.2	0.167	0.143	0.125	0.111	0.1	0.091	0.083	0.077
4	0.2	0.167	0.143	0.125	0.111	0.1	0.091	0.083	0.077	0.071
5	0.167	0.143	0.125	0.111	0.1	0.091	0.083	0.077	0.071	0.067
6	0.143	0.125	0.111	0.1	0.091	0.083	0.077	0.071	0.067	0.063
7	0.125	0.111	0.1	0.091	0.083	0.077	0.071	0.067	0.063	0.059
8	0.111	0.1	0.091	0.083	0.077	0.071	0.067	0.063	0.059	0.056
9	0.1	0.091	0.083	0.077	0.071	0.067	0.063	0.059	0.056	0.053

```
V:=eigenvals(A)
```

V =

	0
0	$1.093 \cdot 10^{-13}$
1	$2.267 \cdot 10^{-11}$
2	$2.147 \cdot 10^{9}$
3	$1.229 \cdot 10^{-7}$
4	$4.73 \cdot 10^{-6}$
5	$1.287 \cdot 10^{-4}$
6	$2.531 \cdot 10^{-3}$
7	0.036
8	0.343
9	1.752

$VC := \mathrm{eigenvec}(A, V_9)$

VC =

	0
0	0.7
1	0.426
2	0.317
3	0.256
4	0.215
5	0.187
6	0.165
7	0.148
8	0.134
9	0.123

◀

> **Note:** Only the first 10 rows and 10 columns of A and the first 10 components of V and VC are shown to save space.

▶Example 4.29

Find the eigenvalues and eigenvectors of the matrix $[A]$ given by Eq. (E1) in Example 4.23.

Solution

$$A4 := \begin{pmatrix} 8 & -1 & -5 \\ -4 & 4 & -2 \\ 18 & -5 & -7 \end{pmatrix}$$

V := eigenvals(A4)

$$V = \begin{pmatrix} 2+4i \\ 2-4i \\ 1 \end{pmatrix}$$

$VC_0 := \text{eigenvec}(A4, V_0)$ $$VC_0 = \begin{pmatrix} -0.362+0.263i \\ -0.625-0.099i \\ -0.099+0.625i \end{pmatrix}$$

$VC_1 := \text{eigenvec}(A4, V_1)$ $$VC_1 = \begin{pmatrix} -0.362-0.263i \\ -0.625+0.099i \\ -0.099-0.625i \end{pmatrix}$$

$VC_2 := \text{eigenvec}(A4, V_2)$ $$VC_2 = \begin{pmatrix} 0.408 \\ 0.816 \\ 0.408 \end{pmatrix}$$

◀

4.14 Computer Programs

4.14.1 Fortran Programs

Program 4.1 Generation of characteristic equation

The following data are required for input:

N = order of the matrix

NP1 = $N + 1$ = number of polynomial coefficients

A(N, N) = matrix[A]

Illustration (Example 4.9); output:

Solution

```
CHARACTERISTIC POLYNOMIAL

GIVEN MATRIX:

 0.120000E+02  0.600000E+01-0.600000E+01
 0.600000E+01  0.160000E+02 0.200000E+01
-0.600000E+01  0.200000E+01 0.160000E+02

COEFFICIENTS OF THE CHARACTERISTIC POLYNOMIAL:
PC(1), PC(2), ..., PC(NP) IN
PC(NP)*X**N+PC(NP-1)*X**N-1+...+PC(1)=0

-0.172800E+04   0.564000E+03-0.440000E+02 0.100000E+01
```

◀

Program 4.2 Jacobi method

The following data are required for input:

n = order of the matrix

$a1(n, n)$ = given matrix [D]

eps = convergence criterion (usual value: 0.001 to 0.00001)

itmax = maximum number of iterations or rotations permitted (usual value: 10 to 50)

Illustration (Example 4.12).

Using $n = 3$, itmax = 10, and eps = 0.001, the output of Program 4.2 is given next:

```
solution by jacobi method

given matrix:

   0.120000E+02   0.600000E+01  -0.600000E+01
   0.600000E+01   0.160000E+02   0.200000E+01
  -0.600000E+01   0.200000E+01   0.160000E+02

method converged in 6 iterations
```

```
eigenvalues:

   0.215380E+02    0.445603E+01  0.179948E+02

eigenvectors (columnwise):

   0.664311E+00    0.747376E+00  0.447966E-06
   0.528399E+00   -0.469807E+00 -0.707005E+00
  -0.528398E+00    0.469807E+00 -0.707005E+00
```

4.14.2 C Programs

Program 4.3 Matrix iteration method

The following are required for input:

n = size of the matrix eigenvalue problem

d(n,n) = matrix [D] in the eigenvalue problem $[D]\vec{X} = \lambda[M]\vec{X}$

xm(n,n) = matrix [M]

nvec = number of eigenvalues and eigenvectors to be found

eps = convergence requirement (small quantity on the order of 10^{-5})

xs(n) = initial guess vector for the eigenvectors

Illustration (Example 4.19)
The following is the program output

Solution

```
Matrix Iteration Method

Matrix d:
    1.200000e+01     6.000000e+00    -6.000000e+00
    6.000000e+00     1.600000e+01     2.000000e+00
   -6.000000e+00     2.000000e+00     1.600000e+01

Matrix xm:
    1.000000e+00     0.000000e+00     0.000000e+00
    0.000000e+00     1.000000e+00     0.000000e+00
    0.000000e+00     0.000000e+00     1.000000e+00

Number of iterations: 75
Eigenvalue 1 2.15440037E+01
Eigenvector:
    1.000000e+00     7.953724e-01    -7.952949e-01

Number of iterations: 16
```

```
Eigenvalue 2 1.80000000e+01
Eigenvector:
  -4.091846e-05    9.999349e-01    1.000000e+00

Number of iterations: 2
Eigenvalue 3 4.45599626e+00
Eigenvector:
   1.000000e+00   -6.286670e-01    6.286670e-01

Summary of results:

Eigenvalues:

   2.154400e+01    1.800000e+01    4.455996e+00

Eigenvectors (Columns):

   1.000000e+00   -4.091846e-05    1.000000e+00
   7.953724e-01    9.999349e-01   -6.286670e-01
  -7.952949e-01    1.000000e+00    6.286670e-01
```

◀

Program 4.4 Jacobi method

The following are required for data: Same as in the case of Program 4.2.

n = size of the matrix eigenvalue problem (size of the matrix [D])

d(n,n) = matrix [D]

itmax = maximum number of iterations or rotations permitted (on the order of 100)

eps = convergence requirement (small quantity on the order of 10^{-5})

Illustration (Example 4.12). Using $n = 3$, itmax = 200, and eps = $1.0e - 5$, the output of Program 4.4 is given here:

Solution

```
Eigenvalue Solution by Jacobi Method:

Given matrix:

   1.200000e+01    6.000000e+00   -6.000000e+00
   6.000000e+00    1.600000e+01    2.000000e+00
  -6.000000e+00    2.000000e+00    1.600000e+01
```

```
Eigenvalues:

   2.154400e+01     1.800000e+01     4.455996e+00

Eigenvectors (Columns):

   6.644392e-01     2.651391e-07     7.473423e-01
   5.284508e-01     7.071066e-01    -4.698297e-01
  -5.284508e-01     7.071069e-01     4.698292e-01
```

◀

REFERENCES AND BIBLIOGRAPHY

4.1. D. K. Faddeev and V. N. Faddeeva, *Computational Methods of Linear Algebra*, (Translated by R. C. Williams), W. H. Freeman & Co., San Francisco, 1963.

4.2. M. Givens, "Computation of Plane Unitary Rotation Transforming a General Matrix to Triangular Form," *Journal of SIAM*, Vol. 6, pp. 26–50, 1958.

4.3. G. H. Hostetter, M. S. Santina, and P. D'Carpio-Montalvo, *Analytical, Numerical, and Computational Methods for Science and Engineering*, Prentice Hall, Englewood Cliffs, NJ, 1991.

4.4. R. S. Martin, C. Reinsch, and J. H. Wilkinson, "Householder's Tridiagonalization of a Symmetric Matrix," *Numerische Mathematik*, Vol. 11, pp. 181–195, 1968.

4.5. A. Jennings and J. J. McKeown, *Matrix Computation*, 2d edition, Wiley, Chichester, 1992.

4.6. M. L. James, G. M. Smith, and J. C. Wolford, *Applied Numerical Methods for Digital Computation*, 4th edition, Harper Collins, NY, 1993.

4.7. S. S. Rao, *Mechanical Vibrations*, 3d edition, Addison-Wesley, Reading, MA, 1995.

4.8. A. Ralston and P. Rabinowitz, *A First Course in Numerical Analysis*, 2d edition, McGraw-Hill, New York, 1978.

4.9. J. H. Wilkinson, *The Algebraic Eigenvalue Problem*, Clarendon Press, Oxford, 1965.

4.10. E. L. Wilson, "The Static Condensation Algorithm," *International Journal for Numerical Methods in Engineering*, Vol. 8, pp. 199–203, 1974.

4.11. R. T. Gregory and D. L. Karney, *A Collection of Matrices for Testing Computational Algorithms*, Wiley, New York, 1969.

4.12. L. Meirovitch, *Analytical Methods in Vibrations*, Macmillan, New York, 1967.

REVIEW QUESTIONS

The following questions along with corresponding answers are available in an interactive format at the Companion Web site at http://www.prenhall.com/rao.

4.1. **Define** the following terms:

Eigenvalue, left eigenvector, right eigenvector, characteristic equation, standard eigenvalue problem, general eigenvalue problem, similarity transformation, Sturm sequence, dominant eigenvalue, Hermitian matrix.

4.2. Answer **true or false** to each of the following:

1. The eigenvalues of a symmetric matrix can be complex.
2. The eigenvalues of a diagonal matrix, $[A] = [a_{ii}]$, are given by a_{ii}.
3. A matrix eigenvalue problem denotes a system of homogeneous equations.
4. A general eigenvalue problem can always be transformed into a special eigenvalue problem.
5. The transformation methods are better suited for large matrices.
6. The iterative methods are better suited for sparse matrices.
7. The Sturm sequence can only find the eigenvalues.
8. Householder's method is the fastest and most accurate transformation method.
9. The power method can be used to find the intermediate eigenvalues directly.
10. The Hotelling deflation procedure is applicable to all types of matrices.
11. The eigenvectors of a matrix are unique.
12. The coordinate transformation matrix is always orthogonal.

4.3. **Give brief answers** to each of the following:

1. What is the difference between left and right eigenvectors?
2. Indicate a method of converting a general eigenvalue problem into a special eigenvalue problem.
3. Name two transformation methods of solving an eigenvalue problem.
4. What is the basic principle used in the transformation methods?
5. State the conditions under which transformation methods are preferred for solving an eigenvalue problem.
6. Indicate the type of problems for which iterative methods are preferred.
7. Name two iterative methods of solving an eigenvalue problem.
8. What is the difference between similarity and orthogonal transformations?
9. What is the role of rotation matrices in the Jacobi method?

(*continued, page 339*)

10. What is the basic difference between the Jacobi and the Given methods?
11. What are the advantages of the power method?
12. What is the use of the deflation procedure?

4.4. In each of the following, **fill in the blanks** with suitable word(s):

1. Transformation methods find ______ the eigenvalues and eigenvectors simultaneously.
2. The characteristic equation represents a ______ equation.
3. The power method finds the ______ eigenvalue and the corresponding eigenvector.
4. Any nonzero vector can be chosen as a ______ vector in the power method.
5. The number of sign changes forms a basis in the ______ ______ method.
6. Householder's method is similar to ______ method in reducing a matrix to a tridiagonal form.
7. The Jacobi method finds the eigenvalues and eigenvectors of ______ matrices.
8. In the Jacobi method, the eigenvalues are identified as the ______ elements of the reduced matrix.
9. The eigenvalue problem is defined only for ______ matrices.
10. The left and right eigenvectors are ______.
11. The power method requires the eigenvalues to be ______.
12. The inverse power method finds the ______ eigenvalue.

4.5. For each of the following, select the **most appropriatte answer** out of the multiple choices given:

1. If the eigenvalues of a 3×3 matrix $[A]$ are known to be 10, 20, and 40, the eigenvalues of $[A]^{-1}$ will be

   (a) 0.1, 0.05 and 0.025; (b) 10, 20, and 40; (c) unrelated to 10, 20, and 40.

2. If the eigenvalues of a 3×3 matrix $[A]$ are known to be 10, 20, and 40, the eigenvalues of $[B] = [P]^T[A][P]$, where $[P]$ is an orthogonal matrix, are given by

   (a) 0.1, 0.05 and 0.025; (b) 10, 20, and 40; (c) unrelated to 10, 20, and 40.

3. If the eigenvalues of a 3×3 matrix $[A]$ are known to be 1, 2, and 3, the eigenvalues of $[A]^T$ are given by

   (a) a, $2a$, and $3a$, where a is an arbitrary constant; (b) −1, −2, and −3; (c) 1, 2, and 3.

4. If the eigenvalues of a 3×3 matrix $[A]$ are known to be 1, 2, and 3, the eigenvalues of $-[A]$ are given by

   (a) a, $2a$, and $3a$, where a is an arbitrary constant; (b) −1, −2, and −3; (c) 1, 2, and 3.

(*continued, page 340*)

5. If the eigenvalues of a 3×3 matrix $[A]$ are known to be $1, 2$, and 3, the eigenvalues of $k[A]$, where $k \neq 0$, are given by
   (a) k, $2k$, and $3k$; (b) $-k$, $-2k$, and $-3k$; (c) 1, 2, and 3.
6. The transformations used in the Jacobi method reduce a symmetric matrix to
   (a) tridiagonal matrix; (b) identity matrix; (c) diagonal matrix.
7. The transformation matrices used in the Householder method are
   (a) orthogonal matrices; (b) rotational matrices; (c) tridiagonal matrices.
8. The power method can be considered
   (a) a transformation method; (b) an iterative method; (c) a root finding method.
9. The determination of the eigenvector corresponding to a known eigenvalue requires the solution of
   (a) a system of nonhomogeneous linear equations; (b) a system of nonlinear equations; (c) a system of homogeneous linear equations.
10. The eigenvalues of a nonsymmetric real matrix are
    (a) always real; (b) always complex; (c) real or complex (or both).

4.6. **Match the following:**

1. Jacobi method	(a) generates a tridiagonal matrix
2. Power method	(b) deflation method
3. Sturm sequence method	(c) transformation method
4. Faddeev–Leverrier method	(d) iterative method
5. Given's method	(e) determinant search method
6. Hotelling method	(f) generates the characteristic polynomial

4.7. **Match the following:**

1. Nonsymmetric matrix	(a) all eigenvalues are zero
2. Zero matrix	(b) all eigenvalues are real
3. Singular matrix	(c) all eigenvalues are equal to one
4. Identity matrix	(d) all eigenvalues are nonzero
5. Hermitian matrix	(e) at least one eigenvalue is zero

PROBLEMS

Section 4.1

4.1. Show that the right and left eigenvectors $\vec{X}^{(i)}$ and $\vec{Y}^{(j)}$ corresponding to the eigenvalues λ_i and λ_j are orthogonal.

4.2.–4.10. Verify the characteristics of eigenvalue problems indicated in Table 4.1 by finding the eigenvalues of the given matrices.

4.2. $[A] = \begin{bmatrix} 2 & 3 \\ 4 & 6 \end{bmatrix}.$

4.3. $[B] = \begin{bmatrix} 1 & 2 \\ 3 & 4 \end{bmatrix}.$

4.4. $[C] = \begin{bmatrix} 2 & 3 \\ 3 & 4 \end{bmatrix}.$

4.5. $[D] = \begin{bmatrix} 0 & 0 \\ 0 & 0 \end{bmatrix}.$

4.6. $[E] = \begin{bmatrix} 1 & 0 \\ 0 & 1 \end{bmatrix}.$

4.7. $[F] = \begin{bmatrix} 7 & 0 \\ 0 & 8 \end{bmatrix}.$

4.8. $[G] = [B]^{-1} = \begin{bmatrix} -2 & 1 \\ 1.5 & -0.5 \end{bmatrix}.$

4.9. $[H] = [P]^T[B][P] = \begin{bmatrix} 5 & -1 \\ -2 & 0 \end{bmatrix}.$

$$\text{with } [P] = \frac{1}{\sqrt{2}}\begin{bmatrix} 1 & 1 \\ 1 & -1 \end{bmatrix}.$$

4.10. $[J] = \begin{bmatrix} 5 & 3+i \\ 3-i & 2 \end{bmatrix}.$

4.11. Determine the right and left eigenvectors of the matrix

$$[A] = \begin{bmatrix} 4 & 1 & 1 \\ 2 & 4 & 1 \\ 0 & 1 & 4 \end{bmatrix}$$

corresponding to the eigenvalue $\lambda_3 = 6$.

4.12. If the eigenvalues of the matrix

$$[A] = \begin{bmatrix} 8 & -1 & -5 \\ -4 & 4 & -2 \\ 18 & -5 & -7 \end{bmatrix}$$

are known to be $\lambda_1 = 2 + 4i$, $\lambda_2 = 2 - 4i$, and $\lambda_3 = 1$, determine the right eigenvectors.

4.13. If the eigenvalues of the matrix

$$[A] = \begin{bmatrix} 33 & 16 & 72 \\ -24 & -10 & -57 \\ -8 & -4 & -17 \end{bmatrix}$$

are known to be $\lambda_1 = 1$, $\lambda_2 = 2$, and $\lambda_3 = 3$, determine the corresponding left eigenvectors.

4.14. Find the left eigenvectors of the matrix $[A]$ given in Problem 4.12. The eigenvalues of the matrix are given in Problem 4.12.

4.15. Find the right and left eigenvectors of the following matrix:

$$[A] = \begin{bmatrix} 2 & 3 \\ 4 & 5 \end{bmatrix}.$$

Section 4.2

4.16. The Rayleigh's quotient, Q, corresponding to the eigenvalue problem $[A]\vec{X} = \lambda\vec{X}$ is given by

$$Q = \frac{\vec{Y}^T[A]\vec{Y}}{\vec{Y}^T\vec{Y}}$$

If $\vec{Y}$ represents an approximation to the eigenvector $\vec{X}$, then Q gives an approximate value of λ. Using this property (also known as Rayleigh's method), estimate the fundamental natural frequency of a shaft carrying three rotors as shown in Fig. 4.6. The governing equation of the system is

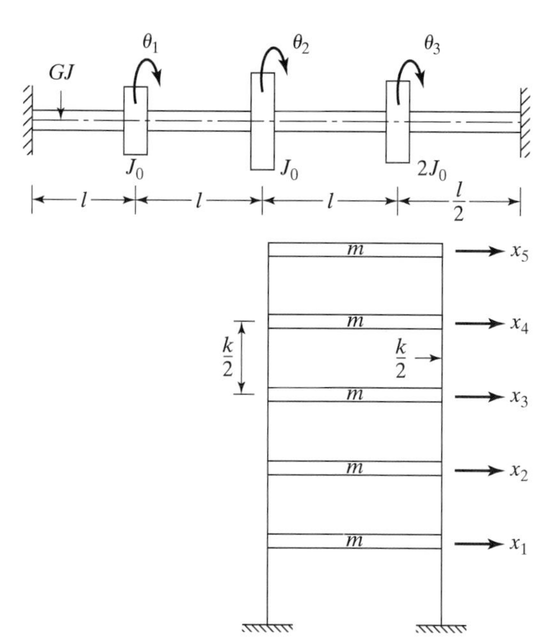

Figure 4.6 Shaft carrying three masses.

Figure 4.7 Building frame ($k = 1.0$ MN/m, $m = 1000$ kg).

given by

$$[D]\vec{\theta} = \lambda\vec{\theta},$$

where $[D]$ is the dynamical matrix

$$[D] = \begin{bmatrix} 10 & 6 & 4 \\ 6 & 12 & 8 \\ 2 & 4 & 12 \end{bmatrix}$$

with $\lambda = \frac{14GJ}{lJ_0\omega^2}$, and GJ is the torsional stiffness of the shaft. Assume $\vec{Y} = \{\ 1\quad 1\quad 1\ \}^T$.

4.17. The equations of motion of a five-story building frame (see Fig. 4.7) are given by

$$10^3 \begin{bmatrix} 1 & 0 & 0 & 0 & 0 \\ 0 & 1 & 0 & 0 & 0 \\ 0 & 0 & 1 & 0 & 0 \\ 0 & 0 & 0 & 1 & 0 \\ 0 & 0 & 0 & 0 & 1 \end{bmatrix} \begin{Bmatrix} \ddot{x}_1 \\ \ddot{x}_2 \\ \ddot{x}_3 \\ \ddot{x}_4 \\ \ddot{x}_5 \end{Bmatrix} + 10^6 \begin{bmatrix} 2 & -1 & 0 & 0 & 0 \\ -1 & 2 & -1 & 0 & 0 \\ 0 & -1 & 2 & -1 & 0 \\ 0 & 0 & -1 & 2 & -1 \\ 0 & 0 & 0 & -1 & 1 \end{bmatrix} \begin{Bmatrix} x_1 \\ x_2 \\ x_3 \\ x_4 \\ x_5 \end{Bmatrix}$$

$$= \begin{Bmatrix} 0 \\ 0 \\ 0 \\ 0 \\ 0 \end{Bmatrix}.$$

Assuming a solution

$$x_i(t) = X_i \cos(\omega t + \phi), i = 1, 2, .., 5,$$

determine the natural frequencies of vibration of the building frame.

4.18. The overhead traveling crane shown in Fig. 4.8(a) can be modeled as a two degree of freedom system as shown in Fig. 4.8(b).

(a) Derive the equations of motion of the system from the free body diagrams of the masses shown in Fig. 4.8(c).

(b) Assuming the harmonic solution

$$x_i(t) = X_i \cos(\omega t + \phi), i = 1, 2$$

derive the eigenvalue problem for finding the natural frequencies, ω.

(c) Find the natural frequencies and mode shapes (eigenvectors) of the system.

4.19. The free vibration equations of a tightly stretched string carrying three masses (shown in Fig. 4.9) are given by

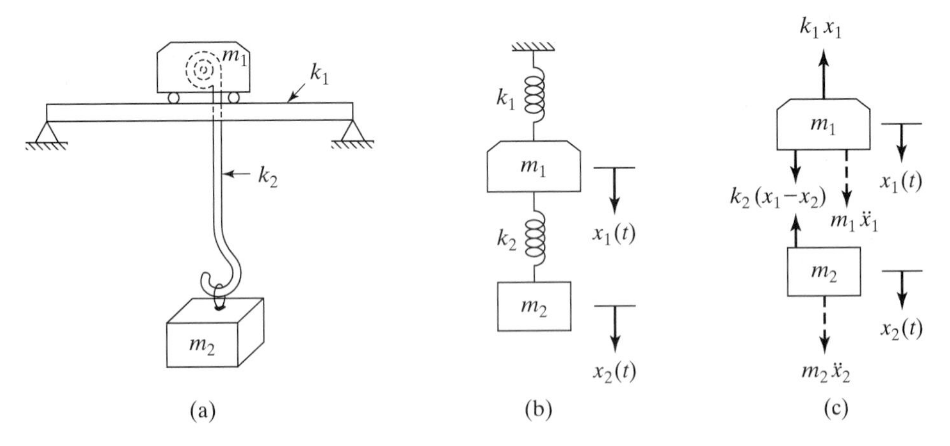

Figure 4.8 Overhead traveling crane.

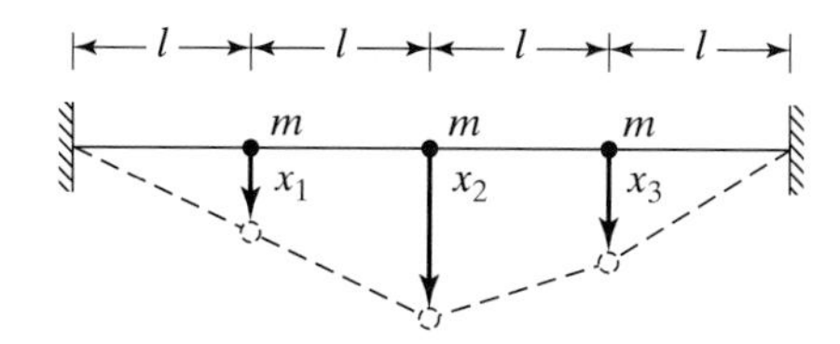

Figure 4.9 String carrying three masses.

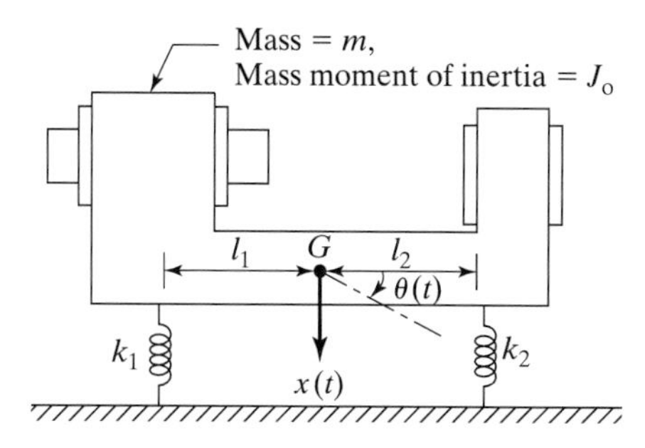

Figure 4.10 Modeling of machine tool and foundation.

$$\frac{l\,m}{4\,P}\begin{bmatrix} 3 & 2 & 1 \\ 2 & 4 & 2 \\ 1 & 2 & 3 \end{bmatrix}\begin{Bmatrix} \ddot{x}_1 \\ \ddot{x}_2 \\ \ddot{x}_3 \end{Bmatrix} + \begin{Bmatrix} x_1 \\ x_2 \\ x_3 \end{Bmatrix} = \begin{Bmatrix} 0 \\ 0 \\ 0 \end{Bmatrix},$$

where P is the tension in the string. Derive the eigenvalue problem assuming harmonic solution and find the natural frequencies and mode shapes of the system.

4.20. A machine tool and its foundation can be modeled as a two degree of freedom system, shown in Fig. 4.10, where $x(t)$ denotes the translatory motion of the c.g. and $\theta(t)$ indicates the rotational motion of the machine tool. The equations of motion of the machine tool can be expressed as

$$m\ddot{x} + k_1(x - l_1\theta) + k_2(x + l_2\theta) = 0$$

and

$$J_o\ddot{\theta} - k_1 l_1(x - l_1\theta) + k_2 l_2(x + l_2\theta) = 0.$$

(i) Assuming harmonic motion

$$x(t) = X\cos(\omega t + \phi),\ \theta(t) = \Theta\cos(\omega t + \phi)$$

derive the free vibration equations of the system.

(ii) Solve the eigenvalue problem derived in part (i) to find the natural frequencies of the system.

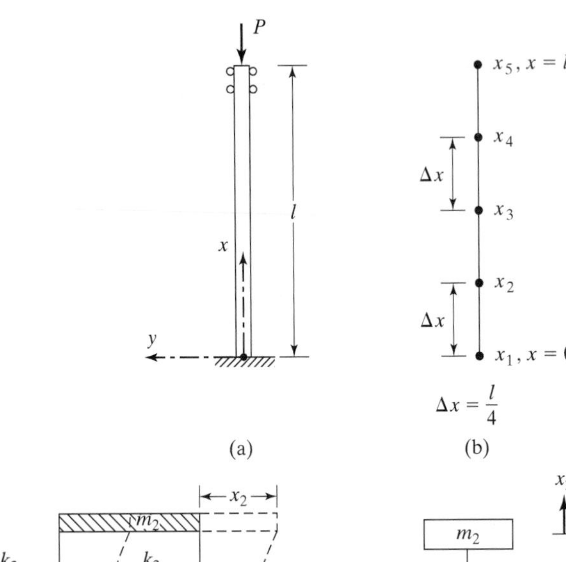

Figure 4.11 Fixed-pinned column.

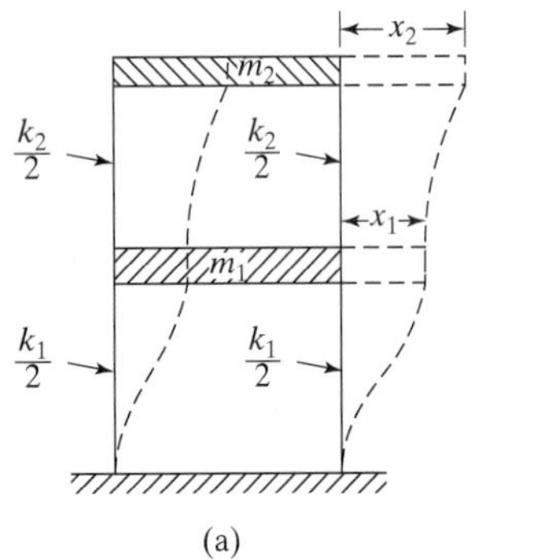

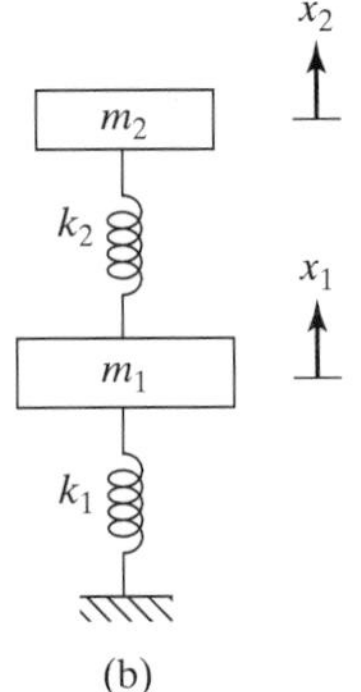

Figure 4.12 Two-story building frame.

4.21. A fixed-pinned column carrys an axial load P shown in Fig. 4.11(a). Divide the column into four segments shown in Fig. 4.11(b), and derive the eigenvalue problem for finding the buckling or critical load using a finite difference approach. The governing equation for finding the buckling load is given by

$$\frac{d^4y}{dx^4} + \frac{P}{EI}\frac{d^2y}{dx^2} = 0.$$

(See Example 4.4.)

4.22. The two-story building frame shown in Fig. 4.12(a) can be modeled as a two degree of freedom system shown in Fig. 4.12(b). The masses m_1 and m_2 represent the masses of the floors and the stiffnesses k_1 and k_2 denote the bending stiffnesses of the columns. The natural frequencies and mode shapes of the frame are given by

$$\omega^2 \begin{bmatrix} m_1 & 0 \\ 0 & m_2 \end{bmatrix} \begin{Bmatrix} x_1 \\ x_2 \end{Bmatrix} = \begin{bmatrix} k_1 + k_2 & -k_2 \\ -k_2 & k_2 \end{bmatrix} \begin{Bmatrix} x_1 \\ x_2 \end{Bmatrix}. \tag{a}$$

Find the natural frequencies and mode shapes of the building frame for $m_1 = m_2 = m$ and $k_1 = k_2 = k$.

Section 4.3

4.23. Using Eqs. (4.25) and (4.26), convert the following eigenvalue problem into a standard eigenvalue problem:

$$\begin{bmatrix} 6 & -2 \\ -2 & 3 \end{bmatrix} \vec{X} = \lambda \begin{bmatrix} 4 & 0 \\ 0 & 8 \end{bmatrix} \vec{X}.$$

4.24. Using Eqs. (4.31) and (4.32), convert the following eigenvalue problem into a standard eigenvalue problem:

$$\begin{bmatrix} 6 & -2 \\ -2 & 3 \end{bmatrix} \vec{X} = \lambda \begin{bmatrix} 8 & 2 \\ 2 & 4 \end{bmatrix} \vec{X}.$$

Section 4.5

4.25. Derive the characteristic polynomial corresponding to the matrix

$$[A] = \begin{bmatrix} 5 & 7 & 6 & 5 \\ 7 & 10 & 8 & 7 \\ 6 & 8 & 10 & 9 \\ 5 & 7 & 9 & 10 \end{bmatrix}$$

by using Faddeev–Leverrier method.

4.26. Find the eigenvalues of the matrix $[A]$ given in Problem 4.25 by finding the roots of the characteristic polynomial equation.

4.27. Find the eigenvalues and eigenvectors of the matrix

$$[A] = \begin{bmatrix} 1 & \frac{1}{2} \\ \frac{1}{2} & \frac{1}{3} \end{bmatrix}.$$

4.28. Find the eigenvalues of the following matrix by finding the roots of the characteristic polynomial equation:

$$[A] = \begin{bmatrix} 1+2i & 3+4i & 21+22i \\ 43+44i & 13+14i & 15+16i \\ 5+6i & 7+8i & 25+26i \end{bmatrix}.$$

4.29. Determine the eigenvalues of the following matrix by finding the roots of the characteristic equation:

$$[A] = \begin{bmatrix} 3 & -2 \\ -2 & 2 \end{bmatrix}.$$

4.30. The principal strains induced in a machine component can be determined by solving the eigenvalue problem

$$\begin{bmatrix} \varepsilon_{xx} & \varepsilon_{xy} & \varepsilon_{xz} \\ \varepsilon_{yx} & \varepsilon_{yy} & \varepsilon_{yz} \\ \varepsilon_{zx} & \varepsilon_{zy} & \varepsilon_{zz} \end{bmatrix} \begin{Bmatrix} l_x \\ l_y \\ l_z \end{Bmatrix} = \lambda \begin{Bmatrix} l_x \\ l_y \\ l_z \end{Bmatrix},$$

where ε_{xx}, ε_{yy}, and ε_{zz} are the normal strains along x, y, and z directions, and $\varepsilon_{xy} = \varepsilon_{yx}$, $\varepsilon_{yz} = \varepsilon_{zy}$, and $\varepsilon_{xz} = \varepsilon_{zx}$ are the shear strains in xy, yz, and xz planes, λ is the eigenvalue (principal strain), and l_x, l_y, and l_z are the direction cosines of the principal strain direction. Determine the principal strains and their directions for the following data:

$$\varepsilon_{xx} = 70,\ \varepsilon_{xy} = -10\sqrt{3},\ \varepsilon_{yy} = 5,\ \varepsilon_{zz} = -20,\ \text{and } \varepsilon_{xz} = \varepsilon_{yz} = 0.$$

4.31. The design of a mechanical component requires the maximum principal stress to be less than the material strength. For a component subjected to arbitrary loads, the principal stresses (σ) are given by the solution of the equations; that is,

$$\begin{bmatrix} \sigma_x & \tau_{xy} & \tau_{xz} \\ \tau_{xy} & \sigma_y & \tau_{yz} \\ \tau_{xz} & \tau_{yz} & \sigma_z \end{bmatrix} \begin{Bmatrix} l_x \\ l_y \\ l_z \end{Bmatrix} = \sigma \begin{Bmatrix} l_x \\ l_y \\ l_z \end{Bmatrix},$$

where σ_x, σ_y, and σ_z denote the normal stresses acting along the x, y, and z directions, τ_{xy}, τ_{yz}, and τ_{xz} indicate the shear stresses acting in the xy, yz, and xz planes, respectively (see Fig. 4.13), and l_x, l_y, and l_z represent the direction cosines that define the principal plane on which σ acts. Determine the principal stresses and principal planes in a machine component for the following stress condition:

$$\begin{bmatrix} \sigma_{xx} & \sigma_{xy} & \sigma_{xz} \\ \sigma_{xy} & \sigma_{yy} & \sigma_{yz} \\ \sigma_{xz} & \sigma_{yz} & \sigma_{zz} \end{bmatrix} = \begin{bmatrix} 10 & 4 & -6 \\ 4 & -6 & 8 \\ -6 & 8 & 14 \end{bmatrix} \text{MPa}.$$

4.32. Find the eigenvalues and eigenvectors of the problem stated in Eq. (b) of Example 4.2.

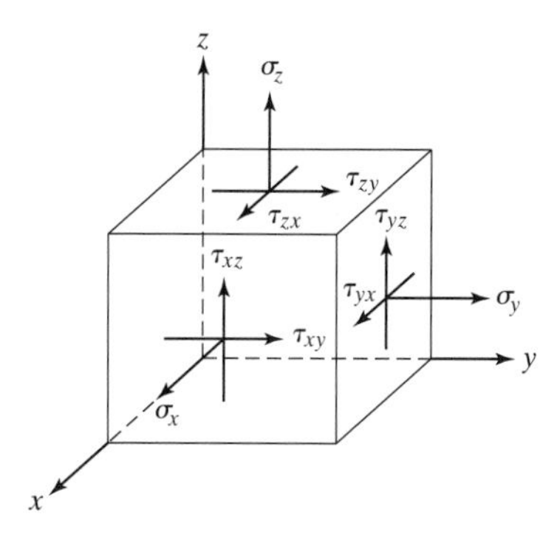

Figure 4.13 State of stress at a point.

Section 4.6

4.33. Prove that the coordinate transformation matrix, given by Eq. (4.50), is orthogonal.

4.34. Find the eigenvalues and eigenvectors of the matrix

$$[A] = \begin{bmatrix} 5 & 4 & 1 & 1 \\ 4 & 5 & 1 & 1 \\ 1 & 1 & 4 & 2 \\ 1 & 1 & 2 & 4 \end{bmatrix}$$

by using the Jacobi method.

4.35. Find the eigenvalues and eigenvectors of the matrix

$$[A] = \begin{bmatrix} 1 & \frac{1}{2} & \frac{1}{3} \\ \frac{1}{2} & \frac{1}{3} & \frac{1}{4} \\ \frac{1}{3} & \frac{1}{4} & \frac{1}{5} \end{bmatrix}$$

by using the Jacobi method.

4.36. Find the eigenvalues and eigenvectors of the Hillbert matrix given by

$$[A] = \begin{bmatrix} 1 & \frac{1}{2} & \frac{1}{3} & \frac{1}{4} \\ \frac{1}{2} & \frac{1}{3} & \frac{1}{4} & \frac{1}{5} \\ \frac{1}{3} & \frac{1}{4} & \frac{1}{5} & \frac{1}{6} \\ \frac{1}{4} & \frac{1}{5} & \frac{1}{6} & \frac{1}{7} \end{bmatrix}$$

by using the subroutine JACOBI.

4.37. Find the eigenvalues and eigenvectors of the matrix given in Problem 4.74 for $n = 4$ using Jacobi method. Compare the eigenvalues found with the exact values

$$\lambda_i = \frac{1}{2}\left[1 - \cos\frac{(2i-1)\pi}{2n+1}\right]^{-1}, \quad i = 1, 2, \ldots, n.$$

Section 4.7

4.38. Reduce the following matrix to tridiagonal form using Given's method:

$$[A] = \begin{bmatrix} 4 & 3 & 2 & 1 \\ 3 & 3 & 2 & 1 \\ 2 & 2 & 2 & 1 \\ 1 & 1 & 1 & 1 \end{bmatrix}.$$

4.39. Find all the real roots of the characteristic equation of Problem 4.38 using the property of Sturm sequence.

4.40. (a) Transform the matrix

$$[A] = \begin{bmatrix} 1 & 2 & 2 \\ 2 & 1 & 2 \\ 2 & 2 & 1 \end{bmatrix}$$

to a tridiagonal form using Given's method.

(b) Generate the Sturm sequence corresponding to the tridiagonal matrix of part (a).

(c) Find the resulting eigenvalues and eigenvectors of the matrix.

Section 4.8

4.41. Reduce the matrix $[A]$ given in Problem 4.38 to tridiagonal form using the Householder method.

4.42. Determine the eigenvalues and eigenvectors of the matrix

$$[A] = \begin{bmatrix} 1 & 2 & -1 \\ 2 & 1 & 2 \\ -1 & 2 & 1 \end{bmatrix}$$

using the Householder method.

Section 4.9

4.43. Generate the functions $f_i(\lambda)$ for the tridiagonal matrix

$$[A] = \begin{bmatrix} \alpha & \beta & 0 & 0 \\ \beta & \alpha & \beta & 0 \\ 0 & \beta & \alpha & \beta \\ 0 & 0 & \beta & \alpha \end{bmatrix},$$

with $\alpha = 1$ and $\beta = 2$.

4.44. Generate the functions $f_i(\lambda)$ for the tridiagonal matrix

$$[A] = \begin{bmatrix} \alpha & 1 & 0 & 0 \\ 1 & \beta & 1 & 0 \\ 0 & 1 & \alpha & 1 \\ 0 & 0 & 1 & \beta \end{bmatrix},$$

for $\alpha = 10$ and $\beta = 20$.

Section 4.10

4.45. Determine the eigenvalues and eigenvectors of the tridiagonal matrix given in Problem 4.44, with $\alpha = 2$ and $\beta = 1$.

4.46. Find the eigenvalues of the matrix

$$[A] = \begin{bmatrix} (\alpha+\beta) & \beta & 0 & 0 \\ \beta & \alpha & \beta & 0 \\ 0 & \beta & \alpha & \beta \\ 0 & 0 & \beta & (\alpha+\beta) \end{bmatrix},$$

for $\alpha = 10$ and $\beta = 10$.

Section 4.11

4.47. Find the largest eigenvalue of the matrix $[A]$ given in Problem 4.34 using the matrix iteration (power) method with the starting vector

$$\vec{X} = \begin{Bmatrix} 1 \\ 1 \\ 1 \\ 1 \end{Bmatrix}.$$

4.48. Using the power method with the starting vector $\vec{X} = \begin{Bmatrix} -3 \\ 2 \\ 1 \end{Bmatrix}$, find the largest eigenvalue and the corresponding eigenvector of the following matrix:

$$[A] = \begin{bmatrix} 33 & 16 & 72 \\ -24 & -10 & -57 \\ -8 & -4 & -17 \end{bmatrix}.$$

4.49. Find the eigenvalues of the matrix

$$[A] = \begin{bmatrix} 4 & 1 & 1 \\ 2 & 4 & 1 \\ 0 & 1 & 4 \end{bmatrix}$$

using the power method.

4.50. Find the largest eigenvalue of the matrix

$$[A] = \begin{bmatrix} -2 & 2 & 2 & 2 \\ -3 & 3 & 2 & 2 \\ -2 & 0 & 4 & 2 \\ -1 & 0 & 0 & 5 \end{bmatrix}$$

using the power method with the starting vector $\vec{X} = \{\ 2 \quad 2 \quad 1 \quad 1\ \}^T$.

4.51. Find the eigenvalues and the eigenvectors of the complex matrix

$$[A] = \begin{bmatrix} 2 & -i \\ i & 1 \end{bmatrix}$$

Section 4.13

4.52. Using (a) MAPLE, (b) MATLAB and (c) MATHCAD, find the eigenvalues and eigenvectors of the following matrix:

$$[A] = \begin{bmatrix} 2 & 1 & 1 & 0 \\ 1 & 1 & 0 & 1 \\ 1 & 0 & 1 & 1 \\ 0 & 1 & 1 & 2 \end{bmatrix}.$$

4.53. Using (a) MAPLE, (b) MATLAB, and (c) MATHCAD, find the eigenvalues and eigenvectors of the following matrix:

$$[A] = \begin{bmatrix} 1 & \frac{1}{2} & \frac{1}{3} & \frac{1}{4} \\ \frac{1}{2} & \frac{1}{3} & \frac{1}{4} & \frac{1}{5} \\ \frac{1}{3} & \frac{1}{4} & \frac{1}{5} & \frac{1}{6} \\ \frac{1}{4} & \frac{1}{5} & \frac{1}{6} & \frac{1}{7} \end{bmatrix}.$$

4.54. Using (a) MAPLE, (b) MATLAB and (c) MATHCAD, find the eigenvalues and eigenvectors of the following matrix:

$$[A] = \begin{bmatrix} 12 & 6 & -6 \\ 6 & 16 & 2 \\ -6 & 2 & 16 \end{bmatrix}.$$

4.55. Using (a) MAPLE, (b) MATLAB, and (c) MATHCAD, determine the eigenvalues and eigenvectors of the matrix $[A]$ given here with $n = 20$:

$$[A] = \begin{bmatrix} n & n-1 & n-2 & \cdots & 2 & 1 \\ n-1 & n-1 & n-2 & \cdots & 2 & 1 \\ n-2 & n-2 & n-2 & \cdots & 2 & 1 \\ \cdot & \cdot & \cdot & \cdots & \cdot & \cdot \\ \cdot & \cdot & \cdot & \cdots & \cdot & \cdot \\ \cdot & \cdot & \cdot & \cdots & \cdot & \cdot \\ 2 & 2 & 2 & \cdots & 2 & 1 \\ 1 & 1 & 1 & \cdots & 1 & 1 \end{bmatrix}.$$

Section 4.14

4.56. Write a Fortran program to find the real roots of a characteristic equation using the property of Sturm sequence. Solve Example 4.17 using this program.

4.57. Write a C program to reduce a symmetric matrix of order n to tridiagonal form using Given's method.

4.58. Find the eigenvalues and eigenvectors of the matrix

$$[A] = \begin{bmatrix} 6 & -1 & -2 \\ -1 & 8 & 1 \\ -6 & 1 & 2 \end{bmatrix}$$

using Program 4.2.

4.59. Generate the characteristic equation corresponding to the matrix $[A]$ given in Problem 4.58 using Program 4.1.

4.60. Find all the eigenvalues and eigenvectors of the matrix $[A]$ given in Problem 4.52 using Program 4.2.

4.61. Find the eigenvalues and eigenvectors of the matrix

$$[A] = \begin{bmatrix} 15 & 3 & 6 \\ 3 & 4 & -3 \\ 6 & -3 & 20 \end{bmatrix}$$

using Program 4.3.

4.62. Find all the eigenvalues and eigenvectors of the matrix $[A]$ given in Problem 4.61 using Program 4.4.

General

4.63. If the eigenvalues of the matrix

$$[A] = \begin{bmatrix} 33 & 16 & 72 \\ -24 & -10 & -57 \\ -8 & -4 & -17 \end{bmatrix}$$

are given by $\lambda_1 = 1$, $\lambda_2 = 2$, and $\lambda_3 = 3$, determine the corresponding right eigenvectors.

4.64. Find the right and left eigenvectors of the following matrix:

$$[A] = \begin{bmatrix} 3 & -2 \\ -2 & 5 \end{bmatrix}.$$

4.65. The equations of motion for the free vibration of the triple pendulum shown in Fig. 4.14 are given by [4.7]

$$ml^2 \begin{bmatrix} 3 & 2 & 1 \\ 2 & 2 & 1 \\ 1 & 1 & 1 \end{bmatrix} \begin{Bmatrix} \ddot{\theta}_1 \\ \ddot{\theta}_2 \\ \ddot{\theta}_3 \end{Bmatrix} + mgl \begin{bmatrix} 3 & 0 & 0 \\ 0 & 2 & 0 \\ 0 & 0 & 1 \end{bmatrix} \begin{Bmatrix} \theta_1 \\ \theta_2 \\ \theta_3 \end{Bmatrix} = \begin{Bmatrix} 0 \\ 0 \\ 0 \end{Bmatrix}.$$

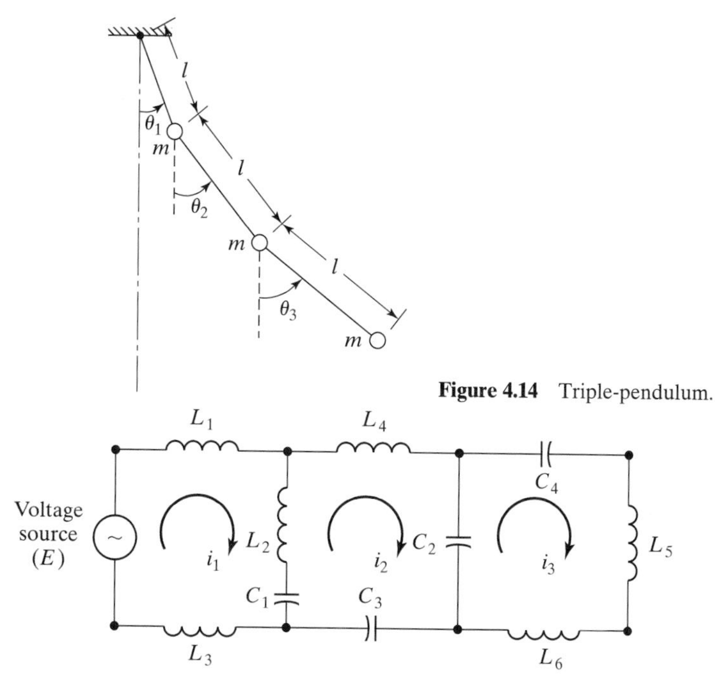

Figure 4.14 Triple-pendulum.

Figure 4.15 Electrical circuit.

(i) Assuming a solution of the form

$$\theta_i(t) = \Theta_i \cos(\omega t + \phi), \quad i = 1, 2, 3,$$

where Θ_i is the amplitude of $\theta_i(t)$, ω is the natural frequency, ϕ is the phase angle, and t is time, derive the eigenvalue problem for finding the natural frequencies of vibration (ω_i).

(ii) Convert the problem into a standard eigenvalue problem.

(iii) Determine the natural frequencies of vibration by solving the characteristic equation.

(iv) Determine the eigenvectors (mode shapes) of the triple pendulum.

4.66. Find the principal strains and their directions by solving the eigenvalue problem stated in Problem 4.30 for the following data:

$$\varepsilon_{xx} = 30,\ \varepsilon_{xy} = 50,\ \varepsilon_{xz} = 80,\ \varepsilon_{yy} = 10,\ \varepsilon_{zz} = 20,\ \varepsilon_{yz} = 0.$$

4.67. (i) Derive the differential equations, in terms of the currents flowing in the various loops (i_1, i_2, i_3) for the electrical circuit shown in Fig. 4.15 by applying Kirchhoff's law.

(ii) Assuming $\frac{dE}{dt} = 0$ and a sinusoidal variation of the currents,

$$i_j = I_j \sin(\omega t + \phi), \quad j = 1, 2, 3,$$

where I_j is the amplitude of current i_j, ω is the natural frequency, and ϕ is the phase angle, derive the eigenvalue problem for determining ω.

(iii) Solve the eigenvalue problem of part (ii) to determine the eigenvalues and eigenvectors of the system. Assume $L_i = 100,\ i = 1, 2, \ldots, 6$ and $C_i = 2, i = 1, 2, 3, 4$.

4.68. Prove that the eigenvectors $\vec{X}^{(i)}$ are orthogonal with respect to the matrices $[A]$ and $[B]$ in a general eigenvalue problem.

4.69. Consider the following matrix:

$$[A] = \begin{bmatrix} 10 & 2 & -2 \\ 2 & 13 & -4 \\ -2 & -4 & 13 \end{bmatrix}.$$

Find the vector closest to the true eigenvector among the following:

$$\vec{Y}_1 = \begin{Bmatrix} 1 \\ 1 \\ 1 \end{Bmatrix}, \vec{Y}_2 = \begin{Bmatrix} 1 \\ 0 \\ 1 \end{Bmatrix}, \text{ and } \vec{Y}_3 = \begin{Bmatrix} 1 \\ 2 \\ -2 \end{Bmatrix}.$$

4.70. Consider the following eigenvalue problem:

$$\begin{bmatrix} 8 & -1 & -5 \\ -4 & 4 & -2 \\ 18 & -5 & -7 \end{bmatrix} \begin{Bmatrix} x_1 \\ x_2 \\ x_3 \end{Bmatrix} = \lambda \begin{Bmatrix} x_1 \\ x_2 \\ x_3 \end{Bmatrix}.$$

Determine the values of λ by finding the roots of the characteristic polynomial equation.

4.71. Find the eigenvalues of the matrix $[A]$ given in Problem 4.63 by finding the roots of the characteristic equation.

4.72. Determine the characteristic equation and find its roots to determine the eigenvalues of the following matrix:

$$[A] = \begin{bmatrix} 1 & 2 & 3 \\ 4 & 5 & 6 \\ 7 & 8 & 9 \end{bmatrix}.$$

4.73. Solve the eigenvalue problem stated in Problem 4.30 and find the principal strains and their directions for the following data:

$$\varepsilon_{xx} = 1200, \varepsilon_{yy} = -400, \text{ and } \varepsilon_{xy} = -1200, \varepsilon_{xz} = \varepsilon_{yz} = \varepsilon_{zz} = 0.$$

4.74. Find the eigenvalues of the matrix $[A]$ defined in Problem 4.55 for $n = 3$ by finding the roots of the characteristic equation.

4.75. Reduce the following matrix to tridiagonal form using Given's method:

$$[A] = \begin{bmatrix} 5 & 4 & 1 & 1 \\ 4 & 5 & 1 & 1 \\ 1 & 1 & 4 & 2 \\ 1 & 1 & 2 & 4 \end{bmatrix}.$$

4.76. Find all the real roots of the characteristic equation of Problem 4.75 using the property of Sturm sequence.

4.77. Reduce the matrix $[A]$ given in Problem 4.75 to tridiagonal form using the Householder method.

4.78. Using power method with the starting vector $\vec{X} = \begin{Bmatrix} 1 \\ 1 \\ 1 \end{Bmatrix}$, find the largest eigenvalue and the corresponding eigenvector of the matrix

$$[A] = \begin{bmatrix} 4 & 1 & 1 \\ 2 & 4 & 1 \\ 0 & 1 & 4 \end{bmatrix}.$$

4.79. Find the eigenvalues and eigenvectors of the complex matrix

$$[A] = \begin{bmatrix} 1 & 1-i \\ 1+i & 1 \end{bmatrix}.$$

4.80. Using Rayleigh's method, find an estimate of the largest eigenvalue of the following matrix

$$[A] = \begin{bmatrix} 6 & 4 & 4 & 1 \\ 4 & 6 & 1 & 4 \\ 4 & 1 & 6 & 4 \\ 1 & 4 & 4 & 6 \end{bmatrix}.$$

(See Problem 4.16.) Use $\vec{Y} = \{\ 1 \quad 2 \quad 1 \quad 2\ \}^T$ as the approximate eigenvector.

4.81. Find the largest eigenvalue of the matrix

$$[A] = \begin{bmatrix} 2 & -1 & 0 \\ -1 & 2 & -1 \\ 0 & -1 & 2 \end{bmatrix}.$$

using Rayleigh's method with $\vec{Y} = \{\ 1 \quad 1 \quad 1\ \}^T$. *Hint*: See Problem 4.16.

4.82. Estimate the largest eigenvalue of the Hillbert matrix $[A]$ given in Problem 4.36 using Rayleigh's method with $\vec{Y} = \{\ 1.00 \quad 0.75 \quad 0.50 \quad 0.25\ \}^T$. *Hint:* See Problem 4.16.

PROJECTS

4.1. The equations of motion of a viscously damped system are given by [4.7]

$$[m]\ddot{\vec{u}} + [c]\dot{\vec{u}} + [k]\vec{u} = \vec{0}, \tag{a}$$

where $[m]$, $[c]$, and $[k]$ are symmetric mass, damping and stiffness matrices, respectively, and $\vec{u}$, $\dot{\vec{u}}$, and $\ddot{\vec{u}}$ are the displacement, velocity, and acceleration vectors, respectively.

(i) Defining

$$\vec{x} = \begin{Bmatrix} \vec{u} \\ \dot{\vec{u}} \end{Bmatrix}, \tag{b}$$

express the equations of motion in the form

$$[A]\dot{\vec{x}} = [B]\vec{x}. \tag{c}$$

When $\vec{x} = \vec{X}e^{\lambda t}$ is assumed, Eq. (c) reduces to the eigenvalue problem

$$\lambda[A]\vec{X} = [B]\vec{X}. \tag{d}$$

(ii) For a system with

$$[m] = \begin{bmatrix} 10 & 0 \\ 0 & 10 \end{bmatrix}, [c] = \begin{bmatrix} 10 & -5 \\ -5 & 5 \end{bmatrix}, \text{ and } [k] = \begin{bmatrix} 4 & -2 \\ -2 & 2 \end{bmatrix}, \tag{e}$$

identify the matrices $[A]$ and $[B]$ and find the eigenvalues λ and the eigenvectors $\vec{X}$ given by Eq. (d).

4.2. Consider the eigenvalue problem

$$[A]\vec{X} = \lambda[B]\vec{X}, \tag{a}$$

where $[A]$ and $[B]$ are symmetric matrices. Corresponding to this problem, a quantity known as Rayleigh's quotient, Q, can be defined as [4.7]

$$Q = \frac{\vec{Y}^T[A]\vec{Y}}{\vec{Y}^T[B]\vec{Y}}, \tag{b}$$

where $\vec{Y}$ is an arbitrary vector.

(i) Show that, if $\vec{Y}$ is an exact eigenvector of the problem stated in Eq. (a), then Q represents the corresponding eigenvalue.

(ii) Show that, for any vector $\vec{Y}$, other than the eigenvector of the problem in Eq. (a), Q gives a lower bound on the largest eigenvalue of the problem in Eq. (a).

(iii) Consider the eigenvalue problem stated in Eq. (a) with

$$[A] = \begin{bmatrix} 2 & -1 & 0 \\ -1 & 2 & -1 \\ 0 & -1 & 1 \end{bmatrix}; [B] = \begin{bmatrix} 1 & 0 & 0 \\ 0 & 1 & 0 \\ 0 & 0 & 1 \end{bmatrix}.$$

Find an eigenvalue of the problem using Rayleigh's quotient.

4.3. The natural frequencies of the stepped cantilever shown in Fig. 4.16 are given by the solution of the equation

$$[A]\vec{X} = \lambda[B]\vec{X}, \tag{a}$$

where

$$[A] = 10^6 \begin{bmatrix} 0.017361 & -0.01736 & -0.10417 & -0.10417 \\ -0.017361 & 0.31399 & 0.10417 & -1.08233 \\ -0.10417 & 0.10417 & 0.83333 & 0.41667 \\ -0.10417 & -1.08233 & 0.41667 & 7.16140 \end{bmatrix}; \tag{b}$$

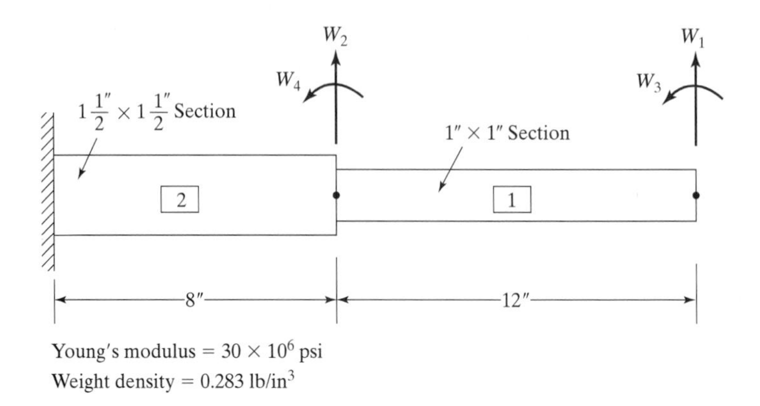

Young's modulus = 30×10^6 psi
Weight density = 0.283 lb/in^3

Figure 4.16 Stepped cantilever.

$$[B] = 10^{-5}\omega^2 \begin{bmatrix} 136.0164 & 47.0826 & -95.9090 & 56.6735 \\ 47.0826 & 442.0572 & -56.6735 & -119.8890 \\ -95.9090 & -56.6735 & 87.1900 & -65.3925 \\ 56.6735 & -119.8890 & -65.3925 & 283.3700 \end{bmatrix}; \tag{c}$$

$$\lambda = 10^{-5}\omega^2; \tag{d}$$

$$\vec{X} = \begin{Bmatrix} W_1 \\ W_2 \\ W_3 \\ W_4 \end{Bmatrix}. \tag{e}$$

(i) Determine the natural frequencies and mode shapes of the stepped beam and formulate the modal matrix

$$[X] = \left[\vec{X}_1\vec{X}_2\vec{X}_3\vec{X}_4\right]. \tag{f}$$

Show that the eigenvalue problem of Eq. (a) can be rewritten as

$$[\bar{A}]\vec{X} = \lambda[\bar{B}]\vec{X}, \tag{g}$$

where

$$[\bar{A}] = [X]^T[A][X] \tag{h}$$

and

$$[\bar{B}] = [X]^T[B][X]. \tag{i}$$

(ii) Reduce the general eigenvalue problem of Eq. (a) into a special eigenvalue problem of the following form:

$$[D]\vec{X} = \lambda\vec{X}. \tag{j}$$

Find the solution of Eq. (j) and compare the results with those found in part (i).

4.4. A complex eigenvalue problem is given by

$$[A]\vec{X} = \lambda\vec{X},$$

where the elements of $[A]$, $\vec{X}$, and λ are complex numbers. The power method can be used by starting with a trial complex vector, $\vec{p}_1 + i\vec{q}_1$, and finding a sequence of improved vectors,

$$\vec{p}_{j+1} + i\vec{q}_{j+1} = [A](\vec{p}_j + i\vec{q}_j) = (\alpha + i\beta)(\vec{p}_j + i\vec{q}_j); \quad j = 1, 2, \ldots.$$

The complex number $(\alpha + i\beta)$ denotes the eigenvalue and the vector $\vec{p}_{j+1} + i\vec{q}_{j+1}$ the eigenvector corresponding to the eigenvalue λ having the largest magnitude, $(\alpha^2 + \beta^2)$ as $(\alpha + i\beta)$ converges. Using this procedure, find the eigenvalues and eigenvectors of the following matrix:

$$[A] = \begin{bmatrix} 2+i & -2-i \\ i & -3+2i \end{bmatrix}.$$

4.5. Many practical systems give rise to eigenvalue problems of large size. For example, in mechanical and structural systems, the displacement degrees of freedom at certain discrete locations constitute the components of the eigenvector. It is not uncommoon to have the total number of degrees of freedom on the order of thousands or more. To reduce the size of the eigenvalue problem to manageable size, a condensation procedure can be used. The procedure involves retaining the degrees of freedom, which are expected to have larger values and eliminate the degrees of freedom that are expected to have smaller values. Let the general eigenvalue problem be

$$\underset{n\times n}{[A]}\ \underset{n\times 1}{\vec{X}} = \lambda\ \underset{n\times n}{[B]}\ \underset{n\times 1,}{\vec{X}} \tag{a}$$

which can be stated in partitioned form as

$$\begin{bmatrix} \underset{p \times p}{[A_{11}]} & \underset{p \times q}{[A_{12}]} \\ \underset{q \times p}{[A_{21}]} & \underset{q \times q}{[A_{22}]} \end{bmatrix} \begin{Bmatrix} \underset{p \times 1}{\vec{X}_1} \\ \underset{q \times 1}{\vec{X}_2} \end{Bmatrix} = \lambda \begin{bmatrix} \underset{p \times p}{[B_{11}]} & \underset{p \times q}{[B_{12}]} \\ \underset{q \times p}{[B_{21}]} & \underset{q \times q}{[B_{22}]} \end{bmatrix} \begin{Bmatrix} \underset{p \times 1}{\vec{X}_1} \\ \underset{q \times 1}{\vec{X}_2} \end{Bmatrix} \tag{b}$$

where $\vec{X}_1$ denotes the vector of degrees of freedom (elements of $\vec{X}$) to be retained, $\vec{X}_2$ indicates the vector of degrees of freedom (elements of $\vec{X}$) to be eliminated, $[A_{11}], [A_{12}], \ldots, [B_{22}]$ are the corresponding partitioned matrices, and $n = p + q$. Consider the special case with $[B_{11}] = [I], [B_{12}] = [0], [B_{21}] = [0]$ and $[B_{22}] = [0]$. Equation (b) can be written as two separate matrix equations as

$$[A_{11}]\vec{X}_1 + [A_{12}]\vec{X}_2 = \lambda[I]\vec{X}_1; \tag{c}$$

$$[A_{21}]\vec{X}_1 + [A_{22}]\vec{X}_2 = \vec{0} \tag{d}$$

(i) Using Eqs. (c) and (d), express the equivalent (reduced) eigenvalue problem as

$$\underset{p \times p}{[\tilde{A}]} \; \underset{p \times 1}{\vec{\tilde{X}}} = \lambda \; \underset{p \times 1}{\vec{\tilde{X}}} \tag{e}$$

and identify $[\tilde{A}]$ and $\vec{\tilde{X}}$.

(ii) Indicate how the solution of Eq. (e) can be used to determine the solution of the original problem stated in Eq. (a).

(iii) Find the solution of the following eigenvalue problem using the condensation procedure using $\vec{X}_1 = \begin{Bmatrix} x_1 \\ x_2 \\ x_3 \end{Bmatrix}$, $\vec{X}_2 = \begin{Bmatrix} x_4 \\ x_5 \end{Bmatrix}$,

Projects

$$\begin{bmatrix} 2 & -1 & 0 & 0 & 0 \\ -1 & 2 & -1 & 0 & 0 \\ 0 & -1 & 2 & -1 & 0 \\ 0 & 0 & -1 & 2 & -1 \\ 0 & 0 & 0 & -1 & 1 \end{bmatrix} \vec{X} = \lambda \begin{bmatrix} 2 & 0 & 0 & 0 & 0 \\ 0 & 2 & 0 & 0 & 0 \\ 0 & 0 & 2 & 0 & 0 \\ 0 & 0 & 0 & 2 & 0 \\ 0 & 0 & 0 & 0 & 1 \end{bmatrix} \vec{X}$$

where

$$\vec{X} = \begin{Bmatrix} x_1 \\ x_2 \\ \cdot \\ \cdot \\ x_5 \end{Bmatrix}.$$

5

Curve Fitting and Interpolation

5.1 Introduction

CHAPTER CONTENTS

In many practical situations, data are available at discrete points, and we are required to fit a smooth and continuous function (curve) to this data. The data may be exact or approximate. To determine the function, we need to find certain coefficients or parameters. The curve fitting to a set of data points can be done using two approaches. In the first approach, the curve is made to pass through every data point. This approach is called *collocation*. This approach is used either when the data are known to be accurate or when the data are generated by evaluating a complicated function at a discrete set of points. Usually, a polynomial, a trigonometric function, or an exponential function is used to approximate a set of data points. If an nth-degree polynomial is used, it can be expressed as

$$f(x) = a_0 + a_1 x + a_2 x^2 + \cdots + a_{n-1} \times x^{n-1} + a_n x^n, \qquad (5.1)$$

where $a_i, i = 0, 1, 2, \ldots, n$, are the unknown coefficients. These $n + 1$ coefficients can be determined uniquely by making the polynomial, Eq. (5.1), pass through the $(n + 1)$ data points. This is illustrated in Fig. 5.1 where a unique polynomial of order one (linear equation) passes through two points $(n + 1 = 2)$, a polynomial of order two (quadratic equation) passes through three points $(n + 1 = 3)$, a polynomial of order three (cubic equation) passes through four points $(n+1 = 4)$, and a polynomial of order four (quartic equation) passes through five points $(n + 1 = 5)$.

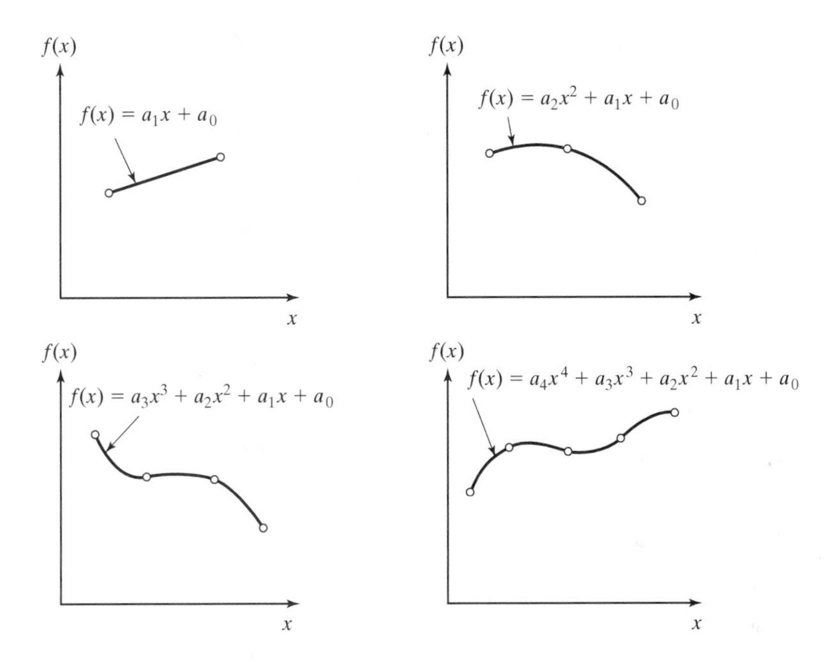

Figure 5.1 Different-degree polynomials.

In some cases, piecewise curve fitting is used in which a specified function (usually a polynomial) is made to pass through subgroups of the data points. Thus, the data are represented by a series of functions (of the same type) in a piecewise manner instead of a single function.

In the second approach, the curve is made to represent the general trend of the data. This approach is useful when there are more data points than the number of unknown coefficients or when the data appear to have significant error or noise. In the latter case, there is a possibility of error in any data point, and hence no attempt is made to pass the curve through every data point; instead, the curve is made to follow the pattern of the data points taken as a group. Such an approach is known as *least-squares regression*. This is illustrated in Fig. 5.2.

Interpolation is the process of estimating the value of a function (or dependent variable) f corresponding to a particular value of the independent variable x when

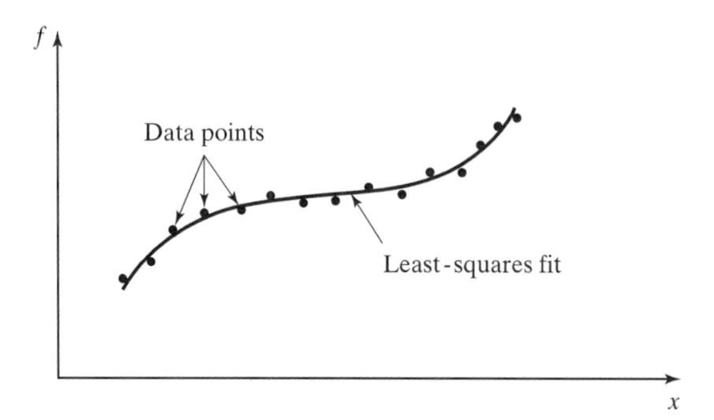

Figure 5.2 Least squares regression.

the function $f(x)$ is described by a set of tabular data. Interpolation can be performed by curve fitting where a suitable approximating function, known as interpolating function, $P(x)$, is used to represent the tabular data. Once the interpolating function $P(x)$ is determined, the common mathematical operations such as determination of roots, differentiation, and integration that are intended for the function $f(x)$ can be performed using $P(x)$ in place of $f(x)$. If $P(x)$ is used to estimate the value of $f(x)$ beyond the range of the tabular data, it is called extrapolation. Usually, interpolation yields better accuracy compared with extrapolation. Inverse interpolation denotes the process of estimating the value of the independent variable x corresponding to a particular value of the function f. Inverse interpolation can be performed either by fitting a curve to the inverse function, $x = x(f)$, or by solving the interpolation function (direct fit) $f(x)$ iteratively for $x(f)$.

5.2 Engineering Applications

The prediction of the behavior of many engineering systems is too complex and sometimes even impossible. In such cases, approximations and empirical relations are to be used to describe the behavior of the system. This invariably involves the use of analytical functions such as polynomials, trigonometric functions, and exponential functions as approximating functions. Sometimes, a missing functional value may have to be determined from a known set of data. This involves using a suitable interpolation. The following examples indicate representative applications of curve fitting and interpolation:

▶Example 5.1

An experiment dealing with the measurement of damping of a solid body immersed in a fluid consists of a solid disk suspended from a fixed support by a thin wire as shown in Fig. 5.3. The disk carries a pointer that rotates with the disk over a fixed scale. The disk is immersed in a bowl of heavy oil, is given an initial angular displacement about the wire axis, and is released to rotate freely about the wire axis. The angular positions of the disk measured at different times are as follows:

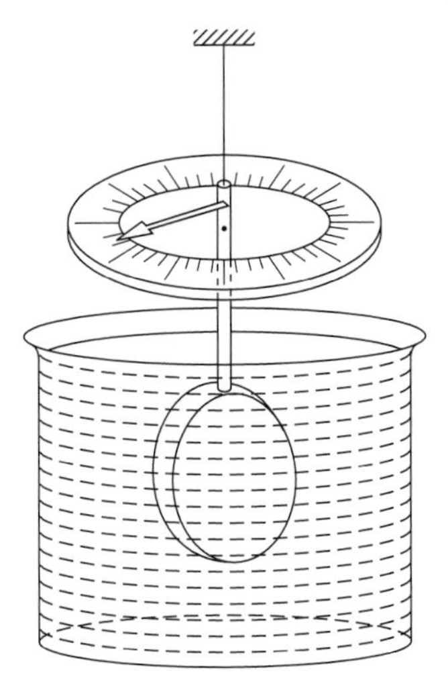

Figure 5.3 Measurement of damping.

Time (sec)	0.0	1.5	3.7	5.2	7.1	9.6	11.8
Angular position (degrees)	110.0	77.5	44.7	31.5	20.1	11.6	7.0

Assuming that the angular displacement, $\theta(t)$, of an overdamped system (in a heavy oil) can be expressed as

$$\theta(t) = a\, e^{bt}, \tag{E1}$$

where t is time, determine the values of the constants a and b.

▶Example 5.2

The machining of low carbon steel with a high-speed steel-cutting tool yielded the following results:

Cutting speed, V (m/min)	Tool life, T (min)
40	40
50	10

If the tool life equation is given by

$$V\, T^a = b, \tag{E1}$$

determine the constants a and b using the foregoing data.

▶Example 5.3

In the finite-element analysis of a fin subjected to an arbitrary temperature distribution, the variation of the unknown temperature in a typical element e is assumed to be a quadratic function [5.1]

$$T^{(e)}(x) = N_i(x)\Phi_i + N_j(x)\Phi_j + N_k(x)\Phi_k, \tag{E1}$$

where $N_i(x)$, $N_j(x)$, and $N_k(x)$ are called the shape functions and Φ_i, Φ_j, and Φ_k are the nodal temperatures (treated as unknowns) at the nodes i, j, and k, respectively. Express the shape functions in terms of the three-station Lagrange interpolation polynomials.

5.3 Collocation-Polynomial Fit

We consider a polynomial of order n and make it pass through $n+1$ data points with no restriction on the spacing of the data. Let the polynomial be of the form

$$f(x) = y = a_0 + a_1\,x + a_2\,x^2 + \cdots + a_{n-1}\,x^{n-1} + a_n\,x^n, \tag{5.2}$$

where x is the independent variable, y is the dependent variable, and a_i are the unknown coefficients. Let the data points be denoted as $(x_i, y_i), i = 0, 1, 2, \ldots, n$, where y_i is the value of the function at $x = x_i$. Since the polynomial passes through the points (x_i, y_i), we have

$$y_i = a_0 + a_1\,x_i + a_2\,x_i^2 + \cdots + a_{n-1}\,x_i^{n-1} + a_n\,x_i^n; i = 0, 1, 2, \ldots, n. \tag{5.3}$$

Equation (5.3) denotes a system of $n+1$ algebraic equations, linear in terms of the coefficients $a_0, a_1, a_2, \ldots, a_n$. These equations can be expressed in matrix form as

$$[B]\vec{a} = \vec{y}, \tag{5.4}$$

where the coefficient matrix, $[B]$, known as a Vandermonde matrix, is given by

$$[B] = \begin{bmatrix} 1 & x_0 & x_0^2 & x_0^3 & \cdot & \cdot & \cdot & x_0^n \\ 1 & x_1 & x_1^2 & x_1^3 & \cdot & \cdot & \cdot & x_1^n \\ \cdot & \cdot & \cdot & \cdot & \cdot & \cdot & \cdot & \cdot \\ \cdot & \cdot & \cdot & \cdot & \cdot & \cdot & \cdot & \cdot \\ \cdot & \cdot & \cdot & \cdot & \cdot & \cdot & \cdot & \cdot \\ 1 & x_n & x_n^2 & x_n^3 & \cdot & \cdot & \cdot & x_n^n \end{bmatrix}, \tag{5.5}$$

$$\vec{a} = \begin{Bmatrix} a_0 \\ a_1 \\ \cdot \\ \cdot \\ \cdot \\ a_n \end{Bmatrix}, \tag{5.6}$$

and

$$\vec{y} = \begin{Bmatrix} y_0 \\ y_1 \\ \cdot \\ \cdot \\ \cdot \\ y_n \end{Bmatrix}. \tag{5.7}$$

Equations (5.4) can be solved using the methods of Chapter 3 to find a unique solution for the coefficients $a_0, a_1, a_2, \ldots, a_n$. Unless some of the data points are duplicated, a unique solution exists to Eq. (5.4). However, note that the coefficient matrix B is prone to ill conditioning, and hence the determination of $f(x)$ through the solution of Eq. (5.4) may not be an efficient procedure. Other polynomial models, known as interpolation models, can be used to represent the function $f(x)$ more efficiently.

▶Example 5.4

The ac voltage applied to an electrical circuit is given by

$$e(t) = E \sin \frac{2\pi t}{T} \equiv 110 \sin 100\pi t, \tag{E1}$$

where $E = 110$ volts and $T = 0.02$ sec. Approximate this voltage over half a cycle using a quadratic equation that matches $e(t)$ exactly at $t = 0$, $t = T/4$, and $t = T/2$.

Solution

The values of e at $t_0 = 0$, $t_1 = \frac{T}{4} = 0.005$, and $t_2 = \frac{T}{2} = 0.01$ are given by 0, 110, and 0, respectively. By denoting t by x, the approximating quadratic can be expressed as $y(x) = a_0 + a_1 x + a_2 x^2$. In this case, Eq. (5.4) can be expressed as

$$[B]\vec{a} = \vec{y}, \tag{E2}$$

where

$$[B] = \begin{bmatrix} 1 & x_0 & x_0^2 \\ 1 & x_1 & x_1^2 \\ 1 & x_2 & x_2^2 \end{bmatrix} = \begin{bmatrix} 1 & 0 & 0 \\ 1 & 0.005 & 25 \times 10^{-6} \\ 1 & 0.01 & 1 \times 10^{-4} \end{bmatrix},$$

$$\vec{a} = \begin{Bmatrix} a_0 \\ a_1 \\ a_2 \end{Bmatrix},$$

and

$$\vec{y} = \begin{Bmatrix} y_0 \\ y_1 \\ y_2 \end{Bmatrix} = \begin{Bmatrix} 0 \\ 110 \\ 0 \end{Bmatrix}.$$

The solution of Eq. (E2) gives $a_0 = 0, a_1 = 4.4 \times 10^4$, $a_2 = -4.4 \times 10^6$, and hence

$$y(x) = 4.4 \times 10^4 x - 4.4 \times 10^6 x^2 \qquad \text{(E3)}$$

◀

5.4 Interpolation

Interpolation is the process of estimating an intermediate value from a set of discrete (or tabulated) values. A collocation function can be used for interpolation, because it passes through all the data points. The collocation function is often called an interpolating function. Interpolation of a set of data points can be done using polynomials, spline functions, or Fourier series. However, polynomial interpolation is most commonly used, and many numerical methods are based on polynomial interpolation. For example, many numerical integration methods are based on polynomial approximation, and numerical differentiation methods make use of differentiating interpolating polynomials. Although there is one and only one polynomial of order n that passes through $n + 1$ data points, there are several alternative forms in which the polynomial can be expressed. Among them are Lagrange interpolation, Newton forward or backward interpolation, and Hermite interpolation methods.

5.5 Lagrange Interpolation Formula

The Lagrange interpolation can be used to obtain an exact fit to a set of data points. As seen in Section 5.3, the solution of a set of linear algebraic equations is required to find the coefficients of a polynomial. For values of n greater than three, this solution needs a computer program and may not be very accurate. In fact, the terms involving powers of x_i in matrix $[B]$ of Eq. (5.4) become very large for large values of n, thereby causing round-off errors in the solution of Eq. (5.4). The Lagrange interpolation is one of the methods that does not require the solution of a set of linear equations to determine the interpolation formula.

5.5.1 Basic Idea

To see the basic idea behind Lagrange interpolation, we consider fitting a quadratic function through three data points (x_i, y_i), $i = 0, 1, 2$. Instead of expressing the quadratic function in the usual manner as

$$f(x) = a_0 + a_1 x + a_2 x^2, \tag{5.8}$$

we express it in a different form, known as the Lagrange polynomial of order two, as

$$f(x) = y = a_0(x - x_1)(x - x_2) + a_1(x - x_0)(x - x_2) + a_2(x - x_0)(x - x_1), \tag{5.9}$$

where a_0, a_1, and a_2 are constants to be determined by making the function $f(x)$ pass through the three data points. Note that each term on the right-hand side of Eq. (5.9) is a quadratic function in x, and hence their sum is also a quadratic function in x. By setting $f(x = x_i) = y_i$ in Eq. (5.9) for $i = 0, 1, 2$, in succession, we obtain

$$\begin{aligned} y_0 &= a_0(x_0 - x_1)(x_0 - x_2), \\ y_1 &= a_1(x_1 - x_0)(x_1 - x_2), \end{aligned} \tag{5.10}$$

and

$$y_2 = a_2(x_2 - x_0)(x_2 - x_1).$$

Equation (5.10) can be solved readily to determine the coefficients a_0, a_1, and a_2 as

$$\begin{aligned} a_0 &= \frac{y_0}{(x_0 - x_1)(x_0 - x_2)}, \\ a_1 &= \frac{y_1}{(x_1 - x_0)(x_1 - x_2)}, \end{aligned} \tag{5.11}$$

and

$$a_2 = \frac{y_2}{(x_2 - x_0)(x_2 - x_1)}.$$

By introducing Eq. (5.11) into Eq. (5.9), we obtain the second-order Lagrange interpolation polynomial

$$f(x) = y = y_0 \frac{(x - x_1)(x - x_2)}{(x_0 - x_1)(x_0 - x_2)} + y_1 \frac{(x - x_0)(x - x_2)}{(x_1 - x_0)(x_1 - x_2)} + y_2 \frac{(x - x_0)(x - x_1)}{(x_2 - x_0)(x_2 - x_1)}, \tag{5.12}$$

which can be expressed in a compact form as

$$f(x) = y = \sum_{i=0}^{2} y_i \prod_{j=0, j \neq i}^{2} \frac{(x - x_j)}{(x_i - x_j)}. \tag{5.13}$$

Note that the ith term on the right-hand side of Eq. (5.12) or Eq. (5.13) does not contain $(x - x_i)$ in the numerator and $(x_i - x_j)$ in the denominator. The just described procedure can be generalized for $n+1$ data points, $(x_i, y_i), i = 0, 1, 2, \ldots, n$, to obtain an nth-order Lagrange interpolation polynomial as

$$f(x) = y = \sum_{i=0}^{n} y_i \prod_{j=0, j \neq i}^{n} \frac{(x - x_j)}{(x_i - x_j)}. \tag{5.14}$$

Equation (5.14) gives an exact fit for any set of $n+1$ data points, $(x_i, y_i), i = 0, 1, 2, \ldots, n$. Note that the points x_i need not be uniformly spaced. Also the determination of the polynomial does not involve the solution of a system of equations. Although Eq. (5.14) appears to be lengthy, it has a very simple structure and hence can be implemented into a computer program in a simple manner.

▶Example 5.5

Develop a Lagrange interpolation polynomial that passes through the points (0, 0), (0.005, 110), and (0.01, 0).

Solution

By denoting $x_0 = 0, x_1 = 0.005, x_2 = 0.01, y_0 = 0, y_1 = 110$, and $y_2 = 0$, the required Lagrange polynomial is given by Eq. (5.12):

$$y(x) = 110\frac{(x-0)(x-0.01)}{(0.005-0)(0.005-0.01)} = -4.4(10^6) \times (x - 0.01).$$ ◀

5.5.2 Error in Lagrange Interpolation

Let $y(x)$ be the function from which the data points (x_i, y_i) are sampled with $y_i = y(x_i); i = 0, 1, 2, \ldots, n$. If $f(x)$ denotes the Lagrange interpolation polynomial of order n to approximate the function $y(x)$, then $f(x)$ will pass through all the $n+1$ data points. If $y(x)$ is a polynomial of order n or less, $f(x)$ becomes identical to $y(x)$. Otherwise, there will be an error, $e(x)$, given by

$$e(x) = y(x) - f(x). \tag{5.15}$$

The error, $e(x)$, depends on the following factors:

1. Range of the interpolation domain, $x_n - x_0$.
2. Spacing of the values of x_i.
3. Order of the interpolating polynomial (or equivalently, the number of data points, $n+1$).

In general, the maximum value of the error, $|e(x)|$, approaches zero as the range, $x_n - x_0$, is reduced. For large values of the range, $x_n - x_0$, sometimes the error can dominate even $|f(x)|$. In most data, the values of x_i are uniformly spaced. For this case, the error $|e(x)|$ usually will be smallest in the middle of the range and largest near the ends, x_0 and x_n. For any specified range of interpolation $(x_n - x_0)$, the maximum error, $|e(x)|$, usually decreases with an increase in the order of the polynomial, n. Beyond a certain value of n, however, the error may start increasing. An estimate of the error in Lagrange interpolation polynomial is given by [5.1]

$$e(x) = g(x)y^{(n+1)}(\xi); \quad x_0 \le \xi \le x_n, \tag{5.16}$$

where

$$g(x) = \frac{(x - x_0)(x - x_1)(x - x_2)\cdots(x - x_n)}{(n+1)!} \tag{5.17}$$

and $y^{(n+1)}(\xi)$ is the $(n+1)$st derivative of y with respect to x evaluated at $x = \xi$. Note that if $y(x)$ is a polynomial of order n or less, then its $(n+1)$st derivative will be zero and hence the error becomes zero.

5.6 Newton's Divided-Difference Interpolating Polynomials

The Lagrange interpolation polynomial considered in the previous section has the following limitations:

(1) If the number of data points is to be increased or decreased, the results of the previous calculations cannot be used.

(2) The computational effort required is more for a single interpolation.

(3) If interpolation is required at another value of x, there will not be any savings in the computational effort since the results of the previous computations can not be used.

(4) The estimation of error in interpolation is not easy.

Newton's divided-difference interpolation polynomial overcomes all these limitations and hence is used more extensively. First we consider the general formulas applicable to arbitrarily spaced data points, and then we derive the simplified formulas for equally spaced data points.

General interpolation formulas

The procedure used for a general nth-order interpolation formula can be seen by considering the derivation of the lowest order interpolation formulas.

5.6.1 Linear Interpolation

If two data points are available as $\{x_0, f(x_0)\}$ and $\{x_1, f(x_1)\}$, the points can be connected by a straight line to obtain the simplest form of interpolation, namely, linear interpolation. (See Fig. 5.4.) The formula for linear interpolation can be derived by considering the similar triangles ADE and ABC, which give

$$\frac{DE}{AE} = \frac{BC}{AC}, \quad \text{or} \quad \frac{f_1(x) - f(x_0)}{x - x_0} = \frac{f(x_1) - f(x_0)}{x_1 - x_0}, \tag{5.18}$$

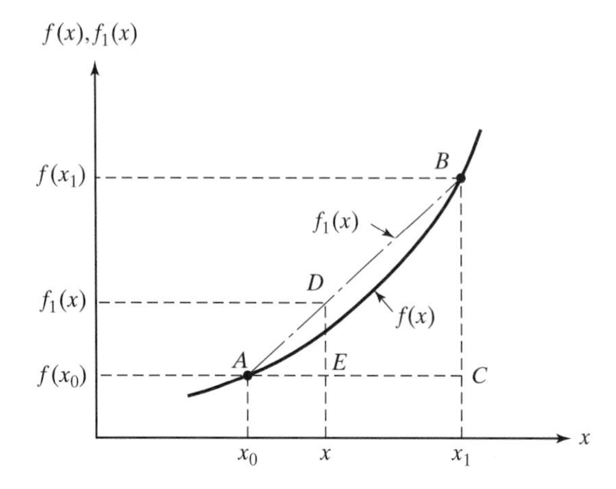

Figure 5.4 Linear interpolation.

where $f_1(x)$ denotes the interpolating polynomial, with the subscript 1 indicating the order of the polynomial. Equation (5.18) can be rewritten as

$$f_1(x) = f(x_0) + \left\{ \frac{f(x_1) - f(x_0)}{x_1 - x_0} \right\} (x - x_0), \tag{5.19}$$

or

$$f_1(x) = a_0 + a_1(x - x_0), \tag{5.20}$$

where a_0 and a_1 are the coefficients of the polynomial with a_1 representing the finite-divided-difference approximation of the first derivative:

$$a_0 = f(x_0), a_1 = \frac{f(x_1) - f(x_0)}{x_1 - x_0}. \tag{5.21}$$

▶Example 5.6

Develop a linear interpolation formula for the function $e^{0.5x}$ using the values at $x_0 = 0$ and $x_1 = 2$, and use it to estimate the value of $e^{0.5x}$ at $x = 1$.

Solution

The values of the function $f(x)$ at x_0 and x_1 are given by $f(x_0) = f(0) = e^0 = 1.0$ and $f(x_1) = f(2) = e^{1.0} = 2.718282$, and hence the linear interpolation formula, Eq. (5.20), can be expressed as

$$f_1(x) = a_0 + a_1(x - x_0), \tag{E1}$$

where $a_0 = f(x_0) = 1.0$ and

$$a_1 = \frac{f(x_1) - f(x_0)}{x_1 - x_0} = \frac{2.718282 - 1.0}{2.0 - 0.0} = 0.859141.$$

Thus, Eq. (E1) can be rewritten as

$$f_1(x) = 1.0 + 0.859141x. \tag{E2}$$

The interpolated value of $e^{0.5x}$ at $x = 1.0$ can be obtained from Eq. (E2) as

$$f_1(1) = 1.0 + 0.859141(1) = 1.859141.$$

When compared with the exact value $f(1) = e^{0.5} = 1.648721$, the linearly interpolated value can be seen to be in error by 12.762620%. ◀

5.6.2 Quadratic Interpolation

If three data points are available as $\{x_0, f(x_0)\}$, $\{x_1, f(x_1)\}$, and $\{x_2, f(x_2)\}$, we can use a second-order or quadratic polynomial to fit the data. (See Fig. 5.5.) It is convenient to assume the quadratic interpolation formula of the form

$$f_2(x) = a_0 + a_1(x - x_0) + a_2(x - x_0)(x - x_1), \tag{5.22}$$

where the subscript of $f_2(x)$ denotes the order of the polynomial, and the coefficients a_0, a_1, and a_2 can be evaluated using the data points. By using the relation $f_2(x = x_0) = f(x_0)$ when $x = x_0$, Eq. (5.22) gives

$$a_0 = f(x_0). \tag{5.23}$$

When $x = x_1$, we use the relation $f_2(x = x_1) = f(x_1)$ and obtain from Eq. (5.22):

$$f_2(x_1) = f(x_1) = a_0 + a_1(x_1 - x_0),$$

or

$$a_1 = \frac{f(x_1) - a_0}{x_1 - x_0} = \frac{f(x_1) - f(x_0)}{x_1 - x_0}. \tag{5.24}$$

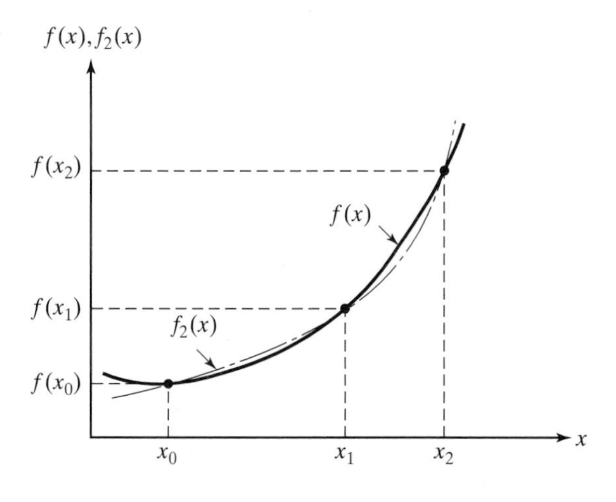

Figure 5.5 Quadratic interpolation.

Finally, Eq. (5.22) is evaluated at $x = x_2$ to obtain

$$f_2(x_2) = f(x_2) = a_0 + a_1(x_2 - x_0) + a_2(x_2 - x_0)(x_2 - x_1),$$

or

$$a_2 = \frac{f(x_2) - a_0 - a_1(x_2 - x_0)}{(x_2 - x_0)(x_2 - x_1)}. \tag{5.25}$$

Substituting Eqs. (5.23) and (5.24) into Eq. (5.25) yields

$$a_2 = \frac{1}{(x_2 - x_0)} \left\{ \frac{f(x_2) - f(x_1)}{x_2 - x_1} - \frac{f(x_1) - f(x_0)}{x_1 - x_0} \right\}. \tag{5.26}$$

Notes

(1) The first two terms of Eq. (5.22) can be seen to be identical to the linear interpolation formula, Eq. (5.20). This shows that all previous computations can still be used when new data points are considered to increase the order of the interpolating polynomial.
(2) The constant a_1 still denotes the finite-difference approximation of the derivative [slope of the line joining the function values $f(x_0)$ and $f(x_1)$], while a_2 can be seen to represent the finite-divided-difference approximation of the second derivative.

▶Example 5.7

Develop a quadratic interpolation formula for the function $e^{0.5x}$ using the values at $x_0 = 0$, $x_1 = 2$, and $x_2 = 4$ and use it to estimate the value of $e^{0.5x}$ at $x = 1$.

Solution

The values of the function $f(x)$ at x_0, x_1, and x_2 are given by

$$f(x_0) = f(0) = e^0 = 1.0,$$

$$f(x_1) = f(2) = e^{1.0} = 2.718282,$$

$$f(x_2) = f(4) = e^{2.0} = 7.389056,$$

and hence the quadratic interpolation formula, Eq. (5.22), can be expressed as

$$f_2(x) = a_0 + a_1(x - x_0) + a_2(x - x_0)(x - x_1), \tag{E1}$$

where

$$a_0 = f(x_0) = 1.0,$$

$$a_1 = \frac{f(x_1) - f(x_0)}{x_1 - x_0} = \frac{2.718282 - 1.0}{2.0 - 0.0} = 0.859141,$$

and

$$a_2 = \frac{1}{x_2 - x_0}\left(\frac{f(x_2) - f(x_1)}{x_2 - x_1} - \frac{f(x_1) - f(x_0)}{x_1 - x_0}\right)$$

$$= \frac{1}{(4-0)}\left(\frac{7.389056 - 2.718282}{4.0 - 2.0} - \frac{2.718282 - 1.0}{1.0 - 0.0}\right) = 0.3690615.$$

Thus, Eq. (E1) becomes

$$f_2(x) = 1.0 + 0.859141x + 0.3690615x(x - 2.0). \tag{E2}$$

The interpolated value of $e^{0.5x}$ at $x = 1$ can be found from Eq. (E2) as

$$f_2(1) = 1.0 + 0.859141(1) + 0.3690615(1)(1 - 2.0) = 1.4900795.$$

When compared with the exact value $f(1) = e^{0.5} = 1.648721$, the quadratically interpolated value can be seen to be in error by 9.622095%. ◀

5.6.3 *n*th-order Polynomial Interpolation

The features exhibited by the linear and quadratic interpolation formulas, Eqs. (5.20) and (5.22), permit us to express an nth-order polynomial as

$$f_n(x) = a_0 + a_1(x - x_0) + a_2(x - x_0)(x - x_1) + a_3(x - x_0)(x - x_1)(x - x_2)$$
$$+ \cdots + a_n(x - x_0)(x - x_1)(x - x_2)\cdots(x - x_{n-1}), \tag{5.27}$$

which is known as the nth-order Newton's divided-difference interpolating polynomial. The subscript n in $f_n(x)$ indicates the order of the polynomial and the $n+1$ coefficients $a_0, a_1, a_2, \ldots, a_n$ can be evaluated using the $n+1$ data points $\{x_i, f(x_i)\}$, $i = 0, 1, 2, \ldots, n$. Also, we introduce a series of functions $g(x_0), g(x_0, x_1), \ldots$, in order to develop a recursive formula for finding the coefficients $a_0, a_1, a_2, \ldots, a_n$. By using the data point $\{x_0, f(x_0)\}$ in Eq. (5.27), we obtain

$$a_0 = f(x_0) \equiv g(x_0). \tag{5.28}$$

When the next data point $\{x_1, f(x_1)\}$ is used, Eq. (5.27) gives

$$a_1 = \left\{\frac{f(x_1) - f(x_0)}{x_1 - x_0}\right\} \equiv g(x_1, x_0). \tag{5.29}$$

When the data point $\{x_2, f(x_2)\}$ is used, Eq. (5.27) yields

$$a_2 = \frac{1}{(x_2 - x_0)}\left\{\frac{f(x_2) - f(x_1)}{x_2 - x_1} - \frac{f(x_1) - f(x_0)}{x_1 - x_0}\right\} \equiv g(x_2, x_1, x_0). \tag{5.30}$$

Similarly, when other data points are used, Eq. (5.27) yields the other constants, which can be expressed as

$$a_3 \equiv g(x_3, x_2, x_1, x_0); \tag{5.31}$$

$$\vdots$$

$$a_n \equiv g(x_n, x_{n-1}, x_{n-2}, \ldots, x_1, x_0). \tag{5.32}$$

In Eqs. (5.28) through (5.32), the functions $g(x_0), g(x_1, x_0), g(x_2, x_1, x_0), \ldots, g(x_n, x_{n-1}, \ldots, x_2, x_1, x_0)$ denote the finite divided differences:

$$g(x_0) = f(x_0) \text{ (zeroth-order forward difference);} \tag{5.33}$$

$$g(x_1, x_0) = \frac{g(x_1) - g(x_0)}{x_1 - x_0} = \frac{f(x_1) - f(x_0)}{x_1 - x_0}$$

$$\text{(first-order forward difference);} \tag{5.34}$$

$$g(x_2, x_1, x_0) = \frac{g(x_2, x_1) - g(x_1, x_0)}{x_2 - x_0}$$

$$= \frac{1}{x_2 - x_0}\left\{\frac{g(x_2) - g(x_1)}{x_2 - x_1} - \frac{g(x_1) - g(x_0)}{x_1 - x_0}\right\}$$

$$\text{(second-order forward difference);} \tag{5.35}$$

$$\vdots$$

$$g(x_n, x_{n-1}, \ldots, x_2, x_1, x_0) = \frac{g(x_n, x_{n-1}, \ldots, x_2, x_1) - g(x_{n-1}, x_{n-2}, \ldots, x_1, x_0)}{x_n - x_0}$$

$$\text{(}n\text{th-order forward difference).} \tag{5.36}$$

Note that Eq. (5.36) indicates a general recursive formula that enables us to compute the higher order forward differences from the lower order forward differences as shown graphically in Fig. 5.6. Thus, the coefficients $a_0, a_1, a_2, \ldots, a_n$ can be evaluated successively. Once the coefficients are determined, the interpolating polynomial can be obtained from Eq. (5.27). In view of the notation in Eqs. (5.28) –(5.32), Eq. (5.27) can be expressed as

$$f_n(x) = g(x_0) + (x - x_0)g(x_1, x_0) + (x - x_0)(x - x_1)g(x_2, x_1, x_0) + \cdots + (x - x_0)$$
$$\times (x - x_1)(x - x_2)\cdots(x - x_{n-1})g(x_n, x_{n-1}, \ldots, x_2, x_1, x_0). \tag{5.37}$$

It can be seen that the general structure of the Newton's intolating polynomial, Eq. (5.37), is similar to the Taylor's series expansion, as terms involving higher order derivatives are added in succession to improve the accuracy of the polynomial. Hence, the expression for the error involved with the Taylor's series expansion can be used to estimate the error of Newton's interpolating polynomial. The remainder term (R_n), and hence the truncation error, in a Taylor's series expansion is given by [5.4]:

$$R_n = \frac{f^{(n+1)}(\xi)}{(n+1)!}(x_{i+1} - x_i)^{n+1}; x_i < \xi < x_{i+1}. \tag{5.38}$$

In a similar manner, the error in an nth-order Newton's interpolating polynomial can be expressed as

$$E = R_n \approx \frac{f^{(n+1)}(\xi)}{(n+1)!}(x - x_0)(x - x_1)(x - x_2)\cdots(x - x_n); x_0 < \xi < x_n, \tag{5.39}$$

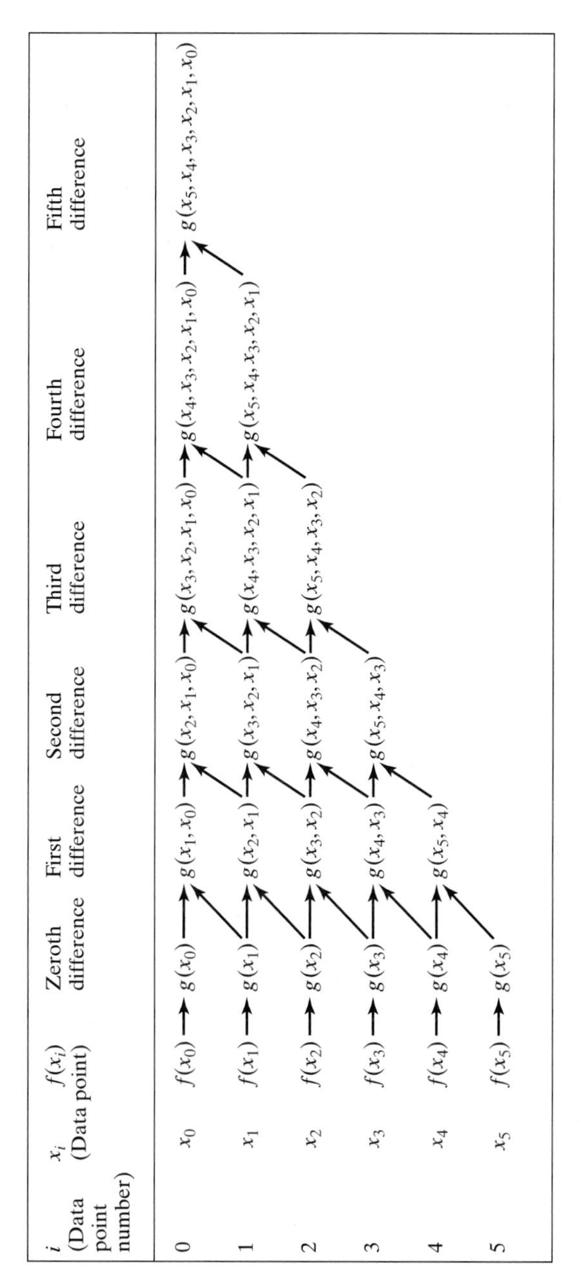

Figure 5.6 Forward divided difference table.

where the $(n+1)$st order derivative is evaluated at some point ξ in the range of integration. Note that Eq. (5.39) is same as the expression for error associated with an nth-order Lagrangian interpolation formula. Also, Eq. (5.39) indicates that the true function $f(x)$ must be known and differentiable. However, $f(x)$ is not usually known. Hence, we approximate the $(n+1)$st derivative using a finite-divided-difference formula and write Eq. (5.39) as

$$E = R_n \approx g(x, x_n, x_{n-1}, \ldots, x_2, x_1, x_0)(x - x_0)(x - x_1) \cdots (x - x_n), \tag{5.40}$$

where $g(x, x_n, x_{n-1}, \ldots, x_2, x_1, x_0)$ denotes the $(n+1)$st-order forward difference. Since the sequential expansion of $g(x, x_n, x_{n-1}, \ldots, x_2, x_1, x_0)$ ultimately requires the unknown function $f(x)$, we cannot evaluate the error using Eq. (5.40). However, if an additional data point is known as $\{x_{n+1}, f(x_{n+1})\}$, the error can be estimated as

$$E = R_n \approx g(x_{n+1}, x_n, x_{n-1}, \ldots, x_2, x_1, x_0)(x - x_0)(x - x_1) \cdots (x - x_n). \tag{5.41}$$

Since the additional data point $\{x_{n+1}, f(x_{n+1})\}$ is usually not available, we use the interpolation polynomial itself to find the additional data point as $\{x_{n+1}, f(x_{n+1}) = f_n(x_{n+1})\}$ for any arbitrary value of $x_{n+1} \neq x_i (i = 0, 1, 2, \ldots, n)$.

▶Example 5.8

Develop a third-order Newton's polynomial for approximating $e^{0.5x}$ using the values at $x_0 = 0, x_1 = 2, x_2 = 4$, and $x_3 = 5$, and use it to estimate the value of $e^{0.5x}$ at $x = 1$.

Solution

The values of the function $f(x)$ at x_0, x_1, x_2, and x_3 are given by

$$f(x_0) = f(0) = e^0 = 1;$$

$$f(x_1) = f(2) = e^{1.0} = 2.718282;$$

$$f(x_2) = f(4) = e^{2.0} = 7.389056;$$

$$f(x_3) = f(5) = e^{2.5} = 12.182494.$$

The coefficients of the polynomial can be evaluated recursively as

$$g(x_0) = f(x_0) = 1.0;$$

$$g(x_1) = f(x_1) = 2.718282;$$

$$g(x_2) = f(x_2) = 7.389056;$$

$$g(x_3) = f(x_3) = 12.182494;$$

$$g(x_1, x_0) = \frac{g(x_1) - g(x_0)}{x_1 - x_0} = \frac{2.718282 - 1.0}{2.0 - 0.0} = 0.859141;$$

$$g(x_2, x_1) = \frac{g(x_2) - g(x_1)}{x_2 - x_1} = \frac{7.389056 - 2.718282}{4 - 2} = 2.335387;$$

$$g(x_3, x_2) = \frac{g(x_3) - g(x_2)}{x_3 - x_2} = \frac{12.182494 - 7.389056}{5 - 4} = 4.793438;$$

$$g(x_2, x_1, x_0) = \frac{g(x_2, x_1) - g(x_1, x_0)}{x_2 - x_0} = \frac{2.335387 - 0.859141}{4 - 0} = 0.3690615;$$

$$g(x_3, x_2, x_1) = \frac{g(x_3, x_2) - g(x_2, x_1)}{x_3 - x_1} = \frac{4.793438 - 2.335387}{5 - 2} = 0.819350;$$

$$g(x_3, x_2, x_1, x_0) = \frac{g(x_3, x_2, x_1) - g(x_2, x_1, x_0)}{x_3 - x_0} = \frac{0.819350 - 0.3690615}{5 - 0} = 0.0900577.$$

Thus, the approximating third-degree Newton's polynomial is given by Eq. (5.37) with $n = 3$:

$$\begin{aligned} f_3(x) &= g(x_0) + (x - x_0)g(x_1, x_0) + (x - x_0)(x - x_1)g(x_2, x_1, x_0) \\ &\quad + (x - x_0)(x - x_1)(x - x_2)g(x_3, x_2, x_1, x_0) \\ &= 1.0 + 0.859141x + 0.3690615x(x - 2) + 0.0900577x(x - 2)(x - 4). \end{aligned}$$

The value of $f_3(x)$ at $x = 1$ can be found as

$$f_3(1) = 1.0 + 0.859141 - 0.3690615 + 0.2701731 = 1.7602526.$$

Since the exact value is $f(1) = e^{0.5} = 1.648721$, the error in $f_3(1)$ can be seen to be 6.764735%.

◀

5.6.4 Newton's Interpolation Polynomials for Uniformly Spaced Data

The general Newton's interpolation formula derived in the previous section can be simplified when the data points are equally spaced along the x-axis. The resulting formulas are also known as Newton–Gregory formulas. If h denotes the interval between any two consecutive values of x_i (i.e., $x_{i+1} - x_i = h$), we can write

$$x_i = x_0 + ih; i = 1, 2, \ldots, n. \tag{5.42}$$

In this case, the coefficients of the interpolation polynomial, $a_0, a_1, \ldots, a_n$, can be expressed, from Eqs. (5.28) through (5.32), as

$$a_0 = f(x_0) \equiv \frac{\Delta^0 f_0}{h^0}, \tag{5.43}$$

$$a_1 = \left\{ \frac{f(x_1) - f(x_0)}{x_1 - x_0} \right\} = \frac{\Delta^1 f_0}{h^1} = \frac{\Delta f_0}{h}, \tag{5.44}$$

$$a_2 = \frac{1}{(x_2 - x_0)} \left\{ \frac{f(x_2) - f(x_1)}{x_2 - x_1} - \frac{f(x_1) - f(x_0)}{x_1 - x_0} \right\}$$

$$= \frac{f(x_2) - 2f(x_1) + f(x_0)}{2h^2} = \frac{\Delta^2 f_0}{2!h^2}, \tag{5.45}$$

$$\vdots$$

and

$$a_n = \frac{\Delta^n f_0}{n!h^n}, \tag{5.46}$$

where $\Delta^j f_0$ is the jth-order forward difference ($j = 0, 1, 2, \dots, n$) evaluated at the base point $x = x_0$ with $\Delta^0 f_0$ denoting the zeroth-order forward difference that corresponds to the value of the function f_0. Using these relations, Eqs. (5.43) through (5.46), we can write the nth-order Newton's interpolation polynomial of Eq. (5.27) as

$$f_n(x) = f(x_0) + \frac{\Delta f_0}{h}(x - x_0) + \frac{\Delta^2 f_0}{2!h^2}(x - x_0)(x - x_0 - h)$$
$$+ \cdots + \frac{\Delta^n f_0}{n!h^n}(x - x_0)(x - x_0 - h)\cdots(x - x_0 - (n-1)h). \tag{5.47}$$

Equation (5.47) is known as the Newton–Gregory forward interpolation formula. Note that the forward differences at the point $x = x_i$ are given by

$$\Delta^j f_i = \Delta^{j-1} f_{i+1} - \Delta^{j-1} f_i;\ j = 1, 2, \dots, n. \tag{5.48}$$

Another frequently used formula, known as the Newton–Gregory backward interpolation formula, makes use of backward differences and can be derived as follows:

Let the various order backward differences be denoted as

$$\nabla^0 f_n = f_n; \tag{5.49}$$

$$\nabla^1 f_n = \frac{f_n - f_{n-1}}{h}; \tag{5.50}$$

$$\nabla^2 f_n = \frac{\nabla f_n - \nabla f_{n-1}}{h} = \frac{f_n - 2f_{n-1} + f_{n-2}}{h^2}; \tag{5.51}$$

$$\vdots$$

$$\nabla^j f_n = \nabla^{j-1} f_n - \nabla^{j-1} f_{n-1};\ j = 1, 2, \dots, n. \tag{5.52}$$

Then the Newton–Gregory backward interpolating polynomial can be written as

$$f_n(x) = f(x_n) + \frac{\nabla f_n}{h}(x - x_n) + \frac{\nabla^2 f_n}{2!h^2}(x - x_n)(x - x_n + h)$$
$$+ \cdots + \frac{\nabla^n f_n}{n!h^n}(x - x_n)(x - x_n + h)\cdots(x - x_n + (n-1)h). \tag{5.53}$$

Once the interpolation polynomial is defined, either the forward difference formula (Eq. (5.47)), or the backward difference formula (Eq. (5.53)), the desired value of the function $f(x)$, approximated by $f_n(x)$, can be found for any specified value of x. The choice of the formula, Eq. (5.47) or Eq. (5.53), depends on the value of x at which the function value, $f(x)$, is to be determined and its closeness to the known data points. For example, if x is close to x_0, then the forward difference formula (Eq. (5.47)) is expected to be better. On the other hand, if x is close to x_n, then the backward difference formula is expected to be better.

5.7 Interpolation using Chebyshev Polynomials

As indicated earlier, the error in polynomial interpolations such as nth-order polynomial, Lagrange formulas or Newton's formulas, is usually small in the middle of the domain, but increases towards the endpoints. However, in the computer implementation of the method, we have no control on where in the domain the interpolation (or approximation) will be used. So, the previously discussed polynomial approximations are not usually appropriate. The use of Chebyshev polynomial interpolation provides a more uniform distribution of error throughout the domain.

5.7.1 Chebyshev Polynomials

Chebyshev polynomials can be expressed in the form of power series, or, equivalently, in the form of cosine functions. The Chebyshev polynomials, in power series, are defined by

$$P_0(x) = 1,$$

$$P_1(x) = x,$$

and

$$P_{i+1}(x) = 2xP_i(x) - P_{i-1}(x); i = 1, 2, \ldots. \tag{5.54}$$

The first five Chebyshev polynomials are given by

$$P_0(x) = 1,$$

$$P_1(x) = x,$$

$$P_2(x) = 2x^2 - 1,$$

$$P_3(x) = 4x^3 - 3x,$$

and

$$P_4(x) = 8x^4 - 8x^2 + 1, \tag{5.55}$$

which are also shown graphically in Fig. 5.7. The Chebyshev polynomials, in the form of cosine functions, are defined by

$$P_i(x) = \cos(i \cos^{-1} x); -1 \leq x \leq 1; i = 0, 1, 2, 3, \ldots, \tag{5.56}$$

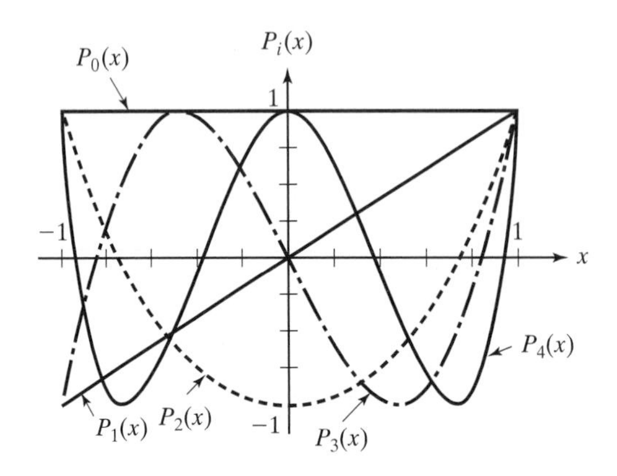

Figure 5.7 First five Chebyshev polynomials in [−1, 1].

or

$$P_i(x) = \cos i\theta, \tag{5.57}$$

where

$$\theta = \cos^{-1} x; -1 \le x \le 1; i = 0, 1, 2, 3, \ldots, \tag{5.58}$$

or

$$\cos\theta = x. \tag{5.59}$$

The equivalence of Eqs. (5.55) and (5.56) can be seen as follows:

$$\text{For } i = 0, \cos(0) = 1, \text{ and hence } P_0(x) = 1. \tag{5.60}$$

$$\text{For } i = 1, \text{ we have } P_1(x) = \cos(\cos^{-1} x) = x. \tag{5.61}$$

For $i = 2$, Eq. (5.56) gives $P_2(x) = \cos(2\cos^{-1} x)$. Noting that $\cos 2\theta = 2\cos^2\theta - 1$, we find that

$$P_2(x) = 2x^2 - 1. \tag{5.62}$$

Similarly, in view of the relation

$$\cos 3\theta = 4\cos^3\theta - 3\cos\theta,$$

we have

$$P_3(x) = 4x^3 - 3x. \tag{5.63}$$

In general, from the trigonometric identity

$$\cos(i+1)\theta + \cos(i-1)\theta = 2\cos\theta\cos i\theta, \tag{5.64}$$

we obtain

$$P_{i+1}(x) + P_{i-1}(x) = 2x P_i(x), \tag{5.65}$$

which can be seen to be same as the recursive relation, Eq. (5.54).

From the relation $P_i(x) = \cos i\theta$, we find that Chebyshev polynomials have their local minimum and maximum of −1 and +1, respectively, in the interval $-1 \le x \le 1$. Since $|\cos i\theta| = 1$, for $i\theta = 0, \pi, 2\pi, \ldots$, and θ varies from 0 to π as

x varies from 1 to -1, $P_i(x)$ attains its maximum value of one $n+1$ times in the interval $[-1, 1]$.

The zeroes of a Chebyshev polynomial of order i can be found by noting that a cosine function becomes zero at $\pm\frac{\pi}{2}, \pm\frac{3\pi}{2}, \ldots$, so that

$$i \cos^{-1}(x_j) = \left(i + \frac{1}{2} - j\right)\pi; j = 1, 2, \ldots, i, \tag{5.66}$$

or

$$x_j = \cos\left(\frac{i + \frac{1}{2} - j}{i}\pi\right); j = 1, 2, \ldots, i. \tag{5.67}$$

For example, for $i = 3$, x_j are given by

$$x_1 = -0.86602, x_2 = 0, x_3 = +0.86602. \tag{5.68}$$

5.7.2 Interpolation

(i) Power series expansion

We can use Chebyshev polynomials to express an nth-order polynomial. The various order powers of x can be written in terms of Chebyshev polynomials as

$$\begin{aligned}
1 &= P_0(x); \\
x &= P_1(x); \\
x^2 &= \frac{1}{2}\{P_0(x) + P_2(x)\}; \\
x^3 &= \frac{1}{4}\{3P_1(x) + P_3(x)\}; \\
x^4 &= \frac{1}{8}\{3P_0(x) + 4P_2(x) + P_4(x)\}; \\
x^5 &= \frac{1}{16}\{10P_1(x) + 5P_3(x) + P_5(x)\}; \\
x^6 &= \frac{1}{32}\{10P_0(x) + 15P_2(x) + 6P_4(x) + P_6(x)\}; \\
&\vdots
\end{aligned} \tag{5.69}$$

By substituting Eqs. (5.69) into the nth-order polynomial and collecting terms in $P_i(x)$, we create a Chebyshev series approximation of the function $f(x)$.

▶Example 5.9

Express the function $f(x) = 4 - 7x + 2x^2 + 5x^3$ in terms of Chebyshev polynomials.

Solution

With the help of Eq. (5.69), the function $f(x)$ can be rewritten as

$$f(x) = 4P_0(x) - 7P_1(x) + 2\left[\frac{1}{2}\{P_0(x) + P_2(x)\}\right] + 5\left[\frac{1}{4}\{3P_1(x) + P_3(x)\}\right],$$

which can be simplified as

$$f(x) = 5P_0(x) - \frac{13}{4}P_1(x) + P_2(x) + \frac{5}{4}P_3(x).$$

◀

(ii) Direct expansion in terms of Chebyshev polynomials

The function $f(x)$ can be expressed directly in terms of Chebyshev polynomials. For this, we make use of the orthogonal property of Chebyshev polynomials:

$$\int_{-1}^{1} \frac{P_n(x)P_m(x)}{\sqrt{1-x^2}}dx = \begin{cases} 0, & n \neq m \\ \pi, & n = m = 0 \\ \frac{\pi}{2}, & n = m \neq 0 \end{cases}. \tag{5.70}$$

The function $f(x)$ is expressed as an infinite series expansion in terms of Chebyshev polynomials as

$$f(x) = \frac{a_0}{2} + \sum_{i=1}^{\infty} a_i P_i(x), \tag{5.71}$$

where the constants a_i can be derived from the orthogonal property as

$$a_i = \frac{2}{\pi}\int_{-1}^{1} \frac{f(x)P_i(x)}{\sqrt{1-x^2}}dx; i = 0, 1, 2, \ldots. \tag{5.72}$$

The series in Eq. (5.71) can be truncated after the $(n+1)$st term, which amounts to approximating $f(x)$ by an nth-order polynomial. Note that Eq. (5.71) is nothing, but the Fourier series expansion of the function $f(\cos\theta)$.

(iii) Interpolation using roots of Chebyshev polynomials

Assuming that the range of integration is $[-1, 1]$, the i roots, x_j $(j = 1, 2, \ldots, i)$, of the Chebyshev polynomial of order i can be used as abscissas in the Lagrange interpolation formula, Eq. (5.14), instead of using equally spaced values of x_i. This gives the following interpolation formula of order $(i-1)$:

$$f(x) = y = \sum_{j=0}^{i-1} y_j \prod_{k=0, k\neq j}^{i-1} \left(\frac{x - x_k}{x_j - x_k}\right). \tag{5.73}$$

Here, (x_k, y_k), $k = 0, 1, 2, \ldots, (i-1)$ denote the i data points, and x_k are the values of the abscissa of the data points, assumed to be the same as the roots of the Chebyshev polynomial of order i.

Notes

(1) The numbering of points in finding the roots of Chebyshev polynomials, Eq. (5.67), and those used in the Lagrange interpolation polynomial, Eq. (5.73), are different. For example, if $i = 3$, then the Lagrange interpolation polynomial will be of order 2, and the roots of Eq. (5.68) are to be renumbered as

$$x_0 = -0.86602, x_1 = 0, x_2 = +0.86602 \tag{5.74}$$

for use in Eq. (5.73).

(2) No data corresponding to the end points ($x = -1$ and $x = 1$) is used in Eq. (5.73). If necessary, Eq. (5.73) will be used as an *extrapolation* polynomial in the intervals $[-1, -0.86602]$ and $[0.86602, 1]$.

(3) The range of Chebyshev polynomial interpolation was assumed to be $[-1, 1]$ in Eq. (5.73). Equation (5.73) can be applied to any other range $[a, b]$ by transforming or mapping $[-1, 1]$ onto the new range $[a, b]$. The necessary transformation is given by

$$x = \frac{2z - b - a}{b - a}; a \le z \le b; -1 \le x \le 1, \tag{5.75}$$

or

$$z = \frac{(b - a)x + a + b}{2}; -1 \le x \le 1; a \le z \le b. \tag{5.76}$$

Thus, by substituting the roots of Chebyshev polynomials x_j in $[-1, 1]$, given by Eq. (5.67), into Eq. (5.76), we obtain the roots of Chebyshev polynomials z_j in $[a, b]$ as

$$z_j = \frac{1}{2}\left[(b - a)\cos\left\{\left(\frac{i + \frac{1}{2} - j}{i}\right)\pi\right\} + a + b\right]; j = 1, 2, \ldots, i. \tag{5.77}$$

(4) The error in the interpolation using Chebyshev roots is given by Eq. (5.16). However, the function $g(x)$ appearing in Eq. (5.16), defined by Eq. (5.17), now becomes a Chebyshev polynomial, because it passes through the roots of the Chebyshev polynomial. Thus, the error of the formula of Eq. (5.73) is more evenly distributed than in the case of equispaced values of x_j.

5.8 Interpolation using Splines

In all the polynomial interpolation methods discussed so far, an nth-order polynomial is used to interpolate between $n + 1$ data points. Although the polynomial passes through all the data points, the errors of a single polynomial tend to increase drastically as its order n becomes large. Often, a high-order polynomial introduces unnecessary oscillations or wiggles as indicated in Fig. 5.8(b).

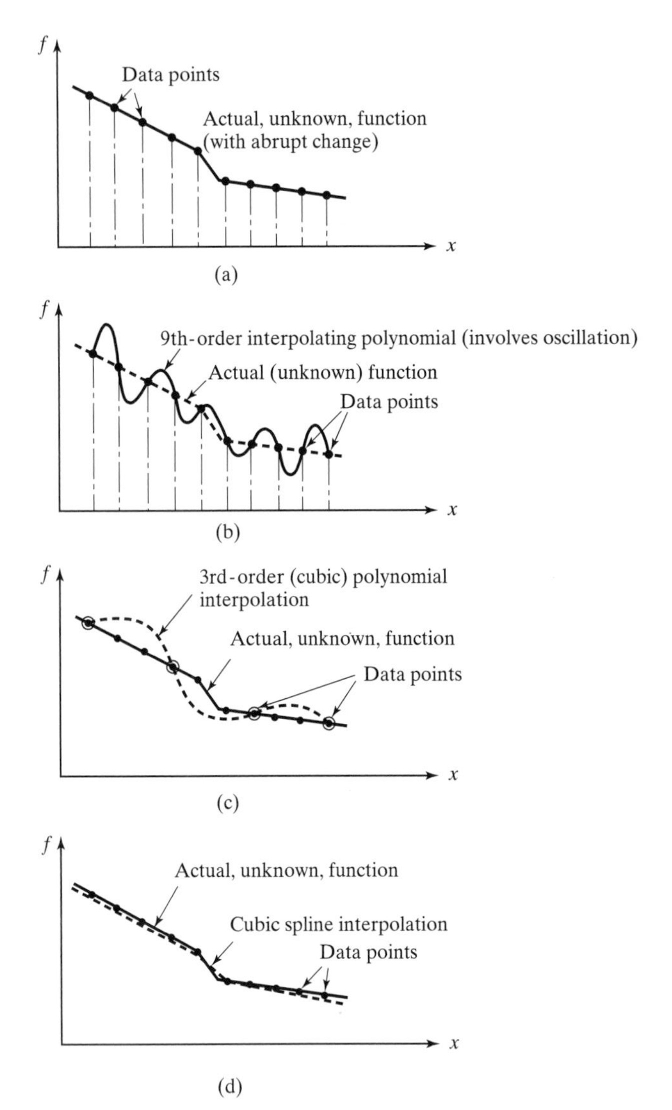

Figure 5.8 Advantage of using low-order polynomials.

In Fig. 5.8(a), the function undergoes an abrupt change in the middle of the range of interpolation. This introduces oscillations in interpolating polynomials as shown in Figs. 5.8(b) and (c). Hence, the polynomial interpolation in this case will not be very accurate. Alternatively, we can use a low-order piecewise polynomial joined together at the data points. When this procedure is applied for the function shown in Fig. 5.8(a), using third-order piecewise polynomials (cubic splines), the interpolation will be much superior and more acceptable as shown in Fig. 5.8(d).

Definition

A spline is defined as a piecewise polynomial of low order. The piecewise polynomials connected at the data points are called knots.

In most practical problems, it is desirable to maintain a continuous second derivative of the function (curvature). For example, let the curve given by the interpolation polynomial be used to machine a curved surface on a numerically controlled milling machine. The program in the milling machine makes the tool follow the path of the interpolation curve to generate the curved shape. If there is a sudden jump in the second derivative of the path (i.e., jump in acceleration), there will be a jump in the force applied to the cutting tool, since force is equal to the product of mass and acceleration. This sudden change in the force will result in a rough surface finish and sometimes may ruin the surface altogether.

First we discuss the basic concepts of splines by considering low-order splines. Then we examine cubic splines, which are most commonly used in engineering applications.

5.8.1 Step Function Spline

The step-function, or zeroth-order, spline is the simplest possible approximation. As shown in Fig. 5.9, it has jumps at the data points (knots) and hence is rarely used.

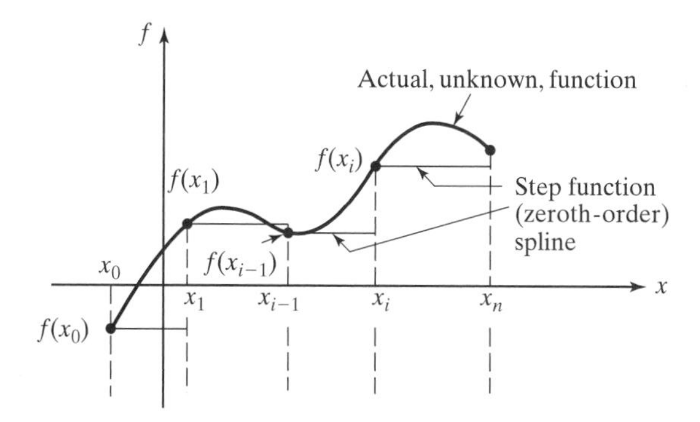

Figure 5.9 Zeroth-order spline.

5.8.2 Linear Spline

The linear or first-order spline represents a straight line between the data points (knots) as shown in Fig. 5.10. Let $n+1$ data points be available as $\{x_i, f(x_i)\}$, $i = 0, 1, 2, \ldots, n$. If we consider two neighboring data points $\{x_{i-1}, f(x_{i-1})\}$ and $\{x_i, f(x_i)\}$, the equation of the line joining the two points is given by

$$f_i(x) = f(x_{i-1}) + \left\{\frac{f(x_i) - f(x_{i-1})}{x_i - x_{i-1}}\right\}(x - x_{i-1}); i = 1, 2, \ldots, n. \tag{5.78}$$

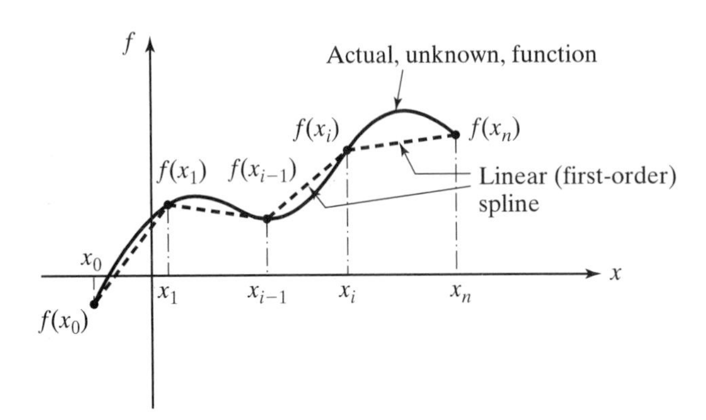

Figure 5.10 First-order spline.

Equation (5.78) represent a set of n piecewise linear equations (splines) using the $n+1$ data points. Thus, the function can be interpolated at any specified value of $x = \bar{x}$ by first locating the interval (x_{i-1}, x_i) in which $\bar{x}$ lies and then using the corresponding equation for $f_i(x)$ from Eq. (5.78). It can be seen that linear splines are not smooth. Although the function is continuous, the first derivative is discontinuous at the knots. In general, an nth-order spline ensures the continuity of the derivatives of the function up to order $n-1$.

▶Example 5.10

Find a linear spline to fit the following data:

i	0	1	2	3	4
x_i	2.0	3.0	6.5	8.0	12.0
$f(x_i)$	14.0	20.0	17.0	16.0	23.0

Use the results to estimate the value of f at $x = 7.0$.

Solution

The data points can be joined by straight lines to represent the linear spline. Since $x = 7.0$ lies in the interval (6.5 to 8.0), Eq. (5.78) is used to find the equation of the straight line between $x = 6.5$ and $x = 8.0$ as

$$f_3(x) = f(x_2) + \left\{ \frac{f(x_3) - f(x_2)}{x_3 - x_2} \right\} (x - x_2) = 17.0 + \left(\frac{16.0 - 17.0}{8.0 - 6.5} \right) (x - 6.5)$$
$$= 21.3333 - 0.6667x.$$

The value of f at $x = 7.0$ can be estimated as

$$f(7.0) = f_3(7.0) = 21.3333 - 0.6667(7.0) = 16.6664.$$

◀

5.8.3 Quadratic Spline

The second-order spline represents a quadratic equation between any two consecutive data points or knots. (See Fig. 5.11). Assuming that $\{x_i, f(x_i)\}, i = 0, 1, 2, \ldots, n$ denote $n + 1$ data points, the equation of the quadratic spline between the points $\{x_{i-1}, f(x_{i-1})\}$ and $\{x_i, f(x_i)\}$ can be expressed as

$$f_i(x) = a_i + b_i x + c_i x^2; \; i = 1, 2, \ldots, n, \tag{5.79}$$

where a_i, b_i, and c_i are the unknown constants (coefficients). Thus, for the n intervals, we need to evaluate $3n$ constants. The following conditions are used to evaluate these constants:

(1) The function value at the interior knot x_i must be equal to $f(x_i)$ whether it is computed using the quadratic $f_i(x)$ or $f_{i+1}(x)$:

$$f_i(x = x_i) = a_i + b_i x_i + c_i x_i^2 = f(x_i); \; i = 1, 2, \ldots, n - 1, \tag{5.80}$$

and

$$f_{i+1}(x = x_i) = a_{i+1} + b_{i+1} x_i + c_{i+1} x_i^2 = f(x_i); \; i = 1, 2, \ldots, n - 1. \tag{5.81}$$

Equations (5.80) and (5.81) give $(2n - 2)$ conditions.

(2) The first and last functions, $f_1(x)$ and $f_n(x)$, must pass through the endpoints x_0 and x_n, respectively:

$$f_1(x = x_0) = a_1 + b_1 x_0 + c_1 x_0^2 = f(x_0), \tag{5.82}$$

and

$$f_n(x = x_n) = a_n + b_n x_n + c_n x_n^2 = f(x_n). \tag{5.83}$$

Equations (5.82) and (5.83) provide two conditions.

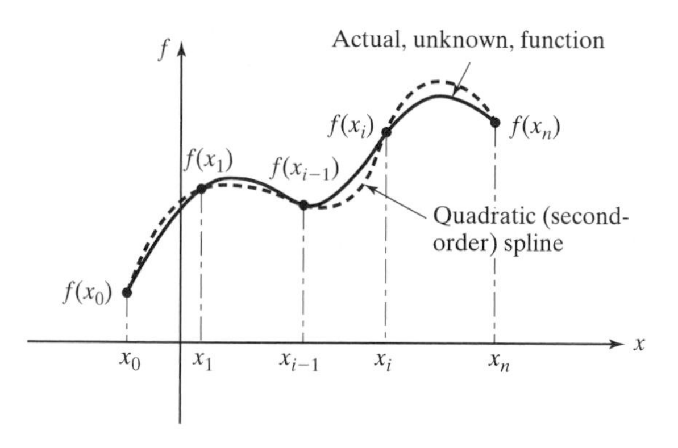

Figure 5.11 Second-order spline.

(3) The first derivative or slope at each of the interior knots must be continuous. Equation (5.79) gives the slope as

$$f_i'(x) = b_i + 2c_i x;\ i = 1, 2, \ldots, n, \tag{5.84}$$

and hence, the continuity of slope leads to

$$f_i'(x = x_i) = f_{i+1}'(x = x_i);$$

that is,

$$b_i + 2c_i x_i = b_{i+1} + 2c_{i+1} x_i;\ i = 1, 2, \ldots, n-1. \tag{5.85}$$

Equation (5.85) yields $(n - 1)$ conditions.

(4) Equations (5.80) through (5.83) and (5.85) give a total of $(3n - 1)$ conditions. Thus, we need one more condition to evaluate all $3n$ constants of Eq. (5.79). There are several possible conditions that can be used. For example, we can assume the second derivative (curvature) to be zero at the final point (x_n). This gives the relation

$$f_n''(x = x_n) = 2c_n = 0, \text{ or } c_n = 0. \tag{5.86}$$

▶Example 5.11

Find a quadratic spline to fit the data given in Example 5.11. Use the results to estimate the value of f at $x = 7.0$.

Solution

There are four intervals and five data points in this example. By equating the function values given by the quadratic functions $f(x_i)$ and $f(x_{i+1})$ at the interior points, we obtain (Eqs. (5.80) and (5.81))

$$f_1(x_1) = a_1 + 3b_1 + 9c_1 = 20, \tag{E1}$$

$$f_2(x_2) = a_2 + 6.5b_2 + 42.25c_2 = 17, \tag{E2}$$

$$f_3(x_3) = a_3 + 8b_3 + 64c_3 = 16, \tag{E3}$$

$$f_2(x_1) = a_2 + 3b_2 + 9c_2 = 20, \tag{E4}$$

$$f_3(x_2) = a_3 + 6.5b_3 + 42.25c_3 = 17, \tag{E5}$$

and

$$f_4(x_3) = a_4 + 8b_4 + 64c_4 = 16. \tag{E6}$$

By making the functions $f_1(x)$ and $f_4(x)$ pass through the endpoints x_0 and x_4, respectively, we obtain (Eqs. (5.82) and (5.83))

$$f_1(x_0) = a_1 + 2b_1 + 4c_1 = 14, \tag{E7}$$

and

$$f_4(x_4) = a_4 + 12b_4 + 144c_4 = 23. \tag{E8}$$

By making the first derivative at each of the interior knots continuous, we obtain (Eq. (5.85)):

$$b_1 + 6c_1 - b_2 - 6c_2 = 0; \tag{E9}$$

$$b_2 + 13c_2 - b_3 - 13c_3 = 0; \tag{E10}$$

$$b_3 + 16c_3 - b_4 - 16c_4 = 0. \tag{E11}$$

By setting the second derivative equal to zero at x_4, we obtain (Eq. (5.86))

$$c_4 = 0. \tag{E12}$$

It can be seen that there are 12 linear equations in 12 unknowns $(a_i, b_i, c_i;\ i = 1, 2, 3, 4)$. The simultaneous solution of Eqs. (E1) through (E12) gives

$$a_1 = -25.7858, a_2 = 10.1683, a_3 = 105.1112, a_4 = 2.0000;$$

$$b_1 = 29.1548, b_2 = 5.1854, b_3 = -24.0278, b_4 = 1.7500;$$

$$c_1 = -4.6309, c_2 = -0.6361, c_3 = 1.6111, c_4 = 0.0000.$$

Thus, the quadratic spline is given by

$$f_1(x) = -25.7858 + 29.1548x - 4.6309x^2;\ 2 \le x \le 3; \tag{E13}$$

$$f_2(x) = 10.1683 + 5.1854x - 0.6361x^2;\ 3 \le x \le 6.5; \tag{E14}$$

$$f_3(x) = 105.1112 - 24.0278x + 1.6111x^2;\ 6.5 \le x \le 8.0; \tag{E15}$$

$$f_4(x) = 2.0 + 1.75x;\ 8.0 \le x \le 12.0. \tag{E16}$$

The value of f at $x = 7.0$ can be estimated from Eq. (E15) as

$$f_3(7.0) = 105.1112 - 24.0278(7.0) + 1.6111(49.0) = 15.8605.$$ ◀

5.8.4 Cubic Spline

The third-order spline denotes a cubic equation (Fig. 5.12) in any of the n intervals corresponding to the $n + 1$ data points, $\{x_i, f(x_i)\}, i = 0, 1, 2, \ldots, n$. The equation of the cubic spline in the ith interval, $[x_{i-1}, x_i]$, can be expressed as

$$f_i(x) = a_i + b_i x + c_i x^2 + d_i x^3;\ i = 1, 2, \ldots, n, \tag{5.87}$$

where the $4n$ coefficients a_i, b_i, c_i, and $d_i (i = 1, 2, \ldots, n)$ can be evaluated using the following conditions:

(1) The function value at the interior knot x_i must be equal to $f(x_i)$ whether it is computed using the cubic equation $f_i(x)$ or $f_{i+1}(x)$. This gives $(2n - 2)$ conditions (similar to Eqs. (5.80) and (5.81)).

(2) The first and last functions, $f_1(x)$ and $f_n(x)$, must pass through the endpoints x_0 and x_n, respectively. This gives two conditions (similar to Eqs. (5.82) and (5.83)).

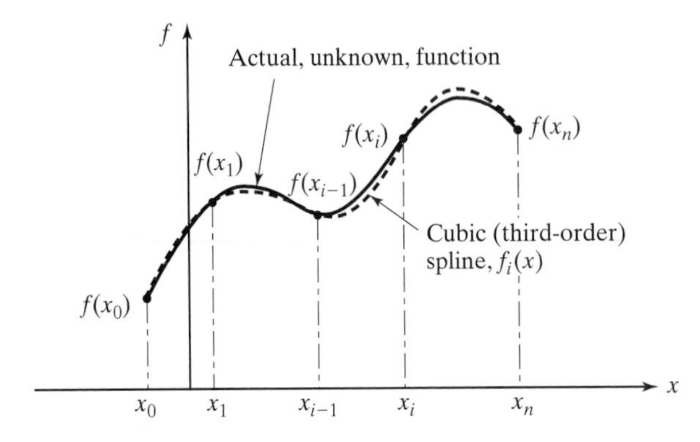

Figure 5.12 Third-order spline.

(3) The first derivative (slope) at each of the interior knots must be continuous. This gives $(n-1)$ conditions (similar to Eq. (5.85)).

(4) The second derivative (curvature) at each of the interior knots must be continuous. Equation (5.87) gives the curvature as

$$f_i''(x) = 2c_i + 6d_i x, \tag{5.88}$$

and hence the continuity of curvature leads to

$$f_i''(x = x_i) = f_{i+1}''(x = x_i);$$

that is,

$$2c_i + 6d_i x_i = 2c_{i+1} + 6d_{i+1} x_i;\ i = 1, 2, \ldots, n-1. \tag{5.89}$$

Equation (5.89) gives $(n-1)$ conditions.

(5) We still need two more conditions in order to have a total of $4n$ conditions. There are several possible relations that can be used. One possibility is to assume the curvature (second derivative) to be zero at the endpoints x_0 and x_n:

$$f_1''(x = x_0) = 2c_1 + 6d_1 x_0 = 0, \tag{5.90}$$

and

$$f_n''(x = x_n) = 2c_n + 6d_n x_n = 0. \tag{5.91}$$

Instead of solving $4n$ simultaneous equations to find the $4n$ constants of Eq. (5.87), a simpler procedure, that requires the solution of a substantially smaller number of equations, can be used. For this, we note that the second derivative in the ith interval (i.e., between the knots x_{i-1} and x_i) varies linearly. Hence, we can represent the variation of the second derivative, $f_i''(x)$, using the first-order Lagrange interpolation formula (see Eq. (5.14)) as

$$f_i''(x) = f''(x_{i-1}) \left(\frac{-x_i + x}{-x_i + x_{i-1}} \right) + f''(x_i) \left(\frac{x - x_{i-1}}{x_i - x_{i-1}} \right);\ x_{i-1} \le x \le x_i;\ i = 1, 2, \ldots, n, \tag{5.92}$$

where $f''(x_{i-1})$ and $f''(x_i)$ denote the values of the second derivative of f at the knot points x_{i-1} and x_i, respectively (unknown at this stage). Equation (5.92), when integrated twice, gives an expression for $f_i(x)$ involving two unknown or integration constants. These constants can be evaluated by using the conditions that $f_i(x)$ must be equal to $f(x_{i-1})$ and $f(x_i)$ at x_{i-1} and x_i, respectively. The resulting cubic equation can be expressed as

$$
\begin{aligned}
f_i(x) = {} & f''(x_{i-1})\frac{(-x_i+x)^3}{6(-x_i+x_{i-1})} + f''(x_i)\frac{(x-x_{i-1})^3}{6(x_i-x_{i-1})} \\
& + \left\{\frac{f(x_{i-1})}{(x_i-x_{i-1})} - f''(x_{i-1})\left(\frac{x_i-x_{i-1}}{6}\right)\right\}(x_i-x) \\
& + \left\{\frac{f(x_i)}{(x_i-x_{i-1})} - f''(x_i)\left(\frac{x_i-x_{i-1}}{6}\right)\right\}(x-x_{i-1}); \\
& x_{i-1} \le x \le x_i;\ i = 1, 2, \ldots, n.
\end{aligned} \tag{5.93}
$$

Notice that the second derivatives at the two ends of the interval, $f''(x_{i-1})$ and $f''(x_i)$, are unknown in Eq. (5.93). To evaluate these constants, we enforce the continuity of slope (first derivative) at each interior knot:

$$f'_i(x = x_i) = f'_{i+1}(x = x_i);\ i = 1, 2, \ldots, n-1. \tag{5.94}$$

By differentiating Eq. (5.93) and using the condition of Eq. (5.94), we get the equation

$$
\begin{aligned}
& f''(x_{i-1})(x_i - x_{i-1}) + 2f''(x_i)(x_{i+1} - x_{i-1}) + f''(x_{i+1})(x_{i+1} - x_i) \\
& = 6\left\{\frac{f(x_{i+1}) - f(x_i)}{x_{i+1} - x_i} - \frac{f(x_i) - f(x_{i-1})}{x_i - x_{i-1}}\right\};\ i = 1, 2, \ldots, n-1.
\end{aligned} \tag{5.95}
$$

There are $n+1$ unknown second derivatives at $n+1$ knot points. Equation (5.95) gives $n-1$ conditions. Hence, two more relations are needed to determine the $n+1$ second derivatives uniquely. Several possibilities exist as indicated next.

(i) Cubic spline with $y''(0) = y''(x_n) = 0$ (Natural cubic spline)

In the most commonly used spline, known as a natural cubic spline, the second derivatives at the end knots are assumed to be zero:

$$f''(x_0) = 0;\ f''(x_n) = 0. \tag{5.96}$$

In practice, this spline results in a curve $f(x)$ that is flatter than the true (desired) curve near the endpoints. Using Eq. (5.96), we can write Eq. (5.95) as the tridiagonal system of equations (see Problem 5.19)

$$[A]\vec{y} = \vec{p}, \tag{5.97}$$

where

$$[A] = \begin{bmatrix} 2(h_1+h_2) & h_2 & \cdot & \cdots & \cdot \\ h_2 & 2(h_2+h_3) & h_3 & \cdots & \cdot \\ \cdot & h_3 & 2(h_3+h_4) & \cdots & \cdot \\ \cdot & \cdot & \cdot & \cdots & \cdot \\ \cdot & \cdot & \cdot & \cdots & \cdot \\ \cdot & \cdot & \cdot & \cdots\ 2(h_{n-2}+h_{n-1}) & h_{n-1} \\ \cdot & \cdot & \cdot & \cdots\ h_{n-1} & 2(h_{n-1}+h_n) \end{bmatrix}, \tag{5.98}$$

$$\vec{y} = \begin{Bmatrix} f_1'' \\ f_2'' \\ \cdot \\ \cdot \\ \cdot \\ f_{n-1}'' \end{Bmatrix}, \tag{5.99}$$

and

$$\vec{p} = 6 \begin{Bmatrix} \dfrac{f_2 - f_1}{h_2} - \dfrac{f_1 - f_0}{h_1} \\ \dfrac{f_3 - f_2}{h_3} - \dfrac{f_2 - f_1}{h_2} \\ \vdots \\ \dfrac{f_n - f_{n-1}}{h_n} - \dfrac{f_{n-1} - f_{n-2}}{h_{n-1}} \end{Bmatrix}, \tag{5.100}$$

with

$$f_i'' = f''(x_i); \tag{5.101}$$

$$f_i = f(x_i); \tag{5.102}$$

and

$$h_i = x_i - x_{i-1}. \tag{5.103}$$

It can be seen that the number of equations to be solved (Eq. (5.97)) is only $n-1$ compared with the $4n$ equations to be solved according to the definition in Eq. (5.87).

(ii) Cubic spline with $y''(x_0) = y''(x_1)$ and $y''(x_n) = y''(x_{n-1})$

In this case, the second derivatives at the endpoints are assumed to be same as those at their adjacent points:

$$f''(x_0) = f''(x_1); \; f''(x_n) = f''(x_{n-1}). \tag{5.104}$$

This assumption states that $f''(x)$ is constant in the first and last intervals. This implies that $f(x)$ is a quadratic in these intervals, $[x_0, x_1]$ and $[x_{n-1}, x_n]$. This results in a curve $f(x)$ that has larger curvature than the true (desired) curve near the endpoints. Using Eq. (5.104), we can express Eq. (5.95) in tridiagonal form as Eq. (5.97) with (see Problem 5.52)

$$[A] = \begin{bmatrix} (3h_1 + 2h_2) & h_2 & \cdot & \cdots & \cdot \\ h_2 & 2(h_2 + h_3) & h_3 & \cdots & \cdot \\ \cdot & h_3 & 2(h_3 + h_4) & \cdots & \cdot \\ \cdot & \cdot & \cdot & \cdots \; 2(h_{n-2} + h_{n-1}) & h_{n-1} \\ \cdot & \cdot & \cdot & \cdots \; h_{n-1} & (2h_{n-1} + 3h_n) \end{bmatrix} \tag{5.105}$$

and $\vec{y}$ and $\vec{p}$ given by Eqs. (5.99) and (5.100), respectively.

(iii) Cubic spline with y'' values at endpoints as linear extrapolations

In this case, the values of $y''(x_0)$ and $y''(x_n)$ are taken as the linear extrapolations of the y'' values of the two nearest data points as

$$\frac{f_1'' - f_0''}{h_1} = \frac{f_2'' - f_1''}{h_2};$$

that is,

$$f_0'' = \left(\frac{h_1 + h_2}{h_2}\right) f_1'' - \left(\frac{h_1}{h_2}\right) f_2'' \tag{5.106}$$

and

$$\frac{f_n'' - f_{n-1}''}{h_n} = \frac{f_{n-1}'' - f_{n-2}''}{h_{n-1}};$$

that is,

$$f_n'' = \left(\frac{h_n + h_{n-1}}{h_{n-1}}\right) f_{n-1}'' - \left(\frac{h_n}{h_{n-1}}\right) f_{n-2}''. \tag{5.107}$$

Using Eqs. (5.106) and (5.107), we can express Eq. (5.95) in tridiagonal form as Eq. (5.97) with (see Problem 5.20)

$$[A] = \begin{bmatrix} \frac{(h_1+h_2)(h_1+2h_2)}{h_2} & \frac{(h_2^2-h_1^2)}{h_2} & \cdots & \cdot \\ h_2 & 2(h_2+h_3) & h_3 \cdots & \cdot \\ \cdot & h_3 & \cdots & \cdot \\ \cdot & \cdot & \cdots & \cdot \\ \cdot & \cdot & \cdots & \cdot \\ \cdot & \cdot & \cdot\; 2(h_{n-2}+h_{n-1}) & h_{n-1} \\ \cdot & \cdot & \cdot\; \frac{h_{n-1}^2-h_n^2}{h_{n-1}} & \frac{(h_{n-1}+h_n)(2h_{n-1}+h_n)}{h_{n-1}} \end{bmatrix} \tag{5.108}$$

and $\vec{y}$ and $\vec{p}$ given by Eqs. (5.99) and (5.100), respectively. After finding the values of $y_1'', y_2'', \ldots, y_{n-1}''$ from the solution of Eq. (5.97), f_0'' and f_n'' can be found from Eqs. (5.106) and (5.107).

(iv) Other types of cubic splines

In some cases, the first derivatives (slopes) of the function at the ends are specified as

$$f'(x_0) = g_0; \; f'(x_n) = g_n. \tag{5.109}$$

In this case, Eq. (5.93) can be differentiated with respect to x and set equal to g_0 and g_n at $x = x_0$ and x_n, respectively. The resulting equations, along with Eq. (5.95), enable us to find the $n+1$ unknowns $f_0'', f_1'', \ldots, f_n''$. On the other hand, in some cases, the data points may correspond to a periodic function. In such a case, the periodic conditions are enforced as

$$f'(x_0) = f'(x_n); \; f''(x_0) = f''(x_n). \tag{5.110}$$

Equations (5.110) and (5.95) then provide the solution of the $n+1$ unknown second derivatives f_i'', $i = 0, 1, 2, \ldots, n$.

▶Example 5.12

Find a cubic spline to fit the data given in Example 5.11. Use the results to estimate the value of f at $x = 7.0$.

Solution

There are four intervals and five data points in this example. By making the second derivative of f continuous at each of the interior knots, we obtain (Eq. (5.95))

$$f''(x_0)(x_1-x_0) + 2f''(x_1)(x_2-x_0) + f''(x_2)(x_2-x_1)$$
$$= 6\left\{\frac{f(x_2)-f(x_1)}{x_2-x_1} - \frac{f(x_1)-f(x_0)}{x_1-x_0}\right\}; \tag{E1}$$

$$f''(x_1)(x_2 - x_1) + 2f''(x_2)(x_3 - x_1) + f''(x_3)(x_3 - x_2)$$
$$= 6\left\{\frac{f(x_3) - f(x_2)}{x_3 - x_2} - \frac{f(x_2) - f(x_1)}{x_2 - x_1}\right\}; \quad \text{(E2)}$$
$$f''(x_2)(x_3 - x_2) + 2f''(x_3)(x_4 - x_2) + f''(x_4)(x_4 - x_3)$$
$$= 6\left\{\frac{f(x_4) - f(x_3)}{x_4 - x_3} - \frac{f(x_3) - f(x_2)}{x_3 - x_2}\right\}. \quad \text{(E3)}$$

With the known data, Eqs. (E1) through (E3) take the following form:

$$f''(2) + 9f''(3) + 3.5f''(6.5) = -41.1429; \quad \text{(E4)}$$
$$3.5f''(3) + 10f''(6.5) + 1.5f''(8) = 1.1427; \quad \text{(E5)}$$
$$1.5f''(6.5) + 11f''(8) + 4.0f''(12) = 14.5002. \quad \text{(E6)}$$

By setting the second derivative of f at x_0 and x_4 equal to zero, we obtain

$$f''(x_0) = f''(2) = 0; \quad \text{(E7)}$$
$$f''(x_4) = f''(12) = 0. \quad \text{(E8)}$$

The five linear simultaneous equations, (E4) through (E8), can be solved to obtain

$$f''(2) = 0,\ f''(3) = -5.270676,\ f''(6.5) = 1.798055,\ f''(8) = 1.073011,\ f''(12) = 0.$$

Using these values, with Eq. (5.93) we can define the cubic spline in each of the four intervals as

$$f_1(x) = -0.878446(x-2)^3 + 14.0(3-x) + 20.878446(x-2),\ 2 \le x \le 3;$$
$$f_2(x) = 0.250985(x-6.5)^3 + 0.0856217(x-3)^3 + 8.788847(6.5-x)$$
$$+3.808278(x-3),\ 3 \le x \le 6.5;$$
$$f_3(x) = -0.199784(x-8)^3 + 0.119223(x-6.5)^3 + 10.883822(8-x)$$
$$+10.3984149(x-6.5),\ 6.5 \le x \le 8;$$
$$f_4(x) = -0.044709(x-12)^3 + 3.284659(12-x) + 5.75(x-8),\ 8 \le x \le 12. \quad \text{(E9)}$$

The value of f at $x = 7.0$ can be estimated from Eq. (E9) as $f_3(7.0) = 16.297714$. ◀

5.9 Least-Squares Regression

5.9.1 Linear Regression

Regression is the method of obtaining the best fit to a given set of data. The procedure of fitting the best straight line (linear equation) to the data is known as linear regression. Let the data points be given by $(x_1, y_1), (x_2, y_2), \ldots, (x_n, y_n)$, where x is the independent variable and y is the dependent variable. The equation of the straight line can be expressed as

$$y = a_0 + a_1 x, \tag{5.111}$$

where a_0 and a_1 are constants to be determined. Since the linear Equation (5.111) is only an approximating function, there will be an error between the model, Eq. (5.111), and the data points (true values). The error or residual at the data point (x_i, y_i) is given by

$$e_i = y_i - a_0 - a_1 x_i. \tag{5.112}$$

The most common approach used to fit the best straight line is by using the method of least squares. For this, we formulate the sum of squares of the error (S) as

$$S = \sum_{i=1}^{n} e_i^2 = \sum_{i=1}^{n} (y_i - a_0 - a_1 x_i)^2 \tag{5.113}$$

and minimize S with respect to the parameters a_0 and a_1. For minimizing S with respect to a_0 and a_1, we differentiate S with respect to a_0 and a_1 and set them equal to zero. This gives

$$\frac{\partial S}{\partial a_0} = -2 \sum_{i=1}^{n} (y_i - a_0 - a_1 x_i) = 0; \tag{5.114}$$

$$\frac{\partial S}{\partial a_1} = -2 \sum_{i=1}^{n} (y_i - a_0 - a_1 x_i) = 0. \tag{5.115}$$

Equations (5.114) and (5.115) can be rewritten as follows:

$$\sum_{i=1}^{n} y_i - \sum_{i=1}^{n} a_0 - a_1 \sum_{i=1}^{n} x_i = 0; \tag{5.116}$$

$$\sum_{i=1}^{n} x_i y_i - a_0 \sum_{i=1}^{n} x_i - a_1 \sum_{i=1}^{n} x_i^2 = 0. \tag{5.117}$$

Noting that $\sum_{i=1}^{n} a_0 = n a_0$, we can express Eqs. (5.116) and (5.117) as two simultaneous linear equations in the unknowns a_0 and a_1 as

$$a_0(n) + a_1 \left(\sum_{i=1}^{n} x_i \right) = \sum_{i=1}^{n} y_i; \tag{5.118}$$

$$a_0 \left(\sum_{i=1}^{n} x_i \right) + a_1 \left(\sum_{i=1}^{n} x_i^2 \right) = \sum_{i=1}^{n} x_i y_i. \tag{5.119}$$

These equations can be solved for the unknowns a_0 and a_1, using Cramer's rule, to obtain

$$a_0 = \frac{\begin{vmatrix} \sum_{i=1}^{n} y_i & \sum_{i=1}^{n} x_i \\ \sum_{i=1}^{n} x_i y_i & \sum_{i=1}^{n} x_i^2 \end{vmatrix}}{\begin{vmatrix} n & \sum_{i=1}^{n} x_i \\ \sum_{i=1}^{n} x_i & \sum_{i=1}^{n} x_i^2 \end{vmatrix}} = \frac{\sum_{i=1}^{n} y_i \sum_{i=1}^{n} x_i^2 - \sum_{i=1}^{n} x_i \sum_{i=1}^{n} x_i y_i}{n \sum_{i=1}^{n} x_i^2 - \left(\sum_{i=1}^{n} x_i\right)^2}, \tag{5.120}$$

$$a_1 = \frac{\begin{vmatrix} n & \sum_{i=1}^{n} y_i \\ \sum_{i=1}^{n} x_i & \sum_{i=1}^{n} x_i y_i \end{vmatrix}}{\begin{vmatrix} n & \sum_{i=1}^{n} x_i \\ \sum_{i=1}^{n} x_i & \sum_{i=1}^{n} x_i^2 \end{vmatrix}} = \frac{n \sum_{i=1}^{n} x_i y_i - \sum_{i=1}^{n} x_i \sum_{i=1}^{n} y_i}{n \sum_{i=1}^{n} x_i^2 - \left(\sum_{i=1}^{n} x_i\right)^2}. \tag{5.121}$$

5.9.2 Accuracy of Linear Regression

To find the accuracy of the straight line obtained from the least-squares fit, we compute the sum of squares of the deviation of the data around the mean (S_0) for the dependent variable (y) before the application of regression. This is given by

$$S_0 = \sum_{i=1}^{n} (y_i - \bar{y})^2, \tag{5.122}$$

where $\bar{y}$ denotes the mean value of y_i:

$$\bar{y} = \frac{1}{n} \sum_{i=1}^{n} y_i. \tag{5.123}$$

The sum of squares of the deviation of the data about the straight line (after the application of regression), S, is given by

$$S = \sum_{i=1}^{n} (y_i - a_0 - a_1 x_i)^2, \tag{5.124}$$

where a_0 and a_1 are given by Eqs. (5.120) and (5.121), respectively. The difference between S_0 and S provides a measure of the accuracy of regression or the extent of

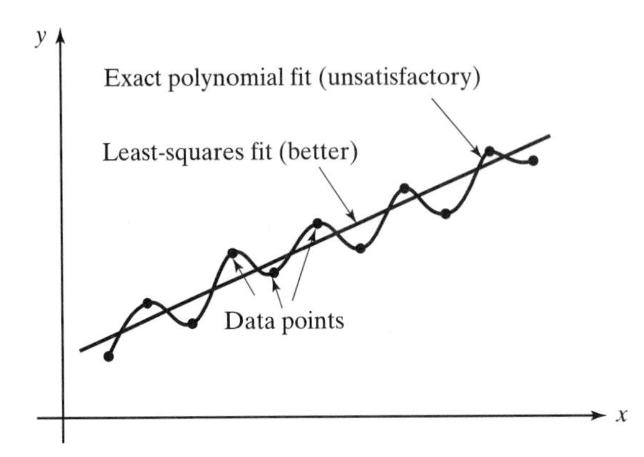

Figure 5.13 Least squares straight-line fit.

improvement achieved by the least squares fit. This difference is usually normalized to define a quantity r, known as the correlation coefficient, as

$$r^2 = \left(\frac{S_0 - S}{S_0}\right). \tag{5.125}$$

For a perfect straight-line fit, all the data points (x_i, y_i) fall on the straight line and S will be zero. Then the correlation coefficient r will be equal to one. If the straight-line fit does not represent any improvement, $S = S_0$ and hence $r = 0$. Thus, a good least-squares fit is indicated by a large value of r, the maximum of which is 1.0. In practice, it is desirable to plot the data points, along with the straight line fitted to see, qualitatively, how good the regression is. (See Fig. 5.13.)

▶Example 5.13

Use linear regression to fit the following data points:

i	1	2	3	4	5
x_i	1	2	3	4	5
y_i	0.7	2.2	2.8	4.4	4.9

Solution

Here $n = 5$, $\sum_{i=1}^{n} x_i = 15$, $\sum_{i=1}^{n} y_i = 15.0$,

$$\sum_{i=1}^{n} x_i^2 = 1 + 4 + 9 + 16 + 25 = 55,$$

$$\sum_{i=1}^{n} x_i y_i = 0.7 + 4.4 + 8.4 + 17.6 + 24.5 = 55.6.$$

Equations (5.120) and (5.121) yield

$$a_0 = \frac{15(55) - 15(55.6)}{5(55) - (15)^2} = -0.18,$$

and

$$a_1 = \frac{5(55.6) - 15(15)}{5(55) - (15)^2} = 1.06.$$

Thus, the regression equation is $y(x) = -0.18 + 1.06x$.

To determine the accuracy of regression, we compute the correlation coefficient r using Eq. (5.125):

$$\bar{y} = \frac{1}{n}\sum_{i=1}^{n} y_i = \frac{1}{5}(15) = 3.0;$$

$$S_0 = \sum_{i=1}^{n}(y_i - \bar{y})^2 = (0.7 - 3.0)^2 + (2.2 - 3.0)^2$$
$$+(2.8 - 3.0)^2 + (4.4 - 3.0)^2 + (4.9 - 3.0)^2 = 11.54;$$

$$S = \sum_{i=1}^{n}(y_i - a_0 - a_1 x_i)^2 = (0.7 + 0.18 - 1.06(1))^2 + (2.2 + 0.18 - 1.06(2))^2$$
$$+(2.8 + 0.18 - 1.06(3))^2 + (4.4 + 0.18 - 1.06(4))^2$$
$$+(4.9 + 0.18 - 1.06(5))^2 = 0.304;$$

$$r = \left(\frac{S_0 - S}{S_0}\right)^{\frac{1}{2}} = \left(\frac{11.54 - 0.304}{11.54}\right)^{\frac{1}{2}} = 0.9867.$$

This denotes a good fit. ◀

5.9.3 Polynomial Regression

As seen in the previous section, linear regression provides the *best* straight-line fit for a given set of data. The use of linear regression implies that the relationship between the independent and dependent variables is linear. In many engineering applications, the independent and dependent variables may be related nonlinearly. In such a case, the data points exhibit a nonlinear trend and fitting a straight line will be inappropriate. (See Fig. 5.14.) Hence, to detect situations such as the one shown in the figure, it is necessary to plot the data and find whether linear regression would be appropriate. For data points exhibiting a nonlinear behavior, we can use a polynomial regression by assuming the following relationship:

$$y = a_0 + a_1 x + a_2 x^2 + \cdots + a_m x^m. \tag{5.126}$$

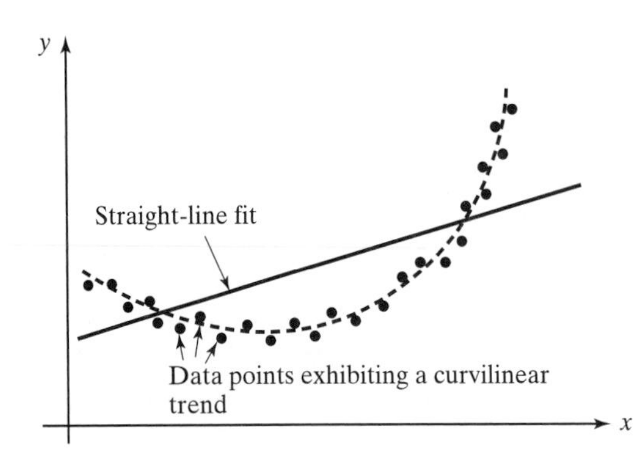

Figure 5.14 Inadequacy of linear fit when trend is nonlinear.

If there are n data points $(x_i, y_i), i = 1, 2, \ldots, n$, the error, e_i, for the ith data point is defined by

$$e_i = y_i - a_0 - a_1 x_i - a_2 x_i^2 - \cdots - a_m x_i^m. \tag{5.127}$$

The sum of squares error (S) is given by

$$S = \sum_{i=1}^{n} e_i^2 = \sum_{i=1}^{n} (y_i - a_0 - a_1 x_i - a_2 x_i^2 - \cdots - a_m x_i^m)^2. \tag{5.128}$$

For the least-squares fit of the polynomial, we minimize S with respect to the $m+1$ unknown coefficients $a_0, a_1, a_2, \ldots, a_m$. For this, we set the partial derivatives of S with respect to the coefficients equal to zero:

$$\frac{\partial S}{\partial a_0} = -2 \sum_{i=1}^{n} (y_i - a_0 - a_1 x_i - a_2 x_i^2 - \cdots - a_m x_i^m) = 0; \tag{5.129}$$

$$\frac{\partial S}{\partial a_1} = -2 \sum_{i=1}^{n} x_i (y_i - a_0 - a_1 x_i - a_2 x_i^2 - \cdots - a_m x_i^m) = 0; \tag{5.130}$$

$$\vdots$$

$$\frac{\partial S}{\partial a_m} = -2 \sum_{i=1}^{n} x_i^m (y_i - a_0 - a_1 x_i - a_2 x_i^2 - \cdots - a_m x_i^m) = 0. \tag{5.131}$$

Equations (5.129) through (5.131) can be rewritten as a system of $m+1$ simultaneous linear equations in the unknowns $a_0, a_1, a_2, \ldots, a_m$ as

$$a_0(n) + a_1\left(\sum_{i=1}^{n} x_i\right) + a_2\left(\sum_{i=1}^{n} x_i^2\right) + \cdots + a_m\left(\sum_{i=1}^{n} x_i^m\right) = \sum_{i=1}^{n} y_i; \tag{5.132}$$

$$a_0\left(\sum_{i=1}^{n} x_i\right) + a_1\left(\sum_{i=1}^{n} x_i^2\right) + a_2\left(\sum_{i=1}^{n} x_i^3\right) + \cdots$$

$$+ a_m \left(\sum_{i=1}^{n} x_i^{m+1} \right) = \sum_{i=1}^{n} x_i y_i; \tag{5.133}$$

$$\vdots$$

$$a_0 \left(\sum_{i=1}^{n} x_i^{m} \right) + a_1 \left(\sum_{i=1}^{n} x_i^{m+1} \right) + a_2 \left(\sum_{i=1}^{n} x_i^{m+2} \right) + \cdots$$

$$+ a_m \left(\sum_{i=1}^{n} x_i^{2m} \right) = \sum_{i=1}^{n} x_i^{m} y_i. \tag{5.134}$$

The solution of Eqs. (5.132) through (5.134) gives the coefficients $a_0, a_1, a_2, \ldots, a_m$ that define the polynomial fit of Eq. (5.126).

Notes

(1) Usually, low-order polynomials are used to obtain a better fit. For $m \geq 4$ or 5, the equations tend to become ill conditioned and the round-off errors make the solution of $a_0, a_1, a_2, \ldots, a_m$ inaccurate.

(2) The ill-conditioning problems associated with the use of high-order polynomials can be reduced by using orthogonal polynomials for y in Eq. (5.126).

(3) As in the case of linear regression, the correlation coefficient (r) can be used as a measure of the accuracy of polynomial regression. It is defined as

$$r^2 = \frac{S_0 - S}{S_0}, \tag{5.135}$$

where S_0 is given by Eq. (5.122) and S by Eq. (5.128). A value of r closer to one is desirable for a better polynomial fit.

▶Example 5.14

The velocities measured at various points along the boundary layer in free convection over a vertical plate are given below (in nondimensional form):

i	1	2	3	4	5	6
x_i (thickness)	0.0	0.2	0.4	0.6	0.8	1.0
y_i (velocity)	0.00	1.05	0.85	0.35	0.10	1.00

Fit a second-order polynomial to the data.

Solution

The various quantities involved in the computations are given by the following:

$$m = 2, n = 6, \sum_{i=1}^{n} x_i = 3.0, \sum_{i=1}^{n} y_i = 3.35;$$

$$\sum_{i=1}^{n} x_i^2 = 0.00 + 0.04 + 0.16 + 0.36 + 0.64 + 1.00 = 2.20;$$

$$\sum_{i=1}^{n} x_i^3 = 0.0 + 0.008 + 0.064 + 0.216 + 0.512 + 1.0 = 1.800;$$

$$\sum_{i=1}^{n} x_i^4 = 0.0 + 0.0016 + 0.0256 + 0.1296 + 0.4096 + 1.0 = 1.5664;$$

$$\sum_{i=1}^{n} x_i y_i = 0.0 + 0.21 + 0.34 + 0.21 + 0.08 + 1.0 = 1.84;$$

$$\sum_{i=1}^{n} x_i^2 y_i = 0.0 + 0.042 + 0.136 + 0.126 + 0.064 + 1.0 = 1.368.$$

Equations (5.132) through (5.134) take the following form:

$$6.0a_0 + 3.0a_1 + 2.2a_2 = 3.35;$$

$$3.0a_0 + 2.2a_1 + 1.8a_2 = 1.84;$$

$$2.2a_0 + 1.8a_1 + 1.5664a_2 = 1.368.$$

The solution of these equations is given by $a_0 = 0.383927$, $a_1 = 0.659830$, and $a_2 = -0.424115$, and hence, the least-squares quadratic equation is given by $y(x) = 0.383927 + 0.659830x - 0.424115x^2$. ◀

5.9.4 Nonlinear Regression

The method of least squares is not limited to linear and polynomial regression. In some situations, it is desirable to use other types of nonlinear functions for regression. As an example, we use trigonometric functions for regression as

$$y = a_1 \sin \omega x + a_2 \cos \omega x, \tag{5.136}$$

where ω is a known constant (also, called the frequency) and a_1 and a_2 are the unknown coefficients. Using the data points $(x_i, y_i), i = 1, 2, \ldots, n$, the sum of squares of the error (S) can be expressed as

$$S = \sum_{i=1}^{n} (y_i - a_1 \sin \omega x_i - a_2 \cos \omega x_i)^2. \tag{5.137}$$

The partial derivatives of S with respect to a_1 and a_2 are set equal to zero for minimizing S:

$$\frac{\partial S}{\partial a_1} = -2\omega \sum_{i=1}^{n} (y_i - a_1 \sin \omega x_i - a_2 \cos \omega x_i) \cos \omega x_i = 0, \tag{5.138}$$

and

$$\frac{\partial S}{\partial a_2} = -2\omega \sum_{i=1}^{n} (y_i - a_1 \sin \omega x_i - a_2 \cos \omega x_i) \sin \omega x_i = 0. \tag{5.139}$$

Equations (5.138) and (5.139) can be expressed as a set of two linear equations:

$$\begin{bmatrix} \sum_{i=1}^{n} (\sin \omega x_i \cos \omega x_i) & \sum_{i=1}^{n} (\cos^2 \omega x_i) \\ \sum_{i=1}^{n} (\sin^2 \omega x_i) & \sum_{i=1}^{n} (\sin \omega x_i \cos \omega x_i) \end{bmatrix} \begin{Bmatrix} a_1 \\ a_2 \end{Bmatrix} = \begin{Bmatrix} \sum_{i=1}^{n} (y_i \cos \omega x_i) \\ \sum_{i=1}^{n} (y_i \sin \omega x_i) \end{Bmatrix} \tag{5.140}$$

The solution of Eq. (5.140) gives the values of a_1 and a_2 that define the regression curve, Eq. (5.136). If necessary, the high-frequency terms such as $a_3 \sin 2\omega x$ and $a_4 \cos 2\omega x$ can also be included in Eq. (5.136). Note that according to Fourier series expansion, any periodic function can be represented exactly by an infinite sum of sine and cosine terms. As another example, consider a least-squares fit using the relation

$$y = ae^{bx}, \tag{5.141}$$

where a and b are the unknown constants. For the data points $(x_i, y_i), i = 1, 2, \ldots, n$, the sum of the squares of error (S) can be expressed as

$$S = \sum_{i=1}^{n} (y_i - ae^{bx_i})^2. \tag{5.142}$$

The conditions for minimizing S are given by

$$\frac{\partial S}{\partial a} = -2 \sum_{i=1}^{n} (y_i - ae^{bx_i}) e^{bx_i} = 0, \tag{5.143}$$

and

$$\frac{\partial S}{\partial b} = -2 \sum_{i=1}^{n} (y_i - ae^{bx_i}) x_i e^{bx_i} = 0. \tag{5.144}$$

Equations (5.143) and (5.144) can be rewritten as two simultaneous nonlinear equations as

$$a \sum_{i=1}^{n} e^{2bx_i} = \sum_{i=1}^{n} y_i e^{bx_i}, \tag{5.145}$$

and

$$a \sum_{i=1}^{n} x_i e^{2bx_i} = \sum_{i=1}^{n} x_i y_i e^{bx_i}. \tag{5.146}$$

The solution of Eqs. (5.145) and (5.146) can be obtained by using the method described in Section 2.12.

5.9.5 Linearization

Certain types of nonlinear relations can be linearized by using suitable transformations. In such cases, linear regression can be used to yield, effectively, a nonlinear curve fit. For example, for the regression considered in Eq. (5.141), we have

$$y = ae^{bx}. \tag{5.147}$$

This equation can be linearized by taking logarithms on both sides:

$$\ln y = \ln a + bx \ln e = \ln a + bx. \tag{5.148}$$

By defining a new variable $Y = \ln y$, we can use a linear regression with the relation

$$Y = a_0 + a_1 x, \tag{5.149}$$

where $a_0 = \ln a$ and $a_1 = b$ are the new unknown coefficients. Some nonlinear relations and the transformations that can be used for linearization are given in Table 5.1.

Table 5.1 Nonlinear relations that can be reduced to linear form.

Original relation	Transformation and linear relation, $\bar{y} = c\bar{x} + d$
1. $y = ax^b$	$\ln y = \ln a + b \ln x$
	$\bar{y} = \ln y,\ \bar{x} = \ln x,\ c = b,\ d = \ln a$
2. $y = ae^{bx}$	$\ln y = \ln a + bx$
	$\bar{y} = \ln y,\ \bar{x} = x,\ c = b,\ d = \ln a$
3. $y = \frac{ax}{b+x}$	$\frac{1}{y} = \frac{b+x}{ax}$
	$\bar{y} = \frac{1}{y},\ \bar{x} = \frac{1}{x},\ c = \frac{b}{a},\ d = \frac{1}{a}$
4. $y = \frac{a}{b+x}$	$\frac{1}{y} = \frac{b+x}{a}$
	$\bar{y} = \frac{1}{y},\ \bar{x} = x,\ c = \frac{1}{a},\ d = \frac{b}{a}$
5. $y = a_0 x_1^{a_1} x_2^{a_2} \ldots x_m^{a_m}$	$\ln y = \ln a_0 + a_1 \ln x_1 + \cdots + a_m \ln x_m$
	$\bar{y} = c_0 + c_1\bar{x}_1 + c_2\bar{x}_2 + \cdots + c_m\bar{x}_m$
	$\bar{y} = \ln y,\ c_0 = \ln a_0,\ c_i = a_i,\ i = 1, 2, \ldots, m,\ \bar{x}_i = \ln x_i, i = 1, 2, \ldots, m$

▶Example 5.15

The vibration amplitude of a machine (x_i) is measured at different instants of time (t_i) and the results are as follows:

i	1	2	3	4
t_i (sec)	0	2	4	6
x_i (mm)	5.0	3.7	2.7	2.0

Fit a curve of the form

$$x(t) = ae^{-bt} \tag{E1}$$

using the data given.

Solution

By taking logarithms of Eq. (E1), we obtain the linear equation

$$\ln x = \ln a - bt. \tag{E2}$$

By defining $y = \ln x$, $c = \ln a$, and $d = -b$, we can write Eq. (E2) as

$$y = c + dt. \tag{E3}$$

The sum of the squares of error, S, can be expressed as

$$S = \sum_{i=1}^{n} (y_i - c - dt_i)^2, \tag{E4}$$

where n is the number of data points. The conditions for the minimum of S yield the simultaneous equations

$$nc + d\left(\sum_{i=1}^{n} t_i\right) = \sum_{i=1}^{n} y_i, \tag{E5}$$

and

$$c\left(\sum_{i=1}^{n} t_i\right) + d\left(\sum_{i=1}^{n} t_i^2\right) = \sum_{i=1}^{n} t_i y_i. \tag{E6}$$

In the present case, $n = 4$, $t_1 = 0$, $t_2 = 2$, $t_3 = 4$, $t_4 = 6$, $y_1 = \ln x_1 = \ln 5.0 = 1.6094$, $y_2 = \ln x_2 = \ln 3.7 = 1.3083$, $y_3 = \ln x_3 = \ln 2.7 = 0.9932$, $y_4 = \ln x_4 = \ln 2.0 = 0.6931$, and $\sum_{i=1}^{n} t_i = 12$, $\sum_{i=1}^{n} t_i^2 = 0 + 4 + 16 + 36 = 56$, $\sum_{i=1}^{n} y_i = 4.6040$, $\sum_{i=1}^{n} t_i y_i = 0 + 2(1.3083) + 4(0.9932) + 6(0.6931) = 10.7480$.

Hence, Eqs. (E5) and (E6) take the following form:

$$4c + 12d = 4.6040; \tag{E7}$$

$$12c + 56d = 10.7480. \tag{E8}$$

The solution of Eqs. (E7) and (E8) gives $c = 0.6914$ and $d = -0.1532$, which imply that $a = e^c = 1.9965$ and $b = 0.1532$. Thus, the least-squares fit leads to the curve

$$x(t) = 1.9965e^{-0.1532t} \text{ mm.} \tag{E9}$$

◀

5.10 Curve Fitting with Multiple Variables

In many practical applications, we encounter variables that are functions of several independent variables. For example, the temperature of a gas depends on the volume and the pressure. If y is known to be a function of the several independent variables, $x_1, x_2, \ldots, x_m$, and if only experimental data are available, we would be interested in finding an approximate relationship of the form

$$y = y(x_1, x_2, \ldots, x_m). \tag{5.150}$$

The curve-fitting procedures outlined in the previous sections for functions of one variable can be extended to this case. In most cases, a linear relationship is used for simplicity:

$$y = a_0 + a_1x_1 + a_2x_2 + \cdots + a_mx_m. \tag{5.151}$$

Here, the coefficients $a_0, a_1, a_2, \ldots, a_m$ are to be determined using the data points

$$(x_{1,i}, x_{2,i}, \ldots, x_{n,i}, y_i); \; i = 1, 2, \ldots, n.$$

In this case, the sum of the squares of error (S) can be expressed as

$$S = \sum_{i=1}^{n} (y_i - a_0 - a_1x_{1,i} - a_2x_{2,i} - \cdots - a_mx_{m,i})^2. \tag{5.152}$$

For the minimization of S, we set the partial derivatives of S with respect to $a_0, a_1, a_2, \ldots, a_m$ equal to zero. This leads to a set of linear algebraic equations that can be expressed in matrix form as follows:

$$\begin{bmatrix} n & \sum x_{1,i} & \sum x_{2,i} & \cdots & \sum x_{m,i} \\ \sum x_{1,i} & \sum x_{1,i}^2 & \sum x_{1,i}x_{2,i} & \cdots & \sum x_{1,i}x_{m,i} \\ \sum x_{2,i} & \sum x_{1,i}x_{2,i} & \sum x_{2,i}^2 & \cdots & \sum x_{2,i}x_{m,i} \\ \vdots & \vdots & \vdots & \vdots\vdots\vdots\vdots & \\ \sum x_{m,i} & \sum x_{1,i}x_{m,i} & \sum x_{2,i}x_{m,i} & \cdots & \sum x_{m,i}^2 \end{bmatrix} \begin{Bmatrix} a_0 \\ a_1 \\ a_2 \\ \vdots \\ a_m \end{Bmatrix} = \begin{Bmatrix} \sum y_i \\ \sum x_{1,i}y_i \\ \sum x_{2,i}y_i \\ \vdots \\ \sum x_{m,i}y_i \end{Bmatrix}. \tag{5.153}$$

The accuracy of multiple linear regression can be measured in terms of the correlation coefficient as in the case of single-variable regression.

▶Example 5.16

The experimental data relating a dependent variable y and two independent variables x_1 and x_2 are given as follows:

i	1	2	3	4	5	6
x_{1i}	10	10	20	50	60	60
x_{2i}	5	45	25	25	5	45
y_i	50	40	36	32	32	19

Use multiple linear regression to fit these data.

Solution

Let the linear relationship be given by

$$y = a_0 + a_1 x_1 + a_2 x_2, \tag{E1}$$

and the sum of the squares of error (S) can be expressed as

$$S = \sum_{i=1}^{n} (y_i - a_0 - a_1 x_{1i} - a_2 x_{2i})^2. \tag{E2}$$

The conditions for the minimum of S can be written as

$$\frac{\partial S}{\partial a_0} = -2 \sum_{i=1}^{n} (y_i - a_0 - a_1 x_{1i} - a_2 x_{2i}) = 0; \tag{E3}$$

$$\frac{\partial S}{\partial a_1} = -2 \sum_{i=1}^{n} x_{1i} (y_i - a_0 - a_1 x_{1i} - a_2 x_{2i}) = 0; \tag{E4}$$

$$\frac{\partial S}{\partial a_2} = -2 \sum_{i=1}^{n} x_{2i} (y_i - a_0 - a_1 x_{1i} - a_2 x_{2i}) = 0. \tag{E5}$$

Equations (E3) through (E5) can be expressed as a set of three simultaneous linear equations:

$$n a_0 + a_1 \left(\sum_{i=1}^{n} x_{1i} \right) + a_2 \left(\sum_{i=1}^{n} x_{2i} \right) = \sum_{i=1}^{n} y_i, \tag{E6}$$

$$a_0 \left(\sum_{i=1}^{n} x_{1i} \right) + a_1 \left(\sum_{i=1}^{n} x_{1i}^2 \right) + a_2 \left(\sum_{i=1}^{n} x_{1i} x_{2i} \right) = \sum_{i=1}^{n} y_i x_{1i}, \tag{E7}$$

and

$$a_0 \left(\sum_{i=1}^{n} x_{2i} \right) + a_1 \left(\sum_{i=1}^{n} x_{1i} x_{2i} \right) + a_2 \left(\sum_{i=1}^{n} x_{2i}^2 \right) = \sum_{i=1}^{n} y_i x_{2i}. \tag{E8}$$

From the given data, we have

$$n = 6, \sum_{i=1}^{n} x_{1i} = 210, \sum_{i=1}^{n} x_{2i} = 150 \sum_{i=1}^{n} y_i = 209,$$

$$\sum_{i=1}^{n} x_{2i}^2 = 25 + 2025 + 625 + 625 + 25 + 2025 = 5350,$$

$$\sum_{i=1}^{n} x_{1i}^2 = 100 + 100 + 400 + 2500 + 3600 + 3600 = 10300,$$

$$\sum_{i=1}^{n} x_{1i} x_{2i} = 50 + 450 + 500 + 1250 + 300 + 2700 = 5250,$$

$$\sum_{i=1}^{n} y_i x_{1i} = 500 + 400 + 720 + 1600 + 1920 + 1140 = 6280,$$

and

$$\sum_{i=1}^{n} y_i x_{2i} = 250 + 1800 + 900 + 800 + 160 + 855 = 4765,$$

and hence Eqs. (E6) through (E8) take the following form:

$$6a_0 + 210a_1 + 150a_2 = 209; \quad \text{(E9)}$$

$$210a_0 + 10300a_1 + 5250a_2 = 6280; \quad \text{(E10)}$$

$$150a_0 + 5250a_1 + 5350a_2 = 4765. \quad \text{(E11)}$$

The solution of Eqs. (E9) through (E11) is given by $a_0 = 52.300507$, $a_1 = -0.350848$, and $a_2 = -0.287500$. Thus, the least squares linear equation is given by

$$y(x_1, x_2) = 52.300507 - 0.350848x_1 - 0.287500x_2.$$

◀

5.11 Choice of Method

In interpolation, the interpolating polynomial (of degree n) and the function $f(x)$ agree in value at a set of points x_i, $i = 0, 1, \ldots, n$, belonging to an interval $[a, b]$. Interpolation is used when the data points are known to be precise and accurate. The interpolating polynomial is used to determine the value of $f(x)$ at any point $x \neq x_i$. If $x \epsilon (a, b)$, the process is called interpolation; otherwise, it is known as extrapolation. Usually, interpolation produces better results. The interpolating polynomial can also be used to find the roots of $f(x)$, derivatives of $f(x)$ at any point in the interval (a, b) and integral of $f(x)$ between specified limits. Interpolating polynomials are easy to derive and implement. If the number of data points is large, a high-degree polynomial is to be used. For a high-degree polynomial, not only do the computations become costly, but the computed results may also become less reliable, due to round-off errors. In some cases, a higher degree polynomial may produce poorer results than lower degree polynomials.

An interpolation polynomial can be derived in several ways: as Newton's divided-difference interpolating polynomial, as Lagrange polynomial, or using orthogonal polynomials (such as Chebyshev polynomials). When the order of the polynomial is known, the Lagrange interpolating polynomial is simpler to use and program. In addition, the procedure does not require the computation and use of the finite divided differences. If a new data point is to be added, the degree of the interpolating polynomial is to be increased by one unit. This involves recomputing the higher degree Lagrange polynomial. The lower degree Lagrange polynomial will be of no use in computing the higher degree Lagrange polynomial. The Newton's interpolating polynomial is to be used when the proper order of the polynomial is not known. In this case, one can try polynomials of different order and compare the results before selecting the order of the polynomial. The procedure can easily be programmed for a computer. If a new data point is to be added either at the beginning or at the end of the tabular data, the previous Newton's divided difference formula can be updated by just adding the term $(x - x_0)(x - x_1)\cdots(x - x_n)g(x_0, x_1, \ldots, x_{n+1})$. Generally, the error with the Lagrange and Newton's interpolating polynomials is small in the middle of the domain, but increases toward the endpoints of the domain. If uniform accuracy is desired throughout the domain, orthogonal polynomials such as Chebyshev polynomials can be used for interpolation.

In the interpolation by splines, a low-order polynomial is developed for each interval between the data points. By enforcing the derivatives of adjacent polynomials to be same at their connecting points (knots), the polynomial fit is made smooth. Splines are to be used to fit data points that exhibit a smooth behavior, except at few local regions that exhibit a rough behavior. The use of a high-degree polynomial is not desired in such cases, as it will unnecessarily induce wild oscillations in the behavior of the dependent variable. Although linear and quadratic splines are possible, the cubic spline is most commonly used.

If the data are not very accurate and precise, the least-squares fit can be used. Although the least-squares curve may not pass through any of the individual data points, it captures the overall trend of the data. The least-squares methods are also known as regression methods. Linear regression can be used when the independent and dependent variables are related linearly. In this method, a straight line is found by minimizing the sum of the squares of differences between the data and the straight line. In some cases, where the relationship between the independent and dependent variables is nonlinear, a suitable transformation (such as logarithmic transformation) can be used to linearize the relationship. In such cases, the least-squares approach can be used to the transformed variables to find the best-fitting straight line. If a suitable transformation cannot be found to obtain a linear relationship between the independent and dependent variables, polynomial regression can be used to fit a nonlinear equation for the data. When the dependent variable is a function of two or more independent variables, multiple regression can be used. If the relationship between the variables is linear—either inherently or through transformation—a least-squares linear equation can be derived. If the relationship between the variables is nonlinear, multiple curvilinear equations can be developed, based on least-squares approach.

5.12 Use of Software Packages

5.12.1 MATLAB

▶Example 5.17

The values of a function $f(x)$ at certain discrete values of x are as follows:

i	1	2	3	4	5
x_i	2	3	6.5	8	12
$f(x_i)$	14	20	17	16	23

Using linear, cubic, spline, and nearest interpolation functions for the data, estimate the value of $f(7.0)$ in each case.

Solution

```
>> x=[2,3,6.5,8,12]'

x =

    2.0000
    3.0000
    6.5000
    8.0000
   12.0000

>> y=[14,20,17,16,23]'

y =

   14
   20
   17
   16
   23
```

```
>> xx=7

xx =
     7

>> yy=interpl(x,y,xx,'linear')

yy =
     16.6667

>> yy=interpl(x,y,xx,'cubic')

yy =

     16.3335

>> yy=interpl(x,y,xx,'spline')

yy =

     16.3335

>> yy=interpl(x,y,xx,'nearest')

yy =

     17
```

◀

Note:

`yy = interp1 (x,y,xx,'linear')`

This statement or command uses the data of x (values of the independent variable) and y (values of the dependent variable), fits a linear equation using least-squares technique, and gives the interpolated values of $y(yy)$ at the specified values of $x(xx)$. The word 'linear' can be replaced by 'cubic', 'spline', or 'nearest' to achieve interpolation based on cubic equation, cubic spline, or nearest value, respectively. If nothing is specified, linear interpolation is implied.

▶Example 5.18

Fit polynomials of degree two, three, and four for the data given in Example 5.17. Also, find the values of f given by each polynomial at x_i, $i = 1, 2, \ldots, 5$.

Solution

```
>> x=[2,3,6.5,8,12]

x =

    2.0000    3.0000    6.5000    8.0000    12.0000

>> y=[14,20,17,16,23]

y =

    14    20    17    16    23

>> z=polyfit(x,y,2)

z =

    0.1062    -0.9121    18.1568

>> yy=polyval (z,x)

yy =

   16.7573    16.3761    16.7136    17.6544    22.4987

>> z=polyfit(x,y,3)

z =

   0.1185    -2.4194    14.3435    -5.4040

>> yy=polyval (z,x)

yy =

   14.5534    19.0513    18.1499    15.1699    23.0755

>> z=polyfit(x,y,4)

z =

   -0.0216    0.6810    -7.3327    31.1271    -24.0260
```

```
>> yy=polyval (z,x)

yy =

   14.0000    20.0000    17.0000    16.0000    23.0000
```

◀

Note:

`p = polyfit(x,y,n)`

This statement or command fits a least-squares polynomial of degree n for the specified set of data x and y. The coefficients of the polynomial are returned through the vector p. The elements of p denote the coefficients of the polynomial starting from that of the highest power of x.

▶Example 5.19

Consider the values of the function $f(x) = \sin^{12} x$ at the discrete values of x, $x_i = \frac{i\pi}{10}$ as indicated in the following table:

i	x_i	$f_i = f(x_i) \times 10^4$
0	0	0
1	0.31416	0.8315
2	0.62832	142.4809
3	0.94248	1835.1291
4	1.25664	6693.5116
5	1.57080	10000.0000
6	1.88496	6693.3858
7	2.19912	1835.0555
8	2.51328	142.4702
9	2.82744	0.8314
10	3.14160,0	0

Use linear, cubic, spline, and nearest neighbor interpolations for this data and find the interpolated value of f at $x_i = (2i - 1)\frac{\pi}{20}$, $i = 1, 2, \ldots, 10$ in each case. Also, compute the error, E, associated with each interpolation using the relation

$$E = \sum_{i=1}^{10} (f_{i,\text{exact}} - f_{i,\text{interpolation}})^2.$$

Solution

```
>> x=0:pi/10:pi;
>> y=(sin(x)).^12;
>> data=[x' y']

data =

         0         0
    0.3142    0.0000
    0.6283    0.0017
    0.9425    0.0786
    1.2566    0.5476
    1.5708    1.0000
    1.8850    0.5476
    2.1991    0.0786
    2.5133    0.0017
    2.8274    0.0000
    3.1416    0.0000

>> xi=pi/20:pi/10:pi-pi/20;
>> yi=interp1(x,y,xi);
>> yy=(sin(xi)).^12;
>> interl=[xi'yy'yi']

interl =

    0.1571    0.0000    0.0000
    0.4712    0.0001    0.0009
    0.7854    0.0156    0.0402
    1.0996    0.2504    0.3131
    1.4137    0.8619    0.7738
    1.7279    0.8619    0.7738
    2.0420    0.2504    0.3131
    2.3562    0.0156    0.0402
    2.6704    0.0001    0.0009
    2.9845    0.0000    0.0000
>> E=FitE(yy',yi');

>> E
E =
    0.0246
>> plot(xi,yi,'--',xi,yi,'+',xi,yy)
>> xlabel('x')
```

```
>> ylabel('y')
>> title('Linear Interpolation')
>> gtext('y=(sin(x))^12')
```

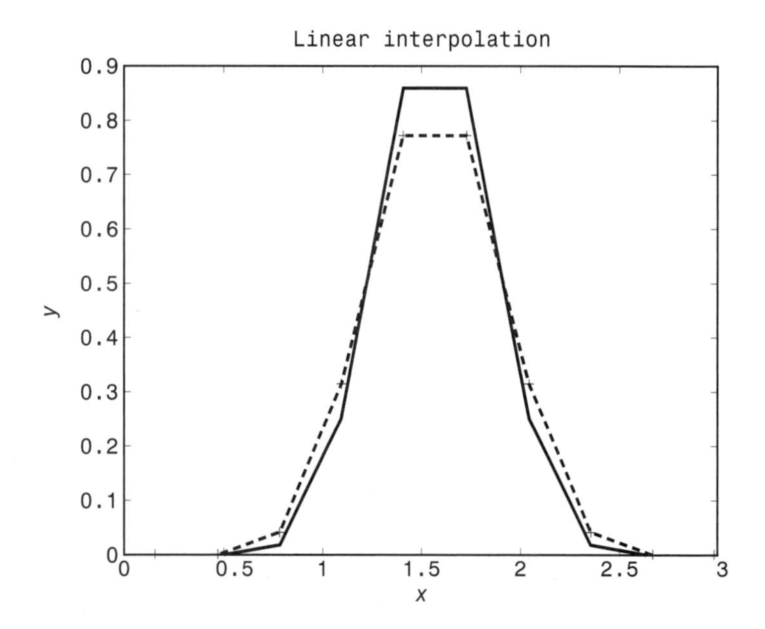


```
>> yc=interpl(x,y,xi,'cubic');
>> interc=[xi' yy' yc']

interc =

    0.1571    0.0000   -0.0002
    0.4712    0.0001   -0.0040
    0.7854    0.0156    0.0110
    1.0996    0.2504    0.2896
    1.4137    0.8619    0.8314
    1.7279    0.8619    0.8314
    2.0420    0.2504    0.2896
    2.3562    0.0156    0.0110
    2.6704    0.0001   -0.0040
    2.9845    0.0000   -0.0002
>> E=FitE(yy', yc');
>> E
E =
    0.0050
>> plot(xi,yc,'--',xi,yc,'+',xi,yy)
>> xlabel('x')
```

```
>> ylabel('y')
>> title('Cubic Interpolation')
```

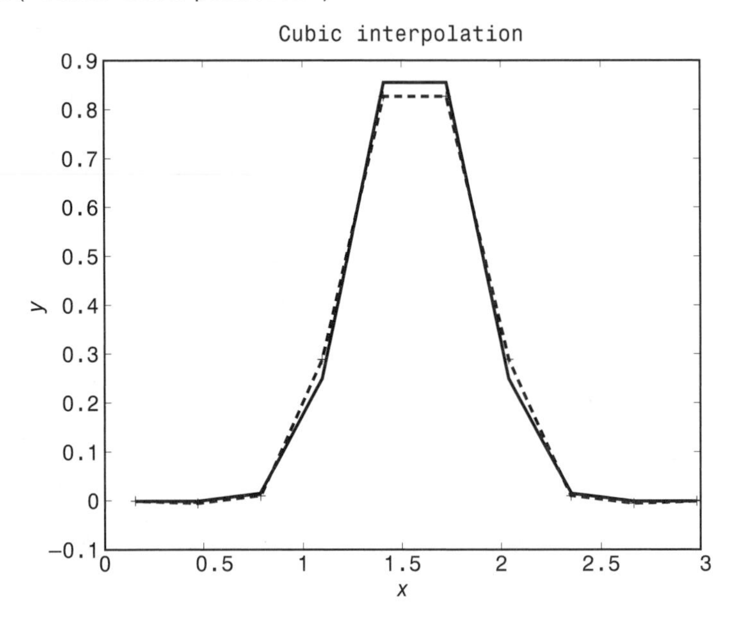


```
>> ys=interpl(x,y,xi,'spline');
>> inters=[xi' yy' ys']

inters =

    0.1571    0.0000    -0.0018
    0.4712    0.0001     0.0022
    0.7854    0.0156     0.0077
    1.0996    0.2504     0.2664
    1.4137    0.8619     0.8523
    1.7279    0.8619     0.8523
    2.0420    0.2504     0.2664
    2.3562    0.0156     0.0077
    2.6704    0.0001     0.0022
    2.9845    0.0000    -0.0018

>> E=FitE(yy', ys')
E =
   8.4216e-004
>> plot(xi,ys,'--',xi,ys,'+',xi,yy)
>> xlabel('x')
>> ylabel('y')
>> title('Spline Interpolation')
```

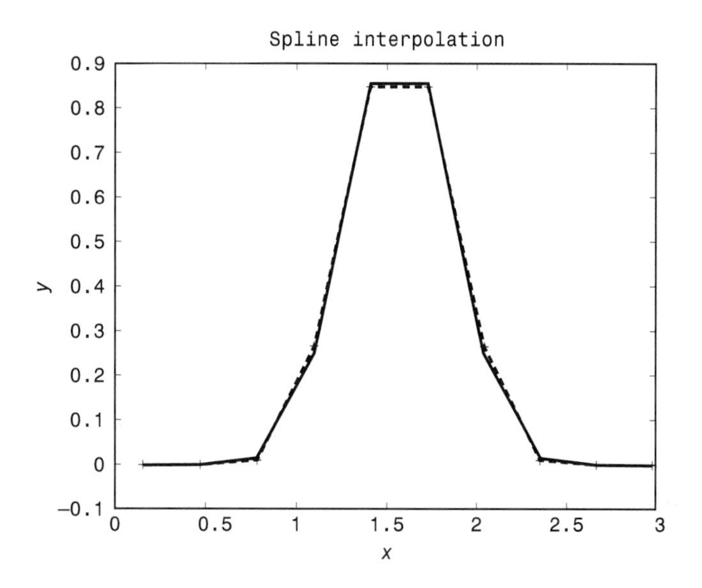


```
>> yn=interp1(x,y,xi,'nearest');
>> intern=[xi' yy' yn']

intern =

    0.1571    0.0000    0.0000
    0.4712    0.0001    0.0017
    0.7854    0.0156    0.0786
    1.0996    0.2504    0.5476
    1.4137    0.8619    1.0000
    1.7279    0.8619    0.5476
    2.0420    0.2504    0.0786
    2.3562    0.0156    0.0017
    2.6704    0.0001    0.0000
    2.9845    0.0000    0.0000

>> yn=interp1(x,y,xi,'nearest');

>> E=FitE(yy', yn')
E =
    0.2399

>> plot(xi,yn,'--',xi,yn,'+',xi,yy)
>> xlabel('x')
>> ylabel('y')
>> title('Nearest Interpolation')
```

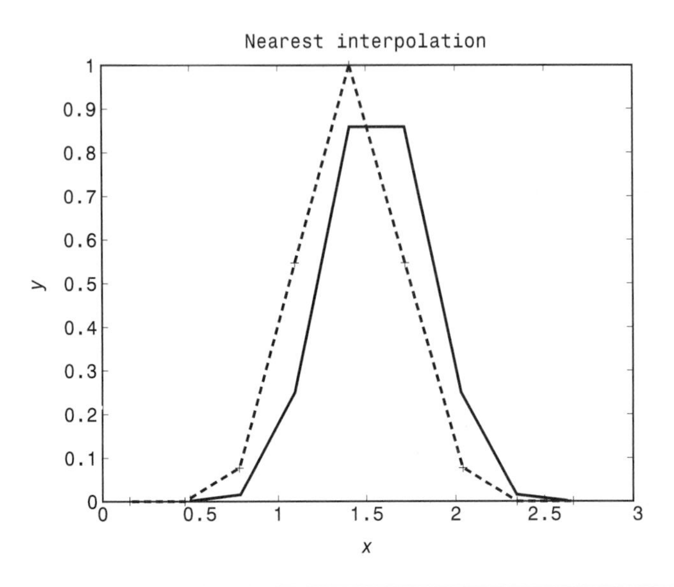

◀

Note:

The vector y (also the vector yy) denotes the values of the function f generated at the specified set of values of x. xi denotes the set of values of x at which interpolated values, yi, are determined.

```
yy = interp(x,y,xi)
```

This statement fits a linear equation and generates the interpolated values yi corresponding to xi. The error (E) is computed using a MATLAB program called FitE. The accuracy of interpolated values (yi), with respect to the input data (yy), is shown through the plots (xi, yi) and (xi, yy).

5.12.2 MAPLE

▶Example 5.20

Fit linear and cubic spline interpolation functions and a fourth-degree polynomial for the data shown in Example 5.17. Estimate the value of $f(7.0)$ in each case.

Solution

```
>

> x:=linalg[vector] (5,[2,3,6.5,8,12]);
```

$$x := [2, 3, 6.5, 8, 12]$$

```
> y:=linalg[vector] (5,[14,20,17,16,23]);
```

$$y := [14, 20, 17, 16, 23]$$

```
> readlib (spline):
```

```
> spline(x,y,z,linear);
```

$$\begin{cases} 2.000000009 + 5.999999996\, z & z < 3 \\ 22.57142858 - .857142858\, z & z < 6.5 \\ 21.33333333 - .666666666\, z & z < 8 \\ 1.999999979 + 1.750000002\, z & otherwise \end{cases}$$

```
> f:=unapply(%,z);
```

$$f := z \to \text{piecewise}(z < 3, 2.000000009 + 5.999999996\, z, z < 6.5, 22.57142858$$
$$-.857142858\, z, z < 8, 21.33333333 - .666666666\, z, 1.999999979 + 1.750000002\, z)$$

```
> f(7);
```

$$16.66666667$$

```
> spline(x,y,z,cubic);
```

$$\begin{cases} 7.270680092 - 3.662913496\, z + 5.270680090\, z^2 - .8784466818\, z^3 & z < 3 \\ -25.53578015 + 29.14354676\, z - 5.66480661\, z^2 + .3366074014\, z^3 & z < 6.5 \\ 89.03020903 - 23.73306357\, z + 2.470056464\, z^2 - .08056506673\, z^3 & z < 8 \\ 70.67134228 - 16.84848854\, z + 1.609484585\, z^2 - .04470790514\, z^3 & otherwise \end{cases}$$

```
> f:=unapply(%,z);
```

$$f := z \to \text{piecewise}(z < 3, 7.270680092 - 3.662913496\, z + 5.270680090\, z^2$$
$$-.8784466818\, z^3, z < 6.5, -25.53578015 + 29.14354676\, z - 5.664806661\, z^2$$
$$+.3366074014\, z^3, z < 8, 89.03020903 - 23.73306357\, z + 2.470056464\, z^2$$
$$-.08056506673\, z^3, 70.67134228 - 16.84848854\, z + 1.609484585\, z^2 - .04470790514\, z^3)$$

```
> f(7);
```

$$16.29771284$$

```
> interp(x,y,z);
```

$$-.02157287158\, z^4 + .6809884562\, z^3 - 7.332720060\, z^2 + 31.12705629\, z - 24.02597404$$

```
> f:=unapply(%,z);
```

$$f := z \to -.02157287158\,z^4 + .6809884562\,z^3 - 7.332720060\,z^2 + 31.12705629\,z$$
$$-24.02597404$$

```
> f(7);
```

$$16.34271286$$

```
>
```

◀

Note:

```
f := unapply(%,z);
```

This statement gives the equation of the interpolation function (f) in terms of the independent variable (z).

```
interp(x,y,z);
```

The statement finds the $n-1$st degree polynomial that passes through n data points specified by (x, y).

▶Example 5.21

(a) Fit linear and cubic spline interpolation functions and a 10-degree polynomial for the data given in Example 5.19.

(b) Determine the values of f at $x_i = (2i-1)\frac{\pi}{20}$, $i = 1, 2, \ldots, 10$ for each interpolation.

(c) Find the error, E, associated with each interpolation using the following relation:

$$E = \sum_{i=1}^{10} (f_{i,\text{exact}} - f_{i,\text{interpolation}})^2$$

Solution

```
> x:=evalf([seq(Pi*i/10,i=0..10)]);
```

$$x := [0, .3141592654, .6283185308, .9424777962, 1.256637062, 1.570796327, 1.884955592,$$
$$2.199114858, 2.513274123, 2.827433389, 3.141592654]$$

```
> y:=evalf([seq((sin(Pi*i/10))^12,i=0..10)]);
```

$$y := [0, .7582080152\;10^{-6}, .001700686339, .07861252319, .5476157200, 1., .5476157200,$$
$$.07861252319, .001700686339, .7582080152\;10^{-6}, 0]$$

```
> interp(x,y,z);
```

$$-1.198109420\,z^{10} + 18.81985878\,z^{9} - 126.0732667\,z^{8} + 469.8160280\,z^{7} - 1065.586503\,z^{6} + 1513.318425\,z^{5} - 1334.994478\,z^{4} + 700.2539445\,z^{3} - 196.2920285\,z^{2} + 22.05233488\,z$$

```
> f:=unapply(%,z);
```

$$f := z \to -1.198109420\,z^{10} + 18.81985878\,z^{9} - 126.0732667\,z^{8} + 469.8160280\,z^{7} - 1065.586503\,z^{6} + 1513.318425\,z^{5} - 1334.994478\,z^{4} + 700.2539445\,z^{3} - 196.2920285\,z^{2} + 22.05233488\,z$$

```
> xl:=evalf([seq(Pi*i/10-Pi/20,i=1..10)]);
```

xl := [.1570796327, .4712388981, .7853981635, 1.099557429, 1.413716694, 1.727875960, 2.042035225, 2.356194491, 2.670353756, 2.984513021]

```
> yy:=evalf([seq((sin(Pi*i/10-Pi/20))^12,i=1..10)]);
```

yy := [.2147777549 10^{-9}, .00007665865214, .01562500000, .2503632667, .8618647624, .8618647624, .2503632667, .01562500000, .00007665865214, .2147777549 10^{-9}]

```
>
>
```

```
> yl:=[seq(f(xl[i]),i=1..10)];
>
>
```

yl := [.651717087, −.11523589, .05064682, .23700674, .86554466, .86555380, .23704912, .05072596, −.11520772, .65205759]

```
> E:=proc(yy,yl)
> local i, error;
> error:=0;
> for i from 1 to 10 do
> error:=error+(yy[i]-yl[i])^2;
> od;
> error;
> end:
```

```
>
```

```
> E(yy,yl);
>
>
```

$$.8793431527$$

```
> plot([sin^12,f],0..pi,linestyle=[1,5]);
```

```
>
```

```
> readlib(spline):
```

```
> spline(x,y,z,linear);
```

$$\begin{cases} .2413451070\ 10^{-5}\ z & z < .3141592654 \\ -.001699169923 + .005411039298\ z & z < .6283185308 \\ -.1521229874 + .2448179803\ z & z < .9424777962 \\ -1.328397066 + 1.492883541\ z & z < 1.256637062 \\ -1.261921404 + 1.439983889\ z & z < 1.570796327 \\ 3.261921403 - 1.439983889\ z & z < 1.884955592 \\ 3.361634893 - 1.492883538\ z & z < 2.199114858 \\ .6169953803 - .2448179799\ z & z < 2.513274123 \\ .01530011139 - .005411039299\ z & z < 2.827433389 \\ .7582080152\ 10^{-5} - .2413451070\ 10^{-5}\ z & \mathit{otherwise} \end{cases}$$

```
> f:=unapply(%,z);
```

$f := z \to \text{piecewise}(z < .3141592654, .2413451070\ 10^{-5} z, z < .6283185308,$
$-.001699169923 + .005411039298\, z, z < .9424777962, -.1521229874 + .2448179803\, z,$
$z < 1.256637062, -1.328397066 + 1.492883541\, z, z < 1.570796327,$
$-1.261921404 + 1.439983889\, z, z < 1.884955592, 3.261921403 - 1.439983889\, z,$
$z < 2.199114858, 3.361634893 - 1.492883538\, z,$
$z < 2.513274123, .6169953803 - .2448179799\, z,$
$z < 2.827433389, .01530011139 - .005411039299\, z, .7582080152\ 10^{-5}$
$-.2413451070\ 10^{-5}\, z)$

```
> ys:=[seq(f(xl[i]),i=1..10)];
```

$ys := [.3791040076\ 10^{-6}, . - 000850722273, .0401566047, .313114122,$
$.773807859, .773807858, .313114122, .0401566048, .00085072227, .379104008\ 10^{-6}]$

```
> E(yy,ys);
```

$.02458817392$

```
> plot([sin^12,f],0..pi,linestyle=[1,5]);
```

```
> spline(x,y,z,cubic);
```

$$\begin{cases} -.004700157568\,z + .04764700618\,z^3 & z < .3141592654 \\ .007164967603 - .07312055826\,z + .2177888996\,z^2 - .1834341933\,z^3 & z < .6283185308 \\ -.7966225333 + 3.764682052\,z - 5.890263653\,z^2 + 3.056988327\,z^3 & z < .9424777962 \\ 3.290153256 - 9.243929318\,z + 7.912301701\,z^2 - 1.824670569\,z^3 & z < 1.256637062 \\ 17.40950672 - 42.95140279\,z + 34.73585680\,z^2 - 8.939839542\,z^3 & z < 1.570796327 \\ -51.88827803 + 89.39761707\,z - 49.52014608\,z^2 + 8.939839579\,z^3 & z < 1.884955592 \\ -4.235461643 + 13.55580430\,z - 9.284814761\,z^2 + 1.824670837\,z^3 & z < 2.199114858 \\ 47.68173646 - 57.26887033\,z + 22.92117683\,z^2 - 3.056988931\,z^3 & z < 2.513274123 \\ -3.760679323 + 4.135990288\,z - 1.511040906\,z^2 + .1834345875\,z^3 & z < 2.827433389 \\ 1.462592903 - 1.406073668\,z + .4490632575\,z^2 - .04764709146\,z^3 & otherwise \end{cases}$$

```
> f:=unapply(%,z);
```

$$f := z \to \text{piecewise}(z < .3141592654, -.004700157568\,z + .04764700618\,z^3,$$
$$z < .6283185308, .007164967603 - .07312055826z + .2177888996z^2 - .1834341933z^3,$$
$$z < .9424777962, -.7966225333 + 3.764682052\,z - 5.890263653\,z^2 + 3.056988327\,z^3,$$
$$z < 1.256637062, 3.290153256 - 9.243929318\,z + 7.912301701\,z^2 - 1.824670569\,z^3,$$
$$z < 1.570796327, 17.40950672 - 42.95140279\,z + 34.73585680\,z^2 - 8.939839542\,z^3,$$
$$z < 1.884955592, -51.88827803 + 89.39761707\,z - 49.52014608\,z^2 + 8.939839579\,z^3,$$
$$z < 2.199114858, -4.235461643 + 13.55580430\,z - 9.284814761\,z^2 + 1.824670837\,z^3,$$
$$z < 2.513274123, 47.68173646 - 57.26887033\,z + 22.92117683\,z^2 - 3.056988931\,z^3,$$
$$z < 2.827433389, -3.760679323 + 4.135990288\,z - 1.511040906\,z^2 + .1834345875\,z^3,$$
$$1.462592903 - 1.406073668\,z + .4490632575\,z^2 - .04764709146\,z^3)$$

```
> yc:=[seq(f(xl[i]),i=1..10)];
```

$$yc := [-.0005536294922, .00187555935, .007769612, .266398541, .85225503, .85225507,$$
$$.26639853, .00776962, .001875556, -.000553631]$$

```
> E(yy,yc);
```

$$.0008294520401$$

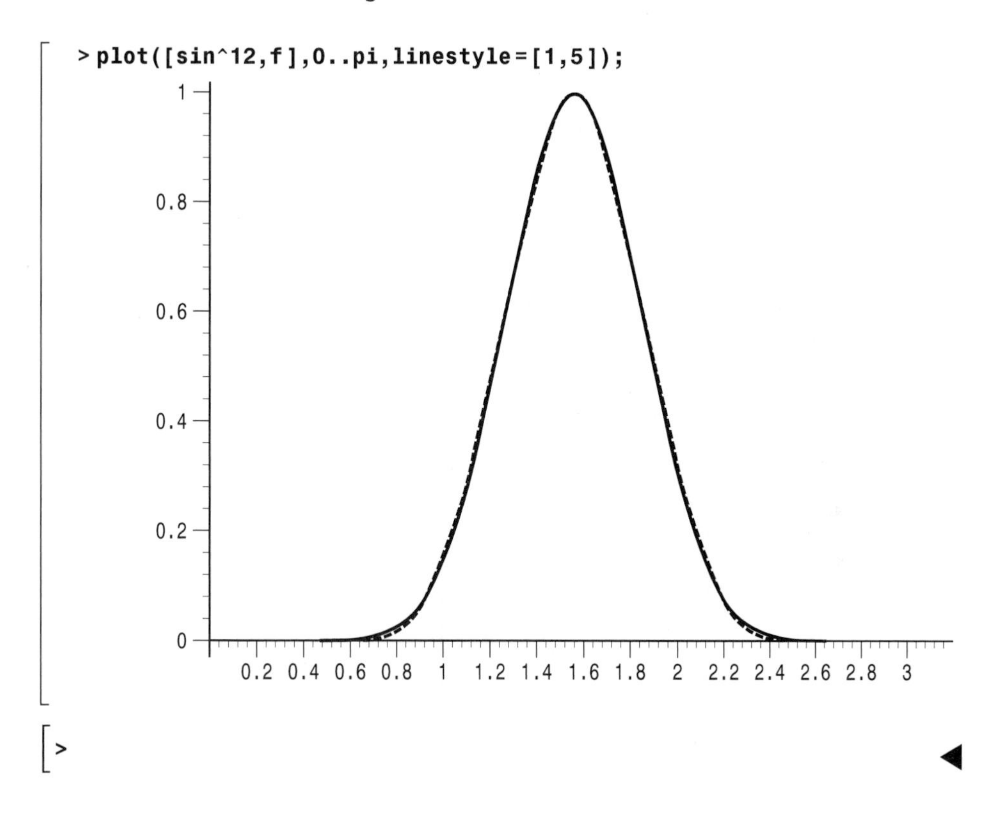

[>

5.12.3 Mathcad

▶Example 5.22

Estimate the value of $f(7.0)$ using linear interpolation, cubic spline with cubic endpoints and cubic spline with parabolic endpoints for the data given in Example 5.17.

Solution

$$x := \begin{pmatrix} 2 \\ 3 \\ 6.5 \\ 8 \\ 12 \end{pmatrix} \qquad y := \begin{pmatrix} 14 \\ 20 \\ 17 \\ 16 \\ 23 \end{pmatrix}$$

linterp(x, y, 7) = 16.667

cs := cspline(x, y)

interp(cs, x, y, 7) = 16.334

ls := lspline(x, y)

interp(ls, x, y, 7) = 16.298

ps := pspline(x, y)

interp(ps, x, y, 7) = 16.35

$$\text{cs} = \begin{pmatrix} 0 \\ 3 \\ 2 \\ -6.184 \\ -4.473 \\ 1.514 \\ 1.106 \\ 0.017 \end{pmatrix} \qquad \text{ls} = \begin{pmatrix} 0 \\ 3 \\ 0 \\ 0 \\ -5.271 \\ 1.798 \\ 1.073 \\ 0 \end{pmatrix} \qquad \text{ps} = \begin{pmatrix} 0 \\ 3 \\ 1 \\ -4.686 \\ -4.686 \\ 1.634 \\ 0.803 \\ 0.803 \end{pmatrix}$$

▶Example 5.23

(a) Estimate the values of $f(x_i)$ at $x_i = (2i - 1)\frac{\pi}{20}$, $i = 1, 2, \ldots, 10$ using linear interpolation, cubic spline with cubic endpoints, and cubic spline with parabolic endpoints for the data given in Example 5.19.

(b) Compute the error, E, associated with each interpolation using the relation

$$E = \sum_{i=1}^{10} (f_{i,\text{exact}} - f_{i,\text{interpolation}})^2.$$

(c) Plot the function $f(x_i)$ in case of each interpolation.

Solution

```
i:=0..10
```

$x_i := \frac{i \cdot \pi}{10}$

$y_i := (\sin(x_i))^{12}$

x =

	0
0	0
1	0.314
2	0.628
3	0.942
4	1.257
5	1.571
6	1.885
7	2.199
8	2.513
9	2.827
10	3.142

y =

	0
0	0
1	$.582 \cdot 10^{-7}$
2	$.701 \cdot 10^{-3}$
3	0.079
4	0.548
5	1
6	0.548
7	0.079
8	$.701 \cdot 10^{-3}$
9	$.582 \cdot 10^{-7}$
10	0

```
i:=0..9
```

$xx_i := \frac{-\pi}{20} + (1+i) \cdot \frac{\pi}{10}$

$yy_i := (\sin(xx_i))^{12}$

$yl_i := \text{linterp}(x, y, xx_i)$

xx =

	0
0	0.157
1	0.471
2	0.785
3	1.1
4	1.414
5	1.728
6	2.042
7	2.356
8	2.67
9	2.985

yy =

	0
0	$.148 \cdot 10^{-10}$
1	$7.666 \cdot 10^{-5}$
2	0.016
3	0.25
4	0.862
5	0.862
6	0.25
7	0.016
8	$7.666 \cdot 10^{-5}$
9	$.148 \cdot 10^{-10}$

yl =

	0
0	$.791 \cdot 10^{-7}$
1	$.507 \cdot 10^{-4}$
2	0.04
3	0.313
4	0.774
5	0.774
6	0.313
7	0.04
8	$.507 \cdot 10^{-4}$
9	$.791 \cdot 10^{-7}$

```
E1:= | error ← 0
     | for i∈ 0..9
     |       error ← error + (yy_i − y1_i)^2
     | error
```

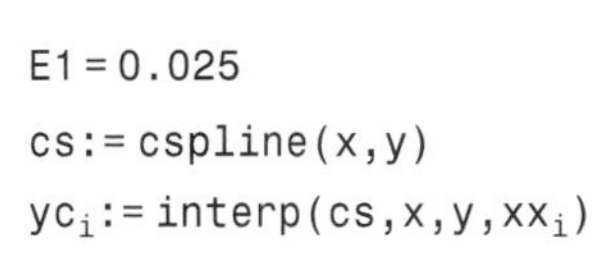

E1 = 0.025

cs:= cspline(x,y)

$yc_i := \text{interp}(cs, x, y, xx_i)$

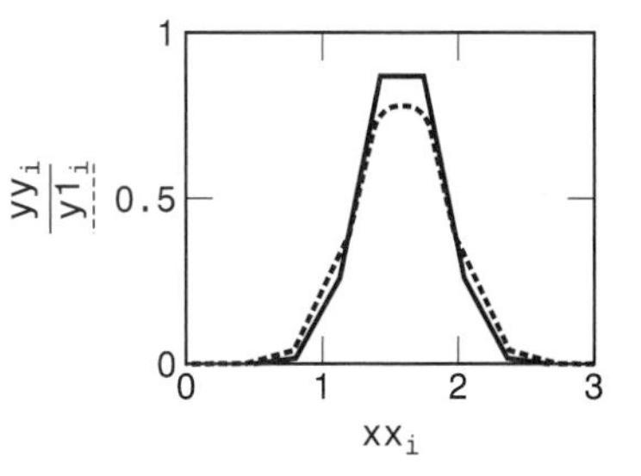

i	xx =	yy =	yc =
0	0.157	$.148 \cdot 10^{-10}$	$1.777 \cdot 10^{-3}$
1	0.471	$7.666 \cdot 10^{-5}$	$2.203 \cdot 10^{-3}$
2	0.785	0.016	$7.682 \cdot 10^{-3}$
3	1.1	0.25	0.266
4	1.414	0.862	0.852
5	1.728	0.862	0.852
6	2.042	0.25	0.266
7	2.356	0.016	$7.682 \cdot 10^{-3}$
8	2.67	$7.666 \cdot 10^{-5}$	$2.203 \cdot 10^{-3}$
9	2.985	$.148 \cdot 10^{-10}$	$1.777 \cdot 10^{-3}$

```
Ec:= | error ← 0
     | for i ∈ 0..9
     |      error ← error + (yy_i − yc_i)^2
     | error
```

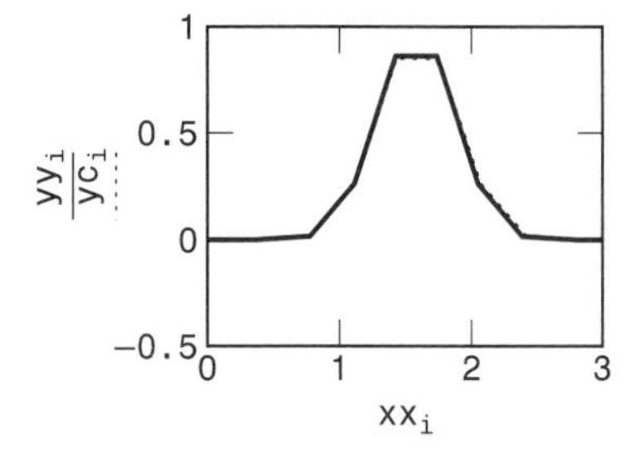

$Ec = 8.422 \times 10^{-4}$

1s:= 1spline(x,y)

$y1s_i$:= interp(1s,x,y,xx_i)

i	xx=	yy=	y1s=
0	0.157	$.148 \cdot 10^{-10}$	$5.536 \cdot 10^{-4}$
1	0.471	$7.666 \cdot 10^{-5}$	$1.876 \cdot 10^{-3}$
2	0.785	0.016	$7.77 \cdot 10^{-3}$
3	1.1	0.25	0.266
4	1.414	0.862	0.852
5	1.728	0.862	0.852
6	2.042	0.25	0.266
7	2.356	0.016	$7.77 \cdot 10^{-3}$
8	2.67	$7.666 \cdot 10^{-5}$	$1.876 \cdot 10^{-3}$
9	2.985	$.148 \cdot 10^{-10}$	$5.536 \cdot 10^{-4}$

```
E1s:= | error ← 0
      | for i ∈ 0..9
      |      error ← error + (yy_i − y1s_i)^2
      | error
```

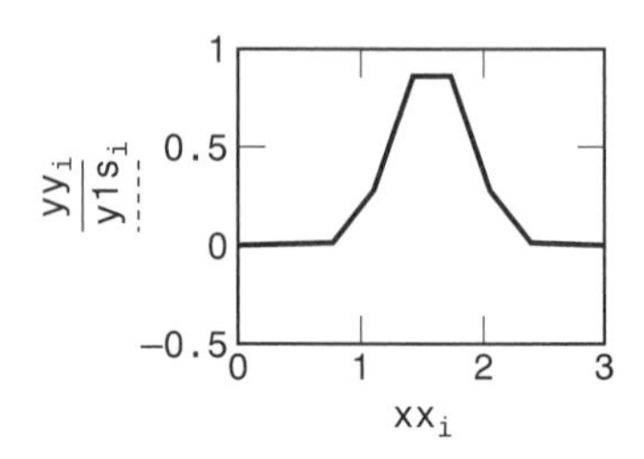

$E1s = 8.295 \times 10^{-4}$

ps:= pspline(x, y)

yps_i:= interp(ps, x, y, xx_i)

xx =

	0
0	0.157
1	0.471
2	0.785
3	1.1
4	1.414
5	1.728
6	2.042
7	2.356
8	2.67
9	2.985

yy =

	0
0	$.148 \cdot 10^{-10}$
1	$7.666 \cdot 10^{-5}$
2	0.016
3	0.25
4	0.862
5	0.862
6	0.25
7	0.016
8	$7.666 \cdot 10^{-5}$
9	$.148 \cdot 10^{-10}$

yps =

	0
0	$8.735 \cdot 10^{-4}$
1	$1.961 \cdot 10^{-3}$
2	$7.747 \cdot 10^{-3}$
3	0.266
4	0.852
5	0.852
6	0.266
7	$7.747 \cdot 10^{-3}$
8	$1.961 \cdot 10^{-3}$
9	$8.735 \cdot 10^{-4}$

```
Eps := | error ← 0
       | for i ∈ 0..9
       |     error ← error + (yy_i − yps_i)^2
       | error
```

Eps = 8.322×10^{-4}

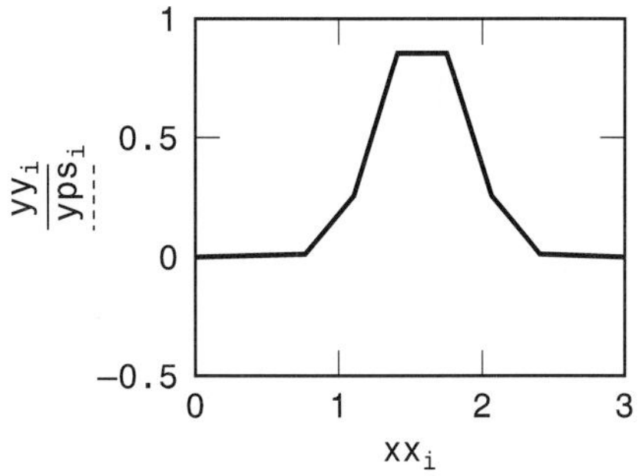

cs =

	0
0	0
1	3
2	2
3	0.271
4	0.017
5	−0.236
6	5.501
7	2.068
8	−14.785
9	2.068
10	5.501
11	−0.236
12	0.017
13	0.271

1s =

	0
0	0
1	3
2	0
3	0
4	0.09
5	−0.256
6	5.506
7	2.067
8	−14.784
9	2.067
10	5.506
11	−0.256
12	0.09
13	0

ps =

	0
0	0
1	3
2	1
3	0.071
4	0.071
5	−0.251
6	5.505
7	2.067
8	−14.784
9	2.067
10	5.505
11	−0.251
12	0.071
13	0.071

◀

5.13 Computer Programs

5.13.1 Fortran Programs

Program 5.1 Lagrange interpolation

Input data required

N = number of data points
$X(1), \ldots, X(N)$ = values of X at which function values are known
$F(1), \ldots, F(N)$ = known function values at $X(1), \ldots, X(N)$
XX = value of X at which the function value F is to be determined

Illustration:

Find the interpolated value of f at $x = 3.5$ using the following data:

$(x_i, f_i) = (1.0, 1.5708), (3.0, 1.5719), (5.0, 1.5738)$

Output of the program:

```
LAGRANGE INTERPOLATION USING 3 DATA POINTS

VALUES OF X(I):
   0.100000E+01   0.300000E+01   0.500000E+01

VALUES OF F(I):
   0.157080E+01   0.157190E+01   0.157380E+01

RESULT OF INTERPOLATION:
VALUE OF X:   0.350000E+01
PREDICTED VALUE OF F:   0.157230E+01
```

Program 5.2 Least squares polynomial regression

Input data required

N = number of data points
$X(1), \ldots, X(N)$ = values of independent variable X at which function values are known
$F(1), \ldots, F(N)$ = values of function F at $X(1), \ldots, X(N)$
M = order of the polynomial to be fitted to the data

Illustration (Example 5.14). Using $N = 6$, $X(I) = 0.0, 0.2, 0.4, 0.6, 0.8, 1.0$, $Y(I) = F(I) = 0.0, 1.05, 0.85, 0.35, 0.1, 1.0$, and $M = 2$, the output of Program 5.2 is as follows:

```
LEAST-SQUARES POLYNOMIAL OF ORDER 2 THROUGH 6 DATA POINTS

VALUES OF X(I):
 0.000000E+00  0.2000E+00  0.4000E+00   0.6000E+00  0.8000E+00  0.1000E+01

VALUES OF Y(I):
 0.000000E+00  0.1050E+01  0.8500E+00   0.3500E+00  0.1000E+00  0.1000E+01

COEFFICIENTS OF POLYNOMIAL, B(I), I=1,..., MP1:
(B(1) = CONSTANT TERM)
 0.383929E+00  0.659821E+00  -0.424106E+00

MEASURES OF GOODNESS OF FIT:
SUMO = 0.108708E+01
SUM = 0.103745E+01
CORRELATION COEFFICIENT (R2) = 0.213684E+00
```

5.13.2 C Programs

Program 5.3 Cubic spline

Input data required

$x(i) = n$ data points (values of x) at which values of f are given
$f(i) =$ known function values of f at $x(i), i = 1, 2, \ldots, n$
$n =$ number of data points
iend = end conditions identifier (1 if sdf(1) $=$ sdf(n) $= 0$
for a natural cubic spline; 2 if
sdf(1) $=$ sdf(2) and sdf($n-1$) $=$ sdf(n))
$xx =$ value of x at which the interpolated value of $f(ff)$ is desired

Illustration

Example 5.12

Program output

```
Cubic spline using 5 data points

value of x[1]:    2.000000e+00,  value of f[1]:    1.400000e+01
value of x[2]:    3.000000e+00,  value of f[2]:    2.000000e+01
value of x[3]:    6.500000e+00,  value of f[3]:    1.700000e+01
value of x[4]:    8.000000e+00,  value of f[4]:    1.600000e+01
value of x[5]:    1.200000e+01,  value of f[5]:    2.300000e+01

b[1] = -4.114286e+01 b[2] = 1.142857e+00 b[3] = 1.450000e+01
d1[1] = 0.000000e+00, d2[1] = 1.028571e+01, d3[1] = 3.500000e+00
d1[2] = 3.214286e+00, d2[2] = 1.000000e+01, d3[2] = -9.166667e+00
d1[3] = 4.000000e+00, d2[3] = 2.566667e+01, d3[3] = 0.000000e+00
b[1] = -4.114286e+01 b[2] = 1.142857e+00 b[3] = 1.450000e+01
x2[1] = -4.389554e+00 x2[2] = 1.246574e+00 x2[3] = 1.010140e+00
```

```
Result of interpolation:
Value of x:    7.000000e+00
Predicted value of f:    1.622876e+01
```

Program 5.4 Least-squares linear regression

Input data required

n = number of data points
$x(0), \ldots, x(n-1)$ = values of x at which function values are known
$y(0), \ldots, y(n-1)$ = known values of the function y at $x(0), .., y(n-1)$

Illustration (Example 5.13). Using $n = 6$, $x[i] = 0, 1, 2, 3, 4, 5$, and $y[i] = 0.0, 0.7, 2.2, 2.8, 4.4, 4.9$, the output of Program 5.4 is as follows:

```
Least squares straight line fit through 5 data points

Data:
values of x[0]:    1.000000e+00  values of f[0]:    7.000000e-01
values of x[1]:    2.000000e+00  values of f[1]:    2.200000e+00
values of x[2]:    3.000000e+00  values of f[2]:    2.800000e+00
values of x[3]:    4.000000e+00  values of f[3]:    4.400000e+00
values of x[4]:    5.000000e+00  values of f[4]:    4.900000e+00

coefficients of linear equation, a0 and a1:    -1.800000e-01    1.060000e+00

Measures of goodness of fit:
sum0= 1.154000e+001 sum= 3.040000e-001

Correlation coefficient (r2) = 9.867405e-001
```

REFERENCES AND BIBLIOGRAPHY

5.1. S. S. Rao, *The Finite Element Method in Engineering*, 3d edition, Butterworth-Heinemann, Boston, 1999.

5.2. B. Irons and N. G. Shrive, *Numerical Methods in Engineering and Applied Science: Numbers are Fun*, Ellis Horwood, Chichester, 1987.

5.3. M. C. Kohn, *Practical Numerical Methods*, Macmillan, New York, 1987.

5.4. B. Carnahan, H. A. Luther, and J. O. Wilkes, *Applied Numerical Methods*, John Wiley, New York, 1969.

5.5. C. Phillips and B. Cornelius, *Computational Numerical Methods*, Ellis Horwood, Chichester, 1986.

REVIEW QUESTIONS

The following questions along with corresponding answers are available in an interactive format at the Companion Web site at http://www.prenhall.com/rao.

5.1. **Define** the following:

Vandermonde matrix, Spline, Knot, Natural cubic spline, Correlation coefficient, Least-squares approach, Curve fitting, Interpolation, Collocation, Multiple regression.

5.2. Answer **true or false** to each of the following:

1. Collocation is used when the data are known to be accurate.
2. The determination of the polynomial passing through n data points always requires the solution of a set of n simultaneous equations.
3. The polynomial of degree n passing through $n + 1$ points is unique.
4. The Vandermonde matrix is prone to ill conditioning.
5. Collocation functions are commonly used for interpolation.
6. Spline functions cannot be used for interpolation.
7. Fourier series can be used for interpolation.
8. The derivation of Lagrange interpolation function does not require the solution of a set of linear equations.
9. Any nth-degree polynomial can be expressed in terms of Chebyshev polynomials.
10. In general, a low-order piecewise polynomial is better than a single high-order polynomial to interpolate n data points.
11. Linear least-squares regression with multiple variables involves the solution of two linear equations.
12. The correlation coefficient is not useful to measure the accuracy of a multiple variable regression.

5.3. **Fill in the blanks** with suitable word(s):

1. A ______ passing through a specific set of data points can be expressed in different forms.
2. Collocation functions are also known as ______ functions.
3. The Lagrange interpolation can be used to obtain an ______ fit to a set of data points.
4. The Lagrange interpolation functions can be used to represent polynomials of any ______ .

(*continued, page 438*)

5. The error in a Lagrange interpolation polynomial depends on the ______ of the domain.
6. The error in a Lagrange interpolation polynomial approaches ______ as the range of interpolation is reduced.
7. For uniformly spaced data points, the error in a Lagrange interpolation polynomial is, in general, ______ in the middle of the range.
8. For uniformly spaced data points, the error in a Lagrange interpolation polynomial is, in general, ______ near the endpoints of the range.
9. For a given range, the maximum error in a Lagrange interpolation polynomial decreases with ______ in the order of the polynomial.
10. The coefficients of the polynomial in a Newton's interpolating polynomial are found using a ______ formula.
11. The general structure of Newton's interpolating polynomial is similar to ______ series expansion where higher order derivative terms are added in succession.
12. The Chebyshev interpolation provides a more ______ distribution of error throughout the domain compared with Lagrange and Newton interpolations.
13. Chebyshev polynomials can be expressed in the form of power series of ______ functions.
14. The expansion of a function $f(x)$ in terms of Chebyshev polynomials is equivalent to the ______ series expansion of the function $f(\cos\theta)$.
15. The Chebyshev polynomial of order n has ______ roots in the range [-1, 1].
16. In general, a higher order polynomial introduces unnecessary ______ .
17. The first derivative of a function represents the ______ .
18. ______ splines are most commonly used in engineering applications.
19. The data points are also known as ______ in the development of splines.
20. With linear splines, the function is continuous but its first derivative is ______ at the knots.
21. In an nth-order spline, in general, the continuity of derivatives up to order ______ is ensured at the knots.
22. The development of a cubic spline involves the solution of a ______ system of linear equations.
23. Regression is the method of obtaining the ______ fit to a given set of data.
24. In regression analysis, a value of ______ is desirable for the correlation coefficient.
25. A ______ -order polynomial usually yields a better fit in regression.

(*continued, page 439*)

5.4. Give **short answers**:

1. What is the difference between curve fitting and interpolation?
2. Name three types of interpolation functions.
3. Are the data points required to be uniformly spaced to define a Lagrange interpolation polynomial?
4. What is the difference between Newton's and Newton–Gregory interpolation formulas?
5. What is least-squares regression?
6. What is the role of correlation coefficient in the context of regression?
7. What is the difference between linear and polynomial regressions?
8. Under what conditions can we use linear regression for nonlinear relations?

5.5. Select the **most appropriate answer** out of the multiple choices given:

1. The following type of functions are most commonly used for interpolation:
   (a) polynomials
   (b) splines
   (c) Fourier series
2. The error in a Lagrange interpolation polynomial depends on
   (a) value of f
   (b) spacing of data points
   (c) value of x_0
3. All previous computations can still be used when new data points are added in the following type of polynomials:
   (a) Lagrange's
   (b) Splines
   (c) Newton's
4. The second derivative of a function represents the
   (a) curvature
   (b) slope
   (c) continuity
5. The development of a quadratic spline with $n+1$ data points requires the use of
   (a) $3n$ conditions
   (b) $3(n+1)$ conditions
   (c) $4n$ conditions
6. The development of a cubic spline with $n+1$ data points requires the use of
   (a) $3n$ conditions
   (b) $3(n+1)$ conditions
   (c) $4n$ conditions

(*continued, page 440*)

7. The ill-conditioning problems associated with high-order polynomials can be reduced by using
   (a) orthogonal polynomials
   (b) high-order splines
   (c) high-order Lagrange polynomials
8. The method of least squares is useful for
   (a) interpolation
   (b) collocation
   (c) linear regression

5.6. **Match the following:**

(1) Collocation	(a) piecewise polynomial
(2) Regression	(b) useful to represent any periodic function
(3) Interpolation	(c) curve passing through every data point
(4) Inverse interpolation	(d) curve passing through $n + 1$ data points
(5) nth-degree polynomial	(e) curve made to represent the general trend of the data
(6) Spline	(f) estimation of the value of a function between known data points
(7) Fourier series	(g) estimation of the value of independent variable for a specified value of the function

PROBLEMS

Section 5.2

5.1. In the finite-element analysis of beams, the transverse displacement, $\phi(x)$, is assumed to be a cubic polynomial as

$$\phi(x) = N_1(x)\Phi_1 + N_2(x)\Phi_2 + N_3(x)\Phi_3 + N_4(x)\Phi_4,$$

where $N_i(x)$ are the shape functions and Φ_i are the nodal displacements and rotations (treated as unknowns), $i = 1, 2, 3, 4$. (See Fig. 5.15.) The shape function $N_i(x)$ is defined such that it will have a value of unity along the degree of freedom Φ_i and zero along other degrees of freedom. Express the shape functions in terms of suitable interpolation polynomials.

5.2. The kinematic viscosity of SAE 30 oil with variation in temperature was found to be as follows:

Value of i	1	2	3	4	5	6	7
Temperature, T_i (°C)	1	20	40	60	80	100	120
Viscosity, μ_i (m^2/sec)	2.5×10^{-3}	5.5×10^{-4}	1×10^{-4}	5×10^{-5}	2×10^{-5}	1.2×10^{-5}	6×10^{-6}

Develop a relationship between the two parameters of the following form:

$$\mu = ae^{(b/T)}.$$

Section 5.3

5.3. The data on the variation of the ratio of stagnation pressure to static pressure (r) with Mach number (M) for air flowing through a duct are as follows:

i	1	2	3	4	5
M_i	0.2	0.4	0.6	0.8	1.0
r_i	1.05	1.1	1.3	1.55	1.9

Fit a fourth-degree polynomial to the data.

5.4. The power developed by a hydraulic impulse turbine (P) by changing the penstock diameter (D) is found to be as follows:

i	1	2	3	4
$D_i(m)$	0.4	0.6	0.8	1.0
$P_i(MW)$	20	50	105	180

Fit a cubic equation to the data.

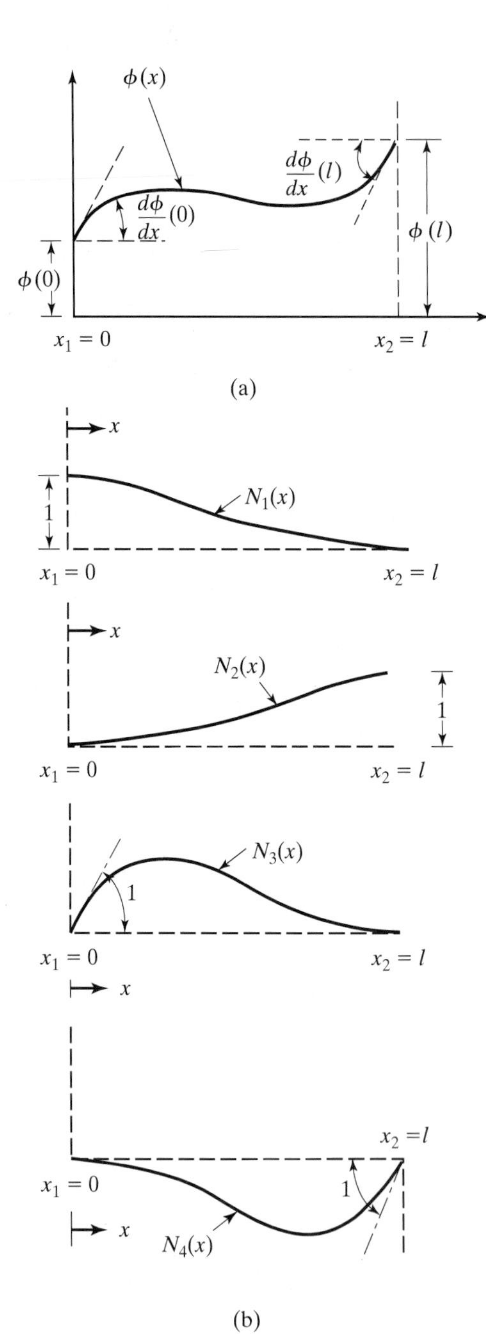

Figure 5.15 Shape functions of a beam element.

Section 5.5

5.5. The amplitude of vibration in the vertical direction of an automobile, after passing over a road bump, is found to be as follows:

Time, t_i (sec)	0	0.64	1.28	1.92
Amplitude, a_i (mm)	5	2	0.75	0.3

Develop a Lagrange interpolation polynomial that passes through the aforementioned data.

5.6. Using the values, $\cos(1.1) = 0.453598$ and $\cos(1.3) = 0.267502$, find an approximate value of $\cos(1.25)$ using Lagrange interpolation. Also, determine a bound on the error involved.

5.7. Derive the Lagrange interpolation polynomial that passes through the following data points:

x	-4	-3	-2	-1	0
y	5	0	3	2	9

Use this polynomial to estimate the value of y at $x = -2.5$.

Section 5.6

5.8. (a) Develop a Newton's interpolation formula for the following data:

x_i	0.5	1.0	2.0	3.5
f_i	0.472366	0.223130	0.0497870	0.00524752

(b) Use this formula to estimate the first and second derivatives of f at $x = 1.5$.

5.9. Consider a second-degree polynomial $f(x)$ passing through three data points; (x_i, f_i), $(x_{i+1} = x_i + h, f_{i+1})$, and $(x_{i+2} = x_i + kh, f_{i+2})$. Derive expressions for the first and second derivatives of f at x, $x_i \le x \le x_{i+2}$, based on Newton's interpolation formula.

5.10. Derive the forward and backward difference polynomials for the following data:

x	0.2	0.3	0.4	0.5	0.6
y	1.60	1.76	1.96	2.20	2.48

Use the result to compute the value of y at $x = 0.35$.

Section 5.7

5.11. Express the function $f(x) = -3x^3 + 8x^2 - 5x + 4$ in terms of Chebyshev polynomials.

5.12. Using the orthogonal property of Chebyshev polynomials,

$$\int_{-1}^{1} \frac{P_i(x)P_j(x)}{\sqrt{1-x^2}} dx = \begin{cases} 0, & i \neq j \\ \frac{\pi}{2}, & i = j \neq 0 \\ \pi, & i = j = 0 \end{cases} \tag{E1}$$

determine the least squares approximation of second degree for $f(x) = x^4$ in the interval $-1 \leq x \leq 1$.

5.13. Develop a Lagrange interpolation polynomial using three Chebyshev points to fit the data given by $y = \ln x$ in the interval [1, 5].

5.14. Express $\sin x$ in terms of Chebyshev polynomials by retaining three terms in its power series expansion.

Section 5.8

5.15. Consider the function $f(x) = \frac{30}{1+3x^2}$ defined over the interval, $-3 \leq x \leq 3$. Fit a linear spline using the values of f at $x_i = -3, -2, -1, 0, 1, 2,$ and 3.

5.16. For the function $f(x) = 9 - 7x + 5x^2 - 3x^3$, fit a quadratic spline using the values of f at $x_i = -3, -2, -1, 0, 1, 2,$ and 3.

5.17. For the function defined in Problem 5.16, fit a cubic spline using the values of f at $x_i = -3, -2, -1, 0, 1, 2,$ and 3 according to the three types of end conditions described in Sections 5.8.4 (i), (ii), and (iii).

5.18. Obtain different types of cubic spline approximations for the following function:

x	0	1	2	3
y	2	3	34	245

Hint: Use cases (i), (ii), and (iii) of Section 5.8.4.

5.19. Derive Eq. (5.97) using Eqs. (5.95) and (5.96).

5.20. Derive Eq. (5.108) using Eqs. (5.106), (5.107), and (5.95).

Section 5.9

5.21. Experiments conducted during the machining of AISI-4140 steel with fixed values of depth of cut and feed rate yielded the following results:

Cutting speed, V (m/min)	Tool life, T (min)
160	7.0
180	5.5
200	5.0
220	3.5
240	2.0

Determine the tool life equation, $VT^a = b$, where a and b are constants, using the method of least squares.

5.22. The relative roughness ($r = \frac{e}{D}$), where e is the roughness and D is the diameter, of cast iron pipes is known to vary with the diameter (D) of the pipe as follows:

i	1	2	3	4	5
D_i (inches)	1	10	20	50	100
r_i	0.01	0.001	0.0005	0.0002	0.0001

Find a relationship between r and D.

5.23. The thermal conductivity of iron (k) is found to vary with temperature (T) as follows:

T_i (°K)	200	600	1000	1400
$k_i \left(\frac{W}{\text{cm} - K}\right)$	1.0	0.4	0.3	0.25

Determine a relationship of the form $Tk^a = b$, where a and b are constants.

5.24. The heat-transfer coefficient (h) in a forced convection heat transfer in cross-flow past a cylinder at room temperature is found to vary with the velocity of the fluid (v) flowing past the cylinder as follows:

v_i (m/s)	2	4	6	8
$h_i \left(\frac{W}{m^2 - K}\right)$	6,000	10,000	13,000	15,000

Fit a linear equation between h and v.

5.25. The vertical displacement of a large electric motor mounted on isolators due to the forced vibration caused by the rotating unbalance in the rotor is shown in the following table:

Speed of motor, v_i (rpm)	100	200	300	400	500
Displacement, d_i (mm)	0.10	0.35	0.70	0.40	0.35

Develop a suitable equation between v and d.

5.26. The number of graduate students enrolled in the Mechanical Engineering department of a university over a period of 10 years is tabulated as follows:

Year, x_i	1	2	3	4	5	6	7	8	9	10
Number, y_i	204	216	208	220	240	257	271	270	285	291

Fit a linear equation to the data.

Section 5.10

5.27. The efficiency (η) of a reaction turbine is found to vary with the output power (p) and the head (h) as follows:

p_i (1000 hp)	22	30	40	32	43	55	52	62	80
h_i (ft)	240	240	240	330	330	330	430	430	430
η_i (%)	75	80	85	75	80	85	75	80	85

Develop a linear relationship between the variables p, h, and η.

5.28. The overall efficiency (η) of a gear pump with volumetric flow rate of the fluid (v) and pressure of the fluid (p) is found to vary as follows:

i	1	2	3	4	5	6	7	8
v_i (gallons / min)	42	53	58	36	46	57	49	65
p_i (psig)	3000	3000	3000	2000	2000	2000	1000	1000
η_i (%)	83	86	88	83	86	88	83	86

Fit a linear equation to the data.

Section 5.12

5.29. Use MAPLE to fit a linear, a quadratic, and a cubic spline for the data given in Problem 5.2 and compare the results corresponding with a temperature of 90°C.

5.30. Use MAPLE to solve Problem 5.26.

5.31. Use MATLAB to fit a cubic spline for the data given in Problem 5.2.

5.32. Use MATLAB to solve Problem 5.26.

5.33. Use MATHCAD to fit a cubic spline for the data given in Problem 5.2.

5.34. Use MATHCAD to solve Problem 5.26.

Section 5.13

5.35. Solve Problem 5.41 using Program 5.1, and estimate the value of heat transfer per unit area at a water pressure of 3 MPa.

5.36. Solve Problem 5.7 using Program 5.1.

5.37. Write a computer program to solve Problem 5.49.

5.38. Use Program 5.2 to solve Problem 5.56 using a third-degree polynomial.

5.39. Fit a cubic spline for the data shown in Problem 5.3 using Program 5.3.

5.40. Solve Problem 5.24 using Program 5.4.

General

5.41. The variation of heat transfer per unit area (q) during the boiling of water under pressure (p) has been found to be as follows:

$q(MW/m^2)$	1.1	2.4	3.4	3.9	4.0	3.8	3.0	1.2
p (MPa)	0	1	2	4	6	10	15	20

Develop a suitable polynomial relation between q and p.

5.42. An experiment conducted to find the variation, with temperature, of kinematic viscosity of air yielded the following results:

i	1	2	3	4	5	6	7
Temperature, T_i (°C)	1	20	40	60	80	100	120
Viscosity, μ_i (10^5 m^2/sec)	1.5	1.6	1.8	2.0	2.1	2.4	3.0

Fit a curve of the form, $\mu = \frac{a\sqrt{T}}{1+\frac{b}{T}}$, where a and b are constants, using the data.

5.43. Find polynomial approximations, $P(x)$, of orders 1, 2, 3, and 4 to the function $f(x) = e^{-2x}$ using Taylor's series expansion about $x_0 = 0$, and determine the maximum error involved in the interval $0 \leq x \leq 1$ in each case.

5.44. In the finite-element analysis of a fin subjected to thermal boundary conditions, the variation of the unknown temperature in a typical element e is assumed to be a quadratic function as

$$T^{(e)}(x) = N_i(x)\Phi_i + N_j(x)\Phi_j + N_k(x)\Phi_k,$$

where N_i, N_j, and N_k are called the shape functions and Φ_i, Φ_j, and Φ_k are the temperatures at nodes i, j, and k, respectively (treated as unknowns). The shape function, $N_j(x)$, for example, has the characteristic of being equal to one, at node j and zero at the other nodes i and k. Express the shape functions in terms of three station Lagrange interpolation polynomials.

5.45. Find the Newton's interpolating polynomial passing through the points $(x_1, y_1) = (-4, 5)$, $(x_2, y_2) = (-2, 3)$ and $(x_3, y_3) = (0, 9)$. Derive the polynomial using a divided-difference table.

5.46. Consider the data shown in Problem 5.22. Derive Newton's interpolation polynomials of orders 1, 2, 3, and 4 using the data. Use the resulting polynomials to estimate the value of relative roughness (r) for a pipe of diameter $D = 5$ in.

5.47. Consider the data given in Problem 5.56. Derive Newton's interpolation polynomials of orders 1, 2, 3, 4, 5, and 6 using the data. Use the resulting polynomials to estimate the torque at the crank angle of $30°$.

5.48. Express the cubic equation $f(x) = 9 - 7x + 5x^2 - 3x^3$ in terms of Chebyshev polynomials.

5.49. The following data corresponds to the force developed in an engine at different locations of the crank:

Location of the crank	0	1	2	3	4	5
Force developed	2	15	16	6	7	20

Fit a linear spline for this data.

5.50. For the data given in Problem 5.49, fit a quadratic spline.

5.51. For the data given in Problem 5.49, fit a natural cubic spline.

5.52. Show that Eqs. (5.104) and (5.95) lead to Eqs. (5.97) and (5.105).

5.53. A wind tunnel test conducted on an airfoil section yielded the following data between the lift coefficient (C_L) and the angle of attack (α):

α (degrees)	0	4	8	12	16	20
C_L	0.11	0.55	0.95	1.40	1.71	1.38

Develop a suitable polynomial relationship between α and C_L.

5.54. The drag coefficient (C) with Reynolds number (R) for a smooth sphere is found to vary according to the following data:

R_i	0.1	1	10	100	1000	10000
C_i	210	30	4	1	0.5	0.4

Develop a suitable relationship between R and C.

5.55. The temperature measured at various points along the thickness of a slab subjected to a sudden change in the wall temperature yielded the following results:

Distance, d_i (% wall thickness)	0	0.25	0.5	0.75	1.0
Temperature, T_i (°C)	100	70	45	25	15

Develop a suitable equation for the variation of temperature along the thickness of the wall.

5.56. The variation of crankshaft torque of a four stroke four cylinder engine with crank angle is as follows:

Crank angle, θ_i (degrees)	0	60	120	180	240	300	360
Torque, T_i (N-m)	0	−300	800	0	−300	800	0

Derive a relationship between the torque (T) and the crank angle (θ).

PROJECTS

5.1. (a) The first-order Hermite interpolation polynomial, $P(x)$, can be used to approximate a function $f(x)$ by satisfying not only the values of f, but also the first derivatives of f, at $x = x_1 = 0$ and $x = x_2 = a$. It is given by

$$P(x) = H_{01}(x)f(0) + H_{02}(x)f(a) + H_{11}(x)\frac{df}{dx}(0) + H_{12}(x)\frac{df}{dx}(a),$$

where H_{01}, H_{02}, H_{11}, and H_{12}—called the Hermite interpolation coefficients—are given by

$$H_{01}(x) = \frac{1}{a^3}(2x^3 - 3ax^2 + a^3);$$

$$H_{02}(x) = -\frac{1}{a^3}(2x^3 - 3ax^2);$$

$$H_{11}(x) = \frac{1}{a^2}(x^3 - 2ax^2 + a^2x);$$

$$H_{12}(x) = \frac{1}{a^2}(x^3 - ax^2).$$

Derive the Hermite interpolation polynomial, $P(x)$, corresponding to the following data:

x	0	2
$f(x)$	1.0	54.5981
$\frac{df}{dx}(x)$	1.0	109.1963

Use $P(x)$ to estimate the value of $f(x)$ and $\frac{df}{dx}(x)$ at $x = 1$.

(b) The second-order Hermite interpolation polynomial, $P(x)$, can be used to approximate a function $f(x)$ by satisfying not only the values of f, but also the first and second derivatives of f, at $x = x_1$ and $x = x_2$. It can be expressed as

$$P(x) = \sum_{i=1}^{2} H_{0i}(x) f(x_i) + \sum_{i=1}^{2} H_{1i}(x) \frac{df}{dx}(x_i) + \sum_{i=1}^{2} H_{2i}(x) \frac{d^2 f}{dx^2}(x_i),$$

where H_{0i}, H_{1i}, and H_{2i} are known as Hermite interpolation coefficients, each of degree five. Using the conditions

$$\frac{d^r H_{ki}}{dx^r}(x_p) = \delta_{ip}\delta_{kr};\ i, p = 1, 2;\ k, r = 0, 1, 2,$$

with $\delta_{ip} = 1$ if $i = p$ and 0 if $i \neq p$. Determine the Hermite interpolation coefficients.

5.2. Consider a two-dimensional function $f(x, y)$ whose values are known at all combinations of $x_0, x_1, \ldots, x_m$ and $y_0, y_1, \ldots, y_n$. Then the function $f(x, y)$ can be approximated by the Lagrangian interpolation polynomial, $P(x, y)$, defined as

$$P(x, y) = \sum_{i=0}^{m} \sum_{j=0}^{n} X_i(x) Y_j(y) f(x_i, y_j),$$

where $X_i(x)$ and $Y_j(y)$ are known as Lagrange interpolation coefficients given by

$$X_i(x) = \prod_{k=0,\ k \neq i}^{m} \left(\frac{x - x_k}{x_i - x_k} \right);\ i = 0, 1, \ldots, m,$$

$$Y_j(y) = \prod_{k=0,\ k \neq j}^{n} \left(\frac{y - y_k}{y_j - y_k} \right);\ j = 0, 1, \ldots, n.$$

Find the Lagrange interpolation polynomial, $P(x, y)$, corresponding to the following data:

y_j \ x_i	0	1	2
0	1	2.7183	7.3891
2	7.3891	20.0855	54.5982
4	54.5982	148.4132	403.4288

Use the result to estimate the value of f at $(x = 1.5, y = 3.0)$.

6

Statistical Methods

CHAPTER CONTENTS

6.1 Introduction

In most engineering applications, data are collected to understand the behavior of systems. Since some of the data may be inexact and the amount of data may be very small, this *raw* data cannot be used directly to describe the performance characteristics of the system. Hence the data are used to develop statistical models, which, in turn, can be used to predict the behavior of the system under any specified set of conditions. Most commonly, the data are used to identify the parameters of an assumed probability distribution. Once the distribution is defined based on the sample data, it is possible to establish the confidence limits for the estimated parameters of the distribution for use on the population of data.

6.2 Engineering Applications

▶Example 6.1

An elevator is supported by a wire rope that can withstand a mean stress of

30,000 psi with a standard deviation of 1000 psi. The load on the elevator is a random variable that causes a stress with a mean value of 25,000 psi and a standard deviation of 2000 psi. Assuming the strength of the rope and the load on the elevator to be normally distributed, determine the probability of failure of the elevator.

Solution

The probability of failure of the rope (P_f) is same as the probability of realizing that the strength of the rope (R) is less than the stress due to load on the elevator (E).

$$P_f = P(R - E < 0). \tag{E1}$$

This can be rewritten as (see Section 6.6.1)

$$P_f = P(X < 0) = P\left(\frac{X - \bar{X}}{\sigma_X} < \frac{0 - \bar{X}}{\sigma_X}\right) = P\left(Z < -\frac{\bar{X}}{\sigma_X}\right), \tag{E2}$$

where $X = R - E =$ reserve strength, $Z = \left(\frac{X - \bar{X}}{\sigma_X}\right)$ is the standard normal variate, and the mean ($\bar{X}$) and standard deviation (σ_X) of X are given by

$$\bar{X} = \bar{R} - \bar{E} = 30,000 - 25,000 = 5000$$

and

$$\sigma_X = \sqrt{\sigma_R^2 + \sigma_E^2} = \sqrt{(1000)^2 + (2000)^2} = 2236.$$

Thus, Eq. (E2) gives

$$P_f = P\left(Z < -\frac{5000}{2236}\right) = P(Z < -2.2361).$$

From standard normal tables (Appendix G), we obtain the probability of failure of the elevator as

$$P_f = 0.004207.$$ ◀

▶Example 6.2

The outer diameters of 10 piston rings are found to be 121.5, 119.4, 126.7, 117.9, 120.2, 124.3, 122.5, 120.8, 121.9, and 123.6 mm. If the diameters follow normal distribution, find the 95% confidence interval for the outer diameters of the entire population of piston rings.

Solution

The lower and upper bounds on the mean value, corresponding to the significance level α, are given by (see Section 6.7.2)

$$L = \bar{X} - t_{\frac{\alpha}{2},\, n-1} \frac{s_X}{\sqrt{n}}$$

and

$$U = \bar{X} + t_{\frac{\alpha}{2},\, n-1} \frac{s_X}{\sqrt{n}},$$

where $\bar{X}$ is the sample mean, s_X is the sample standard deviation of the diameters of piston rings (X), and $t_{\frac{\alpha}{2},\, n-1}$ is the 100 $\left(\frac{\alpha}{2}\right)$ percent point of the t-distribution with $n-1$ degrees of freedom. For the given data,

$$\bar{X} = \frac{1}{n} \sum_{i=1}^{n} = \frac{1}{10}(121.5 + 119.4 + \cdots + 123.6) = 121.88 \text{ mm}$$

and

$$s_X = \left\{ \frac{1}{n-1} \sum_{i=1}^{n} (x_i - \bar{X})^2 \right\}^{\frac{1}{2}}$$

$$= \left\{ \frac{1}{9} \left[(121.5 - 121.88)^2 + (119.4 - 121.88)^2 + \cdots + (123.6 - 121.88)^2 \right] \right\}^{\frac{1}{2}}$$

$$= 2.5507 \text{ mm.}$$

From Appendix G, we find

$$t_{\frac{\alpha}{2},\, n-1} = t_{0.0025,9} = 2.262,$$

and hence, the lower and upper bounds on the population mean diameter are given by

$$L = 121.88 - \left(\frac{2.262 \times 2.5507}{\sqrt{10}} \right) = 120.06 \text{ mm}$$

and

$$U = 121.88 + \left(\frac{2.262 \times 2.5507}{\sqrt{10}} \right) = 123.70 \text{ mm.}$$

◀

6.3 Basic Definitions

1. **Statistics.** It is a branch of mathematics that deals with the scientific method of collecting, analyzing, and displaying data and developing models for describing physical phenomena.
2. **Population.** It is the collection of all possible objects that have a common measurable or observable feature or characteristic.
3. **Sample.** It is a group of individual members drawn, at random, from the population for the purpose of experimentation.
4. **Sample size.** It is the number of individual members in the sample.

5. **Experiment.** It is the act of performing some thing, the exact outcome of which is not known beforehand. For example, tossing a coin and measuring the diameter of a machined shaft can be called experiments.

6. **Event.** It is the outcome of an experiment. The number of outcomes of an experiment can be finite or infinite. For example, in the case of tossing a coin, the outcome can be a head or a tail. On the other hand, the outcome may be any real number, say, between 1.95 in and 2.05 in the case of measuring the diameter of a shaft. Realizing a head on tossing a coin or observing a value larger than 1.99 in for the diameter of a shaft can be called events.

7. **Random variable.** It is the characteristic of the individual member that is observed or measured. The outcome of a coin toss and the diameter of a shaft can be considered as random variables.

8. **Discrete random variable.** It is a random variable that can assume only a discrete set (finite number) of values. For example, the tossing of a coin is a discrete random variable since it has only two possible outcomes (head and tail).

9. **Continuous random variable.** It is a random variable that can have an infinite number of possible values. For example, the diameter of a shaft is a continuous random variable since it can have any real value, for example, in the interval 1.95 to 2.05 inches.

10. **Probability.** It is the chance of occurrence of a well-defined event. The probability of occurrence of an event E is denoted as $P(E)$. If an experiment is repeated N times and the event E is observed to occur n times, then $P(E)$ is given by

$$P(E) = \lim_{N\to\infty} \frac{n}{N}. \tag{6.1}$$

6.4 Histogram and Probability Density Function

6.4.1 Histogram

Consider a discrete random variable X that can take any of the m values, $x_1, x_2, \ldots, x_m$. Let an experiment to find the value of X be repeated n times. Assume that the values $x_1, x_2, \ldots, x_m$ are observed $n_1, n_2, \ldots, n_m$ times, respectively, with $n_1+n_2+\cdots+n_m = n$. The graph between X (values x_i) and the probability of realizing X $\left(P(x_i) = \frac{n_i}{n}\right)$ is known as the histogram or the probability mass function (Fig. 6.1). It can be seen that the sum of the ordinates in Fig. 6.1 is one since

$$\frac{n_1}{n} + \frac{n_2}{n} + \cdots + \frac{n_m}{n} = 1.$$

6.4.2 Probability Density Function

The probability density function of a random variable, X, denoted $f_X(x)$, is defined such that $f_X(x)\Delta x$ represents the probability of occurrence of X in the interval,

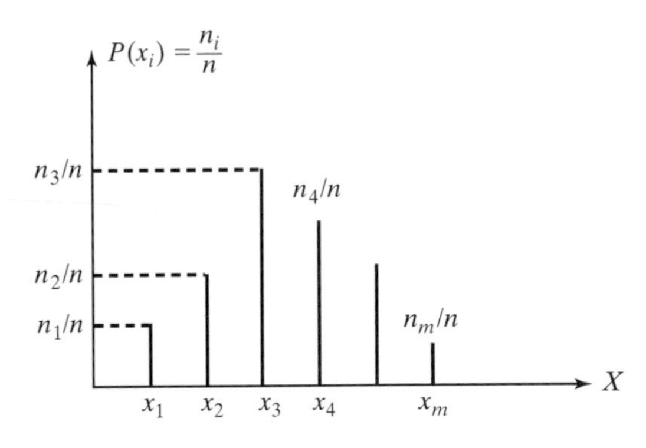

Figure 6.1 Probability mass function (histogram).

$x - \frac{\Delta x}{2}$ to $x + \frac{\Delta x}{2}$. Consider a continuous random variable, X, that can take any value in the range (x_L, x_R). Divide the range (x_L, x_R) into m parts, $\Delta x_1, \Delta x_2, \ldots, \Delta x_m$, so that $x_R - x_L = \Delta x_1 + \Delta x_2 + \cdots + \Delta x_m$. The ranges $\Delta x_1, \Delta x_2, \ldots, \Delta x_m$ are called *class intervals*. For convenience, we label the midpoints of the various intervals as $x_1, x_2, \ldots, x_m$ called *class midpoints* and usually choose whole numbers. Assume that $n_1, n_2, \ldots, n_m$ experimental values fall in the intervals $\Delta x_1, \Delta x_2, \ldots, \Delta x_m$, respectively. Usually, all the intervals are taken to be of equal length so that $\Delta x_1 = \Delta x_2 = \ldots = \Delta x_m = \Delta x$ and $x_i = x_L + (i - \frac{1}{2})\Delta x; i = 1, 2, \ldots, m$. Then, from the definition of probability density function, the probability of realizing a value in the interval Δx around x_i is given by

$$f(x_i)\Delta x = \frac{n_i}{n} = P(x_i). \tag{6.2}$$

The graph between X (values x_i) and the probability of realizing $X(P(x_i) = \frac{n_i}{n})$ is known as the histogram for the data. As $n \to \infty$ and $\Delta x \to 0$, the bar chart of Fig. 6.2(a) reduces to a continuous curve as shown by dotted lines in Fig. 6.2(a). The continuous curve is known as the probability density function. It can be seen that the area under the probability density function is equal to one.

6.4.3 Probability Distribution Function

The distribution function or the cumulative distribution function of a random variable, denoted $F_X(x)$, is defined as the probability of realizing the value of X less than or equal to x. For a discrete random variable X that can assume values $x_1, x_2, \ldots, x_m$, $F_X(x_i)$ is given by

$$F_X(x_i) = P(X \le x_i) = \sum_{j=1}^{i} f_X(x_j). \tag{6.3}$$

The distribution function corresponding to the probability mass function of Fig. 6.2(a) is shown in Fig. 6.2(b). For a continuous random variable X that can

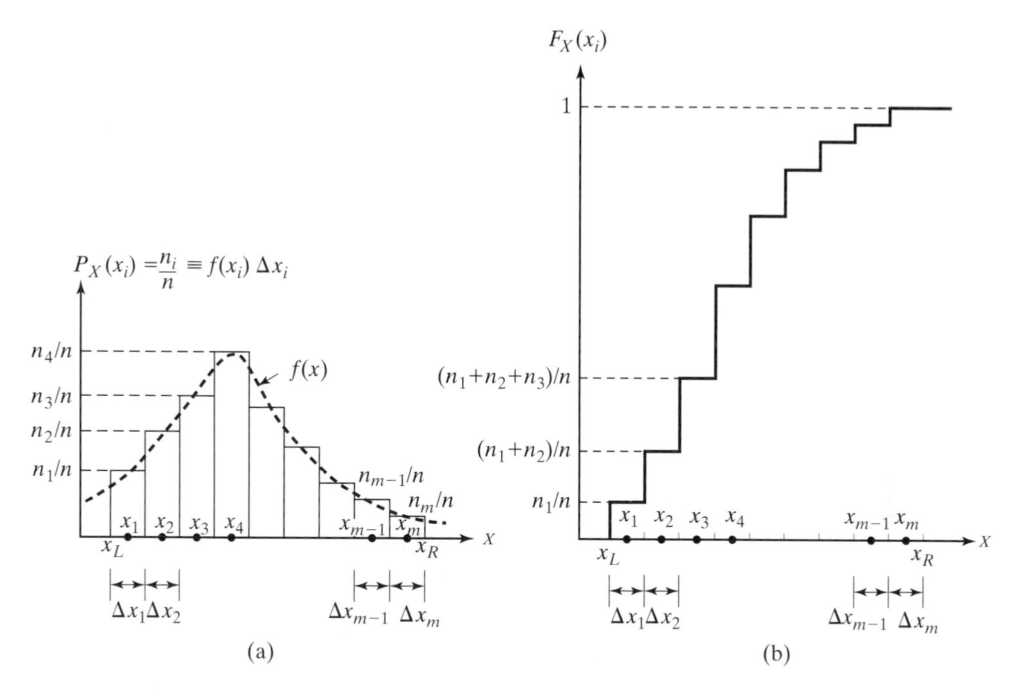

Figure 6.2 Probability mass and distribution functions.

assume any real value in the range x_L to x_R, $F_X(x)$ is given by

$$F_X(x) = P(X \leq x) = \int_{x'=x_L}^{x} f_X(x')dx', \tag{6.4}$$

which can be rewritten as

$$F_X(x) = \int_{x'=-\infty}^{x} f_X(x')dx' \tag{6.5}$$

since $f_X(x) = 0$ for $x < x_L$. A typical probability density function and the corresponding distribution function of a continuous random variable are shown in Fig. 6.3.

▶Example 6.3

The data on the yield strength of steel, from a sample of size 50, are shown in Table 6.1. Plot the histogram and the cumulative distribution of the data.

Solution

The yield strength (X) data are arranged with increasing magnitudes in Table 6.2. It can be seen that some of the values, x_i, are repeated. By inspecting the data, we

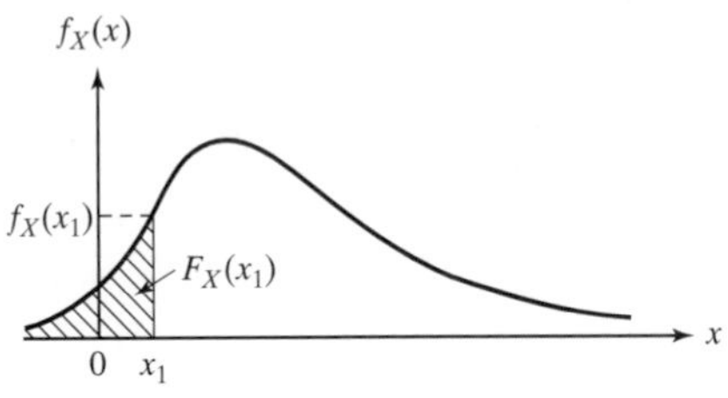

(a) Probability density function.

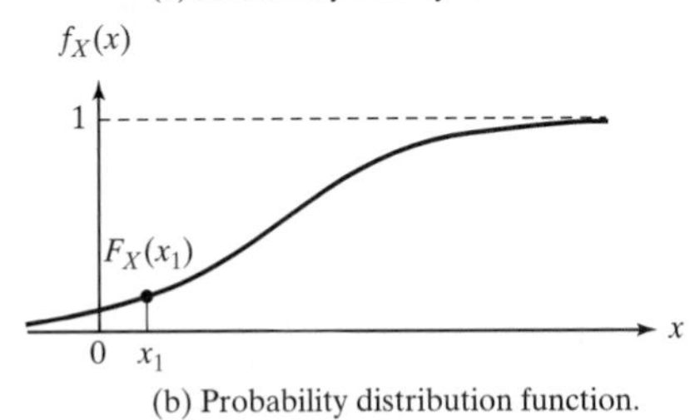

(b) Probability distribution function.

Figure 6.3 Probability density and distribution functions.

Table 6.1 Yield strength of steel for a sample of 50 (in kpsi).

Sample number	Yield strength	Sample number	Yield strength
1	29.4	26	32.1
2	30.5	27	30.1
3	30.5	28	32.2
4	28.3	29	29.3
5	33.0	30	30.1
6	28.2	31	31.3
7	31.4	32	30.4
8	29.7	33	31.9
9	29.9	34	31.2
10	30.9	35	27.6
11	29.3	36	29.5
12	29.8	37	28.4
13	30.3	38	31.3
14	28.1	39	32.3
15	30.7	40	29.9
16	32.8	41	29.7
17	29.4	42	29.2
18	31.6	43	27.8
19	30.8	44	31.7
20	29.8	45	30.6
21	28.9	46	29.1
22	31.2	47	30.2
23	29.3	48	29.4
24	28.8	49	30.3
25	31.2	50	27.2

Table 6.2 Yield strength of steel in increasing order of magnitude (in kpsi).

Serial number	Yield strength	Serial number	Yield strength
1	27.2	26	30.1
2	27.6	27	30.2
3	27.8	28	30.3
4	28.1	29	30.3
5	28.2	30	30.4
6	28.3	31	30.5
7	28.4	32	30.5
8	28.8	33	30.6
9	28.9	34	30.7
10	29.1	35	30.8
11	29.2	36	30.9
12	29.3	37	31.2
13	29.3	38	31.2
14	29.3	39	31.2
15	29.4	40	31.3
16	29.4	41	31.3
17	29.4	42	31.4
18	29.5	43	31.6
19	29.7	44	31.7
20	29.7	45	31.9
21	29.8	46	32.1
22	29.8	47	32.2
23	29.9	48	32.3
24	29.9	49	32.8
25	30.1	50	33.0

notice that the range of the data is (27.2, 33.0), and hence we divide the range into seven equal class intervals, (26.6–27.5), (27.6–28.5), (28.6–29.5), ..., and (32.6–33.5) with 27.0, 28.0, 29.0, ..., and 33.0 denoting the respective class midpoints. Thus, the members of class i are determined by rounding the values of x while placing the values of x that lie at the boundary between two classes in the lower class as

$$\left(\text{class midpoint} - \frac{\text{class interval}}{2}\right) < x_i < \left(\text{class midpoint} + \frac{\text{class interval}}{2}\right). \tag{6.6}$$

When the data of Table 6.2 are grouped according to the scheme of Eq. (6.6), the resulting frequencies and cumulative frequencies are indicated in Table 6.3. The cumulative frequency at class interval i is determined by summing the frequencies n_1 through n_i. The corresponding histogram and the cumulative frequency distribution can be represented as in Figs. 6.2(a) and (b), respectively. ◀

Table 6.3 Frequency table for the yield strength of steel.

Interval number, i	Class midpoint, x_i (kpsi)	Frequency in the class (n_i)	Cumulative frequency
1	27	1	1
2	28	6	7
3	29	11	18
4	30	14	32
5	31	10	42
6	32	6	48
7	33	2	50

6.5 Statistical Characteristics

Several characteristics can be used to describe statistical data. For example, the central tendency of the data can be described through the mean, mode, or median. The spread or variability of the data can be described using the characteristics, range, variance, or standard deviation. Similarly, the symmetry of the data can be described using the characteristic, or skewness coefficient. The various statistical characteristics are described in this section.

6.5.1 Mean

The mean value (also called the average or expected value) of a random variable, X, denotes the arithmetic mean. If the data consists of n values, $x_1, x_2, \ldots, x_n$, then the mean of X, denoted $\bar{X}$, is given by

$$\bar{X} = \frac{1}{n} \sum_{i=1}^{n} x_i. \tag{6.7}$$

If the data are separated into m classes, the mean can be computed as

$$\bar{X} = \frac{1}{n} \sum_{i=1}^{m} x_{im} n_i, \tag{6.8}$$

where x_{im} is the midpoint of class i and n_i is the number of values of X in class i.

6.5.2 Mode

It is the value of a random variable, X, that is most likely to occur. If the data are separated into m classes, the mode of X can be determined as

$$X_{\text{mod}} = x_i \text{ with largest value of } n_i \ (i = 1, 2, \ldots, m). \tag{6.9}$$

The distribution is called unimodal if it has only one peak; it is called multimodal if it has several peaks. (See Fig. 6.4.) Note that the two peaks in Fig. 6.4(b) denote relative maxima of the bimodal distribution.

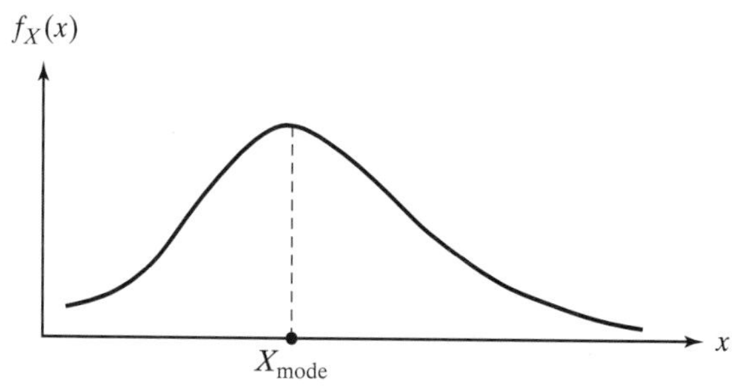

(a) Unimodal distribution.

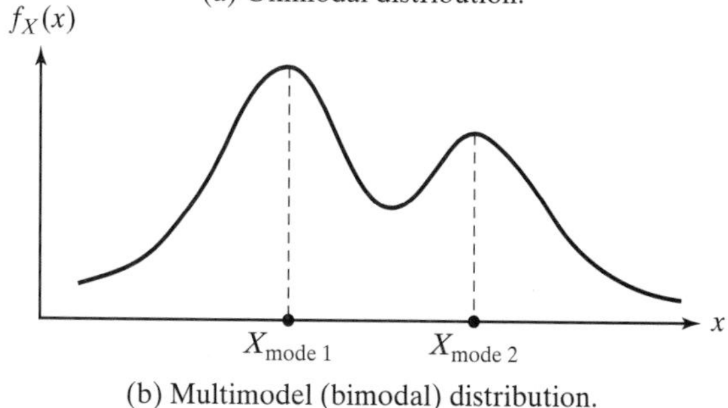

(b) Multimodel (bimodal) distribution.

Figure 6.4 Unimodal and bimodal distributions.

6.5.3 Median

The median is the value of the random variable at which values below and above it can occur with equal probability. If n sorted data points are available (n values of X in increasing order), then the median, X_{med}, can be determined as

$$X_{\text{med}} = \frac{1}{2}(x_{n/2} + x_{n/2+1}) \text{ if } n \text{ is even} \tag{6.10}$$

or

$$x_{n/2+\frac{1}{2}} \text{ if } n \text{ is odd.}$$

If the data are grouped into m classes, first we need to find the class that contains the median. If X_{med} belongs to class j, then

$$X_{\text{med}} = x_{j-1} + \left(\frac{\frac{n}{2} \sum_{i=1}^{j-1} n_i}{n_j} \right) w, \tag{6.11}$$

where x_{j-1} denotes the lower bound on class j in which the median lies, n is the total number of data points, n_i is the frequency in class j, and w is the width of the class interval ($w = x_2 - x_1 = x_3 - x_2 = \ldots$). Figure 6.5 shows the locations of mean, mode, and median in three different distributions.

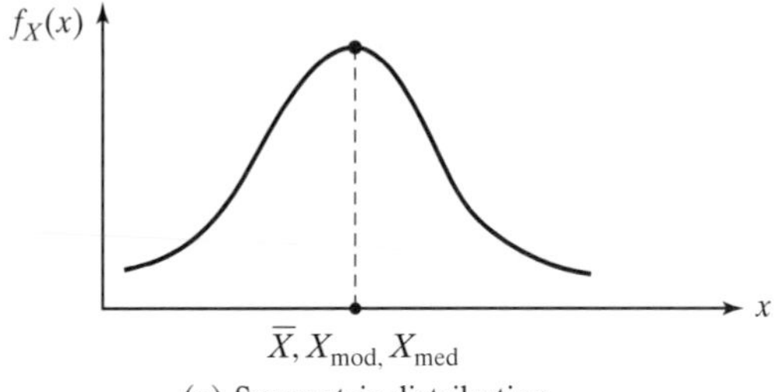

(a) Symmetric distribution.

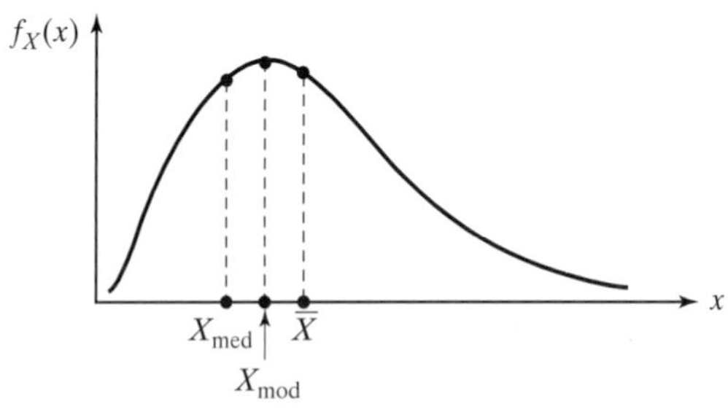

(b) Distribution with positive skewness.

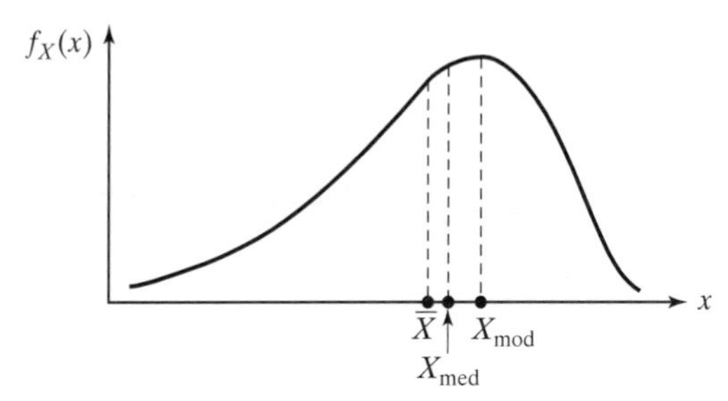

(c) Distribution with negative skewness.

Figure 6.5 Mean, mode and median of a distribution.

6.5.4 Variance and Standard Deviation

The variance of a random variable X, denoted s^2, is defined as the weighted average of the squares of the deviations of x_i from the arithmetic mean, $\bar{X}$:

$$s^2 = \frac{1}{n-1}\sum_{i=1}^{n}(x_i - \bar{X})^2 \equiv \frac{1}{n-1}\left\{\sum_{i=1}^{n} x_i^2 - \mathrm{n}\,\bar{X}^2\right\}. \tag{6.12}$$

The positive square root of the variance is called the standard deviation (s) of the random variable X. Thus,

$$s = \sqrt{\frac{1}{n-1}\sum_{i=1}^{n}(x_i - \bar{X})^2}. \tag{6.13}$$

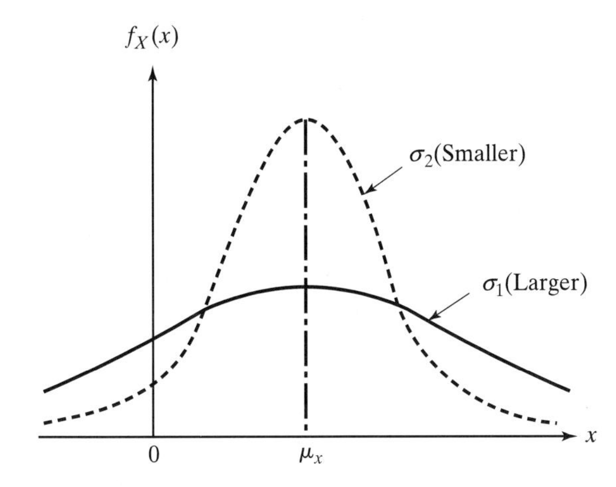

Figure 6.6 Distributions with different standard deviations.

Sometimes, the coefficient of variation (V) is used to describe the variability of X. It is defined as

$$V = \frac{s}{\bar{X}}. \tag{6.14}$$

Figure 6.6 denotes typical distributions with smaller and larger standard deviations.

The variance and standard deviation computed from Eqs. (6.12) and (6.13) are known as sample variance and sample standard deviation. When the data represents the population (and not a sample), the population variance, denoted σ^2, and the population standard deviation, σ, are defined as

$$\sigma^2 = \frac{1}{n}\sum_{i=1}^{n}(x_i - \bar{X})^2 \tag{6.15}$$

and

$$\sigma = \sqrt{\sigma^2} = \sqrt{\frac{1}{n}\sum_{i=1}^{n}(x_i - \bar{X})^2}. \tag{6.16}$$

6.5.5 Skewness

A skewed distribution is one that is not symmetric. The skewness of a random variable is defined as the expected value of the cube of its deviation from the arithmetic mean:

$$\text{Skewness} = \frac{1}{n-1}\sum_{i=1}^{n}(x_i - \bar{X})^3. \tag{6.17}$$

The skewness coefficient, γ, is defined as

$$\gamma = \frac{\text{skewness}}{s^3}, \tag{6.18}$$

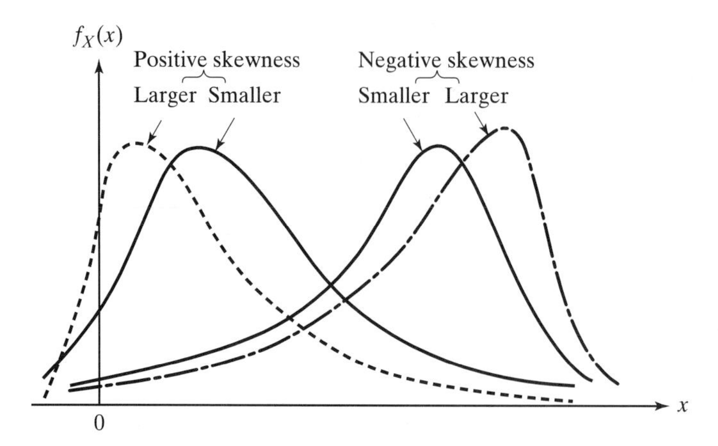

Figure 6.7 Distributions with different skewnesses.

where s is the standard deviation defined by Eq. (6.13). Figures 6.5(a), (b), and (c) denote distributions with zero, positive, and negative values of skewness. Figure 6.7 represents typical distributions with different values of skewness.

▶Example 6.4

Find the mean, mode, median, variance, standard deviation, and skewness coefficient of the yield strength of steel corresponding to the sample data of Table 6.1.

Solution

The mean value of the yield strength, computed from Eq. (6.7), is 30.1320. The median of yield strength from Eq. (6.10) is given by 30.1. The mode can be found from Table 6.3 as 30.0. The variance and standard deviation can be determined from Eqs. (6.12) and (6.13) as 1.84385 and 1.35789, respectively. The skewness coefficient given by Eq. (6.18) is 0.00982057. ◀

6.6 Normal Distribution

The normal or Gaussian distribution is one of the most important distributions in statistics. Many random phenomena encountered in engineering applications follow normal distribution. Normal distribution can be used to describe the probability characteristics of a continuous random variable that can take on any value between minus infinity and plus infinity. Although the random variables corresponding to most physical phenomena are bounded in magnitude, the normal distribution, which is not bounded, can be used as a good model to describe bounded variables. The normal probability density function is defined as

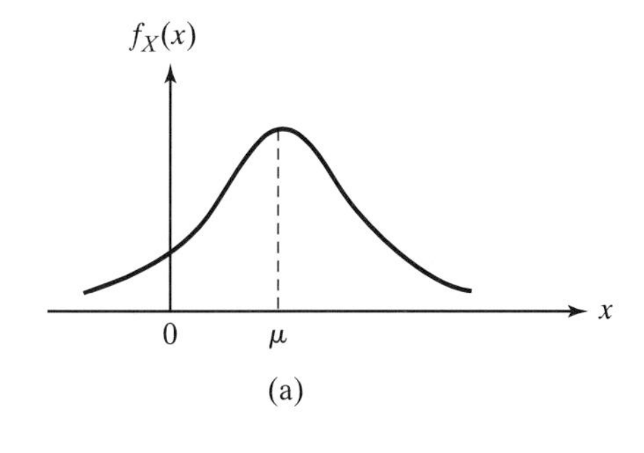

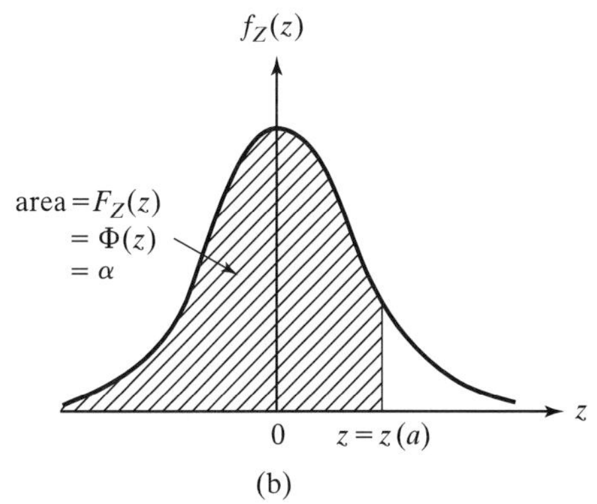

Figure 6.8 Normal distribution.

$$f_X(x) = \frac{1}{\sqrt{2\pi}} e^{-\frac{1}{2}\left(\frac{x-\mu}{\sigma}\right)^2}, \tag{6.19}$$

where μ and σ denote, respectively, the population mean and standard deviations of the random variable, X. If the values of μ and σ are not available, the sample mean ($\bar{X}$) and sample standard deviation (s) can be used in their place in Eq. (6.19). The density function given by Eq. (6.19) is a symmetric bell-shaped curve as shown in Fig. 6.8(a).

6.6.1 Standard Normal Distribution

A normally distributed random variable having $\mu = 0$ and $\sigma = 1$ is called a standard normal variate, denoted by Z, and is described by the standard normal distribution given by

$$f_Z(z) = \frac{1}{\sqrt{2\pi}} e^{-\frac{1}{2}z^2}. \tag{6.20}$$

The graph of Eq. (6.20) is shown in Fig. 6.8(b). A normal random variable with mean μ and standard deviation σ can be converted into a standard normal variable by using the transformation

$$Z = \frac{X - \mu}{\sigma}. \tag{6.21}$$

The cumulative distribution function of Z, denoted as $F_Z(z^*)$ or more commonly as $\Phi(z^*)$, is given by

$$\Phi(z^*) = F_Z(z^*) = \int_{-\infty}^{z^*} f_Z(z)dz = \frac{1}{\sqrt{2\pi}} \int_{-\infty}^{z^*} e^{-\frac{1}{2}z^2} dz, \qquad (6.22)$$

where $F_Z(z^*)$ represents the area under the probability density curve from $-\infty$ to z^*. The integral in Eq. (6.22) cannot be evaluated in closed form. Hence, an approximate method is to be used to evaluate the integral. The values of $f_Z(z)$ and $\Phi(z)$ corresponding to different values of z are usually tabulated in the form of standard normal tables in most books on statistics and probability. A typical standard normal table is given in Table G.1 of Appendix G.

The probabilities and percentiles for standard normal distribution can be determined using the following relations:

(a) The probability of realizing Z less than or equal to a is given by $\Phi(a)$.

(b) The probability of realizing Z between a and b (with $b > a$) is given by $\Phi(b) - \Phi(a)$.

(c) The probability of realizing Z greater than a is given by $1 - \Phi(a)$.

(d) The probability of realizing Z less than or equal to $-a$ is given by $\Phi(-a) = 1 - \Phi(+a)$.

(e) The value of Z corresponding to an area a defines the percentile of the standard normal distribution. The 100 a percentile, denoted by $z(a)$, is defined as [see Fig. 6.8(b)]

$$P[Z \leq z(a)] = a. \qquad (6.23)$$

(f) Due to the symmetry of normal distribution, the percentiles below 50 ($a = 0.5$) are related to those above 50 ($a = 0.5$) as

$$z(a) = -z(1 - a). \qquad (6.24)$$

6.6.2 Probabilities and Percentiles for Any Normal Distribution

The standard normal tables can be used to compute the probabilities and percentiles for any normal random variable X by transforming X into the standard normal variable Z as indicated in Eq. (6.21). The procedure is illustrated through the following examples.

▶Example 6.5

Consider the yield strength of steel (X) with mean $\mu = 30.25$ kpsi and standard deviation $\sigma = 2.00$ kpsi. Determine the following:

(a) Probability of finding the yield strength less than or equal to 30.50 kpsi,

(b) Probability of finding the yield strength greater than or equal to 30.00 kpsi,

(c) Probability of finding the yield strength between 29.00 and 31.00 kpsi, and
(d) Central 97 percentile probability limits of the yield strength of steel.

Solution

(a) We need to find the value of $P(X \leq 30.50)$. By subtracting μ and dividing the result by σ on both sides of the inequality, we obtain

$$P(X \leq 30.50) = P\left(\frac{X-\mu}{\sigma} \leq \frac{30.50-\mu}{\sigma}\right). \tag{a}$$

Noting that $\frac{X-\mu}{\sigma} = Z$, $\mu = 30.25$, and $\sigma = 2.00$, Eq. (a) can be expressed as

$$P\left(Z \leq \frac{30.50-30.25}{2.00}\right) = P(Z \leq 0.125). \tag{b}$$

From standard normal tables, the probability indicated in Eq. (b) can be found as $\Phi(0.125) = 0.5498$.

(b) We need to determine the value of $P(X \geq 30.00)$. This value can be found by using the following equivalence:

$$P(X \geq 30.00) = 1 - P(X \leq 30.00). \tag{c}$$

As before, we transform the variable X to Z and obtain

$$P(X \leq 30.00) = P\left(\frac{X-\mu}{\sigma} \leq \frac{30.00-\mu}{\sigma}\right)$$

or

$$P\left(Z \leq \frac{30.00-30.25}{2.00}\right) = P(Z \leq -0.125). \tag{d}$$

Since standard normal tables usually do not contain values of $\Phi(z)$ corresponding to negative values of z, we rewrite Eq. (d) as

$$P(Z \leq -0.125) = 1 - P(Z \leq +0.125) = 1 - \Phi(0.125). \tag{e}$$

Since the value of $\Phi(0.125) = 0.5498$, we obtain

$$P(X \leq 30.00) = P(Z \leq -0.125) = 1 - 0.5498 = 0.4502. \tag{f}$$

Thus, the probability of finding the yield strength greater than or equal to 30.00 kpsi is given by

$$P(X \geq 30.00) = 1 - P(X \leq 30.00) = 1 - 0.4502 = 0.5498. \tag{g}$$

(c) The required probability can be determined as

$$P(29.00 \leq X \leq 31.00) = P\left(\frac{29.00-\mu}{\sigma} \leq \frac{X-\mu}{\sigma} \leq \frac{31.00-\mu}{\sigma}\right). \tag{h}$$

Since $\frac{29.00-\mu}{\sigma} = \frac{29.00-30.25}{2.00} = -0.625$ and $\frac{31.00-\mu}{\sigma} = \frac{31.00-30.25}{2.00} = 0.375$, the probability indicated by Eq. (h) can be evaluated as

$$P(-0.625 \le Z \le 0.375) = \Phi(0.375) - \Phi(-0.625).$$

The values of $\Phi(0.375)$ and $\Phi(-0.625)$ can be found as $\Phi(0.375) = 0.6462$ and $\Phi(-0.625) = 1 - \Phi(0.625) = 1 - 0.7341 = 0.2659$. Thus, the required probability is given by

$$P(29.0 \le X \le 31.00) = \Phi(0.375) - \Phi(-0.625)$$
$$= 0.6462 - 0.2659 = 0.3803. \quad \text{(i)}$$

(d) The central 97th percentile probability limits of the yield strength can be determined by finding the 2nd and 98th percentiles $\left(97 = \frac{2+98}{2}\right)$ of the yield strength. From standard normal tables, we find the 2nd and 98th percentiles of standard normal distribution as $z(0.02) = -z(0.98) = -2.054$ and $z(0.98) = +2.054$. Thus, the limits are located 2.054 standard deviations on either side of the mean. Hence, the required limits of the yield strength can be found as

$$\mu + z(0.02)\,\sigma = 30.25 - 2.054(2.00) = 26.142 \text{ kpsi}. \quad \text{(j)}$$

and

$$\mu + z(0.98)\,\sigma = 30.25 + 2.054(2.00) = 34.358 \text{ kpsi}. \quad \text{(k)}$$

This also implies that $P(26.142 \le X \le 34.358) = 0.98$. ◀

6.7 Statistical Tests

When sample data are used to draw conclusions about a population parameter, we need to use procedures, known as statistical tests, to determine the validity of the conclusions drawn. For example, a manufacturer of gears may claim that a maximum of only 2% of the gears are defective. This claim can be verified by using a sampling procedure. Similarly, let the average SAT score of seniors in one high school be different from another high school. If one wants to find whether this difference is a coincidence or due to other reasons, one needs to conduct a statistical test.

In a statistical test, we start with a hypothesis, known as the null hypothesis (H_0), and either accept it or reject it depending on the outcome of the test. A probability of error, α, is specified for accepting or rejecting the null hypothesis. Usually, the value of α is specified as 0.01 (1%) or 0.05 (5%). The value, $(1-\alpha)100$, is also known as the confidence level at which the null hypothesis is accepted or rejected. For testing the null hypothesis, we use random samples generated from a specific distribution. Some of the common distributions used for generating samples are normal, t- and χ^2-distributions.

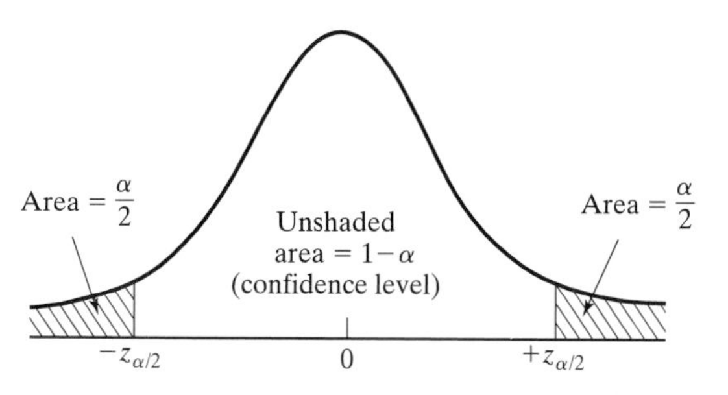

Figure 6.9 Significance of probability of error.

6.7.1 Normal Distribution

The significance of α, the probability of error, is indicated in Fig. 6.9. For example, the values of z corresponding to $\alpha = \pm 0.01$ or 1% are given by ± 2.576, and for $\alpha = 0.05$ or 5%, they are given by ± 1.96. In order to test the mean using normal distribution, we need to have either a large number of samples or a knowledge that the population is normally distributed with a known value of the standard deviation (σ). Then the lower and upper bounds on the mean can be computed as

$$\bar{X} \pm z_{\frac{\alpha}{2}} \sigma_s, \tag{6.25}$$

where $\bar{X}$ is the sample mean, $\sigma_s = \frac{\sigma}{\sqrt{n}}$ is the sample standard deviation, σ is the known standard deviation of the population, n is the number of samples, and $z_{\frac{\alpha}{2}}$ is the value indicated in Fig. 6.9. Equation (6.25) implies that

$$P\left(-z_{\frac{\alpha}{2}} \le z \le z_{\frac{\alpha}{2}}\right), \tag{6.26}$$

where

$$z = \frac{\bar{X} - \mu}{\frac{\sigma}{\sqrt{n}}}. \tag{6.27}$$

▶Example 6.6

Steel shafts, with a nominal diameter of 10 mm, are produced on a lathe. The machine is adjusted to achieve a mean diameter of 10 mm with a standard deviation of $\sigma = 0.005$ mm. During production, 20 samples are inspected and the sample mean is found to be $\bar{X} = 10.0015$ mm. Determine whether this value of $\bar{X}$ is within the bounds of $\mu = 10$ corresponding to an error probability of 0.01.

Solution

Although the number of samples is small ($n = 20$), we assume normal distribution with a known value of population standard deviation, $\sigma = 0.005$. The null hypothesis (H_0) is that $\mu = 10.0$ mm. Since the value of $z_{\frac{\alpha}{2}}$ corresponding to $\alpha = 0.01$ is ± 2.576, the bounds on the mean are given by

$$\mu \pm z_{\frac{\alpha}{2}}\, \sigma_S = \mu \pm z_{\frac{\alpha}{2}} \frac{\sigma}{\sqrt{n}}. \tag{a}$$

With $\mu = 10.0$, $z_{\frac{\alpha}{2}} = 2.576$, $\sigma = 0.005$ and $n = 20$, Eq. (a) gives the bounds on the mean as

$$10.0 \pm 2.576 \left(\frac{0.005}{\sqrt{20}} \right) = (9.99712,\ 10.00288)\ \text{mm}. \tag{b}$$

This shows that the sample mean, $\bar{X} = 10.0015$ mm, falls within the bounds for $\alpha = 0.01$. Hence, the null hypothesis ($\mu = 10.0$ mm) is accepted. ◀

6.7.2 *t*-Distribution

The normal distribution cannot be used for testing when μ and σ are not known and are to be replaced, respectively, by $\bar{X}$ and s based on a small number of samples. In such cases, the t-distribution, also known as Student t-distribution, can be used. The t-distribution takes the sample size into account and uses a term, known as the degree of freedom (f), which is defined as the difference between the sample size (n) and the number of parameters to be estimated (m):

$$f = n - m. \tag{6.28}$$

The probability density function of the t-distribution, with f degrees of freedom, is defined as

$$f_T(t) = \frac{\Gamma\left(\frac{f+1}{2}\right)}{\sqrt{\pi f}\,\Gamma\left(\frac{f}{2}\right)} \left(1 + \frac{t^2}{f}\right)^{-\frac{f+1}{2}}; \ -\infty < t < \infty, \tag{6.29}$$

where $\Gamma(\alpha)$ is the gamma function defined by

$$\Gamma(\alpha) = \int_0^\infty e^{-t} t^{\alpha-1} dt \tag{6.30}$$

The gamma function can also be considered as a generalized factorial function. $\Gamma(\alpha)$ has the following characteristics:

$$\Gamma(\alpha+1) = \alpha\Gamma(\alpha),$$
$$\Gamma(1) = 1, \tag{6.31}$$

and

$$\Gamma\left(\frac{1}{2}\right) = \sqrt{\pi}.$$

Equations (6.31) lead to the following relations:

$$\Gamma\left(\frac{x}{2}\right) = \begin{cases} \left(\frac{x}{2} - 1\right)! \text{ for } x = 4, 6, 8, \ldots \\ \left(\frac{x}{2} - 1\right)\left(\frac{x}{2} - 2\right)\cdots\left(\frac{1}{2}\right)\sqrt{\pi} \text{ for } x = 3, 5, 7, \ldots. \end{cases} \tag{6.32}$$

The density function given by Eq. (6.29) has a bell shape with symmetry about the origin. For large values of f, it approaches the standard normal distribution. The values of the cumulative distribution function corresponding to Eq. (6.29) for different values of t and f are given in Table G.2 in Appendix G.

Similar to the case of normal distribution, the bounds on the mean, corresponding to a probability of error α, are given by

$$\bar{X} \pm t_{\frac{\alpha}{2},n-1} \frac{s}{\sqrt{n}}, \tag{6.33}$$

where $t_{\frac{\alpha}{2},n-1}$ is the $100(\frac{\alpha}{2})$ percent point of the t-distribution with $n-1$ degrees of freedom. This implies that

$$P\left[-t_{\frac{\alpha}{2},n-1} \leq t \leq +t_{\frac{\alpha}{2},n-1}\right] = 1 - \alpha, \tag{6.34}$$

where

$$t = \frac{\bar{X} - \mu}{\left(\frac{s}{\sqrt{n}}\right)}. \tag{6.35}$$

▶Example 6.7

The yield strength of steel is known to follow normal distribution. Ten samples are used to find the mean as $\bar{X} = 27.23$ kpsi and the standard deviation as $s = 1.12$ kpsi. Determine the bounds on the population mean of the yield strength corresponding to an error, $\alpha = 0.01$ (or confidence level of 0.99).

Solution

Here, $\frac{\alpha}{2} = 0.005$, $f = n - m = 10 - 1 = 9$ (sample size $= n = 10$ and the number of parameters estimated $= m = 1$, the mean value), $\bar{X} = 27.23$, and $s = 1.12$. The value of $t_{\frac{\alpha}{2},\, n-1} = t_{0.005,\, 9} = 3.250$. Equation (6.33) gives the bounds on the population mean as

$$\bar{X} \pm t_{0.005,\, 9} \frac{s}{\sqrt{n}} = 27.23 \pm 3.250 \left(\frac{1.12}{\sqrt{10}}\right) = (26.078932, 28.381068) \text{ kpsi.}$$ ◀

▶Example 6.8

The yield strength of steel is known to follow normal distribution. Ten samples are used to find the mean and standard deviations as $\bar{X} = 27.23$ kpsi and $s = 1.12$ kpsi. If the nominal value of the yield strength is assumed as $\mu = 29.0$ kpsi, determine whether the deviation in the observed mean is significant.

Solution

We assume the null hypothesis (H_0) to be "The deviation in the observed mean is insignificant at 95% confidence level ($\alpha = 0.05$)." Here, $\frac{\alpha}{2} = 0.025$, $f = n - m =$

$10 - 1 = 9$, $\mu = 29.0$, and $s = 1.12$. The value of $t_{\frac{\alpha}{2},n-1} = t_{0.025,\,9} = 2.262$. If the nominal value of the yield strength is $\mu = 29.0$ kpsi, the range of yield strength corresponding to an error of $\alpha = 0.05$ is given by

$$\mu \pm t_{\frac{\alpha}{2},\,n-1}\frac{s}{\sqrt{n}} = 29.0 \pm 2.262\left(\frac{1.12}{\sqrt{10}}\right) = (28.198856, 29.801143)\text{ kpsi.} \quad \text{(a)}$$

Since the observed mean $\bar{X} = 27.23$ kpsi is outside the range given in Eq. (a), the hypothesis is to be rejected. This implies that the observed deviation of $\bar{X}$ from the nominal value is significant. ◀

6.7.3 χ^2-Distribution

If we consider samples of size n drawn from a normal population with standard deviation σ and if for each sample we compute χ^2 as

$$\chi^2 = y = \frac{\sum_{i=1}^{n}(x_i - \bar{X})^2}{\sigma^2} \equiv \frac{(n-1)s^2}{\sigma^2}, \quad (6.36)$$

then we obtain the sampling distribution of χ^2 (with $n - 1$ degrees of freedom). In Eq. (6.36), s denotes the sample standard deviation given by

$$s^2 = \frac{1}{n-1}\sum_{i=1}^{n}(x_i - \bar{X})^2. \quad (6.37)$$

The probability density function of chi-squared (χ^2) distribution, with f degrees of freedom, is given by

$$f_{\chi^2}(y) = \frac{y^{\frac{f}{2}-1}e^{-\frac{y}{2}}}{2^{\frac{f}{2}}\Gamma\left(\frac{f}{2}\right)};\; y \ge 0, \quad (6.38)$$

where $\Gamma(\alpha)$ is the gamma function defined by Eq. (6.30). The density function of Eq. (6.38) is shown in Fig. 6.10 for $f = 5$. The values of the cumulative distribution function corresponding to Eq. (6.38) for different values of y and f are given in Table G.3 of Appendix G.

Using the notation

$$P\left[Y > \chi^2_{\frac{\alpha}{2},\,n}\right] = \frac{\alpha}{2}, \quad (6.39)$$

the bounds on y, corresponding to a probability of error α [at confidence level, 100 $(1 - \alpha)$ %], can be expressed as

$$P\left[-y_{(1-\frac{\alpha}{2}),\,n} \le y \le +y_{\frac{\alpha}{2},\,n}\right] = 1 - \alpha, \quad (6.40)$$

which can be rewritten, by substituting the expression of y from Eq. (6.36), as

$$P\left[\frac{(n-1)s^2}{\chi^2_{\frac{\alpha}{2},\,n-1}} \le \sigma^2 \le \frac{(n-1)s^2}{\chi^2_{(1-\frac{\alpha}{2}),\,n-1}}\right] = 1 - \alpha. \quad (6.41)$$

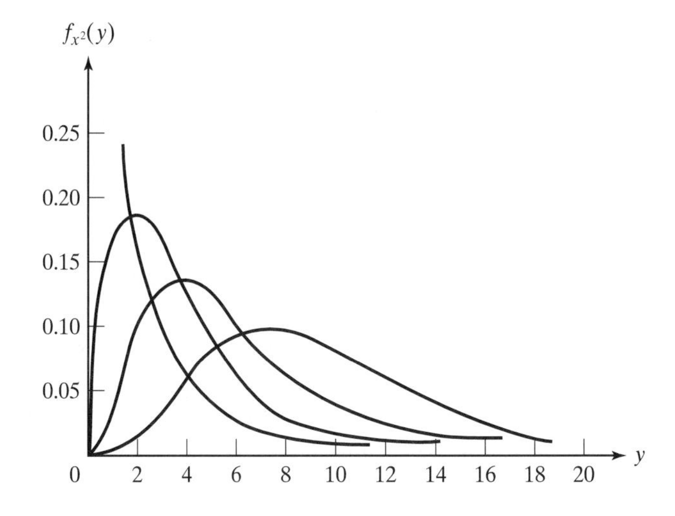

Figure 6.10 Chi-squared distribution.

▶Example 6.9

The standard deviation of yield strength of steel in a random sample of size 16 is found to be 0.24 kpsi. Determine the 95% confidence limits of the (population) standard deviation of the yield strength of steel.

Solution

For $n - 1 = 15$, 95% confidence limits correspond to $\chi^2_{0.025,15}$ and $\chi^2_{0.975,\,15}$. Table G.3 of Appendix G gives the values $\chi^2_{0.975,\,15} = 6.26214$ and $\chi^2_{0.025,\,15} = 27.4884$. Thus, $\chi_{0.975,\,15} = 2.502427$ and $\chi_{0.025,\,15} = 5.242938$. The 95% confidence limits on the population standard deviation are given by

$$\frac{\sqrt{(n-1)}s}{\chi_{0.025,\,15}} = \frac{\sqrt{15}(0.24)}{5.242938} = 0.177289 \text{ kpsi}$$

and

$$\frac{\sqrt{(n-1)}s}{\chi_{0.975,\,15}} = \frac{\sqrt{15}(0.24)}{2.502427} = 0.371446 \text{ kpsi.}$$ ◀

▶Example 6.10

In the past, the standard deviation of the diameter of forged shafts with a nominal diameter of 40.0 mm was found to be 0.25 mm. A random sample of 18 shafts indicated a standard deviation of 0.33 mm. Is the observed increase in the standard deviation significant at (a) 0.01 and (b) 0.05 level of significance?

Solution

Let the null hypothesis (H_0) be "The observed value of $s = 0.33$ mm compared to $\sigma = 0.25$ mm is due to changes in the forging process." From the distribution given in Eq. (6.36), we find the sample value of χ^2 as

$$\chi^2 = \frac{(n-1)s^2}{\sigma^2} = \frac{(18-1)(0.33)^2}{(0.25)^2} = 29.6208.$$

(a) The null hypothesis H_0 will be rejected at 0.01 level of significance if the sample value of χ^2 is larger than ${\chi^2}_{0.01}$ (using only one tail) for 17 degrees of freedom. Since the value of ${\chi^2}_{0.01,17}$ is 33.4087, we accept the null hypothesis H_0.

(b) The null hypothesis H_0 will be rejected at 0.05 level of significance if the sample value of χ^2 is larger than ${\chi^2}_{0.05}$ (using only one tail) for 17 degrees of freedom. Since the value of ${\chi^2}_{0.05,17}$ is 27.5871, we reject the null hypothesis H_0. This implies that the variability of the forging process has increased, and hence, an examination of the forging process is to be made to find the exact cause. ◀

6.8 Chi-Square Test for Distribution

Usually a theoretical distribution is assumed for a given set of data either on the basis of the general shape of the histogram or on the basis of the data plotted on a given probability paper. The validity of the assumed distribution can be verified or disproved using statistical tests, known as goodness-of-fit tests. The chi-squares and Kolmogorov–Smirnov tests are generally used for this purpose. The chi-squares test is described in this section.

Consider a sample of n observed values of a random variable. Let the range of the random variable be divided into m intervals with $n_1, n_2, \ldots, n_m$ representing the number of values or frequencies observed in the various intervals $(n_1 + n_2 + \cdots + n_m = n)$. Let the frequencies corresponding to the various intervals given by an assumed theoretical distribution be $e_1, e_2, \ldots, e_m$. Then the following quantity is computed:

$$y = \sum_{i=1}^{m} \frac{(n_i - e_i)^2}{e_i}. \tag{6.42}$$

It can be proved that the distribution of y defined by Eq. (6.42) approaches the chi-squared distribution with $f = m - 1$ degrees of freedom as $n \to \infty$. Note that, if the parameters of the theoretical distribution are unknown and are to be estimated from the data, the number of degrees of freedom is given by $f = m - k - 1$, where k denotes the number of parameters of the distribution estimated from the data.

The assumed theoretical distribution is considered acceptable, at the significance level α [or at the confidence level, $100(1 - \alpha)$], if

$$P\left[y \le \chi^2_{(1-\alpha),f}\right] \le (1-\alpha). \tag{6.43}$$

Otherwise, the assumed theoretical distribution is considered unacceptable for the given data at the significance level α.

▶Example 6.11

Sixty castings of engine blocks were inspected for internal cracks. It was found that 21, 18, 14, 5, and 2 castings have 0, 1, 2, 3, and 4 cracks, respectively. From the shape of the histogram, shown in Fig. 6.11, it was felt that Poisson distribution might be an appropriate model to describe the number of cracks in castings (X). Determine whether Poisson distribution is a suitable model at 5% significance level.

Solution

The average number of internal cracks per casting can be determined from the data as

$$\bar{X} = \lambda = \frac{21(0) + 18(1) + 14(2) + 5(3) + 2(4)}{60} = \frac{69}{60} = 1.15.$$

The Poisson distribution of a discrete random variable (X) is given by

$$P(X = x) = \frac{\lambda^x e^{-\lambda}}{x!};\ x = 0, 1, 2, \ldots,$$

where λ is the average value of X. In the present case, the Poisson distribution becomes

$$P(X = x) = \frac{1.15^x e^{-1.15}}{x!};\ x = 0, 1, 2, \ldots.$$

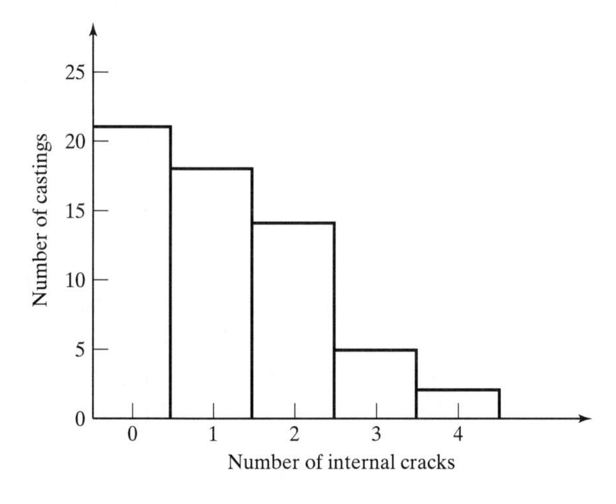

Figure 6.11 Histogram for number of cracks in castings.

Table 6.4 χ^2-test for number of cracks in castings.

Number of cracks (i)	Observed frequency (n_i)	Theoretical frequency (e_i)	$(n_i - e_i)^2$	$\frac{(n_i-e_i)^2}{e_i}$
0	21	18.998206	4.007179	0.210924
1	18	21.847953	14.806742	0.677718
2	14	12.562573	2.066196	0.164472
3	5	4.815653	0.033984	0.007057
≥ 4	2	1.775640	0.050337	0.028349
Σ	60	60	—	1.08852

Thus, we have the following:

$P(X = 0) = 0.316637$; number of castings with zero cracks $= e_0 = 0.316637(60) = 18.998206$.

$P(X = 1) = 1.15\, e^{-1.15} = 0.364132$; number of castings with one crack $= e_1 = 21.847953$.

$P(X = 2) = \frac{1}{2}1.15^2\, e^{-1.15} = 0.209376$; number of castings with two cracks $= e_2 = 12.562573$.

$P(X = 3) = \frac{1}{6}1.15^3\, e^{-1.15} = 0.080261$; number of castings with three cracks $= e_3 = 4.815653$.

$P(X \geq 4) = 1.0 - (0.316637 + 0.364132 + 0.209376 + 0.080261) = 0.029594$; number of castings with four cracks or more $= e_4 = 1.775640$.

The subsequent computational details are shown in Table 6.4. For simplicity, the number of cracks four or larger is entered in Table 6.4 since the theoretical distribution predicts the probability of realizing more than four cracks. Because the number of intervals (groups) of data is five and one parameter of the distribution (mean, λ) is estimated from the data, the number of degrees of freedom is $f = 5 - 1 - 1 = 3$. As the computed value of $\sum_i \frac{(n_i-e_i)^2}{e_i} = 1.08852$ is less than the value, $\chi^2_{0.05,3} = 7.81473$, we can assume Poisson distribution as a valid model for describing the number of cracks in castings at the 5% significance level. ◀

6.9 Choice of Method

When several values of a random parameter are known, the central tendency (value) and the variability of the parameter are characterized by the mean (average, median, or mode) and the standard deviation (variance), respectively. The statistical distribution of the parameter is described by the histogram or the discrete probability distribution function corresponding to the data. Most parameters used in engineering practice can be described by normal distribution. The mean and standard deviation of the parameter, computed from available data, can be used to define the normal distribution function of the parameter. However, if the data are collected only from a small number of samples, normal distribution cannot be used accurately. In such cases, t-distribution can be used. The t-distribution accounts for the sample size and makes use of a term known as the degrees of freedom. Usually a theoretical distribution such as normal, exponential, or Weibull distribution is assumed to describe a random variable either on the basis of the general shape of the histogram or on the basis of the data plotted on a specific probability paper. The validity of the assumed distribution can be verified or disproved using goodness-of-fit tests. The chi-square test is commonly used for this purpose.

6.10 Use of Software Packages

6.10.1 MATLAB

▶Example 6.12

Find the mean, median, standard deviation, and histogram corresponding to the following data:

29.4, 30.5, 30.5, 28.3, 33.0, 28.2, 31.4, 29.7, 29.9, 30.9, 29.3, 29.8, 30.3, 28.1, 30.7, 32.8, 29.4, 31.6, 30.8, 29.8, 28.9, 31.2, 29.3, 28.8, 31.2, 32.1, 30.1, 32.2, 29.3, 30.1, 31.3, 30.4, 31.9, 31.2, 27.6, 29.5, 28.4, 31.3, 32.3, 29.9, 29.7, 29.2, 29.2, 27.8, 31.7, 30.6, 29.1, 30.2, 29.4, 30.3, 27.2.

Solution

```
>>  q=[29.4 30.5 30.5 28.3 33.0 28.2 31.4 29.7 29.9 30.9
       29.3 29.8 30.3 28.1 30.7 32.8 29.4 31.6 30.8 29.8
       31.2 31.2 29.3 28.8 31.2 32.1 30.1 32.2 29.3 30.1
       31.3 30.4 31.9 31.2 27.6 29.5 28.4 31.3 32.3 29.9
       29.7 29.2 27.8 31.7 30.6 29.1 30.2 29.4 30.3 27.2];

>> mean (q)

ans =

   30.1320
>> median (q)

ans =

   30.1000
```

```
>> std (q)

ans =

    1.3579

>> hist (q)
>> title ('Histogram of q')
```

Note:

`hist(q)`

This function plots a 10-bin (or 10-interval) histogram for the data in the vector q. The bins are equally spaced between the minimum and maximum values in q.

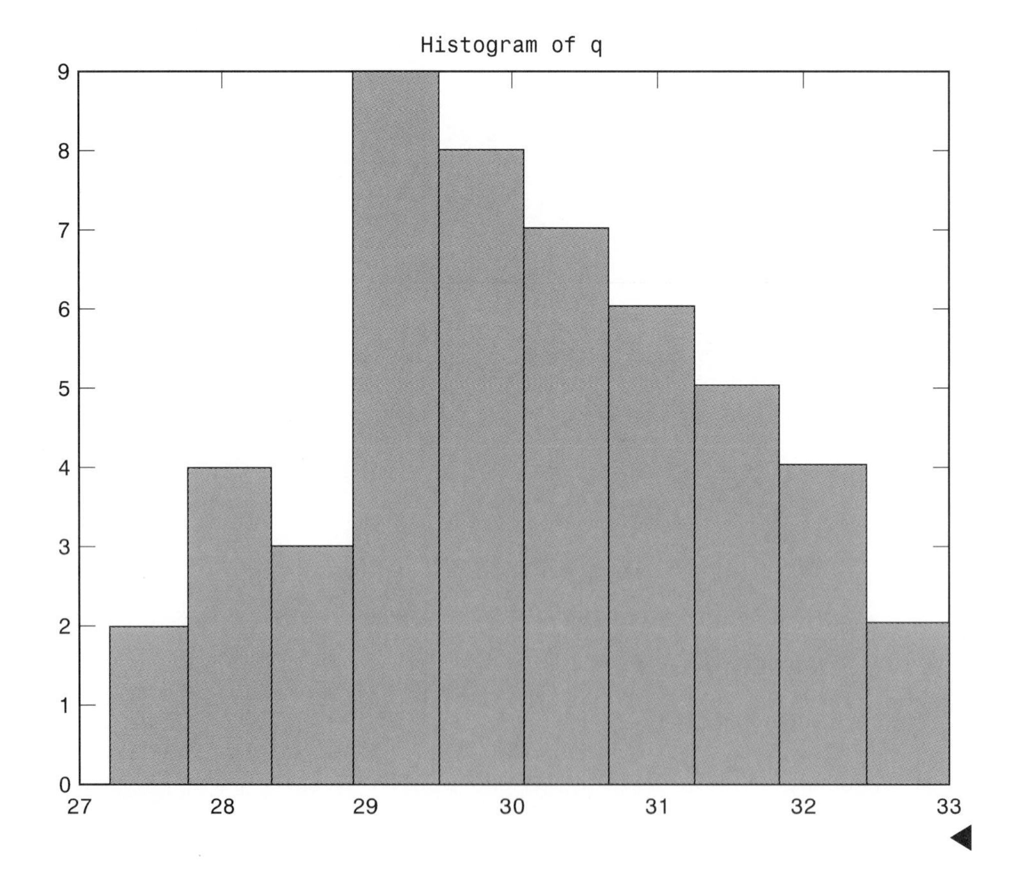

▶Example 6.13

Determine the value of the normal distribution function

$$\Phi(z) = \int_{-\infty}^{z} \frac{1}{\sqrt{2\pi}} e^{-0.5\,y^2}\, dy$$

for a specified value of z using a suitable numerical integration procedure.

Solution

The function $\frac{1}{\sqrt{2\pi}} e^{-0.5\,y^2}$ is defined through an m-file labeled *normp.m.*

The following is a listing of *normp.m*:

```
function pdf=normp(x)

pdf=1./sqrt(2.0*pi).*exp(-0.5*x.^2);

>> q=quad('normp', -7,1.8)

q =

   0.96407008801677

>> q=quad ('normp', -7,2.3)

q =

   0.98927521657472

>> q=quad8('normp', -7,1.8)

q =

   0.96407021262010

>> q=quad8('normp', -7,2.3)

q =

   0.98927588990392
>> q=cdf('norm', 1.8,0,1) - cdf ('norm', -7,0,1)

q =

   0.96406968088579
```

```
>> q=cdf('norm', 2.3,0,1) - cdf('norm', -7,0,1)
q =
   0.98927588997704
```

◀

Note:

q = quad ('function',a,b)

and

quad8 ('function',a,b)

can be used for the numerical integration of 'function' between the limits a and b. q denotes the value of the integral evaluated. *quad* implements a low order method (Simpson's rule) and *quad8* implements a higher order method (Newton-Cotes 8 panel rule).

The function "cdf" for finding the cumulative distribution is available only with "Statistics Toolbox."

6.10.2 Maple

▶Example 6.14

Find the mean, median, standard deviation, variance, and histogram corresponding to the data given in Example 6.12 using Maple.

Solution

```
>
> with (stats):
> q: = [29.4, 30.5, 30.5, 28.3, 33.0, 28.2, 31.4, 29.7, 29.9, 30.9,
> 29.3, 29.8, 30.3, 28.1, 30.7, 32.8, 29.4, 31.6, 30.8, 29.8, 28.9,
> 31.2, 29.3, 28.8, 31.2, 32.1, 30.1, 32.2, 29.3, 30.1, 31.3, 30.4,
> 31.9, 31.2, 27.6, 29.5, 28.4, 31.3, 32.3, 29.9, 29.7, 29.2, 27.8,
> 31.7, 30.6, 29.1, 30.2, 29.4, 30.3, 27.2];
```

$q := [29.4, 30.5, 30.5, 28.3, 33.0, 28.2, 31.4, 29.7, 29.9, 30.9,$
$29.3, 29.8, 30.3, 28.1, 30.7, 32.8, 29.4, 31.6, 30.8, 29.8,$
$28.9, 31.2, 29.3, 28.8, 31.2, 32.1, 30.1, 32.2, 29.3, 30.1,$
$31.3, 30.4, 31.9, 31.2, 27.6, 29.5, 28.4, 31.3, 32.3, 29.9,$
$29.7, 29.2, 27.8, 31.7, 30.6, 29.1, 30.2, 29.4, 30.3, 27.2]$

```
> describe [mean] (q);
```

30.13200000

```
> describe [median] (q);
```

30.1

```
> describe [standarddeviation] (q);
```

1.344238074

```
> describe [variance] (q);
```

1.806976000

```
> statplots[histogram] (q);
```

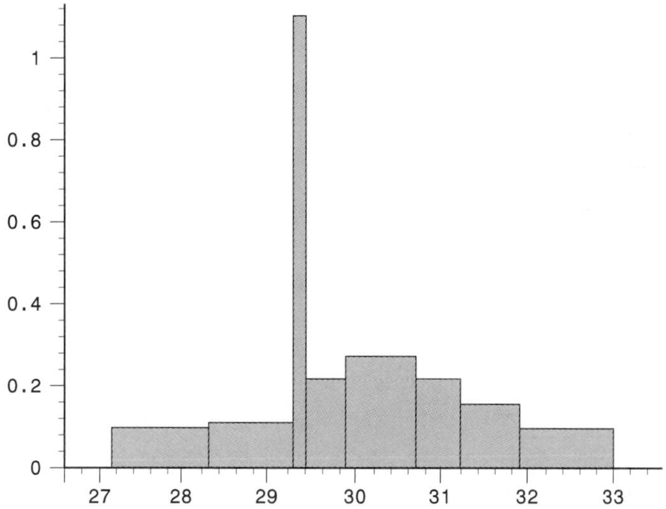

```
>
```

▶Example 6.15

Find the value of the cumulative normal distribution function, $\Phi(z)$, corresponding to a specified value of z.

Solution

```
> q1: = statevalf[cdf, normald] (1.8) -statevalf [cdf, normald] (-7);
```

$$q1 := .9640696809$$

```
> q2: = statevalf[cdf, normald] (2.3) -statevalf [cdf, normald] (-7);
```

$$q2 := .9892758900$$

```
> q3: = 1-statevalf[cdf, normald] (-8);
```

$$q3 := 1$$

```
> q4: = statevalf[icdf, normald] (0.9);
```

$$q4 := 1.281551566$$

```
>
```

◀

6.10.3 Mathcad

▶Example 6.16

Compute the mean, median, standard deviation, variance, and histogram corresponding to the data given in Example 6.12.

Solution

```
q:=(29.4 30.5 30.5 28.3 33.0 28.2 31.4 29.7 29.9 30.9 29.3 29.8 30.3 28.1 30.7 32.
Q:=q^T

Mean(q) = 30.132    Median(q) = 30.1    Stdev(q) = 1.344    Var(q) = 1.807

i:= 0.49

j:=0.10

int_j := 27.2 + 0.58·j

f := hist(int,Q)

n := 0.9
```

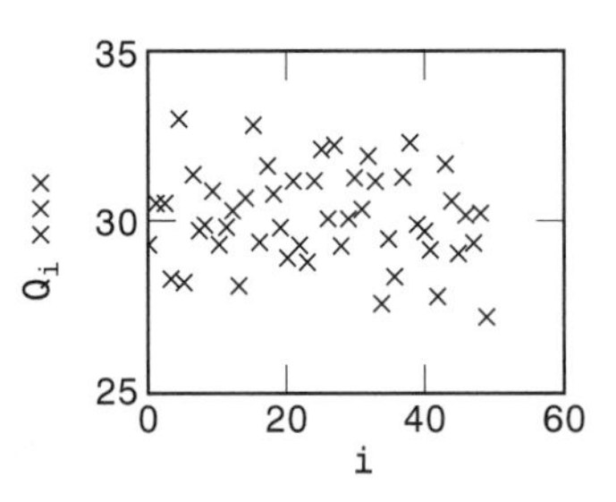

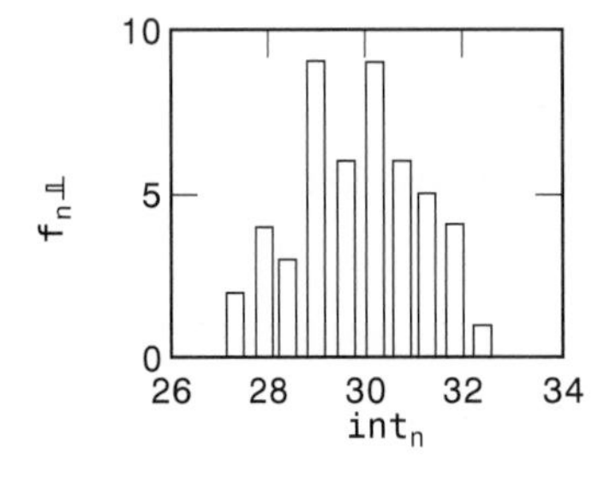


```
Min(q)=27.2

Max(q)=33
```

◀

▶Example 6.17

Find the value of the cumulative normal distribution function, $\Phi(z)$, corresponding to a specified value of z using MATHCAD.

Solution

dnorm(−8, 0, 1) = 5.052×10^{-15}

dnorm(−7, 0, 1) = 9.135×10^{-12}

dnorm(1.8, 0, 1) = 0.079 dnorm(2.3, 0, 1) = 0.028

cnorm(−8) = 0 pnorm(−8, 0, 1) = 0

cnorm(1.8) − cnorm(−7) = 0.964

pnorm(1.8, 0, 1) − pnorm(−7, 0, 1) = 0.964

pnorm(2.3, 0, 1) − pnorm(−7, 0, 1) = 0.989

qnorm(0.9, 0, 1) = 1.282

◀

Note:

The functions

dnorm(x, μ, σ), pnorm(x, μ, σ), and qnorm(x, μ, σ) give, respectively, the values of the probability density function, cumulative distribution function and the inverse cumulative distribution function corresponding to x for a normally distributed variable with mean μ and standard deviation σ.

The function

cnorm(x) ≡ pnorm(x, 0, 1)

gives the value of the cumulative distribution function corresponding to x of a standard normal variate.

6.11 Computer Programs

6.11.1 Fortran Programs

Program 6.1 Histogram

The following input data are required:

N = size of vector X (data)
X(1), ..., X(N) = sample values of X (data)

INT = number of intervals to be considered in the range of data
XMIN, XMAX = minimum and maximum values for the data $X(1), \ldots, X(N)$

Illustration (Example 6.3)

The following is a program output:

```
HISTOGRAM GENERATION

DATA:

5.90  6.20  5.80  7.80  6.50  6.30  8.90  5.30  3.70  1.40
2.10  6.80  9.10  4.30  3.20  7.20  6.10  5.70  4.90  2.60
3.40  6.80  8.30  5.10  7.30  8.20  7.70  5.40  3.70  4.50
4.10  5.60  6.40  6.70  7.90  6.90  7.50  5.20  4.30  6.60
5.40  6.40  7.20  8.10  3.90  4.60  7.20  8.00  3.30  6.90

DATA IN ASCENDING ORDER:

1.40  2.10  2.60  3.20  3.30  3.40  3.70  3.70  3.90  4.10
4.30  4.30  4.50  4.60  4.90  5.10  5.20  5.30  5.40  5.40
5.60  5.70  5.80  5.90  6.10  6.20  6.30  6.40  6.40  6.50
6.60  6.70  6.80  6.80  6.90  6.90  7.20  7.20  7.20  7.30
7.50  7.70  7.80  7.90  8.00  8.10  8.20  8.30  8.90  9.10

POINTS          P(X)          F(X)        HISTOGRAM

      2        0.0400        0.0400         1.8278   **
      1        0.0201        0.0600         2.6833   *
      6        0.1200        0.1800         3.5389   ******
      5        0.1001        0.2801         4.3945   *****
      7        0.1400        0.4201         5.2500   *******
      9        0.1801        0.6001         6.1056   *********
     10        0.2000        0.8001         6.9611   **********
      7        0.1400        0.9401         7.8167   *******
      3        0.0600        1.0000         8.6722   ***
```

Program 6.2 t-distribution

The required input data are

NA = number of values of $\alpha/2$ to be considered
NDOF = number of degrees of freedom to be considered starting from 1
ALF(NA) = values of $\alpha/2$
TINF(NA) = values of t-distribution for ∞ number of degrees of freedom

Illustration: Generation of a sample t-distribution table. Using NA = 5, NDOF = 25, ALF(I) = 0.1, 0.05, 0.025, 0.01, 0.005, and TINF(I) = 1.28, 1.84, 1.96, 2.33, 2.58, the output of Program 6.2 is as follows:

```
STUDENTS T-DISTRIBUTION

 DOF                  VALUE OF ALPHA

          0.1000   0.0500    0.0250    0.0100    0.0050

   1      3.0827   6.3188   12.7111   31.8265   63.6561
   2      1.8906   2.9250    4.3077    6.9696    9.9299
   3      1.6450   2.3550    3.1850    4.5449    5.8449
   4      1.5350   2.1350    2.7850    3.7549    4.6049
   5      1.4850   2.0250    2.5750    3.3749    4.0349
   6      1.4450   1.9450    2.4550    3.1449    3.7149
   7      1.2750   1.6550    1.9750    2.3049    2.4849
   8      1.2850   1.6550    1.9750    2.3149    2.4949
   9      1.1250   1.4050    1.6150    1.7749    1.8549
  10      1.1350   1.4150    1.6150    1.7849    1.8549
  11      1.0150   1.2350    1.3850    1.4949    1.5349
  12      1.0150   1.2450    1.3850    1.4949    1.5349
  13      0.9250   1.1150    1.2350    1.3149    1.3449
  14      0.9350   1.1250    1.2350    1.3149    1.3449
  15      0.8650   1.0250    1.1150    1.1849    1.2049
  16      0.8650   1.0250    1.1250    1.1849    1.2149
  17      0.8050   0.9550    1.0350    1.0849    1.1049
  18      0.8150   0.9550    1.0350    1.0949    1.1149
  19      0.7650   0.8950    0.9650    1.0149    1.0249
  20      0.7650   0.8950    0.9650    1.0149    1.0349
  21      0.7250   0.8450    0.9050    0.9549    0.9649
  22      0.7250   0.8450    0.9150    0.9549    0.9749
  23      0.6850   0.7950    0.8650    0.8949    0.9149
  24      0.6950   0.8050    0.8650    0.9049    0.9149
  25      0.6550   0.7650    0.8250    0.8549    0.8649

 INF      1.2800   1.6400    1.9600    2.3300    2.5800
```

6.11.2 C Programs

Program 6.3 Statistics

The input data required are as follows:

N = size of sample data (dimension of data vector X)
X(1), ..., X(N) = sample values of X (data)

Illustration (Example 6.4)

The following is the program output:

```
Computation of statistics for given data

Given data:

5.90  6.20   5.80   7.80   6.50   6.30   8.90   5.30   3.70   1.40
2.10  6.80   9.10   4.30   3.20   7.20   6.10   5.70   4.90   2.60
3.40  6.80   8.30   5.10   7.30   8.20   7.70   5.40   3.70   4.50
4.10  5.60   6.40   6.70   7.90   6.90   7.50   5.20   4.30   6.60
5.40  6.40   7.20   8.10   3.90   4.60   7.20   8.00   3.30   6.90

Given data in ascending order:

1.40  2.10   2.60   3.20   3.30   3.40   3.70   3.70   3.90   4.10
4.30  4.30   4.50   4.60   4.90   5.10   5.20   5.30   5.40   5.40
5.60  5.70   5.80   5.90   6.10   6.20   6.30   6.40   6.40   6.50
6.60  6.70   6.80   6.80   6.90   6.90   7.20   7.20   7.20   7.30
7.50  7.70   7.80   7.90   8.00   8.10   8.20   8.30   8.90   9.10

xmean = 0.584800E+01
var   = 0.327438E+01
stdev = 0.180953E+01
skew  = -0.398621E+00
```

Program 6.4 Normal Distribution

The required input data are

NUM = 1 if $F(Z)$ is to be found with known value of Z; = 2 if Z is to be found with known value of $F(Z)$
ZCON = value of Z if NUM is set equal to 1; = value of $F(Z)$ if NUM is set equal to 2

Illustration (Example 6.5). Using NUM = 2 and ZCON = $F(Z)$ = 0.9990, the output of Program 6.4 is as follows:

```
Computations with normal distribution

num = 1 (num = 1 if z is given; = 2 if F(z) is given)
value of z = 2.5000
value of F(z) = 0.9938
```

```
Computations with normal distribution

num = 2 (num = 1 if z is given; = 2 if F(z) is given)
value of F(z) = 0.9000
value of z = 1.2817
```

REFERENCES AND BIBLIOGRAPHY

6.1. S. S. Rao, *Reliability-Based Design*, McGraw-Hill, New York, 1992.

6.2. M. R. Spiegel, *Theory and Problems of Statistics*, Schaum Publishing Co., New York, 1961.

6.3. T. J. Akai, *Applied Numerical Methods for Engineers*, Wiley, New York, 1994.

6.4. K. S. Krishnamoorthi, *Reliability Methods for Engineers*, ASQC Quality Press, Milwaukee, WI, 1992.

6.5. A. H. S. Ang and W. H. Tang, *Probability Concepts in Engineering Planning and Design, Vol. 1—Basic Principles*, Wiley, New York, 1975.

REVIEW QUESTIONS

The following questions along with corresponding answers are available in an interactive format at the Companion Web site at http://www.prenhall.com/rao.

6.1. Define the following terms:

Statistics, Population, Sample, Probability, Random variable, Experiment, Event, Histogram, Coefficient of variation, Degrees of freedom, Percentile, and Standard normal distribution.

6.2. Give **brief answers** to the following:

1. What is the difference between a discrete and a continuous random variable?
2. What is the difference between a histogram and a probability density function?
3. What is the difference between population and sample?
4. What is the difference between a probability density function and a probability (cumulative) distribution function?
5. Under what condition(s) do(es) the mean, mode, and median of a distribution coincide?
6. What is the difference between a unimodal and a multimodal distribution?
7. What is the relationship between variance and standard deviation?
8. What is the difference between normal distribution and standard normal distribution?
9. What are the two measures of the spread of a random variable?
10. What is a statistical test?
11. What is a null hypothesis?
12. What is the purpose of a chi-square test?
13. What are five engineering examples involving uncertainty?
14. What is the relationship between an experiment and an event?
15. What are the three measures of central tendency of a random variable?

6.3. Answer **true or false** for the following:

1. The number of outcomes of an experiment are always finite.
2. The number realized on rolling a die is a discrete random variable.
3. A normally distributed random variable can take negative values.
4. Normal distribution is a multimodal distribution.

(*continued, page 489*)

5. A continuous random variable is one that can take an infinite number of values.
6. Most physical phenomena can be described using normal distribution.
7. A larger value of standard deviation implies a narrower probability density function.
8. The mean of a random variable is always smaller than its mode.
9. A random variable with zero standard deviation is a real number.
10. The probability density function of a random variable can have a value greater than one.

6.4. **Match the following:**

Term		Meaning	
1.	Mean	(a)	Number in the middle
2.	Mode	(b)	Average value
3.	Median	(c)	Peak value
4.	Standard deviation	(d)	Measure of asymmetry
5.	Variance	(e)	Measure of variability
6.	Skewness coefficient	(f)	Mean square deviation

6.5. **Fill in the blanks** with suitable words:

1. An event is a possible ______ of an experiment.
2. A standard normal variable has ______ mean and ______ standard deviation.
3. The skewness coefficient must be equal to zero for the probability function to be ______ .
4. A random variable with zero standard deviation is a ______ ______ .
5. The area under a probability density function is equal to ______ .
6. The t-distribution is also called ______ t-distribution.
7. The normal distribution is also known as ______ distribution.
8. The chi-square test is used for testing the validity of a ______ .
9. The gamma function can be considered a generalized ______ function.
10. A statistical test can be used to validate or reject a ______ hypothesis.

6.6. Select the **most appropriate answer** out of the multiple choices given:

1. The standard deviation of a sample of size n is given by

(a) $\frac{1}{n}\sum_{i=1}^{n}(x_i - \bar{X})^2$

(*continued, page 490*)

(b) $\frac{1}{n-1}\sum_{i=1}^{n}(x_i - \bar{X})^2$

(c) $\sqrt{\frac{1}{n-1}\sum_{i=1}^{n}(x_i - \bar{X})^2}$.

2. If X is a random variable with mean μ and standard deviation σ, the corresponding standard normal variable (Z) is defined as
   (a) $\frac{X-\mu}{\sigma}$ (b) $\frac{X}{\mu}$ (c) $\frac{X-\sigma}{\mu}$.
3. The skewness coefficient of a random variable with zero standard deviation is
   (a) zero (b) positive (c) negative.
4. The value of $\Gamma\left(\frac{1}{2}\right)$ is
   (a) $\frac{1}{2}$ (b) $\sqrt{\pi}$ (c) 1.
5. If $\Gamma(\alpha+1) = \alpha\Gamma(\alpha)$ with $\Gamma(1) = 1$ and $\Gamma(\frac{1}{2}) = \sqrt{\pi}$, the value of $\Gamma(3.5)$ is
   (a) $\frac{15}{4}\sqrt{\pi}$ (b) $\frac{7}{2}\sqrt{\pi}$ (c) $\frac{15}{8}\sqrt{\pi}$.
6. If 95th percentile of a standard normal variate (Z) is denoted as y, it implies that
   (a) $P(Z \leq y) = 0.05$ (b) $P(Z \leq y) = 0.95$ (c) $P(Z \geq y) = 0.95$.
7. If $\Phi(z)$ denotes the cumulative distribution function of a standard normal variate, then:
   (a) $\Phi(-z) = \Phi(+z)$ (b) $\Phi(-z) = 0.5 + \Phi(+z)$ (c) $\Phi(-z) = 1 - \Phi(+z)$
8. If 20 sample values are used to estimate two parameters of an assumed distribution, the number of degrees of freedom in a χ^2-test is
   (a) 17 (b) 18 (c) 19.
9. An inspection of 200 welded joints revealed that 10 joints were defective. The probability of finding a randomly chosen welded joint to be good is
   (a) 0.05 (b) 0.95 (c) 0.10.
10. If the yield strength of a particular type of steel has a mean value of 30 kpsi and a standard deviation of 1.5 kpsi, then its coefficient of variation is given by
    (a) 1.5% (b) 20.0% (c) 5.0%.

PROBLEMS

Section 6.2

6.1. The number of automobiles arriving at a toll booth in any one minute (X) is described by the following distribution [$p(x)$]:

x	$p(x)$
0	0.05
5	0.12
10	0.21
15	0.26
20	0.19
25	0.14
30	0.03

(a) Plot the distribution of X.

(b) Find the mean and standard deviation of X.

6.2. The mean and standard deviation of shear strength of a bolted joint are 30 kpsi and 3.9 kpsi, respectively. The joint is loaded such that the stress induced has a mean value of 27 kpsi and a standard deviation of 4.2 kpsi. If both the shear strength and the induced stress are independent and normally distributed, find the probability of failure of the bolted joint.

Section 6.4

6.3. The maximum load carried by 30 nominally identical welded beams (in kiloNewtons) before failure is given here:

176.2	157.2	131.2	143.3	154.1
164.6	124.6	151.1	158.7	129.4
168.1	140.8	123.6	164.3	138.2
171.0	151.5	141.2	152.9	127.9
186.7	129.7	151.7	164.6	137.5
147.8	156.2	137.2	147.2	132.3

Construct the histogram and the corresponding cumulative distribution function.

6.4. The eccentricity of the applied load (in inches) in a sample of 40 nominally identical columns is as follows:

0.411	0.051	0.091	0.196	0.346
0.156	0.321	0.121	0.291	0.064
0.276	0.232	0.142	0.266	0.217
0.071	0.156	0.304	0.434	0.132
0.536	0.111	0.206	0.085	0.135
0.126	0.186	0.480	0.177	0.146
0.381	0.166	0.256	0.182	0.239
0.221	0.106	0.209	0.203	0.210

Plot the histogram and the corresponding cumulative distribution function.

Section 6.5

6.5. Find the mean, median, mode, variance, standard deviation, and skewness coefficient of the load carried by the welded beams corresponding to the data given in Problem 6.3.

6.6. Find the following for the data given in Problem 6.4:

Mean, mode, median, standard deviation, coefficient of variation, and skewness coefficient.

Section 6.6

6.7. The life of an automobile battery is known to follow normal distribution with a mean and standard deviation of 2500 and 220 days, respectively. Determine the following: (a) The percentage of batteries that fail before 2000 days. (b) The warranty period to be specified by the manufacturer so that the manufacturer will encounter only 10% failures of the battery.

6.8. The width of a slot in a machine is specified as 0.25 ± 0.02 inch. The widths of the slot measured on a random sample of machines gave a mean of 0.26 in and a standard deviation of 0.01 in. Assuming normal distribution for the width of the slot, determine the following:

(a) the proportion of the slots that are outside the specs and

(b) the proportion of the defective slots if the process mean is shifted to coincide with the center of the specs.

6.9. A fair coin is tossed six times. (a) Find the probability density and distribution functions of the number of heads realized. (b) Find the probability of realizing heads at least four times out of the six trials.

Section 6.7

6.10. A machine for production of bolts was adjusted so that the mean nominal diameter is 40 mm and the standard deviation is 0.03 mm. During production, 25 samples are inspected and the sample mean diameter is found to be $\bar{X} = 40.015$ mm. Determine whether this value of $\bar{X}$ is within the bounds of $\mu = 40$ mm corresponding to an error probability of 0.05.

6.11. A random sample of five resistors showed the following resistance values: 12.8, 13.5, 12.7, 13.3, and 13.1 ohms. Assuming that resistances of resistors follow normal distribution with a (population) standard deviation of 0.3, determine a 99% confidence interval for the average resistance for the population of resistors.

6.12. A random sample of five resistors yielded the following resistance values: 12.8, 13.5, 12.7, 13.3, and 13.1 ohms. Assuming that resistances of resistors follow normal distribution, determine a 99% confidence interval for the standard deviation of resistance for the population of resistors.

6.13. A manufacturer of light bulbs claims that their bulbs have an average life not less than 1000 hours with standard deviation of 50 hours. A sample of 50 bulbs were found to have an average life of 980 hours. Is the null hypothesis, $H_0 : \mu = 1,000$ hours, valid at $\alpha = 0.01$?.

6.14. The measurement of the internal diameters of a random sample of 200 piston rings produced in one week showed a mean of 8.24 in and a standard deviation of 0.42 in. Find (a) 99% and (b) 95% confidence limits for the mean internal diameter of all the piston rings.

6.15. In the past, the standard deviation of the diameter of ball bearings, having a nominal diameter of 0.5 in, was found to be 0.025 in. A random sample of 25 ball bearings showed a standard deviation of 0.032 in. Determine whether the observed increase in the standard deviation is significant at (a) 0.01 and (b) 0.05 level of significance.

Section 6.8

6.16. Twenty-five observations of the number of vehicles arriving at an intersection (per minute) gave the following values:

1, 4, 2, 0, 3, 1, 2, 3, 1, 1, 0, 4, 3, 1, 1, 2, 3, 1, 0, 2, 4, 2, 0, 0, 1.

Determine whether the arrival rate of vehicles can be described by the Poisson distribution at the 1% significance level using chi-square test.

6.17. Sixty identical electronic components are tested until failure and the following data are obtained:

Life to failure (Hours)	Number of components failed
0 –199	19
200 –399	13
400 –599	11
600 –799	10
800 –999	7

Determine whether the failure life follows exponential distribution with constant failure rate at 10% significance.

6.18. One hundred units of a mechanical component are tested for life and the following data are obtained:

Life (Hours)	Number of units failed
0 –999	22
1000 –1999	25
2000 –2999	18
3000 –3999	12
4000 –4999	23

Determine whether the life of the component follows exponential distribution with constant failure rate at 10% significance.

Section 6.10

6.19. Find the standard deviation, median, mean, and histogram corresponding to the data of Problem 6.4 using MAPLE.

6.20. Use MAPLE to find the values of the cumulative normal distribution, $\Phi(z)$, corresponding to $z = 1.0, 2.0, 3.0, 4.0$, and 5.0.

6.21. Solve Problem 6.19 using MATLAB.

6.22. Solve Problem 6.20 using MATLAB.

6.23. Solve Problem 6.19 using MATHCAD.

6.24. Solve Problem 6.20 using MATHCAD.

6.25. Use MAPLE to generate the values of t-distribution corresponding to degrees of freedom 1 to 25.

6.26. Use MATHCAD to solve Problem 6.25.

6.27. Use MAPLE to generate the values of x that correspond to a probability of 0.8 in a chi-square distribution for degrees of freedom ranging from 1 to 10.

Section 6.11

6.28. Plot the histogram corresponding to the data given in Problem 6.4 using Program 6.1.

6.29. Generate the values of a t-distribution for $\alpha/2 = 0.25, 0.1, 0.0025, 0.001$, and 0.0005 corresponding to degrees of freedom 1 through 25 using Program 6.2.

6.30. Find the statistics corresponding to the data of Problem 6.4 using Program 6.3.

6.31. Generate the values of $f(z)$ for $z = 0.5, 1.5, 2.5$, and 3.5 using Program 6.4.

General

6.32. The point reached by the end effector of a robot manipulator along a line is given by 15 ± 0.25 in. What is the probability of the end effector reaching a point beyond 15.5 in?

6.33. The number of automobiles passing a toll booth during 9 am–10 am on 50 different days were found as follows:

259	275	320	279	255
316	311	364	309	232
326	269	321	323	252
391	314	267	251	275
316	271	238	239	306
302	258	274	282	259
341	259	250	283	269
310	247	251	318	323
381	275	338	288	252
308	285	271	331	280

Plot the histogram and the cumulative distribution function.

6.34. Determine the mean, mode, median, variance, standard deviation, coefficient of variation, and skewness coefficient of the data given in Problem 6.33.

6.35. The error in the altitude of a helicopter predicted by an altimeter can be approximated as a normal variate with mean μ and standard deviation 150 ft. Determine the following probabilities for the error:

(a) less than or equal to -250 ft,

(b) greater than or equal to 350 ft, or

(c) between −200 ft and 150 ft.

6.36. Consider the diameters of ball bearings that have a mean value of 4.0 in and a standard deviation of 0.1 in. Assuming that the diameters of ball bearings follow normal distribution, determine the following:

(a) The probability of finding the diameter less than or equal to 4.05 in,

(b) The probability of finding the diameter between 3.9 in and 4.2 in, and

(c) The central 98th percentile probability limits of the diameter.

6.37. The average life of an electric motor is stated to be 10,000 hours. The standard deviation is $\sigma = 500$ hours. A sample of 30 motors showed an average life of 9600 hours. Can the null hypothesis, $H_0 = 10{,}000$ hours, be accepted with an error probability of 0.01?

6.38. A random sample of five resistors gave the following resistance values: 12.8, 13.5, 12.7, 13.3, and 13.1 ohms. Assuming that resistances of resistors follow normal distribution, determine a 99% confidence interval for the average resistance for the population of resistors.

6.39. The nth moment of a continuous random variable about a, denoted as $E[(X-a)^n]$, is given by

$$E[(X-a)^n] = \int_{-\infty}^{\infty} (x-a)^n f_X(x)dx,$$

where $f_X(x)$ is the probability density function of X. Using this relation, determine the first and second moments of a random variable X with

$$f_X(x) = 100e^{-100x};\ x \geq 0$$

for (a) $a = 0$ and (b) $a = 100$.

6.40. The bursting strengths (pressures) of a sample of nominally identical metal tubes are found to be as follows:

Bursting strength (kpsi)	Observed frequency
< 7.00	8
7.00–7.25	16
7.25–7.50	23
7.50–7.75	30
7.75–8.00	29
8.00–8.25	22
8.25–8.50	11
8.50–8.75	5
> 8.75	1

The sample mean and sample standard deviation are observed to be 7.76 and 0.78 kpsi, respectively. Determine whether the bursting strengths can be described by normal distribution at the 1% significance level using a chi-square test.

PROJECTS

6.1. The magnitude of wind load, in kpsi, measured on a multistory building at different times are

$$1250, 360, 1190, -730, 110, -1340, -660, 280, \text{ and } -460.$$

If the building is modeled as a uniform bar subjected to an axial (wind) load as shown in Fig. 6.12, determine the cross section of the bar (A) using a Chebyshev inequality.

The Chebyshev inequality states that the probability of the absolute value of the difference between a random variable (x) and its mean (μ) exceeding a constant t times the standard deviation (σ) is not greater than t^{-2}, that is,

$$P[|x - \mu| \geq t\sigma] \leq \frac{1}{t^2}.$$

Assume the permissible stress of the bar (s) to be deterministic with $s =$ 2000 kpsi.

6.2. If the value of a function $f = f(x_1, x_2, \ldots, x_n)$ is to be determined from the measured values of $x_1, x_2, \ldots, x_n$, the confidence interval for f is defined by

$$P\left[-z_{\frac{\alpha}{2}} < \frac{f - \bar{f}}{s_f} < z_{\frac{\alpha}{2}}\right] = 1 - \alpha,$$

where

$$\bar{f} = f(\bar{x}_1, \bar{x}_2, \ldots, \bar{x}_n) = \text{mean value of } f$$

and

$$s_f = \left\{ \sum_{i=1}^{n} \left(\left. \frac{\partial f}{\partial x_i} \right|_{(\bar{x}_1, \bar{x}_2, \ldots, \bar{x}_n)} \right)^2 s_{x_i}^2 \right\}^{\frac{1}{2}}$$

$$= \text{standard deviation of } f$$

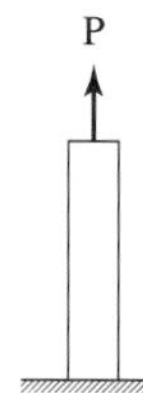

Figure 6.12 Building modeled as a bar subjected to axial load P.

and $(\bar{x}_i, s_{x_i})$ are computed from the measured values of $x_i, i = 1, 2, \ldots, n$. Equation (E1) gives the $(1-\alpha)$ confidence interval of f as

$$\left(\bar{f} - z_{\frac{\alpha}{2}} s_f, \bar{f} + z_{\frac{\alpha}{2}} s_f\right).$$

The dimensions of a casting, which is in the form of a hollow circular prism, are measured independently eight times and the following results are obtained:

Inner diameter (d_i) in inches: 6.5, 6.4, 6.6, 6.8, 6.7, 6.2, 6.3, 6.5.

Outer diameter (d_o) in inches: 11.9, 12.3, 12.1, 12.4, 11.8, 11.7, 12.5, 12.0

Height (h) in inches: 20.1, 19.7, 20.2, 20.3, 20.0, 19.9, 19.8, 20.2

Determine the following:

(a) the mean and standard deviations of x_i, $i = 1, 2, 3$;

(b) the mean and standard deviation of the volume of the casting;

(c) the 99% confidence interval on the true volume of the casting; and

(d) the 95% confidence interval on the true volume of the casting.

6.3. The partial derivative rule [6.1] can be used to compute the approximate mean and standard deviation of a function of several random variables. If $f = f(x_1, x_2, \ldots, x_n)$ denotes a function in terms of the independent random variables $x_1, x_2, \ldots, x_n$, then Taylor's series expansion of f about the mean values of the random variables, $\bar{x}_1, \bar{x}_2, \ldots, \bar{x}_n$, gives

$$f(x_1, x_2, \ldots, x_n) = f(\bar{x}_1, \bar{x}_2, \ldots, \bar{x}_n) + \sum_{i=1}^{n} \left.\frac{\partial f}{\partial x_i}\right|_{\bar{x}_1, \bar{x}_2, \ldots, \bar{x}_n} (x_i - \bar{x}_i)$$

$$+ \frac{1}{2} \sum_{i=1}^{n} \sum_{j=1}^{n} \left.\frac{\partial^2 f}{\partial x_i \partial x_j}\right|_{\bar{x}_1, \bar{x}_2, \ldots, \bar{x}_n} (x_i - \bar{x}_i)(x_j - \bar{x}_j) + \cdots. \quad \text{(E1)}$$

(a) By neglecting terms involving the second derivatives of f, show that the mean and variance of f are given by

$$\bar{f} = f(\bar{x}_1, \bar{x}_2, \ldots, \bar{x}_n) \quad \text{(E2)}$$

and

$$\sigma_f^2 = \sum_{i=1}^{n} \left(\left.\frac{\partial f}{\partial x_i}\right|_{\bar{x}_1, \bar{x}_2, \ldots, \bar{x}_n} \right)^2 \sigma_{x_i}^2, \quad \text{(E3)}$$

where $\bar{x}_i$ and $\sigma_{x_i}^2$ denote the mean and variance of x_i, respectively.

(b) Using Eqs. (E2) and (E3), determine the mean and standard deviation of the length of a belt (l) required to connect the two pulleys of a belt drive (shown in Fig. 6.13). The length of the belt is given by

$$l = \sqrt{4c^2 - (d_2 - d_1)^2} + \frac{1}{2}(d_2\,\theta_2 + d_1\,\theta_1), \quad \text{(E4)}$$

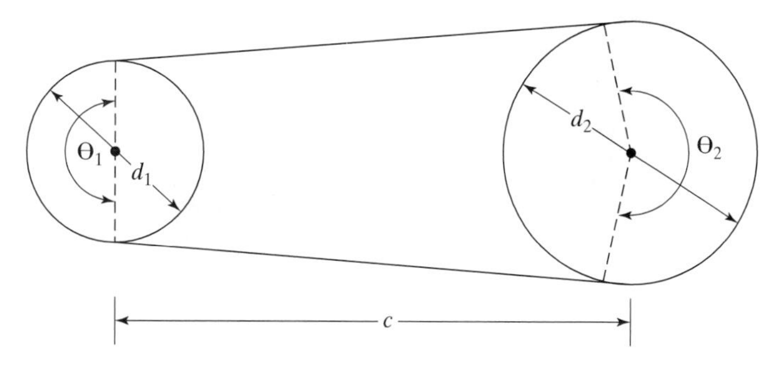

Figure 6.13 A belt drive.

where

$$\theta_1 = \pi - 2\sin^{-1}\left(\frac{d_2 - d_1}{2\,c}\right) \tag{E5}$$

and

$$\theta_2 = \pi + 2\sin^{-1}\left(\frac{d_2 - d_1}{2\,c}\right), \tag{E6}$$

where d_1, d_2 = diameters of smaller, larger pulley, respectively, and c = center distance bet- ween the pulleys.

Data: $\bar{d}_1 = 20$ cm, $\sigma_{d_1} = 0.2$ cm, $\bar{d}_2 = 40$ cm, $\sigma_{d_2} = 0.5$ cm, $\bar{c} = 200$ cm, and $\sigma_c = 1.0$ cm.

7

Numerical Differentiation

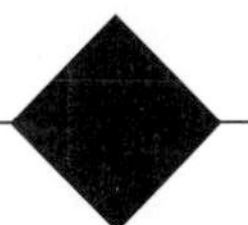

7.1 Introduction

CHAPTER CONTENTS

Numerical differentiation is the procedure by which the derivatives of a function are computed when the function is given either as an analytical expression or as a series of numbers at discrete points in the region of interest. The mathematical models of most engineering and science problems require numerical differentiation. For example, in mechanics of rigid bodies, the velocity of a body is given by the time derivative of its displacement and the acceleration by the second derivative of the displacement. In robotics, the force required to move an object along a specified (displacement) trajectory is given by the product of its mass and acceleration. If only a set of discrete points is specified along the path of the object (instead of trajectory being specified as a continuous curve), we need to compute the acceleration of the object using numerical differentiation. In an ac electrical circuit, the computation of the voltage across an inductance coil requires the time derivative of the current flowing through the coil. The computation of heat flow along a rod requires the derivative of temperature with respect to the length coordinate. In the case of fluid flow in a channel, the evaluation of shear stress involves the derivative of velocity with respect to the spatial coordinate along the depth of the channel.

The differential equations in many engineering applications require numerical solutions due to the presence of complex geometry, varying material properties, or complicated boundary conditions. Sometimes nonlinear differential equations or a large number of simultaneous differential equations may have to be solved. Often, such equations can only be solved using numerical methods.

In all these situations, the derivatives are replaced by their discrete forms, known as finite-difference approximations. The use of finite-differences transforms an ordinary differential equation into an algebraic equation, which will be relatively simpler to solve. The numerical analysis that deals with the discretization of derivatives is known as *finite-difference calculus*. The calculus of finite differences enables the numerical integration of differential equations. Also, if the values of the function at a discrete set of points (such as experimental data) are given, these may be differentiated or integrated using the calculus of finite differences. However, it is to be noted that numerical differentiation is inherently less accurate than numerical integration. The calculus of finite differences is also useful in the derivation of interpolation or extrapolation polynomials. (See Section 5.6.)

7.2 Engineering Applications

▶Example 7.1

The radial temperature distribution, $T(r)$, in a cylinder, initially at temperature T_i, suddenly immersed in a liquid of temperature T_0 is given by [7.1]

$$\theta = \frac{T - T_0}{T_i - T_0} = \sum_{i=1}^{\infty} c_i e^{-\left(\frac{\beta_i^2 \alpha t}{r_0^2}\right)} J_0\left(\frac{\beta_i r}{r_0}\right), \tag{a}$$

where

$$c_i = \frac{2}{\beta_i} \frac{J_1(\beta_i)}{J_0^2(\beta_i) + J_1^2(\beta_i)}, \tag{b}$$

β_i is the root of the equation

$$\beta_i \frac{J_1(\beta_i)}{J_0(\beta_i)} = \frac{h_0 r_0}{k}, \tag{c}$$

α = thermal diffusivity, k = thermal conductivity, h_0 = heat transfer coefficient, r_0 = radius of the cylinder, and J_0 and J_1 = Bessel functions of the first kind, defined by

$$J_0(x) = 1 - \frac{x^2}{2^2(1!)^2} + \frac{x^4}{2^4(2!)^2} - \frac{x^6}{2^6(3!)^2} + \cdots \tag{d}$$

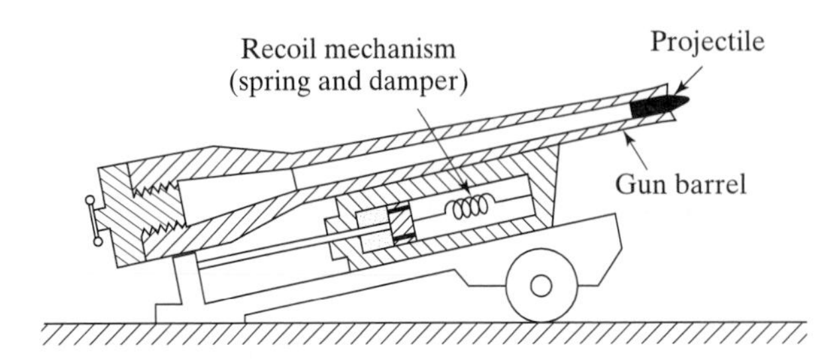

Figure 7.1 Recoil mechanism of a cannon.

and

$$J_1(x) = \frac{x}{2} - \frac{x^3}{2^3(1!\,2!)} + \frac{x^5}{2^5(2!\,3!)} - \frac{x^7}{2^7(3!\,4!)} + \cdots. \tag{e}$$

Determine the rate of heat transfer in the radial direction, $k\frac{\partial T}{\partial r}$, for a brass cylinder of radius 10 cm with $k = 110$ W/m-K and $\alpha = 3.4 \times 10^{-5}$ m^2/s. Assume $h_0 = 575$ W/m^2-K, $t_i = 293.15$ K, and $T_0 = 373.15$ K.

▶Example 7.2

The barrel and the recoil mechanism of a cannon (Fig. 7.1) have a mass of 500 kg with a recoil spring stiffness of 10,000 N/m. The response of the cannon, with a critically damped recoil mechanism, is given by [7.2]

$$x(t) = (c_1 + c_2 t)e^{-\omega_n t},$$

where $c_1 = x_0$, $c_2 = \dot{x}_0 + \omega_n x_0$, x_0 is the initial displacement, $\dot{x}_0$ is the initial velocity, and $\omega_n = \sqrt{\frac{k}{m}}$ is the undamped natural frequency of the system. Find the maximum recoil velocity of the cannon using numerical differentiation with $x_0 = 0$ and $\dot{x}_0 = 5$ m/s.

▶Example 7.3

The load carrying capacity of a tilt thrust bearing (P), shown in Fig. 7.2, is given by [7.3]

$$P = \frac{6\mu(U_1 - U_2)lb^2}{h_2^2}\left\{\frac{1}{(m-1)^2}\left[\ln m - \frac{2(m-1)}{m+1}\right]\right\},$$

where μ = viscosity of the lubricant, U_1 = surface velocity of the supporting plate, U_2 = surface velocity of the pad, l = length of the bearing, b = length of lubricant flow path, h_2 = film thickness at exit, h_1 = film thickness at entrance, and $m = \frac{h_1}{h_2}$ = tilt ratio. The change in the load due to a small change in the tilt ratio, given by $\frac{dP}{dm}$, will be of interest from a design point of view. Find the value of $\frac{dP}{dm}$ for the following data: $\mu = 1.95 \times 10^{-6}$ reyn, $P = 5000$ lb, $b = 3$ in, $l = 1.5$ in, $U_1 - U_2 = 150$ in/sec, $h_1 = 0.1$ in, and $h_2 = 0.05$ in.

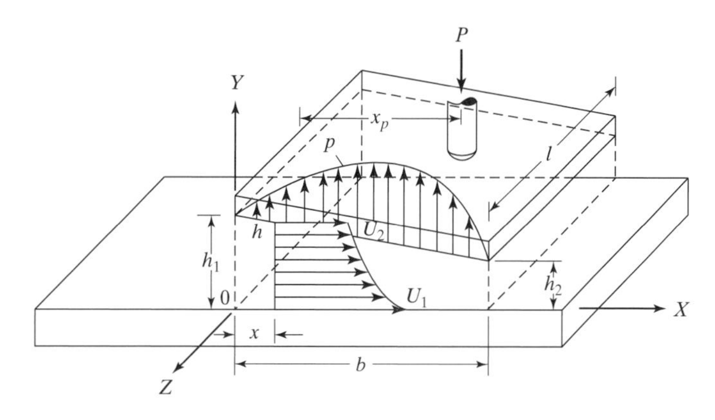

Figure 7.2 Tilt thrust bearing.

7.3 Definition of the Derivative

The derivative of a function $f(x)$ is defined as

$$\left.\frac{df(x)}{dx}\right|_{x_0} = f'(x_0) = \lim_{x \to x_0} \frac{f(x) - f(x_0)}{x - x_0}. \tag{7.1}$$

The value of $x - x_0$ does not approach zero but remains a finite quantity in calculus of finite differences. Thus, the derivative of Eq. (7.1) can be approximated as

$$f'(x_0) \approx \frac{f(x) - f(x_0)}{\Delta x}, \tag{7.2}$$

where

$$\Delta x = x - x_0. \tag{7.3}$$

Although Eq. (7.2) is an approximate expression, according to the mean-value theorem, there is a point, ξ, in the interval of interest (a, b) at which Eq. (7.2) gives the exact value of the derivative. The mean-value theorem states that if $f(x)$ is continuous in the interval $[a, b]$ and differentiable in $a < x < b$, then a point $\xi, a < \xi < b$ exists such that

$$f'(\xi) = \frac{f(b) - f(a)}{b - a}. \tag{7.4}$$

This theorem forms a basis for the development of not only differential calculus, but also finite difference calculus.

7.4 Basic Finite-Difference Approximations

Consider the function $f(x)$ shown in Fig. 7.3. The values of the function at the discrete points x_{i-1}, x_i, and x_{i+1} can be used to find an approximate expression for the derivative of f using the concept of Eq. (7.2). The derivative of f at x_i, $\left.\frac{df}{dx}\right|_{x_i} = \left.\frac{df}{dx}\right|_i$ can be approximated as $\frac{\Delta f}{\Delta x}$, where Δx denotes the discrete change in f over the discrete interval Δx. The discrete changes in f and x can be found in three different ways, each giving a different approximation for $\left.\frac{df}{dx}\right|_i$:

$$\left.\frac{df}{dx}\right|_i \approx \frac{\Delta f}{\Delta x} = \frac{f_{i+1} - f_i}{x_{i+1} - x_i} \tag{7.5}$$

(two point forward-difference approximation; Fig. 7.3(a)),

$$\left.\frac{df}{dx}\right|_i \approx \frac{\Delta f}{\Delta x} = \frac{f_i - f_{i-1}}{x_i - x_{i-1}} \tag{7.6}$$

(two point backward-difference approximation; Fig. 7.3(b)),

$$\left.\frac{df}{dx}\right|_i \approx \frac{\Delta f}{\Delta x} = \frac{f_{i+1} - f_{i-1}}{x_{i+1} - x_{i-1}} \tag{7.7}$$

(two point central-difference approximation; Fig. 7.3(c)),

where

$$f_i = f(x_i),\ f_{i-1} = f(x_{i-1}),\ \text{and}\ f_{i+1} = f(x_{i+1}).$$

The finite-difference calculus, including Eqs. (7.5) through (7.7), can be developed using three different approaches—Taylor's series expansions of the function, the difference operators, and the differentiation of the interpolating polynomials. These approaches are described briefly in the following sections.

7.5 Using Taylor's Series Expansions

Taylor's series expansion of a function $f(x)$ about a point $x = x_i$ is given by

$$f(x) = f(x_i) + (x - x_i)f'(x_i) + \frac{(x - x_i)^2}{2!}f''(x_i) + \frac{(x - x_i)^3}{3!}f'''(x_i) + \ldots + \frac{(x - x_i)^n}{n!}f^{(n)}(x_i) + \cdots, \tag{7.8}$$

where the expansion is valid for values of x sufficiently close to $x = x_i$, a prime denotes differentiation with respect to x, and all the derivatives are evaluated at x_i, for example,

$$f'(x_i) = \frac{df}{dx}(x = x_i).$$

Denoting $\Delta x = x - x_i$, we can write Eq. (7.8) as

$$f(x_i + \Delta x) = f(x_i) + \Delta x f'(x_i) + \frac{(\Delta x)^2}{2!}f''(x_i) + \frac{(\Delta x)^3}{3!}f'''(x_i) + \ldots + \frac{(\Delta x)^n}{n!}f^{(n)}(x_i) + \cdots. \tag{7.9}$$

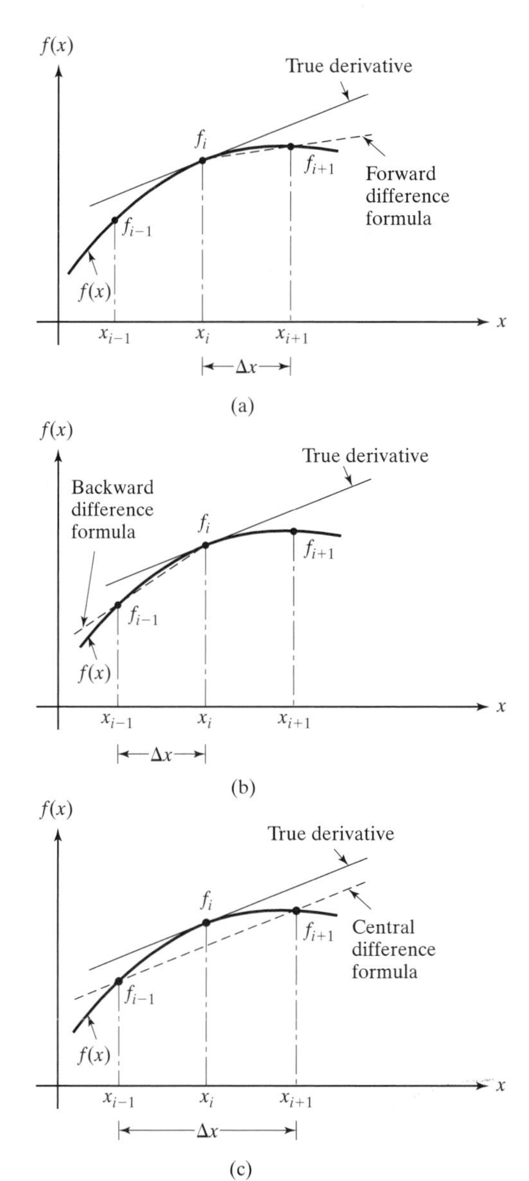

Figure 7.3 Graphical interpretation of finite difference approximations.

By using $-\Delta x = x - x_i$, we can express Eq. (7.8) as

$$f(x_i - \Delta x) = f(x_i) - \Delta x f'(x_i) + \frac{(\Delta x)^2}{2!} f''(x_i) - \frac{(\Delta x)^3}{3!} f'''(x_i) + \ldots + (-1)^n \frac{(\Delta x)^n}{n!} f^{(n)}(x_i) + \cdots. \tag{7.10}$$

The exact values of $f(x_i + \Delta x)$ and $f(x_i - \Delta x)$ can be determined only if an infinite number of terms are considered in Eqs. (7.9) and (7.10). Since the consideration of an infinite number of terms is not possible, we compute the approximate values of $f(x_i + \Delta x)$ and $f(x_i - \Delta x)$ by including only the first few terms in the series of Eqs. (7.9) and (7.10). The approximations achieved by retaining one, two, or three terms in the series (Eq. (7.8) or (7.9)) are shown graphically in Fig. 7.4. It can be observed that the one-term, the two-term, or the three-term approximation becomes the exact solution if $f(x)$ is a constant, a linear equation, or a quadratic equation, respectively. When terms up to nth-order derivative term (a total of $n+1$ terms) are retained in the series of Eqs. (7.8) through (7.10), the error or remainder (to account for the contribution of all terms from $(n+2)$ to infinity) is given by

$$R_n = \frac{(x - x_i)^{n+1}}{(n+1)!} f^{(n+1)}(\xi) = \frac{(\Delta x)^{n+1}}{(n+1)!} f^{(n+1)}(\xi);\ x_i < \xi < x = x_i + \Delta x, \quad (7.11)$$

where the subscript n to R indicates that the error or remainder is for the nth-order approximation, $f^{(n+1)}$ is the $(n+1)$th-order derivative of f evaluated at ξ, which lies somewhere between x_i and $x = x_i + \Delta x$. It can be seen that the nth-order Taylor's series expansion will be exact if $f(x)$ is an nth-order polynomial since the $(n+1)$th-derivative of $f(x)$ and hence R_n will be zero. Since the quantity $f^{(n+1)}(\xi)$ is constant for a given interval, the truncation error or remainder, Eq. (7.11), is usually written as

$$R_n = O\left((\Delta x)^{n+1}\right), \quad (7.12)$$

indicating that the error is on the order of $(\Delta x)^{n+1}$. Equation (7.12) shows that the error will be reduced if the interval Δx is reduced. For example, if Δx is halved, the error becomes $\left(\frac{1}{2^{n+1}}\right)$ of the previous error. The graphical interpretation of Eq. (7.11) is given in Fig. 7.5 for the case of a one-term (or zeroth-order) expansion.

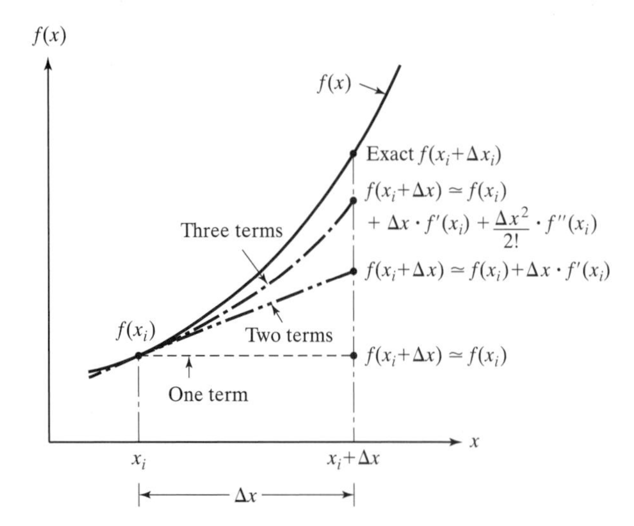

Figure 7.4 Approximations using different number of terms.

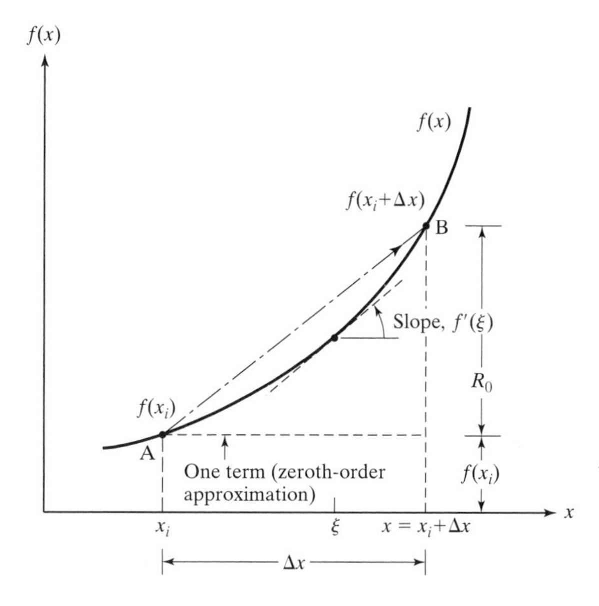

Figure 7.5 Error with one-term approximation.

For this case, the error or remainder, R_0, is given by the sum of all the neglected terms in Eq. (7.8) or Eq. (7.9):

$$R_0 = (\Delta x) f'(x_i) + \frac{(\Delta x)^2}{2!} f''(x_i) + \frac{(\Delta x)^3}{3!} f'''(x_i) + \cdots. \tag{7.13}$$

Usually the lower order derivative terms contribute more to the remainder than the higher order terms. Hence, instead of considering infinite terms in Eq. (7.13), we approximate it by the first-order derivative term as

$$R_0 \approx \Delta x f'(x_i). \tag{7.14}$$

According to the mean-value theorem, Eq. (7.4), there exists a value, $x = \xi$, in the range of our interest $(x_i, x_i + \Delta x)$, where the slope $f'(\xi)$ will be parallel to the line joining the exact function values $f(x_i)$ and $f(x_i + \Delta x)$. Using the slope at ξ, $f'(\xi)$, the error or remainder, R_0, can be expressed as

$$R_0 = \Delta x f'(\xi). \tag{7.15}$$

(See Fig. 7.5.) Notice that the *approximation sign* in Eq. (7.14) is replaced by the *equality sign* in Eq. (7.15). Equation (7.15) can be seen to be the zeroth-order version of Eq. (7.11). The higher order versions of Eq. (7.11) can be derived by logically extending the reasoning used in deriving Eq. (7.15).

▶Example 7.4

The pressure (p)–specific volume (v) relationship of superheated water vapor at 350° C is given by van der Waals equation [7.4]

$$p = \frac{RT}{v - b} - \frac{a}{v^2},$$

where $R =$ specific gas constant $= 0.461889$ kJ/kg-K, $T =$ temperature in kelvin $= 623.15$ K, $a = 1.7048$, and $b = 0.0016895$. Expand the pressure in Taylor's series and estimate the value of p at $v = 0.051, 0.054$, and 0.060 assuming that the value of p and its derivatives are known at $v = 0.05$.

Solution

Using the known values of R, T, a, and b, we find that the value of p and its derivatives at $v = 0.05$ can be computed as

$$p = RT(v-b)^{-1} - av^{-2} = 5275.917952, \tag{a}$$

$$p' = \frac{dp}{dv} = -RT(v-b)^{-2} + 2av^{-3} = -96047.0727, \tag{b}$$

$$p'' = \frac{d^2p}{dv^2} = 2RT(v-b)^{-3} - 6av^{-4} = 3,468,860.696, \tag{c}$$

and

$$p''' = \frac{d^3p}{dv^3} = -6RT(v-b)^{-4} + 24av^{-5} = -161,560,045.4. \tag{d}$$

Taylor's series expansion of p at $v + dv$ is given by

$$p(v+dv) = p(v) + p'(v)dv + p''(v)\frac{(dv)^2}{2} + p'''(v)\frac{(dv)^3}{6} + \cdots. \tag{e}$$

The values of $p(v + dv)$ given by Eq. (e), by retaining different number of terms on the right-hand side, at different values of dv are as follows:

Value of v	Two-term solution	Three-term solution	Four-term solution	Exact solution
0.051	5179.871582	5181.605957	5181.579102	5181.575195
0.054	4891.729980	4919.480957	4917.757812	4917.626953
0.060	4315.447754	4488.890625	4461.963867	4462.539062

◀

7.5.1 Finite-Difference Approximation of First Derivatives

Taylor's series expansions of a function $f(x)$ in the neighborhood of $x = x_i$ (Fig. 7.3) can be expressed as

$$f_{i+1} = f_i + \Delta x f_i' + \frac{(\Delta x)^2}{2!} f_i'' + \frac{(\Delta x)^3}{3!} f_i''' + \frac{(\Delta x)^4}{4!} f_i'''' + \frac{(\Delta x)^5}{5!} f_i^{(5)} + \cdots, \tag{7.16}$$

where $\Delta x = x_{i+1} - x_i$, and

$$f_{i-1} = f_i - \Delta x f_i' + \frac{(\Delta x)^2}{2!} f_i'' - \frac{(\Delta x)^3}{3!} f_i''' + \frac{(\Delta x)^4}{4!} f_i'''' - \frac{(\Delta x)^5}{5!} f_i^{(5)} + \cdots, \tag{7.17}$$

where $\Delta x = x_i - x_{i-1}$. Solving Eq. (7.16) for f_i' yields

$$f_i' = \frac{f_{i+1} - f_i}{\Delta x} - \frac{(\Delta x)}{2!} f_i'' - \frac{(\Delta x)^2}{3!} f_i''' - \cdots, \tag{7.18}$$

which, upon the application of the mean-value theorem, becomes

$$f_i' = \frac{f_{i+1} - f_i}{\Delta x} - \frac{\Delta x}{2} f''(\xi); x_i < \xi < x_{i+1} \tag{7.19}$$

or

$$f_i' = \frac{f_{i+1} - f_i}{\Delta x} + O(\Delta x), \tag{7.20}$$

where the term

$$O\,(\Delta x) = -\frac{\Delta x}{2} f''(\xi)$$

indicates that the error is proportional to the step length (Δx) and also to the second derivative of f. Equation (7.20) becomes the forward-difference approximation [given earlier in Eq. (7.5)] when the error term is truncated. The solution of Eq. (7.17) for f_i' gives

$$f_i' = \frac{f_i - f_{i-1}}{\Delta x} + \frac{\Delta x}{2!} f_i'' - \frac{(\Delta x)^2}{3!} f_i''' + \cdots. \tag{7.21}$$

The application of the mean-value theorem reduces Eq. (7.21) to

$$f_i' = \frac{f_i - f_{i-1}}{\Delta x} + \frac{\Delta x}{2} f''(\xi); x_{i-1} < \xi < x_i \tag{7.22}$$

or

$$f_i' = \frac{f_i - f_{i-1}}{\Delta x} + O(\Delta x), \tag{7.23}$$

where the error

$$O\,(\Delta x) = \frac{\Delta x}{2} f''(\xi)$$

is, as in the case of Eq. (7.20), proportional to the step length and the second derivative of f. Equation (7.23) becomes the backward-difference approximation [given earlier in Eq. (7.6)] when the error term is neglected. Subtracting Eq. (7.17) from Eq. (7.16) yields

$$f_{i+1} - f_i = 2\Delta x f_i' + \frac{(\Delta x)^3}{3} f_i''' + \frac{(\Delta x)^5}{60} f_i^{(5)} + \cdots, \tag{7.24}$$

which can be solved for f_i' as

$$f_i' = \frac{f_{i+1} - f_{i-1}}{2\Delta x} - \frac{(\Delta x)^2}{6} f_i''' - \frac{(\Delta x)^4}{120} f_i^{(5)} - \cdots, \tag{7.25}$$

or

$$f_i' = \frac{f_{i+1} - f_{i-1}}{2\Delta x} - \frac{(\Delta x)^2}{6} f'''(\xi); x_{i-1} < \xi < x_{i+1}, \tag{7.26}$$

or

$$f_i' = \frac{f_{i+1} - f_{i-1}}{2\Delta x} + O(\Delta x^2), \tag{7.27}$$

where the error

$$O(\Delta x^2) = -\frac{(\Delta x)^2}{2} f'''(\xi)$$

is proportional to $(\Delta x)^2$ and the third derivative of f. Equation (7.27) becomes the central-difference approximation [given earlier in Eq. (7.7)] upon the truncation of the error term. Notice that the error of the central-difference approximation is proportional to $(\Delta x)^2$ rather than Δx and hence is more accurate than the forward- and backward-difference approximations. Also when Δx is decreased, the error decreased more rapidly than in the other two approximations. For example, when Δx is halved, the error is also approximately halved in the forward- and backward-difference approximations, whereas the error is reduced to approximately one-fourth for the central-difference approximation.

7.5.2 Finite-Difference Approximation of Second Derivatives

The basic principle in deriving an expression for the second derivative is to eliminate the first derivative from Taylor's series expansion. For this, we consider Taylor's series expansion for $f_{i+2} = f(x_{i+2})$ using Eq. (7.8) with $x - x_i = x_{i+2} - x_i = 2\Delta x$ (Fig. 7.6):

$$f_{i+2} = f(x_i) + (2\Delta x) f_i' + \frac{(2\Delta x)^2}{2!} f_i'' + \frac{(2\Delta x)^3}{3!} f_i''' + \cdots. \tag{7.28}$$

Equation (7.16), when multiplied by two and subtracted from Eq. (7.28), gives

$$f_{i+2} - 2f_{i+1} = -f_i + (\Delta x)^2 f_i'' + (\Delta x)^3 f_i''' - \cdots, \tag{7.29}$$

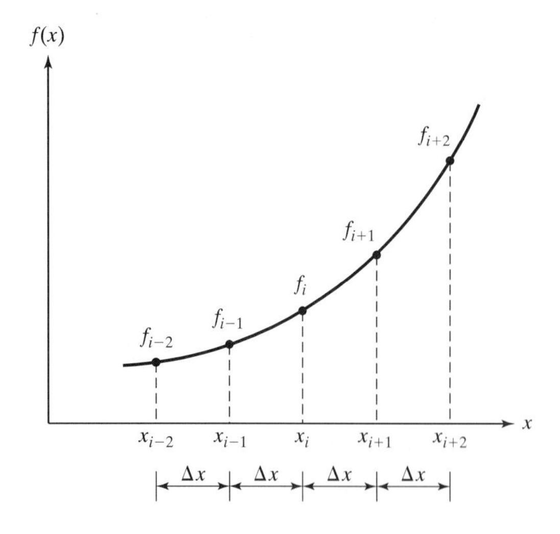

Figure 7.6 Grid for Taylor's series expansion of f_{i+2}.

from which the second derivative can be expressed as

$$f_i'' = \frac{f_{i+2} - 2f_{i+1} + f_i}{(\Delta x)^2} - (\Delta x) f_i''' + \cdots \tag{7.30}$$

or

$$f_i'' = \frac{f_{i+2} - 2f_{i+1} + f_i}{(\Delta x)^2} + O(\Delta x). \tag{7.31}$$

This equation, without the error term $O(\Delta x)$, represents the forward-difference approximation for f_i''. To derive the backward-difference approximation for f_i'', we consider $f_{i-2} = f(x_{i-2})$ using Eq. (7.8) with $x - x_i = -2\Delta x$:

$$f_{i-2} = f_i - (2\Delta x) f_i' + \frac{(2\Delta x)^2}{2!} f_i'' - \frac{(2\Delta x)^3}{3!} f_i''' + \cdots. \tag{7.32}$$

Equation (7.32) minus twice Eq. (7.17) gives

$$f_{i-2} - 2f_{i-1} = -f_i + (\Delta x)^2 f_i'' - (\Delta x)^3 f_i''' + \cdots, \tag{7.33}$$

from which we can obtain

$$f_i'' = \frac{f_i - 2f_{i-1} + f_{i-2}}{(\Delta x)^2} + O(\Delta x). \tag{7.34}$$

Equation (7.34), with the truncation of the error term $O(\Delta x)$, represents the backward-difference approximation for f_i''. Finally, the addition of Eq. (7.16) and (7.17) yields

$$f_{i+1} + f_{i-1} = 2f_i + (\Delta x)^2 f_i'' + \frac{(\Delta x)^4}{12} f_i'''' + \cdots \tag{7.35}$$

from which the second derivative, f_i'', can be expressed as

$$f_i'' = \frac{f_{i+1} - 2f_i + f_{i-1}}{(\Delta x)^2} + O((\Delta x)^2). \tag{7.36}$$

This equation, without the error term $O\left((\Delta x)^2\right)$, denotes the central-difference approximation for f_i''.

▶Example 7.5

Consider the pressure (p)–specific volume (v) relationship given in Example 7.4. Determine the first and second derivatives of p, $\frac{dp}{dv}$ and $\frac{d^2p}{dv^2}$, at $v = 0.05$ using backward-, forward-, and central-difference formulas. Compare the results with the exact derivatives.

Solution

Let us consider the step length, Δv, as 0.005, which is 10% of v. Noting that

$$p|_v = \frac{RT}{v-b} - \frac{a}{v^2} = \frac{287.8261304}{v - 0.0016895} - \frac{1.7048}{v^2},$$

the finite-difference derivatives can be computed as follows:

Backward differences

$$\left.\frac{dp}{dv}\right|_{v=0.05} = \frac{1}{\Delta v}\left(p|_v - p|_{v-\Delta v}\right) = \frac{1}{0.005}(p(0.05) - p(0.045))$$

$$= (5275.918457 - 5803.767090)/0.005$$

$$= -0.1055697266 \times 10^6$$

$$\left.\frac{d^2p}{dv^2}\right|_{v=0.05} = \frac{1}{(\Delta v)^2}\left(p|_v - 2p|_{v-\Delta v} + p|_{v-2\Delta v}\right)$$

$$= \frac{1}{(0.005)^2}(p(0.05) - 2p(0.045) + p(0.04))$$

$$= \frac{1}{(0.005)^2}(5275.918457 - 2(5803.767090) + 6447.483398)$$

$$= 0.4634707000 \times 10^7$$

Forward differences

$$\left.\frac{dp}{dv}\right|_{v=0.05} = \frac{1}{\Delta v}\left(p|_{v+\Delta v} - p|_v\right) = \frac{1}{0.005}(p(0.055) - p(0.05))$$

$$= (4835.481445 - 5275.918457)/0.005 = -0.8808740625 \times 10^5$$

$$\left.\frac{d^2p}{dv^2}\right|_{v=0.05} = \frac{1}{(\Delta v)^2}\left(p|_{v+2\Delta v} - 2p|_{v+\Delta v} + p|_v\right)$$

$$= \frac{1}{(0.005)^2}(p(0.06) - 2p(0.055) + p(0.05))$$

$$= \frac{1}{(0.005)^2}(4462.539062 - 2(4835.481445) + 5275.918457)$$

$$= 0.2699785250 \times 10^7$$

Central differences

$$\left.\frac{dp}{dv}\right|_{v=0.05} = \frac{1}{2\Delta v}\left(p|_{v+\Delta v} - p|_{v-\Delta v}\right) = \frac{1}{0.01}(p(0.055) - p(0.045))$$

$$= \frac{1}{0.01}(4835.481445 - 5803.767090) = -0.9682857031 \times 10^5$$

$$\left.\frac{d^2p}{dv^2}\right|_{v=0.05} = \frac{1}{(\Delta v)^2}\left(p|_{v+\Delta v} - 2p|_v + p|_{v-\Delta v}\right)$$

$$= \frac{1}{(0.005)^2}(p(0.055) - 2p(0.05) + p(0.045))$$

$$= \frac{1}{(0.005)^2}(4835.481445 - 2(5275.918457) + 5803.767090)$$

$$= 0.3496465000 \times 10^7$$

Exact derivatives

$$\left.\frac{dp}{dv}\right|_{v=0.05} = -0.9204707031 \times 10^5 \text{(from Example 7.4)}$$

$$\left.\frac{d^2p}{dv^2}\right|_{v=0.05} = 0.3468860750 \times 10^7 \text{(from Example 7.4)}$$

◀

7.5.3 Finite-Difference Approximation of Higher Order Derivatives

The procedure used for deriving the finite-difference approximations for first and second derivatives can be extended easily for deriving higher order derivatives. This involves finding Taylor's series expansions for f_{i+3} (with $x - x_i = 3\,\Delta x$), f_{i-3} (with $x - x_i = -3\,\Delta x$), and f_{i+4} (with $x - x_i = 4\,\Delta x$), ... using Eq. (7.8) and then manipulating the expressions of $f_{i+1}, f_{i-1}, f_{i+2}, f_{i-2}, f_{i+3}, f_{i-3}, \ldots$ properly. For example, we consider the forward-finite-difference approximation for f_i'''. This requires f_{i+3}:

$$f_{i+3} = f_i + (3\,\Delta x) f_i' + \frac{(3\,\Delta x)^2}{2!} f_i'' + \frac{(3\,\Delta x)^3}{3!} f_i''' + \frac{(3\,\Delta x)^4}{4!} f_i'''' + \cdots \tag{7.37}$$

Equations (7.37), (7.28), and (7.16) can now be used to find

$$f_{i+3} - 3f_{i+2} + 3f_{i+1} = f_i + (\Delta x)^3 f_i''' + \frac{3}{2}(\Delta x)^4 f_i'''' + \cdots, \tag{7.38}$$

$$f_i''' = \frac{f_{i+3} - 3f_{i+2} + 3f_{i+1} - f_i}{(\Delta x)^3} - \frac{3}{2}(\Delta x) f_i'''', \tag{7.39}$$

or

$$f_i''' = \frac{f_{i+3} - 3f_{i+2} + 3f_{i+1} - f_i}{(\Delta x)^3} + O\,(\Delta x)\,. \tag{7.40}$$

Similarly, for the central-difference approximation of f_i''', we subtract Eq. (7.32) from Eq. (7.28) to obtain

$$f_{i+2} - f_{i-2} = 4\,\Delta x f_i' + \frac{8(\Delta x)^3}{3} f_i''' + \frac{8(\Delta x)^5}{15} f_i^{(5)} + \cdots \tag{7.41}$$

Substituting Eq. (7.25) into Eq. (7.41) for f_i' yields

$$f_{i+2} - f_{i-2} = 4\Delta x \left\{ \frac{f_{i+1} - f_{i-1}}{2\Delta x} - \frac{(\Delta x)^2}{6} f_i''' - \frac{(\Delta x)^5}{120} f_i^{(5)} - \cdots \right\}$$

$$+ \frac{8}{3}(\Delta x)^3 f_i''' + \frac{8}{15}(\Delta x)^5 f_i^{(5)} + \cdots$$

Table 7.1 Common finite-difference formulas.

Type of approximation	Formula	Truncation error
Forward differences	$f_i' = (f_{i+1} - f_i)/(\Delta x)$	$O(\Delta x)$
	$f_i'' = (f_{i+2} - 2f_{i+1} + f_i)/(\Delta x)^2$	
	$f_i''' = (f_{i+3} - 3f_{i+2} + 3f_{i+1} - f_i)/(\Delta x)^3$	
	$f_i'''' = (f_{i+4} - 4f_{i+3} + 6f_{i+2} - 4f_{i+1} + f_i)/(\Delta x)^4$	
Backward differences	$f_i' = (f_i - f_{i-1})/(\Delta x)$	$O(\Delta x)$
	$f_i'' = (f_i - 2f_{i-1} + f_{i-2})/(\Delta x)^2$	
	$f_i''' = (f_i - 3f_{i-1} + 3f_{i-2} - f_{i-3})/(\Delta x)^3$	
	$f_i'''' = (f_i - 4f_{i-1} + 6f_{i-2} - 4f_{i-3} + f_{i-4})/(\Delta x)^4$	
Central differences	$f_i' = (f_{i+1} - f_{i-1})/(2\,\Delta x)$	$O(\Delta x^2)$
	$f_i'' = (f_{i+1} - 2f_i + f_{i-1})/(\Delta x)^2$	
	$f_i''' = (f_{i+2} - 2f_{i+1} + 2f_{i-1} - f_{i-2})/(2(\Delta x)^3)$	
	$f_i'''' = (f_{i+2} - 4f_{i+1} + 6f_i - 4f_{i-1} + f_{i-2})/(\Delta x)^4$	

$$= 2f_{i+1} - 2f_{i-1} - \frac{2}{3}(\Delta x)^3 f_i''' - \frac{1}{30}(\Delta x)^5 f_i^{(5)} - \cdots$$
$$+\frac{8}{3}(\Delta x)^3 f_i''' + \frac{8}{15}(\Delta x)^5 f_i^{(5)} + \cdots \tag{7.42}$$

Equation (7.42) can be rearranged to obtain the central-difference approximation for f_i''' as

$$f_i''' = \frac{f_{i+2} - 2f_{i+1} + 2f_{i-1} - f_{i-2}}{2(\Delta x)^3} - \frac{(\Delta x)^2}{4} f_i^{(5)} - \cdots \tag{7.43}$$

or

$$f_i''' = \frac{f_{i+2} - 2f_{i+1} + 2f_{i-1} - f_{i-2}}{2(\Delta x)^3} + O((\Delta x)^2). \tag{7.44}$$

It can be seen that the required manipulations become more involved with increasing order of the derivatives. Some of the commonly used finite-difference approximations are given in Table 7.1.

7.5.4 Higher Accuracy Finite-Difference Approximations

It can be seen that all the forward- and backward-finite-difference approximations as previously derived have an error, $O(\Delta x)$, while all the central-difference approximations have an error, $O\left((\Delta x)^2\right)$. The finite-difference approximations with higher accuracy can be derived by including additional terms in Taylor's series expansions. As an example, consider the forward-difference approximation for f_i' with a higher accuracy. For this, we use Eq. (7.18),

$$f_i' = \frac{f_{i+1} - f_i}{\Delta x} - \frac{\Delta x}{2} f_i'' - \frac{(\Delta x)^2}{6} f_i''' - \cdots \tag{7.45}$$

and, instead of truncating all terms beyond the first one (as was done in Eq. (7.19)), we retain the second-derivative term also in Eq. (7.45) to yield

$$f_i' = \frac{f_{i+1} - f_i}{\Delta x} - \frac{\Delta x}{2} f_i'' - \frac{(\Delta x)^2}{6} f_i''' - \cdots \tag{7.46}$$

Substituting Eq. (7.30) into Eq. (7.46) results in

$$f_i' = \frac{f_{i+1} - f_i}{\Delta x} - \frac{\Delta x}{2}\left[\frac{f_{i+2} - 2f_{i+1} + f_i}{(\Delta x)^2} - (\Delta x) f_i''' + \cdots\right] - \frac{(\Delta x)^2}{6} f_i''' - \cdots$$

$$= \frac{-f_{i+2} + 4f_{i+1} - 3f_i}{2\,\Delta x} + \frac{(\Delta x)^2}{3} f_i''' - \cdots, \tag{7.47}$$

which can be expressed as

$$f_i' = \frac{-f_{i+2} + 4f_{i+1} - 3f_i}{2\,\Delta x} + O((\Delta x)^2). \tag{7.48}$$

Thus, the accuracy of f_i' is improved from $O(\Delta x)$ in Eq. (7.20) to $O\left((\Delta x)^2\right)$ in Eq. (7.48) by including the second-derivative term. A similar procedure can be adopted to derive higher accuracy backward- and central-difference approximations for f_i'. Table 7.2 gives some of the higher accuracy finite-difference approximations.

The formulation of still higher accuracy finite-difference approximations is possible by retaining additional terms in Taylor's series expansions. For example, the forward- and backward-difference approximations with accuracy of $O\left((\Delta x)^3\right)$ and the central-difference approximation with accuracy of $O\left((\Delta x)^4\right)$ can be derived

Table 7.2 Higher order finite-difference formulas.

Type of formula	Formula	Truncation error
Forward differences	$f_i' = (-f_{i+2} + 4f_{i+1} - 3f_i)/(2(\Delta x))$	$O(\Delta x)^2$
	$f_i'' = (-f_{i+3} + 4f_{i+2} - 5f_{i+1} + 2f_i)/(\Delta x)^2$	
	$f_i''' = (-3f_{i+4} + 14f_{i+3} - 24f_{i+2} + 18f_{i+1} - 5f_i)/(2(\Delta x)^3)$	
	$f_i'''' = (-2f_{i+5} + 11f_{i+4} - 24f_{i+3} + 26f_{i+2} - 14f_{i+1} + 3f_i)/(\Delta x)^4$	
Backward differences	$f_i' = (3f_i - 4f_{i-1} + f_{i-2})/(2(\Delta x))$	$O(\Delta x^2)$
	$f_i'' = (2f_i - 5f_{i-1} + 4f_{i-2} - f_{i-3})/(\Delta x)^2$	
	$f_i''' = (5f_i - 18f_{i-1} + 24f_{i-2} - 14f_{i-3} + 3f_{i-4})/(2(\Delta x)^3)$	
	$f_i'''' = (3f_i - 14f_{i-1} + 26f_{i-2} - 24f_{i-3} + 11f_{i-4} - 2f_{i-5})/(\Delta x)^4$	
Central differences	$f_i' = (-f_{i+2} + 8f_{i+1} - 8f_{i-1} + f_{i-2})/(12(\Delta x))$	$O(\Delta x^4)$
	$f_i'' = (-f_{i+2} + 16f_{i+1} - 30f_i + 16f_{i-1} - f_{i-2})/(12(\Delta x)^2)$	
	$f_i''' = (-f_{i+3} + 8f_{i+2} - 13f_{i+1} + 13f_{i-1} - 8f_{i-2} + f_{i-3})/(8(\Delta x)^3)$	
	$f_i'''' = (-f_{i+3} + 12f_{i+2} - 39f_{i+1} + 56f_i - 39f_{i-1} + 12f_{i-2} - f_{i-3})/(6(\Delta x)^4)$	

as (See Problem 7.36)

$$f_i' = \frac{-11f_i + 18f_{i+1} - 9f_{i+2} + 2f_{i+3}}{6\,\Delta x} + O((\Delta x)^3), \tag{7.49}$$

$$f_i' = \frac{11f_i - 18f_{i-1} + 9f_{i-2} - 2f_{i-3}}{6\,\Delta x} + O((\Delta x)^3), \tag{7.50}$$

and

$$f_i' = \frac{-f_{i+2} + 8f_{i+1} - 8f_{i-1} + f_{i-2}}{12\,\Delta x} + O((\Delta x)^4). \tag{7.51}$$

Notes

1. In all the finite-difference formulas, the sum of all the coefficients of the function values (f_i) appearing in the numerator can be seen to be zero. Physically this implies that the derivative becomes zero if $f(x)$ is a constant.
2. The accuracy of the computed derivatives can be improved either by using a smaller step size (Δx) or by using a higher accuracy formula.

▶Example 7.6

The displacement of an instrument subjected to a random vibration test, at different instants of time, is found to be as follows:

Station, i	Time, t_i (sec)	Displacement, y_i (inch)
1	0.05	0.144
2	0.10	0.172
3	0.15	0.213
4	0.20	0.296
5	0.25	0.070
6	0.30	0.085
7	0.35	0.525
8	0.40	0.110
9	0.45	0.062
10	0.50	0.055
11	0.55	0.042
12	0.60	0.035

Determine the velocity $\left(\frac{dy}{dt}\right)$, acceleration $\left(\frac{d^2y}{dt^2}\right)$, and jerk $\left(\frac{d^3y}{dt^3}\right)$ at $t = 0.05$, 0.20, and 0.60 sec using suitable finite-difference formulas with a step size, Δt, of 0.05 sec.

Solution

Since the data are known for $0.05 \le t \le 0.60$ sec, we can find the required derivatives at $t = 0.05$ sec and $t = 0.60$ sec using forward- and backward-difference formulas, respectively. We use the central-difference formulas to evaluate the derivatives at $t = 0.20$ sec.

At $t = 0.05$ sec

The forward-difference formulas (see Table 7.1) are given by

$$y_i' = \frac{dy}{dt}(t_i) = \frac{y_{i+1} - y_i}{\Delta t}, \tag{a}$$

$$y_i'' = \frac{d^2y}{dt^2}(t_i) = \frac{y_{i+2} - 2y_{i+1} + y_i}{(\Delta t)^2}, \tag{b}$$

$$y_i''' = \frac{d^3y}{dt^3}(t_i) = \frac{y_{i+3} - 3y_{i+2} + 3y_{i+1} - y_i}{(\Delta t)^3}. \tag{c}$$

Using $i = 1$ and $\Delta t = 0.05$, Eqs. (a) through (c) yield

$$y_1' = y'(0.05) = \frac{y_2 - y_1}{\Delta t} = \frac{0.172 - 0.144}{0.05} = 0.56 \text{ in/sec},$$

$$y_1'' = y''(0.05) = \frac{y_3 - 2y_2 + y_1}{(\Delta t)^2} = \frac{0.213 - 2(0.172) + 0.144}{(0.05)^2} = 5.20 \text{ in/sec}^2,$$

and

$$\begin{aligned} y_1''' = y'''(0.05) &= \frac{y_4 - 3y_3 + 3y_2 - y_1}{(\Delta t)^3} \\ &= \frac{0.296 - 3(0.213) + 3(0.172) - 0.144}{(0.05)^3} \\ &= 232.0 \text{ in/sec}^3. \end{aligned}$$

At $t = 0.20$ sec

The central-difference formulas (see Table 7.1) are given by

$$y_i' = \frac{dy}{dt}(t_i) = \frac{y_{i+1} - y_{i-1}}{2\,\Delta t}, \tag{d}$$

$$y_i'' = \frac{d^2y}{dt^2}(t_i) = \frac{y_{i+1} - 2y_i + y_{i-1}}{(\Delta t)^2}, \tag{e}$$

and

$$y_i''' = \frac{d^3y}{dt^3}(t_i) = \frac{y_{i+2} - 2y_{i+1} + 2y_{i-1} - y_{i-2}}{2(\Delta t)^3}. \tag{f}$$

For $i = 4$ and $\Delta t = 0.05$, Eqs. (d) through (f) give

$$y_4' = y'(0.20) = \frac{y_5 - y_3}{2\,\Delta t} = \frac{0.070 - 0.213}{2(0.05)}$$
$$= -1.43 \text{ in/sec},$$
$$y_4'' = y''(0.20) = \frac{y_5 - 2y_4 + y_3}{(\Delta t)^2} = \frac{0.070 - 2(0.296) + 0.213}{(0.05)^2}$$
$$= -123.60 \text{ in/sec}^2,$$

and

$$y_4''' = y'''(0.20) = \frac{y_6 - 2y_5 + 2y_3 - y_2}{2(\Delta t)^3}$$
$$= \frac{0.085 - 2(0.070) + 2(0.213) - 0.172}{2(0.05)^3} = 796.0 \text{ in/sec}^3.$$

At $t = 0.60$ sec

The backward-difference formulas (see Table 7.1) are given by

$$y_i' = \frac{dy}{dt}(t_i) = \frac{y_i - y_{i-1}}{\Delta t}, \tag{g}$$
$$y_i'' = \frac{d^2y}{dt^2}(t_i)$$
$$= \frac{y_i - 2y_{i-1} + y_{i-2}}{(\Delta t)^2}, \tag{h}$$

and

$$y_i''' = \frac{d^3y}{dt^3}(t_i) = \frac{y_i - 3y_{i-1} + 3y_{i-2} - y_{i-3}}{(\Delta t)^3}. \tag{i}$$

For $i = 12$, Eqs. (g), (h), and (i) result in

$$y_{12}' = y'(0.60) = \frac{y_{12} - y_{11}}{\Delta t} = \frac{0.035 - 0.042}{0.05} = -0.14 \text{ in/sec},$$
$$y_{12}'' = y''(0.60) = \frac{y_{12} - 2y_{11} + y_{10}}{(\Delta t)^2} = \frac{0.035 - 2(0.042) + 0.055}{(0.05)^2}$$
$$= 2.4 \text{ in/sec}^2,$$

and

$$y_{12}''' = y'''(0.60) = \frac{y_{12} - 3y_{11} + 3y_{10} - y_9}{(\Delta t)^3} = \frac{0.035 - 3(0.042) + 3(0.055) - 0.062}{(0.05)^3}$$
$$= 96.0 \text{ in/sec}^3.$$

◀

7.6 Using Difference Operators

The calculus of finite differences is used in conjunction with a series of discrete values, which may denote either experimental data, such as $f_{i-2}, f_{i-1}, f_i, f_{i+1}, f_{i+2}$, or discrete values of a continuous function $f(x)$, such as* $f(x-2h), f(x-h), f(x), f(x+h)$, and $f(x+2h)$, where the values of the dependent variable f_i or $f(x)$ correspond to equally spaced values of the independent variable, x.

The following difference operators are defined to deal with the discrete values of f such as those indicated previously

$$\Delta = \text{forward-difference operator:}$$

$$\Delta f_i = f_{i+1} - f_i; \tag{7.52}$$

$$\nabla = \text{backward-difference operator:}$$

$$\nabla f_i = f_i - f_{i-1}; \tag{7.53}$$

and

$$\delta = \text{central-difference operator:}$$

$$\delta f_i = f_{i+\frac{1}{2}} - f_{i-\frac{1}{2}} \tag{7.54}$$

or

$$\delta f_{i+\frac{1}{2}} = f_{i+1} - f_i, \tag{7.55}$$

where

$$f_i = f(x_i),\ f_{i+1} = f(x_{i+1}) = f(x_i + \Delta x),\ f_{i+\frac{1}{2}} = f\left(x_i + \frac{\Delta x}{2}\right).$$

These operators can be treated as algebraic variables since they satisfy the associative, commutative, and distributive laws of algebra [7.5]. The higher order difference operators can be written as powers of the difference operators. For example, the kth-order difference operators are written as Δ^k, ∇^k, and δ^k. The second-order forward difference of f at x_i can be derived, using the definition of the first-order difference operator given in Eq. (7.52), as

$$\begin{aligned}\Delta^2 f_i &= \Delta(\Delta f_i) = \Delta(f_{i+1} - f_i)\\ &= \Delta f_{i+1} - \Delta f_i = (f_{i+2} - f_{i+1}) - (f_{i+1} - f_i)\\ &= f_{i+2} - 2f_{i+1} + f_i.\end{aligned} \tag{7.56}$$

Similarly, the second-order backward- and central-differences of f at x_i can be derived, using the definitions of Eqs. (7.53) and (7.54), as

$$\begin{aligned}\nabla^2 f_i &= \nabla(\nabla f_i) = \nabla(f_i - f_{i-1}) = \nabla f_i - \nabla f_{i-1}\\ &= (f_i - f_{i-1}) - (f_{i-1} - f_{i-2}) = f_i - 2f_{i-1} + f_{i-2}\end{aligned} \tag{7.57}$$

*The step size is denoted as h (instead of Δx) in this section to avoid confusion with the forward-difference operator, Δ.

and

$$\delta^2 f_i = \delta(\delta f_i) = \delta(f_{i+\frac{1}{2}} - f_{i-\frac{1}{2}}) = \delta f_{i+\frac{1}{2}} - \delta f_{i-\frac{1}{2}}$$
$$= (f_{i+1} - f_i) - (f_i - f_{i-1}) = f_{i+1} - 2f_i + f_{i-1}. \tag{7.58}$$

The difference operators can also be mixed, for example, as

$$\Delta\nabla f_i = \Delta(\nabla f_i) = \Delta(f_i - f_{i-1}) = \Delta f_i - \Delta f_{i-1} = (f_{i+1} - f_i) - (f_i - f_{i-1})$$
$$= f_{i+1} - 2f_i + f_{i-1} \tag{7.59}$$

and

$$\nabla\Delta f_i = \nabla(\Delta f_i) = \nabla(f_{i+1} - f_i) = \nabla f_{i+1} - \nabla f_i$$
$$= (f_{i+1} - f_i) - (f_i - f_{i-1}) = f_{i+1} - 2f_i + f_{i-1}. \tag{7.60}$$

The higher order difference operators of f at x_i can also be derived using the same procedure. For example, $\Delta^3 f_i$, $\Delta^4 f_i$, and $\Delta^5 f_i$ (see Problem 7.14) are given by

$$\Delta^3 f_i = f_{i+3} - 3f_{i+2} + 3f_{i+1} - f_i, \tag{7.61}$$

$$\Delta^4 f_i = f_{i+4} - 4f_{i+3} + 6f_{i+2} - 4f_{i+1} + f_i, \tag{7.62}$$

and

$$\Delta^5 f_i = f_{i+5} - 5f_{i+4} + 10f_{i+3} - 10f_{i+2} + 5f_{i+1} - f_i. \tag{7.63}$$

7.7 Approximation of Derivatives Using Difference Operators

In order to express the derivatives of a function $f(x)$ in terms of difference operators, the differential operators are approximated as follows:

$$\frac{d}{dx} \approx \frac{\Delta}{\Delta x} \text{ (approximation by forward-difference operators)}, \tag{7.64}$$

$$\frac{d}{dx} \approx \frac{\nabla}{\nabla x} \text{ (approximation by backward-difference operators)}, \tag{7.65}$$

and

$$\frac{d}{dx} \approx \frac{\delta}{\delta x} \text{ (approximation by central-difference operators)}. \tag{7.66}$$

By applying the operators of Eqs. (7.64) through (7.66) to the function $f(x)$, the various types of finite-difference approximations can be derived for the derivative of $f(x)$:

$$\left.\frac{df(x)}{dx}\right|_{x_i} \approx \left.\frac{\Delta}{\Delta x} f(x)\right|_{x_i} = \frac{\Delta f_i}{\Delta x_i} = \frac{f_{i+1} - f_i}{x_{i+1} - x_i} = \frac{f_{i+1} - f_i}{h} \tag{7.67}$$

(forward-difference formula),

$$\left.\frac{df(x)}{dx}\right|_{x_i} \approx \left.\frac{\nabla}{\nabla x} f(x)\right|_{x_i} = \frac{\nabla f_i}{\nabla x_i} = \frac{f_i - f_{i-1}}{x_i - x_{i-1}} = \frac{f_i - f_{i-1}}{h} \tag{7.68}$$

(backward-difference formula),

and

$$\left.\frac{df(x)}{dx}\right|_{x_i} \approx \left.\frac{\delta}{\delta x} f(x)\right|_{x_i} = \frac{\delta f_i}{\delta x_i} = \frac{f_{i+1} - f_{i-1}}{x_{i+1} - x_{i-1}} = \frac{f_{i+1} - f_{i-1}}{2h} \tag{7.69}$$

(central-difference formula).

Note that the central-difference formula, Eq. (7.69), is based on a step size of $2h$ in order to avoid the values of the function at the midpoints of the grid, $f_{i+\frac{1}{2}}$ and $f_{i-\frac{1}{2}}$. The central-difference formula can also be obtained as the average of the forward- and backward-difference formulas as

$$\left.\frac{df(x)}{dx}\right|_{x_i} \approx \frac{1}{2}\left(\frac{\Delta}{\Delta x} + \frac{\nabla}{\nabla x}\right) f_i = \frac{1}{2}\left(\frac{\Delta f_i}{\Delta x_i} + \frac{\nabla f_i}{\nabla x_i}\right) = \frac{f_{i+1} - f_{i-1}}{2h}. \tag{7.70}$$

The higher order differential operators can be approximated by finite-difference operators by applying the first-order approximations of Eqs. (7.64) through (7.66) the required number of times. For example, the second derivative of a function $f(x)$ at x_i can be approximated as follows:

$$\begin{aligned}\left.\frac{d^2 f(x)}{dx^2}\right|_{x_i} &= \left.\frac{d}{dx}\left(\frac{df(x)}{dx}\right)\right|_{x_i} \approx \left.\frac{\Delta}{\Delta x}\left(\frac{\Delta f}{\Delta x}\right)\right|_{x_i} \approx \frac{\Delta^2 f_i}{(\Delta x_i)^2} \\ &= \frac{f_{i+2} - 2f_{i+1} + f_i}{h^2}\end{aligned} \tag{7.71}$$

(forward-difference approximation),

$$\begin{aligned}\left.\frac{d^2 f(x)}{dx^2}\right|_{x_i} &= \left.\frac{d}{dx}\left(\frac{df(x)}{dx}\right)\right|_{x_i} \approx \left.\frac{\nabla}{\nabla x}\left(\frac{\nabla f}{\nabla x}\right)\right|_{x_i} \approx \frac{\nabla^2 f_i}{(\nabla x_i)^2} \\ &= \frac{f_i - 2f_{i-1} + f_{i-2}}{h^2}\end{aligned} \tag{7.72}$$

(backward-difference formula),

and

$$\begin{aligned}\left.\frac{d^2 f(x)}{dx^2}\right|_{x_i} &= \left.\frac{d}{dx}\left(\frac{df(x)}{dx}\right)\right|_{x_i} \approx \left.\frac{\delta}{\delta x}\left(\frac{\delta f}{\delta x}\right)\right|_{x_i} \approx \frac{\delta^2 f_i}{(\delta x_i)^2} \\ &= \frac{f_{i+1} - 2f_i + f_{i-1}}{h^2}\end{aligned} \tag{7.73}$$

(central-difference formula).

The relationships between the differential operator and the difference operators given in Eqs. (7.64) through (7.66) are only approximations. Exact relationships between $\frac{d}{dx}$ and Δ, ∇, or δ can also be derived [7.6].

7.8 Using Differentiation of Interpolating Polynomials

The interpolation polynomials discussed in Section 5.6.4 can be used to derive the finite difference approximations of functions. Although any interpolating polynomial can be used for this purpose, the power series type and Newton's interpolating polynomials are used in this section for illustration.

7.8.1 Power Series Type Interpolating Polynomial

Consider an nth order interpolation polynomial passing through $(n + 1)$ data (grid) points. The $(n+1)$ coefficients of the polynomial are uniquely determined by the $(n+1)$ data points. The derivation of interpolating polynomials was considered in detail in Chapter 5. For simplicity of illustration, we consider a second-order polynomial $f(x)$ passing through three data points (x_i, f_i), (x_{i+1}, f_{i+1}), and (x_{i+2}, f_{i+2}) as shown in Fig. 7.7 by assuming the interpolation polynomial $f(x)$ as

$$f(x) = a_0 + a_1 x + a_2 x^2, \tag{7.74}$$

where a_0, a_1, and a_2 are unknown coefficients. By making this polynomial pass through the three data points, we obtain the relations

$$f(x = x_i) = f_i = a_0 + a_1 x_i + a_2 x_i^2, \tag{7.75}$$

$$f(x = x_{i+1} = x_i + h) = f_{i+1} = a_0 + a_1(x_i + h) + a_2(x_i + h)^2, \tag{7.76}$$

and

$$f(x = x_{i+2} = x_i + 2h) = f_{i+2} = a_0 + a_1(x_i + 2h) + a_2(x_i + 2h)^2. \tag{7.77}$$

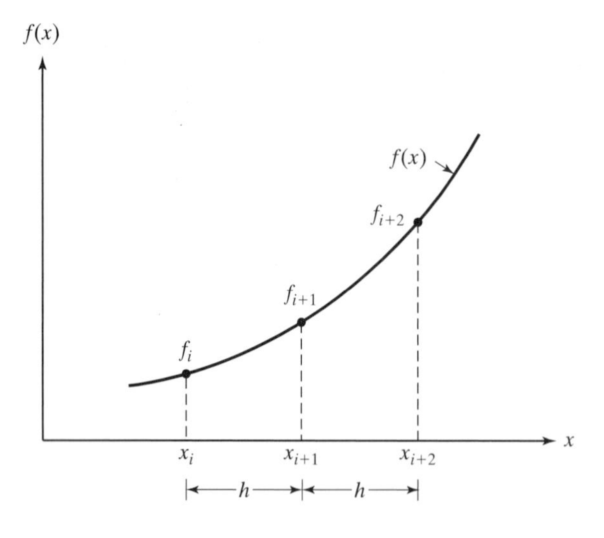

Figure 7.7 Second-order polynomial.

Equations (7.75) through (7.77) can be solved for the coefficients a_0, a_1, and a_2. Since the finite-difference approximations are expected to be independent of the location of the points, we can use any value of x_i. The algebra can be simplified by taking $x_i = 0, x_{i+1} = h$, and $x_{i+2} = 2h$. For this choice, Eqs. (7.75) through (7.77) simplify as

$$f_i = a_0, \tag{7.78}$$

$$f_{i+1} = a_0 + a_1 h + a_2 h^2, \tag{7.79}$$

and

$$f_{i+2} = a_0 + 2a_1 h + 4a_2 h^2. \tag{7.80}$$

The solution of Eqs. (7.78) through (7.80) can be found as

$$a_0 = f_i, \tag{7.81}$$

$$a_1 = \frac{-f_{i+2} + 4f_{i+1} - 3f_i}{2h}, \tag{7.82}$$

and

$$a_2 = \frac{f_{i+2} - 2f_{i+1} + f_i}{2h^2}. \tag{7.83}$$

Differentiating Eq. (7.74) and using Eqs. (7.82) and (7.83) yields

$$f'(x_i) = f_i' = f'(x_i = 0) = a_1 + 2a_2 x_i = a_1 = \frac{-f_{i+2} + 4f_{i+1} - 3f_i}{2h} \tag{7.84}$$

and

$$f''(x_i) = f_i'' = f''(x_i = 0) = 2a_2 = \frac{f_{i+2} - 2f_{i+1} + f_i}{h^2}. \tag{7.85}$$

Equations (7.84) and (7.85) denote forward-difference approximations and can be seen to be the same as Eqs. (7.48) and (7.36) without the error terms, respectively. The derivation of finite-difference approximations from interpolating polynomials is particularly useful when the data (base) points have nonuniform spacing. As an example, consider the data points shown in Fig. 7.8, where $x_{i+1} = x_i + h$ and $x_{i+2} = x_i + 4h$. By using $x_i = 0, x_{i+1} = h$, and $x_{i+2} = 4h$ in Eq. (7.74), we obtain the relations

$$f_i = a_0, \tag{7.86}$$

$$f_{i+1} = a_0 + a_1 h + a_2 h^2, \tag{7.87}$$

and

$$f_{i+2} = a_0 + 4h\, a_1 + 16h^2 a_2. \tag{7.88}$$

The solution of Eqs. (7.86) through (7.88) gives

$$a_0 = f_i, \tag{7.89}$$

$$a_1 = \frac{-f_{i+2} + 16f_{i+1} - 15f_i}{12h}, \tag{7.90}$$

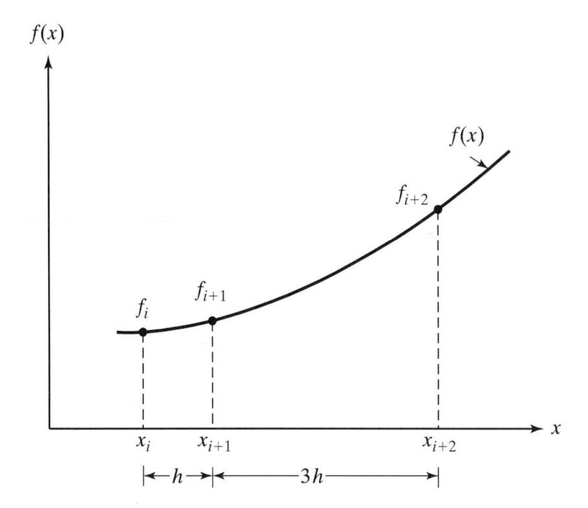

Figure 7.8 Nonuniformly-spaced data points.

and

$$a_2 = \frac{a_{i+2} - 4f_{i+1} + 3f_i}{12h^2}. \tag{7.91}$$

By differentiating Eq. (7.74) and using Eqs. (7.90) and (7.91), we obtain

$$f_i'(x_i) = f_i' = f'(x_i = 0) = a_1 + 2a_2 x_i = a_1 = \frac{-f_{i+2} + 16f_{i+1} - 15f_i}{12h} \tag{7.92}$$

and

$$f_i''(x_i) = f_i'' = f''(x_i = 0) = 2a_2 = \frac{f_{i+2} - 4f_{i+1} + 3f_i}{6h^2}. \tag{7.93}$$

This procedure can be used to find the finite-difference approximations to higher order derivatives, also.

7.8.2 Newton's Interpolation Polynomial

Although either the forward or the backward Newton–Gregory interpolation formula can be used, we consider the forward interpolation formula for illustration. The nth-order Newton–Gregory forward interpolation formula, $p(x)$, fitted to $(n+1)$ data points is given by Eq. (5.47):

$$p(x) = f_n(x) = f(x_0) + \frac{\Delta f_0}{h}(x - x_0) + \frac{\Delta^2 f_0}{2!\,h^2}(x - x_0)(x - x_0 - h)$$
$$+ \cdots + \frac{\Delta^n f_0}{n!\,h^n}(x - x_0)(x - x_0 - h)\cdots(x - x_0 - (n-1)h). \tag{7.94}$$

To derive the second-order difference approximations, for example, we consider $n = 2$ for which Eq. (7.94) reduces to

$$p(x) = f(x_0) + \frac{\Delta f_0}{h}(x - x_0) + \frac{\Delta^2 f_0}{2h^2}(x - x_0)(x - x_0 - h). \tag{7.95}$$

Differentiation of Eq. (7.95) yields

$$p'(x) = \frac{1}{h}\Delta f_0 + \frac{\Delta^2 f_0}{2h^2}(2x - 2x_0 - h). \tag{7.96}$$

By setting, in sequence, $x = x_0, x_0 + h$, and $x_0 + 2h$ in Eq. (7.96) and using Eqs. (7.52) and (7.56), we obtain

$$\begin{aligned} p'(x_0) &= \frac{1}{2h}[2\Delta f_0 - \Delta^2 f_0] = \frac{1}{2h}[2(f_1 - f_0) - (f_2 - 2f_1 + f_0)] \\ &= \frac{1}{2h}(-f_2 + 4f_1 - 3f_0), \end{aligned} \tag{7.97}$$

$$\begin{aligned} p'(x_0 + h) = p'(x_1) &= \frac{1}{2h}[2\Delta f_0 + \Delta^2 f_0] = \frac{1}{2h}[2(f_1 - f_0) + (f_2 - 2f_1 + f_0)] \\ &= \frac{1}{2h}(f_2 - f_0), \end{aligned} \tag{7.98}$$

and

$$\begin{aligned} p'(x_0 + 2h) = p'(x_2) &= \frac{1}{2h}[2\Delta f_0 + 3\Delta^2 f_0] = \frac{1}{2h}[2(f_1 - f_0) + 3(f_2 - 2f_1 + f_0)] \\ &= \frac{1}{2h}(3f_2 - 4f_1 + f_0). \end{aligned} \tag{7.99}$$

Equations (7.97), (7.98), and (7.99) denote, respectively, the forward-difference approximation at the grid point 0, the central-difference approximation at the grid point 1, and the backward-difference approximation at grid point 2. Equations (7.97) through (7.99) can be written in general form by replacing the subscripts 0, 1, and 2 by i, $i+1$, and $i+2$. The error in the difference approximations as described earlier can be found by including one more term in the polynomial. (i.e., using $n = 3$ instead of $n = 2$ in deriving Eq. (7.95)). In this case, Eqs. (7.97) to (7.99) can be found to have an accuracy of $O(h^2)$. For deriving mth-order difference approximation, we need to consider the Newton–Gregory interpolation formula of order m or greater.

7.9 Finite-Difference Approximations for Partial Derivatives

The finite difference approximations considered in the previous sections are valid for total or ordinary derivatives where the function f depends on a single independent variable x. In many engineering problems, the function f depends on two or more independent variables. In such cases, the finite-difference approximations of partial derivatives will be of interest. By noting that a partial derivative denotes the local variation of the function with respect to a particular independent variable while all other independent variables are held constant, we can readily adapt the finite-difference approximations of ordinary derivatives for partial derivatives. If there are two independent variables, we use the notation (i, j) to designate the pivot point, and if there are three independent variables, we use the notation (i, j, k), where i, j, and k are the counters in the x, y, and z directions. The step sizes along x, y, and z directions are denoted as Δx, Δy, and Δz, respectively. Typical two- and three-dimensional finite-difference grids are shown as in Figs. 7.9 and 7.10. Consider a two-dimensional function $f(x, y)$. The finite-difference approximation for the partial derivative $\frac{\partial f(x,y)}{\partial x}$ at $(x = x_i, y = y_j)$ can be found by fixing the value of y at y_j and treating $f(x, y_j)$ as a one-variable function. Then the forward-, backward-, and central-difference approximations of $\frac{\partial f}{\partial x}$ can be expressed as follows:

$$\left.\frac{\partial f}{\partial x}\right|_{i,j} \approx \frac{f(x_i + \Delta x, y_j) - f(x_i, y_j)}{\Delta x}, \tag{7.100}$$

$$\left.\frac{\partial f}{\partial y}\right|_{i,j} \approx \frac{f(x_i, y_j) - f(x_i - \Delta x, y_j)}{\Delta x}, \tag{7.101}$$

and

$$\left.\frac{\partial f}{\partial x}\right|_{i,j} \approx \frac{f(x_i + \Delta x, y_j) - f(x_i - \Delta x, y_j)}{2\Delta x}. \tag{7.102}$$

The central-difference approximations of second partial derivatives at (x_i, y_j) can be derived as

$$\left.\frac{\partial^2 f}{\partial x^2}\right|_{i,j} \approx \frac{f(x_i + \Delta x, y_j) - 2f(x_i, y_j) + f(x_i - \Delta x, y_j)}{(\Delta x)^2}, \tag{7.103}$$

$$\frac{\partial^2 f}{\partial y^2} \approx \frac{f(x_i, y_j + \Delta y) - 2f(x_i, y_j) + f(x_i, y_j - \Delta y)}{(\Delta y)^2}, \tag{7.104}$$

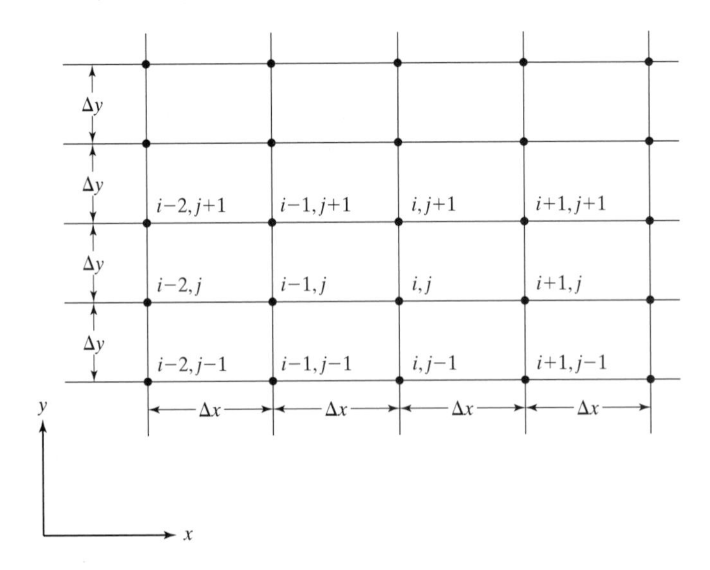

Figure 7.9 Two-dimensional grid.

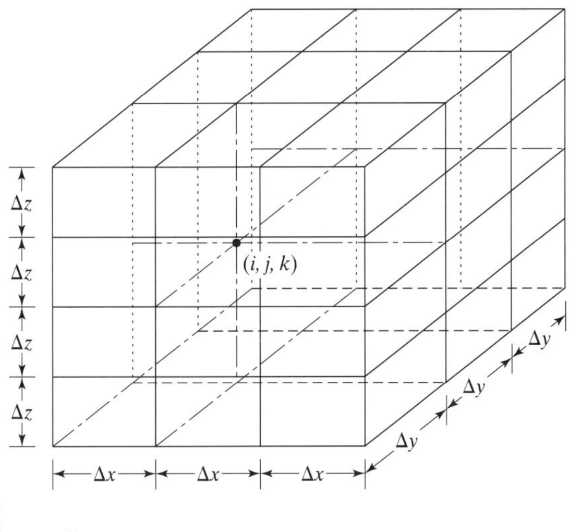

Figure 7.10 Three-dimensional grid.

and

$$\frac{\partial^2 f}{\partial x \partial y} \approx \frac{\left\{ \begin{array}{l} f(x_i + \Delta x, y_j + \Delta y) - f(x_i + \Delta x, y_j - \Delta y) \\ \quad - f(x_i - \Delta x, y_j + \Delta y) + f(x_i - \Delta x, y_j - \Delta y) \end{array} \right\}}{(4 \Delta x \Delta y)}. \tag{7.105}$$

The errors associated with the finite-difference approximations of partial derivatives can be found using Taylor's series expansions of $f(x, y)$ around the point (x_i, y_j) as

$$f_{i \pm 1, j} = f_{i,j} \pm \Delta x \left. \frac{\partial f}{\partial x} \right|_{i,j} + \frac{(\Delta x)^2}{2!} \left. \frac{\partial^2 f}{\partial x^2} \right|_{i,j} \pm \frac{(\Delta x)^3}{3!} \left. \frac{\partial^3 f}{\partial x^3} \right|_{i,j} + \cdots \tag{7.106}$$

and

$$f_{i, j \pm 1} = f_{i,j} \pm \Delta y \left. \frac{\partial f}{\partial y} \right|_{i,j} + \frac{(\Delta y)^2}{2!} \left. \frac{\partial^2 f}{\partial y^2} \right|_{i,j} \pm \frac{(\Delta y)^3}{3!} \left. \frac{\partial^3 f}{\partial y^3} \right|_{i,j} + \cdots \tag{7.107}$$

If Eq. (7.106), for example, is truncated after the nth order derivative term, the error or remainder can be expressed as

$$R_{x,n} \simeq (-1)^{n+1} \frac{(\Delta x)^{n+1}}{(n+1)!} \left. \frac{\partial^{n+1} f(x, y)}{\partial x^{n+1}} \right|_{i,j}. \tag{7.108}$$

▶Example 7.7

Evaluate the mixed partial derivative $\frac{\partial^4 f}{\partial x^2 \partial y^2}$ of the function $f(x, y) = 2x^4 y^3$ using central differences at $x = 1$ and $y = 1$ with a step size $\Delta x = \Delta y = 0.1$.

Solution

Denoting the constant step size as $\Delta x = \Delta y = h$, the mixed partial derivative, $\frac{\partial^4 f}{\partial x^2 \partial y^2}$, at $(x = x_i, y = y_j)$ can be expressed as

$$\frac{\partial^4 f}{\partial x^2 \partial y^2}(x_i, y_j) = \frac{\partial^2}{\partial y^2}\left(\left. \frac{\partial^2 f}{\partial x^2} \right|_{x_i} \right) \Bigg|_{y_j}. \tag{a}$$

The use of the central-difference formula for $\frac{\partial^2 f}{\partial x^2}|_{x_i}$ yields

$$\begin{aligned}
\frac{\partial^4 f}{\partial x^2 \partial y^2}(x_i, y_j) &= \frac{\partial^2}{\partial y^2}\left\{ \frac{1}{h^2}(f_{x_{i-1}, y_j} - 2 f_{x_i, y_j} + f_{x_{i+1}, y_j}) \right\} \\
&= \frac{1}{h^2}\left\{ \frac{\partial^2 f}{\partial y^2}(x_{i-1}, y_j) - 2\frac{\partial^2 f}{\partial y^2}(x_i, y_j) + \frac{\partial^2 f}{\partial y^2}(x_{i+1}, y_j) \right\} \\
&= \frac{1}{h^2}\left[\frac{1}{h^2}\{f_{i-1,j-1} - 2 f_{i-1,j} + f_{i-1,j+1}\} - \frac{2}{h^2}\{f_{i,j-1} \right. \\
&\qquad \left. -2 f_{i,j} + f_{i,j+1}\} + \frac{1}{h^2}\{f_{i+1,j-1} - 2 f_{i+1,j} + f_{i+1,j+1}\} \right] \\
&= \frac{1}{h^4}\{f_{i-1,j-1} - 2 f_{i-1,j} + f_{i-1,j+1} - 2 f_{i,j-1} + 4 f_{i,j} \\
&\qquad -2 f_{i,j+1} + f_{i+1,j-1} - 2 f_{i+1,j} + f_{i+1,j+1}\}.
\end{aligned} \tag{b}$$

The various function values, $f(x, y)$, involved in Eq. (b) can be evaluated as

$$\begin{aligned}
f_{i-1,j-1} &= f(0.9, 0.9) = 0.95659357, \\
f_{i-1,j} &= f(0.9, 1.0) = 1.31219983, \\
f_{i-1,j+1} &= f(0.9, 1.1) = 1.74653804, \\
f_{i,j-1} &= f(1.0, 0.9) = 1.45799982, \\
f_{i,j} &= f(1.0, 1.0) = 2.0, \\
f_{i,j+1} &= f(1.0, 1.1) = 2.66200018, \\
f_{i+1,j-1} &= f(1.1, 0.9) = 2.13465762, \\
f_{i+1,j} &= f(1.1, 1.0) = 2.92820024,
\end{aligned}$$

and

$$f_{i+1,j+1} = f(1.1, 1.1) = 3.89743471.$$

With these, Eq. (b) gives the value of the numerical mixed partial derivative as $\frac{\partial^4 f}{\partial x^2 \partial y^2}(x = 1.0, y = 1.0) = 144.23606873$. This value can be compared with the exact (analytical) value of $\frac{\partial^4 f}{\partial x^2 \partial y^2}(1.0, 1.0) = 144x^2 y|_{x=1.0, y=1.0} = 144.0$. ◀

7.10 Choice of Method

Although the central-difference method yields the most accurate results, the forward- and backward-difference methods are most useful in solving initial value problems. (See Chapter 9.) When finite-difference approximations are used for the solution of ordinary and partial differential equations, it is most desirable to use expressions having the same order of error for the various order derivatives involved in the equation in order to reduce instability problems that might be caused by the propagation of error. In general, the evaluation of derivatives of functions whose values are known in tabular form (from experiments) is to be avoided as far as possible to minimize the errors. If tabular data are to be used for the computation of derivatives, it is desirable to approximate the functional relationship based on least-squares approximation before differentiation.

When function values are known only at some discrete points (as in the case of experimental data), it is not desirable to use the finite-difference formulas directly. This is particularly true when the data are available at unequally spaced points. In these cases, it is desirable to fit an interpolation polynomial for the data and differentiate the resulting polynomial for evaluating the required derivatives.

7.11 Use of Software Packages

7.11.1 MATLAB

▶Example 7.8

Find the first three derivatives of $f(x) = x^2 e^{-4x}$ and express Taylor's series expansion of $f(x)$ at $x = 1$.

Solution

The first three derivatives of $f(x)$ are labeled y1, y2 and y3 and are evaluated at x = 1.0 to express Taylor's series expansion of $f(x)$.

```
>> x=1.0;
>> y1=2*x*exp(-4*x)-4*x^2*exp(-4*x)
y1  =
    -0.03663127777747
>> y2 = 2*x*exp(-4*x)-16*x*exp(-4*x)+16*x^2*exp(-4*x)
```

```
y2  =
    0.03663127777747
>> y3=-24*exp(-4*x)+96*x*exp(-4*x)-64*x^2*exp(-4*x)
y3  =
    0.1452511110987
```

Taylor's series expansion of $f(x)$ at x_0:

$$f(x) = f(x_0) + (x - x_0)f'(x_0) + \frac{1}{2!}(x - x_0)^2 f''(x_0) + \frac{1}{3!}(x - x_0)^3 f'''(x_0) + \cdots$$

$$= e^{-4} + (x-1)(-0.03663128) + \frac{1}{2}(x-1)^2(0.03663128)$$

$$+\frac{1}{6}(x-1)^3(0.14652511) + \cdots .$$

◀

▶Example 7.9

Consider the displacement (y) data given in Example 7.6.

(a) Find $\frac{dy}{dt}$, $\frac{d^2y}{dt^2}$, and $\frac{d^3y}{dt^3}$ using forward-difference formula with a step size, Δt, of 0.05.

(b) Find $\frac{dy}{dt}$, $\frac{d^2y}{dt^2}$, and $\frac{d^3y}{dt^3}$ by fitting a fourth-degree polynomial for the data.

Solution

(a)
```
>>h=0.05;
>>y=[0.144 0.172 0.213 0.296 0.070 0.085 0.525 0.110 0.062 0.055 0.042 0.035];
>>y1=diff(y)/h
y1 =
  Columns 1 through 4
   0.56000000000000  0.82000000000000  1.66000000000000 -4.52000000000000
  Columns 5 through 8
   0.30000000000000  8.80000000000000 -8.30000000000000 -0.96000000000000
  Columns 9 through 11
  -0.14000000000000 -0.26000000000000 -0.14000000000000
 >>y2=diff(y1)/h
  y2 =
  1.0e+0.02 *
  Columns 1 through 4
   0.05200000000000  0.16800000000000 -1.23600000000000   0.96400000000000
  Columns 5 through 8
   1.70000000000000 -3.42000000000000  1.46800000000000   0.16400000000000
  Columns 9 through 10
  -0.02400000000000  0.02400000000000
>> y3=diff(y2)/h
y3 =
  1.0e+0.04 *
```

```
Columns 1 through 4
 0.02320000000000 -0.28080000000000  0.44000000000000   0.14720000000000
Columns 5 through 8
-1.02400000000000  0.97760000000000 -0.26080000000000  -0.03760000000000
Column 9
 0.00960000000000
```

Note:

The function
z = diff(y)
calculates differences between adjacent elements of y (forward differences). The output vector z is one element shorter than the input vector y.

(b) The discrete values of t are defined before fitting a least squares 4th order polynomial using the command z = polyfit(t,y,4). The function y1 = diff(f(t))/h first computes the values of the polynomial $f(t)$ at the discrete values of t and then finds the values of differences between adjacent elements of $f(t)$ divided by h.

```
>> t=0.05 : 0.05 : 0.60;
>> t
t  =
   Columns 1 through 4
    0.05000000000000   0.10000000000000  0.15000000000000  0.20000000000000
   Columns 5 through 8
    0.25000000000000   0.30000000000000  0.35000000000000  0.40000000000000
   Columns 9 through 12
    0.45000000000000   0.50000000000000  0.55000000000000  0.60000000000000

>> z=polyfit(t,y,4)
z  =
   Columns 1 through 4
    34.21328671328872 -42.27991452991719 15.26277680652800 -1.62366792929313
   Column 5
    0.20928030303031

>> y1=diff(34.21328671328872*t.^4-42.27991452991719*t.^3+15.26277680652800
     *t.^2-1.62366792929313*t+0.20928030303031)/h
y1 =
   Columns 1 through 4
    -0.01000000000006  0.46171328671327  0.55582750582752  0.37498251748254
   Columns 5 through 8
    0.02181818181820  -0.40102564102563 -0.79090909090910 -1.04519230769232
   Columns 9 through 11
    -1.06123543123544 -0.73639860139859   0.03195804195809
```

```
>> y2=diff(y1)/h
   y2 =
   Columns 1 through 4
    9.43426573426676    1.88228438228494   -3.61689976689963   -7.06328671328680
   Columns 5 through 8
   -8.45687645687658 -7.79766899766945   -5.08566433566435   -0.32086247086238
   Columns 9 through 10
    6.49673659673700   15.36713286713360

>> y3=diff(y2)/h
   y3 =
   1.0e+002*
   Columns 1 through 4
    -1.51039627039636 -1.09983682983691   -0.68927738927743   -0.27871794871796
   Columns 5 through 8
    0.13184149184143  0.54240093240102    0.95296037296039    1.36351981351988
   Column 9
    1.77407925407932
```

◀

7.11.2 MAPLE

▶Example 7.10

Find the first three derivatives of $f(x) = x^2e^{-4x}$ and determine Taylor's series expansion of $f(x)$ at $x_0 = 1$.

Solution

The command

diff (f,x);

is used to differentiate the given function with respect to x in symbolic form.

g: = unapply (%,x);

defines the derivative as g(x) and

evalf(g(1));

returns the value of g(x) at x = 1. The command

taylor(f,x = 1);

returns 6 terms of the Taylor's series expansion of the given function f(x) at x = 1. The command

taylor(f, x = 1, 4);

returns 4 terms of Taylor's function f(x) at x = 1.

```
> diff(x^2* exp(-4*x), x);
```

$$2x\,e^{(-4x)} - 4x^2\,e^{(-4x)}$$

```
>
>
> g := unapply(%, x);
```

$$g := x \to 2xe^{(-4x)} - 4x^2e^{(-4x)}$$

```
> evalf(g(1));
```

$$-.03663127778$$

```
>
> diff(x^2* exp(-4*x),x, x);
```

$$2e^{(-4x)} - 16xe^{(-4x)} + 16x^2e^{(-4x)}$$

```
> gg := unapply(%,x);
```

$$gg := x \to 2e^{(-4x)} - 16xe^{(-4x)} + 16x^2e^{(-4x)}$$

```
> evalf(gg(1));
```

$$.03663127778$$

```
>
> diff(x^2* exp(-4*x), x, x,x);
```

$$-24e^{(-4x)} + 96xe^{(-4x)} - 64x^2e^{(-4x)}$$

```
> ggg := unapply(%, x);
```

$$ggg := x \to -24e^{(-4x)} + 96xe^{(-4x)} - 64x^2e^{(-4x)}$$

```
> evalf(ggg(1));
```

$$.1465251111$$

```
>
> taylor(x^2* exp(-4*x), x = 1);
```

$$e^{(-4)} - 2e^{(-4)}(x-1) + e^{(-4)}(x-1)^2 + \frac{4}{3}e^{(-4)}(x-1)^3 - \frac{8}{3}e^{(-4)}(x-1)^4 + \frac{32}{15}e^{(-4)}(x-1)^5 + O((x-1)^6)$$

```
>
> taylor(x^2* exp(-4*x), x = 1, 4);
```

$$e^{(-4)} - 2e^{(-4)}(x-1) + e^{(-4)}(x-1)^2 + \frac{4}{3}e^{(-4)}(x-1)^3 + O((x-1)^4)$$

```
> evalf(%);
```

$$.01831563889 - .03663127778(x - 1.) + .01831563889(x - 1.)^2 + .02442085185(x - 1.)^3 + O((x - 1.)^4)$$

```
>
```

▶Example 7.11

Consider the displacement data given in Example 7.6. Fit a polynomial through the data and find the first, second, and third derivatives of the displacement at $t_i, i = 1, 2, \ldots, 12$.

Solution

The vector t (having n components $t_1, t_2, \ldots, t_n$) is defined using the statement

t := linalg[vector](n, [$t_1, \ldots, t_n$]);

The statement

interp(t,y,z);

finds the polynomial of order n − 1 through n data points (t_i, y_i) given by the vectors t and y. The statement

f := unapply(%, z);

stores the polynomial as f(z). The statement

yl := [seq(f(t[i]), i = 1..12)];

defines 12 values of f at the discrete values of t[i] and stores them as vector yl. The statement

diff(f(z),z);

differentiates the function $f(z)$ with respect to z symbolically.

```
>
```

```
> t := linalg [vector](12, [0.05, 0.10, 0.15, 0.20, 0.25, 0.30, 0.35, 0.40, 0.45,
  0.50, 0.55, 0.60]);
```

$$t := [.05, .10, .15, .20, .25, .30, .35, .40, .45, .50, .55, .60]$$

```
> y := linalg [vector](12, [0.144, 0.172, 0.213, 0.296, 0.070, 0.085, 0.525, 0.110,
  0.062, 0.055,0.042, 0.035]);
```

$$y := [.144, .172, .213, .296, .070, .085, .525, .110, .062, .055, .042, .035]$$

```
> interp(t, y, z);
```

$$-.8149507996\ 10^9 z^{11} + .2873044167\ 10^{10} z^{10} - .4455578243\ 10^{10} z^9 + .3999647507\ 10^{10} z^8 - .2300941981\ 10^{10} z^7 + .8871454633\ 10^9 z^6 - .2328470799\ 10^9 z^5 + .4138347339\ 10^8 z^4 - .4849759291\ 10^7 z^3 + 354017.6607 z^2 - 14324.38178 z + 239.4450051$$

```
> f := unapply (%, z);
```

$$f := z \rightarrow -.8149507996\ 10^9 z^{11} + .2873044167\ 10^{10} z^{10} - .4455578243\ 10^{10} z^9 + .3999647507\ 10^{10} z^8 - .2300941981\ 10^{10} z^7 + .8871454633\ 10^9 z^6 - .2328470799\ 10^9 z^5 + .4138347339\ 10^8 z^4 - .4849759291\ 10^7 z^3 + 354017.6607 z^2 - 14324.38178 z + 239.4450051$$

```
> yl := [seq(f(t[i]), i = 1..12)];
```

$$yl := [.1440002, .1720001, .2129841, .2959791, .0699301, .0848311, .5258221, .1094031, .0610941, .0408951, .0574261, .0378371]$$

```
> diff(f(z), z);
```

$$.2873044167\ 10^{11} z^9 - .4010020419\ 10^{11} z^8 - .8964458796\ 10^{10} z^{10} + .3199718006\ 10^{11} z^7 - .1610659387\ 10^{11} z^6 + .5322872780\ 10^{10} z^5 - .1164235400\ 10^{10} z^4 + .1655338936\ 10^9 z^3 - .1454927787\ 10^8 z^2 + 708035.3214 z - 14324.38178$$

```
> g := unapply(%, z);
```

$$g := z \rightarrow .2873044167\ 10^{11} z^9 - .4010020419\ 10^{11} z^8 - .8964458796\ 10^{10} z^{10} + .3199718006\ 10^{11} z^7 - .1610659387\ 10^{11} z^6 + .5322872780\ 10^{10} z^5 - .1164235400\ 10^{10} z^4 + .1655338936\ 10^9 z^3 - .1454927787\ 10^8 z^2 + 708035.3214 z - 14324.38178$$

```
> yl := [seq(g(t[i]), i = 1..12)];
```

$$yl := [-445.32610, 45.40915, -7.89598, 2.76642, -7.71528, 9.45562, 1.32972, -11.72218, 9.80382, -20.58908, 75.73902, -669.30198]$$

```
>
>
> diff(f(z), z, z);
```

$$.2585739750\ 10^{12} z^8 - .3208016335\ 10^{12} z^7 - .8964458796\ 10^{11} z^9 + .2239802604\ 10^{12} z^6 - .9663956322\ 10^{11} z^5 + .2661436390\ 10^{11} z^4 - .4656941600\ 10^{10} z^3 + .4966016808\ 10^9 z^2 - .2909855574\ 10^8 z + 708035.3214$$

```
> gg := unapply(%, z);
```

$gg := z \rightarrow .2585739750\ 10^{12} z^8 - .3208016335\ 10^{12} z^7 - .8964458796\ 10^{11} z^9$
$+ .2239802604\ 10^{12} z^6 - .9663956322\ 10^{11} z^5 + .2661436390\ 10^{11} z^4 - .4656941600\ 10^{10} z^3$
$+ .4966016808\ 10^9 z^2 - .2909855574\ 10^8 z + 708035.3214$

```
> yl := [seq(gg(t[i]), i = 1..12)];
```

$yl := [51892.9380, -3308.0946, 545.6994, -292.2766, 136.1864, 262.2694, -492.3886,$
$207.6514, 177.3414, -818.3486, 5120.3614, -75382.0186]$

```
>
>
> diff(f(z), z, z, z);
```

$.2068591800\ 10^{13} z^7 - .2245611435\ 10^{13} z^6 - .8068012916\ 10^{12} z^8 + .1343881562\ 10^{13} z^5$
$- .4831978161\ 10^{12} z^4 + .1064574556\ 10^{12} z^3 - .1397082480\ 10^{11} z^2 + .9932033616\ 10^9 z$
$- .2909855574\ 10^8$

```
> ggg := unapply(%, z);
```

$ggg := z \rightarrow .2068591800\ 10^{13} z^7 - .2245611435\ 10^{13} z^6 - .8068012916\ 10^{12} z^8$
$+ .1343881562\ 10^{13} z^5 - .4831978161\ 10^{12} z^4 + .1064574556\ 10^{12} z^3 - .1397082480\ 10^{11} z^2$
$+ .9932033616\ 10^9 z - .2909855574\ 10^8$

```
> yl := [seq(ggg(t[i]), i = 1..12)];
```

$yl := [-.369179418\ 10^7, 43201.76, 13105.76, -12204.74, 17381.66, -14534.24,$
$-3781.14, 22150.86, -31243.04, 41395.06, 29371.16, -.514899674\ 10^7]$

```
>
```

◀

7.11.3 MATHCAD

▶Example 7.12

Find Taylor's series expansion of the function $f(x) = x^2 e^{-4}$x about $x = 1$ using the first three derivative terms.

Solution

The function

series, $x = z, m$

expands the given expression in terms of the variable x around the point z and gives m terms symbolically. By defining $x := 1$ and using the expression $\frac{d}{dx}(x^2 e^{-4} x)$ yields the derivative of $(x^2 e^{-4} x)$ evaluated at $x = 1$.

$x^2 \cdot e^{-4x}$series,

$x = 1, 4 \rightarrow \exp(-4) - 2\exp(-4)\cdot(x-1) + \exp(-4)\cdot(x-1)^2$

$+\frac{4}{3}\cdot\exp(-4)\cdot(x-1)^3$

$x := 1$

$$\frac{d}{dx}(x^2 \cdot e^{-4\cdot x}) \rightarrow -2\cdot\exp(-4) = -0.037$$

$$\frac{d^2}{dx^2}(x^2 \cdot e^{-4\cdot x}) \rightarrow 2\cdot\exp(-4) = 0.037$$

$$\frac{d^3}{dx^3}(x^2 \cdot e^{-4\cdot x}) \rightarrow 8\cdot\exp(-4) = 0.147$$

◀

▶Example 7.13

Consider the displacement data given in Example 7.6. Fit polynomials of degrees 4 and 5 and evaluate the first, second, and third derivatives of displacement at $t_i, i = 1, 2, \ldots, 12$ in each case.

Solution

S := linfit(x,y,F)

returns a vector S that contains the coefficients to be used to create a linear combination of the functions in F that best approximates the data points in the vectors x and y. F contains the vector of functions to be linearly combined. The elements of x should be in ascending order.

g(x) := F(x) · S

defines the approximating function as a dot product of F and S. The given data (y) and the approximating function (g) are plotted on the same graph for comparison.

$$t := \begin{pmatrix} 0.05 \\ 0.10 \\ 0.15 \\ 0.20 \\ 0.25 \\ 0.30 \\ 0.35 \\ 0.40 \\ 0.45 \\ 0.50 \\ 0.55 \\ 0.60 \end{pmatrix} \quad y := \begin{pmatrix} 0.144 \\ 0.172 \\ 0.213 \\ 0.296 \\ 0.070 \\ 0.085 \\ 0.525 \\ 0.110 \\ 0.062 \\ 0.055 \\ 0.042 \\ 0.035 \end{pmatrix} \quad F(x) := \begin{pmatrix} 1 \\ x \\ x^2 \\ x^3 \\ x^4 \end{pmatrix}$$

$$j := 0..11 \qquad S := \text{linfit}(t,y,F)$$

$$S = \begin{pmatrix} 0.209 \\ -1.624 \\ 15.263 \\ -42.28 \\ 34.213 \end{pmatrix} \quad r := 00.005..0.8$$

$$g(x) := F(x) \cdot S$$

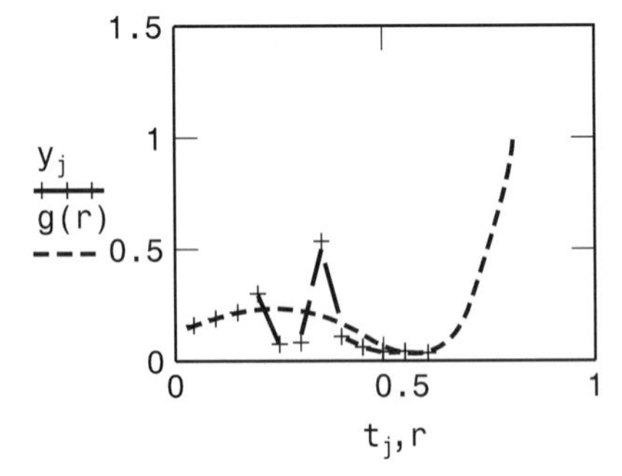

$i := 0.05, 0.10..0.60$

$f1(x) := \frac{d}{dx} g(x)$ $\quad f2(x) := \frac{d^2}{dx^2} g(x)$ $\quad f3(x) := \frac{d^3}{dx^3} g(x)$

g(i) =	f1(i) =	f2(i) =	f3(i) =
0.161	−0.397	18.868	−212.624
0.161	0.297	9.263	−171.568
0.184	0.563	1.711	−130.512
0.212	0.503	−3.788	−89.456
0.23	0.219	−7.234	−48.4
0.231	−0.187	−8.628	−7.344
0.211	−0.61	−7.969	33.712
0.172	−0.949	−5.257	74.768
0.12	−1.101	−0.492	115.824
0.066	−0.964	6.326	156.88
0.03	−0.435	15.196	197.936
0.031	0.59	26.119	238.992

$$F(x) := \begin{pmatrix} 1 \\ x \\ x^2 \\ x^3 \\ x^4 \\ x^5 \end{pmatrix}$$

$j := 0..11$ $\quad S := \text{linfit}(t,y,F)$

$$S = \begin{pmatrix} -0.094 \\ 7.377 \\ -67.148 \\ 273.887 \\ -500.346 \\ 328.959 \end{pmatrix} \qquad r := 0, 0.005..0.8$$

$g(x) := F(x) \cdot S$

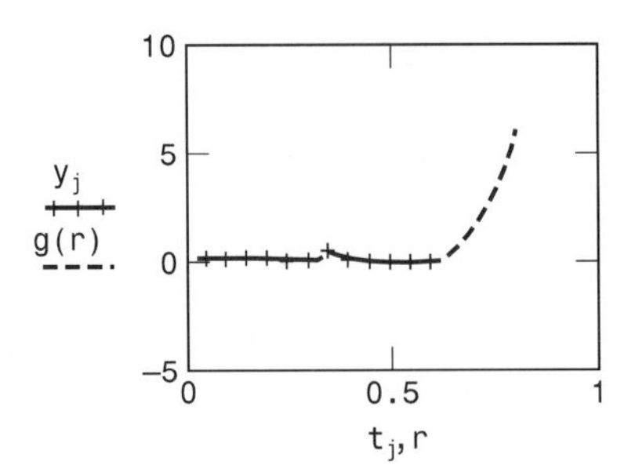

i := 0.05, 0.10..0.60

$f1(x) := \frac{d}{dx}g(x)$ $f2(x) := \frac{d^2}{dx^2}g(x)$ $f3(x) := \frac{d^3}{dx^3}g(x)$

g(i) =	f1(i) =	f2(i) =	f3(i) =
0.139	2.477	−66.319	$1.092 \cdot 10^3$
0.2	0.327	−23.427	639.865
0.198	−0.202	−0.687	286.17
0.192	$.899 \cdot 10^{-3}$	6.835	31.163
0.2	0.31	4.074	−125.157
0.218	0.323	−4.036	−182.789
0.225	−0.1	−12.56	−141.733
0.202	−0.858	−16.565	−1.989
0.139	−1.599	−11.115	236.442
0.052	−1.729	8.724	573.562
$9.402 \cdot 10^{-3}$	−0.405	47.886	$1.009 \cdot 10^3$
0.054	3.464	111.306	$1.544 \cdot 10^3$

◀

7.12 Computer Programs

7.12.1 Fortran Programs

Program 7.1 Finite difference derivatives

The following input data are required:

X = value of x at which derivatives are to be evaluated
DX = increment of x to be used in the finite difference formulas

F(X) = function to be differentiated in the form of the subroutine

```
SUBROUTINE FUN(X,F)
F = ...
RETURN
END
```

Illustration

Evaluate the derivatives of $f(x) = e^x$ at $x = 2$.

Program output

```
FINITE DIFFERENCE DERIVATIVES:
EVALUATED AT X = 0.200000E+01 STEP LENGTH (DX) = 0.200000E-01

FORWARD DIFFERENCES:
FIRST, SECOND, THIRD DERIVATIVES:
  0.746343E+01   0.753880E+01   0.762939E+01

BACKWARD DIFFERENCES:
FIRST, SECOND, THIRD DERIVATIVES:
  0.731566E+01   0.724435E+01   0.733137E+01

CENTRAL DIFFERENCES:
FIRST, SECOND, THIRD DERIVATIVES:
  0.738955E+01   0.738859E+01   0.736117E+01

EXACT VALUES:
FIRST, SECOND, THIRD DERIVATIVES:
  0.738906E+01   0.738906E+01   0.738906E+01
```

Program 7.2 Derivatives based on Newton's interpolation

The Input data required are

N = number of data points
X(1), ..., X(N) = values of x at which function values are known
F(1), ..., F(N) = known values of f at X(1), ..., X(N)
XX = value of x at which derivatives are to be evaluated

Illustration

Find the derivatives of $f(x)$ at $x = 3.5$ using the following data:

x_i	0	1	3	4.5
$f_i = f(x_i)$	1.0	2.718282	20.085537	90.017131

The output of Program 7.2 is as follows:

```
RESULTS FROM NEWTONS INTERPOLATION FORMULA:
FUNCTION VALUE AT 2.0:     0.356105E+02
FIRST DERIVATIVE AT 2.0:     0.378891E+02
SECOND DERIVATIVE AT 2.0:     0.292496E+02

EXACT VALUES:
FUNCTION VALUE:     0.331155E+02
FIRST DERIVATIVE:     0.331155E+02
SECOND DERIVATIVE:     0.331155E+02
```

7.12.2 C Programs

Program 7.3 Higher accuracy finite difference formulas

The required input data are

X = value of x at which derivatives are to be evaluated.
DX = increment of x to be used in finite difference formulas
F(X) = function to be differentiated in the form of the subroutine:
SUBROUTINE FUN(X,F)
F = . . .
RETURN
END

Illustration
Evaluate the derivatives of the function $f(x) = e^x$ at $x = 2$.

Program output

```
Higher accuracy finite difference derivatives:

evaluated at x = 0.200000E+01
step length (dx) = 0.200000E-01

Forward differences:
First, Second, Third derivatives:
   0.738807E+01 0.738621E+01 0.715256E+01

Backward differences:
First, Second, Third derivatives:
   0.738808E+01 0.739217E+01 0.858307E+01

Central differences:
First, Second, Third derivatives:
   0.738905E+01 0.738770E+01 0.738353E+01
```

```
Exact derivatives:
First, Second, Third derivatives:
   0.738906E+01 0.738906E+01 0.738906E+01
```

Program 7.4 Taylor's series approximation

The Following input data are required:

X = base value of x at which function and its derivatives are known
XN = value of x at which function value is to be estimated
F(X) = function to be used in the form of the subroutine:
SUBROUTINE FUN(X,F)
F = ...
RETURN
END

Illustration

Estimate the value of $f(x) = e^x$ at $x = 2.2$ using Taylor's series expansion at $x = 2.0$

Using $x = 2.0$ and $xn = 2.2$, the output of Program 7.4 is

```
Approximation of f using Taylors series

at xn = 0.220000E+01
base point = 0.200000E+01

Central differences:
step length, dx = 0.200000E-01
First, Second, Third derivatives:
   0.738955E+01   0.738859E+01    0.736117E+01

Using first derivative term, fn1 = 0.886697E+01
Using first two derivative terms, fn2 = 0.901474E+01
Using first three derivative terms, fn3 = 0.902455E+01

Exact function value at xn = 0.220000E+01
fn = 0.902501E+01
```

REFERENCES AND BIBLIOGRAPHY

7.1. F. M. White, *Heat and Mass Transfer*, Addison-Wesley, Reading, MA, 1988.

7.2. S. S. Rao, *Mechanical Vibrations*, 3d edition, Addison-Wesley, Reading, MA, 1995.

7.3. A. H. Burr and J. B. Cheatham, *Mechanical Analysis and Design*, 2d edition, Prentice Hall, Upper Saddle River, NJ, 1995.

7.4. W. F. Stoecker, *Design of Thermal Systems*, 3d edition, McGraw-Hill, New York, 1989.

7.5. A. Ralston, *A First Course in Numerical Analysis*, McGraw-Hill, New York, 1965.
7.6. A. Constantinides, *Applied Numerical Methods with Personal Computers*, McGraw-Hill, New York, 1987.
7.7. R. C. Juvinall and K. M. Marshek, *Fundamentals of Machine Component Design*, 2d edition, Wiley, New York, 1991.
7.8. J. C. Lange, *Kinematics: A Graphical Approach*, Prentice Hall, Upper Saddle River, NJ, 1995.
7.9. J. E. Shigley and C. R. Mischke, *Mechanical Engineering Design*, 5th edition, McGraw-Hill, New York, 1989.
7.10. I. Cochin, *Analysis and Design of Dynamic Systems*, Harper & Row, New York, 1980.
7.11. S. S. Rao, *Engineering Optimization: Theory and Practice*, 3d edition, Wiley, New York, 1996.

REVIEW QUESTIONS

The following questions along with corresponding answers are available in an interactive format at the Companion Web site at http://www.prenhall.com/rao.

7.1. Give **brief answers:**

1. What is finite-difference calculus?
2. What is mean-value theorem?
3. Name three different approaches that can be used for deriving finite-difference formulas.
4. Indicate the order of truncation error involved in Taylor's series expansion of a function.
5. Give Taylor's series expansion of the function $f(x)$ about x_i.
6. How many terms are needed in Taylor's series expansion for an exact representation of an nth-order polynomial?
7. What is the significance of mean-value theorem in the context of remainder term in Taylor's series expansion?
8. What is the basis for deriving higher accuracy finite-difference formulas?
9. What is the physical justification for having the sum of all the coefficients of the function values in the numerator of any finite-difference formula equal to zero?
10. What is the difference between an ordinary derivative and a partial derivative?

7.2. **Fill in the blanks** with suitable words:

1. The use of finite-differences transforms an ordinary differential equation into an ______ equation.
2. The three-term Taylor's series approximation of a function denotes the exact representation of the function if the function is a ______ .
3. The sum of all the coefficients of the function values (f_i) appearing in the numerator of all the finite-difference formulas is ______ .
4. The accuracy of the computed derivative using finite differences can be improved by using either a smaller ______ or a higher ______ formula.
5. The difference operators can be treated as ______ variables.

7.3. Answer **true or false:**

1. Numerical differentiation is inherently more accurate than numerical integration.
2. Any interpolation polynomial can be used to derive the finite-difference formulas.
3. Usually the lower order derivative terms contribute less to the remainder term in Taylor's series expansion.

(*continued, page 546*)

4. Higher accuracy finite-difference formulas can be derived by retaining more terms in Taylor's series expansions.
5. The relationship indicated between a differential operator and the corresponding difference operator is exact.
6. The sum of coefficients in the numerator of certain finite-difference formulas can be nonzero.

7.4. **Define** the following:

Derivative of a function $f(x)$; partial derivatives of a function $f(x, y)$; remainder term of Taylor's series expansion; difference operators; basic finite difference formulas.

7.5. **Match the following:**

1. Two-point forward differences	(a) $(f_{i+2} - 2f_{i+1} + f_i)/(\Delta x)^2$
2. Two-point backward differences	(b) $(f_{i+1} - f_i)/(\Delta x)$
3. Two-point central differences	(c) $(f_i - 2f_{i-1} + f_{i-2})/(\Delta x)^2$
4. Three-point forward differences	(d) $(f_i - f_{i-1})/(\Delta x)$
5. Three-point backward differences	(e) $(f_{i+1} - 2f_i + f_{i-1})/(\Delta x)^2$
6. Three-point central differences	(f) $(f_{i+1} - f_{i-1})/2(\Delta x)$

7.6. **Match the following:**

Formula	Error proportional to
1. $f_i' = (f_{i+1} - f_i)/(\Delta x)$	(a) $(\Delta x)^2$ and f'''
2. $f_i' = (f_{i+1} - f_{i-1})/(2\Delta x)$	(b) Δx and f''
3. $f_i'' = (f_{i+2} - 2f_{i+1} + f_i)/(\Delta x)^2$	(c) $(\Delta x)^2$ and f'''
4. $f_i'' = (f_{i+1} - 2f_i + f_{i-1})/(\Delta x^2)$	(d) Δx and f''''
5. $f_i''' = (f_{i+3} - 3f_{i+2} - 3f_{i+1} - f_i)/(\Delta x^3)$	(e) Δx and f'''
6. $f_i' = (-f_{i+2} + 4f_{i+1} - 3f_i)/(2\Delta x)$	(f) $(\Delta x)^2$ and f''''

7.7. **Match the following:**

Notation	Meaning
1. Δf_i	(a) $f_{i+1} - 2f_i + f_{i-1}$
2. ∇f_i	(b) $f_{i+1} - 2f_i + f_{i-1}$
3. δf_i	(c) $f_{i+1} - f_i$
4. $\nabla\Delta f_i$	(d) $f_i - 2f_{i-1} + f_{i+2}$
5. $\Delta^2 f_i$	(e) $f_i - f_{i-1}$
6. $\nabla^2 f_i$	(f) $f_{i+\frac{1}{2}} - f_{i-\frac{1}{2}}$
7. $\delta^2 f_i$	(g) $f_{i+2} - 2f_{i+1} + f_i$

PROBLEMS

Section 7.2

7.1. The contact (shear) stress developed between the two spheres shown in Fig. 7.11 is given by [7.9]

$$\tau_{xz} = \frac{1}{2} p_{\max} \left[-(1+\nu)\left(1 - \tilde{z}\tan^{-1}\frac{1}{\tilde{z}}\right) + \frac{1.5}{1+\tilde{z}^2} \right], \tag{a}$$

where

$$p_{\max} = \frac{3P}{2\pi a^2} \tag{b}$$

and

$$a = \left[\frac{3P}{8} \left\{ \frac{1-\nu_1^2}{E_1} + \frac{1-\nu_2^2}{E_2} \right\} \frac{d_1 d_2}{d_1 + d_2} \right]^{\frac{3}{2}}. \tag{c}$$

P = load applied, $\tilde{z} = z/a$, z = distance from the contact surface, ν_1 and ν_2 = Poisson's ratio of spheres 1 and 2, E_1 and E_2 = Young's moduli of spheres 1 and 2, d_1 and d_2 = diameters of spheres 1 and 2, ν = Poisson's ratio of the sphere under consideration, $p_{\max}$ = maximum pressure on the contact surface, and a = radius of the contact area. Determine the rate of change of τ_{xz} with respect to $\tilde{z}$, $\frac{d\tau_{xz}}{d\tilde{z}}$, for the following data:

$P = 1000$ lb, $\nu_1 = \nu_2 = \nu = 0.3$, $E_1 = E_2 = 30 \times 10^6$ psi, $d_1 = 2$ in, $d_2 = 1$ in, and $\bar{z} = 2$.

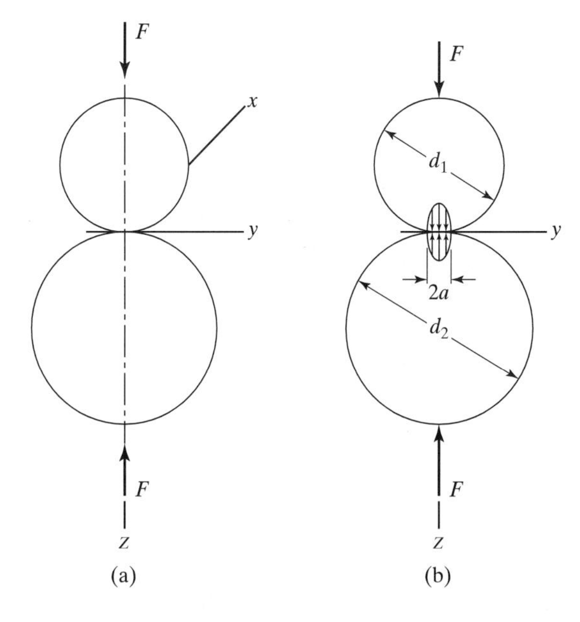

Figure 7.11 Contact of two spheres.

7.2. The length of a belt connecting two pulleys, x, is given by [7.8]

$$x = 2\sqrt{L^2 - (R_1 - R_2)^2} + \pi(R_1 + R_2) + 2(R_1 - R_2)\sin^{-1}\left(\frac{R_1 - R_2}{L}\right),$$

where L is the distance between the centers of the pulleys and R_1 and R_2 are the radii of the pulleys. Find the values of $\frac{dx}{dR_1}$ and $\frac{dx}{dR_2}$ when $L = 36$ in, $R_1 = 8$ in, and $R_2 = 6$ in.

7.3. The efficiency of a power screw, η, is given by [7.7]

$$\eta = \frac{L}{\pi d}\left(\frac{\pi d \cos\alpha_n - fL}{\pi f d + L\cos\alpha_n}\right),$$

where L = lead of the screw, d = mean diameter of thread contact, f = coefficient of friction, and α_n = thread angle measured in the normal plane, given by

$$\alpha_n = \tan^{-1}(\tan\alpha\cos\lambda),$$

where λ = helix angle and α = thread angle measured in the axial plane. Determine the value of $\frac{d\eta}{d\lambda}$ for the following data:
$L = 0.4''$, $d = 0.9''$, $f = 0.16$, $\alpha = 14.5°$, and $\lambda = 8.0°$.

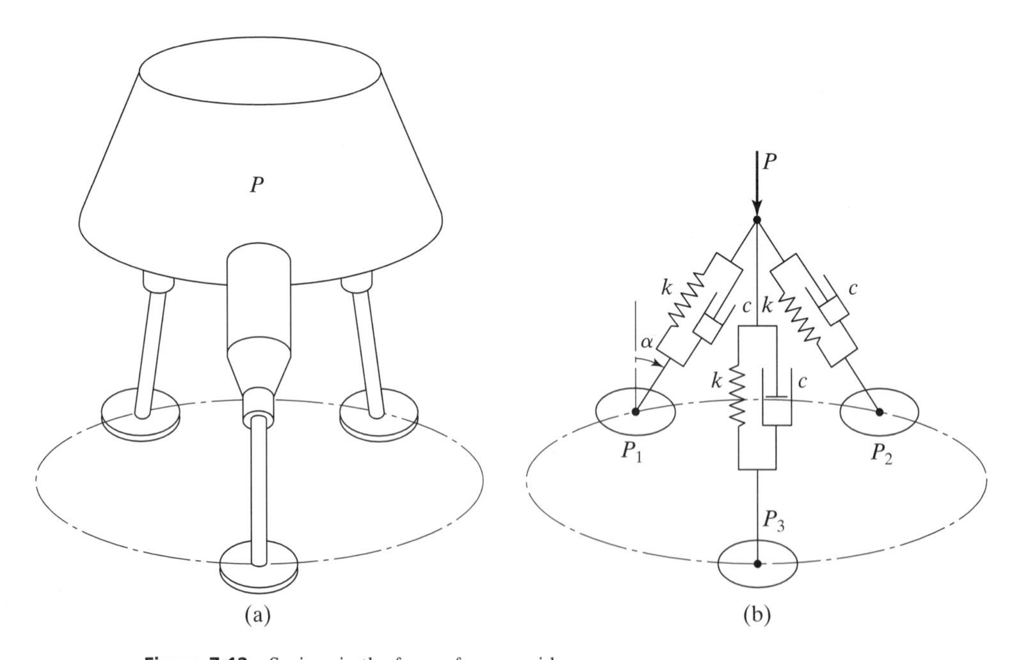

Figure 7.12 Springs in the form of a pyramid.

7.4. Three identical springs, of stiffness k_s each, are used to form a triangular pyramid whose apex supports a load P (Fig. 7.12). If y_s denotes the deformation of each spring, the load (P) can be expressed as [7.10]

$$P = 3k_s \cos\theta y_s, \tag{E1}$$

where y_s is related to the vertical deflection of the load, y, by

$$y = l\cos\theta\left[1 - \sqrt{1 - \left(\frac{2y_s + y_s^2}{l\cos^2\theta}\right)}\right]. \tag{E2}$$

The equivalent spring constant of the system can be determined as

$$k = \frac{dP}{dy} = \frac{dP}{dy_s}\frac{dy_s}{dy}. \tag{E3}$$

Determine the spring constant by evaluating the derivative, $\frac{dy}{dy_s}$, numerically for the following data:

$$P = 500 \text{ lb}, k_s = 1000 \text{ lb/in}, l = 30 \text{ in, and } \theta = 50^\circ.$$

7.5. The capacitance, C, of two eccentric cylinders, shown in Fig. 7.13, is given by [7.10]

$$C = \frac{2\pi\varepsilon l}{\cosh^{-1}\left(\frac{a^2 + b^2 - x^2}{2ab}\right)},$$

where ε = permitivity of the medium (water), a and b = radii of the cylinders, l = length of the cylinders, and x = eccentricity. Evaluate the derivative, $\frac{dC}{dx}$, numerically for the following data:

$$\varepsilon = 71 \times 10^{-11} \text{ F/m}, l = 0.25 \text{ m}, a = 0.1 \text{ m}, b = 0.3 \text{ m, and } x = 0.05 \text{ m}.$$

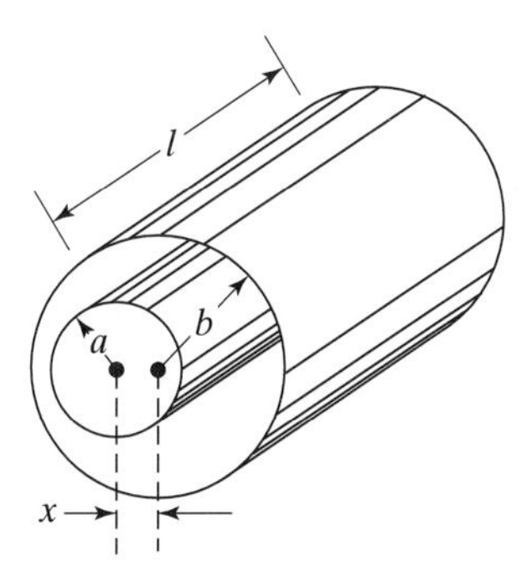

Figure 7.13 Two eccentric cylinders.

Section 7.4

7.6. The following table gives the values of a function $f(x)$:

x_i	$f(x_i)$
2.00	7.3891
2.05	7.7679
2.10	8.1662
2.15	8.5849
2.20	9.0250
2.25	9.4877
2.30	9.9742
2.35	10.486
2.40	11.023
2.45	11.588
2.50	12.182

Prepare tables for $f'(x_i)$ and $f''(x_i)$ in the range $2.00 \leq x \leq 2.50$ using suitable finite-difference formulas. Compare your results with the exact results given by $f(x) = e^x$.

7.7. Derive expressions for the first, second, and third derivatives of a function based on unequally spaced data.

Section 7.5

7.8. Find the value of $f = \sqrt{x}$ at $x = 27$ using Taylor's series expansion with known values of f and its derivatives at $x = 25$.

7.9. The altitude of a helicopter at three different instants is found to be as follows:

Time, t_i (sec)	Altitude, h_i (ft)
$t_1 = 0.20$	$h_1 = 445.98$
$t_2 = 0.30$	$h_2 = 471.85$
$t_3 = 0.41$	$h_3 = 503.46$

Determine the rate of climb, $\frac{dh}{dt}$, at $t = 0.30$ sec using a finite-difference approach.

Hint: Expand $h(t)$ in Taylor's series about t_2 twice by first using the step length -0.10 and then by using the step length $+0.11$. Derive an expression for $\frac{dh(t_2)}{dt}$ using a central-difference type of approach.

7.10. Using the displacement data given in Example 7.6, determine the velocity, acceleration, and jerk at $t = 0.05, 0.30$, and 0.60 sec using the higher accuracy finite-difference formulas given in Table 7.2.

7.11. Compute the central-difference approximations of f', f'', and f''' of the following function at $x = 2.0$ using formulas with truncation errors of $O(\Delta x^2)$ and $O(\Delta x^4)$:

$$f(x) = e^{2x};\ \Delta x = 0.1.$$

7.12. Solve Example 7.1.

7.13. Solve Example 7.3.

Section 7.6

7.14. Derive the following higher order difference operators of f at x_i : $\nabla^3 f_i$, $\nabla^4 f_i$, and $\nabla^5 f_i$.

7.15. Derive Eqs. (7.61), (7.62), and (7.63).

Section 7.7

7.16. Show the equivalence of the following expressions:

(a) $\nabla^3 f_{i+3} = \Delta^3 f_i$;

(b) $\delta^2 f_{i+1} = \Delta^2 f_i$.

7.17. Using the difference operators Δ, ∇, and δ, derive the forward-, backward-, and central-difference formulas for the fourth derivative of a function $f(x)$ at x_i.

Section 7.8

7.18. Using a third-order polynomial $f(x)$ passing through four uniformly spaced data points, derive the central difference formulas for $\frac{df}{dx}$ and $\frac{d^2 f}{dx^2}$.

7.19. Prove that Eqs. (7.97) through (7.99) have an error of $O(h^2)$.

7.20. Derive the forward-, backward-, and central-difference formulas for the first and second derivatives of $f(x)$ based on the polynomial $f(x) = a_0 + a_1 x + a_2 x^2$

Hint: Use the method of Section 7.8.1.

Section 7.9

7.21. Evaluate the following partial derivatives of the function $f(x, y) = 2x^4y^3$ using central differences at $x = 1$ and $y = 1$ with the step size $\Delta x = \Delta y = 0.1$:

$$\text{(a) } \frac{\partial^2 f}{\partial x^2}, \text{ (b) } \frac{\partial^2 f}{\partial y^2}, \quad \text{and} \quad \text{(c) } \frac{\partial^2 f}{\partial x \partial y}$$

Section 7.11

7.22. Use MAPLE to find Taylor's series expansion of the following function at $x = 0.5$:

$$f(x) = \frac{x^4}{(x-1)(x-3)^3}.$$

7.23. Use MATLAB to find the forward-, backward-, and central-difference values of the first derivative of the function given in Problem 7.22 at $x = 0.5$.

7.24. Use MATHCAD to find Taylor's series expansion of the function given in Problem 7.22 at $x = 0.5$.

7.25. Consider the data given in Problem 7.6. Use MATLAB to find the first, second, and third derivatives of the function $f(x)$

(a) using forward-difference formula with a step size of $\Delta x = 0.05$,

(b) by fitting a fourth-degree polynomial to the data.

7.26. Use MAPLE to fit a suitable polynomial for the data given in Problem 7.6 and find the first three derivatives of the function $f(x)$ at $x_i, i = 1, 2, \ldots, 11$.

7.27. Use MATHCAD to fit polynomials of degree 4, 5, and 6 for the data given in Problem 7.6 and determine the first three derivatives of the function $f(x)$ at $x_i, i = 1, 2, \ldots, 11$ in each case.

Section 7.12

7.28. Use Program 7.1 to find the first, second, and third derivatives of the following function:

$$f(x) = \frac{x^4}{(x-1)(x-3)^3} \text{ at } x = 0.5.$$

7.29. Use Program 7.2 to find the first and second derivatives of the function $f(x)$ at $x = 1.8$ using the following data:

x_i	0.0	0.5	1.5	4.0
$f_i = f(x_i)$	0.62	0.95	0.24	0.19

7.30. Use Program 7.3 to find the higher accuracy first, second, and third derivatives of the function given in Problem 7.28 at $x = 0.5$.

7.31. Use Program 7.4 to find Taylor's series approximation of the function given in Problem 7.28 at $x = 0.5$.

General

7.32. The follower of a cam is subjected to the following displacement [7.8]:

$$s(\theta) = L\left\{1.0 - 2.65\left(\frac{\theta}{\beta}\right)^2 + 2.80\left(\frac{\theta}{\beta}\right)^5 + 3.25\left(\frac{\theta}{\beta}\right)^6 - 6.85\left(\frac{\theta}{\beta}\right)^7 + 2.60\left(\frac{\theta}{\beta}\right)^8\right\}, \quad \text{(a)}$$

Here, L is the total lift, β is the total angular displacement, and θ is the angular displacement of the cam at time t. Assuming that $\theta(t) = \sin 120\pi t$, $L = 1$ in and $\beta = \frac{\pi}{2}$, determine the velocity $\left(\frac{ds}{dt}\right)$, acceleration $\left(\frac{d^2s}{dt^2}\right)$, and jerk $\left(\frac{d^3s}{dt^3}\right)$ at $t = \frac{1}{240}$.

7.33. The stress distribution across section A–A of the crane hook shown in Fig. 7.14 is given by [7.9]

$$\sigma = \frac{F}{A} + \frac{M(r_n - r)}{Aer}, \quad \text{(a)}$$

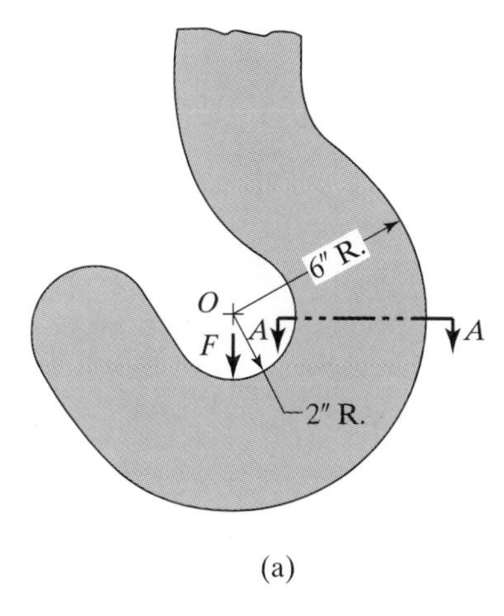

(a)

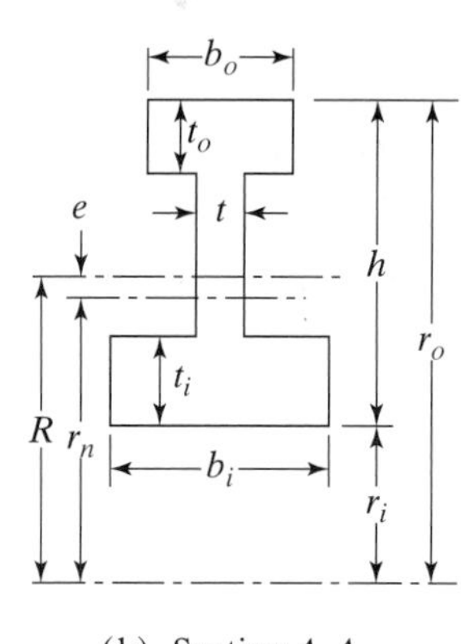

(b) Section A-A

Figure 7.14 Crane hook.

where F = load lifted, A = area of cross section at section A–A, M = moment = $F\,R$, R = radius of the centroidal axis, $e = R - r_n$ = eccentricity = distance from the centroidal axis to the neutral axis, r = radial distance from the center (O) of the hook, and r_n = radius of the neutral axis with

$$r_n = \frac{t_i(b_i - t) + t_0(b_0 - t) + ht_0}{b_i \ln \dfrac{r_i + t}{r_i} + t \ln \dfrac{r_0 - t_0}{r_i + t_i} + b_0 \ln \dfrac{r_0}{r_0 - t_0}} \tag{b}$$

and

$$R = r_i + \frac{\frac{1}{2}h^2 t + \frac{1}{2}t_i^2(b_i - t) + t_0(b_0 - t)\left(h - \frac{t_0}{2}\right)}{t_i(b_i - t) + t_0(b_0 - t) + ht}. \tag{c}$$

Determine the values of $\frac{\partial \sigma}{\partial t}$ and $\frac{\partial \sigma}{\partial h}$ numerically for the following data:

$$F = 5000 \text{ lb}, h = 4'', r_i = 2'', r_0 = 6'', b_i = b_0 = 1'', r = 4'',$$

$$t = 0.5'', \text{ and } t_i = t_0 = 0.5''.$$

7.34. The load (P)–deflection (δ) relation for the belleville spring, shown in Fig. 7.15, is given by [7.3]

$$P = \frac{C_1 C_2 E t^4}{(1 - \nu^2) b^2}, \tag{a}$$

where t = thickness of disk, E = Young's modulus, ν = Poisson's ratio, $\alpha = b/a$, b = outer radius, a = inner radius, h = unloaded disk height, δ = deflection,

$$C_1 = \frac{1}{2}\left(\frac{\delta}{t}\right)^3 - \frac{3h}{2t}\left(\frac{\delta}{t}\right)^2 + \left(1 + \frac{h^2}{t^2}\right)\frac{\delta}{t} \tag{b}$$

and

$$C_2 = \pi\left(\frac{\alpha + 1}{\alpha - 1} - \frac{2}{\ln \alpha}\right)\left(\frac{\alpha}{\alpha - 1}\right)^2. \tag{c}$$

Determine the spring rate, $k = \frac{dP}{d\delta}$, numerically for the following data:

$$E = 30 \times 10^6 \text{ psi}, \nu = 0.3, b = 8 \text{ in}, a = 4 \text{ in}, t = 0.1 \text{ in},$$

$$h = 0.25 \text{ in, and } \delta = 0.1 \text{ in}.$$

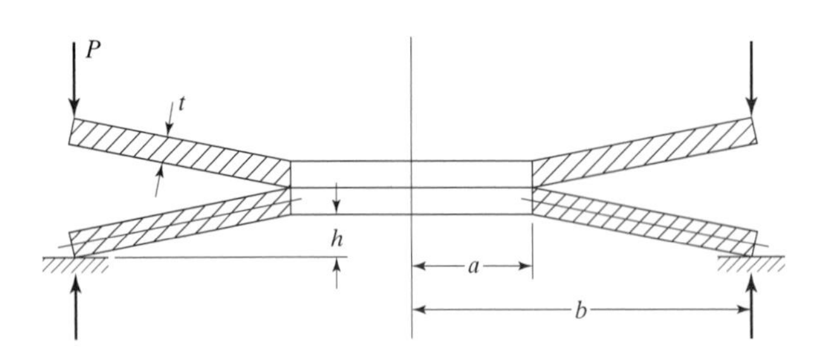

Figure 7.15 Belleville spring.

7.35. The values of the gamma function, $\Gamma(x_i)$, at unequally spaced points (x_i) are as follows:

x_i	$\Gamma(x_i)$
1.00	1.00000
1.02	0.98884
1.05	0.97350
1.09	0.95546
1.11	0.94739
1.14	0.93642
1.15	0.93304
1.20	0.91817
1.23	0.91075
1.27	0.90250
1.30	0.89747

Compute the values of $\frac{d\Gamma(x_i)}{dx}$ at each of the locations x_i indicated in the previous table.

7.36. Derive the high-accuracy forward-, backward-, and central-difference formulas given by Eqs. (7.49), (7.50), and (7.51).

7.37. Solve Example 7.2 using higher accuracy formulas.

7.38. Derive expressions for the following difference operators:

(a) $\Delta^2 \nabla f_i$

(b) $\Delta \nabla^2 f_i$

7.39. The values of a function $f(x)$ corresponding to three different values of x are

x_i	2.0	2.3	2.8
$f_i = f(x_i)$	0.69315	0.83291	1.02962

Find the approximate values of $f'(2.0)$ and $f''(2.0)$ based on linear and quadratic interpolation and compare the results with those given by the exact functional relation, $f(x) = \log x$. Also, determine an upper bound on the error in each case.

7.40. Lagrange interpolation polynomial of order n that fits the data (x_i, f_i), $i = 0, 1, 2, \ldots, n$ is given by

$$L(x) = \sum_{k=0}^{n} f_k L_k(x),$$

where $L_k(x)$ are called the Lagrange interpolation coefficients defined as

$$L_k(x) = \prod_{i=0, i \neq k}^{n} \left(\frac{x - x_i}{x_k - x_i} \right).$$

Using the following known values of $f(x)$, determine the approximate values of $f'(x)$ and $f''(x)$ using different order Lagrange interpolation polynomials at $x = 2.0$:

i	0	1	2	3
x_i	2.0	2.2	2.6	3.1
f_i	7.3890561	9.0250135	13.4637380	22.1979513

Compare the results with those given by the exact function $f(x) = e^x$.

7.41. If $f(h)$ denotes the approximate value of f, obtained by using a step length h and a method of order p, and $f(kh)$ indicates the value of f, obtained by using a step length kh and the same method of order p, we can express

$$f(h) = f + ch^p + O(h^{p+1}) \tag{a}$$

and

$$f(kh) = f + c(kh)^p + O(h^{p+1}) \tag{b}$$

Eliminating c from Eqs. (a) and (b) gives

$$f^{(1)}(h) = \frac{k^p f(h) - f(kh)}{k^p - 1} + O(h^{p+1}). \tag{c}$$

Equation (c) gives a higher order (order $p+1$ instead of p) approximation of f and is known as Richardson's extrapolation. Using Richardson's extrapolation method, in conjunction with a central difference formula, find $f'(2)$ from the following data:

x_i	−1	1	2	3	4	5	7
$f_i = f(x_i)$	−2	2	16	54	128	250	686

PROJECTS

7.1. The nodal displacements of the crane shown in Fig. 7.16 are given by the solution of the equations [7.11]

$$[K]\vec{y} = \vec{p}, \tag{a}$$

where $[K] = [k_{ij}]$ with

$$k_{11} = 30 \times 10^6 \left(\frac{0.8A_1}{55.9017} + \frac{0.8A_2}{55.9017} + \frac{0.8A_3}{167.7051} \right),$$

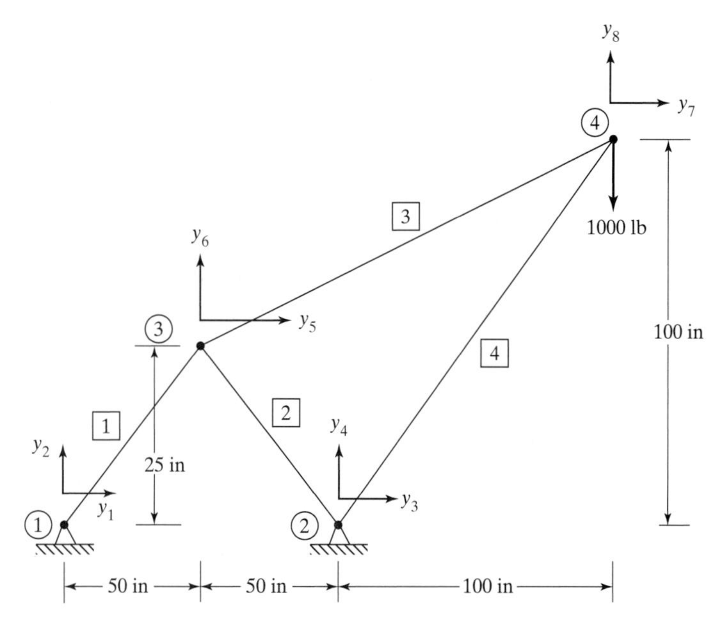

Figure 7.16 A crane (planar truss) structure.

$$k_{12} = k_{21} = 30 \times 10^6 \left(\frac{0.4A_1}{55.9017} - \frac{0.4A_2}{55.9017} + \frac{0.4A_3}{167.7051} \right),$$

$$k_{13} = k_{31} = 30 \times 10^6 \left(-\frac{0.8A_3}{167.7051} \right),$$

$$k_{14} = k_{41} = 30 \times 10^6 \left(-\frac{0.4A_3}{167.7051} \right),$$

$$k_{22} = 30 \times 10^6 \left(\frac{0.2A_1}{55.9017} + \frac{0.2A_2}{55.9017} + \frac{0.2A_3}{167.7051} \right),$$

$$k_{23} = k_{32} = 30 \times 10^6 \left(-\frac{0.4A_3}{167.7051} \right),$$

$$k_{24} = k_{42} = 30 \times 10^6 \left(-\frac{0.2A_3}{167.7051} \right),$$

$$k_{33} = 30 \times 10^6 \left(\frac{0.8A_3}{167.7051} + \frac{0.5A_4}{141.4214} \right),$$

$$k_{34} = k_{43} = 30 \times 10^6 \left(\frac{0.4A_3}{167.7051} + \frac{0.5A_4}{141.4214} \right),$$

$$k_{44} = 30 \times 10^6 \left(\frac{0.2A_3}{167.7051} + \frac{0.5A_4}{141.4214} \right),$$

$$\vec{y} = \begin{Bmatrix} y_5 \\ y_6 \\ y_7 \\ y_8 \end{Bmatrix}, \tag{b}$$

and

$$\vec{p} = \begin{Bmatrix} 0 \\ 0 \\ 0 \\ -1000 \end{Bmatrix}, \tag{c}$$

where A_i is the cross-sectional area of member i and y_i is the ith-displacement component ($i = 5, 6, 7, 8$). Find the value of $\frac{\partial y_8}{\partial A_i}$; $i = 1, 2, 3$, and 4, using the relation

$$\frac{\partial \vec{y}}{\partial A_i} = [K]^{-1}\left(-\frac{\partial [K]}{\partial A_i}\vec{y}\right). \tag{d}$$

Assume the data as $A_1 = A_2 = 2$ in^2 and $A_3 = A_4 = 1$ in^2.

7.2. An eigenvalue problem is given by

$$[K]\vec{Y} = \lambda[M]\vec{Y}, \tag{a}$$

where $[K]$ is the stiffness matrix, $[M]$ is the mass matrix, λ is the eigenvalue, and $\vec{Y}$ is the eigenvector. The stiffness and mass matrices of the system, shown in Fig. 7.17, are given by

$$[K] = k\begin{bmatrix} 2 & -1 \\ -1 & 2 \end{bmatrix} \text{ and } [M] = m\begin{bmatrix} 1 & 0 \\ 0 & 1 \end{bmatrix},$$

where $k = \frac{d^4G}{8D^3n}$, d = wire diameter, $D = x_1$ = coil diameter, G = shear modulus and n = number of turns of the spring. The eigenvalue solution of the system is given by

$$\lambda_1 = \omega_1^2 = \frac{k}{m}, \lambda_2 = \omega_2^2 = \frac{3k}{m} \tag{b}$$

and

$$\vec{Y}_1 = c_1\begin{Bmatrix} 1 \\ 1 \end{Bmatrix}, \vec{Y}_2 = c_2\begin{Bmatrix} 1 \\ -1 \end{Bmatrix}, \tag{c}$$

where ω_1 and ω_2 are the natural frequencies of vibration of the system and c_1 and c_2 are constants.

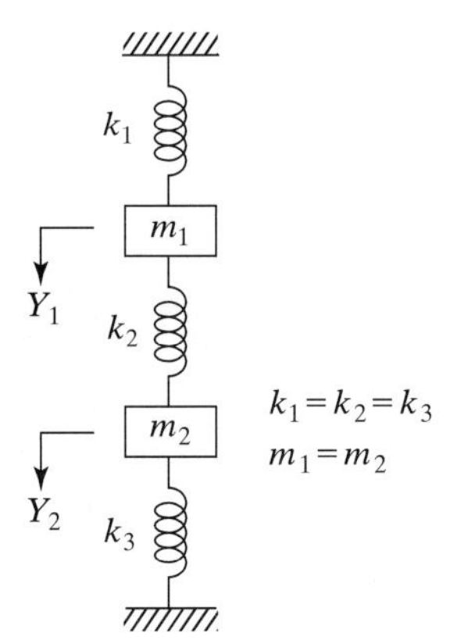

Figure 7.17 A two degree-of-freedom spring-mass system.

Determine the values of $\frac{\partial \lambda_i}{\partial x_j}$ for the spring–mass system shown in Fig. 7.17 using the relation

$$\frac{\partial \lambda_i}{\partial x_j} = \vec{Y}_i^T \left[\frac{\partial [K]}{\partial x_j} - \lambda_i \frac{\partial [M]}{\partial x_j} \right] \vec{Y}_i, \quad i = 1, 2; j = 1.$$

Assume the data as $d = 0.04$ in, $G = 11.5 \times 10^6$ psi, $x_1 = D = 0.4$ in, $n = 10$, and $m = 32.2$ lb-sec^2/in.

8

Numerical Integration

8.1 Introduction

CHAPTER CONTENTS

Frequently, many engineering problems require the evaluation of the integral

$$I = \int_a^b f(x)\,dx, \tag{8.1}$$

where the function $f(x)$ is called the integrand and a and b are called the limits of integration. If the function $f(x)$ is continuous, finite, and well behaved over the range of integration $a \leq x \leq b$, the integral (I) can be evaluated using the available mathematical techniques. If $f(x)$ denotes a simple function such as a polynomial, an exponential, or a trigonometric function, the integrals are well known from calculus. If $f(x)$ involves more complicated functions, often, standard tables of integrals can be used to evaluate the integral (I) in closed form. The analytical or closed form expressions for the integrals, if available, are very valuable, since they are exact and no error is involved in their evaluation. In addition, the influence of changing some physical parameter of the engineering problem on the integral can be studied easily. Finally, the closed form

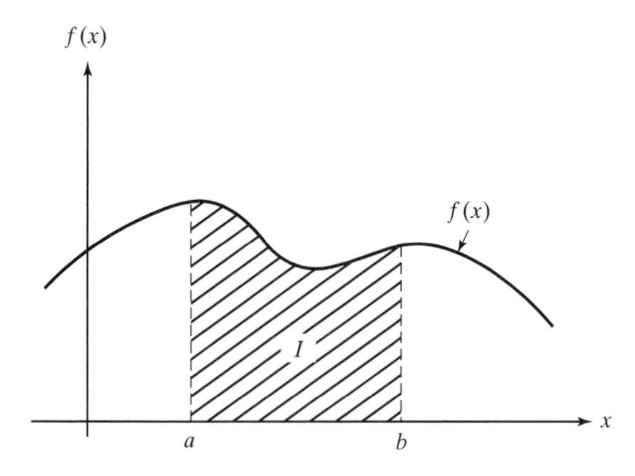

Figure 8.1 Integral as area under the curve.

expressions of the integral (I) can be used to verify the accuracy of numerical integration.

On the other hand, the function $f(x)$ may be a complicated continuous function that is difficult or impossible to integrate in closed form. In some cases, $f(x)$ may not be known in analytical form; it may be known only in a tabular form, where the values of x and $f(x)$ are available at a number of discrete points in the interval a to b (may be, from an experimental study). The limits of integration may be infinite or the function $f(x)$ may be discontinuous or may become infinite at some point in the interval a to b. In all these cases, the integral (I) can be evaluated only numerically.

The integral of a function $f(x)$ between the limits a and b basically denotes the area under the curve of $f(x)$ between a and b as shown in Fig. 8.1. Integration is also known as quadrature. A simple, intuitive approach to evaluate the integral in Eq. (8.1) is to plot the function $f(x)$ on a grid or graph paper and count the number of boxes or rectangles that approximate the area under the curve of $f(x)$. (See Fig. 8.2.) The product of the number of boxes and the area of each box gives

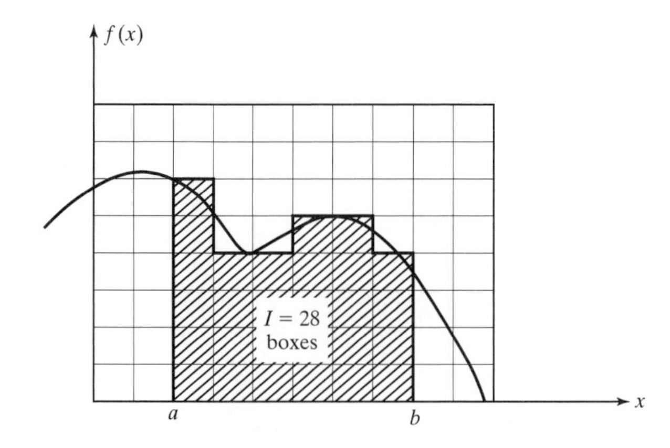

Figure 8.2 Evaluation of an integral using a grid or graph paper.

an estimate of the total area under the curve (i.e., the integral, I). This estimate can be refined, if necessary, by using a finer grid. However, the method used is very impractical and inaccurate in many cases.

8.2 Engineering Applications

▶Example 8.1

A semiinfinite solid body, initially at temperature T_i, is suddenly exposed to a fluid at temperature T_0 at the face $x = 0$ as shown in Fig. 8.3. If the diffusivity of the material (α) is constant, the unsteady state temperature distribution in the body, $T(x, t)$, is governed by the equation

$$\frac{d^2\theta}{d\eta^2} + 2\eta\frac{d\theta}{d\eta} = 0 \tag{a}$$

subject to

$$\theta(\eta) \to 0 \quad \text{as} \quad \eta \to \infty \tag{b}$$

and

$$\theta(0) = 1, \tag{c}$$

where

$$\theta = \frac{T - T_i}{T_0 - T_i} \tag{d}$$

and

$$\eta = \frac{x}{2\sqrt{\alpha t}}. \tag{e}$$

Determine the temperature distribution in the body.

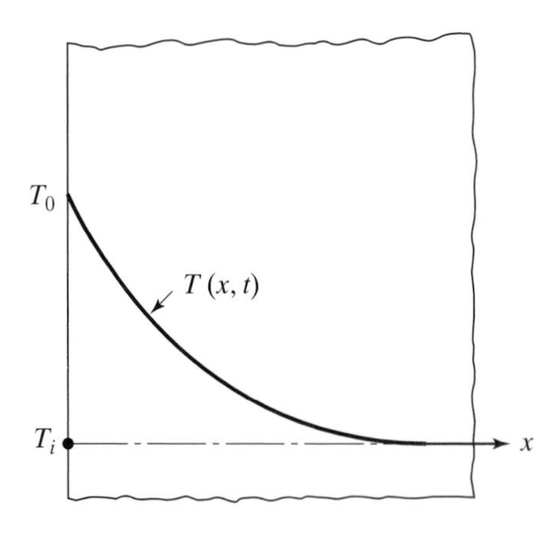

Figure 8.3 Semi-infinite solid.

Solution

The solution of Eq. (a) is given by

$$\theta(\eta) = c_1 + c_2 \int e^{-\eta^2} d\eta, \tag{f}$$

where the constants c_1 and c_2 can be evaluated, using Eqs. (b) and (c), as $c_1 = 1$ and $c_2 = -\frac{2}{\sqrt{\pi}}$. Thus, the solution becomes

$$\theta(\eta) = \frac{T - T_i}{T_0 - T_i} = 1 - \frac{2}{\sqrt{\pi}} \int_0^{\eta} e^{-z^2} dz. \tag{g}$$

The right-hand side of Eq. (g) is known as the complementary error function and the integral of Eq. (g) cannot be evaluated in closed form. ◀

▶Example 8.2

The axial displacement (du) of an elemental length (dx) of the bar, shown in Fig. 8.4, under a load P is given by

$$\frac{du}{dx} = \frac{\sigma}{E} = \frac{P}{EA}, \tag{a}$$

where E is Young's modulus and A is the cross-sectional area. Determine the axial displacement of the bar for the following data: $P = 5000$ lb, $l = 10$ in, $E = 30 \times 10^6(1 - 0.01x - 0.0005x^2)$ psi, and $A = A_0 e^{-0.1x} = 2e^{-0.1x}$ in^2.

Solution

The axial displacement of the bar at $x = l$ can be determined by integrating Eq. (a) as

$$u = \int du = \int_0^l \frac{P}{EA} dx, \tag{b}$$

where the integral can be conveniently evaluated using a numerical integration procedure. ◀

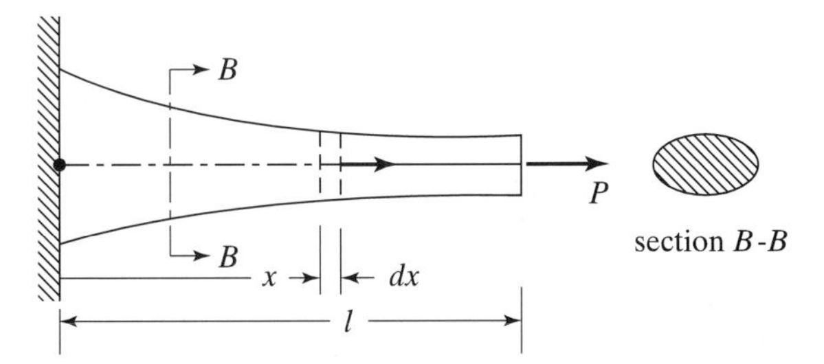

Figure 8.4 Nonuniform bar under axial load.

▶Example 8.3

The turning moment developed at various positions of the crank from the inner dead center in a multicylinder internal combustion engine is shown in Fig. 8.5. If the speed of the engine is 1,500 rpm, determine the power developed.

Solution

The power developed per cycle (one revolution of the crank) can be found as the area under the turning-moment–crank-angle diagram. The integral of the function $f(x)$ can be evaluated analytically as follows:

$$I = \int_0^{1.5} f(x)\,dx = 5.5606613. \tag{a}$$

Thus, the area under the turning moment diagram per one cycle of the engine is given by

$$5.5606613\ (100)\left(\frac{4\pi}{3}\right) = 2{,}329.249805 \text{ lb-ft/rev.}$$

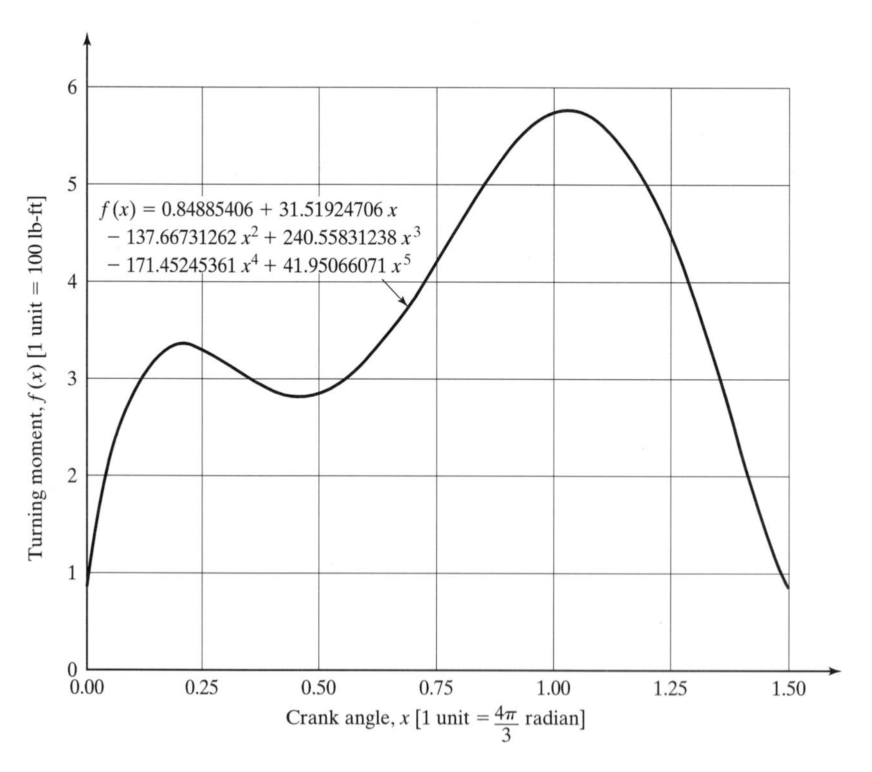

Figure 8.5 Turning moment diagram.

The power developed by the engine (P) can be determined by multiplying the power per cycle and the speed of the engine. This gives

$$P = 2329.249805 \times 1500 = 3.4938747 \times 10^6 \text{ lb-ft/min}$$
$$= \frac{3.4938747 \times 10^6}{33,000} = 105.874991 \text{ hp}$$

◀

8.3 Newton–Cotes Formulas

The Newton–Cotes formulas are the most commonly used numerical integration methods. They are based on replacing a complicated function or tabular data by some approximating function that can be integrated easily; that is,

$$I = \int_a^b f(x)\,dx \approx \int_a^b p_m(x)\,dx, \tag{8.2}$$

where $p_m(x)$ is the approximating function, usually taken as an mth-degree polynomial, viz.,

$$p_m(x) = a_m x^m + a_{m-1}x^{m-1} + \cdots + a_2x^2 + a_1x + a_0, \tag{8.3}$$

where the coefficients of the polynomial (constants) $a_m, a_{m-1}, \ldots a_1, a_0$ are determined such that $f(x)$ and $p_m(x)$ have the same values at a finite number of points. (See Section 5.3.) Figure 8.6 shows the approximation of $f(x)$ using three of the simplest polynomials, namely, a constant, a straight line, and a parabola.

8.3.1 Rectangular Rule

The function or data of $f(x)$ can also be approximated using a series of piecewise polynomials shown in Fig. 8.7. In this approach, the range of integration $a \le x \le b$ is first divided into a finite number (n) of intervals or strips such that the width of each interval is given by

$$h = \Delta x = \frac{b-a}{n}. \tag{8.4}$$

The discrete points in the range of integration are then defined as $x_0 = a, x_1, x_2, \ldots,$ x_{n-1}, and $x_n = b$ with

$$x_i = a + ih; i = 0, 1, 2, \ldots, n. \tag{8.5}$$

The values of the function $f(x)$ at the discrete point x_i is assumed to be known as $f_i (i = 0, 1, 2, \ldots, n)$. As shown in Fig. 8.7(a), the simplest approximation to the function $f(x)$ is a piecewise polynomial of order zero (i.e., a series of constants) Clearly, from Fig. 8.7(a), the function $f(x)$ can be approximated over the interval $x_i \le x \le x_{i+1}$ either by the value of f_i or f_{i+1}. If the values of f_i are used (i.e., $f(x)$ is approximated by its values at the beginning of each interval), the area under the curve $f(x)$ in the interval $x_i \le x \le x_{i+1}$ is taken as $(f_i h)$ and hence the integral (I)

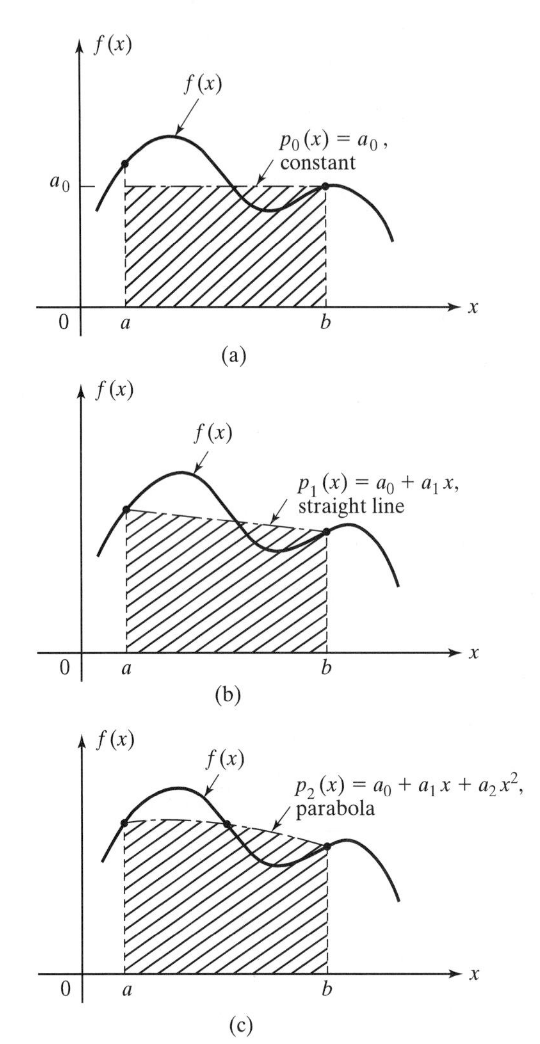

Figure 8.6 Different types of approximation of $f(x)$.

is evaluated as

$$I = \int_a^b f(x)\,dx \approx h\left(\sum_{i=0}^{n-1} f_i\right) \tag{8.6}$$

On the other hand, if the values of f_{i+1} are used (i.e., $f(x)$ is approximated by its values at the end of each interval), the area under the curve $f(x)$ in the interval $x_i \le x \le x_{i+1}$ is taken as $(f_{i+1}h)$ and hence the integral (I) is evaluated as

$$I = \int_a^b f(x)\,dx \approx h\left(\sum_{i=0}^{n-1} f_{i+1}\right) \equiv h\left(\sum_{i=1}^{n} f_i\right). \tag{8.7}$$

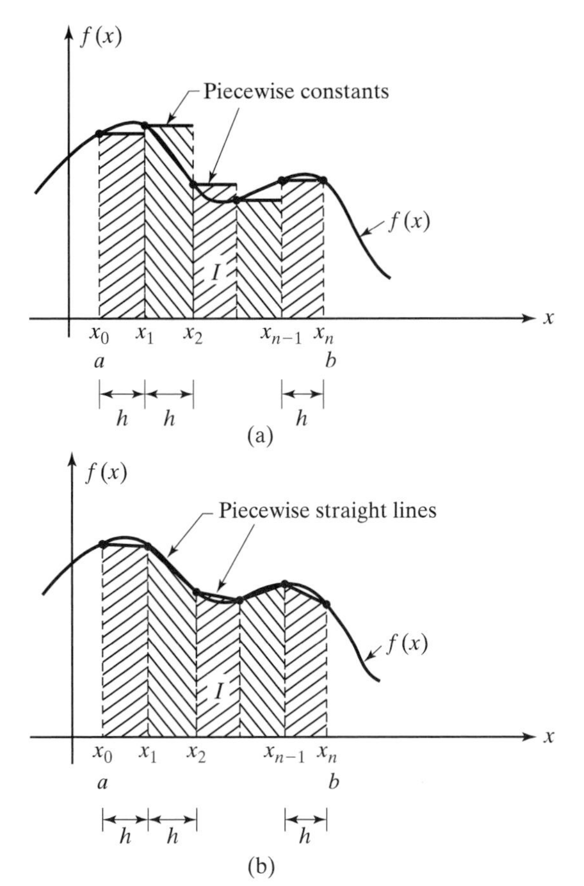

Figure 8.7 Approximation of $f(x)$ by piecewise polynomials of degree 0 and 1.

For a monotonically increasing function, Eq. (8.6) underestimates and Eq. (8.7) overestimates the actual value of the integral. (See Fig. 8.8.) On the other hand, for a monotonically decreasing function, Eq. (8.6) overestimates and Eq. (8.7) underestimates the true value of the integral. In practice, the rectangular rule leads to large truncation errors for general nonlinear functions $f(x)$ and, hence, is not commonly used. However, the method serves to illustrate the basic concepts used in numerical integration and Newton–Cotes formulas. An improvement in accuracy of the piecewise-constant approximation (rectangular rule) can be achieved by using the average value of f_i and f_{i+1} in the interval $x_i \leq x \leq x_{i+1}$ as shown in Fig. 8.9. In this case, the integral (I) is evaluated as

$$I = \int_a^b f(x)\,dx \approx h \sum_{i=0}^{n-1} \left(\frac{f_i + f_{i+1}}{2} \right). \tag{8.8}$$

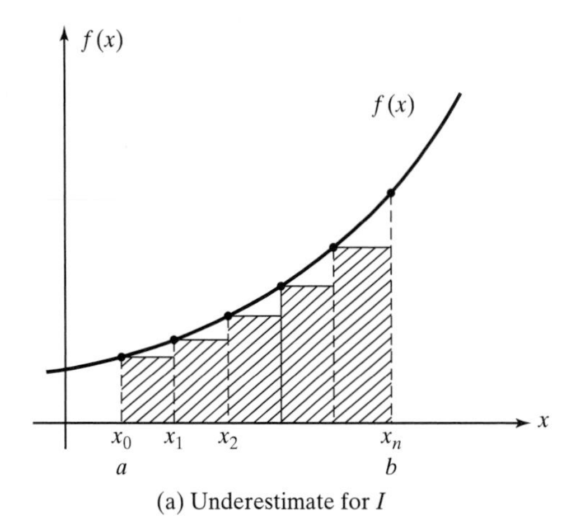

(a) Underestimate for I

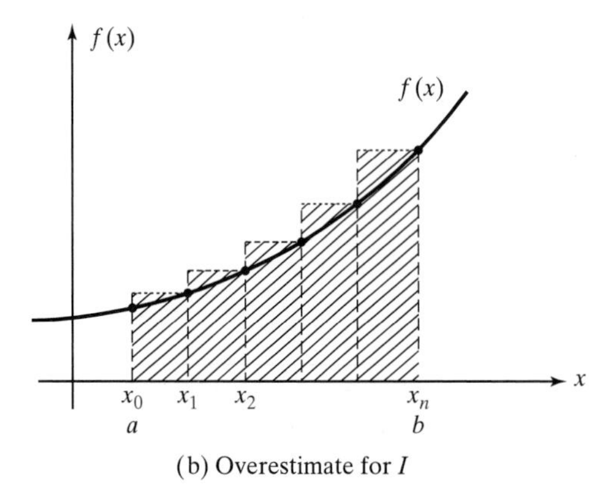

(b) Overestimate for I

Figure 8.8 Under- and over-estimation of I.

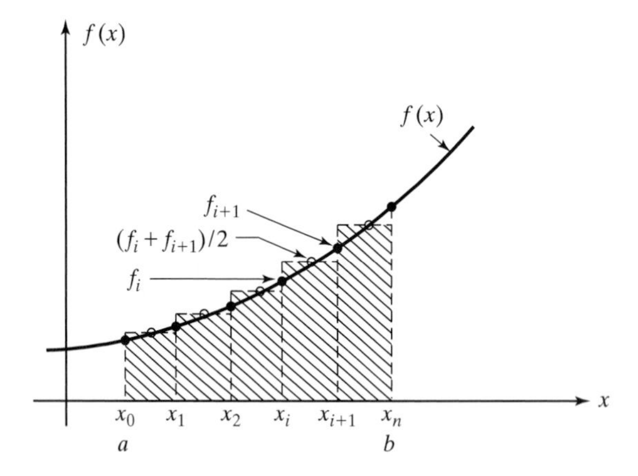

Figure 8.9 Approximation of $f(x)$ by $(f_i + f_{i+1})/2$ in $x_i \le x \le x_{i+1}$.

8.3.2 Trapezoidal Rule

The trapezoidal rule is extensively used in engineering applications because of its simplicity in developing a computer program. The method corresponds to the approximation of $f(x)$ by piecewise polynomials of order one $[p_1(x) = c_1 x + c_0]$, that is, by straight-line segments as shown in Fig. 8.7(b). In this case, the area under the curve $f(x)$ in the interval $x_i \le x \le x_{i+1}$ is equal to the area of the trapezoid; hence the name trapezoidal rule. Denoting the areas of the trapezoids as $I_1, I_2, \ldots, I_n$, we have (Fig. 8.10)

$$I_1 = \left(\frac{f_0 + f_1}{2}\right)h,\ I_2 = \left(\frac{f_1 + f_2}{2}\right)h, \ldots,$$

$$I_i = \left(\frac{f_{i-1} + f_i}{2}\right)h, \ldots, \text{ and } I_n = \left(\frac{f_{n-1} + f_n}{2}\right)h. \tag{8.9}$$

The integral can be evaluated as

$$I = \int_a^b f(x)\,dx \approx \sum_{i=1}^{n} I_i = \frac{h}{2}\left(f_0 + 2f_1 + 2f_2 + \cdots + 2f_{n-1} + f_n\right). \tag{8.10}$$

8.3.3 Truncation Error in Trapezoidal Rule

The basic truncation error (E) of the trapezoidal rule is given by

$$E = \int_a^b f(x)\,dx - \left[\frac{f(a) + f(b)}{2}\right](b - a), \tag{8.11}$$

where the first term on the right-hand side of Eq. (8.11) denotes the exact integral and the second term represents the approximate integral given by the trapezoidal rule. Note that only one segment is considered in the interval for simplicity. (See Fig. 8.11.) To derive a more convenient expression for the error, we use Taylor's

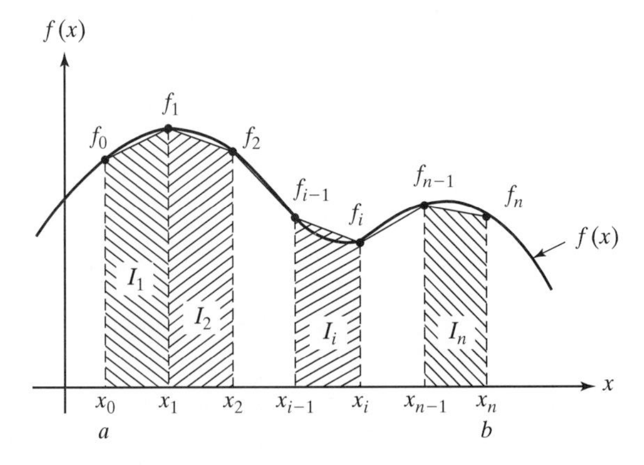

Figure 8.10 Trapezoidal rule.

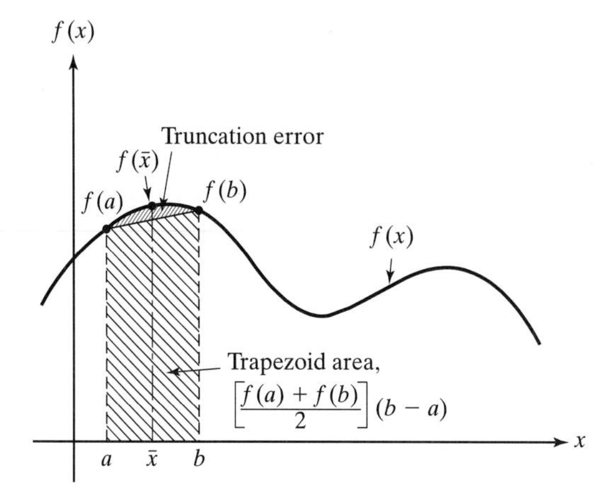

Figure 8.11 Truncation error.

series expansion of $f(x)$ about the midpoint of the range, $\bar{x} = \frac{a+b}{2}$:

$$f(x) = f(\bar{x}) + yf'(\bar{x}) + \frac{y^2}{2!}f''(\bar{x}) + \cdots. \tag{8.12}$$

Here $y = x - \bar{x}$, a prime indicates a derivative, and the function $f(x)$ is assumed to be analytical in the interval $a \leq x \leq b$. Equation (8.12) can be used to express

$$\int_a^b f(x)\,dx = \int_{-h/2}^{h/2} \left\{ f(\bar{x}) + yf'(\bar{x}) + \frac{y^2}{2!}f''(\bar{x}) + \cdots \right\} dy, \tag{8.13}$$

where $y = -h/2$ and $y = +h/2$ can be seen to correspond to $x = a$ and $x = b$, respectively. By carrying out the integration in Eq. (8.13), we obtain

$$\begin{aligned} \int_a^b f(x)\,dx &= f(\bar{x})\,(y)\Big|_{-h/2}^{h/2} + f'(\bar{x}) \left(\frac{y^2}{2}\right)\Bigg|_{-h/2}^{h/2} + \frac{1}{2}f''(\bar{x}) \left(\frac{y^3}{3}\right)\Bigg|_{-h/2}^{h/2} + \cdots \\ &= hf(\bar{x}) + \frac{1}{24}h^3 f''(\bar{x}) + \cdots. \end{aligned} \tag{8.14}$$

Substituting $x = a$ and $x = b$ into Eq. (8.12) yields

$$f(a) = f(\bar{x}) - \frac{h}{2}f'(\bar{x}) + \frac{1}{2}\left(\frac{h}{2}\right)^2 f''(\bar{x}) - \cdots; \tag{8.15}$$

$$f(b) = f(\bar{x}) + \frac{h}{2}f'(\bar{x}) + \frac{1}{2}\left(\frac{h}{2}\right)^2 f''(\bar{x}) + \cdots, \tag{8.16}$$

where the values of y at $x = a$ and $x = b$ are taken as $x - \bar{x} = a - \bar{x} = -\frac{h}{2}$ and $x - \bar{x} = b - \bar{x} = +\frac{h}{2}$, respectively. Noting that $(b - a) = h$, the second term on the right-hand side of Eq. (8.11) can be expressed as

$$(b-a)\left[\frac{f(a)+f(b)}{2}\right]=\frac{h}{2}\left[f(\bar{x})-\frac{h}{2}f'(\bar{x})+\frac{1}{8}h^2f''(\bar{x})-\cdots+f(\bar{x})+\frac{h}{2}f'(\bar{x})\right.$$
$$\left.+\frac{1}{8}h^2f''(\bar{x})+\cdots\right]=hf(\bar{x})+\frac{1}{8}h^3f''(\bar{x})+\cdots. \tag{8.17}$$

Substituting Eqs. (8.14) and (8.17) into Eq. (8.11) and truncating the higher order derivative terms yields

$$E=\left[hf(\bar{x})+\frac{1}{24}h^3f''(\bar{x})+\cdots\right]-\left[hf(\bar{x})+\frac{1}{8}h^3f''(\bar{x})+\cdots\right]$$
$$\approx-\frac{1}{12}h^3f''(\bar{x}). \tag{8.18}$$

This shows that the error of the trapezoidal rule (per segment or step) is proportional to $f''(\bar{x})$ and h^3. Thus, the error can be reduced by reducing the value of $h=b-a$.

The error in the multisegmented trapezoidal rule, Eq. (8.10), can be determined by summing the errors of the individual segments $(x_0,x_1),(x_1,x_2),\ldots,(x_{n-1},x_n)$. Since the range of integration is divided into n equal segments, we have $h=\frac{b-a}{n}$ and hence

$$E\approx-\frac{1}{12}\left(\frac{b-a}{n}\right)^3\sum_{i=1}^{n}f''(\bar{x}_i), \tag{8.19}$$

where $\bar{x}_i$ is the midpoint between x_i and x_{i+1}. By defining an average value of the second derivative

$$\bar{f}''=\frac{1}{n}\sum_{i=1}^{n}f''(\bar{x}_i), \tag{8.20}$$

Eq. (8.19) can be written as

$$E\approx-\frac{1}{12}(b-a)\left(\frac{b-a}{n}\right)^2\bar{f}''=-\frac{1}{12}(b-a)h^2\bar{f}''=O(h^2). \tag{8.21}$$

This indicates that the error of the multisegment trapezoidal rule, Eq. (8.10), is proportional to h^2 (since $(b-a)$ is fixed).

8.3.4 Truncation Error in Rectangular Rule

The foregoing procedure can be used to evaluate the truncation error in rectangular rule. The error can be expressed, for a single segment $a\le x\le b$, as

$$E=\int_a^b f(x)\,dx-f(a)h,\ \text{ for Eq. (8.6)}, \tag{8.22}$$

and

$$E=\int_a^b f(x)\,dx-f(b)h,\ \text{ for Eq. (8.7)}, \tag{8.23}$$

where the first term on the right-hand sides of Eqs. (8.22) and (8.23) denotes the exact integral and the second term represents the approximate integral given by the particular rectangular rule. Taylor's series expansion of $f(x)$ about a is given by

$$f(x) = f(a) + (x-a)f'(a) + \frac{(x-a)^2}{2!}f''(a) + \cdots. \tag{8.24}$$

The integration of Eq. (8.24) yields

$$\begin{aligned}\int_a^b f(x)\,dx &= \int_0^h \left[f(a) + yf'(a) + \frac{y^2}{2!}f''(a) + \cdots\right] dy \\ &= f(a)y|_0^h + f'(a)\left.\frac{y^2}{2}\right|_0^h + f''(a)\left.\frac{y^3}{6}\right|_0^h + \cdots \\ &= f(a)h + f'(a)\frac{h^2}{2} + f''(a)\frac{h^3}{6} + \cdots,\end{aligned} \tag{8.25}$$

where $y = x - a$ and $h = b - a$. Thus, Eq. (8.22) gives

$$E = f'(a)\frac{h^2}{2} + f''(a)\frac{h^3}{6} + \cdots. \tag{8.26}$$

Similarly, Taylor's series expansion of $f(x)$ about b can be expressed as

$$f(x) = f(b) - (b-x)f'(b) + \frac{(b-x)^2}{2!}f''(b) - \cdots. \tag{8.27}$$

The integration of Eq. (8.27) gives

$$\begin{aligned}\int_a^b f(x)\,dx &= \int_a^b \left[f(b) - yf'(b) + \frac{y^2}{2!}f''(b) - \cdots\right] dy \\ &= f(b)\,y|_0^h - f'(b)\left.\frac{y^2}{2}\right|_0^h + f''(b)\left.\frac{y^3}{6}\right|_0^h - \cdots \\ &= f(b)h - \frac{h^2}{2}f'(b) + \frac{h^3}{6}f''(b) - \cdots,\end{aligned} \tag{8.28}$$

where $y = b - x$ and $h = b - a$. Thus, Eq. (8.23) yields

$$E = -\frac{h^2}{2}f'(b) + \frac{h^3}{6}f''(b) - \cdots. \tag{8.29}$$

Equations (8.26) and (8.29) indicate that the error of the rectangular rule per step is proportional to h^2 and $f'(a)$ or $f'(b)$. By proceeding as in the case of the trapezoidal rule, the error in a multistep rectangular rule can be expressed as (see Problem 8.5)

$$E = \frac{1}{2}(b-a)\left(\frac{b-a}{n}\right)\bar{f}' = \frac{1}{2}(b-a)h\bar{f}', \text{ for Eq. (8.6)}, \tag{8.30}$$

and

$$E = -\frac{1}{2}(b-a)\left(\frac{b-a}{n}\right)\bar{f}' = -\frac{1}{2}(b-a)h\bar{f}', \text{ for Eq. (8.7)}, \tag{8.31}$$

where $\bar{f}'$ in Eqs. (8.30) and (8.31) denotes the average value of the first derivative at the discrete points $a, x_1, x_2, \ldots, x_{n-1}$ and $x_1, x_2, \ldots, x_{n-1}, b$, respectively. This shows that the error in a multistep rectangular rule, Eq. (8.6) or (8.7), is proportional to h since $(b-a)$ is fixed.

▶Example 8.4

Determine the value of the integral $I = \int_a^b f(x)\,dx$, where

$$f(x) = 0.84885406 + 31.51924706x - 137.66731262x^2 + 240.55831238x^3 - 171.45245361x^4 + 41.95066071x^5 \tag{a}$$

with $a = 0.0$ and $b = 1.5$ using trapezoidal rule with different step lengths.

Solution

The value of the integral given by the trapezoidal rule with n steps is

$$I = \frac{h}{2}\left(f_0 + 2f_1 + 2f_2 + \cdots + 2f_{n-1} + f_n\right), \tag{b}$$

where $h = (b-a)/n$ and $f_i = f(x = a + ih)$. For $n = 1$ with $h = 1.5$, Eq. (b) becomes

$$I = \left(\frac{f(a) + f(b)}{2}\right)(b-a). \tag{c}$$

Since $a = 0.0$, $b = 1.5$, $f(a) = f(0.0) = 0.84885406$ and $f(b) = f(1.5) = 0.84542847$, Eq. (c) gives the value of the integral as $I = 1.2707119$. The exact value of the integral, determined analytically, is 5.5606613. Thus, the error in the one-step trapezoidal rule is 77.148186%. Using $n = 3$ with $h = 0.5$, the values of the function $f(x)$ are given by

$$f_0 = f(0.0) = 0.84885406,\ f_1 = f(a+h) = f(0.5) = 2.8566201,\ f_2 = f(a+2h) = f(1.0) = 5.7573166, \text{ and } f_3 = f(a+3h) = f(1.5) = 0.84542847.$$

The trapezoidal rule with $n = 3$ gives the value of the integral as

$$I = \frac{h}{2}(f_0 + 2f_1 + 2f_2 + f_3) = 4.7305388. \tag{d}$$

Compared with the exact value of the integral, 5.5606613, the error in the trapezoidal rule is 14.928484%. The significance of I, for $n = 1$ and 3, is shown in Fig. 8.12. The results given by the trapezoidal rule with $n = 1, 2, 3, 4, 5, 6, 9, 12$, and 15 are shown in Table 8.1. ◀

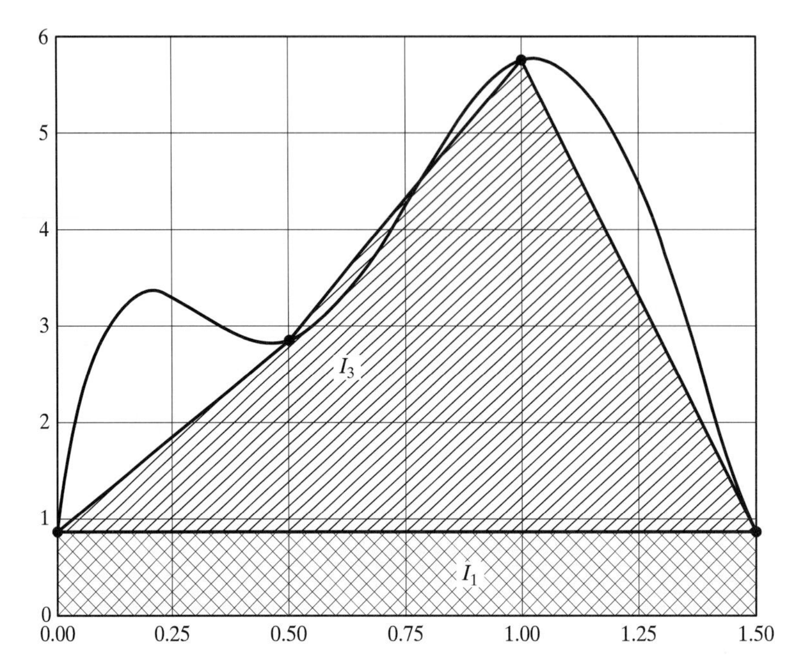

Figure 8.12 Significance of trapezoidal rule.

Table 8.1

Number of steps (n)	Step length (h)	Value of the integral (I)	Percent error
1	1.5	1.2707119	77.148186
2	0.75	3.8171761	31.353920
3	0.5	4.7305388	14.928484
4	0.375	5.0828342	8.5929899
5	0.3	5.2516432	5.5572190
6	0.25	5.3448267	3.8814559
9	0.16666667	5.4640293	1.7377790
12	0.125	5.5061684	0.97997248
15	0.1	5.5257416	0.62797815

Exact value of the integral: 5.5606613

8.4 Simpson's Rule

The accuracy of the trapezoidal rule can be improved by reducing the step size h (or increasing the number of segments n). However, the round-off error increases with a reduction in the step size h. Another way of obtaining a more accurate estimate of an integral is to use higher order polynomials for approximating the function $f(x)$. For example, we can use piecewise quadratic functions to approximate $f(x)$, which corresponds to the case of $m = 2$ in the Newton–Cotes formulas (Fig. 8.6). This method is also known as Simpson's one-third rule. As a further improvement, we can use piecewise cubic functions to approximate $f(x)$, corresponding to the case of $m = 3$ in the Newton–Cotes formulas. The method is also known as Simpson's three-eighths rule.

8.4.1 Simpson's One-Third Rule

As just stated, the integral

$$I = \int_a^b f(x)\,dx \tag{8.32}$$

is evaluated using a parabola or second-order polynomial for approximating $f(x)$. Assuming that $a \le x_{i-1} < x_i < x_{i+1} \le b$, the three points (x_{i-1}, f_{i-1}), (x_i, f_i) and (x_{i+1}, f_{i+1}), as shown in Fig. 8.13, are used to define a second-degree polynomial, $p_2(x)$. By making the polynomial

$$p_2(x) = c_2x^2 + c_1x + c_0 \tag{8.33}$$

pass through the three points shown in Fig. 8.13, the constants c_0, c_1, and c_2 can be determined. For this, we take the origin at $x_i\,(x = 0$ at $x_i)$ so that x_{i-1} and x_{i+1} correspond to $-h$ and $+h$, respectively. Such a choice of the origin does not influence the final result. By using the relations

For x_{i-1},

$$p_2(x = -h) = f_{i-1} = c_2(-h)^2 + c_1(-h) + c_0 = c_2h^2 - c_1h + c_0; \tag{8.34}$$

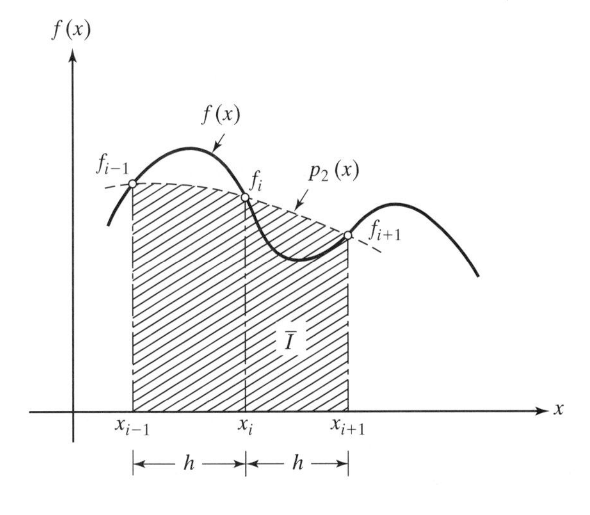

Figure 8.13 Simpson's one-third rule.

For x_i,

$$p_2(x=0) = f_i = c_2(0)^2 + c_1(0) + c_0 = c_0; \tag{8.35}$$

For x_{i+1},

$$p_2(x=h) = f_{i+1} = c_2(h)^2 + c_1(h) + c_0 = c_2h^2 + c_1h + c_0 \tag{8.36}$$

the solution of Eqs. (8.34) through (8.36) can be found as

$$c_2 = \frac{f_{i-1} - 2f_i + f_{i+1}}{2h^2}, c_1 = \frac{f_{i+1} - f_{i-1}}{2h}, \text{ and } c_0 = f_i. \tag{8.37}$$

(See Problem 8.46.) The area $(\bar{I})$ under the second-degree polynomial $p_2(x)$ between x_{i-1} and x_{i+1} can be determined as follows:

$$\begin{aligned}\bar{I} &= \int_{x_{i-1}}^{x_{i+1}} p_2(x)\,dx = \int_{-h}^{h} \left(c_2x^2 + c_1x + c_0\right) dx \\ &= \frac{c_2}{3}\,(x^3)\Big|_{-h}^{h} + \frac{c_1}{2}\,(x^2)\Big|_{-h}^{h} + c_0\,(x)|_{-h}^{h} \\ &= \frac{2}{3}c_2\,h^3 + 2\,c_0\,h.\end{aligned} \tag{8.38}$$

By substituting for c_2 and c_0 from Eqs. (8.37), Eq. (8.38) gives

$$\bar{I} = \frac{2}{3}h^3\left(\frac{f_{i-1} - 2\,f_i + f_{i+1}}{2\,h^2}\right) + 2\,h\,f_i = \frac{h}{3}\left(f_{i-1} + 4\,f_i + f_{i+1}\right). \tag{8.39}$$

The term "$\frac{1}{3}$" in Simpson's one-third rule refers to the presence of the factor "$\frac{1}{3}$" in Eq. (8.39). Note that two segments are used in deriving Eq. (8.39). Thus, for a multistage application of Simpson's one-third rule, we need to divide the range $a \le x \le b$ into n segments of equal width $h = \frac{b-a}{n}$. The number of segments must be an even number so that Eq. (8.39) can be applied for groups of two segments. The integral in Eq. (8.32) can be evaluated as

$$I = \int_a^b f(x)\,dx \approx \sum_{j=1}^{n/2} \left(\bar{I}\right)_j, \tag{8.40}$$

where $\left(\bar{I}\right)_j$ denotes the value of $\bar{I}$ corresponding to the jth pair of segments and is given by Eq. (8.39) with $i = 2j - 1$. Equation (8.39) and (8.40) lead to

$$I \approx \frac{h}{3}\left[f_0 + 4 \sum_{i=1,3,5,\ldots}^{n-1} f_i + 2 \sum_{i=2,4,6,\ldots}^{n-2} f_i + f_n\right]. \tag{8.41}$$

▶Example 8.5

Determine the value of the integral described in Example 8.4 with $a = 0.0$ and $b = 1.5$ using Simpson's $\frac{1}{3}$ rule with different step sizes.

Solution

Simpson's $\frac{1}{3}$ rule gives the value of the integral, with n steps, as

$$I = \frac{h}{3}\left(f_0 + 4 \sum_{i=1,3,5,\ldots}^{n-1} f_i + 2 \sum_{i=2,4,6,\ldots}^{n-2} f_i + f_n \right). \tag{a}$$

For $n = 2$ and $h = 0.75$, Eq. (a) gives

$$I = \frac{h}{3}\left(f_0 + 4\, f_1 + f_2\right), \tag{b}$$

where $f_0 = f(0.0) = 0.84885406$, $f_1 = f(0.75) = 4.2424269$, and $f_2 = f(1.5) = 0.84542847$. Thus, Eq. (b) gives $I = 4.66599753$. Noting that the exact value of the integral is 5.5606613, the error in Simpson's $\frac{1}{3}$ rule is 23.7064321%. For $n = 4$ and $h = 0.375$, Eq. (a) gives

$$I = \frac{h}{3}\left(f_0 + 4\, f_1 + 2\, f_2 + 4\, f_3 + f_4\right), \tag{c}$$

where $f_0 = f(0.0) = 0.84885406$, $f_1 = f(0.375) = 2.9153550$, $f_2 = f(0.75) = 4.2424269$, $f_3 = f(1.125) = 5.5493011$, and $f_4 = f(1.5) = 0.84542847$. Thus, Eq. (c) gives $I = 5.50472009$ with an error of 1.0060172%. By proceeding in a similar manner, the value of the integral is computed for $n = 6, 8, 10$, and 12. The results are given in Table 8.2. ◀

8.4.2 Simpson's Three-Eighth's Rule

In this method, the integral is evaluated by approximating the function $f(x)$ by a third-degree polynomial, $p_3(x)$, as shown in Fig. 8.14. By assuming the polynomial $p_3(x)$ as

$$p_3(x) = c_3x^3 + c_2x^2 + c_1x + c_0, \tag{8.42}$$

Table 8.2

Number of steps (n)	Step length (h)	Value of the integral (I)	Percent error
2	0.75	4.6659975	16.089163
4	0.375	5.5047202	1.0060153
6	0.25	5.5495892	0.19911586
8	0.1875	5.5571246	0.063602172
10	0.15	5.5592151	0.026008544
12	0.125	5.5599494	0.012802755
14	0.10714286	5.5602551	0.0073060603
16	0.09375	5.5604043	0.0046220263

Exact value of the integral: 5.5606613

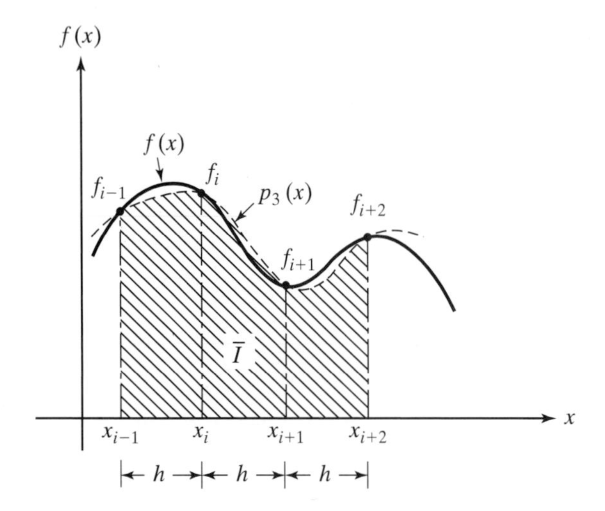

Figure 8.14 Simpson's three-eighth's rule.

the constants c_0, c_1, c_2, and c_3 are determined by making the polynomial pass through the four points (x_{i-1}, f_{i-1}), (x_i, f_i), (x_{i+1}, f_{i+1}), and (x_{i+2}, f_{i+2}). By taking the origin at x_i ($x = 0$ at x_i), x_{i-1}, x_{i+1}, and x_{i+2} can be assumed to correspond to $x = -h, h$, and $2\,h$, respectively. Such a choice of the origin does not influence the final result. By using the relations

For x_{i-1},

$$p_3(x = -h) = f_{i-1} = -h^3\,c_3 + h^2 c_2 - h\,c_1 + c_0; \tag{8.43}$$

For x_i,

$$p_3(x = 0) = f_i = c_0; \tag{8.44}$$

For x_{i+1},

$$p_3(x = h) = f_{i+1} = h^3\,c_3 + h^2\,c_2 + h\,c_1 + c_0; \tag{8.45}$$

For x_{i+2},

$$p_3(x = 2\,h) = f_{i+2} = 8\,h^3\,c_3 + 4\,h^2\,c_2 + 2\,h\,c_1 + c_0 \tag{8.46}$$

the solution of Eqs. (8.43) through (8.46) can be determined as

$$c_0 = f_i \tag{8.47}$$

$$c_1 = \frac{1}{6\,h}(-f_{i+2} + 6\,f_{i+1} - 3f_i - 2\,f_{i-1}); \tag{8.48}$$

$$c_2 = \frac{1}{2\,h^2}(f_{i-1} - 2\,f_i + f_{i+1}); \tag{8.49}$$

$$c_3 = \frac{1}{6\,h^3}(f_{i+2} - 3\,f_{i+1} + 3\,f_i - f_{i-1}). \tag{8.50}$$

(See Problem 8.47.) The area ($\bar{I}$) under the third-degree polynomial $p_3(x)$ between x_{i-1} to x_{i+2} can be found as

$$\bar{I} = \int_{x_{i-1}}^{x_{i+2}} p_3(x)\,dx = \int_{-h}^{2h} \left(c_3 x^3 + c_2 x^2 + c_1 x + c_0\right) dx$$

$$= \frac{c_3}{4}\left(x^4\right)\Big|_{-h}^{2h} + \frac{c_2}{3}\left(x^3\right)\Big|_{-h}^{2h} + \frac{c_1}{2}\left(x^2\right)\Big|_{-h}^{2h} + c_0\left(x\right)\Big|_{-h}^{2h}$$

$$= \frac{c_3}{4}(15h^4) + \frac{c_2}{3}(9h^3) + \frac{c_1}{2}(3h^2) + c_0\,(3h)\,. \tag{8.51}$$

By substituting for c_0 to c_3 from Eqs. (8.47) through (8.50), Eq. (8.51) gives

$$\bar{I} = \frac{15h^4}{4}\left(\frac{f_{i+2} - 3f_{i+1} + 3f_i - f_{i-1}}{6h^3}\right) + 3h^3\left(\frac{f_{i+1} - 2f_i + f_{i-1}}{2h^2}\right)$$

$$+\frac{3h^2}{2}\left(\frac{-f_{i+2} + 6f_{i+1} - 3f_i - 2f_{i-1}}{6h}\right) + 3hf_i$$

$$= \frac{3h}{8}\left[f_{i+2} + 3f_{i+1} + 3f_i + f_{i-1}\right]. \tag{8.52}$$

The term "$\frac{3}{8}$" in Simpson's three-eighths rule refers to the presence of the factor "$\frac{3}{8}$" in Eq. (8.52). Note that three segments are used in deriving Eq. (8.52). Thus, for a multistage application of Simpson's three-eighths rule, we need to divide the range $a \leq x \leq b$ into n segments of equal width $h = \frac{b-a}{n}$. The number of segments n must be a multiple of 3 so that Eq. (8.52) can be applied for groups of three segments. The integral in Eq. (8.32) can be evaluated as

$$I = \int_a^b f(x)\,dx \approx \sum_{j=1}^{n/3} \left(\bar{I}\right)_j, \tag{8.53}$$

where $\left(\bar{I}\right)_j$ represents the value of $\bar{I}$ corresponding to the jth group of three segments and is given by Eq. (8.52) with $i = 3j - 2$. The use of Eqs. (8.52) and (8.53) yields

$$I \approx \frac{3h}{8}\left[f_0 + 3\sum_{i=1,4,7,\ldots}^{n-2}\left(f_i + f_{i+1}\right) + 2\sum_{i=3,6,9,\ldots}^{n-3} f_i + f_n\right]. \tag{8.54}$$

It can be shown that the truncation error in using Eq. (8.54) is of the same order as that of Simpson's one-third rule. But the use of Eq. (8.54) requires the number of segments to be a multiple of 3. Hence, Eq. (8.54) is rarely used by itself. Often both Simpson's one-third and three-eighths rules are used together so that the number of segments n need not be constrained in any way. If the number of segments is even, Simpson's one-third rule can be used. On the other hand, if the number of segments is odd, Simpson's three-eighths rule can be applied, for instance, for the first three segments and Simpson's one-third rule can be used for the remaining even number of segments.

Table 8.3

Number of steps (n)	Step length (h)	Value of the integral (I)	Percent error
3	0.5	5.1630173	7.1510205
6	0.25	5.5357828	0.44740185
9	0.16666667	5.5557156	0.088941850
12	0.125	5.5590858	0.028332422
15	0.1	5.5600042	0.011816609
18	0.08333333	5.5603180	0.0061741355
21	0.071428575	5.5604572	0.003670180
24	0.0625	5.5605288	0.0023839022

Exact value of the integral: 5.5606613

▶Example 8.6

Determine the value of the integral described in Example 8.4 with $a = 0.0$ and $b = 1.5$ using Simpson's $\frac{3}{8}$ rule with different step sizes.

Solution

Equation (8.54) gives the value of the integral according to Simpson's $\frac{3}{8}$ rule for n steps. For $n = 3$ and $h = 0.5$, Eq. (8.54) gives

$$I = \frac{3h}{8}\left[f_0 + 3f_1 + 3f_2 + f_3\right], \tag{a}$$

where $f_0 = f(0.0) = 0.84885406$, $f_1 = f(0.5) = 2.8566201$, $f_2 = f(1.0) = 5.7573166$, and $f_3 = f(1.5) = 0.84542847$. Equation (a) gives $I = 5.16301737$ with an error of 7.1510187%. For $n = 6$ and $h = 0.25$, Eq. (8.54) gives

$$I = \frac{3h}{8}\left[f_0 + 3f_1 + 3f_2 + 2f_3 + 3f_4 + 3f_5 + f_6\right], \tag{b}$$

where $f_0 = f(0.0) = 0.84885406$, $f_1 = f(0.25) = 3.2544143$, $f_2 = f(0.5) = 2.8566201$, $f_3 = f(0.75) = 4.2424269$, $f_4 = f(1.0) = 5.7573166$, $f_5 = f(1.25) = 4.4213867$, and $f_6 = f(1.5) = 0.84542847$. Thus, Eq. (b) gives $I = 5.5357828$ with an error of 0.44740185%. By proceeding in a similar manner, the value of the integral is computed for $n = 9$, 12, 15, and 18. The results are shown in Table 8.3. ◀

8.4.3 Truncation Error

As in the case of trapezoidal rule, the basic truncation error (E) in Simpson's one-third rule, considering only two segments in the interval a to b, is given by

$$E = \int_a^b f(x)\,dx - \left(\frac{b-a}{6}\right)\left[f(a) + 4f(x_1) + f(b)\right], \tag{8.55}$$

where the first term on the right-hand side of Eq. (8.55) denotes the exact integral, while the second term represents the approximate integral given by Simpson's one-third rule. (See Fig. 8.15.) We expand $f(x)$ using Taylor's series about the midpoint of the range, x_1, that is,

$$f(x) = f(x_1) + yf'(x_1) + \frac{y^2}{2!}f''(x_1) + \frac{y^3}{3!}f'''(x_1) + \frac{y^4}{4!}f''''(x_1) + \frac{y^5}{5!}f'''''(x_1) + \cdots, \tag{8.56}$$

where $y = x - x_1$. Equation (8.56) can be used to express the integral of $f(x)$ as

$$\int_a^b f(x)\,dx = \int_{-h}^{h} \left\{ f(x_1) + yf'(x_1) + \frac{y^2}{2}f''(x_1) + \frac{y^3}{6}f'''(x_1) + \frac{y^4}{24}f''''(x_1) + \frac{y^5}{120}f'''''(x_1) + \cdots \right\} dy, \tag{8.57}$$

where $y = -h$ and $y = h$ correspond to $x = a$ and $x = b$, respectively. By carrying out the integration in Eq. (8.57), we obtain

$$\begin{aligned}\int_a^b f(x)\,dx &= f(x_1)\,(y)\big|_{-h}^{h} + f'(x_1)\left(\frac{y^2}{2}\right)\bigg|_{-h}^{h} + f''(x_1)\left(\frac{y^3}{6}\right)\bigg|_{-h}^{h} \\ &\quad + f'''(x_1)\left(\frac{y^4}{24}\right)\bigg|_{-h}^{h} + f''''(x_1)\left(\frac{y^5}{120}\right)\bigg|_{-h}^{h} + f'''''(x_1)\left(\frac{y^6}{720}\right)\bigg|_{-h}^{h} + \cdots \\ &= 2hf(x_1) + \frac{h^3}{3}f''(x_1) + \frac{h^5}{60}f''''(x_1) + \cdots. \end{aligned} \tag{8.58}$$

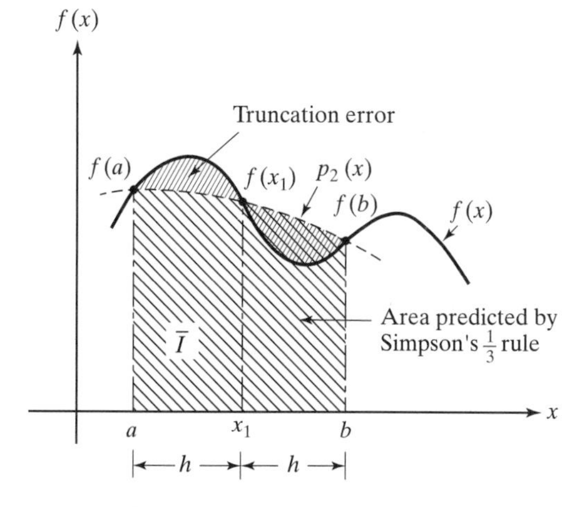

Figure 8.15 Truncation error.

Substituting $x = a(y = -h)$, $x = x_1(y = 0)$, and $x = b(y = h)$ into Eq. (8.56) yields

$$f(a) = f(x_1) - hf'(x_1) + \frac{h^2}{2}f''(x_1) - \frac{h^3}{6}f'''(x_1) + \frac{h^4}{24}f''''(x_1) - \frac{h^5}{120}f'''''(x_1) + \cdots; \tag{8.59}$$

$$f(x_1) = f(x_1); \tag{8.60}$$

$$f(b) = f(x_1) + hf'(x_1) + \frac{h^2}{2}f''(x_1) + \frac{h^3}{6}f'''(x_1) + \frac{h^4}{24}f''''(x_1) + \frac{h^5}{120}f'''''(x_1) + \cdots. \tag{8.61}$$

Now the second term on the right-hand side of Eq. (8.55) can be expressed, using Eqs. (8.59) through (8.61), as

$$\begin{aligned}&\left(\frac{b-a}{6}\right)\left[f(x_1) - hf'(x_1) + \frac{h^2}{2}f''(x_1) - \frac{h^3}{6}f'''(x_1)\right.\\ &+\frac{h^4}{24}f''''(x_1) - \frac{h^5}{120}f'''''(x_1) + \cdots + 4f(x_1)\\ &+f(x_1) + hf'(x_1) + \frac{h^2}{2}f''(x_1) + \frac{h^3}{6}f'''(x_1)\\ &\left.+\frac{h^4}{24}f''''(x_1) + \frac{h^5}{120}f'''''(x_1) + \cdots\right]\\ &= \left(\frac{b-a}{6}\right)\left[6f(x_1) + h^2f''(x_1) + \frac{h^4}{12}f''''(x_1) + \cdots\right].\end{aligned} \tag{8.62}$$

Substituting Eqs. (8.58) and (8.62) into Eq. (8.55) and truncating terms involving derivatives higher than the fifth gives

$$\begin{aligned}E &\approx \left[2\left(\frac{b-a}{2}\right)f(x_1) + \frac{1}{3}\left(\frac{b-a}{2}\right)^3 f''(x_1) + \frac{1}{60}\left(\frac{b-a}{2}\right)^5 f''''(x_1)\right]\\ &\quad - \left[(b-a)f(x_1) + \left(\frac{b-a}{6}\right)\left(\frac{b-a}{2}\right)^2 f''(x_1) + \left(\frac{b-a}{6}\right)\frac{1}{12}\left(\frac{b-a}{2}\right)^4 f''''(x_1)\right]\\ &\approx -\frac{1}{2880}(b-a)^5 f''''(x_1)\\ &\approx -\frac{1}{90}h^5 f''''(x_1).\end{aligned} \tag{8.63}$$

This indicates that the error of Simpson's one-third rule (per each pair of segments) is proportional to h^5 and $f''''(x_1)$. Thus, the error will be zero if $f(x)$ is a third-order polynomial, since $f'''' = 0$.

The error in a multisegmented Simpson's one-third rule, Eq. (8.54), can be found by summing the errors of the individual pairs of segments (x_0, x_2), (x_2, x_4), ..., (x_{n-2}, x_n):

$$E \approx -\frac{h^5}{90} \sum_{j=1,3,5,\ldots}^{n-1} f''''(x_j). \tag{8.64}$$

By defining an average value of the fourth derivative, $\overline{f''''}$, as

$$\overline{f''''} = \frac{2}{n} \left(\sum_{j=1,3,5,\ldots}^{n-1} f''''(x_j) \right). \tag{8.65}$$

Eq. (8.64) can be expressed as follows:

$$\begin{aligned} E &\approx -\frac{1}{90} h^5 \frac{n}{2} \overline{f''''} \\ &\approx -\frac{1}{180} h^4 (b-a) \overline{f''''} \\ &= O\left(h^4\right). \end{aligned} \tag{8.66}$$

This indicates that the error in a multisegment Simpson's one-third rule, Eq. (8.41), is proportional to h^4, since $(b - a)$ is fixed.

By following a similar approach, the truncation error in a multisegment Simpson's three-eighths rule can also be shown to be proportional to h^4. (See Problem 8.8.)

8.5 General Newton–Cotes Formulas

As stated earlier, the Newton–Cotes formulas are derived by using a polynomial of order m to approximate the function $f(x)$. That is,

$$\int_a^b f(x)\,dx \approx \int_a^b p_m(x)\,dx, \tag{8.67}$$

where

$$p_m(x) = c_m x^m + c_{m-1} x^{m-1} + \cdots + c_2 x^2 + c_1 x + c_0. \tag{8.68}$$

Numerical integration formulas corresponding to $m = 0$ (rectangular rule), $m = 1$ (trapezoidal rule), $m = 2$ (Simpson's one-third rule), and $m = 3$ (Simpson's three-eighths rule) have been derived in Sections 8.3 and 8.4. Formulas corresponding to higher order polynomials can also be derived. An estimate of error associated with any formula can also be derived as outlined earlier. A summary of some of

Table 8.4

Value of m	Name of formula	Formula $h = (b-a)/n$	Estimate of truncation error	Number of segments of width h (in each group)
0	Rectangular	hf_i or hf_{i+1}	$\frac{1}{2}h^2f'$ or $-\frac{1}{2}h^2f'$	0
1	Trapezoidal	$\frac{h}{2}(f_{i-1}+f_i)$	$-\frac{1}{12}h^3f''$	1
2	Simpson's one-third	$\frac{h}{3}(f_{i-1}+4f_i+f_{i+1})$	$-\frac{1}{90}h^5f''''$	2
3	Simpson's three-eighth	$\frac{3h}{8}(f_{i-1}+3f_i+3f_{i+1}+f_{i+2})$	$-\frac{3}{80}h^5f''''$	3
4	Boole's	$\frac{2h}{45}(7f_{i-2}+32f_{i-1}+12f_i$ $+32f_{i+1}+7f_{i+2})$	$-\frac{8}{945}h^7f^{vi}$	4
5	—	$\frac{5h}{288}(19f_{i-2}+75f_{i-1}+50f_i$ $+50f_{i+1}+75f_{i+2}+19f_{i+3})$	$-\frac{275}{12096}h^7f^{vi}$	5

the Newton–Cotes formulas, along with the associated error estimates, is given in Table 8.4.

8.6 Richardson's Extrapolation

In many engineering problems, the integrals are to be evaluated very accurately. One possibility is to use a large number of segments (n) in the trapezoidal or Simpson's method to reduce the truncation error. However, beyond a certain number of segments, the round-off error begins to dominate and the accuracy of the result may suffer as shown in Fig. 8.16. Also, the computational effort required will be

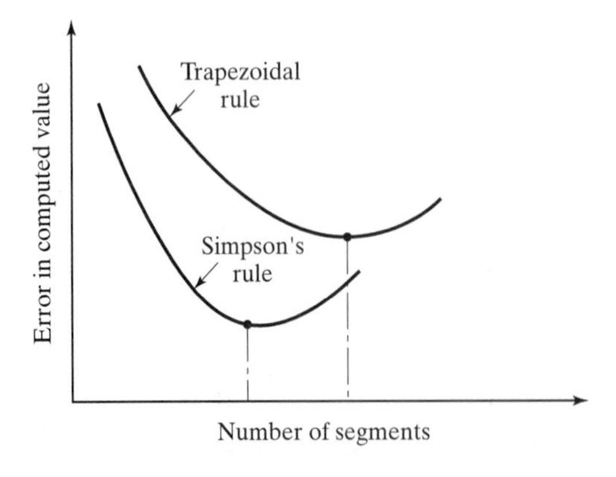

Figure 8.16 Variation of accuracy with increasing number of segments.

more with larger number of segments. Another possibility to improve the accuracy is to use a higher order Newton–Cotes formula. Alternatively, the accuracy of the estimated integral can be improved by using a scheme known as Richardson's extrapolation, in which two numerical integral estimates are combined to obtain a third, more accurate value. The computational algorithm, which implements Richardson's extrapolation in an efficient manner, is known as Romberg integration. This is a recursive procedure that can be used to generate the value of the integral to within a prespecified error tolerance.

Richardson's extrapolation is a numerical procedure that can be used to improve the accuracy of the results obtained from another numerical method, provided an estimate of the error is available. The procedure can be used not only in numerical integration of functions, but also in other methods such as numerical integration of differential equations. In this section, we consider the application of Richardson's extrapolation procedure to trapezoidal and Simpson's rules.

8.6.1 Trapezoidal Rule

The truncation error in multisegmented trapezoidal rule is given by Eq. (8.21):

$$E \approx -\frac{1}{12}(b-a)h^2\overline{f''}. \tag{8.69}$$

If $I_1(h_1)$ denotes the value of the integral (approximate value) given by the trapezoidal rule and $E_1(h_1)$ indicates the truncation error with a step size h_1, the exact value of the integral can be expressed as

$$I \approx I_1(h_1) + E_1(h_1) \approx I_1(h_1) + ch_1^2, \tag{8.70}$$

where $c = -\frac{1}{12}(b-a)\bar{f}''$ is a constant. Similarly, if $I_2(h_2)$ denotes the value of the integral given by the trapezoidal rule with a step size h_2 and $E_2(h_2)$ represents the associated truncation error, we can write

$$I \approx I_2(h_2) + E_2(h_2) \approx I_2(h_2) + ch_2^2 \tag{8.71}$$

by assuming that $\overline{f''}$ is constant regardless of the step size. Equations (8.70) and (8.71) can be used to obtain

$$I_1(h_1) + ch_1^2 \approx I_2(h_2) + ch_2^2,$$

or

$$c \approx \frac{I_2(h_2) - I_1(h_1)}{h_1^2 - h_2^2}. \tag{8.72}$$

Substituting this expression of c into Eq. (8.71) yields an improved estimate of the integral (I) as

$$I \approx I_2(h_2) + \frac{I_2(h_2) - I_1(h_1)}{\left\{\left(\frac{h_1}{h_2}\right)^2 - 1\right\}}. \tag{8.73}$$

It can be shown [8.3] that the error of this estimate is $O(h^4)$, which means that we combined two estimates given by the trapezoidal rule, which has an error of $O(h^2)$, to yield a new estimate having an error of $O(h^4)$. This can also be seen for the special case where the interval is halved $\left(h_2 = \frac{h_1}{2}\right)$. Using $h_2 = \frac{h_1}{2}$, we find that Eq. (8.73) gives

$$I \approx I_2(h_2) + \frac{I_2(h_2) - I_1(h_1)}{3}$$
$$\approx \frac{4}{3}I_2(h_2) - \frac{1}{3}I_1(h_1). \tag{8.74}$$

It can be verified that this expression is identical to the one given by Simpson's one-third rule with a step size of h_2. Note that the estimate given by Simpson's one-third rule has an error of $O(h^4)$.

▶Example 8.7

Using the results of the trapezoidal rule given in Table 8.1, find an improved estimate of the value of the integral using Richardson's extrapolation.

Solution

Using $h_1 = 1.5, I_1 = 1.2707119, h_2 = 0.75$, and $I_2 = 3.8171761$, we find that Eq. (8.74) gives the improved estimate of

$$I = \frac{4}{3}(3.8171761) - \frac{1}{3}(1.2707119) = 4.66599750,$$

which corresponds to an error of 16.0891619%. Similarly, by using $h_1 = 0.5$, $I_1 = 4.7305388$, $h_2 = 0.375$, and $I_2 = 5.0828342$, Eq. (8.74) yields the improved estimate of I as

$$I = \frac{4}{3}(5.0828342) - \frac{1}{3}(4.7305388) = 5.20026600,$$

which corresponds to an error of 6.48115899%. It can be seen that, in both the cases, the value of the integral predicted by Richardson's extrapolation is superior to the original estimates. ◀

8.6.2 Simpson's One-Third Rule

The truncation error in a multisegmented Simpson's one-third rule is given by Eq. (8.66):

$$E \approx -\frac{1}{180}(b-a)h^4\overline{f''''}. \tag{8.75}$$

If $I_1(h_1)$ and $I_2(h_2)$ denote the values of the integral given by Simpson's one-third rule with step sizes h_1 and h_2, and the corresponding error estimates are given by

$E_1(h_1)$ and $E_2(h_2)$, respectively, we have

$$I \approx I_1(h_1) + E_1(h_1) \approx I_1(h_1) + ch_1^4 \tag{8.76}$$

and

$$I \approx I_2(h_2) + E_2(h_2) \approx I_2(h_2) + ch_2^4. \tag{8.77}$$

These equations yield

$$c \approx \frac{I_2(h_2) - I_1(h_1)}{h_1^4 - h_2^4} \tag{8.78}$$

Substituting this expression into Eq. (8.77) gives an improved estimate of the integral (I) as

$$I \approx I_2(h_2) + \frac{I_2(h_2) - I_1(h_1)}{\left\{\left(\frac{h_1}{h_2}\right)^4 - 1\right\}}. \tag{8.79}$$

It can be shown that the error of this estimate is $O(h^6)$. This implies that we combined two estimates given by Simpson's one-third rule, which has an error of $O(h^4)$, to obtain a new estimate having an error of $O(h^6)$. When h_2 is taken as $\frac{1}{2}h_1$, Eq. (8.79) gives

$$I \approx \frac{16}{15}I_2(h_2) - \frac{1}{15}I_1(h_1). \tag{8.80}$$

8.7 Romberg Integration

We saw in the previous section that the results of numerical integration given by the trapezoidal rule, for example, can be improved considerably by computing the integral twice using two different steps and then applying Eq. (8.73). Although the computational effort is doubled, the method is very simple to use; we just need to apply the numerical method twice. As stated earlier, we obtain an improved estimate of the integral of $O(h^4)$ by combining two numerical integrals of $O(h^2)$. In fact, if we have three numerical values of the integral, given by the trapezoidal rule with three different step sizes, each with $O(h^2)$, we can combine the first and second numerical integrals (I_1 and I_2) to obtain a better estimate (I_4) with error $O(h^4)$. Similarly, the second and third numerical integrals (I_2 and I_3) can be combined to find an improved estimate (I_5) with error $O(h^4)$. The two improved estimates (I_4 and I_5), in turn, can be combined to yield an even better estimate of the integral (I_6) with error $O(h^6)$.

The general concept can be explained by reconsidering the truncation error in the trapezoidal rule. The error estimate given by Eq. (8.21) was derived by neglecting terms involving derivatives higher than two in Taylor's series expansion of Eq. (8.12). If the higher derivative terms are also included, the error estimate (E) in the trapezoidal rule can be expressed as

$$E = a_1h^2 + a_2h^4 + a_3h^6 + a_4h^8 + a_5h^{10} + \cdots, \tag{8.81}$$

where $a_1, a_2, \ldots$ are constants. It can be proved that each successive application of Richardson's extrapolation eliminates the leading term in Eq. (8.81). Thus, the application of Richardson's extrapolation once (using I_1 and I_2) produces a result with error of $O(h^4)$, the next application produces a result with error of $O(h^6)$, etc. In fact, Richardson's extrapolation can be applied in succession (assuming that we have applied the basic numerical integration formula, such as trapezoidal rule, enough number of times using different step sizes to obtain the numerical values of the integral) to obtain a sequence of improved estimates of the integral. The general formulation of the procedure, which is well suited for computer implementation, is known as Romberg integration. The general formula used in Romberg integration can be derived as follows:

Let $I_{0,n}$ and $I_{0,\frac{n}{2}}$ indicate the values of the integral computed by the trapezoidal rule with n and $\frac{n}{2}$ segments (that is, with step sizes $\frac{h}{2}$ and h), respectively. Then the improved value of the integral given by Richardson's extrapolation, Eq. (8.74), can be written as

$$I_{1,n} = \frac{4 I_{0,n} - I_{0,\frac{n}{2}}}{4^1 - 1}, \tag{8.82}$$

where the first subscript 1 in $I_{1,n}$ indicates the number of times the Richardson's extrapolation is applied. When Richardson's extrapolation is applied again, it produces the second extrapolation for the integral $(I_{2,n})$ as

$$I_{2,n} = \frac{4^2 I_{1,n} - I_{1,\frac{n}{2}}}{4^2 - 1} \tag{8.83}$$

When the extrapolation process is repeated k times, it gives the value of the integral as

$$I_{k,n} = \frac{4^k I_{k-1,n} - I_{k-1,\frac{n}{2}}}{4^k - 1}. \tag{8.84}$$

As stated earlier, the value given by Eq. (8.82) has an error of $O(h^4)$, the value given by Eq. (8.83) has $O(h^6)$, and so on. The Romberg integration procedure can be represented schematically as shown in Fig. 8.17.

Notice that we can obtain any specified accuracy using a proper number of Richardson's extrapolations. The process can be stopped when the convergence criterion

$$\left| I_{k,n} - I_{k-1,n} \right| \leq \varepsilon, \tag{8.85}$$

where ε is a small number, is satisfied.

8.8 Gauss Quadrature

While the methods considered so far required the evaluation of the integrand at equal intervals, the Gauss quadrature requires the evaluation of the integrand at specified, but unequal, intervals. Gauss quadrature is a powerful method of

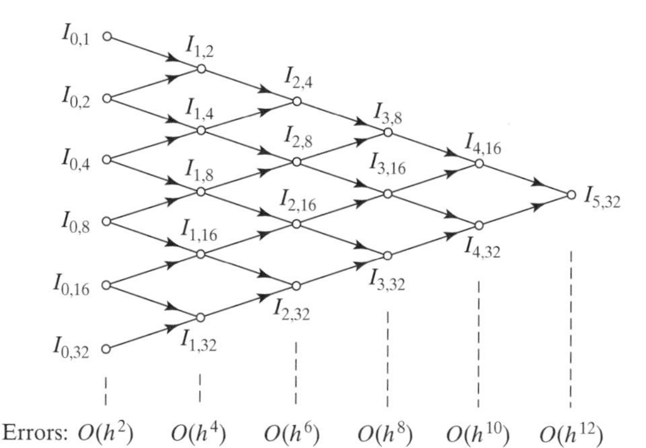

Figure 8.17 Romberg integration procedure.

numerical integration and its accuracy is much higher than the Newton–Cotes formulas. As such, Gauss quadrature is not useful to integrate functions that are given in tabular form with equispaced intervals. The most popular form of Gauss quadrature, also known as Gauss–Legendre quadrature, uses Legendre polynomials to approximate the function $f(x)$. The method uses the roots of Legendre polynomials to locate the points at which the integrand is evaluated. In Gauss integration, the integral is evaluated by using the formula

$$\int_{-1}^{1} f(x)\,dx = \sum_{i=1}^{n} w_i f(x_i), \tag{8.86}$$

where n is called the number of Gauss points, w_i are the unknown coefficients, also called weights, and x_i are the specific values of x, also called Gauss points, at which the integrand is evaluated. For any specified n, the values of w_i and x_i are chosen so that the formula will be exact for polynomials up to and, including degree $(2n-1)$. For example, when $n = 2$, the values of w_1, w_2, x_1, and x_2 are selected so that the formula will give the exact value of the integral for polynomials up to degree three.

8.8.1 Coordinate Transformation

As can be seen from Eq. (8.86), Gauss integration requires the range of integration from -1 to $+1$. For convenience of notation, let the original coordinate be y and the range of integration of $f(y)$ be from a to b. Then the transformation

$$x = \frac{2y - a - b}{b - a} \tag{8.87}$$

gives the normalized coordinate $x = -1$, when $y = a$ and $x = +1$, when $y = b$. The transformation from x to y is given by

$$y = \frac{(b-a)x + a + b}{2}. \tag{8.88}$$

By noting that $dy = \left(\frac{b-a}{2}\right) dx$, the original integral $\int_a^b f(y)\, dy$ can be rewritten as

$$\int_a^b f(y)\, dy = \int_{-1}^{1} f(y) \frac{dy}{dx} dx = \frac{b-a}{2} \sum_{i=1}^{n} w_i f(y_i). \tag{8.89}$$

If x_i is the Gauss point of the normalized coordinate, the corresponding value of y_i can be determined, using Eq. (8.88), as

$$y_i = \frac{(b-a)x_i + a + b}{2}. \tag{8.90}$$

Since the weights w_i remain the same, the integral can be evaluated using the right-hand side expression of Eq. (8.89).

8.8.2 Derivation of Two-Point Gauss Formula

The two-point Gauss integration formula is given by [Eq. (8.86) with $n = 2$]

$$\int_{-1}^{1} f(x)\, dx = w_1 f(x_1) + w_2 f(x_2), \tag{8.91}$$

where the evaluation of the four unknowns w_1, w_2, x_1, and x_2 requires the use of four conditions. Since $n = 2$, the formula should give the exact value for polynomials of order three and below. By enforcing the formula to be exact for the polynomials $f(x) = 1, x, x^2$, and x^3 in turn, we obtain the following equations:

When $f(x) = 1$,

$$\int_{-1}^{1} f(x)\, dx = \int_{-1}^{1} 1\, dx = 2$$
$$= w_1 f(x_1) + w_2 f(x_2) = w_1 + w_2. \tag{8.92}$$

When $f(x) = x$,

$$\int_{-1}^{1} f(x)\, dx = \int_{-1}^{1} x\, dx = \left(\frac{x^2}{2}\right)_{-1}^{1} = 0$$
$$= w_1 f(x_1) + w_2 f(x_2) = w_1 x_1 + w_2 x_2. \tag{8.93}$$

When $f(x) = x^2$,

$$\int_{-1}^{1} f(x)\, dx = \int_{-1}^{1} x^2\, dx = \left(\frac{x^3}{3}\right)_{-1}^{1} = \frac{2}{3}$$
$$= w_1 f(x_1) + w_2 f(x_2) = w_1 x_1^2 + w_2 x_2^2. \tag{8.94}$$

When $f(x) = x^3$,

$$\int_{-1}^{1} f(x)\,dx = \int_{-1}^{1} x^3\,dx = \left(\frac{x^4}{4}\right)_{-1}^{1} = 0$$

$$= w_1 f(x_1) + w_2 f(x_2) = w_1 x_1^3 + w_2 x_2^3. \tag{8.95}$$

Since the limits of integration, -1 and $+1$, are symmetric about $x = 0$, we expect x_1 and x_2 also to be symmetric about $x = 0$. By setting $x_2 = -x_1$, we obtain the following from Eqs. (8.93) and (8.92):

$$w_1 = w_2 = 1.$$

These values automatically satisfy Eq. (8.95), and Eq. (8.94) gives

$$x_1^2 = \frac{1}{3},$$

from which we obtain

$$x_1 = \frac{1}{\sqrt{3}} = 0.577350269189626$$

and

$$x_2 = -x_1 = -\frac{1}{\sqrt{3}} = -0.577350269189626.$$

Note: The two-point Gauss formula, like Simpson's rule, is exact for polynomials up to order three, but it requires much smaller number of function evaluations.

8.8.3 General Procedure

Although the derivation of the two-point Gauss integration formula is not complex, the derivation of a formula using more than two Gauss points is quite involved. However, the general procedure of finding w_i and x_i involves the following steps:

1. The Gauss points $x_1, x_2, \ldots, x_n$ are the roots of the Legendre polynomial of degree n, $P_n(x)$. The Legendre polynomials are orthogonal on the interval $[-1, 1]$, so that

   $$\int_{-1}^{1} P_n(x) P_m(x)\,dx = 0; \quad n \neq m$$

   and

   $$\int_{-1}^{1} \{P_n(x)\}^2\,dx = c(n) \neq 0. \tag{8.96}$$

   where $c(n)$ is a constant whose value depends on n. The Legendre polynomials are defined by

   $$P_0(x) = 1,$$

   $$P_1(x) = x,$$

and

$$P_n(x) = \left(\frac{2n-1}{n}\right) x P_{n-1}(x) - \left(\frac{n-1}{n}\right) P_{n-2}(x); n = 2, 3, 4, \ldots. \quad (8.97)$$

In general, any arbitrary nth-degree polynomial, $p_n(x)$, can be represented by a linear combination of Legendre polynomials as

$$p_n(x) = \sum_{i=0}^{n} \alpha_i P_i(x), \quad (8.98)$$

where α_i are constants.

▶Example 8.8

Express a fifth-degree polynomial, $p_5(x)$, in terms of Legendre polynomials.

Solution

The Legendre polynomials $P_i(x)$ for $i = 0, 1, 2, 3, 4$, and 5 can be obtained from Eq. (8.97) as

$$P_0(x) = 1,\ P_1(x) = x,\ P_2(x) = \frac{1}{2}(3x^2 - 1),$$

$$P_3(x) = \frac{1}{2}(5x^3 - 3x),\ P_4(x) = \frac{1}{8}(35x^4 - 30x^2 + 3),$$

and

$$P_5(x) = \frac{1}{8}(63x^5 - 70x^3 + 15x).$$

Thus, the polynomial, $p_5(x)$, can be written as

$$p_5(x) = \alpha_0 P_0(x) + \alpha_1 P_1(x) + \alpha_2 P_2(x) + \alpha_3 P_3(x) + \alpha_4 P_4(x) + \alpha_5 P_5(x). \quad \text{(a)}$$

Equation (a) can be rewritten as

$$p_5(x) = \beta_0 + \beta_1 x + \beta_2 x^2 + \beta_3 x^3 + \beta_4 x^4 + \beta_5 x^5, \quad \text{(b)}$$

where β_i can be expressed in terms of α_i by equating the coefficients of like powers of x in Eqs. (a) and (b). (See Problem 8.22.) ◀

2. The weights w_i can be computed as

$$w_i = \frac{2(1 - x_i^2)}{\{n P_{n-1}(x_i)\}^2}. \quad (8.99)$$

Although the routine computation of x_i and w_i is quite complex, the values of x_i and w_i for various values of n have been generated and tabulated [8.6]. Values of x_i and w_i corresponding to $n = 1, 2, \ldots, 6$ are given in Table 8.5.

Table 8.5 Gauss points, weights, and error estimates.

Number (n)	Locations (x_i)	Weights (w_i)	Error proportional to
1	0.00000 00000 00000	2.00000 00000 00000	$f^{(2)}(\xi)$
2	$\pm$ 0.57735 02691 89626	1.00000 00000 00000	$f^{(4)}(\xi)$
3	$\pm$ 0.77459 66692 41483	0.55555 55555 55555	$f^{(6)}(\xi)$
	0.00000 00000 00000	0.88888 88888 88889	
4	$\pm$ 0.86113 63115 94053	0.34785 48451 47454	$f^{(8)}(\xi)$
	$\pm$ 0.33998 10435 84856	0.65214 51548 62546	
5	$\pm$ 0.90617 98459 38664	0.23692 68850 56189	$f^{(10)}(\xi)$
	$\pm$ 0.53846 93101 05683	0.47862 86704 99366	
	0.00000 00000 00000	0.56888 88888 88889	
6	$\pm$ 0.93246 95142 03152	0.17132 44923 79170	$f^{(12)}(\xi)$
	$\pm$ 0.66120 93864 66265	0.36076 15730 48139	
	$\pm$ 0.23861 91860 83197	0.46791 39345 72691	

$-1 < \xi < 1$

8.8.4 Error Estimate

The error (E) in the n-point Gauss formula (more accurately, the Gauss–Legendre formula) is given by

$$E \approx \frac{2^{2n+1}\{(n)!\}^4}{(2n+1)\{(2n)!\}^3} f^{(2n)}(\xi); \quad -1 < \xi < 1. \tag{8.100}$$

As stated earlier, the n-point formula integrates a polynomial of degree $(2n-1)$ exactly, since the derivative of order $(2n)$, $f^{(2n)}$, is zero in this case. Assuming that the magnitudes of higher order derivatives decrease (or increase only slowly) with increasing values of n, the Gauss formulas are significantly more accurate than the Newton–Cotes formulas. The truncation errors for different Gauss formulas are also indicated in Table 8.5.

▶Example 8.9

Evaluate the integral

$$I = \int_0^2 ye^{2y}\,dy \tag{a}$$

using the Gauss–Legendre quadrature.

Solution

To obtain the limits of integration as -1 to 1, the variable y is transformed as

$$y = \frac{(b-a)x + a + b}{2} = \frac{(2-0)x + 0 + 2}{2} = x + 1. \tag{b}$$

Table 8.6

Number of Gauss points (n)	Value of the integral (I)
2	37.966793
3	41.058926
4	41.195549
5	41.198578
6	41.198608

Thus, Eq. (a) can be rewritten as

$$I = \int_{-1}^{1} f(x)\,dx \equiv \int_{-1}^{1} (x+1)e^{2(x+1)}\,dx. \tag{c}$$

The n-point Gauss–Legendre quadrature gives the value of the integral by Eq. (8.86). Using the weights (w_i) and the Gauss points (x_i) given in Table 8.5, results are obtained as indicated in Table 8.6. ◀

8.8.5 Other Gauss Quadrature Formulas

The Gauss quadrature formulas considered in the previous section are based on Legendre orthogonal polynomials. Gauss quadrature formulas, based on other types of orthogonal polynomials such as Chebyshev, Hermite, and Laguerre polynomials, are also available.

The derivations of all these formulas are very similar. All n-point Gauss quadrature formulas give exact values when the function involved, $f(x)$, is a polynomial of degree $(2n - 1)$ or less.

8.8.6 Gauss–Chebyshev Quadrature Formulas

The Gauss–Chebyshev quadrature is based on Chebyshev polynomials and can be used to evaluate integrals of the type

$$\int_{-1}^{1} \frac{1}{\sqrt{1-x^2}} f(x)\,dx.$$

The formula is given by

$$\int_{-1}^{1} \frac{1}{\sqrt{1-x^2}} f(x)\,dx = \sum_{i=1}^{n} w_i f(x_i), \tag{8.101}$$

where x_i are the roots of the Chebyshev polynomial of order n, $P_n(x)$. The Chebyshev polynomials are defined as

$$P_0(x) = 1;$$

$$P_1(x) = x;$$

$$P_2(x) = 2x^2 - 1;$$

$$P_n(x) = 2xP_{n-1}(x) - P_{n-2}(x); n = 3, 4, 5, \ldots. \quad (8.102)$$

The roots of the nth-degree Chebyshev polynomial are given by [8.1]

$$x_i = \cos\frac{\left(i - \frac{1}{2}\right)\pi}{n}; i = 1, 2, \ldots, n. \quad (8.103)$$

The weights are given by

$$w_i = \frac{\pi}{n}; i = 1, 2, \ldots, n. \quad (8.104)$$

The error (E) in the n-point Gauss–Chebyshev formula is given by [8.1]

$$E \approx \frac{2\pi}{2^{2n}(2n)!} f^{(2n)}(\xi); -1 < \xi < 1. \quad (8.105)$$

Any arbitrary integration limits $[a, b]$ can be changed to $[-1, 1]$ by using the transformation indicated in Eqs. (8.87) and (8.88).

8.8.7 Gauss–Hermite Quadrature Formulas

The Gauss–Hermite quadrature is based on Hermite polynomials, which are defined as

$$P_0(x) = 1;$$

$$P_1(x) = 2x;$$

$$P_2(x) = 4x^2 - 2;$$

$$P_3(x) = 8x^3 - 12x;$$

$$P_n(x) = 2xP_{n-1}(x) - 2(n-1)P_{n-2}(x); n = 2, 3, 4, \ldots. \quad (8.106)$$

The Gauss–Hermite formula can be used to evaluate integrals of the form

$$\int_{-\infty}^{\infty} e^{-x^2} f(x)\, dx$$

and is given by

$$\int_{-\infty}^{\infty} e^{-x^2} f(x)\, dx = \sum_{i=1}^{n} w_i f(x_i) \quad (8.107)$$

The values of x_i (same as the roots of the Hermite polynomial of order n) and the weights w_i for $n = 2, 3, 4$, and 5 are given in Table 8.7. The error (E) in the n-point Gauss–Hermite quadrature formula is given by [8.1]

$$E \approx \frac{n!\sqrt{\pi}}{2^n(2n)!} f^{(2n)}(\xi) \text{ for } -\infty < \xi < \infty \quad (8.108)$$

Note that the limits of integration are $-\infty$ and ∞ in Eq. (8.107).

Table 8.7 Hermite points, weights, and error estimates.

n	x_i	w_i	Error estimate
2	± 0.70710 67811	0.88622 69255	$f^{(4)}(\xi)$
3	0.00000 00000	1.18163 59006	$f^{(6)}(\xi)$
	± 1.22474 48714	0.29540 89752	
4	± 0.52464 76233	0.80491 40900	$f^{(8)}(\xi)$
	± 1.65068 01239	0.08131 28354	
5	0.00000 00000	0.94530 87205	$f^{(10)}(\xi)$
	± 0.95857 24646	0.39361 93232	
	± 2.02018 28705	0.01995 32421	

$-1 < \xi < 1$

8.8.8 Gauss–Laguerre Quadrature Formulas

The Gauss–Laguerre formulas are based on the use of following Laguerre polynomials:

$$P_0(x) = 1;$$

$$P_1(x) = -x + 1;$$

$$P_2(x) = x^2 - 4x + 2;$$

$$P_3(x) = -x^3 + 9x^2 - 18x + 6;$$

$$P_n(x) = (2n - x - 1)P_{n-1}(x) - (n-1)^2\, P_{n-2}(x); n = 2, 3, 4, \ldots. \tag{8.109}$$

The Gauss–Laguerre quadrature formula can be used to evaluate integrals of the form

$$\int_0^\infty e^{-x} f(x)\, dx$$

and is given by

$$\int_0^\infty e^{-x} f(x)\, dx = \sum_{i=1}^{n} w_i f(x_i). \tag{8.110}$$

The values of x_i, given by the roots of the nth-order Laguerre polynomial and the weights w_i for $n = 2, 3$, and 4 are given in Table 8.8.

The error (E) associated with the n-point Gauss–Laguerre quadrature formula is given by

$$E \approx \frac{(n!)^2}{(2n)!} f^{(2n)}(\xi) \text{ for } 0 < \xi < \infty. \tag{8.111}$$

The Gauss–Laguerre formula can also be applied to evaluate the integral

$$\int_a^\infty e^{-y} f(y)\, dy,$$

Table 8.8 Laguerre points, weights, and error estimates.

n	x_i	w_i	Error proportional to
2	0.58578 64376	0.85355 33906	$f^{(4)}(\xi)$
	3.41421 35624	0.14644 66094	
3	0.41577 45568	0.71109 30099	$f^{(6)}(\xi)$
	2.29428 03603	0.27851 77336	
	6.28994 50829	0.01038 92565	
4	0.32254 76896	0.60315 41043	$f^{(8)}(\xi)$
	1.74576 11012	0.35741 86924	
	4.53662 02969	0.03888 79085	
	9.39507 09123	0.00053 92947	

$0 < \xi < \infty$

where a is an arbitrary constant, by using the transformation

$$y = x + a. \tag{8.112}$$

This gives

$$\int_a^\infty e^{-y} f(y)\,dy = \int_0^\infty e^{-(x+a)} f(x+a)\,dx = e^{-a}\int_0^\infty e^{-x} f(x+a)\,dx. \tag{8.113}$$

By combining Eqs. (8.110) and (8.113), we obtain

$$\int_a^\infty e^{-y} f(y)\,dy = e^{-a}\sum_{i=1}^{n} w_i f(x_i + a), \tag{8.114}$$

where w_i and x_i are those given in Table 8.8.

8.9 Integration with Unequal Segments

All methods of numerical integration considered so far in this chapter, except Gauss quadrature formulas, have been based on equally spaced data points. In practice, we encounter many situations where the segments have unequal widths. For example, the function value may vary very slowly (or very abruptly), and we may like to use a smaller (or larger) number of function values without significantly affecting the accuracy of the integral. In some cases, experimental or numerical values of the function may be available only at specific values of x, which may not be equally spaced. Hence, we need to consider a method of numerical integration with unequal segments. Basically, we can use two approaches for this purpose.

In the first approach, which is commonly used in engineering, we employ a curve-fitting technique to obtain a continuous function, $\bar{f}(x)$, for the discrete set of data points. Then, we divide the range of integration into equally spaced segments and use some form of Newton–Cotes formula (such as trapezoidal or Simpson's rule). In this procedure, the values of the function required at equally distributed segments are obtained from curve fitting (that is, computed from $\bar{f}(x)$).

In the second approach, we try to use the unequally spaced data directly. For example, we can apply the trapezoidal rule to each segment and sum the results to obtain the value of the integral (I) as

$$I = \left(\frac{f_0 + f_1}{2}\right) h_1 + \left(\frac{f_1 + f_2}{2}\right) h_2 + \cdots + \left(\frac{f_{n-1} + f_n}{2}\right) h_n, \tag{8.115}$$

where h_i denotes the width of segment i. Note that this approach is same as the one used in the case of the multisegment trapezoidal rule, Eq. (8.10), the only difference being that Eq. (8.10) assumes all the segment widths (h_i) to be equal (h). Although Eq. (8.115) appears to be more involved compared with Eq. (8.10), a computer program can be easily developed to implement Eq. (8.115). In some cases, although the segments have unequal widths, we may find groups of two or three segments with equal widths in the data. In such a case, we can use a combination of trapezoidal, Simpson's one-third, and Simpson's three-eighths rules. If adjacent segments are of unequal width, we use the trapezoidal rule. If there are two adjacent segments with the same width, we use Simpson's one-third rule. If there are three adjacent segments of the same width, we apply Simpson's three-eighths rule. Note that the use of Simpson's rules, where applicable, will increase the accuracy of the integral. A computer program can be developed to implement this combination strategy. The program can be developed to check the widths of adjacent segments and apply the appropriate formula. Finally, the program can be made to sum all the values computed for the various segments to yield the total value of the integral.

▶Example 8.10

The turning-moment–crank-angle data of an internal combustion engine are shown in Table 8.9. Determine the area under the turning-moment diagram using the trapezoidal rule.

Solution

By denoting the jth term on the right-hand side of Eq. (8.115) as $I_j = \left(\frac{f_{j-1}+f_j}{2}\right) h_j$, the values of $I_1, I_2, \ldots, I_9$ can be found as $I_1 = 0.182500005$, $I_2 = 0.453750044$, $I_3 = 0.250000030$, $I_4 = 0.497249961$, $I_5 = 0.669999957$, $I_6 = 1.44000006$, $I_7 = 1.06500018$, $I_8 = 0.592499912$, and $I_9 = 0.287999958$. Thus, the integral is given by $I = \sum_{j=1}^{9} I_j = 5.43900013$. ◀

8.10 Numerical Integration of Improper Integrals

In most of the methods discussed so far (except in the Gauss–Laguerre and Gauss–Hermite formulas), the limits of integration a and b were assumed to be finite

Table 8.9

i	Crank angle, x_i	Turning moment, $f(x_i)$	Step length, h
0	0.0	0.85	
			0.10
1	0.1	2.80	
			0.15
2	0.25	3.25	
			0.08
3	0.33	3.00	
			0.17
4	0.50	2.85	
			0.20
5	0.70	3.85	
			0.30
6	1.00	5.75	
			0.20
7	1.20	4.90	
			0.15
8	1.35	3.00	
			0.15
9	1.50	0.84	

and the integrand $f(x)$ was assumed to be finite and continuous over the range of integration. Many engineering applications require the evaluation of improper integrals in which either the limit(s) of integration are infinite or the integrand is singular at some point(s) in the range of integration. Assuming that the integral exists and has a finite value, we consider some methods of numerical integration of improper integrals in this section.

8.10.1 Numerical Integration with Infinite Limits

The integrals of the form

$$I_1 = \int_a^{\infty} f(x)\, dx, \tag{8.116}$$

$$I_2 = \int_{-\infty}^{b} f(x)\, dx, \tag{8.117}$$

and

$$I_3 = \int_{-\infty}^{\infty} f(x)\, dx \tag{8.118}$$

involve infinity either as a lower limit, upper limit, or both lower and upper limits. If $f(x)$ is integrable in a semiinfinite or infinite range, then the function $f(x)$ usually will have a value close to zero everywhere, except over a small part of the range where it will have a value significantly different from zero. In such cases, we can use a modified trapezoidal rule to evaluate the integral in Eq. (8.116) as

$$I_1 = \int_a^{\infty} f(x)\,dx \approx h \sum_{i=1}^{n} f_i, \tag{8.119}$$

where n is a sufficiently large number; $f_i = f(x_i), i = 1, 2, \ldots, n; x_i = a+(i-1)h, i = 1, 2, \ldots, n$; and h is the step size. When n is very large, the value of x_n approaches infinity. Similarly, the integral in Eq. (8.118) can be evaluated as

$$I_3 = \int_{-\infty}^{\infty} f(x)\,dx \approx h \sum_{i=-n}^{n} f_i, \tag{8.120}$$

where n is a sufficiently large number, h is the step size, and $f_i = f(x_i)$ with

$$x_i = ih; i = -n, -(n-1), -(n-2), \ldots, 1, 0, 1, 2, \ldots, (n-1), n. \tag{8.121}$$

It can be seen that when n is very large, the values of x_{-n} and x_n approach minus and plus infinity, respectively. In practice, the value of n to be used in Eqs. (8.120) or (8.121) is not known. Hence, the integral is evaluated with successively increasing values of n until any further increase of n does not improve the value of the integral significantly.

8.10.2 Numerical Integration with Singularities

In some applications, we need to evaluate an improper integral in which the integrand, $f(x)$, has a singularity or infinite discontinuity at some point in the range of integration. In spite of the singularity, the integral may have a finite value. The following are some examples of such integrals:

$$I_1 = \int_0^4 \frac{1}{\sqrt{4-x^2}}\,dx, \tag{8.122}$$

$$I_2 = \int_{-2}^0 \frac{1}{\sqrt{4-x^2}}\,dx, \tag{8.123}$$

and

$$I_3 = \int_{-2}^{26} \frac{1}{(x+1)^{2/3}}\,dx. \tag{8.124}$$

In Eqs. (8.122) and (8.123), the integrand becomes singular at the upper and lower limits of integration. On the other hand, in the case of Eq. (8.124), the integrand becomes singular at an intermediate point, $x = -1$, in the range of integration. In general, if the integrand is singular at some point c in the range of integration

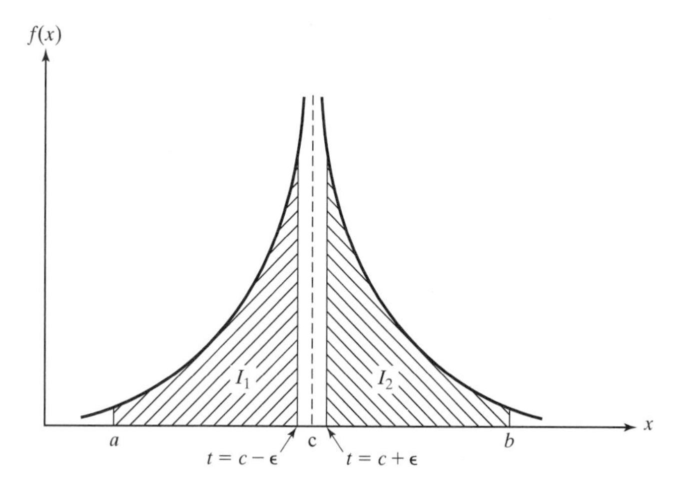

Figure 8.18 Integration involving singularity.

a to b, we can write the integral as

$$I = \int_a^b f(x)\,dx = I_1 + I_2 \equiv \int_a^c f(x)\,dx + \int_c^b f(x)\,dx$$
$$= \lim_{t \to c^-} \int_a^t f(x)\,dx + \lim_{t \to c^+} \int_t^b f(x)\,dx, \tag{8.125}$$

where the limits are found as t approaches c from left ($t \to c^-$) or from the right ($t \to c^+$). The numerical implementation of this procedure is shown in Fig. 8.18. Here, starting with a selected small value of ε, the integrals I_1 and I_2 are evaluated with successively smaller values of ε. Whenever the computed values of I_1 and I_2 are not significantly affected by a further reduction in the value of ε, the process can be assumed to have converged.

Based on the nature of singularity involved, several other methods are available to evaluate integrals involving singularities. For example, the integration by parts can sometimes avoid the singularity as indicated next:

$$I = \int_0^5 \frac{e^{2x}}{\sqrt{x}}\,dx = 2e^{2x}\sqrt{x}\Big|_0^5 - \int_0^5 4\sqrt{x}e^{2x}\,dx. \tag{8.126}$$

In Eq. (8.126), the first term on the right-hand side can be evaluated easily, and the value of the second term can be found either analytically or numerically as it does not involve any singularity. In some cases, the transformation of the variable can eliminate the singularity. For illustration, consider the integral

$$I = \int_0^1 \frac{1 + x^3 + x^5}{\sqrt{x^2 - 1}}\,dx. \tag{8.127}$$

Here the use of the transformation $x = \sin y$ or $dx = \cos y\, dy$ gives $\sqrt{x^2 - 1} = \cos y$, and hence, Eq. (8.127) reduces to

$$I = \int_0^{\pi/2} (1 + sin^3 y + sin^5 y)\, dy, \tag{8.128}$$

which does not involve any singularity.

8.11 Numerical Integration in Two- and Three-Dimensional Domains

In many engineering and other practical situations, we need to evaluate integrals over a two- or three-dimensional domain. If a double integral is to be evaluated over the domain shown in Fig. 8.19, we have

$$I = \int\int_A f(x, y)\, dx\, dy = \int_a^b \left\{ \int_{p(x)}^{q(x)} f(x, y)\, dy \right\} dx, \tag{8.129}$$

where A denotes the area (domain) of integration shown in Fig. 8.19. Equation (8.129) can be expressed as

$$I = \int_a^b X(x)\, dx, \tag{8.130}$$

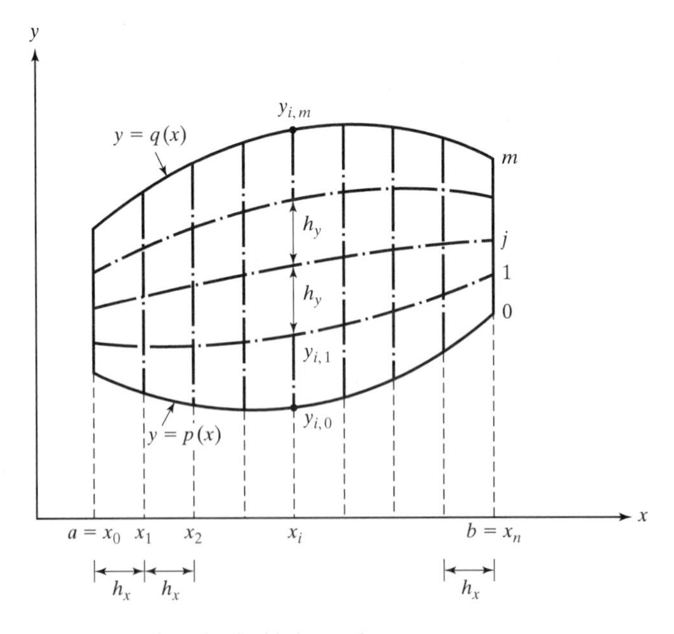

Figure 8.19 Evaluation of a double integral.

where the function, $X(x)$, is given by

$$X(x) = \int_{p(x)}^{q(x)} f(x, y)\,dy. \tag{8.131}$$

To evaluate the integral in Eq. (8.130), we first divide the domain of integration into n segments of equal width (h_x) along the x-axis, and into m segments of equal width along y-axis at any specified value of x_i. (See Fig. 8.19.) We can then use any of the integration methods described in the previous sections. For example, if the trapezoidal rule is used, the integral in Eq. (8.130) can be evaluated as

$$I = \frac{h_x}{2}\left[X_0 + 2X_1 + 2X_2 + \cdots + 2X_i + \cdots + 2X_{n-1} + X_n\right], \tag{8.132}$$

where h_x is the width of segment along the x-axis. The width of segment (h_y) along the y-axis at x_i is given by

$$h_y = \frac{q(x_i) - p(x_i)}{m}. \tag{8.133}$$

By denoting the grid values of y at x_i as $y_{i,0}, y_{i,1}, y_{i,2}, \ldots, y_{i,m}$, X_i appearing in Eq. (8.132) can be evaluated using the trapezoidal rule, for the y-direction, as

$$X_i = \int_{p(x_i)}^{q(x_i)} f(x_i, y)\,dy = \frac{h_y}{2}\left[f_{i,0} + 2f_{i,1} + \cdots + 2f_{i,j} + \cdots + 2f_{i,m-1} + f_{i,m}\right], \tag{8.134}$$

where

$$f_{i,j} = f(x_i, y_{i,j});\; j = 0, 1, 2, \ldots, m. \tag{8.135}$$

Note that we can use any other numerical integration scheme instead of the trapezoidal rule to evaluate the double integral of Eq. (8.129). If Gauss–Legendre quadrature is used, we consider the double integral of the form

$$I = \int_{-1}^{1}\int_{-1}^{1} f(x, y)\,dx\,dy \tag{8.136}$$

and evaluate it as follows:

$$\begin{aligned} I &= \int_{-1}^{1}\int_{-1}^{1} f(x, y)\,dx\,dy = \int_{-1}^{1}\left[\sum_{i=1}^{n} w_i f(x_i, y)\right] dy \\ &= \sum_{j=1}^{n} w_j \left[\sum_{i=1}^{n} w_i f(x_i, y_j)\right] \\ &= \sum_{i=1}^{n}\sum_{j=1}^{n} w_i w_j f(x_i, y_j). \end{aligned} \tag{8.137}$$

Here, the weights $w_i (w_j)$ and Gauss points x_i (and y_j) are same as those given in Table 8.5. Note that the number of integration points in each direction was assumed to be the same. Clearly, it is not necessary, and sometimes, it may be advantageous to use different numbers of integration points in each direction. The integral of

Eq. (8.129), in which the limits of integration are not -1 and $+1$, we can use coordinate transformations similar to the one described in Eq. (8.87) and (8.88).

The procedure described can be extended to triple integrals also. For example, by using Gauss–Legendre quadrature formula, we can evaluate a triple integral with limits of -1 and $+1$ as

$$\int_{-1}^{1}\int_{-1}^{1}\int_{-1}^{1} f(x, y, z)\, dx\, dy\, dz = \sum_{i=1}^{n}\sum_{j=1}^{n}\sum_{k=1}^{n} w_i w_j w_k f(x_i, y_j, z_k), \tag{8.138}$$

where the weights w_i, w_j, and w_k and the Gauss points x_i, x_j, and x_k are same as those given in Table 8.5.

▶Example 8.11

Evaluate the integral

$$I = \int_{z=0}^{2}\int_{y=1}^{4}\int_{x=-1}^{3} 5xy^3z^2\, dx\, dy\, dz \tag{a}$$

using the two-point Gauss–Legendre quadrature rule.

Solution

First we transform the variables x, y, and z to $\bar{x}$, $\bar{y}$, and $\bar{z}$ to obtain the limits of integration as -1 to 1 for each variable:

For x,

$$x = \frac{(b-a)\bar{x} + a + b}{2} = \frac{(3+1)\bar{x} - 1 + 3}{2} = 2\bar{x} + 1. \tag{b}$$

For y,

$$y = \frac{(b-a)\bar{y} + a + b}{2} = \frac{(4-1)\bar{y} + 1 + 4}{2} = 1.5\bar{y} + 2.5. \tag{c}$$

For z,

$$z = \frac{(b-a)\bar{z} + a + b}{2} = \frac{(2-0)\bar{z} + 0 + 2}{2} = \bar{z} + 1. \tag{d}$$

Using Eqs. (b), (c), and (d) with $dx = 2d\bar{x}$, $dy = 1.5d\bar{y}$ and $dz = d\bar{z}$, the integral in Eq. (a) can be rewritten as

$$I = \int_{-1}^{1}\int_{-1}^{1}\int_{-1}^{1} 15(2\bar{x}+1)(1.5\bar{y}+2.5)^3(\bar{z}+1)^2 d\bar{x}\, d\bar{y}\, d\bar{z}. \tag{e}$$

The evaluation of the integral in Eq. (e) using the two-point Gauss–Legendre rule leads to

$$I = \sum_{i=1}^{2}\sum_{j=1}^{2}\sum_{k=1}^{2} w_i\, w_j\, w_k f(\bar{x}_i, \bar{y}_j, \bar{z}_k), \tag{f}$$

where the weights and the values of the variables are given in Table 8.5. The results are indicated in Table 8.10. ◀

Table 8.10

i	j	k	$f(\bar{x}_i, \bar{y}_j, \bar{z}_k)$
1	1	1	3066.80835
1	1	2	220.186966
1	2	1	350.808624
1	2	2	25.1869297
2	1	1	−220.186905
2	1	2	−15.8087111
2	2	1	−25.1869240
2	2	2	−1.80833995

Value of integral: 3400.00000

8.12 Choice of Method

If the function to be integrated is available in tabular form at equal intervals, the clear choice is a Newton–Cotes formula in order to avoid interpolation. If the function to be integrated is available in analytical (equation) form, clearly the Gauss quadrature is to be used. It requires, in general, a smaller number of function evaluations compared with Newton–Cotes formulas for the same accuracy. One minor limitation with Gaussian quadrature is that the limits of integration need to be converted to −1 to 1. Although a higher order Gaussian quadrature is expected to result in higher accuracy, we need to store the Gaussian abscissas and weights in the computer.

In some engineering applications, the function to be integrated may not be available in analytical or tabular form. For example, in the generation of element stiffness matrices and load vectors in finite element analysis, the function to be integrated is defined only implicitly. In such cases, the Gaussian quadrature is used most frequently to achieve accurate results.

8.13 Use of Software Packages

8.13.1 MATLAB

▶Example 8.12

Evaluate the integral

$$I = \int_0^{10} e^{-x^2}\, dx. \tag{E1}$$

Solution

The integrand is defined as an m.file labeled *funb.m.*, and Simpson's method is used for integration.

Listing of *funb.m*:

```
function y=funb (x)

y=exp(-x.^2);


>> quad('funb', 0,10)

ans =
    0.88622705339348

>> quad8('funb', 0,10)

ans =
    0.88622707431169
```

◀

▶Example 8.13

Evaluate the integral

$$I = \int_{-1}^{1} (x+1)e^{2(x+1)}\, dx. \tag{E1}$$

Solution

The integrand is defined as an m.file labeled *func.m*.

Listing of *func.m*

```
function  y=func (x)

y=(x+1).*exp(2*(x+1));


>> quad('func', -1,1)

ans =
   41.19873991682360

>> quad8('func', -1,1)

ans =
   41.19861252495721
```

◀

▶Example 8.14

Evaluate the integral

$$I = \int_0^4 10(1 - e^{-2t}) \sin\left(e^t - 1\right) dt. \quad \text{(E1)}$$

Solution

The integrand is defined as an m.file labeled *fune.f.*

Listing of *fune.m*

```
function y=fune (x)

y=10*(1-exp(-2*x)).*sin(exp(x)-1);


>> quad('fune', 0,4)

ans =
   4.50201737935197

>> quad8('fune', 0,4)

ans =
   4.50202232290777
```

◀

8.13.2 Maple

▶Example 8.15

Evaluate the integral

$$I = \int \frac{1}{1+x^2}\, dx \quad \text{(E1)}$$

(a) as an indefinite integral and (b) as a definite integral with limits 0 and 1.

Solution

```
[>
[> int (1/(1+x^2), x=0..1);
```

$$\frac{1}{4}\pi$$

```
> evalf(%);
```

$$.7853981635$$

```
>
> int(1/(1+x^2),x);
```

$$\arctan(x)$$

```
>
```

◀

▶Example 8.16

Evaluate the integral in Eq. (E1) of Example 8.12 part (a) as an indefinite integral and part (b) as a definite integral with limits 0 and 10.

Solution

```
>
> int(exp(-x^2), x=0..10);
```

$$\frac{1}{2}\operatorname{erf}(10)\sqrt{\pi}$$

```
> evalf(%);
```

$$.8862269255$$

```
>
```

◀

▶Example 8.17

Evaluate the integral

$$I = \int_{z=0}^{2}\int_{y=1}^{4}\int_{x=-1}^{3} 5xy^3z^2\,dx\,dy\,dz. \tag{E1}$$

Solution

```
> with(student):
> Tripleint (5*x*y^3*z^2, x=-1..3, y=1..4, z=0..2);
```

$$\int_{0}^{2}\int_{1}^{4}\int_{-1}^{3} 5xy^3z^2\,dx\,dy\,dz$$

```
> evalf(%);
```

$$3400.000000$$

◀

▶Example 8.18

Evaluate the integral of Eq. (E1) in Example 8.14.

Solution

```
> 
> int(10*(1-exp (-2*t))* sin(exp(t)-1), t=0..4);
```

$$-\frac{5}{2}\left(I e^{(-I)} e^{(-8)}\left((-1)^{\left(\frac{e^4}{\pi}\right)}\right)^2 - e^{(-I)} e^{(-4)}\left((-1)^{\left(\frac{e^4}{\pi}\right)}\right)^2 - 3I e^{(-I)} \mathrm{Ei}(1, -I e^4)(-1)^{\left(\frac{e^4}{\pi}\right)} - I e^{(-8)} e^{I}\right.$$
$$- e^{(-4)} e^{I} + 3I e^{I} \mathrm{Ei}(1, I e^4)(-1)^{\left(\frac{e^4}{\pi}\right)} + 2(-1)^{\left(\frac{e^4}{\pi}\right)} + 3I e^{(-I)} \mathrm{Ei}(1, -I)(-1)^{\left(\frac{e^4}{\pi}\right)}$$
$$\left.- 3I e^{I} \mathrm{Ei}(1, I)(-1)^{\left(\frac{e^4}{\pi}\right)}\right) \Big/ (-1)^{\left(\frac{e^4}{\pi}\right)}$$

```
> evalf(%);
```

$$4.502022358 + .1063410279\,10^{-9} I$$

```
> 
```

◀

8.13.3 MATHCAD

▶Example 8.19

Evaluate the integral of Eq. (E1) in Example 8.15 part (a) as an indefinite integral and part (b) as a definite integral with limits 0 and 1.

Solution

$$\int_0^1 \frac{1}{1+x^2}\,dx = 0.785,$$

$$\int \frac{1}{1+x^2}\,dx \rightarrow \operatorname{atan}(x).$$

◀

▶Example 8.20

Evaluate the integral of Eq. (E1) in Example 8.17.

Solution

$$\int_0^2 \int_1^4 \int_{-1}^3 5 \cdot x \cdot y^3 \cdot z^2\,dx\,dy\,dz = 3.4 \times 10^3.$$

◀

▶Example 8.21

Evaluate the integral of Eq. (E1) in Example 8.14.

Solution

$$\int_0^4 10 \cdot (1 - e^{-2 \cdot t}) \cdot \sin(e^t - 1)dt = 4.502.$$

◀

8.14 Computer Programs

8.14.1 Fortran Programs

Program 8.1 Simpson method

Input data required

A,B = limits of integration
ITMAX = maximum number of iterations permitted (usual value: 20)
EPS = convergence requirement (usual value: 10^{-6})
Function to be integrated; to be given in the form of a subroutine as:

```
FUNCTION F(X)
F=...
RETURN
END
```

Illustration

Evaluation of the integral $I = \int_0^3 \frac{dx}{(3x+1)^3}$.

Program output

```
INPUT DATA TO SIMPSON INTEGRATION METHOD:

A = 0.00000000E+00
B = 0.30000000E+01
ITMAX =    20
EPS = 0.10000000E-05

OUTPUT FROM PROGRAM:

     ITER              H                      INTEGRAL
       1         0.15000000E+01         0.51252109E+00
       2         0.75000000E+00         0.28453422E+00
       3         0.37500000E+00         0.19495969E+00
       4         0.18750000E+00         0.16988339E+00
       5         0.93750000E-01         0.16551381E+00
       6         0.46875000E-01         0.16503969E+00
       7         0.23437500E-01         0.16500266E+00
       8         0.11718750E-01         0.16500020E+00
       9         0.58593750E-02         0.16499999E+00
      10         0.29296875E-02         0.16500008E+00
```

```
NUMBER OF ITERATIONS TAKEN FOR CONVERGENCE =        10
CONVERGED VALUE OF THE INTEGRAL =      0.16500008E+00
```

Program 8.2 Gauss–Legendre quadrature

Input data required

NG = number of Gauss points to be used (between 2 and 6)
Subroutine FUN (XX,FF) to evaluate the function to be integrated, FF(XX)

Illustration

Evaluation of the integral

$$I = \int_{-1}^{1} (x+1)e^{2(x+1)}\, dx$$

The output of Program 8.2 is given next:

```
GAUSS-LEGENDRE QUADRATURE

LIMITS OF INTEGRATION:    A  =  -1.00      B  =    1.00

NUMBER OF GAUSS POINTS:    2     INTEGRAL:     0.37966793E+02
NUMBER OF GAUSS POINTS:    3     INTEGRAL:     0.41058926E+02
NUMBER OF GAUSS POINTS:    4     INTEGRAL:     0.41195549E+02
NUMBER OF GAUSS POINTS:    5     INTEGRAL:     0.41198578E+02
NUMBER OF GAUSS POINTS:    6     INTEGRAL:     0.41198608E+02
```

Program 8.3 Triple integral by Gauss–Legendre quadrature

Input data required

A1,B1 = limits of integration for x
A2,B2 = limits of integration for y
A3,B3 = limits of integration for z
ITMAX = maximum number of iterations permitted (usual value: 20)
EPS = convergence criterion (usual value: 10^{-6})
Function to be integrated, in the form of FUNCTION F(X,Y,Z)

Illustration

Evaluation of the integral

$$I = \int_{-1}^{2} \int_{4}^{8} \int_{3}^{6} x^2 y^3 z^5\, dx\, dy\, dz$$

The output of Program 8.3 is given next:

```
TRIPLE INTEGRATION BY GAUSS-LEGENDRE QUADRATURE

LIMITS OF INTEGRATION:
FOR X:  -0.1000E+01   0.2000E+01
FOR Y:   0.4000E+01   0.8000E+01
FOR Z:   0.3000E+01   0.6000E+01

TRIPLE INTEGRAL = 0.22044964E+08
```

8.14.2 C Programs

Program 8.4 Trapezoidal rule

Input data required

a, b = limits of integration
itmax = maximum number of iterations permitted (usual value: 20)
eps = convergence requirement (usual value: 10^{-6})
function subprogram to evaluate the function f(x) to be integrated

Illustration

Evaluation of the integral

$$\int_0^3 \frac{dx}{(3x+1)^3}.$$

Program output.

```
Integration by Trapezoidal Rule

Input Data::
a = 0.00000000E+00
b = 0.30000000E+01
itmax = 20
eps = 0.99999997E-05

Output from program:

 iter              h                      integral

   1         0.15000000E+01          0.75976586E+00
   2         0.75000000E+00          0.40334210E+00
   3         0.37500000E+00          0.24705528E+00
   4         0.18750000E+00          0.18917638E+00
   5         0.93750000E-01          0.17142946E+00
   6         0.46875000E-01          0.16663711E+00
   7         0.23437500E-01          0.16541126E+00
```

```
     8          0.11718750E-01          0.16510296E+00
     9          0.58593750E-02          0.16502571E+00
    10          0.29296875E-02          0.16500647E+00
    11          0.14648438E-02          0.16500174E+00
    12          0.73242188E-03          0.16500062E+00

Number of iterations taken for convergence =     12
Converged value of the integral =     0.16500062E+00
```

Program 8.5 Gauss–Chebyshev quadrature

Input data required

a, b = limits of integration
itmax = maximum number of iterations permitted (usual value: 20)
eps = convergence criterion (usual value: 10^{-6})
subprogram to evaluate the function $f(x)$ to be integrated

Illustration

Evaluation of

$$\int_0^3 \frac{dx}{(3x+1)^3}.$$

The output of Program 8.5 is given next:

```
Result of Gauss-Chebyshev Quadrature

Lower limit of integration, a =    0.0000E+00
Upper limit of integration, b =    0.3000E+01
Convergence criterion, eps =   0.1000E-05
Maximum number of iterations permitted, itmax =   20

     1      0.13631159E+00
     2      0.20988578E+00
     3      0.17750710E+00
     4      0.16753747E+00
     5      0.16561028E+00
     6      0.16515115E+00
     7      0.16503772E+00
     8      0.16500941E+00
     9      0.16500244E+00
    10      0.16500063E+00
    11      0.16500020E+00
    12      0.16499998E+00
    13      0.16499984E+00
```

```
Number of iterations required, iter = 13
Value of the integral, xint = 0.16499984E+00
```

REFERENCES AND BIBLIOGRAPHY

8.1. B. Carnahan, H. A. Luther, and J. O. Wilkes, *Applied Numerical Methods*, Wiley, New York, 1969.

8.2. E. W. Swokowski, *Calculus with Analytic Geometry*, alternate edition, PWS Publishers, Boston, 1983.

8.3. A. Ralston and P. Rabinowitz, *A First Course in Numerical Analysis*, 2d edition, McGraw-Hill, New York, 1978.

8.4. Y. Jaluria, *Computer Methods for Engineering*, Allyn and Bacon, Boston, 1988.

8.5. R. W. Fox and A. T. McDonald, *Introduction to Fluid Mechanics*, 4th edition, Wiley, New York, 1992.

8.6. *Tables of Functions and Zeros of Functions*, National Bureau of Standards Applied Mathematics Series, Vol. 37, Washington, DC, 1954.

8.7. A. H. Burr and J. B. Cheatham, *Mechanical Analysis and Design*, 2d edition, Prentice Hall, Upper Saddle River, NJ, 1995.

8.8. S. S. Rao, *Mechanical Vibrations*, 3d edition, Addison-Wesley, Reading, MA, 1995.

REVIEW QUESTIONS

The following questions along with corresponding answers are available in an interactive format at the Companion Web site at http://www.prenhall.com/rao.

8.1. **Define** the following terms:

Integrand; quadrature, Gauss points; trapezoidal rule; improper integral; Richardson extrapolation.

8.2. **Give short answers** to the following questions:

1. What is the advantage of evaluating an integral in closed form?
2. What is the most commonly used numerical integration method?
3. What is Romberg integration?
4. Which integration method, out of Simpson's rule and Gauss formula, is more accurate?
5. What is the relationship between Gauss formula and Legendre polynomials?
6. How do you evaluate multiple integrals numerically?
7. How do you handle singularities during numerical integration?
8. How do you deal with infinite limits during numerical integration?
9. Indicate two methods of evaluating integrals involving unequal segments.
10. How are finite limits of integration (a and b) handled in Gauss–Legendre quadrature?
11. What is the difference between Simpson's $\frac{1}{3}$ and $\frac{3}{8}$ rules?
12. Explain the physical significance of the trapezoidal rule.
13. What is rectangular rule?
14. Describe a possible method of converting an infinite limit of integration to a finite limit?

8.3. **Fill in the blanks** with suitable word(s):

1. The rectangular rule is not generally used for numerical integration of nonlinear functions as it leads to large _______ errors.
2. In the trapezoidal rule, the integrand $f(x)$ is approximated by _______ _______ segments.
3. The Romberg integration implements the _______ extrapolation.
4. The accuracy of Gauss quadrature is _______ compared with that of Newton–Cotes formulas.

(*continued, page 616*)

5. The most popular form of Gauss quadrature is known as Gauss ______ quadrature.
6. Gauss quadrature with n-Gauss points represents exact integration of polynomials of order ______ .
7. The range of integration is ______ to ______ for Gauss (Gauss-Legendre) quadrature.
8. The truncation error in rectangular rule is ______ than that of the trapezoidal rule.
9. The integration by parts can sometimes avoid ______ in an indefinite integral.
10. The integrand is evaluated at the ______ of the Legendre polynomials in the Gauss–Legendre quadrature.

8.4. Answer **True or False:**

1. Numerical integration is also known as quadrature.
2. By increasing the number of trapezoids, the trapezoidal rule always gives a more accurate value of the integral.
3. In general, Simpson's method is less accurate than the trapezoidal method.
4. The trapezoidal and Simpson's methods are special cases of Newton–Cotes methods.
5. Simpson's one-third rule can be considered as a Newton–Cotes method.
6. Richardson's extrapolation can be used to improve the accuracy of results obtained by other methods.
7. The error in trapezoidal rule is proportional to h^2.
8. The error in Simpson's rule is proportional to h^4.
9. Gauss quadrature requires the evaluation of the integrand at equal intervals.
10. Any desired accuracy can be obtained by using Richardson's extrapolation.
11. Gauss quadrature can be used when the integrand is available in tabular form with equal intervals.
12. Newton–Cotes formulas are nearly as accurate as Gauss formulas for numerical integration.
13. If $f(x)$ is available in tabular form at unequally spaced intervals, the Gauss quadrature formulas cannot be directly used.
14. If $f(x)$ is available in tabular form at unequally spaced intervals, a curve-fitting technique is used before applying Newton–Cotes formulas.
15. Numerical integration with infinite limits can be performed using modified quadrature rules.
16. The Gauss–Legendre quadrature formula is applicable to integrals having arbitrary, but finite, limits of integration.

(continued, page 617)

17. The number of Gauss points must be same in each coordinate direction in two- and three-dimensional integrations.
18. An improper integral has always a value of infinity.
19. Different Gauss quadrature formulas predict different values for the same integral.
20. The transformation of variable helps, in some cases, the evaluation of an indefinite integral.

8.5. **Match the following:**

1. Simpson's rule	(a) evaluates $\int_{-1}^{1} f(x)\,dx$
2. Newton–Cotes formulas	(b) evaluates $\int_{0}^{\infty} e^{-x} f(x)\,dx$
3. Romberg integration	(c) evaluates $\int_{-\infty}^{\infty} e^{-x^2} f(x)\,dx$
4. Gauss–Legendre formula	(d) $f(x)$ is approximated by linear segments
5. Trapezoidal rule	(e) $f(x)$ is approximated by second- or third-degree polynomials
6. Gauss–Chebyshev formula	(f) $f(x)$ is approximated by polynomials
7. Gauss–Hermite formula	(g) evaluates $\int_{-1}^{1} \frac{1}{\sqrt{1-x^2}} f(x)\,dx$
8. Gauss–Laguerre formula	(h) implements Richardson's extrapolation

PROBLEMS

Section 8.2

8.1. The variation of torque of an engine with the angular displacement of the crank is shown in Table 8.11:

Use a graphical method to determine the following:

(i) The energy generated during a cycle by integrating the torque-displacement function over a cycle.

(ii) The mean torque generated over a cycle.

8.2. The equation of a catenary (cable hanging under its own weight as shown in Fig. 8.20) is given by

$$y(x) = \frac{4bx}{a^2}(a - x) + (y_2 - y_1)\frac{x}{a}$$

Table 8.11 Variation of engine torque with crank angle.

Crank angle, θ (deg)	Torque, T (lb-in)
0	0
15	4200
30	3150
60	3240
80	2400
110	1950
150	800
180	0
190	−150
240	−475
270	−360
310	0
330	190
360	0
390	−200
410	0
450	370
480	475
515	270
540	0
570	−315
610	−510
630	−520
650	−420
680	−750
700	−1100
720	0

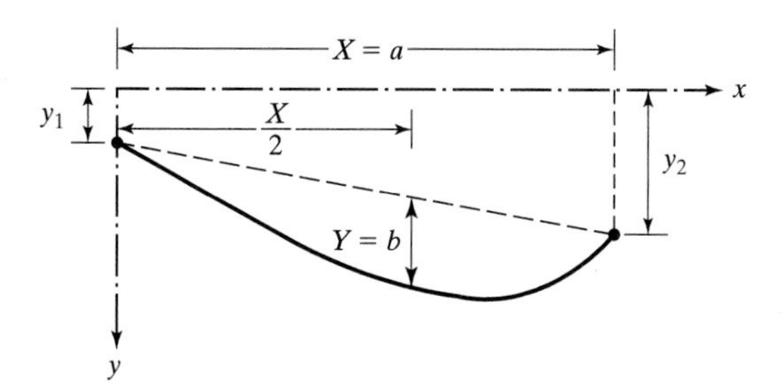

Figure 8.20 Catenary.

and the length of the cable, L, can be determined as

$$L = \int_0^a \sqrt{1 + \left(\frac{dy}{dx}\right)^2}\, dx.$$

Find the value of b for a cable of length $L = 120$ in with $a = 90$ in, $y_2 = 70$ in, and $y_1 = 10$ in.

8.3. The velocity profile for turbulent flow of a fluid in a smooth pipe can be represented by the power-law equation [8.5]

$$\frac{u}{U} = \left(1 - \frac{r}{R}\right)^{\frac{1}{n}}, \tag{a}$$

where u = velocity at radius r, U = center line velocity, and R = radius of the pipe. In contrast, for a fully developed laminar flow, the velocity profile will be a parabolic profile:

$$\frac{u}{U} = 1 - \left(\frac{r}{R}\right)^2. \tag{b}$$

Determine the flow rates (Q) according to Eqs. (a) and (b) by integrating the velocity over the area of the pipe (A):

$$Q = \int_A u\, dA. \tag{c}$$

Use $U = 10$ ft/sec, $n = 5$, $R = 0.5$ ft, and two-point Gauss quadrature rule.

Section 8.3

8.4. Show that the integral given by the trapezoidal rule (Eq. (8.10)) is the average of the integrals given by the two rectangular rules, Eqs. (8.6) and (8.7). Indicate the physical significance of the result.

8.5. Derive Eqs. (8.30) and (8.31) for a multistep rectangular rule.

8.6. Find the axial displacement of the bar described in Example 8.2 using trapezoidal rule with different step lengths.

8.7. Find the area under the turning-moment–crank-angle diagram shown in Fig. 8.5 using different rectangular rules and determine the power developed by the internal combustion engine in each case. Assume the speed of the engine as 1500 rpm.

Section 8.4

8.8. Prove that the truncation error in a multistep Simpson's $\frac{3}{8}$ rule is proportional to h^4.

8.9. Find the axial displacement of the bar described in Example 8.2 using Simpson's $\frac{1}{3}$ rule with different step lengths.

8.10. Find the axial displacement of the bar described in Example 8.2 using Simpson's $\frac{3}{8}$ rule with different step lengths.

Section 8.5

8.11. Evaluate the integral

$$I = \int_0^{2\pi} \cos^2 x \, dx$$

using the following methods with 6 function evaluations:
(a) Trapezoidal rule
(b) Simpson's $\frac{1}{3}$ rule
(c) Simpson's $\frac{3}{8}$ rule
(d) Gauss–Legendre quadrature formula
Discuss the results in comparison to the exact value of the integral, $I = \pi$.

8.12. Evaluate the integral

$$I = \int_0^1 \frac{dx}{1+x^2}$$

using the six Newton–Cotes formulas indicated in Table 8.4 and compare the results.

Section 8.6

8.13. Using the results of Simpson's $\frac{1}{3}$ rule given in Table 8.2, find an improved estimate of the value of the integral using Richardson's extrapolation.

8.14. Derive the general formula for the improved values of the integral for the following methods when Richardson's extrapolation is applied k times:
(a) Simpson's $\frac{1}{3}$ rule and (b) Simpson's $\frac{3}{8}$ rule.

Section 8.7

8.15. The trapezoidal rule gave the values of $I = \int_0^1 (1 + e^{-x^2}) \, dx$ as 1.7313702 for $n = 2$, 1.7429841 for $n = 4$, 1.7458656 for $n = 8$, 1.7465847 for $n = 16$, and 1.7467644 for $n = 32$. Determine the value of I, using Romberg procedure, to satisfy the convergence criterion:

$$|I_{k,n} - I_{k-1,n}| \leq 10^{-5}.$$

8.16. Simpson's $\frac{1}{3}$ rule gave the values of $I = \int_0^1 (1 + e^{-x^2})\,dx$ as 1.7471803 for $n = 2$, 1.7468554 for $n = 4$, 1.7468261 for $n = 8$, 1.7468241 for $n = 16$, and 1.7468241 for $n = 32$. Determine the value of I, using Romberg procedure, to satisfy the convergence criterion:

$$|I_{k,n} - I_{k-1,n}| \leq 10^{-5}.$$

Section 8.8

8.17. Evaluate the following integral using Gauss–Legendre quadrature formula with $n = 2$, 4, and 6:

$$I = \int_{-1}^{1} \frac{dx}{x^2\sqrt{x^2+1}}.$$

8.18. Evaluate the following integral using Gauss–Laguerre quadrature formula with $n = 2$, 3, and 4:

$$I = \int_0^{\infty} x^2 e^{-x^2}\,dx.$$

Hint: Use the relation $I = \int_0^{\infty} g(x)\,dx = \int_0^{\infty} e^{-x} f(x)\,dx$, where $f(x) = e^x g(x)$.

8.19. Evaluate the integral

$$I = \int_{-\infty}^{\infty} \frac{dx}{1+x^2}$$

using Gauss–Hermite quadrature formula with $n = 2$, 3, 4, and 5.

8.20. Evaluate the integral

$$I = \int_{-1}^{1} \frac{dx}{x^2\sqrt{x^2+1}}$$

using Gauss–Chebyshev quadrature formula with $n = 2$, 3, 4, and 5.

8.21. Evaluate the integral

$$I = \int_1^3 \frac{dx}{x^2(100 - x^2)^{3/2}}$$

using Gauss–Legendre quadrature with different number of Gauss points. Compare the results with the exact value of the integral given by

$$I = \frac{1}{10,000}\left[-\frac{\sqrt{100-x^2}}{x} + \frac{x}{\sqrt{100-x^2}}\right]_1^3$$

and explain the errors involved with the help of Eq. (8.100).

8.22. Derive expressions for the coefficients of a fifth-degree polynomial, β_i (Eq. (b) of Example 8.8), in terms of Legendre polynomials, $P_i(x)$.

8.23. Express an arbitrary fifth-degree polynomial in terms of Laguerre polynomials, $P_i(x)$.

Section 8.9

8.24. Solve Problem 8.1 using a suitable numerical procedure.

8.25. The pressure acting on a piston, of diameter 8 inch, in an internal combustion engine at different positions is given here:

Position from top dead center (x), in	0	1	2	2.5	4	5	7	8	11	12	13	14	15
Pressure (p), psi	110	130	140	170	185	210	215	185	135	100	90	70	40

Find the work done during the 15 inch travel of the piston using the relation, Work $= A\int_0^{15} p(x)\,dx$, where A is the area of the piston.

Section 8.10

8.26. Using Simpson's $\frac{1}{3}$ rule, evaluate the integral

$$I = \int_2^\infty \frac{dx}{(x-1)^2}$$

and compare the numerical value with the exact value of 1.

8.27. An improper integral of the form $\int_a^\infty f(x)\,dx$ can be evaluated conveniently, in some cases, by transforming the variable so that the infinite limit of integration becomes a finite limit for the new variable. Suggest a suitable variable transformation for the following cases:

(a)
$$I_1 = \int_1^\infty \frac{1}{1+x^{-1}+x^3}\,dx;$$

(b)
$$I_2 = \int_{-\infty}^0 \frac{1}{e^{-2x}+e^x}\,dx.$$

8.28. An improper integral of the form $\int_a^\infty f(x)\,dx$ can be expressed, in some cases, as

$$\int_a^\infty f(x)\,dx = \int_a^c f(x)\,dx + \int_c^\infty \bar{f}(x)\,dx, \tag{a}$$

where c is a sufficiently large quantity so that the function $f(x)$ can be approximated by a simpler function $\bar{f}(x)$ that can be integrated in closed form. Thus, the first and second terms on the right-hand side of Eq. (a) can be evaluated numerically and analytically, respectively. Use this technique to evaluate the integral

$$I = \int_0^\infty \frac{1}{e^{2x} + e^{-5x} + 4e^{-3x}}\, dx. \tag{b}$$

Section 8.11

8.29. Evaluate the integral

$$I = \int_{x=1}^{3} \int_{y=-1}^{4} (xy^2 - 8x^3 + 5x^2 y)\, dx\, dy$$

using the two-point Gauss-quadrature rule.

8.30. The volume of a solid (I) bounded by the graphs of $x^2 + y^2 = 25$ and $y^2 + z^2 = 25$ is given by

$$I = 8 \int_{y=0}^{5} \int_{x=0}^{\sqrt{25-y^2}} \sqrt{25 - y^2}\, dx\, dy. \tag{a}$$

Evaluate the integral in Eq. (a) using the two-point Gauss quadrature rule.

Section 8.13

8.31. Use MATLAB to find the value of the complete elliptic integral of the first kind given by

$$K_1(m) = \int_0^{\frac{\pi}{2}} \frac{dx}{\sqrt{1 - m \sin^2 x}}$$

for $m = 0.5$.

8.32. Use MATLAB to find the value of the integral I given in Problem 8.29.

8.33. Use MATLAB to evaluate the triple integral in Eq. (E1) of Example 8.17.

8.34. Use MAPLE to find the value of the complete elliptic integral of the second kind given by

$$K_2(m) = \int_0^{\frac{\pi}{2}} \sqrt{1 - m \sin^2 x}\, dx$$

for $m = 0.5$.

8.35. Use MAPLE to evaluate the following integrals:

(a) $I = \int \frac{dx}{x^2\sqrt{1+x^2}}$, and (b) $I = \int_{-1}^{1} \frac{dx}{x^2\sqrt{1+x^2}}$.

8.36. Use MAPLE to evaluate the integral

$$I = \int_0^3 \frac{dx}{(3x+1)^3}.$$

8.37. Use MATHCAD to evaluate the integral of Problem 8.29.

8.38. Use MATHCAD to find the values of $K_1(m)$ and $K_2(m)$ of Problems 8.31 and 8.34 for $m = 0.5$.

Section 8.14

8.39. Use Program 8.1 to evaluate the integral of Problem 8.12.

8.40. Use Program 8.2 to solve Problem 8.12.

8.41. Use Program 8.3 to evaluate the following integral:

$$I = \int_{x=0}^{1} \int_{y=-1}^{0} \int_{z=-1}^{1} e^x y^2 z \, dx \, dy \, dz.$$

8.42. Use Program 8.4 to evaluate the integral of Problem 8.12.

8.43. Use Program 8.5 to solve Problem 8.12.

General

8.44. The vibration of a mass, m, connected to a nonlinear spring of stiffness, $k(x)$, is governed by the equation [8.8]

$$m\ddot{x} + k(x) = 0 \text{ or } \ddot{x} + F(x) = 0, \tag{a}$$

where $F(x) = k(x)/m$, $\ddot{x} = \frac{d^2x}{dt^2}$, and $x(t)$ = displacement of the mass at time t. The integration of Eq. (a) twice yields

$$t = \frac{1}{\sqrt{2}} \int_0^x \frac{d\xi}{\left\{ \int_\xi^{x_0} F(\eta) d\eta \right\}^{\frac{1}{2}}}, \tag{b}$$

where the initial displacement and initial velocity of the mass are assumed to be x_0 and 0, respectively. Assuming that $F(x) = x^5$ and $x_0 = 1$, find the value of t corresponding to $x = 1$ using a suitable numerical method.

8.45. A closed cylindrical barrel, of radius R and length L, is half full with oil of weight density w and lies on the ground on the edge AB as shown in Fig. 8.21.

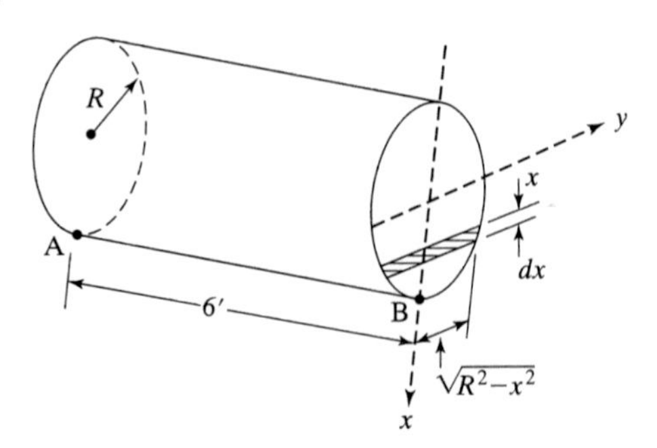

Figure 8.21 Closed cylindrical barrel.

The force exerted by the oil on the circular side (F) is given by

$$F = \int_0^R 2w\sqrt{R^2 - x^2}x\,dx. \tag{a}$$

Find the value of F for $R = 2$ ft and $w = 90$ lb/ft^3 using the following methods:
(a) Analytical integration.
(b) Trapezoidal rule using 12 steps.
(c) Simpson's $\frac{1}{3}$ rule using 12 steps.
(d) Simpson's $\frac{3}{8}$ rule using 12 steps.

8.46. Solve Eqs. (8.34) through (8.36) for c_0, c_1, and c_2.

8.47. Show that the solution of Eqs. (8.43) through (8.46) is given by Eqs. (8.47) through (8.50).

8.48. Using the results of Simpson's $\frac{3}{8}$ rule given in Table 8.3, find an improved estimate of the value of the integral using Richardson's extrapolation.

8.49. Evaluate the integral $I = \int_a^b f(x)\,dx$, where $a = 0$, $b = 1.5$ and

$$f(x) = 0.84885406 + 31.51924706x - 137.66731262x^2$$
$$+240.55831238x^3 - 171.45245361x^4 + 41.95066071x^5$$

using Gauss–Legendre quadrature rule with
(a) Two Gauss points;
(b) Three Gauss points;
(c) Four Gauss points;
(d) Five Gauss points;
(e) Six Gauss points;

8.50. Express the fifth-degree polynomial

$$p_5(x) = \beta_0 + \beta_1 x + \beta_2 x^2 + \beta_3 x^3 + \beta_4 x^4 + \beta_5 x^5$$

in terms of Chebyshev polynomials, $P_i(x)$.

8.51. Evaluate the value of the improper integral

$$I = \int_0^\infty \frac{e^{-2x}}{\sqrt{x}}\,dx$$

as accurately as possible and compare the numerical value with the exact value of $\sqrt{\frac{\pi}{2}}$.

8.52. The gravitational force, $f(x)$, exerted on a body is given by $f(x) = \frac{c}{x^2}$, where c is a constant. If the body weighs 500 lb on the surface of the earth, determine the work to be done to send the body to infinity in a radial direction (Fig. 8.22). Assume that the earth is spherical with a diameter of 8500 miles.

8.53. A solid body, in the form of a partial ellipsoid, is shown in Fig. 8.23. If the density of the solid is ρ, its mass moment of inertia about the y-axis (I) can be

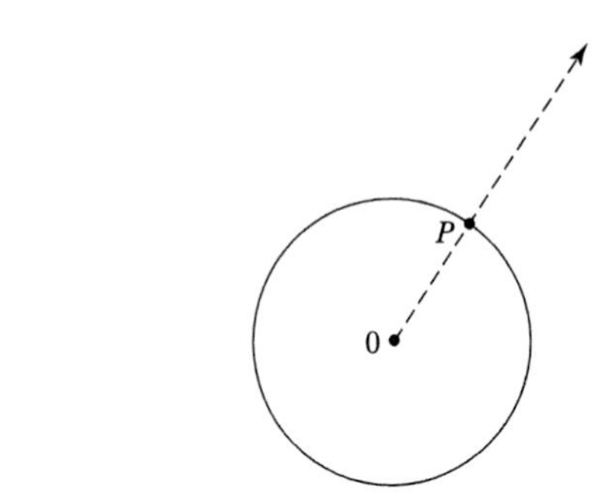

Figure 8.22 Body moving away from earth.

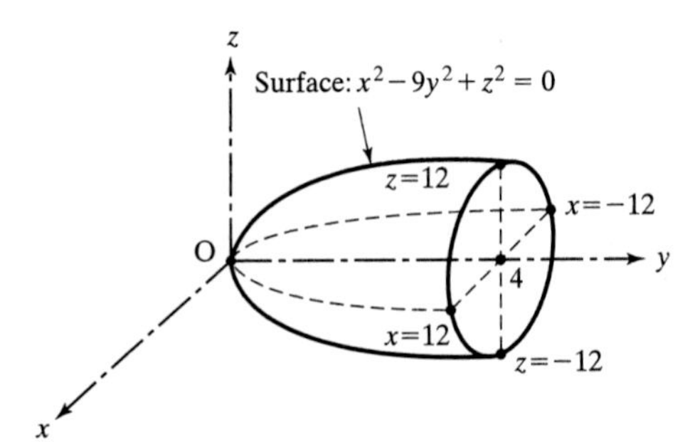

Figure 8.23 Solid body in the form of partial ellipsoid.

expressed as

$$I = \rho \int_{y=0}^{4} \int_{x=-3y}^{3y} \int_{z=-\sqrt{x^2-9y^2}}^{\sqrt{x^2-9y^2}} (x^2+z^2)\,dx\,dy\,dz. \tag{a}$$

Suggest a method of evaluating the integral in Eq. (a) using the trapezoidal rule.

PROJECTS

8.1. The large deflection theory of beams is based on the relation

$$\frac{1}{\rho} = \frac{d\theta}{ds} = \frac{M}{EI} = \frac{P(l-x)}{EI}, \tag{a}$$

where ρ = radius of curvature, M = bending moment, E = Young's modulus, I = area moment of inertia of beam cross section, P = end load, and l = distance of load from the fixed end (Fig. 8.24). According to this theory, the end deflection of the beam (w_0) is given by

$$\frac{w_0}{L} = \frac{1}{\beta}\left[\int_{\phi_0}^{\pi/2} \frac{d\phi}{\sqrt{1-k^2\sin^2\phi}} - 2\int_{\phi_0}^{\pi/2} \sqrt{1-k^2\sin^2\phi}\,d\phi\right], \tag{b}$$

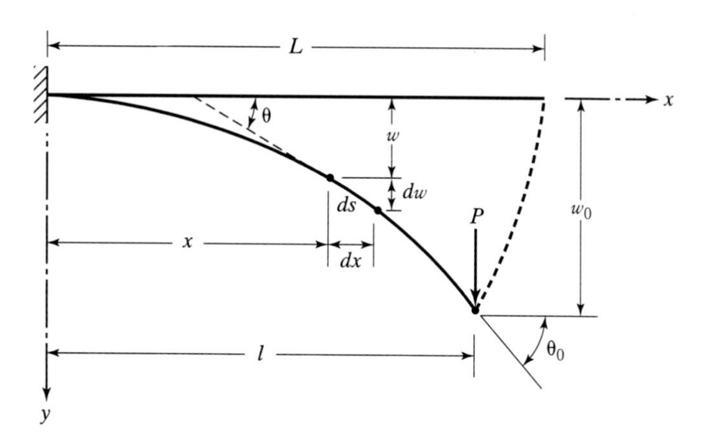

Figure 8.24 Large deflection of a beam.

where

$$\beta = \sqrt{\frac{PL^2}{EI}}, \tag{c}$$

$$k^2 = \frac{1}{2}(1 + \sin\theta_0), \tag{d}$$

and

$$\sin\phi_0 = \frac{1}{\sqrt{2}k}, \tag{e}$$

where θ_0 is the slope at the free end of the beam. It can be shown that [8.7]

$$\beta = \int_{\phi_0}^{\pi/2} \frac{d\phi}{\sqrt{1 - k^2 \sin^2\phi}}. \tag{f}$$

The first and second integrals in Eq. (b) are known as the complete and incomplete elliptic integrals of the second kind, respectively. Determine the value of $\frac{w_0}{L}$ corresponding to $\theta_0 = 60°$.

Hint: Find k, ϕ_0, β and $\frac{w_0}{L}$ using Eqs. (d), (e), (f), and (b), respectively.

8.2. The motion of a mass (m) supported by a spring (Fig. 8.25) is governed by the equation

$$\ddot{x} + a^2 F(x) = 0, \tag{a}$$

where $\ddot{x} = \frac{d^2x}{dt^2}$, $t =$ time, $F(x) =$ spring or restoring force corresponding to the deflection x, and $a^2 = \frac{1}{m}$. Equation (a) can be rewritten as

$$\frac{d}{dx}\left(\dot{x}^2\right) + 2a^2 F(x) = 0. \tag{b}$$

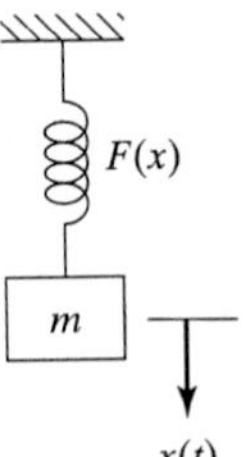

Figure 8.25 Spring-mass system.

Assuming that the initial displacement and velocity of the mass as x_0 and 0, respectively, Eq. (b) can be integrated to obtain

$$t - t_0 = \frac{1}{\sqrt{2a}} \int_0^x \frac{dy}{\left\{ \int_y^{x_0} F(\eta) d\eta \right\}^{1/2}}, \tag{c}$$

where t_0 is the value of t at $x = 0$. When Eq. (c) is considered from zero displacement to maximum displacement, the period of vibration (τ) can be obtained as

$$\tau = \frac{4}{\sqrt{2a}} \int_0^{x_0} \frac{dy}{\left\{ \int_y^{x_0} F(\eta) d\eta \right\}^{1/2}}. \tag{d}$$

Determine the period of vibration for the following cases:
(a) When $m = 10$, $x_0 = 2$, and $F(x) = 1000x^3$
(corresponds to the motion of a mass supported by a cubic spring).
(b) When $m = 1$, $x_0 = 2$, and $F(x) = \frac{mg}{l} \sin x = 4.905 \sin x$
(corresponds to the motion of a pendulum of length 2 m with mass 1 kg and initial angular displacement of 2 radians; see Fig. 8.26).

8.3. The probability of failure of a mechanical or structural component (P_f), for which the strength of the component and the load acting on it follow normal distribution, is given by

$$P_f = \frac{1}{2\pi} \int_{z_1}^{\infty} e^{-\frac{1}{2}z^2} \, dz, \tag{a}$$

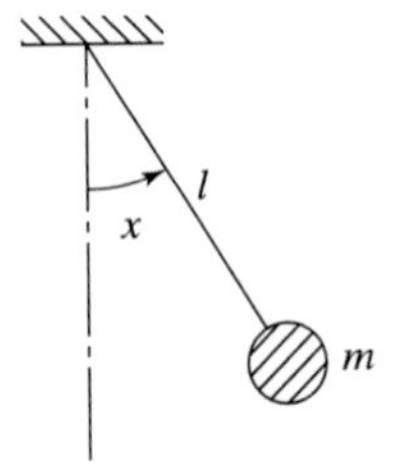

Figure 8.26 Pendulum.

where

$$z_1 = \left(\frac{\mu_L - \mu_S}{\sqrt{\sigma_L^2 + \sigma_S^2}} \right), \tag{b}$$

$\mu_L(\mu_S)$ = mean value of load (strength), and $\sigma_L(\sigma_S)$ = standard deviation of load (strength).

(a) Since the integral in Eq. (a) cannot be evaluated analytically, it is often approximated as

$$P_f = f(z_1)(0.31938153\,t - 0.356563782\,t^2 + 1.781477937\,t^3 - 1.821255978\,t^4 + 1.330274429\,t^5), \tag{c}$$

where

$$f(z_1) = \frac{1}{\sqrt{2\pi}} e^{-\frac{1}{2}z_1^2} \tag{d}$$

and

$$t = \frac{1}{1 + 0.2316419 z_1}.$$

(b) Determine the value of P_f for different values of z_1 ranging from 0.0 to 5.0 in increments of 0.1 using a suitable Gauss quadrature formula. Compare this value of P_f with the one given by Eq. (b) for different values of z_1.

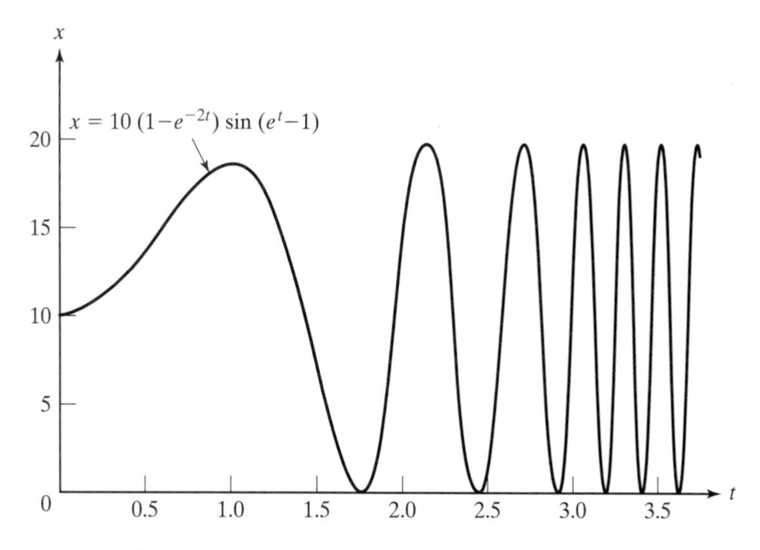

Figure 8.27 Graph of $x(t)$.

8.4. The probability density function of strength of steel, x in kpsi, is given by the Weibull distribution:

$$f(x) = \frac{m}{v-\varepsilon}\left(\frac{x-\varepsilon}{v-\varepsilon}\right)^{m-1} e^{-\left(\frac{x-\varepsilon}{v-\varepsilon}\right)^m}; \quad x \geq \varepsilon; \quad m > 0; \quad v > \varepsilon. \qquad \text{(a)}$$

The mean and standard deviation of x are given by

$$\mu = \varepsilon + (v-\varepsilon)\Gamma\left(1 + \frac{1}{m}\right); \qquad \text{(b)}$$

$$\sigma^2 = (v-\varepsilon)^2\left\{\Gamma\left(1+\frac{2}{m}\right) - \Gamma^2\left(1+\frac{1}{m}\right)\right\}, \qquad \text{(c)}$$

where $\Gamma(\cdot)$ is the gamma function defined as

$$\Gamma(x) = \int_0^\infty y^{x-1}e^{-y}\,dy; \quad x > 0. \qquad \text{(d)}$$

It can be proved that $\Gamma(x+1) = x\Gamma(x)$ for any x and $\Gamma(x+1) = x!$ when x is an integer with $\Gamma(1) = 1$.

(a) Find the values of $\Gamma(x)$ in the range 1 to 2 in increments of 0.05.

(b) Determine the values of μ and σ when $\varepsilon = 100$, $m = 5$, and $v = 150$.

8.5. Develop a procedure for evaluating the integral of a function whose graph is shown in Fig. 8.27 using the trapezoidal rule.

Hint: We need to use an adaptive refinement of the step length.

9

Ordinary Differential Equations: Initial-Value Problems

CHAPTER CONTENTS

9.1 Introduction

An equation involving the derivatives or differentials of the dependent variable is called a differential equation. A differential equation involving only one independent variable is called an ordinary differential equation. If a differential equation involves two or more independent variables (with partial derivatives), it is called a partial differential equation.

Ordinary differential equations are classified according to their order, linearity, and boundary conditions. The order of an ordinary differential equation is defined to be the order of the highest derivative present in that equation. Some examples of first-, second-, and third-order differential equations are

$$\frac{dy}{dx} + ay = f(x)$$

$$\text{(First-order equation),} \qquad (9.1)$$

$$x^2 \frac{d^2y}{dx^2} + \frac{dy}{dx} + 2y = \sin x$$

$$\text{(Second-order equation),} \qquad (9.2)$$

and

$$\frac{d^3y}{dx^3}+a\frac{d^2y}{dx^2}+by\frac{dy}{dx}=c\sin x \quad \text{(Third-order equation)}, \tag{9.3}$$

where x is the independent variable; y is the dependent variable ($y = y(x)$); $f(x)$ is a known function of x; and a, b, and c are constants. Ordinary differential equations can be classified as linear and nonlinear equations. The ordinary differential equation

$$F\left(x, y, \frac{dy}{dx}, \frac{d^2y}{dx^2}, \ldots, \frac{d^ny}{dx^n}\right)=0 \tag{9.4}$$

is said to be linear if F is a linear function of the variables y, $\frac{dy}{dx}$, $\frac{d^2y}{dx^2}$, ..., $\frac{d^ny}{dx^n}$. Thus, a general linear differential equation of order n will have the form

$$a_n(x)y^{(n)}+a_{n-1}(x)y^{(n-1)}+\cdots+a_1y^{(1)}+a_0y=g(x), \tag{9.5}$$

where $a_i(x)$, $i = 0, 1, 2, \ldots, n$, and $g(x)$ are known functions of x and $y^{(i)} = \frac{d^iy}{dx^i}$, $i = 1, 2, \ldots, n$. An ordinary differential equation is said to be nonlinear if it is not of the form of Eq. (9.5), or equivalently, an equation will be nonlinear if it contains products of dependent variable, its derivatives, or both. It can be seen that Eqs. (9.1) and (9.2) are linear, while Eq. (9.3) is nonlinear because of the presence of the term $y\frac{dy}{dx}$.

An ordinary differential equation can be homogeneous or nonhomogeneous. If $g(x) = 0$ in Eq. (9.5), the equation is called homogeneous. On the other hand, if $g(x) \neq 0$, the equation is said to be nonhomogeneous. The coefficients $a_n(x), a_{n-1}(x), \ldots, a_1(x)$, and $a_0(x)$ in Eq. (9.5) are called variable coefficients when any of them is a function of x and are called constant coefficients when they are all constants. An ordinary differential equation is said to be autonomous if the independent variable does not appear explicitly in that equation; otherwise, the equation is called nonautonomous.

Ordinary differential equations can be classified as initial value problems or boundary value problems. An equation is called an initial value problem if the values of the dependent variables or their derivatives are known at the initial value of the independent variables. An equation for which the values of the dependent variables or their derivatives are known at the final value of the independent variable is called a final value problem. If the dependent variable or its derivatives are known at more than one point of the independent variable, the differential equation is a boundary-value problem. In particular, if some of the dependent variable or its derivatives are specified at the initial value of the independent variable, and the remaining ones are specified at the final value of the independent variable, the differential equation is known as a two-point boundary-value problem.

9.2 Engineering Applications

▶Example 9.1

During taxiing, an airplane experiences vibration in the vertical direction due to the unevenness of the runway. By idealizing the airplane and the landing gear as a simple mass–spring–damper system as shown in Fig. 9.1, the equation of motion of the airplane can be expressed as [9.13]

$$m\frac{d^2x}{dt^2} + c\frac{dx}{dt} + kx = -m\frac{d^2y}{dt^2}, \tag{E1}$$

where m = mass of the airplane, c = damping constant of the landing gear, k = stiffness of the landing gear, x = displacement of the airplane relative to the runway, y = runway roughness, and t = time. By idealizing the runway roughness to be sinusoidal with an amplitude y_0 and wavelength l and the airplane to be initially at rest, state the resulting boundary-value problem. Assume the velocity of the airplane as v.

Solution

The frequency of the ground excitation (ω) can be found by dividing the speed of the airplane by the length of one cycle of runway roughness:

$$\omega = \frac{2\pi v}{l}. \tag{E2}$$

Thus, the runway roughness can be expressed as a sinusoidal wave as

$$y(t) = y_0 \sin \omega t = y_0 \sin \frac{2\pi vt}{l}. \tag{E3}$$

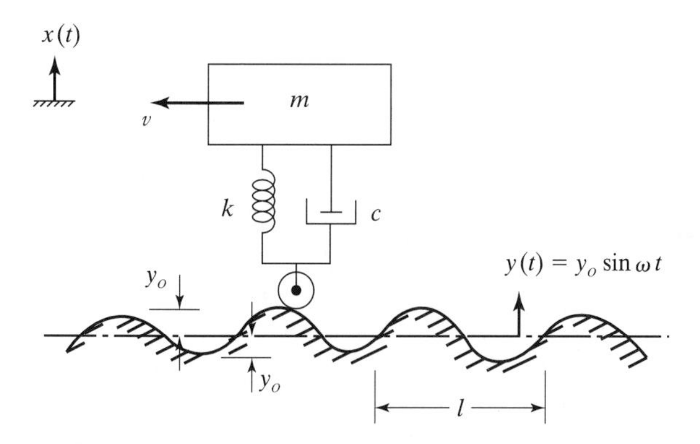

Figure 9.1 Motion along an uneven runway.

The equation of motion, Eq. (E1), becomes

$$m\frac{d^2x}{dt^2} + c\frac{dx}{dt} + kx = m\left(\frac{2\pi v}{l}\right)^2 y_0 \sin\frac{2\pi vt}{l}. \tag{E4}$$

Since the airplane is initially at rest, the initial conditions are given by

$$x(0) = 0; \quad \frac{dx(0)}{dt} = 0. \tag{E5}$$

◀

▶Example 9.2

An electrical circuit consists of an inductance L, capacitance C, resistance R, and an applied voltage $V(t)$ as shown in Fig. 9.2. Derive the differential equation governing the variation of the current and the charge in the circuit.

Solution

The application of Kirchhoff's voltage law gives the following:

Applied voltage = Sum of voltage drops across L, C, and R, that is,

$$V(t) = V_L + V_C + V_R = L\frac{di}{dt} + \frac{1}{C}\int_0^t i\,dt + Ri, \tag{E1}$$

where V_L, V_C, and V_R denote the voltage drops across the inductance, capacitance, and resistance, respectively, i is the electric current flowing in the circuit and t is time. Denoting $i = \frac{dq}{dt}$ where $q(t)$ is the change on the condenser (capacitance C) at time t, Eq. (E1) can be rewritten as

$$L\frac{d^2q(t)}{dt^2} + R\frac{dq(t)}{dt} + \frac{q(t)}{C} = V(t). \tag{E2}$$

The initial conditions can be stated as

$$q(0) = q_0 \tag{E3}$$

and

$$\frac{dq}{dt}(0) = i(0) = i_0, \tag{E4}$$

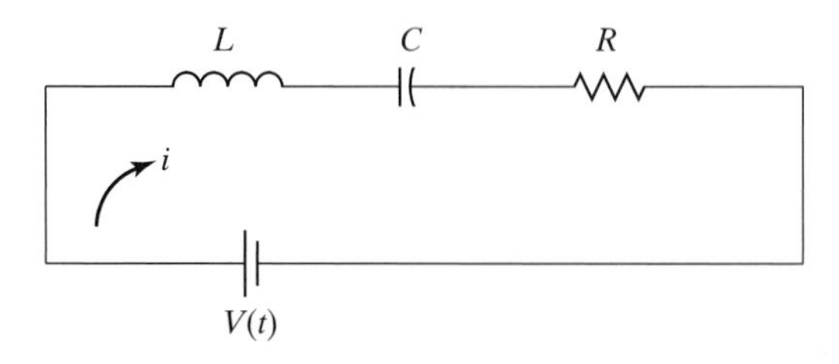

Figure 9.2 Electrical circuit.

where q_0 and i_0 indicate the values of change and current at time 0. It can be seen that Eq. (E2) represents a second-order nonhomogeneous differential equation with two initial conditions given by Eqs. (E3) and (E4). ◀

▶Example 9.3

A simple pendulum of length l, having a bob of mass m, undergoes large angular displacements as shown in Fig. 9.3. Derive the equation of motion and the initial conditions of the system.

Solution

When the pendulum undergoes an angular displacement $\theta(t)$, the inertia moment is given by $I_0\ddot{\theta}$, where $I_0 = ml^2$ and $\ddot{\theta} = \frac{d^2\theta}{dt^2}$, and the restoring moment due to gravity can be expressed as $mgl\sin\theta$. The application of Newton's second law of motion yields

$$ml^2\ddot{\theta} = -mgl\sin\theta,$$

that is,

$$ml^2\ddot{\theta} + mgl\sin\theta = 0. \tag{E1}$$

The initial conditions can be stated in terms of the angular displacement and angular velocity of the pendulum at time 0 as

$$\theta(0) = \theta_0 \text{ and } \dot{\theta}(0) = \dot{\theta}_0, \tag{E2}$$

where θ_0 and $\dot{\theta}_0$ are the values of $\theta(t)$ and $\dot{\theta}(t)$ at $t = 0$, respectively. It can be observed that Eq. (E1) denotes a nonlinear second-order homogeneous differential equation with the initial conditions given by Eq. (E2). ◀

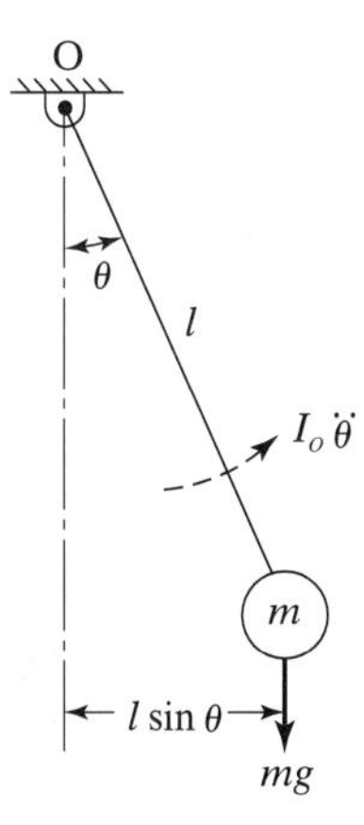

Figure 9.3 Simple pendulum.

9.3 Simultaneous Differential Equations

The analysis of many engineering systems involves the solution of systems of simultaneous differential equations. The numerical solution of one or more high-order ordinary differential equations is most conveniently obtained if the system of equations is represented in the form of a set of n simultaneous first-order differential equations of the form

$$\begin{aligned} \frac{dy_1}{dx} &= f_1(y_1, y_2, \ldots, y_n, x); \\ \frac{dy_2}{dx} &= f_2(y_1, y_2, \ldots, y_n, x); \\ &\vdots; \\ \frac{dy_n}{dx} &= f_n(y_1, y_2, \ldots, y_n, x). \end{aligned} \tag{9.6}$$

Equations (9.6) are said to be in canonical form. A differential equation of high order or a system of differential equations of same or mixed order can be transformed to the canonical form of Eq. (9.6) by using a series of transformations. For illustration, consider the nth-order differential equation:

$$a_n(x)\frac{d^n y}{dx^n} + a_{n-1}(x)\frac{d^{n-1} y}{dx^{n-1}} + \cdots + a_1(x)\frac{dy}{dx} + a_0(x)y = g(x). \tag{9.7}$$

This can be rewritten as

$$\frac{d^n y}{dx^n} = F\left(y, \frac{dy}{dx}, \frac{d^2 y}{dx^2}, \ldots, \frac{d^{n-1} y}{dx^{n-1}}, x\right). \tag{9.8}$$

By using the transformations

$$\begin{aligned} y &= y_1, \\ \frac{dy}{dx} &= \frac{dy_1}{dx} = y_2, \\ \frac{d^2 y}{dx^2} &= \frac{dy_2}{dx} = y_3, \\ &\vdots, \\ \frac{d^{n-1} y}{dx^{n-1}} &= \frac{dy_{n-1}}{dx} = y_n, \end{aligned}$$

and

$$\frac{d^n y}{dx^n} = \frac{dy_n}{dx}. \tag{9.9}$$

in the nth-order equation, Eq. (9.8), the following equivalent set of first-order equations can be obtained:

$$\frac{dy_1}{dx} = f_1(y_1, y_2, \ldots, y_n, x) \equiv y_2,$$

$$\frac{dy_2}{dx} = f_2(y_1, y_2, \ldots, y_n, x) \equiv y_3,$$

$$\vdots,$$

$$\frac{dy_{n-1}}{dx} = f_{n-1}(y_1, y_2, \ldots, y_n, x) \equiv y_n,$$

and

$$\frac{dy_n}{dx} = f_n(y_1, y_2, \ldots, y_n, x) \equiv F(y_1, y_2, \ldots, y_n, x). \tag{9.10}$$

Equation (9.10) can be stated, using vector notation, as

$$\frac{d\vec{y}}{dx} = \vec{f}(\vec{y}, x), \tag{9.11}$$

where

$$\vec{y} = \begin{Bmatrix} y_1 \\ y_2 \\ \cdot \\ \cdot \\ \cdot \\ y_n \end{Bmatrix}, \quad \vec{f} = \begin{Bmatrix} f_1 \\ f_2 \\ \cdot \\ \cdot \\ \cdot \\ f_n \end{Bmatrix}. \tag{9.12}$$

▶Example 9.4

The equations of motion of a body moving in a plane in earth's gravitational field are given by

$$\frac{d^2x}{dt^2} = -G\frac{x}{(x^2 + y^2)^{3/2}} \tag{E1}$$

and

$$\frac{d^2y}{dt^2} = -G\frac{y}{(x^2 + y^2)^{3/2}}, \tag{E2}$$

where G is the gravitational constant and x and y are the rectangular coordinates defining the position of the body at time t. Convert Eqs. (E1) and (E2) into a system of first-order differential equations.

Solution

By defining

$$x_1 = x, x_2 = \frac{dx}{dt}, x_3 = y, x_4 = \frac{dy}{dt},$$

Eqs. (E1) and (E2) can be stated as

$$\frac{d\vec{X}}{dt} = \vec{f}(\vec{X}, t), \tag{E3}$$

where

$$\vec{X}(t) = \begin{Bmatrix} x_1(t) \\ x_2(t) \\ x_3(t) \\ x_4(t) \end{Bmatrix} \text{ and } \vec{f}(t) = \begin{Bmatrix} f_1(t) \\ f_2(t) \\ f_3(t) \\ f_4(t) \end{Bmatrix} = \begin{Bmatrix} x_2 \\ -G\dfrac{x_1}{(x_1^2 + x_3^2)^{3/2}} \\ x_4 \\ -G\dfrac{x_3}{(x_1^2 + x_3^2)^{3/2}} \end{Bmatrix}.$$

The initial conditions are usually stated as

$$\vec{X}(0) = \vec{X}_0 = \begin{Bmatrix} x_{10} \\ x_{20} \\ x_{30} \\ x_{40} \end{Bmatrix}, \tag{E4}$$

where

$$x_{10} = x(t = 0), x_{20} = \frac{dx}{dt}(t = 0), x_{30} = y(t = 0), \text{ and } x_{40} = \frac{dy}{dt}(t = 0).$$ ◀

9.4 Solution Concept

For simplicity of illustration, consider the numerical solution of the single-first-order differential equation

$$\frac{dy}{dx} = f(x, y) \tag{9.13}$$

with the initial condition

$$y(x_0) = y_0. \tag{9.14}$$

Basically, two approaches can be used to solve the differential equation stated in Eqs. (9.13) and (9.14). The first approach involves integrating the function $f(x, y)$ using a numerical integration technique, and the second approach involves using a finite-difference approximation to the derivatives. In the first approach, Eq. (9.13) is rewritten as

$$dy = f(x, y)dx \tag{9.15}$$

and integrated between the limits of $x_i \leq x \leq x_{i+1}$ and $y_i \leq y \leq y_{i+1}$:

$$\int_{y_i}^{y_{i+1}} dy = \int_{x_i}^{x_{i+1}} f(x, y)\, dx. \tag{9.16}$$

This gives

$$y_{i+1} - y_i = \int_{x_i}^{x_{i+1}} f(x, y)\, dx. \tag{9.17}$$

Now the function $f(x, y)$ on the right-hand side of Eq. (9.17) can be replaced by a suitable interpolation polynomial and the integral evaluated between the appropriate limits. In the second approach, first the derivative of y on the left-hand side of Eq. (9.13) is replaced by a suitable finite difference approximation and then a solution is sought. Although both the approaches yield the desired solution, the second approach is more elegant since it finds the tangential trajectories of y, rather than the areas under the function $f(x, y)$. Hence, the second approach is used in the following sections. The procedures and formulas developed for the solution of a single differential equation can be readily generalized to a set of differential equations that can be stated as

$$\frac{d\vec{y}}{dx} = \vec{f}(x, \vec{y}), \tag{9.18}$$

with initial conditions

$$\vec{y}(x_0) = \vec{y}_0, \text{ that is, } \begin{Bmatrix} y_1(x_0) \\ y_2(x_0) \\ \vdots \\ y_n(x_0) \end{Bmatrix} = \begin{Bmatrix} y_{1,0} \\ y_{2,0} \\ \vdots \\ y_{n,0} \end{Bmatrix}. \tag{9.19}$$

9.5 Euler's Method

Euler's method is one of the simplest and also the earliest techniques developed for the solution of ordinary differential equations. It is not commonly used since there are several other methods that are more efficient. However, the method is considered here as it is very simple and provides most of the basic features of numerical methods used for the solution of ordinary differential equations. As with all other methods of solution, Euler's method finds the solution of Eq. (9.13) in terms of the values of the function $y(x)$ at discrete values of x, called the node or grid points, for $x > x_0$. The value of y at x_0 is taken as y_0 (as given by the initial condition, Eq. (9.14)). If the grid points are uniformly spaced, the grid point x_i is given by

$$x_i = x_0 + ih, \tag{9.20}$$

where h denotes the uniform step size. Since the value of y_0 is known at the outset, the general numerical procedure must be able to find y_{i+1} from the value of y at the previous grid point, y_i. On the other hand, if x_0 denotes the final value of x, we need to find the value of y for $x < x_0$. In such a case, the value of x_i is taken as

$$x_i = x_0 - ih \tag{9.21}$$

and the value of y_{i+1} is to be found from the known value, y_i.

Although Euler's method can be derived in several ways, the derivation based on Taylor's series expansion is considered here. The value of y_{i+1} can be expressed, using Taylor's series expansion about x_i, as

$$y_{i+1} = y_i + hy_i' + \frac{h^2}{2}y''(\xi);\ x_i < \xi < x_{i+1}, \tag{9.22}$$

where the third term on the right-hand side of Eq. (9.22) denotes the error or remainder term, $y_i = y(x_i)$, $y_{i+1} = y(x_{i+1})$, and $y_i' = \frac{dy}{dx}(x_i)$. Since $y(x)$ satisfies the differential Equation (9.13), $y_i' = f(x_i, y_i)$ and Eq. (9.22) can be expressed as

$$y_{i+1} = y_i + hf(x_i, y_i) + \frac{h^2}{2}y''(\xi). \tag{9.23}$$

If h is small, the error term can be neglected, and Eq. (9.23) yields

$$y_{i+1} = y_i + hf(x_i, y_i);\ i = 0, 1, 2, \ldots. \tag{9.24}$$

This equation is referred to as Euler's or Euler–Cauchy or the point slope method.

The geometric interpretation of Euler's method is shown in Fig. 9.4. At the starting point, x_0, the tangent line to $y(x)$ is used to approximate the graph $y(x)$ in the interval $0 \le x \le h$. For larger values of x, the tangent to the curve at the point x_i, which has a slope of $f(x_i, y_i)$, is used to predict x_{i+1} according to Eq. (9.24). Hence, the method is called the point–slope method. In this method, the errors accumulate and the numerical solution deviates more and more from the exact solution as the value of x_i increases as indicated in Fig. 9.5.

▶Example 9.5

Find the solution of the initial value problem

$$y' = y + 2x - 1, \tag{E1}$$

with $y(x = 0) = 1$ over the interval $0 \le x \le 1$.

Solution

We divide the interval $0 \le x \le 1$ into 10 parts such that $h = 1/10 = 0.1$, $x_0 = 0$, $x_1 = 0.1$, $x_2 = 0.2, \ldots, x_{10} = 1.0$, and $y_0 = 1$. Noting that $f(x, y) = y + 2x - 1$, we can express the iterative process as

$$y_{i+1} = y_i + hf(x_i, y_i);\ i = 0, 1, 2, \ldots, 10. \tag{E2}$$

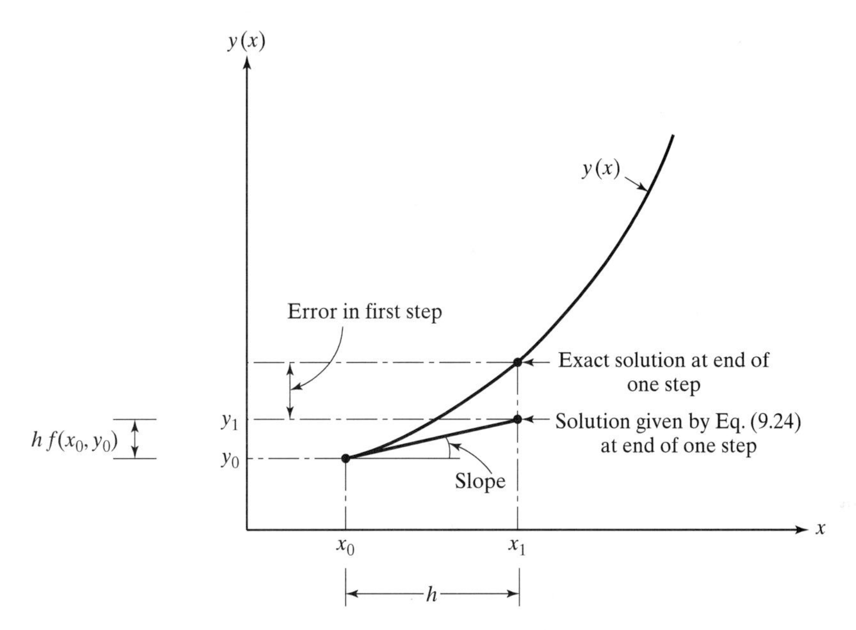

Figure 9.4 Geometric interpretation of Euler's method.

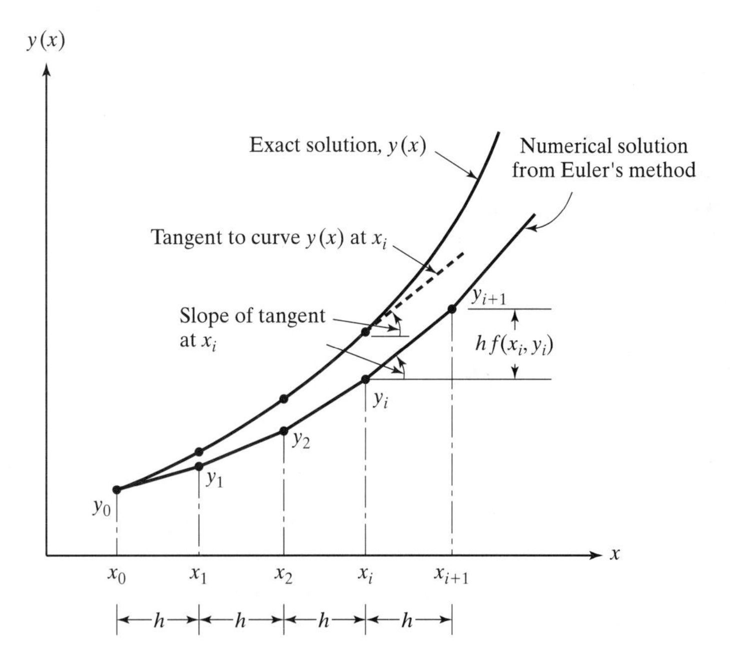

Figure 9.5 Accumulation of error in Euler's method.

Table 9.1 Results of Euler's method.

x_i	Value of y_i with						Exact solution, y_i
	$h = 0.1$	Error	$h = 0.05$	Error	$h = 0.01$	Error	
0.00	—	—	—	—	—	—	1.0
0.10	1.000 0000	0.010 3426	1.005 0001	0.005 3425	1.009 2459	0.001 0967	1.010 3426
0.20	1.020 0005	0.022 8062	1.031 0125	0.011 7941	1.040 3824	0.002 4242	1.042 8066
0.30	1.062 0003	0.037 7178	1.080 1907	0.019 5274	1.095 7012	0.004 0169	1.099 7181
0.40	1.128 2005	0.055 4495	1.154 9110	0.028 7390	1.177 7315	0.005 9185	1.183 6500
0.50	1.221 0207	0.076 4227	1.257 7896	0.039 6538	1.289 2666	0.008 1768	1.297 4434
0.60	1.343 1225	0.101 1143	1.391 7131	0.052 5246	1.433 3963	0.010 8414	1.444 2368
0.70	1.497 4346	0.130 0707	1.559 8640	0.067 6422	1.613 5292	0.013 9761	1.627 5053
0.80	1.687 1786	0.163 9032	1.765 7499	0.085 3319	1.833 4312	0.017 6506	1.851 0818
0.90	1.915 8964	0.203 3100	2.013 2389	0.105 9675	2.097 2662	0.021 9402	2.119 2064
1.00	2.187 4857	0.249 0778	2.306 5958	0.129 9677	2.409 6279	0.026 9356	2.436 5635

Using $f(x_i, y_i) = y_i + 2x_i - 1$, Eq. (E2) gives

$$y_{i+1} = y_i + h(y_i + 2x_i - 1);\ i = 0, 1, 2, \ldots, 10. \tag{E3}$$

Thus, we have the following results:

For $i = 0$;

$$y_1 = y_0 + hf(x_0, y_0) = y_0 + h(y_0 + 2x_0 - 1) = 1 + 0.1\{1 + 2(0) - 1\} = 1.0;$$

for $i = 1$;

$$y_2 = y_1 + hf(x_1, y_1) = y_1 + h(y_1 + 2x_1 - 1) = 1 + 0.1\{1 + 2(0.1) - 1\} = 1.02;$$

$$\vdots$$

These results, as well as the results obtained with the reduced step lengths $h = 0.05$ and $h = 0.01$, are shown in Table 9.1. The exact solution of Eq. (E1) is known to be (can be verified by substitution)

$$y(x) = -1 - 2x + 2e^x. \tag{E4}$$

The values of y at $x_0, x_1, x_2, \ldots, x_{10}$ found from Eq. (E4) are also indicated in Table 9.1. ◀

9.5.1 Solution Using Numerical Integration

As indicated by Eq. (9.17), the solution of the differential equation can be found by numerically integrating the right-hand side of Eq. (9.17):

$$y_{i+1} = y_i + \int_{x_i}^{x_{i+1}} f(x, y)\, dx. \tag{9.25}$$

Thus, the area under the $f(x, y)$ curve from $x = x_i$ to $x = x_{i+1}$ is to be added to y_i to obtain the solution y_{i+1}. For simplicity, the value of $f(x, y)$ can be assumed to be a constant, equal to $f(x_i, y_i)$, in the interval x_i to x_{i+1} to find the area under the curve (i.e., using the rectangular rule) as

$$\int_{x_i}^{x_{i+1}} f(x, y)\, dx \approx f(x_i, y_i)(x_{i+1} - x_i) = hf(x_i, y_i). \tag{9.26}$$

This leads to

$$y_{i+1} = y_i + hf(x_i, y_i). \tag{9.27}$$

The solution obtained according to Eq. (9.27) is shown qualitatively in Fig. 9.6. As stated earlier, the numerical solution procedure indicated by Eq. (9.25) is not as elegant as the solution procedure implied by Eq. (9.24).

9.5.2 Solution of a System of Differential Equations

As stated earlier, the method of solving a single differential equation can easily be extended to solve a system of differential equations. To illustrate the procedure, we consider Euler's method for the solution of a set of n first-order differential equations given by Eq. (9.18):

$$\frac{dy_1}{dx} = f_1(x, y_1, y_2, \ldots, y_n);\; y_1(x_0) = y_{1,0}; \tag{9.28}$$

$$\frac{dy_2}{dx} = f_2(x, y_1, y_2, \ldots, y_n);\; y_2(x_0) = y_{2,0}; \tag{9.29}$$

$$\vdots;$$

$$\frac{dy_n}{dx} = f_n(x, y_1, y_2, \ldots, y_n);\; y_n(x_0) = y_{n,0}. \tag{9.30}$$

Euler's method, Eq. (9.24) can be extended to generate the solution at x_{i+1} from the known solution at x_i as

$$y_{1,i+1} = y_{1,i} + hf_1(x_i, y_{1,i}, y_{2,i}, \ldots, y_{n,i}), \tag{9.31}$$

$$y_{2,i+1} = y_{2,i} + hf_2(x_i, y_{1,i}, y_{2,i}, \ldots, y_{n,i}), \tag{9.32}$$

$$\vdots,$$

and

$$y_{n,i+1} = y_{n,i} + hf_n(x_i, y_{1,i}, y_{2,i}, \ldots, y_{n,i}). \tag{9.33}$$

These equations indicate that, from the known solution $y_{1,i}, y_{2,i}, \ldots, y_{n,i}$ at x_i ($i = 0$ at the beginning), the values of $y_{1,i+1}, y_{2,i+1}, \ldots, y_{n,i+1}$ can be determined using Eqs. (9.31) through (9.33).

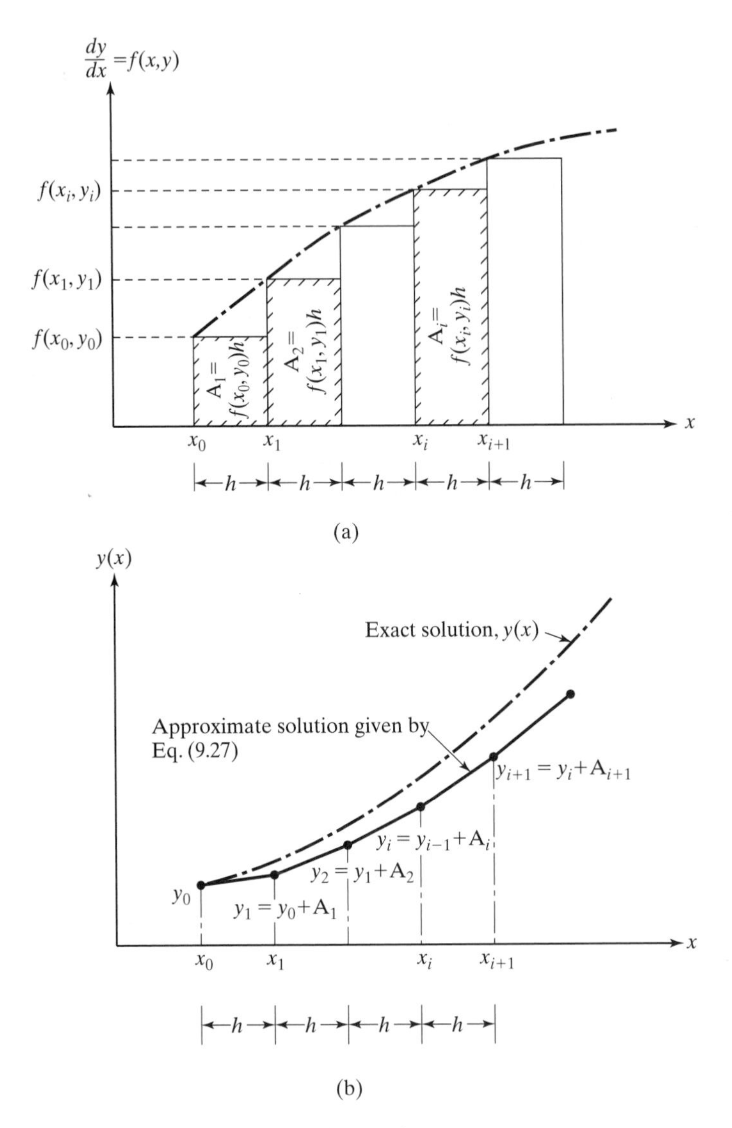

Figure 9.6 Solution using numerical integration.

9.5.3 Error Analysis of Euler's Method

There are two types of errors in the numerical solution of ordinary differential equations—round-off error and truncation error. The round-off error is caused by the limited number of significant digits used in the computation by a computer. For a given computer, the round-off error can be reduced by using double precision for the computations. The truncation error is caused by the approximate procedure used in computing the values of y. In the case of Euler's method, the truncation error is caused by the truncation of higher order derivative terms in Taylor's series expansion of Eq. (9.22). The truncation errors are of two types—one is the local truncation error caused by the application of the method in one step, and the other is the propagated truncation error caused by the accumulation of errors of the previous steps. The sum of the local and propagated errors is known as the global or total error. The propagated error is important for the stability of the numerical procedure.

To estimate the truncation error in Euler's method, consider Taylor's series expansion of $y(x)$ about x_i to find $y(x_{i+1})$ as

$$y(x_{i+1}) = y(x_i) + h \left.\frac{dy}{dx}\right|_{x_i} + \left.\frac{h^2}{2!}\frac{d^2y}{dx^2}\right|_{x_i} + \left.\frac{h^3}{3!}\frac{d^3y}{dx^3}\right|_{x_i} + \cdots, \tag{9.34}$$

where $y(x_i)$, $y(x_{i+1})$, and the derivatives of y computed at x_i are assumed to be exact. Next, assuming that the exact solution of the differential equation at x_i, $y(x_i)$, is known, Euler's method can be used to compute the approximate solution at x_{i+1}, $\bar{y}_{i+1}$, as

$$\bar{y}_{i+1} = y(x_i) + hf(x_i, y(x_i)). \tag{9.35}$$

Subtracting Eq. (9.35) from Eq. (9.34), and noting that $\left.\frac{dy}{dx}\right|_{x_i} = f(x_i, y(x_i))$, we obtain the truncation error from x_i to x_{i+1} as

$$y(x_{i+1}) - \bar{y}_{i+1} = \left.\frac{h^2}{2}\frac{d^2y}{dx^2}\right|_{x_i} + \left.\frac{h^3}{6}\frac{d^3y}{dx^3}\right|_{x_i} + \cdots. \tag{9.36}$$

Equation (9.36) denotes that the local truncation error or the truncation error per step is of order $O(h^2)$. Although the solution y_0 is exact at x_0, the local truncation errors will accumulate as the numerical solutions $y_1, y_2, \ldots, y_i$ are generated. The total truncation error at x_{i+1} will be equal to the product of the truncation error per step and the number of steps. The truncation error per step is of order $O(h^2)$, and the number of steps, $i + 1$, is equal to $\left(\frac{x_{i+1}}{h}\right)$. Therefore, the total truncation error will be of order $O(h)$. Because the total truncation error is of the first order of h, Euler's method is called a *first-order method*.

The truncation error in Euler's method can be reduced by using a smaller step size h; the reduction in error being linear with h. As the value of h is reduced, the number of steps to be used increases, thereby, increasing the round-off error. Thus, the round-off error increases as the truncation error decreases, as shown qualitatively in Fig. 9.7. Although this figure shows that there is an optimum step size to minimize the total error, usually its value is not known. Hence, in practice, for a given differential equation, a series of solutions, each with a smaller step size, can be generated until two successive step sizes give essentially the same solution. At that stage, the solution can be assumed to have converged to the exact solution. However, it is to be noted that, due to the presence of round-off error, the numerical solution will always be different from the exact solution. In the absence of round-off errors, if a numerical solution approaches the exact solution as the step size (h)

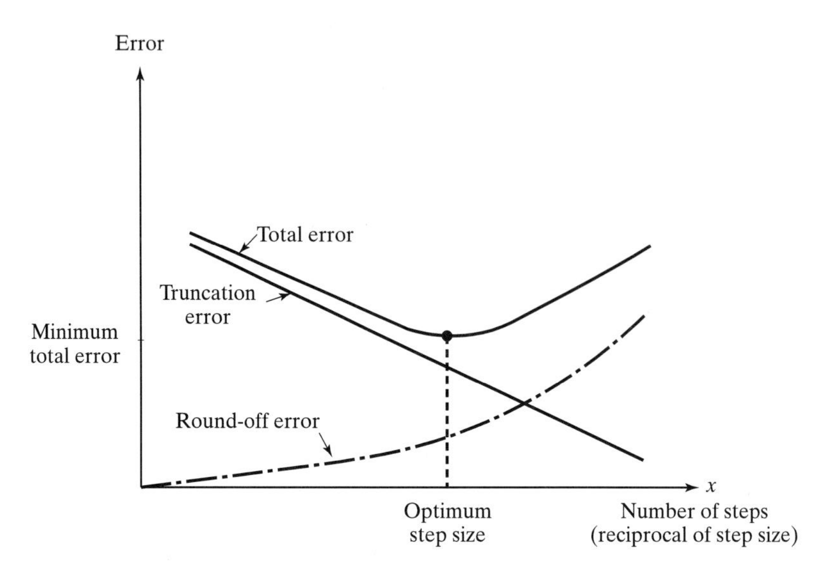

Figure 9.7 Round-off and truncation errors.

approaches zero, the numerical method is said to be *convergent.* All the numerical methods considered in this chapter are convergent and hence the numerical solutions can be assumed to approach the exact solutions as the step size is reduced.

Another important consideration in the numerical solution of differential equations is the numerical stability of the method used. In general, a numerical method is considered to be *unstable* if it gives an unbounded solution (errors grow at an exponential rate) for a problem for which the exact solution is bounded. The numerical instability occurs whenever the error propagates without bounds. The numerical stability of a method depends on the method as well as on the differential equation. As an example, consider the differential equation

$$\frac{dy}{dx} = \lambda y$$

$$\text{Initial condition; } y(x_0) = y_0 \tag{9.37}$$

$$\lambda \text{ real, } y_0 \text{ finite.}$$

The exact solution of this problem is given by

$$y(x) = y_0 e^{\lambda x}. \tag{9.38}$$

The numerical solution of this problem can be expressed, using Euler's method, as

$$y_{i+1} = y_i + h\lambda y_i = (1 + h\lambda)y_i. \tag{9.39}$$

Repeated application of Eq. (9.39), after n steps, yields

$$y_n = y_0(1 + h\lambda)^n. \tag{9.40}$$

This solution can be seen to be stable only if $|1 + h\lambda| \leq 1$, which implies that Euler's method is conditionally stable for

$$\lambda < 0 \text{ and } h \leq \frac{2}{|\lambda|}$$

and unstable for

$$|1 + h\lambda| > 1.$$

In general, a numerical method is said to be unconditionally stable, if it is stable for any values of the step size and other parameters of the differential equation.

9.6 Improvements and Modifications of Euler's Method

Euler's method discussed in the previous section is simple, but it requires a small step h to provide acceptable levels of accuracy. If the solution of the differential equation is required over a large range of x, then the method becomes computationally expensive. Also, the local errors of various steps may accumulate in an unpredictable manner. Hence, several modifications are suggested to make the method more accurate and stable.

9.6.1 Higher Order Taylor's Series Methods

Since Euler's method was derived from Taylor's series expansion by truncating all terms involving derivatives of order two and higher, our first attempt to improve Euler's method should consider retaining more terms in Taylor's series expansion. If terms up to the nth order derivative are included, Taylor's series expansion of $y(x)$ about x_i can be written as

$$y(x_{i+1}) = y(x_i) + hy'(x_i) + \frac{h^2}{2!}y''(x_i) + \frac{h^3}{3!}y'''(x_i)$$
$$+ \cdots + \frac{h^n}{n!}y^{(n)}(x_i) + \frac{h^{n+1}}{(n+1)!}y^{(n+1)}(\xi);\ x_i < \xi < x_{i+1}, \tag{9.41}$$

where the last term represents the truncation error. Since the function $f(x, y)$ in the original differential equation, Eq. (9.13), involves both the independent and dependent variables, the various order derivatives appearing in Eq. (9.41) can be evaluated using the chain rule of differentiation as follows:

$$y'(x) = f(x, y) \quad \text{(from Eq. (9.13))}, \tag{9.42}$$

$$y''(x) = \frac{d[y'(x)]}{dx} = \frac{df(x, y)}{dx} = \frac{\partial f(x, y)}{\partial x} + \frac{\partial f(x, y)}{\partial y}\frac{dy}{dx}$$
$$= \frac{\partial f(x, y)}{\partial x} + \frac{\partial f(x, y)}{\partial y} f(x, y) \equiv f'(x, y), \tag{9.43}$$

$$y'''(x) = \frac{df'(x, y)}{dx} \equiv f''(x, y), \tag{9.44}$$

$$\vdots,$$

and

$$y^{(n)}(x) = f^{(n-1)}(x, y). \tag{9.45}$$

Substituting these results into Eq. (9.41) yields

$$y(x_{i+1}) = y(x_i) + hf(x_i, y_i) + \frac{h^2}{2!}f'(x_i, y_i) + \frac{h^3}{3!}f''(x_i, y_i)$$
$$+ \cdots + \frac{h^n}{n!}f^{(n-1)}(x_i, y_i) + \frac{h^{n+1}}{(n+1)!}y^{(n+1)}(\xi);\ x_i < \xi < x_{i+1}. \tag{9.46}$$

Using Eq. (9.46), Taylor's method of order n can be implemented to solve the differential equation as follows:

$$y(x_0) = y_0 \text{ (initial condition)};$$

$$y_{i+1} = y_i + hf(x_i, y_i) + \frac{h^2}{2!}f'(x_i, y_i) + \frac{h^3}{3!}f''(x_i, y_i)$$
$$+ \cdots + \frac{h^n}{n!}f^{(n-1)}(x_i, y_i). \tag{9.47}$$

According to this terminology, Euler's method can be considered as Taylor's method of order 1. The local truncation error in Taylor's method of order n can be seen to be $O(h^{n+1})$ from the remainder term in Eq. (9.46).

▶Example 9.6

Find the solution of the problem

$$y' = y + 2x - 1, \tag{E1}$$

with $y(x = 0) = 1$ over the interval $0 \le x \le 1$ using Taylor's series methods of order 1, 2, 3, and 4.

Solution

By dividing the interval of x into 10 parts with $h = 0.1$, $x_0 = 0.0$, $x_1 = 0.1, \ldots, x_{10} = 1.0$, and $y_0 = 1.0$, the iterative process indicated by Eq. (9.47) can be used to find the values of $y_1, y_2, \ldots$, and y_{10}. In the present case, Eq. (E1) gives

$$f(x, y) = y + 2x - 1, \tag{E2}$$

and hence,

$$f' = \frac{df}{dx} = \frac{\partial f}{\partial x} + \frac{\partial f}{\partial y}y' = 2 + (y + 2x - 1) = y + 2x + 1, \tag{E3}$$

$$f'' = \frac{df'}{dx} = \frac{\partial f'}{\partial x} + \frac{\partial f'}{\partial y} y' = \frac{\partial f'}{\partial x} + \frac{\partial f'}{\partial y} f$$
$$= 2 + (y + 2x - 1) = y + 2x + 1, \tag{E4}$$

and

$$f''' = \frac{\partial f''}{\partial x} + \frac{\partial f''}{\partial y} y' = \frac{\partial f''}{\partial x} + \frac{\partial f''}{\partial y} f$$
$$= 2 + (y + 2x - 1) = y + 2x + 1. \tag{E5}$$

Thus, the iterative process can be expressed as follows:
Using the first-order method,

$$y_{i+1} = y_i + hf(x_i, y_i) = y_i + h(y_i + 2x_i - 1). \tag{E6}$$

With the second-order method

$$y_{i+1} = y_i + hf(x_i, y_i) + \frac{h^2}{2!} f'(x_i, y_i)$$
$$= y_i + h(y_i + 2x_i - 1) + \frac{h^2}{2}(y_i + 2x_i + 1). \tag{E7}$$

Third-order method is exemplified in

$$y_{i+1} = y_i + hf(x_i, y_i) + \frac{h^2}{2!} f'(x_i, y_i) + \frac{h^3}{3!} f''(x_i, y_i)$$
$$= y_i + h(y_i + 2x_i - 1) + \frac{h^2}{2}(y_i + 2x_i + 1) + \frac{h^3}{6}(y_i + 2x_i + 1). \tag{E8}$$

Using the fourth-order method,

$$y_{i+1} = y_i + hf(x_i, y_i) + \frac{h^2}{2!} f'(x_i, y_i) + \frac{h^3}{3!} f''(x_i, y_i) + \frac{h^4}{4!} f'''(x_i, y_i)$$
$$= y_i + h(y_i + 2x_i - 1) + \frac{h^2}{2}(y_i + 2x_i + 1)$$
$$+ \frac{h^3}{6}(y_i + 2x_i + 1) + \frac{h^4}{24}(y_i + 2x_i + 1). \tag{E9}$$

With $h = 0.1$, $x_i = i(0.1)$, and $y_0 = 1.0$, Eqs. (E6) through (E9) yield the results shown in Table 9.2.

◀

9.6.2 Heun's Method

In Euler's method, the value of the function $f(x, y)$, which denotes the derivative $\frac{dy}{dx}$, is computed at the beginning of the interval h and is assumed to be a constant over the entire interval. This assumption is a major source of error since the derivative, $\frac{dy}{dx}$, changes from point to point over the interval h. In Heun's method, the derivative

Table 9.2 Results given by Taylor's series methods.

x_i	Value of y_i given by Taylor's series method				Exact solution, y_i
	Order 1	Order 2	Order 3	Order 4	
0.0	1.0	1.0	1.0	1.0	1.0
0.1	1.000 0000	1.010 0002	1.010 3340	1.010 3426	1.010 3426
0.2	1.020 0005	1.042 0504	1.042 7876	1.042 8066	1.042 8066
0.3	1.062 0003	1.098 4650	1.099 6876	1.099 7190	1.099 7181
0.4	1.128 2005	1.181 8037	1.183 6052	1.183 6510	1.183 6500
0.5	1.221 0207	1.294 8933	1.297 3814	1.297 4434	1.297 4434
0.6	1.343 1225	1.440 8569	1.444 1557	1.444 2377	1.444 2368
0.7	1.497 4346	1.623 1470	1.627 3994	1.627 5053	1.627 5053
0.8	1.687 1786	1.845 5772	1.850 9474	1.851 0818	1.851 0818
0.9	1.915 8964	2.112 3629	2.119 0395	2.119 2055	2.119 2064
1.0	2.187 4857	2.428 1607	2.436 3594	2.436 5625	2.436 5635

or slope, $\frac{dy}{dx}$, is computed at two points—one at the beginning and the other at the end of the interval h—and their average value is used to achieve an improvement. Thus, the computational procedure of Heun's method can be stated as follows:

1. Start with the known initial condition, $y(x_0) = y_0$.
2. For $i = 0, 1, 2, \ldots$, determine the slope at beginning of step as $f(x_i, y_i)$.
3. Find the value of $y_{i+1}^{(0)}$ as

$$y_{i+1}^{(0)} = y_i + hf(x_i, y_i). \tag{9.48}$$

4. Find the slope at the end of step as

$$f(x_{i+1}, y_{i+1}^{(0)}). \tag{9.49}$$

5. Determine the value of y at x_{i+1} as

$$y_{i+1} = y_i + h\left[\frac{f(x_i, y_i) + f(x_{i+1}, y_{i+1}^{(0)})}{2}\right]. \tag{9.50}$$

Heun's method can be seen to be self-starting, since only the values at x_i are used in finding y_{i+1}. The method can be considered as a simple form of the predictor–corrector methods that are discussed in Section 9.10. The reason is that the value of $y_{i+1}^{(0)}$ computed using Eq. (9.48) is not the final value of y_{i+1}, but only an intermediate prediction that enables us to estimate the value of the slope, $\frac{dy}{dx}$, at the end of the interval, using Eq. (9.49). Thus, Eq. (9.48) can be called a predictor equation. Equation (9.50), which uses the average of the slopes at x_i and x_{i+1} to estimate the value of y at x_{i+1}, can be called the corrector equation. The geometric interpretation of Heun's method is indicated in Fig. 9.8.

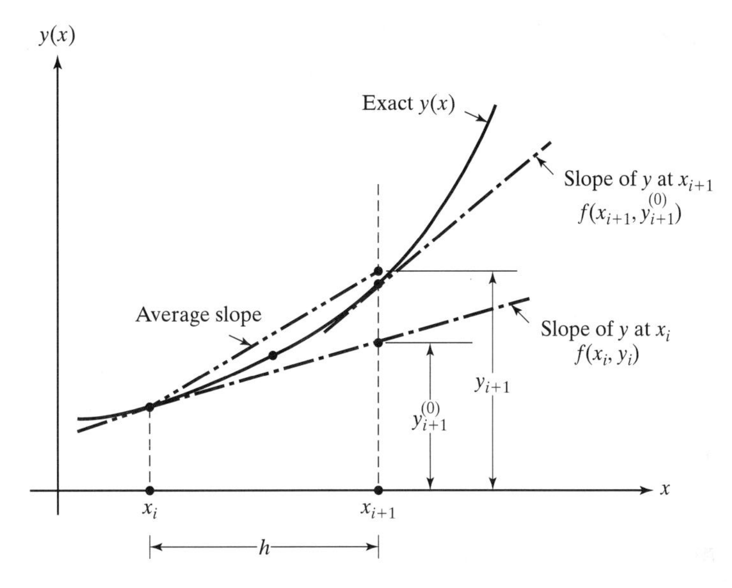

Figure 9.8 Graphical interpretation of Heun's method

Further improvement in Heun's method can be achieved by modifying the corrector Equation (9.50) as

$$y_{i+1}^{(k+1)} = y_i + h\left[\frac{f(x_i, y_i) + f(x_{i+1}, y_{i+1}^{(k)})}{2}\right]; \; k = 0, 1, 2, \ldots. \tag{9.51}$$

This indicates that *corrective* solution is found iteratively. The iterative process of Eq. (9.51) can be stopped when the difference between successive values of $y_{i+1}^{(k)}$ becomes sufficiently small, that is, when

$$\left|\frac{y_{i+1}^{(k+1)} - y_{i+1}^{(k)}}{y_{i+1}^{(k)}}\right| \le \varepsilon, \tag{9.52}$$

where ε is a small number used for convergence.

Error Analysis

In order to determine the error in Heun's method, consider Taylor's series expansion of y_{i+1}:

$$y_{i+1} = y_i + hy_i' + \frac{h^2}{2!}y_i'' + \frac{h^3}{3!}y'''(\xi); \; x_i < \xi < x_{i+1}. \tag{9.53}$$

Using a forward difference approximation for y_i'' as

$$y_i'' = y''(x_i) = \frac{y'(x_{i+1}) - y'(x_i)}{h}, \tag{9.54}$$

Eq. (9.53) can be written as

$$y_{i+1} = y_i + hy_i' + \frac{h^2}{2}\left(\frac{y_{i+1}' - y_i'}{h}\right) + O(h^3)$$

or

$$y_{i+1} = y_i + h\left(\frac{y_i' + y_{i+1}'}{2}\right) + O(h^3). \tag{9.55}$$

Noting that $y_i' = f(x_i, y_i)$ and $y_{i+1}' = f(x_{i+1}, y_{i+1})$, we can see that Eq. (9.55), without the error term, is the same as Eq. (9.50) or Eq. (9.51). This shows that the local truncation error for Heun's method is of order $O(h^3)$, and hence, there is a significant improvement in accuracy compared to the basic Euler's method, which has a local truncation error of order $O(h^2)$. The global error for Heun's method will be of order $O(h^2)$; hence, the method is considered to be a second-order method.

▶Example 9.7

Find the solution of the problem

$$y' = -5y + e^{-2x};\ y(0) = 1.0 \tag{E1}$$

in the interval $0 \le x \le 1$ using Heun's method.

Solution

The interval $0 \le x \le 1$ is divided into 10 parts with $h = 0.1$ so that $x_0 = 0.0, x_1 = 0.1, \ldots, x_{10} = 1.0$, and $y_0 = 1.0$. By using the iterative process,

$$y_{i+1} = y_i + \frac{h}{2}\left\{f_i + f(x_i + h, y_i + hf_i)\right\}, \tag{E2}$$

Heun's method yields the results shown in Table 9.3. The exact solution of Eq. (E1) is known to be

$$y(x) = \tfrac{1}{3}(e^{-2x} + 2e^{-5x}). \tag{E3}$$

The exact values of y_i are also shown in Table 9.3. ◀

9.6.3 Modified Euler's Method (Improved Polygon Method)

Another simple modification of Euler's method involves using the derivative, $\frac{dy}{dx}$, at the midpoint of the interval $(x = x_i + \frac{h}{2})$. In this procedure, first Euler's method is

Table 9.3 Results given by Heun's method.

x_i	y_i given by Heun's method	Exact solution, y_i
0.0	1.000 0000	1.000 0000
0.1	0.690 9364	0.677 2639
0.2	0.485 8195	0.468 6931
0.3	0.347 8358	0.331 6906
0.4	0.253 5841	0.239 9997
0.5	0.188 1173	0.177 3498
0.6	0.141 8300	0.133 5893
0.7	0.108 5035	0.102 3305
0.8	0.084 0744	0.079 5092
0.9	0.065 8588	0.062 5056
1.0	0.052 0610	0.049 6037

used to find the value of y at the midpoint of the interval as

$$y_{i+\frac{1}{2}} = y\left(x_i + \frac{h}{2}\right) = y_i + \frac{h}{2} f(x_i, y_i). \tag{9.56}$$

Next, this value of $y_{i+\frac{1}{2}}$ is used to find the slope, $\frac{dy}{dx}$, at the midpoint as

$$y'_{i+\frac{1}{2}} = f\left(x_{i+\frac{1}{2}}, y_{i+\frac{1}{2}}\right). \tag{9.57}$$

Finally, this slope (Eq. (9.57)) is assumed to be a reasonable approximation for the average slope for the entire interval, and Euler's method is once again applied to find the value of y_{i+1} as

$$y_{i+1} = y_i + hf\left(x_{i+\frac{1}{2}}, y_{i+\frac{1}{2}}\right). \tag{9.58}$$

Note that since y_{i+1} is not appearing on both sides of Eq. (9.58) (as in the case of Eq. (9.50)), the solution (y_{i+1}) cannot be improved iteratively. The modified Euler's method is shown graphically in Fig. 9.9. As can be seen from this figure, the modified Euler's method is superior to the original Euler's method. The improvement essentially comes from the use of the slope at the midpoint instead of at the beginning of the interval. As in the case of Heun's method, the local truncation error of the modified Euler's method can be shown to be $O(h^3)$.

▶Example 9.8

Find the solution of the problem

$$y' = -5y + e^{-2x}; \; y(0) = 1.0 \tag{E1}$$

in the interval $0 \le x \le 1$ using the modified Euler's method.

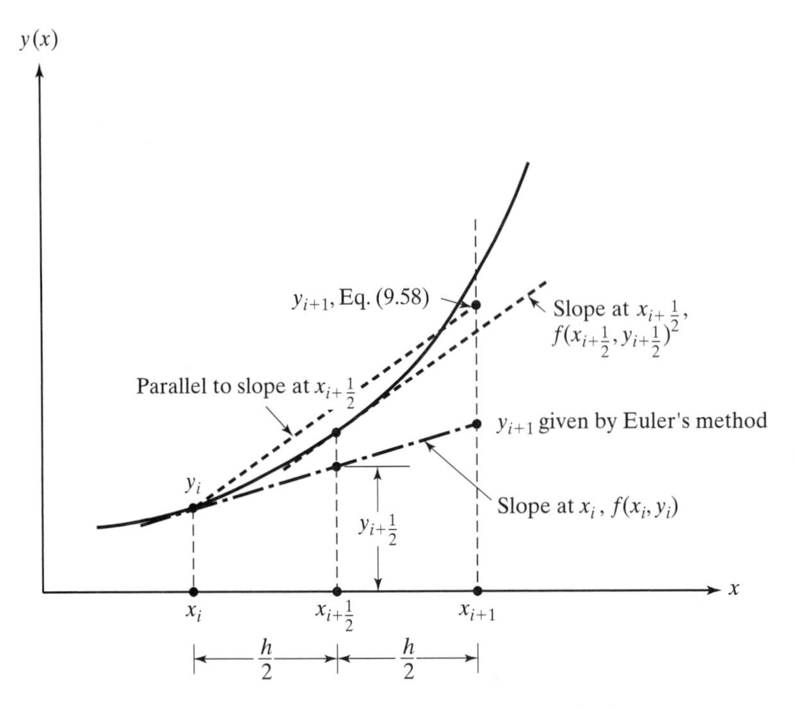

Figure 9.9 Graphical representation of modified Euler's method

Solution

In this case, the iterative process,

$$y_{i+1} = y_i + hf\left(x_i + \tfrac{1}{2}h,\, y_i + \tfrac{1}{2}hf_i\right), \tag{E2}$$

is used with $y_0 = 1.0$, $h = 0.1$, and $x_i = i(0.1)$. The numerical results are given in Table 9.4. The exact solution is also indicated in Table 9.4 for comparison (from Eq. (E3) of Example 9.7).

9.7 Runge–Kutta Methods

The high-order Taylor's series approaches outlined in Section 9.6.1 have a smaller local truncation error, but they require the computation and evaluation of the derivatives of $f(x, y)$. This can be a very complicated and time-consuming procedure for many problems and as such Taylor's series methods are not commonly used in practice. The Runge–Kutta methods do not require the derivatives of $f(x, y)$, but produce results equivalent in accuracy to the higher order Taylor formulas. The

Table 9.4 Results given by the modified Euler's method.

x_i	y_i given by modified Euler's method	Exact value, y_i
0.0	1.000 0000	1.000 0000
0.1	0.690 4836	0.677 2639
0.2	0.485 1658	0.468 6931
0.3	0.347 1236	0.331 6906
0.4	0.252 8905	0.239 9997
0.5	0.187 4802	0.177 3498
0.6	0.141 2652	0.133 5893
0.7	0.108 0140	0.102 3305
0.8	0.083 6568	0.079 5092
0.9	0.065 5065	0.062 5056
1.0	0.051 7659	0.049 6037

Runge–Kutta methods require only one initial point to start the procedure. The only disadvantage of Runge–Kutta methods is that they require several evaluations of $f(x, y)$ for each step of integration, which make them somewhat slower than the other methods. Although many variations exist, all the Runge–Kutta methods can be stated in the form

$$y_{i+1} = y_i + h\alpha(x_i, y_i, h), \tag{9.59}$$

where $\alpha(x_i, y_i, h)$ is called the increment function, which is chosen to represent the average slope over the interval $x_i \leq x \leq x_{i+1}$. The increment function can be expressed as

$$\alpha(x_i, y_i, h) = c_1 k_1 + c_2 k_2 + \cdots + c_n k_n, \tag{9.60}$$

where n denotes the order of the Runge–Kutta method; $c_1, c_2, \ldots, c_n$ are constants; and $k_1, k_2, \ldots$, and k_n are recurrence relations given by

$$k_1 = f(x_i, y_i), \tag{9.61}$$

$$k_2 = f(x_i + p_2 h, y_i + a_{21} h k_1), \tag{9.62}$$

$$k_3 = f(x_i + p_3 h, y_i + a_{31} h k_1 + a_{32} h k_2), \tag{9.63}$$

$$\vdots,$$

and

$$k_n = f(x_i + p_n h, y_i + a_{n1} h k_1 + a_{n2} h k_2 + \cdots + a_{n,n-1} h k_{n-1}). \tag{9.64}$$

Equations (9.61) through (9.64) denote recurrence relations, because k_1 appears in the equation for k_2, k_2 appears in the equation for k_3, and so forth. Equations (9.59) through (9.64) can be written in compact form as

$$y_{i+1} = y_i + h \sum_{j=1}^{n} c_j k_j, \tag{9.65}$$

where

$$k_j = f\left(x_i + p_j h,\, y_i + \sum_{l=1}^{j-1} a_{jl} k_l\right). \tag{9.66}$$

9.7.1 First-Order Runge–Kutta Method

For $n = 1$, Eqs. (9.65) and (9.66) reduce to

$$y_{i+1} = y_i + hc_1 k_1$$

and

$$k_1 = f(x_i, y_i), \tag{9.67}$$

which can be rewritten as

$$y_{i+1} = y_i + hf(x_i, y_i) \tag{9.68}$$

with $c_1 = 1$. The first order Runge–Kutta method (Eq. (9.68)) can be seen to be same as Euler's method.

9.7.2 Second-Order Runge–Kutta Method

The general procedure used for deriving a Runge–Kutta method can be described in six steps. These steps are described in the context of derivation of the second-order Runge–Kutta method.

Step 1 Select the value of n depending on the accuracy requirements of the numerical solution. For the second-order Runge–Kutta method, $n = 2$. Use Taylor's series expansion to express y_{i+1} using $(n + 1)$ terms as follows:

$$y_{i+1} = y_i + hy_i' + \frac{h^2}{2} y_i'' + O(h^3). \tag{9.69}$$

Step 2 Express each of the derivatives of y appearing in Eq. (9.69) by its equivalent in f, using the chain rule of differentiation for differentiating $f(x, y)$:

$$y_i' = \left.\frac{dy}{dx}\right|_{x_i} = f(x_i, y_i) \equiv f_i \tag{9.70}$$

and

$$y_i'' = \left.\frac{d^2y}{dx^2}\right|_{x_i} = \left.\frac{df}{dx}\right|_{x_i} = \left.\left(\frac{\partial f}{\partial x} + \frac{\partial f}{\partial y}\frac{dy}{dx}\right)\right|_{x_i} = (f_x + f_y f)_i. \tag{9.71}$$

Step 3 Use Eqs. (9.70) and (9.71) to express Eq. (9.69) as follows:

$$y_{i+1} = y_i + hf_i + \frac{h^2}{2} f_{x_i} + \frac{h^2}{2} f_i f_{y_i} + O(h^3). \tag{9.72}$$

Step 4 Express Eq. (9.65) using n terms in the summation:

$$y_{i+1} = y_i + hc_1 k_1 + hc_2 k_2, \tag{9.73}$$

where

$$k_1 = f(x_i, y_i) \tag{9.74}$$

and

$$k_2 = f(x_i + p_2 h, y_i + a_{21} h k_1). \tag{9.75}$$

Step 5 Use Taylor's series expansion for functions of two variables to expand the function(s) f appearing in Eq. (9.75):

$$f(x_i + p_2 h, y_i + a_{21} h k_1) = k_2 = f_i + p_2 h f_{x_i} + a_{21} h f_{y_i} f_i + O(h^2), \tag{9.76}$$

that is,

$$y_{i+1} = y_i + h f_i (c_1 + c_2) + h^2 f_{x_i} (c_2 p_2) + h^2 f_i f_{y_i} (c_2 a_{21}) + O(h^3). \tag{9.77}$$

Step 6 Since both Eqs. (9.72) and (9.77) express the same quantity, y_{i+1}, the coefficients of the corresponding terms must be same. This gives a system of nonlinear algebraic equations in terms of the unknown constants c_j, p_j, and a_{jl}. In the present case, we obtain

$$c_1 + c_2 = 1,$$

$$c_2 p_2 = \frac{1}{2}, \tag{9.78}$$

and

$$c_2 a_{21} = \frac{1}{2}.$$

It can be seen that there are four unknowns (c_1, c_2, p_2, and a_{21}) and three equations. Hence, this method is said to have one degree of freedom. The third- and fourth-order Runge–Kutta methods have two degrees of freedom. For fifth-order Runge–Kutta method, there are at least five degrees of freedom. Since there are more unknowns than the number of equations, there is no unique solution to Eq. (9.78). The degrees of freedom permit us to choose some of the constants (parameters) and determine the others. Thus, there is a family of nth-order Runge–Kutta methods instead of a single version.

In the present case (second-order Runge–Kutta method), let us specify the value of c_2. Then Eq. (9.78) yield

$$c_1 = 1 - c_2, \; p_2 = \frac{1}{2c_2}, \; a_{21} = \frac{1}{2c_2}. \tag{9.79}$$

We now consider three of the most commonly used versions of second-order Runge–Kutta method.

Modified Euler's Method

Let $c_2 = 1$. Equation (9.79) gives $c_1 = 0$, $p_2 = \frac{1}{2}$, and $a_{21} = \frac{1}{2}$. Thus, the solution, Eqs. (9.73) through (9.75), is given by

$$y_{i+1} = y_i + h k_2$$

with

$$k_1 = f(x_i, y_i) \tag{9.80}$$

and

$$k_2 = f\left(x_i + \frac{h}{2}, y_i + \frac{1}{2}hk_1\right).$$

This method, Eq. (9.80), can be seen to be the modified Euler's method or the improved polygon method.

Heun's Method

Let $c_2 = \frac{1}{2}$. Equations (9.79) give $c_1 = \frac{1}{2}$, $p_2 = 1$ and $a_{21} = 1$. The solution, Eqs. (9.73) to (9.75), becomes

$$y_{i+1} = y_i + \tfrac{1}{2}h(k_1 + k_2)$$

with

$$k_1 = f(x_i, y_i) \tag{9.81}$$

and

$$k_2 = f(x_i + h, y_i + hk_1),$$

which can be seen to be same as Heun's method or Euler's predictor–corrector method.

Ralston's Method

Let $c_2 = \frac{2}{3}$. Equations (9.79) yield $c_1 = \frac{1}{3}$, $p_2 = \frac{3}{4}$, and $a_{21} = \frac{3}{4}$. The solution, Eqs. (9.73) through (9.75), becomes

$$y_{i+1} = y_i + \tfrac{1}{3}h(k_1 + 2k_2)$$

with

$$k_1 = f(x_i, y_i) \tag{9.82}$$

and

$$k_2 = f\left(x_i + \tfrac{3}{4}h, y_i + \tfrac{3}{4}hk_1\right).$$

9.7.3 Third-Order Runge–Kutta Method

For the third-order Runge–Kutta method ($n = 3$), the derivation is similar to the one described for the second-order method. The procedure yields six equations and eight unknowns. Thus, the values of two constants are selected, and the remaining ones determined. One of the popular versions of the third-order Runge–Kutta method is as follows:

$$y_{i+1} = y_i + \frac{h}{6}(k_1 + 4k_2 + k_3),$$

where

$$k_1 = f(x_i, y_i), \tag{9.83}$$
$$k_2 = f\left(x_i + \tfrac{1}{2}h, y_i + \tfrac{1}{2}hk_1\right),$$

and

$$k_3 = f(x_i + h, y_i - hk_1 + 2hk_2).$$

In cases where $f(x, y)$ is a function of x only, the third-order Runge–Kutta method reduces to Simpson's one-third rule. It can be noted that Eq. (9.83) will yield exact results when $f(x, y) = f(x)$ is a cubic and the solution $y(x)$ is a quartic equation in x. The reason for this is that Simpson's rule gives the exact result when used for integration of cubic polynomials.

9.7.4 Fourth-Order Runge–Kutta Methods

The fourth-order Runge–Kutta methods are accurate to the fourth-order term in Taylor's series expansion, and hence, the local truncation error is $O(h^5)$. The fourth-order Runge–Kutta methods are most popularly used. There are several versions of the method and the following are commonly used in practice;

Method Attributed to Runge

The iterative process is given by

$$y_{i+1} = y_i + \frac{h}{6}(k_1 + 2k_2 + 2k_3 + k_4),$$

where

$$k_1 = f(x_i, y_i),$$
$$k_2 = f\left(x_i + \tfrac{1}{2}h, y_i + \tfrac{1}{2}hk_1\right), \tag{9.84}$$
$$k_3 = f\left(x_i + \tfrac{1}{2}h, y_i + \tfrac{1}{2}hk_2\right),$$

and

$$k_4 = f(x_i + h, y_i + hk_3).$$

This method reduces to Simpson's one-third rule when $f(x, y) = f(x)$.

Method Attributed to Kutta

The following describes the method attributed to Kutta:

$$y_{i+1} = y_i + \frac{h}{8}(k_1 + 3k_2 + 3k_3 + k_4),$$

where

$$
\begin{aligned}
k_1 &= f(x_i, y_i), \\
k_2 &= f\left(x_i + \tfrac{1}{3}h,\, y_i + \tfrac{1}{3}hk_1\right), \\
k_3 &= f\left(x_i + \tfrac{2}{3}h,\, y_i - \tfrac{1}{3}hk_1 + hk_2\right),
\end{aligned}
\tag{9.85}
$$

and

$$k_4 = f(x_i + h,\, y_i + hk_1 - hk_2 + hk_3).$$

This method reduces to Simpson's 3/8 rule when $f(x, y) = f(x)$.

Runge–Kutta–Gill Method

This is the most widely used fourth-order method, and the constants are selected to reduce the amount of storage required in the solution of a large number of simultaneous first-order differential equations. The method can be described as follows:

$$y_{i+1} = y_i + \frac{h}{6}\left[k_1 + 2\left(1 - \frac{1}{\sqrt{2}}\right)k_2 + 2\left(1 + \frac{1}{\sqrt{2}}\right)k_3 + k_4\right],$$

where

$$
\begin{aligned}
k_1 &= f(x_i, y_i), \\
k_2 &= f\left(x_i + \frac{1}{2}h,\, y_i + \frac{1}{2}hk_1\right), \\
k_3 &= f\left(x_i + \frac{1}{2}h,\, y_i + \left(-\frac{1}{2} + \frac{1}{\sqrt{2}}\right)hk_1 + \left(1 - \frac{1}{\sqrt{2}}\right)hk_2\right),
\end{aligned}
\tag{9.86}
$$

and

$$k_4 = f\left(x_i + h,\, y_i - \frac{1}{\sqrt{2}}hk_2 + \left(1 + \frac{1}{\sqrt{2}}\right)hk_3\right).$$

▶Example 9.9

Find the solution of the initial value problem

$$y' = y + 2x - 1;\ y(0) = 1 \tag{E1}$$

in the interval $0 \le x \le 1$ using the fourth-order Runge–Kutta methods.

Solution

The interval $0 \le x \le 1$ is divided into 10 steps so that $x_0 = 0.0, x_1 = 0.1, \ldots,$ and $x_{10} = 1.0$ with $h = 0.1$ and $y_0 = y(x = 0) = 1.0$. The iterative formulas given by Eqs. (9.84) to (9.86) are used to obtain the results shown in Table 9.5. ◀

Table 9.5 Results given by the fourth-order Runge–Kutta methods.

x_i	y_i given by the method of			Exact value, y_i
	Runge (Eq. 9.84)	Kutta (Eq. 9.85)	Runge–Kutta–Gill (Eq. 9.86)	
0.0	1.0	1.0	1.0	1.0
0.1	1.0103 4164	1.0103 4164	1.0103 4164	1.0103 4260
0.2	1.0428 0472	1.0428 0472	1.0428 0472	1.0428 0663
0.3	1.0997 1619	1.0997 1619	1.0997 1619	1.0997 1809
0.4	1.1836 4716	1.1836 4811	1.1836 4811	1.1836 5002
0.5	1.2974 3958	1.2974 4053	1.2974 4148	1.2974 4339
0.6	1.4442 3389	1.4442 3485	1.4442 3580	1.4442 3676
0.7	1.6275 0149	1.6275 0244	1.6275 0340	1.6275 0530
0.8	1.8510 7708	1.8510 7803	1.8510 7994	1.8510 8185
0.9	2.1192 0071	2.1192 0166	2.1192 0357	2.1192 0643
1.0	2.4365 5777	2.4365 5777	2.4365 6063	2.4365 6349

9.7.5 Error Analysis of Runge–Kutta Methods

According to the procedure outlined earlier, the nth-order Runge–Kutta method was generated by requiring it to be the same as Taylor's series expansion of the solution, $y(x)$, through terms of order h^n. Thus, the local truncation error (e) can be expressed as

$$e = Ch^{n+1} + O(h^{n+2}), \tag{9.87}$$

where C is a constant whose value depends on the nature of the function $f(x, y)$ and its higher order derivatives. It is desirable to use a small step size (h) to make the truncation error small, but the step size should be large to make the round-off error small. In addition, it is not desirable to use a method that requires the evaluation of a large number of derivatives because they require substantial computational time. Note that the nth-order Runge–Kutta method requires the evaluation of the derivative n times for each integration step.

9.7.6 Error Estimation Using Richardson's Extrapolation

Richardson's extrapolation described earlier (see Section 8.6) can be used to derive an estimate of the error in Runge–Kutta methods. For this, let the values of y be found as $y_{i+1}^{(1)}$ and $y_{i+1}^{(2)}$ by integrating the differential equation between two points, x_i and x_{i+1}, using two different step sizes $h_1 = h$ and $h_2 = \frac{h}{2}$, respectively. Let the exact solution of the differential equation at $x = x_{i+1}$ be $\bar{y}_{i+1}$. Then, by considering only the dominant term in Eq. (9.87), we can write

$$\bar{y}_{i+1} - y_{i+1}^{(1)} = Ch_1^{n+1} \tag{9.88}$$

and

$$\bar{y}_{i+1} - y_{i+1}^{(2)} = Ch_2^{n+1}\left(\frac{h_1}{h_2}\right) = 2Ch_2^{n+1}, \tag{9.89}$$

where the factor $\frac{h_1}{h_2} = 2$ is included in Eq. (9.89) due to the accumulation of error in two steps of size h_2 (from x_i to x_{i+1}). Dividing Eq. (9.88) by Eq. (9.89) yields

$$\frac{\bar{y}_{i+1} - y_{i+1}^{(1)}}{\bar{y}_{i+1} - y_{i+1}^{(2)}} = \frac{1}{2}\left(\frac{h_1}{h_2}\right)^{n+1} = 2^n$$

or

$$\bar{y}_{i+1} = \frac{y_{i+1}^{(1)} - 2^n y_{i+1}^{(2)}}{1 - 2^n}. \tag{9.90}$$

Equation (9.90) gives a more accurate approximation to the solution and can be used to find the local truncation error as

$$e = \bar{y}_{i+1} - y_{i+1}^{(1)} = \frac{2^n\left(y_{i+1}^{(2)} - y_{i+1}^{(1)}\right)}{2^n - 1}. \tag{9.91}$$

For the second-, third-, and fourth-order Runge–Kutta methods, Eq. (9.91) gives the error as

$$e = \begin{cases} \frac{4}{3}\left(y_{i+1}^{(2)} - y_{i+1}^{(1)}\right) & \text{for second-order method} \\ \frac{8}{7}\left(y_{i+1}^{(2)} - y_{i+1}^{(1)}\right) & \text{for third-order method} \\ \frac{16}{15}\left(y_{i+1}^{(2)} - y_{i+1}^{(1)}\right) & \text{for fourth-order method} \end{cases}. \tag{9.92}$$

Equation (9.92) can be used to determine the maximum step size (h_{max}) that can be used to satisfy a specified accuracy requirement, ε. For this, we use Eqs. (9.91) and (9.88) to express C as

$$C = \frac{2^n}{(2^n - 1)} \frac{\left(y_{i+1}^{(2)} - y_{i+1}^{(1)}\right)}{h_1^{(n+1)}}. \tag{9.93}$$

Since the error with the step size, h_{max}, is required to be less than or equal to ε, we have

$$Ch_{max}^{(n+1)} = \frac{2^n}{(2^n - 1)} \frac{\left(y_{i+1}^{(2)} - y_{i+1}^{(1)}\right)}{h_1^{(n+1)}} h_{max}^{(n+1)} \leq \varepsilon,$$

and hence,

$$h_{max} \leq h_1 \left\{\frac{\varepsilon(2^n - 1)}{2^n\left(y_{i+1}^{(2)} - y_{i+1}^{(1)}\right)}\right\}^{\frac{1}{n+1}}. \tag{9.94}$$

Equation (9.94) gives the step size to achieve any specified accuracy. However, the use of Eq. (9.94) at each step increases the total number of calculations nearly three times compared to the number of calculations required with the use of just one step size h_1. One possible compromise is to adjust the step size after every m steps where m is chosen arbitrarily.

9.7.7 Runge–Kutta–Fehlberg Method

Fehlberg used two Runge–Kutta methods of different orders to find two different solutions and then subtracted the results to find an estimate of error [9.2–9.4]. In particular, the method finds solutions using the following fourth- and fifth-order Runge–Kutta methods:

Fourth-order Method

$$y_{i+1} = y_i + h\left(\frac{25}{216}k_1 + \frac{1408}{2565}k_3 + \frac{2197}{4104}k_4 - \frac{1}{5}k_5\right). \tag{9.95}$$

Fifth-order Method

$$y_{i+1} = y_i + h\left(\frac{16}{135}k_1 + \frac{6656}{12825}k_3 + \frac{28561}{56430}k_4 - \frac{9}{50}k_5 + \frac{2}{55}k_6\right), \tag{9.96}$$

where

$$\begin{aligned}
k_1 &= f(x_i, y_i),\\
k_2 &= f\left(x_i + \frac{1}{4}h,\ y_i + \frac{1}{4}hk_1\right),\\
k_3 &= f\left(x_i + \frac{3}{8}h,\ y_i + \frac{3}{32}hk_1 + \frac{9}{32}hk_2\right),\\
k_4 &= f\left(x_i + \frac{12}{13}h,\ y_i + \frac{1932}{2197}hk_1 - \frac{7200}{2197}hk_2 + \frac{7296}{2197}hk_3\right),\\
k_5 &= f\left(x_i + h,\ y_i + \frac{439}{216}hk_1 - 8hk_2 + \frac{3680}{513}hk_3 - \frac{845}{4104}hk_4\right),
\end{aligned} \tag{9.97}$$

and

$$k_6 = f\left(x_i + \frac{1}{2}h,\ y_i - \frac{8}{27}hk_1 + 2hk_2 - \frac{3544}{2565}hk_3 + \frac{1859}{4104}hk_4 - \frac{11}{40}hk_5\right).$$

The error estimate (e) is found by subtracting Eq. (9.95) from Eq. (9.96) to yield

$$e = h\left(\frac{1}{360}k_1 - \frac{128}{4275}k_2 - \frac{2197}{75240}k_3 + \frac{1}{50}k_5 + \frac{2}{55}k_6\right). \tag{9.98}$$

It can be seen that the Runge–Kutta–Fehlberg method appears to be more complex than the classical fourth-order Runge–Kutta method. However, the method is useful in finding good error estimates.

Another procedure to adjust the step size was given by Collatz [9.10]. According to this procedure, the ratio,

$$r = \left|\frac{k_3 - k_2}{k_2 - k_1}\right|$$

is computed using Eq. (9.84) after each step. A large value of r on the order of few hundredths indicates that the error is too large, and hence, the step size is to be reduced. Some other procedures that can be used to control the error in Runge–Kutta methods are given in Ref. [9.11].

▶Example 9.10

Find the solution of the initial value problem

$$y' = y + 2x - 1; \; y(0) = 1.0 \tag{E1}$$

in the interval $0 \le x \le 1$ using the Runge–Kutta–Fehlberg method.

Solution

The given interval $0 \le x \le 1$ is divided into 10 uniform steps of size $h = 0.1$ with $x_0 = 0.0, x_1 = 0.1, \ldots, x_{10} = 1.0$. Using the initial value $y_0 = 1.0$, the fourth- and fifth-order Runge–Kutta–Fehlberg iterative algorithms, Eqs. (9.95) and (9.96), yield the results indicated in Table 9.6. ◀

9.7.8 Stability of Runge–Kutta Methods

A numerical solution is considered to be unstable if the errors introduced in each step are propagated without a bound. Certain equations with specified initial conditions are known to be inherently unstable implying that no numerical procedure can generate a stable solution. Thus, inherent instability is independent of the procedure used for solution. Another type of instability, known as partial instability, may make the solution unstable even when the equation is not inherently unstable. Hence, partial instability depends on the method of solution used. It has been shown [9.11] that for the differential equation given by Eqs. (9.13) and (9.14), the solution will be unstable

Table 9.6 Results given by Runge–Kutta–Fehlberg method.

x_i	y_i given by Runge–Kutta–Fehlberg method		Exact solution, y_i
	Fourth order, Eq. (9.95)	Fifth order, Eq. (9.96)	
0.0	1.0	1.0	1.0
0.1	1.0103 4164	1.0103 4164	1.0103 4260
0.2	1.0428 0567	1.0428 0567	1.0428 0663
0.3	1.0997 1809	1.0997 1809	1.0997 1809
0.4	1.1836 5002	1.1836 5002	1.1836 5002
0.5	1.2974 4339	1.2974 4339	1.2974 4339
0.6	1.4442 3866	1.4442 3866	1.4442 3676
0.7	1.6275 0721	1.6275 0626	1.6275 0530
0.8	1.8510 8376	1.8510 8280	1.8510 8185
0.9	2.1192 0834	2.1192 0738	2.1192 0643
1.0	2.4365 6635	2.4365 6540	2.4365 6349

if $\frac{\partial f(x,y)}{\partial y}$ is negative. On the other hand, if $\frac{\partial f(x,y)}{\partial y}$ is positive, the solution may become unstable. However, by choosing a very small value of the step size (h), the instability can be avoided even when $\frac{\partial f(x,y)}{\partial y}$ is positive. In practice, we can find the solution of the differential equation using different step sizes and if the resulting solutions are sufficiently close to one another, the procedure can be assumed to be stable.

9.8 Multistep Methods

The methods considered in the previous sections require information about the solution at a single point $x = x_i$ to find the solution y_{i+1} at the next point $x = x_{i+1}$. As such, those methods are called *single-step methods*. The implication is that they are self-starting and the step size (h) can be changed between iterations. There are some methods, known as *multistep methods*, which require information about the solution at more than one point to find the solution y_{i+1} at the next point x_{i+1}. Let us assume that the solution has already been found at the points $x_0, x_1, x_2, \ldots, x_n$ as $y_0, y_1, y_2, \ldots, y_n$, respectively. In the multistep methods, first an interpolating polynomial passing through $(k + 1)$ known points $y'_n, y'_{n-1}, \ldots, y'_{n-k}$ is found and then the resulting polynomial is extrapolated over the next interval of integration, x_i to $x_{i+1}$to find y'_{n+1}. A major disadvantage of multistep methods is that they are not self-starting. The reason is that only the initial condition $y(x_0) = y_0$ is known and the solution at the earlier points, $x_{-1}, x_{-2}, \ldots$ is not known while starting at x_0. Hence, usually, a self-starting method that has the same order of accuracy as that of the multistep method, such as a Runge–Kutta method, is used in the first few steps to generate the solution needed for starting the multistep method. Also, since most interpolation polynomials are based on a fixed-step size, it is not possible to change the step size during the iterative process of a multistep method.

As will be shown later, the multistep formulas require only one derivative evaluation per step, compared to four evaluations per step with the fourth-order Runge–Kutta methods, and hence, they require less computational effort and are faster. There are two types of multistep methods, namely, the open or explicit methods and the closed or implicit methods. If y_{i+1} is expressed explicitly in terms of the known values of the dependent variable y and the function $f(x, y)$ at $x_i, x_{i-1}, x_{i-2}, \ldots$, the method is known as the open or explicit method. On the other hand, if y_{i+1} is expressed in terms of a function that involves the unknown solution y_{i+1} along with the known values $y_i, y_{i-1}, \ldots$, the solution y_{i+1} has to be found iteratively, and the method is known as the closed or implicit method. Both the open and closed methods are considered in the following section.

9.9 Adams Methods

Adams methods can be derived in a variety of ways. We use Taylor's series expansion of the dependent variable y around x_i to derive Adams methods.

9.9.1 Adams–Bashforth Open Formulas

Taylor's series expansion of y around x_i can be used to express y_{i+1} as

$$y_{i+1} = y_i + hf_i + \frac{h^2}{2!} f_i' + \frac{h^3}{3!} f_i'' + \frac{h^4}{4!} f_i''' + \cdots, \tag{9.99}$$

where

$$f_i = \left.\frac{dy}{dx}\right|_{x_i}, \; f_i' = \left.\frac{d^2y}{dx^2}\right|_{x_i}, \; f_i'' = \left.\frac{d^3y}{dx^3}\right|_{x_i}, \ldots . \tag{9.100}$$

A series of formulas can be developed by retaining different number of terms in Eq. (9.99) and using finite difference approximations for the derivatives of f_i. The backward differences are used to derive the Adams–Bashforth or Adams open formulas. When only the first three terms are retained in Eq. (9.99), we obtain the first-order Adams–Bashforth open formula as

$$y_{i+1} = y_i + hf_i + \frac{h^2}{2} f_i' + O(h^3). \tag{9.101}$$

When the first four terms are retained, Eq. (9.99) gives

$$y_{i+1} = y_i + hf_i + \frac{h^2}{2} f_i' + \frac{h^3}{6} f_i'' + O(h^4). \tag{9.102}$$

By using the backward difference formula for f_i' (Eq. (7.23))

$$f_i' = \frac{f_i - f_{i-1}}{h} + \frac{h}{2} f_i'' + O(h^2), \tag{9.103}$$

Eq. (9.102) yields

$$y_{i+1} = y_i + hf_i + \frac{h^2}{2}\left[\frac{f_i - f_{i-1}}{h} + \frac{h}{2} f_i'' + O(h^2)\right] + \frac{h^3}{6} f_i'' + O(h^4), \tag{9.104}$$

which can be rearranged to obtain the second-order open Adams–Bashforth formula as

$$y_{i+1} = y_i + h\left(\frac{3}{2} f_i - \frac{1}{2} f_{i-1}\right) + \frac{5}{12} h^3 f_i'' + O(h^4). \tag{9.105}$$

When the first four terms are retained in Eq. (9.99), we obtain

$$y_{i+1} = y_i + hf_i + \frac{h^2}{2} f_i' + \frac{h^3}{6} f_i'' + O(h^4). \tag{9.106}$$

By using backward difference formulas for f_i' (from Table 7.2) and f_i'' (from Eq. (7.34)), Eq. (9.106) gives

$$y_{i+1} = y_i + hf_i + \frac{h^2}{2}\left[\frac{3f_i - 4f_{i-1} + f_{i-2}}{2h} + O(h^2)\right]$$

$$+ \frac{h^3}{6}\left[\frac{f_i - 2f_{i-1} + f_{i-2}}{h^2} + O(h)\right] + O(h^4),$$

Table 9.7 Coefficients α_{nk} and local truncation error for Adams–Bashforth Open Formulas.

Order of the formula (n)	$k=1$ α_{n1}	$k=2$ α_{n2}	$k=3$ α_{n3}	$k=4$ α_{n4}	$k=5$ α_{n5}	$k=6$ α_{n6}	Local truncation error, $O(h^{n+1})$
1	1						$\frac{1}{2}h^2 f'(\xi)$
2	$\frac{3}{2}$	$-\frac{1}{2}$					$\frac{5}{12}h^3 f''(\xi)$
3	$\frac{23}{12}$	$-\frac{16}{12}$	$\frac{5}{12}$				$\frac{9}{24}h^4 f'''(\xi)$
4	$\frac{55}{24}$	$-\frac{59}{24}$	$\frac{37}{24}$	$-\frac{9}{24}$			$\frac{251}{720}h^5 f^{(4)}(\xi)$
5	$\frac{1901}{720}$	$-\frac{2774}{720}$	$\frac{2616}{720}$	$-\frac{1274}{720}$	$\frac{251}{720}$		$\frac{475}{1440}h^6 f^{(5)}(\xi)$
6	$\frac{4277}{1440}$	$-\frac{7923}{1440}$	$\frac{9982}{1440}$	$-\frac{7298}{1440}$	$\frac{2877}{1440}$	$-\frac{475}{1440}$	$\frac{19087}{60480}h^7 f^{(6)}(\xi)$

that is,

$$y_{i+1} = y_i + h\left(\frac{23}{12}f_i - \frac{16}{12}f_{i-1} + \frac{5}{12}f_{i-2}\right) + O(h^4). \tag{9.107}$$

Note that a higher accuracy formula was used for f_i' compared with f_i'' in order to have the same order of error in each of the terms involving derivatives of f_i in Eq. (9.106). Equation (9.107) denotes the third-order Adams–Bashforth open formula. The higher order Adams–Bashforth formulas can be developed by using the same procedure. The general expression for the nth-order Adams–Bashforth open formula can be represented as

$$y_{i+1} = y_i + h\sum_{k=1}^{n} \alpha_{nk} f_{i-k+1} + O(h^{n+1}), \tag{9.108}$$

where the coefficients α_{nk} for different values of n are given in Table 9.7.

▶Example 9.11

Find the solution of the initial value problem

$$y' = y + 2x - 1;\ y(0) = 1 \tag{E1}$$

in the interval $0 \le x \le 1$ using Adams–Bashforth open formulas of orders 2 through 6.

Solution

Using a step size of $h = 0.1$ with $x_0 = 0.0, x_1 = 0.1, \ldots,$ and $x_{10} = 1.0$ and $y_0 = 1.0$, the iterative processes of Adams–Bashforth formulas (Eq. (9.108)) are applied as follows:

when $n = 2$,

$$y_{i+1} = y_i + \frac{h}{2}(3f_i - f_{i-1}). \tag{E2}$$

when $n = 3$,

$$y_{i+1} = y_i + \frac{h}{12}(23f_i - 16f_{i-1} + 5f_{i-2}). \tag{E3}$$

when $n = 4$,

$$y_{i+1} = y_i + \frac{h}{24}(55f_i - 59f_{i-1} + 37f_{i-2} - 9f_{i-3}). \tag{E4}$$

when $n = 5$,

$$y_{i+1} = y_i + \frac{h}{720}(1901f_i - 2774f_{i-1} + 2616f_{i-2} - 1274f_{i-3} + 251f_{i-4}). \tag{E5}$$

when $n = 6$,

$$y_{i+1} = y_i + \frac{h}{1440}(4277f_i - 7923f_{i-1} + 9982f_{i-2} - 7298f_{i-3} + 2877f_{i-4} - 475f_{i-5}). \tag{E6}$$

The values computed from the exact solution,

$$y(x) = -1 - 2x + 2e^x, \tag{E7}$$

are given as the starting values in the iterative process. For example, the formula corresponding to $n = 5$ requires the values of y at $x = 0, 0.1, 0.2, 0.3,$ and 0.4 to use Eq. (E5). These values are computed from Eq. (E7). Also it is to be noted that the Adams–Bashforth formula of order $n = 1$ is same as Euler's method. The numerical results given by Eqs. (E2) through (E6) are shown in Table 9.8. ◀

Table 9.8 Results given by Adams–Bashforth formulas.

x_i	y_i given by Adams–Bashforth formula					Exact solution, y_i
	$n = 2$	$n = 3$	$n = 4$	$n = 5$	$n = 6$	
0.0	1.0	1.0	1.0	1.0	1.0	1.0
0.1	1.010 3426	1.010 3426	1.010 3426	1.010 3426	1.010 3426	1.010 3426
0.2	1.011 8942	1.042 8066	1.042 8066	1.042 8066	1.042 8066	1.042 8066
0.3	1.073 1611	1.099 6323	1.099 7181	1.099 7181	1.099 7181	1.099 7181
0.4	1.153 5406	1.183 4517	1.183 6414	1.183 6500	1.183 6500	1.183 6500
0.5	1.262 9137	1.297 1125	1.297 4234	1.297 4424	1.297 4434	1.297 4434
0.6	1.404 6736	1.443 7504	1.444 2053	1.444 2368	1.444 2387	1.444 2368
0.7	1.582 2287	1.626 8320	1.627 4576	1.627 5034	1.627 5072	1.627 5053
0.8	1.799 3298	1.850 1883	1.851 0160	1.851 0780	1.851 0838	1.851 0818
0.9	2.060 1177	2.118 0534	2.119 1187	2.119 2007	2.119 2083	2.119 2064
1.0	2.369 1692	2.435 1063	2.436 4510	2.436 5559	2.436 5664	2.436 5635

9.9.2 Adams–Moulton Closed Formulas

Taylor's series expansion of y around x_{i+1} can be used to express y_i (with a negative step length) as

$$y_i = y_{i+1} - hy'_{i+1} + \frac{h^2}{2!}y''_{i+1} - \frac{h^3}{3!}y'''_{i+1} + \cdots, \tag{9.109}$$

which can be rewritten as

$$y_i = y_{i+1} - hf_{i+1} + \frac{h^2}{2!}f'_{i+1} - \frac{h^3}{3!}f''_{i+1} + \cdots \tag{9.110}$$

Equation (9.110) gives, after rearrangement of terms,

$$y_{i+1} = y_i + hf_{i+1} - \frac{h^2}{2!}f'_{i+1} + \frac{h^3}{3!}f''_{i+1} - \cdots \tag{9.111}$$

As in the case of open formulas, a series of formulas can be developed by retaining different number of terms in Eq. (9.111) and using finite-difference approximations for the derivatives of f_i. By using backward differences, we obtain the Adams–Moulton or Adams closed formulas. When only the first two terms are retained in Eq. (9.111), we obtain the first-order formula

$$y_{i+1} = y_i + hf_{i+1}, \tag{9.112}$$

where the local truncation error can be seen by comparing Eqs. (9.111) and (9.112), as

$$-\frac{h^2}{2}f'(\xi);\ x_i < \xi < x_{i+1} \text{ or } O(h^2).$$

Since $f_{i+1} = f(x_{i+1}, y_{i+1})$, y_{i+1} appears on the right-hand side of Eq. (9.112), and hence, the formula is an implicit one. In general, when $f(x, y)$ is a nonlinear function, Eq. (9.112) cannot be solved for y_{i+1} exactly. An iterative procedure is to be used to find y_{i+1}. To start the iterative procedure, x_i is fixed and an estimate of y_{i+1} is obtained as

$$y_{i+1}^{(1)} = y_i + h\left[f(x_i, y_i)\right]. \tag{9.113}$$

Subsequently, the improved values of y_{i+1} are determined using the iterative procedure as

$$y_{i+1}^{(j+1)} = y_i + \frac{h}{2}\left[f(x_i, y_i) + f(x_{i+1}, y_{i+1}^{(j)})\right];\ j = 1, 2, \ldots. \tag{9.114}$$

This iterative process is terminated when the following convergence criterion is satisfied:

$$\left|\frac{y_{i+1}^{(j+1)} - y_{i+1}^{(j)}}{y_{i+1}^{(j)}}\right| \le \varepsilon. \tag{9.115}$$

Here, ε is a small number. When the first three terms are retained, Eq. (9.111) gives

$$y_{i+1} = y_i + hf_{i+1} - \frac{h^2}{2}f'_{i+1} + \frac{h^3}{6}f''_{i+1} + O(h^4). \tag{9.116}$$

By substituting the backward-difference approximation for f'_{i+1}, when we use Eq. (7.23), Eq. (9.116) yields

$$y_{i+1} = y_i + hf_{i+1} - \frac{h^2}{2}\left[\frac{f_{i+1} - f_i}{h} + \frac{h}{2}f''_{i+1} + O(h^2)\right] + \frac{h^3}{6}f''_{i+1} + O(h^4), \quad (9.117)$$

that is,

$$y_{i+1} = y_i + h\left(\frac{1}{2}f_i + \frac{1}{2}f_{i+1}\right) - \frac{1}{12}h^3 f''_{i+1} + O(h^4). \quad (9.118)$$

Thus, the second-order Adams–Moulton closed formula is given by

$$y_{i+1} = y_i + h\left(\frac{1}{2}f_i + \frac{1}{2}f_{i+1}\right) \quad (9.119)$$

with a local truncation error of

$$-\frac{1}{2}h^3 f''_{i+1}(\xi);\ x_i < \xi < x_{i+1} \quad \text{or} \quad O(h^3).$$

Similarly, when the first four terms are retained, Eq. (9.111) gives

$$y_{i+1} = y_i + hf_{i+1} - \frac{h^2}{2}f'_{i+1} + \frac{h^3}{6}f''_{i+1} - \frac{h^4}{24}f'''_{i+1} + \cdots. \quad (9.120)$$

By substituting a higher accuracy backward-difference formula for f'_{i+1} (from Table 7.2) and the common backward-difference formula for f''_{i+1} (from Table 7.1) into Eq. (9.120) yields

$$y_{i+1} = y_i + hf_{i+1} - \frac{h^2}{2}\left[\frac{3f_{i+1} - 4f_i + f_{i-1}}{2h} + O(h^2)\right]$$
$$+\frac{h^3}{6}\left[\frac{f_{i+1} - 2f_i + f_{i-1}}{h^2} + O(h)\right] + O(h^4), \quad (9.121)$$

that is,

$$y_{i+1} = y_i + h\left[\frac{5}{12}f_{i+1} + \frac{8}{12}f_i - \frac{1}{12}f_{i-1}\right] + O(h^4). \quad (9.122)$$

Equation (9.122), without the error term, denotes the third-order Adams–Moulton closed formula. A similar procedure can be used to derive the higher order Adams–Moulton formulas. The general expression for the nth-order Adams–Moulton formula can be written as

$$y_{i+1} = y_i + h\sum_{k=1}^{n}\alpha_{nk} f_{i-k+2} + O(h^{n+1}), \quad (9.123)$$

where the coefficients α_{nk} for different values of n are given in Table 9.9. Note that all the Adams–Moulton formulas are implicit, and hence, an iterative procedure similar to the one outlined in Eqs. (9.113) and (9.114) is to be used with any formula.

Table 9.9 Coefficients α_{nk} and local truncation error for Adams–Moulton closed formulas.

Order of the formula (n)	$k=1$ α_{n1}	$k=2$ α_{n2}	$k=3$ α_{n3}	$k=4$ α_{n4}	$k=5$ α_{n5}	$k=6$ α_{n6}	Local truncation error, $O(h^{n+1})$
1	1						$-\frac{1}{2}h^2 f'(\xi)$
2	$\frac{1}{2}$	$\frac{1}{2}$					$-\frac{1}{12}h^3 f''(\xi)$
3	$\frac{5}{12}$	$\frac{8}{12}$	$-\frac{1}{12}$				$-\frac{1}{24}h^4 f'''(\xi)$
4	$\frac{9}{24}$	$\frac{19}{24}$	$-\frac{5}{24}$	$\frac{1}{24}$			$-\frac{19}{720}h^5 f^{(4)}(\xi)$
5	$\frac{251}{720}$	$\frac{646}{720}$	$-\frac{264}{720}$	$\frac{106}{720}$	$-\frac{19}{720}$		$-\frac{27}{1440}h^6 f^{(5)}(\xi)$
6	$\frac{475}{1440}$	$\frac{1427}{1440}$	$-\frac{798}{1440}$	$\frac{482}{1440}$	$-\frac{173}{1440}$	$\frac{27}{1440}$	$-\frac{863}{60480}h^7 f^{(6)}(\xi)$

▶Example 9.12

Indicate the method of solving the initial value problem

$$y' = y + 2x - 1;\ y(0) = 1 \tag{E1}$$

in the interval $0 \le x \le 1$ using Adams–Moulton closed formulas of orders 1 to 6.

Solution

Using a step size of $h = 0.1$ with $x_0 = 0.0, x_1 = 0.1, \ldots, x_{10} = 1.0$ and $y_0 = 1.0$, the iterative processes of the Adams–Moulton formulas (Eqs. (9.123)) are applied as follows:

when $n = 1$,

$$y_{i+1} = y_i + hf_{i+1}. \tag{E2}$$

when $n = 2$,

$$y_{i+1} = y_i + \frac{h}{2}(f_{i+1} - f_i). \tag{E3}$$

when $n = 3$,

$$y_{i+1} = y_i + \frac{h}{12}(5f_{i+1} + 8f_i - f_{i-1}). \tag{E4}$$

when $n = 4$,

$$y_{i+1} = y_i + \frac{h}{24}(9f_{i+1} + 19f_i - 5f_{i-1} + f_{i-2}). \tag{E5}$$

when $n = 5$,

$$y_{i+1} = y_i + \frac{h}{720}(251f_{i+1} + 646f_i - 264f_{i-1} + 106f_{i-2} - 19f_{i-3}). \tag{E6}$$

when $n = 6$,

$$y_{i+1} = y_i + \frac{h}{1440}(475 f_{i+1} + 1427 f_i - 798 f_{i-1} + 482 f_{i-2} - 173 f_{i-3} + 27 f_{i-4}). \quad \text{(E7)}$$

The values computed from the exact solution,

$$y(x) = -1 - 2x + 2e^x, \quad \text{(E8)}$$

are given as the starting values in the iterative process. For example, the Adams–Moulton formula of order $n = 5$ requires the values of y at $x = 0.0, 0.1, 0.2,$ and 0.3 to use Eq. (E6). These values are computed from Eq. (E8). It is to be noted that the Adams–Moulton method of order m is comparable to the Adams–Bashforth method of order $m + 1$, since both require $m + 1$ evaluations of the function f per step and both have the term $h^m f^{(m+1)}(\xi)$i n their local truncation errors. ◀

9.10 Predictor–Corrector Methods

As stated earlier, the explicit formulas, such as Adams–Bashforth formulas, are also called open formulas, and the implicit formulas, such as Adams–Moulton formulas, are also known as closed formulas. Sometimes they are used in combination as a pair of formulas, and the procedure is known as a predictor–corrector method. In such a case, the procedure consists of applying the open-type formula, called a predictor step and then applying the closed-type formula, known as a corrector step, in each interval. This means that the predictor estimates the solution at a new point x_{i+1}, and then the corrector improves its accuracy. Some of the common predictor–corrector methods are described in the following sections.

9.10.1 Fourth-Order Adams Predictor–Corrector Method

The fourth-order Adams–Bashforth open formula is used as a predictor, that is,

$$y_{i+1}^{(1)} = y_i + h\left(\frac{55}{24} f_i - \frac{59}{24} f_{i-1} + \frac{37}{24} f_{i-2} - \frac{9}{24} f_{i-3}\right), \quad (9.124)$$

where $y_{i+1}^{(1)}$ is the predicted value of y at x_{i+1}. The fourth-order Adams–Moulton closed formula is used as a corrector to improve the solution y_{i+1} as

$$y_{i+1} = y_i + h\left(\frac{9}{24} f_{i+1}^{(j)} + \frac{19}{24} f_i - \frac{5}{24} f_{i-1} + \frac{1}{24} f_{i-2}\right); \; j = 1, 2, 3, \ldots. \quad (9.125)$$

It can be shown that the error in both the predictor and corrector formulas is $O(h^5)$.

The computational procedure to solve the differential equation

$$y' = f(x, y) \text{ with } y(x_0) = y_0$$

involves the following steps:

Step 1 Select the values of h and ε, and compute

$$x_i = x_0 + ih; \; i = 1, 2, 3, \ldots.$$

Step 2 From the known initial condition $y(x_0) = y_0$, use the fourth-order Runge–Kutta method (because it has the same level of accuracy as the fourth-order Adams formulas), find the solution $y_1 = y(x_1)$, $y_2 = y(x_2)$, and $y_3 = y(x_3)$.

Step 3 Compute the following for each fixed $i = 3, 4, \ldots$.

Step 4 Find $y_{i+1}^{(1)}$, using Eq. (9.124).

Step 5 Find $y_{i+1}^{(j+1)}$, using Eq. (9.125).

Step 6 Iterate on j until

$$\left| \frac{y_{i+1}^{(j+1)} - y_{i+1}^{(j)}}{y_{i+1}^{(j)}} \right| \leq \varepsilon. \tag{9.126}$$

If the convergence criterion of Eq. (9.126) is not satisfied in a reasonable number of iterations, repeat the whole procedure with a reduced step size, h.

▶Example 9.13

Find the solution of the initial value problem

$$y' = -1 + 2x + y; \; y(0) = 1 \tag{E1}$$

at $x = 0.4$ using the fourth-order Adams predictor–corrector method with $h = 0.1$.

Solution

To start the procedure, we need the values of y_0, y_1, y_2, and y_3. The values of y_1, y_2, and y_3 can be determined using the fourth-order Runge–Kutta–Gill method. From the solution of Example 9.9, we obtain

$$x_0 = 0.0, \; y_0 = 1.0;$$

$$x_1 = 0.1, \; y_1 = 1.01034164;$$

$$x_2 = 0.2, \; y_2 = 1.04280472;$$

and

$$x_3 = 0.3, \; y_3 = 1.09971619. \tag{E2}$$

Noting that

$$f(x) = y' = -1 + 2x + y, \tag{E3}$$

we obtain, using Eq. (E2),

$$f_0 = -1 + 0 + 1 = 0.0,$$

$$f_1 = -1 + 2(0.1) + 1.01034164 = 0.21034164,$$

$$f_2 = -1 + 2(0.2) + 1.04280472 = 0.44280472,$$

and

$$f_3 = -1 + 2(0.3) + 1.09971619 = 0.69971619.$$

The application of Eq. (9.124) gives the predicted value of y_4 as

$$\begin{aligned} y_4^{(1)} &= y_3 + \frac{h}{24}(55f_3 - 59f_2 + 37f_1 - 9f_0) \\ &= 1.09971619 + \frac{0.1}{24}\{55(0.69971619) - 59(0.44280472) \\ &\quad + 37(0.21034164) - 9(0.0)\} = 1.18363933 \end{aligned} \tag{E4}$$

Corresponding to the predicted value of y_4 in Eq. (E4), Eq. (E3) gives

$$f_4^{(1)} = -1 + 2x_4 + y_4^{(1)} = -1 + 2(0.4) + 1.18363933 = 0.98363933. \tag{E5}$$

Then the corrected value of y_4 is given by Eq. (9.125) as

$$\begin{aligned} y_4 &= y_3 + \frac{h}{24}(9f_4^{(1)} + 19f_3 - 5f_2 + f_1) \\ &= 1.09971619 + \frac{0.1}{24}\{9(0.98363933) + 19(0.69971619) \\ &\quad -5(0.44280472) + 0.21034164\} = 1.18364819. \end{aligned} \tag{E6}$$

This value can be compared with the exact value, $y_4 = 1.1836\,5002$. Notice that using the correction formula reduced the error in y_4 from 0.0000 1069 to 0.0000 0183. ◀

9.10.2 Milne's Method

The fourth-order Milne's method uses the three-point Newton–Cotes open formula,

$$y_{i+1}^{(1)} = y_{i-3} + \frac{4h}{3}(2f_i - f_{i-1} + 2f_{i-2}), \tag{9.127}$$

as a predictor and the three-point Newton–Cotes closed formula,

$$y_{i+1}^{(j+1)} = y_{i-1} + \frac{h}{3}(f_{i+1}^{(j)} + 4f_i + f_{i-1});\ j = 1, 2, 3, \ldots \tag{9.128}$$

as a corrector. The local truncation error of each of these formulas is $O(h^5)$. The detailed step-by-step procedure is similar to the one described in the previous section.

Although both Adams and Milne's predictor–corrector methods have the same order of error, in general, Milne's method is more accurate. The reason is that the coefficients of the error terms in Milne's method are smaller compared with those of Adams method as shown in Table 9.10. In addition, Milne's method is simpler to implement and uses fewer function evaluations. In spite of these advantages, Milne's method is subject to an instability problem in certain cases. Because of the possible instability, Milne's method is usually not used. Instead, the Adams method is commonly used.

Table 9.10 Errors in Adams and Milne's methods.

Error in Predictor		Error in Corrector	
Adams	Milne's	Adams	Milne's
$\frac{251}{720}h^5 f^{(4)}(\xi)$	$\frac{14}{45}h^5 f^{(4)}(\xi)$ $= \frac{224}{720}h^5 f^{(4)}(\xi)$	$-\frac{19}{720}h^5 f^{(4)}(\xi)$	$-\frac{1}{90}h^5 f^{(4)}(\xi)$ $-\frac{8}{720}h^5 f^{(4)}(\xi)$

▶Example 9.14

Find the solution of the initial value problem

$$y' = -1 + 2x + y; \; y(0) = 1 \tag{E1}$$

at $x = 0.4$ using Milne's method with $h = 0.1$.

Solution

To start the Milne's method, we need the values of y_0, y_1, y_2, and y_3. The fourth-order Runge–Kutta–Gill method can be used to find the values of y_1, y_2, and y_3. From the solution of Example 9.9, we have

$$x_0 = 0.0, \; y_0 = 1.0;$$

$$x_1 = 0.1, \; y_1 = 1.01034164;$$

$$x_2 = 0.2, \; y_2 = 1.04280472;$$

and

$$x_3 = 0.3, \; y_3 = 1.09971619. \tag{E2}$$

Since

$$f(x) = y' = -1 + 2x + y, \tag{E3}$$

we find that

$$x_1 = 0.1, \; f_1 = -1 + 2(0.1) + 1.01034164 = 0.21034164;$$

$$x_2 = 0.2, \; f_2 = -1 + 2(0.2) + 1.04280472 = 0.44280472;$$

and

$$x_3 = 0.3, \; f_3 = -1 + 2(0.3) + 1.09971619 = 0.69971619. \tag{E4}$$

The application of Eq. (9.127) gives the predicted value of y_4 as

$$\begin{aligned} y_4^{(1)} &= y_0 + \frac{4h}{3}(2f_3 - f_2 + 2f_1) \\ &= 1.0 + \frac{0.4}{3}\{2(0.69971619) - 0.44280472 + 2(0.21034164)\} \\ &= 1.18364146. \end{aligned} \tag{E5}$$

Equations (E3) and (E5) yield

$$f_4^{(1)} = f(x_4, y_4^{(1)}) = -1 + 2(0.4) + 1.18364146 = 0.98364146. \tag{E6}$$

Then the corrected value of y_4 is given by Eq. (9.128) as

$$\begin{aligned} y_4 &= y_2 + \frac{h}{3}(f_4^{(1)} + 4f_3 + f_2) \\ &= 1.04280472 + \frac{0.1}{3}\{0.98364146 + 4(0.69971619) + 0.44280472\} \\ &= 1.18364842. \end{aligned} \tag{E7}$$

This value can be compared with the exact value, $y_4 = 1.1836\,5002$. Notice that using the correction formula reduced the error in y_4 from 0.0000 0956 to 0.0000 0160. Also note that Milne's method is slightly better than the Adams method considered in Example 9.13.

◀

9.10.3 Hamming's Method

The instability of Milne's method has been found to be due to the corrector. Attempts have been made to develop stable correctors that can be used with Milne's predictor. Hamming's method is one such method [9.12]. Hamming's method considers a general class of corrector formulas of the form

$$\begin{aligned} y_{i+1} &= a_i\, y_i + a_{i-1}\, y_{i-1} + a_{i-2}\, y_{i-2} \\ &\quad + h(b_{i+1}\, y'_{i+1} + b_i\, y'_i + b_{i-1}\, y'_{i-1} + b_{i-2}\, y'_{i-2}) + O(h^5), \end{aligned} \tag{9.129}$$

where

$$y'_j = f_j;\ j = i+1, i, i-1, i-2. \tag{9.130}$$

Equation (9.129) includes Milne's as well as Adams corrector formulas as special cases. Taylor's series expansions about x_i, including terms up to $y_i^{(4)}$, can be obtained as

$$y_{i-1} = y_i - hy'_i + \frac{h^2}{2!}y''_i - \frac{h^3}{3!}y'''_i + \frac{h^4}{4!}y_i^{(4)} \tag{9.131}$$

(expansion of y, step size $-h$);

$$y_{i-2} = y_i - (2h)y'_i + \frac{(2h)^2}{2!}y''_i - \frac{(2h)^3}{3!}y'''_i + \frac{(2h)^4}{4!}y_i^{(4)} \tag{9.132}$$

(expansion of y, step size $-2h$);

$$y'_{i+1} = y'_i + hy''_i + \frac{h^2}{2!}y'''_i + \frac{h^3}{3!}y_i^{(4)} \tag{9.133}$$

(expansion of y', step size $+h$);

$$y'_{i-1} = y'_i - hy''_i + \frac{h^2}{2!}y'''_i - \frac{h^3}{3!}y_i^{(4)} \tag{9.134}$$

(expansion of y', step size $-h$);

$$y'_{i-2} = y'_i - (2h)y''_i + \frac{(2h)^2}{2!}y'''_i - \frac{(2h)^3}{3!}y_i^{(4)} \tag{9.135}$$

(expansion of y', step size $-2h$);

$$y_{i+1} = y_i + hy'_i + \frac{h^2}{2!}y''_i + \frac{h^3}{3!}y'''_i + \frac{h^4}{4!}y_i^{(4)} \tag{9.136}$$

(expansion of y, step size $+h$).

Substituting Eq. (9.136) in the left-hand side and Eqs. (9.131) to (9.135) in the right-hand side of Eq. (9.129), and equating the coefficients of y_i and its derivatives on both sides of the resulting equation yields

$$\begin{aligned} a_i + a_{i-1} + a_{i-2} &= 1, \\ -a_{i-1} - 2a_{i-2} + b_{i+1} + b_i + b_{i-1} + b_{i-2} &= 1, \\ \frac{1}{2}a_{i-1} + 2a_{i-2} + b_{i+1} - b_{i-1} - 2b_{i-2} &= \frac{1}{2}, \\ -\frac{1}{6}a_{i-1} - \frac{4}{3}a_{i-2} + \frac{1}{2}b_{i+1} + \frac{1}{2}b_{i-1} + 2b_{i-2} &= \frac{1}{6}, \end{aligned} \tag{9.137}$$

and

$$\frac{1}{24}a_{i-1} + \frac{2}{3}a_{i-2} + \frac{1}{6}b_{i+1} - \frac{1}{6}b_{i-1} - \frac{4}{3}b_{i-2} = \frac{1}{24}.$$

It can be seen that there are seven unknowns and five equations (Eq. (9.137)). In one solution, Hamming assumed that

$$a_{i-1} = b_{i-2} = 0. \tag{9.138}$$

With these values, Eq. (9.137) can be solved to find

$$a_i = \frac{9}{8},\ a_{i-2} = -\frac{1}{8},\ b_{i+1} = \frac{3}{8},\ b_i = \frac{3}{4},\ \text{and } b_{i-1} = -\frac{3}{8}. \tag{9.139}$$

Substituting these values into Eq. (9.129), the corrector formula can be written as

$$y_{i+1} = \frac{9}{8}y_i - \frac{1}{8}y_{i-2} + \frac{3}{8}h(y'_{i+1} + 2y'_i - y'_{i-1}). \tag{9.140}$$

As seen earlier, the error in the predictor formula, Eq. (9.127), is given by

$$e_p = \frac{14}{45}h^5 y^{(4)}(\xi_p), \tag{9.141}$$

where $x_{i-3} < \xi_p < x_{i+1}$. The error in the corrector formula, Eq. (9.140), can be shown to be

$$e_c = -\frac{1}{40}h^5 y^{(4)}(\xi_c) \tag{9.142}$$

where $x_{i-3} < \xi_c < x_{i+1}$. (See Problem 9.44.) Assuming that $y^{(4)}$ does not vary drastically in the interval (x_{i-3}, x_{i+1}), we use the relation

$$y^{(4)}(\xi_p) = y^{(4)}(\xi_c) \equiv y^{(4)}(\xi). \tag{9.143}$$

Thus, the errors in predictor and corrector formulas, given by Eqs. (9.157) and (9.155), can be expressed as

$$e^p = \frac{c_1}{c_1 + c_2}\left(y_i^{(m+1)} - y_i^{(1)}\right) = \frac{112}{121}\left(y_i^{(m+1)} - y_i^{(1)}\right) \tag{9.144}$$

and

$$e_c = -\frac{c_2}{c_1 + c_2}\left(y_{i+1}^{(m+1)} - y_{i+1}^{(1)}\right) = -\frac{9}{121}\left(y_{i+1}^{(m+1)} - y_{i+1}^{(1)}\right), \tag{9.145}$$

noting that $c_1 = \frac{14}{45}$ from Eq. (9.141) and $c_2 = \frac{1}{40}$ from Eq. (9.142).

Equation (9.140) can be used along with a suitable predictor formula such as Milne's formula, Eq. (9.127). Instead of just using Eqs. (9.127) and (9.140) as predictor and corrector, Hamming suggested an alternative procedure to eliminate the iterations associated with the usual corrector formulas. This involves the use of predictor and corrector modifiers. The complete Hamming procedure can be stated as follows:

$$\bar{y}_{i+1} = y_{i-3} + \frac{4}{3}h(2y_i' - y_{i-1}' + 2y_{i-2}'): \text{ predictor,} \tag{9.146}$$

$$y_{i+1}^{(1)} = \bar{y}_{i+1} - \frac{112}{121}(\bar{y}_i - \bar{\bar{y}}_i): \text{ predictor modifier,} \tag{9.147}$$

$$\bar{\bar{y}}_{i+1} = \frac{9}{8}y_i - \frac{1}{8}y_{i-2} + \frac{3}{8}h[f(x_{i+1}, y_{i+1}^{(1)}) + 2y_i' - y_{i-1}']: \text{ corrector,} \tag{9.148}$$

and

$$y_{i+1} = \bar{\bar{y}}_{i+1} + \frac{9}{121}(\bar{y}_{i+1} - \bar{\bar{y}}_{i+1}): \text{ corrector modifier.} \tag{9.149}$$

It can be seen that Eqs. (9.146) through (9.149) do not involve any iterative procedure, and Eq. (9.149) gives the final value of y_{i+1}. Also note that the method is not self-starting; as such, a self-starting method such as the fourth-order Runge–Kutta method can be used in the beginning. In the first step in which Hamming's method (Eqs. (9.146) through (9.149)) is used, the quantity $(\bar{y}_i - \bar{\bar{y}}_i)$ is to be taken as zero in Eq. (9.147). The fourth-order Hamming's method, just given is one of the most popularly used multistep methods.

▶Example 9.15

Find the solution of the initial value problem

$$y' = -1 + 2x + y; \; y(0) = 1 \tag{E1}$$

at $x = 0.4$ using Hamming's method with $h = 0.1$.

Solution

To initiate Hamming's method, the values of y_0, y_1, y_2, and y_3 are required. The values of y_1, y_2, and y_3 can be determined using the fourth-order Runge–Kutta–Gill method. From the solution of Example 9.9, we have

$$x_0 = 0.0,\ y_0 = 1.0;$$

$$x_1 = 0.1,\ y_1 = 1.01034164;$$

$$x_2 = 0.2,\ y_2 = 1.04280472;$$

and

$$x_3 = 0.3,\ y_3 = 1.09971619. \tag{E2}$$

Since

$$f(x) = y' = -1 + 2x + y, \tag{E3}$$

we obtain

$$x_0 = 0.0,\ f_0 = -1 + 2(0) + 1 = 0.0;$$

$$x_1 = 0.1,\ f_1 = -1 + 2(0.1) + 1.01034164 = 0.21034164;$$

$$x_2 = 0.2,\ f_2 = -1 + 2(0.2) + 1.04280472 = 0.44280472;$$

and

$$x_3 = 0.3,\ f_3 = -1 + 2(0.3) + 1.09971619 = 0.69971619. \tag{E4}$$

Equation (9.148) gives the predicted value of y_4 as

$$\begin{aligned}\bar{y}_4 &= y_0 + \frac{4h}{3}(2f_3 - f_2 + 2f_1)\\ &= 1.0 + \frac{0.4}{3}\{2(0.69971619) - 0.44280472 + 2(0.21034164)\}\\ &= 1.18364146.\end{aligned} \tag{E5}$$

Since $\bar{\bar{y}}_4$ is not available, we assume $(\bar{y}_4 - \bar{\bar{y}}_4)$ to be zero in Eq. (9.147) so that the predictor modifier gives

$$y_4^{(1)} = \bar{y}_4 = 1.18364146. \tag{E6}$$

Equations (E6) and (E3) yield

$$f(x_4, y_4^{(1)}) = -1 + 2x_4 + y_4^{(1)} = -1 + 2(0.4) + 1.18364146 = 0.98364146. \tag{E7}$$

The corrector Eq. (9.148) gives

$$\begin{aligned}\bar{\bar{y}}_4 &= \frac{9}{8}y_3 - \frac{1}{8}y_1 + \frac{3h}{8}(f(x_4, y_4^{(1)}) + 2f_3 - f_2)\\ &= \frac{9}{8}(1.09971619) - \frac{1}{8}(1.01034146) + \frac{0.3}{8}(0.98364146\\ &\quad + 2(0.69971619) - 0.44280472) = 1.18364812.\end{aligned} \tag{E8}$$

Finally the corrector modifier, Eq. (9.149), yields

$$y_4 = \bar{\bar{y}}_4 + \frac{9}{121}(\bar{y}_4 - \bar{\bar{y}}_4) = 1.18364812 + \frac{9}{121}(1.18364146 - 1.18364812)$$
$$= 1.18364763. \tag{E9}$$

This value can be compared with the exact value, $y_4 = 1.1836\,5002$. Thus, the error in the value given by Eq. (E9) is 0.0000 0239.

◀

9.10.4 Error Analysis of Predictor–Corrector Methods

Let the formulas used as predictor and corrector have the same order of accuracy. This assumption permits the estimation of the local truncation error during the computations and also the derivation of a criterion for adjusting the step size. The local truncation errors of the predictor and corrector formulas are given in Tables 9.7 and 9.9, respectively. When an nth-order predictor–corrector method is used, the errors given in Tables 9.7 and 9.9 can be used to express the exact value of y_{i+1} as

$$y_{i+1}|_{\text{exact}} = y_{i+1}^{(1)} + c_1(h)^{n+1} f^{(n)}(\xi_1) \tag{9.150}$$

and

$$y_{i+1}|_{\text{exact}} = y_{i+1}^{(m+1)} - c_2(h)^{(n+1)} f^{(n)}(\xi_2), \tag{9.151}$$

where $y_{i+1}^{(1)}$ denotes the value of y_{i+1} given by the predictor formula, $y_{i+1}^{(m+1)}$ indicates the value of y_{i+1} given by the corrector formula (converged solution after m iterations of the corrector formula), the values of the constants c_1 and c_2 are given in Tables 9.7 and 9.9, and ξ_1 and ξ_2 are, respectively, the values of x at which the nth-order derivative of f or the $(n+1)$st derivative of y with respect to x is evaluated for estimating the error in the predictor and corrector formulas. If the nth derivative of f does not vary by a large amount in the interval over which the predictor–corrector formula is defined, we can assume that

$$f^{(n)}(\xi_1) = f^{(n)}(\xi_2) \equiv f^{(n)}(\xi). \tag{9.152}$$

Subtracting Eq. (9.150) from Eq. (9.151) yields

$$y_{i+1}^{(m+1)} - y_{i+1}^{(1)} - (c_1 + c_2)h^{n+1} f^{(n)}(\xi) = 0, \tag{9.153}$$

which can be rearranged to obtain

$$-\frac{c_2\left(y_{i+1}^{(m+1)} - y_{i+1}^{(1)}\right)}{(c_1 + c_2)} = -c_2 h^{(n+1)} f^{(n)}(\xi). \tag{9.154}$$

Notice that, with this rearrangement, the right-hand side of Eq. (9.154) is made the same as the error term of the corrector formula, Eq. (9.151). Thus, Eq. (9.154) gives the local truncation error (e_c) of the corrector formula as

$$e_c = -\frac{c_2}{(c_1 + c_2)}\left(y_{i+1}^{(m+1)} - y_{i+1}^{(1)}\right). \tag{9.155}$$

Equation (9.153) can also be rearranged to obtain

$$\frac{c_1\left(y_{i+1}^{(m+1)} - y_{i+1}^{(1)}\right)}{(c_1 + c_2)} = c_1 h^{n+1} f^{(n)}(\xi). \tag{9.156}$$

Noting that the right-hand side of Eq. (9.156) is the same as the error term of the predictor formula, and using the subscript i in place of $i+1$, Eq. (9.156) gives the local truncation error (e_p) of the predictor formula as

$$e_p = \frac{c_1\left(y_i^{(m+1)} - y_i^{(1)}\right)}{(c_1 + c_2)}. \tag{9.157}$$

For example, in the case of fourth-order Adams predictor–corrector method $(n = 4)$, the error associated with the predictor, Eq. (9.124), is given by

$$\frac{251}{720} h^5 f^{(4)}(\xi) = c_1 h^5 f^{(4)}(\xi) \tag{9.158}$$

(from Table 9.7) and that of the corrector, Eq. (9.125), by

$$-\frac{19}{720} h^5 f^{(4)}(\xi) = -c_2 h^5 f^{(4)}(\xi) \tag{9.159}$$

(from Table 9.9). Thus, the constants c_1 and c_2 can be seen to be

$$c_1 = \frac{251}{720} \text{ and } c_2 = \frac{19}{720},$$

and hence, the errors of predictor and corrector, given by Eqs. (9.157) and (9.155), can be expressed as

$$e_p = \frac{251}{720}\left(y_i^{(m+1)} - y_i^{(1)}\right) \tag{9.160}$$

and

$$e_c = -\frac{19}{720}\left(y_{i+1}^{(m+1)} - y_{i+1}^{(1)}\right). \tag{9.161}$$

Similarly, in the case of Milne's method $(n = 4)$, the predictor, Eq. (9.127), has an error of

$$\frac{14}{45} h^5 f^{(4)}(\xi) = c_1 h^5 f^{(4)}(\xi), \tag{9.162}$$

and the corrector, Eq. (9.128), has an error of

$$-\frac{1}{90} h^5 f^{(4)}(\xi) = -c_2 h^5 f^{(4)}(\xi). \tag{9.163}$$

The constants are given by $c_1 = \frac{14}{45}$ and $c_2 = \frac{1}{90}$, and hence, the errors of predictor and corrector, Eqs. (9.157) and (9.155), can be expressed as

$$e_p = \frac{28}{29}\left(y_i^{(m+1)} - y_i^{(1)}\right) \tag{9.164}$$

and

$$e_c = -\frac{1}{29}\left(y_{i+1}^{(m+1)} - y_{i+1}^{(1)}\right). \tag{9.165}$$

Adjusting the Step Size

Equations (9.155) and (9.157) can be used to adjust the step size during the course of a computation. If Eqs. (9.155) or (9.157) indicates that the error e_c or e_p is greater than the maximum permissible value, then the step size could be reduced as indicated next. Equation (9.154) gives the absolute value of the error in the corrector solution (with step size h) as

$$\frac{c_2}{(c_1 + c_2)}\left(y_{i+1}^{(m+1)} - y_{i+1}^{(1)}\right) = c_2 h^{n+1} f^{(n)}(\xi). \tag{9.166}$$

If the step size is reduced to (rh), Eq. (9.154) gives the error as

$$\frac{c_2}{(c_1 + c_2)}(\bar{y}_{i+1}^{(m+1)} - \bar{y}_{i+1}^{(1)}) = c_2 (rh)^{n+1} f^{(n)}(\bar{\xi}), \tag{9.167}$$

where $\bar{y}_{i+1}^{(1)}$ and $\bar{y}_{i+1}^{(m+1)}$ denote, respectively, the values given by the predictor and the corrector formulas when a step size of (rh) is used. If the nth derivative of f does not vary abruptly, we can assume that

$$f^{(n)}(\xi) = f^{(n)}(\bar{\xi}). \tag{9.168}$$

Thus, in order to have the new error (when the step sizer rh is used) less than a specified maximum value (ε), we need to have

$$-c_2 (rh)^{n+1} f^{(n)}(\bar{\xi}) = -c_2 (rh)^{n+1} f^{(n)}(\xi) \le \varepsilon. \tag{9.169}$$

When Eq. (9.166) is used for $c_2\, h^{n+1} f^{(n)}(\xi)$, Eq. (9.169) yields

$$r^{n+1}\left|\frac{c_2}{(c_1 + c_2)}\left(y_{i+1}^{(m+1)} - y_{i+1}^{(1)}\right)\right| \le \varepsilon$$

or

$$r \le \left\{\frac{\varepsilon(c_1 + c_2)}{c_2\left(y_{i+1}^{(m+1)} - y_{i+1}^{(1)}\right)}\right\}^{\frac{1}{n+1}}. \tag{9.170}$$

Thus, the new step size to control the error, rh, is given by

$$rh \le h\left\{\frac{\varepsilon(c_1 + c_2)}{c_2\left(y_{i+1}^{(m+1)} - y_{i+1}^{(1)}\right)}\right\}^{\frac{1}{n+1}}. \tag{9.171}$$

Modification of Solution

Equations (9.155) and (9.157) can also be used to make the solution more accurate and efficient. It can be seen that the number of iterations needed (m) to find the converged solution, $y_{i+1}^{(m+1)}$, depends on the accuracy of the initial value, $y_{i+1}^{(1)}$, given by the predictor formula. Since Eq. (9.157) gives the error in the predictor formula, the value given by the predictor formula can be modified so that it is closer to the final converged value of the corrector as

$$y_{i+1}^{(1)}\Big|_{cf} = y_{i+1}^{(1)}\Big|_{pf} + \frac{c_1\left(y_i^{(m+1)} - y_i^{(1)}\right)}{(c_1 + c_2)}, \tag{9.172}$$

where $y_{i+1}^{(1)}\Big|_{cf}$ denotes the initial value used for the corrector formula and $y_{i+1}^{(1)}\Big|_{pf}$ indicates the value given by the predictor formula. Note that, in Eq. (9.172), the error between the predicted value and the final converged value of corrector in the ith step is used to modify the predicted value in the $(i + 1)$st step. This greatly reduces the number of iterations required in the corrector formula in the $(i + 1)$st step, if the nth derivative of f is relatively constant from step to step.

Similarly, Eq. (9.155) gives the error or discrepancy between the final converged value of the corrector and the exact solution. Hence, Eq. (9.155) can be added to the converged value of the corrector, $y_{i+1}^{(m+1)}$, to obtain a better or refined value of y_{i+1} as

$$y_{i+1\text{ refined value}}^{(m+1)} = y_{i+1}^{(m+1)} - \frac{c_2}{(c_1 + c_2)}(y_{i+1}^{(m+1)} - y_{i+1}^{(1)}). \tag{9.173}$$

Since Eqs. (9.155) and (9.157) are used to modify the solutions, they are also known as corrector and predictor modifiers.

9.11 Simultaneous Differential Equations

As indicated in Section 9.3, any nth-order linear differential equation can be expressed as a system of n simultaneous first-order differential equations

$$\frac{d\vec{y}}{dx} = \vec{f}(\vec{y}, x), \tag{9.174}$$

where

$$\vec{y} = \begin{Bmatrix} y_1 \\ y_2 \\ \cdot \\ \cdot \\ \cdot \\ y_n \end{Bmatrix}, \quad \vec{f} = \begin{Bmatrix} f_1 \\ f_2 \\ \cdot \\ \cdot \\ \cdot \\ f_n \end{Bmatrix}, \quad \text{and } \vec{y}_0 = \begin{Bmatrix} y_{1,0} \\ y_{2,0} \\ \cdot \\ \cdot \\ \cdot \\ y_{n,0} \end{Bmatrix}.$$

Similarly, a set of simultaneous higher order differential equations can also be expressed as an equivalent set of first-order differential equations of the form of

Eq. (9.174). All the solution methods discussed in the previous sections of this chapter are applicable for the solution of Eq. (9.174) provided the vectors $\vec{y}$ and $\vec{f}$ are used in place of y and f, respectively.

9.12 Stiff Equations

A stiff differential equation is one whose general solution contains an exponential term such as $e^{\lambda x}$ for some constant λ. When λ is a large negative quantity, such equations are particularly troublesome, because it causes the solution to decay rapidly to zero. To illustrate the difficulties associated with stiff equations, consider the (stiff) differential equation

$$\frac{dy}{dx} = \lambda y \tag{9.175}$$

with

$$y(x_0) = y_0.$$

The exact solution of Eq. (9.175) is given by

$$y(x) = y_0 e^{\lambda x}, \tag{9.176}$$

where λ, called the eigenvalue of the equation, remains constant throughout the integration. If Euler's method is used, the solution is given by

$$y_{i+1} = y_i + h\lambda y_i = (1 + h\lambda)y_i. \tag{9.177}$$

Let an error e_i be introduced (due to round-off) in the ith step. Then the subsequent computations yield the values of y as

$$y_{i+1} = (1 + h\lambda)(y_i + e_i),$$

$$y_{i+2} = (1 + h\lambda))^2(y_i + e_i),$$

$$\vdots,$$

and

$$y_{i+m} = (1 + h\lambda)^m(y_i + e_i).$$

This indicates that the error (E) caused in y_{i+m} due to the error in y_i is given by

$$E = (1 + h\lambda)^m e_i. \tag{9.178}$$

If $\lambda > 0$, the exact solution (given by Eq. (9.176)) grows exponentially and the error E may not be significant. However, if $\lambda < 0$, the exact solution (given by Eq. (9.176)) approaches zero. If $\lambda < 0$ and also $|1 + h\lambda| > 1$, that is, if $h > -\frac{2}{\lambda}$, the error grows exponentially and may dominate the approximate solution also. This indicates that it is possible to choose a sufficiently small step length to solve a stiff differential equation with reasonable accuracy. For a general initial value problem,

$$\frac{dy}{dx} = f(x, y) \tag{9.179}$$

and

$$y(x_0) = y_0,$$

the nonlinear function $f(x, y)$ can be linearized at each step so that λ can be obtained as $\lambda = \frac{\partial f}{\partial y}\big|_{x_i}$. Here, the value of λ varies in magnitude at each step of integration. Thus, Eq. (9.179) can be expected to be stiff when $\frac{\partial f}{\partial y}$ is negative at some point in the range of integration. In such cases, a solution method, which would have the largest region of absolute stability, is to be used. For example, Heun's (trapezoidal) method is a second-order method that is known to be absolutely stable for $\lambda < 0$ and that can be used to solve stiff differential equations. Heun's method is given by [Eq. (9.51)]

$$y_{i+1} = y_i + \frac{h}{2}[f(x_i, y_i) + f(x_{i+1}, y_{i+1})]. \tag{9.180}$$

The stability regions of some solution methods are given in Table 9.11. An extensive treatment of stiff differential equations can be found in Ref. [9.6].

Stiff equations are also encountered in systems of differential equations. For this, consider the following system of n simultaneous differential equations:

$$\frac{dy_1}{dx} = f_1(x, y_1, y_2, \ldots, y_n),$$

$$\frac{dy_2}{dx} = f_2(x, y_1, y_2, \ldots, y_n), \tag{9.181}$$

$$\vdots,$$

and

$$\frac{dy_n}{dx} = f_n(x, y_1, y_2, \ldots, y_n).$$

Table 9.11 Real stability regions of some solution methods [9.14].

Method	Stability boundary
Explicit Euler	$-2 \le h\lambda \le 0$
Implicit Euler	$0 < h < \infty$, for $\lambda < 0$, $-2 \le h\lambda \le 0$ for $\lambda > 0$
Second-order Runge–Kutta	$-2 < h\lambda \le 0$
Third-order Runge–Kutta	$-2.5 \le h\lambda \le 0$
Fourth-order Runge–Kutta	$-2.785 \le h\lambda \le 0$

By linearizing the functions $f_i, i = 1, 2, \ldots, n$, we obtain the Jacobian matrix,

$$[A] = \begin{bmatrix} \frac{\partial f_1}{\partial y_1} & \frac{\partial f_1}{\partial y_2} & \cdots & \frac{\partial f_1}{\partial y_n} \\ \frac{\partial f_2}{\partial y_1} & \frac{\partial f_2}{\partial y_2} & \cdots & \frac{\partial f_2}{\partial y_n} \\ \cdot & \cdot & \cdots & \cdot \\ \cdot & \cdot & \cdots & \cdot \\ \cdot & \cdot & \cdots & \cdot \\ \frac{\partial f_n}{\partial y_1} & \frac{\partial f_n}{\partial y_2} & \cdots & \frac{\partial f_n}{\partial y_n} \end{bmatrix}. \tag{9.182}$$

The stability of the numerical solution of Eqs. (9.181) is determined by the eigenvalues ($\lambda_i, i = 1, 2, \ldots, n$) of the Jacobian matrix, $[A]$. The step size is to be selected based on the maximum eigenvalue and the stability boundary of the method. If all the eigenvalues of the matrix $[A]$ have the same order of magnitude, then there will not be any problem during the numerical solution (integration) of Eq. (9.181). However, if the orders of magnitude of the maximum and the minimum eigenvalues are different by several orders of magnitude, the equations are said to be stiff. A measure of stiffness (M) is defined as

$$M = \frac{\max\limits_{1 \le i \le n} |\mathrm{Re}(\lambda_i)|}{\min\limits_{1 \le i \le n} |\mathrm{Re}(\lambda_i)|}. \tag{9.183}$$

The largest eigenvalue determines the step size, while the smallest eigenvalue determines the final time of integration. Hence, the solution of differential equations using explicit methods may be time intensive. Thus, implicit methods are recommended for solving stiff differential equations [9.5].

9.13 Choice of Method

The Runge–Kutta and predictor–corrector methods are found to be very efficient and reliable compared with all other methods. As such, we compare Runge–Kutta methods with predictor–corrector methods of corresponding order in this section.

The Runge–Kutta methods are self-starting and are easy to program for solution on a digital computer. Although the predictor–corrector methods are not self-starting, a fourth-order Runge–Kutta method can be used to start the solution. The interval between steps can be changed at will in Runge–Kutta methods.

Although the accuracies of the two methods are comparable, the Runge–Kutta methods might require a smaller step size in order to achieve a specified accuracy.

The monitoring of local truncation error does not involve additional function evaluations in predictor–corrector methods, while Runge–Kutta methods require additional computations. The number of evaluations of the function, $f(x, y)$, required in each step is two for predictor–corrector methods and equal to the order of the method in Runge–Kutta methods. Since the evaluation of $f(x, y)$ is computationally expensive in most practical problems, the predictor–corrector methods yield the results faster in most cases. For example, a fourth-order predictor–corrector method requires only half the computational time compared with a fourth-order Runge–Kutta method. On the other hand, if the function $f(x, y)$ is easy to evaluate and the accuracy needed is small, on the order of about 10^{-4}, the Runge–Kutta methods can be used.

For large systems of differential equations, a predictor–corrector method is preferred not only because of smaller computational time, but also due to the simplicity to estimate the error at any given stage.

9.14 Use of Software Packages

9.14.1 MATLAB

▶Example 9.16

Solve the equation

$$\frac{d^2y}{dx^2} + 0.2\frac{dy}{dx} + \sin\frac{x}{2} + x - 1 = 0;\ 0 \le x \le 4;\ y(0) = 0,\ \frac{dy}{dx}(0) = 0. \tag{E1}$$

Solution

By defining $t = x$, $x(1) = y$, and $x(2) = \frac{dy}{dx}$, we can rewrite Eq. (E1) as a set of two first-order equations as

$$\frac{d}{dt}\begin{Bmatrix} x(1) \\ x(2) \end{Bmatrix} = \begin{Bmatrix} f_1 \\ f_2 \end{Bmatrix} = \begin{Bmatrix} x(2) \\ -0.2\,x(2) - \sin\frac{t}{2} - t + 1 \end{Bmatrix}.$$

The equations are defined in an m.file labeled *dfuc.m.*

Listing of *dfuc.m*:

```
function f=dfuc(t,x)

f=zeros(2,1);

f(1)=x(2);
f(2)=1-sin(t./2)-0.2*x(2)-t;
```

Solution using the MATLAB function ode23

```
>> tspan=0:0.1:4;
>> x0=[0 0];
>> [t x]=ode23('dfuc',tspan,x0);
>> [t x]
ans =

                 0                 0                 0
  0.10000000000000  0.00471806177692  0.09155690728341
  0.20000000000000  0.01775596034598  0.16645712926249
  0.30000000000000  0.03746524228324  0.22504908467896
  0.40000000000000  0.06223276877909  0.26768655722444
  0.50000000000000  0.09048073143808  0.29472861102453
  0.60000000000000  0.12066750703390  0.30653932399784
  0.70000000000000  0.15128834266012  0.30348747136999
  0.80000000000000  0.18087628134233  0.28594628922655
  0.90000000000000  0.20800149210137  0.25429342752677
  1.00000000000000  0.23126921360146  0.20891062630203
  1.10000000000000  0.24932904582983  0.15018230823094
  1.20000000000000  0.26086541536194  0.07849724790953
  1.30000000000000  0.26459861935127 -0.00575341366087
  1.40000000000000  0.25929836169649 -0.10217663188833
  1.50000000000000  0.24376012851779 -0.21037584175088
  1.60000000000000  0.21683245755399 -0.32995484498797
  1.70000000000000  0.17739311545766 -0.46051413853479
  1.80000000000000  0.12436603885925 -0.60165538088468
  1.90000000000000  0.05671411740475 -0.75297870047535
  2.00000000000000 -0.02656117212823 -0.91408527157928
  2.10000000000000 -0.12641939580722 -1.08457621007875
  2.20000000000000 -0.24378028181640 -1.26405400610325
  2.30000000000000 -0.37952270531268 -1.45212290167909
  2.40000000000000 -0.53449038335318 -1.64838915646657
  2.50000000000000 -0.70947819271804 -1.85246241808275
  2.60000000000000 -0.90524830799563 -2.06395576638079
  2.70000000000000 -1.12253788968649 -2.28248274771690
  2.80000000000000 -1.36200273687692 -2.50766869102308
  2.90000000000000 -1.62430921255568 -2.73913633257798
  3.00000000000000 -1.91009076974809 -2.97651159516495
  3.10000000000000 -2.21987716505931 -3.21943911529512
  3.20000000000000 -2.55419370500374 -3.46756439252351
  3.30000000000000 -2.91358422040505 -3.72052832275887
  3.40000000000000 -3.29851893748813 -3.97798753860276
  3.50000000000000 -3.70939331113431 -4.23961566122852
```

```
    3.60000000000000  -4.14660252734614  -4.50508637435310
    3.70000000000000  -4.61055122785474  -4.77407123467219
    3.80000000000000  -5.10157456292913  -5.04626283842222
    3.90000000000000  -5.61994961106334  -5.32137021699821
    4.00000000000000  -6.16595326006659  -5.59910245374169
>> plot(t,x)
>> xlabel('t')
>> gtext('x(2)')
>> title('Problem 1 using ODE23(t=0:0.1:4.0)')
```

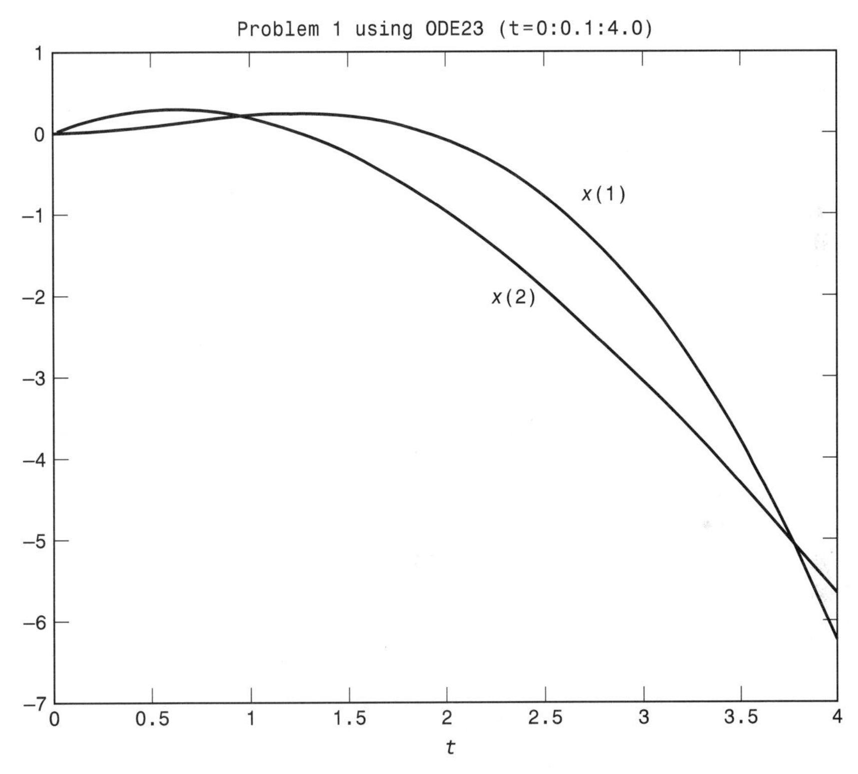

Note:

The statement

$$\texttt{f} = \texttt{zeros(m, n);}$$

gives the m by n matrix f with zero elements.

The statement

$$\texttt{tspan} = \texttt{0:0.1:4;}$$

gives the discrete values of tspan as 0.0, 0.1, 0.2, . . . , 4.0.

The statement

[t, y]=ode23(func, tspan, x0); or [t, y]=ode45(func, tspan, x0);

integrates a set of first-order differential equations described in the function *func* from an initial time to a final time given in *tspan*, using the initial conditions $x0$. The result is stored in the vector y at discrete values of t.

▶Example 9.17

Solve the following equations:

$$9\frac{dx}{dt} = 490x - 1996y$$

and

$$90\frac{dy}{dt} = 12475x - 49990y \tag{E1}$$

with the initial conditions $x(0) = 10$, $y(0) = -20$.

Solution

The equations are defined as an m.file labeled *ddfu.m*.

Listing of *ddfu.m*:

```
function f=ddfu(t,x)

f=zeros(2,1);

f(1)=490/9*x(1)-1996/9*x(2);
f(2)=12475/90*x(1)-49990/90*x(2);


>> tspan=0:0.01:1.0;
>> [t x]=ode23('ddfu',tspan,x0);

>> plot(t,x)
>> xlabel('t')

>> gtext('x(t)')
>> gtext('y(t)')
>> title('Problem 2 using ODE23(t=0:0.01:1.0)')
```

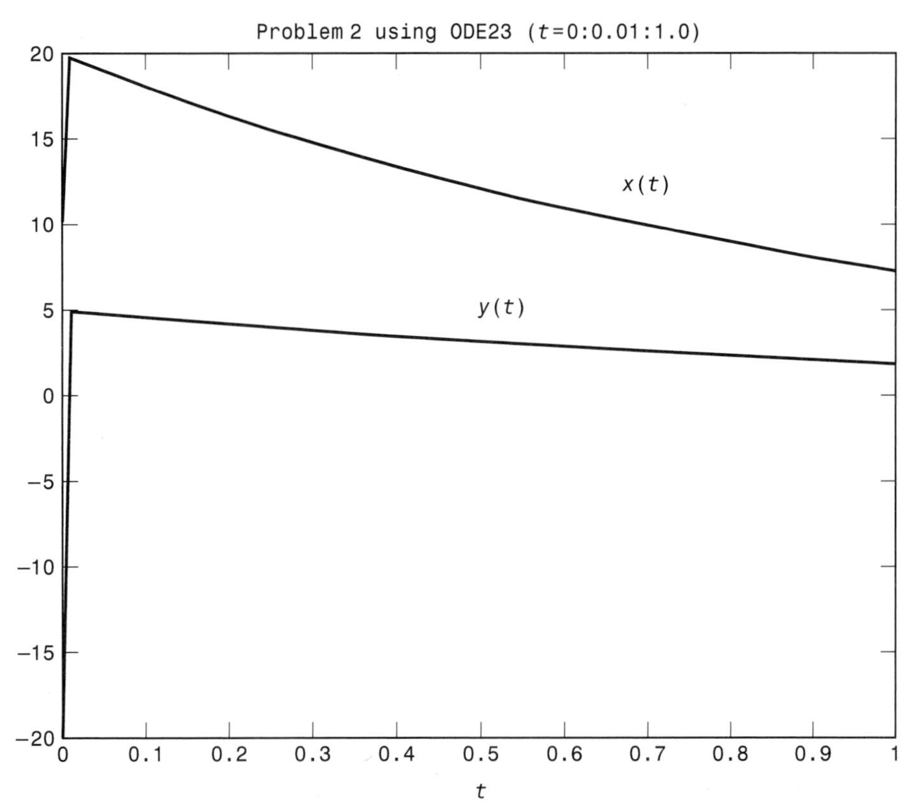

The solutions given by the MATLAB functions ode23, ode45, ode15s, and ode23s are compared with the Euler and exact solutions in the following table;

Comparison of solutions.

MATLAB function ($\Delta t = 0.1$)	$x = x(1)$ at $t = 0.1$	$y = x(2)$ at $t = 0.1$	$x = x(1)$ at $t = 1.0$	$y = x(2)$ at $t = 1.0$
ode23	18.09731551	4.52560499	7.35757806	1.83937038
ode45	18.09734243	4.52567227	7.35738294	1.83888250
ode15s	18.10786334	4.52696586	7.36007759	1.84001940
ode23s	18.09666964	4.52417076	7.35485529	1.83871382
Euler solution	508.0	1229.5	−0.79792315E + 18	−0.19948078E + 19
Exact solution	18.09674836	4.52418709	7.35758781	1.83939799

9.14.2 MAPLE

▶Example 9.18

Solve Eq. (E1) of Example 9.16 using MAPLE.

Solution

Note:

`diff(y(x),x)` and `diff(y(x),x,x)`

return the expressions $\frac{\partial y(x)}{\partial x}$ and $\frac{\partial^2 y(x)}{\partial x^2}$, respectively.

`dsolve(deqs union dic, var)`

solves the differential equation or the set of differential equations defined in deqs using the initial conditions of `dic` for the variables given in `var`.

```
>
> ode1:={diff(y(x),x,x)+0.2*diff(y(x),x)+x+sin(x/2)-1=0};
```

$$ode1 := \left\{\left(\frac{\partial^2}{\partial x^2}y(x)\right) + .2\left(\frac{\partial}{\partial x}y(x)\right) + x + \sin\left(\frac{1}{2}x\right) - 1 = 0\right\}$$

```
> ic1:={y(0)=0,D(y)(0)=0};
```

$$icl := \{y(0) = 0, \mathrm{D}(y)(0) = 0\}$$

```
> sol1:=dsolve(ode1 union ic1,{y(x)});
```

$$sol1 := y(x) = 30.x - 2.500000000\,x^2 + 1.379310345\cos(.5000000000\,x) - 160.0000000 + 3.448275862\sin(.5000000000\,x) + 158.6206897\,e^{(-.2000000000\,x)}$$

```
>
> eval(y(x),sol1);
```

$$30.x - 2.500000000\,x^2 + 1.379310345\cos(.5000000000\,x) - 160.0000000 + 3.448275862\sin(.5000000000\,x) + 158.6206897\,e^{(-.2000000000\,x)}$$

```
> y1:=unapply(%,x);
```

$$y1 := x \to 30.x - 2.500000000\,x^2 + 1.379310345\cos(.5000000000\,x) - 160.0000000 + 3.448275862\sin(.5000000000\,x) + 158.6206897\,e^{(-.2000000000x)}$$

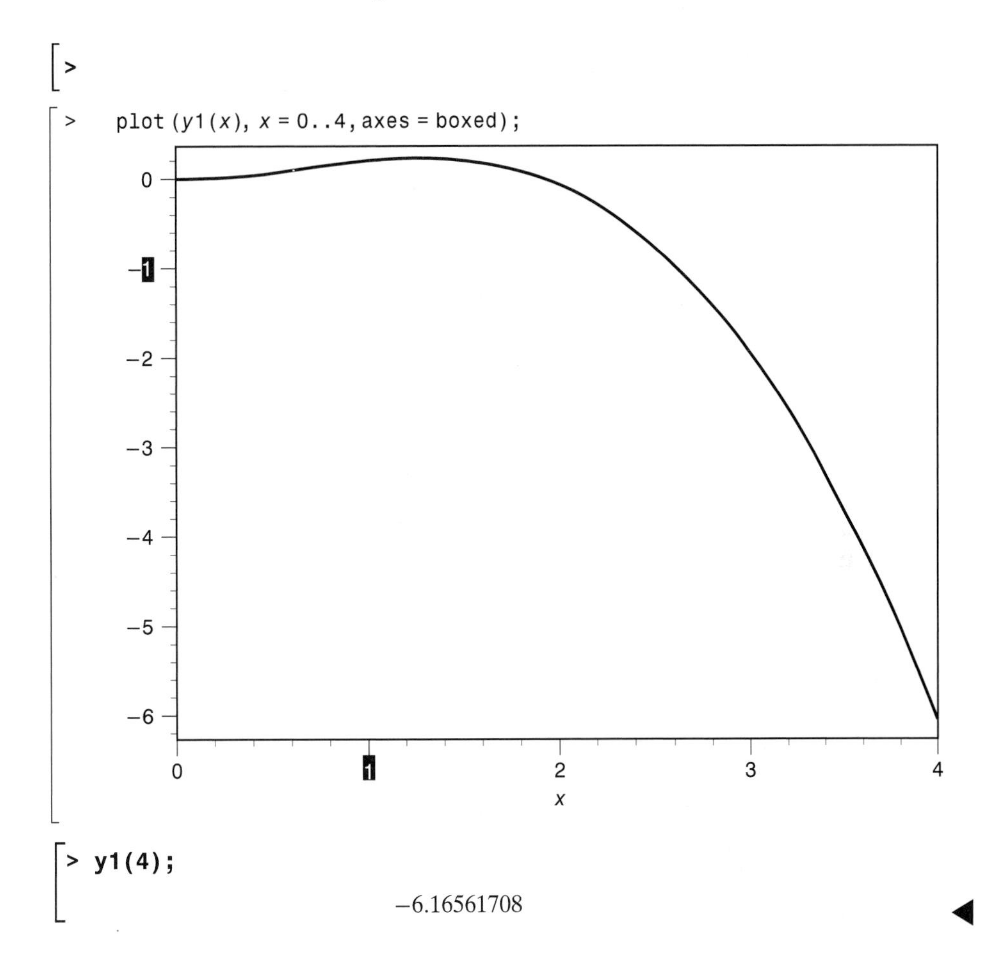


```
> plot(y1(x), x = 0..4, axes = boxed);
```

```
> y1(4);
```

$$-6.16561708$$

◀

▶Example 9.19

Solve Eq. (E1) of Example 9.17 using MAPLE.

Solution

```
>
> ode2:={9*diff(x(t),t)=490*x(t)-1996*y(t),
  90*diff(y(t),t)=12475*x(t)-49990*y(t)};
```

$$ode2 := \{90\left(\frac{\partial}{\partial t}\mathrm{y}(t)\right) = 12475\mathrm{x}(t) - 49990\,\mathrm{y}(t),\ 9\left(\frac{\partial}{\partial t}\mathrm{x}(t)\right) = 490\,\mathrm{x}(t) - 1996\,\mathrm{y}(t)\}$$

```
> ic2:={x(0)=10,y(0)=-20};
```

$$ic2 := \{x(0) = 10,\ y(0) = -20\}$$

```
> sol2:=dsolve(ode2 union ic2,{x(t),y(t)});
```

$$sol2 := \{x(t) = -10e^{(-500\,t)} + 20e^{(-t)},\ y(t) = 5e^{(-t)} - 25e^{(-500t)}\}$$

```
>

>

> x2:=unapply(eval(x(t),sol2),t);
```

$$x2 := t \to -10e(-500\,t) + 20e^{(-t)}$$

```
> plot(x2(t), t = 0..1,axes = boxed);
```

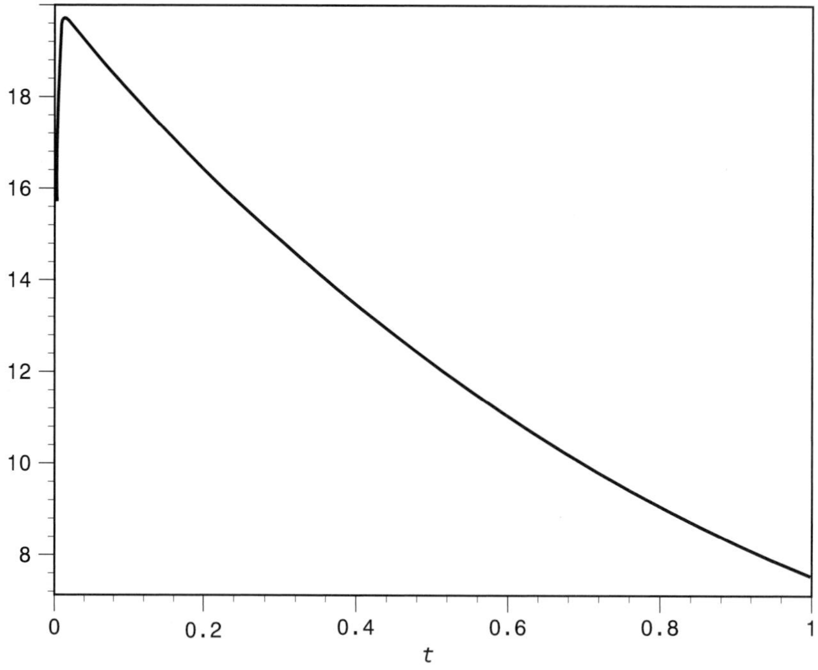

18
16
14
12
10
8
0
0.2
0.4
0.6
0.8
1
t

```
>

> y2:=unapply(eval(y(t),sol2),t);
```

$$y2 := t \to 5e^{(-t)} - 25e^{(-500\,t)}$$

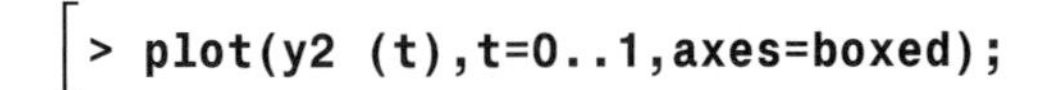

```
> plot(y2 (t),t=0..1,axes=boxed);
```

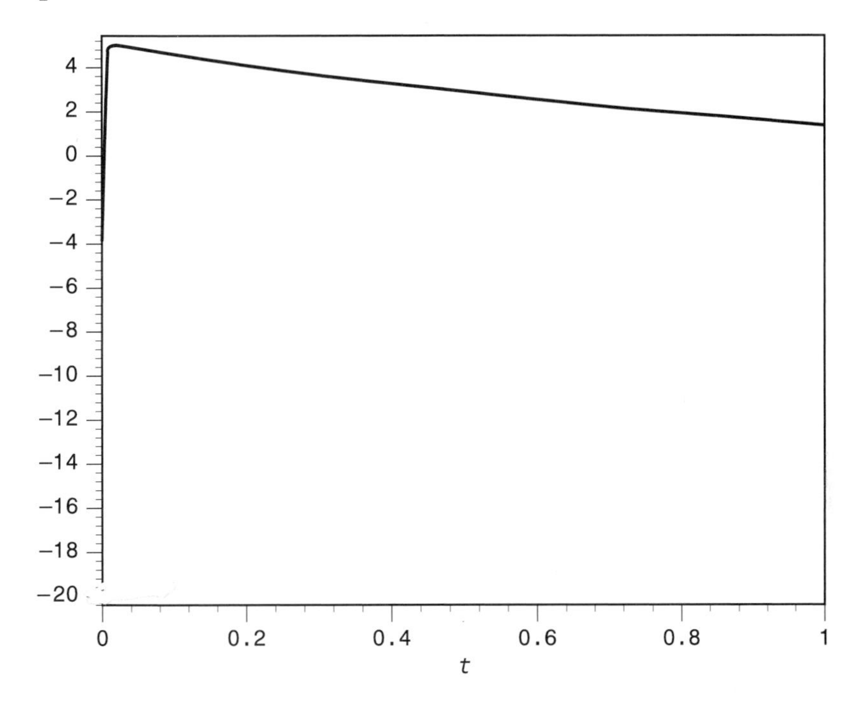


```
> evalf(y2 (0.1));
```

4.524187090

```
> evalf(y2 (0.2));
```

4.093653766

```
>
```

◀

9.14.3 Mathcad

▶Example 9.20

Solve equation in Example 9.16 using Mathcad.

Solution

Intial Condition

Solve y"=1−0.2*y'−sin(0.5*x)−x y(0)=0 y'(0)=0

$$y := \begin{pmatrix} 0 \\ 0 \end{pmatrix}$$

$$D(x,y) := \begin{pmatrix} y_1 \\ 1-0.2\cdot y_1-\sin(0.5\cdot x)-x \end{pmatrix}$$

Z := rkfixed(y,0,4,100,D) i := 0..100

Z =

	0	1	2
0	0	0	0
1	0.04	$7.819\cdot10^{-4}$	0.039
2	0.08	$3.056\cdot10^{-3}$	0.075
3	0.12	$6.713\cdot10^{-3}$	0.108
4	0.16	0.012	0.138
5	0.2	0.018	0.166
6	0.24	0.025	0.192
7	0.28	0.033	0.215
8	0.32	0.042	0.235
9	0.36	0.052	0.253
10	0.4	0.062	0.268
11	0.44	0.073	0.28
12	0.48	0.085	0.291
13	0.52	0.096	0.298
14	0.56	0.108	0.304
15	0.6	0.121	0.307

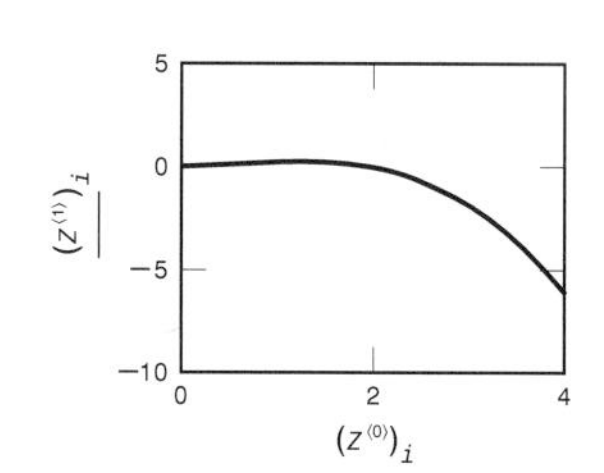

$Z_{100,1} = -6.166$

Z := Rkadapt(y,0,4,100,D) i := 0..100

Z =

	0	1	2
0	0	0	0
1	0.04	$7.819\cdot10^{-4}$	0.039
2	0.08	$3.056\cdot10^{-3}$	0.075
3	0.12	$6.713\cdot10^{-3}$	0.108
4	0.16	0.012	0.138
5	0.2	0.018	0.166
6	0.24	0.025	0.192
7	0.28	0.033	0.215
8	0.32	0.042	0.235
9	0.36	0.052	0.253
10	0.4	0.062	0.268
11	0.44	0.073	0.28
12	0.48	0.085	0.291
13	0.52	0.096	0.298
14	0.56	0.108	0.304
15	0.6	0.121	0.307

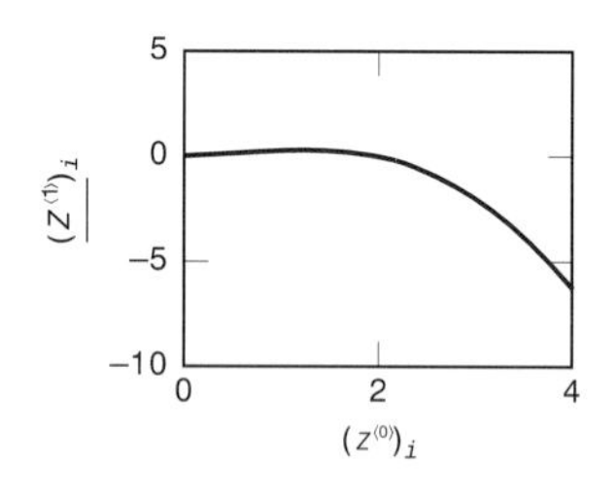

$Z_{100,1} = -6.166$

Note:

The statement

rkfixed(y, x1, x2, np, D)

solves a system of first-order differential equations by using a fourth-order Runge–Kutta method and returns a two-column matrix in which the left-hand column contains the points at which the solution to the differential equation is evaluated and the right-hand column contains the corresponding values of the solution:

y = vector of n initial values, where n denotes the number of first-order differential equations

x1, x2 = end points of the interval (initial values in y are the values at $x1$)

np = number of points beyond the initial point at which the solution is to be found

D(x, y) = n-element vector-valued function containing the first n derivatives of the unknown function

The statement

rkadapt(y, x1, x2, np, D)

is similar to "rkfixed" function, except that "rkadapt" yields a more accurate solution for a fixed number of points (np) by using frequent evaluations of the solution when the solution changes fast and infrequent evaluations when the solution changes more slowly.

◀

▶Example 9.21

Solve equations in Example 9.17 using MATHCAD.

Solution

```
Solve 9*x'(t)=490*x(t)-1996*y(t)   90*y'(t)=12475*x(t)-49990*y(t)


Intial Conditon  x(0)=10  y(0)=-20
```

$$x := \begin{pmatrix} 10 \\ -20 \end{pmatrix}$$

$$D(t,x) := \begin{pmatrix} 54.44444 \cdot x_0 - 221.77778 \cdot x_1 \\ 138.61111 \cdot x_0 - 555.44444 \cdot x_1 \end{pmatrix}$$

```
Z := rkfixed(x,0,1,10,D)
```

Z =

	0	1	2
0	0	10	−20
1	0.1	$-2.408\cdot10^{6}$	$-6.02\cdot10^{6}$
2	0.2	$-5.798\cdot10^{11}$	$-1.449\cdot10^{12}$
3	0.3	$-1.396\cdot10^{17}$	$-3.49\cdot10^{17}$
4	0.4	$-3.361\cdot10^{22}$	$-8.403\cdot10^{22}$
5	0.5	$-8.094\cdot10^{27}$	$-2.023\cdot10^{28}$
6	0.6	$-1.949\cdot10^{33}$	$-4.872\cdot10^{33}$
7	0.7	$-4.692\cdot10^{38}$	$-1.173\cdot10^{39}$
8	0.8	$-1.13\cdot10^{44}$	$-2.825\cdot10^{44}$
9	0.9	$-2.721\cdot10^{49}$	$-6.801\cdot10^{49}$
10	1	$-6.551\cdot10^{54}$	$-1.638\cdot10^{55}$

i := 0 ..10

$Z_{1,2} = -6.02 \times 10^{6}$

$Z_{2,2} = -1.449 \times 10^{12}$

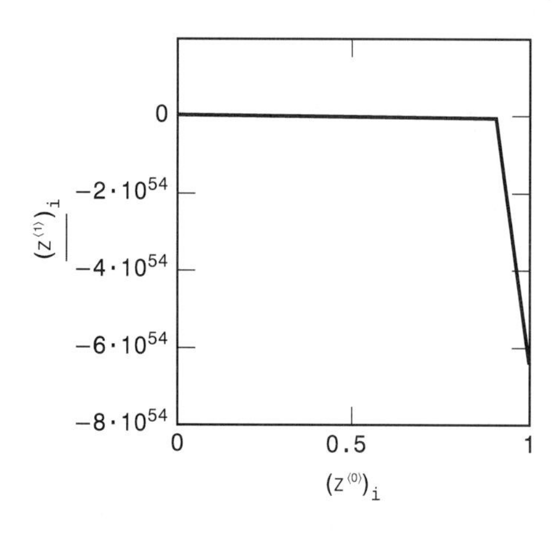

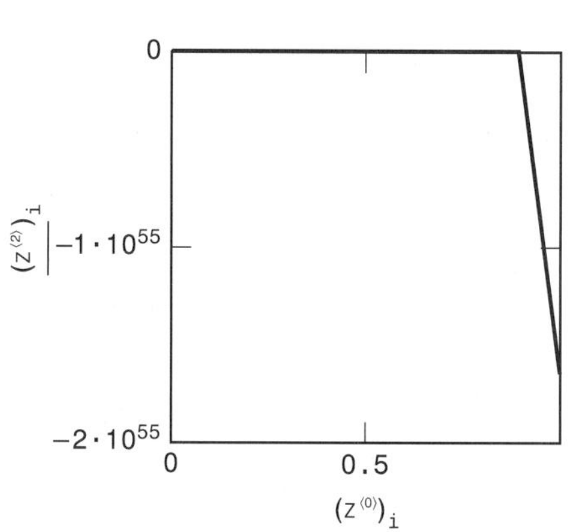


```
Z := Rkadapt(x,0,1,10,D)
```

Z =

	0	1	2
0	0	10	−20
1	0.1	18.096	4.523
2	0.2	16.374	4.093
3	0.3	14.816	3.702
4	0.4	13.406	3.351
5	0.5	12.13	3.032
6	0.6	10.976	2.744
7	0.7	9.931	2.481
8	0.8	8.986	2.246
9	0.9	8.131	2.032
10	1	7.357	1.838

i := 0..10

$Z_{1,2} = 4.523$

$Z_{2,2} = 4.093$

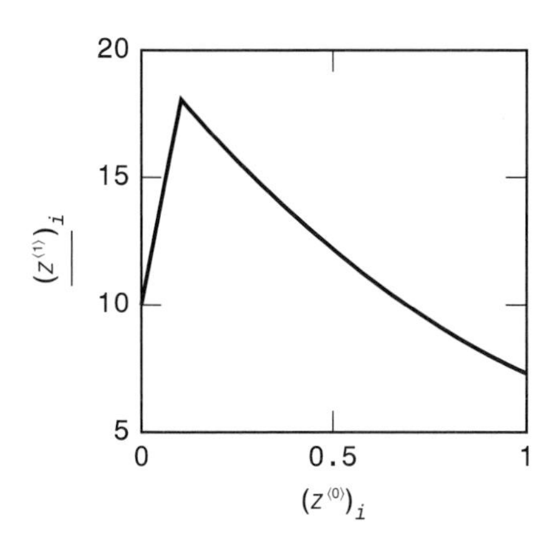

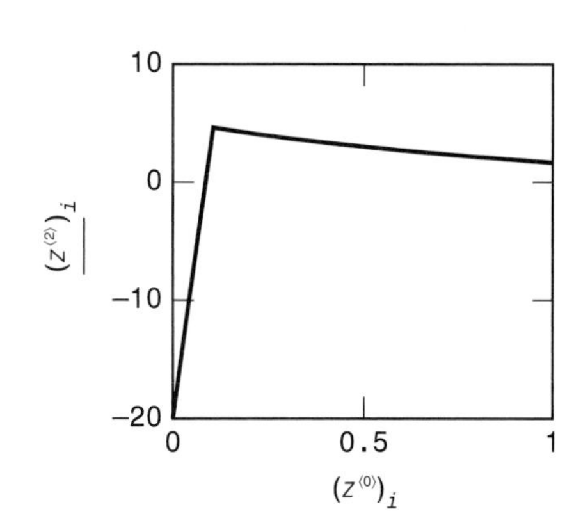

◀

9.15 Computer Programs

9.15.1 Fortran Programs

Program 9.1 Hamming's predictor–corrector method

The following input data are required:

N = number of steps to be used
H = step length to be used (so that the final value of the independent variable = X(1) + N*H)
X(1) = initial value of x
Y(1) = initial value of the dependent variable, y (initial condition)
Function $f(x, y)$ in the form of subroutine
(Differential equation is of the form $y'(x) = f(x, y)$)

```
SUBROUTINE FUN1(XM, YM1, YI1)
YI1 = defined in terms of y (YM1) and x (XM)
RETURN
END
```

Illustration (Example 9.15)

Program output:

```
SOLUTION BY HAMMING METHOD

STEP LENGTH = 0.100
NUMBER OF STEPS =  10
INITIAL CONDITIONS:  X(1)   =   0.0000E+00   Y(1)   =   0.1000E+01
```

STEP (I)	XX	Y(I)	YYT	ERR
2	0.100000E+00	0.101034E+01	0.101034E+01	0.238419E-06
3	0.200000E+00	0.104419E+01	0.104281E+01	-0.138474E-02
4	0.300000E+00	0.110398E+01	0.109972E+01	-0.425923E-02
5	0.400000E+00	0.118844E+01	0.118365E+01	-0.478637E-02
6	0.500000E+00	0.130257E+01	0.129744E+01	-0.512588E-02
7	0.600000E+00	0.144963E+01	0.144424E+01	-0.538862E-02
8	0.700000E+00	0.163345E+01	0.162751E+01	-0.593972E-02
9	0.800000E+00	0.185766E+01	0.185108E+01	-0.658250E-02
10	0.900000E+00	0.212651E+01	0.211921E+01	-0.730228E-02
11	0.100000E+01	0.244464E+01	0.243656E+01	-0.807238E-02

Program 9.2 Runge–Kutta methods

The following input data are required:

NN = order of the Runge–Kutta method to be used
= 1 for first-order Runge–Kutta method
= 2 for second-order Runge–Kutta method
= 3 for third-order Runge–Kutta method
= 4 for fourth-order Runge–Kutta method (Runge's)
= 5 for fourth-order Runge–Kutta method (Kutta's)
= 6 for fourth-order Runge–Kutta method (Runge–Kutta–Gill)
= 7 for fourth-order Runge–Kutta method (Runge–Kutta–Fehlberg)
= 8 for fifth-order Runge–Kutta–Fehlberg method
N = number of steps to be used
H = step length (so that the final value of the independent variable X(1) + N*H)
X(1) = initial value of x
Y1(1), ..., Y8(1) = initial value of y (initial condition)
Function $f(x, y)$ in the form of subroutine

```
SUBROUTINE FUN1(XM, YM1, XI, YI1)
YI1 = defined in terms of y (YM1) and x (XM)
RETURN
END
```

Illustration (Example 9.9).

Using N = 10, H = 0.1, X(1) = 0.0, Y1(1) = Y2(1) = Y3(1) = Y4(1) = Y5(1) = Y6(1) = Y7(1) = Y8(1) = 1.0, the output of Program 9.2 is as follows:

```
SOLUTION BY DIFFERENT ORDER RUNGE--KUTTA METHODS

STEP LENGTH = 0.100
NUMBER OF STEPS =  10
```

```
INITIAL CONDITIONS:
X(1) =  0.000000E+00
Y1(1)=Y2(1)=Y3(1)=Y4(1)=Y5(1)=Y6(1)=Y7(1)=Y8(1)=   0.100000E+01

STEP (I)    X(I)        YI4            YI5            YI6
                        YI7            YI8            YIT

YI4 = 4TH ORDER RUNGE--KUTTA (RUNGE)    YI5 = 4TH ORDER RUNGE--KUTTA (KUTTA)
YI6 = 4TH ORDER RUNGE--KUTTA-GILL       YI7 = 4TH ORDER R--K--FEHLBERG
YI8 = 5TH ORDER R--K--FEHLBERG          YIT = EXACT SOLUTION

   1      0.1000      1.01034164     1.01034164     1.01034164
                      1.01034188     1.01034188     1.01034188

   2      0.2000      1.04280508     1.04280508     1.04280508
                      1.04280555     1.04280555     1.04280555

   3      0.3000      1.09971690     1.09971690     1.09971690
                      1.09971774     1.09971762     1.09971774

   4      0.4000      1.18364835     1.18364835     1.18364835
                      1.18364954     1.18364942     1.18364954

   5      0.5000      1.29744112     1.29744112     1.29744112
                      1.29744279     1.29744256     1.29744244

   6      0.6000      1.44423580     1.44423580     1.44423580
                      1.44423795     1.44423759     1.44423771

   7      0.7000      1.62750316     1.62750316     1.62750316
                      1.62750578     1.62750542     1.62750530

   8      0.8000      1.85107899     1.85107899     1.85107899
                      1.85108232     1.85108185     1.85108185

   9      0.9000      2.11920261     2.11920261     2.11920261
                      2.11920691     2.11920619     2.11920643

  10      1.0000      2.43655920     2.43655920     2.43655920
                      2.43656445     2.43656349     2.43656349
```

9.15.2 C Programs

Program 9.3 Taylor's methods of order 1, 2, 3, and 4

The following input data are required:

n = number of steps to be used
h = step length (so that the final value of the independent variable is $x[1] + n * h$)
$x[1]$ = initial value of x

$y1[1], y2[1], y3[1], y4[1]$ = initial value of y (initial condition) for Taylor's method of order 1, 2, 3, and 4
Function $f(x, y)$ in the form of subprogram
(Differential equation is of the form $y'(x) = f(x, y)$)
fun (xm, ym, yi)
yi = defined in terms of $y(ym)$ and $x(xm)$
return
Derivatives f', f'' and f''' are also to be evaluated in terms of $y(ym)$ and $x(xm)$

Illustration (Example 9.6)

Program output:

```
Results of Taylors methods of order 1, 2, 3 and 4

Step length = 0.100
Number of steps =  10
Initial conditions: x(1)  =  0.0000E+00 y1(1)  =  0.1000E+01
```

i	x(i)	y1(i) first order	y2(i) second order	y3(i) third order	y4(i) fourth order	yit exact sol	err error in y1(i)
1	0.100	1.0000	1.0100	1.0103	1.0103	1.0103	0.0103
2	0.200	1.0200	1.0421	1.0428	1.0428	1.0428	0.0228
3	0.300	1.0620	1.0985	1.0997	1.0997	1.0997	0.0377
4	0.400	1.1282	1.1818	1.1836	1.1836	1.1836	0.0554
5	0.500	1.2210	1.2949	1.2974	1.2974	1.2974	0.0764
6	0.600	1.3431	1.4409	1.4442	1.4442	1.4442	0.1011
7	0.700	1.4974	1.6231	1.6274	1.6275	1.6275	0.1301
8	0.800	1.6872	1.8456	1.8509	1.8511	1.8511	0.1639
9	0.900	1.9159	2.1124	2.1190	2.1192	2.1192	0.2033
10	1.000	2.1875	2.4282	2.4364	2.4366	2.4366	0.2491

Program 9.4 Adams–Bashforth method (6th order method)

The following input data are required:

n = number of steps to be used
h = step length (so that the final value of independent variable is $x[1] + n * h$)
x[1], ..., x[6] = values of x, $x[i] = x[1] + (i - 1) * h; i = 2, 3, \ldots, 6$
$y6[1], y6[2], \ldots, y6[6]$ = values of dependent variable y at $x[1], x[2], \ldots, x[6]$, respectively
Function $f(x, y)$ in the form of subprogram
(Differential equation is of the form $y'(x) = f(x, y)$)
fun (xm,ym,yi)

yi = defined in terms of $y(ym)$ and $x(xm)$
return

Illustration (Example 9.11).

Using $n = 10$, h = 0.1, x[1] = 0.0, x[2] = 0.1, x[3] = 0.2, x[4] = 0.3, x[5] = 0.4, x[6] = 0.5, y6[1] = 1.0, y6[2] = 1.01034260, y6[3] = 1.04280663, y6[4] = 1.09971809, y6[5] = 1.18365002, and y6[6] = 1.29744339, the output of Program 9.4 is given as follows:

```
Adams-Bashforth Method (N=6)

Step length  =  0.100
Number of steps  =  10
Initial conditions: x(1) = 0.0000E+00 y6(1)  =  0.1000E+01

step (i)    xi          y6(i)           yit             error

   2      0.1000   0.10103419E+01   0.10103419E+01   0.00000000E+00
   3      0.2000   0.10428056E+01   0.10428056E+01   0.00000000E+00
   4      0.3000   0.10997177E+01   0.10997177E+01   0.00000000E+00
   5      0.4000   0.11836495E+01   0.11836495E+01   0.00000000E+00
   6      0.5000   0.12974424E+01   0.12974424E+01   0.00000000E+00
   7      0.6000   0.14442375E+01   0.14442377E+01   0.23841858E-06
   8      0.7000   0.16275052E+01   0.16275053E+01   0.11920929E-06
   9      0.8000   0.18510814E+01   0.18510818E+01   0.47683716E-06
  10      0.9000   0.21192055E+01   0.21192064E+01   0.95367432E-06
  11      1.0000   0.24365628E+01   0.24365635E+01   0.71525574E-06
```

REFERENCES AND BIBLIOGRAPHY

9.1. W. H. Enright, T. E. Hull, and B. Lindberg, "Comparing Numerical Methods for Stiff Systems of Ordinary Differential Equations," *BIT*, Vol. 15, pp. 10–48, 1975.

9.2. E. Fehlberg, "New High-Order Runge–Kutta Formulas with Step-Size Control for Systems of First- and Second-Order Differential Equations," *ZAMM*, Vol. 44, pp. 17–29, 1964.

9.3. E. Fehlberg, "New High-Order Runge–Kutta Formulas with an Arbitrary Small Truncation Error," *Zeitschrift fur Angewandte Mathematik und Mechanik*, Vol. 46, pp. 1–16, 1966.

9.4. E. Fehlberg, "Klassische Runge–Kutta formeln vierter und niedrigerer Ordnung mit Schrittweiteen–Kontrolle und ihre Anwendung auf Warmeleitungs-probleme," *Computing*, Vol. 6, pp. 61–71, 1970.

9.5. B. A. Finlayson, *Nonlinear Analysis in Chemical Engineering*, McGraw-Hill, New York, 1980.

9.6. C. W. Gear, *Numerical Initial-Value Problems in Ordinary Differential Equations*, Prentice–Hall, Englewood Cliffs, NJ, 1971.

9.7. T. E. Hull, W. H. Enright, B. M. Fellen, and A. E. Sedgewick, "Comparing Numerical Methods for Ordinary Differential Equations," *SIAM Journal of Numerical Analysis*, Vol. 9, No. 4, pp. 603–637, 1972.

9.8. T. E. Hull and W. H. Enright, "Test Results on Initial-Value Methods for Nonstiff Ordinary Differential Equations," *SIAM Journal of Numerical Analysis*, Vol. 13, pp. 944–961, 1976.

9.9. J. M. A. Danby, *Computing Applications to Differential Equations*, Reston Publishing, Reston, VA, 1985.

9.10. L. Collatz, *The Numerical Treatment of Differential Equations*, 3d edition, Springer–Verlag, Berlin, 1960.

9.11. B. Carnahan, H. A. Luther, and J. O. Wilkes, *Applied Numerical Methods*, Wiley, New York, 1969.

9.12. R. W. Hamming, *Numerical Methods for Scientists and Engineers*, 2d edition, McGraw-Hill, New York, 1973.

9.13. S. S. Rao, *Mechanical Vibrations*, 3d edition, Addison-Wesley, Reading, MA, 1995.

9.14. A. Constantinedes, *Applied Numerical Methods with Personal Computers*, McGraw-Hill, New York, 1987.

9.15. I. Cochin, *Analysis and Design of Dynamic Systems*, Harper & Row, New York, 1980.

REVIEW QUESTIONS

The following questions along with corresponding answers are available in an interactive format at the Companion Web site at http://www.prenhall.com/rao.

9.1. **Indicate the order** of the following differential equations:

(a)

$$x^3 y'' + x^2 y' + xy = \cos x.$$

(b)

$$y'' + 2xy^2 = 0.$$

(c)

$$y''' + x^2 y' + \sin^2 xy = x^4.$$

(d)

$$c_1 y^{\mathtt{iv}} + c_2 y^{\mathtt{iii}} + c_3 y^{\mathtt{ii}} + c_4 y^{\mathtt{i}} + c_5 y = c_6.$$

$(c_1, c_2, \ldots, c_6$ are constants).

(e)

$$y' - 2xy = 0.$$

9.2. State whether the following differential equations are **linear or nonlinear**:

(a)

$$x(y'')^3 + (y')^5 - y = 0.$$

(b)

$$yy'' = x.$$

(c)

$$y'' + y + x\cos x = 0.$$

(d)

$$y' = -\frac{2xy^3}{3x^2 + y^2}.$$

(continued, page 706)

(e)

$$y' = y^{\frac{2}{3}}.$$

9.3. **Determine** whether the following problems are **initial- or boundary-value** problems:

(a)

$$A\frac{d^4w}{dx^4} = B;\ w(0) = w(L) = 0;\ \frac{d^2w}{dx^2}(0) = \frac{d^2w}{dx^2}(L) = 0.$$

(b)

$$y'' + \lambda y = 0;\ y(0) = 0,\ y(1) = 0.$$

(c)

$$y'' + \sin xy' + \cos xy = 0;\ y(0) = 1,\ y'(0) = 0.$$

(d)

$$y'' + \lambda y = 0;\ y(0) + y'(0) = 0,\ y(1) = 0.$$

(e)

$$\frac{d^4w}{dx^4} - \lambda w = 0;\ w(0) = \frac{dw}{dx}(0) = w(1) = \frac{dw}{dx}(1) = 0.$$

9.4. **Find** whether the following **differential equations** are **autonomous**:

(a)

$$y'' + y + y^3 = 0;\ y(0) = c,\ y'(0) = 0$$

$$(c = \text{constant}).$$

(b)

$$\frac{d^2i}{dt^2} + \frac{R}{L}\frac{di}{dt} + \frac{1}{LC}i = 0$$

$$(R, L, \text{ and } C \text{ are constants}).$$

(c)

$$\left(\frac{d^3y}{dx^3}\right)^4 - \frac{dy}{dx} - 2xy = 0.$$

(*continued, page 707*)

(d)

$$\left(\frac{d^2y}{dx^2}\right)^3 - \left(\frac{dy}{dx}\right)^4 = y.$$

(e)

$$\frac{dy}{dx} + y = x^2.$$

9.5. **Identify** whether the following **equations** are **homogeneous or nonhomogeneous**:

(a)

$$\frac{dy}{dx} = e^x.$$

(b)

$$\frac{d^2y}{dx^2} - 20\frac{dy}{dx} + 50y = 0.$$

(c)

$$\frac{d^2y}{dx^2} - 20\frac{dy}{dx} + 100y = 10.$$

(d)

$$\frac{dy}{dx} = e^{x+y}.$$

(e)

$$\frac{d^4w}{dx^4} - p(x) = 0.$$

9.6. **Define** the following:

Ordinary differential equation, autonomous equation, first-order differential equation, homogeneous differential equation, stiff equation, stable method, convergent method, single-step method, multistep method, and predictor–corrector method.

9.7. Answer **true or false**:

1. A set of differential equations of different orders cannot be expressed in a canonical form.

(*continued, page 708*)

2. Heun's method can be considered a simple form of the predictor–corrector method.
3. It is possible to have Runge–Kutta method of any order.
4. Single step methods are not self-starting.
5. Multistep methods are not self-starting.
6. Hamming's method can be considered a generalization of Adams and Milne's methods.
7. The equation $\frac{dy}{dx} + 5y^2 = \sin x$ is linear.
8. We can have a differential equation of zeroth order.
9. A nonlinear differential equation is characterized by the appearance of nonlinear terms in terms of the independent variable.
10. Any nth order linear differential equation can be expressed as a set of n first-order linear differential equations.
11. An nth order nonlinear differential equation can be expressed as a set of n first-order nonlinear differential equations.
12. An autonomous differential equation is one in which the independent variable does not appear explicitly in the equation.
13. Any solution method that is applicable for a single-differential equation is also applicable for the solution of a set of differential equations.
14. Euler's method is an improvement on the point–slope method.
15. The numerical solution deviates more and more from the exact solution with increasing values of the independent variable, x_i.
16. The global error represents the sum of the local and propagated errors.
17. Euler's method is self-starting.
18. Heun's method is a second-order method.
19. All Runge–Kutta methods are not self-starting.
20. Euler's method is same as the first-order Runge–Kutta method.
21. Modified Euler's, Heun's, and Ralston's methods are all second-order Runge–Kutta methods.

9.8. **Fill in the blanks**:

1. The boundary-value problem is one in which the dependent variable or its derivatives are known at more than one point of the ________ variable.
2. Euler's method is________ efficient compared with most other numerical methods.
3. Euler's method is also known as Euler–________ method.
4. The round-off error can be reduced by using________ precision in the computations.

(*continued, page 709*)

5. The________ error is caused by the approximate procedure used in the computations.
6. The propagated error is important for the________ of the numerical method.
7. Euler's method is considered a first-order method, since the total truncation error is of the order of the ________ ________.
8. Even when the step size is reduced sufficiently, the numerical solution will not be exact due to the presence of________ error.
9. A numerical method is considered unstable if it gives an________ solution for a problem whose exact solution is bounded.
10. Euler's method uses the value of the derivative $\frac{dy}{dx}$ at ________ to be valid over the interval, $x_{i-1} \leq x \leq x_i$.
11. Heun's method uses the average value of the derivative $\frac{dy}{dx}$ at________ and________ to be valid over the interval, $x_{i-1} \leq x \leq x_i$.
12. Heun's method can be considered as a________ -order method.
13. The first-order Runge–Kutta method is the same as the________ method.
14. Although both Adams and Milne's methods have the same order of error, Milne's method is not commonly used due to the possibility of ________.

9.9. **Give short answers** to the following:

1. Indicate two different approaches for solving a differential equation numerically.
2. When is a numerical method said to be convergent?
3. What is the difference between a general boundary-value problem and a two-point boundary-value problem?
4. When is a set of differential equations said to be in canonical form?
5. What is the difference between an autonomous and a nonautonomous differential equation?
6. How many initial conditions are needed for the solution of a second-order differential equation?
7. Why is Euler's method called a first-order method?
8. When is a numerical method considered to be unstable?
9. When is a numerical method considered to be self-starting?

9.10. **Indicate** the **most appropriate answer** out of the multiple choices given:

1. Euler's method can be considered as Taylor's method of order
   (a) 0 (b) 1 (c) 2.
2. The numerical stability of a method depends on
   (a) only the numerical method.
   (b) only the differential equation.
   (c) both the numerical method and the differential equation.

(*continued, page 710*)

3. The order of the most popularly used Runge–Kutta–Gill method is
   (a) 4 (b) 3 (c) 2.
4. A stiff differential equation is characterized by
   (a) large values of eigenvalues.
   (b) large difference between the minimum and maximum eigenvalues.
   (c) small values of eigenvalues.
5. Stiff differential equations can be solved efficiently using
   (a) implicit methods.
   (b) explicit methods.
   (c) combination of implicit and explicit methods.

9.11. **Match the following:**

Equation	Nature of the equation
1. $c^2 \frac{\partial^2 y}{\partial x^2} = \frac{\partial y}{\partial t}$	(a) Ordinary differential equation
2. $c^2 \frac{d^2 y}{dx^2} = x + y$	(b) Nonlinear differential equation
3. $\frac{dy}{dx} + cy = 0$	(c) Second-order linear homogeneous equation
4. $\frac{d^2 y}{dt^2} + c^2 \sin y = 0$	(d) Initial value problem
5. $\frac{d^2 y}{dt^2} + c^2 y = 0$	(e) Partial differential equation
6. $\frac{d^3 y}{dx^3} + x \frac{dy}{dx} = x^3$	(f) Boundary-value problem
7. $y'' + cy = 0,\ y(0) = 1,\ y'(0) = 0$	(g) First-order equation
8. $y'' = f(x, y, y'),\ y(x_1) = y_1,\ y(x_2) = y_2$	(h) Third-order equation

PROBLEMS

Section 9.1

9.1. Find the analytical solution of the equation

$$y'' - 8y' + 15y = 225x$$

satisfying the initial conditions $y(0) = 1$ and $y'(0) = 0$.

9.2. Find the analytical solution of the equation

$$y'' - 3y' + 3y = 0.$$

Section 9.2

9.3. A sum of P dollars is deposited in a bank that pays an interest of r percent. If additional funds are deposited continuously at the rate of c dollars per unit period, derive the differential equation for determining the value of the investment at the end of k periods. Assume that the interest is compounded continuously.

9.4. A person opens an individual retirement account (IRA) at age 30 by depositing $2000 and subsequently deposits $2000 every year. Determine the amount accumulated at age 65 if the interest rate is (a) 6%, (b) 8%, and (c) 10%.

9.5. A charged particle of mass m is subjected to a steady force $\vec{F}$. If a resisting force, proportional to the speed, acts on the particle, derive the equation of motion of the particle.

9.6. A railway wagon, moving with velocity v, strikes a viscously damped buffer. The equation of motion is given by

$$m\frac{dv}{dt} = -cv; \; v(0) = v_0 = \text{known}.$$

Formulate the problem of finding the velocity of the truck, $v(t)$, at time $t = 1$ sec. Assume the values of the mass, damping, and initial velocity as $m = 5,000$ kg, $c = 15,000$ N-s/m, and $v_0 = 3$ m/s.

9.7. The deflection equation of a cantilever beam, of length l, is given by

$$EI(x)y'' = M(x); \; y(0) = y'(0) = 0,$$

where $y(x)$ is the deflection and $M(x)$ is the bending moment of the beam at x. Formulate the problem of finding $y(x)$ for (a) an end-loaded beam with $EI = w = 1$(w = uniformly distributed load per unit length) and (b) a tapered cantilever beam with loading as shown in Fig. 9.10.

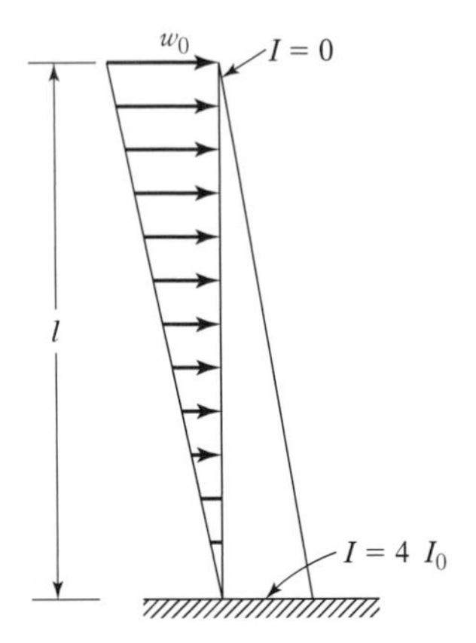

Figure 9.10 Tapered cantilever beam.

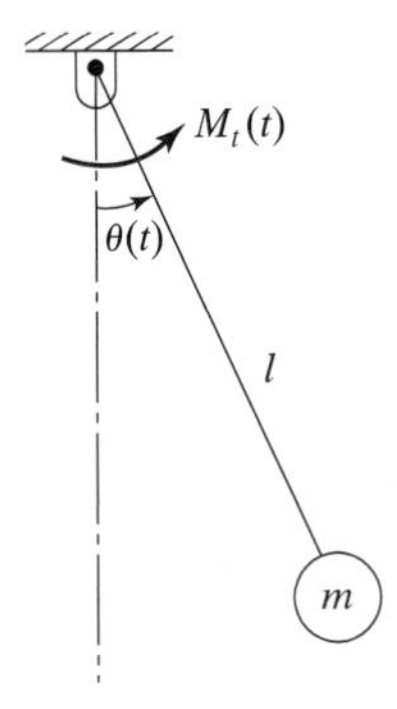

Figure 9.11 Simple pendulum.

Section 9.3

9.8. The equation of motion of a simple pendulum, subject to damping and external torque (Fig. 9.11), is given by

$$ml^2 \frac{d^2\theta}{dt^2} + c\frac{d\theta}{dt} + mgl\sin\theta = M_t(t),$$

where m = mass of the bob, l = length, c = damping constant, g = acceleration due to gravity, $M_t(t)$ = external torque, θ = angular deflection, and t = time. Express this equation as a system of linear first-order equations.

9.9. The equations of motion of the spring–mass–damper system shown in Fig. 9.12 are given by

$$[m]\ddot{\vec{x}} + [c]\dot{\vec{x}} + [k]\vec{x} = \vec{F}, \tag{E1}$$

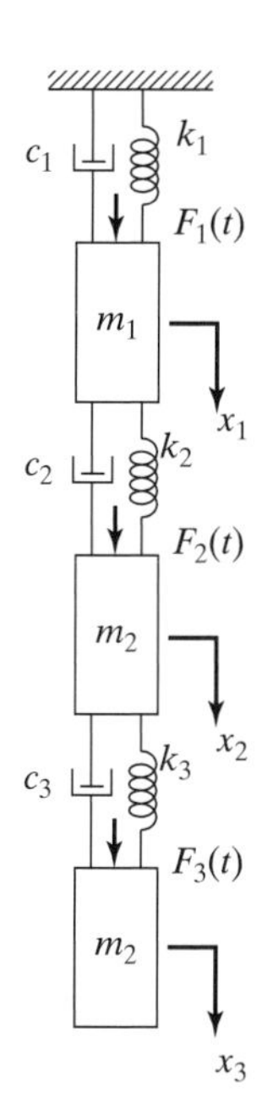

Figure 9.12 Three-degree of freedom spring-mass-damper system.

where

$$[m] = \begin{bmatrix} m_1 & 0 & 0 \\ 0 & m_2 & 0 \\ 0 & 0 & m_3 \end{bmatrix}, [c] = \begin{bmatrix} c_1 + c_2 & -c_2 & 0 \\ -c_2 & c_2 + c_3 & -c_3 \\ 0 & -c_3 & c_3 \end{bmatrix},$$

$$[k] = \begin{bmatrix} k_1 + k_2 & -k_2 & 0 \\ -k_2 & k_2 + k_3 & -k_3 \\ 0 & -k_3 & k_3 \end{bmatrix},$$

and

$$\vec{x} = \begin{Bmatrix} x_1(t) \\ x_2(t) \\ x_3(t) \end{Bmatrix}, \dot{\vec{x}} = \begin{Bmatrix} \frac{dx_1(t)}{dt} \\ \frac{dx_2(t)}{dt} \\ \frac{dx_3(t)}{dt} \end{Bmatrix}, \ddot{\vec{x}} = \begin{Bmatrix} \frac{d^2x_1(t)}{dt^2} \\ \frac{d^2x_2(t)}{dt^2} \\ \frac{d^2x_3(t)}{dt^2} \end{Bmatrix}, \text{ and } \vec{F} = \begin{Bmatrix} F_1(t) \\ F_2(t) \\ F_3(t) \end{Bmatrix}.$$

Convert Eqs. (E1) into a system of first-order differential equations.

Section 9.4

9.10. In the mirror used in a solar heater, all the incident light rays are required to reflect through a single point (focus) as shown in Fig. 9.13. For this, the shape of the mirror is governed by the equation

$$x\left(\frac{dy}{dx}\right)^2 - 2y\frac{dy}{dx} - x = 0.$$

Propose a method of solving this equation.

9.11. The base of a spring–mass system with Coulomb damping is connected to a slider crank mechanism as shown in Fig. 9.14. The equation of motion of the mass, m, is given by [9.13]

$$m\ddot{x} + kx \pm \mu N = ky, \tag{E1}$$

where

$$y(x) = r - r\cos\omega t + \frac{1}{4}\left(\frac{r}{l}\right)^2 - \frac{l}{4}\left(\frac{r}{l}\right)^2 \cos 2\omega t + \ldots \tag{E2}$$

is the base displacement, r = crank length, l = connecting rod length, ω = angular velocity of the crank, μ = coefficient of friction between the mass (m)

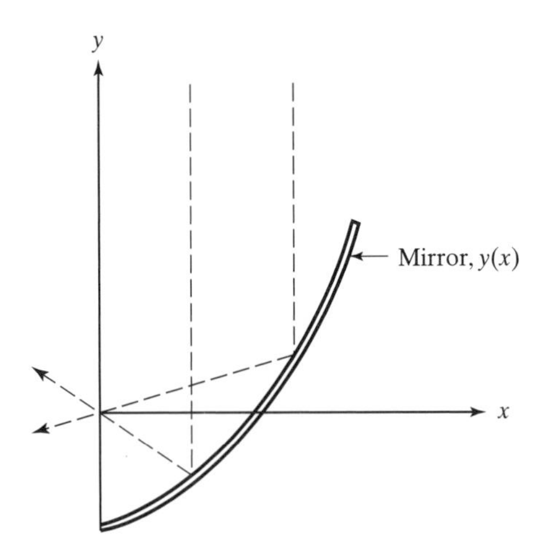

Figure 9.13 Mirror in a solar heater.

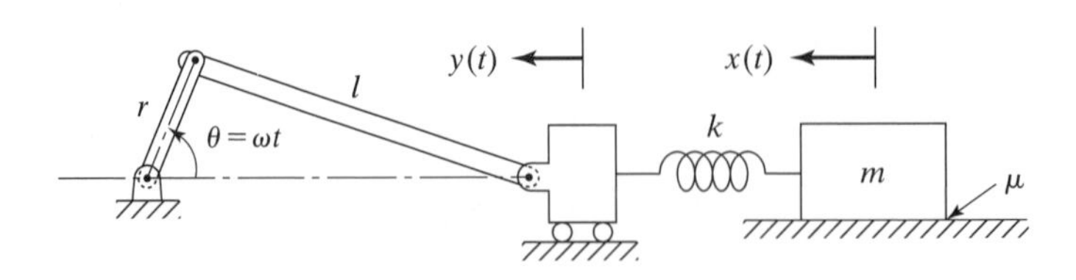

Figure 9.14 Spring-mass system connected to slider crank mechanism.

and the horizontal surface, and N = weight of the mass. In Eq. (E1), the term μN should have a positive sign when $\frac{dx}{dt}$ is positive, and a negative sign when $\frac{dx}{dt}$ is negative. Indicate a procedure for the numerical solution of Eq. (E1).

Section 9.5

9.12. Determine the solution of the equation

$$y' = 2 - 3x + 4y; \; y(0) = 1$$

using Euler's method with a step size of (a) $h = 0.1$, (b) $h = 0.05$, and (c) $h = 0.01$. Plot the solutions over the interval $0 \le x \le 1$.

9.13. Find the solution of Example 9.7 using Euler's method with a step size of $h = \Delta x = 0.1$.

Section 9.6

9.14. Find the solution of the equation

$$y' = 2 - 3x + 4y; \; y(0) = 1$$

over the interval $0 \le x \le 1$ using Taylor's series methods of order 1, 2, 3, and 4. Use a step size of $h = 0.1$.

9.15. Find the solution of the problem

$$y' = 2 - 3x + 4y; \; y(0) = 1$$

in the interval $0 \le x \le 1$ using Heun's method. Use a step size of $h = 0.1$.

9.16. Find the solution of the problem

$$y' = 2 - 3x + 4y; \; y(0) = 1$$

in the interval $0 \le x \le 1$ with $h = 0.1$ using the modified Euler's method.

Section 9.7

9.17. Find the solution of Problem 9.16 using Ralston's method.

9.18. Find the solution of Problem 9.16 using the fourth-order Runge–Kutta methods.

9.19. Find the solution of Problem 9.16 using the Runge–Kutta–Fehlberg method.

9.20. Find the solution of Problem 9.16 using the Runge–Kutta–Fehlberg method with the step size $h = 0.05$.

Section 9.9

9.21. Find the solution of the problem

$$y' = 2 - 3x + 4y; \; y(0) = 1$$

in the interval $0 \leq x \leq 1$ with $h = 0.1$ using Adams–Bashforth open formulas of order 2 through 6.

9.22. Find the solution of the problem

$$y' = 2 - 3x + 4y;\ y(0) = 1$$

in the interval $0 \leq x \leq 1$ with $h = 0.1$ using Adams–Moulton closed formulas of order 1 through 6.

Section 9.10

9.23. Find the solution of the problem

$$y' = 2 - 3x + 4y;\ y(0) = 1$$

at $x = 0.4$ using the fourth-order Adams predictor–corrector method with $h = 0.05$.

9.24. Solve Problem 9.23 using Milne's method.

9.25. Solve Problem 9.23 using Hamming's method.

Section 9.11

9.26. Each of the two tanks shown in Fig. 9.15 contain 100 gallons of water. The water in tank 1 contains 1000 grams of dissolved salt, while the water in tank 2 contains 100 grams of dissolved salt. At time $t = 0$, pipe 1 starts pumping pure water at the rate of 10 gallons/minute, while pipes 2, 3, and 4 start pumping mixed water at rates of 15, 5, and 10 gallons/minute, respectively, in the directions indicated in Fig. 9.16. To determine the amounts (concentrations) of salt in the two tanks at any instant, the following equations are to be solved:

$$\frac{dx}{dt} = \frac{5}{100}y - \frac{15}{100}x$$

and

$$\frac{dy}{dt} = \frac{15}{100}x - \frac{15}{100}y$$

with $x(t = 0) = 1,000$, $y(t = 0) = 100$, $x(t) =$ amount of salt in tank 1 at time t, and $y(t) =$ amount of salt in tank 2 at time t. Determine the concentration of salt in the tanks at $t = 3$.

Section 9.12

9.27. (a) Solve the following system of equations over $0 \leq t \leq 1$ using Euler's method:

$$x_1'(t) = -7x_1(t) + 8x_2(t);\ x_1(0) = 5$$

and

$$x_2'(t) = -6x_1(t) + 7x_2(t);\ x_2(t) = 0.$$

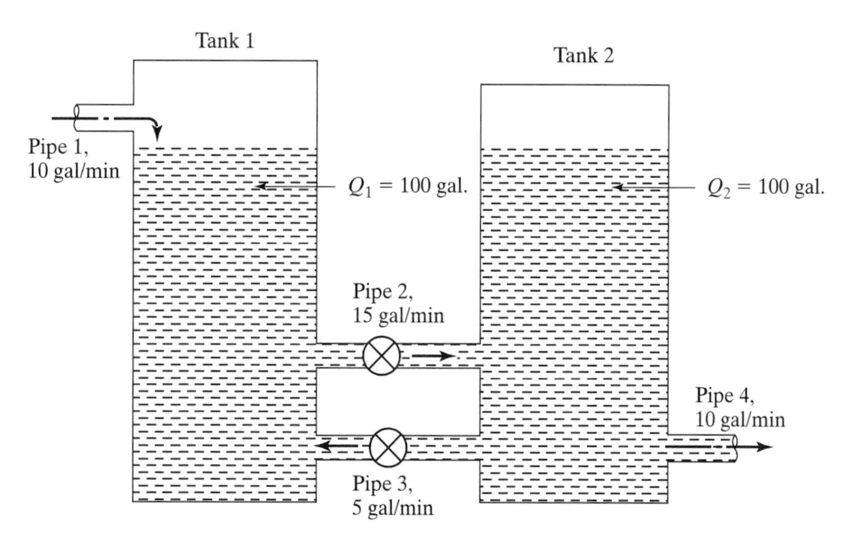

Figure 9.15 Two connected tanks with dissolved salt.

(b) Compare the solution with the exact solution given by

$$x_1(t) = 20e^{-t} - 15e^{-500t};\ x_2(t) = 15e^{-t} - 15e^{-500t}.$$

Section 9.14

9.28. Solve the following equation using MAPLE:

$$y'' = 4y + \sin 3x$$

with $y(0) = 0$ and $y'(0) = 0$.

9.29. Solve the equation given in Problem 9.28 using MATLAB.

9.30. Solve the equation given in Problem 9.28 using MATHCAD.

9.31. Solve the following equation using (a) MATLAB, (b) MAPLE, and (c) MATHCAD:

$$y' = 20 - 4y;\ 0 \le t \le 5$$
$$y(0) = 1.$$

9.32. Solve Example 9.16 using MATLAB functions ode23, ode45, ode15s, and ode23s with a time step $\Delta t = 0.01$.

Section 9.15

9.33. Use Program 9.1 to find the solution of the equation, $y' = 0.5x - 0.5y$, with $y(0) = 1$ over the interval $0 \le x \le 2$.

9.34. Use Program 9.2 to solve the differential equation given in Problem 9.33.

9.35. Find the solution of the differential equation given in Problem 9.33 using Program 9.3

9.36. Find the solution of the differential equation given in Problem 9.33 using Program 9.4

General

9.37. If the rate of change of the population of elephants at any time is directly proportional to the population at that time, determine the population at time t.

9.38. A sum of \$100 is deposited in a bank that pays 10% interest. Find the value of the investment at the end of 5 years if the interest is compounded (a) once every year, (b) once every month, (c) once every day, and (d) continuously.

9.39. The angular deflection of the pointer of an electrical meter is given by

$$A\frac{d^2\phi}{dt^2} + B\frac{d\phi}{dt} + C\phi = 0,$$

where A is a function of the electrical inductance and inertia of the pointer and coil, B is a function of the friction and circuit resistance, and C is a function of the stiffness of the spring and the capacitance. Find the analytical solution of the equation for $A = 1$, $B = -20$, and $C = 50$.

9.40. A ball of mass m is thrown vertically up with an initial velocity, v_0. Determine the equation for finding the height reached by the ball.

9.41. One end of a spring of stiffness k is attached to a pivot point O and the other end to a mass m, and the resulting spring–mass system is made to oscillate as a simple pendulum as shown in Fig. 9.16. If the unstretched length of the spring is l_0, the equations of motion along the directions of l and θ yield

$$\frac{d^2 l}{dt^2} - l\left(\frac{d\theta}{dt}\right)^2 - g\cos\theta + \frac{k}{m}(l - l_0) = 0$$

and

$$l^2\frac{d^2\theta}{dt^2} + 2l\frac{dl}{dt}\frac{d\theta}{dt} + gl\sin\theta = 0.$$

State these equations of motion as a system of first-order equations.

9.42. Solve Problem 9.15 using a step size of $h = 0.05$.

9.43. Find the solution of Problem 9.16 using the fourth-order Runge–Kutta methods with the step size $h = 0.05$.

9.44. Show that the error in the corrector formula, Eq. (9.140), is given by Eq. (9.142).

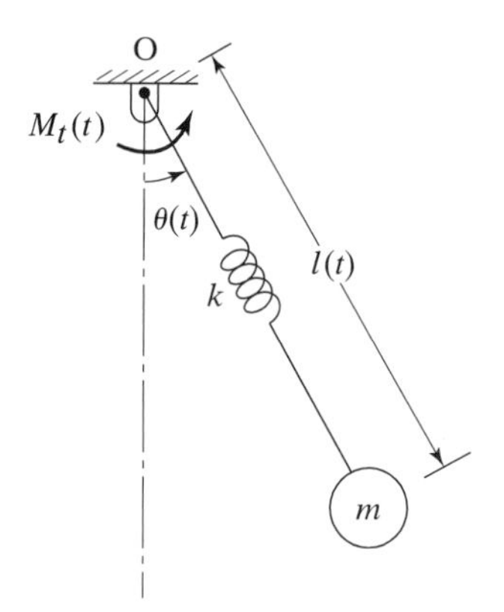

Figure 9.16 Oscillation of spring-mass system.

PROJECTS

9.1. Consider the equation of motion governing large deflections of a simple pendulum:

$$ml^2 \frac{d^2\theta}{dt^2} + c\frac{d\theta}{dt} + mgl \sin\theta = M_t(t), \tag{E1}$$

where m = mass of the bob, l = length, c = damping constant, g = acceleration due to gravity, $M_t(t)$ = external torque, θ = angular deflection, and t = time.

(a) Linearize the equation for small angular displacements using the approximation, $\sin\theta \approx \theta$. Find the solution of the linearized equation for the following conditions:

$$l = 1 \text{ m}, g = 9.81 \text{ m/s}^2, c = M_t(t) = 0, \theta(t=0) = 10^\circ, \text{ and } \frac{d\theta}{dt}(t=0) = 0.$$

(b) Find the solution of the linearized equation for the following conditions:

$$l = 1 \text{ m}, g = 9.81 \text{ m/s}^2, c = M_t(t) = 0, \theta(t=0) = 45^\circ, \text{ and } \frac{d\theta}{dt}(t=0) = 0.$$

(c) Find the solution of the nonlinear equation, Eq. (E1), for the conditions indicated in part (a).

(d) Find the solution of the nonlinear equation, Eq. (E1), for the conditions indicated in part (b).

(e) Find the solution of the linearized equation for the following conditions:

$$l = 1 \text{ m}, g = 9.81 \text{ m/s}^2, c = 10 \text{ N-m-s}, M_t(t) = e^{-5t} \text{ N-m}, \theta(t=0) = 10^\circ,$$

$$\text{and } \frac{d\theta}{dt}(t=0) = 0.$$

(f) Find the solution of the linearized equation for the following conditions:

$$l = 1 \text{ m}, \; g = 9.81 \text{ m/s}^2, \; c = 10 \text{ N-m-s}, \; M_t(t) = e^{-5t} \text{ N-m}, \; \theta(t = 0) = 45^\circ,$$

$$\text{and } \frac{d\theta}{dt}(t = 0) = 0.$$

(g) Find the solution of the nonlinear equation, Eq. (E1), for the conditions indicated in Eq. (e).

(h) Find the solution of the nonlinear equation, Eq. (E1), for the conditions indicated in Eq. (f).

9.2. A milling machine is supported on an elastic foundation as shown in Fig. 9.17(a). An external force, $F(t)$, is applied to the machine during the metal cutting operation. In addition, the floor on which the machine is mounted is subjected to a harmonic disturbance, $y(t)$, due to the operation of an unbalanced engine in the vicinity of the milling machine. The equation

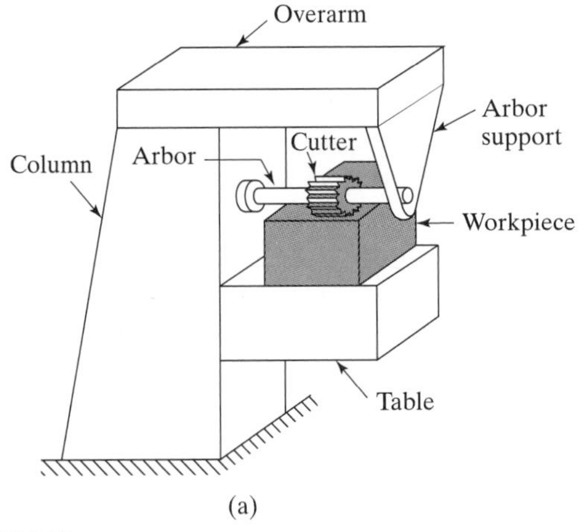

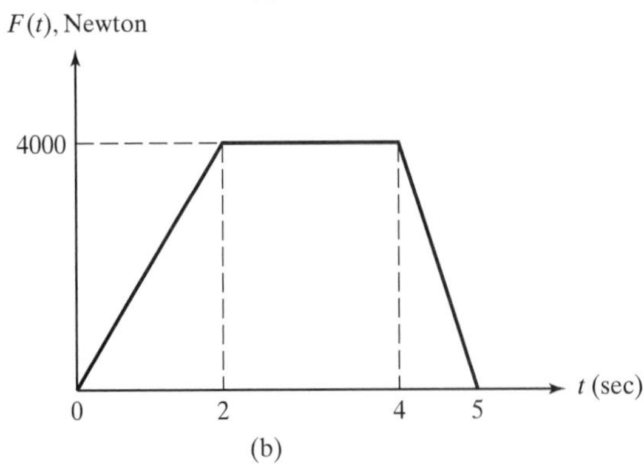

Figure 9.17 Milling machine and external force applied.

governing the vibration of the machine is given by [9.13]

$$m\ddot{x} + c\dot{x} + kx = c\dot{y} + ky + F(t). \qquad \text{(E1)}$$

If $m = 5000$ N, $c = 1$ kN-s/m, $k = 1$ MN/m, $y(t) = 2\sin 200\pi t$ mm, and $F(t)$ is as indicated in Fig. 9.17(b), determine the following:

(a) The displacement of the milling machine as a function of time, t, for $0 \le t \le 10$s.

(b) The variation of velocity and acceleration of the milling machine over the period, $0 \le t \le 10$s.

9.3. An automobile, traveling on a rough road, can be modeled for pitch and bounce motions as shown in Fig. 9.18(a). The free-body diagram of the automobile, Fig. 9.18(b), leads to the following equations:

$$m\ddot{x} + (k_f + k_r)x + (k_r l_2 - k_f l_1)\theta = k_f y_f + k_r y_r \qquad \text{(E1)}$$

For motion along θ (pitch),

$$J_0\ddot{\theta} + (l_2 k_r - l_1 k_f)x + (k_r l_2^2 + k_f l_1^2)\theta = k_r l_2 y_r - k_f l_1 y_f, \qquad \text{(E2)}$$

where m = mass of the automobile, J_0 = mass moment of inertia of the automobile, $k_f(k_r)$ = stiffness of the front (rear) suspension and tires, $l_1(l_2)$ = distance of the front (rear) wheels from the center of gravity, and $y_f(y_r)$ = downward ground displacement of the front (rear) wheels. Solve Eqs. (E1)

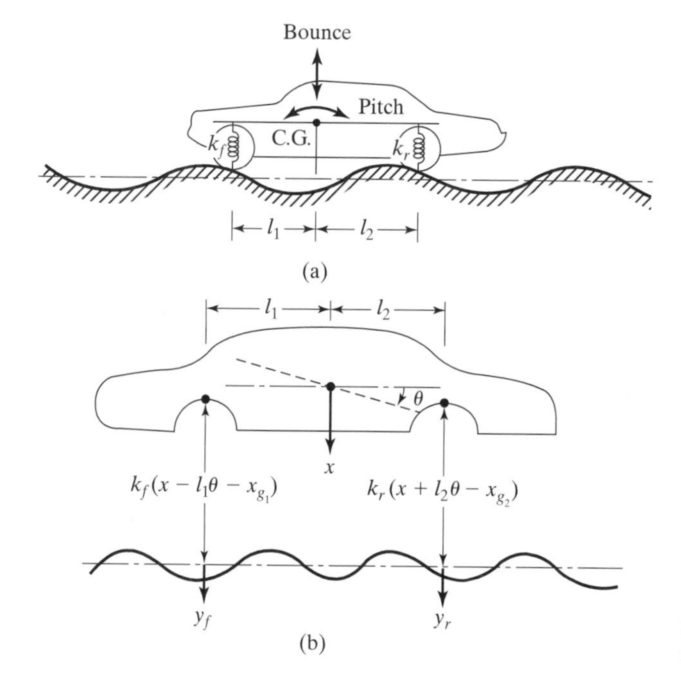

Figure 9.18 Automobile traveling on rough road.

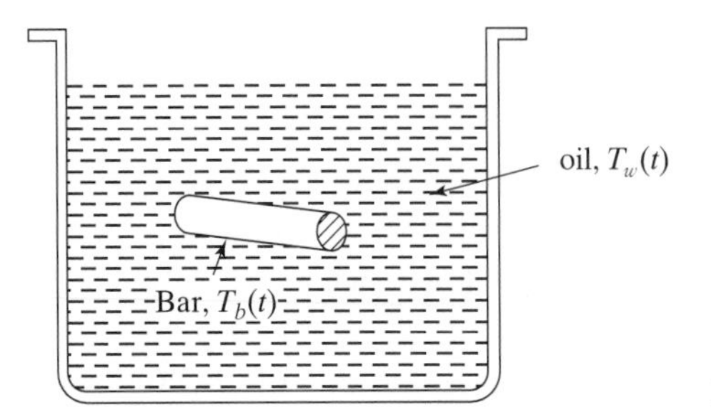

Figure 9.19 Quenching of a heated bar.

and (E2) for the following data:

$$m = 1000 \text{ kg}, \; J_0 = 800 \text{ kg} - \text{m}^2, \; k_f = 20 \text{ kN/m}, \; k_r = 25 \text{ kN/m},$$

$$l_1 = 1.0 \text{ m}, \; l_2 = 1.5 \text{ m}, \; y_f(t) = 0.05 \sin \frac{2\pi v}{10} t \text{ m},$$

$$y_r(t) = 0.05 \sin \left[\frac{2\pi v}{10} t - \frac{2\pi (l_1 + l_2)}{d} \right] \text{ m},$$

$$v = \text{velocity} = 15 \text{ m/s, and } d = 10 \text{ m}.$$

9.4. In the quenching operation used to harden a metal component, the component or bar is heated to a high temperature T_b and immersed in an oil (or water) bath at temperature T_w as shown in Fig. 9.19. The consideration of thermal equilibrium of the bar and the oil lead to the equations [9.15]

$$\frac{m_b c_b}{hA} \frac{dT_b}{dt} + T_b = T_w \tag{E1}$$

and

$$\frac{m_w c_w}{hA} \frac{dT_w}{dt} + T_w = T_b \tag{E2}$$

with initial conditions $T_b(0) = T_{10}$ and $T_w(0) = T_{20}$, where $T_b(T_w) =$ temperature of the bar (oil), $m_b(m_w) =$ mass of the bar (oil), $c_b(c_w) =$ specific heat of the bar (oil), $h =$ convection heat transfer coefficient and $A =$ surface area of the bar.

Determine the variations of T_b and T_w as functions of time, t, for $0 \le t \le 10$ sec for the following data:

$$h = 500 \text{ BTU/hr-ft}^2 - ^\circ\text{F}, \; A = 0.1 \text{ ft}^2, \; m_b = 0.02 \text{ lb}, \; c = 0.22 \text{ BTU/lb} - ^\circ\text{F},$$

$$T_{10} = 1200^\circ\text{F}, \; T_{20} = 65^\circ\text{F}, \; m_w = 2.5 \text{ lb}, \text{ and } c_w = 1.00 \text{ BTU/lb} - ^\circ\text{F}.$$

10

Ordinary Differential Equations: Boundary-Value Problems

10.1 Introduction

In the previous chapter, we considered the numerical solution of initial-value problems in which all the conditions to be satisfied by the solution are specified at one value of the independent variable. In many engineering applications, we need to solve differential equations for which the conditions are specified at more than one point. An ordinary differential equation for which the conditions are specified at two or more points of the independent variable is known as a boundary-value problem. For a boundary-value problem, the differential equation may be linear or nonlinear, and the boundary conditions may be linear or nonlinear, separated or mixed, and specified at two points or more points. A boundary value problem is considered nonhomogeneous if the differential equation or any of the boundary

CHAPTER CONTENTS

conditions contain at least one nonzero term that is independent of y and its derivatives. Since a boundary-value problem has at least two conditions specified on the solution at two different values of the independent variable, the differential equation will be of second or higher order. A simple example of a second-order boundary-value problem is

$$\frac{d^2y(x)}{dx^2} = f\left(x, y, \frac{dy}{dx}\right); \; a \le x \le b \tag{10.1}$$

with the boundary conditions

$$y(x=a) = \alpha; \; y(x=b) = \beta.$$

Sometimes the boundary conditions may be specified in mixed form as

$$c_1 y(x=a) + c_2 \frac{dy}{dx}(x=a) = c_3 \tag{10.2}$$

and

$$c_4 y(x=b) + c_5 \frac{dy}{dx}(x=b) = c_6, \tag{10.3}$$

where c_1 to c_6 are constants. A simple example of a fourth-order boundary value problem is

$$\frac{d^4y(x)}{dx^4} + ky(x) = f; \; 0 \le x \le 1 \tag{10.4}$$

with the boundary conditions

$$y(x=0) = \frac{dy}{dx}(x=0) = 0$$

and

$$y(x=1) = \frac{dy}{dx}(x=l) = 0.$$

In the boundary-value problem of Eq. (10.1), two conditions are specified on y at two different values of the independent variable x. Thus, the solution must satisfy the specified conditions at $x = a$ and $x = b$. We cannot start the numerical solution at $x = a$ and proceed along x with a step size h since the value of $\frac{dy}{dx}$ and hence that of f is not known at $x = a$. Several methods, such as shooting methods, finite-difference methods, and collocation methods, are available for solving boundary-value problems. These methods are discussed in this chapter.

10.2 Engineering Applications

▶Example 10.1

The differential equation governing the transverse deflection of a beam, $w(x)$, subjected to a distributed load, $p(x)$, as shown in Fig. 10.1 is given by [10.1]

$$\frac{d^2}{dx^2}\left(EI\frac{d^2w}{dx^2}\right) = p(x), \tag{E1}$$

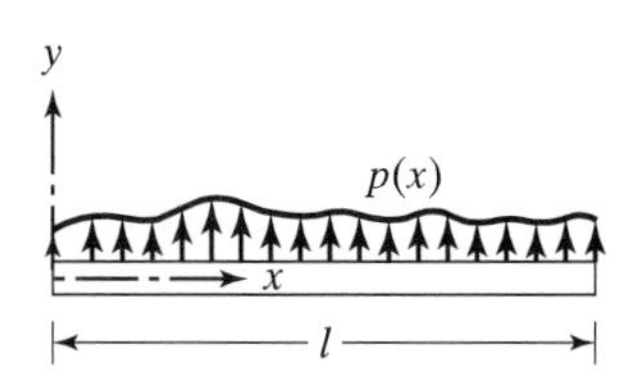

Figure 10.1 Beam under distributed load.

where E = Young's modulus and I = area moment of inertia of the beam. Formulate the boundary-value problem for a uniform beam (a) fixed at both ends and (b) simply supported at both ends.

Solution

For a uniform beam, E and I are constants and Eq. (E1) reduces to

$$\frac{d^4w}{dx^4} = \frac{p(x)}{EI}. \tag{E2}$$

(a) For a fixed–fixed beam, the deflection and slope are zero, and hence, the boundary conditions become

$$w(0) = 0; \quad \frac{dw(0)}{dx} = 0$$

and

$$w(1) = 0; \quad \frac{dw(1)}{dx} = 0. \tag{E3}$$

(b) For a simply supported beam, the deflection and bending moment are zero, and hence, the boundary conditions become

$$w(0) = 0; \quad \frac{d^2w(0)}{dx^2} = 0$$

and

$$w(1) = 0; \quad \frac{d^2w(1)}{dx^2} = 0. \tag{E4}$$

◀

▶Example 10.2

A cooling fin extends from a hot furnace wall as shown in Fig. 10.2. Assuming that heat flows only in the x-direction, the thermal equilibrium equation leads to [10.2]

$$kA\frac{d^2T}{dx^2} - hP(T - T_\infty) = 0, \tag{E1}$$

where k = thermal conductivity, A = cross-sectional area, h = convection heat transfer coefficient, T_∞ = surrounding temperature, P = perimeter, and $T(x)$ = temperature at x. State the boundary conditions of the problem.

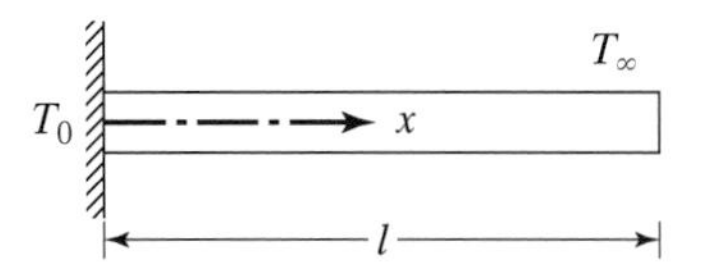

Figure 10.2 Cooling fin.

Solution

The condition at the root of the fin gives

$$T(0) = T_0. \tag{E2}$$

The convection loss at the end of the fin must be equal to the heat transfer by conduction. This gives

$$-k\frac{dT(l)}{dx} = h(T(l) - T_\infty). \tag{E3}$$

◀

▶Example 10.3

A pin-ended column of length l is subjected to an axial load P as shown in Fig. 10.3. The transverse deflection of the column, $w(x)$, can be found by solving the equation [10.1]

$$EI\frac{d^2w}{dx^2} + Pw = 0. \tag{E1}$$

State the boundary conditions of the problem.

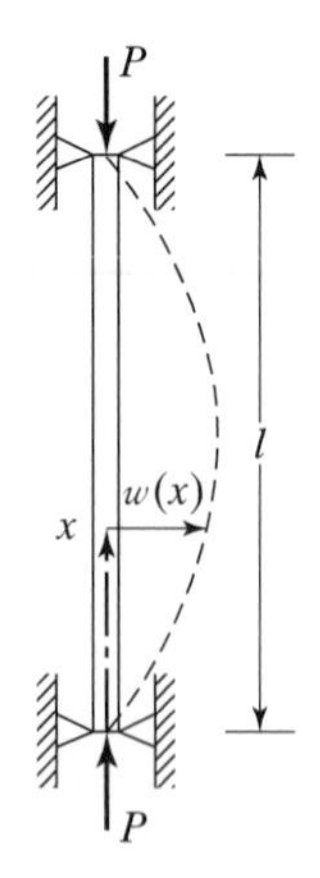

Figure 10.3 Pin-ended column.

Solution

Since the column is pin-connected at the ends, the transverse deflection is zero:

$$w(0) = 0;\ w(l) = 0. \tag{E2}$$

◀

10.3 Shooting Methods

The shooting methods are applicable to both linear and nonlinear boundary-value problems. These methods are easy to implement and, although there is no guarantee of convergence, when they converge, they usually converge faster. In shooting methods, the basic approach is to convert a boundary-value problem to an initial-value problem and solve the resulting problem iteratively using the methods described in Chapter 9. The general procedure, applicable to both linear and nonlinear boundary-value problems, can be stated by the following steps:

1. The unspecified initial conditions of the differential equation are guessed so that the problem can be solved as an initial-value problem.
2. The variational equations denoting the sensitivity of the dependent variables with respect to the guessed initial conditions are derived.
3. The differential equation and the variational equations are integrated along the x-direction as a set of simultaneous initial-value equations.
4. The results of sensitivities found in step 3 are used to correct the guessed initial conditions.
5. With the new (corrected) initial conditions found in step 4, steps 2 through 4 are repeated until the specified second (terminal) boundary conditions are satisfied.

This procedure is similar to the Newton–Raphson method used for finding the roots of equations in several variables. Hence, the method is often called the *Newton–Raphson* method. The method is also called the *shooting method* because a trial and error approach is used to solve the problem as a series of initial-value problems.

The detailed procedure can be developed by considering the second-order boundary-value problem

$$\frac{d^2y(x)}{dx^2} = f\left(x, y, \frac{dy}{dx}\right);\ a \le x \le b \tag{10.5}$$

with the boundary conditions

$$y(x = a) = y_a$$

and

$$\frac{dy}{dx}(x = b) = y_b',$$

where a and b denote the starting and final values of x. By defining the variables

$$y_1(x) = y(x) \text{ and } y_2(x) = \frac{dy_1(x)}{dx} = \frac{dy(x)}{dx}$$

the problem of Eq. (10.5) can be restated as a set of two first-order ordinary differential equations as

$$\frac{dy_1(x)}{dx} = f_1(x, y_1, y_2) \equiv y_2(x), \tag{10.6}$$

$$\frac{dy_2(x)}{dx} = f_2(x, y_1, y_2) \equiv f(x, y_1, y_2), \tag{10.7}$$

$$y_1(a) = y_{1,s} \equiv y_a, \tag{10.8}$$

and

$$y_2(b) = y_{2,f} \equiv y_b'. \tag{10.9}$$

To start the integration of Eqs. (10.6) and (10.7), we need the values of $y_1(x)$ and $y_2(x)$ at the starting point a. Since only $y_1(a) = y_{1,s}$ is known from Eq. (10.8), we guess the initial value, $y_2(a)$, as

$$y_2(x = a) = Y. \tag{10.10}$$

With the value of $y_2(a)$, the two Equations (10.6) and (10.7) are integrated from a to b to find the solutions $y_1(x)$ and $y_2(x)$ using any of the methods described in Chapter 9, such as the fourth-order Runge–Kutta method. Since the initial value, $y_2(a)$, used was only a guess, the final value of $y_2(x)$ at b, denoted as $y_2(b, Y)$, may not satisfy the boundary condition of Eq. (10.9) and may overshoot or undershoot the required solution as shown in Fig. 10.4. We now can try to make another guess for Y (by making suitable correction to the earlier value of Y) to be used as the initial condition of $y_2(x)$. By defining a function, $h(Y)$, as

$$h(Y) = y_2(b, Y) - y_{2,f} \tag{10.11}$$

and expanding $h(Y)$ in Taylor's series about the current value of Y gives

$$h(Y + \Delta Y) = h(Y) + \frac{\partial h}{\partial Y}\Delta Y + \dots. \tag{10.12}$$

Assuming that the use of $Y + \Delta Y$ instead of Y as the initial condition in Eq. (10.10) satisfies the final boundary condition on $y_2(x)$, we set Eq. (10.12) equal to zero by retaining only the first two terms:

$$h(Y) + \frac{\partial h}{\partial Y}\Delta Y = 0$$

or

$$\Delta Y = -\frac{h(Y)}{\left(\frac{\partial h}{\partial Y}\right)}. \tag{10.13}$$

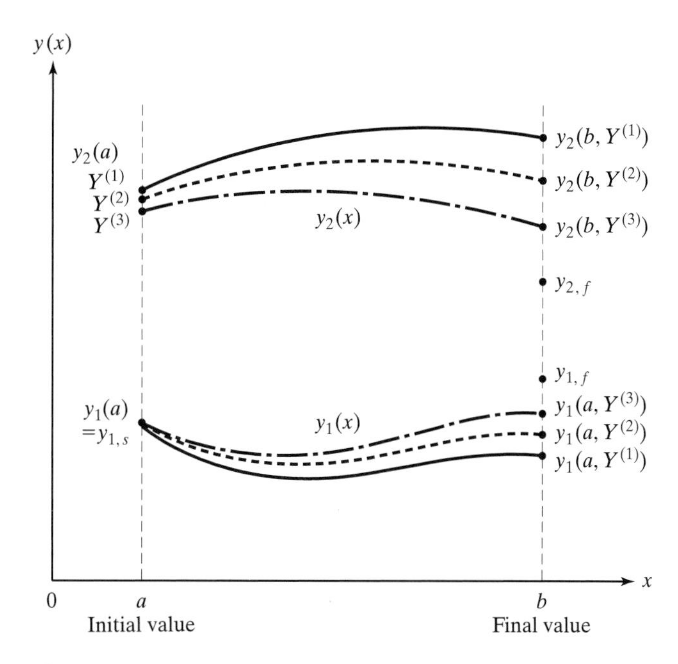

Figure 10.4 Interpretation of the shooting method.

Noting that Eq. (10.11) gives

$$\frac{\partial h(Y)}{\partial Y} = \frac{\partial y_2(b, Y)}{\partial Y}, \tag{10.14}$$

Eq. (10.13) can be rewritten as

$$\Delta Y = -\frac{(y_2(b, Y) - y_{2,f})}{\left\{\frac{\partial y_2(b, Y)}{\partial Y}\right\}}. \tag{10.15}$$

Thus, the new guess value of the initial condition on $y_2(x)$ becomes

$$y_2(a) = Y + \Delta Y, \tag{10.16}$$

where Y is the old guess value (used in Eq. (10.10)) and ΔY is given by Eq. (10.15).

The partial derivative, $\frac{\partial y_2(b,Y)}{\partial Y}$, needed for the computation of ΔY in Eq. (10.15) can be obtained by partially differentiating the differential Equations (10.6) and (10.7) as

$$\frac{\partial}{\partial Y}\left(\frac{dy_1}{dx}\right) = \frac{d}{dx}\left(\frac{\partial y_1}{\partial Y}\right) = \frac{\partial f_1}{\partial y_1}\frac{\partial y_1}{\partial Y} + \frac{\partial f_1}{\partial y_2}\frac{\partial y_2}{\partial Y} \tag{10.17}$$

and

$$\frac{\partial}{\partial Y}\left(\frac{dy_2}{dx}\right) = \frac{d}{dx}\left(\frac{\partial y_2}{\partial Y}\right) = \frac{\partial f_2}{\partial y_1}\frac{\partial y_1}{\partial Y} + \frac{\partial f_2}{\partial y_2}\frac{\partial y_2}{\partial Y}. \tag{10.18}$$

By defining the sensitivity functions,

$$g_1 = \frac{\partial y_1}{\partial Y} \text{ and } g_2 = \frac{\partial y_2}{\partial Y}, \tag{10.19}$$

Eqs. (10.17) and (10.18) can be expressed as

$$\frac{dg_1}{dx} = \frac{\partial f_1}{\partial y_1} g_1 + \frac{\partial f_1}{\partial y_2} g_2 \tag{10.20}$$

and

$$\frac{dg_2}{dx} = \frac{\partial f_2}{\partial y_1} g_1 + \frac{\partial f_2}{\partial y_2} g_2. \tag{10.21}$$

The initial conditions for g_1 and g_2 can be obtained from the definitions given in Eqs. (10.8), (10.10), and (10.19):

$$g_1(a) = \left.\frac{\partial y_1}{\partial Y}\right|_a = 0 \tag{10.22}$$

and

$$g_2(a) = \left.\frac{\partial y_2}{\partial Y}\right|_a = 1. \tag{10.23}$$

Equations (10.6), (10.7), (10.20), and (10.21) are to be solved (integrated from a to b) as a system of four simultaneous first-order differential equations using the initial conditions of Eqs. (10.8), (10.9), (10.22), and (10.23). After integrating these equations, the values of $y_2(a, Y)$ and $\frac{\partial y_2}{\partial Y}|_a$ can be determined, from which the correction needed, ΔY, for the next iteration can be computed using Eq. (10.16).

The whole procedure is to be repeated the required number of times, each time using the new boundary condition ($Y_{\text{new}} = Y_{\text{old}} + \Delta Y$), until the difference, $h(Y) = y_2(b, Y) - y_{2,f}$, satisfies the convergence criterion

$$|h(Y)| = \left|y_2(b, Y) - y_{2,f}\right| \leq \varepsilon, \tag{10.24}$$

where ε is a specified small number.

▶Example 10.4

Find the radial temperature distribution in a cylinder, with inner radius 5″ and outer radius 10″, when its inner and outer surfaces are maintained at 120° F and 60° F, respectively.

Solution

Assuming that the temperature distribution is one dimensional, the governing equation is given by

$$\frac{d^2T}{dr^2} + \frac{1}{r}\frac{dT}{dr} = 0 \tag{E1}$$

with boundary conditions

$$T(r=5) = 120 \text{ and } T(r=10) = 60. \tag{E2}$$

Equation (E1) can be written as a set of two first-order equations as

$$\frac{dT_1}{dr} \equiv \frac{dT}{dr} = T_2 \tag{E3}$$

and

$$\frac{dT_2}{dr} = \frac{1}{r} T_2. \tag{E4}$$

We assume the value of $\frac{dT}{dr}$ at $r = 5$ as $Y^{(1)}$ and find the resulting $T(r)$. If the value of temperature at the outer radius of the cylinder, $T(r = 10)$, coincides with the specified value, 60, the corresponding $T(r)$ will be the correct solution of the problem, Eqs. (E1) and (E2). For a linear variation of temperature in the cylinder, the rate of change of temperature is given by

$$\frac{dT}{dr}(r=5) = \frac{60-120}{5} = -12^\circ \text{ F/in.}$$

We assume, arbitrarily, the value of $Y^{(1)} = \frac{dT}{dr}(r = 5)$ as -11.0 and use the fourth-order Runge–Kutta method to find the solution $T(r = 10) = 81.8769^\circ$ F and $\frac{dT}{dr}(r = 10) = -5.5^\circ$ F/in. The function $h(Y)$ of Eq. (10.11) is given by

$$h(Y^{(1)}) = y_2(b, Y^{(1)}) - y_{2,f} = 81.8769 - 60 = 21.8769. \tag{E5}$$

Next, we assume the value of $Y^{(2)} = \frac{dT}{dr}(r = 5)$ as -13.0 and apply the fourth-order Runge–Kutta method to find the solution $T(r = 10) = 74.9455^\circ$ F and $\frac{dT}{dr}(r = 10) = -6.5^\circ$ F/in. This gives the value of the function $h(Y)$ as

$$h(Y^{(2)}) = y_2(b, Y^{(2)}) - y_{2,f} = 74.9455 - 60.0 = 14.9455. \tag{E6}$$

Equation (E5) and (E6) can be used to find the value of $\frac{\partial h}{\partial Y}$ at $Y^{(1)}$, by forward-finite differences, as

$$\left.\frac{\partial h}{\partial Y}\right|_{Y^{(1)}} = \frac{h(Y^{(2)}) - h(Y^{(1)})}{Y^{(2)} - Y^{(1)}} = \frac{74.9455 - 81.8769}{-13.0 + 11.0} = 3.4657. \tag{E7}$$

By setting the linear expansion of $h(Y)$ equal to zero, we obtain

$$h(Y) = h(Y^{(1)}) + \frac{\partial h}{\partial Y}(Y^{(1)})\Delta Y = 0, \tag{E8}$$

that is,

$$\Delta Y = -\frac{h(Y^{(1)})}{\frac{\partial h}{\partial Y}(Y^{(1)})} = -\frac{21.8769}{3.4657} = -6.3124. \tag{E9}$$

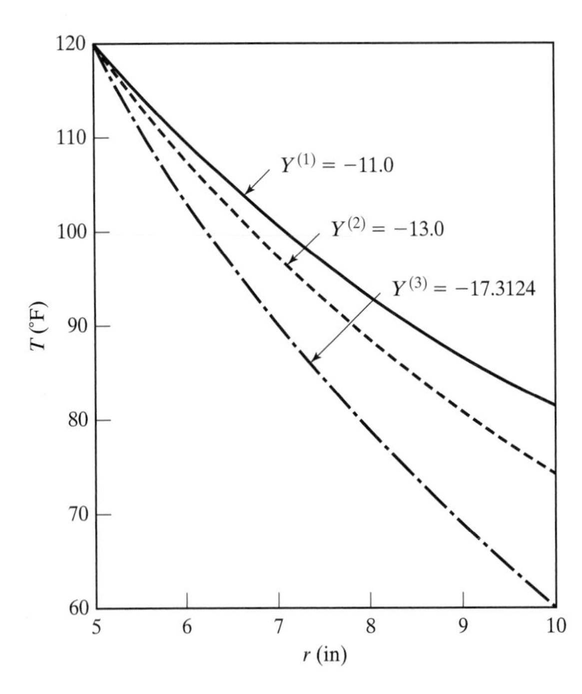

Figure 10.5 Temperature distribution given by $Y^{(i)}$.

This gives the next estimate, $Y^{(3)}$, as $Y^{(1)} + \Delta Y = -11.0 - 6.3124 = -17.3124$. The corresponding solution given by the fourth-order Runge–Kutta method is

$$T(r = 10) = 60.0^\circ \text{ F}, \ \frac{dT}{dr}(r = 10) = -8.6562^\circ \text{ F/in.} \tag{E10}$$

Since this solution satisfies the boundary condition at $r = 10$ (namely, $T(r = 10) = 60^\circ$ F), it is taken as the final solution. The temperature distributions given by $Y^{(1)}$, $Y^{(2)}$, and $Y^{(3)}$ are shown in Fig. 10.5.

10.4 Generalization to n Equations

Let the set of differential equations be expressed as

$$\frac{dy_1}{dx} = f_1(x, y_1, y_2, \ldots, y_n),$$

$$\frac{dy_2}{dx} = f_2(x, y_1, y_2, \ldots, y_n),$$

and

$$\frac{dy_n}{dx} = f_n(x, y_1, y_2, \ldots, y_n), \ a \leq x \leq b \tag{10.25}$$

with boundary conditions specified for the first k variables at the final point b, and for the remaining $(n-k)$ variables at the initial point a:

$$y_i(b) = y_{i,f};\ i = 1, 2, \ldots, k \tag{10.26}$$

and

$$y_i(a) = y_{i,s};\ i = k+1, k+2, \ldots, n. \tag{10.27}$$

Note that the boundary conditions are specified at the final point, b, for the first k variables y_i, $i = 1, 2, \ldots, k$, to simplify the notation.

10.4.1 General Procedure

To solve the Eqs. (10.25) as an initial-value problem, we start with the guess initial conditions for $y_i(x)$, $i = 1, 2, \ldots, k$:

$$y_i(a) = Y_i;\ i = 1, 2, \ldots, k. \tag{10.28}$$

When the solution of the differential equations is found with the guessed initial conditions, Eqs. (10.28), the discrepancies in the final values are denoted by

$$h_i(Y_1, Y_2, \ldots, Y_k) = y_i(b, Y_1, Y_2, \ldots, Y_k) - y_{i,f}; i = 1, 2, \ldots, k. \tag{10.29}$$

Assuming that $(Y_i + \Delta Y_i)$, $i = 1, 2, \ldots, k$ satisfy the specified final boundary conditions, we set

$$h_i(Y_1 + \Delta Y_1, \ldots, Y_k + \Delta Y_k) = h_i(Y_1, \ldots, Y_k) + \sum_{j=1}^{k} \frac{\partial h_i}{\partial Y_j} \Delta Y_j = 0; i = 1, 2, \ldots, k. \tag{10.30}$$

Equation (10.30) can be rewritten as

$$\left(\frac{\partial h_1}{\partial Y_1}\right) \Delta Y_1 + \left(\frac{\partial h_1}{\partial Y_2}\right) \Delta Y_2 + \cdots + \left(\frac{\partial h_1}{\partial Y_k}\right) \Delta Y_k = -h_1(Y_1, \ldots, Y_k),$$

$$\vdots,$$

$$\left(\frac{\partial h_k}{\partial Y_1}\right) \Delta Y_1 + \left(\frac{\partial h_k}{\partial Y_2}\right) \Delta Y_2 + \cdots + \left(\frac{\partial h_k}{\partial Y_k}\right) \Delta Y_k = -h_k(Y_1, \ldots, Y_k), \tag{10.31}$$

where the partial derivatives are evaluated at $x = b$. Noting that $\frac{\partial h_i}{\partial Y_j} = \frac{\partial y_i}{\partial Y_j}$ from Eq. (10.29), the solution of Eqs. (10.31), which gives the corrections to be made to the guessed values to start the next iteration, can be expressed as

$$\Delta Y = -\left[J(\vec{Y}, b)\right]^{-1} \vec{h}(\vec{Y}), \tag{10.32}$$

where

$$\vec{\Delta Y} = \begin{Bmatrix} \Delta Y_1 \\ \cdot \\ \cdot \\ \cdot \\ \Delta Y_k \end{Bmatrix}, \tag{10.33}$$

$$\vec{h}(\vec{Y}) = \begin{Bmatrix} h_1(Y_1, \ldots, Y_k) \\ \cdot \\ \cdot \\ \cdot \\ h_k(Y_1, \ldots, Y_k) \end{Bmatrix}, \tag{10.34}$$

and

$$\left[J(\vec{Y}, b)\right] = \begin{bmatrix} \frac{\partial y_1}{\partial Y_1} & \cdots & \frac{\partial y_1}{\partial Y_k} \\ \cdot & \cdots & \cdot \\ \cdot & \cdots & \cdot \\ \cdot & \cdots & \cdot \\ \frac{\partial y_k}{\partial Y_1} & \cdots & \frac{\partial y_k}{\partial Y_k} \end{bmatrix}_{x=b} . \tag{10.35}$$

As in the case of two differential equations, the partial derivatives, $\frac{\partial y_i}{\partial Y_k}$, in Eq. (10.35) can be found by differentiating Eq. (10.25) as

$$\frac{\partial}{\partial Y_j}\left(\frac{dy_i}{dx}\right) = \frac{d}{dx}\left(\frac{\partial y_i}{\partial Y_j}\right) = \sum_{k=1}^{n} \frac{\partial f_i}{\partial y_k}\frac{\partial y_k}{\partial Y_j}; i = 1, 2, \ldots, n; j = 1, 2, \ldots, k. \tag{10.36}$$

By defining the sensitivity functions

$$g_{ij} = \frac{\partial y_i}{\partial Y_j};\ i = 1, 2, \ldots, n;\ j = 1, 2, \ldots, k, \tag{10.37}$$

Eqs. (10.36) can be expressed as a set of $n \times k$ ordinary differential equations

$$\frac{dp_{ij}}{dx} = \sum_{k=1}^{n} \frac{\partial f_i}{\partial y_k} p_{ij};\ i = 1, 2, \ldots, n;\ j = 1, 2, \ldots, k \tag{10.38}$$

with the initial conditions (found from the definition of Eq. (10.37) and Eqs. (10.27) and (10.28)) of

$$g_{ij}(a) = \begin{cases} 1 \text{ if } i = j \\ 0 \text{ if } i \neq j \end{cases} ; \; i = 1, 2, \ldots, n; \; j = 1, 2, \ldots, k. \tag{10.39}$$

Thus, the differential equations, Eqs. (10.25) and (10.38), are solved (integrated from a to b) using the initial conditions given by Eqs. (10.27), (10.28), and (10.39). After integrating these equations, the values of $y_i(b, Y_1, \ldots, Y_k)$ and $\frac{\partial y_i}{\partial Y_j}|_b$ can be used to determine $\Delta\vec{Y}$ according to Eq. (10.32). Finally the corrected guessed initial values to start the next iteration are computed as

$$\vec{Y}_{\text{new}} = \vec{Y}_{\text{old}} + \Delta\vec{Y} = \vec{Y}_{\text{old}} - [J(\vec{Y}, b)]^{-1}\vec{h}(\vec{Y}) \tag{10.40}$$

or

$$y_1(a) = Y_1|_{\text{new}} = Y_1|_{\text{old}} + \Delta Y_1,$$

$$\vdots,$$

$$y_k(a) = Y_k|_{\text{new}} = Y_k|_{\text{old}} + \Delta Y_k.$$

The entire procedure is to be repeated until each of the functions $h_i(\vec{Y})$ satisfy the convergence criterion

$$|h_i(Y_1, \ldots, Y_k)| = \left|y_i(b, Y_1, \ldots, Y_k) - y_{i,f}\right| \le \varepsilon; \; i = 1, 2, \ldots, k, \tag{10.41}$$

where ε is a specified small number.

10.4.2 Shooting Method for Linear Differential Equations

The procedure described earlier can be used to find the numerical solution of any linear or nonlinear set of ordinary differential equations. However, if the differential equations are linear, the principle of superposition can be applied to find the solution in a simpler manner without using an iterative procedure. To illustrate the procedure for a set of linear differential equations, consider a system of two linear ordinary differential equations:

$$\frac{dy_1(x)}{dx} = f_1(x, y_1, y_2) = g_1(x) + y_1 g_2(x) + y_2 g_3(x) \tag{10.42}$$

and

$$\frac{dy_2(x)}{dx} = f_2(x, y_1, y_2) = g_4(x) + y_1 g_5(x) + y_2 g_6(x) \tag{10.43}$$

with the boundary conditions

$$y_1(a) = y_{1,s} \tag{10.44}$$

and

$$y_2(b) = y_{2,f}. \tag{10.45}$$

As in the previous case, we can convert the problem into an equivalent initial-value problem by using the initial guess for $y_2(a)$ as

$$y_2(a) = Y. \tag{10.46}$$

In fact, the initial-value problem defined by Eqs. (10.42), (10.44), and (10.46) is solved twice; first by using the initial guess $y_2(a) = Y_1$ and next by using the initial guess $y_2(a) = Y_2$. Let the resulting solutions be denoted as $y_1^{(1)}(x)$, $y_2^{(1)}(x)$, and $y_1^{(2)}(x)$, $y_2^{(2)}(x)$, where the superscripts 1 and 2 denote the solutions corresponding to the initial guesses Y_1 and Y_2, respectively. For linear differential equations, the principle of superposition is applicable, and hence, in the present case, a linear combination of the two solutions found will also be a solution of Eq. (10.42). Thus, the solution

$$y_1(x) = c_1 y_1^{(1)}(x) + c_2 y_1^{(2)}(x), \tag{10.47}$$

where c_1 and c_2 are constants, also satisfies the boundary condition $y_1(a) = y_{1,s}$. We set the solution given by

$$y_2(x) = c_1 y_2^{(1)}(x) + c_2 y_2^{(2)}(x) \tag{10.48}$$

at $x = b$ equal to the specified boundary value of

$$y_2(b) = c_1 y_2^{(1)}(b) + c_2 y_2^{(2)}(b) = y_{2,f}. \tag{10.49}$$

Because the solution given by Eq. (10.48) satisfies the differential Equation (10.43), by substituting we obtain

$$\left[c_1 \frac{dy_2^{(1)}(x)}{dx} + c_2 \frac{dy_2^{(2)}(x)}{dx}\right] = g_4(x) + \left[c_1 y_1^{(1)}(x) + c_2 y_1^{(2)}(x)\right] g_5(x)$$
$$+ \left[c_1 y_2^{(1)}(x) + c_2 y_2^{(2)}(x)\right] g_6(x), \tag{10.50}$$

which can be rearranged in the form

$$c_1 \left[\frac{dy_2^{(1)}(x)}{dx} - g_4(x) - y_1^{(1)}(x) g_5(x) - y_2^{(1)}(x) g_6(x)\right]$$
$$+ c_2 \left[\frac{dy_2^{(2)}}{dx} - g_4(x) - y_1^{(2)}(x) g_5(x) - y_2^{(2)}(x) g_6(x)\right] + (c_1 + c_2 - 1) g_4(x) = 0. \tag{10.51}$$

Since the two solutions $\left(y_1^{(1)}(x), y_2^{(1)}(x)\right)$ and $\left(y_1^{(2)}(x), y_2^{(2)}(x)\right)$ independently satisfy the differential Equation (10.43), the coefficients of c_1 and c_2 in the first two terms of Eq. (10.51) are zero, so that

$$(c_1 + c_2 - 1) g_4(x) = 0$$

or

$$c_1 + c_2 = 1. \tag{10.52}$$

Equations (10.49) and (10.52) can be solved to find the constants c_1 and c_2 as

$$c_1 = \frac{y_{2,f} - y_2^{(2)}(b)}{y_2^{(1)}(b) - y_2^{(2)}(b)} \tag{10.53}$$

and

$$c_2 = \frac{y_2^{(1)}(b) - y_{2,f}}{y_2^{(1)}(b) - y_2^{(2)}(b)}. \tag{10.54}$$

Equations (10.47), (10.48), (10.53), and (10.54) can be used to find the desired solution (which satisfies both the boundary conditions, Eqs. (10.44) and (10.45)) as

$$y_1(x) = \left[\frac{y_{2,f} - y_2^{(2)}(b)}{y_2^{(1)}(b) - y_2^{(2)}(b)}\right] y_1^{(1)}(x) + \left[\frac{y_2^{(1)}(b) - y_{2,f}}{y_2^{(1)}(b) - y_2^{(2)}(b)}\right] y_1^{(2)}(x) \tag{10.55}$$

and

$$y_2(x) = \left[\frac{y_{2,f} - y_2^{(2)}(b)}{y_2^{(1)}(b) - y_2^{(2)}(b)}\right] y_2^{(1)}(x) + \left[\frac{y_2^{(1)}(b) - y_{2,f}}{y_2^{(1)}(b) - y_2^{(2)}(b)}\right] y_2^{(2)}(x). \tag{10.56}$$

Although only two differential equations were considered in Eqs. (10.42) and (10.43), the procedure can be used even for a general system of n linear differential equations. If the number of missing initial conditions is k, then we need to find $(k+1)$ solutions to apply the procedure.

▶Example 10.5

Find the solution of the boundary-value problem

$$\begin{aligned} &y'' - y' + y = 3e^{2x} - 2\sin x, \\ &y(1) = 6.308447,\ y(2) = 55.430436. \end{aligned} \tag{a}$$

Solution

Defining $y_1 = x$, $y_2 = y$, and $y_3 = y'$, we can write the problem of Eq. (a)

$$\begin{Bmatrix} y_1' \\ y_2' \\ y_3' \end{Bmatrix} = \begin{Bmatrix} 1 \\ y' \\ y'' \end{Bmatrix} = \begin{Bmatrix} 1 \\ y_3 \\ 3e^{2y_1} - 2\sin y_1 + y_3 - y_2 \end{Bmatrix}. \tag{b}$$

Equation (b) is solved twice by assuming two different trial values for $y'(1)$. By assuming the value of $y'(1) = y_3(1)$ as 0, the initial conditions can be expressed as

$$\begin{Bmatrix} y_1(1) \\ y_2(1) \\ y_3(1) \end{Bmatrix} = \begin{Bmatrix} 1 \\ 6.308447 \\ 0 \end{Bmatrix}. \tag{c}$$

Next, we define $z_1 = x$, $z_2 = y$, and $z_3 = y'$, and the problem is solved by assuming a value of 1 for $y'(1) = y_3(1)$. Thus, the problem becomes

$$\begin{Bmatrix} z_1' \\ z_2' \\ z_3' \end{Bmatrix} = \begin{Bmatrix} 1 \\ y' \\ y'' \end{Bmatrix} = \begin{Bmatrix} 1 \\ z_3 \\ 3e^{2z_1} - 2\sin z_1 + z_3 - z_2 \end{Bmatrix} \tag{d}$$

with initial conditions

$$\begin{Bmatrix} z_1(1) \\ z_2(1) \\ z_3(1) \end{Bmatrix} = \begin{Bmatrix} 1 \\ 6.308447 \\ 1 \end{Bmatrix}. \tag{e}$$

For simplicity, the two problems, defined by Eqs. (b) and (c) and Eqs. (d) and (e), can be combined and solved together. For this, we define $y_4 = z_2 = y$ and $y_5 = z_3 = y'$ and the combined problem is expressed as

$$\begin{Bmatrix} y_1' \\ y_2' \\ y_3' \\ y_4' \\ y_5' \end{Bmatrix} = \begin{Bmatrix} 1 \\ y_3 \\ 3e^{2y_1} - 2\sin y_1 + y_3 - y_2 \\ y_5 \\ 3e^{2y_1} - 2\sin y_1 + y_5 - y_4 \end{Bmatrix} \tag{f}$$

with initial conditions

$$\begin{Bmatrix} y_1(1) \\ y_2(1) \\ y_3(1) \\ y_4(1) \\ y_5(1) \end{Bmatrix} = \begin{Bmatrix} 1 \\ 6.308447 \\ 0 \\ 6.308447 \\ 1 \end{Bmatrix}. \tag{g}$$

The system of first-order differential equations defined by Eqs. (f) and (g) are solved using the fourth-order Runge–Kutta method. The solution of Eqs. (b) and (c) is given by $y_2(x)$, while that of Eqs. (d) and (e) is given by $y_4(x)$. The exact solution can be expressed as

$$y(x_i) = \gamma y_2(x_i) + (1 - \gamma) y_4(x_i);\ i = 1, 2, \ldots, N, \tag{h}$$

where x_i denotes the ith discrete value of x, N indicates the number of equal divisions into which the range of x is divided to define the discrete values of x with

$x_1 = 1$ and $x_{N+1} = 2$, and

$$\gamma = \frac{55.430436 - y_4(2)}{y_2(2) - y_4(2)}. \tag{i}$$

The results are as follows.

VALUE OF X	VALUE OF $Y(X)$	EXACT $Y(X)$	ERROR IN $Y(X)$
1.0000	0.63084469E+01	0.63084517E+01	0.47683716E−05
1.0100	0.64745941E+01	0.64746037E+01	0.95367432E−05
1.0200	0.66438675E+01	0.66438770E+01	0.95367432E−05
1.0300	0.68163223E+01	0.68163314E+01	0.90599060E−05
1.0400	0.69920273E+01	0.69920273E+01	0.00000000E+00
1.0500	0.71710205E+01	0.71710272E+01	0.66757202E−05
1.0600	0.73533859E+01	0.73533926E+01	0.66757202E−05
1.0700	0.75391846E+01	0.75391898E+01	0.52452087E−05
1.0800	0.77284775E+01	0.77284822E+01	0.47683716E−05
1.0900	0.79213333E+01	0.79213362E+01	0.28610229E−05
1.1000	0.81178207E+01	0.81178217E+01	0.95367432E−06
⋮			
1.9000	0.45347778E+02	0.45347763E+02	−.15258789E−04
1.9100	0.46269714E+02	0.46269680E+02	−.34332275E−04
1.9200	0.47209778E+02	0.47209770E+02	−.76293945E−05
1.9300	0.48168396E+02	0.48168415E+02	0.19073486E−04
1.9400	0.49145966E+02	0.49145966E+02	0.00000000E+00
1.9500	0.50142822E+02	0.50142815E+02	−.76293945E−05
1.9600	0.51159363E+02	0.51159351E+02	−.11444092E−04
1.9700	0.52195953E+02	0.52195976E+02	0.22888184E−04
1.9800	0.53253052E+02	0.53253086E+02	0.34332275E−04
1.9900	0.54331085E+02	0.54331104E+02	0.19073486E−04
2.0000	0.55430420E+02	0.55430443E+02	0.22888184E−04

◀

10.5 Finite-Difference Methods

In these methods, the finite-difference approximation of the differential equation is obtained at a number of mesh points in the interval of integration thereby converting the differential equation to a set of simultaneous algebraic equations. The set of algebraic equations will be linear (or nonlinear) if the differential equation is linear (or nonlinear). The finite-difference method of solution is presented for a variety of boundary-value problems in this section. It is not necessary to convert an nth-order differential equation into an equivalent system of n first-order differential equations, since finite-difference approximations can be used for derivatives of all order.

10.5.1 Solution of Second-Order Equations

Consider the linear second-order differential equation:

$$y''(x) + p(x)y'(x) + q(x)y(x) = r(x);\ a \leq x \leq b \tag{10.57}$$

with the boundary conditions

$$y(a) = \alpha \tag{10.58}$$

and

$$y(b) = \beta. \tag{10.59}$$

Equation (10.57) requires finite-difference approximations for y' and y''. To derive the finite-difference equations, the interval of integration, $[a, b]$, is divided into N equal parts of width $h = \left(\frac{b-a}{N}\right)$. This implies that $x_0 = a$ and $x_N = b$ are the boundary mesh points, and

$$x_i = a + ih \text{ and } i = 1, 2, \ldots, N-1 \tag{10.60}$$

are the interior mesh points. The central-difference formulas are usually used for the derivatives, because they lead to greater accuracy. The formulas for y' and y'' are

$$y'(x_i) = \frac{y_{i+1} - y_{i-1}}{2h} \tag{10.61}$$

and

$$y''(x_i) = \frac{y_{i+1} - 2y_i + y_{i-1}}{h^2}, \tag{10.62}$$

where $y_i = y(x_i), i = 0, 1, 2, \ldots, N$. Note that both Eqs. (10.61) and (10.62) have an error of $O(h^2)$. The finite-difference approximation of Eq. (10.57) at the interior node i is given by

$$\frac{y_{i+1} - 2y_i + y_{i-1}}{h^2} + p(x_i)\frac{y_{i+1} - y_{i-1}}{2h} + q(x_i)y_i = r(x_i);\ i = 1, 2, \ldots, N-1. \tag{10.63}$$

By defining

$$p(x_i) = p_i,\ q(x_i) = q_i, \text{ and } r(x_i) = r_i \tag{10.64}$$

and multiplying Eq. (10.63) throughout by h^2, we can rewrite Eq. (10.63) as

$$\left(1+\frac{h}{2}p_i\right)y_{i+1}+\left(-2+h^2q_i\right)y_i+\left(1-\frac{h}{2}p_i\right)y_{i-1}=h^2r_i;\ i=1,2,\ldots,N-1. \tag{10.65}$$

It can be seen that for $i = 1$, y_0 appears on the left-hand side of Eq. (10.65), and for $i = N$, y_N appears on the left-hand side of Eq. (10.65). Since the values of y_0 and y_N are specified as α and β (Eqs. (10.58) and (10.59)), the terms involving y_0 and y_N can be taken to the right-hand side of Eq. (10.65). By using $i = 1, 2, \ldots, N-1$ in Eq. (10.65), we obtain

$$\begin{gathered}
(-2+h^2q_1)y_1+\left(1+\frac{h}{2}p_1\right)y_2=h^2r_1-\left(1-\frac{h}{2}p_1\right)\alpha,\\
\left(1-\frac{h}{2}p_2\right)y_1+(-2+h^2q_2)y_2+\left(1+\frac{h}{2}p_2\right)y_3=h^2r_2,\\
\left(1-\frac{h}{2}p_3\right)y_2+(-2+h^2q_3)y_3+\left(1+\frac{h}{2}p_3\right)y_4=h^2r_3,\\
\vdots\\
\left(1-\frac{h}{2}p_{N-2}\right)y_{N-3}+(-2+h^2q_{N-2})y_{N-2}+\left(1+\frac{h}{2}p_{N-2}\right)y_{N-1}=h^2r_{N-2},
\end{gathered} \tag{10.66}$$

and

$$\left(1-\frac{h}{2}p_{N-1}\right)y_{N-2}+(-2+h^2q_{N-1})y_{N-1}=h^2r_{N-1}-\left(1+\frac{h}{2}p_{N-1}\right)\beta.$$

Equations (10.66) represent a system of $(N-1)$ linear equations in the $(N-1)$ unknowns $y_1, y_2, \ldots, y_{N-1}$. The coefficients of y_i and the right-hand side constants in Eqs. (10.66) can be computed since the functions $p(x)$, $q(x)$, and $r(x)$ are known functions of x. Thus, Eqs. (10.66) can be solved by using the methods described in Chapter 3. Equations (10.66) can be expressed in matrix form as

$$[A]\vec{y}=\vec{b}, \tag{10.67}$$

where

$$\vec{y}=\begin{Bmatrix} y_1\\ y_2\\ \cdot\\ \cdot\\ \cdot\\ y_{N-2}\\ y_{N-1} \end{Bmatrix}, \tag{10.68}$$

$$\vec{b} = \begin{Bmatrix} h^2 r_1 - \left(1 - \frac{h}{2} p_1\right) \alpha \\ h^2 r_2 \\ \cdot \\ \cdot \\ \cdot \\ h^2 r_{N-2} \\ h^2 r_{N-1} - \left(1 + \frac{h}{2} p_{N-1}\right) \beta \end{Bmatrix}, \tag{10.69}$$

and

$$[A] = \begin{bmatrix} Q_1 & P_1 & & & & & \bigcirc \\ R_2 & Q_2 & P_2 & & & & \\ & R_3 & Q_3 & P_3 & & & \\ & & \ddots & \ddots & \ddots & & \\ & & & & & & \\ & \bigcirc & & & R_{N-2} & Q_{N-2} & P_{N-2} \\ & & & & & R_{N-1} & Q_{N-1} \end{bmatrix} \tag{10.70}$$

with

$$P_i = 1 + \frac{h}{2} p_i;\ i = 1, 2, \ldots, N-2, \tag{10.71}$$

$$Q_i = -2 + h^2 q_i;\ i = 1, 2, \ldots, N-1, \tag{10.72}$$

and

$$R_i = 1 - \frac{h}{2} p_i;\ i = 2, 3, \ldots, N-1. \tag{10.73}$$

Since Eqs. (10.67) denote a tridiagonal system of equations, they can be solved very efficiently. (See Section 3.17.)

▶Example 10.6

Find the solution of the boundary-value problem

$$y'' - y' + y = 3e^{2x} - 2\sin x;$$
$$y(1) = 6.308447,\ y(2) = 55.430436. \quad \text{(a)}$$

Solution

The interval $1 \le x \le 2$ is divided into $N = 99$ equal divisions so that $h = 1/99$, $x_1 = 1.0$, and $x_i = x_1 + (i-1)h$, $i = 2, 3, \ldots, 100$ with $x_{100} = 2.0$. The finite-difference equations are given by Eq. (10.65),

$$\left(1 + \frac{h}{2}p_i\right)y_{i+1} + (-2 + h^2 q_i)y_i + \left(1 - \frac{h}{2}p_i\right)y_{i-1} = h^2 r_i;\ i = 2, 3, \ldots, N, \quad \text{(b)}$$

where $p_i = -1$, $q_i = 1$, and $r_i = 3e^{2x_i} - 2\sin x_i$. Equation (b) leads to a tridiagonal system of equations whose solution can be determined easily. The results are as follows:

X:	VALUE OF $Y(X)$:	EXACT $Y(X)$:	ERROR:
1.0000	0.63084469E+01	0.63084517E+01	0.47683716E−05
1.0200	0.66439004E+01	0.66438770E+01	−.23365021E−04
1.0300	0.68163695E+01	0.68163314E+01	−.38146973E−04
1.0400	0.69920797E+01	0.69920273E+01	−.52452087E−04
1.0500	0.71710930E+01	0.71710272E+01	−.65803528E−04
1.0600	0.73534737E+01	0.73533926E+01	−.81062317E−04
1.0700	0.75392838E+01	0.75391898E+01	−.93936920E−04
1.0800	0.77285900E+01	0.77284822E+01	−.10776520E−03
1.0900	0.79214592E+01	0.79213362E+01	−.12302399E−03
1.1000	0.81179590E+01	0.81178217E+01	−.13732910E−03
⋮			
1.9000	0.45348049E+02	0.45347763E+02	−.28610229E−03
1.9100	0.46269947E+02	0.46269680E+02	−.26702881E−03
1.9200	0.47210014E+02	0.47209770E+02	−.24414063E−03
1.9300	0.48168629E+02	0.48168415E+02	−.21362305E−03
1.9400	0.49146156E+02	0.49145966E+02	−.19073486E−03

1.9500	0.50142979E+02	0.50142815E+02	−.16403198E−03
1.9600	0.51159489E+02	0.51159351E+02	−.13732910E−03
1.9700	0.52196075E+02	0.52195976E+02	−.99182129E−04
1.9800	0.53253151E+02	0.53253086E+02	−.64849854E−04
1.9900	0.54331131E+02	0.54331104E+02	−.26702881E−04
2.0000	0.55430435E+02	0.55430443E+02	0.76293945E−05

◀

10.5.2 Mixed-Boundary Conditions

In some applications, the boundary conditions may be stated in a more complicated manner. Frequently, the value of a linear combination of the dependent variable and its derivative is prescribed at a boundary as

$$c_1 y(a) + c_2 y'(a) = c_3. \tag{10.74}$$

In this case, Eq. (10.74) replaces Eq. (10.58). When a forward-difference formula is used for $y'(a)$, Eq. (10.74) becomes

$$c_1 y(a) + c_2 \frac{y(a+h) - y(a)}{h} = c_3, \tag{10.75}$$

which can be rewritten as

$$y_0(c_1 h - c_2) + c_2 y_1 = c_3 h$$

or

$$y_0 = \left(\frac{c_3 h - c_2 y_1}{c_1 h - c_2}\right), \tag{10.76}$$

where $y_0 = y(a)$ and $y_1 = y(a+h)$. Equation (10.65), when applied at node $i = 1$, gives

$$\left(1 + \frac{h}{2}p_1\right) y_2 + (-2 + h^2 q_1) y_1 + \left(1 - \frac{h}{2}p_1\right) y_0 = h^2 r_1. \tag{10.77}$$

Substituting Eq. (10.76) for y_0 and rearranging Eq. (10.77) yields

$$\left(1 + \frac{h}{2}p_1\right) y_2 + \left\{-2 + h^2 q_1 - \frac{c_2}{(c_1 h - c_2)} + \frac{h c_2 p_1}{2(c_1 h - c_2)}\right\} y_1$$
$$+ \left\{h^2 r_1 - \frac{c_3 h}{(c_1 h - c_2)} + \frac{h^2 c_3 p_1}{2(c_1 h - c_2)}\right\}. \tag{10.78}$$

In this case, Eq. (10.78) replaces the first equation of Eqs. (10.66). Since all other equations of Eqs. (10.66) remain unaltered, the resulting tridiagonal system of equations can readily be solved as in the previous case.

If the mixed-boundary condition is specified at the other end as

$$c_1 y(b) + c_2 y'(b) = c_3, \tag{10.79}$$

this condition replaces Eq. (10.59). In this case, a backward-difference formula is used for $y'(b)$ so that Eq. (10.79) gives

$$c_1 y(b) + c_2 \frac{y(b) - y(b-h)}{h} = c_3, \tag{10.80}$$

which can be rearranged to obtain

$$(c_1 h + c_2) y_N - c_2 y_{N-1} = c_3 h \tag{10.81}$$

or

$$y_N = \frac{c_3 h + c_2 y_{N-1}}{(c_1 h + c_2)}. \tag{10.82}$$

Equation (10.65), when applied at node $N-1$, gives

$$\left(1 + \frac{h}{2} p_{N-1}\right) y_N + (-2 + h^2 q_{N-1}) y_{N-1} + \left(1 - \frac{h}{2} p_{N-1}\right) y_{N-2} = h^2 r_{N-1}. \tag{10.83}$$

Substituting Eq. (10.82) into Eq. (10.83) results in

$$\left(1 - \frac{h}{2} p_{N-1}\right) y_{N-2} + \left[-2 + h^2 q_{N-1} + \frac{c_2}{(c_1 h + c_2)} + \frac{h c_2 p_{N-1}}{2(c_1 h + c_2)}\right] y_{N-1}$$
$$= \left[h^2 r_{N-1} - \frac{c_3 h}{(c_1 h + c_2)} - \frac{c_3 h^2 p_{N-1}}{2(c_1 h + c_2)}\right]. \tag{10.84}$$

In this case, Eq. (10.84) replaces the last equation of Eqs. (10.66). Since all other equations of Eqs. (10.66) remain unchanged, the resulting tridiagonal system of equations can be solved as in the earlier case.

Note that the finite-difference formulas used in Eqs. (10.74) and (10.79) have an error of $O(h)$ only, and hence, the error in the final solution will also be $O(h)$. To obtain a solution that has an error of $O\left(h^2\right)$ throughout, we can use a central-difference formula in Eqs. (10.74) and (10.79). When a central-difference formula is used in the boundary condition of Eq. (10.74) or (10.79), we need to introduce a fictitious or exterior mesh point as shown in Fig. 10.6. For example, the mesh point $x_{-1} = a - h$ will be needed for Eq. (10.74) and $x_{N+1} = b + h$ for Eq. (10.79).

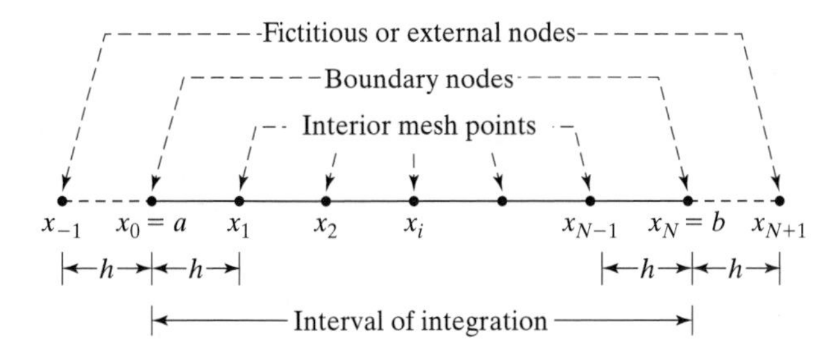

Figure 10.6 Fictitious or exterior node points.

When a central-difference formula, having an error of $O\left(h^2\right)$, is used for $y'(a)$ in Eq. (10.74), we obtain

$$c_1 y(a) + c_2 \left\{ \frac{y(a+h) - y(a-h)}{2h} \right\} = c_3, \tag{10.85}$$

$$-\frac{c_2}{2h} y_{-1} + c_1 y_0 + \frac{c_2}{2h} y_1 = c_3, \tag{10.86}$$

or

$$y_{-1} = \left(\frac{2hc_1}{c_2} \right) y_0 + y_1 - \frac{2hc_3}{c_2}, \tag{10.87}$$

where $y_{-1} = y(a-h)$. Since y_{-1} appears in Eq. (10.87), we must consider y_0 as an unknown in addition to $y_1, y_2, \ldots, y_{N-1}$. The application of Eq. (10.65) at node point $i = 0$ gives

$$\left(1 + \frac{h}{2} p_0\right) y_1 + \left(-2 + h^2 q_0\right) y_0 + \left(1 - \frac{h}{2} p_0\right) y_{-1} = h^2 r_0. \tag{10.88}$$

The use of Eq. (10.87) in Eq. (10.88) yields

$$\left(-2 + h^2 q_0 + \frac{2hc_1}{c_2} - \frac{h^2 c_1 p_0}{c_2}\right) y_0 + \left(-1 + h^2 q_1 - \frac{h}{2} p_0\right) y_1$$
$$= h^2 r_0 + \frac{2hc_3}{c_2} - \frac{h^2 c_3 p_0}{c_2}. \tag{10.89}$$

The application of Eq. (10.65) at node point $i = 1$ yields

$$\left(1 - \frac{h}{2} p_1\right) y_0 + \left(-2 + h^2 q_1\right) y_1 + \left(1 + \frac{h}{2} p_1\right) y_2 = h^2 r_1. \tag{10.90}$$

Thus, the new system of equations is the same as Eqs. (10.66), except that the first equation in (10.66) is now replaced by Eqs. (10.89) and (10.90). Hence, the system consists of N simultaneous equations in tridiagonal form in the N unknowns $y_0, y_1, y_2, \ldots, y_{N-1}$. A similar procedure can be adopted when a central-difference formula is used in the boundary condition of Eq. (10.79). (See Problem 10.29.)

10.5.3 Solution of Fourth-Order Equations

To illustrate the applicability of the finite-difference method to the solution of higher order differential equations, we consider a beam deflection problem for which the differential equation is of order four. For this, consider a beam of length l with a nonuniform cross section as shown in Fig. 10.7(a). When the beam is subjected to a load, $p(x)$, the deflection of the beam, $y(x)$, is governed by the differential equation (derived by considering the equilibrium of vertical forces for a small element of the

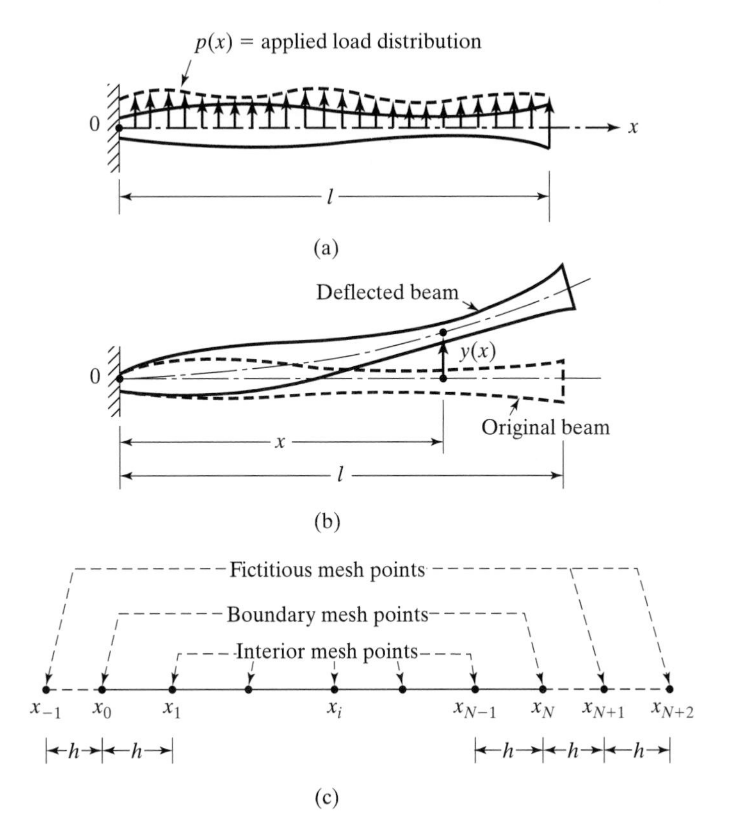

Figure 10.7 Nonuniform beam and finite difference grid.

beam located at x in Fig. 10.7(b))

$$\frac{d^2}{dx^2}\left[EI(x)\frac{d^2y(x)}{dx^2}\right] = p(x) \tag{10.91}$$

or

$$EI(x)\frac{d^4y(x)}{dx^4} + 2\frac{d}{dx}(EI(x))\frac{d^3y(x)}{dx^3} + \frac{d^2}{dx^2}(EI(x))\frac{d^2y(x)}{dx^2} = p(x), \tag{10.92}$$

where $EI(x)$ is the flexural rigidity, E is Young's modulus, $I(x)$ is the area moment of inertia of the cross section of the beam at x, and $p(x)$ is the load per unit length of the beam applied at a distance x. The beam is assumed to be fixed at $x = 0$ and free at $x = l$ (cantilever) so that the boundary conditions can be stated as

$$y(0) = 0\text{: deflection is zero at } x = 0, \tag{10.93}$$

$$y'(0) = 0\text{: slope is zero at } x = 0, \tag{10.94}$$

$$y''(l) = 0\text{: bending moment is zero at } x = l, \tag{10.95}$$

and

$$y'''(l) = 0\text{: shear force is zero at } x = l. \tag{10.96}$$

The beam is divided into N equal parts of width $h = \frac{l}{N}$ and the mesh points are labeled as (Fig. 10.7(c)).

$$x_0 = 0, x_1 = h, x_2 = 2h, \ldots, x_i = ih, \ldots, x_N = Nh = l \tag{10.97}$$

The differential Equation (10.92), when applied at the mesh point i gives

$$EI_i y_i'''' + 2EI_i' y_i''' + EI_i'' y_i'' = p_i, \tag{10.98}$$

where

$$I_i = I(x_i),\ I_i' = \frac{dI(x_i)}{dx},\ y_i'' = \frac{d^2y(x_i)}{dx^2},\ p_i = p(x_i).$$

The following central-difference formulas, each having an accuracy of $O\left(h^2\right)$, can be used for the derivatives of y appearing in Eq. (10.98):

$$y_i'' = \frac{y_{i-1} - 2y_i + y_{i+1}}{h^2}, \tag{10.99}$$

$$y_i''' = \frac{-y_{i-2} + 2y_{i-1} - 2y_{i+1} + y_{i+2}}{2h^3}, \tag{10.100}$$

and

$$y_i'''' = \frac{y_{i-2} - 4y_{i-1} + 6y_i - 4y_{i+1} + y_{i+2}}{h^4}. \tag{10.101}$$

The derivatives of y appearing in the boundary conditions of Eqs. (10.94) through (10.96) are replaced by central-difference formulas as

$$y'(0) = y_0' = \frac{y_1 - y_{-1}}{2h} = 0, \tag{10.102}$$

$$y''(l) = y_N'' = \frac{y_{N-1} - 2y_N + y_{N+1}}{h^2} = 0, \tag{10.103}$$

and

$$y'''(l) = y_N''' = \frac{-y_{N-2} + 2y_{N-1} - 2y_{N+1} + y_{N+2}}{2h^3} = 0. \tag{10.104}$$

The derivatives of the area moment of inertia are computed as follows:

$$I_i' = \frac{I_{i+1} - I_{i-1}}{2h}; i = 1, 2, \ldots, N-1 \tag{10.105}$$

(central difference formula),

$$I_i'' = \frac{I_{i-1} - 2I_i + I_{i+1}}{h^2}; i = 1, 2, \ldots, N-1 \tag{10.106}$$

(central-difference formula),

$$I_N' = \frac{3I_N - 4I_{N-1} + I_{N-2}}{2h} \tag{10.107}$$

(backward-difference formula of accuracy $O(h^2)$),

and

$$I_N'' = \frac{2I_N - 5I_{N-1} + 4I_{N-2} - I_{N-3}}{h^2} \tag{10.108}$$

(backward-difference formula of accuracy $O(h^2)$).

Note that the derivatives of the area moment of inertia at the mesh point N are expressed in terms of backward-difference formulas of accuracy $O\left(h^2\right)$. The reason is that the moment of inertia of the beam at a fictitious node (as required by a central-difference formula) does not make sense. However, the deflection of the beam at fictitious nodes, introduced by the central-difference formulas, can be eliminated (or expressed) in terms of the deflections of the interior mesh points using the boundary conditions. By substituting Eqs. (10.99) to (10.101), (10.105), and (10.106) into Eq. (10.98), we obtain the finite-difference equation for the mesh point i as

$$\begin{aligned}&\left(EI_i - \tfrac{1}{2}EI_{i+1} + \tfrac{1}{2}EI_{i-1}\right) y_{i-2} + \left(-6EI_i + 2EI_{i+1}\right) y_{i-1}\\ &+ \left(10EI_i - 2EI_{i-1} - 2EI_{i+1}\right) y_i + \left(-6EI_i + 2EI_{i-1}\right) y_{i+1}\\ &+ \left(EI_i + \tfrac{1}{2}EI_{i+1} - \tfrac{1}{2}EI_{i-1}\right) y_{i+2} = h^4 p_i; i = 1, 2, \ldots, N,\end{aligned} \tag{10.109}$$

which can be written in a compact form as

$$a_i y_{i-2} + b_i y_{i-1} + c_i y_i + d_i y_{i+1} + e_i y_{i+2} = h^4 p_i; i = 1, 2, \ldots, N, \tag{10.110}$$

where the coefficients $a_i, b_i, \ldots, e_i$ can be identified by comparing Eqs. (10.109) and (10.110). To implement the boundary conditions at $x = 0$, we have from Eqs. (10.93) and (10.94)

$$y(0) = y_0 = 0 \tag{10.111}$$

and

$$y'(0) = y_0' = \frac{y_1 - y_{-1}}{2h} = 0 \text{ or } y_{-1} = y_1. \tag{10.112}$$

Equation (10.110) gives, for $i = 1$,

$$a_1 y_{-1} + b_1 y_0 + c_1 y_1 + d_1 y_2 + e_1 y_3 = h^4 p_1, \tag{10.113}$$

which can be expressed, by using the boundary conditions of Eqs. (10.111) and (10.112), as

$$(a_1 + c_1) y_1 + d_1 y_2 + e_1 y_3 = h^4 p_1. \tag{10.114}$$

Similarly, Eq. (10.110) can be written for $i = 2$ (using $y_0 = 0$ from Eq. (10.111)) as

$$b_2 y_1 + c_2 y_2 + d_2 y_3 + e_2 y_4 = h^4 p_2. \tag{10.115}$$

In order to apply the boundary conditions at $x = l$, we express Eqs. (10.95) and (10.96), using Eqs. (10.99) and (10.100), as

$$y''(l) = y_N'' = \left(\frac{y_{N-1} - 2y_N + y_{N+1}}{h^2}\right) = 0,$$

or

$$y_{N+1} = 2y_N - y_{N-1}, \tag{10.116}$$

$$y'''(l) = y_N''' = \left(\frac{-y_{N-2} + 2y_{N-1} - 2y_{N+1} + y_{N+2}}{2h^3} \right) = 0,$$

$y_{N+2} = y_{N-2} - 2y_{N-1} + 2y_{N+1} = y_{N-2} - 2y_{N-1} + 2(2y_N - y_{N-1})$, using Eq. (10.116), or

$$y_{N+2} = 4y_N - 4y_{N-1} + y_{N-2}. \tag{10.117}$$

When Eq. (10.110) is written for $i = N - 1$ and $i = N$, we obtain

$$a_{N-1}y_{N-3} + b_{N-1}y_{N-2} + c_{N-1}y_{N-1} + d_{N-1}y_N + e_{N-1}y_{N+1} = h^4 p_{N-1} \tag{10.118}$$

and

$$a_N y_{N-2} + b_N y_{N-1} + c_N y_N + d_N y_{N+1} + e_N y_{N+2} = h^4 p_N. \tag{10.119}$$

By substituting Eqs. (10.116) and (10.117) for y_{N+1} and y_{N+2}, Eqs. (10.118) and (10.119) take the form

$$a_{N-1}y_{N-3} + b_{N-1}y_{N-2} + (c_{N-1} - e_{N-1})y_{N-1} + (d_{N-1} + 2e_{N-1})y_N = h^4 p_{N-1} \tag{10.120}$$

and

$$(a_N + e_N)y_{N-2} + (b_N - d_N - 4e_N)y_{N-1} + (c_N + 2d_N + 4e_N)y_N = h^4 p_N. \tag{10.121}$$

Finally, the finite-difference equations for the mesh-point deflections of the beam are given by Eq. (10.114) for $i = 1$, Eq. (10.115) for $i = 2$, Eq. (10.110) for $i = 3, 4, \ldots, N - 2$, Eq. (10.120) for $i = N - 1$, and Eq. (10.121) for $i = N$. These represent N simultaneous linear equations in the N unknowns $y_1, y_2, \ldots, y_N$. These equations, when written in matrix form, appear in a band form. The solution techniques discussed in Chapter 3 can be used for the solution of these equations.

10.5.4 Accuracy

The accuracy of the finite-difference solution depends on the type of finite-difference formula used and the number of mesh points used in the interval of integration. The use of higher order finite-difference formulas increases the accuracy, but might introduce difficulties in implementing the boundary conditions, especially by requiring fictitious mesh points where the values of y are unknown. With an increase in the number of mesh points, the accuracy improves, but the number of equations to be solved, and hence, the computational time also increases. In order to obtain a reasonable solution, the boundary-value problem has to be solved a number of times with different step lengths (h). The convergence of solution at the common mesh points can be used to stop the procedure.

▶Example 10.7

Find the deflection of the uniform fixed–fixed beam shown in Fig. 10.8(a) using the finite-difference method.

The data are $h = 9''$, $E = 30 \times 10^6$ psi, $l = 36''$, and $I = 2$ in^4.

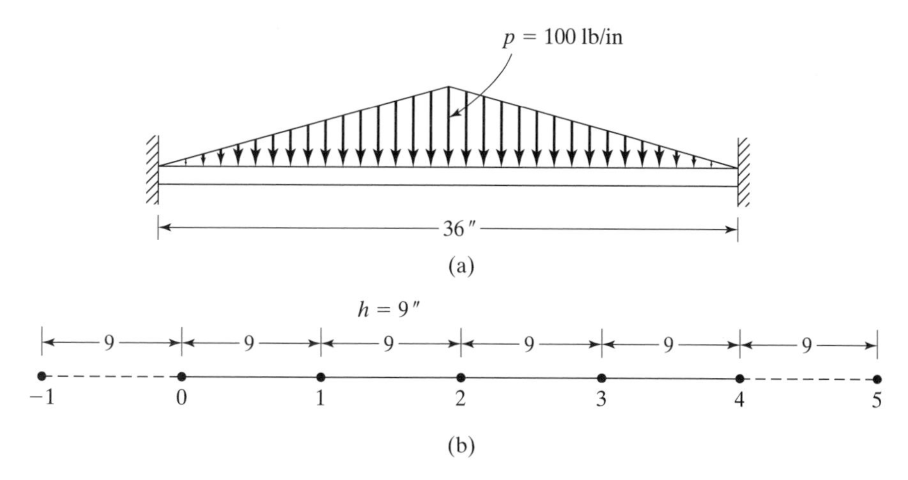

Figure 10.8 Fixed-fixed beam and grid points.

Solution

Since the beam is fixed at $x = 0$ and $x = l = 36''$, we introduce one imaginary node on each of the left and right sides of the beam as shown in Fig. 10.8(b). For a uniform beam, Eqs. (10.98) and (10.101) yield

$$EIY_i'''' = p_i \tag{E1}$$

or

$$y_{i-2} - 4y_{i-1} + 6y_i - 4y_{i+1} + y_{i+2} = \frac{p_i h^4}{EI}. \tag{E2}$$

Applying Eq. (E2) at nodes 1, 2, and 3, we obtain

$$y_{-1} - 4y_0 + 6y_1 - 4y_2 + y_3 = \frac{p_1 h^4}{EI}, \tag{E3}$$

$$y_0 - 4y_1 + 6y_2 - 4y_3 + y_4 = \frac{p_2 h^4}{EI}, \tag{E4}$$

and

$$y_1 - 4y_2 + 6y_3 - 4y_4 + y_5 = \frac{p_3 h^4}{EI}. \tag{E5}$$

The boundary conditions can be expressed as

$$y_0 = 0, \left.\frac{dy}{dx}\right|_0 = \frac{y_1 - y_{-1}}{2h} = 0 \text{ or } y_{-1} = y_1 \tag{E6}$$

and

$$y_4 = 0, \left.\frac{dy}{dx}\right|_4 = \frac{y_5 - y_3}{2h} = 0 \text{ or } y_5 = y_3. \tag{E7}$$

By introducing the boundary conditions of Eqs. (E6) and (E7) and the values $p_1 = p_3 = 50$, $p_2 = 100$, $E = 30 \times 10^6$ psi, and $I = 2$ into Eqs. (E3) to (E5), we obtain

$$7y_1 - 4y_2 + y_3 = 5.4675 \times 10^{-3}, \tag{E8}$$

$$-4y_1 + 6y_2 - 4y_3 = 10.935 \times 10^{-3}, \tag{E9}$$

and

$$y_1 - 4y_2 + 7y_3 = 5.4675 \times 10^{-3}. \tag{E10}$$

The solution of Eqs. (E8) to (E10) gives the beam deflections as

$$y_1 = y_3 = 4.78406 \times 10^{-3} \text{ in and } y_2 = 8.20125 \times 10^{-3} \text{ in.}$$ ◀

10.6 Solution of Nonlinear Boundary-Value Problems

The finite-difference method can be applied to nonlinear boundary-value problems also. However, the resulting algebraic equations will be nonlinear, and hence, an iterative process has to be used to solve the problem. To illustrate the procedure, consider the problem

$$y''(x) = f(x, y, y');\ a \le x \le b \tag{10.122}$$

with the boundary conditions

$$y(a) = \alpha \tag{10.123}$$

and

$$y(b) = \beta, \tag{10.124}$$

where $f(x, y, y')$ is a nonlinear function of x, y, and y'. As in the linear case, we divide the interval $[a, b]$ into N equal parts of width $h = \left(\frac{b-a}{N}\right)$. The finite-difference approximation (using central-difference formulas for y' and y'') of Eq. (10.122) at the interior node point i is given by

$$\frac{y_{i+1} - 2y_i + y_{i-1}}{h^2} = f\left(x_i, y_i, \frac{y_{i+1} - y_{i-1}}{2h}\right);\ i = 1, 2, \ldots, N-1. \tag{10.125}$$

Equation (10.125) can be written for $i = 1$ and $i = N - 1$ as

$$y_2 - 2y_1 + y_0 - h^2 f\left(x_1, y_1, \frac{y_2 - y_0}{2h}\right) = 0 \tag{10.126}$$

and

$$y_N - 2y_{N-1} + y_{N-2} - h^2 f\left(x_{N-1}, y_{N-1}, \frac{y_N - y_{N-2}}{2h}\right) = 0. \tag{10.127}$$

By using the boundary conditions

$$y_0 = \alpha,\ y_N = \beta \tag{10.128}$$

in Eqs. (10.126) and (10.127), we can use Eq. (10.126), Eq. (10.125) for $i = 2, 3, \ldots, N-2$, and Eq. (10.127) to yield the following system of $(N-1)$ nonlinear algebraic equations in the $(N-1)$ unknowns $y_1, y_2, \ldots, y_{N-1}$:

$$y_2 - 2y_1 + \alpha - h^2 f\left(x_1, y_1, \frac{y_2 - \alpha}{2h}\right) = 0,$$

$$y_3 - 2y_2 + y_1 - h^2 f\left(x_2, y_2, \frac{y_3 - y_1}{2h}\right) = 0,$$

$$y_4 - 2y_3 + y_2 - h^2 f\left(x_3, y_3, \frac{y_4 - y_2}{2h}\right) = 0,$$

$$\vdots$$

$$y_{N-1} - 2y_{N-2} + y_{N-3} - h^2 f\left(x_{N-2}, y_{N-2}, \frac{y_{N-1} - y_{N-3}}{2h}\right) = 0, \quad (10.129)$$

and

$$\beta - 2y_{N-1} + y_{N-2} - h^2 f\left(x_{N-1}, y_{N-1}, \frac{\beta - y_{N-2}}{2h}\right) = 0.$$

These $(N-1)$ nonlinear equations can be solved by using the Newton–Raphson method described in Section 2.12.

▶Example 10.8

The temperature distribution in a rectangular fin, considering conduction and radiation heat transfers, is given by

$$\frac{d^2T}{dx^2} = \frac{\sigma \varepsilon P}{kA}(T^4 - T_\infty^4), \quad \text{(a)}$$

where T = temperature, k = thermal conductivity, $A = bd$ = cross-sectional area, $P = 2(b+d)$ = perimeter, T_∞ = surrounding temperature, σ = Stefan–Boltzman constant, and ε = emissivity. (See Fig. 10.9(a).) The data are given by

$$k = 42 \text{ W/m} -^\circ \text{ K}, b = 0.5 \text{ m}, d = 0.2 \text{ m}, T_\infty = 500^\circ \text{ K},$$

$$l = 2 \text{ m}, \varepsilon = 0.1, \sigma = 5.7 \times 10^{-8} \text{ W/m}^2 -^\circ \text{ K}^4,$$

$$T(x=0) = 1000^\circ \text{ K}, \text{ and } T(x=l) = 350^\circ \text{ K}.$$

Find the solution of this nonlinear boundary-value problem.

Solution

We divide the length of the fin into four equal parts as shown in Fig. 10.9(b). By approximating the second derivative using central differences, we can apply Eq. (a)

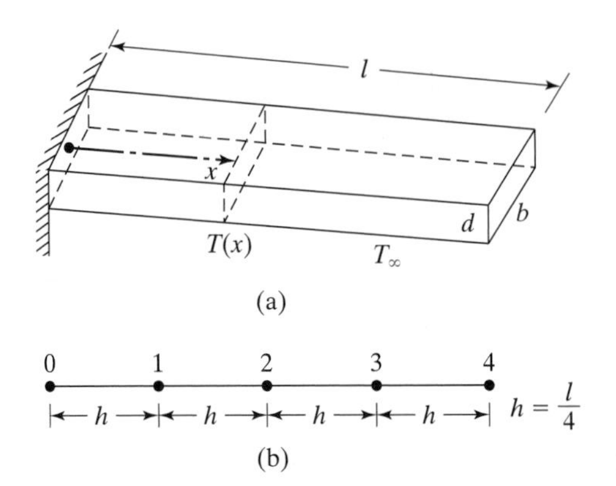

Figure 10.9 Rectangular fin.

to the interior nodes 1, 2, and 3 to yield

$$\frac{T_0 - 2T_1 + T_2}{h^2} - cT_1^4 + cT_\infty^4 = 0, \tag{b}$$

$$\frac{T_1 - 2T_2 + T_3}{h^2} - cT_2^4 + cT_\infty^4 = 0, \tag{c}$$

and

$$\frac{T_2 - 2T_3 + T_4}{h^2} - cT_3^4 + cT_\infty^4 = 0, \tag{d}$$

where

$$c = \frac{\sigma\varepsilon P}{kA} = \frac{2(b+d)\sigma\varepsilon}{kbd} \tag{e}$$

with known values of $\varepsilon = 0.1$, $\sigma = 5.7 \times 10^{-8}$, $b = 0.5$, $d = 0.2$, $h = 0.5$, $k = 42$, $T_0 = 1000$, and $T_4 = 350$. Eqs. (b) to (d) become

$$-2T_1 - 0.0475 \times 10^{-8} T_1^4 + T_2 + 1029.6875 = 0, \tag{f}$$

$$T_1 - 2T_2 - 0.0475 \times 10^{-8} T_2^4 + T_3 + 29.6875 = 0, \tag{g}$$

and

$$T_2 - 2T_3 - 0.0475 \times 10^{-8} T_3^4 + 379.6875 = 0. \tag{h}$$

Equations (f) to (h) are solved using Newton–Raphson method with starting values, $T_1 = 800$, $T_2 = 700$, and $T_3 = 600$. The results are given by $T_1 = 739.945°$ K, $T_2 = 592.597°$ K, and $T_3 = 474.139°$ K. ◀

10.7 Solution of Eigenvalue Problems

Certain problems in vibration analysis, stability analysis, and elasticity theory give rise to eigenvalue problems described in the form of homogeneous ordinary differential equations. A boundary-value problem becomes an eigenvalue problem, when the differential equation is homogeneous (that is, the forcing term is zero) and the boundary conditions are also homogeneous (that is, the boundary conditions do not contain any nonzero term that is independent of y and its derivatives). For example, the following boundary-value problem represents an eigenvalue problem

$$y'' + \lambda f(x) y = 0;\ 0 \le x \le 1 \tag{10.130}$$

with the boundary conditions

$$c_1 y(0) + c_2 y'(0) = 0 \tag{10.131}$$

and

$$c_3 y(1) + c_4 y'(1) = 0, \tag{10.132}$$

where λ is an arbitrary parameter, known as the eigenvalue, and c_1 to c_4 are constants. The solution of an eigenvalue problem is not unique. If $y(x) = \phi(x)$ is a solution, then $y(x) = k\phi(x)$ will also be a solution for any value of the constant k. (This can be verified by substituting the solution into the differential Equation (10.130).) The eigenvalue problems of ordinary differential equations are closely related to the matrix eigenvalue problems considered in Chapter 4, because the application of the finite-difference technique converts the former into the latter. To illustrate the application of finite-difference technique to an eigenvalue problem, we consider the longitudinal vibration of an elastic bar with a variable cross section (Fig. 10.10). The equilibrium of forces of an element of the bar during free vibration (including the inertia force) gives the differential equation

$$\frac{\partial}{\partial x}\left[EA(x)\frac{\partial u(x,t)}{\partial x}\right] = \rho A(x)\frac{\partial^2 u(x,t)}{\partial t^2}, \tag{10.133}$$

where E is Young's modulus, ρ is the density, $A(x)$ is the area of cross section of the bar, and $u(x,t)$ is the axial displacement of the bar at cross-section x at time t. The motion of the bar is assumed to be harmonic so that

$$u(x,t) = U(x)\cos\omega t, \tag{10.134}$$

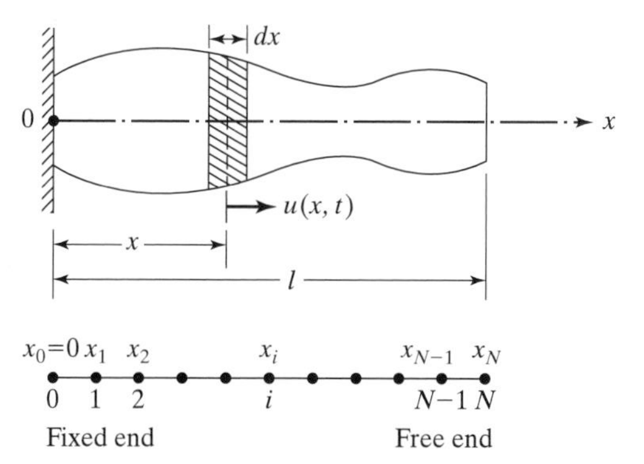

Figure 10.10 Longitudinal vibration of a bar.

where $U(x)$ is the axial displacement of the bar (amplitude) at the cross-section x (independent of time) and ω is called the natural frequency of vibration. By substituting Eq. (10.134) into Eq. (10.133), and dividing throughout by $\cos\omega t$, we obtain

$$\frac{d}{dx}\left[EA(x)\frac{dU(x)}{dx}\right] = -\omega^2\rho A(x)U(x). \tag{10.135}$$

We now assume the bar to be uniform so that Eq. (10.135) reduces to

$$\frac{d^2U(x)}{dx^2} + \alpha^2 U(x) = 0, \tag{10.136}$$

where

$$\alpha^2 = \frac{\rho\omega^2}{E}. \tag{10.137}$$

The bar is assumed to be fixed at $x = 0$ and free at $x = l$ so that the boundary conditions can be expressed as

$$U(0) = 0\text{: fixed end} \tag{10.138}$$

$$\text{(displacement is zero at } x = 0)$$

and

$$\frac{dU}{dx}(l) = 0\text{: free end} \tag{10.139}$$

$$\text{(stress is zero at } x = l\text{; hence strain is zero).}$$

▶Example 10.9

Derive the eigenvalue problem corresponding to the free longitudinal vibration of a fixed-free bar using central differences.

Solution

The length of the bar is divided into N equal parts of width $h = \frac{l}{N}$. This gives

$$x_0 = 0, x_1 = h, x_2 = 2h, \ldots, x_i = ih, \ldots, x_N = l. \tag{a}$$

The application of Eq. (10.136) at node i using a central-difference formula for $\frac{d^2U(x)}{dx^2}$ gives

$$\frac{U_{i+1} - 2U_i + U_{i-1}}{h^2} + \alpha^2 U_i = 0, \tag{b}$$

where $U_{i-1} = U(x_{i-1})$, $U_i = U(x_i)$ and $U_{i+1} = U(x_{i+1})$. Equation (b) can be rewritten as

$$U_{i-1} + (-2 + h^2\alpha^2)U_i + U_{i+1} = 0, \tag{c}$$

where $\lambda = h^2\alpha^2$. The application of Eq. (c) to mesh points $1, 2, \ldots, N$ yields the following equations:

$$\begin{aligned} U_0 + (-2+\lambda)U_1 + U_2 &= 0, \\ U_1 + (-2+\lambda)U_2 + U_3 &= 0 \\ &\vdots, \\ U_{N-2} + (-2+\lambda)U_{N-1} + U_N &= 0, \end{aligned} \tag{d}$$

and

$$U_{N-1} + (-2+\lambda)U_N + U_{N+1} = 0.$$

The boundary condition of Eq. (10.138) gives

$$U_0 = 0, \tag{e}$$

and the boundary condition of Eq. (10.139) can be expressed, using a central-difference formula, as

$$\frac{dU}{dx}(l) = \frac{dU}{dx}(x_N) = \frac{U_{N+1} - U_{N-1}}{2h} = 0$$

or

$$U_{N+1} = U_{N-1}. \tag{f}$$

When Eq. (e) is used, the first formula of Eq. (d) reduces to

$$(-2+\lambda)U_1 + U_2 = 0. \tag{g}$$

Similarly, Eq. (f) can be used to reduce the last formula of Eq. (d) to

$$2U_{N-1} + (-2+\lambda)U_N = 0. \tag{h}$$

When the first and the last formulas of Eq. (d) are replaced, respectively, by Eqs. (g) and (h), the resulting system of N linear equations in the N unknowns $U_1, U_2, \ldots, U_N$ can be expressed in matrix form as

$$\begin{bmatrix} -2+\lambda & 1 & & & & \bigcirc \\ 1 & -2+\lambda & 1 & & & \\ & 1 & -2+\lambda & 1 & & \\ & & \ddots & \ddots & \ddots & \\ \bigcirc & & & 1 & -2+\lambda & 1 \\ & & & & 2 & -2+\lambda \end{bmatrix} \begin{Bmatrix} U_1 \\ U_2 \\ U_3 \\ \cdot \\ \cdot \\ \cdot \\ U_{N-1} \\ U_N \end{Bmatrix} = \begin{Bmatrix} 0 \\ 0 \\ 0 \\ \cdot \\ \cdot \\ \cdot \\ 0 \\ 0 \end{Bmatrix}. \tag{i}$$

Equation (i) denotes a matrix eigenvalue problem that can be expressed in a compact form as

$$[[A] - \lambda[I]]\,\vec{U} = \vec{0}, \tag{j}$$

where the tridiagonal matrix $[A]$ is given by

$$[A] = \begin{bmatrix} 2 & -1 & & & & \\ -1 & 2 & -1 & & \bigcirc & \\ & -1 & 2 & -1 & & \\ & & \ddots & \ddots & \ddots & \\ & \bigcirc & & -1 & 2 & -1 \\ & & & & -2 & 2 \end{bmatrix}. \tag{k}$$

$\vec{U}$ is called the eigenvector,

$$\vec{U} = \begin{Bmatrix} U_1 \\ U_2 \\ \cdot \\ \cdot \\ \cdot \\ U_N \end{Bmatrix}, \tag{l}$$

and $[I]$ is an identity matrix of order N. The matrix eigenvalue problem of Eq. (j) can be solved by using the methods discussed in Chapter 4. ◀

▶Example 10.10

The equation governing the free transverse vibration of a uniform beam is given by [10.3]

$$\frac{d^4 w}{dx^4} - \beta^4 w = 0, \tag{E1}$$

where

$$\beta^4 = \frac{\rho A \omega^2}{EI}, \tag{E2}$$

w = transverse deflection, ρ = density, A = area of cross section, E = Young's modulus, I = area moment of inertia, and ω = natural frequency of vibration. Using Eqs. (E1) and (E2), determine the natural frequencies and mode shapes of a beam simply supported at both ends.

Use the following data:

$$\text{unit weight} = \rho g = 0.29 \text{ lb/in}^3,\ A = 3 \text{ in}^2,\ I = 1 \text{ in}^4,$$
$$E = 30 \times 10^6 \text{ psi},\ l = 36'',\ \text{and } h = 9''(N = 4).$$

Solution

The finite-difference approximation of Eq. (E1) for a uniform beam at node i is given by

$$\frac{w_{i-2} - 4w_{i-1} + 6w_i - 4w_{i+1} + w_{i+2}}{h^4} - \beta^4 w_i = 0, \tag{E3}$$

which can be rewritten as

$$w_{i-2} - 4w_{i-1} + (6 - \lambda)w_i - 4w_{i+1} + w_{i+2} = 0. \tag{E4}$$

The application of Eq. (E4) at the interior nodes 1, 2, and 3 gives

$$w_{-1} - 4w_0 + (6 - \lambda)w_1 - 4w_2 + w_3 = 0,$$
$$w_0 - 4w_1 + (6 - \lambda)w_2 - 4w_3 + w_4 = 0, \tag{E5}$$

and

$$w_1 - 4w_2 + (6 - \lambda)w_3 - 4w_4 + w_5 = 0.$$

The boundary conditions can be expressed as

$$w_0 = 0,\ \left.\frac{d^2w}{dx^2}\right|_0 = \frac{w_{-1} - 2w_0 + w_1}{h^2} = 0 \text{ or } w_{-1} = -w_1 \tag{E6}$$

and

$$w_4 = 0,\ \left.\frac{d^2w}{dx^2}\right|_4 = \frac{w_5 - 2w_4 + w_3}{h^2} = 0 \text{ or } w_5 = -w_3. \tag{E7}$$

By introducing the boundary conditions of Eqs. (E6) and (E7) into Eqs. (E5), we obtain

$$(5 - \lambda)w_1 - 4w_2 + w_3 = 0,$$
$$-4w_1 + (6 - \lambda)w_2 - 4w_3 = 0, \tag{E8}$$

and

$$w_1 - 4w_2 + (5 - \lambda)w_3 = 0,$$

where

$$\lambda = h^4\beta^4. \tag{E9}$$

Equation (E9) can be written in matrix form as

$$\begin{bmatrix} (5-\lambda) & -4 & 1 \\ -4 & (6-\lambda) & -4 \\ 1 & -4 & (5-\lambda) \end{bmatrix} \begin{Bmatrix} w_1 \\ w_2 \\ w_3 \end{Bmatrix} = \begin{Bmatrix} 0 \\ 0 \\ 0 \end{Bmatrix}. \tag{E10}$$

For a nontrivial solution of Eqs. (E10), the determinant of the coefficient matrix must be zero:

$$\begin{vmatrix} (5-\lambda) & -4 & 1 \\ -4 & (6-\lambda) & -4 \\ 1 & -4 & (5-\lambda) \end{vmatrix} = 0$$

or

$$\lambda^3 - 16\lambda^2 + 52\lambda - 16 = 0. \tag{E11}$$

The roots of Eq. (E11) yield the natural frequencies of vibration as

$$\lambda_1 = 0.343124,\ \omega_1 = \sqrt{\frac{\lambda_1 EI}{\rho A h^4}} = 834.757 \text{ rad/sec}; \tag{E12}$$

$$\lambda_2 = 4.0,\ \omega_2 = \sqrt{\frac{\lambda_2 EI}{\rho A h^4}} = 1688.233 \text{ rad/sec}; \tag{E13}$$

and

$$\lambda_3 = 11.6569,\ \omega_3 = \sqrt{\frac{\lambda_3 EI}{\rho A h^4}} = 4865.484 \text{ rad/sec}. \tag{E14}$$

The eigenvectors or mode shapes can be found by solving Eqs. (E8). For λ_1, we obtain

$$\begin{aligned} 4.656876w_1 - 4w_2 + w_3 &= 0, \\ -4w_1 + 5.656876w_2 - 4w_3 &= 0, \end{aligned} \tag{E15}$$

and

$$w_1 - 4w_2 + 4.656876w_3 = 0.$$

Letting $w_1 = 1$ arbitrarily, Eqs. (E15) can be solved to find the first mode shape as

$$\vec{w}^{(1)} = \begin{Bmatrix} w_1 \\ w_2 \\ w_3 \end{Bmatrix}^{(1)} = \begin{Bmatrix} 1.0 \\ 1.414225 \\ 1.000024 \end{Bmatrix}. \tag{E16}$$

For λ_2, Eqs. (E8) yield

$$\begin{aligned} w_1 - 4w_2 + w_3 &= 0, \\ -4w_1 + 2w_2 - 4w_3 &= 0, \end{aligned} \tag{E17}$$

and

$$w_1 - 4w_2 + w_3 = 0.$$

If $w_1 = 1.0$ is assumed, Eqs. (E17) give the second mode shape as

$$\vec{w}^{(2)} = \begin{Bmatrix} w_1 \\ w_2 \\ w_3 \end{Bmatrix}^{(2)} = \begin{Bmatrix} 1.0 \\ 0 \\ -1.0 \end{Bmatrix}. \tag{E18}$$

Finally, for λ_3, Eqs. (E8) give

$$-6.6569w_1 - 4w_2 + w_3 = 0,$$
$$-4w_1 - 5.6569w_2 - 4w_3 = 0, \tag{E19}$$

and

$$w_1 - 4w_2 - 6.6569w_3 = 0.$$

If $w_1 = 1.0$ is assumed, Eqs. (E19) give the third mode shape as

$$\vec{w}^{(3)} = \begin{Bmatrix} w_1 \\ w_2 \\ w_3 \end{Bmatrix}^{(3)} = \begin{Bmatrix} 1.0 \\ -1.414219 \\ -1.000024 \end{Bmatrix}. \tag{E20}$$

◀

10.8 Choice of Method

A boundary-value problem in ordinary differential equations is characterized by the specification of conditions at more than one point. The shooting and finite-difference methods can be used for the solution of boundary-value problems. When the finite-difference method is applied to a linear boundary-value problem, it results in a tridiagonal system of equations. These equations can be solved without any need for a trial-and-error procedure. The application of the finite-difference method to nonlinear boundary-value problems results in a set of nonlinear algebraic equations that require an iterative procedure for their solution. An advantage with the finite-difference method is that it is not necessary to convert an nth-order differential equation into an equivalent system of n first-order differential equations. The accuracy of the solution attainable with the finite-difference method depends on the mesh size and the order of the finite-difference approximation used. As the mesh size is reduced, the number of equations to be solved, and hence, the computational time increases. In practice, it is desirable to solve the differential equation with different step sizes and compare the solutions at specific mesh points to observe the convergence and accuracy of the solution. Although higher order finite-difference approximations lead to higher accuracy for the same mesh size, they lead to complications, especially near the boundaries of the interval where the exterior values are not known.

The shooting methods can be used to solve both linear and nonlinear boundary-value problems. Although they are easy to implement, there is no guarantee of convergence of the shooting methods. However, when they converge, they usually converge faster. An advantage of the shooting methods is that an existing program for initial-value problems can be used.

The eigenvalue problems are governed by homogeneous boundary-value problems. The application of finite-difference approximation leads to algebraic eigenvalue problems. The standard methods of solution such as power, Jacobi, and Householder methods, discussed in Chapter 4, can be used to solve the resulting algebraic eigenvalue problems.

10.9 Use of Software Packages

10.9.1 MATLAB

▶Example 10.11

Find the solution of the boundary-value problem (Example 10.5)

$$y'' = f(x, y, y') \equiv y' - y + 3e^{2x} - 2\sin x; a \le x \le b$$
$$y(a) = y_a = 6.308447, \ y(b) = y_b = 55.430436 \tag{E1}$$

with $a = 1$ and $b = 2$.

Solution

We solve this problem using the shooting method. We start by solving the initial-value problem (y is replaced by u in Eq. (E1)):

$$u'' = f(x, u, u')$$
$$u(a) = y_a, \ u'(a) = t_k, \tag{E2}$$

where t_k is the assumed or trial value of u' at $x = a$ in iteration k ($k = 1$ to start with). The error in the solution of Eq. (E2) is given by

$$m(t_k) = u(b, t_k) - y_b. \tag{E3}$$

Next, we assume a new value, t_{k+1}, and solve the problem in Eq. (E2) to find a new solution $u(x, t_{k+1})$. This procedure is continued until the error satisfies the convergence criterion

$$|m(t_{k+1})| = |u(b, t_{k+1}) - y_b| \le \varepsilon, \tag{E4}$$

where ε is a small number. Newton's method is used to update the value of t_k. The computational procedure can be summarized by the following steps:

Step 1 Assume a value of t_k, and solve the ordinary differential equations given by

$$u'' = f(x, u, u'); u(a) = y_a, u'(a) = t_k$$

$$v'' = v\frac{\partial f}{\partial u}(x, u, u') + v'\frac{\partial f}{\partial u'}(x, u, u'); v(a) = 0, v'(a) = 1. \quad \text{(E5)}$$

Step 2 Stop the procedure if

$$|m(t_k)| = |u(b, t_k) - y_b| \le \varepsilon. \quad \text{(E6)}$$

Otherwise, take

$$t_{k+1} = t_k - \frac{m(t_k)}{v(b, t_k)}, \quad \text{(E7)}$$

and go to Step 1.

The MATLAB program used to implement this procedure for the solution of Eq. (E1) along with the results is given below. Note that $z(1) = y$, $z(2) = y'$, $z(3) = v$, $z(4) = v'$ and $z(1)$ and $z(2)$ are plotted in the figure.

```
%P1.m: this program is used to solve a differential equation
ya = 6.308447;
yb = 55.430436;
a = 1;
b = 2;
maxiter = 10;
eps = 0.0001;
t(1) = 1.0; %assumed b.c.
tol = 1.0;
i = 1;
tspan = [a: 0.01: b];
while ((tol > eps)&(i < maxiter))
   z0 = [ya t(i) 0 1]';
   [x,z] = ode23 ('func', tspan, z0);
   [n nn] = size(z);
   m(i) = z(n,1) - yb;
   tol = abs (m(i));
   t(i+1) = t(i) - m(i)/z(n, 3);
   i = i+1;
end
disp ('Iteration number:');
i
zz = [x, z];
disp('    x              Solution z');
```

```
disp(zz)
plot(x,z(:, 1));
hold on;
plot(x, z(:,2));
title ('Solution of differential equation');
xlabel ('x');
ylabel ('z');
text(1.5, 50, 'z(2)');
text(1.5, 25, 'z(1)');

function dz = func(x,z)
dz = [z(2)
   z(2)-z(1)+3*exp(2*x)-2*sin(x)
   z(4)
   -z(3) + z(4)];

>> p1
Iteration number:

i =
     4
    x          Solution z
    1.0000     6.3084    16.4627          0     1.0000
    1.0100     6.4746    16.7720     0.0100     1.0100
    1.0200     6.6439    17.0871     0.0202     1.0200
    1.0300     6.8164    17.4083     0.0305     1.0300
    1.0400     6.9921    17.7355     0.0408     1.0400
    1.0500     7.1711    18.0689     0.0512     1.0500
    1.0600     7.3535    18.4088     0.0618     1.0600
    1.0700     7.5393    18.7551     0.0724     1.0699
    1.0800     7.7286    19.1080     0.0832     1.0799
    1.0900     7.9214    19.4677     0.0940     1.0899
    1.1000     8.1179    19.8342     0.1050     1.0998
    .
    .
    .
    1.9000     45.3475    91.3001     1.2729     1.7520
    1.9100     46.2694    93.0996     1.2905     1.7567
    1.9200     47.2094    94.9355     1.3080     1.7613
    1.9300     48.1680    96.8085     1.3257     1.7658
    1.9400     49.1456    98.7193     1.3434     1.7701
    1.9500     50.1425   100.6687     1.3611     1.7743
    1.9600     51.1591   102.6573     1.3788     1.7784
    1.9700     52.1958   104.6860     1.3967     1.7823
```

1.9800	53.2530	106.7556	1.4145	1.7861
1.9900	54.3310	108.8671	1.4324	1.7898
2.0000	55.4304	111.0211	1.4503	1.7933

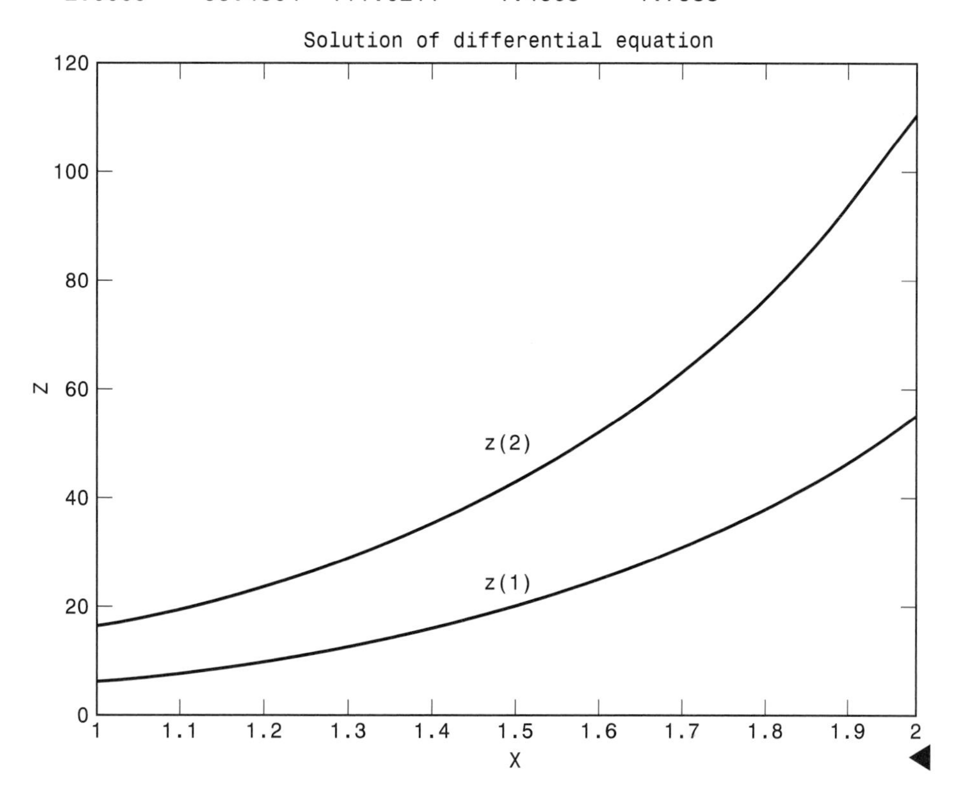

▶Example 10.12

Solve Example 10.11 using the finite-difference method.

Solution

For the general boundary-value problem

$$y''(x) + p(x)y'(x) + q(x)y(x) = r(x);\ a \le x \le b$$

$$y(a) = \alpha,\ y(b) = \beta, \tag{E1}$$

the tridiagonal system of Equations (10.67) are to be solved. In the present case, $p(x) = -1$, $q(x) = 1$, $r(x) = 3e^{2x} - 2\sin x$, $a = 1$, $b = 2$, $\alpha = 6.308447$, and $\beta = 55.430436$. The interval $a \le x \le b$ is divided into $n = 99$ equal divisions so that $h = 1/99$, $x_1 = 1.0$, $x_i = x_1 + (i-1)h$, $i = 2, 3, \ldots, 100$, with $x_{100} = 2.0$. The MATLAB program to solve Eqs. (10.67) along with the results is given next. Note that the matrix zz contains the values of x (array xx) in the first column and the values of y (array yy) in the second column. The variation of y with x (array yy) is plotted in the figure.

```
%fdm.m
%Finite difference method for solving a two-point boundary
%value problem
%Start of problem data
a = 1;
b = 2;
n = 99;
p = ones (1, n-1);
q = -ones (1, n-1);
h = (b-a)/n;
x = linspace (a+h, b, n);
r = 3 * exp(2*x) - 2*sin(x);
alp = 6.308447;
bet = 55.430436;
%End of problem data
h2 = h/2;
hh = h*h;
pp = zeros(1, n-1);
rr = pp;
pp(1:n-2) = 1-p(1,1:n-2) * h2;
qq = -(2 + hh * q);
rr(2: n-1) = 1 + p(1, 2: n-1) * h2;
rhs(1) = hh * r(1) - (1 + p(1)*h2)*alp;
rhs(2: n-2) = hh * r(2: n-2);
rhs(n-1) = hh* r(n-1) - (1 - p(n-1)*h2)*bet;
y = tridiag (pp, qq, rr, rhs);
xx = [a x];
yy = [alp y bet];
zz = [xx' yy'];
zz
plot (xx, yy);
xlabel ('xx');
ylabel ('yy');

function x = tridiag (p, q, r, b)
n = length (q);
p(1) = p(1)/q(1);
b(1) = b(1)/q(1);
for i = 2: n-1
   dr = q(i) - r(i)*p (i - 1);
   if (dr == 0)
```

```
        error ('denominator zero')
    end
    p(i) = p(i)/dr;
    b(i) = (b(i) - r(i)*b(i-1))/dr;
end
b(n) = (b(n) - r(n)*b (n-1))/(q(n) - r(n)*p (n-1));
x(n) = b(n);
for i = n-1: -1 : 1
    x(i) = b(i) - p(i)*x (i + 1);
end

>> fdm

zz =

     1.0000     6.3084
     1.0101     6.4763
     1.0202     6.6473
     1.0303     6.8216
     1.0404     6.9992
     1.0505     7.1802
     1.0606     7.3646
     1.0707     7.5525
     1.0808     7.7439
     1.0909     7.9390
     1.1010     8.1379
     1.1111     8.3405
     1.1212     8.5469
     1.1313     8.7573
     1.1414     8.9717
     1.1515     9.1902
     .
     .
     .
     1.9091    46.1851
     1.9192    47.1331
     1.9293    48.1000
     1.9394    49.0862
     1.9495    50.0920
     1.9596    51.1179
     1.9697    52.1643
     1.9798    53.2315
     1.9899    54.3201
     2.0000    55.4304
```

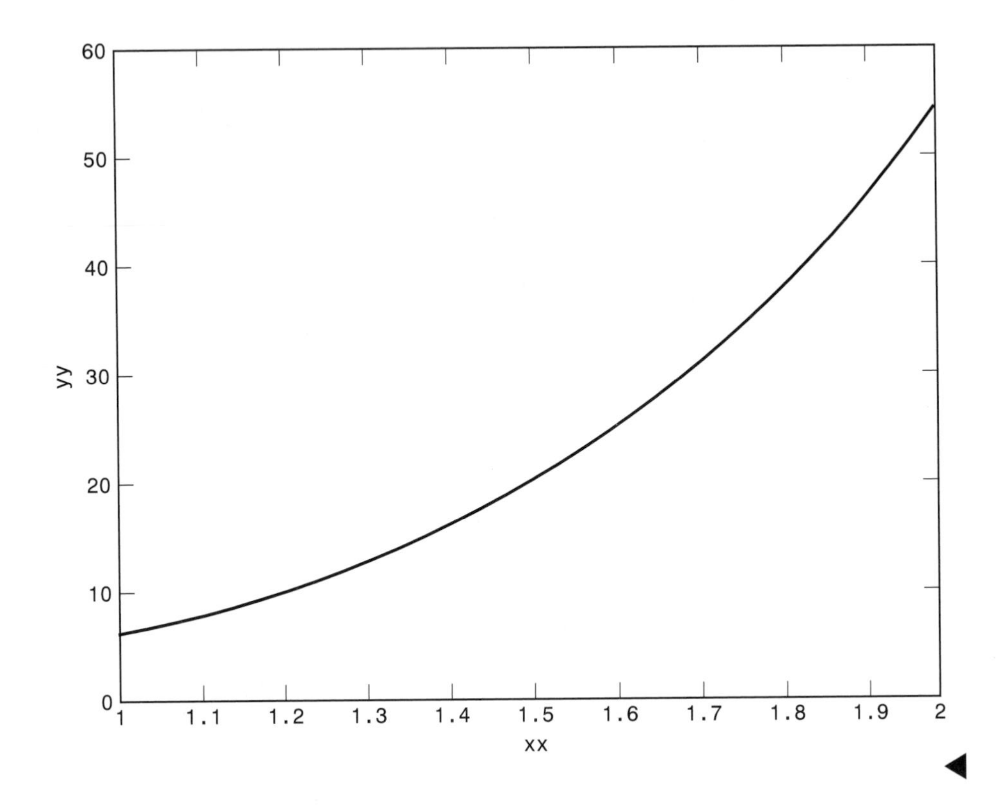

10.9.2 MAPLE

▶Example 10.13

Solve the boundary-value problem

$$\frac{d^2y}{dx^2} + \lambda y = 0; y(0) = 0, y(1) = 0.$$

Solution

The following MAPLE functions are used in finding the exact solution of the boundary-value problem.

The statement

```
dsolve({ode1,in1}, {y(x)});
```

solves the differential equation or the set of differential equations defined in *odel* by using the initial conditions stated in *in1* for the variables given in $y(x)$.

The statement

```
fsolve(detA,K = 8..12);
```

finds the root of the equation, $detA = 0$, in the range 8 through 12.

The statement

```
subs(s1, s2, ...,expr);
```

returns the expression that is generated by substituting $s1, s2, \ldots$ into the expression *expr*.

The statement

```
rhs(expr);
```

yields the right-hand side of the equation, inequality or relation, given in *expr*.

The statement

```
solve(eqs,vars);
```

determines the solution of the equation or the set of equations given in *eqs* in terms of the variables *vars*.

```
>
> ode1:=diff(y(x),x,x) +K*y(x) = 0;
```

$$ode1 := \left(\frac{\partial^2}{\partial x^2} y(x)\right) + K y(x) = 0$$

```
> in1:=y(0) = 0, y(1) = 0;
```

$$in1 := y(0) = 0, y(1) = 0$$

```
> tem:=dsolve(ode1,y(x));
```

$$tem := y(x) = \_C1 \sin(\sqrt{K} x) + \_C2 \cos(\sqrt{K} x)$$

```
> t1:=coeff(rhs(tem), _ C1);
> t2:=coeff(rhs(tem), _ C2);
```

$$t1 := \sin(\sqrt{K} x)$$

$$t2 := \cos(\sqrt{K} x)$$

```
>
> with (linalg):
```

Warning, new definition for norm

Warning, new definition for trace

```
> A:=array([[subs(x=0, t1),      subs(x=0, t2)],
[subs(x=1, t1), subs(x=1, t2)]]);
```

$$A := \begin{bmatrix} \sin(0) & \cos(0) \\ \sin(\sqrt{K}) & \cos(\sqrt{K}) \end{bmatrix}$$

```
> detA:=det(A);
```

$$detA := -\sin(\sqrt{K})$$

```
> plot(detA, K=0..100);
```

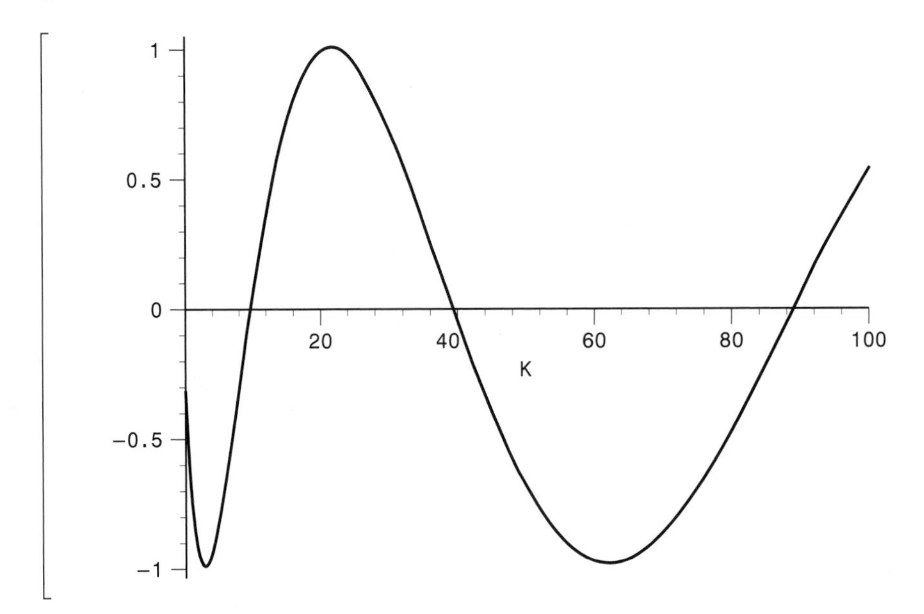


```
> eig1:=fsolve(detA, K=8..12);
```

$$eig1 := 9.869604401$$

```
> eig2:=fsolve(detA,K=36..44);
```

$$eig2 := 39.47841760$$

```
> eig3:=fsolve(detA,K=84..92);
```

$$eig3 := 88.82643961$$

```
>
```

```
> y1:=subs(K=eig1,rhs(tem));
```

$$y1 := \_C1\sin(3.141592654x) + \_C2\cos(3.141592654x)$$

```
> tem1:=subs(x=0,y1);
```

$$tem1 := \_C1\sin(0) + \_C2\cos(0)$$

```
> solve(tem1, _C2);
```

$$0$$

```
> y(x):=subs(_C2=%,y1);
```

$$y(x) := \_C1\sin(3.141592654x)$$

```
>
```

```
> plot(subs(_C1 = 1,y(x)), x = 0..1);
```

1
0.8
0.6
0.4
0.2
0
0 0.2 0.4 0.6 0.8 1
x

```
>
> y2:=subs(K=eig2,rhs(tem));
```

$$y2 := \_C1\sin(9.424777961x) + \_C2\cos(9.424777961x)$$

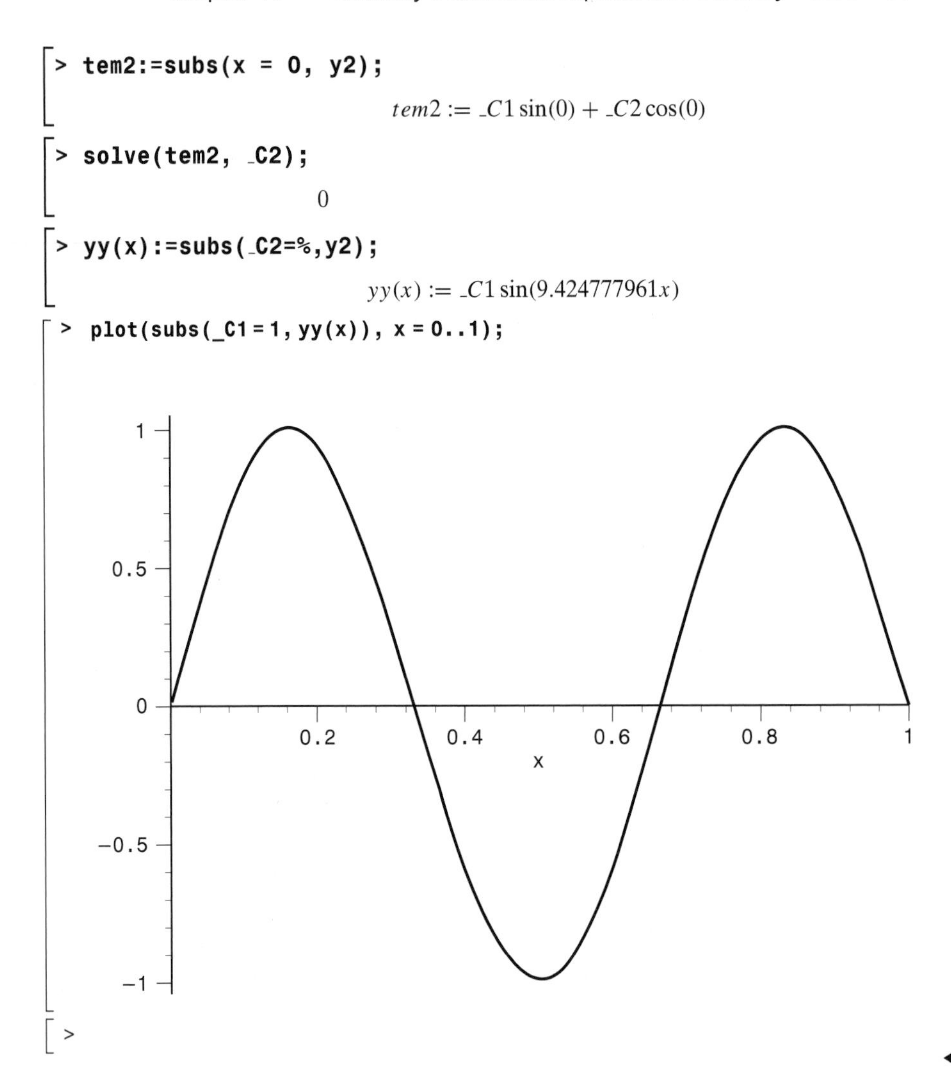

▶Example 10.14

The equation governing the bending of a beam, subjected to axial tension (T) and a distributed transverse load (w), is given by

$$\frac{d^2y}{dx^2} = \frac{T}{EI}y + \frac{wx(x-L)}{2EI}; 0 \leq x \leq L, \tag{E1}$$

where L is the length, E = Young's modulus, and I is the area moment of inertia of the beam. Solve Eq. (E1) for $L = 50$, $T = 2000$, $w = 400$, $E = 2 \times 10^7$, and $I = 100$ for a beam with $y(0) = y(50) = 0$.

Solution

For the given data, Eq. (E1) takes the form

$$\frac{d^2y}{dx^2} = 10^{-6}y + 10^{-7}(x^2 - 50x); 0 \le x \le 50. \tag{E2}$$

```
>
```

```
> ode3:=diff(q(x),x,x) = 0.000001*q(x) + 0.0000001*(x^2-50*x);
```

$$ode3 := \frac{\partial^2}{\partial x^2}q(x) = .110^{-5}q(x) + .110^{-6}x^2 - .5010^{-5}x$$

```
> in3:=q(0)=0, q(50)=0;
```

$$in3 := q(0) = 0, q(50) = 0$$

```
> dsolve({ode3, in3}, q(x));
```

$$q(x) = -200000.0000 + 5.0x - .1000000000x^2 - 4998.956865\,\sinh(.001000000000x)$$
$$+200000.\,\cosh(.001000000000x)$$

```
> Q(x):=rhs(%);
```

$$Q(x) := -200000.0000 + 5.0x - .1000000000x^2 - 4998.956865\,\sinh(.001000000000x)$$
$$+200000.\,\cosh(.001000000000x)$$

```
> plot (Q(x), x = 0..100);
```

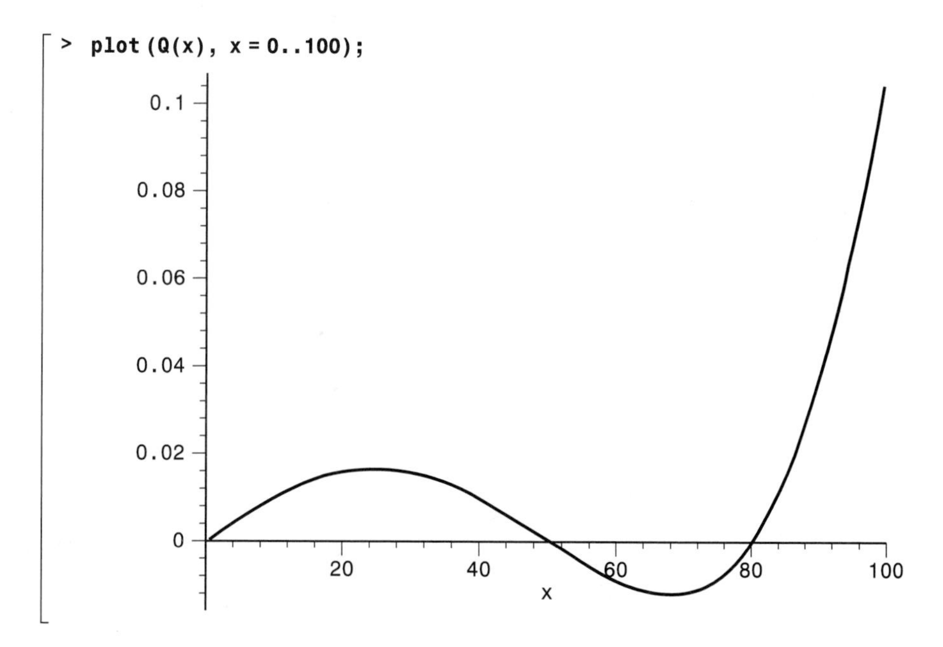

```
> diff(Q(x),x);
```

$5.0 - .200000000000x - 4.998956865\cosh(.001000000000x)$
$+200.0000000\sinh(.001000000000x)$

```
>evalf(5.0-4.998956865*cosh(0)+200.0*sinh(0));;
```

.001043135

◀

10.9.3 MATHCAD

▶Example 10.15

Find the eigenvectors corresponding to the first two eigenvalues $\lambda_1 = 9.869604401$ and $\lambda_2 = 39.47841760$ of the boundary-value problem

$$\frac{d^2y}{dx^2} + \lambda y = 0; y(0) = 0, y(1) = 0 \tag{E1}$$

Solution

The solution of the two-point boundary-value problem can be found when using the MATHCAD function

sbval(v,x1,x2,D,load,score)

where v = vector of guesses for the initial values of y and y' for a second-order differential equation, $x1$, $x2$ = end points of the interval on which the solution of the differential equation is to be found, $D(x, y)$ = an n-element vector-valued function containing the first derivatives of the unknown function(s), n = order of the differential equation, load$(x1, v)$ = a vector-valued function whose n elements correspond to the values of the n unknown functions at $x1$ (some of these values are specified by the known initial conditions and others are unknowns at the start, but will be found by "sbval"), and score$(x2, y)$ = a vector-valued function having the same number of elements as v (each element of v is the difference between an initial condition at $x2$ as specified originally and the corresponding estimate from the solution).

The solution of Eq. (E1) is indicated here:

Solve y"+λ*y=0   B.C: y(0)=0 y(1)=0

λ:=9.8698604401

$$v := \begin{pmatrix} 0 \\ 1 \end{pmatrix} \qquad \text{load}(x1, v) := \begin{pmatrix} v_0 \\ v_1 \end{pmatrix}$$

$$D(x,y) := \begin{pmatrix} y_1 \\ -\lambda \cdot y_0 \end{pmatrix}$$

score(x2,y) := y_0

S:=sbval(v,0,1,D,load,score)

$$S = \begin{pmatrix} 0 \\ 1 \end{pmatrix}$$

Z:=rkfixed(S,0,1,100,D)

n:=0..100

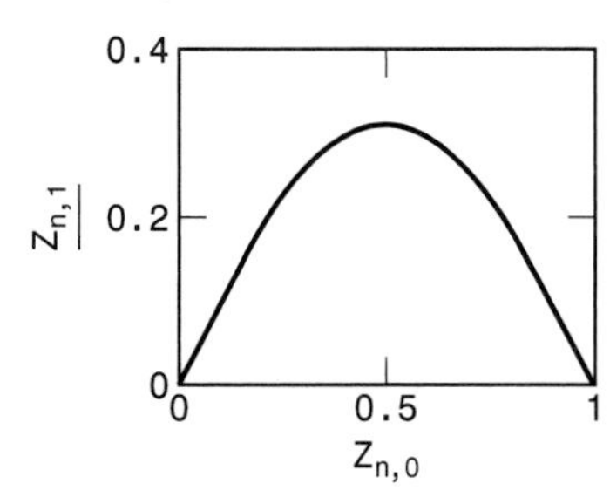

Z =

	0	1	2
0	0	0	1
1	0.01	$.998 \cdot 10^{-3}$	1
2	0.02	0.02	0.998
3	0.03	0.03	0.996
4	0.04	0.04	0.992
5	0.05	0.05	0.988
6	0.06	0.06	0.982
7	0.07	0.069	0.976
8	0.08	0.079	0.969
9	0.09	0.089	0.96
10	0.1	0.098	0.951
11	0.11	0.108	0.941
12	0.12	0.117	0.93
13	0.13	0.126	0.918
14	0.14	0.136	0.905
15	0.15	0.145	0.891

λ:=39.47841760

$$v := \begin{pmatrix} 0 \\ 1 \end{pmatrix} \qquad \text{load}(x1,v) := \begin{pmatrix} v_0 \\ v_1 \end{pmatrix}$$

$$D(x,y) := \begin{pmatrix} y_1 \\ -\lambda \cdot y_0 \end{pmatrix}$$

score(x2,y) := y_0

S:=sbval(v,0,1,D,load,score)

$$S = \begin{pmatrix} 0 \\ 1 \end{pmatrix}$$

Z:=rkfixed(S,0,1,100,D)

n:=0..100

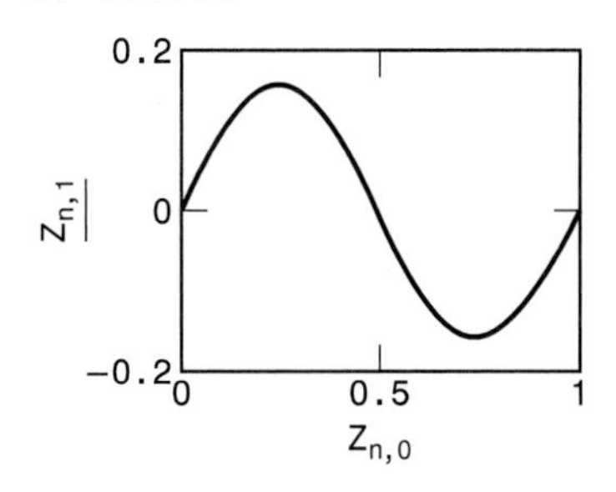

Z =

	0	1	2
0	0	0	1
1	0.01	$.993 \cdot 10^{-3}$	0.998
2	0.02	0.02	0.992
3	0.03	0.03	0.982
4	0.04	0.04	0.969
5	0.05	0.049	0.951
6	0.06	0.059	0.93
7	0.07	0.068	0.905
8	0.08	0.077	0.876
9	0.09	0.085	0.844
10	0.1	0.094	0.809
11	0.11	0.101	0.771
12	0.12	0.109	0.729
13	0.13	0.116	0.685
14	0.14	0.123	0.637
15	0.15	0.129	0.588

◀

▶Example 10.16

Solve the boundary-value problem $y'' = y + x$, $y(0) = 0$, $y(1) = 0$ using MATHCAD.

Solution

```
Solve y"=y+x
B.C: y(0)=0  y(1)=0
```

$$v := \begin{pmatrix} 0 \\ -.1490818720 \end{pmatrix}$$

$$\mathrm{load}(x1, v) := \begin{pmatrix} v_0 \\ v_1 \end{pmatrix}$$

$$D(x, y) := \begin{pmatrix} y_1 \\ y_0 + x \end{pmatrix}$$

$\mathrm{score}(x2, y) := y_0$

S:=sbval(v,0,1,D,load,score)

$$S = \begin{pmatrix} 2.056 \times 10^{-10} \\ -0.149 \end{pmatrix}$$

Z:=rkfixed(S,0,1,10,D)

Z =

	0	1	2
0	0	$.056 \cdot 10^{-10}$	−0.149
1	0.1	−0.015	−0.145
2	0.2	−0.029	−0.132
3	0.3	−0.041	−0.111
4	0.4	−0.05	−0.08
5	0.5	−0.057	−0.04
6	0.6	−0.058	$8.734 \cdot 10^{-3}$
7	0.7	−0.055	0.068
8	0.8	−0.044	0.138
9	0.9	−0.027	0.219
10	1	$-1.029 \cdot 10^{-6}$	0.313

n:=0..10

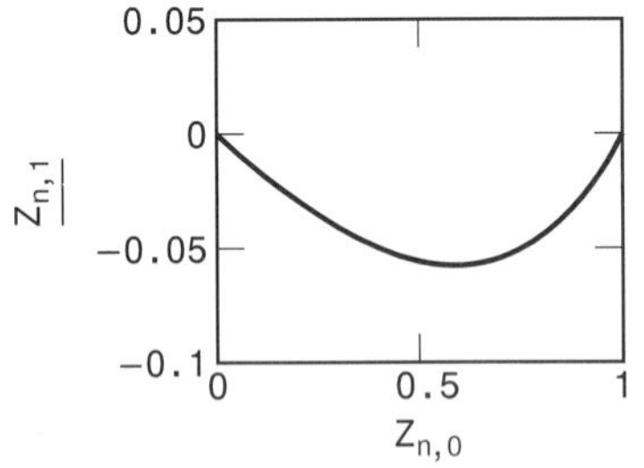

$$Z = \begin{pmatrix} 0 & 2.056 \times 10^{-10} & -0.149 \\ 0.2 & -0.029 & -0.132 \\ 0.4 & -0.05 & -0.08 \\ 0.6 & -0.058 & 8.731 \times 10^{-3} \\ 0.8 & -0.044 & 0.138 \\ 1 & -1.552 \times 10^{-5} & 0.313 \end{pmatrix}$$

Z:=rkfixed(S,0,1,5,D)

n:=0..5

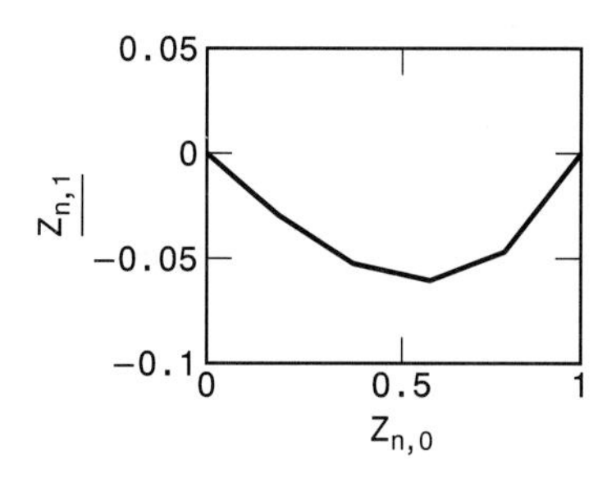

Z:=rkfixed(S,0,1,20,D)

n:=0..20

Z =

	0	1	2
0	0	$.056\cdot10^{-10}$	−0.149
1	0.05	$-7.436\cdot10^{-3}$	−0.148
2	0.1	−0.015	−0.145
3	0.15	−0.022	−0.139
4	0.2	−0.029	−0.132
5	0.25	−0.035	−0.122
6	0.3	−0.041	−0.111
7	0.35	−0.046	−0.096
8	0.4	−0.05	−0.08
9	0.45	−0.054	−0.061
10	0.5	−0.057	−0.04
11	0.55	−0.058	−0.017
12	0.6	−0.058	$8.734\cdot10^{-3}$
13	0.65	−0.057	0.037
14	0.7	−0.055	0.068
15	0.75	−0.05	0.102

◀

10.10 Computer Programs

10.10.1 Fortran Programs

Program 10.1 Shooting method

Solution of the boundary-value problem (Example 10.5):

$y'' - y' + y = g(x);\ y(a) = \alpha,\ y(b) = \beta.$

The following input data are required:

NE = number of first-order equations involved = 5
NDX = number of steps (divisions) in the interval (A, B)
A, B = initial and final values of X
ALP, BET = α, β (boundary values of Y at A, B)

Subroutine to evaluate the function $g(X)$ in the following form:

```
SUBROUTINE FUN (Y,F,NE)
DIMENSION Y(NE),F(NE)
```

```
F(1)=1.0
F(2)=Y(3)
F(3)=g(Y(1))+Y(3)-Y(2)
F(4)=Y(5)
F(5)=g(Y(1))+Y(5)-Y(4)
RETURN
END
```

Function program to evaluate the exact solution (if available):

```
FUNCTION FUNEX(X)
FUNEX=in terms of X
RETURN
END
```

Illustration (Example 10.5)

Program output:

```
SOLUTION OF BOUNDARY VALUE PROBLEM
USING SHOOTING METHOD

VALUE OF X     VALUE OF Y(X)       EXACT Y(X)          ERROR IN Y(X)

  1.0000       0.63084469E+01      0.63084517E+01      0.47683716E-05
  1.0100       0.64745941E+01      0.64746037E+01      0.95367432E-05
  1.0200       0.66438675E+01      0.66438770E+01      0.95367432E-05
  1.0300       0.68163223E+01      0.68163314E+01      0.90599060E-05
  1.0400       0.69920273E+01      0.69920273E+01      0.00000000E+00
  1.0500       0.71710205E+01      0.71710272E+01      0.66757202E-05
  1.0600       0.73533859E+01      0.73533926E+01      0.66757202E-05
  1.0700       0.75391846E+01      0.75391898E+01      0.52452087E-05
  1.0800       0.77284775E+01      0.77284822E+01      0.47683716E-05
  1.0900       0.79213333E+01      0.79213362E+01      0.28610229E-05
  1.1000       0.81178207E+01      0.81178217E+01      0.95367432E-06
    .
    .
    .
  1.9000       0.45347778E+02      0.45347763E+02      -.15258789E-04
  1.9100       0.46269714E+02      0.46269680E+02      -.34332275E-04
  1.9200       0.47209778E+02      0.47209770E+02      -.76293945E-05
  1.9300       0.48168396E+02      0.48168415E+02      0.19073486E-04
  1.9400       0.49145966E+02      0.49145966E+02      0.00000000E+00
  1.9500       0.50142822E+02      0.50142815E+02      -.76293945E-05
  1.9600       0.51159363E+02      0.51159351E+02      -.11444092E-04
  1.9700       0.52195953E+02      0.52195976E+02      0.22888184E-04
```

```
   1.9800     0.53253052E+02     0.53253086E+02     0.34332275E-04
   1.9900     0.54331085E+02     0.54331104E+02     0.19073486E-04
   2.0000     0.55430420E+02     0.55430443E+02     0.22888184E-04
```

Program 10.2 Finite-difference method

Solution of the boundary-value problem (Example 10.6):

$y'' - y' + y = g(x); y(a) = \alpha, y(b) = \beta$

The following input data are required:

N = number of steps (divisions) in the interval $(a, b) + 1$
A1, B1 = initial and final values of X (a, b)
ALP, BET = α, β (boundary values of y at a, b)

Function program to evaluate the function $g(x)$ in the following form:

```
FUNCTION FR(X)
FR=g(X)
RETURN
END
```

Function program to evaluate the exact solution (if available) in the form:

```
FUNCTION FUNEX (X)
FUNEX=exact solution in terms of X
RETURN
END
```

Illustration (Example 10.6). Using $N = 101$, A1 = 1.0, B1 = 2.0, ALP = 6.308447, and BET = 55.430436, the output of Program 10.2 is as follows:

```
SOLUTION OF BOUNDARY VALUE PROBLEM
USING FINITE DIFFERENCE METHOD

VALUE OF X    VALUE OF Y(X)      EXACT Y(X)         ERROR IN Y(X)

   1.0000     0.63084469E+01     0.63084517E+01     0.47683716E-05
   1.0200     0.66439004E+01     0.66438770E+01     -.23365021E-04
   1.0300     0.68163695E+01     0.68163314E+01     -.38146973E-04
   1.0400     0.69920797E+01     0.69920273E+01     -.52452087E-04
   1.0500     0.71710930E+01     0.71710272E+01     -.65803528E-04
   1.0600     0.73534737E+01     0.73533926E+01     -.81062317E-04
   1.0700     0.75392838E+01     0.75391898E+01     -.93936920E-04
   1.0800     0.77285900E+01     0.77284822E+01     -.10776520E-03
   1.0900     0.79214592E+01     0.79213362E+01     -.12302399E-03
```

```
1.1000    0.81179590E+01   0.81178217E+01   -.13732910E-03
:
1.9000    0.45348049E+02   0.45347763E+02   -.28610229E-03
1.9100    0.46269947E+02   0.46269680E+02   -.26702881E-03
1.9200    0.47210014E+02   0.47209770E+02   -.24414063E-03
1.9300    0.48168629E+02   0.48168415E+02   -.21362305E-03
1.9400    0.49146156E+02   0.49145966E+02   -.19073486E-03
1.9500    0.50142979E+02   0.50142815E+02   -.16403198E-03
1.9600    0.51159489E+02   0.51159351E+02   -.13732910E-03
1.9700    0.52196075E+02   0.52195976E+02   -.99182129E-04
1.9800    0.53253151E+02   0.53253086E+02   -.64849854E-04
1.9900    0.54331131E+02   0.54331104E+02   -.26702881E-04
2.0000    0.55430435E+02   0.55430443E+02   0.76293945E-05
```

10.10.2 C Program

Program 10.3 Nonlinear boundary-value problem

One-dimensional temperature distribution, including radiation (Example 10.8):
The following input data are required:

SIG = σ = Stefan–Boltzman constant
EPS = ε = emissivity
XL = length of rod
P = perimeter of rod
A = cross-sectional area of rod
XK = thermal conductivity
TINF = T_∞ = surrounding temperature
T_0 = temperature at $x = 0$
T_4 = temperature at x = XL

Illustration (Example 10.8):
Program output:

```
Solution of nonlinear boundary value problem

Data:
Number of equations = 3
Maximum number of iterations permitted (miter) = 50
x(i):   0.800000E+03   0.700000E+03   0.600000E+03

Iter =  1  x(i) =  0.800000E+03   0.700000E+03   0.600000E+03
           f(i) = -0.648726E+02  -0.843600E+02  -0.181872E+03

Iter =  2  x(i) =  0.746028E+03   0.604425E+03   0.484896E+03
           f(i) = -0.507825E+01  -0.116350E+02  -0.119383E+02
```

```
Iter =  3  x(i) =  0.740011E+03   0.592721E+03   0.474230E+03
           f(i) = -0.568848E-01  -0.140869E+00  -0.750732E-01

Solution converged in 4 iterations
Solution is: x(i):  0.739945E+03   0.592597E+03   0.474139E+03
             f(i):  0.000000E+00  -0.610352E-04   0.305176E-04
```

REFERENCES AND BIBLIOGRAPHY

10.1. E. P. Popov, *Engineering Mechanics of Solids*, 2d edition, Prentice Hall, Upper Saddle River, NJ, 1999.

10.2. F. M. White, *Heat and Mass Transfer*, Addison-Wesley, Reading, MA, 1988.

10.3. S. S. Rao, *Mechanical Vibrations*, 3d edition, Addison-Wesley, Reading, MA, 1995.

10.4. W. E. Boyce and R. C. DiPrima, *Elementary Differential Equations and Boundary Value Problems*, 4th edition, Wiley, New York, 1986.

10.5. R. G. Budynas, *Advanced Strength and Applied Stress Analysis*, McGraw-Hill, New York, 1977.

REVIEW QUESTIONS

The following questions along with corresponding answers are available in an interactive format at the Companion Web site at http://www.prenhall.com/rao.

10.1. **State** whether the following **boundary-value problem** is **homogeneous or nonhomogeneous**:

1.
$$f'' + 2f = \sin x; f(0) = 0, f(1) = 0.$$

2.
$$f'' + (\lambda - x^2)f = 0; f'(0) = f(0), f'(1) = f(1).$$

3.
$$A(x)f'' + B(x)f' + C(x)f = 0, 0 < x < 1,$$
$$a_1 f(0) + a_2 f'(0) = a_3, b_1 f(1) + b_2 f'(1) = 0.$$

10.2. **Give short answers**:

1. What is a boundary-value problem?
2. State the form of mixed-boundary conditions for a second-order boundary value problem.
3. State two methods of solving boundary-value problems.
4. Why is the shooting method called the Newton–Raphson method?
5. What is the relationship between the eigenvalue problem of an ordinary differential equation and a matrix eigenvalue problem?

10.3. Answer **true or false**:

1. A boundary-value problem can be nonlinear.
2. The boundary conditions of a boundary-value problem are always linear.
3. The shooting method is applicable to both linear and nonlinear boundary-value problems.
4. The shooting method requires an iterative procedure for linear boundary-value problems.
5. The shooting method requires the conversion of an nth-order differential equation into a set of n first-order equations.
6. The finite difference method requires the conversion of an nth-order differential equation into a set of n first-order equations.

(*continued, page 783*)

7. The use of higher order finite-difference formulas requires more fictitious mesh points.
8. The solution of a differential equation denoting a boundary-value problem (for specified boundary conditions) is unique.
9. The boundary conditions can be specified at more than two points in a boundary-value problem.

10.4. **Fill in the blanks** with suitable word(s):

1. The differential equation must be at least of order ______ for a boundary-value problem.
2. In shooting methods, a boundary-value problem is first converted to ______ value problem.
3. A ______ and ______ approach is used in the shooting method.
4. The central-difference formula for the second derivative, with a step size h, involves error of order ______ .
5. In finite-difference method, the domain of integration is replaced by a set of ______ points.
6. The finite-difference method converts a boundary-value problem into a ______ system of linear simultaneous equations.
7. A larger number of mesh points in the interval of integration leads to ______ computational effort.

10.5. **Verify** whether the given function $y(x)$ represents the solution of the **stated boundary value** problem:

1.
$$y'' - y = 0; 0 \leq < x \leq 1$$
subject to
$$y'(0) = 0,\ y(1) = 1.$$
$$y(x) = \frac{\cosh x}{\cosh 1}.$$

2.
$$y'' = y' + y - (2x - 1)e^x; 0 \leq x \leq 1$$
subject to
$$y(0) = 1,\ y(1) = 1.$$
$$y(x) = (2x + 1)e^x.$$

3.
$$y'' + y + x = 0; 0 \leq x \leq 1$$

(*continued, page 784*)

subject to

$$y(0) = 0, \; y(1) = 0.$$

$$y(x) = \frac{\sin x}{\sin 1} - x$$

10.6. **Match the following**:

1. Nonlinear boundary-value problem with finite differences	(a) solution of a set of algebraic equations
2. Shooting method	(b) solution of eigenvalue problem
3. General finite-difference method	(c) solution of a series of initial-value problems
4. Homogeneous ordinary differential equation with homogeneous boundary conditions	(d) iterative solution of nonlinear set of algebraic equations

PROBLEMS

Section 10.2

10.1. The radial heat transfer in a thick hollow cylinder is given by

$$\frac{d^2T}{dr^2} + \frac{1}{r}\frac{dT}{dr} = 0. \tag{E1}$$

Formulate the boundary-value problem for an infinitely long hollow cylinder whose inner surface ($r = 5''$) is maintained at 200° F, while the outer surface ($r = 10''$) is maintained at 65° F.

10.2. The equation governing the heat transfer in a fin (described in Example 10.2) can be expressed as

$$\frac{d^2\theta}{dx^2} = \frac{hp}{kA}\theta, \tag{E1}$$

where $\theta = \frac{T-T_\infty}{T_0-T_\infty}$. If the fin is insulated at $x = l$, state the boundary conditions. Verify that the solution

$$\frac{T - T_\infty}{T_0 - T_\infty} = \frac{\cosh\left(\frac{l-x}{\lambda}\right)}{\cosh\left(\frac{l}{\lambda}\right)}, \tag{E2}$$

where $\lambda = \left(\frac{kA}{hp}\right)^{\frac{1}{2}}$, is exact.

Section 10.3

10.3. Find the radial temperature distribution in the hollow cylinder described in Problem 10.1 using the shooting method.

10.4. Find the temperature distribution in a fin insulated at $x = l$ using the shooting method for the following data:

$$l = 0.1 \text{ m}, A = 3.2 \times 10^{-4} \text{ m}^2, k = 240\frac{W}{m - {}^\circ C}, P = 7.1554 \times 10^{-2} \text{ m},$$

$$h = 9\frac{W}{m^2 - {}^\circ C}, T_0 = 200^\circ \text{ C, and } T_\infty = 20^\circ \text{ C.}$$

10.5. The transverse deflection of a flexible cable or string (Fig. 10.11) is governed by the equation [10.3]

$$\frac{d^2w}{dx^2} = \frac{Tw}{R} + \frac{px(x-l)}{2R}. \tag{E1}$$

where p is the distributed transverse load, T is the tension in the cable, and R is the flexural stiffness. Find the deflection of a cable using the shooting method with the following data:

$$l = 100 \text{ in}, T = 1000 \text{ lb}, p = 100 \text{ lb/in, and } R = 50 \times 10^6 \text{ lb-in}^2.$$

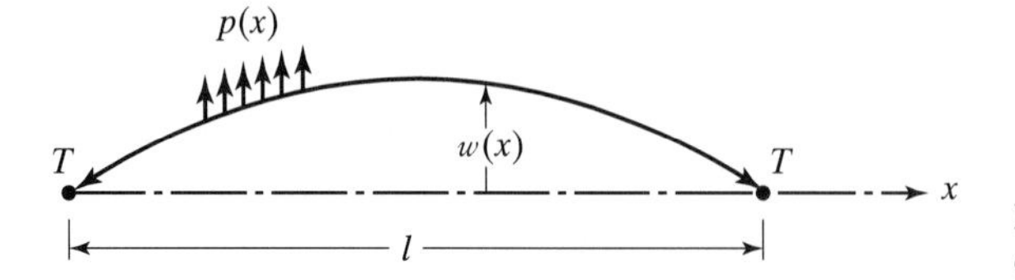

Figure 10.11 Transverse deflection of a cable.

10.6. Consider the boundary-value problem

$$\frac{d^2y}{dx^2} = y + x; \; 0 \le x \le 1$$

with the conditions $y(0) = 0$ and $y(1) = 0$. Find the solution of the problem using the shooting method.

Section 10.5

10.7. Find the radial temperature distribution in the hollow cylinder described in Problem 10.1 using the finite-difference method.

10.8. Solve Problem 10.4 using the finite-difference method.

10.9. The radial stress (σ_r) induced in a rotating thick-walled cylinder (Fig. 10.12) is governed by the equation [10.5]

$$r\frac{d^2\sigma_r}{dr^2} + 3\frac{d\sigma_r}{dr} + (3 + \nu)\rho\omega^2 r = 0 \tag{E1}$$

where r is the radial distance, ν is Poisson's ratio, ρ is the mass density, and ω is the angular velocity of the cylinder.

(a) State the boundary conditions for a cylinder rotating at 500 rpm and subjected to an internal pressure of 100 psi with no external pressure.

(b) Find the radial stress distribution in a cylinder using the finite-difference method with the following data:

$$r_i = 1 \text{ in}, \, r_o = 6 \text{ in}, \, \nu = 0.3, \, \omega = 104.72 \text{ rad/sec, and}$$

$$\rho = 724.638 \times 10^{-6} \text{ lb-sec}^2/\text{in}^4.$$

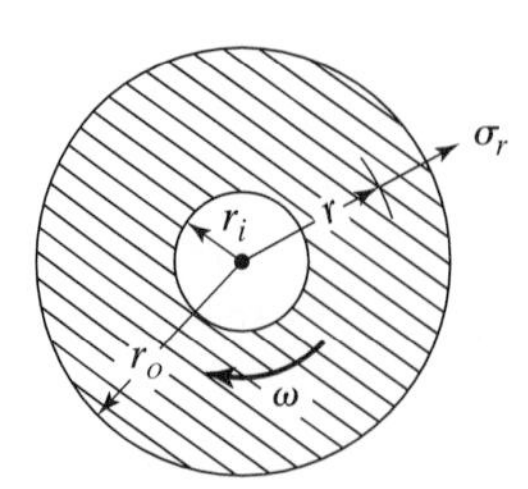

Figure 10.12 Cross section of a thick-walled cylinder.

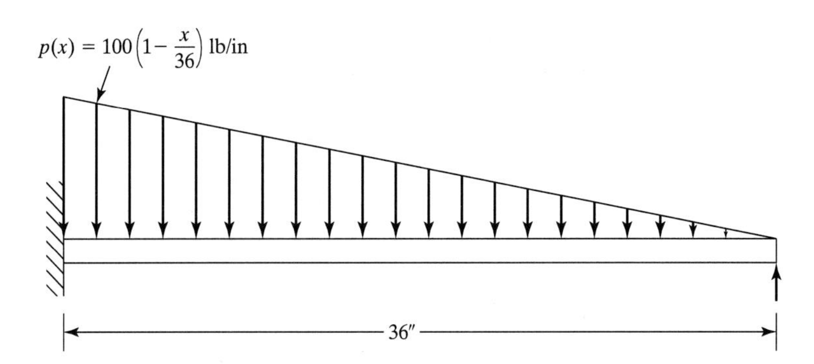

Figure 10.13 Fixed-simply supported beam.

10.10. The deflection of a beam, $w(x)$, lying on an elastic foundation and subjected to a distributed load, $p(x)$, is governed by the equation,

$$EI\frac{d^4w}{dx^4} = p(x) - kw, \tag{E1}$$

where E = Young's modulus, I = moment of inertia of the cross section, and k = foundation modulus (stiffness per unit length).

(a) Specify the boundary conditions if the beam is simply supported at both ends.

(b) Find the deflection of the beam using the finite-difference method for the following data:

$$E = 30 \times 10^6 \text{ psi}, I = 1 \text{ in}^4, p(x) = 500 \text{ lb/in},$$

$$k = 1000 \text{ lb/in}^2, \text{ and } l = 36 \text{ in}.$$

10.11. Find the deflection of the uniform fixed simply supported beam shown in Fig. 10.13 using the finite-difference method.

Use the following data: $E = 30 \times 10^6$ psi, $l = 36''$, $I = 2$ in^4,

$$p(x) = 100\left(1 - \frac{x}{36}\right) \text{ lb/in, and } h = 9''(N = 4).$$

Section 10.6

10.12. Consider the nonlinear two-point boundary-value problem

$$y'' = e^y;\ 0 \le x \le 1$$

subject to the conditions $y(0) = a$ and $y(1) = b$. Solve this problem using shooting method for $a = b = 0$. Compare your solution with the exact solution

$$y(x) = \ln \alpha - 2\ln\left\{\cos\left(\sqrt{\frac{\alpha}{2}}x + \beta\right)\right\},$$

where α and β are determined from the boundary conditions

$$a = \ln \alpha - 2 \ln \{\cos \beta\}$$

and

$$b = \ln \alpha - 2 \ln \left\{ \cos \left(\sqrt{\frac{\alpha}{2}} + \beta \right) \right\}.$$

Section 10.7

10.13. Find the natural frequencies of longitudinal vibration of a fixed-free bar using a total of five mesh points in the length of the bar.

10.14. Find the natural frequencies and mode shapes of a uniform simply supported-fixed beam of length 36″ using the finite-difference method with $h = 9''$ ($N = 4$). Assume the material of the beam as steel with $E = 30 \times 10^6$ psi and $\rho g = 0.29$ lb/in^3, and the cross section as circular with diameter 2 in.

Section 10.9

10.15. Use MAPLE to solve the following boundary-value problem that describes the free vibration of a uniform beam of length L with both ends clamped:

$$w'''' - \beta^4 w = 0, 0 \le x \le L$$

with

$$w(0) = 0, w'(0) = 0;$$

$$w(L) = 0, w'(L) = 0;$$

and

$$\beta^4 = \frac{m\omega^2}{EI},$$

where m = mass per unit length of the beam, E = Young's modulus, I = area moment of inertia of the cross section of the beam, and ω = natural frequency of vibration of the beam.

10.16. Use MATLAB to solve the nonlinear boundary-value problem stated in Example 10.8.

10.17. Solve Example 10.14 using MATLAB.

10.18. Find the eigenvector corresponding to the given eigenvalue of the following boundary-value problems using MATLAB:

(a)

$$\frac{d^2y}{dx^2} + \lambda y = 0; y(0) = y(1) = 0,$$

$\lambda = 39.47841760.$

(b)

$$\frac{d^4y}{dx^4} + \beta^4 y = 0; y(0) = y(1) = 0, \frac{d^2y}{dx^2}(0) = \frac{d^2y}{dx^2}(1) = 0,$$

$\beta = 0.0872664626$

10.19. Solve Example 10.12 using MAPLE.

10.20. Solve the following boundary-value problem using MAPLE:

$$\frac{d^4y}{dx^4} + \beta^4 y = 0; y(0) = y(1) = 0, \frac{d^2y}{dx^2}(0) = \frac{d^2y}{dx^2}(1) = 0.$$

10.21. Solve Example 10.14 using MATHCAD.

10.22. Solve Problem 10.20 using MATHCAD.

Section 10.10

10.23. Solve the boundary-value problem

$$y'' = y' - y + e^x + 4\cos x; y(0) = 3.161217, y(2) = 5.724486$$

using Program 10.1

10.24. Solve the boundary-value problem stated in Problem 10.23 using Program 10.2.

10.25. Find the temperature distribution in a fin, considering conduction and radiation heat transfer (similar to Example 10.8), using Program 10.3 for the following data:
Fin's cross section is circular with radius 2 cm,

$$\varepsilon = 0.1, \sigma = 5.7 \times 10^{-4}\ \text{W/m}^2 - {}^\circ\ \text{K}^4, T(x=0) = 800^\circ\ \text{K},$$

$$T(x=l) = 350^\circ\ \text{K}, T_\infty = 400^\circ\ \text{K}, k = 50\ \text{W/m} - {}^\circ\ \text{K, and } l = 1\ \text{m}.$$

General

10.26. The heat transfer in a solid cylinder, in radial direction, with a uniform heat generation is governed by the equation [10.2]

$$\frac{1}{r}\frac{d}{dr}\left(r\frac{dT}{dr}\right) + \frac{\dot{q}}{k} = 0, \tag{E1}$$

where $\dot{q}$ is the heat generated per unit volume and k is the thermal conductivity. State the boundary conditions of the problem.

10.27. The differential equation for the transverse deflection of a beam is given by

$$EI\frac{d^2w}{dx^2} = M, \tag{E1}$$

where E is Young's modulus, I is the area moment of inertia, and M is the bending moment at x. If a beam is fixed at $x = 0$, supported on a spring, of

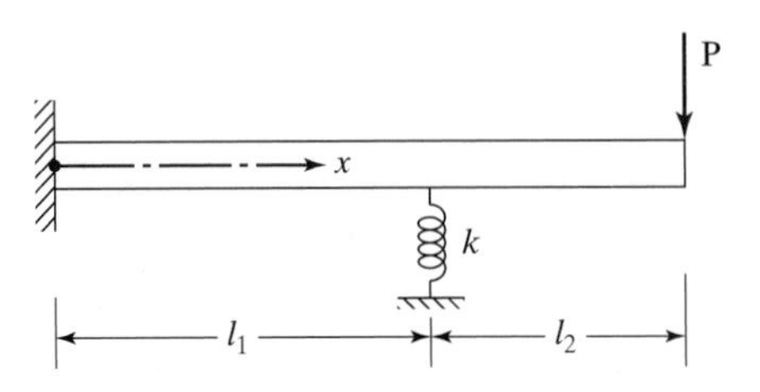

Figure 10.14 Beam with spring support.

stiffness k, at $x = l_1$ and subjected to a load P at $x = l_1 + l_2$ as shown in Fig. 10.14, determine the deflection of the beam using the shooting method Use the following data:

$E = 30 \times 10^6$ psi, $I = 2$ in^4, $l_1 = 12$ in, $l_2 = 18$ in, $k = 500$ lb/in, and $P = 1000$ lb.

10.28. Consider the boundary-value problem

$$\begin{aligned} y'' &= f(x) + g(x)y + h(x)y'; \; a \le x \le b; \\ y(a) &= \alpha; \; y(b) = \beta. \end{aligned} \tag{E1}$$

In the shooting method, the following initial-value problem is solved using two trial values of $y'(a)$:

$$\begin{aligned} y'' &= f(x) + g(x)y + h(x)y'; \; a \le x \le b; \\ y(a) &= \alpha; \; y'(a) = \gamma. \end{aligned} \tag{E2}$$

Let the solutions obtained with the trial values γ_1 and γ_2 be denoted $y_1(x)$ and $y_2(x)$, respectively.

(a) Prove that the function

$$z(x) = \lambda y_1(x) + (1 - \lambda) y_2(x) \tag{E3}$$

is a solution of the following problem:

$$\begin{aligned} z'' &= f(x) + g(x)z + h(x)z'; \; a \le x \le b, \\ z(a) &= \alpha. \end{aligned} \tag{E4}$$

(b) Find the value of λ in Eq. (E3) that satisfies the condition, $z(b) = \beta$.

10.29. Derive the system of equations resulting from the boundary condition of Eq. (10.79) in a tridiagonal form using the central difference method.

10.30. The lateral deflection (w) of a circular membrane, subjected to a uniform lateral pressure p and tension T, is governed by the equation [10.3]:

$$\frac{d^2 w}{dr^2} + \frac{1}{r}\frac{dw}{dr} = -\frac{p}{T}. \tag{E1}$$

Determine the deflection of an annular membrane with the following data:

Inner radius = 10 in, outer radius = 20 in, $p = 50$ psi, and $T = 300$ lb/in.

Use the finite-difference method for solution.

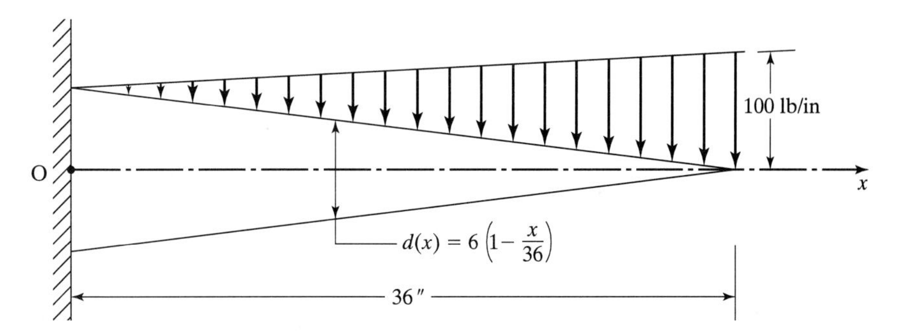

Figure 10.15 Tapered beam subjected to varying load.

10.31. Find the deflection of the tapered beam shown in Fig. 10.15 using the finite-difference method. The diameter of the beam, $d(x)$, varies linearly along the length of the beam as $d(x) = 6\left(1 - \frac{x}{36}\right)$, while the intensity of the loading varies as $p(x) = \frac{25}{9}x$ lb/in.
Use the following data:

$$E = 30 \times 10^6 \text{ psi and } h = 9''(N = 4).$$

PROJECTS

10.1. For a fin with a variable cross section, the governing equation is given by

$$\frac{d}{dx}\left(kA\frac{d\theta}{dx}\right) - hp\theta = 0 \tag{E1}$$

with $\theta = T - T_\infty$. (Fig. 10.16.)

(a) Divide the fin into n equal parts and derive the finite-difference form of Eq. (E1) that relates the temperature at node i, T_i, to the temperatures at the neighboring nodes using the forward-finite-difference formula.

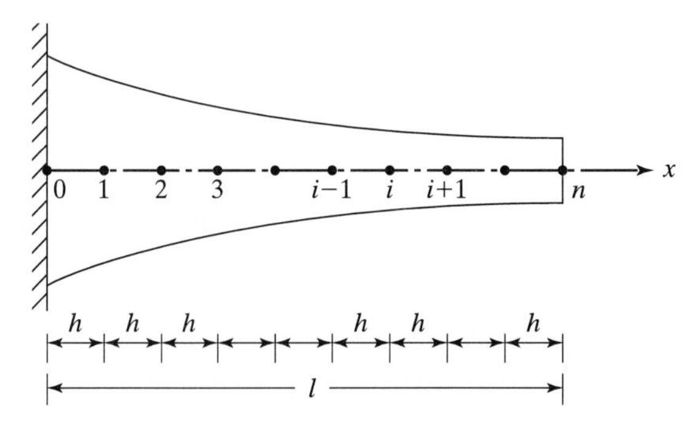

Figure 10.16 A fin with a variable cross section.

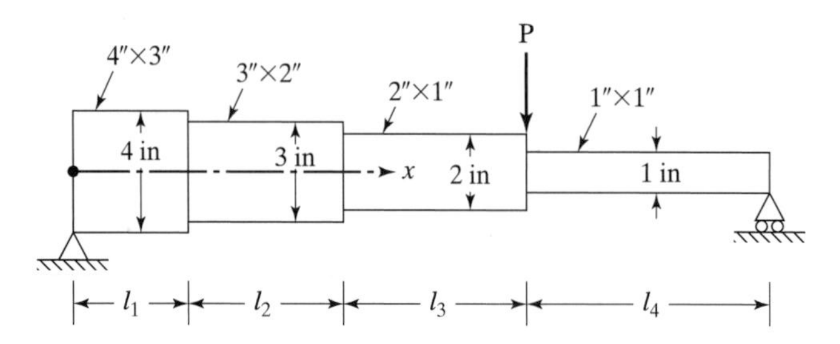

Figure 10.17 A stepped beam.

(b) Express the forward-difference approximations for a fin with (i) insulation at $x = l$ and (ii) convection at $x = l$.

(c) Formulate the forward-finite-difference equations for a linearly tapered fin that is insulated at $x = l$ with the following data:

$$A(x) = A_0\left(1 - \frac{x}{l}\right),\ p(x) = p_0\left(1 - \frac{x}{l}\right),\ A_0 = 1\ \text{cm}^2,\ p_0 = 4\ \text{cm},$$

$$l = 10\ \text{cm},\ k = 40\frac{W}{m - K},\ h = 10\frac{W}{m^2 - K},\ \theta_0 = 100^\circ\ \text{C, and}\ n = 5.$$

(d) Solve the equations formulated in part (c) and find the temperatures at nodes $i, i = 1, 2, \ldots, 5$.

10.2. Solve Project 10.1 using the backward-difference method.

10.3. Solve Project 10.1 using the central-difference method.

10.4. The transverse deflection of a beam, $w(x)$, is governed by the equation

$$EI\frac{d^2w}{dx^2} = M, \tag{E1}$$

where E is Young's modulus, I is the area moment of inertia, and M is the bending moment at x. Find the variation of the deflection of the stepped beam shown in Fig. 10.17 using the central-difference method.
Use the following data:

$l_1 = 12$ in, $l_2 = 18$ in, $l_3 = 24$ in, $l_4 = 30$ in, $P = 1000$ lb, and $E = 30 \times 10^6$ psi.

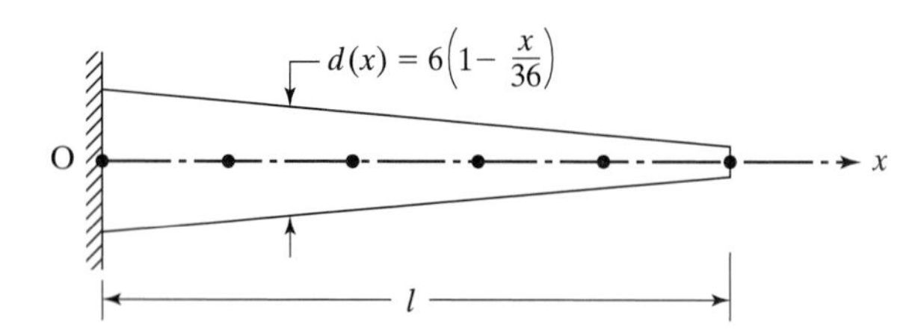

Figure 10.18 Tapered cantilever beam.

10.5. Consider a linearly tapered cantilever beam with circular cross section whose diameter varies as $d(x) = 6\left(1 - \frac{x}{36}\right)$ in Fig. 10.18.

(a) Formulate the eigenvalue problem using the finite-difference method.

(b) Solve the problem formulated in part (a) and find the natural frequencies and mode shapes.

Use the following data:

$$E = 30 \times 10^6 \text{ psi}, \rho g = 0.29 \text{ lb/in}^3, l = 36'', \text{ and } h = 9''(N = 4).$$

11

Partial Differential Equations

CHAPTER CONTENTS

11.1 Introduction

Differential equations involving more than one independent variable are called partial differential equations. Partial differential equations can be classified in several ways; in fact, many of the classifications indicated for ordinary differential equations are also valid for partial differential equations. Thus, a partial differential equation may be linear or nonlinear, of the first order or of a higher order, homogeneous or nonhomogeneous, and may involve two or more independent variables. The linear second-order partial differential equations in two independent variables are further classified according to their mathematical form. For example, consider a general second-order partial differential equation in two variables:

$$A\frac{\partial^2\phi}{\partial x^2}+B\frac{\partial^2\phi}{\partial x\partial y}+C\frac{\partial^2\phi}{\partial y^2}=G\left(x,y,\phi,\frac{\partial\phi}{\partial x},\frac{\partial\phi}{\partial y}\right). \quad (11.1)$$

If A, B, and C are functions of x, y, ϕ, $\frac{\partial \phi}{\partial x}$, and $\frac{\partial \phi}{\partial y}$; ϕ is the dependent variable; and x and y are the independent variables, then Eq. (11.1) is called a quasilinear equation. When A, B, and C are functions of x and y and G is a linear function of ϕ, $\frac{\partial \phi}{\partial x}$, and $\frac{\partial \phi}{\partial y}$, as in

$$A\frac{\partial^2 \phi}{\partial x^2} + B\frac{\partial^2 \phi}{\partial x \partial y} + C\frac{\partial^2 \phi}{\partial y^2} + D\frac{\partial \phi}{\partial x} + E\frac{\partial \phi}{\partial y} + F\phi = H, \tag{11.2}$$

the equation is said to be linear. Equation (11.2) is homogeneous if $H = 0$; otherwise, it is nonhomogeneous. In many engineering applications, the parameters $A, B, \ldots, G$ are constants. In general, the solution of Eq. (11.2) is to be found over a region (or domain) R as indicated in Fig. 11.1. Equation (11.2) is classified as follows [11.1]:

$$\text{Elliptic equation}: \quad \text{if } B^2 - 4AC < 0; \tag{11.3}$$

$$\text{Parabolic equation}: \quad \text{if } B^2 - 4AC = 0; \tag{11.4}$$

$$\text{Hyperbolic equation}: \quad \text{if } B^2 - 4AC > 0. \tag{11.5}$$

This classification into elliptic, parabolic, and hyperbolic equations has also been applied to equations not of order two. The classification is also extended to second-order equations involving three independent variables; this classification includes elliptic, parabolic, hyperbolic, and ultrahyperbolic equations [11.2, 11.3]. Usually, elliptic equations are associated with equilibrium states, and parabolic equations are related to diffusion states. The hyperbolic equations are associated with oscillating or vibrating systems. Unfortunately, analytical solutions can be obtained to very few partial differential equations. So, most partial differential equations are solved using numerical methods.

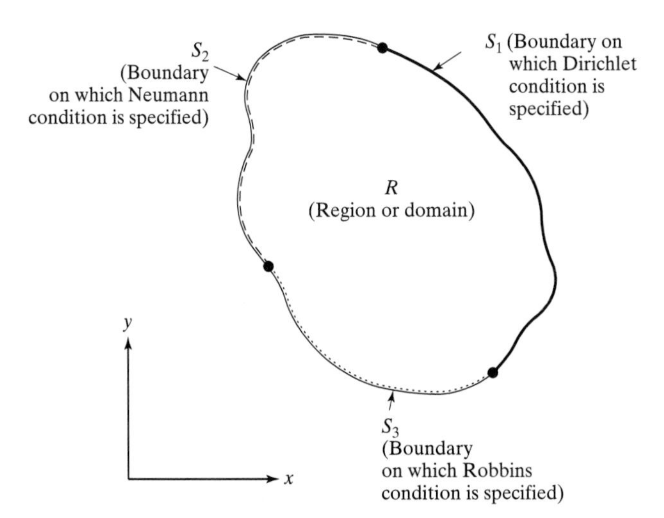

Figure 11.1 Arbitrary domain and boundary.

11.2 Engineering Applications

▶Example 11.1

Derive the equation governing the temperature distribution in a one-dimensional fin (Fig. 11.2).

Solution

Considering an element of the fin, we can write the thermal equilibrium or energy balance equation, in time dt, as

$$\text{Heat inflow } (Q_{\text{in}}) + \text{Heat generated } (Q_g) = \text{Heat outflow } (Q_{\text{out}}) + \text{Change in internal energy } (Q_c), \tag{E1}$$

where the heat inflow can be expressed, from the heat conduction relation, as

$$Q_{\text{in}} = q_x\, dt = -k_x A_x \frac{\partial T}{\partial x}\, dt, \tag{E2}$$

where q_x is the rate at which heat enters the element, k_x is the thermal conductivity in the x-direction, A_x is the area of cross section through which heat flows, and $\frac{\partial T}{\partial x}$ is the temperature gradient in the x-direction. If $\dot{q}$ denotes the rate of heat generation per unit volume,

$$Q_g = \dot{q} A_x\, dx\, dt. \tag{E3}$$

The heat outflow can be expressed as

$$Q_{\text{out}} = \left(q_x + \frac{\partial q_x}{\partial x}\, dx \right) dt = -k_x A_x \frac{\partial T}{\partial x}\, dt - \frac{\partial}{\partial x}\left(k_x A_x \frac{\partial T}{\partial x} \right) dx\, dt. \tag{E4}$$

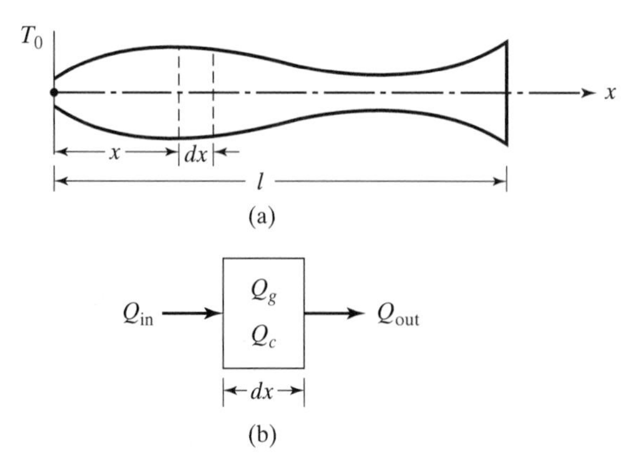

Figure 11.2 One-dimensional fin.

The change in internal energy is given by

$$Q_c = \rho c \frac{\partial T}{\partial t} A_x \, dx \, dt, \tag{E5}$$

where ρ is the density of the material, c is the specific heat, and $\frac{\partial T}{\partial t} dt = dT$ is the temperature change in the element in time dt. Equations (E1) through (E5) lead to the heat conduction (parabolic) equation

$$\frac{\partial}{\partial x}\left(k_x A_x \frac{\partial T}{\partial x}\right) + \dot{q} A_x = \rho c A_x \frac{\partial T}{\partial t}. \tag{E6}$$

The boundary and initial conditions can be expressed as follows:

(i) $T(0, t) = T_0(t)$ for $t > 0$ on S_1 (Temperature specified at $x = 0$);

(ii) $\frac{\partial T}{\partial x}(l, t) = q(t)$ for $t > 0$ on S_2 (Gradient of temperature specified as $q(t)$ at $x = 1$; $q = 0$ if insulated at $x = 1$);

(iii) $T(x, 0) = f(x)$ for $0 \leq x \leq 1$ (Initial temperature distribution specified as $f(x)$); ◀

▶Example 11.2

Derive the equation for the seepage flow through a porous medium such as the flow through an earth dam and a foundation (Fig. 11.3).

Solution

The flow of water through a porous medium can be assumed to be an irrotational flow, since there will be no rotation or distortion of the fluid particles during their movement. The seepage flow is governed by Darcy's law, which is

$$u = -k_x \frac{\partial \phi}{\partial x}, \quad v = -k_y \frac{\partial \phi}{\partial y}, \quad w = -k_z \frac{\partial \phi}{\partial z}, \tag{E1}$$

where u, v, and w are the x-, y-, and z-components, respectively, of seepage velocity (of the fluid), k_x, k_y, and k_z are the coefficients of permeability in x-, y-, and z-directions respectively, and ϕ is the fluid potential ($\phi = \frac{p}{\gamma} + z$, where p is the fluid pressure, γ is the specific weight of the fluid, and z is the elevation head). The

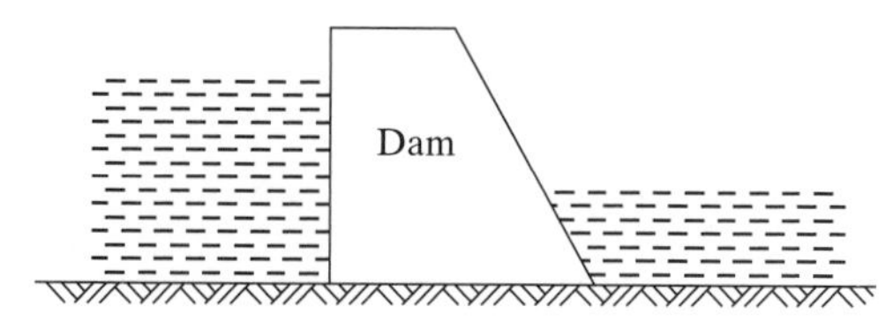

Figure 11.3 Seepage flow through a dam.

continuity equation for an incompressible fluid can be expressed as

$$\frac{\partial u}{\partial x} + \frac{\partial v}{\partial y} + \frac{\partial w}{\partial z} = 0. \tag{E2}$$

Equations (E1) and (E2) lead to the governing equation for a saturated flow

$$\frac{\partial}{\partial x}\left(k_x \frac{\partial \phi}{\partial x}\right) + \frac{\partial}{\partial y}\left(k_y \frac{\partial \phi}{\partial y}\right) + \frac{\partial}{\partial z}\left(k_z \frac{\partial \phi}{\partial z}\right) = 0, \tag{E3}$$

which can be seen to be an elliptic equation. If there is a recharge (quantity of fluid added per unit time) of $\dot{q}$ and the flow is unsteady, Eq. (E3) becomes

$$\frac{\partial}{\partial x}\left(k_x \frac{\partial \phi}{\partial x}\right) + \frac{\partial}{\partial y}\left(k_y \frac{\partial \phi}{\partial y}\right) + \frac{\partial}{\partial z}\left(k_z \frac{\partial \phi}{\partial z}\right) + \dot{q} = \alpha \frac{\partial \phi}{\partial t}, \tag{E4}$$

where α denotes the effective porosity divided by the acquifer thickness for unconfined flow and the specific storage for confined flow. Equation (E4) is a parabolic equation.

The boundary and initial conditions can be stated as follows:

(i) $\phi(x,t) = \phi_0(t)$ on S_1 (potential or head is prescribed on S_1);

(ii) $k_x \frac{\partial \phi}{\partial x} l_x + k_y \frac{\partial \phi}{\partial y} l_y + k_z \frac{\partial \phi}{\partial z} l_z + q(t) = 0$ on S_2 (velocity of the fluid moving out of the domain is prescribed as $q(t)$ on S_2, l_x, l_y, l_z are the direction cosines of the outward normal to the boundary S_2);

(iii) $\phi(x,0) = \bar{\phi}(x)$ (initial potential distribution is specified). ◀

▶Example 11.3

Derive the equation governing the free vibration of a beam. (See Fig. 11.4.)

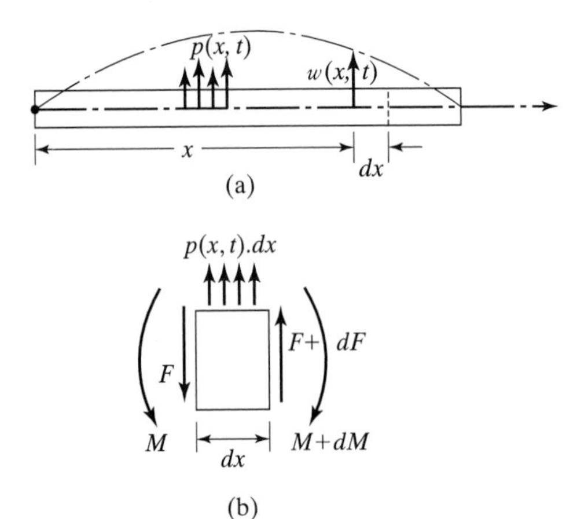

Figure 11.4 Vibration of a beam.

Solution

By considering the dynamic equilibrium of an element of the beam as shown in Fig. 11.4(b), we obtain

$$\frac{\partial F}{\partial x} = -p(x,t), \tag{E1}$$

where F is the shear force and p is the load acting on the beam per unit length. The moment equilibrium equation can be derived as

$$\frac{\partial M}{\partial x} = F(x,t), \tag{E2}$$

where $M(x,t)$ is the bending moment. The curvature of the deflected center line of the beam, for small deflections, can be expressed as

$$\frac{1}{R} = -\frac{\partial^2 w}{\partial x^2}, \tag{E3}$$

where R is the radius of curvature and w is the deflection of the beam. The moment–curvature relation is given by

$$\frac{1}{R} = \frac{M}{EI}, \tag{E4}$$

where E is Young's modulus and I is the area moment of inertia of the beam. Combining Eqs. (E1)–(E4) leads to the free-vibration equation

$$EI\frac{\partial^4 w}{\partial x^4} + m\frac{\partial^2 w}{\partial t^2} = 0, \tag{E5}$$

where $p(x,t) = -m\frac{\partial^2 w}{\partial t^2}$ was used to denote the inertia force and m is the mass of the beam per unit length.

Assuming that the beam is fixed at $x = 0$ and free at $x = 1$, we can express the boundary conditions as follows:

(i) Deflection $= w(0,t) = 0$, Slope $= \frac{\partial w(0,t)}{\partial x} = 0$ (Fixed end);

(ii) Bending moment $= \frac{\partial^2 w(l,t)}{\partial x^2} = 0$, Shear force $= \frac{\partial^3 w(l,t)}{\partial x^3} = 0$ (Free end);

Typical initial conditions are given by

(iii) $w(x,t=0) = 0$, $\frac{\partial w(x,t=0)}{\partial t} = 0$ (Initial deflection and velocity of all points are zero); ◀

11.3 Initial and Boundary Conditions

In general, a partial differential equation can have both initial and boundary conditions. The initial and boundary conditions are specified to obtain a unique solution to the partial differential equation. The partial differential equations can be classified as initial-value or boundary-value problems. In the case of an initial-value problem, at least one of the independent variables will have an open region. In the case of a boundary-value problem, conditions are specified at all boundaries, and the region will be closed with respect to all independent variables. In general, three types of boundary conditions can be specified for a partial differential equation as follows, (Fig. 11.1):

(1) *Dirichlet condition*:

The value of the dependent variable is specified on part of the boundary (S_1) as

$$\phi(x, y) = u(x, y), \text{ for all } (x, y)\varepsilon S_1. \quad (11.6)$$

In many physical problems, $u(x, y)$ will be a constant and independent of the coordinates x and y.

(2) *Neumann condition*:

The gradient of the dependent variable (i.e., the derivative of the dependent variable in a direction normal to the boundary) is specified on part of the boundary (S_2) as

$$\frac{\partial \phi}{\partial n}(x, y) = v(x, y), \text{ for all } (x, y)\varepsilon S_2, \quad (11.7)$$

where $v(x, y)$ will be a constant in many practical problems.

(3) *Robbins, or mixed, condition*:

A linear combination of the dependent variable and its gradient is specified on part of the boundary (S_3) as

$$\frac{\partial \phi}{\partial n}(x, y) + r\phi(x, y) = w(x, y), \text{ for all } (x, y)\varepsilon S_3, \quad (11.8)$$

where r is a constant and $w(x, y)$ will be a constant in many engineering applications.

11.4 Elliptic Partial Differential Equations

We consider the solution of the following general elliptic partial differential equation using a finite-difference approach in this section:

$$\nabla[g(x, y)\nabla\phi(x, y)] + h(x, y)\phi(x, y) = f(x, y), \text{ for all } (x, y)\varepsilon R. \quad (11.9)$$

Here R is the region of integration (see Fig. 11.1), $g(x, y)$, $h(x, y)$, and $f(x, y)$ are known functions of (x, y). The function $f(x, y)$ is called the forcing, source, or nonhomogeneous term. For simplicity of derivation of the finite-difference equations, we assume a special type of elliptic equation, known as the Poisson equation, which is given by

$$\nabla^2\phi(x, y) = \frac{\partial^2\phi(x, y)}{\partial x^2} + \frac{\partial^2\phi(x, y)}{\partial y^2} = f(x, y). \tag{11.10}$$

The region R is assumed to be rectangular, as shown in Fig. 11.5 with the following boundary conditions:

$$\phi(0, y) = 0 \quad \text{(left boundary);} \tag{11.11}$$

$$\frac{\partial\phi}{\partial x}(a, y) = 0 \quad \text{(right boundary);} \tag{11.12}$$

$$\phi(x, 0) = 0 \quad \text{(bottom boundary);} \tag{11.13}$$

$$\frac{\partial\phi}{\partial y}(x, b) = 0 \quad \text{(top boundary).} \tag{11.14}$$

The rectangular domain of integration is divided into m equal parts along the x-direction and n equal parts along the y-direction, so that the step sizes are given by $\Delta x = \frac{a}{m}$ and $\Delta y = \frac{b}{n}$. The coordinates of the mesh points are denoted by (x_i, y_j); $i = 0, 1, 2, \ldots, m$; $j = 0, 1, 2, \ldots, n$, with (see Fig. 11.6)

$$x_0 = 0, x_1 = \Delta x, \ldots, x_i = i\Delta x (i = 1, 2, \ldots, m), x_m = m\Delta x = a \tag{11.15}$$

and

$$y_0 = 0, y_1 = \Delta y, \ldots, y_j = j\Delta y (j = 1, 2, \ldots, n), y_n = n\Delta y = b. \tag{11.16}$$

To derive the finite-difference equation for an interior grid point (i, j), we use the central-difference formulas

$$\left.\frac{\partial^2\phi}{\partial x^2}\right|_{i,j} = \frac{\phi_{i-1,j} - 2\phi_{i,j} + \phi_{i+1,j}}{\Delta x^2}, \tag{11.17}$$

$$\left.\frac{\partial^2\phi}{\partial y^2}\right|_{i,j} = \frac{\phi_{i,j-1} - 2\phi_{i,j} + \phi_{i,j+1}}{\Delta y^2}, \tag{11.18}$$

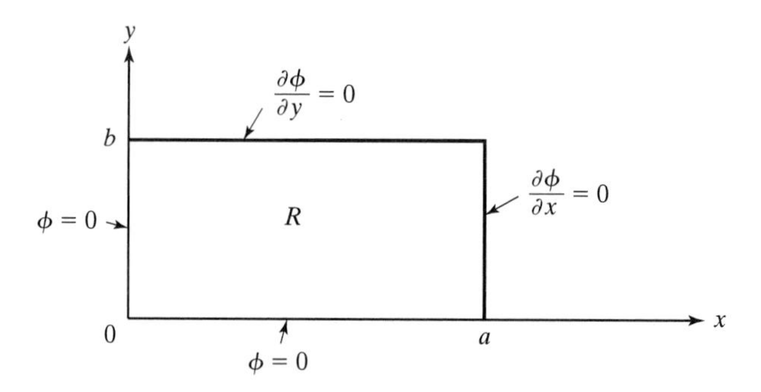

Figure 11.5 Rectangular region.

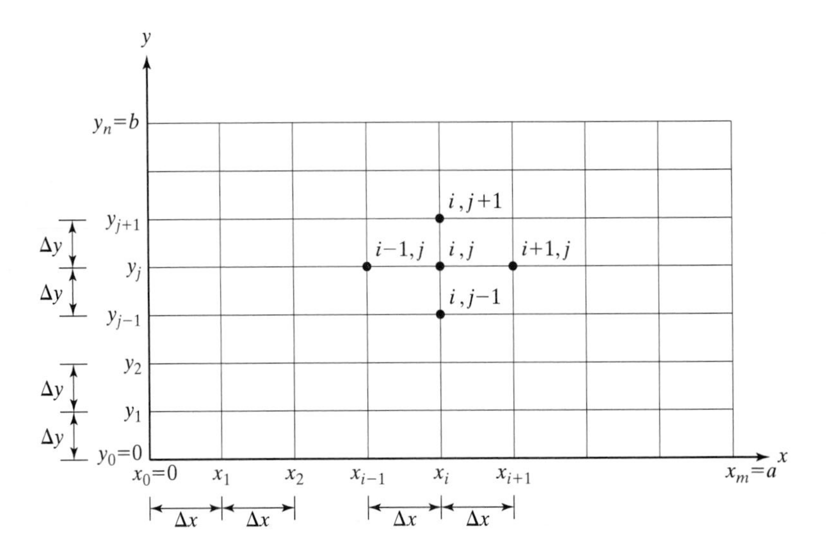

Figure 11.6 Finite-difference grid.

where $\phi_{i,j} = \phi(x_i, y_j)$. By substituting Eqs. (11.17) and (11.18) into Eq. (11.10), we obtain the finite-difference equation for an interior grid point (i, j):

$$\frac{\phi_{i-1,j} - 2\phi_{i,j} + \phi_{i+1,j}}{\Delta x^2} + \frac{\phi_{i,j-1} - 2\phi_{i,j} + \phi_{i,j+1}}{\Delta y^2} = f_{i,j},$$

$$i = 1, 2, \ldots, m-1; j = 1, 2, \ldots, n-1. \tag{11.19}$$

Here $f_{i,j} = f(x_i, y_j)$. The boundary conditions given by Eqs. (11.11) through (11.14) can be expressed as follows:

$$\phi(0, y) = \phi_{0,j} = 0; j = 0, 1, 2, \ldots, n, \tag{11.20}$$

$$\frac{\partial \phi}{\partial x}(a, y) = \left.\frac{\partial \phi}{\partial x}\right|_{m,j} = 0; j = 0, 1, 2, \ldots, n, \tag{11.21}$$

$$\phi(x, 0) = \phi_{i,0} = 0; i = 0, 1, 2, \ldots, m, \tag{11.22}$$

and

$$\frac{\partial \phi}{\partial y}(x, b) = \left.\frac{\partial \phi}{\partial y}\right|_{i,n} = 0; i = 0, 1, 2, \ldots, m. \tag{11.23}$$

Since the values of ϕ are specified as zero along the left and bottom boundaries (Eqs. (11.20) and (11.22)), the derivation of the finite-difference equations is not needed for all grid points lying on these boundaries. In addition, the condition $\phi_{0,j} = 0$ (Eq. (11.20)) is to be used in the finite-difference equation (Eq. (11.19)) corresponding to $i = 1$ and the condition $\phi_{i,0} = 0$ (Eq. (11.22)) is to be used in Eq. (11.19) corresponding to $j = 1$. The boundary conditions of Eqs. (11.21) and (11.23) can be implemented in several ways.

Approach 1

One possibility is to use the backward-difference formulas:

$$\left.\frac{\partial \phi}{\partial x}\right|_{m,j} = \frac{\phi_{m,j} - \phi_{m-1,j}}{\Delta x} = 0;\ j = 0, 1, 2, \ldots, n, \tag{11.24}$$

$$\left.\frac{\partial \phi}{\partial y}\right|_{i,n} = \frac{\phi_{i,n} - \phi_{i,n-1}}{\Delta y} = 0;\ i = 0, 1, 2, \ldots, m. \tag{11.25}$$

The equivalences

$$\phi_{m,j} = \phi_{m-1,j},\ j = 0, 1, 2, \ldots, n, \tag{11.26}$$

and

$$\phi_{i,n} = \phi_{i,n-1},\ i = 0, 1, 2, \ldots, m, \tag{11.27}$$

given by Eqs. (11.24) and (11.25) are to be used in the finite-difference equations (Eq. (11.19)) corresponding to $i = m$ and $j = n$, respectively.

Approach 2

Another possibility is to derive suitable finite-difference equivalence of Eq. (11.10) for the grid points lying on the right and top boundaries and enforce the conditions given by Eqs. (11.21) and (11.23). For the grid points lying on the right boundary, the second partial derivative $\frac{\partial^2 \phi}{\partial x^2}$ of Eq. (11.10) is approximated using a backward-difference formula in terms of $\frac{\partial \phi}{\partial x}$ with half the stepsize, $\frac{\Delta x}{2}$:

$$\left.\frac{\partial^2 \phi}{\partial x^2}\right|_{m,j} = \frac{\left.\frac{\partial \phi}{\partial x}\right|_{m,j} - \left.\frac{\partial \phi}{\partial x}\right|_{m-\frac{1}{2},j}}{\frac{\Delta x}{2}};\ j = 1, 2, \ldots, n-1. \tag{11.28}$$

By setting $\left.\frac{\partial \phi}{\partial x}\right|_{m,j}$ equal to zero (from Eq. (11.21)) and using a central-difference formula with a step size $\frac{\Delta x}{2}$ for the derivative $\left.\frac{\partial \phi}{\partial x}\right|_{m-\frac{1}{2},j}$, we see that Eq. (11.28) yields

$$\left.\frac{\partial^2 \phi}{\partial x^2}\right|_{m,j} = -\frac{2}{\Delta x}\left.\frac{\partial \phi}{\partial x}\right|_{m-\frac{1}{2},j} = -\frac{2}{\Delta x}\left[\frac{\phi_{m,j} - \phi_{m-1,j}}{\Delta x}\right];\ j = 1, 2, \ldots, n-1. \tag{11.29}$$

Similarly, for the grid points lying on the top boundary, the second partial derivative $\frac{\partial^2 \phi}{\partial y^2}$ of Eq. (11.9) is approximated using a backward difference formula in terms of $\frac{\partial \phi}{\partial y}$ with half the stepsize $\frac{\Delta y}{2}$:

$$\left.\frac{\partial^2 \phi}{\partial y^2}\right|_{i,n} = \frac{\left.\frac{\partial \phi}{\partial y}\right|_{i,n} - \left.\frac{\partial \phi}{\partial y}\right|_{i,n-\frac{1}{2}}}{\frac{\Delta y}{2}};\ i = 1, 2, \ldots, m-1. \tag{11.30}$$

By substituting $\frac{\partial \phi}{\partial y}\big|_{i,n} = 0$ (from Eq. (11.23)), and using a central-difference formula with a step size $\frac{\Delta y}{2}$ for the derivative $\frac{\partial \phi}{\partial y}\big|_{i,n-\frac{1}{2}}$, Eq. (11.30) can be expressed as

$$\left.\frac{\partial^2 \phi}{\partial y^2}\right|_{i,n} = -\frac{2}{\Delta y}\left.\frac{\partial \phi}{\partial y}\right|_{i,n-\frac{1}{2}} = -\frac{2}{\Delta y}\left[\frac{\phi_{i,n} - \phi_{i,n-1}}{\Delta y}\right]; i = 1, 2, \ldots, m-1. \quad (11.31)$$

By combining Eqs. (11.29) and (11.19), we find that the finite-difference equations for the grid point lying on the right boundary can be expressed as

$$-\frac{2}{\Delta x^2}(\phi_{m,j} - \phi_{m-1,j}) + \left(\frac{\phi_{m,j-1} - 2\phi_{m,j} + \phi_{m,j+1}}{\Delta y^2}\right) = f_{m,j}; j = 1, 2, \ldots, n-1. \quad (11.32)$$

Similarly, the finite-difference equations for the grid points lying on the top boundary can be written, using Eqs. (11.31) and (11.19), as

$$\left(\frac{\phi_{i-1,n} - 2\phi_{i,n} + \phi_{i+1,n}}{\Delta x^2}\right) - \frac{2}{\Delta y^2}\left(\phi_{i,n} - \phi_{i,n-1}\right) = f_{i,n}; i = 1, 2, \ldots, m-1. \quad (11.33)$$

For the corner grid point (m, n), Eqs. (11.29) and (11.31) can be used for $\frac{\partial^2 \phi}{\partial x^2}$ and $\frac{\partial^2 \phi}{\partial y^2}$, respectively, and the finite-difference equation can be expressed as

$$-\frac{2}{\Delta x^2}(\phi_{m,n} - \phi_{m-1,n}) - \frac{2}{\Delta y^2}(\phi_{m,n} - \phi_{m,n-1}) = f_{m,n}. \quad (11.34)$$

Thus, Eqs. (11.19), (11.32), (11.33), and (11.34) constitute a complete system of finite-difference equations for the grid shown in Fig. 11.6. Of course, the boundary conditions given by Eqs. (11.20) and (11.22) are to be incorporated before solving the equations.

11.4.1 Incorporation of Robbins (Mixed-) Boundary Condition

Let the Robbins or mixed-boundary condition be specified as

$$\frac{\partial \phi}{\partial n}(x, y) + r\phi(x, y) = w; r, w \text{ constants}, \quad (11.35)$$

along the left and top boundaries of a rectangular region (similar to the region shown in Fig. 11.5). It can be seen that the outward normal direction points along $-x$- and $+y$-directions for the left and top boundaries, respectively. Hence, Eq. (11.35) becomes

$$-\frac{\partial \phi}{\partial x}(x, y) + r\phi(x, y) = w; \quad \text{for } x = 0 \text{ and } 0 \le y \le b, \quad (11.36)$$

$$\frac{\partial \phi}{\partial y}(x, y) + r\phi(x, y) = w; \quad \text{for } y = b \text{ and } 0 \le x \le a. \quad (11.37)$$

As indicated earlier for the case of Neumann condition, two different approaches can be adopted to incorporate the boundary conditions of Eqs. (11.36) and (11.37).

Approach 1

Use a forward-difference formula for $\frac{\partial \phi}{\partial x}$, and express Eq. (11.36) as

$$-\left(\frac{\phi_{1,j}-\phi_{0,j}}{\Delta x}\right)+r\phi_{0,j}=w;\ j=0,1,2,\ldots,n,$$

or

$$\phi_{0,j}=\frac{\Delta x w+\phi_{1,j}}{1+r\Delta x};\ j=0,1,2,\ldots,n. \tag{11.38}$$

Use a backward-difference formula for $\frac{\partial \phi}{\partial y}$, and write Eq. (11.37) as

$$\left(\frac{\phi_{i,n}-\phi_{i,n-1}}{\Delta y}\right)+r\phi_{i,n}=w;\ i=0,1,2,\ldots,m,$$

or

$$\phi_{i,n}=\frac{\Delta y w+\phi_{i,n-1}}{1+r\Delta y};\ i=0,1,2,\ldots,m. \tag{11.39}$$

Equations (11.38) and (11.39) can be used to eliminate $\phi_{0,j}$ and $\phi_{i,n}$, respectively, from the finite-difference form of Eq. (11.19) corresponding to $i=0$ and $j=n$.

Approach 2

The finite-difference form of Eq. (11.10) for the grid points lying on the left and top boundaries are derived, and Eqs. (11.36) and (11.37) are substituted in them. For the grid points lying on the left boundary, $\frac{\partial^2 \phi}{\partial x^2}$ can be approximated using a forward-difference formula in terms of $\frac{\partial \phi}{\partial x}$ with half the step size $\frac{\Delta x}{2}$:

$$\left.\frac{\partial^2 \phi}{\partial x^2}\right|_{0,j}=\frac{\left.\frac{\partial \phi}{\partial x}\right|_{0,j}-\left.\frac{\partial \phi}{\partial x}\right|_{\frac{1}{2},j}}{\left(\frac{\Delta x}{2}\right)};\ j=1,2,\ldots,n-1. \tag{11.40}$$

By using

$$\left.\frac{\partial \phi}{\partial x}\right|_{0,j}=(r\phi_{0,j}-w) \tag{11.41}$$

from Eq. (11.36) and a central-difference formula with a step size $\frac{\Delta x}{2}$ for the derivative $\left.\frac{\partial \phi}{\partial x}\right|_{\frac{1}{2},j}$, we see that Eq. (11.40) results in

$$\left.\frac{\partial^2 \phi}{\partial x^2}\right|_{0,j}=\frac{(r\phi_{0,j}-w)-\left(\frac{\phi_{1,j}-\phi_{0,j}}{\Delta x}\right)}{\left(\frac{\Delta x}{2}\right)}=\frac{2}{\Delta x^2}[\phi_{0,j}(1+r\Delta x)-\phi_{1,j}-w\Delta x];$$

$$j=1,2,\ldots,n-1. \tag{11.42}$$

Similarly, for the grid points lying on the top boundary, the second partial derivative $\frac{\partial^2\phi}{\partial y^2}$ of Eq. (11.10) is approximated using a backward-difference formula in terms of $\frac{\partial\phi}{\partial y}$ with half the step size $\frac{\Delta y}{2}$:

$$\left.\frac{\partial^2\phi}{\partial y^2}\right|_{i,n} = \frac{\left.\frac{\partial\phi}{\partial y}\right|_{i,n} - \left.\frac{\partial\phi}{\partial y}\right|_{i,n-\frac{1}{2}}}{\left(\frac{\Delta y}{2}\right)}; i = 1, 2, \ldots, m-1. \tag{11.43}$$

By substituting

$$\left.\frac{\partial\phi}{\partial y}\right|_{i,n} = (-r\phi_{i,n} + w) \tag{11.44}$$

from Eq. (11.37) and using a central-difference formula with a step size $\frac{\Delta y}{2}$ for the derivative $\left.\frac{\partial\phi}{\partial y}\right|_{i,n-\frac{1}{2}}$, we find that Eq. (11.43) can be rewritten as

$$\left.\frac{\partial^2\phi}{\partial y^2}\right|_{i,n} = \frac{(-r\phi_{i,n} + w) - \left(\frac{\phi_{i,n} - \phi_{i,n-1}}{\Delta y}\right)}{\left(\frac{\Delta y}{2}\right)}$$

$$= \frac{2}{\Delta y^2}[-(r\Delta y + 1)\phi_{i,n} + \phi_{i,n-1} + w]; i = 1, 2, \ldots, m-1. \tag{11.45}$$

By combining Eqs. (11.42) and (11.19), we can express the finite-difference approximation of the grid points lying on the left boundary as

$$\frac{2}{\Delta x^2}[\phi_{0,j}(1 + r\Delta x) - \phi_{1,j} - w\Delta x] + \left(\frac{\phi_{1,j-1} - 2\phi_{1,j} + \phi_{1,j+1}}{\Delta y^2}\right) = f_{1,j};$$

$$j = 1, 2, \ldots, n-1. \tag{11.46}$$

Similarly, the finite-difference equivalence of Eq. (11.10) for the grid points lying on the top boundary can be written, using Eqs. (11.45) and (11.19), as

$$\left(\frac{\phi_{i-1,n} - 2\phi_{i,n} + \phi_{i+1,n}}{\Delta x^2}\right) + \frac{2}{\Delta y^2}[-(1 + r\Delta y)\phi_{i,n} + \phi_{i.n-1} + w]$$

$$+ f_{i,n}; i = 1, 2, \ldots, m-1. \tag{11.47}$$

For the corner grid point (1,n), Eqs. (11.42) and (11.45) can be used for $\frac{\partial^2\phi}{\partial x^2}$ and $\frac{\partial^2\phi}{\partial y^2}$, respectively, and the finite-difference approximation of Eq. (11.9) can be expressed as

$$\frac{2}{\Delta x^2}[(1 + r\Delta x)\phi_{0,n} - \phi_{1,n} - w\Delta x]$$

$$+ \frac{2}{\Delta y^2}[-(1 + r\Delta y)\phi_{1,n} + \phi_{1,n-1} + w] = f_{1,n}. \tag{11.48}$$

11.4.2 Irregular Boundaries

In some problems, the domain for the partial-differential equation may be irregular. In such cases, the mesh points of a rectangular grid may not fall on the boundary of the irregular domain, as shown in Fig. 11.7. Then a different approach is to be adopted to derive the finite-difference approximation of the partial differential equation. For this, consider the general case of grid points surrounding a base grid point (i, j) as shown in Fig. 11.8. The distances of the grid points on the right and left sides of the base grid point (i, j) are assumed to be $\alpha_1 \Delta x$ and $\alpha_2 \Delta x$, where α_1 and α_2 denote fractions of the standard spacing along the x-direction, Δx. Similarly, the distances of the grid points lying above and below the base grid point (i, j) are taken to be $\beta_1 \Delta y$ and $\beta_2 \Delta y$, where β_1 and β_2 indicate fractions of the standard spacing along the y-direction, Δy. The first derivatives, $\frac{\partial \phi}{\partial x}$, can be evaluated at the grid points $(i+1, j)$ and $(i-1, j)$ using, respectively, backward- and forward-difference formulas:

$$\left.\frac{\partial \phi}{\partial x}\right|_{i+1,j} = \frac{\phi_{i+1,j} - \phi_{i,j}}{\alpha_1 \Delta x}, \tag{11.49}$$

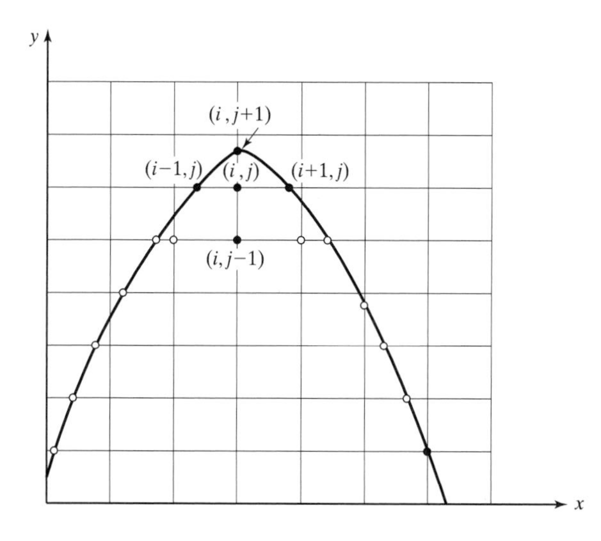

Figure 11.7 A domain with irregular boundary.

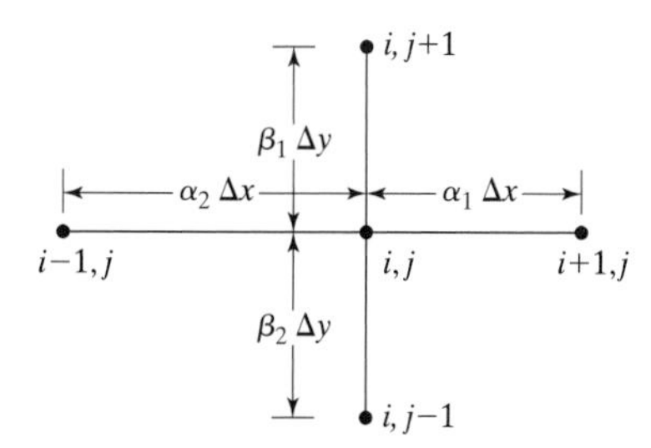

Figure 11.8 Grid points at an irregular boundary.

$$\left.\frac{\partial \phi}{\partial x}\right|_{i-1,j} = \frac{\phi_{i,j} - \phi_{i-1,j}}{\alpha_2 \Delta x}. \tag{11.50}$$

Next, the second derivative $\frac{\partial^2 \phi}{\partial x^2}$ at the base node (i, j) can be expressed using a formula similar to the central-difference formula as

$$\left.\frac{\partial^2 \phi}{\partial x^2}\right|_{i,j} = \frac{\left.\frac{\partial \phi}{\partial x}\right|_{i+1,j} - \left.\frac{\partial \phi}{\partial x}\right|_{i-1,j}}{\alpha_1 \Delta x + \alpha_2 \Delta x}. \tag{11.51}$$

Substituting Eqs. (11.49) and (11.50) into Eq. (11.51) yields

$$\left.\frac{\partial^2 \phi}{\partial x^2}\right|_{i,j} = \left\{ \frac{\frac{\phi_{i+1,j} - \phi_{i,j}}{\alpha_1 \Delta x} - \frac{\phi_{i,j} - \phi_{i-1,j}}{\alpha_2 \Delta x}}{(\alpha_1 + \alpha_2)\Delta x} \right\}. \tag{11.52}$$

Similarly, we can obtain

$$\left.\frac{\partial^2 \phi}{\partial y^2}\right|_{i,j} = \left\{ \frac{\frac{\phi_{i,j+1} - \phi_{i,j}}{\beta_1 \Delta y} - \frac{\phi_{i,j} - \phi_{i,j-1}}{\beta_2 \Delta y}}{(\beta_1 + \beta_2)\Delta y} \right\}. \tag{11.53}$$

Finally, the finite-difference approximation of the Poisson equation (Eq. (11.10)) at grid point (i, j) is given by

$$\left\{ \frac{\left(\frac{\phi_{i+1,j} - \phi_{i,j}}{\alpha_1 \Delta x}\right) - \left(\frac{\phi_{i,j} - \phi_{i-1,j}}{\alpha_2 \Delta x}\right)}{(\alpha_1 + \alpha_2)\Delta x} \right\} + \left\{ \frac{\left(\frac{\phi_{i,j+1} - \phi_{i,j}}{\beta_1 \Delta y}\right) - \left(\frac{\phi_{i,j} - \phi_{i,j-1}}{\beta_2 \Delta y}\right)}{(\beta_1 + \beta_2)\Delta y} \right\} = f_{i,j}. \tag{11.54}$$

Note that if the boundary values of $\phi(x, y)$ are specified (Dirichlet condition), those specified values are to be substituted into Eq. (11.54).

▶Example 11.4

Determine the steady-state temperature distribution in a rectangular plate of size $15'' \times 20''$ by solving the Laplace equation using $\Delta x = \Delta y = 5''$. The temperatures on the four sides of the plate are specified as indicated in Fig. 11.9.

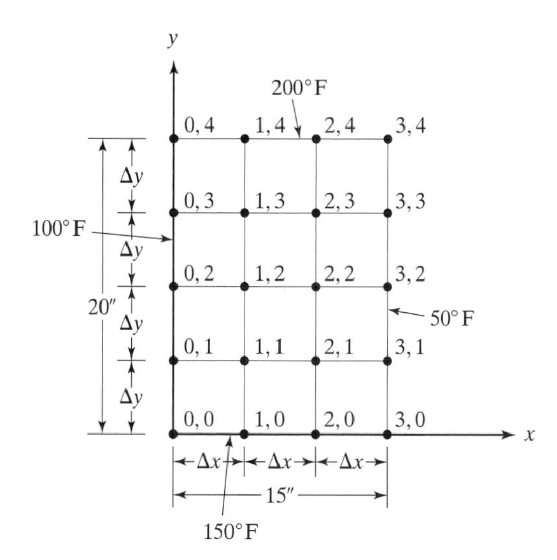

Figure 11.9 Finite difference grid of rectangular plate.

Solution

The finite-difference grid, with $\Delta x = \Delta y = 5''$, is shown in Fig. 11.9. Since the temperature is specified on all the boundaries, we approximate the governing equation

$$\frac{\partial^2 T}{\partial x^2} + \frac{\partial^2 T}{\partial y^2} = 0 \tag{E1}$$

only at the interior grid points. The central-difference approximation of Eq. (E1) at the interior node (1,1) yields the equation

$$\frac{T_{01} - 2T_{11} + T_{21}}{(\Delta x)^2} + \frac{T_{12} - T_{11} + T_{10}}{(\Delta y)^2} = 0. \tag{E2}$$

Since $\Delta x = \Delta y = 5$, $T_{01} = 100$, and $T_{10} = 150$, Eq. (E2) reduces to

$$-4T_{11} + T_{12} + T_{21} + 250 = 0. \tag{E3}$$

Similarly, the central-difference approximation of Eq. (E1) at the other interior nodes (2,1), (1,2), (2,2), (1,3), and (2,3) results in the following equations:

$$T_{11} - 4T_{21} + T_{22} + 200 = 0, \tag{E4}$$

$$-4T_{12} + T_{22} + T_{13} + T_{11} + 100 = 0, \tag{E5}$$

$$T_{12} + 4T_{22} + T_{23} + T_{21} + 50 = 0, \tag{E6}$$

$$-4T_{13} + T_{23} + T_{12} + 300 = 0, \tag{E7}$$

$$T_{13} - 4T_{23} + T_{22} + 250 = 0. \tag{E8}$$

Equations (E3) through (E8) can be expressed in matrix form as

$$\begin{bmatrix} -4 & 1 & 1 & 0 & 0 & 0 \\ 1 & -4 & 0 & 1 & 0 & 0 \\ 1 & 0 & -4 & 1 & 1 & 0 \\ 0 & 1 & 1 & -4 & 0 & 1 \\ 0 & 0 & 1 & 0 & -4 & 1 \\ 0 & 0 & 0 & 1 & 1 & -4 \end{bmatrix} \begin{Bmatrix} T_{11} \\ T_{21} \\ T_{12} \\ T_{22} \\ T_{13} \\ T_{23} \end{Bmatrix} = \begin{Bmatrix} -250 \\ -200 \\ -100 \\ -50 \\ -300 \\ -250 \end{Bmatrix}. \tag{E9}$$

The solution of Eqs. (E9) gives the steady-state temperatures at the interior nodes (in °F) as $T_{11} = 116.04555$, $T_{21} = 103.00208$, $T_{12} = 111.18013$, $T_{22} = 95.962730$, $T_{13} = 132.71222$, and $T_{23} = 119.66874$.

Note that the accuracy of the result can be improved by using smaller step sizes Δx and Δy. ◀

▶Example 11.5

Derive the finite-difference equations for solving the Poisson equation

$$\frac{\partial^2 \phi}{\partial x^2} + \frac{\partial^2 \phi}{\partial y^2} = f(x, y) \tag{E1}$$

over a rectangular region of size $10'' \times 12''$ using $\Delta x = 5''$ and $\Delta y = 6''$ with the following boundary conditions:

$$\frac{\partial \phi}{\partial x} - \phi = 2 \text{ at } x = 0, \tag{E2}$$

$$\frac{\partial \phi}{\partial y} - 2\phi = -1 \text{ at } y = 0, \tag{E3}$$

$$\phi = 3 \text{ at } x = 10 \text{ in}, \tag{E4}$$

$$\phi = 4 \text{ at } y = 12 \text{ in}. \tag{E5}$$

Solution

The finite-difference grid is established with $\Delta x = 5''$ and $\Delta y = 6''$ as shown in Fig. 11.10. Since the values of ϕ are prescribed on the top and right-hand-side boundaries, we approximate Eq. (E1) using finite differences only at the grid points (0,1), (1,1), (0,0), and (1,0).

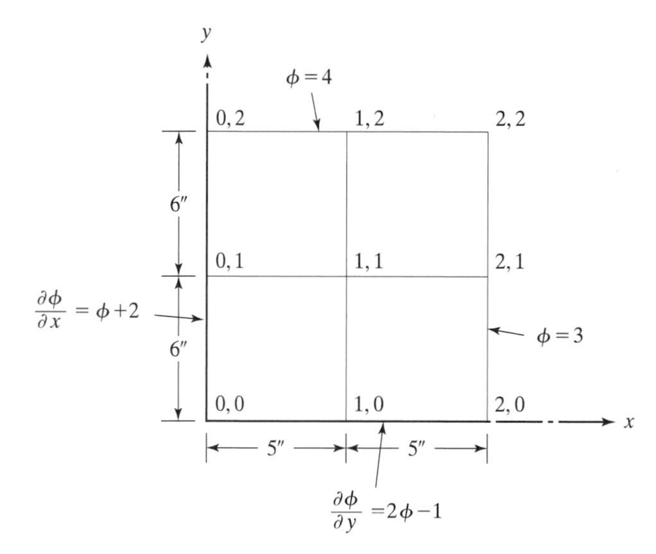

Figure 11.10 Finite difference grid of rectangular region.

Equation (E1) at grid point (0,0)

Using

$$\left.\frac{\partial^2\phi}{\partial x^2}\right|_{0,0} = \frac{\left.\dfrac{\partial\phi}{\partial x}\right|_{\frac{1}{2},0} - \left.\dfrac{\partial\phi}{\partial x}\right|_{0,0}}{\dfrac{\Delta x}{2}} = \frac{\left(\dfrac{\phi_{10}-\phi_{00}}{\Delta x}\right) - (\phi_{00}+2)}{\dfrac{\Delta x}{2}}$$

$$= \frac{2}{25}(\phi_{10} - 6\phi_{00} - 10) \tag{E6}$$

and

$$\left.\frac{\partial^2\phi}{\partial y^2}\right|_{0,0} = \frac{\left.\dfrac{\partial\phi}{\partial y}\right|_{0,\frac{1}{2}} - \left.\dfrac{\partial\phi}{\partial y}\right|_{0,0}}{\dfrac{\Delta y}{2}} = \frac{\dfrac{\phi_{01}-\phi_{00}}{\Delta y} - (2\phi_{00}-1)}{\dfrac{\Delta y}{2}}$$

$$= \frac{1}{18}(\phi_{01} - 13\phi_{00} + 6), \tag{E7}$$

we see that Eq. (E1) can be expressed as

$$\frac{2}{25}(\phi_{10} - 6\phi_{00} - 10) + \frac{1}{18}(\phi_{01} - 13\phi_{00} + 6) = f(0, 0). \tag{E8}$$

Note that the values $\frac{\partial\phi}{\partial x}\big|_{0,0} = \phi_{00} + 2$ and $\frac{\partial\phi}{\partial y}\big|_{0,0} = 2\phi_{00} - 1$ from Eqs. (E2) and (E3) were used in deriving Eqs. (E6) and (E7), respectively.

Equation (E1) at grid point (1,0)

We use the relations

$$\left.\frac{\partial^2\phi}{\partial x^2}\right|_{1,0} = \frac{\phi_{20} - 2\phi_{10} + \phi_{00}}{(\Delta x)^2} = \frac{1}{25}(\phi_{20} - 2\phi_{10} + \phi_{00}) \tag{E9}$$

and

$$\left.\frac{\partial^2\phi}{\partial y^2}\right|_{1,0} = \frac{\left.\frac{\partial\phi}{\partial y}\right|_{1,\frac{1}{2}} - \left.\frac{\partial\phi}{\partial y}\right|_{1,0}}{\frac{\Delta y}{2}} = \frac{\left(\frac{\phi_{11} - \phi_{10}}{\Delta y}\right) - (2\phi_{10} - 1)}{\frac{\Delta y}{2}}$$

$$= \frac{1}{18}(\phi_{11} - 13\phi_{10} + 6) \tag{E10}$$

to obtain

$$\frac{1}{25}(\phi_{20} - 2\phi_{10} + \phi_{00}) + \frac{1}{18}(\phi_{11} - 13\phi_{10} + 6) = f(5, 0). \tag{E11}$$

Equation (E1) at grid point (0,1)

Using

$$\left.\frac{\partial^2\phi}{\partial x^2}\right|_{0,1} = \frac{\left.\frac{\partial\phi}{\partial x}\right|_{\frac{1}{2},1} - \left.\frac{\partial\phi}{\partial x}\right|_{0,1}}{\frac{\Delta x}{2}} = \frac{\left(\frac{\phi_{11} - \phi_{01}}{\Delta x}\right) - (\phi_{01} + 2)}{\frac{\Delta x}{2}}$$

$$= \frac{2}{25}(\phi_{11} - 6\phi_{01} - 10) \tag{E12}$$

and

$$\left.\frac{\partial^2\phi}{\partial y^2}\right|_{0,1} = \frac{\phi_{02} - 2\phi_{01} + \phi_{00}}{(\Delta y)^2} = \frac{1}{36}(\phi_{02} - 2\phi_{01} + \phi_{00}), \tag{E13}$$

we can write Eq. (E1) as

$$\frac{2}{25}(\phi_{11} - 6\phi_{01} - 10) + \frac{1}{36}(\phi_{02} - 2\phi_{01} + \phi_{00}) = f(0, 6). \tag{E14}$$

Equation (E1) at grid point (1,1)

We use the central-difference relations

$$\left.\frac{\partial^2\phi}{\partial x^2}\right|_{1,1} = \frac{\phi_{21} - 2\phi_{11} + \phi_{01}}{(\Delta x)^2} \tag{E15}$$

and

$$\left.\frac{\partial^2\phi}{\partial y^2}\right|_{1,1} = \frac{\phi_{12} - \phi_{11} + \phi_{10}}{(\Delta y)^2} \tag{E16}$$

to obtain

$$\frac{1}{25}(\phi_{21} - 2\phi_{11} + \phi_{01}) + \frac{1}{36}(\phi_{12} - 2\phi_{11} + \phi_{10}) = f(5,6). \tag{E17}$$

Using the boundary conditions of Eqs. (E4) and (E5), we can write Eqs. (E8), (E11), (E14), and (E17) as

$$\frac{2}{25}\phi_{10} - \frac{541}{450}\phi_{00} + \frac{1}{18}\phi_{01} = f(0,0) + \frac{7}{15}, \tag{E18}$$

$$\frac{1}{25}\phi_{00} - \frac{361}{450}\phi_{10} + \frac{1}{18}\phi_{11} = f(5,0) - \frac{34}{75}, \tag{E19}$$

$$\frac{1}{36}\phi_{00} - \frac{241}{450}\phi_{01} + \frac{2}{25}\phi_{11} = f(0,6) + \frac{31}{45}, \tag{E20}$$

and

$$-\frac{61}{450}\phi_{11} + \frac{1}{25}\phi_{01} + \frac{1}{36}\phi_{10} = f(5,6) - \frac{52}{225}. \tag{E21}$$

Thus, Eqs. (E18) through (E21) represent the finite-difference equations corresponding to Eqs. (E1) through (E5) for the grid shown in Fig. 11.10. ◀

▶Example 11.6

Derive the central-difference equations corresponding to Poisson's equation

$$\frac{\partial^2\phi}{\partial x^2} + \frac{\partial^2\phi}{\partial y^2} = 2xy \tag{E1}$$

at nodes (2,1), (2,2), and (1,2) in the quarter circle shown in Fig. 11.11.

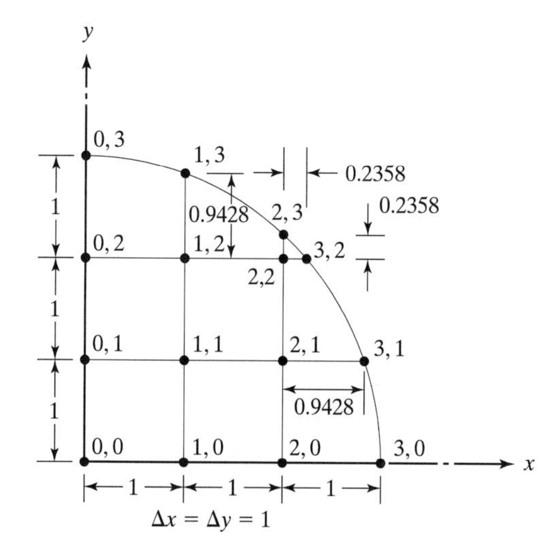

Figure 11.11 Grid for a quarter-circle.

Solution

Equation (E1) at node (2,1)

Equation (11.54) is applicable with $i = 2$, $j = 1$; $x_2 = 2$, $y_1 = 1$; $\alpha_1 = 0.9428$, $\alpha_2 = 1$, $\beta_1 = 1$, $\beta_2 = 1$:

$$\frac{\dfrac{\phi_{31}-\phi_{21}}{0.9428}-\dfrac{\phi_{21}-\phi_{11}}{1.0}}{1.9428}+\frac{\dfrac{\phi_{22}-\phi_{21}}{1.0}-\dfrac{\phi_{21}-\phi_{20}}{1.0}}{2.0}=2(2)(1)=4,$$

or

$$2\phi_{31}-7.5490\,\phi_{21}+1.8856\,\phi_{11}+1.8317\,\phi_{22}+1.8317\,\phi_{20}=14.6536. \quad \text{(E2)}$$

Equation (E1) at node (2,2)

Equation (11.54) is applicable with $i = 2$, $j = 2$; $x_2 = 2$, $y_2 = 2$; $\alpha_1 = 0.2358$, $\alpha_2 = 1$, $\beta_1 = 0.2358$, $\beta_2 = 1$:

$$\frac{\dfrac{\phi_{32}-\phi_{22}}{0.2358}-\dfrac{\phi_{22}-\phi_{12}}{1.0}}{1.2358}+\frac{\dfrac{\phi_{23}-\phi_{22}}{0.2358}-\dfrac{\phi_{22}-\phi_{21}}{1.0}}{1.2358}=2(2)(2)=8,$$

or

$$\phi_{32}-2.5716\,\phi_{22}+0.2358\,\phi_{12}+\phi_{23}+0.2358\,\phi_{21}=2.3312. \quad \text{(E3)}$$

Equation (E1) at node (1,2)

Equation (11.54) is applicable with $i = 1$, $j = 2$; $x_1 = 1$, $y_2 = 2$; $\alpha_1 = 1$, $\alpha_2 = 1$, $\beta_1 = 0.9428$, $\beta_2 = 1$:

$$\frac{\dfrac{\phi_{22}-\phi_{12}}{1.0}-\dfrac{\phi_{12}-\phi_{02}}{1.0}}{2.0}+\frac{\dfrac{\phi_{13}-\phi_{12}}{0.9428}-\dfrac{\phi_{12}-\phi_{11}}{1.0}}{1.9428}=2(1)(2)=4,$$

or

$$1.8317\,\phi_{22}-7.5490\,\phi_{12}+1.8317\,\phi_{02}+2\,\phi_{13}+1.8856\,\phi_{11}=14.6536. \quad \text{(E4)}$$

◀

11.5 Parabolic Partial Differential Equations

The general form of a parabolic partial differential equation is given by

$$(c_1\phi_x)_x+(c_2\phi_y)_y+(c_3\phi_z)_z+c_4\phi_x+c_5\phi_y+c_6\phi_z+c_7\phi=\phi_t, \quad (11.55)$$

where c_i are functions of x, y, z, and t with $c_j \geq 0$ for $j = 1, 2, 3$ and a subscript x, y, z, or t denotes partial derivative with respect to x, y, z, or t, respectively. For specificity, we consider the parabolic equation

$$\alpha^2\frac{\partial^2\phi}{\partial x^2}(x,t)=\frac{\partial\phi}{\partial t}(x,t); 0 \leq x \leq 1, \quad (11.56)$$

subject to the boundary conditions

$$\phi(0, t) = f(t); t > 0,$$
$$\phi(1, t) = g(t); t > 0 \tag{11.57}$$

and initial condition

$$\phi(x, 0) = h(x); 0 \leq x \leq 1, \tag{11.58}$$

where α^2 is a constant. This problem can be solved using finite-difference approximation in two ways: one known as the explicit method and the other called the implicit method.

11.5.1 Explicit Method

To derive the finite-difference equations, the grid shown in Fig. 11.12 is considered. The length along the x-axis (l) is divided into m equal parts of width $\Delta x = \frac{l}{m}$ each, and the time axis is divided into n equal parts of width Δt each, so that

$$x_0 = 0, x_1 = \Delta x, \ldots, x_i = i\Delta x, \ldots, x_m = m\Delta x = l, \tag{11.59}$$

$$t_0 = 0, t_1 = \Delta t, \ldots, t_j = j\Delta t, \ldots, t_n = n\Delta t = T, \tag{11.60}$$

where T is the maximum time used to find the solution. The grid points are denoted as (i, j) with coordinates (x_i, t_j). The derivatives appearing in Eq. (11.56) can be approximated at grid point (i, j) by central-difference formulas as

$$\left.\frac{\partial^2\phi}{\partial x^2}\right|_{i,j} = \frac{\partial^2\phi}{\partial x^2}(x_i, t_j) = \frac{\phi_{i-1,j} - 2\phi_{i,j} + \phi_{i+1,j}}{\Delta x^2} \tag{11.61}$$

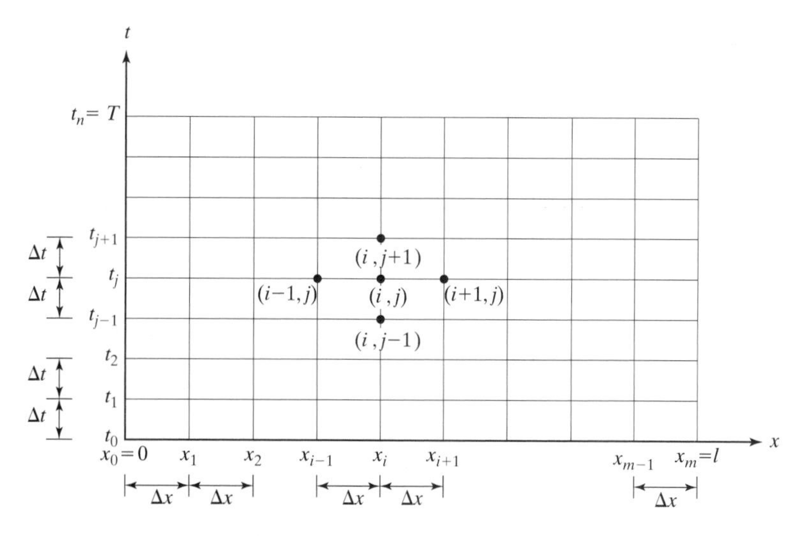

Figure 11.12 Finite-difference grid for Eq. (11.56).

and

$$\left.\frac{\partial \phi}{\partial t}\right|_{i,j} = \frac{\partial \phi}{\partial t}(x_i, t_j) = \frac{\phi_{i,j+1} - \phi_{i,j-1}}{2\Delta t}, \tag{11.62}$$

where Eqs. (11.61) and (11.62) have accuracy of $O(\Delta x^2)$ and $O(\Delta t^2)$, respectively. Substituting Eqs. (11.61) and (11.62) into Eq. (11.56) and rearranging terms yields

$$\phi_{i,j+1} = \phi_{i,j-1} + \frac{2\alpha^2 \Delta t}{\Delta x^2}(\phi_{i+1,j} - 2\phi_{i,j} + \phi_{i-1,j}), \tag{11.63}$$

which has an accuracy $O\left(\Delta x^2 + \Delta t^2\right)$. Equation (11.63) permits the calculation of the dependent variable at the next time step from the known values at the current and previous time steps. Thus, once the boundary and the initial conditions are specified, we can compute the following:

$$\phi_{0,j} = f_j = f(t_j); j = 1, 2, \ldots, n, \tag{11.64}$$

$$\phi_{m,j} = g_j = g(t_j); j = 1, 2, \ldots, n, \tag{11.65}$$

and

$$\phi_{i,0} = h_i = h(x_i); i = 1, 2, \ldots, m. \tag{11.66}$$

Thus, the solution can be generated using Eq. (11.63). If the initial and boundary conditions given by Eqs. (11.64) through (11.66) do not match at the corner grid points (0,0) and (m,0), $\phi(x, t)$ will be discontinuous at these corners. We can use the average value as an approximation. Equation (11.63) is known as the explicit formula and is found to be unstable, because of the negative term on the right-hand side [11.4]. In general, when the finite-difference equation is written with all the known values on the right-hand side as

$$\phi_{i,j+1} = p\phi_{i+1,j} + q\phi_{i,j} + r\phi_{i-1,j}, \tag{11.67}$$

the numerical procedure can be shown to be stable if p, q, and r are positive and $p+q+r \leq 1$. The instability problem associated with Eq. (11.63) can be eliminated by using a forward-difference formula for $\frac{\partial \phi}{\partial t}$ in Eq. (11.56) as

$$\left.\frac{\partial \phi}{\partial t}\right|_{i,j} = \frac{\partial \phi}{\partial t}(x_i, t_j) = \frac{\phi_{i,j+1} - \phi_{i,j}}{\Delta t}, \tag{11.68}$$

which has an accuracy of only $O(\Delta t)$. Substituting Eqs. (11.61) and (11.68) into Eq. (11.56) gives

$$\phi_{i,j+1} = \left(\frac{\alpha^2 \Delta t}{\Delta x^2}\right)\phi_{i+1,j} + \left(1 - \frac{2\alpha^2 \Delta t}{\Delta x^2}\right)\phi_{i,j} + \left(\frac{\alpha^2 \Delta t}{\Delta x^2}\right)\phi_{i-1,j}, \tag{11.69}$$

which has an accuracy of $O(\Delta x^2 + \Delta t)$, or an accuracy of $O(\Delta x^2)$ in the x-direction and $O(\Delta t)$ in t-direction. It can be seen from Eq. (11.69) that the solution will be stable if

$$1 - \frac{2\alpha^2 \Delta t}{\Delta x^2} \geq 0,$$

or

$$\frac{\alpha^2 \Delta t}{\Delta x^2} \leq \frac{1}{2}. \tag{11.70}$$

Thus, Equation (11.70) determines the relationship between Δx and Δt for a stable solution. If the equality relation is used, Eq. (11.70) gives

$$\Delta t = \frac{1}{2\alpha^2}\Delta x^2 \tag{11.71}$$

and Eq. (11.69) reduces to

$$\phi_{i,j+1} = \frac{1}{2}(\phi_{i+1,j} + \phi_{i-1,j}). \tag{11.72}$$

Note that the accuracy of Eq. (11.69) or Eq. (11.72) in the t-direction is reduced compared with that of Eq. (11.63). However, the advantage of ensuring the stability of solution outweighs the loss of accuracy.

11.5.2 Implicit Method

The explicit method has the following disadvantages:

(1) To ensure the stability of solution, the requirement

$$0 < \frac{\alpha^2 \Delta t}{\Delta x^2} \leq \frac{1}{2} \tag{11.73}$$

imposes a restriction on the step sizes Δt and Δx. This might require an excessive computational effort for problems extending over large values of time.

(2) Since the expression for $\phi_{i,j+1}$ involves only $\phi_{i+1,j}$, $\phi_{i,j}$, and $\phi_{i-1,j}$, a typical value at the final time, $\phi_{i,n}$, depends only on $\phi_{i+1,n-1}$, $\phi_{i,n-1}$, and $\phi_{i-1,n-1}$. This implies that only the values of ϕ at the grid points shown by the dark dots in Fig. 11.13 influence the value of $\phi_{i,n}$, but not the values of ϕ at the grid points lying in the triangles DEF and DGH. However, theoretically, the solution of the partial differential equation, $\phi(x, y)$, should depend on all values of ϕ, including the values of ϕ at the grid points lying in the triangles DEF and DGH.

Next, we consider the implicit method, which overcomes the aforementioned disadvantages, to solve parabolic partial differential equations. The method involves approximating the second derivative, $\frac{\partial^2 \phi}{\partial x^2}\big|_{i,j}$, by a finite-difference formula involving the value of ϕ at an advanced time (t_{j+1}). Using the grid shown in Fig. 11.14, the partial derivative at the middle grid point, $\frac{\partial \phi}{\partial t}\big|_{i,j+\frac{1}{2}}$, is computed using a central-difference formula:

$$\left.\frac{\partial \phi}{\partial t}\right|_{i,j+\frac{1}{2}} = \frac{\phi_{i,j+1} - \phi_{i,j}}{\Delta t}. \tag{11.74}$$

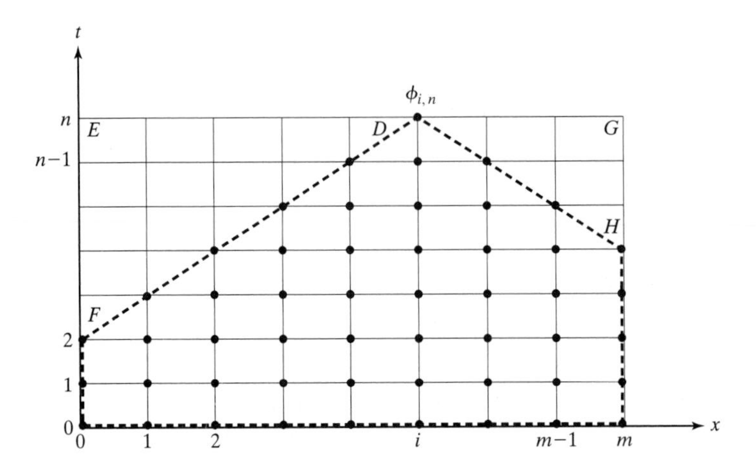

Figure 11.13 Grid points for Eq. (11.72).

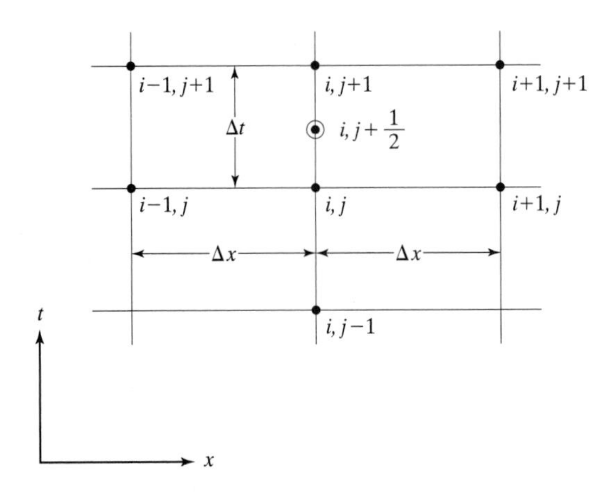

Figure 11.14 Grid points for Eq. (11.74).

Further, the second partial derivative of ϕ with respect to x at the middle grid point $(i, j + \frac{1}{2})$ is computed as a weighted average of the central-difference values of $\frac{\partial^2\phi}{\partial x^2}$ at the grid points (i, j) and $(i, j + 1)$ by

$$\left.\frac{\partial^2\phi}{\partial x^2}\right|_{i,j+\frac{1}{2}} = \theta\left.\frac{\partial^2\phi}{\partial x^2}\right|_{i,j+1} + (1-\theta)\left.\frac{\partial^2\phi}{\partial x^2}\right|_{i,j}, \tag{11.75}$$

where

$$\left.\frac{\partial^2\phi}{\partial x^2}\right|_{i,j} = \frac{\phi_{i+1,j} - 2\phi_{i,j} + \phi_{i-1,j}}{\Delta x^2}, \tag{11.76}$$

and

$$\left.\frac{\partial^2\phi}{\partial x^2}\right|_{i,j+1} = \frac{\phi_{i+1,j+1} - 2\phi_{i,j+1} + \phi_{i-1,j+1}}{\Delta x^2}, \tag{11.77}$$

and θ denotes a weighting parameter in the range, $0 \le \theta \le 1$. The use of Eqs. (11.74) through (11.77) in Eq. (11.56) gives the weighted implicit finite-difference approximation of Eq. (11.56) at the middle grid point $(i, j + \frac{1}{2})$ as

$$\theta\alpha^2 \left\{ \frac{1}{\Delta x^2} \left(\phi_{i+1,j+1} - 2[1 + \frac{\Delta x^2}{\theta\alpha^2 \Delta t}]\phi_{i,j+1} + \phi_{i-1,j+1} \right) \right\}$$

$$= -(1-\theta)\alpha^2 \left\{ \frac{1}{\Delta x^2} \left(\phi_{i+1,j} - 2[1 - \frac{\Delta x^2}{(1-\theta)\alpha^2 \Delta t}]\phi_{i,j} + \phi_{i-1,j} \right) \right\}. \quad (11.78)$$

This equation is called the *variable-weighted implicit formula*, because it involves the variable-weighting factor θ and also more than one unknown at the time step $j+1$. It can be seen that Eq. (11.78) reduces to the explicit formula, Eq. (11.69), for $\theta = 0$. Equation (11.78) gives the backward-implicit formula for $\theta = 1$ as

$$-\lambda\phi_{i-1,j+1} + (1+2\lambda)\phi_{i,j+1} - \lambda\phi_{i+1,j+1} = \phi_{i,j}, \quad (11.79)$$

where

$$\lambda = \frac{\alpha^2 \Delta t}{\Delta x^2}. \quad (11.80)$$

It can be shown that this method is stable for all values of λ. The boundary conditions given by Eqs. (11.64) through (11.66) are still valid. At any time step, Eq. (11.79) can be written for each grid point $i (1 \le i \le m-1)$, which result in $m-1$ simultaneous equations in the $m-1$ unknowns, $\phi_{i,j+1}$. Noting that the boundary conditions give $\phi_{0,j+1} = f(t_{j+1})$ and $\phi_{m,j+1} = g(t_{j+1})$, we see that Eq. (11.79) can be written for $i = 1, 2, \ldots, m-1$ to obtain

$$(1+2\lambda)\phi_{1,j+1} - \lambda\phi_{2,j+1} = \phi_{1,j} + \lambda f(t_{j+1}),$$

$$-\lambda\phi_{i-1,j+1} + (1+2\lambda)\phi_{i.j+1} - \lambda\phi_{i+1,j+1} = \phi_{i,j}; i = 2, 3, \ldots, m-2, \quad (11.81)$$

and

$$-\lambda\phi_{m-2,j+1} + (1+2\lambda)\phi_{m-1,j+1} = \phi_{m-1,j} + \lambda g(t_{j+1}).$$

These equations can be written in a compact form as

$$b_1\beta_1 + c_1\beta_2 = d_1$$

$$a_2\beta_1 + b_2\beta_2 + c_2\beta_3 = d_2$$

$$a_3\beta_2 + b_3\beta_3 + c_3\beta_4 = d_3$$

$$\ldots$$

$$\ldots$$

$$a_i\beta_{i-1} + b_i\beta_i + c_i\beta_{i+1} = d_i$$

$$\ldots$$

$$\ldots$$

$$a_{m-2}\beta_{m-3} + b_{m-2}\beta_{m-2} + c_{m-2}\beta_{m-1} = d_{m-2}$$
$$a_{m-1}\beta_{m-2} + b_{m-1}\beta_{m-1} = d_{m-1}, \tag{11.82}$$

where

$$a_2 = a_3 = \cdots = a_{m-1} = -\lambda, \tag{11.83}$$

$$b_1 = b_2 = \cdots = b_{m-1} = (1 + 2\lambda), \tag{11.84}$$

$$c_1 = c_2 = \cdots = c_{m-2} = -\lambda, \tag{11.85}$$

$$d_1 = \phi_{1,j} + \lambda f(t_{j+1}), \tag{11.86}$$

$$d_k = \phi_{k,j}; k = 2, 3, \ldots, m-2, \tag{11.87}$$

$$d_{m-1} = \phi_{m-1,j} + \lambda g(t_{j+1}), \tag{11.88}$$

and

$$\beta_k = \phi_{k,j+1}; k = 1, 2, \ldots, m-1. \tag{11.89}$$

Since Eqs. (11.82) represent a system of $m-1$ linear algebraic equations in tridiagonal form, their solution can be readily found using the methods described in Chapter 3.

▶Example 11.7

A metal rod of length 1 m is initially at 70° C. The steady-state temperatures of the left and right ends of the rod are given by 50° C and 20° C, respectively. Using $\alpha^2 = 0.1$ m^2/ min, $\Delta x = 0.2$ m and $\Delta t = 0.1$ min, determine the temperature distribution in the rod for $0 \le t \le 0.3$ min.

Solution

With $\Delta x = 0.2$ m and $\Delta t = 0.1$ min, the finite-difference grid is set up over the solution region as shown in Fig. 11.15. Since the temperature, $T(x, t)$, is 70° C at $t = 0$ and the boundary conditions state the temperatures at the left and right ends of the rod as 50° C and 20° C, respectively, we assume the values of T_{00} and T_{50} as the averages of the respective initial and steady-state values:

$$T_{00} = \frac{70 + 50}{2} = 60^\circ \text{ C}, T_{50} = \frac{70 + 20}{2} = 45^\circ \text{ C}.$$

We assume that the boundary conditions are valid for $t > 0$, so that $T_0 j = 50^\circ$ C and $T_{5j} = 20^\circ$ C for $j = 1, 2, 3$, and 4. The value of

$$\frac{\alpha^2 \Delta t}{(\Delta x)^2} = \frac{(0.1)(0.1)}{(0.2)^2} = \frac{1}{4} < \frac{1}{2}$$

indicates that the stability of the solution is assured. To find the desired solution, Eq. (11.69) is used:

$$T_{i,j+1} = 0.25T_{i,j} + 0.25T_{i-1,j}. \tag{E1}$$

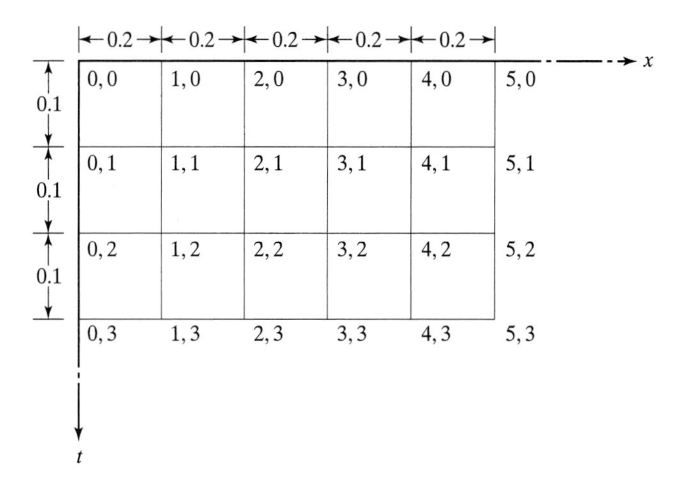

Figure 11.15 Solution region of metal rod.

For $j = 0$, Eq. (E1) becomes

$$T_{i,0} = 0.25T_{i+1,0} + 0.5T_{i,0} + 0.25T_{i-1,0}, \tag{E2}$$

with $T_{00} = 60$, $T_{50} = 45$, $T_{10} = T_{20} = T_{30} = T_{40} = 70$, $T_{0j} = 50$, and $T_{5j} = 20 (j = 1, 2, 3, 4)$. With $i = 1$, Eq. (E2) becomes

$$T_{11} = 0.25T_{20} + 0.5T_{10} + 0.25T_{00} = 0.25(70) + 0.5(70) + 0.25(60) = 67.5^\circ \text{ C}.$$

With $i = 2$, Eq. (E2) gives

$$T_{21} = 0.25T_{30} + 0.5T_{20} + 0.25T_{10} = 0.25(70) + 0.5(70) + 0.25(70) = 70^\circ \text{ C}.$$

With $i = 3$, Eq. (E2) yields

$$T_{31} = 0.25T_{40} + 0.5T_{30} + 0.25T_{20} = 0.25(70) + 0.5(70) + 0.25(70) = 70^\circ \text{ C}.$$

With $i = 4$, Eq. (E2) gives

$$T_{41} = 0.25T_{50} + 0.5T_{40} + 0.25T_{30} = 0.25(45) + 0.5(70) + 0.25(70) = 63.75^\circ \text{ C}.$$

For $j = 1$, Eq. (E1) becomes

$$T_{i,2} = 0.25T_{i+1,1} + 0.5T_{i,1} + 0.25T_{i-1,1}, \tag{E3}$$

with $T_{02} = 50$ and $T_{52} = 20$. Using $i = 1, 2, 3$, and 4 in Eq. (E3), we obtain $T_{12} = 63.75^\circ$ C, $T_{22} = 69.375^\circ$ C, $T_{32} = 68.4375^\circ$ C, and $T_{42} = 54.375^\circ$ C, respectively. For $j = 2$, Eq. (E1) yields

$$T_{i,3} = 0.25T_{i+1,2} + 0.5T_{i,2} + 0.25T_{i-1,2}, \tag{E4}$$

with $T_{03} = 50$ and $T_{53} = 20$. Using $i = 1, 2, 3$, and 4 in Eq. (E4), we obtain $T_{13} = 61.71875^\circ$ C, $T_{23} = 67.734375^\circ$ C, $T_{33} = 65.15625^\circ$ C, and $T_{43} = 49.296875^\circ$ C, respectively. ◀

11.6 Crank–Nicholson Method

For $\theta = \frac{1}{2}$, Eq. (11.78) yields the following formula, known as the Crank–Nicholson implicit formula:

$$-\lambda\phi_{i-1,j+1} + 2(1+\lambda)\phi_{i,j+1} - \lambda\phi_{i+1,j+1} = \lambda\phi_{i-1,j} + 2(1-\lambda)\phi_{i,j} + \lambda\phi_{i+1,j}. \quad (11.90)$$

Here λ is given by Eq. (11.80). It can be shown that this method is unconditionally stable (regardless of the time step Δt) and has an accuracy of $O(\Delta x^2 + \Delta t^2)$. Equation (11.90) can be seen to be similar to Eq. (11.79) and hence results in a set of linear equations in tridiagonal form. It has been observed [11.5] that if $\theta = (\frac{6\lambda-1}{12\lambda})$ is used in Eq. (11.78), the accuracy of the method improves to $O(\Delta x^4)$ and further if $\lambda = \frac{1}{\sqrt{20}}$ is used, the accuracy of the method improves to $O(\Delta x^6)$. However, it is to be noted that these improvements may not be realized if the partial differential equation is other than Eq. (11.56).

▶Example 11.8

Solve Example 11.7 using the Crank–Nicholson method with $\Delta x = 0.2$ m and $\Delta t = 0.3$ min.

Solution

With $\Delta x = 0.2$ and $\Delta t = 0.3$, the value of λ can be seen to be

$$\lambda = \frac{\alpha^2 \Delta t}{(\Delta x)^2} = \frac{(0.1)(0.3)}{(0.2)^2} = 0.75,$$

which should not effect the stability of the Crank–Nicholson (implicit) method. By using the initial conditions, $T_{i0} = 70$ for $i = 0, 1, 2, 3, 4, 5$ and the boundary conditions as $T_{01} = 50$ and $T_{51} = 20$, Eq. (11.90) can be written, for $j = 0$, as

$$0.75T_{i-1,0} + 0.5T_{i,0} + 0.75T_{i+1,0} + 0.75T_{i-1,1} - 3.5T_{i,1} + 0.75T_{i+1,1}. \quad \text{(E1)}$$

For $i = 1$, Eq. (E1) gives

$$0.75T_{00} + 0.5T_{10} + 0.75T_{20} + 0.75T_{01} - 3.5T_{11} + 0.75T_{21} = 0,$$

or

$$177.5 - 3.5T_{11} + 0.75T_{21} = 0. \quad \text{(E2)}$$

For $i = 2$, Eq. (E1) yields

$$0.75T_{10} + 0.5T_{20} + 0.75T_{30} + 0.75T_{11} - 3.5T_{21} + 0.75T_{31} = 0,$$

or

$$140.0 + 0.75T_{11} - 3.5T_{21} + 0.75T_{31} = 0. \quad \text{(E3)}$$

For $i = 3$, Eq. (E1) becomes

$$0.75T_{20} + 0.5T_{30} + 0.75T_{40} + 0.75T_{21} - 3.5T_{31} + 0.75T_{41} = 0,$$

or

$$140.0 + 0.75T_{21} - 3.5T_{31} + 0.75T_{41} = 0. \quad \text{(E4)}$$

For $i = 4$, Eq. (E1) gives

$$0.75T_{30} + 0.5T_{40} + 0.75T_{50} + 0.75T_{31} - 3.5T_{41} + 0.75T_{51} = 0,$$

or

$$155.0 + 0.75T_{31} - 3.5T_{41} = 0. \quad \text{(E5)}$$

Equations (E2) through (E5) can be expressed in matrix form as

$$\begin{bmatrix} 3.5 & -0.75 & 0 & 0 \\ -0.75 & 3.5 & -0.75 & 0 \\ 0 & -0.75 & 3.5 & -0.75 \\ 0 & 0 & -0.75 & 3.5 \end{bmatrix} \begin{Bmatrix} T_{11} \\ T_{21} \\ T_{31} \\ T_{41} \end{Bmatrix} = \begin{Bmatrix} 177.5 \\ 140.0 \\ 140.0 \\ 155.0 \end{Bmatrix}. \quad \text{(E6)}$$

The solution of Eq. (E6) gives the desired solution as $T_{11} = 65.3751$, $T_{21} = 68.4171$, $T_{31} = 67.2381$, and $T_{41} = 58.6939$. ◀

11.7 Method of Lines

In the method of lines, the partial differential equation is converted to a system of first-order ordinary differential equations by approximating only the spatial derivative, $\frac{\partial^2 \phi}{\partial x^2}(x, t)$, using a finite-difference formula. Thus, the time derivative, $\frac{\partial \phi}{\partial t}(x, t)$, remains unchanged in the equation

$$\alpha^2 \frac{\partial^2 \phi}{\partial x^2}(x, t) = \frac{\partial \phi}{\partial t}(x, t). \quad (11.91)$$

Using a central-difference formula at the grid point $i(x = x_i)$, Eq. (11.91) can be expressed as

$$\frac{\partial \phi}{\partial t}(x_i, t) = \frac{d\phi_i(t)}{dt} = \frac{\alpha^2}{\Delta x^2}[\phi_{i-1}(t) - 2\phi_i(t) + \phi_{i+1}(t)]; i = 1, 2, \ldots, m-1. \quad (11.92)$$

The complete system of $m - 1$ ordinary differential equations given by Eq. (11.92) can be written as follows

$$\frac{d\phi_1}{dt} = \frac{\alpha^2}{\Delta x^2}(\phi_0 - 2\phi_1 + \phi_2), \quad (11.93)$$

$$\frac{d\phi_2}{dt} = \frac{\alpha^2}{\Delta x^2}(\phi_1 - 2\phi_2 + \phi_3), \quad (11.94)$$

$$\vdots$$

$$\frac{d\phi_i}{dt} = \frac{\alpha^2}{\Delta x^2}(\phi_{i-1} - 2\phi_i + \phi_{i+1}), \tag{11.95}$$

$$\vdots$$

$$\frac{\phi_{m-1}}{dt} = \frac{\alpha^2}{\Delta x^2}(\phi_{m-2} - 2\phi_{m-1} + \phi_m). \tag{11.96}$$

The known boundary conditions are to be used to modify the equations corresponding to the boundary grid points, namely, Eqs. (11.93) and (11.96). If the boundary condition at $x = 0$ is specified as

$$\phi_0 = \phi(x = 0) = u_0 = \text{constant (Dirichlet condition); for } t > 0, \tag{11.97}$$

Eq. (11.93) is modified to

$$\frac{d\phi_1}{dt} = \frac{\alpha^2}{\Delta x^2}(u_0 - 2\phi_1 + \phi_2). \tag{11.98}$$

If the boundary condition at $x = 0$ is specified as

$$\frac{\partial \phi_0}{\partial x} = \frac{\partial \phi}{\partial x}(x = 0, t) = v_0 = \text{ constant (Neumann condition); for } t > 0, \tag{11.99}$$

the partial derivative of Eq. (11.99) is replaced by a forward-difference approximation as

$$\frac{\phi_1 - \phi_0}{\Delta x} = v_0,$$

or

$$\phi_0 = \phi_1 - v_0 \Delta x. \tag{11.100}$$

Equation (11.93) is now modified, using Eq. (11.100), to

$$\frac{d\phi_1}{dt} = \frac{\alpha^2}{\Delta x^2}(-v_0 \Delta x - \phi_1 + \phi_2). \tag{11.101}$$

Similarly, if the boundary condition at $x = 0$ is specified as

$$\frac{\partial \phi}{\partial x}(x = 0, t) + r\phi(x = 0, t) = \frac{\partial \phi_0}{\partial x} + r\phi_0 = w_0$$

$$= \text{constant (mixed condition); for } t > 0, \tag{11.102}$$

Eq. (11.102) is rewritten as

$$\frac{\phi_1 - \phi_0}{\Delta x} + r\phi_0 = w_0,$$

or

$$\phi_0 = \left(\frac{w_0 \Delta x - \phi_1}{r \Delta x - 1} \right). \tag{11.103}$$

This equation can be used to modify Eq. (11.93) as follows:

$$\frac{d\phi_1}{dt} = \frac{\alpha^2}{\Delta x^2} \left[\frac{w_0 \Delta x}{r \Delta x - 1} - \left(2 + \frac{1}{r \Delta x - 1} \right) \phi_1 + \phi_2 \right]. \tag{11.104}$$

The specified boundary conditions at $x = x_m = 1$ can be used to modify Eq. (11.96) in a similar manner. Once the boundary conditions are incorporated, the resulting system of $m - 1$ simultaneous first-order ordinary differential equations can be integrated along the time-axis using the specified initial conditions. The methods described in Chapter 9 can be used for this purpose. This integration is shown schematically in Fig. 11.16.

▶Example 11.9

A metal rod of length 1 m has an initial temperature of 70° C. The steady state temperatures of left and right ends of the rod are given by 50° C and 20° C, respectively. Using $\alpha^2 = 0.1\ \text{m}^2/\text{min}$ and $\Delta x = 0.2$ m, derive the equations for determining the temperature distribution in the rod for $0 \leq t \leq 0.3$ min using the method of lines.

Solution

The finite-difference grid along the x-axis is set up as shown in Fig. 11.17. Using the central-difference formula for the spatial derivative at the grid point i, the governing equation can be expressed as

$$\frac{\alpha^2}{(\Delta x)^2}(T_{i-1} - 2T_i + T_{i+1}) = \frac{dT_i}{dt}; i = 1, 2, 3, 4, \tag{a}$$

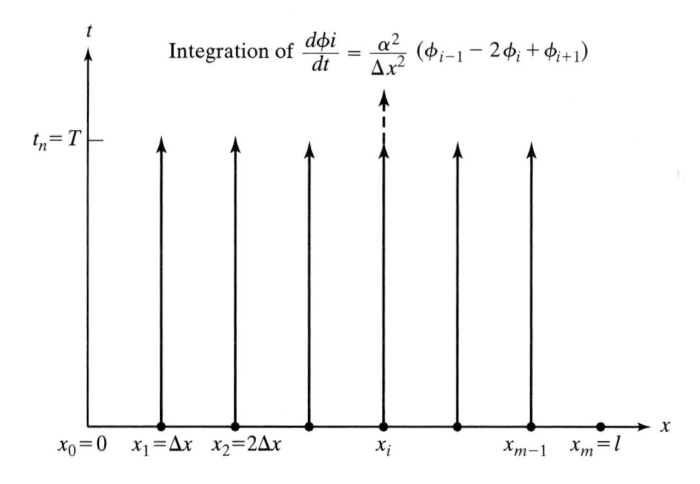

Figure 11.16 Integration in method of lines.

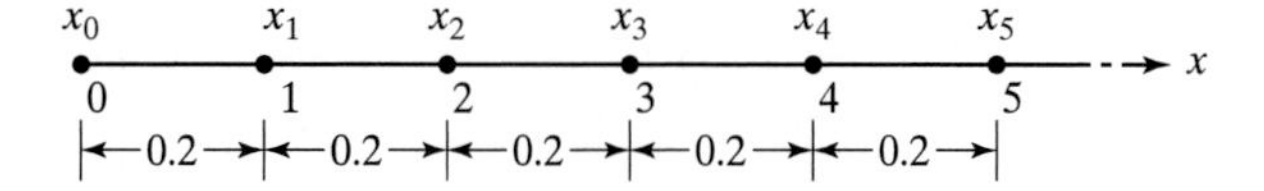

Figure 11.17 Finite difference grid along x-axis.

which can be rewritten as

$$\frac{dT_1}{dt} = 2.5(T_0 - 2T_1 + T_2), \tag{b}$$

$$\frac{dT_2}{dt} = 2.5(T_1 - 2T_2 + T_3), \tag{c}$$

$$\frac{dT_3}{dt} = 2.5(T_2 - 2T_3 + T_4), \tag{d}$$

$$\frac{dT_4}{dt} = 2.5(T_3 - 2T_4 + T_5). \tag{e}$$

When the boundary conditions $T_0 = 50$ and $T_5 = 20$ are substituted, Eqs. (b) through (e) reduce to

$$\frac{dT_1}{dt} = 2.5(50 - 2T_1 + T_2), \tag{f}$$

$$\frac{dT_2}{dt} = 2.5(T_1 - 2T_2 + T_3), \tag{g}$$

$$\frac{dT_3}{dt} = 2.5(T_2 - 2T_3 + T_4), \tag{h}$$

and

$$\frac{dT_4}{dt} = 2.5(T_3 - 2T_4 + T_5). \tag{i}$$

The system of Eqs. (f) through (i) can be solved using the initial conditions, $T_1(0) = T_2(0) = T_3(0) = T_4(0) = 70$. ◀

11.8 Two-Dimensional Parabolic Problems

The methods discussed in Section 11.5 can be extended to two-dimensional parabolic partial differential equations, but each has certain limitations. For example, consider the following equation:

$$\frac{\partial^2\phi(x, y, t)}{\partial x^2} + \frac{\partial^2\phi(x, y, t)}{\partial y^2} = \frac{\partial\phi(x, y, t)}{\partial t}. \tag{11.105}$$

11.8.1 Explicit Formula

By denoting the step sizes along the x-, y-, and t-directions as Δx, Δy, and Δt, and a general grid point with coordinates $(x_i, y_j, t_k) = (i\Delta x, j\Delta y, k\Delta t)$ as (i, j, k), the space derivatives are approximated by central-difference formulas and the time derivative by a forward-difference formula. This yields the finite-difference approximation of Eq. (11.105) at the grid point (i, j, k):

$$\frac{\phi_{i+1,j,k} - 2\phi_{i,j,k} + \phi_{i-1,j,k}}{\Delta x^2} + \frac{\phi_{i,j+1,k} - 2\phi_{i,j,k} + \phi_{i,j-1,k}}{\Delta y^2} = \frac{\phi_{i,j,k+1} - \phi_{i,j,k}}{\Delta t}. \tag{11.106}$$

Although the solution of this explicit formula (Eq. (11.106)) is straight forward, the step sizes have to satisfy the relation (similar to Eq. (11.70)):

$$\Delta t\left(\frac{1}{\Delta x^2}+\frac{1}{\Delta y^2}\right)\leq\frac{1}{2}. \tag{11.107}$$

11.8.2 Implicit Formula

The backward-implicit formula for the solution of Eq. (11.105) can be derived by proceeding as in the case of Eq. (11.79). The basic approximation is given by

$$\frac{\phi_{i,j,k+1}-\phi_{i,j,k}}{\Delta t}=\left.\frac{\partial^2\phi}{\partial x^2}\right|_{i,j,k+1}+\left.\frac{\partial^2\phi}{\partial y^2}\right|_{i,j,k+1}. \tag{11.108}$$

Using $\Delta x=\Delta y$, and central-difference formulas for the second partial derivatives of ϕ in Eq. (11.108), we can obtain an equation, similar to Eq. (11.79):

$$-\lambda\phi_{i-1,j,k+1}-\lambda\phi_{i,j-1,k+1}+(1+4\lambda)\phi_{i,j,k+1}-\lambda\phi_{i+1,j,k+1}-\lambda\phi_{i,j+1,k+1}=\phi_{i,j,k}. \tag{11.109}$$

Here

$$\lambda=\frac{\Delta t}{\Delta x^2}. \tag{11.110}$$

This method can be shown to be conditionally stable (for all values of λ). It can be seen that there are five unknowns per equation in Eq. (11.109), so the efficient solution techniques of tridiagonal systems cannot be applied. Thus, the method requires the simultaneous solution of Eq. (11.109) for all the grid points in the whole domain in every time step, which requires considerable amount of computation in every time step.

The Crank–Nicholson implicit formula, similar to Eq. (11.90), can also be derived for the solution of Eq. (11.105). However, the method has limitations similar to those of the backward-implicit formula already discussed. The following method, known as the alternating-direction implicit method, avoids the disadvantages of the explicit and implicit methods.

11.8.3 Alternating-Direction Implicit Method

The alternating-direction implicit method (ADI), proposed by Peaceman and Rachford [11.6], is a widely used method for the solution of parabolic partial differential equations in two spatial variables. It is unconditionally stable, yet the solution in each step involves only a tridiagonal system of equations along each grid line. In this method, a central-difference approximation for $\frac{\partial^2\phi}{\partial x^2}$ is used at the beginning of the time step, and a similar expression for $\frac{\partial^2\phi}{\partial y^2}$ is used at the end of the time step. The bias in the procedure is balanced by reversing the order of approximation for $\frac{\partial^2\phi}{\partial x^2}$ and $\frac{\partial^2\phi}{\partial y^2}$ in the next time step. Thus, the finite-difference equations for Eq. (11.105) can be stated as follows:

$$\frac{\phi_{i,j,k+1} - \phi_{i,j,k}}{\Delta t} = \left\{ \underbrace{\frac{\phi_{i+1,j,k} - 2\phi_{i,j,k} + \phi_{i-1,j,k}}{\Delta x^2}}_{\text{at beginning of time step, } t_k} + \underbrace{\frac{\phi_{i,j+1,k+1} - 2\phi_{i,j,k+1} + \phi_{i,j-1,k+1}}{\Delta y^2}}_{\text{at end of time step, } t_{k+1}} \right\} \tag{11.111}$$

and

$$\frac{\phi_{i,j,k+2} - \phi_{i,j,k+1}}{\Delta t} = \left\{ \underbrace{\frac{\phi_{i+1,j,k+2} - 2\phi_{i,j,k+2} + \phi_{i-1,j,k+2}}{\Delta x^2}}_{\text{at end of time step, } t_{k+2}} + \underbrace{\frac{\phi_{i,j+1,k+1} - 2\phi_{i,j,k+1} + \phi_{i,j-1,k+1}}{\Delta y^2}}_{\text{at beginning of time step, } t_{k+1}} \right\}. \tag{11.112}$$

The following example is considered to illustrate the ADI method:

▶Example 11.10

An aluminum plate of size 0.3m × 0.3 m is initially at the temperature 30° C. If two adjacent sides of the plate are suddenly brought to 120° C and maintained at that temperature, derive the equations necessary for the determination of the time variation of temperature in the plate using the ADI method.

Data: $k = 236$ W/m-K, $c = 900$ J/kg-K, $\rho = 2700$ kg/m^3, $\Delta x = \Delta y = 0.1$ m.

Solution

The finite-difference grid corresponding to $\Delta x = \Delta y = 0.1$ is shown in Fig. 11.18. The governing equation for the transient heat conduction in two dimensions is given by

$$\frac{\partial T}{\partial t} = \frac{k}{\rho c}\left(\frac{\partial^2 T}{\partial x^2} + \frac{\partial^2 T}{\partial y^2}\right). \tag{a}$$

We use a forward-difference approximation for $\frac{\partial T}{\partial t}$ and central-difference approximations for $\frac{\partial^2 T}{\partial x^2}$ and $\frac{\partial^2 T}{\partial y^2}$ at the beginning and end of the time step, respectively, in time step k. In the next time step $k+1$, we use the central-difference approximations of $\frac{\partial^2 T}{\partial x^2}$ and $\frac{\partial^2 T}{\partial y^2}$ at the end and beginning. Using

$$s = \frac{k\Delta t}{\rho c(\Delta x)^2}, \tag{b}$$

we can write Eq. (a)

$$T_{i,j,k+1} - T_{i,j,k} = s\{T_{i+1,j,k} - 2T_{i,j,k} + T_{i-1,j,k} + T_{i,j+1,k+1} - 2T_{i,j,k+1} + T_{i,j-1,k+1}\} \tag{c}$$

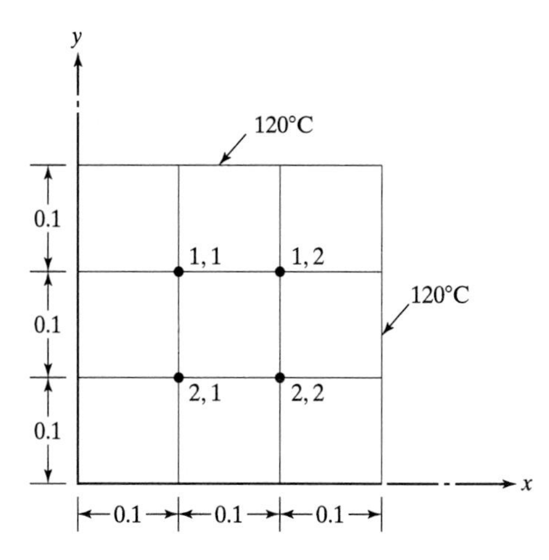

Figure 11.18 Finite difference grid in xy-plane.

and

$$T_{i,j,k+2} - T_{i,j,k+1} = s\{T_{i+1,j,k+2} \\ -2T_{i,j,k+2} + T_{i-1,j,k+2} + T_{i,j+1,k+1} - 2T_{i,j,k+1} + T_{i,j-1,k+1}\}. \quad \text{(d)}$$

Equations (c) and (d) can be rewritten as

$$T_{i,j,k}(-1+2s) + T_{i,j,k+1}(1+2s) \\ +T_{i+1,j,k}(-s) + T_{i-1,j,k}(-s) + T_{i,j+1,k+1}(-s) + T_{i,j-1,k+1}(-s) = 0 \quad \text{(e)}$$

and

$$T_{i,j,k+2}(1+2s) + T_{i,j,k+1}(-1+2s) \\ +T_{i+1,j,k+2}(-s) + T_{i-1,j,k+2}(-s) + T_{i,j+1,k+1}(-s) + T_{i,j-1,k+1}(-s) = 0. \quad \text{(f)}$$

Equations (e) and (f) can be used at different time stations ($k = 0, 1, 2, \ldots$) and different spatial grid points ($i, j = 1, 2, \ldots$), and the resulting system of linear algebraic equations can be solved using the known boundary and initial conditions. ◀

11.9 Hyperbolic Equations

Several engineering problems, such as those involving the convective transport of matter, and elastic, electromagnetic and acoustic waves, are governed by hyperbolic partial differential equations. The hyperbolic partial differential equations appear in both first- and second-order forms. For example, the convection equation

$$\frac{\partial \phi}{\partial t} + c\frac{\partial \phi}{\partial x} = 0, \quad (11.113)$$

where c is a constant, indicating the convective velocity denotes a first-order hyperbolic equation. The following equation, known as the general wave equation, represents a second-order hyperbolic equation:

$$\frac{\partial^2 \phi}{\partial t^2} = c^2 \frac{\partial^2 \phi}{\partial x^2}. \quad (11.114)$$

c is a constant, denoting the velocity of wave propagation. Equation (11.114) can be obtained by differentiating Eq. (11.113). To see this, consider the partial differentiation of Eq. (11.113) with respect to time, t:

$$\frac{\partial}{\partial t}\left(\frac{\partial \phi}{\partial t}\right) = \frac{\partial}{\partial t}\left(-c\frac{\partial \phi}{\partial x}\right). \quad (11.115)$$

Interchanging the order of differentiation on the right-hand side of Eq. (11.115) gives

$$\frac{\partial^2 \phi}{\partial t^2} = -c\frac{\partial}{\partial x}\left(\frac{\partial \phi}{\partial t}\right). \quad (11.116)$$

Substituting $\frac{\partial \phi}{\partial t} = -c\frac{\partial \phi}{\partial x}$ from Eq. (11.113), we see that Eq. (11.116) yields

$$\frac{\partial^2 \phi}{\partial t^2} = -c\frac{\partial}{\partial x}\left(-c\frac{\partial \phi}{\partial x}\right) = c^2\frac{\partial^2 \phi}{\partial x^2}, \quad (11.117)$$

which can be seen to be same as Eq. (11.114). This indicates that the methods of solving Eq. (11.113) can also be used for solving Eq. (11.114).

The finite-difference solutions can be used to solve hyperbolic partial differential equations. However, the finite-difference approximations will be good only when the solution is smooth. It can be shown that the solution of hyperbolic partial differential equations can have jumps, and hence the finite difference methods have to be applied with care. Note that the solution of elliptic or parabolic partial differential equations is always continuous in its spatial and time domains.

11.9.1 Finite-Difference Solution of First-Order Equations

Consider the first-order hyperbolic equation

$$\frac{\partial \phi(x,t)}{\partial t} + c\frac{\partial \phi(x,t)}{\partial x} = 0 \quad (11.118)$$

subject to

$$\phi(0,t) = f(t)(\text{boundary condition}) \quad (11.119)$$

and

$$\phi(x,0) = g(x)(\text{initial condition}), \quad (11.120)$$

where c is a constant with $c > 0$. The finite-difference grid is shown in Fig. 11.19. At the grid point (i, j), the time and spatial derivatives of ϕ can be approximated using different finite-difference schemes. Some of the schemes are indicated next.

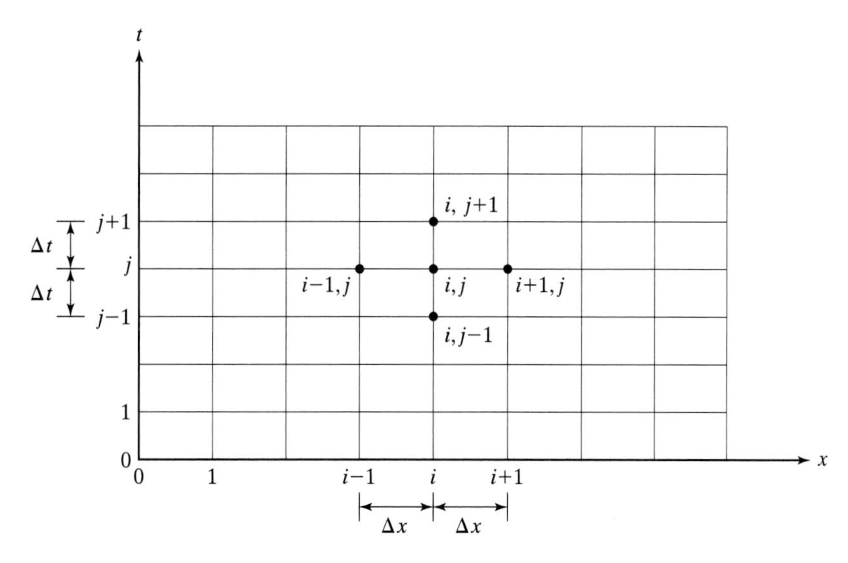

Figure 11.19 Finite-difference grid.

(i) Backward Differences in Space and Forward Differences in Time

The finite-difference approximations of the derivatives of ϕ, in this case, are given by

$$\left.\frac{\partial\phi}{\partial x}\right|_{i,j} = \frac{\phi_{i,j} - \phi_{i-1,j}}{\Delta x} \quad \text{(backward-difference formula)} \tag{11.121}$$

and

$$\left.\frac{\partial\phi}{\partial t}\right|_{i,j} = \frac{\phi_{i,j+1} - \phi_{i,j}}{\Delta t} \quad \text{(forward-difference formula).} \tag{11.122}$$

With these approximations, Eq. (11.118) becomes

$$\frac{\phi_{i,j+1} - \phi_{i,j}}{\Delta t} + c\left(\frac{\phi_{i,j} - \phi_{i-1,j}}{\Delta x}\right) = 0, \tag{11.123}$$

which can be solved for $\phi_{i,j+1}$ as

$$\phi_{i,j+1} = \alpha\phi_{i-1,j} + (1-\alpha)\phi_{i,j}, \tag{11.124}$$

where

$$\alpha = \frac{c\Delta t}{\Delta x}$$

is known as the Courant number. The boundary and initial conditions of Eqs. (11.119) and (11.120) can be incorporated as follows:

When $i = 1$, $\phi_{0,j}$ are found from the boundary condition, Eq. (11.119), as

$$\phi_{0,j} = \phi(0, t_j) = f(t_j). \tag{11.125}$$

Similarly, when $j = 0$, $\phi_{i,0}$ are determined from the initial condition, Eq. (11.120), as

$$\phi_{i,0} = \phi(x_i, 0) = g(x_i). \tag{11.126}$$

Thus, Eq. (11.124) can be used to generate the solution for all the grid points directly.

If $c \le 0$, the derivative $\frac{\partial \phi}{\partial x}$ is approximated by a forward-difference formula instead of a backward-difference formula, so that the finite difference equivalence of Eq. (11.118) becomes:

$$\frac{\phi_{i,j+1} - \phi_{i,j}}{\Delta t} + c\left(\frac{\phi_{i+1,j} - \phi_{i,j}}{\Delta x}\right) = 0, \tag{11.127}$$

or

$$\phi_{i,j+1} = -\alpha\phi_{i+1,j} + (1+\alpha)\phi_{i,j}. \tag{11.128}$$

The boundary and initial conditions can be used to evaluate the values of $\phi_{0,j}$ and $\phi_{i,0}$ in Eq. (11.127). Equation (11.124) or Eq. (11.128) has to be used, depending on the sign of c. Both the equations can be combined into a single equation (which takes care of the sign of c) as

$$\frac{\phi_{i,j+1} - \phi_{i,j}}{\Delta t} + c\left(\frac{\phi_{i+1,j} - \phi_{i-1,j}}{2\Delta x}\right) - \frac{|c|\Delta x}{2}\left(\frac{\phi_{i+1,j} - 2\phi_{i,j} + \phi_{i-1,j}}{\Delta x^2}\right) = 0 \tag{11.129}$$

or

$$\phi_{i,j+1} = (1 - |\alpha|)\phi_{i,j} + \left(-\frac{\alpha}{2} + \frac{|\alpha|}{2}\right)\phi_{i+1,j} + \left(\frac{\alpha}{2} + \frac{|\alpha|}{2}\right)\phi_{i-1,j}. \tag{11.130}$$

The first term in Eq. (11.129) is the forward-difference formula for $\frac{\partial \phi}{\partial t}$, the second term is c times the central-difference formula for $\frac{\partial \phi}{\partial x}$, and the third term is $(\frac{-|c|\Delta x}{2})$ times the central-difference formula for $\frac{\partial^2 \phi}{\partial x^2}$. Thus, although no central-difference formula was used in its derivation, Eq. (11.129) or Eq. (11.130) can be interpreted as using the central-difference formula for $\frac{\partial \phi}{\partial x}$ and artificially adding the third term in Eq. (11.129), which is often termed the numerical viscosity term [11.7]. This method can be shown to be stable for $\alpha \le 1$ [11.7].

(ii) Central Differences in Space and Backward Differences in Time:

According to this scheme, the partial derivatives of ϕ are expressed as

$$\left.\frac{\partial \phi}{\partial x}\right|_{i,j+1} = \frac{\phi_{i+1,j+1} - \phi_{i-1,j+1}}{2\Delta x}. \tag{11.131}$$

and

$$\left.\frac{\partial \phi}{\partial t}\right|_{i,j+1} = \frac{\phi_{i,j+1} - \phi_{i,j}}{\Delta t}. \tag{11.132}$$

Thus, the finite difference equivalence of Eq. (11.118) is given by

$$\frac{\phi_{i,j+1} - \phi_{i,j}}{\Delta t} + c\left(\frac{\phi_{i+1,j+1} - \phi_{i-1,j+1}}{2\Delta x}\right) = 0, \tag{11.133}$$

which, upon rearrangement of terms, yields

$$-\frac{\alpha}{2}\phi_{i-1,j+1} + \phi_{i,j+1} + \frac{\alpha}{2}\phi_{i+1,j+1} = \phi_{i,j}. \tag{11.134}$$

For $i = 1$, the boundary condition, Eq. (11.119), can be used in Eq. (11.134) to obtain

$$\phi_{1,j+1} + \frac{\alpha}{2}\phi_{2,j+1} = \phi_{1,j} + \frac{\alpha}{2}\phi_{0,j+1} = \phi_{1,j} + \frac{\alpha}{2}f(t_{j+1}). \tag{11.135}$$

Note that $\phi_{2,j+1}$ in Eq. (11.135) and $\phi_{i+1,j+1}$ in Eq. (11.134), in general, are not known. They are computed using an extrapolation scheme as [11.8]

$$\phi_{i+1,j+1} = 2\phi_{i,j+1} - \phi_{i-1,j+1}. \tag{11.136}$$

With this, Eq. (11.134) becomes

$$-\frac{\alpha}{2}\phi_{i-1,j+1} + (1+\alpha)\phi_{i,j+1} = \phi_{i,j}; i = 1, 2, \ldots. \tag{11.137}$$

Equation (11.137) yields a tridiagonal system of equations, which can be solved readily. This method can be shown to be unconditionally stable [11.7].

11.9.2 Solution of Second-Order Equations

Consider the second-order hyperbolic partial differential equation

$$\frac{\partial^2\phi(x,t)}{\partial t^2} - \alpha^2\frac{\partial^2\phi(x,t)}{\partial x^2} = 0;\ 0 \le x \le 1;\ 0 \le t \le T \tag{11.138}$$

subject to the conditions

$$\phi(0,t) = 0;\ 0 \le t \le T, \tag{11.139}$$

$$\phi(1,t) = 0;\ 0 \le t \le T, \tag{11.140}$$

$$\phi(x,0) = f(x);\ 0 \le x \le 1, \tag{11.141}$$

and

$$\frac{\partial\phi(x,0)}{\partial t} = g(x); 0 \le x \le 1, \tag{11.142}$$

where α^2 is a constant. A two-dimensional finite-difference grid is assumed as shown in Fig. 11.19, so that a general grid point is assumed to be (i, j) with coordinates $(x_i, t_j) = (i\Delta x, j\Delta t)$, where Δx and Δt are step sizes along the x and t directions, respectively:

$$x_0 = 0, x_1 = \Delta x, \ldots, x_i = i\Delta x, \ldots, x_m = m\Delta x = 1,$$

and

$$t_0 = 0, t_1 = \Delta t, \ldots, t_j = j\Delta t, \ldots.$$

At grid point (i, j), the central-difference formulas are used for $\frac{\partial^2 \phi}{\partial t^2}$ and $\frac{\partial^2 \phi}{\partial x^2}$, so that the finite-difference equivalence of Eq. (11.138) becomes

$$\left(\frac{\phi_{i,j+1} - 2\phi_{i,j} + \phi_{i,j-1}}{\Delta t^2}\right) - \alpha^2 \left(\frac{\phi_{i+1,j} - 2\phi_{i,j} + \phi_{i-1,j}}{\Delta x^2}\right) = 0, \quad (11.143)$$

which has an accuracy of $O(\Delta t^2 + \Delta x^2)$. Equation (11.143) can be restated as

$$\phi_{i,j+1} = 2(1-\lambda)\phi_{i,j} + \lambda(\phi_{i+1,j} + \phi_{i-1,j}) - \phi_{i,j-1}; i = 1, 2, \ldots, m-1; j = 1, 2, \ldots, \quad (11.144)$$

where

$$\lambda = \frac{\alpha^2 \Delta t^2}{\Delta x^2}. \quad (11.145)$$

If $\lambda = 1$ is chosen, Eq. (11.144) simplifies to

$$\phi_{i,j+1} = \phi_{i+1,j} + \phi_{i-1,j} - \phi_{i,j-1}; i = 1, 2, \ldots, m-1; \; j = 1, 2, \ldots. \quad (11.146)$$

It has been shown that the stability of the solution requires that $\lambda \leq 1$ [11.7]. Equation (11.144) can be written in matrix form as

$$\begin{Bmatrix} \phi_{1,j+1} \\ \phi_{2,j+1} \\ \cdot \\ \vdots \\ \cdot \\ \phi_{m-1,j+1} \end{Bmatrix} = \begin{bmatrix} 2(1-\lambda) & \lambda & 0 & 0 & 0 & \cdot & \cdot & 0 \\ \lambda & 2(1-\lambda) & \lambda & 0 & 0 & \cdot & \cdot & 0 \\ 0 & \ddots & \ddots & \ddots & & & & \vdots \\ \vdots & & \ddots & \ddots & \ddots & & & 0 \\ \vdots & & & & \lambda & 2(1-\lambda) & & \lambda \\ 0 & \cdots & \cdots & & 0 & \lambda & & 2(1-\lambda) \end{bmatrix}$$

$$\times \begin{Bmatrix} \phi_{1,j} \\ \phi_{2,j} \\ \vdots \\ \vdots \\ \phi_{m-1,j} \end{Bmatrix} - \begin{Bmatrix} \phi_{1,j-1} \\ \phi_{2,j-1} \\ \vdots \\ \vdots \\ \phi_{m-1,j-1} \end{Bmatrix}. \quad (11.147)$$

The boundary conditions, Eqs. (11.139) and (11.140), give

$$\phi_{0,j} = \phi(0, t_j) = 0; \; j = 1, 2, \ldots, \quad (11.148)$$

$$\phi_{m,j} = \phi(l, t_j) = 0; \; j = 1, 2, \ldots, \quad (11.149)$$

and the initial conditions yield

$$\phi_{i,0} = \phi(x_i, 0) = f(x_i);\ i = 1, 2, \ldots, m-1, \tag{11.150}$$

$$\frac{\partial \phi(x_i, 0)}{\partial t} = g(x_i);\ i = 1, 2, \ldots, m-1. \tag{11.151}$$

The initial condition of Eq. (11.151) can be incorporated in two different ways. In the first approach, the derivative is approximated by a backward-difference formula at $t = t_1$ as

$$\frac{\partial \phi_{i,1}}{\partial t} = \frac{\phi_{i,1} - \phi_{i,0}}{\Delta t} = g(x_i);\ i = 1, 2, \ldots, m-1, \tag{11.152}$$

which can be rewritten, using Eq. (11.150), as

$$\phi_{i,1} = \phi_{i,0} + \Delta t g(x_i) = f(x_i) + \Delta t g(x_i);\ i = 1, 2, \ldots, m-1. \tag{11.153}$$

It is to be noted that Eq. (11.153) has only an accuracy of $O(\Delta t)$. In the second approach, the function $\phi(x_i, t)$ is expanded in Taylor's series around $t = 0$ as

$$\phi(x_i, t_1) = \phi(x_i, 0) + \Delta t \frac{\partial \phi(x_i, 0)}{\partial t} + \frac{\Delta t^2}{2} \frac{\partial^2 \phi(x_i, 0)}{\partial t^2}$$
$$+ \frac{\Delta t^3}{6} \frac{\partial^3 \phi(x_i, \xi)}{\partial t^3}; 0 < \xi < t_1. \tag{11.154}$$

This equation can be rewritten as

$$\frac{\phi(x_i, t_1) - \phi(x_i, 0)}{\Delta t} = \frac{\partial \phi(x_i, 0)}{\partial t} + \frac{\Delta t}{2} \frac{\partial^2 \phi(x_i, 0)}{\partial t^2} + O(\Delta t^2). \tag{11.155}$$

Assuming that the partial differential equation, Eq. (11.138), is valid at $t = 0$, we can use Eq. (11.141) to express

$$\frac{\partial^2 \phi(x_i, 0)}{\partial t^2} = \alpha^2 \frac{\partial^2 \phi(x_i, 0)}{\partial x^2} = \alpha^2 \frac{\partial^2}{\partial x^2}[\phi(x_i, 0)] = \alpha^2 \frac{\partial^2 f(x_i)}{\partial x^2}. \tag{11.156}$$

Substituting Eqs. (11.151) and (11.156) into Eq. (11.155) yields

$$\frac{\phi(x_i, t_1) - \phi(x_i, 0)}{\Delta t} = g(x_i) + \frac{\Delta t \alpha^2}{2} \frac{\partial^2 f(x_i)}{\partial x^2}, \tag{11.157}$$

which has an accuracy of $O(\Delta t^2)$. By using a central difference formula for $\frac{\partial^2 f}{\partial x^2}(x_i)$, we can rewrite Eq. (11.157) as

$$\phi_{i,1} = \phi_{i,0} + \Delta t g(x_i) + \frac{\alpha^2 \Delta t^2}{2 \Delta x^2}[f(x_{i+1}) - 2f(x_i) + f(x_{i-1})]. \tag{11.158}$$

Using Eqs. (11.150) and (11.145), we can express Eq. (11.158) as

$$\phi_{i,1} = (1 - \lambda) f(x_i) + \frac{\lambda}{2} f(x_{i+1}) + \frac{\lambda}{2} f(x_{i-1}) + \Delta t g(x_i);\ i = 1, 2, \ldots, m-1, \tag{11.159}$$

which has an accuracy of $O(\Delta t^2)$. The step-by-step computational procedure is as follows:

(1) Choose the values of m and n, and find $\Delta x = \frac{1}{m}$ and $\Delta t = \frac{T}{n}$.

(2) Find $x_i = i\Delta x; i = 0, 1, 2, \ldots, m; t_j = j\Delta t; j = 0, 1, 2, \ldots, n$; and $\lambda = \frac{\alpha^2 \Delta t^2}{\Delta x^2}$.

(3) Set the boundary conditions

$$\phi_{0,j} = \phi_{m,j} = 0; j = 1, 2, \ldots, n.$$

Set the initial conditions

$$\phi_{i,0} = f(x_i); i = 0, 1, 2, \ldots, m,$$

$$\phi_{i,1} = (1 - \lambda) f(x_i) + \frac{\lambda}{2}[f(x_{i+1}) + f(x_{i-1})] + \Delta t g(x_i); i = 1, 2, \ldots, m - 1.$$

(4) Take the time step $j = 1$.

(5) Find $\phi_{i,j+1}$ using Eq. (11.144) for $i = 1, 2, \ldots, m - 1$.

(6) Increment the value of j by 1. If $j \le n - 1$, go to step 5. If $j = n$, the computations are completed and hence stop the procedure.

▶Example 11.11

A string of length 36 inches, fixed at both ends, is plucked at a point 9 inches from one end and displaced by 1 inch from its stationary position. The string weighs 0.1 lb/in. If the tension in the string is 100 lb, determine the following:

(a) The frequency of vibration of the string.

(b) The time it takes to complete one cycle of motion.

(c) The displacement history of the various points of the string as functions of time.

Assume the initial velocity of the string to be zero.

Solution

(a) The initial displacement of the string is linear as shown in Fig. 11.20:

$$w(x) \equiv f(x) = \left\{ \begin{array}{l} x/9; 0 \le x \le 9 \text{ in} \\ -x/27 + 4/5; 9 \text{ in} \le x \le 36 \text{ in} \end{array} \right\}. \tag{a}$$

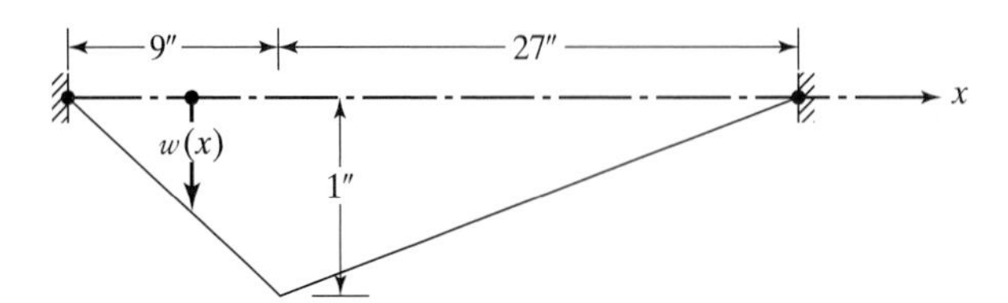

Figure 11.20 Initial displacement of string.

The equation governing the free vibration of a string is given by Eq. (11.138) with $\alpha^2 = \frac{P}{\rho}$, where P is the tension in the string and ρ is the mass of the string per unit length. Here,

$$\alpha = \sqrt{\frac{P}{\rho}} = \sqrt{\frac{100}{(0.1/386.4)}} = 621.6108 \text{ in/sec.}$$

The fundamental frequency of vibration of the string is given by [11.9]

$$\omega = \frac{\alpha\pi}{l} = \frac{621.6108\pi}{36} = 54.2659 \text{ rad/sec.}$$

(b) The time period to complete one cycle of motion is given by

$$\tau = \frac{2\pi}{\omega} = \frac{2\pi}{54.2659} = 0.1158 \text{ sec.}$$

(c) By using $\Delta x = 4.5''$ and $\lambda = 1$, we obtain

$$\lambda = \frac{\alpha^2(\Delta t)^2}{(\Delta x)^2} = 1, \text{ or } \Delta t = \frac{\Delta x}{\alpha} = \frac{4.5}{621.6108} = 0.007239 \text{ sec.} \tag{b}$$

To cover the length of string and one time period, we use

$$m = 1/4.5 = 36/4.5 = 8 \text{ and } n = \frac{\tau}{\Delta t} = \frac{0.1158}{0.007239} \approx 16.$$

The boundary conditions are given by

$$w_{0j} = w_{8j} = 0; j = 1, 2, \ldots, 16,$$

and the initial conditions by

$$f(x_0) = w_{00} = 0,\ f(x_1) = w_{10} = 0.5,\ f(x_2) = w_{20} = 1.0,$$
$$f(x_3) = w_{30} = 5/6,$$
$$f(x_4) = w_{40} = 4/6,\ f(x_5) = w_{50} = 3/6,\ f(x_6) = w_{60} = 2/6,$$
$$f(x_7) = w_{70} = 1/6,\ f(x_8) = w_{80} = 0$$

and

$$w_{i1} = (1-\lambda) f(x_i) + \frac{\lambda}{2}\{f(x_{i+1}) + f(x_{i-1})\} + \Delta t g(x_i); i = 1, 2, \ldots, m-1. \tag{c}$$

Since $\lambda = 1$ and $g(x_i) = 0$, Eq. (c) reduces to

$$w_{i1} = \frac{1}{2}\left\{f(x_{i+1}) + f(x_{i-1})\right\}, \tag{d}$$

which yields

$$w_{11} = 0.5,\ w_{21} = 2/3,\ w_{31} = 5/6,\ w_{41} = 2/3,$$
$$w_{51} = 1/2,\ w_{61} = 1/3,\ w_{71} = 1/6.$$

Equation (11.144) gives, for $\lambda = 1$,

$$w_{i,j+1} = w_{i+1,j} + w_{i-1,j} - w_{i,j-1}; i = 1, 2, \ldots, 7; j = 1, 2, \ldots, 16. \quad \text{(e)}$$

This equation can be used to generate the desired displacement history of the string. ◀

▶Example 11.12

Find the solution of the equation

$$\frac{\partial^2 \phi}{\partial x^2} = \frac{\partial^2 \phi}{\partial t^2}, 0 \le x \le 1, \quad \text{(a)}$$

subject to the boundary conditions

$$\phi(0, t) = 0, t > 0, \quad \text{(b)}$$

$$\phi(1, t) = 0, t > 0, \quad \text{(c)}$$

and the initial conditions

$$\phi(x, 0) = \sin \pi x, 0 \le x \le 1, \quad \text{(d)}$$

and

$$\frac{\partial \phi}{\partial t}(x, 0) = 0, 0 \le x \le 1. \quad \text{(e)}$$

Solution

Using central-difference approximations for $\frac{\partial^2 \phi}{\partial x^2}$ and $\frac{\partial^2 \phi}{\partial t^2}$ with step sizes Δx and Δt, respectively, Eq. (a) can be expressed as

$$\frac{\phi_{i+1,j} - 2\phi_{i,j} + \phi_{i-1,j}}{(\Delta x)^2} - \frac{\phi_{i,j+1} - 2\phi_{i,j} + \phi_{i,j-1}}{(\Delta t)^2} = 0, \quad \text{(f)}$$

or

$$\phi_{i,j+1} = 2(1 - r^2)\phi_{i,j} + r^2(\phi_{i+1,j} + \phi_{i-1,j}) - \phi_{i,j-1}, \quad \text{(g)}$$

where $r = \frac{\Delta t}{\Delta x}$. Using $\Delta x = 1/4$ and $r = 1/2$, we have $\Delta t = 1/8$. (See Fig. 11.21.) The boundary conditions of Eqs. (b) and (c) can be expressed as

$$\phi_{0,j} = 0, j = 0, 1, 2, \ldots, \quad \text{(h)}$$

$$\phi_{4,j} = 0, j = 0, 1, 2, \ldots. \quad \text{(i)}$$

The initial conditions of Eqs. (d) and (e) can be expressed as

$$\phi_{i,0} = \sin\left(\frac{i\pi}{4}\right), i = 0, 1, \ldots, 4, \quad \text{(j)}$$

$$\phi_{i,-1} = \phi_{i,1}, i = 0, 1, \ldots, 4. \quad \text{(k)}$$

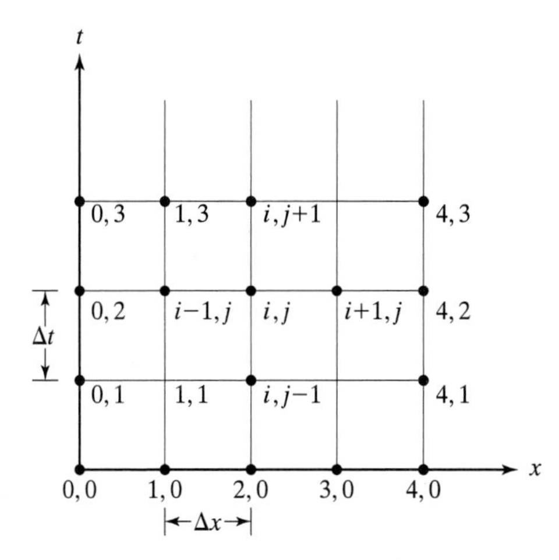

Figure 11.21 Grid in xt-space.

Equation (g) gives, for $i = 1, 2, 3$,

$$\phi_{i,j+1} = \frac{3}{2}\phi_{i,j} + \frac{1}{4}(\phi_{i+1,j} + \phi_{i-1,j}) - \phi_{i,j-1}; i = 1, 2, 3. \qquad (1)$$

For $j = 0$, Eq. (l) becomes, using Eq. (k),

$$\phi_{i,1} = \frac{3}{4}\phi_{i,0} + \frac{1}{8}(\phi_{i+1,0} + \phi_{i-1,0}); i = 1, 2, 3. \qquad (m)$$

Using Eqs. (j) and (m), we get

$$\phi_{1,1} = \frac{3}{4}\phi_{1,0} + \frac{1}{8}(\phi_{2,0} + \phi_{0,0}) = \frac{3}{4}\sin\frac{\pi}{4} + \frac{1}{8}\left(\sin\frac{\pi}{2} + \sin 0\right) = 0.6553,$$

$$\phi_{2,1} = \frac{3}{4}\phi_{2,0} + \frac{1}{8}(\phi_{3,0} + \phi_{1,0}) = \frac{3}{4}\sin\frac{\pi}{2} + \frac{1}{8}\left(\sin\frac{3\pi}{4} + \sin\frac{\pi}{4}\right) = 0.9268,$$

$$\phi_{3,1} = \frac{3}{4}\phi_{3,0} + \frac{1}{8}(\phi_{4,0} + \phi_{2,0}) = \frac{3}{4}\sin\frac{3\pi}{4} + \frac{1}{8}\left(\sin\pi + \sin\frac{\pi}{2}\right) = 0.6553.$$

For $j = 1$, Eq. (l) gives

$$\phi_{i,2} = \frac{3}{2}\phi_{i,1} + \frac{1}{4}(\phi_{i+1,1} + \phi_{i-1,1}) - \phi_{i,0},$$

which yields

$$\phi_{1,2} = \frac{3}{2}\phi_{1,1} + \frac{1}{4}(\phi_{2,1} + \phi_{0,1}) - \phi_{1,0} = 0.50755,$$

$$\phi_{2,2} = \frac{3}{2}\phi_{2,1} + \frac{1}{4}(\phi_{3,1} + \phi_{1,1}) - \phi_{2,0} = 0.71785,$$

$$\phi_{3,2} = \frac{3}{2}\phi_{3,1} + \frac{1}{4}(\phi_{4,1} + \phi_{2,1}) - \phi_{3,0} = 0.50755.$$

Similarly, for $j = 2$, Eq. (l) gives

$$\phi_{i,3} = \frac{3}{2}\phi_{i,2} + \frac{1}{4}(\phi_{i+2,2} + \phi_{i-1,2}) - \phi_{i,1},$$

which results in

$$\phi_{1,3} = \frac{3}{2}\phi_{1,2} + \frac{1}{4}(\phi_{2,2} + \phi_{0,2}) - \phi_{1,1} = 0.28549,$$

$$\phi_{2,3} = \frac{3}{2}\phi_{2,2} + \frac{1}{4}(\phi_{3,2} + \phi_{1,2}) - \phi_{2,1} = 0.40375,$$

$$\phi_{3,3} = \frac{3}{2}\phi_{3,2} + \frac{1}{4}(\phi_{4,2} + \phi_{2,2}) - \phi_{3,1} = 0.28549.$$

This process can be continued to find $\phi_{i,j}$ for all $j \geq 1$. The exact solution of Eqs. (a) through (e) is known to be

$$\phi(x, t) = \sin \pi x \cos \pi t, \tag{n}$$

which gives

$$\phi_{1,3} = \phi(0.25, 0.375) = 0.27059,$$

$$\phi_{2,3} = \phi(0.5, 0.375) = 0.38268,$$

and

$$\phi_{3,3} = \phi(0.75, 0.375) = 0.27059.$$

The accuracy of the finite-difference solution can be improved by using smaller values of Δx and Δt. ◀

11.10 Method of Characteristics

11.10.1 Solution of a Single First-Order Equation

Consider the general first-order hyperbolic partial differential equation

$$a\frac{\partial \phi(x,t)}{\partial x} + b\frac{\partial \phi(x,t)}{\partial t} = c, \tag{11.160}$$

where x and t are the independent variables; ϕ is the dependent variable; and a, b, and c are constants or functions of x, t, and ϕ (but not of $\frac{\partial \phi}{\partial x}$ or $\frac{\partial \phi}{\partial t}$). Let the solution of Eq. (11.160) be represented by the surface shown in Fig. 11.22. The change in the value of ϕ between two neighboring points on the surface, P and Q, is assumed to be $d\phi$. Here, $d\phi$ can be written in terms of the rates of change, $\frac{\partial \phi}{\partial x}$ and $\frac{\partial \phi}{\partial t}$;

$$d\phi = \frac{\partial \phi}{\partial x}\,dx + \frac{\partial \phi}{\partial t}\,dt. \tag{11.161}$$

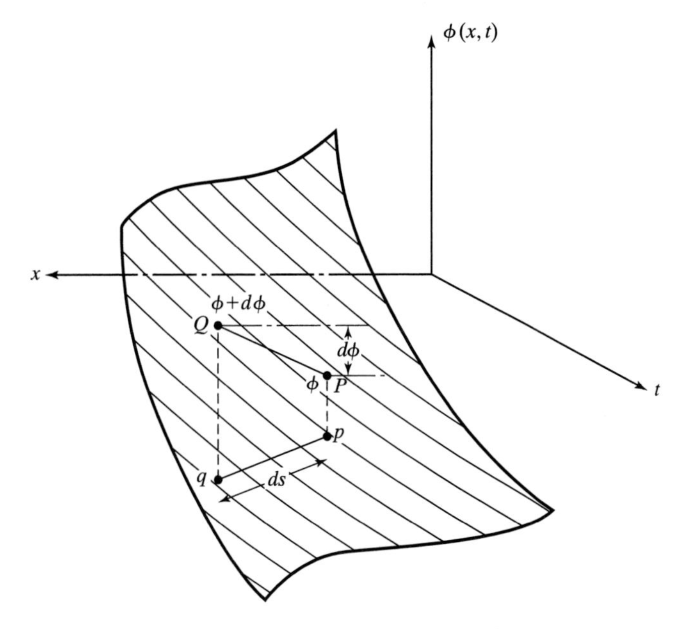

Figure 11.22 Surface denoting the solution of Eq. (11.160).

Equations (11.160) and (11.161) can be expressed in matrix form as

$$\begin{bmatrix} a & b \\ dx & dt \end{bmatrix} \begin{Bmatrix} \dfrac{\partial \phi}{\partial x} \\ \dfrac{\partial \phi}{\partial t} \end{Bmatrix} = \begin{Bmatrix} c \\ d\phi \end{Bmatrix}, \tag{11.162}$$

or

$$[D] \begin{Bmatrix} \dfrac{\partial \phi}{\partial x} \\ \dfrac{\partial \phi}{\partial t} \end{Bmatrix} = \begin{Bmatrix} c \\ d\phi \end{Bmatrix}, \tag{11.163}$$

where

$$[D] = \begin{bmatrix} a & b \\ dx & dt \end{bmatrix}.$$

This system will have a unique solution if the determinant of the matrix, $[D]$ is nonzero. On the other hand, the system will have an infinite number of solutions (i.e., an infinite number of surfaces containing the line PQ) if the determinant of the matrix $[D]$ is zero:

$$|[D]| = \begin{vmatrix} a & b \\ dx & dt \end{vmatrix} = 0. \tag{11.164}$$

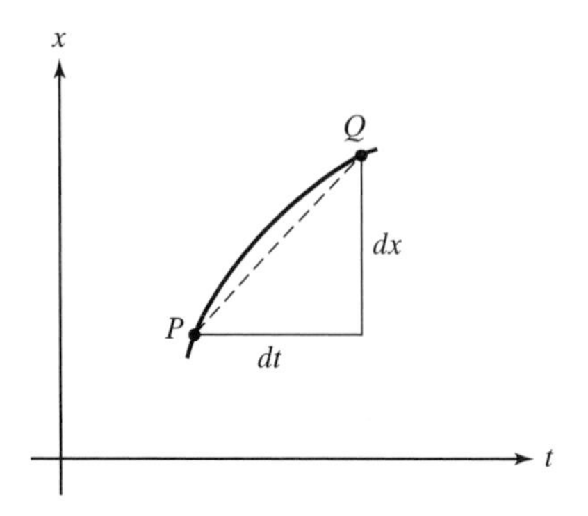

Figure 11.23 Slope of the characteristic.

A *characteristic* is defined as a curve along which the determinant of the matrix $[D]$ is zero. Thus, the direction of the characteristic can be found from Eq. (11.164):

$$a\,dt - b\,dx = 0, \text{ or } a\,dt = b\,dx. \tag{11.165}$$

The slope of the characteristic is defined by

$$\frac{dx}{dt} = \frac{a}{b}. \tag{11.166}$$

(See Fig. 11.23). When Eq. (11.164) holds, Cramer's rule implies that a solution of Eq. (11.163) cannot be obtained unless the determinants of the matrices obtained by substituting the right-hand-side column of Eq. (11.163) into the first or second column of the matrix $[D]$ are equal to zero:

$$\begin{vmatrix} c & b \\ d\phi & dt \end{vmatrix} = 0, \text{ or } c dt = b d\phi, \tag{11.167}$$

and

$$\begin{vmatrix} a & c \\ dx & d\phi \end{vmatrix} = 0, \text{ or } a d\phi = c dx. \tag{11.168}$$

Equations (11.167) and (11.168) yield

$$\frac{d\phi}{dt} = \frac{c}{b}, \tag{11.169}$$

and

$$\frac{d\phi}{dx} = \frac{c}{a}. \tag{11.170}$$

It can be seen that the hyperbolic partial differential equation, Eq. (11.160), is now replaced by the ordinary differential equations, Eqs. (11.166), (11.169), and (11.170). The method of characteristics involves the determination of the characteristic curve as $x = x(t)$ by integrating Eq. (11.166) and the solution of Eq. (11.169) or (11.170) by integration along the curve $x = x(t)$.

11.10.2 Solution of a Set of Two First-Order Equations

Consider the following two simultaneous first-order hyperbolic partial differential equations:

$$a_1 \frac{\partial \phi}{\partial x} + b_1 \frac{\partial \phi}{\partial y} + c_1 \frac{\partial \psi}{\partial x} + d_1 \frac{\partial \psi}{\partial y} = f_1, \tag{11.171}$$

and

$$a_2 \frac{\partial \phi}{\partial x} + b_2 \frac{\partial \phi}{\partial y} + c_2 \frac{\partial \psi}{\partial x} + d_2 \frac{\partial \psi}{\partial y} = f_2, \tag{11.172}$$

where a_i, b_i, c_i, d_i, and $f_i (i = 1, 2)$ are constants or functions of x, y, ϕ, or ψ (but not of $\frac{\partial \phi}{\partial x}$, $\frac{\partial \phi}{\partial y}$, $\frac{\partial \psi}{\partial x}$, or $\frac{\partial \psi}{\partial y}$). The differentials of ϕ and ψ can be expressed as

$$\frac{\partial \phi}{\partial x} dx + \frac{\partial \phi}{\partial y} dy = d\phi, \tag{11.173}$$

$$\frac{\partial \psi}{\partial x} dx + \frac{\partial \psi}{\partial y} dy = d\psi. \tag{11.174}$$

Equations (11.171) through (11.174) can be expressed in matrix form as

$$\begin{bmatrix} a_1 & b_1 & c_1 & d_1 \\ a_2 & b_2 & c_2 & d_2 \\ dx & dy & 0 & 0 \\ 0 & 0 & dx & dy \end{bmatrix} \begin{Bmatrix} \frac{\partial \phi}{\partial x} \\ \frac{\partial \phi}{\partial y} \\ \frac{\partial \psi}{\partial x} \\ \frac{\partial \psi}{\partial y} \end{Bmatrix} = \begin{Bmatrix} f_1 \\ f_2 \\ d\phi \\ d\psi \end{Bmatrix}. \tag{11.175}$$

The direction(s) of the characteristic curve(s) can be obtained by setting the determinant of the coefficient matrix in Eq. (11.175) equal to zero:

$$\begin{vmatrix} a_1 & b_1 & c_1 & d_1 \\ a_2 & b_2 & c_2 & d_2 \\ dx & dy & 0 & 0 \\ 0 & 0 & dx & dy \end{vmatrix} = 0. \tag{11.176}$$

Expanding the determinant, we obtain

$$\left(\frac{dy}{dx}\right)^2 (a_1 c_2 - a_2 c_1) - \left(\frac{dy}{dx}\right)(b_1 c_2 - c_1 b_2 + a_1 d_2 - a_2 d_1) + (b_1 d_2 - d_1 b_2) = 0. \tag{11.177}$$

This equation is a quadratic in $\frac{dy}{dx}$, and, hence there can be no real root (if the system of equations—Eqs. (11.171) and (11.172)—is elliptic), one independent real root (if the system of equations is parabolic), or two real roots (if the system is hyperbolic). For the case of a hyperbolic system of equations, the two roots of

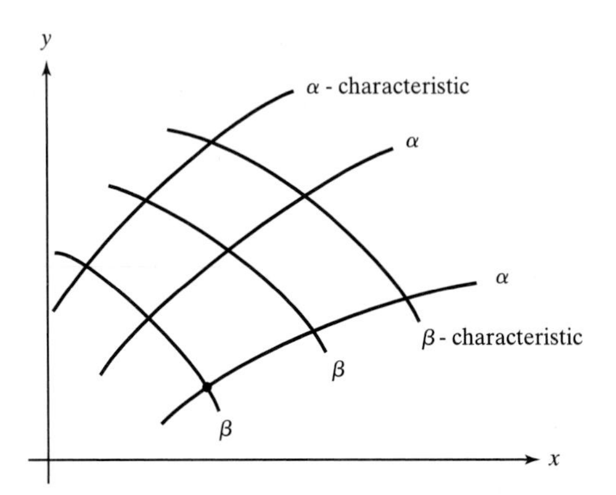

Figure 11.24 Two farnilies of characteristics.

Eq. (11.177) are denoted as $(\frac{dy}{dx})_\alpha$ and $(\frac{dy}{dx})_\beta$. The curve along which the slope is equal to $\frac{dy}{dx}\big|_\alpha$ (or $\frac{dy}{dx}\big|_\beta$) is termed as the $\alpha -$ (or $\beta -$) characteristic. Thus, there will be two families of characteristic curves in the domain (x, y-plane) of the problem. (See Fig. 11.24.)

As in the case of previous section (Eq. (11.160)), if the determinant of the coefficient matrix is zero in Eq. (11.175), the right-hand side must be compatible with this in order to have a solution to Eq. (11.175). This implies that when the right-hand side is substituted for any of the four columns in the matrix on left, the determinant of the resulting matrix must be zero. For example, when the fourth column of the matrix on the left side is replaced by the right-hand-side column of Eq. (11.175) and the determinant is set equal to zero, we obtain

$$\begin{vmatrix} a_1 & b_1 & c_1 & f_1 \\ a_2 & b_2 & c_2 & f_2 \\ dx & dy & 0 & d\phi \\ 0 & 0 & dx & d\psi \end{vmatrix} = 0, \tag{11.178}$$

which yields, upon expansion,

$$d\phi(a_1b_2 - a_2b_1) + d\psi\left\{(a_1c_2 - a_2c_1)\frac{dy}{dx} - (b_1c_2 - b_2c_1)\right\} + \left\{(b_2f_1 - b_1f_2) + \frac{dy}{dx}(f_2a_1 - f_1a_2)\right\} = 0. \tag{11.179}$$

When $\frac{dy}{dx}\big|_\alpha$ and $\frac{dy}{dx}\big|_\beta$ are substituted, one at a time, for $\frac{dy}{dx}$ in Eq. (11.179), we obtain two ordinary differential equations to determine ϕ and ψ along the α - and β - characteristics. Thus, the system of hyperbolic partial differential equations are solved by solving two ordinary differential equations.

11.10.3 Solution of a Second-Order Equation

Consider the second-order quasilinear hyperbolic differential equation

$$a\frac{\partial^2\phi}{\partial x^2}+b\frac{\partial^2\phi}{\partial x\partial y}+c\frac{\partial^2\phi}{\partial y^2}=f, \tag{11.180}$$

where a, b, c, and f are constants or functions of x, y, $\frac{\partial\phi}{\partial x}$, or $\frac{\partial\phi}{\partial y}$. The differentials $d(\frac{\partial\phi}{\partial x})$ and $d(\frac{\partial\phi}{\partial y})$ can be expressed as

$$d\left(\frac{\partial\phi}{\partial x}\right)=\frac{\partial}{\partial x}\left(\frac{\partial\phi}{\partial x}\right)dx+\frac{\partial}{\partial y}\left(\frac{\partial\phi}{\partial x}\right)dy=\frac{\partial^2\phi}{\partial x^2}dx+\frac{\partial^2\phi}{\partial x\partial y}dy, \tag{11.181}$$

$$d\left(\frac{\partial\phi}{\partial y}\right)=\frac{\partial}{\partial x}\left(\frac{\partial\phi}{\partial y}\right)dx+\frac{\partial}{\partial y}\left(\frac{\partial\phi}{\partial y}\right)dy=\frac{\partial^2\phi}{\partial x\partial y}dx+\frac{\partial^2\phi}{\partial y^2}dy. \tag{11.182}$$

Equations (11.180) through (11.182) can be expressed in matrix form as

$$\begin{bmatrix} a & b & c \\ dx & dy & 0 \\ 0 & dx & dy \end{bmatrix}\begin{Bmatrix} \dfrac{\partial^2\phi}{\partial x^2} \\ \dfrac{\partial^2\phi}{\partial x\partial y} \\ \dfrac{\partial^2\phi}{\partial y^2} \end{Bmatrix}=\begin{Bmatrix} f \\ d\left(\dfrac{\partial\phi}{\partial x}\right) \\ d\left(\dfrac{\partial\phi}{\partial y}\right) \end{Bmatrix}. \tag{11.183}$$

As in the previous cases, Eqs. (11.183) can be solved uniquely unless the determinant of the coefficient matrix is zero:

$$\begin{vmatrix} a & b & c \\ dx & dy & 0 \\ 0 & dx & dy \end{vmatrix}=a(dy)^2-b(dxdy)+c(dx)^2=0. \tag{11.184}$$

This yields

$$a\left(\frac{dy}{dx}\right)^2-b\left(\frac{dy}{dx}\right)+c=0,$$

or

$$\frac{dy}{dx}=\frac{b\pm\sqrt{b^2-4ac}}{2a}. \tag{11.185}$$

In Eq. (11.185), the descriminant, b^2-4ac, will be negative, zero, or positive if Eq. (11.180) is elliptic, parabolic, or hyperbolic, respectively. For the case of a hyperbolic equation, Eq. (11.185) gives the slopes $\frac{dy}{dx}\big|_\alpha$ and $\frac{dy}{dx}\big|_\beta$ of the two characteristic curves, labeled as α-curve and β-curve. By substituting the right-hand side for the third column of the coefficient matrix on the left of Eq. (11.183) and setting the determinant of the resulting matrix equal to zero, we obtain

$$\begin{vmatrix} a & b & f \\ dx & dy & d\left(\frac{\partial \phi}{\partial x}\right) \\ 0 & dx & d\left(\frac{\partial \phi}{\partial y}\right) \end{vmatrix} = \left(b - a\frac{dy}{dx}\right) d\left(\frac{\partial \phi}{\partial y}\right) + ad\left(\frac{\partial \phi}{\partial x}\right) - f\,dx = 0, \quad (11.186)$$

Note that

$$d\phi = \frac{\partial \phi}{\partial x}\,dx + \frac{\partial \phi}{\partial y}\,dy. \quad (11.187)$$

The ordinary differential equation that results by substituting $\frac{dy}{dx}\big|_\alpha$ given by Eq. (11.185) for $\frac{dy}{dx}$ in Eq. (11.186) and Eq. (11.187) are solved as a system of two ordinary differential equations to find the solution along the α-characteristic. Similarly, the ordinary differential equation that results by substituting $\frac{dy}{dx}\big|_\beta$ given by Eq. (11.185) for $\frac{dy}{dx}$ in Eq. (11.186) and Eq. (11.187) are integrated to find the solution along the β-characteristic.

11.11 Finite-Difference Formulas in Polar Coordinate System

In many engineering applications, the domain of the partial differential equations is a cylinder for which the polar coordinates best describe the geometry. The transformation between Cartesian coordinates (x, y) and polar coordinates (r, θ) is given by

$$x = r\cos\theta,\ y = r\sin\theta, \quad (11.188)$$

or

$$r = \sqrt{x^2 + y^2},\ \theta = \tan^{-1}\left(\frac{y}{x}\right). \quad (11.189)$$

(See Fig. 11.25). The Laplace operator can be expressed in the polar coordinate system as

$$\nabla^2 \phi = \frac{\partial^2 \phi}{\partial x^2} + \frac{\partial^2 \phi}{\partial y^2} = \frac{\partial^2 \phi}{\partial r^2} + \frac{1}{r}\frac{\partial \phi}{\partial r} + \frac{1}{r^2}\frac{\partial^2 \phi}{\partial \theta^2}. \quad (11.190)$$

To derive a finite-difference approximation of the Laplace operator in polar coordinates, the finite-difference grid is set up as shown in Fig. 11.26. A general grid point is denoted as (i, j) with coordinates (θ_i, r_j), where

$$\theta_0 = 0, \theta_1 = \Delta\theta, \theta_2 = 2\Delta\theta, \ldots, \theta_i = i\Delta\theta, \ldots, \theta_m = m\Delta\theta = 2\pi$$

and

$$r_0 = 0, r_1 = \Delta r, r_2 = 2\Delta r, \ldots, r_j = j\Delta r, \ldots, r_n = n\Delta r = R,$$

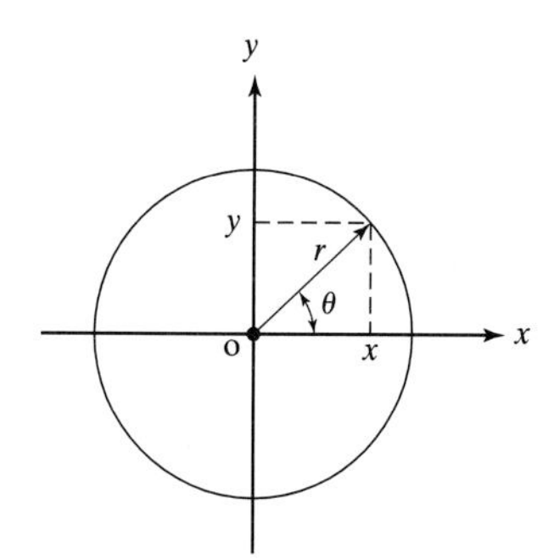

Figure 11.25 Coordinate transformation.

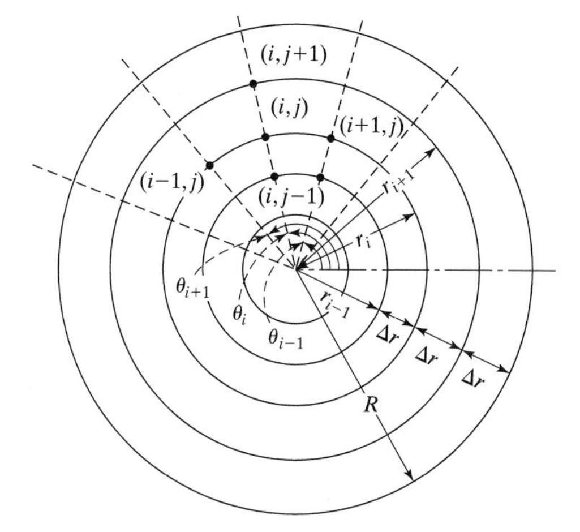

Figure 11.26 Finite-difference grid in polar coordinates.

with $\Delta\theta$ and Δr denoting the step sizes in θ and r directions. The finite-difference approximations of the derivatives of the ϕ are given by

$$\left.\frac{\partial^2\phi}{\partial\theta^2}\right|_{i,j} = \frac{1}{\Delta\theta^2}(\phi_{i+1,j} - 2\phi_{i,j} + \phi_{i-1,j}); \tag{11.191}$$

$$\left.\frac{\partial^2\phi}{\partial r^2}\right|_{i,j} = \frac{1}{\Delta r^2}(\phi_{i,j+1} - 2\phi_{i,j} + \phi_{i,j-1}); \tag{11.192}$$

$$\left.\frac{\partial\phi}{\partial r}\right|_{i,j} = \frac{1}{2\Delta r}(\phi_{i,j+1} - \phi_{i,j-1}). \tag{11.193}$$

Thus, the finite-difference approximations of Eq. (11.190) is given by

$$\nabla^2\phi = \frac{1}{\Delta r^2}(\phi_{i,j+1} - 2\phi_{i,j} + \phi_{i,j-1}) + \frac{1}{2\Delta r r_i}(\phi_{i,j+1} - \phi_{i,j-1})$$
$$+ \frac{1}{r_i^2\Delta\theta^2}(\phi_{i+1j} - 2\phi_{i,j} + \phi_{i-1j}). \tag{11.194}$$

11.12 Choice of Method

The classification of linear second-order partial differential equations into parabolic, elliptic, and hyperbolic types is useful in solving different types of engineering problems using efficient and specialized solution techniques. In engineering mechanics, the equilibrium problems are governed by elliptic equations that involve only the boundary conditions. Problems such as steady-state temperature distribution, steady fluid flow, and stress analysis of solid bodies under static loads lead to elliptic equations. The eigenvalue problems are also governed by elliptic equations, but involve two parts that are related by an eigenvalue. Only the boundary conditions are to be considered in these problems. The stability of structures and natural frequency analysis of solids lead to eigenvalue problems. The propagation problems are typically governed by parabolic and hyperbolic equations and involve both boundary and initial conditions. These problems are also known as initial-boundary-value problems or simply as initial-value problems. Diffusion problems lead to parabolic equations, while oscillatory systems lead to hyperbolic equations. Examples of propagation problems include those related to unsteady-state heat transfer, viscous flow, wave motion, and stress analysis under dynamic loads.

Basically, the finite-difference method (considered in this chapter) and finite-element method (considered in chapter 13) are available for the numerical solution of partial-differential equations. Depending on the type of partial-differential equation and the choice of the finite-difference approximation used (among backward-, central-, and forward-difference techniques), the resulting equations can be solved either directly or by using an iterative method. If the domain or region of interest occupied by the independent variables is regular, a uniform finite-difference grid can be used. On the other hand, if the domain is irregular, a nonuniform grid is to be used. The finite-difference method leads to a set of linear algebraic equations for equilibrium problems.

For the solution of parabolic equations, explicit or implicit methods can be used. Although the explicit methods are simple to use, they impose restrictions on step sizes to ensure stability of the solution. The implicit methods are somewhat more difficult to solve, but do not place any restrictions on the mesh spacing. The Crank–Nicholson (implicit) method, which basically averages the second-derivative difference formula at two adjacent time steps, is a stable method with better truncation error than the simple implicit method. The ADI method is unconditionally stable and is recommended for parabolic equations involving two spatial variables.

For the solution of hyperbolic equations using the finite difference method, different finite difference schemes can be used for approximating time and spatial derivatives. For these problems also, explicit and implicit methods can be used with implicit methods exhibiting unconditional stability characteristics.

11.13 Use of Software Packages

11.13.1 MATLAB

▶Example 11.13

Solve the following hyperbolic equation using MATLAB:

$$u_{tt} = c^2 u_{xx}, 0 \le x \le l, 0 \le t \le T.$$

Boundary conditions:

$$u(0,t) = f_1(t), u(1.t) = f_2(t), 0 \le t \le T.$$

Initial conditions:

$$u(x,0) = g_1(x), \frac{\partial u}{\partial t}(x,0) = g_2(x), 0 \le x \le l.$$

Solution

Input data:

$l = 1, T = 2, c = 1, f_1(t) = 0, f_2(t) = 0,$
$g_1(x) = x(1-x), g_2(x) = 0,$
$n =$ number of grid points along x-axis $= 10,$
$m =$ number of grid points along t-axis $= 10,$

The following is a listing of the m.file to implement Eq. (11.144) or Eq. (11.147) along with the results.

```
% p20.m
l = 1;
t = 2;
n = 10;
n1 = n + 1;
m = 20;
m1 = m + 1;
h = 1/n;
f = t/m;
c = 1.0;
x(1) = 0;

for i = 2: n1
   x(i) = x(1) + (i-1)*h;
end
```

```
t(1) = 0;
for i = 2: m1
   t(i) = t(1) + (i-1)*f;
end
for i = 1: nl
   y(i, 1) = feval ('f1', x(i));
   dy(i, 1) = 0;
end

y(1, 1:m1) = 0;
y(n1, 1:m1) = 0;
p = (c*f/h)^2;
p2 = p/2;
p22 = 2*(1-p);
y(2,2) = (1-p) * y(2,1) + p2*y(3,1) + f*dy (2,1);
y(3:n-1,2)=p2*y(2:n-2,1)+(1-p)*y(3:n-1,1)+p2*y(4:n,1)
+f*dy(3:n-1,1);
y(n, 2) = p2*y (n-1, 1) + (1-p) *y(n,1) + f*dy(n,1);
y(2,3) = p22*y(2,2) + p*y(3,2) - y (2,1);
y(3:n-1,3)=p*y(2:n-2,2)+p22*y(3:n-1,2)+p*y(4:n,2)+y(3:n-1,1);
y(n,3) = p*y(n-1,2) + p22*y(n,2) - y(n,1);

for j = 4: ml
   y(2, j) = p22*y(2, j-1) + p*y(3, j-1) - y(2, j-2);
   y(3:n-1,j)=p*y(2:n-2,j-1)+p22*y(3:n-1,j-1)+p*y(4:n,j-1)
   -y(3:n-1,j-2);
   y(n, j) = p*y(n-1, j-1) + p22*y(n, j-1) - y(n, j-2);
end
% y1 = y';
disp('Solution of hyperbolic equation');
disp('    ');
disp('x-axis:horizontal,delta_x=0.1;t-axis:vertical,delta_t=0.1');
disp('   ');
disp('Column 1 through 6');
disp(y(1:n1, 1:6));
disp('Column 7 through 12');
disp(y(1:n1, 7:12))
disp('Column 13 through 18');
disp(y(1:n1, 13:18))
disp('Column 18 through 21');
disp(y(1:n1, 18:21))
mesh(t, x, y)
xlabel('x');
ylabel('t');
```

```
zlabel ('y');
title('Hyperbolic curve for y = sin(pi*x)');

function y = f1(x)
y = sin (pi*x);

>> P20
solution of hyperbolic equation

x-axis: horizontal, delta_x = 0.1; t-axis: vertical, delta_t = 0.1

Column 1 through 6

          0         0         0         0         0         0
     0.3090    0.2939    0.2500    1.3572    1.7135    1.9021
     0.5878    0.5590    1.6511    1.9635    3.2593    3.6180
     0.8090    0.7694    2.2725    3.5532    3.8680    4.9798
     0.9511    0.9045    2.6715    4.1771    5.2737    5.2361
     1.0000    0.9511    2.8090    4.3920    5.5451    6.1554
     0.9511    0.9045    2.6715    4.1771    5.2737    5.2361
     0.8090    0.7694    2.2725    3.5532    3.8680    4.9798
     0.5878    0.5590    1.6511    1.9635    3.2593    3.6180
     0.3090    0.2939    0.2500    1.3572    1.7135    1.9021
          0         0         0         0         0         0

Column 7 through 12

          0         0         0         0         0         0
     1.9045    1.7205    1.3680    0.8817   -0.3090   -0.2939
     3.6226    3.2725    2.6022    1.0590    0.5878   -0.5590
     4.9861    4.5043    2.9635    2.3083    0.8090   -0.7694
     5.8615    4.6771    4.2104    2.7135    0.9511   -0.9045
     4.9271    5.5676    4.4271    2.8532    1.0000   -0.9511
     5.8615    4.6771    4.2104    2.7135    0.9511   -0.9045
     4.9861    4.5043    2.9635    2.3083    0.8090   -0.7694
     3.6226    3.2725    2.6022    1.0590    0.5878   -0.5590
     1.9045    1.7205    1.3680    0.8817   -0.3090   -0.2939
          0         0         0         0         0         0

Column 13 through 18

          0         0         0         0         0         0
    -0.2500   -1.3572   -1.7135   -1.9021   -1.9045   -1.7205
    -1.6511   -1.9635   -3.2593   -3.6180   -3.6226   -3.2725
```

```
   -2.2725   -3.5532   -3.8680   -4.9798   -4.9861   -4.5043
   -2.6715   -4.1771   -5.2737   -5.2361   -5.8615   -4.6771
   -2.8090   -4.3920   -5.5451   -6.1554   -4.9271   -5.5676
   -2.6715   -4.1771   -5.2737   -5.2361   -5.8615   -4.6771
   -2.2725   -3.5532   -3.8680   -4.9798   -4.9861   -4.5043
   -1.6511   -1.9635   -3.2593   -3.6180   -3.6226   -3.2725
   -0.2500   -1.3572   -1.7135   -1.9021   -1.9045   -1.7205
         0         0         0         0         0         0

Column 18 through 21

         0         0         0         0
   -1.7205   -1.3680   -0.8817    0.3090
   -3.2725   -2.6022   -1.0590   -0.5878
   -4.5043   -2.9635   -2.3083   -0.8090
   -4.6771   -4.2104   -2.7135   -0.9511
   -5.5676   -4.4271   -2.8532   -1.0000
   -4.6771   -4.2104   -2.7135   -0.9511
   -4.5043   -2.9635   -2.3083   -0.8090
   -3.2725   -2.6022   -1.0590   -0.5878
   -1.7205   -1.3680   -0.8817    0.3090
         0         0         0         0
```

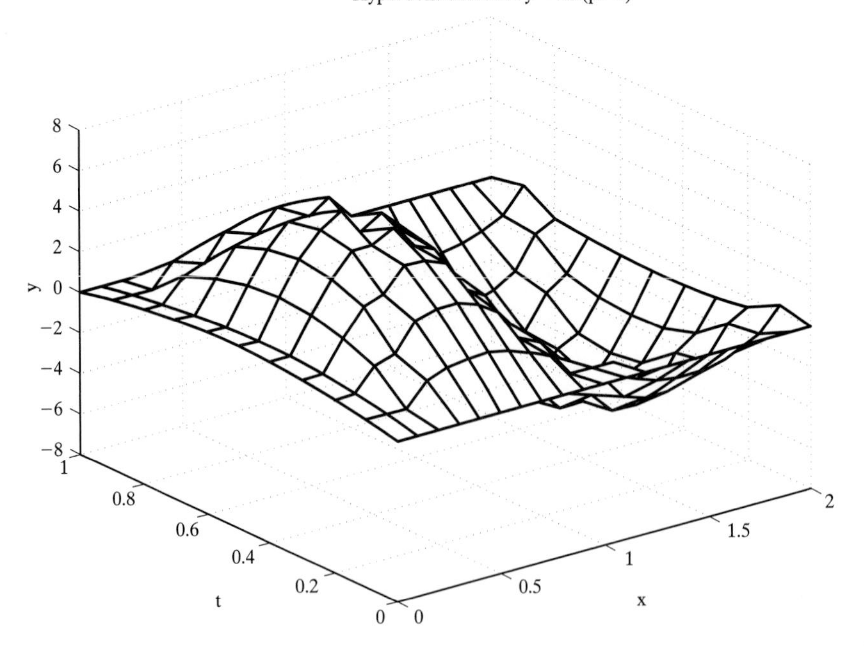

▶Example 11.14

Find the temperature distribution in a metal rod of length 1 m that is initially at 70° C. The steady-state temperatures at the left and right ends of the rod are given by 50° C and 20° C, respectively. Use the implicit method (Section 11.5.2) with $\alpha^2 = 0.1$, $\Delta x = 0.1$, and (i) $\Delta t = 0.025$ and (ii) $\Delta t = 0.075$.

Solution

The MATLAB program to implement the implicit method is given below along with the results. The temperature is denoted as u in the Program. It can be seen that both the time steps $\Delta t = 0.025$ and $\Delta t = 0.075$ gave similar (stable) results.

```
function [u, A] = HEATi (t, x, time, length)
n=time/t;                % number of steps in time
m=length/x;              % number of steps in length
u=zeros(m+1, n+1)        % initialization of matrix u

for i=1: n + 1
u(1, i)=50;              % boundary condition on left side
u(m + 1, i) = 20;        % boundary condition on right side
end

s = 0.1*t /(x)^2;

for k = 2:m               % loop for length
   u(k, 1) = 70;          % initial condition
end

A = zeros(m-1,m-1);
for i=1:m-1
   for j=1:m-1
      if j==i
         A(i,j)=1+2*s;
      elseif j==i+1
         A(i, j)=-s;
      elseif j==i-1
         A(i, j)=-s;
      end
   end
end
```

```
AA=inv(A);

w = zeros(m-1,n+1);
for i=1:m-1
   for j=1:n+1
      w(i, j)=u(i+1,j);
   end
end

for j=2:n+1
   w(1, j-1)= w(1, j-1) + s*50;
   w(m-1, j-1) = w(m-1, j-1) + s*20;
   w(:, j) = AA*w(:, j-1);
   w(1, j-1) = w(1, j-1)-s*50;
   w(m-1, j-1) = w(m-1, j-1) - s*20;
end

for i=2:m
   for j=1:n + 1
      u(i, j)=w(i-1, j);
   end
end

u = u';          %  transpose of matrix u

>> t = 0.025;
  time = 1.5;
  x = 0.1;
  length = 1.0;
  [u, A] = HEATi (t, x, time, length)

u =

Columns 1 through 4

50.00000000000000  70.00000000000000  70.00000000000000  70.00000000000000
50.00000000000000  66.56853624274655  69.41121745647922  69.89876849612878
50.00000000000000  64.14208985111762  68.57839413571951  69.68340513728258
50.00000000000000  62.37418336769267  67.67674080168561  69.37268489954293
50.00000000000000  61.04799495212743  66.79123624199384  68.99245929309319
50.00000000000000  60.02522700381587  65.95938221438554  68.56612131452205
50.00000000000000  59.21585140642546  65.19420042328927  68.11182227576803
50.00000000000000  58.55999154929866  64.49654367009016  67.64246877808516
```

```
50.00000000000000  58.01693267464370  63.86162985066756  67.16667174900100
50.00000000000000  57.55836463117650  63.28245708848423  66.68985849705861
.
.
.
50.00000000000000  50.41273308844326  50.49973594573781  49.96406241253666
50.00000000000000  50.33307227337028  50.34750128644868  49.75299166237056
50.00000000000000  50.25515535351619  50.19864302761600  49.54669766638509
50.00000000000000  50.17895399488611  50.05310255525190  49.34508922616124

Columns 5 through 8

70.00000000000000  70.00000000000000  70.00000000000000  70.00000000000000
69.98139352029344  69.98959262563187  69.95616223349784  69.74738077535503
69.92696270346079  69.95279700230836  69.83144880786189  69.21124691087157
69.82574804644167  69.87395256526391  69.60677933590836  68.44092821873865
69.67277991839357  69.74122803150151  69.27877800955982  67.50432268222393
69.46750850037402  69.54781001414781  68.85443945850683  66.46371469865392
69.21224797323072  69.29163156212013  68.34630134289887  65.36841866124573
68.91097989534870  68.97441870108422  67.76900606267606  64.25441230337655
68.56852553099769  68.60056185559036  67.13717079820755  63.14643868295067
68.19000689786350  68.17608076613161  66.46423027656471  62.06061770042639
.
.
.
48.56463391271574  46.13971697659368  42.62459611507527  38.06108364789255
48.31419903762802  45.87366691253463  42.36893453120506  37.84155581439462
48.06957632121237  45.61396411037706  42.11954069091146  37.62754191027143
47.83064213617519  45.36045830604032  41.87625125855852  37.41888648166492

Columns 9 through 11

70.00000000000000  70.00000000000000  20.00000000000000
68.52812241863229  61.42135373643870  20.00000000000000
66.44650955594742  55.35532075028370  20.00000000000000
64.19380233303728  50.93584755569535  20.00000000000000
61.98344520882914  47.62113923860175  20.00000000000000
59.91055800452085  45.06585249315464  20.00000000000000
58.00935182995981  43.04546030042972  20.00000000000000
56.28379311260040  41.41093905238655  20.00000000000000
54.72381208599028  40.06126138258941  20.00000000000000
53.31372119419094  38.92646112075809  20.00000000000000
.
.
.
```

```
32.59675761853946   26.47269599860490   20.00000000000000
32.43606576359241   26.38780829300200   20.00000000000000
32.27948751313870   26.30512011419111   20.00000000000000
32.12689999034522   26.22456340785161   20.00000000000000

>> mesh (u)
>> xlabel ('Length')
>> ylabel ('Time Step t = 0.025')
>> zlable ('Temperature')
>> title ('Heat Problem (Implicit Method)')
```

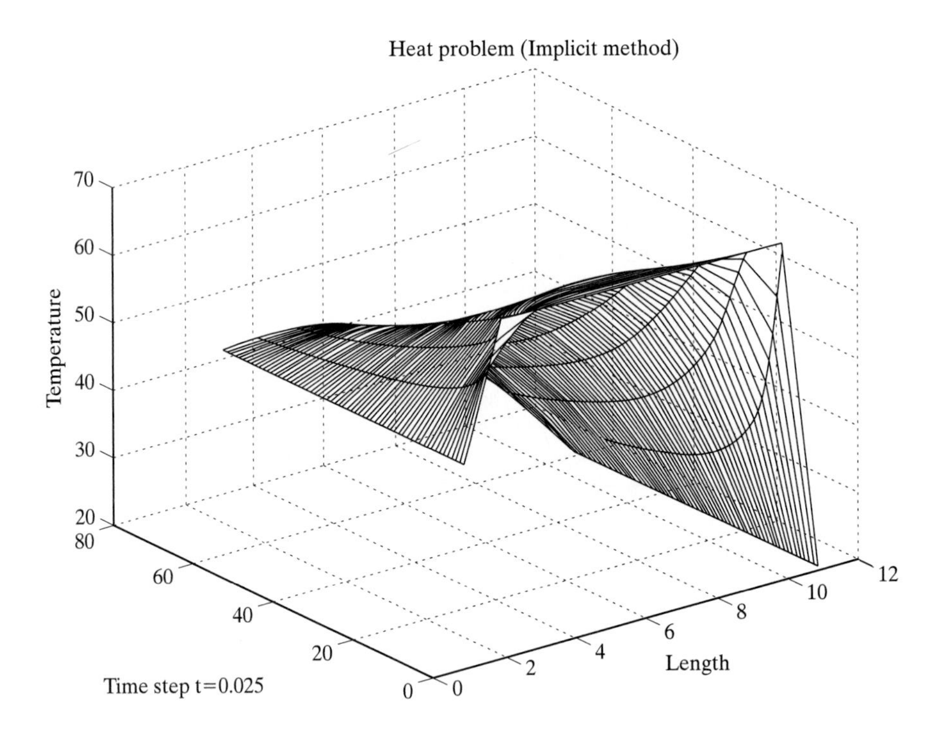


```
>> t = 0.075;

  time = 1.5;
  x = 0.1;
  length = 1.0;

>>  [u,A] = HEATi (t, x, time, length)
```

```
u =

Columns 1 through 4

50.0000000000000  70.0000000000000  70.0000000000000  70.0000000000000
50.0000000000000  63.33107533692076 67.77025112306917 69.23642840664310
50.0000000000000  59.98786334575513 65.51811070328944 68.04550416778403
50.0000000000000  58.08832582499711 63.64393495565020 66.70064308945096
50.0000000000000  56.87077492594001 62.11814865313725 65.33114064365053
50.0000000000000  56.00135431306218 60.84348114228728 63.98605129037905
50.0000000000000  55.32206552341746 59.73841266064196 62.68133515567266
50.0000000000000  54.75319625891142 58.74790016514807 61.42192074405954
50.0000000000000  54.25265924466133 57.83793580365589 60.20992654732751
50.0000000000000  53.79763478487486 56.98857029003439 59.04701844369859
.
.
.
50.0000000000000  51.02372365597235 51.67263913230986 51.60231892083372
50.0000000000000  50.75912370014268 51.16544745917913 50.89551565404128
50.0000000000000  50.51041324619071 50.68921255377877 50.23303198749967
50.0000000000000  50.27693912488966 50.24257942137793 49.61270887466504

Columns 5 through 8

70.0000000000000  70.0000000000000  70.0000000000000  70.0000000000000
69.68451023240783 69.71193903471637 69.35528654997998 68.13901613188358
68.98499864713320 68.99181101278288 68.03845268252115 65.32931586231435
67.96420311880759 67.86670244373011 66.26905700991556 62.31221734596020
66.71812937309653 66.44368644159447 64.27188884057819 59.43720034711207
65.33516896744216 64.84033943696610 62.20771390031887 56.82285510999252
63.88463613776150 63.15389334694284 60.17455576942645 54.48434068405351
62.41672210748686 61.45430476388131 58.22576930952703 52.39885190864019
60.96592502868974 59.78752740498914 56.38675996943230 50.53398008041582
59.55492245919086 58.18148971535135 54.66667338532806 48.85840827649913
.
.
.
50.52774201616641 48.24976251771918 44.67758836105758 39.84481938344217
49.68317949318017 47.34475996833742 43.79967037765229 39.08402347576011
48.89353986583172 46.50119490769919 42.98409653538232 38.37956637337217
48.15574084417263 45.71504078480147 42.22613522823331 37.72661459546647

Columns 9 through 11

70.0000000000000  70.0000000000000  20.0000000000000
64.44143388963195 53.33243016688959 20.0000000000000
```

```
58.87391201601520  44.99514567156041  20.00000000000000
54.33257966019933  40.29783216668396  20.00000000000000
50.76915585518179  37.34987962322812  20.00000000000000
47.95220267017347  35.32561265034329  20.00000000000000
45.67610636409521  33.83307696936588  20.00000000000000
43.79128280720224  32.67061562990703  20.00000000000000
42.19470442043350  31.72665757809287  20.00000000000000
40.81604742911460  30.93547725997153  20.00000000000000
.
.
.
33.91519406903225  27.17365814353858  20.00000000000000
33.35398203029183  26.87565786650298  20.00000000000000
32.83576007484478  26.60099116905463  20.00000000000000
32.35649159215869  26.34734394526946  20.00000000000000

>> mesh (u)
>> xlabel ('Length')
>> ylabel ('Time Step t=0.075')
>> zlabel ('Temperature')
>> title ('Heat Problem (Implicit Method)')
```

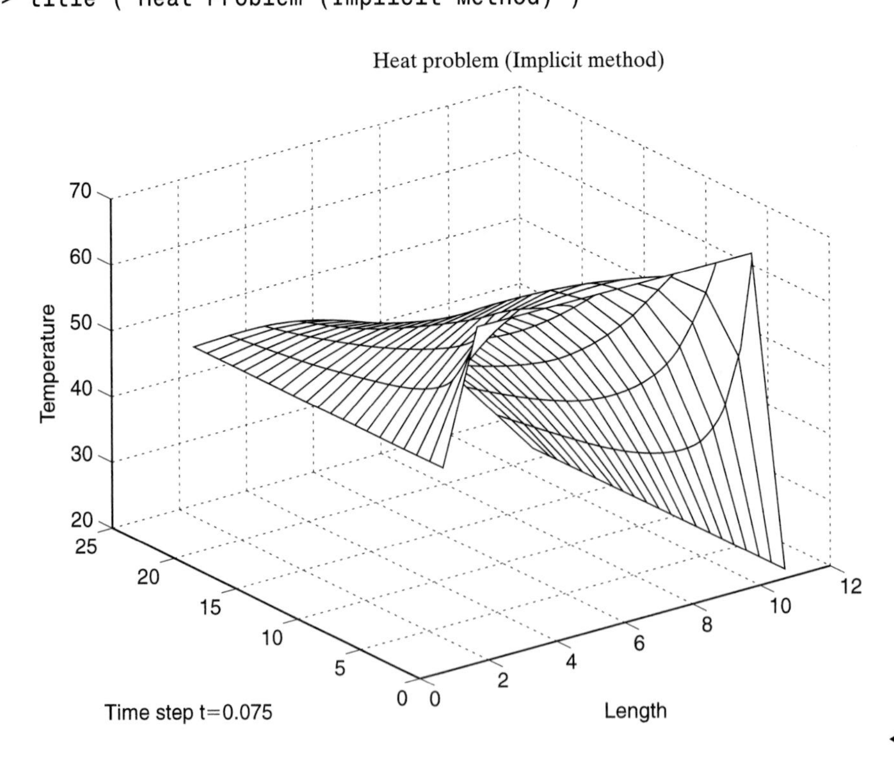

11.13.2 Maple

▶Example 11.15

Find the solution of the heat equation

$$c\frac{\partial^2 u}{\partial x^2} - \frac{\partial u}{\partial t} = 0.$$

Solution

```
> with(PDEtools):
> heat:=(C* diff(u(x,t),x,x)-diff(u(x,t),t));
```

$$heat := C\left(\frac{\partial^2}{\partial x^2}\mathrm{u}(x,t)\right) - \left(\frac{\partial}{\partial t}\mathrm{u}(x,t)\right)$$

```
>sol:=pdsolve(heat,u(x,t));
```

$$sol := (\mathrm{u}(x,t) = \_\mathrm{F1}(x)\_\mathrm{F2}(t)) \text{ where } \left[\left\{\frac{\partial}{\partial t}\_\mathrm{F2}(t) = C\_c_1\_\mathrm{F2}(t), \frac{\partial^2}{\partial x^2}\_\mathrm{F1}(x) = \_c_1\_\mathrm{F1}(x)\right\}\right]$$

```
> sol1:=pdsolve(heat,u(x,t),HINT=X(x)*T(t));
```

$$sol1 := (\mathrm{u}(x,t) = \mathrm{X}(x)\mathrm{T}(t))\&\text{where} \left[\left\{\frac{\partial}{\partial t}\mathrm{T}(t) = C\_c_1\mathrm{T}(t), \frac{\partial^2}{\partial x^2}\mathrm{X}(x) = \_c_1\mathrm{X}(x)\right\}\right]$$

```
> sol1:=pdsolve(heat,u(x,t),'build');
```

$$sol1 := \mathrm{u}(x,t) = \_C3\mathrm{e}^{(C\_c_1 t)}\_C1\cosh(\sqrt{\_c_1}x) + \_C3\mathrm{e}^{(C\_c_1 t)}\_C2\sinh(\sqrt{\_c_1}x)$$ ◀

11.13.3 Mathcad

▶Example 11.16

Use Mathcad to find the solution of the elliptic equation

$$\frac{\partial^2 \phi}{\partial x^2} + \frac{\partial^2 \phi}{\partial y^2} = f(x,y) \tag{E1}$$

over the region $0 \le x \le$ a and $0 \le y \le b$ with $a = b = 1$. The boundary conditions are given by the following:

$\phi(0,y) = c_3 = 150, \phi(a,y) = c_4 = 300,$
$\phi(x,0) = c_1 = 50, \phi(x,b) = c_2 = 400,$
Use (i) $f(x,y) = 0$ and (ii) $f(x,y) = 25$.

Solution

Using central-difference approximation for the second derivatives of ϕ, we find that Eq. (E1) yields

$$\phi_{ij}^{k+1} = \frac{1}{4}[\phi_{i+1j}^{k} + \phi_{i-1j}^{k} + \phi_{ij+1}^{k} + \phi_{i.j-1}^{k} - h^2 f_{ij}^{k}], \qquad \text{(E2)}$$

where $\Delta x = \Delta y = h$ and superscript k denotes the kth iterative value. Assuming the initial values of ϕ as $\phi_{ij}^{1} = 200$ (for $i = 1, 2, \ldots, m-1$ and $j = 1, 2, \ldots, n-1$), we can write a Mathcad program to solve the problem:

h := 0.1

$m := \frac{1}{h}$ $n := \frac{1}{h}$

C1 := 50 C2 := 400 C3 := 150 C4 := 300

C := 200

```
X(h, m, n, C1, C2, C3, C4, C) := | for i ∈ 0..m − 1
                                 |   for j ∈ 0..n − 1
                                 |     X_i,j ← C1 if j = 0
                                 |     X_i,j ← C2 if j = n − 1
                                 |     X_i,j ← C3 if i = 0
                                 |     X_i,j ← C4 if i = m − 1
                                 |     X_i,j ← C otherwise
                                 | X
```

X := X(h, m, n, C1, C2, C3, C4, C)

X =

	0	1	2	3	4	5	6	7	8	9
0	150	150	150	150	150	150	150	150	150	150
1	50	200	200	200	200	200	200	200	200	400
2	50	200	200	200	200	200	200	200	200	400
3	50	200	200	200	200	200	200	200	200	400
4	50	200	200	200	200	200	200	200	200	400
5	50	200	200	200	200	200	200	200	200	400
6	50	200	200	200	200	200	200	200	200	400
7	50	200	200	200	200	200	200	200	200	400
8	50	200	200	200	200	200	200	200	200	400
9	300	300	300	300	300	300	300	300	300	300

(i) With f(x,y) = 0:

```
Y(X) := | for i ∈ 0..m − 1
        |   for j ∈ 0..n − 1
        |     | Y_{i,j} ← C1 if j = 0
        |     | Y_{i,j} ← C2 if j = n − 1
        |     | Y_{i,j} ← C3 if i = 0
        |     | Y_{i,j} ← C4 if i = m − 1
        |     | Y_{i,j} ← 1·(X_{i+1,j} + X_{i−1,j} + X_{i,j+1} + X_{i,j−1}) / 4  otherwise
        | k ← 0
        | while k < 100
        |     | ee ← 0
        |     | for i ∈ 0..m − 1
        |     |   for j ∈ 0..n − 1
        |     |     | et ← |Y_{i,j} − X_{i,j}|
        |     |     | ee ← et if ee < et
        |     | break if ee ≤ 0.00001
        |     | X ← Y if ee > 0.00001
        |     | for i ∈ 0..m − 1
        |     |   for j ∈ 0..n − 1
        |     |     | Y_{i,j} ← C1 if j = 0
        |     |     | Y_{i,j} ← C2 if j = n − 1
        |     |     | Y_{i,j} ← C3 if i = 0
        |     |     | Y_{i,j} ← C4 if i = m − 1
        |     |     | Y_{i,j} ← 1·(X_{i+1,j} + X_{i−1,j} + X_{i,j+1} + X_{i,j−1}) / 4  otherwise
        |     | k ← k + 1
        | Y
```

$Y := Y(X)$

Y =

	0	1	2	3	4	5	6	7	8
0	150	150	150	150	150	150	150	150	150
1	50	106.846	133.526	150.043	163.235	176.541	193.357	219.918	272.249
2	50	93.86	127.218	153.415	176.364	199.577	226.974	264.07	319.082
3	50	91.381	128.08	160.044	189.241	218.438	250.905	290.312	340.012
4	50	93.588	133.686	169.454	202.133	234.046	267.907	306.272	350.66
5	50	99.29	143.6341	181.967	215.81	247.723	280.42	316.22	356.362
6	50	109.944	159.605	198.984	231.435	260.632	289.845	321.838	358.576
7	50	130.885	185.869	222.943	250.33	273.542	296.503	322.72	356.107
8	50	177.733	230.049	256.599	273.409	286.715	299.913	316.442	343.137
9	300	300	300	300	300	300	300	300	300

(ii) With f(x,y) = 25:

```
YY(X) := | for i ∈ 0..m − 1
         | for j ∈ 0..n − 1
         |     | Y[i,j] ← C1 if j = 0
         |     | Y[i,j] ← C2 if j = n − 1
         |     | Y[i,j] ← C3 if i = 0
         |     | Y[i,j] ← C4 if i = m − 1
         |     | Y[i,j] ← 1·(X[i+1,j] + X[i−1,j] + X[i,j+1] + X[i,j−1] − 0.1^2·25)/4 otherwise
         | k ← 0
         | while k < 100
         |     | ee ← 0
         |     | for i ∈ 0..m − 1
         |     |   for j ∈ 0..n − 1
         |     |       | et ← |Y[i,j] − X[i,j]|
         |     |       | ee ← et if ee < et
         |     | break if ee ≤ 0.00001
         |     | X ← Y if ee > 0.00001
         |     | for i ∈ 0..m − 1
         |     |   for j ∈ 0..n − 1
         |     |       | Y[i,j] ← C1 if j = 0
         |     |       | Y[i,j] ← C2 if j = n − 1
         |     |       | Y[i,j] ← C3 if i = 0
         |     |       | Y[i,j] ← C4 if i = m − 1
         |     |       | Y[i,j] ← 1·(X[i+1,j] + X[i−1,j] + X[i,j+1] + X[i,j−1] − 0.1^2·25)/4 otherwise
         |     | k ← k + 1
         | Y
```

Y := YY(X)

Y =

	0	1	2	3	4	5	6	7	8
0	150	150	150	150	150	150	150	150	150
1	50	106.543	133.045	149.46	162.606	175.911	192.774	219.437	271.946
2	50	93.379	126.43	152.443	175.307	198.52	226.003	263.281	318.601
3	50	90.798	127.108	158.836	187.922	217.119	249.697	289.341	339.429
4	50	92.958	132.629	168.135	200.69	232.602	266.588	305.215	350.03
5	50	98.66	142.578	180.648	214.366	246.279	279.101	315.164	355.733
6	50	109.361	158.634	197.777	230.116	259.313	288.637	320.866	357.992
7	50	130.404	185.08	221.972	249.273	272.485	295.531	321.932	355.626
8	50	177.43	229.568	256.016	272.78	286.085	299.33	315.961	342.834
9	300	300	300	300	300	300	300	300	300

◀

11.14 Computer Programs

11.14.1 Fortran Programs

Program 11.1 Solution of elliptic equation

Problem
Steady-state temperature distribution in a rectangular plate. Governing equation (Laplace equation):

$$\frac{\partial^2 T}{\partial x^2} + \frac{\partial^2 T}{\partial y^2} = 0.$$

Temperature is initially zero. Part of boundary at $y = 0$ is suddenly heated to linearly varying temperature. Boundary at $y = b$ is insulated.
Input date required:

M = number of grid points along x-axis,
N = number of grid points along y-axis,
$M1 = M - 1$,
$N1 = N - 1$,
MB = number of grid points with specified temperature,
EPS = small number for testing convergence,
IRMAX = maximum number of iterations in relaxation method,
PHI(I,1) = specified temperature at grid point I (for I $= 1, 2, \ldots,$ MB).

The program listing and output are as follows:

```
SOLUTION OF ELLIPTIC PARTIAL DIFFERENTIAL EQUATION

DATA:
NUMBER OF GRID POINTS ALONG X AND Y = 10 AND 15
MAXIMUM NUMBER OF ITERATIONS IN RELAXATION METHOD = 150

TEMPERATURES AT VARIOUS GRID POINTS:
(X-AXIS: HORIZONTAL, ROWS, Y-AXIS: VERTICAL, COLUMNS)

ITERATION NUMBER:        1
 0.0  0.0  0.0  0.0  0.0  0.0  0.0  0.0  0.0  0.0  0.0  0.0  0.0  0.0  0.0
15.0  0.0  0.0  0.0  0.0  0.0  0.0  0.0  0.0  0.0  0.0  0.0  0.0  0.0  0.0
30.0  0.0  0.0  0.0  0.0  0.0  0.0  0.0  0.0  0.0  0.0  0.0  0.0  0.0  0.0
45.0  0.0  0.0  0.0  0.0  0.0  0.0  0.0  0.0  0.0  0.0  0.0  0.0  0.0  0.0
60.0  0.0  0.0  0.0  0.0  0.0  0.0  0.0  0.0  0.0  0.0  0.0  0.0  0.0  0.0
 0.0  0.0  0.0  0.0  0.0  0.0  0.0  0.0  0.0  0.0  0.0  0.0  0.0  0.0  0.0
 0.0  0.0  0.0  0.0  0.0  0.0  0.0  0.0  0.0  0.0  0.0  0.0  0.0  0.0  0.0
 0.0  0.0  0.0  0.0  0.0  0.0  0.0  0.0  0.0  0.0  0.0  0.0  0.0  0.0  0.0
 0.0  0.0  0.0  0.0  0.0  0.0  0.0  0.0  0.0  0.0  0.0  0.0  0.0  0.0  0.0
 0.0  0.0  0.0  0.0  0.0  0.0  0.0  0.0  0.0  0.0  0.0  0.0  0.0  0.0  0.0
```

```
ITERATION NUMBER:      2
 0.0  0.0  0.0  0.0  0.0  0.0  0.0  0.0  0.0  0.0  0.0  0.0  0.0  0.0  0.0
15.0  3.8  0.9  0.2  0.1  0.0  0.0  0.0  0.0  0.0  0.0  0.0  0.0  0.0  0.0
30.0  8.4  2.3  0.6  0.2  0.0  0.0  0.0  0.0  0.0  0.0  0.0  0.0  0.0  0.0
45.0 13.4  3.9  1.1  0.3  0.1  0.0  0.0  0.0  0.0  0.0  0.0  0.0  0.0  0.0
60.0 18.3  5.6  1.7  0.5  0.1  0.0  0.0  0.0  0.0  0.0  0.0  0.0  0.0  0.0
15.0  8.3  3.5  1.3  0.4  0.1  0.0  0.0  0.0  0.0  0.0  0.0  0.0  0.0  0.0
 3.8  3.0  1.6  0.7  0.3  0.1  0.0  0.0  0.0  0.0  0.0  0.0  0.0  0.0  0.0
 0.9  1.0  0.7  0.3  0.2  0.1  0.0  0.0  0.0  0.0  0.0  0.0  0.0  0.0  0.0
 0.2  0.3  0.2  0.1  0.1  0.0  0.0  0.0  0.0  0.0  0.0  0.0  0.0  0.0  0.0
 0.1  0.2  0.2  0.1  0.1  0.0  0.0  0.0  0.0  0.0  0.0  0.0  0.0  0.0  0.0

ITERATION NUMBER:      3
 0.0  0.0  0.0  0.0  0.0  0.0  0.0  0.0  0.0  0.0  0.0  0.0  0.0  0.0  0.0
15.0  6.1  2.2  0.7  0.2  0.1  0.0  0.0  0.0  0.0  0.0  0.0  0.0  0.0  0.0
30.0 12.9  4.9  1.7  0.6  0.2  0.1  0.0  0.0  0.0  0.0  0.0  0.0  0.0  0.0
45.0 20.1  7.9  2.9  1.0  0.3  0.1  0.0  0.0  0.0  0.0  0.0  0.0  0.0  0.0
60.0 23.5  9.1  3.5  1.3  0.5  0.2  0.1  0.0  0.0  0.0  0.0  0.0  0.0  0.0
20.1 12.5  6.1  2.7  1.1  0.4  0.2  0.1  0.0  0.0  0.0  0.0  0.0  0.0  0.0
 6.8  5.5  3.3  1.6  0.8  0.3  0.1  0.1  0.0  0.0  0.0  0.0  0.0  0.0  0.0
 2.2  2.2  1.5  0.9  0.4  0.2  0.1  0.0  0.0  0.0  0.0  0.0  0.0  0.0  0.0
 0.7  0.8  0.7  0.4  0.2  0.1  0.1  0.0  0.0  0.0  0.0  0.0  0.0  0.0  0.0
 0.5  0.6  0.5  0.4  0.2  0.1  0.1  0.0  0.0  0.0  0.0  0.0  0.0  0.0  0.0
 .
 .
 .

ITERATION NUMBER:      50
 0.0  0.0  0.0  0.0  0.0  0.0  0.0  0.0  0.0  0.0  0.0  0.0  0.0  0.0  0.0
15.0 11.4  8.3  6.0  4.4  3.2  2.3  1.7  1.2  0.9  0.6  0.4  0.3  0.1  0.0
30.0 22.2 16.0 11.5  8.3  6.0  4.4  3.2  2.4  1.7  1.2  0.9  0.6  0.3  0.0
45.0 31.5 22.1 15.8 11.4  8.4  6.2  4.6  3.4  2.5  1.8  1.3  0.8  0.4  0.0
60.0 36.8 25.2 18.1 13.4 10.0  7.5  5.6  4.2  3.1  2.3  1.6  1.1  0.6  0.0
37.2 30.8 23.9 18.4 14.2 10.9  8.3  6.4  4.8  3.6  2.7  2.0  1.4  0.8  0.0
27.6 25.4 21.6 17.7 14.2 11.2  8.8  6.9  5.3  4.0  3.1  2.3  1.7  1.2  1.0
22.9 21.9 19.5 16.7 13.9 11.3  9.0  7.1  5.6  4.3  3.3  2.5  2.0  1.6  1.5
20.6 20.0 18.3 16.1 13.6 11.3  9.2  7.3  5.8  4.5  3.5  2.7  2.1  1.8  1.7
19.9 19.4 17.9 15.9 13.6 11.3  9.2  7.4  5.9  4.6  3.6  2.8  2.2  1.9  1.8
 .
 .
 .

ITERATION NUMBER:      75
 0.0  0.0  0.0  0.0  0.0  0.0  0.0  0.0  0.0  0.0  0.0  0.0  0.0  0.0  0.0
15.0 11.5  8.6  6.4  4.7  3.5  2.7  2.0  1.6  1.2  0.9  0.6  0.4  0.2  0.0
30.0 22.5 16.5 12.2  9.0  6.8  5.2  4.0  3.0  2.3  1.7  1.3  0.8  0.4  0.0
45.0 32.0 22.9 16.8 12.5  9.5  7.3  5.6  4.3  3.3  2.5  1.8  1.2  0.6  0.0
```

```
60.0 37.7 26.4 19.5 14.8 11.5  8.9  7.0  5.5  4.2  3.2  2.4  1.6  0.9  0.0
38.5 32.3 25.6 20.2 16.0 12.7 10.1  8.0  6.3  5.0  3.9  2.9  2.1  1.2  0.0
29.6 27.4 23.7 19.8 16.3 13.3 10.8  8.7  7.0  5.6  4.4  3.4  2.6  2.0  1.6
25.3 24.3 21.9 19.1 16.2 13.6 11.2  9.2  7.4  6.0  4.8  3.8  3.1  2.6  2.4
23.2 22.5 20.9 18.6 16.1 13.7 11.4  9.4  7.7  6.2  5.0  4.1  3.4  2.9  2.8
22.6 22.0 20.5 18.4 16.1 13.7 11.5  9.5  7.8  6.3  5.1  4.2  3.5  3.1  2.9
  .
  .
  .

ITERATION NUMBER:   100
 0.0  0.0  0.0  0.0  0.0  0.0  0.0  0.0  0.0  0.0  0.0  0.0  0.0  0.0  0.0
15.0 11.6  8.7  6.5  4.9  3.7  2.9  2.2  1.7  1.3  1.0  0.7  0.5  0.2  0.0
30.0 22.7 16.8 12.5  9.4  7.2  5.6  4.3  3.4  2.6  2.0  1.5  1.0  0.5  0.0
45.0 32.2 23.3 17.3 13.1 10.1  7.9  6.2  4.9  3.8  3.0  2.2  1.5  0.8  0.0
60.0 38.0 27.0 20.2 15.6 12.2  9.7  7.7  6.1  4.9  3.8  2.9  2.0  1.1  0.0
39.1 33.0 26.4 21.1 16.9 13.6 11.0  8.9  7.2  5.7  4.5  3.5  2.6  1.5  0.0
30.6 28.4 24.7 20.8 17.4 14.4 11.8  9.7  7.9  6.4  5.2  4.1  3.2  2.5  2.0
26.4 25.4 23.1 20.3 17.4 14.7 12.3 10.2  8.4  6.9  5.7  4.6  3.8  3.2  3.0
24.5 23.8 22.1 19.8 17.3 14.9 12.6 10.5  8.8  7.2  6.0  4.9  4.1  3.6  3.5
23.9 23.3 21.8 19.7 17.3 14.9 12.7 10.7  8.9  7.3  6.1  5.0  4.3  3.8  3.6
 .
 .
 .

ITERATION NUMBER:   109
 0.0  0.0  0.0  0.0  0.0  0.0  0.0  0.0  0.0  0.0  0.0  0.0  0.0  0.0  0.0
15.0 11.6  8.8  6.6  5.0  3.8  2.9  2.3  1.8  1.4  1.1  0.8  0.5  0.3  0.0
30.0 22.7 16.9 12.6  9.5  7.3  5.7  4.4  3.5  2.7  2.1  1.5  1.0  0.5  0.0
45.0 32.3 23.4 17.4 13.2 10.2  8.0  6.3  5.0  3.9  3.1  2.3  1.5  0.8  0.0
60.0 38.1 27.1 20.4 15.8 12.4  9.9  7.9  6.3  5.0  3.9  3.0  2.1  1.1  0.0
39.2 33.1 26.6 21.3 17.1 13.8 11.2  9.1  7.4  5.9  4.7  3.7  2.7  1.6  0.0
30.8 28.6 24.9 21.1 17.6 14.6 12.1  9.9  8.1  6.6  5.4  4.3  3.4  2.6  2.1
26.7 25.7 23.4 20.5 17.7 15.0 12.6 10.5  8.7  7.2  5.9  4.8  4.0  3.4  3.1
24.7 24.1 22.4 20.1 17.6 15.2 12.9 10.8  9.0  7.5  6.2  5.1  4.3  3.8  3.6
24.2 23.6 22.1 20.0 17.6 15.2 13.0 10.9  9.1  7.6  6.3  5.2  4.5  4.0  3.8

ITERATION NUMBER:   110
 0.0  0.0  0.0  0.0  0.0  0.0  0.0  0.0  0.0  0.0  0.0  0.0  0.0  0.0  0.0
15.0 11.6  8.8  6.6  5.0  3.8  2.9  2.3  1.8  1.4  1.1  0.8  0.5  0.3  0.0
30.0 22.7 16.9 12.6  9.5  7.3  5.7  4.4  3.5  2.7  2.1  1.5  1.0  0.5  0.0
45.0 32.3 23.4 17.4 13.2 10.2  8.0  6.3  5.0  4.0  3.1  2.3  1.5  0.8  0.0
60.0 38.1 27.1 20.4 15.8 12.4  9.9  7.9  6.3  5.0  3.9  3.0  2.1  1.1  0.0
39.3 33.2 26.6 21.3 17.1 13.8 11.2  9.1  7.4  5.9  4.7  3.7  2.7  1.6  0.0
30.8 28.6 24.9 21.1 17.6 14.6 12.1 10.0  8.2  6.7  5.4  4.3  3.4  2.6  2.1
26.7 25.7 23.4 20.6 17.7 15.0 12.6 10.5  8.7  7.2  5.9  4.8  4.0  3.4  3.1
24.8 24.1 22.4 20.2 17.7 15.2 12.9 10.8  9.0  7.5  6.2  5.1  4.3  3.8  3.6
24.2 23.6 22.1 20.0 17.6 15.2 13.0 11.0  9.2  7.6  6.3  5.3  4.5  4.0  3.8
```

Program 11.2 Solution of parabolic equation

Problem

Transient heat equation

$$\frac{\partial^2 u}{\partial x^2} = \frac{\partial u}{\partial t}, 0 \le x \le 1.$$

Boundary conditions:

$$u(0, t) = u(1, t) = 0.$$

Initial condition:

$$u(x, 0) = \sin \pi x.$$

Input data:
N = number of interior grid points
DX = step length along x-axis
XK = step length along t-axis
MAX = maximum number of time steps
Using N = 9, DX = 0.1, XK = 0.005 and MAX = 25, the output of Program 11.2 is as follows:

```
solution OF PARABOLIC PARTIAL DIFFERENTIAL EQUATION

NUMBER OF GRID POINTS (N) = 9
STEP LENGTHS ALONG X AND T AXES: DELTA X = 0.100 DELTA T = 0.005
MAXIMUM NUMBER OF TIME STEPS (MAX) = 25

INITIAL SOLUTION

0.00000000E+00 0.30901769E+00 0.58778644E+00 0.80901831E+00 0.95105743E+00
0.10000000E+01 0.95105511E+00 0.80901396E+00 0.58778059E+00 0.30901068E+00
0.00000000E+00

NUMERICAL SOLUTION AT TIME STEP    1

0.00000000E+00 0.29459894E+00 0.56036037E+00 0.77126962E+00 0.90668124E+00
0.95334017E+00 0.90667909E+00 0.77126563E+00 0.56035525E+00 0.29459417E+00
0.00000000E+00

EXACT SOLUTION AT TIME STEP      1

0.00000000E+00 0.29413834E+00 0.55948424E+00 0.77006376E+00 0.90526360E+00
0.95184958E+00 0.90526140E+00 0.77005959E+00 0.55947870E+00 0.29413170E+00
0.00000000E+00

ERROR AT TIME STEP      1

 0.00000000E+00-0.46059489E-03-0.87612867E-03-0.12058616E-02-0.14176369E-02
-0.14905930E-02-0.14176965E-02-0.12060404E-02-0.87654591E-03-0.46247244E-03
 0.00000000E+00
```

```
NUMERICAL SOLUTION AT TIME STEP    2

0.00000000E+00 0.28085297E+00 0.53421402E+00 0.73528224E+00 0.86437553E+00
0.90885746E+00 0.86437356E+00 0.73527867E+00 0.53420967E+00 0.28084952E+00
0.00000000E+00

EXACT SOLUTION AT TIME STEP      2

0.00000000E+00 0.27997547E+00 0.53254491E+00 0.73298490E+00 0.86167485E+00
0.90601766E+00 0.86167270E+00 0.73298097E+00 0.53253961E+00 0.27996913E+00
0.00000000E+00

ERROR AT TIME STEP      2

 0.00000000E+00-0.87749958E-03-0.16691089E-02-0.22973418E-02-0.27006865E-02
-0.28398037E-02-0.27008653E-02-0.22976995E-02-0.16700625E-02-0.88039041E-03
 0.00000000E+00
.
.
.
NUMERICAL SOLUTION AT TIME STEP   25

0.00000000E+00 0.93579523E-01 0.17799881E+00 0.24499434E+00 0.28800806E+00
0.30282962E+00 0.28800803E+00 0.24499427E+00 0.17799875E+00 0.93579486E-01
0.00000000E+00

EXACT SOLUTION AT TIME STEP     25

0.00000000E+00 0.89989439E-01 0.17117004E+00 0.23559526E+00 0.27695864E+00
0.29121128E+00 0.27695796E+00 0.23559399E+00 0.17116834E+00 0.89987397E-01
0.00000000E+00

ERROR AT TIME STEP     25

 0.00000000E+00-0.35900846E-02-0.68287700E-02-0.93990862E-02-0.11049420E-01
-0.11618346E-01-0.11050075E-01-0.94002783E-02-0.68304092E-02-0.35920888E-02
 0.00000000E+00
```

11.14.2 C Program

Program 11.3 Solution of hyperbolic equation

Problem

Solution of the equation:

$$\frac{\partial^2 \phi}{\partial x^2} = \frac{\partial^2 \phi}{\partial t^2}, 0 \leq x \leq 1, 0 \leq t \leq 0.5.$$

Boundary conditions:

$$\phi(0, t) = \phi(1, t) = 0, 0 \leq t \leq 0.5.$$

Initial conditions:

$$\phi(x, 0) = f(x) = \sin \pi x, \frac{\partial \phi}{\partial t}(x, 0) = 0, 0 \leq x \leq 1.$$

Input data:
M = number of grid points along x-axis,
N = number of grid points along t-axis,
DX = Δx = step length along x-axis,
DT = Δt = step length along t-axis.

The following shows the output of Program 11.3:

```
Solution of Hyperbolic equation

Number of grid points along x and t axes: 10 and 10
Step lengths along x and t axes: 0.100 and 0.050

Values of phi at various grid points
(x-axis: horizontal (rows), t-axis: vertical (columns)

0.000 0.000 0.000 0.000 0.000 0.000 0.000 0.000 0.000 0.000 0.000
0.309 0.305 0.294 0.276 0.250 0.219 0.182 0.141 0.097 0.050 0.001
0.588 0.581 0.559 0.524 0.476 0.417 0.347 0.269 0.184 0.094 0.003
0.809 0.799 0.770 0.721 0.655 0.573 0.477 0.370 0.253 0.130 0.004
0.951 0.939 0.905 0.848 0.771 0.674 0.561 0.435 0.297 0.153 0.005
1.000 0.988 0.951 0.892 0.810 0.709 0.590 0.457 0.313 0.161 0.005
0.951 0.939 0.905 0.848 0.771 0.674 0.561 0.435 0.297 0.153 0.005
0.809 0.799 0.770 0.721 0.655 0.573 0.477 0.370 0.253 0.130 0.004
0.588 0.581 0.559 0.524 0.476 0.417 0.347 0.269 0.184 0.094 0.003
0.309 0.305 0.294 0.276 0.250 0.219 0.182 0.141 0.097 0.050 0.001
0.000 0.000 0.000 0.000 0.000 0.000 0.000 0.000 0.000 0.000 0.000
```

REFERENCES AND BIBLIOGRAPHY

11.1. R. Haberman, *Elementary Applied Partial Differential Equations with Fourier Series and Boundary Value Problems*, Prentice-Hall, Englewood Cliffs, NJ, 1983.

11.2. A. N. Tychonov and A. A. Samarski, *Partial Differential Equations of Mathematical Physics*, Holden-Day, San Francisco, 1964.

11.3. R. Vichnevetsky, *Computer Methods for Partial Differential Equations*, Vol. 1, Prentice-Hall, Englewood Cliffs, NJ, 1981.

11.4. B. Carnahan, H. A. Luther, and J. O. Wilkes, *Applied Numerical Methods*, Wiley, New York, 1969.

11.5. G. E. Forsythe and W. R. Wasow, *Finite-Difference Methods for Partial Differential Equations*, Wiley, New York, 1960.

11.6. D. W. Peaceman and H. H. Rachford, "The Numerical Solution of Parabolic and Elliptic Differential Equations," *J. Soc. Ind. Appl. Math.,* Vol. 3, pp. 28–41, 1955.

11.7. S. Nakamura, *Applied Numerical Methods with Software*, Prentice Hall, Englewood Cliffs, NJ, 1991.

11.8. H. C. Yee, R. M. Beam, and R. F. Warming, "Stable Boundary Approximations for a Class of Implicit Schemes for the One-Dimensional Inviscid Equations of Gas Dynamics," *AIAA Computational Fluid Dynamics Conference*, Palo Alto, CA, June 22–23, 1981.

11.9. S. S. Rao, *Mechanical Vibrations*, 3d edition, Addison-Wesley, Reading, MA, 1995.

11.10. S. S. Rao, *The Finite Element Method in Engineering*, 3d edition, Butterworth-Heinemann, Boston, 1999.

REVIEW QUESTIONS

The following questions along with corresponding answers are available in an interactive format at the Companion Web site at http://www.prenhall.com/rao.

11.1. **Classify** the following **equations** as **hyperbolic**, **parabolic**, or **elliptic**:
1. $\phi_{xx} + 2\phi_{xy} + \phi_{yy} = \phi_x - x\phi_y$.
2. $2\phi_{xx} + 5\phi_{xy} + 2\phi_{yy} = 0$.
3. $\phi_{xx} + 4\phi_{xy} + 20\phi_{yy} - 5\phi_x + y\phi = 0$.
4. $\phi_{xx} + 4\phi_{xy} + \phi_{yy} = 2\phi\phi_x$.
5. $y^2\phi_{xx} - 2xy\,\phi_{xy} + x^2\,\phi_{yy} = 0$.
6. $\phi_{xx} + x^2\phi_{xy} - (\frac{x^2}{2} + \frac{1}{4})\phi_{yy} = 0$.

11.2. **Find the region** in which the following **equation behaves** as (a) a **hyperbolic equation**, (b) a **parabolic equation**, and (c) an **elliptic equation**:
1. $x^2\phi_{xx} + 4\phi_{yy} = \phi$.
2. $y\phi_{xx} + \phi_{yy} = 0$.

11.3. **Define the following terms:**
Dirichlet condition, Neumann condition, Robbins condition, Courant number, Characteristic, Laplace operator in Cartesian coordinates, Laplace operator in polar coordinates.

11.4. **Give short answers** to each of the following:
1. How many initial and boundary conditions are needed to solve a partial differential equation?
2. How many types of boundary conditions can be specified for a partial differential equation?
3. What is the difference between an ordinary differential equation and a partial differential equation?
4. How are partial differential equations classified?
5. For a rectangular region, the boundary conditions are stated as $\phi(0, y) = 0$, $\frac{\partial\phi(a,y)}{\partial x} = 0$, $\phi(x, 0) = 0$, and $\frac{\partial\phi(x,b)}{\partial y} = 0$. Identify each of these conditions as a Dirichlet, Neumann, or Robbins condition.
6. Give two practical applications governed by hyperbolic equations.
7. What is the significance of the numerical viscosity term in the finite-difference solution of a hyperbolic equation?

(*continued, page 871*)

8. If the Cartesian and polar coordinates are related as $x = r\cos\theta$ and $y = r\sin\theta$, express r and θ in terms of x and y.

11.5. Determine whether each of the following statements is **true or false**:

1. The classification into elliptic, parabolic, and hyperbolic equations is applicable only to second-order partial differential equations.
2. Partial differential equations can be classified as initial- and boundary-value problems.
3. Analytical solutions can be found to most partial differential equations.
4. All partial differential equations need boundary and initial conditions for their solution.
5. Three types of boundary conditions can be specified for a partial differential equation.
6. The Robbins boundary condition can be considered as a mixed-boundary condition.
7. The Crank–Nicholson method is a conditionally stable method.
8. The ADI method is an unconditionally stable method.
9. The solutions of elliptic and parabolic equations are not always continuous in their spatial and time domains.

11.6. **Fill in the blanks** with suitable word(s):

1. Second-order partial differential equations involving three independent variables are classified as elliptic, parabolic, hyperbolic, and ______ equations.
2. The grid points at an irregular boundary are defined using ______ of standard grid-point spacings.
3. A parabolic differential equation can be solved using both ______ and ______ methods.
4. The step sizes along x- and t-axes in a parabolic equation need to satisfy certain relation to ensure the ______ of the explicit method of solution.
5. The Crank–Nicholson method is a ______ method.
6. In the method of lines, the partial differential equation is converted into a system of ______ ______ equations.
7. The alternating-direction implicit (ADI) method is widely used for the solution of parabolic equations involving ______ spatial variables.
8. The solution of a hyperbolic equation can have ______ .
9. The method of characteristics involves the determination of the ______ curve.

(*continued, page 872*)

11.7. **Match the following:**

1. $\phi(x, y) = c_1$.	(a) Neumann condition.
2. $\dfrac{\partial \phi(x, y)}{\partial n} = c_2$.	(b) Robbins condition.
3. $\dfrac{\partial \phi(x, y)}{\partial n} + r\phi(x, y) = c_3$.	(c) Dirichlet condition.

11.8. **Consider the following equation:**

$$A(x, y)\phi_{xx} + B(x, y)\phi_{xy} + C(x, y)\phi_{yy} = D(x, y).$$

Match each of the following conditions with its corresponding determinant for the equation:

1. Elliptic equation.	(a) $B^2 - 4AC = 0$.
2. Parabolic equation.	(b) $B^2 - 4AC > 0$.
3. Hyperbolic equation.	(c) $B^2 - 4AC < 0$.

PROBLEMS

Section 11.2

11.1. The electric potential (ϕ) in an isotropic dielectric medium with a permittivity of ε and a charge density ρ is governed by the equation

$$\frac{\partial^2 \phi}{\partial x^2} + \frac{\partial^2 \phi}{\partial y^2} + \frac{\rho}{\varepsilon} = 0 \tag{E1}$$

with values of ϕ specified on the boundary of the medium. For the cross section (hollow rectangular section) of a coaxial cable shown in Fig. 11.27, suggest a method of determining the distribution of the electric potential with the following data:

$$\varepsilon = 22.15 \times 10^{-12}\ \text{F/m}, \rho = 0, c_1 = 100V, c_2 = 0.$$

Use symmetry conditions and model only one-quarter of the cross section for the solution.

Section 11.3

11.2. The forced vibration of a membrane, subjected to a tension P and transverse loading $f(x, y, t)$, is governed by the equation

$$P\left(\frac{\partial^2 w}{\partial x^2} + \frac{\partial^2 w}{\partial y^2}\right) + f = \rho \frac{\partial^2 w}{\partial t^2}, \tag{E1}$$

where ρ is the mass per unit area and $w(x, y, t)$ is the transverse deflection. If the membrane is fixed at (x_1, y_1), the boundary condition can be expressed as

$$w(x_1, y_1, t) = 0, t > 0. \tag{E2}$$

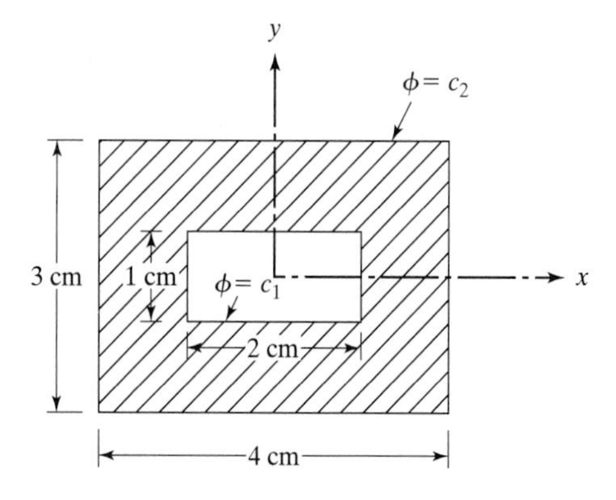

Figure 11.27 Cross-section of coaxial cable.

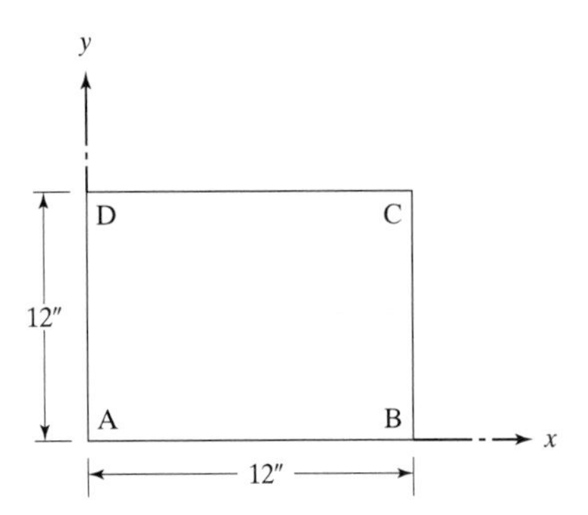

Figure 11.28 Square membrane.

If the membrane is free to deflect transversely in z-direction at (x_1, y_1), the boundary condition can be expressed as

$$P\frac{\partial w}{\partial n}(x_1, y_1, t) = 0, t > 0, \tag{E3}$$

where n is the normal direction to the boundary of the membrane at (x_1, y_1). If the membrane shown in Fig. 11.28 is fixed along the edges AB and CD and free along the edges BC and DA, state the boundary and initial conditions of the problem. Assume that the membrane is initially at rest.

Section 11.4

11.3. The steady-state irrotational flow of an incompressible inviscid fluid is governed by the equation

$$\frac{\partial^2\phi}{\partial x^2} + \frac{\partial^2\phi}{\partial y^2} = 0, \tag{E1}$$

where ϕ is the potential function from which the components of velocity along x- and y-directions, u and v determined by

$$u = \frac{\partial\phi}{\partial x}, v = \frac{\partial\phi}{\partial y}. \tag{E2}$$

For a confined flow around a cylinder, as shown in Fig. 11.29, indicate a method of determining the potential function and the velocity components of the fluid using a finite-difference scheme.

Make use of the double symmetry of the problem, and use a finite-difference mesh only in one-quarter of the problem.

11.4. Derive the equations for determining the steady-state temperature distribution in the plate shown in Fig. 11.30 using the finite-difference grid indicated by the dotted lines.

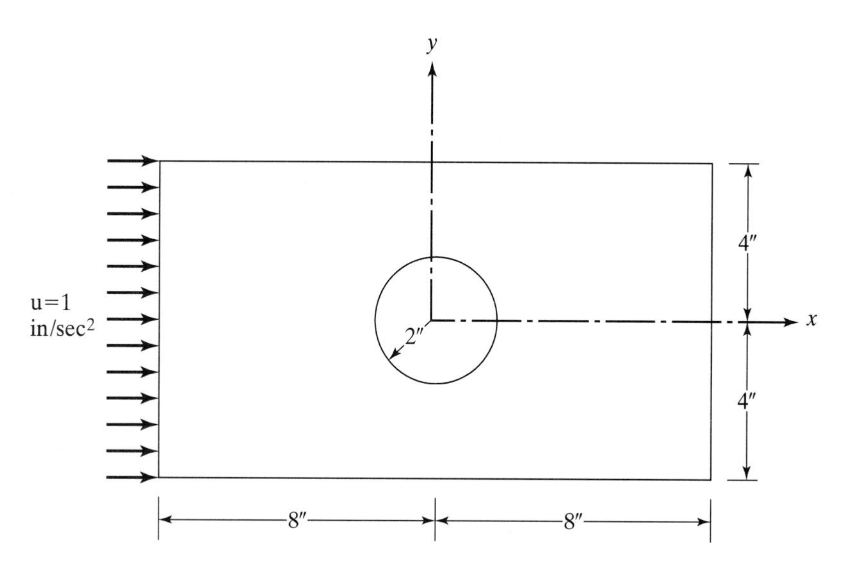

Figure 11.29 Confined flow around a cylinder.

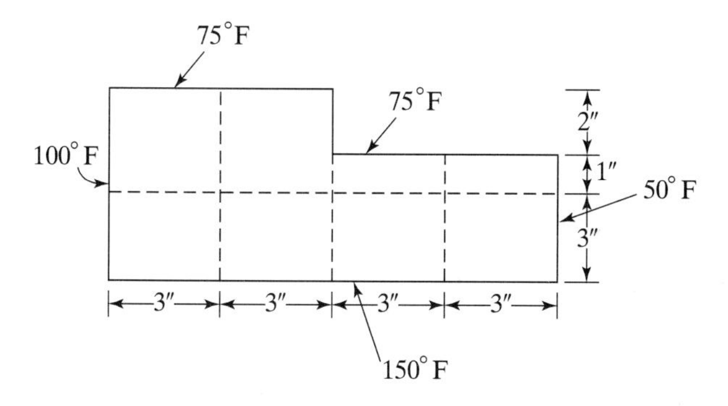

Figure 11.30 Finite difference grid of plate.

11.5. Determine the steady-state temperature distribution in a rectangular plate of size 15" × 20" using $\Delta x = \Delta y = 2.5$". The temperatures on the four sides of the plate are specified in Fig. 11.9. Compare the results with those of Example 11.4.

Section 11.5

11.6. Investigate the stability condition when the equation $\frac{\partial^2\phi}{\partial x^2} = \frac{\partial\phi}{\partial t}$ is approximated as

$$\frac{\phi_{i+1,j} - 2\phi_{i,j} + \phi_{i-1,j}}{(\Delta x)^2} = \frac{\phi_{i,j+1} - \phi_{i,j-1}}{2\Delta t}.$$

11.7. The initial temperature of a metal rod of length 1 m is given by

$$T(x, 0) = x^2, 0 < x < 1.$$

The temperatures of the rod at $x = 0$ and $x = 1$ are given by

$$T(0, t) = 0, T(1, t) = 1, t > 0.$$

Using $\alpha^2 = 0.75$, $\Delta x = 0.2$ and $\Delta t = 0.02$, determine the temperature distribution in the rod for $0 \le t \le 1$ using the explicit method.

11.8. Solve Problem 11.7 using the implicit method with $\Delta x = 0.2$ and $\Delta t = 0.04$.

Section 11.6

11.9. Solve Problem 11.7 using the Crank–Nicholson method with $\Delta x = 0.2$ and $\Delta t = 0.04$.

11.10. Solve the following one-dimensional heat equation using the Crank–Nicholson method:

$$T_{xx} = T_t, 0 \le x \le 1, 0 \le t \le 0.6.$$

The initial conditions are

$$T(x, 0) = \sin \pi x, 0 \le x \le 1$$

The boundary conditions (with both ends insulated) are

$$\frac{\partial T}{\partial x}(0, t) = 0, \frac{\partial T}{\partial x}(1, t) = 0, 0 \le t \le 0.6.$$

Use $\Delta x = 0.2$ and $\Delta t = 0.1$.

Section 11.7

11.11. Derive the equations for determining the temperature distribution in the metal rod described in Problem 11.7 using the method of lines.

11.12. A metal rod of unit length is subjected to the initial conditions

$$T(x, 0) = \sin \pi x, 0 \le x \le 1.$$

Both the ends of the rod are insulated so that

$$\frac{\partial T}{\partial x}(0, t) = 0, \frac{\partial T}{\partial x}(1, t) = 0, 0 \le t \le 0.6.$$

Derive the equations governing the temperature distribution in the rod using the method of lines. Use $\Delta x = 0.2$ and $\Delta t = 0.1$.

Section 11.8

11.13. Consider the two-dimensional heat conduction equation

$$\frac{\partial^2 \phi}{\partial x^2} + \frac{\partial^2 \phi}{\partial y^2} = \frac{\partial \phi}{\partial t}, \quad 0 \le x, y \le 1, t > 0 \qquad \text{(a)}$$

subject to the boundary conditions

$$\phi(x, y, t) = 0, \text{ on the boundary, for } t > 0, \tag{b}$$

and the initial conditions

$$\phi(x, y, 0) = \sin \pi x \sin \pi y, 0 \leq x, y \leq 1. \tag{c}$$

Find the solution of Eqs. (a) through (c) using a finite-difference method with $\Delta x = \Delta y = 1/3$ and $\Delta t = (1/8)(\Delta x)^2$. Compare the solution with the exact result

$$\phi(x, y, t) = e^{-\pi^2 t} \sin \pi x \sin \pi y. \tag{d}$$

Section 11.9

11.14. A vibrating string of unit length is fixed at both ends and is subjected to the initial conditions

$$u(x, 0) = x(1 - x), \frac{\partial u}{\partial t}(x, 0) = 0, 0 \leq x \leq 1.$$

Solve the governing equation

$$\frac{\partial^2 u}{\partial x^2} = \frac{\partial^2 u}{\partial t^2}, 0 \leq x \leq 1, t \geq 0,$$

for $0 \leq t \leq 1$ using $\Delta x = 0.2$ and $\Delta t = 0.2$.

11.15. A rectangular membrane is subjected to the following boundary and initial conditions:

$$w(x, y, t) = 0; x = 0, x = a; y = 0, y = b; t \geq 0;$$

$$w(x, y, 0) = w_0 \sin \frac{\pi x}{a} \sin \frac{\pi y}{b}; 0 \leq x \leq a; 0 \leq y \leq b;$$

$$\frac{\partial w}{\partial t}(x, y, 0) = 0; 0 \leq x \leq a; 0 \leq y \leq b.$$

Determine the deflection of the membrane, $w(x, y, t)$, using a finite-difference scheme with $f = 0$ and $P = 2\rho$ in Eq. (E1) of Problem 11.2. Assume $a = b = w_0 = 1$.

11.16. The longitudinal vibration of a bar is governed by

$$c^2 \frac{\partial^2 u}{\partial x^2} = \frac{\partial^2 u}{\partial t^2},$$

with $c = \sqrt{\frac{E}{\rho}}$, where $u = u(x, t)$ is the axial displacement, E is Young's modulus, and ρ is the mass density of the bar. The boundary and initial conditions are given by

$$u(0, t) = 0; t \geq 0;$$

$$\frac{\partial u}{\partial x}(l, t) = 0; t \geq 0;$$

$$u(x, 0) = u_0(x); 0 \leq x \leq 1;$$

$$\frac{\partial u}{\partial t}(x, 0) = \dot{u}_0; 0 \leq x \leq 1.$$

Determine the variation of the axial displacement of the bar using a finite-difference method with the following data:

$$E = 30 \times 10^6, \rho = 0.283, l = 20, u_0 = 1, \dot{u}_0 = 0.$$

Section 11.10

11.17. Consider the partial differential equation

$$a\frac{\partial \phi}{\partial x} + b\frac{\partial \phi}{\partial t} = c, 0 \leq x \leq 1, 0 \leq t \leq 1,$$

with $a = x$, $b = \phi$, and $c = -\phi^2$. The initial and boundary conditions are given by

$$\phi(x, 0) = 1, \phi(0, t) = 0.$$

Suggest a procedure for solving this equation by coupling the method of characteristics and numerical integration scheme.

Section 11.11

11.18. Consider the Poisson equation

$$\nabla^2 \phi(x, y) = xy \text{ with } \phi(x, y) = 0 \text{ on the boundary.} \tag{E1}$$

Express this equation in polar coordinates. Indicate a procedure for solving the problem over a circular region of radius R using finite differences.

Section 11.13

11.19. Consider the Poisson equation

$$\frac{\partial^2 T}{\partial x^2} + \frac{\partial^2 T}{\partial y^2} = f(x, y); 0 \leq x \leq a; 0 \leq y \leq b$$

with

$$T(0, y) = c_1; \frac{\partial T}{\partial x}(a, y) = c_2;$$

$$T(x, 0) = c_3; \frac{\partial T}{\partial y}(x, b) = c_4.$$

Write a MATLAB program for finding $T(x, y)$ using central-difference approximations for $\frac{\partial^2 T}{\partial x^2}$ and $\frac{\partial^2 T}{\partial y^2}$ and forward- and backward-difference approximations with $O(h^2)$ for $\frac{\partial T}{\partial x}$ and $\frac{\partial T}{\partial y}$ at the boundaries. Use the program to find $T(x, y)$ for $c_1 = 200$, $c_2 = 0$, $c_3 = 400$, $c_4 = 0$, $f(x, y) = -5000$, and $a = b = 1$.

11.20. Develop a Maple program to solve Problem 11.19.

11.21. Develop a Mathcad program to solve Problem 11.19.

11.22. Solve Example 11.16 using MATLAB.

11.23. Use MATLAB to solve Example 11.14 with an explicit method (Section 11.5.1) with $\alpha^2 = 0.1$, $\Delta x = 0.1$, and (i) $\Delta t = 0.025$, and (ii) $\Delta t = 0.075$.

Section 11.14

11.24. Use Program 11.1 to find the solution of Laplace equation over the region $0 \le x, y \le 5$ when the temperature along the edge $y = 0$ is specified as x^3.

11.25. Use Program 11.2 to solve the transient heat equation

$$T_{xx} = T_t, 0 \le x \le 1, 0 \le t \le 2$$

with boundary conditions

$$T(0, t) = T(1, t) = 0$$

and initial condition

$$T(x, 0) = \sin^2 \pi x.$$

11.26. Use Program 11.3 to solve the partial differential equation

$$\phi_{xx} = \phi_{tt}, 0 \le x \le 1, 0 \le t \le 1$$

with boundary conditions

$$\phi(0, t) = \phi(1, t) = 0, 0 \le t \le 1$$

and initial conditions

$$\phi(x, 0) = x(1 - x), \frac{\partial \phi}{\partial t}(x, 0) = 0, 0 \le x \le 1.$$

11.27. Write a Fortran computer program to solve Problem 11.19.

General

11.28. The torsion of a prismatic shaft is governed by the equation

$$\frac{\partial^2 \phi}{\partial x^2} + \frac{\partial^2 \phi}{\partial y^2} = -2, \quad \text{(E1)}$$

where $\phi = \frac{\tilde{\phi}}{G\theta l^2}$, $x = \frac{\tilde{x}}{l}$, $y = \frac{\tilde{y}}{l}$, $\tilde{\phi}$ is the stress function, θ is the angle of twist per unit length, G is the shear modulus, and l is the length of the shaft. The shear stresses developed in the cross section can be expressed, in terms of $\tilde{\phi}$, as

$$\sigma_{\tilde{x}\tilde{z}} = -\frac{\partial \tilde{\phi}}{\partial \tilde{x}}, \sigma_{\tilde{y}\tilde{z}} = \frac{\partial \tilde{\phi}}{\partial \tilde{y}}. \quad \text{(E2)}$$

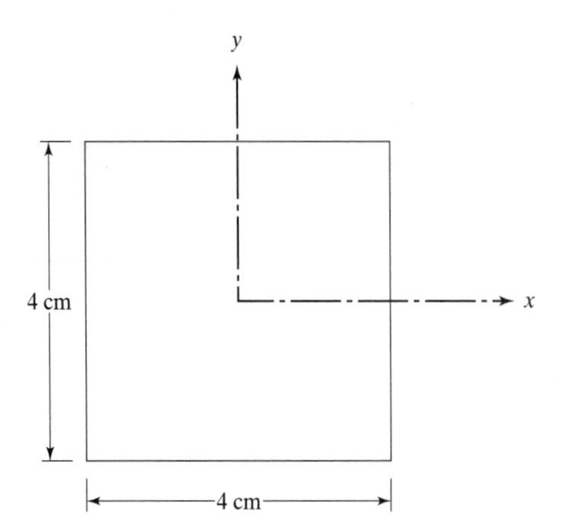

Figure 11.31 Square cross-section of prismatic shaft.

Since the shear stress is zero on an external boundary, the boundary condition on the external boundary becomes $\tilde{\phi} = 0$.

For the square cross section shown in Fig. 11.31, set up the equations for determining the distributions of $\tilde{\phi}$, $\sigma_{\tilde{x}\tilde{z}}$ and $\sigma_{\tilde{y}\tilde{z}}$ using a finite-difference method. Assume that $G = 0.8 \times 10^6$ kgf/ cm^2 and $\theta = 2°$ /m.

11.29. The variation of velocity, within the cross section, of a fluid flowing in a long, straight, uniform duct or pipe is governed by the equation

$$\frac{\partial^2 \phi}{\partial x^2} + \frac{\partial^2 \phi}{\partial y^2} + 1 = 0, \tag{E1}$$

where

$$\phi = \frac{u}{2u_0 f R_e}. \tag{E2}$$

Here u is the axial velocity of the fluid, u_0 is the mean velocity of the fluid, f is the Fanning friction factor, R_e is Reynold's number—$\frac{u_0 D \rho}{\mu}$, where $D = \frac{4(\text{area})}{\text{perimeter}}$ = hydraulic diameter of the duct, μ = absolute viscosity and ρ = density. The boundary condition is that u (and hence ϕ) is zero on the duct boundary. Suggest a method of finding the velocity distribution of a fluid flowing in the triangular duct shown in Fig. 11.32.

11.30. Derive the equations for determining the steady-state temperature distribution in the plate shown in Fig. 11.33 using the finite-difference grid indicated by the dotted lines.

11.31. Find the solution of the equation

$$\nabla^2 \phi = 0; 0 \leq x^2 + y^2 \leq 1, x \geq 0, y \geq 0; \tag{a}$$

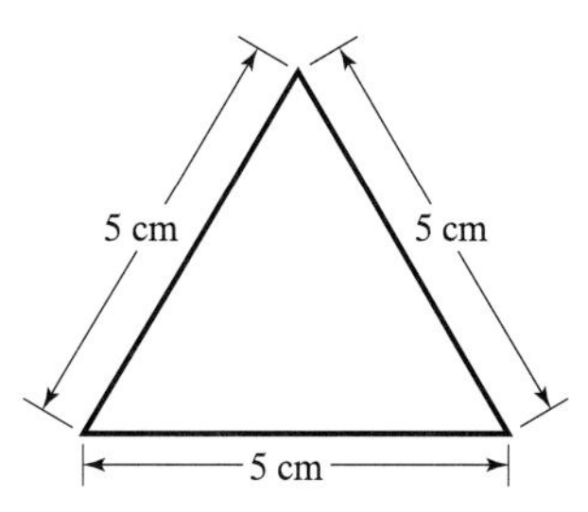

Figure 11.32 Triangular duct.

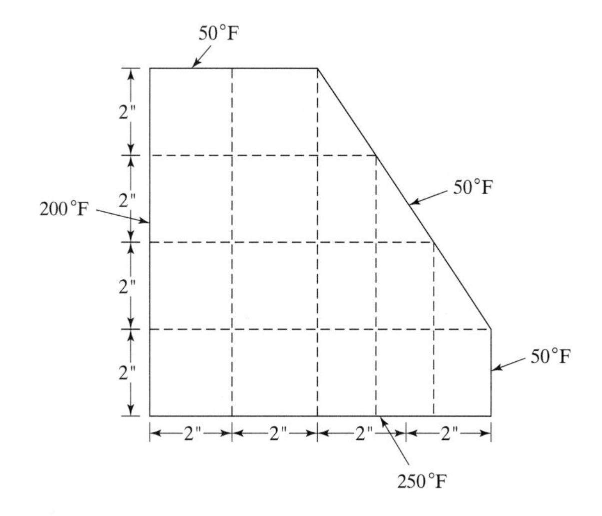

Figure 11.33 Finite difference grid of irregular plate.

subject to the boundary conditions

$$\phi(x, y) = 0; x = 0, y = 0; \tag{b}$$

$$\frac{\partial \phi}{\partial \vec{n}} = x - y; x^2 + y^2 = 1; \tag{c}$$

using a finite-difference method with $\Delta x = \Delta y = 1/2$.

11.32. Investigate the stability condition when the equation $\frac{\partial^2 \phi}{\partial x^2} = \frac{\partial \phi}{\partial t}$ is approximated as

$$\frac{\phi_{i+1,j} - 2\phi_{i,j} + \phi_{i-1,j}}{(\Delta x)^2} = \frac{\phi_{i,j} - \phi_{i,j-1}}{\Delta t}.$$

11.33. The vibration of a string or cable is governed by the equation

$$\frac{\partial^2 w}{\partial x^2} = \frac{\rho}{P} \frac{\partial^2 w}{\partial t^2},$$

where P is the tension in the string and ρ is the mass per unit length of the string. A cable is fixed at both ends and is subjected to the initial conditions

$$w(x, 0) = 0,$$

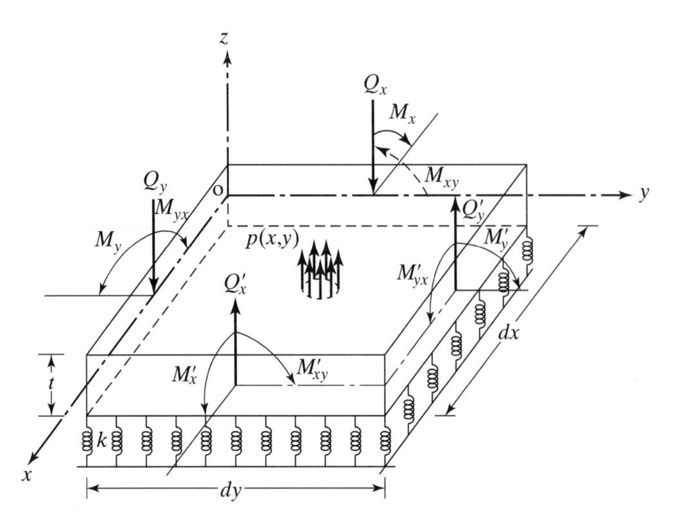

Figure 11.34 Forces and moments on a plate element.

$$\frac{\partial w}{\partial t}(x, 0) = \begin{cases} \dfrac{2ax}{1}; 0 \leq x \leq \dfrac{1}{2}; \\ 2a(1 - \frac{x}{l}); \frac{1}{2} \leq x \leq 1. \end{cases}$$

Determine the variation of the displacement $w(x, t)$ using a finite-difference method. Data: $l = 90$, $a = 1$, $P = 100$, $P/\rho = 1$.

11.34. Consider the boundary-value problem

$$\frac{\partial^2 \phi}{\partial x^2} + \frac{\partial^2 \phi}{\partial y^2} = \frac{\partial^2 \phi}{\partial t^2} \tag{a}$$

over a square region subject to the boundary conditions

$$\phi(x, y, t) = 0 \tag{b}$$

over the boundary and the initial conditions

$$\phi(x, y, 0) = \sin \pi x \sin \pi y; 0 \leq x, y \leq 1; \tag{c}$$

$$\frac{\partial \phi}{\partial t}(x, y, 0) = 0; 0 \leq x, y \leq 1. \tag{d}$$

Find the solution of the problem using a finite-difference method with $\Delta x = \Delta y = 1/3$ and $\Delta t = 1/2$.

PROJECTS

11.1. For determining the equation governing the transverse deflection (w) of a plate on an elastic foundation, the equilibrium of an element of size $dx \times dy$

is considered as shown in Fig. 11.34. The equilibrium of vertical forces and moments about the x- and y-axes and the strain–displacement relations can be combined to obtain the final plate deflection equation [11.10] to give

$$\frac{\partial^4 w}{\partial x^4} + 2\frac{\partial^4 w}{\partial^2 x \partial^2 y} + \frac{\partial^4 w}{\partial y^4} = \frac{p(x, y)}{D} - \frac{kw(x, y)}{D}, \tag{E1}$$

where p is the distributed load acting on the plate, k is the stiffness of the foundation (force per unit area per unit deflection), and D is the flexural rigidity

$$D = \frac{Et^3}{12(1 - \nu^2)}, \tag{E2}$$

where E and ν denote Young's modulus and Poisson's ratio of the plate, respectively.

The boundary conditions can be expressed as follows:

(a) Simply supported edge (y is constant):

$$w(x, y) = 0, M_y = -D\left(\frac{\partial^2 w}{\partial y^2} + \nu\frac{\partial^2 w}{\partial x^2}\right) = 0; 0 \le x \le a. \tag{E3}$$

(b) Fixed edge (y is constant):

$$w(x, y) = 0, \frac{\partial w(x, y)}{\partial y} = 0; 0 \le x \le a. \tag{E4}$$

(c) Free edge (y is constant):

$$M_y = -D\left(\frac{\partial^2 w}{\partial y^2} + \nu\frac{\partial^2 w}{\partial x^2}\right) = 0,$$

$$Q_y = -(2 - \nu)D\frac{\partial^3 w}{\partial x^2 \partial y} - D\frac{\partial^3 w}{\partial y^3} = 0; 0 \le x \le a. \tag{E5}$$

(i) State the boundary conditions if the edge x = constant is simply supported, fixed, or free.

(ii) A rectangular plate of size 30″ × 30″ × 0.5″ is simply supported on all the edges and rests on an elastic foundation. It is subjected to a uniformly distributed transverse load of 100 lb/in^2. Find the deflection of the plate using the central-difference method with step sizes $\Delta x = \Delta y = 10$ in. Assume that the foundation stiffness is $k = 50$ lb/in^3, $E = 30 \times 10^6$ psi, and $\nu = 0.3$.

(iii) Find the delection of the plate described in part (ii) by assuming that all the edges of the plate are fixed.

11.2. Show that the equation for the lateral vibration of a circular membrane is given by

$$\frac{\partial^2 w}{\partial r^2} + \frac{1}{r}\frac{\partial w}{\partial r} + \frac{1}{r^2}\frac{\partial^2 w}{\partial \theta^2} = \frac{\rho}{P}\frac{\partial^2 w}{\partial t^2},$$

where w is the lateral displacement, ρ is the mass per unit area, P is the tension, t is the time and (r, θ) are the polar coordinates. If a drum of diameter 30″ is made to vibrate by giving an initial displacement of 1″ at the center, determine the displacement of the drum using a finite-difference method for the following data:

$$P = 20 \text{ lb/in}, \ \rho = 0.001 \text{ lb/in}^2.$$

12

Optimization

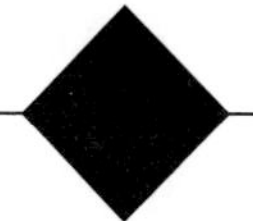

12.1 Introduction

CHAPTER CONTENTS

Optimization is the process by which the best solution is selected from among several possible solutions. Most engineering problems, including those associated with the analysis, design, construction, operation, and maintenance, involve decision making. Usually, there will be a criterion that is to be minimized or maximized while satisfying several social, economical, physical, and technological constraints. There will be a number of parameters that can be varied in the decision-making process. As the number of parameters increases, it becomes necessary to use systematic procedures for solving the decision-making problems. In this chapter, the methods of optimization are presented for solving a variety of decision-making problems.

12.1.1 Definition of an Optimization Problem

The formulation of an optimization problem involves the development of a mathematical model for the physical or engineering problem. In practice, several assumptions have to be made to develop a reasonably simple mathematical model that can predict the behavior of the system fairly accurately. The results of optimization will be different with different mathematical models of the same physical system. Hence, it is necessary to have a good mathematical model of the system, so that the results of optimization can be used to improve the performance of the system.

A general optimization problem can be stated in mathematical form as

$$\text{Find } \vec{X} = \begin{Bmatrix} x_1 \\ x_2 \\ \cdot \\ \cdot \\ \cdot \\ x_n \end{Bmatrix} \tag{12.1}$$

$$\text{that minimizes } f(\vec{X})$$

subject to

$$g_j(\vec{X}) \leq 0; \; j = 1, 2, \ldots, m;$$

and

$$h_k(\vec{X}) = 0; \; k = 1, 2, \ldots, p;$$

where $x_i (i = 1, 2, \ldots, n)$ are the decision variables, $\vec{X}$ is the vector of decision variables, $f(\vec{X})$ is the merit, or criterion, or objective, function, $g_j(\vec{X})$ is the jth inequality constraint function that is required to be less than or equal to zero, $h_k(\vec{X})$ is the kth equality constraint function that is required to be equal to zero, n is the number of decision variables, m is the number of inequality constraints, and p is the number of equality constraints. Each of the quantities appearing in Eq. (12.1) is described next.

12.1.2 Terminology

Decision variables

The formulation of an optimization problem begins with the identification of a set of variables that can be varied to change the performance of the system. These are called the decision, or design, variables, and their values are freely controlled by the decision maker. A set of numerical values, one for each decision variable, constitutes a solution (acceptable or unacceptable) to the optimization problem.

Objective function

When different solutions are obtained by changing the decision variables, a criterion is needed to judge whether one solution is better than another. This criterion, when expressed in terms of the decision variables, is called the objective, merit, or cost function. The interest of the decision maker is to select suitable values for the decision variables so as to minimize or maximize the objective function.

Inequality constraints

In any decision-making problem, there will be limitations or conditions imposed on the decision variables that may include economical, physical, or functional limitations. In many cases, the validity of the mathematical model used for the physical system imposes restrictions on the decision variables. These limitations, or conditions, when expressed in terms of the decision variables, are known as constraint functions. When the constraint function is restricted to have only negative or zero values, the resulting constraint is known as the inequality constraint. On the other hand, if the constraint function is required to be equal to zero, the resulting constraint is called the equality constraint.

Feasible solution

Any set of decision variables that satisfy the constraints of the problem (both inequality and equality constraints) is known as a feasible solution. A feasible solution is an acceptable solution to the decision maker in terms of the constraints, but may or may not minimize the objective function.

Optimum solution

A feasible solution that minimizes the objective function is known as the optimum solution.

▶Example 12.1

A beam of length 30 inches with a rectangular cross section is to be designed for minimum weight to carry a load of 1000 lb at the end as shown in Fig. 12.1. The maximum deflection and the maximum stress induced under the load are to be limited to 1 in and 20,000 lb/in^2, respectively. Formulate the optimization problem

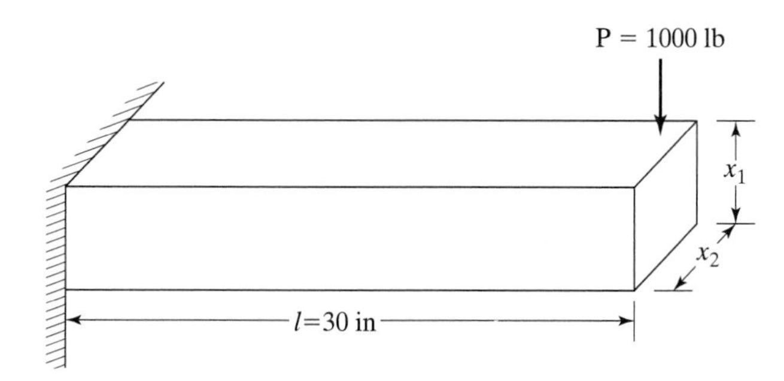

Figure 12.1 Design of minimum weight beam.

by identifying the design variables, objective function, and constraints. Assume Young's modulus (E) as 30×10^6 lb/in^2 and the unit weight (ρ) as 0.282 lb/in^3 for the material.

Solution

Since the cross-sectional dimensions of the beam are not specified, we choose the design variables as the depth (x_1) and the width (x_2) of the beam, in inches. The weight of the beam, in lb, to be minimized can be expressed as

$$f(x_1, x_2) = \rho l x_1 x_2 = (0.282)(30)x_1 x_2 = 8.46 x_1 x_2. \tag{a}$$

The maximum deflection δ in inches, at the tip of the beam, is given by [12.1]

$$\delta = \frac{Pl^3}{3EI} = \frac{4Pl^3}{Ex_1^3 x_2}, \tag{b}$$

where $I = \frac{1}{12}x_1^3 x_2$ is the area moment of inertia of the cross section. Using the known data, the constraint on deflection can be expressed as

$$g_1(x_1, x_2) = \delta - 1.0 \leq 0,$$

or

$$g_1(x_1, x_2) = \frac{3.6}{x_1^3 x_2} - 1.0 \leq 0. \tag{c}$$

The maximum bending stress induced (σ) in lb/in^2 (at the fixed end) of the beam is given by [12.1]

$$\sigma = \frac{Pl}{I}\frac{(x_1)}{2} = \frac{6Pl}{x_1^2 x_2}. \tag{d}$$

Using the known data, the constraint on stress can be expressed as

$$g_2(x_1, x_2) = \sigma - 20000 \leq 0,$$

$$g_2(x_1, x_2) = \frac{6(1000)(30)}{x_1^2 x_2} - 20000 \leq 0,$$

or

$$g_2(x_1, x_2) = \frac{9}{x_1^2 x_2} - 1 \leq 0. \tag{e}$$

Thus, the optimization problem is to find x_1 and x_2 to minimize the objective function given by Eq. (a) subject to the inequality constraints given by Eqs. (c) and (e). ◀

12.2 Types of Optimization Problems

The problem stated in Eq. (12.1) is called a constrained optimization problem. Optimization problems are also known as mathematical programming problems. In some practical situations, the constraints may be absent, and the problem reduces to

$$\text{find } \vec{X} = \begin{Bmatrix} x_1 \\ x_2 \\ \cdot \\ \cdot \\ \cdot \\ x_n \end{Bmatrix} \tag{12.2}$$

$$\text{that minimizes } f(\vec{X}).$$

This problem is known as an unconstrained optimization problem. Depending on the nature of functions, optimization problems can be classified as linear and nonlinear programming problems. A linear programming problem is one in which all the functions involved, namely, $f(\vec{X})$, $g_j(\vec{X})$, and $h_k(\vec{X})$ in Eq. (12.1), are linear in terms of the design variables. A nonlinear programming problem is one in which at least one of the functions is nonlinear in terms of the decision variables.

Some practical problems require the maximization of a function $f(\vec{X})$ instead of minimization. However, the maximum of a function $f(\vec{X})$ can be found by seeking the minimum of the negative of the same function. This situation is illustrated in Fig. 12.2 for the case of a function of a single variable. In this figure, $x = x^*$ corresponds to the maximum of the function $f(x)$. The same solution, $x = x^*$, also corresponds to

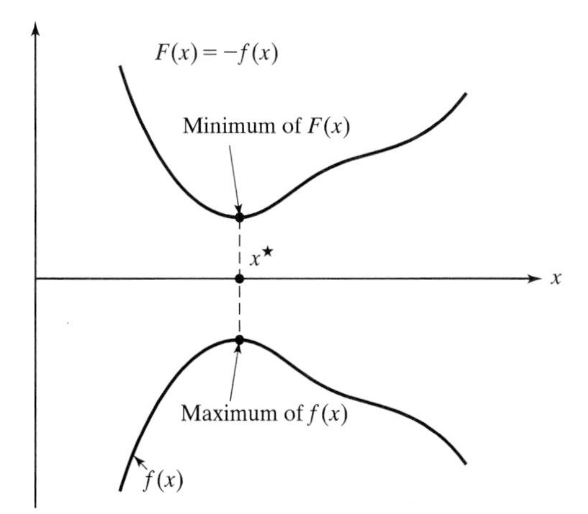

Figure 12.2 Maximum of $f(x)$ same as minimum of $-f(x)$.

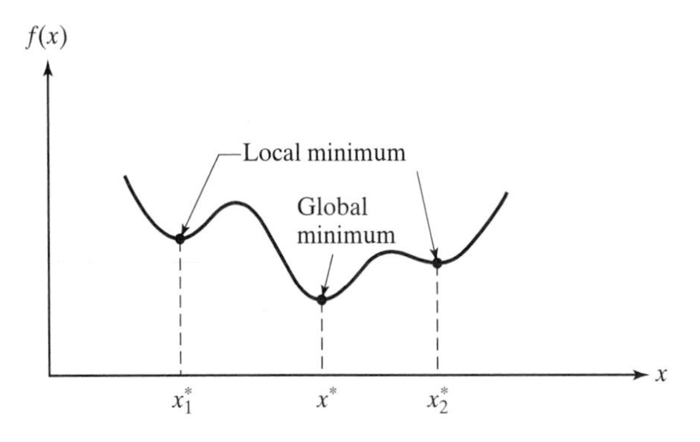

Figure 12.3 Local and global minima.

the minimum of the function $F(x) = -f(x)$. Hence, with out loss of generality, all optimization problems will be assumed to be of minimization type in this chapter.

The solution of an optimization problem may be local or global. In Fig. 12.3, three solutions are indicated for the function $f(x)$ in the absence of constraints. A local minimum x^* indicates the smallest value of the function $f(x)$ in the vicinity of x^*. A global minimum x^*, on the other hand, denotes the smallest value of the function in the entire range of x. For a linear programming problem, the optimum solution can be proved to be always a global minimum. However, for a nonlinear programming problem—constrained or unconstrained—the solution can be a local or a global minimum. Unfortunately, most numerical solutions can guarantee only a local minimum of the problem. In spite of this limitation, optimization methods are quite useful in solving several practical (and complex) decision-making problems.

12.3 Engineering Applications

▶Example 12.2

A manufacturer produces three components—A, B, and C—using two types of materials. The amount of each type of material required to produce one unit of each of the components is as follows:

For each unit of component type	Amount of material required (lb) Material 1	Material 2
A	8	6
B	3	5
C	5	4

The maximum amounts of materials of type 1 and 2 available per day are 2000 lb and 2500 lb, respectively. The profit for each unit of the components A, B, and C

(in dollars) is given by $500 - 2a - c$, $1200 - a - 2b$, and $800 - 2b - 3c$, respectively, where a, b, and c denote the number of components A, B, and C produced per day. Formulate the problem of maximizing the daily profit of the company, assuming that the company can sell all the components it manufactures.

Solution

Let a, b, and c denote the number of components A, B, and C produced per day, respectively. The design vector is given by

$$\vec{X} = \begin{Bmatrix} x_1 \\ x_2 \\ x_3 \end{Bmatrix} \equiv \begin{Bmatrix} a \\ b \\ c \end{Bmatrix}.$$

The constraint on a type-1 material is

$$8a + 3b + 5c \leq 2000.$$

The constraint on a type-2 material is

$$6a + 5b + 4c \leq 2500.$$

The bounds on design variables are as follows:
The design variables are expected to be nonnegative so that

$$a \geq 0, \ b \geq 0, \ c \geq 0.$$

The objective function for maximization is

$$\text{Total profit} = f(\vec{X}) = a(500 - 2a - c) + b(1200 - a - 2b) + c(800 - 2b - 3c). \quad ◀$$

▶Example 12.3

A power screw, having double square threads, is to be designed to lift a load of 1500 lb with maximum efficiency. (See Fig. 12.4.) Formulate the optimization problem by treating the major diameter (d), the number of threads per inch (N), and the mean diameter of the collar (d_c) as design variables.

Data: Load $= P = 1500$ lb, μ is the coefficient of friction of threads in nut $= 0.08$, μ_c is the coefficient of friction of thrust collar in bearing $= 0.08$, h is the height of the screw $= 6$ in. The material is steel with Young's modulus (E) of 30×10^6 psi, permissible shear stress ($\tau_{\max}$) of 12,000 psi and permissible compressive stress ($\sigma_{\max}$) of 20,000 psi.

The following are the requirements:

(1) Direct compressive stress (σ) must be less than $\sigma_{\max}$.

(2) Direct compressive stress (σ) must be less than buckling stress (σ_b).

(3) Shearing of screw threads at minor diameter (d_r) in nut must be avoided.

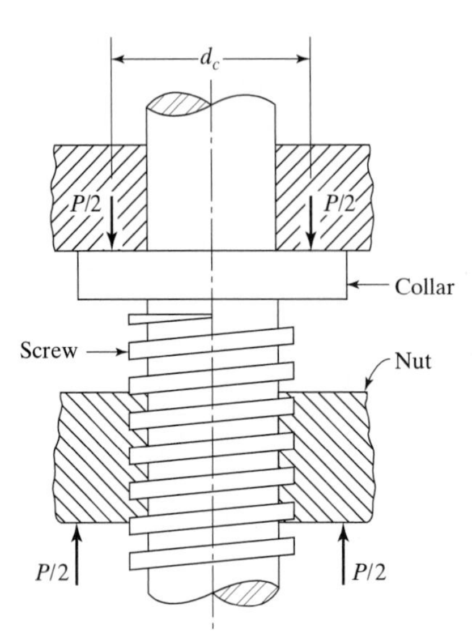

Figure 12.4 Power screw.

(4) Shearing of nut at major diameter of the screw must be avoided.

(5) Bearing stress in threads must be less than $(\sigma_{\max})$.

(6) Shear stress in screw due to applied torque (τ) must be less than $\tau_{\max}$.

The equations are as follows [12.2]:

(1) The efficiency of the screw (e) is

$$e = \frac{\text{Work out}}{\text{Work in}},$$

where Work out $= PL$ and

$$\text{Work in} = 2\pi T = 2\pi \left[\frac{P d_p}{2} \left(\frac{\pi d_p \mu + L}{\pi d_p - \mu L} \right) + \frac{1}{2} \mu_c P d_c \right],$$

where T is the torque, L is the lead, d_p is the pitch diameter, and d_c is the collar diameter.

(2) The tensile area is

$$A_t = \frac{\pi}{4} \left(\frac{d_p + d_r}{2} \right)^2,$$

where the pitch diameter (d_p) and root diameter (d_r) are given by

$$d_p = d - (0.649519/N),\ d_r = d - (1.299038/N),$$

where d denotes the major diameter of the screw.

The direct compressive stress in screw (σ) is P/A_t.

(3) The self-locking condition is $\mu \pi d_p \geq L$.

Solution

The design vector is

$$\vec{X} = \begin{Bmatrix} x_1 \\ x_2 \\ x_3 \end{Bmatrix} \equiv \begin{Bmatrix} d \\ N \\ d_c \end{Bmatrix}.$$

The objective function to be maximized (efficiency) is $f(\vec{X}) = e$.
The following are the constraints.
The direct compressive stress on the screw is

$$g_1(\vec{X}) = \sigma(\vec{X}) - \sigma_{\max} \leq 0.$$

The buckling constraint is

$$g_2(\vec{X}) = \sigma(\vec{X}) - \sigma_b \leq 0,$$

where the buckling stress (σ_b) is given by, assuming pin ends for the screw,

$$\sigma_b = \frac{\pi^2 EI}{h^2 A_t}$$

and

$$I = \frac{\pi d_p^4}{64}.$$

The constraint on the shearing of screw threads is given by

$$g_3(\vec{X}) = \frac{P}{\left(\pi d_r \frac{h}{2}\right)} - \tau_{\max} \leq 0.$$

The constraint on the shearing of nut threads is

$$g_4(\vec{X}) = \frac{P}{\left(\pi d \frac{h}{2}\right)} - \tau_{\max} \leq 0.$$

The constraint on the bearing stress in threads is

$$g_5(\vec{X}) = \frac{P}{\frac{\pi}{4}(d^2 - d_r^2)\frac{h}{p}} - \sigma_{\max} \leq 0,$$

where p is the pitch of the threads.
The constraint on the shear stress in the screw due to applied torque is

$$g_6(\vec{X}) = \frac{Tr}{J} = \frac{16T}{\pi d_r^3} - \tau_{\max} \leq 0.$$

The constraint for self-locking is

$$g_7(\vec{X}) = L - \mu\pi d_p \leq 0.$$

In addition, each of the design variables is restricted to be nonnegative. ◀

12.4 Optimization Methods from Differential Calculus

The theory of maxima and minima from differential calculus can be used to find the solution of certain types of optimization problems. These methods assume that the functions involved are continuous and differentiable.

12.4.1 Functions of One Variable

The simplest possible optimization problem seeks to find $x = x^*$ that minimizes the function $f(x)$. The necessary condition for a local minimum is given by

$$f'(x^*) = \frac{df(x^*)}{dx} = 0. \tag{12.3}$$

The point (solution) x^* given by Eq. (12.3) is guaranteed to be a local minimum (sufficient condition) if

$$f''(x^*) = \frac{d^2 f(x^*)}{dx^2} > 0. \tag{12.4}$$

Similarly, the point x^*, given by Eq. (12.3), is guaranteed to be a local maximum if

$$f''(x^*) = \frac{d^2 f(x^*)}{dx^2} < 0. \tag{12.5}$$

Note that the equation, $\frac{df(x)}{dx} = 0$, has to be solved to find the optimum solution, $x = x^*$. If the function $f(x)$ is extremely difficult to differentiate or if the function $f(x)$ is not available in explicit form (as is the case in many engineering applications), then an approximation procedure, such as numerical differentiation, is to be used to find the derivative of f. In some problems, both the first and the second derivatives will be zero at x^*. In such cases, we need to investigate the sign of the higher derivatives of f at x^* to establish the nature of the point x^*. If $f'(x^*) = f''(x^*) = \cdots = f^{(k-1)}(x^*) = 0$, but $f^{(k)}(x^*) \neq 0$, then $f(x^*)$ will be (i) a local minimum if $f^{(k)}(x^*) > 0$ and k is even, (ii) a local maximum if $f^{(k)}(x^*) < 0$, and k is even, and (iii) neither a minimum nor a maximum if k is odd. Here, $f^{(k)}(x^*)$ denotes the kth derivative of f evaluated at x^*.

▶Example 12.4

Determine the length of the shortest ladder AB that can be made to lean against the wall of a building without interfering an adjoining room $PQRS$ of height 10 ft and width 10 ft shown in Fig. 12.5.

Solution

The length of the ladder, x, is given by

$$x = AQ + QB = 10 \sec\theta + 10 \csc\theta, \tag{a}$$

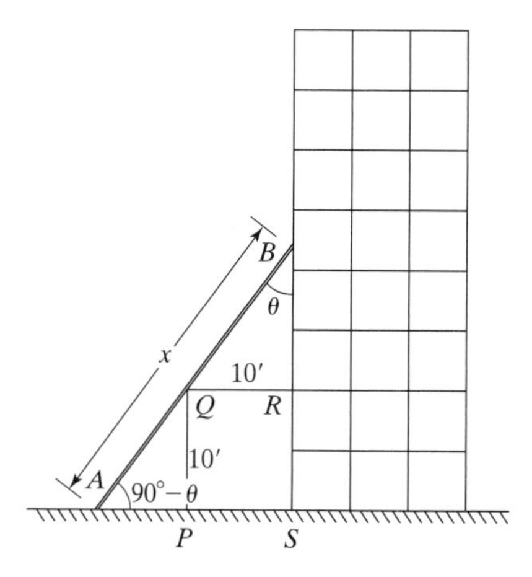

Figure 12.5 Shortest ladder problem.

where θ is the angle between the ladder and the wall. For the minimum of x, the necessary condition is

$$\frac{dx}{d\theta} = 0. \tag{b}$$

Equations (a) and (b) yield

$$\frac{dx}{d\theta} = 10(\sec\theta\tan\theta - \csc\theta\cot\theta) = 0. \tag{c}$$

The simplification of Eq. (c) yields

$$\tan^3\theta = 1, \text{ or } \tan\theta = \tan\frac{\pi}{4}. \tag{d}$$

This gives $\theta^* = \frac{\pi}{4}$, and hence, Eq. (a) yields

$$x^* = 10(\sqrt{2} + \sqrt{2}) = 20\sqrt{2} \text{ ft}. \tag{e}$$

The second derivative of x with respect to θ is

$$\begin{aligned}\frac{d^2x}{d\theta^2} &= \frac{d}{d\theta}(10\sec\theta\tan\theta - 10\csc\theta\cot\theta) \\ &= \frac{10}{\cos\theta}(1 + 2\tan^2\theta) + \frac{10}{\sin\theta}(1 + 2\cot^2\theta).\end{aligned} \tag{f}$$

Using $\tan\theta^* = \cot\theta^* = 1$, Eq. (f) gives

$$\left.\frac{d^2x}{d\theta^2}\right|_{\theta^*} = 10\sqrt{2}(3) + 10\sqrt{2}(3) = 60\sqrt{2} > 0, \tag{g}$$

which proves that x^* is a relative minimum. ◀

12.4.2 Function of Several Variables

Consider a multivariable function

$$f(\vec{X}) = f(x_1, x_2, \ldots, x_n). \tag{12.6}$$

The necessary conditions for the minimum of $f(\vec{X})$ are given by

$$\frac{\partial f}{\partial x_i}(x_1, x_2, \ldots, x_n) = 0;\ i = 1, 2, \ldots, n. \tag{12.7}$$

For a nonlinear function f, Eq. (12.7) represents a set of n simultaneous nonlinear equations. The solution of these equations gives the point $\vec{X}^*$:

$$\vec{X}^* = \begin{Bmatrix} x_1^* \\ x_2^* \\ \cdot \\ \cdot \\ \cdot \\ x_n^* \end{Bmatrix}. \tag{12.8}$$

This solution is guaranteed to be a local minimum solution of $f(\vec{X})$ only if the matrix of second partial derivatives of f, $[A]$, evaluated at $\vec{X}^*$, is positive definite. The matrix $[A]$ is given by

$$\begin{bmatrix} \dfrac{\partial^2 f}{\partial x_1^2} & \dfrac{\partial^2 f}{\partial x_1 \partial x_2} & \cdots & \dfrac{\partial^2 f}{\partial x_1 \partial x_n} \\ \dfrac{\partial^2 f}{\partial x_1 \partial x_2} & \dfrac{\partial^2 f}{\partial x_2^2} & \cdots & \dfrac{\partial^2 f}{\partial x_2 \partial x_n} \\ \cdot & \cdot & \cdots & \cdot \\ \cdot & \cdot & \cdots & \cdot \\ \cdot & \cdot & \cdots & \cdot \\ \dfrac{\partial^2 f}{\partial x_1 \partial x_n} & \dfrac{\partial^2 f}{\partial x_2 \partial x_n} & \cdots & \dfrac{\partial^2 f}{\partial x_n^2} \end{bmatrix}_{\vec{X}=\vec{X}^*}. \tag{12.9}$$

The matrix $[A]$ is called the Hessian matrix. A matrix $[A] = [a_{ij}]$ is said to be positive definite if all its eigenvalues are positive. Since the matrix $[A]$ is symmetric, the methods described in Sections 4.5 through 4.8 can be used to determine its eigenvalues. Alternatively, the positive definiteness of the matrix $[A]$ can be determined by evaluating the following determinants:

$$A_1 = |a_{11}|, \tag{12.10}$$

$$A_2 = \begin{vmatrix} a_{11} & a_{12} \\ a_{21} & a_{22} \end{vmatrix}, \tag{12.11}$$

$$A_3 = \begin{vmatrix} a_{11} & a_{12} & a_{13} \\ a_{21} & a_{22} & a_{23} \\ a_{31} & a_{32} & a_{33} \end{vmatrix}, \tag{12.12}$$

$$\vdots$$

and

$$A_n = \begin{vmatrix} a_{11} & a_{12} & \cdot & \cdot & \cdot & a_{1n} \\ a_{21} & a_{22} & \cdot & \cdot & \cdot & a_{2n} \\ \cdot & \cdot & \cdot & \cdot & \cdot & \cdot \\ \cdot & \cdot & \cdot & \cdot & \cdot & \cdot \\ \cdot & \cdot & \cdot & \cdot & \cdot & \cdot \\ a_{n1} & a_{n2} & \cdot & \cdot & \cdot & a_{nn} \end{vmatrix}. \tag{12.13}$$

The matrix $[A]$ will be positive definite if and only if the values of $A_1, A_2, \ldots, A_n$ are all positive.

▶Example 12.5

The displacement components x_1 and x_2 of node R of the two-bar truss shown in Fig. 12.6 can be found by minimizing the potential energy (f) given by

$$f(x_1, x_2) = \frac{EA}{s}\left(\frac{l}{2s}\right)^2 x_1^2 + \frac{EA}{s}\left(\frac{h}{s}\right)^2 x_2^2 - P_1 x_1 - P_2 x_2, \tag{a}$$

where E is Young's modulus, A is the cross-sectional area of each member, s is the length of each member, l is the base length, and P_1 and P_2 are the loads applied along the horizontal and vertical directions (as shown in Fig. 12.6).

Determine the values of x_1 and x_2 for the following data: $h = 30$ in, $E = 30 \times 10^6$ psi, $A = 1$ in^2, and $l = s = 20\sqrt{3}$ in.

Solution

For the given data, the potential energy can be expressed as

$$f(x_1, x_2) = 0.21650625 \times 10^6\, x_1^2 + 0.866025 \times 10^6\, x_2^2 - 1000\, x_1 - 500\, x_2. \tag{b}$$

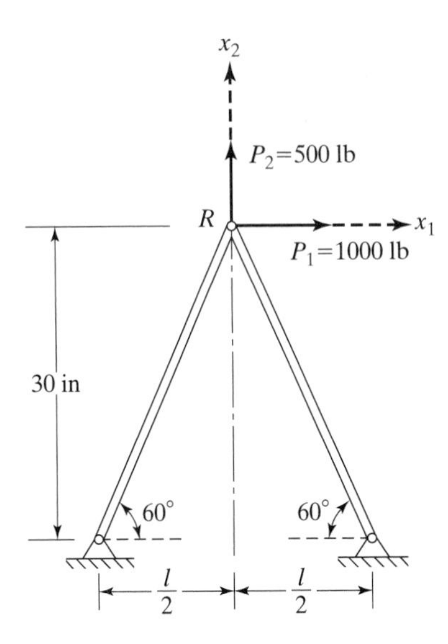

Figure 12.6 Two-bar truss.

The necessary conditions for the minimum of f are given by

$$\frac{\partial f}{\partial x_1} = 0.4330125 \times 10^6 x_1 - 1000 = 0, \tag{c}$$

and

$$\frac{\partial f}{\partial x_2} = 1.2990375 \times 10^6 x_2 - 500 = 0. \tag{d}$$

The solution of Eqs.(c) and (d) is given by

$$x_1^* = 2.309402 \times 10^{-3} \text{ in}, x_2^* = 0.384900 \times 10^{-3} \text{ in}. \tag{e}$$

The Hessian matrix at (x_1^*, x_2^*) is given by

$$[A] = \begin{bmatrix} \dfrac{\partial^2 f}{\partial x_1^2} & \dfrac{\partial^2 f}{\partial x_1 \partial x_2} \\ \dfrac{\partial^2 f}{\partial x_1 \partial x_2} & \dfrac{\partial^2 f}{\partial x_2^2} \end{bmatrix} = \begin{bmatrix} 0.4330125 \times 10^6 & 0 \\ 0 & 1.2990375 \times 10^6 \end{bmatrix}. \tag{f}$$

Since $A_1 = |0.4330125 \times 10^6| > 0$ and

$$A_2 = \begin{vmatrix} 0.4330125 \times 10^6 & 0 \\ 0 & 1.2990375 \times 10^6 \end{vmatrix} = 0.562499 \times 10^{12} > 0,$$

the matrix $[A]$ is positive definite. Hence, the solution (x_1^*, x_2^*) corresponds to the minimum of f. ◀

12.4.3 Function of Several Variables with Equality Constraints

Certain optimization problems involve only equality constraints; for example,

$$\text{find } \vec{X} = \begin{Bmatrix} x_1 \\ x_2 \\ \cdot \\ \cdot \\ \cdot \\ x_n \end{Bmatrix}$$

$$\text{that minimizes } f(\vec{X})$$

subject to

$$h_k(\vec{X}) = 0;\ k = 1, 2, \ldots, p. \tag{12.14}$$

In Eq. (12.14), the number of equality constraints, p is less than the number of design variables, n. If $p > n$, there will be more equations than the number of unknowns, and hence, there may not be a solution to the problem. If $p = n$, then the number of equations will be equal to the number of unknowns. In this case, there will be a unique solution, and there is no possibility for optimization. Thus, the only case of practical significance is when $p < n$. A method, known as the Lagrange multiplier method, can be used to solve the problem in Eq. (12.14).

In the Lagrange multiplier method, the Lagrangian function

$$L(\vec{X}, \vec{\lambda}) = f(\vec{X}) + \sum_{k=1}^{p} \lambda_k h_k(\vec{X}), \tag{12.15}$$

where λ_k are constants (unknowns at this stage) called Lagrange multipliers, and $\vec{\lambda}$ is the vector of Lagrange multipliers

$$\vec{\lambda} = \begin{Bmatrix} \lambda_1 \\ \lambda_2 \\ \cdot \\ \cdot \\ \cdot \\ \lambda_p \end{Bmatrix}, \tag{12.16}$$

is constructed. It can be proved [12.3] that the solution of the original problem, Eq. (12.14), can be obtained by extremizing the function L, given by Eq. (12.15). By treating L as a function of $n + p$ variables, namely, $x_1, x_2, \ldots, x_n, \lambda_1, \lambda_2, \ldots, \lambda_p$, the necessary conditions for the extremum of L can be obtained by

$$\frac{\partial L}{\partial x_i} = \frac{\partial f}{\partial x_i} + \sum_{k=1}^{p} \lambda_k \frac{\partial h_k}{\partial x_i} = 0;\ i = 1, 2, \ldots, n;$$

and

$$\frac{\partial L}{\partial \lambda_k} = h_k = 0;\ k = 1, 2, \ldots, p. \tag{12.17}$$

Equation (12.17) denotes $n + p$ equations in the $n + p$ unknowns, and hence, their solution yields

$$\vec{X}^* = \begin{Bmatrix} x_1^* \\ x_2^* \\ \cdot \\ \cdot \\ \cdot \\ x_n^* \end{Bmatrix} \text{ and } \vec{\lambda}^* = \begin{Bmatrix} \lambda_1^* \\ \lambda_2^* \\ \cdot \\ \cdot \\ \cdot \\ \lambda_p^* \end{Bmatrix}. \tag{12.18}$$

The solution $\vec{X}^*$ will also be the solution (local minimum) of the original problem stated in Eq. (12.14). The sufficiency conditions for the minimum of $f(\vec{X}^*)$ can also be verified, but they are quite involved [12.4]. The following example illustrates the Lagrange multiplier method:

▶Example 12.6

The maximum bending stress induced in a beam of rectangular cross section is given by

$$f = \frac{M\,y}{I}, \tag{a}$$

where M is the maximum bending moment, $y = \frac{x_2}{2}$ is the distance of the outermost fiber from the neutral axis, and $I = \frac{1}{12}x_1x_2^3$ is the moment of inertia of the cross section about the neutral axis. (See Fig. 12.7.) Find the cross-sectional dimensions of the beam that can be cut from a circular log of radius 10 in to minimize the induced stress when $M = 2000$ lb-in.

Solution

The optimization problem can be stated as follows:
Find x_1 and x_2 to minimize

$$f(x_1, x_2) = \frac{(2000)x_2}{2\left(\frac{1}{12}x_1x_2^3\right)} = \frac{12000}{x_1x_2^2} \tag{b}$$

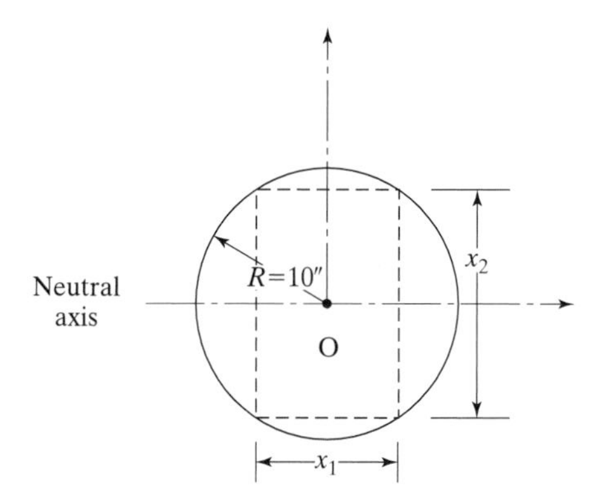

Figure 12.7 Rectangular beam from circular log.

subject to

$$g(x_1, x_2) = \left(\frac{x_1}{2}\right)^2 + \left(\frac{x_2}{2}\right)^2 - R^2 = 0,$$

or

$$g(x_1, x_2) = x_1^2 + x_2^2 - 400 = 0. \tag{c}$$

The Lagrange function is defined as

$$L(x_1, x_2, \lambda) = \frac{12000}{x_1 x_2^2} + \lambda(x_1^2 + x_2^2 - 400). \tag{d}$$

The necessary conditions for the stationariness of L are given by

$$\frac{\partial L}{\partial x_1} = -\frac{12000}{x_1^2 x_2^2} + 2\lambda x_1 = 0, \tag{e}$$

$$\frac{\partial L}{\partial x_2} = -\frac{24000}{x_1 x_2^3} + 2\lambda x_2 = 0, \tag{f}$$

and

$$\frac{\partial L}{\partial \lambda} = x_1^2 + x_2^2 - 400 = 0. \tag{g}$$

Equations (e) and (f) yield

$$2\lambda = \frac{12000}{x_1^3 x_2^2} = \frac{24000}{x_1 x_2^4}, \text{ or } x_2 = \sqrt{2} x_1. \tag{h}$$

Substituting Eq. (h) into Eq. (g) gives

$$x_1^2 + 2x_1^2 = 400, \text{ or } x_1^* = \frac{20}{\sqrt{3}}, \tag{i}$$

and hence, from Eq. (h), we obtain

$$x_2^* = \sqrt{2} x_1^* = \sqrt{\frac{2}{3}}\, 20. \tag{j}$$

Although it is not required, the value of the Lagrange multiplier can be found:

$$\lambda^* = \frac{12000}{x_1^* x_2^{*4}} = \frac{27\sqrt{3}}{3200}. \tag{k}$$

◀

12.4.4 Function of Several Variables with Mixed Equality and Inequality Constraints

Next, consider the general constrained optimization problem involving both inequality and equality constraints indicated in Eq. (12.1). The inequality constraints of Eq. (12.1) can be converted to equality constraints with the equation

$$G_j(\vec{X}, y_j) = g_j(\vec{X}) + y_j^2 = 0; \; j = 1, 2, \ldots, m, \tag{12.19}$$

where y_j are new variables (y_j^2 is used to ensure nonnegative values), known as slack variables, introduced to convert the jth inequality constraint to an equality constraint. With the use of Eq. (12.19), the optimization problem of Eq. (12.1) can be restated as a problem involving only equality constraints as follows:

Find $(x_1, x_2, \ldots, x_n, y_1, y_2, \ldots, y_m)$ that minimizes $f(\vec{X})$ subject to

$$G_j(\vec{X}, y_j) = g_j(\vec{X}) + y_j^2 = 0, \; j = 1, 2, \ldots, m, \tag{12.20}$$

and

$$h_k(\vec{X}) = 0, k = 1, 2, \ldots, p.$$

This equality constrained optimization problem, Eq. (12.20), can be solved by using the Lagrange multiplier method described in the previous section.

12.5 Linear-programming Problem

As stated earlier, a linear programming problem is one in which all the functions are linear in terms of the decision variables. Linear-programming (LP) problems represent the most widely used optimization problems in practice.

12.5.1 Standard Form

An LP problem is generally expressed in the following standard form:

Minimize $f = c_1x_1 + c_2x_2 + \cdots + c_nx_n$ subject to

$$\begin{aligned} a_{11}x_1 + a_{12}x_2 + \cdots + a_{1n}x_n &= b_1, \\ a_{21}x_1 + a_{22}x_2 + \cdots + a_{2n}x_n &= b_2, \\ &\vdots \\ a_{m1}x_1 + a_{m2}x_2 + \cdots + a_{mn}x_n &= b_m, \end{aligned} \tag{12.21}$$

with

$$x_i \geq 0, \; i = 1, 2, \ldots, n,$$

where the constants $c_1, c_2, \ldots, c_n, a_{11}, a_{12}, \ldots, a_{mn}, b_1, b_2, \ldots, b_m$ are assumed to be known. The constants c_i, $i = 1, 2, \ldots, n$, are also known as the cost coefficients. If a constraint is originally stated as a "less than or equal to" type of inequality,

$$a_{i1}x_1 + a_{i2}x_2 + \cdots + a_{in}x_n \leq b_i, \tag{12.22}$$

then a new unknown, called a slack variable, $y_i \geq 0$ is introduced, and the inequality is converted into an equality:

$$a_{i1}x_1 + a_{i2}x_2 + \cdots + a_{in}x_n + y_i = b_i. \tag{12.23}$$

On the other hand, if the original constraint is of the form

$$a_{i1}x_1 + a_{i2}x_2 + \cdots + a_{in}x_n \geq b_i, \tag{12.24}$$

it is converted into an equality by introducing a new unknown, called a surplus variable, $y_i \geq 0$:

$$a_{i1}x_1 + a_{i2}x_2 + \cdots + a_{in}x_n - y_i = b_i. \tag{12.25}$$

The standard form of the LP problem, Eq. (12.21), requires that all the decision variables be nonnegative. In some applications, a decision variable x_i may be unrestricted in sign (i.e., x_i can assume a negative, zero, or positive value in the final solution). Such an unrestricted variable x_i can be replaced by two new variables $x_i' \geq 0$ and $x_i'' \geq 0$ with

$$x_i = x_i' - x_i''. \tag{12.26}$$

It can be seen that x_i can take a negative, zero, or positive value, depending on the relative values of x_i' and x_i''.

▶Example 12.7

Convert to standard form the LP problem of

$$\text{maximizing } F = 2x_1 - 7x_2$$

subject to

$$\begin{aligned} 2x_1 + x_2 &\geq 3, \\ 5x_1 + 3x_2 &\leq 15, \\ x_1 - 4x_2 &\geq -20, \end{aligned} \tag{a}$$

and

$$x_1, x_2 \text{ unrestricted in sign.}$$

Solution

The objective function can be stated as

$$\text{minimize } f = -F = -2x_1 + 7x_2. \tag{b}$$

The inequality constraints can be stated by the equality constraints

$$g_1(x_1, x_2, x_3) = 2x_1 + x_2 - x_3 = 3, \tag{c}$$

$$g_2(x_1, x_2, x_4) = 5x_1 + 3x_2 + x_4 = 15, \tag{d}$$

and

$$g_3(x_1, x_2, x_5) = -x_1 + 4x_2 + x_5 = 20, \tag{e}$$

where x_3 is a nonnegative surplus variable and x_4 and x_5 are the nonnegative slack variables. The unrestricted variables can be replaced by nonnegative variables as

$$x_1 = x_1^+ - x_1^-, \; x_2 = x_2^+ - x_2^-. \tag{f}$$

Thus, the optimization problem can be stated in standard form as find $(x_1^+, x_1^-, x_2^+, x_2^-, x_3, x_4, x_5)$ that minimizes

$$f = -2x_1^+ + 2x_1^- + 7x_2^+ - 7x_2^- \tag{g}$$

subject to

$$g_1 = 2x_1^+ - 2x_1^- + x_2^+ - x_2^- - x_3 = 3, \tag{h}$$

$$g_2 = 5x_1^+ - 5x_1^- + 3x_2^+ - 3x_2^- + x_4 = 15, \tag{i}$$

$$g_3 = -x_1^+ + x_1^- + 4x_2^+ - 4x_2^- + x_5 = 20, \tag{j}$$

and

$$x_1^+ \geq 0, x_1^- \geq 0, x_2^+ \geq 0, x_2^- \geq 0, x_3 \geq 0, x_4 \geq 0, x_5 \geq 0. \tag{k}$$

◀

12.5.2 Graphical Solution

Optimization problems in two variables can be solved using a graphical procedure. In the graphical method, the constraints are plotted in the decision variable space, and the feasible region, in which all the constraints are satisfied, is identified. Then by plotting the objective function contours, we identify the optimum solution. Although the graphical procedure is not applicable for most practical problems involving many decision variables, the method provides a graphical picture of the general characteristics of linear-programming problems. The following example illustrates the graphical optimization procedure:

▶Example 12.8

A manufacturer produces two types of products (products A and B) using lathes, milling machines, and grinding machines. The amount of machining time required for each product and the profit of each product are as follows:

Product	On lathe	On milling machine	On grinding machine	Profit per unit (dollars)
	(machining time required (hours))			
A	16	8	10	90
B	8	14	9	110

The maximum machining times available per day on lathes, milling machines, and grinding machines are, respectively, 128, 112, and 90 hours. Determine the number of units of products A and B to be manufactured per day for maximum profit.

Solution

Let x_1 and x_2 denote the number of units of products A and B manufactured per day. The constraints imposed by the maximum times available on different machines can be stated as follows:

$$\text{Lathes: } 16x_1 + 8x_2 \le 128, \tag{a}$$

$$\text{Milling machines: } 8x_1 + 14x_2 \le 112, \tag{b}$$

$$\text{Grinding machines: } 10x_1 + 9x_2 \le 90. \tag{c}$$

Since x_1 and x_2 cannot have negative values, we have

$$x_1 \ge 0, \tag{d}$$

and

$$x_2 \ge 0. \tag{e}$$

The profit per day is given by

$$\text{Maximize: } f = 90x_1 + 110x_2. \tag{f}$$

The complete optimization problem can be stated as follows: Find x_1 and x_2 to maximize f given by Eq. (f) subject to the constraints of Eqs. (a) through (e).

For the graphical solution, we consider a two-dimensional graph with the decision variables x_1 and x_2 denoting the two coordinates. (See Figure 12.8.) Inequalities (d) and (e) indicate that the solution must lie in the first quadrant. The constraint boundaries of Eqs. (a), (b), and (c) are plotted (by considering equality signs in Eqs. (a), (b), and (c)) as straight lines P_1Q_1, P_2Q_2, and P_3Q_3. It can be seen that all the constraints of Eqs. (a) through (e) will be satisfied by all the points in the hatched (crossed) area. This closed area is called the feasible space. Any point in the two-dimensional space outside the feasible area violates one or more of the constraints Eqs. (a) through (e). Some of the contours of the objective function are also shown in the figure. It can be seen that the maximum value of f that can be achieved without violating any of the constraints is 980.5882. The optimum solution

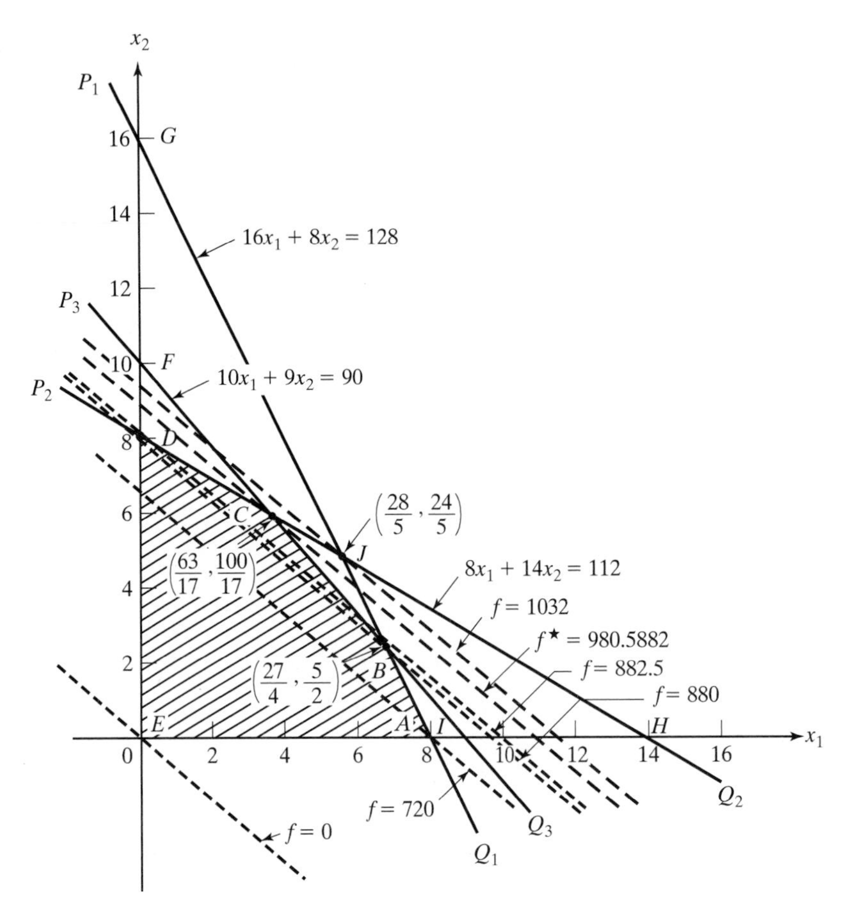

Figure 12.8 Graphical solution of Example 12.8

is given by point C with coordinates ($x_1 = \frac{63}{17}$, $x_2 = \frac{100}{17}$). Thus, the number of products A and B to be manufactured per day for maximum profit are $\frac{63}{17}$ and $\frac{100}{17}$, respectively, and the maximum profit is \$980.5882 per day. ◀

12.6 Simplex Method

The standard LP problem is stated in Eq. (12.21). If the problem has an optimal solution, then it must occur at an extreme point. In Example 12.8, the number of extreme points is equal to 10 and are indicated by the letters $A, B, C, \ldots, J$. The characteristics of these points are indicated in Table 12.1. By finding the basic feasible solutions, the optimum solution of the problem can be identified from among the basic feasible solutions (extreme points). For a problem with n decision

Table 12.1

Extreme point in Fig. 12.8 (Basic Solution)[1]	Coordinates		Basic feasible solution[2] ?	Optimum solution ?	Value of f	Values of		
	x_1	x_2				x_3	x_4	x_5
A	8	0	yes	no	720	0	48	10
B	$\frac{27}{4}$	$\frac{5}{2}$	yes	no	882.5	0	23	0
C	$\frac{63}{17}$	$\frac{100}{17}$	yes	yes	980.5882	$\frac{368}{17}$	0	0
D	0	8	yes	no	880	64	0	18
E	0	0	yes	no	0	128	112	90
F	0	10	no	no	1100	48	−28	0
G	0	16	no	no	1760	0	−112	−54
H	14	0	no	no	1260	−96	0	−50
I	9	0	no	no	810	−16	40	0
J	$\frac{28}{5}$	$\frac{24}{5}$	no	no	1032	0	0	$-\frac{46}{5}$

(1) A basic solution is one in which $(n - m)$ variables are zero.

(2) A basic feasible solution or an extreme point is a basic solution that satisfies the nonnegativity conditions, $x_i \geq 0, i = 1, 2, \ldots, n$.

variables and m equality constraints, the number of basic solutions is given by

$$\text{Number of basic solutions} = \frac{n!}{m!(n-m)!}. \tag{12.27}$$

For most practical problems, this represents a fairly large number of basic solutions. For example, if $n = 25$ and $m = 10$, the number is approximately 3,268,000. Hence, an efficient and systematic procedure is needed to identify the optimum solution from among the basic solutions. The simplex method is a scheme that moves from one extreme point to an adjacent one (with an improved objective function value) until the optimum solution is identified [12.4].

12.7 Search Methods for Nonlinear Optimization

If the optimization problem is nonlinear, the differential calculus based methods discussed in Section 12.4 can be used. However, these methods have the following limitations:

(1) The differential calculus based methods require the solution of a set of simultaneous nonlinear equations derived from the necessary conditions of optimality. For simple objective and constraint functions, $f(\vec{X})$, $g_j(\vec{X})$,

$j = 1, 2, \ldots, m$, and $h_k(\vec{X})$, $k = 1, 2, \ldots, p$, these equations can be solved using the methods described in Chapter 2. However, in many practical problems, the functions $f(\vec{X})$, $g_j(\vec{X})$, and $h_k(\vec{X})$ will be highly nonlinear. In such cases, the solution of the resulting nonlinear equations requires considerable effort.

(2) In many engineering applications, the objective and constraint functions $f(\vec{X})$, $g_j(\vec{X})$, $j = 1, 2, \ldots, m$, and $h_k(\vec{X})$, $k = 1, 2, \ldots, p$, may not be available as explicit functions of $\vec{X}$. Usually, one can only hope to find the numerical values of f, g_j, and h_k by conducting a lengthy computer analysis, such as a finite-element analysis, to find the thermal stress in the tile of a space shuttle, for any given decision variable vector $\vec{X}$. For such cases, the application of the differential calculus methods of optimization become very difficult.

12.8 Optimization of a Function of a Single Variable

The problem can be stated as follows:

$$\text{Find } x \text{ that minimizes } f(x). \tag{12.28}$$

Several numerical methods are available for solving the problem of Eq. (12.28). Three methods, namely, the exhaustive search method, the Fibonacci method, and the cubic interpolation method, are considered in this section. The concept of a *unimodal function*, which is required in the exhaustive search and Fibonacci methods, is introduced first.

12.8.1 Unimodal Function

A unimodal function is one that has only one minimum or maximum point within the region of interest. Such a function is shown in Fig. 12.9. If the values of a unimodal

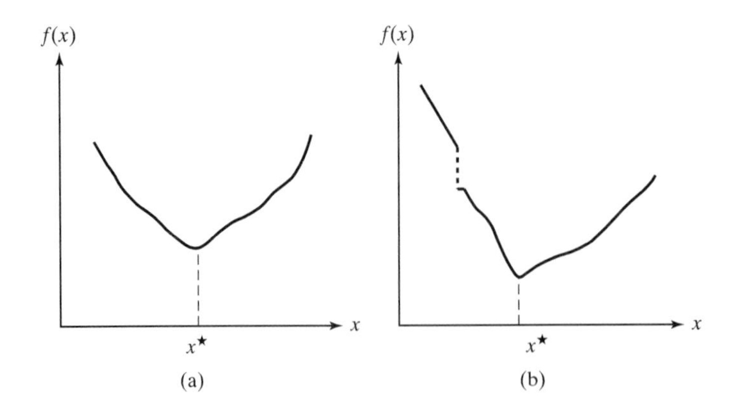

Figure 12.9 Unimodal functions.

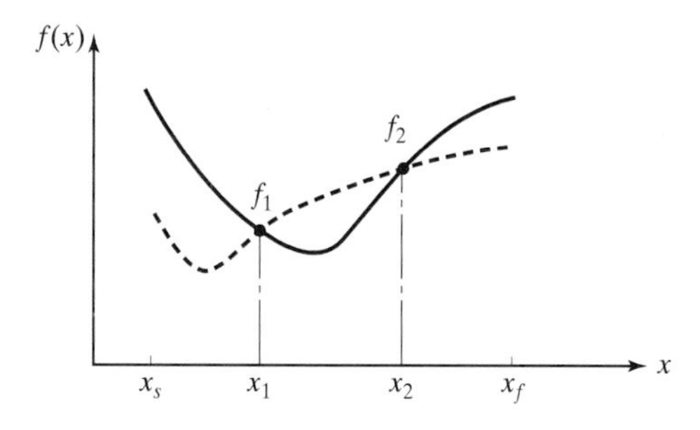

Figure 12.10 Minimum of unimodal function.

function are available at two different locations in a given interval, we can identify the subinterval in which its minimum lies. Consider the case in which two function evaluations are made at x_1 and x_2 for a unimodal function as shown in Fig. 12.10. If the function value $f_1 < f_2$, then we can see that the region x_s to x_2 must contain the minimum. If this were not true [i.e., if the minimum lies in the region (x_2, x_f)], then the function will have a lower value than f_2 on either side of x_2, which contradicts the assumption of unimodality for the function f.

12.8.2 Exhaustive Search Method

Let the minimum of the function $f(x)$ be known to lie between x_s and x_f. The interval $L_0 = (x_s, x_f)$ is known as the initial interval of uncertainty. Furthermore, the function is assumed to be unimodal. In the exhaustive search method, the objective function, f, is evaluated at a predetermined number of equally spaced points in the interval (x_s, x_f), and the interval of uncertainty is reduced using the assumption of unimodality of f. The evaluation of the function $f(x)$ at any specific value of x is termed as an experiment. Suppose that the function is evaluated at six equally spaced points x_1 through x_6 (six experiments are conducted), and the function values appear as shown in Fig. 12.11. Since the function is unimodal, it will have only one minimum; the minimum must lie in between the points x_2 and x_4. Thus, the interval (x_2, x_4) can be considered as the final interval of uncertainty. The final interval can be reduced if the function f is evaluated at more number of points in (x_s, x_f). In general, if the function is evaluated at n equally spaced points in the interval $L_0 = x_f - x_s$, and if the minimum value out of $f_1, f_2, \ldots, f_n$ is found to be f_j, then the final interval of uncertainty is given by

$$L_n = x_{j+1} - x_{j-1} = \left(\frac{2}{n+1}\right) L_0. \tag{12.29}$$

The following example illustrates the exhaustive search method.

▶Example 12.9

A rectangular room of size $20' \times 30' \times h'$ has a light bulb at the middle of the ceiling. (See Fig. 12.12.) The light intensity at a point B on the floor is directly proportional

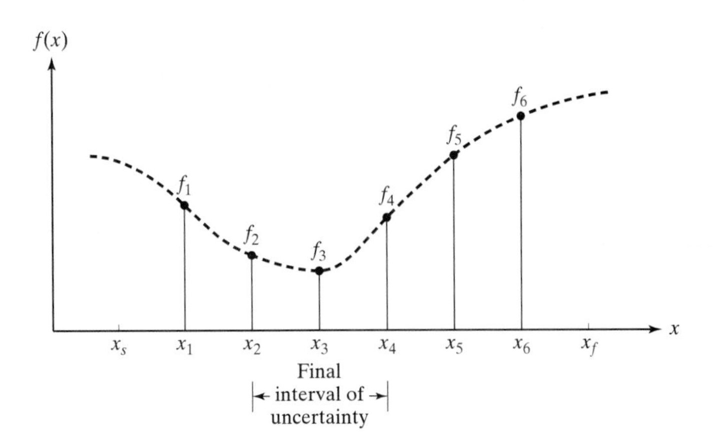

Figure 12.11 Exhaustive search.

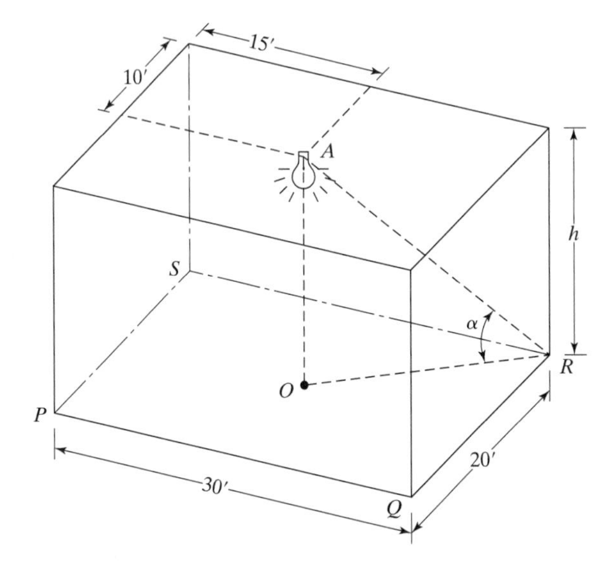

Figure 12.12 Room for maximum light intensity.

to $\sin\theta$ and inversely proportional to the square of the distance from the source, A. Determine the height of the room to maximize the light intensity at corners P, Q, R, and S. Use the exhaustive search method for the solution.

Solution

The light intensity at P can be expressed as

$$I = k\frac{\sin\alpha}{d^2} = k\frac{(OA)}{d^3} = \frac{h}{(h^2 + 325)^{1.5}}, \tag{a}$$

where k is assumed to be unity for simplicity. For maximizing the light intensity using the calculus approach, we set $\frac{dI}{dh}$ equal to zero and obtain h as 12.7475 ft.

For the exhaustive search method, we choose an initial interval of uncertainty $L_0 = (8', 15')$ and the number of experiments $n = 6$. Then experiments are conducted at six equally spaced points within the initial interval of uncertainty to obtain the following objective function values:

$$h_1 = 9,\ I_1 = \frac{9}{(406)^{1.5}} = 0.00110015,$$

$$h_2 = 10,\ I_2 = \frac{10}{(425)^{1.5}} = 0.00114134,$$

$$h_3 = 11,\ I_3 = \frac{11}{(446)^{1.5}} = 0.00116786,$$

$$h_4 = 12,\ I_4 = \frac{12}{(469)^{1.5}} = 0.00118147,$$

$$h_5 = 13,\ I_5 = \frac{13}{(494)^{1.5}} = 0.00118400,$$

and

$$h_6 = 14,\ I_6 = \frac{14}{(521)^{1.5}} = 0.00117726.$$

Since the maximum value of light intensity is observed to be at h_5, the final interval of uncertainty is given by (see Fig. 12.13)

$$L_6 = (h_4, h_6) = (12, 14) \text{ ft.} \tag{b}$$

This interval can be compared with the actual optimum value of $h^* = 12.7475$ ft. ◀

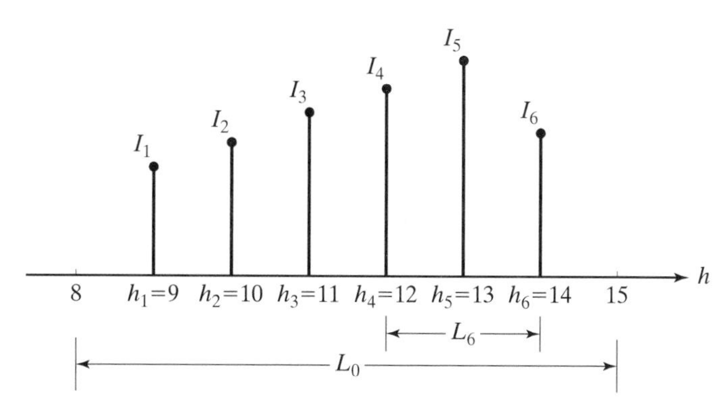

Figure 12.13 Solution of Example 12.9.

12.8.3 Fibonacci Method

This method makes use of certain numbers known as Fibonacci numbers for placing the experiments. The Fibonacci numbers (F_i) are defined as follows:

$$F_0 = F_1 = 1,$$

and

$$F_n = F_{n-1} + F_{n-2};\ n = 2, 3, 4, \ldots. \tag{12.30}$$

Equation (12.30) gives the sequence of Fibonacci numbers as 1, 1, 2, 3, 5, 8, 13, 21, 34, 55, ... The Fibonacci method also assumes the function $f(x)$ to be unimodal. As before, let $L_0 = (x_s, x_f)$ denote the initial interval of uncertainty. In this method, the total number of experiments to be conducted (n) in finding the final interval of uncertainty is to be specified. The larger the value of n, the smaller will be the final interval of uncertainty. Once n is known, a quantity, denoted as L_2^*, is computed:

$$L_2^* = \left(\frac{F_{n-2}}{F_n}\right) L_0. \tag{12.31}$$

The first two experiments (function evaluations) are conducted at the points x_1 and x_2 given by

$$x_1 = x_s + L_2^*, x_2 = x_f - L_2^*. \tag{12.32}$$

Note that x_1 and x_2 are located at the same distance from the two ends. Then a part of the interval is discarded by using the assumption of unimodality of the function $f(x)$. Then there remains a smaller interval of uncertainty with one experiment left in between. Then the next experiment is placed at x_3 such that x_3 and the existing experiment are located symmetrically from the endpoints of the current interval of uncertainty. Again, the unimodality assumption will allow us to discard part of the interval of uncertainty. This process of discarding a certain part of the interval and placing a new experiment is continued until all the predetermined (n) experiments are conducted. The middle point of the final interval of uncertainty is taken as the (approximate) optimum solution of $f(x)$. The following example illustrates the Fibonacci method:

▶Example 12.10

Find the solution of the problem in Example 12.9 using Fibonacci method with $n = 6$ and $L_0 = (8, 15)$ ft.

Solution

The function to be maximized can be expressed as

$$f(x) = \frac{x}{(x^2 + 325)^{1.5}}. \tag{a}$$

The initial interval of uncertainty, L_0, is given by $L_0 = x_f - x_s = 15 - 8 = 7$. The location of the first two experiments are defined by

$$L_2^* = \frac{F_4}{F_6} L_0 = \frac{5}{13}(7) = 2.6923, \tag{b}$$

so that

$$x_1 = x_s + L_2^* = 8 + 2.6923 = 10.6923,\ f_1 = f(x_1) = \frac{10.6923}{(439.3253)^{1.5}} = 0.00116116,$$

$$x_2 = x_f - L_2^* = 15 - 2.6923 = 12.3077,\ f_2 = f(x_2) = \frac{12.3077}{(476.4795)^{1.5}} = 0.00118334.$$

Since $f_2 > f_1$, the maximum of f cannot lie in the interval (x_s, x_1). The new interval of uncertainty is given by [see Fig. 12.14(a)]

$$L_2 = (x_1, x_f) = (10.6923, 15).$$

The location of the third experiment, x_3, should ensure that x_2 and x_3 are at the same distance from the endpoints, x_1 and x_f, respectively. This gives

$$x_3 = 15 - (12.3077 - 10.6923) = 13.3846,\ f_3 = \frac{13.3946}{(504.1475)^{1.5}} = 0.00118330.$$

Since $f_3 < f_2$, the maximum of f cannot lie in the interval (x_3, x_f). Thus, the new interval of uncertainty is given by [see Fig. 12.14(b)]

$$L_3 = (x_1, x_3) = (10.6923, 13.3846).$$

The location of the next experiment, x_4, should ensure that x_2 and x_4 are symmetrically located with respect to the ends. Thus,

$$x_4 = x_3 - (x_2 - x_1) = 11.7692,\ f_4 = \frac{11.7692}{(463.5141)^{1.5}} = 0.00117938.$$

Since $f_4 < f_2$, the maximum of f cannot lie in the interval (x_1, x_4). Thus, the new interval of uncertainty is given by [see Fig. 12.14(c)]

$$L_4 = (x_4, x_3) = (11.7692, 13.3846).$$

The next experiment, x_5, should be located so that its distance from x_3 is same as that of x_2 from x_4:

$$x_5 = x_3 - (x_2 - x_4) = 12.8461,\ f_5 = \frac{12.8461}{(490.0223)^{1.5}} = 0.00118426.$$

Since $f_2 < f_5$, the maximum of f cannot lie in the interval (x_2, x_3). Thus, the new interval of uncertainty is given by [see Fig. 12.14(d)]

$$L_5 = (x_2, x_3) = (12.3077, 13.3846).$$

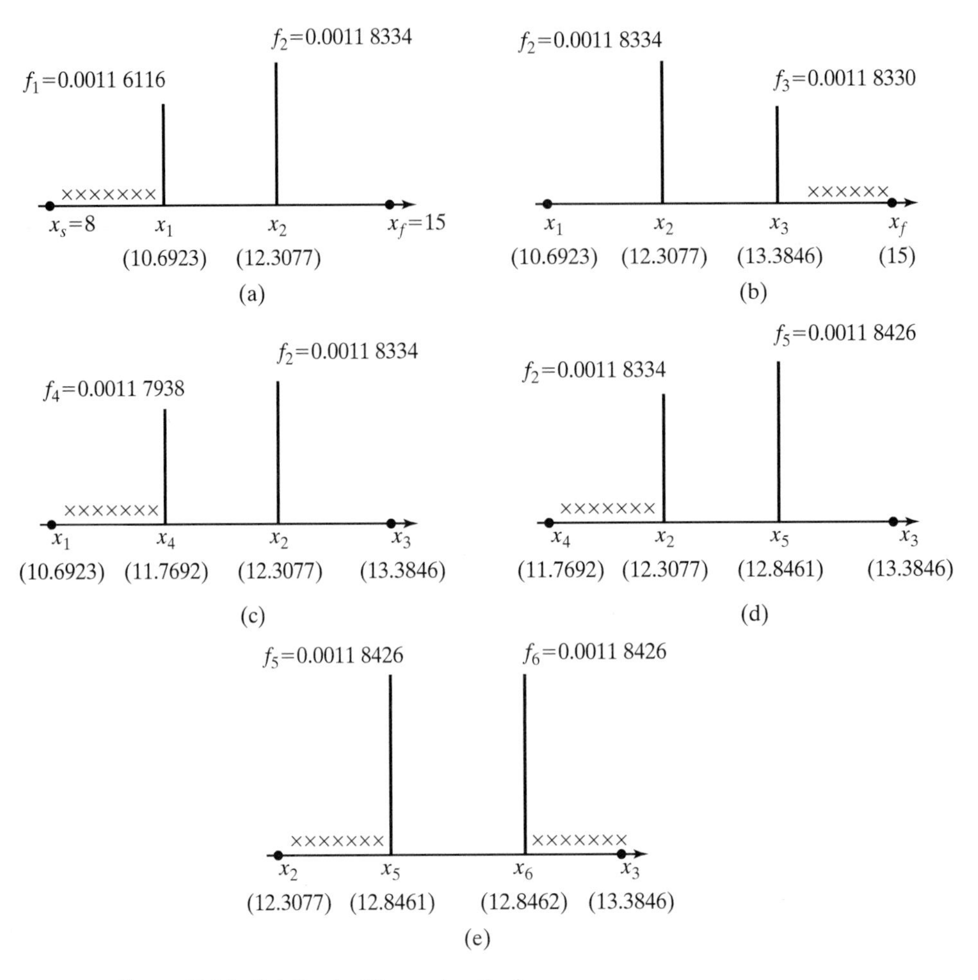

Figure 12.14 Solution by Fibonacci method.

The next experiment, x_6, is to be located so that x_5 and x_6 are symmetrically located with respect to the endpoints x_2 and x_3, respectively. This gives

$$x_6 = x_3 - (x_5 - x_2) = 12.8462, \; f_6 = \frac{12.8462}{(490.0248)^{1.5}} = 0.00118426.$$

Since $f_5 = f_6$, the maximum cannot lie in the intervals (x_2, x_5) and (x_6, x_3). Thus, the final interval of uncertainty is given by (see Fig. 12.14(e))

$$L_6 = (x_5, x_6) = (12.8461, 12.8462).$$

This can be compared with the true optimum value, $x^* = 12.7475$. ◀

12.8.4 Cubic Interpolation Method

The cubic interpolation method assumes that the function $f(x)$ can be approximated by a cubic function $h(x)$ and that the minimum of the cubic function can be used to approximate the minimum of $f(x)$. The cubic equation, $h(x)$, which is used to approximate the function $f(x)$ can be expressed as

$$h(x) = a + bx + cx^2 + dx^3, \tag{12.33}$$

where the constants a, b, c, and d are to be determined from the values of the function, $f(x)$, and its derivatives, $f'(x) = \frac{df(x)}{dx}$, at two locations x_1 and x_2. By equating the values of f and f' to h and h' at x_1 and x_2, we have the following

$$\text{at } x = x_1,\ f(x_1) \equiv h(x_1) = a + bx_1 + cx_1^2 + dx_1^3, \tag{12.34}$$

$$f'(x_1) = \frac{df(x_1)}{dx} \equiv \frac{dh(x_1)}{dx} = b + 2cx_1 + 3dx_1^2, \tag{12.35}$$

$$\text{at } x = x_2,\ f(x_2) \equiv h(x_2) = a + bx_2 + cx_2^2 + dx_2^3, \tag{12.36}$$

and

$$f'(x_2) = \frac{df(x_2)}{dx} \equiv \frac{dh(x_2)}{dx} = b + 2cx_2 + 3dx_2^2. \tag{12.37}$$

Equations (12.34) through (12.37) can be solved to find the four constants a, b, c, and d of Eq. (12.33). Once $h(x)$ is defined, its minimum can be found from

$$\left.\frac{dh}{dx}\right|_{\tilde{x}^*} = b + 2c\tilde{x}^* + 3d(\tilde{x}^*)^2 = 0, \tag{12.38}$$

or

$$\tilde{x}^* = \frac{-2c \pm \sqrt{4c^2 - 12bd}}{6d}. \tag{12.39}$$

To ensure that $\tilde{x}^*$ minimizes $h(x)$, the second derivative of h at $\tilde{x}^*$ must be positive:

$$h''(\tilde{x}^*) = \left.\frac{d^2h}{dx^2}\right|_{\tilde{x}^*} = 2c + 6d\tilde{x}^* = \sqrt{4c^2 - 12bd} > 0. \tag{12.40}$$

This indicates that the positive sign must be used in Eq. (12.39) to ensure that $\tilde{x}^*$ minimizes $h(x)$. The tilde in $\tilde{x}^*$ indicates that $\tilde{x}^*$ minimizes $h(x)$, but not necessarily $f(x)$. Hence, a suitable convergence check is to be made before accepting the point $\tilde{x}^*$ as the minimum point of $f(x)$, namely, x^*. Computationally, the cubic interpolation is implemented by the following step-by-step procedure:

(1) Assume that $x_1 = 0$, and evaluate $f_1 = f(x_1 = 0)$ and $f_1' = \frac{df(x_1=0)}{dx}$. When this method is used in the context of minimization of a multivariable function, it can be shown that $f_1' < 0$.

(2) Assume an initial step size t.

(3) Evaluate $f(t)$ and $f'(t)$.

(4) If $f(t) > f_1$, or $f'(t) > 0$, go to step 6. Otherwise, go to step 5.

(5) Replace t by twice its current value, and go to step 3.

(6) Set $x_2 = t$, $f_2 = f(t)$, and $f_2' = f'(t)$.

(7) Compute $\tilde{x}^*$ using Eq. (12.39) with a positive sign.

(8) Test $\tilde{x}^*$ for convergence. Since the derivative f' will be zero at its minimum, we use the criterion

$$|f'(\tilde{x}^*)| \leq \epsilon, \tag{12.41}$$

where ϵ is a specified small quantity. If Eq. (12.41) is satisfied, we take the desired minimum of $f(x)$ as $x^* = \tilde{x}^*$ and stop the procedure. Otherwise, go to step 9.

(9) If $f'(\tilde{x}^*) < 0$, set $x_1 = \tilde{x}^*$, $f_1 = f(\tilde{x}^*)$, $f_1' = f'(\tilde{x}^*)$, and go to step 7. On the other hand, if $f'(\tilde{x}^*) > 0$, set $x_2 = \tilde{x}^*$, $f_2 = f(\tilde{x}^*)$, $f_2' = f'(\tilde{x}_2)$, and go to step 7.

Note that the cubic interpolation method requires the first derivatives of the function $f(x)$, while the exhaustive search and Fibonacci methods do not require the derivatives of $f(x)$. This is the reason why the cubic interpolation method is considered as a first-order method, while the other two methods are considered zeroth-order methods.

▶Example 12.11

Find the solution of the problem in Example 12.9 using the cubic interpolation method.

Solution

The function to be minimized and its derivative are given by

$$f(x) = \frac{-x}{(x^2 + 325)^{1.5}}, \tag{a}$$

and

$$f'(x) = \frac{df}{dx}(x) = \frac{-325 + 2x^2}{(x^2 + 325)^{2.5}}. \tag{b}$$

The computational steps are given as follows:

(1) $x_1 = 0$, $f_1 = f(0) = 0$, $f_1' = f'(0) = \frac{-1}{325^{1.5}} = -0.00017068$.

(2) Let the initial step be $t = 5.0$.

(3) $f(t) = f(5.0) = \frac{-5.0}{350^{1.5}} = -0.0007636$.

$f'(t) = f'(5.0) = \frac{-325+50}{350^{2.5}} = -0.00011999$.

(4) Since $f(t) < f_1$ and $f'(t) < 0$, we go to step 5.

(5) Set new $t = 2(5.0) = 10.0$ and go to step 3.

(3) $f(t) = f(10.0) = \frac{-10}{425^{1.5}} = -0.00114134.$

$f'(t) = f'(10.0) = \frac{-325+200}{425^{2.5}} = -0.00003357.$

(4) Since $f(t) < f_1$ and $f'(t) < 0$, we go to step 5.

(5) Set new $t = 2(10) = 20$, and go to step 3.

(3) $f(t) = f(20) = \frac{-20}{725^{1.5}} = -0.00102453.$

$f'(t) = f'(20) = \frac{-325+800}{725^{2.5}} = 0.00003356.$

(4) Since $f'(t) > 0$, we go to step 6.

(6) Set $x_2 = t = 20.0$, $f_2 = f(t) = -0.00102453$, $f_2' = f'(t) = 0.00003356$.

(7) Using $x_1 = 0$, $f_1 = f(x_1) = 0$, $f_1' = f'(x_1) = -0.00017068$,

$x_2 = 20.0$, $f_2 = f(x_2) = -0.00102453$, and $f_2' = f'(x_2) = 0.00003356$, solve Eqs. (12.34) through (12.37) to find

$a = 0.0$, $b = -0.00017068$, $c = 0.77060231 \times 10^{-5}$, and
$d = -0.86667512 \times 10^{-7}$.

Equation (12.39) gives $\tilde{x}^* = 14.739540$.

(8) Assuming that

$|f'(\tilde{x}^*)| = |0.15993342 \times 10^{-4}|$

is not sufficiently close to zero, we go to step 9.

(9) Since $f'(\tilde{x}^*) > 0$, we set the new values of

$$x_2 = \tilde{x}^* = 14.739540, \quad f_2 = f(\tilde{x}^*) = -0.0011672927,$$
$$f_2' = f'(\tilde{x}^*) = 0.15993342 \times 10^{-4},$$

and go to step 7.

(7) Solve Eqs. (12.34) through (12.37) with $x_1 = 0.0$, $f_1 = 0.0$, $f_1' = -0.00017068$, $x_2 = 14.739540$, $f_2 = -0.0011672927$, and $f_2' = 0.15993342 \times 10^{-4}$ to find

$a = 0.0$, $b = -0.00017068$, $c = 0.59555896 \times 10^{-5}$, and
$d = 0.17043135 \times 10^{-5}$.

Equation (12.39) gives $\tilde{x}^* = 13.542183$.

(8) Assuming that

$$|f'(\tilde{x}^*)| = |0.71695072 \times 10^{-5}|$$

is sufficiently close to zero, we stop the procedure by taking $x^* = \tilde{x}^* = 13.542183$, which corresponds to $f(x^*) = -0.0011813872$. ◀

12.9 Unconstrained Minimization of a Function of Several Variables

The problem can be stated as

$$\text{find } \vec{X} = \begin{Bmatrix} x_1 \\ x_2 \\ \cdot \\ \cdot \\ \cdot \\ x_n \end{Bmatrix} \tag{12.42}$$

that minimizes $f(\vec{X})$.

There are numerous methods available for the solution of the problem stated in Eq. (12.42). We consider a method known as the Davidon–Fletcher–Powell method in this section. In this method, as in most other methods, starting with an initial guess vector $\vec{X}_1$, an improved vector, which gives a reduced value of the objective function, is found by

$$\vec{X}_{i+1} = \vec{X}_i + \alpha_i^* \vec{S}_i, \tag{12.43}$$

where i denotes the iteration number, $\vec{X}_i$ is the starting point of the ith iteration, $\vec{X}_{i+1}$ is the final point of the ith iteration, $\vec{S}_i$ is the search direction along which the function value can be reduced in the ith iteration, and α_i^* is a scalar (called the step size) that minimizes the function f along the direction $\vec{S}_i$:

$$f(\vec{X}_{i+1}) = \min_{\alpha_i} f(\vec{X}_i + \alpha_i \vec{S}_i) = \min_{\alpha_i} f(\alpha_i). \tag{12.44}$$

To indicate the development of the Davidon–Fletcher–Powell method, consider Taylor's series expansion of the gradient of f around the point $\vec{X}_{i+1}$:

$$\nabla f(\vec{X}_{i+1}) \approx \nabla f(\vec{X}_i) + [H_i](\vec{X}_{i+1} - \vec{X}_i). \tag{12.45}$$

The gradient of f is given by

$$\nabla f(\vec{X}) = \begin{Bmatrix} \dfrac{\partial f(\vec{X})}{\partial x_1} \\ \dfrac{\partial f(\vec{X})}{\partial x_2} \\ \cdot \\ \cdot \\ \cdot \\ \dfrac{\partial f(\vec{X})}{\partial x_n} \end{Bmatrix}, \tag{12.46}$$

and $[H_i]$ denotes the matrix of second partial derivatives of f, also known as the Hessian matrix of f:

$$[H_i] = \begin{bmatrix} \frac{\partial^2 f}{\partial x_1^2} & \frac{\partial^2 f}{\partial x_1 \partial x_2} & \cdots & \frac{\partial^2 f}{\partial x_1 \partial x_n} \\ \frac{\partial^2 f}{\partial x_1 \partial x_2} & \frac{\partial^2 f}{\partial x_2^2} & \cdots & \frac{\partial^2 f}{\partial x_2 \partial x_n} \\ \cdot & \cdot & \cdots & \cdot \\ \cdot & \cdot & \cdots & \cdot \\ \cdot & \cdot & \cdots & \cdot \\ \frac{\partial^2 f}{\partial x_1 \partial x_n} & \frac{\partial^2 f}{\partial x_2 \partial x_n} & \cdots & \frac{\partial^2 f}{\partial x_n^2} \end{bmatrix}_{\vec{X}_i} . \tag{12.47}$$

If $[P_i] = [P(\vec{X}_i)]$ denotes an $n \times n$ matrix that approximates $[H_i]$ in the ith iteration, Eq. (12.45) can be written as

$$\vec{y}_i = [P_i]\vec{z}_i, \tag{12.48}$$

where

$$\vec{y}_i = \nabla f(\vec{X}_{i+1}) - \nabla f(\vec{X}_i) \tag{12.49}$$

and

$$\vec{z}_i = \vec{X}_{i+1} - \vec{X}_i. \tag{12.50}$$

The solution of Eq. (12.48) gives

$$\vec{z}_i = [Q_{i+1}]\vec{y}_i, \tag{12.51}$$

where $[Q_{i+1}]$ is an approximate inverse of the Hessian matrix. Equation (12.51) is known as the quasi-Newton formula. The basis of the Davidon–Fletcher–Powell method is that it updates the matrix $[Q_i]$ in every iteration such that it eventually converges to the inverse of the Hessian matrix. Also, the matrix $[Q_i]$ is maintained positive definite throughout the iterative process.

Equation (12.51) is used as a basis for the Davidon–Fletcher–Powell iterative process with

$$\vec{X}_{i+1} = \vec{X}_i - \alpha_i \vec{S}_i, \tag{12.52}$$

where

$$\vec{S}_i = -[Q_i]\nabla f(\vec{X}_i) \tag{12.53}$$

and $[Q_i]$ is a positive-definite symmetric matrix. The matrix $[Q_i]$ is updated as follows:

$$[Q_{i+1}] = [Q_i] - \frac{[Q_i]\vec{y}_i\vec{y}_i^T[Q_i]}{\vec{y}_i^T[Q_i]\vec{y}_i} + \frac{\vec{z}_i\vec{z}_i^T}{\vec{z}_i^T\vec{y}_i}. \tag{12.54}$$

The step-by-step computational procedure of the Davidon–Fletcher–Powell method is as follows:

(1) Start with an arbitrary n-component vector $\vec{X}_1$ and a $n \times n$ positive definite symmetric matrix, $[Q_1]$. Usually, an identity matrix is used as $[Q_1] = [I]$. Set the iteration number as $i = 1$.

(2) Compute the gradient of the function, $\nabla f(\vec{X}_i)$, and set

$$\vec{S}_i = -[Q_i]\nabla f(\vec{X}_i). \tag{12.55}$$

(3) Minimize f along $\vec{S}_i$ and find the step size α_i^*:

$$f(\vec{X}_i + \alpha_i^* \vec{S}_i) = \min_{\alpha_i} f(\vec{X}_i + \alpha_i \vec{S}_i). \tag{12.56}$$

The methods of minimizing a function of a single variable, discussed in Section 12.8, can be used for this purpose.

(4) Find the new point:

$$\vec{X}_{i+1} = \vec{X}_i + \alpha_i^* \vec{S}_i. \tag{12.57}$$

(5) Test the point $\vec{X}_{i+1}$ for convergence using the relation

$$|\nabla f(\vec{X}_{i+1})| \le \epsilon, \tag{12.58}$$

where ϵ is a prescribed small quantity. If convergence of Eq. (12.58) is satisfied, the optimum point is taken as $\vec{X}^* = \vec{X}_{i+1}$, and the process is stopped. Otherwise, go to step 6.

(6) Update the $[Q_i]$ matrix using Eq. (12.54). Set the new iteration number i to be $i + 1$, and go to step 2.

The following example illustrates the Davidon–Fletcher–Powell method:

▶Example 12.12

The steady-state temperature at nodes 1 and 2 ($x_1 = T_1$ and $x_2 = T_2$) of the one-dimensional fin shown in Fig. 12.15 can be found by minimizing the function

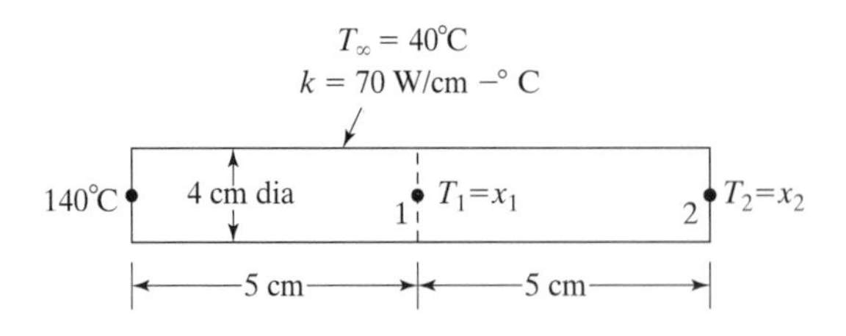

Figure 12.15 One-dimensional fin.

$$f(x_1, x_2) = 0.4380x_1^2 - 0.0810x_1x_2 + 0.2190x_2^2 - 39.9114x_1 - 14.2857x_2 + 2292.4020. \tag{a}$$

Determine the values of x_1 and x_2 by minimizing the function $f(x_1, x_2)$.

Solution

The Davidon–Fletcher–Powell method is used to minimize the function $f(x_1, x_2)$ using the following steps:

(1) The starting vector $\vec{X}_1$ and the matrix $[Q_1]$ are chosen as

$$\vec{X}_1 = \begin{Bmatrix} x_1 \\ x_2 \end{Bmatrix}_1 = \begin{Bmatrix} T_1 \\ T_2 \end{Bmatrix}_1 = \begin{Bmatrix} 100 \\ 50 \end{Bmatrix} \text{ and } [Q_1] = \begin{bmatrix} 1 & 0 \\ 0 & 1 \end{bmatrix}.$$

Set $i = 1$.

(2) The gradient of the function f is given by

$$\nabla f = \begin{Bmatrix} \dfrac{\partial f}{\partial x_1} \\ \dfrac{\partial f}{\partial x_2} \end{Bmatrix} = \begin{Bmatrix} 0.9760x_1 - 0.0810x_2 - 39.9114 \\ -0.0810x_1 + 0.4380x_2 - 14.2857 \end{Bmatrix}, \tag{b}$$

and hence,

$$\nabla f(\vec{X}_1) = \nabla f \begin{pmatrix} x_1 = 100 \\ x_2 = 50 \end{pmatrix} = \begin{Bmatrix} 43.6386 \\ -0.4857 \end{Bmatrix}.$$

The search direction, $\vec{S}_1$, is given by

$$\vec{S}_1 = -[Q_1]\nabla f(\vec{X}_1) = \begin{Bmatrix} -43.6386 \\ 0.4857 \end{Bmatrix}.$$

(3)

$$f(\vec{X}_1 + \alpha_1 \vec{S}_1) = f \begin{pmatrix} x_1 = 100 - 43.6386\,\alpha_1 \\ x_2 = 50 + 0.4857\,\alpha_1 \end{pmatrix}$$

$$= 835.8639\,\alpha_1^2 - 1904.5635\,\alpha_1 + 2109.4770.$$

For the minimum of $f(\alpha_1)$, we use the calculus method, set

$$\frac{df}{d\alpha_1} = 1671.7278\,\alpha_1 - 1904.5635 = 0,$$

and obtain $\alpha_1^* = 1.1393$.

(4) The new design vector $\vec{X}_2$ is given by

$$\vec{X}_2 = \vec{X}_1 + \alpha_1^* \vec{S}_1 = \begin{Bmatrix} 100 \\ 50 \end{Bmatrix} + 1.1393 \begin{Bmatrix} -43.6386 \\ 0.4857 \end{Bmatrix} = \begin{Bmatrix} 50.2825 \\ 50.5534 \end{Bmatrix}.$$

(5) To test the convergence of the vector $\vec{X}_2$, we evaluate $\nabla f(\vec{X}_2)$ using Eq. (b):

$$\nabla f(\vec{X}_2) = \begin{Bmatrix} 0.0413 \\ 3.7938 \end{Bmatrix}.$$

Since $|\nabla f(\vec{X}_2)|$ is not close to zero, we go to step 6.

(6) The matrix $[Q_1]$ is updated using Eq. (12.54). Noting that

$$\vec{z}_1 = \vec{X}_2 - \vec{X}_1 = \begin{Bmatrix} 50.2825 \\ 50.5534 \end{Bmatrix} - \begin{Bmatrix} 100 \\ 50 \end{Bmatrix} = \begin{Bmatrix} -49.7175 \\ 0.5534 \end{Bmatrix},$$

$$\vec{y}_1 = \nabla f(\vec{X}_2) - \nabla f(\vec{X}_1) = \begin{Bmatrix} 0.0413 \\ 3.7938 \end{Bmatrix} - \begin{Bmatrix} 43.6386 \\ -0.4857 \end{Bmatrix} = \begin{Bmatrix} -43.5973 \\ 4.2795 \end{Bmatrix},$$

$$\vec{z}_1^T \vec{y}_1 = (-49.7175 \quad 0.5534) \begin{Bmatrix} -43.5973 \\ 4.2795 \end{Bmatrix} = 2169.9171,$$

$$\vec{y}_1^T [Q_1] \vec{y}_1 = \vec{y}_1^T \vec{y}_1 = (-43.5973 \quad 4.2795) \begin{Bmatrix} -43.5973 \\ 4.2795 \end{Bmatrix} = 1919.0387,$$

$$\vec{z}_1 \vec{z}_1^T = \begin{Bmatrix} -49.7175 \\ 0.5534 \end{Bmatrix} (-49.7175 \quad 0.5534) = \begin{bmatrix} 2471.8298 & -27.5137 \\ -27.5137 & 0.3063 \end{bmatrix},$$

$$\frac{\vec{z}_1 \vec{z}_1^T}{\vec{z}_1^T \vec{y}_1} = \begin{bmatrix} 1.1391 & -0.0127 \\ -0.0127 & 0.0001 \end{bmatrix},$$

$$[Q_1]\vec{y}_1 \vec{y}_1^T [Q_1] = \begin{Bmatrix} -43.5973 \\ 4.2795 \end{Bmatrix} (-43.5973 \quad 4.2795) = \begin{bmatrix} 1900.7246 & -186.5746 \\ -186.5746 & 18.3141 \end{bmatrix},$$

$$\frac{[Q_1]\vec{y}_1 \vec{y}_1^T [Q_1]}{\vec{y}_1^T [Q_1] \vec{y}_1} = \begin{bmatrix} 0.9905 & -0.0972 \\ -0.0972 & 0.0095 \end{bmatrix},$$

Eq. (12.54) gives

$$[Q_2] = \begin{bmatrix} 1 & 0 \\ 0 & 1 \end{bmatrix} - \begin{bmatrix} 0.9905 & -0.0972 \\ -0.0972 & 0.0095 \end{bmatrix} + \begin{bmatrix} 1.1391 & -0.0127 \\ -0.0127 & 0.0001 \end{bmatrix}$$

$$= \begin{bmatrix} 1.1486 & 0.0845 \\ 0.0845 & 0.9906 \end{bmatrix}.$$

Set $i = 2$ and go to step 2.

(2)

$$\vec{S}_2 = -[Q_2]\nabla f(\vec{X}_2) = \begin{bmatrix} 1.1486 & 0.0845 \\ 0.0845 & 0.9906 \end{bmatrix} \begin{Bmatrix} -0.0413 \\ -3.7938 \end{Bmatrix} = \begin{Bmatrix} -0.3681 \\ -3.7616 \end{Bmatrix}.$$

(3)

$$f(\vec{X}_2 + \alpha_2 \vec{S}_2) = f \begin{pmatrix} x_1 = 50.2825 - 0.3681\,\alpha_2 \\ x_2 = 50.5534 - 3.7616\,\alpha_2 \end{pmatrix}$$

$$= 3.0459\,\alpha_2^2 - 14.2483\,\alpha_2 + 1024.5632.$$

For the minimum of $f(\alpha_2)$, we use calculus method and set

$$\frac{df}{d\alpha_2} = 6.0918\,\alpha_2 - 14.2483 = 0$$

and find $\alpha_2^* = 2.3389$.

(4) The new design vector $\vec{X}_3$ is given by

$$\vec{X}_3 = \vec{X}_2 + \alpha_2^* \vec{S}_2 = \begin{Bmatrix} 50.2825 \\ 50.5534 \end{Bmatrix} + 2.3389 \begin{Bmatrix} -0.3681 \\ -3.7616 \end{Bmatrix} = \begin{Bmatrix} 49.4216 \\ 41.7554 \end{Bmatrix}.$$

(5) To test the convergence of the vector $\vec{X}_3$, we evaluate $\nabla f(\vec{X}_3)$ using Eq. (b):

$$\nabla f(\vec{X}_3) = \begin{Bmatrix} -0.0003 \\ 0.0001 \end{Bmatrix}.$$

Since $|\nabla f(\vec{X}_3)| \approx 0$, the optimum solution is given by $\vec{X}^* = \vec{X}_3 = \begin{Bmatrix} 49.4216 \\ 41.7554 \end{Bmatrix}$.

◀

12.10 Constrained Minimization of a Function of Several Variables

A constrained minimization problem can be stated as

$$\text{find } \vec{X} = \begin{Bmatrix} x_1 \\ x_2 \\ \cdot \\ \cdot \\ \cdot \\ x_n \end{Bmatrix}$$

$$\text{that minimizes } f(\vec{X})$$

subject to

$$g_j(\vec{X}) \le 0;\ j = 1, 2, \ldots, m. \tag{12.59}$$

Note that only inequality constraints are considered in Eq. (12.59). The equality constraints can be included without much difficulty [12.6]. One of the popular methods of solving the constrained optimization problem of Eq. (12.59) is known as the interior penalty function method. In this method, the constraints are included in defining a new objective function whose unconstrained minimization yields the solution of the original constrained problem of Eq. (12.59). This allows us to use the unconstrained minimization method discussed earlier. The new objective function, denoted as ϕ, is defined as

$$\phi(\vec{X}, r_i) = f(\vec{X}) - r_i \sum_{j=1}^{m} \frac{1}{g_j(\vec{X})}, \tag{12.60}$$

where r_i is a constant known as the penalty parameter. It has been proved [12.6] that the unconstrained minimization of the function $\phi(\vec{X}, r_i)$ for a series of constants $r_1 > r_2 > \cdots > r_i > r_{i+1} > \cdots$ will converge to the solution of problem of Eq. (12.59). The convergence of the procedure for the simple problem given in Eq. (12.61) is indicated graphically in Fig. 12.16. Consider the problem of

$$\text{finding the value of } x \text{ that minimizes } f(x) = x$$

subject to

$$g_1(x) = x - 2 \le 0. \tag{12.61}$$

The ϕ- function corresponding to the problem of Eq. (12.61) is given by

$$\phi(x, r_i) = x - r_i \left(\frac{1}{x - 2}\right). \tag{12.62}$$

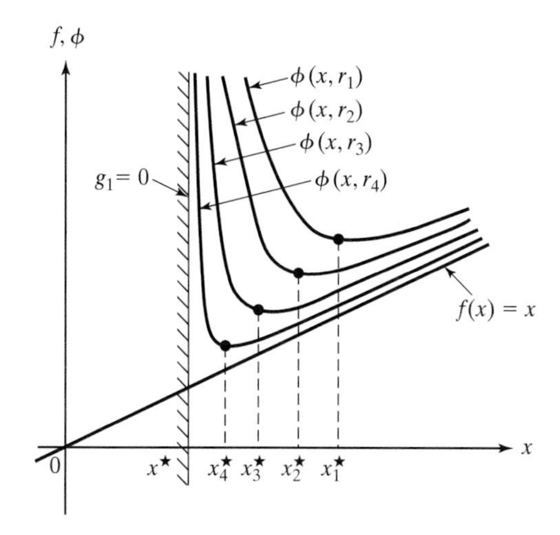

Figure 12.16 Interior penalty function approach.

The contours of ϕ for different values of the constant r_i are shown in Fig. 12.16. The minimum of $\phi(x, r_i)$ is indicated as x_i^*. Figure 12.16 shows that the minima $x_1^*, x_2^*, \ldots$ of $\phi(x, r_i)$ corresponding to a decreasing series of constants $r_1, r_2, \ldots$, converge to the true optimum solution, x^*, of the problem of Eq. (12.61). It can be seen from Eqs. (12.60) and (12.61) that as the constraint $g_j(\vec{X})$ approaches its limiting value of zero, its contribution to ϕ will tend to increase rapidly to infinity. It is to be noted that an infinite number of minimizations of the ϕ- function, with decreasing series of values of r_i, are needed to find the exact constrained minimum point $\vec{X}^*$. However, in practice, even a small number of minimizations (on the order of 5 or 6) yield a vector $\vec{X}_i^*$ that is reasonably close to the true optimum point, $\vec{X}^*$. For a multivariable constrained minimization problem stated in Eq. (12.59), the following step-by-step procedure is used to implement the interior penalty function method:

(1) Start with an initial point $\vec{X}_1$ that satisfies all the constraints with strict inequality sign:

$$g_j(\vec{X}) < 0; \; j = 1, 2, \ldots, m. \tag{12.63}$$

(2) Choose a value for the initial penalty parameter, r_1. Any positive constant, such as 100, can be used for r_1, although some empirical recommendations are available in the literature [12.6]. Set the iteration number $i = 1$.

(3) Formulate the ϕ-function using Eq. (12.60).

(4) Using $\vec{X}_i$ as a starting point, minimize $\phi(\vec{X}, r_i)$ as an unconstrained function and find the solution as $\vec{X}_i^*$.

(5) Test $\vec{X}_i^*$ for the convergence of the procedure. A practical way of testing $\vec{X}_i^*$ for convergence involves first generating the vectors $\vec{Y}_j$ and $\vec{Z}_j$, $j = 1, 2, \ldots, n$ by perturbing each component of $\vec{X}_i^*$ by a small amount on positive and

negative sides:

$$\vec{Y}_j = \vec{X}_i^* - \Delta x_j; \; \vec{Z}_j = \vec{X}_i^* + \Delta x_j; \; j = 1, 2, \ldots, n. \tag{12.64}$$

Then the objective and all constraints are evaluated at each of the vectors $\vec{Y}_j$ and $\vec{Z}_j$:

$$f(\vec{Y}_j), \; g_k(\vec{Y}_j); \; k = 1, 2, \ldots, m, \tag{12.65}$$

$$f(\vec{Z}_j), \; g_k(\vec{Z}_j); \; k = 1, 2, \ldots, m.$$

If $\vec{X}_i^*$ is the true optimum solution (or very close to the optimum), then one of the following conditions will be satisfied:

(i) Neighboring points of $\vec{X}_i^*$ have larger objective function value

$$f(\vec{Y}_j) > f(\vec{X}_i^*); \; j = 1, 2, \ldots, n,$$

$$f(\vec{Z}_j) > f(\vec{X}_i^*); \; j = 1, 2, \ldots, n. \tag{12.66}$$

(ii) Neighboring points of $\vec{X}_i^*$ violate some constraint:

$$g_k(\vec{Y}_j) > 0; \; k = 1, 2, \ldots, m; \; j = 1, 2, \ldots, n,$$

$$g_k(\vec{Z}_j) > 0; \; k = 1, 2, \ldots, m; \; j = 1, 2, \ldots, n. \tag{12.67}$$

(6) If $\vec{X}_i^*$ is not found to be the optimum solution, find the new value of the penalty parameter:

$$r_{i+1} = c r_i; \; c < 1. \tag{12.68}$$

Then set the new iteration number as $i + 1$, and go to step 3.

The following example is given to illustrate the procedure:

▶Example 12.13

Solve the following constrained optimization problem using the interior penalty function method coupled with the calculus method of unconstrained minimization:

$$\text{Minimize } f(x_1, x_2) = (x_1 + 2)^3 + 4x_2$$

subject to

$$x_1 \geq 0, \; x_2 \geq 1.$$

Solution

The problem can be restated as

$$\text{minimize } f(x_1, x_2) = (x_1 + 2)^3 + 4x_2$$

subject to

$$2 - x_1 \leq 0, \tag{a}$$

$$1 - x_2 \leq 0.$$

The ϕ function can be constructed as

$$\phi(\vec{X}, r) = (x_1 + 2)^2 + 4x_2 - \frac{r}{2 - x_1} - \frac{r}{1 - x_2}. \tag{b}$$

To find the unconstrained minimum of ϕ, we use the calculus method:

$$\frac{\partial \phi}{\partial x_1} = 3(x_1 + 2)^2 - \frac{r}{(2 - x_1)^2} = 0, \tag{c}$$

$$\frac{\partial \phi}{\partial x_2} = 4 - \frac{r}{(1 - x_2)^2} = 0. \tag{d}$$

Equations (c) and (d) can be solved to find

$$x_1^*(r) = \left(4 + \left(\frac{r}{3}\right)^{\frac{1}{2}}\right)^{\frac{1}{2}}, \tag{e}$$

$$x_2^*(r) = 1 + \frac{\sqrt{r}}{2}. \tag{f}$$

To find the solution of the original constrained optimization problem, we make $r \to 0$ and find, from Eqs. (e) and (f), that

$$x_1^* = x_1^*(r \to 0) = 2, x_2^* = x_2^*(r \to 0) = 1, f^* = \phi_{\min}(r \to 0) = 68. \tag{g}$$

The values of $x_1^*(r)$, $x_2^*(r)$, $\phi_{\min}(r)$, and $f(r)$ for different values of r, in a decreasing sequence, are given in Table 12.2. ◀

Table 12.2 Convergence of results of Example 12.13

Value of r	$x_1^*(r)$	$x_2^*(r)$	$\phi_{\min}(r)$	$f(r)$
100	3.1263	6.0	267.5001	158.7138
10	2.4137	2.5811	126.8034	96.3066
1	2.1395	1.5	86.1007	76.9322
0.1	2.0451	1.1581	73.6715	70.8217
0.01	2.0144	1.05	69.7881	68.8937
0.001	2.0046	1.0158	68.5649	68.2842
0.0001	2.0014	1.005	68.1786	68.0872
0.00001	2.0005	1.0016	68.0567	68.0304
Exact solution ($r = 0$)	2.0	1.0	68.0	68.0

12.11 Choice of Method

If all the objective and constraint functions are linear in terms of design variables, the solution can be obtained using LP techniques. Any of the available commercial LP software packages can be used for this purpose. On the other hand, if any of the functions involved is nonlinear, the problem is to be solved using a suitable nonlinear-programming (NLP) technique. For the one-dimensional search, the Fibonacci method is very efficient if the initial interval of uncertainty is known, and the gradient of the function cannot be evaluated either accurately or in a simple manner. In the absence of the initial interval of uncertainty, the cubic interpolation method is preferred.

In general, the efficiency and reliability of the various methods are problem dependent, and any efficient computer program must include many heuristic ideas not indicated explicitly by the method. The heuristic considerations are needed to take care of functions with multiple extreme points, rapid variations in the first and second derivatives of the function, and the influence of round-off and truncation errors during the computation. If the problem is unconstrained and the objective function is available in an explicit form that can be readily differentiated with respect to the design variables, the classical approach can be used where the simultaneous nonlinear equations denoting the necessary conditions of optimality are solved using the Newton–Raphson method. On the other hand, if the objective function is not available in explicit form, the Davidon–Fletcher–Powell method, with a finite-difference gradient, is to be used.

Finally, if the optimization problem is constrained, the penalty function approach can be used to find the solution. Although the penalty function method is robust and reliable, there are more efficient methods available in the literature. An efficient method that can readily be applied using available commercial software is called the sequential quadratic programming (SQP) method.

12.12 Use of Software Packages

12.12.1 MATLAB

The *Optimization Toolbox* is used to solve Examples 12.14 through 12.17. In Example 12.14, the function

```
x=fmins('fmin1',x0)
```

is used to find the optimum design vector x. The objective function is defined in *fmin1.m* and *x0* is the initial guess for the vector x. If there is only one design variable, x will be a scalar. In this case, instead of $x0$, the lower and upper bounds on x can be given as indicated in Example 12.15. The function *fmins* can be used for unconstrained optimization. For solving constrained optimization problems, the function *constr* can be used as illustrated in examples 12.16 and 12.17.

▶Example 12.14

Minimize $f(x_1, x_2) = 100(x_2 - x_1^2)^2 + (1 - x_1)^2$ using the starting point $(x_1 = -1.2, x_2 = 1.0)$.

Solution

The function is defined as an m.file labeled *fmin1.m*, whose listing is as follows:

```
function f=fmin1(x)

f=100*(x(2)-x(1)^2)^2+(1-x(1))^2;

>> x0=[-1.2,1.0];

>> x=fmins('fmin1',x0)
x =
       0.99998747911171  0.9999780482028

>> fmin1(x)
ans =
      1.111473033117937e-009
```

◀

▶Example 12.15

Minimize $f(x) = 0.65 - \frac{0.75}{1+x^2} - 0.65x \tan^{-1}\left(\frac{1}{x}\right)$ in the range 0.1 to 1.0.

Solution

The function is defined as an m.file labeled *fmin3.m*, whose listing is as follows:

```
function f=fmin3(x)

f=0.65-0.75/(1+x^2)-0.65*x*atan(1/x);

>> x=fmins('fmin3',0.01,1.0);
>> x
x =
   0.48087500000000
>> fmin3(x)
ans =
  -0.31002050189351
```

◀

▶Example 12.16

Maximize $f = x_1 + x_2 + x_3$

subject to

$$6x_1 + 6x_2 + 3x_3 \leq 24,100,$$
$$3x_1 + 3x_2 + 4x_3 \leq 15,500,$$
$$4x_2 + x_3 \leq 14,800,$$
$$2x_3 \leq 3,600,$$
$$x_3 \leq 3,800,$$
$$x_1 \geq 0, i = 1, 2, 3.$$

Solution

The problem is defined as an m.file labeled *fmin5.m*:

```
function [f,g]=fmin5(x)

f=-(x(1)+x(2)+x(3));

g(1,1)=6*x(1)+6*x(2)+3*x(3)-24100;
g(2,1)=3*x(1)+3*x(2)+4*x(3)-15500;
g(3,1)=4*x(2)+x(3)-14800;
g(4,1)=2*x(3)-3600;
g(5,1)=x(3)-3800;

>> x0=[0,0,0];
>> options=[];
>> vlb=[0,0,0];
>> vub=[];
>> x=constr('fmin5',x0,options,vlb,vub)
x  =

   1.0e+003 *

   1.66333335099390    1.66333331567277    1.38000000000000

>> [f,g]=fmin5(x)

f  =

   -4.706666666666666e+003

g  =

   1.0e+003 *
```

```
                    0
                    0
   -6.76666673730893
   -0.84000000000000
   -2.42000000000000
```

◀

▶Example 12.17

Minimize $f = x_1^3 - 6x_1^2 + 11x_1 + x_3$

subject to

$$g_1 = x_1^2 + x_2^2 - x_3^2 \leq 0,$$

$$g_2 = 4 - x_1^2 - x_2^2 - x_3^2 \leq 0,$$

$$g_3 = x_3 - 5 \leq 0,$$

$$g_4, g_5, g_6 : -x_i \leq 0, i = 1, 2, 3.$$

Solution

The problem is defined as an m.file labeled *fmin7.m*:

```
function [f,g]=fmin7(x)

f=x(1)^3-6*x(1)^2+11*x(1)+x(3);

g(1,1)=x(1)^2+x(2)^2-x(3)^2;
g(2,1)=4-x(1)^2-x(2)^2-x(3)^2;
g(3,1)=x(3)-5;

>> x0=[0.1, 0.1, 3.0];
>> options=[];
>> vlb=[0,0,0];
>> vub=[];
>> x=constr('fmin7',x0,options,vlb,vub)

x  =

          0    1.4142    1.4142

>> [f,g]=fmin7(x)

f  =
      1.4142
```

```
g =

     0.0000
     0.0000
    -3.5858
```
◀

12.12.2 MAPLE

In MAPLE, the functions

Maximize(f(x)) and Minimize(f(x))

can be used for the unconstrained optimization of the function $f(x)$. For constrained optimization, the function

maximize(obj, consts union {x >= 0, y >= 0, z >= 0})

can be used to find the analytical solution with the objective function defined in *obj*, the constraints defined in *consts* and with nonnegativity requirements on x, y and z.

▶Example 12.18

Maximize $f(x) = \frac{1}{10\sqrt{2\pi}} e^{-\frac{1}{2}\left(\frac{x-100}{10}\right)^2}$.

Solution

```
> 
> maximize(exp(-0.5*((x-100)/10)^2))/(10*sqrt(2*Pi));
```

$$.05000000000\frac{\sqrt{2}}{\sqrt{\pi}}$$

```
> evalf(%);
```

$$.03989422803$$
◀

▶Example 12.19

Minimize the function f defined in Example 12.14.

Solution

```
> 
> minimize(100*(y-x^2)^2+(1-x)^2);
```

$$0$$
◀

▶Example 12.20

Redo Example 12.16

Solution

```
> cnsts:={6*x+6*y+3*z<=24100, 3*x+3*y+4*z<=15500, 4*y+z<=14800,
  2*z<=3600, z<=3800};
```

$$cnsts := \{6x + 6y + 3z \le 24100, 3x + 3y + 4z \le 15500, 4y + z \le 14800, 2z \le 3600, z \le 3800\}$$

```
> obj:=x+y+z;
```

$$obj := x + y + z$$

```
> maximize(obj,cnsts union{x>=0,y>=0,z>=0});
```

$$\{y = 0, z = 1380, x = \frac{9980}{3}\}$$

```
> evalf(%);
```

$$\{z = 1380., x = 3326.666667, y = 0\}$$

```
> eval(x+y+z,{x=9980/3,y=0,z=1380});
```

$$\frac{14120}{3}$$

```
> evalf (%);
```

$$4706.666667$$

```
>
```

◀

12.12.3 MATHCAD

The MATHCAD functions

`Maximize(f, x, y, ...)` and `Minimize(f, x, y, ...)`

can be used for the unconstrained optimization of the function f in terms of the design variables $x, y, \ldots$. For constrained optimization, a *Solve Block* is to be used. Four steps can be used to create a *Solve Block*:

(a) Give an initial guess for each unknown.

(b) Type the words "Given" in a separate math region below the guess values. This tells MATHCAD that what follows is a system of constraint equations.

(c) Enter the constraints below the word "Given".

(d) Enter any equation that involves the functions Minimize or Maximize below the constraints.

The procedure is shown in Examples 12.23 and 12.24.

▶Example 12.21

Redo Example 12.14 with MATHCAD.

Solution

$f(x,y) := 100 \cdot (y - x^2)^2 + (1 - x)^2$

$x := -1.2$

$y := 1.0$

$P := \text{Minimize}(f,x,y) \qquad P = \begin{pmatrix} 1 \\ 1 \end{pmatrix}$

$f(P_0, P_1) = 4.257 \times 10^{-11}$ ◀

▶Example 12.22

Redo Example 12.15 with MATHCAD.

Solution

$f(x) := 0.65 - \left(\frac{0.75}{1 + x^2}\right) - 0.65 \cdot x \cdot \text{atan}\left(\frac{1}{x}\right)$

Given

$0.1 \le x \le 1.0$

$P := \text{Minimize}(f,x) \qquad P = 0.481$

$f(P) = -0.31$ ◀

▶Example 12.23

Redo Example 12.16 with MATHCAD.

Solution

$f(x,y,z) := x + y + z$

$x := 0 \qquad y := 0 \qquad z := 0$

Given

$$6 \cdot x + 6 \cdot y + 3 \cdot z \le 24100$$

$$3 \cdot x + 3 \cdot y + 4 \cdot z \le 15500$$

$$4 \cdot y + z \le 14800$$

$$2 \cdot z \le 3600$$

$$z \le 3800$$

$$x \ge 0$$

$$y \ge 0$$

$$z \ge 0$$

$$P := \text{Maximize}(f,x,y,z) \qquad P = \begin{pmatrix} 3.327 \times 10^3 \\ 0 \\ 1.38 \times 10^3 \end{pmatrix}$$

$$f(P_0, P_1, P_2) = 4.707 \times 10^3$$

◀

▶Example 12.24

Redo Example 12.17 with MATHCAD.

Solution

$$f(x,y,z) := x^3 - 6 \cdot x^2 + 11 \cdot x + z$$

$x := 0.1$ $\qquad$ $y := 0.1$ $\qquad$ $z := 3.0$

Given

$$x^2 + y^2 - z^2 \le 0$$

$$4 - x^2 - y^2 - z^2 \le 0$$

$$z - 5 \le 0$$

$$-x \le 0$$

$$-y \le 0$$

$$-z \le 0$$

P := Maximize(f,x,y,z) $\quad P = \begin{pmatrix} 0 \\ 1.414 \\ 1.414 \end{pmatrix}$

$f(P_0, P_1, P_2) = 1.414$

◀

12.13 Computer Programs

12.13.1 Fortran Programs

Program 12.1 Cubic interpolation method

The following data are required for input:

MAXFIT = maximum number of refits permitted (usual value: 5 to 10).
EPS1, EPS2 = convergence requirements (usual values: 0.01 for each).
STEP = initial step length (usual value: 1.0).
F(X) = funtion to be minimized; to be given as a subroutine as follows:

```
      SUBROUTINE FUN (X,S,XL,F).
C     F is to be defined in terms of XL.
      DIMENSION X(*),S(*).
      F = . . .
      RETURN
      END
```

Illustration:

Minimize $f(x) = x^5 - 5x^3 - 20x + 5$.

Solution

$x^* = 2$.

The following is the program output:

```
RESULTS OF CUBIC INTERPOLATION:

LOWER BOUND: 0.000000E+00  FUNCTION: 0.500000E+01  DERIVATIVE: -.200005E+02
UPPER BOUND: 0.100000E+01  FUNCTION: -.190000E+02  DERIVATIVE: -.299996E+02

ADJUSTING STEP SIZE TO: 0.200000E+01

LOWER BOUND: 0.100000E+01  FUNCTION: -.190000E+02  DERIVATIVE: -.299996E+02
UPPER BOUND: 0.300000E+01  FUNCTION: 0.530000E+02  DERIVATIVE: 0.250077E+03

ADJUSTING STEP SIZE TO: 0.200000E+01

LANBDA: 0.200724E+01  FUNCTION: -.429974E+02  DERIVATIVE: 0.743368E+00
```

```
FUNCTION VALUE BY CUBIC:    -0.530227E+02
ACTUAL FUNCTION VALUE:    -0.429974E+02

LOWER BOUND: 0.100000E+01  FUNCTION: -.190000E+02  DERIVATIVE: -.299996E+02
UPPER BOUND: 0.200724E+01  FUNCTION: -.429974E+02  DERIVATIVE: 0.743368E+00

LANBDA: 0.199847E+01  FUNCTION: -.429999E+02  DERIVATIVE: -.138388E+00

LOWER BOUND: 0.199847E+01  FUNCTION: -.429999E+02  DERIVATIVE: -.138388E+00
UPPER BOUND: 0.200724E+01  FUNCTION: -.429974E+02  DERIVATIVE: 0.743368E+00

LANBDA: 0.200003E+01  FUNCTION: -.430000E+02  DERIVATIVE: 0.164984E-01

LOWER BOUND: 0.199847E+01  FUNCTION: -.429999E+02  DERIVATIVE: -.138388E+00
UPPER BOUND: 0.200003E+01  FUNCTION: -.430000E+02  DERIVATIVE: 0.164984E-01

CONVERGED IN     3 ITERATIONS
LAMBASTAR:       0.199982E+01
F (LAMBDASTAR):   -0.430000E+02
DERIVATIVE (LAMBDASTAR):   -0.438730E-02
```

Program 12.2 The Davidon–Fletcher–Powell method of unconstrained minimization

The following are required for input:

ITLIM = maximum number of Davidon–Fletcher–Powell iterations (usual value: 10 to 20).
MAXIN = maximum number of interpolations in one-dimensional search (usual value: 5 to 10).
MAXGR = maximum number of gradient evaluations permitted in each one-dimensional search (usual value: 5 to 10).
N = number of design variables.
X(1), . . . , X(N) = initial (guess) values of design variables.
EPS, EPSS = convergence requirements in unconstrained and one-dimensional searches (usual values: 0.005 for each).
STEPO = initial step length in one-dimensional search (usual value: 1.0).
F(X) = function to be minimized; to be given as a subroutine as follows: SUBROUTINE FUN (X,F,N),
DIMENSION X(N).

```
C     NFUN = number of times the function F(X) is evaluated.
      COMMON/COUNT/NFUN, NGRAD.
      NFUN = NFUN + 1.
      F = . . . .
      RETURN
      END
```

Consider the problem of

minimizing $f(\vec{X}) = 100(x_2 - x_1^2)^2 + (1 - x_1)^2$.

Solution

$x_1^* = 1, x_2^* = 1.$

Using ITLIM = 5, MAXIN = 5, MAXGR = 3, N = 2, EPS = 0.005, EPSS = 0.005, STEPO = 1.0, and X(1) = −1.0, X(2) = 1.0, we get the following output of Program 12.2.

```
STARTING VALUES FOR UNCONSTRAINED MINIMIZATION
OBJ  =    0.400000E+01
X(I) :
   -0.100000E+01    0.100000E+01
S(I) :
   0.999741E+00  -0.227460E-01

ITER   =  1 STEPO = 0.19773E+01
OBJ   = 0.60269E-03 NFUN   =  48 NGRAD = 16
X(I) :
     0.976837E+00    0.955023E+00
S(I) :
  -0.363242E+00    -0.931695E+00

ITER   =  2 STEPO   = 0.22207E-01
OBJ   = 0.27248E-02 NFUN   =  69 NGRAD = 23
X(I) :
     0.968770E+00    0.934333E+00
S(I) :
  -0.932858E+00    0.360245E+00

ITER   =  3 STEPO = 0.21889E-02
OBJ   =  0.11381E-02 NFUN   = 81 NGRAD = 27
X(I) :
     0.966728E+00    0.935121E+00
S(I) :
  -0.398530E-01    0.999206E+00
```

12.13.2 C Programs

Program 12.3 Golden section method (Modified Fibanacci method)

The following input data are required:

XLOW, XUP = lower and upper bounds on the optimum value of X.
EPS = convergence requirement (usual value: 0.001).
F(X) = funtion to be minimized; to be given as a subprogram as follows:
SUBROUTINE FUN (X,S,XLAM,F).

```
C       F is to be defined in terms of XLAM.
        F = . . . .
        RETURN
        END
```

Consider the problem of

minimizing $f(x) = x^5 - 5x^3 - 20x + 5.$

Solution

$x^* = 2.$

The following is the program output:

```
Iteration number=1
Interval of uncertainty=   7.638480e+000

Lambda1=    7.638480e+000  Lambda2=   1.236152e+001

F1=    2.362752e+004  F2=  2.789550e+005

Lower bound=    0.000000e+000  Upper bound= 1.236152e+001

Iteration number=2
Interval of uncertainty=   4.721161e+000

Lambda1=  4.721161e+000    Lambda2=   7.640359e+000

F1=  1.729965e+003    F2=    2.365783e+004

Lower bound=  0.000000e+000    Upper bound= 7.640359e+000

Iteration number=3
Interval of uncertainty=  2.918036e+000

Lambda1=  2.918036e+000  Lambda2=  4.722322e+000

F1= 3.397453e+001    F2=  1.732440e+003

Lower bound= 0.000000e+000   Upper bound=  4.722322e+000
.
.
.
Iteration number=9
Interval of uncertainty= 1.626829e-001
```

```
Lambda1=  1.966485e+000 Lambda2=  2.067076e+000

F1= -4.294514e+001  F2= -4.276428e+001

Lower bound=  1.803802e+000   Upper bound= 2.067076e+000

Iteration number=10
Interval of uncertainty=  1.005504e-001

Lambda1=  1.904353e+000  Lambda2=  1.966525e+000

F1= -4.257238e+001  F2= -4.294527e+001

Lower bound=  1.904353e+000  Upper bound=  2.067076e+000

Result of one-dimensional minimization
by golden-section method
Iteration number=10
Lower bound= 1.904353e+000
Upper bound= 2.067076e+000

Final interval of uncertainty= 1.627229e-001
Final value of lambda= 1.966525e+000
final value of objective function= -4.294527e+001

xlam= 1.966525e+000 f= -4.294527e+001
```

Program 12.4 Interior penalty function method

The following input data are required:

ITLIM = maximum number of iterations permitted in the Davidon–Fletcher–Powell (usual value: 5 to 10).
MAXIN = maximum number of cubic interpolations permitted in each one-dimensional search (usual value: 5).
MAXGR = maximum number of gradient evaluations permitted in each one-dimensional search
(usual value: 3 to 5).
N = number of design variables.
M = number of inequality constraints.
MAXPI = maximum number of penalty parameters considered
(usual value: 5 to 10).
C = reduction factor for penalty parameter (usual value: 0.1).
R = initial value of penalty parameter (usual value: 1.0).
EPS, EPSS = convergence criteria specified (usual values: 0.005 for each).

STEPO = initial step length in one-dimensional search (usual value: 1.0).
X(1), ... X(N) = initial (guess) values of design variables.
F(X),G(X) = objective and constraint functions; to be given in a subprogram as follows:

```
SUBROUTINE FUN (G,X,F,OBJ,N,M,R),
DIMENSION X(N),G(M),
G(1) = ...,
G(2) = ....
....
G(M) = ....
OBJ = ....
F = ....
RETURN
END
```

Illustration (Example 12.17).

Using itlim = 5, maxin = 5, maxgr = 3, n = 3, m = 6, maxpi = 6, eps = 0.005, epss = 0.005, c = 0.1, r = 1.0, and x[1] = 0.0, x[2] = 0.1, x[3] = 0.1, x[4] = 3.0, we get the output of Program 12.4:

```
solution of constrained optimization problem
by interior penalty function method

Data:
Number of design variables (n) = 3
Number of inequality constraints (m) = 6
Reduction factor for penalty parameter (c) = 0.10
Initial value of penalty parameter (r) = 1.00

Starting values for unconstrained minimization
r =  0.10000E+01 pf = 0.251849E+02 obj =     0.404100E+01
x(i) :
     0.100000E+00    0.100000E+00    0.300000E+01
g(i) :
    -0.898000E+01   -0.502000E+01   -0.200000E+01
    -0.100000E+00   -0.100000E+00   -0.300000E+01

iter = 1 stepo  = 0.86444E+00
pf = 0.11950E+02 obj = 0.80026E+01 nfun = 20 ngrad = 5
x(i) :
     0.677236E+00    0.743453E+00    0.299433E+01
g(i) :
    -0.795463E+01   -0.597737E+01   -0.200567E+01
    -0.677236E+00   -0.743453E+00   -0.299433E+01
```

```
.
.
.
iter  =  5 stepo = 0.45253E+00
pf = 0.10366E+02 obj = 0.56677E+01 nfun = 132 ngrad =  33
x(i) :
     0.381986E+00   0.163657E+01   0.228558E+01
g(i) :
    -0.239959E+01  -0.404812E+01  -0.271442E+01
    -0.381986E+00  -0.163657E+01  -0.228558E+01

Starting values for unconstrained minimization

r = 0.10000E+00 pf =   0.613754E+01 obj =   0.566768E+01
x(i) :
     0.381986E+00   0.163657E+01  0.228558E+01
g(i) :
    -0.239959E+01  -0.404812E+01  -0.271442E+01
    -0.381986E+00  -0.163657E+01  -0.228558E+01

iter = 1 stepo = 0.28501E+00
pf = 0.44971E+01 obj = 0.32849E+01 nfun = 162 ngrad = 40
x(i) :
     0.999081E-01   0.163649E+01   0.224481E+01
g(i) :
    -0.235111E+01  -0.372727E+01  -0.275519E+01
    -0.999081E-01  -0.163649E+01  -0.224481E+01
.
.
.
iter = 5 stepo = 0.24185E-01
pf = 0.41259E+01 obj = 0.27261E+01 nfun = 270 ngrad = 67
x(i) :
     0.102026E+00   0.142379E+01   0.166520E+01
g(i) :
    -0.735312E+00  -0.810493E+00  -0.333480E+01
    -0.102026E+00  -0.142379E+01  -0.166520E+01

Starting values for unconstrained minimization

r = 0.10000E-01 pf = 0.286607E+01 obj = 0.272609E+01
```

```
x(i) :
     0.102026E+00   0.142379E+01   0.166520E+01
g(i) :
    -0.735312E+00  -0.810493E+00  -0.333480E+01
    -0.102026E+00  -0.142379E+01  -0.166520E+01

iter = 1 stepo = 0.73940E-01
pf = 0.23602E+01 obj = 0.19660E+01 nfun = 326 ngrad = 81
x(i) :
     0.284585E-01   0.142371E+01   0.165779E+01
g(i) :
    -0.720498E+00  -0.776023E+00  -0.334221E+01
    -0.284585E-01  -0.142371E+01  -0.165779E+01
.
.
.
Starting values for unconstrained minimization
r = 0.10000E-04 pf = 0.144839E+01 obj = 0.143995E+01
x(i) :
     0.148750E-02   0.140716E+01   0.142360E+01
g(i) :
    -0.465473E-01  -0.672626E-02  -0.357640E+01
    -0.148750E-02  -0.140716E+01  -0.142360E+01

iter = 1 stepo = 0.22563E-03
pf = 0.14471E+01 obj = 0.14375E+01 nfun = 1093 ngrad = 267
x(i) :
     0.126280E-02   0.140716E+01   0.142358E+01
g(i) :
    -0.464896E-01  -0.666738E-02  -0.357642E+01
    -0.126280E-02  -0.140716E+01  -0.142358E+01
.
.
.
iter = 5 stepo = 0.38073E-07
pf = 0.14678E+01 obj = 0.14250E+01 nfun = 1314 ngrad = 318
x(i) :
     0.314806E-03   0.140718E+01   0.142154E+01
g(i) :
    -0.406171E-01  -0.928164E-03  -0.357846E+01
    -0.314806E-03  -0.140718E+01  -0.142154E+01
```

REFERENCES AND BIBLIOGRAPHY

12.1. F. P. Beer and E. R. Johnston, *Mechanics of Materials*, 2d edition, McGraw-Hill, New York, 1992.

12.2. R. L. Norton, *Machine Design An Integrated Approach*, Prentice Hall, Upper Saddle River, NJ, 1998.

12.3. M. J. Panik, *Classical Optimization: Foundations and Extensions*, North-Holland, Amsterdam, 1976.

12.4. H. Hancock, *Theory of Maxima and Minima*, Dover, NY, 1960.

12.5. G. B. Dantzig, *Linear Programming and Extensions*, Princeton University Press, Princeton, NJ, 1963.

12.6. S. S. Rao, *Engineering Optimization: Theory and Practice*, 3d edition, Wiley, New York, 1996.

REVIEW QUESTIONS

The following questions along with corresponding answers are available in an interactive format at the Companion Web site at http://www.prenhall.com/rao.

12.1. **Define** the following:
  (a) Objective function.
  (b) Design vector.
  (c) Constraint function.
  (d) Local minimum.
  (e) Global minimum.
  (f) Hessian matrix.
  (g) Unimodal function.
  (h) Fibonacci numbers.
  (i) Experiment.
  (j) Interval of uncertainty.

12.2. Answer **true or false** to each of the following:
  (a) The problem of minimizing $f(\vec{X})$ is called a nonlinear-programming problem.
  (b) The solution x^* at which $\frac{df}{dx} = 0$ is a local optimum solution.
  (c) An optimization problem is same as a decision-making problem.
  (d) An optimum solution is a feasible solution.
  (e) A unimodal function can be discontinuous.
  (f) A unimodal function can be nondifferentiable.
  (g) A nondifferentiable function is called a unimodal function.
  (h) The Fibonacci method can be used for the minimization of any type of functions.
  (i) An optimization problem can have any number of inequality constraints.
  (j) An optimization problem can have any number of equality constraints.
  (k) Most nonlinear programming techniques find the global solution.
  (l) The solution of a nonlinear programming problem can be determined by solving a system of simultaneous nonlinear equations.
  (m) The cubic interpolation method requires the derivative of the function.
  (n) The Davidon–Fletcher–Powell method requires the Hessian matrix of the function.

(*continued, page 946*)

(o) The interior penalty function method can be used to solve any constrained optimization problem.

12.3. Select the **most appropriate answer** out of the multiple choices indicated in each case:

(a) Consider the minimization of the function $f(x_1, x_2) = 2x_1^2 + x_2^2 - 3x_1$ from the point $\vec{X}_i = \begin{Bmatrix} 1 \\ -1 \end{Bmatrix}$ along the direction $\vec{S}_i = \begin{Bmatrix} -1 \\ 2 \end{Bmatrix}$. The resulting one-dimensional minimization problem can be stated as

(a) minimize $f(\lambda) = 6\lambda^2 - 4\lambda - 3$,

(b) minimize $f(\lambda) = 6\lambda^2 - 5\lambda$, and

(c) minimize $f(\lambda) = 2\lambda^2 + \lambda - 3$.

(b) The following values were found during a cubic interpolation process:

$$\lambda = 1.0,\ f(\lambda = 1.0) = 4023.05,\ f'(\lambda = 1.0) = -5611.10,$$

$$\lambda = 2.0,\ f(\lambda = 2.0) = 4101.73,\ f'(\lambda = 2.0) = 2170.89,$$

$$\tilde{\lambda}^* = 1.414,\ f(\tilde{\lambda}^* = 1.414) = 4973.22,\ f'(\tilde{\lambda}^* = 1.414) = 148.70.$$

If $\tilde{\lambda}^*$ does not satisfy the convergence criterion, the values of A and B to be selected for refitting are

(a) $A = 1.0, B = 2.0$, (b) $A = 1.414, B = 2.0$, (c) $A = 1.0, B = 1.414$.

(c) In Davidon–Fletcher–Powell method, the positive definite matrix $[Q_i]$ converges to

(a) the inverse of the hessian matrix of the objective function,

(b) the hessian matrix of the objective function,

(c) the identity matrix.

(d) The solution of a constrained optimization problem

(a) can be a free infeasible point.

(b) is always a bound feasible point.

(c) can be an interior feasible point.

(e) Consider the problem of minimizing

$$f(\vec{X}) = (x_1 - 2)^2 + (x_2 - 1)^2$$

subject to

$$x_1 + x_2 - 2 \le 0 \text{ and } x_1^2 - x_2 \le 0,$$

and the points $\vec{X}_1 = \begin{Bmatrix} 2 \\ 2 \end{Bmatrix}$ and $\vec{X}_2 = \begin{Bmatrix} 1 \\ 1 \end{Bmatrix}$. Then

(*continued, page 947*)

(a) Both $\vec{X}_1$ and $\vec{X}_2$ are infeasible.
(b) Both $\vec{X}_1$ and $\vec{X}_2$ are feasible.
(c) $\vec{X}_1$ is infeasible, but $\vec{X}_2$ is feasible.

(f) The gradient of $f = x_1^3 - 2x_2^2$ at the point $\vec{X}_i = \begin{Bmatrix} -1 \\ 1 \end{Bmatrix}$ is given by

(a) $\begin{Bmatrix} -3 \\ -4 \end{Bmatrix}$, (b) $\begin{Bmatrix} 3 \\ -4 \end{Bmatrix}$, (c) $\begin{Bmatrix} -3 \\ 4 \end{Bmatrix}$.

(g) For the function $f = 12x^5 - 45x^4 + 40x^3 + 7$, the value $x = 1$ is a
(a) local minimum, (b) local maximum, or (c) neither minimum nor maximum.

(h) Consider the problem of maximizing

$$f = 2x_1x_2 - x_1^2 - 2x_2^2 + 4x_1$$

subject to $2x_1 + x_2 \geq 6$.
The corresponding ϕ-function for unconstrained minimization (according to the interior penalty function approach) is
(a) $\phi = -2x_1x_2 + x_1^2 + 2x_2^2 - 4x_1 - \frac{r_k}{2x_1+x_2-6}$,
(b) $\phi = 2x_1x_2 - x_1^2 - 2x_2^2 + 4x_1 - \frac{r_k}{2x_1+x_2-6}$, and
(c) $\phi = -2x_1x_2 + x_1^2 + 2x_2^2 - 4x_1 + \frac{r_k}{2x_1+x_2-6}$.

(i) Consider the minimization of the function $f = x^2 - 1.5x$ and the interval of uncertainty $L_0 = (0, 1)$. If $f(0.4995) = -0.49975$ and $f(0.5005) = -0.50025$, the new interval of uncertainty is
(a) (0.5005, 1), (b) (0, 0.4995), or (c) (0.4995, 1).

(j) When the number of experiments is six in the Fibonacci method, the value of L_2^* to locate the first two experiments is
(a) $\frac{8}{21}$, (b) $\frac{5}{13}$, or (c) $\frac{5}{8}$.

12.4. **Fill in the blanks** with suitable word(s) to each of the following:
(a) A unimodal function has a ______ peak or valley.
(b) A feasible solution satisfies all the ______ .
(c) In Lagrange multiplier method, the number of Lagrange multipliers is same as the number of ______ constraints.
(d) The constraint $a_{11}x_1 + a_{12}x_2 + \ldots + a_{1n}x_n \geq b_1$ can be converted to an equality constraint by introducing a ______ variable.
(e) The solution of an LP problem always occurs at a(n) ______ point of the feasible space.
(f) The cubic interpolation method can be considered as a ______ order method.
(g) The identity matrix is a ______ definite matrix.

(*continued, page 948*)

(h) In the interior penalty function method, the ϕ - function is minimized for a(n) _______ sequence of values of the parameter, r_k.

(i) The starting design vector needs to be an interior _______ point in the interior penalty function method.

(j) When six experiments are conducted in the exhaustive search method, the final interval of uncertainty will be _______ of the initial interval of uncertainty.

12.5. **Match the following** matrices with their characteristics:

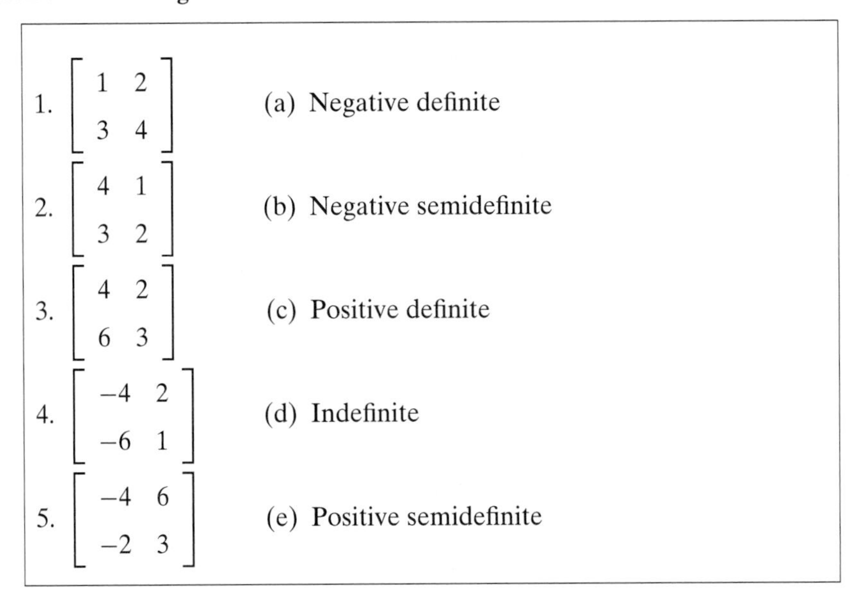

1. $\begin{bmatrix} 1 & 2 \\ 3 & 4 \end{bmatrix}$	(a) Negative definite
2. $\begin{bmatrix} 4 & 1 \\ 3 & 2 \end{bmatrix}$	(b) Negative semidefinite
3. $\begin{bmatrix} 4 & 2 \\ 6 & 3 \end{bmatrix}$	(c) Positive definite
4. $\begin{bmatrix} -4 & 2 \\ -6 & 1 \end{bmatrix}$	(d) Indefinite
5. $\begin{bmatrix} -4 & 6 \\ -2 & 3 \end{bmatrix}$	(e) Positive semidefinite

12.6. Give **short answers** to the following:

(a) What is a nonlinear programming problem?

(b) What is a linear programming problem?

(c) How can you convert an unrestricted variable into an equivalent nonnegative variable(s)?

(d) What is the significance of a basic variable in an LP problem?

(e) How many basic solutions exist for an LP problem?

(f) What are the practical limitations in using calculus methods of optimization?

(g) What is the difference between a constrained and an unconstrained optimization problem?

(h) What is the basic assumption made in using the exhaustive search method?

(i) What is a penalty parameter?

(j) What are Lagrange multipliers?

PROBLEMS

Section 12.3

12.1. A cylindrical vessel with a lid, is to be constructed to transport 200 m^3 of oil from a cargo ship to a storage tank. The sheet metal used for the lid, bottom, and sides costs \$ 20, \$ 50, and \$ 10 per square meter, respectively. If it costs \$ 5 for each round trip of the vessel, formulate the problem of determining the dimensions of the vessel for minimum transportation cost. Assume that the vessel has no salvage value upon completion of the operation.

12.2. A manufacturer produces three products—A, B and C. Each of the products can be produced on either of the machines 1 and 2. The times required to produce one unit of each of the products on machines 1 and 2 are as follows:

Product	Time required, in hours, to produce one unit	
	On machine 1	On machine 2
A	1.0	1.2
B	1.4	1.6
C	1.8	2.0

The operating cost is \$ 7 per hour for machine 1 and \$ 6 per hour for machine 2. It is required that at least 100 units of product A, 90 units of product B, and 80 units of product C be produced per week. Assuming that each machine can be operated for 80 hours per week, formulate the problem of minimizing the overall manufacturing cost per week.

12.3. A flywheel is to be designed for a forging press to store an energy of at least 10,000 J. The minimum and maximum operating speeds of the flywheel are 600 and 800 rpm, respectively. The kinetic energy of the flywheel is given by $E = \frac{1}{2}I\omega^2$, where I is the mass moment of inertia and ω is the rotational speed of the flywheel. By idealizing the flywheel as a rotating annular disk with inner radius r_i, outer radius r_0, and thickness t, we express the mass moment of inertia as

$$I = \frac{1}{2}\pi(r_0^4 - r_i^4)t\rho,$$

where ρ is the density of the material. The tangential and radial stresses developed in the flywheel at a radius r, $r_i \le r \le r_0$, are given by

$$\sigma_t = \rho\omega^2\left(\frac{3+\nu}{8}\right)\left(r_i^2 + r_0^2 + \frac{r_i^2 r_0^2}{r^2} - \frac{1+3\nu}{3+\nu}r^2\right),$$

$$\sigma_r = \rho\omega^2\left(\frac{3+\nu}{8}\right)\left(r_i^2 + r_0^2 - \frac{r_i^2 r_0^2}{r^2} - r^2\right).$$

where ν is Poisson's ratio. Formulate the problem of determining r_i, r_0, and t of the flywheel to minimize the mass without its tangential or radial stress exceeding a value of 2×10^8 Pa. Assume the material of the flywheel as steel with $\rho = 7,700$ kg/m^3 and $\nu = 0.3$.

Section 12.4

12.4. Consider the following functions:
(a) $f(x) = (2x - 3)^4$,
(b) $f(x) = (2x - 3)^3$.
Determine the nature of the optimum solution in each case.

12.5. Determine whether the following matrix is positive definite, negative definite, or indefinite:

$$\text{(a)}\ [A] = \begin{bmatrix} 2 & 1 & 2 \\ 1 & 3 & 0 \\ 2 & 0 & 4 \end{bmatrix};$$

$$\text{(b)}\ [A] = \begin{bmatrix} 1 & -2 & 3 \\ -2 & 5 & -5 \\ 3 & -5 & 7 \end{bmatrix}.$$

12.6. The drag force acting on an airplane is given by $F = c_d v^2$, where v is the velocity of the airplane and c_d is the drag coefficient, which can be expressed as the sum of parasitic drag coefficient (c_1) and induced drag coefficient (c_2/v^4) as

$$c_d = c_1 + \frac{c_2}{v^4},$$

where c_1 and c_2 are constants. If the power required to overcome the drag force is given by $P = F\ v$, determine the velocity of the airplane corresponding to minimum power.

12.7. The values of stress and the corresponding values of strain observed during a material testing experiment are as follows

Stress (ksi)	20	22	24	26	28	30	32
Strain (in/in)	0.005	0.020	0.030	0.040	0.050	0.065	0.075

Find the equation of a straight line between stress and strain that fits the data using the least-squares approach.

12.8. A fence is to be constructed on the bank of a straight canal to enclose land in the shape of a semicircle adjacent to a rectangle as shown in Fig. 12.17. If the length of the fence is restricted to 1000 ft, formulate the problem of

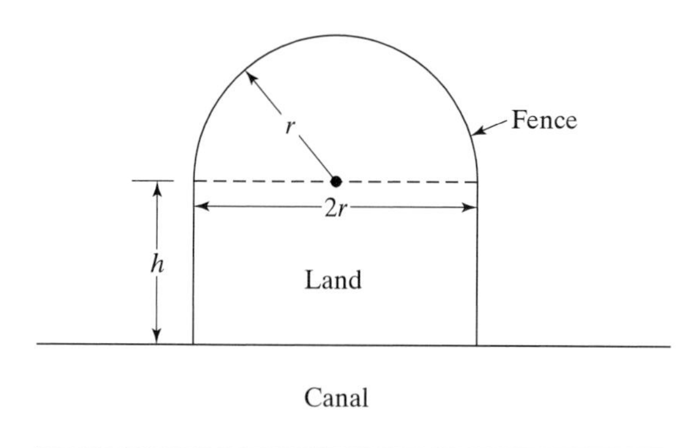

Figure 12.17 Fence to enclose maximum land area.

determining the dimensions h and r that correspond to maximum enclosed land area.

12.9. Find the dimensions of the rectangular box with largest volume with faces parallel to the coordinate planes that can be inscribed in the ellipsoid given by $16x_1^2 + 9x_2^2 + 4x_3^2 - 288 = 0$. Use the Lagrange multiplier method.

12.10. Solve the following nonlinear-programming problem using the Lagrange multiplier method:
Minimize $f = 5x_1^2 + 4x_2^2$ subject to $x_1 + 4x_2 = 5$ and $x_i \geq 0, i = 1, 2$.

12.11. A package needs to have its height plus girth (perimeter of the base) less than or equal to 96 inches in order to be sent by parcel post. Find the dimensions of a package, with a square base, that will have a maximum volume.

Section 12.5

12.12. Solve the following LP problem using a graphical method:
Minimize $f = 4x_1 + 3x_2$ subject to $x_1 + 2x_2 \geq 2, 3x_1 + x_2 \geq 3$, and $5x_1 + 3x_2 \geq 6$, $x_i \geq 0, i = 1, 2$.

12.13. Find the solution of the following problem using a graphical procedure:
Minimize $f = x_1^2 + x_2^2$ subject to $x_1 + x_2 \geq 5$ and $3x_1 + x_2 \geq 6$, where $x_1 x_2 \geq 10$ and $x_i \geq 0, i = 1, 2$.

Section 12.8

12.14. The efficiency of a screw jack is given by

$$e = \frac{1 - \mu \tan \lambda}{1 + \mu \cot \lambda},$$

where λ is the lead angle and μ is the coefficient of friction between the screw threads and the nut. For $\mu = 0.15$, determine the value of the lead angle that maximizes the efficiency. Use an exhaustive search procedure.

12.15. Solve Problem 12.14 using the Fibonacci method.

12.16. Solve Problem 12.14 using the cubic interpolation method.

Section 12.9

12.17. The profit per acre of a farm is given by

$$f(x_1, x_2) = -300 + 20x_1 + 5x_2 - 5x_1x_2 - 15x_1^2 - 10x$$

where x_1 and x_2 denote, respectively, the labor cost and the fertilizer cost in thousands of dollars. Find the values of x_1 and x_2 to maximize the profit using the Davidon–Fletcher–Powell method.

12.18. Find the solution of Problem 12.1 using the Davidon–Fletcher–Powell method.

Section 12.10

12.19. Solve the following problem using the interior penalty function approach with a calculus-based method for unconstrained minimization:

Minimize $f = x_1^2 - 2x_1 - x_2$ subject to $2x_1^2 + 3x_2^2 \leq 8$.

12.20. Consider the following optimization problem:

Minimize $f(x_1, x_2) = (3x_1 - x_2)^2 + (x_2 + 1)^2$ subject to $x_1 + x_2 \leq 8$.

Solve this problem using the interior penalty function method coupled with the classical method of unconstrained minimization.

Section 12.12

12.21. Use MAPLE to solve Problem 12.17.

12.22. Use MATLAB to solve Problem 12.19.

12.23. Use MATHCAD to solve Problem 12.19.

12.24. Use MAPLE to solve Problem 12.8.

12.25. Use MATLAB to solve Problem 12.12.

12.26. Use MATHCAD to solve Problem 12.12.

Section 12.13

12.27. Use Program 12.1 to solve Problem 12.14.

12.28. Use Program 12.2 to solve Problem 12.17.

12.29. Use Program 12.3 to solve Problem 12.14.

12.30. Use Program 12.4 to solve Problem 12.19.

General

12.31. A hollow circular shaft is subjected to a bending moment (M) of 2000 lb-in, a twisting moment (T) of 3000 lb-in, and an axial compressive load (F) of 1000

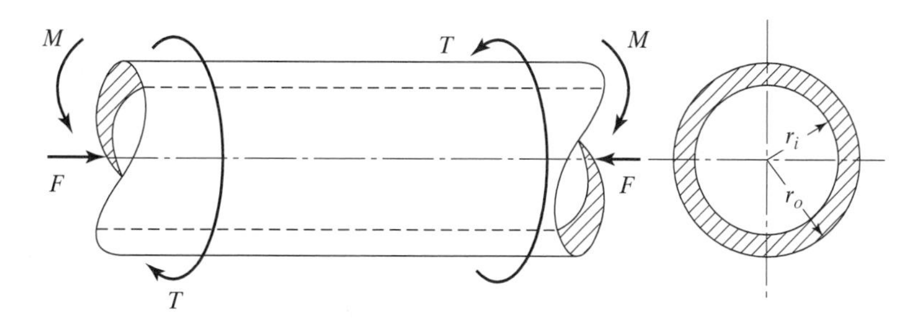

Figure 12.18 Hollow circular shaft for minimum weight.

lb simultaneously. (See Fig. 12.18.) Formulate the problem of designing the shaft for minimum weight when the maximum shear stress is restricted to be less than 20,000 psi.

Note:

The maximum shear stress (τ) induced in the shaft is given by

$$\tau = \left\{ \left(\frac{\sigma_x}{2} \right)^2 + \tau_{xy}^2 \right\}^{\frac{1}{2}},$$

where σ_x and τ_{xy} denote the axial stress in x-direction and shear stress in the xy-plane, respectively, and are given by

$$\sigma_x = \frac{4Mr_0}{\pi(r_0^4 - r_i^4)} + \frac{F}{\pi(r_0^2 - r_i^2)},$$

$$\tau_{xy} = \frac{2Tr_0}{\pi(r_0^4 - r_i^4)}.$$

12.32. Automobile doors are often reinforced by aluminum beams to help absorb side impact energy and reduce injury to passengers from side impacts. Formulate the problem of designing a minimum weight beam of rectangular cross section with the following constraints:

(a) The stress induced in the beam, assumed to be simply supported, during the impact should not exceed 20,000 psi. The impact energy absorbed by the beam can be approximated as

$$E = \frac{1}{2} \frac{m_2(v_1 - v_2)^2}{(1 + \frac{m_2}{m_1})},$$

where m_1 and m_2 are the masses of the two impacting vehicles and v_1 and v_2 are their respective velocities at the time of impact. The impact force induced (F_i) at the middle of the beam is given by $F_i = \frac{2E}{s}$, where s is the net crush distance during impact. The bending stress induced in a simply supported beam due to central concentrated load (F_i) can be expressed as

$$\sigma_b = \frac{3F_i l}{2bh^2},$$

where $l, b,$ and h are the length, width, and depth of the beam, respectively. Assume that $v_1 = 0$, $v_2 = 14.7$ ft/sec (10 mph), $m_1 = m_2 = 2750$ lbm, $s = \frac{2}{3}$ ft, and beam length $= 2$ ft.

(b) The width (b) and depth (h) of the beam should not be less than 0.25″.

(c) The depth of the beam should not exceed the width of the beam.

12.33. A pipe is to be connected to a boiler to convey water at a specified flow rate. The initial cost of installing the pipe is given by 200 d, while the cost of pumping water over the expected life of the pipe is estimated to be $\frac{20\times10^6}{d^4}$, where d is the diameter of the pipe in inches and the costs are in dollars. Determine the diameter of the pipe for minimum overall cost.

12.34. Find the extrema of the following functions:

(a) $f = x_1^2 + 4x_2^2 - x_1 + 2x_2 + 5$;

(b) $f = x_1^2 x_2 + x_2^3 - 4x_1^2 + 1$;

(c) $f = 4x_1^3 - 2x_1^2 x_2 + x_2^2 + 2$.

12.35. A wire of length 100 in is to be cut into two pieces. One of the pieces will be bent into the shape of a square and the other into the shape of a circle. How should the wire be cut so that the sum of the areas of the square and the circle is a maximum?

12.36. Solve Problem 12.35 by replacing the square by an equilateral triangle.

12.37. Convert the following LP problem to the standard form:

Maximize $f = 3x_1 + 4x_2$ subject to $-x_1 + 3x_2 \leq 6$, $x_1 + x_2 \leq 6$, and $2x_1 + 3x_2 \geq 6$, with x_1, x_2 unrestricted in sign.

12.38. Find the solution of the following problem using a graphical procedure:

Maximize $f = 2x_1 + 5x_2$ subject to $x_1 x_2 \leq 10$, $x_1^2 + x_2^2 \leq 25$, and $x_i \geq 0$, $i = 1, 2$.

12.39. Find the dimensions of an open rectangular box of volume V for which the amount (area) of sheet metal required for manufacture is a minimum. Use the Davidon–Fletcher–Powell method.

12.40. Construct the ϕ-function according to the interior penalty function approach, and plot its contours for different values of the penalty parameter for the following problem:

Minimize $f(x) = (x - 1)^2$ subject to $3 \leq x \leq 5$.

12.41. An exterior penalty function method can be used to solve an equality-constrained problem:

Minimize $f(\vec{X})$ subject to $h_j(\vec{X}) = 0;\ j = 1, 2, \ldots, p$.

For this, the ϕ-function is constructed as

$$\phi(\vec{X}, r_k) = f(\vec{X}) + r_k \sum_{j=1}^{p} h_j^2(\vec{X})$$

and the ϕ-function is minimized for an increasing sequence of values of r_k (i.e., $r_k \rightarrow \infty$).

Use the procedure to solve the following problem:

Minimize $f(x_1, x_2) = (3x_1 - x_2)^2 + (x_2 + 1)^2$ subject to $x_1 + x_2 = 8$.

Use a calculus method for the unconstrained minimization of the ϕ-function.

PROJECTS

12.1. A spherical pressure vessel with inner radius r_i and outer radius r_o is to be designed for minimum volume such that the stress does not exceed a value of $\frac{S_y}{n}$ when the vessel is used to store an ideal gas following the relation, $p = \frac{mRT}{V}$, where p is the pressure, m is the mass, V is the volume, R is the gas constant, T is the temperature of the gas, S_y is the yield stress of the material, and n is the factor of safety of the vessel.

(i) Formulate the optimization problem.

(ii) Find the solution of the problem using a graphical method for the following data:

$S_y = 30,000$ psi, $n = 2$, $m = 10$ kg, $R = 0.3464$ kN m/kg – K, $T = 308.2$ K.

Note:

The stress induced in a spherical shell (σ) subjected to an internal pressure (p) is given by $\sigma = \frac{pr_i}{2t}$, where t is the thickness of the shell wall.)

12.2. The actuating force (F) necessary to apply the brake shown in Fig. 12.19 is given by

$$F = \frac{M_n - M_f}{a},$$

$$M_n = p_a trb\left(\frac{\theta_2}{2} - \frac{1}{4}\sin 2\theta_2\right),$$

$$M_f = p_a trf\left(r - r\cos\theta_2 - \frac{b}{2}\sin^2\theta_2\right),$$

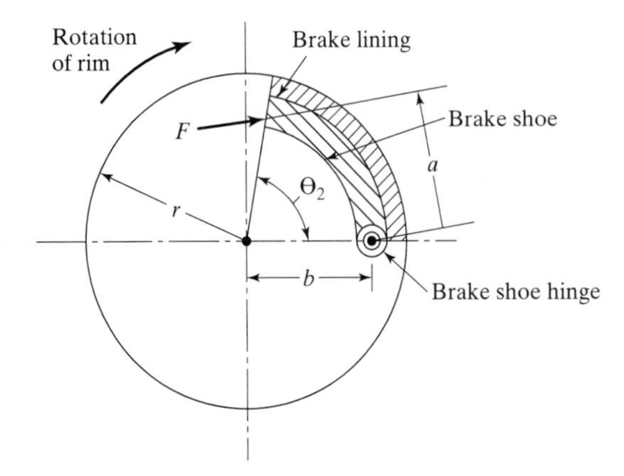

Figure 12.19 Brake design for minimum volume.

where p_a is the maximum pressure exerted on the brake lining, t is the face width of the brake shoes, r is the inner radius of the brake drum, b is the radius of the brake shoe, f is the coefficient of friction, a is the linear distance between the brake shoe hinge and the line of application of the actuating force, and θ_2 is the angle between the brake—shoe hinge and the point of application of the actuating force. (See Fig. 12.19.) Formulate the problem of determining the values of r, b, t, and θ_2 for minimizing the volume of the brake lining with the maximum actuating force limited to a value of $F_{\max}$.

The data are as follows: $p_a = 10^5$ Pa, $f = 0.35$, $a = 20$ cm, $F_{\max} = 80$ N.

Note:

The radial thickness of the brake lining can be assumed to be $(r - b.)$

12.3. A helical spring, subjected to an axial compressive load of $P = 100$ lb, is to be designed for minimum weight. (See Fig. 12.20.) Formulate the optimization problem by using the wire diameter (d), mean coil diameter (D), and the number of active coils (N) as design variables. Assume the following data:

The material is steel (shear modulus of $G = 11.5 \times 10^6$ psi, Young's modulus of $E = 30 \times 10^6$ psi, weight density of $\gamma = 0.283$ lb/in^3).

Number of inactive coils (ends are squared and ground) are 2

The maximum permissible axial deflection is 0.4 in

The maximum permissible shear stress is 80,000 psi

The range of spring index $(C = D/d)$ is 6 to 10

The pitch (p) is 3 d

The lower limit on the surge frequency is 100 Hz

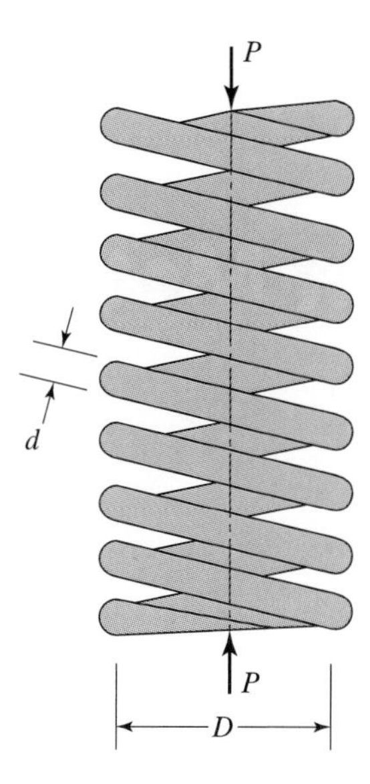

Figure 12.20 Helical spring design.

The Equations are as follows:
Axial deflection (δ):

$$\delta = \frac{8PD^3N}{d^4G}\left(1 + \frac{1}{2C^2}\right).$$

Shear stress (τ):

$$\tau = \left(\frac{2C+1}{2C}\right)\frac{8PD}{\pi d^3}.$$

Stability (buckling) condition:

$$L_0 < \frac{\pi D}{\alpha}\left(\frac{2(E-G)}{2G+E}\right)^{\frac{1}{2}},$$

where L_0 is the free length $= pN + 2d$, and $\alpha = 0.5$ when ends are supported between flat plates.
Fundamental natural frequency in Hz (f):

$$f = \frac{1}{2}\sqrt{\frac{kg}{W}},$$

where k is the spring rate $= \frac{d^4\,G}{8D^3\,N}$, g is the gravitational constant, and W is the weight of the spring.

13

Finite Element Method

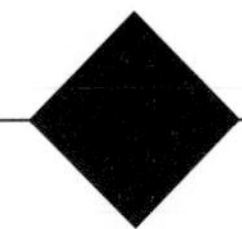

13.1 Introduction

CHAPTER CONTENTS

The finite-element method is a numerical procedure for finding approximate solutions to boundary-value problems. Because of the generality of the ideas underlying the method, it has been used with remarkable success in solving a variety of problems in virtually all areas of engineering and mathematical physics [13.1, 13.2]. The basic concept of the finite-element method is that any solution domain can be divided into several simple subdomains known as finite elements. By assuming a simple form of solution in each finite element, the approximate solution of the problem in the complete domain is determined. The finite-element method has all the advantages of the finite-difference method. In fact, in some respects the finite-element method is superior to the finite-difference method. For example, irregular boundaries can be approximated the same way as regular boundaries. In the case of physical problems, the finite-element method permits different material properties for adjacent elements.

The finite-element method essentially involves five steps:

(1) Discretization of the domain.

(2) Selection of an interpolation or shape function.

(3) Derivation of element characteristic matrices and vectors.

(4) Assemblage of element characteristic matrices and vectors.

(5) Solution of the system equations.

The details of these steps will be discussed in subsequent sections.

13.2 Engineering Applications

The finite-element method can be used to analyze most engineering science problems, including structural, mechanical, biomechanical, and several other nonstructural problems. Typical engineering analysis problems that can be solved using the finite-element method include the stress analysis of trusses, frames, plates and shells, temperature distribution in heated bodies, fluid flow in turbomachinery blades, and distribution of magnetic and electric potential in electromagnetic devices. The following examples represent selected applications of the finite-element method:

▶Example 13.1

Many automobile accidents result in head injuries. In spite of the lower speed limits, safer car designs, and the mandatory use of seat belts in many states, the overall decrease in the number of head injuries is not significant [13.3]. The development of mathematical models, based on finite-element analysis, can provide an effective means of studying the dynamics of impact to the head. A better understanding of the dynamics of impact can help in bringing improvements in the prevention, diagnosis, and treatment of head trauma. Figure 13.1 shows a finite-element model of the head and neck [13.3].

▶Example 13.2

Figure 13.2 shows the front view of an electric transmission tower. It is a three-dimensional frame comprising several beam-type elements. Each node or joint can have six displacement components, also known as degrees of freedom—three translations parallel to the three coordinate axes and three rotations about the three coordinate axes. The loading on the tower includes the tensions in the cables, wind and snow loads, and earthquake loads. The finite-element analysis of the tower helps the analyst designer to quickly obtain the displacements of the nodes and stresses in the members so that it can be designed to meet the specified safety standards and design codes.

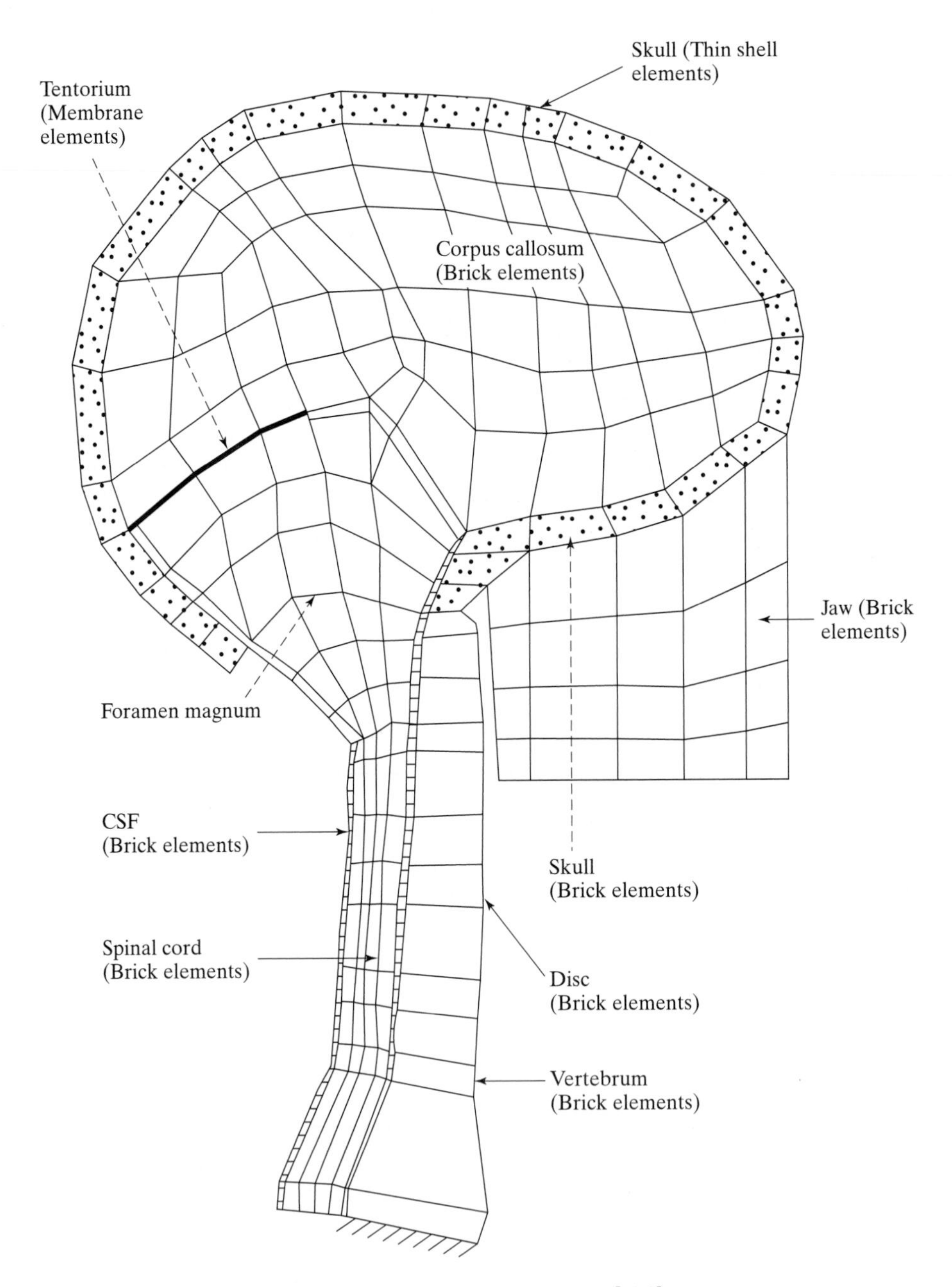

Figure 13.1 Finite element model of the head and neck [13.3].

Figure 13.2 An electric transmission tower.

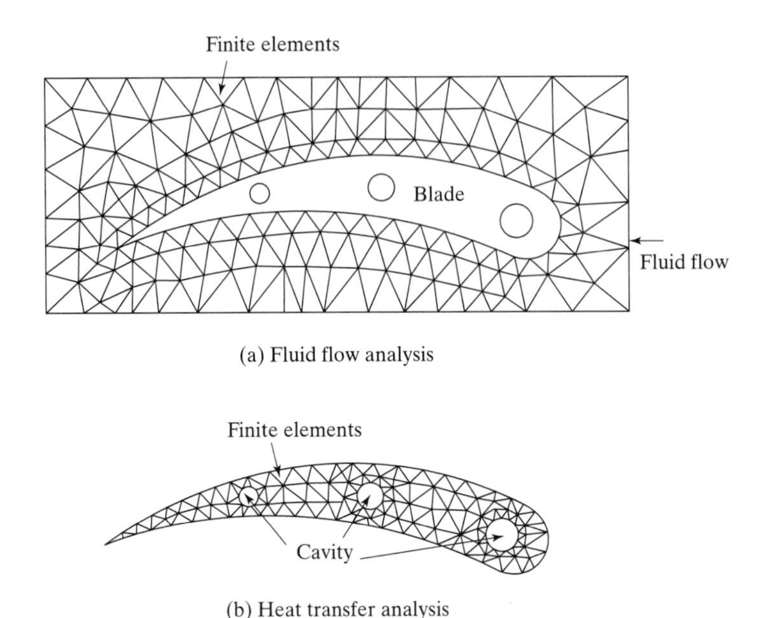

Figure 13.3 Gas turbine blade.

▶Example 13.3

The heat transfer and fluid flow analysis of turbine blades play a major role in the efficient design of gas turbines. Figure 13.3(a) shows the finite-element grid used to study the fluid flow around a blade, while Figure 13.3(b) indicates the finite-element modeling of the blade for the temperature distribution in the blade. The fluid flow analysis helps in designing the blade for maximizing the power developed in the engine. The heat transfer analysis is necessary in order to limit the temperature developed in the blade due to hot gases. Cavities are provided so that the blade can be cooled by passing cooler air through them. The heat transfer analysis can also be used in selecting the sizes and locations of the cavities for maximum thermal efficiency.

13.3 Discretization of the Domain

The first step in the finite-element method is to subdivide the domain or solution region into subdomains, or finite elements. The shapes, sizes, number, and orientations of the elements have to be selected properly so that the original domain is represented as closely as possible without unnecessarily increasing the computational effort required in finding the solution. The shape of the elements is decided mostly by the type of problem and the shape of the solution domain. If the solution domain

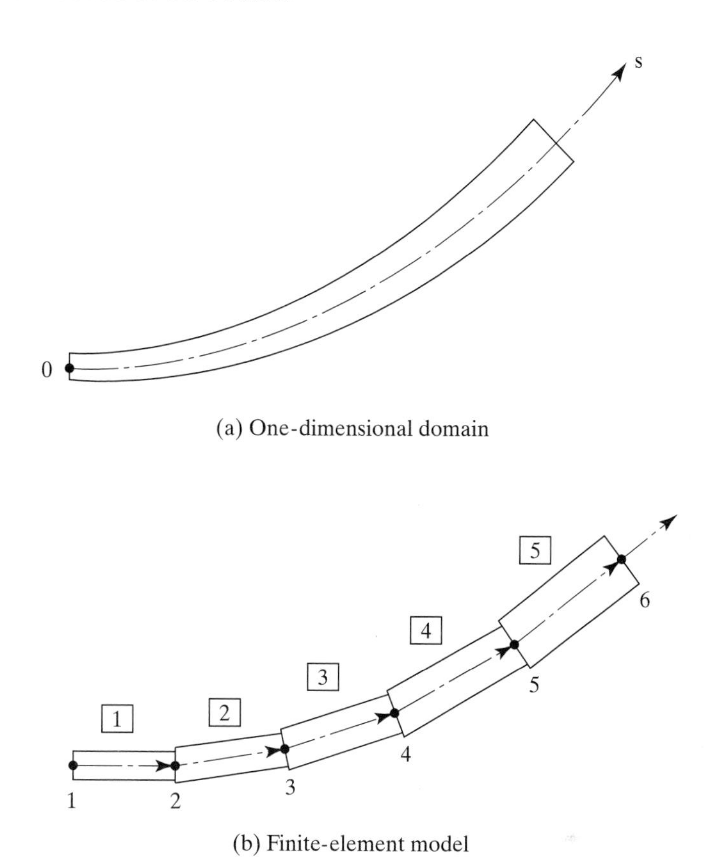

Figure 13.4 Finite element model for one-dimensional domain.

can be described by one spatial coordinate (such as a straight or curved bar), then one-dimensional, or line, elements can be used as shown in Fig. 13.4. If the solution domain is a two-dimensional region (such as a rectangular, triangular, or circular plate), then two-dimensional, or planar, elements can be used as shown in Fig. 13.5. It can be seen that a given domain can be modeled in a variety of ways using finite elements. Even with a particular type of elements, such as triangular or rectangular elements, the modeling can be done using different sizes or orientations. Also note that when straight-sided elements, such as triangular elements, are used to model a curved or irregular boundary, the original domain is not represented completely. For example, in the case of the circular domain shown in Fig. 13.5(a), the finite-element model approximates the circular boundary as a polygon as indicated in Fig. 13.5(b). To represent the geometry or solution domain accurately, sometimes elements with curved sides are used as shown in Fig. 13.6. To avoid numerical problems, the two- and three-dimensional elements should have all sides of approximately the same length. This implies that elongated elements are to be avoided. (See

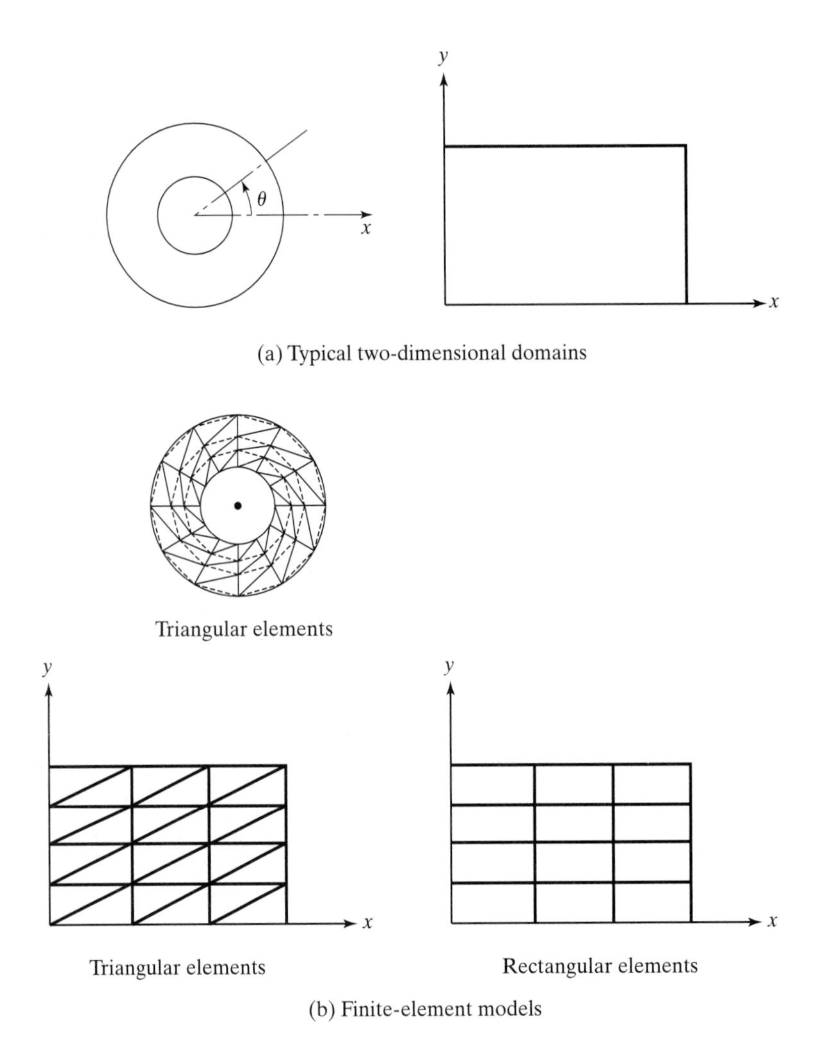

Figure 13.5 Modeling by two-dimensional elements.

Fig. 13.7.) In addition, the nodes at which the corner points of the elements meet must be numbered so as to reduce the bandwidth of the resulting set of algebraic equations that must be solved. The general guideline in this respect is to number the nodes along the shorter side of the domain first. The bandwidth—to be called semibandwidth—more accurately is defined as

$$B = (D + 1)f, \tag{13.1}$$

where D is the maximum difference between the largest and the smallest node numbers for any element and f is the number of degrees of freedom considered per node.

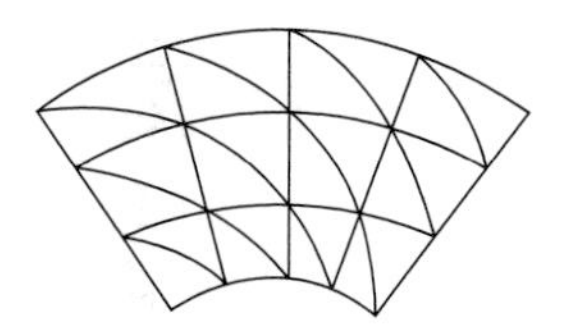

Figure 13.6 Use of curved-sided triangular elements.

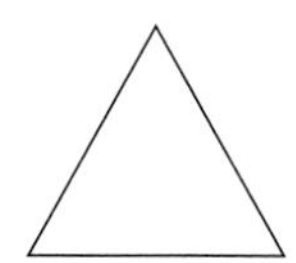

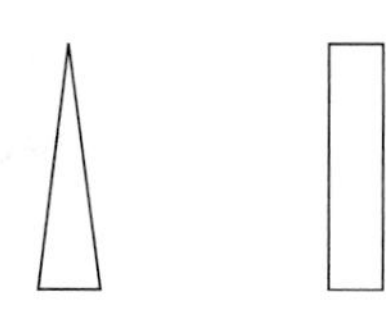

(a) Desirable shapes

(b) Undesirable shapes

Figure 13.7 Element shapes.

▶Example 13.4

Determine the bandwidth of the frame structures shown in Figs. 13.8(a) and (b), assuming one degree of freedom per node.

Solution

In Fig. 13.8(a), the maximum difference between the largest and smallest node numbers for any element (oriented in the vertical direction) is $D = 3$. Hence, the bandwidth is $(D+1)f = (3+1)1 = 4$. On the other hand, in the case of Fig. 13.8(b), the value of D (for elements oriented in the horizontal direction) is 9. Hence, the bandwidth is $(9+1)1 = 10$. ◀

13.4 Interpolation Functions

The basic idea of the finite-element method is piecewise approximation; that is, the solution of the complex problem is approximated by a simple function in each element. The overall accuracy of the finite-element solution depends on the choice of the approximating function. The function used to approximate the solution in each element is called the interpolation or shape function. Polynomial type of interpolation functions are most widely used in the finite-element literature. Since a higher order polynomial usually approximates an arbitrary function better, as shown in Fig. 13.9, it is desirable to use a higher order polynomial. However, a lower order polynomial simplifies the computations. Hence, a compromise has to be made between accuracy and computational effort in selecting the order of the polynomial. Finite elements are classified into three groups according to the order of the polynomial used for interpolation:

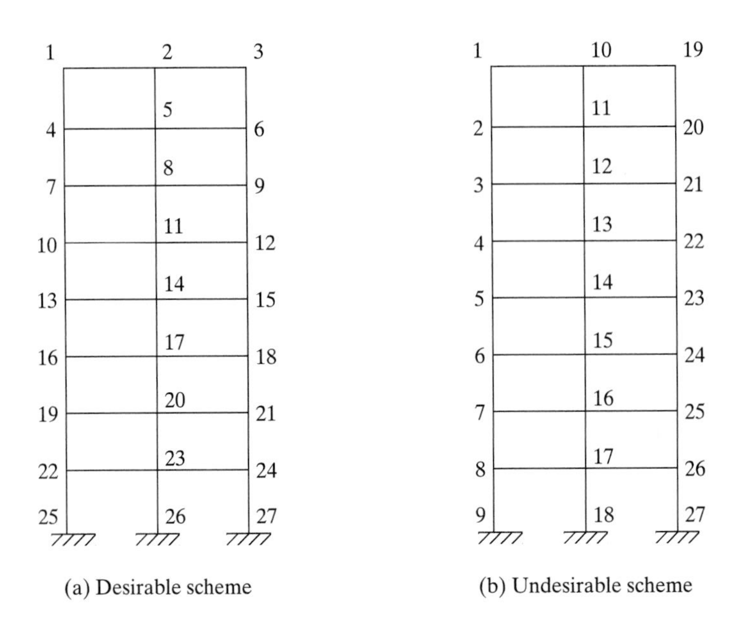

Figure 13.8 Frame structure (node numbering schemes).

(1) Simplex elements,

(2) Complex elements,

(3) Multiplex elements.

The simplex element is one for which the interpolation function is linear. Thus, the interpolation function of a simplex element can be represented as

$$\phi = \begin{cases} a_1 + a_2 x & \text{(for a one-dimensional element)}, \\ a_1 + a_2 x + a_3 y & \text{(for a two-dimensional triangular element)}, \\ a_1 + a_2 x + a_3 y + a_4 z & \text{(for a three-dimensional tetrahedron element)}, \end{cases} \tag{13.2}$$

where ϕ is the field variable and a_1, a_2, a_3, and a_4 are constants. The corners of simplex elements constitute the node points. For example, in two dimensions, the simplex element is a triangle with three corners or nodes. Hence, the values of $\phi(x, y)$ at the three corners can be used to evaluate the three constants a_1, a_2, and a_3 of $\phi(x, y)$.

The complex element is one for which the interpolation function is a higher order polynomial, including a constant, linear, and higher order terms. For example, a second-order polynomial (quadratic) in two dimensions is given by

$$\phi(x, y) = a_1 + a_2 x + a_3 y + a_4 x^2 + a_5 xy + a_6 y^2. \tag{13.3}$$

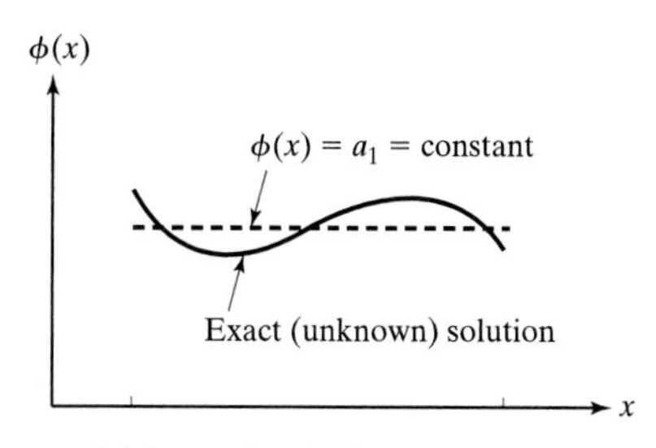

(a) Approximation by a constant

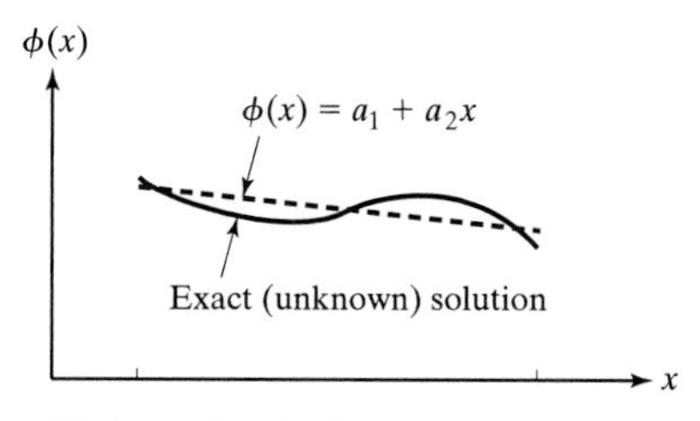

(b) Approximation by a linear equation

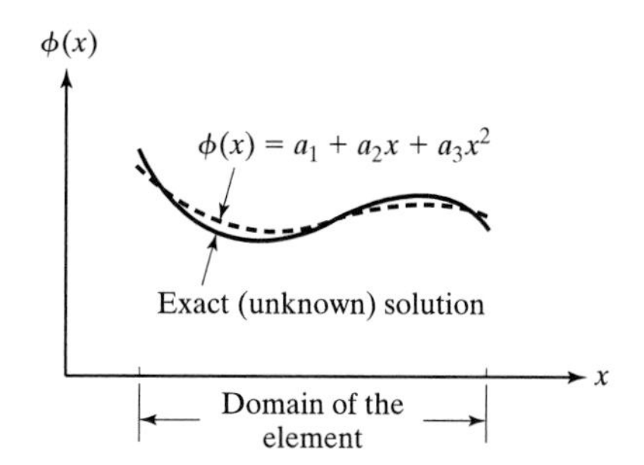

(c) Approximation by a quadratic equation

Figure 13.9 Approximation by different-degree polynomials.

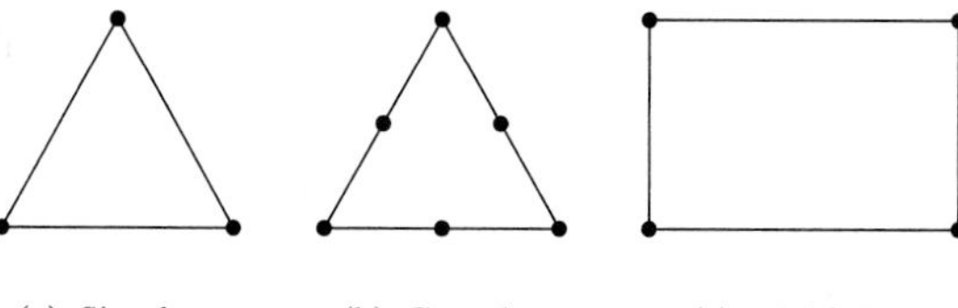

(a) Simplex element (b) Complex element (c) Multiplex element

Figure 13.10 Simplex, complex, and multiplex elements.

Since this interpolation consists of six constants, the midside points along with the corner points are considered as nodes for a triangle. (See Fig. 13.10.) The six constants in Eq. (13.3) can be expressed in terms of the nodal values of the variable ϕ.

The multiplex element is one for which the boundaries are parallel to the coordinate axes (such as a rectangle) and the interpolation function is a higher order polynomial.

13.4.1 Interpolation Function in Terms of Nodal Values of the Solution

The constants appearing in an interpolation function (polynomial) can be expressed in terms of the nodal values of the unknown solution. We consider simplex elements in one dimension to illustrate the procedure.

13.4.2 One-Dimensional Simplex Element

The simplex element in one dimension is a line element of length l with two nodes as shown in Fig. 13.11. The nodes of the element are denoted as i and j with coordinates x_i and x_j. The unknown solution $\phi(x)$, when evaluated at x_i and x_j, gives the nodal values as ϕ_i and ϕ_j. (These are treated as unknowns in the problem.) The solution $\phi(x)$ is assumed to be linear inside the element giving

$$\phi(x) = a_1 + a_2\, x = \left\{ \begin{matrix} 1 & x \end{matrix} \right\} \left\{ \begin{matrix} a_1 \\ a_2 \end{matrix} \right\}, \tag{13.4}$$

where a_1 and a_2 are unknown constants. The nodal values of $\phi(x)$ are given by

$$\phi(x = x_i) = \phi_i = a_1 + a_2 x_i,$$

$$\phi(x = x_j) = \phi_j = a_1 + a_2 x_j,$$

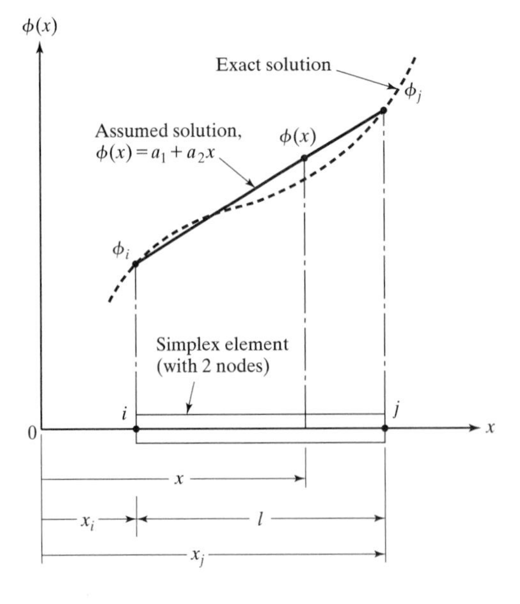

Figure 13.11 Simplex element in one dimension.

or

$$\vec{\phi} = [A]\vec{a}, \tag{13.5}$$

where

$$\vec{\phi} = \begin{Bmatrix} \phi_i \\ \phi_j \end{Bmatrix},$$

$$[A] = \begin{bmatrix} 1 & x_i \\ 1 & x_j \end{bmatrix}$$

and

$$\vec{a} = \begin{Bmatrix} a_1 \\ a_2 \end{Bmatrix}.$$

The solution of Eq. (13.5) can be expressed as

$$\vec{a} = [A]^{-1}\vec{\phi}, \tag{13.6}$$

where

$$[A]^{-1} = \frac{1}{(x_j - x_i)} \begin{bmatrix} x_j & -x_i \\ -1 & 1 \end{bmatrix}.$$

Noting that $x_j - x_i = l$, we see that Eq. (13.6) can be rewritten as

$$\begin{Bmatrix} a_1 \\ a_2 \end{Bmatrix} = \frac{1}{l} \begin{bmatrix} x_j & -x_i \\ -1 & 1 \end{bmatrix} \begin{Bmatrix} \phi_i \\ \phi_j \end{Bmatrix}. \tag{13.7}$$

By using Eq. (13.7), we can write Eq. (13.4) as

$$\phi(x) = \frac{1}{l} \begin{Bmatrix} 1 & x \end{Bmatrix} \begin{bmatrix} x_j & -x_i \\ -1 & 1 \end{bmatrix} \begin{Bmatrix} \phi_i \\ \phi_j \end{Bmatrix},$$

or

$$\phi(x) = N_i(x)\phi_i + N_j(x)\phi_j = \begin{Bmatrix} N_i(x) & N_j(x) \end{Bmatrix} \begin{Bmatrix} \phi_i \\ \phi_j \end{Bmatrix} \equiv [N(x)]\vec{\phi}, \tag{13.8}$$

where $[N(x)]$ is the matrix of shape functions, $N_i(x)$ and $N_j(x)$ are called the shape functions corresponding to nodes i and j, respectively, and are given by

$$N_i(x) = \frac{1}{l}(x_j - x), \tag{13.9}$$

and

$$N_j(x) = \frac{1}{l}(x - x_i). \tag{13.10}$$

The shape functions have an important characteristic; the value of $N_i(x)$ is equal to unity at node i and zero at node j; similarly, the value of $N_j(x)$ is equal to unity at node j and zero at node i.

▶Example 13.5

The pressure distribution of a fluid flowing in a straight pipe is approximated by linear elements. The pressures at nodes i and j of an element, located at 2.5 in and 7.5 in from the origin, are found to be 100 psi and 55 psi, respectively. Determine the pressure distribution and its gradient in the element. Also, find the pressure at a point located 3.5 in from the origin.

Solution

The pressure inside the element, $\phi(x)$, is given by Eq. (13.8):

$$\phi(x) = \frac{1}{l}(x_j - x)\phi_i + \frac{1}{l}(x - x_i)\phi_j. \tag{E1}$$

From given data, we have $\phi_i = 100, \phi_j = 55, x_i = 2.5, x_j = 7.5$, and $l = x_j - x_i = 5.0$. (See Fig. 13.12.) Hence, Eq. (E1) gives the pressure distribution in the element as

$$\phi(x) = \frac{1}{5}(7.5 - x)(100) + \frac{1}{5}(x - 2.5)(55) = 122.5 - 9x \text{ psi}; \ 2.5 \le x \le 7.5. \tag{E2}$$

The gradient of pressure in the element is given by

$$\frac{d\phi(x)}{dx} = \frac{-\phi_i + \phi_j}{l} = -9 \text{ psi/in.} \tag{E3}$$

The pressure at a point located at 3.5 in from the origin is given by

$$\phi(3.5) = 122.5 - 9(3.5) = 81.0 \text{ psi.} \tag{E4}$$

◀

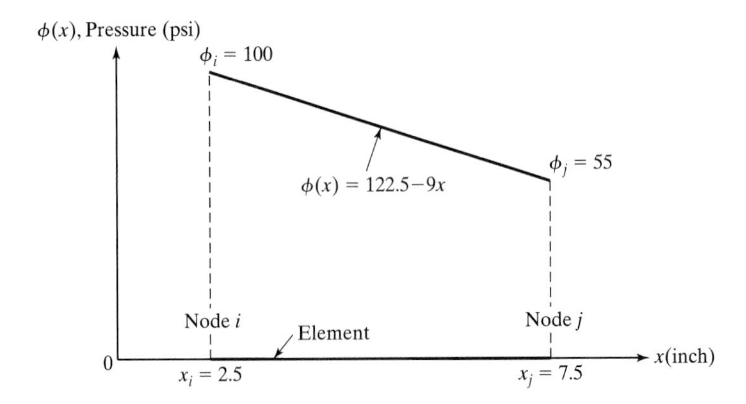

Figure 13.12 Fluid pressure distribution in pipe element.

13.4.3 Two-Dimensional Simplex Element

The simplex element in two dimensions is a triangular element with three corner nodes as shown in Fig. 13.13. The nodes are denoted as i, j, and k with coordinates (x_i, y_i), (x_j, y_j), and (x_k, y_k). The unknown solution $\phi(x, y)$, when evaluated at the node points, gives the nodal values ϕ_i, ϕ_j, and ϕ_k. (These are considered as unknowns in the problem.) The solution

$$\phi(x, y) = a_1 + a_2 x + a_3 y = \{\, 1 \quad x \quad y \,\} \begin{Bmatrix} a_1 \\ a_2 \\ a_3 \end{Bmatrix}, \tag{13.11}$$

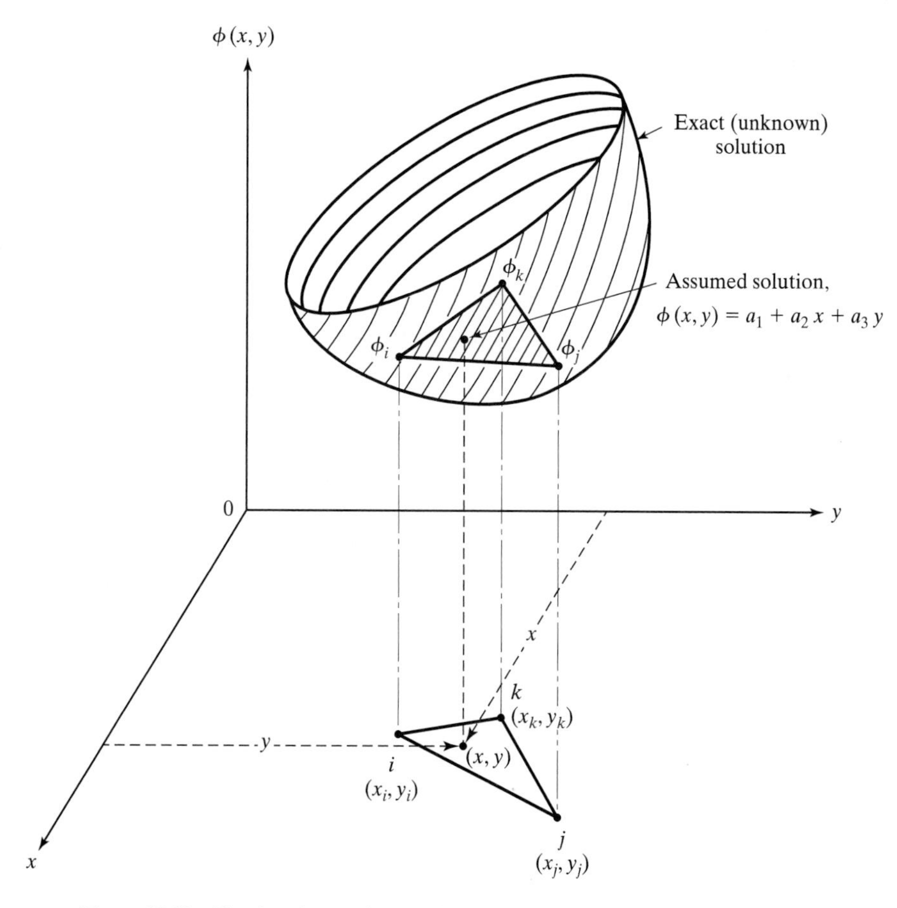

Figure 13.13 Simplex element in two dimensions.

where a_1, a_2, and a_3 are unknown constants is assumed to be linear inside the element. The nodal values of $\phi(x, y)$ are given by

$$\phi(x = x_i, y = y_i) = \phi_i = a_1 + a_2 x_i + a_3 y_i,$$

$$\phi(x = x_j, y = y_j) = \phi_j = a_1 + a_2 x_j + a_3 y_j,$$

$$\phi(x = x_k, y = y_k) = \phi_k = a_1 + a_2 x_k + a_3 y_k,$$

which can be expressed in matrix form as

$$\vec{\phi} = [A]\vec{a}, \tag{13.12}$$

where

$$\vec{\phi} = \begin{Bmatrix} \phi_i \\ \phi_j \\ \phi_k \end{Bmatrix},$$

$$[A] = \begin{bmatrix} 1 & x_i & y_i \\ 1 & x_j & y_j \\ 1 & x_k & y_k \end{bmatrix},$$

and

$$\vec{a} = \begin{Bmatrix} a_1 \\ a_2 \\ a_3 \end{Bmatrix}.$$

Equation (13.12) can be solved to obtain

$$\vec{a} = [A]^{-1}\vec{\phi}, \tag{13.13}$$

where

$$[A]^{-1} = \frac{1}{2\tilde{A}} \begin{bmatrix} (x_j y_k - x_k y_j) & (x_k y_i - x_i y_k) & (x_i y_j - x_j y_i) \\ (y_j - y_k) & (y_k - y_i) & (y_i - y_j) \\ (x_k - x_j) & (x_i - x_k) & (x_j - x_i) \end{bmatrix} \tag{13.14}$$

and $\tilde{A}$ denotes the area of the triangle ijk, which is given by

$$\tilde{A} = \frac{1}{2} \begin{vmatrix} 1 & x_i & y_i \\ 1 & x_j & y_j \\ 1 & x_k & y_k \end{vmatrix}. \tag{13.15}$$

By substituting Eq. (13.13) into Eq. (13.12), we obtain

$$\phi(x, y) = N_i(x, y)\phi_i + N_j(x, y)\phi_j + N_k(x, y)\phi_k$$

$$= \left\{ N_i(x, y) \quad N_j(x, y) \quad N_k(x, y) \right\} \begin{Bmatrix} \phi_i \\ \phi_j \\ \phi_k \end{Bmatrix} \equiv [N(x, y)]\vec{\phi}, \tag{13.16}$$

where the shape functions N_i, N_j, and N_k are given by

$$N_i(x, y) = \frac{1}{2\tilde{A}}\{(x_j y_k - x_k y_j) + (y_j - y_k)x + (x_k - x_j)y\}$$

$$\equiv \frac{1}{2\tilde{A}}(a_i + b_i x + c_i y), \tag{13.17}$$

$$N_j(x, y) = \frac{1}{2\tilde{A}}\{(x_k y_i - x_i y_k) + (y_k - y_i)x + (x_i - x_k)y\}$$

$$\equiv \frac{1}{2\tilde{A}}(a_j + b_j x + c_j y), \tag{13.18}$$

$$N_k(x, y) = \frac{1}{2\tilde{A}}\{(x_i y_j - x_j y_i) + (y_i - y_j)x + (x_j - x_i)y\}$$

$$\equiv \frac{1}{2\tilde{A}}(a_k + b_k x + c_k y) \tag{13.19}$$

and $[N(x, y)]$ denotes the matrix of shape functions. Note that, as before, $N_i(x, y)$ is equal to unity at node i and zero at nodes j and k. Similarly, the shape functions N_j and N_k have a value of unity at nodes j and k, respectively, and zero at other nodes.

▶Example 13.6

The nodal temperatures of a linear triangular element, shown in Fig. 13.14, are given by $\phi_i = 120°$ F, $\phi_j = 90°$ F, and $\phi_k = 80°$ F. Determine the temperature distribution and its gradient in the element. Also, find the temperature at point P whose coordinates are $x = 20$ in and $y = 20$ in.

Solution

The temperature distribution in the element is given by Eq. (13.16). From the given data, we have $\phi_i = 120$, $\phi_j = 90$, $\phi_k = 80$, $x_i = 20$, $y_i = 40$, $x_j = 10$, $y_j = 10$, $x_k = 30$, and $y_k = 20$. Hence, the area of the triangular element is given by

$$\tilde{A} = \frac{1}{2}\begin{vmatrix} 1 & 20 & 40 \\ 1 & 10 & 10 \\ 1 & 30 & 20 \end{vmatrix} = 250 \text{ in}^2.$$

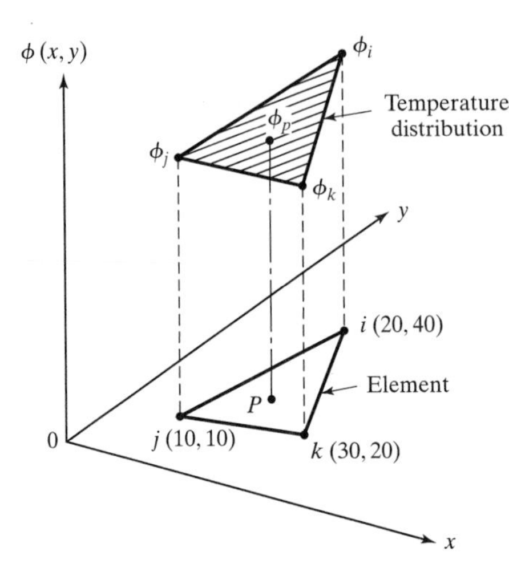

Figure 13.14 Temperature distribution in triangular element.

The shape functions are given by Eqs. (13.17) through (13.19):

$$N_i(x, y) = \frac{1}{2(250)}\{10(20) - 30(10) + (10 - 20)x + (30 - 10)y\}$$

$$= \frac{1}{500}(-100 - 10x + 20y),$$

$$N_j(x, y) = \frac{1}{2(250)}\{30(40) - 20(20) + (20 - 40)x + (20 - 30)y\}$$

$$= \frac{1}{500}(800 - 20x - 10y),$$

$$N_k(x, y) = \frac{1}{2(250)}\{20(10) - 10(40) + (40 - 10)x + (10 - 20)y\}$$

$$= \frac{1}{500}(-200 + 30x - 10y).$$

Thus, the temperature distribution in the element is given by Eq. (13.16):

$$\phi(x, y) = \frac{1}{500}(-100 - 10x + 20y)(120) + \frac{1}{500}(800 - 20x - 10y)(90)$$

$$+\frac{1}{500}(-200 + 30x - 10y)(8) = 88 - 1.2x + 1.4y\,^\circ\text{F}.$$

The gradient of temperature in the element is given by

$$\nabla\phi = \begin{Bmatrix} \dfrac{\partial\phi}{\partial x} \\ \dfrac{\partial\phi}{\partial y} \end{Bmatrix} = \begin{Bmatrix} -1.2 \\ 1.4 \end{Bmatrix} {}^\circ\text{F/inch}.$$

The temperature at point P (20, 20) is given by

$$\phi_P = \phi(20, 20) = 88 - 1.2(20) + 1.4(20) = 92^\circ \text{ F.}$$ ◀

13.5 Derivation of Element Characteristic Matrices and Vectors

The characteristic matrices and vectors of finite elements can be derived by using either a variational approach or a weighted residual approach. In the variational approach, a variational principle valid over the whole domain of the problem is postulated, and an integral I is defined in terms of the unknown solution and its derivatives. The solution, which minimizes the integral I, will be the correct solution of the problem. The solution that minimizes I will also satisfy the governing differential equation and the associated boundary conditions of the problem. In the weighted residual approach such as the Galerkin method and the least-squares method, the governing differential equation is used to define a residual, and the integral of the weighted residual is used to derive the element characteristic matrices and vectors.

13.5.1 Variational Approach

Let the functional I (to be extremized) for the problem be given by

$$\iiint_V F\left\{\phi, \frac{\partial \phi}{\partial x}, \frac{\partial \phi}{\partial y}, \frac{\partial \phi}{\partial z}\right\} dV + \iint_S g\left\{\phi, \frac{\partial \phi}{\partial x}, \frac{\partial \phi}{\partial y}, \frac{\partial \phi}{\partial z}\right\} dS, \tag{13.20}$$

where V and S denote the volume (domain) and surface area (boundary) of the solution domain and F and g are known functionals. When the solution domain is divided into E finite elements, the solution in a typical element e is expressed as

$$\phi(x, y, z) = [N(x, y, z)]\vec{\phi}^{(e)}, \tag{13.21}$$

where $[N]$ is the matrix of shape functions and $\vec{\phi}^{(e)}$ is the vector of nodal values of the solution for element e. If $\vec{\Phi}$ denotes the vector of nodal values of $\phi(x, y, z)$ for the whole domain, the extremization of the functional I with respect to $\vec{\Phi}$ gives the conditions

$$\frac{\partial I}{\partial \vec{\Phi}} = \begin{Bmatrix} \dfrac{\partial I}{\partial \Phi_1} \\ \dfrac{\partial I}{\partial \Phi_2} \\ \cdot \\ \cdot \\ \cdot \\ \dfrac{\partial I}{\partial \Phi_P} \end{Bmatrix} = \vec{0}, \tag{13.22}$$

where $\Phi_1, \Phi_2, \ldots, \Phi_P$ are the components of the vector $\vec{\Phi}$ and P is the total number of nodal unknowns of the complete domain. The functional I can be expressed as a summation of elemental contributions $I^{(e)}$ as

$$I = \sum_{e=1}^{E} I^{(e)}. \tag{13.23}$$

Thus, Eq. (13.22) can be expressed as

$$\frac{\partial I}{\partial \Phi_i} = \sum_{e=1}^{E} \frac{\partial I^{(e)}}{\partial \Phi_i} = 0;\ i = 1, 2, \ldots, P. \tag{13.24}$$

If I is a quadratic function of ϕ and its derivatives, Eq. (13.24) can be used to obtain the elemental equations by

$$\sum_{e=1}^{E} \frac{\partial I^{(e)}}{\partial \vec{\phi}^{(e)}} = \sum_{e=1}^{E} \left(\left[S^{(e)}\right] \vec{\phi}^{(e)} - \vec{Q}^{(e)} \right), \tag{13.25}$$

where $[S^{(e)}]$ and $\vec{\phi}^{(e)}$ are called the element characteristic matrix and vector, respectively.

▶Example 13.7

Consider the differential equation

$$\frac{d^2y}{dx^2} + y + x = 0;\ 0 \le x \le 1, \tag{E1}$$

with the boundary conditions

$$y(0) = y(1) = 0. \tag{E2}$$

The functional I corresponding to the differential equation (E1) is given by

$$I = \frac{1}{2} \int_0^1 \left\{ -\left(\frac{dy}{dx}\right)^2 + y^2 + 2yx \right\} dx. \tag{E3}$$

By dividing the solution domain ($x = 0$ to 1) into four finite elements of equal length and using linear interpolation functions, derive the system equations.

Solution

The solution domain, given by $x = 0$ to $x = 1$, is divided into four finite elements of equal length as shown in Fig. 13.15. Using a linear interpolation polynomial, Eq. (13.8), to approximate the unknown function $y(x)$ inside an element, we have

$$y^{(e)}(x) = [N(x)]\vec{y}^{(e)} = \left\{ N_i(x) \quad N_j(x) \right\} \left\{ \begin{matrix} y_i \\ y_j \end{matrix} \right\}^{(e)}, \tag{E4}$$

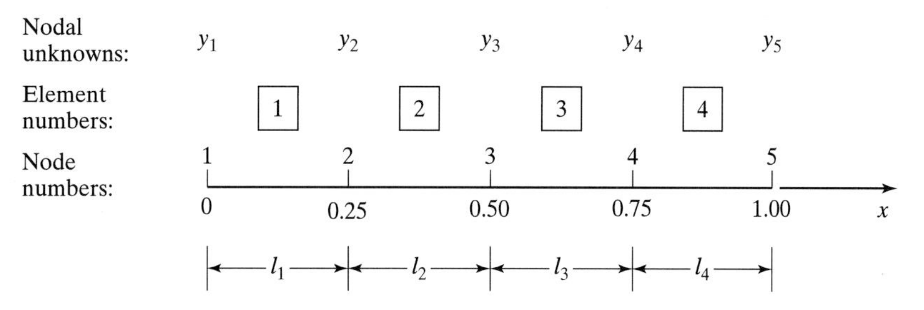

Figure 13.15 Finite element model of Eq. (E1).

where the shape functions $N_i(x)$ and $N_j(x)$ are given by Eqs. (13.9) and (13.10). The functional I is expressed as a sum of elemental quantities $I^{(e)}$ as

$$I = \sum_{e=1}^{E=4} I^{(e)}, \tag{E5}$$

where

$$I^{(e)} = \frac{1}{2}\int_{x_i}^{x_j}\left[-\left(\frac{dy}{dx}\right)^2 + y^2 + 2yx\right]dx. \tag{E6}$$

By substituting Eq. (E4) into Eq. (E6), we obtain

$$I^{(e)} = \frac{1}{2}\int_{x_i}^{x_j}\left\{-\vec{y}^{(e)T}\left[\frac{dN(x)}{dx}\right]^T\left[\frac{dN(x)}{dx}\right]\vec{y}^{(e)} + \vec{y}^{(e)T}[N(x)]^T[N(x)]\vec{y}^{(e)} + \left[2x[N(x)]\vec{y}^{(e)}\right]\right\}dx. \tag{E7}$$

The conditions for the extremization of I are given by

$$\frac{\partial I}{\partial y_m} = \sum_{e=1}^{E}\frac{\partial I^{(e)}}{\partial y_m} = 0; \quad m = 1, 2, \ldots, P, \tag{E8}$$

where y_m is a typical nodal unknown, the number of elements E is equal to 4, and the number of nodal unknowns P is equal to 5. Equation (E8) can be expressed in matrix notation as

$$\sum_{e=1}^{E}\frac{\partial I^{(e)}}{\partial \vec{y}^{(e)}} = \sum_{e=1}^{E}\int_{x_i}^{x_j}\left\{-\left[\frac{dN(x)}{dx}\right]^T\left[\frac{dN(x)}{dx}\right]\vec{y}^{(e)} + [N(x)]^T[N(x)]\vec{y}^{(e)} + x[N(x)]^T\right\}dx = \vec{0}, \tag{E9}$$

which can be rewritten as

$$\sum_{e=1}^{E}[S^{(e)}]\vec{y}^{(e)} = \sum_{e=1}^{E}\vec{Q}^{(e)}, \tag{E10}$$

where the element characteristic matrix, $[S^{(e)}]$, is given by

$$\int_{x_i}^{x_j}\left\{\left[\frac{dN(x)}{dx}\right]^T\left[\frac{dN(x)}{dx}\right] - [N(x)]^T[N(x)]\right\}dx \tag{E11}$$

and the element characteristic vector, $\vec{Q}^{(e)}$, by

$$\vec{Q}^{(e)} = \int_{x_i}^{x_j} x[N(x)]^T dx. \tag{E12}$$

Equations (13.9) and (13.10) give

$$[N(x)] = \left[\left(\frac{x_j - x}{l_e}\right)\left(\frac{x - x_i}{l_e}\right)\right], \tag{E13}$$

$$\left[\frac{dN(x)}{dx}\right] = \left[\left(-\frac{1}{l_e}\right)\left(\frac{1}{l_e}\right)\right]. \tag{E14}$$

Equations (E11) and (E14) yield

$$[S^{(e)}] = \int_{x_i}^{x_j}\left[\left\{\begin{matrix}-\dfrac{1}{l_e}\\ \dfrac{1}{l_e}\end{matrix}\right\}\left\{-\frac{1}{l_e} \quad \frac{1}{l_e}\right\} - \left\{\begin{matrix}\left(\dfrac{x_j - x}{l_e}\right)\\ \left(\dfrac{x - x_i}{l_e}\right)\end{matrix}\right\}\left\{\left(\frac{x_j - x}{l_e}\right)\left(\frac{x - x_i}{l_e}\right)\right\}\right]dx$$

$$= \frac{1}{l_e}\begin{bmatrix} 1 & -1 \\ -1 & 1 \end{bmatrix} - \frac{l_e}{6}\begin{bmatrix} 2 & 1 \\ 1 & 2 \end{bmatrix} \tag{E15}$$

and

$$\vec{Q}^{(e)} = \int_{x_i}^{x_j}\left\{\begin{matrix}\left(\dfrac{x_j - x}{l_e}\right)\\ \left(\dfrac{x - x_i}{l_e}\right)\end{matrix}\right\}x\,dx = \frac{1}{6}\left\{\begin{matrix}(x_j^2 + x_i x_j - 2x_i^2)\\ (2x_j^2 - x_i x_j - x_i^2)\end{matrix}\right\}. \tag{E16}$$

From Fig. 13.15, we obtain $l_1 = l_2 = l_3 = l_4 = \frac{1}{4}$; $x_i = 0$ and $x_j = \frac{1}{4}$ for $e = 1$; $x_i = \frac{1}{4}$ and $x_j = \frac{1}{2}$ for $e = 2$; $x_i = \frac{1}{2}$ and $x_j = \frac{3}{4}$ for $e = 3$; $x_i = \frac{3}{4}$ and $x_j = 1$ for $e = 4$. Thus, the characteristic matrices and vectors of the four elements can be determined from Eqs. (E15) and (E16):

$$[S^{(e)}] = \begin{bmatrix} 4 & -4 \\ -4 & 4 \end{bmatrix} - \begin{bmatrix} \dfrac{1}{12} & \dfrac{1}{24} \\ \dfrac{1}{24} & \dfrac{1}{12} \end{bmatrix}$$

$$= \frac{1}{24}\begin{bmatrix} 94 & -97 \\ -97 & 94 \end{bmatrix}; \; e = 1, 2, 3, 4; \tag{E17}$$

$$\vec{Q}^{(1)} = \frac{1}{96}\begin{Bmatrix} 1 \\ 2 \end{Bmatrix}; \tag{E18}$$

$$\vec{Q}^{(2)} = \frac{1}{96}\begin{Bmatrix} 4 \\ 13 \end{Bmatrix}; \tag{E19}$$

$$\vec{Q}^{(3)} = \frac{1}{96}\begin{Bmatrix} 7 \\ 8 \end{Bmatrix}; \tag{E20}$$

$$\vec{Q}^{(4)} = \frac{1}{96}\begin{Bmatrix} 10 \\ 11 \end{Bmatrix}; \tag{E21}$$

◀

13.5.2 Weighted Residual Approach

The use of the variational approach to derive the element characteristic matrices and vectors requires the availability of a functional for the given problem. For some problems, the variational principle may not be known readily. Hence, the weighted residual approach is more often used to derive the element characteristic matrices and vectors. To indicate the weighted residual approach, let the problem be stated by a differential equation, and the associated boundary conditions as

$$F(\phi) = f(\phi) \text{ in } V, \tag{13.26}$$

$$B_j(\phi) = b_j, \; j = 1, 2, \ldots, p \text{ on } S, \tag{13.27}$$

where F and B_j are differential operators, f is a function of ϕ, b_j are functions of the independent variables, p is the number of boundary conditions, $\phi(x, y, z)$ is the unknown solution, V is the domain, and S is the boundary of the domain. In the weighted residual method, the solution $\phi(x, y, z)$ is approximated by

$$\bar{\phi}(x, y, z) = \sum_{i=1}^{n} c_i f_i(x, y, z), \tag{13.28}$$

where c_i are constants and f_i are linearly independent functions (also called trial functions) of x, y, and z, which are chosen to satisfy all the boundary conditions, Eq. (13.27), and n is the total number of trial functions used. Since the approximate solution $\bar{\phi}$ given by Eq. (13.28) does not satisfy the governing differential equation,

Eq. (13.26), exactly, a residual is defined as

$$R = F(\bar{\phi}) - f(\bar{\phi}). \tag{13.29}$$

Usually, a weighted function of the residual, $w\, g(R)$, is made small or minimum in some sense, with w denoting the weight or weighting function and where $g(R)$ is a function of R chosen to satisfy the smallness criterion. A common form of smallness criterion used in several applications is given by

$$\iiint_V w\, g(R)\, dV = 0. \tag{13.30}$$

According to the most popular type of weighted residual method, namely, the Galerkin method, the weight w is taken to be the trial functions $f_i(x, y, z)$ used in the approximate solution of Eq. (13.28) and the function of the residue is taken as $g(R) = R$. Thus, Eq. (13.30) becomes, for the Galerkin method,

$$\begin{aligned}\iiint_V f_i\, R\, dV &= \iiint_V f_i(x, y, z)\; R(\bar{\phi})\, dV \\ &= \iiint_V f_i(x, y, z) R(c_1, \ldots, c_n, f_1(x, y, z), \ldots, f_n(x, y, z))\, dV = 0; \\ & i = 1, 2, \ldots, n. \end{aligned} \tag{13.31}$$

Equation (13.31) represents n simultaneous equations in the n unknowns $c_1, c_2, \ldots, c_n$. Thus, the solution of Eq. (13.31) gives the approximation solution through Eq. (13.28). It is to be noted that the trial functions $f_i(x, y, z)$ in Eq. (13.28) are defined over the entire domain of the problem.

To derive the finite-element equations, Eqs. (13.31) and (13.28) are interpreted to be valid for a typical element e also. Then the unknowns c_i can be recognized as the nodal unknowns $\Phi_i^{(e)}$, and the functions $f_i(x, y, z)$ as the shape functions $N_i^{(e)}(x, y, z)$. Equation (13.31), when applied to element e, gives

$$\begin{aligned}&\iiint_V [F(\phi^{(e)}(x, y, z)) - f(\phi^{(e)}(x, y, z))] N_i^{(e)}(x, y, z)\, dV = 0; \\ & i = 1, 2, \ldots, n,\end{aligned} \tag{13.32}$$

where the element interpolation function is assumed to be of the form

$$\phi^{(e)}(x, y, z) = \sum_{i=1}^{n} N_i(x, y, z) \Phi_i^{(e)} \equiv [N(x, y, z)] \vec{\Phi}^{(e)}. \tag{13.33}$$

Equation (13.32) yields the required equations for finite element e from which the element characteristic matrix and vector can be readily identified as illustrated in Example 13.9.

13.6 Assemblage of Element Characteristic Matrices and Vectors

Once the element characteristic matrices and vectors are derived for all the elements, the overall or system matrix and vector can be derived as (from Eq. (13.25))

$$[S] = \sum_{e=1}^{E} [S^{(e)}] \tag{13.34}$$

and

$$\vec{Q} = \sum_{e=1}^{E} \vec{Q}^{(e)}, \tag{13.35}$$

where the summation sign indicates assembly over all finite elements. The summation does not indicate the usual algebraic summation, but is based on the requirement of compatibility at the element nodes. This means that on the nodes where the elements are connected, the value of the unknown $\Phi_i^{(e)}$ is the same for all the elements joining at the node. The order of the system matrix $[S]$ and system vector $\vec{Q}$ will be $P \times P$ and $P \times 1$, respectively, where P denotes the total number of unknowns (nodes) considered in the complete domain.

13.7 Solution of System Equations

The system equations can be expressed as

$$[S]\vec{\Phi} = \vec{Q}. \tag{13.36}$$

The validity of Eq. (13.36) can be seen, for example, from Eqs. (13.24) and (13.25). Equation (13.25) can be rewritten as:

$$\sum_{e=1}^{E} \frac{\partial I^{(e)}}{\partial \vec{\Phi}^{(e)}} = \sum_{e=1}^{E} [S^{(e)}]\vec{\Phi}^{(e)} - \sum_{e=1}^{E} \vec{Q}^{(e)} = [S]\vec{\Phi} - \vec{Q} = \vec{0}, \tag{13.37}$$

where

$$\vec{\Phi} = \begin{Bmatrix} \Phi_1 \\ \Phi_2 \\ \cdot \\ \cdot \\ \cdot \\ \Phi_P \end{Bmatrix}. \tag{13.38}$$

The specified boundary conditions of the problem have to be incorporated into Eq. (13.36) before solving them. The solution of Eq. (13.36) yields the nodal values of $\phi(x, y, z)$ from which the solution in any finite element e can be found using Eq. (13.33).

▶Example 13.8

Derive the system equations, and find the solution of the problem considered in Example 13.7.

Solution

To assemble $[S^{(e)}]$ and $\vec{Q}^{(e)}$, we consider one element at a time. For element 1 ($e = 1$), the nodal degrees of freedom are y_1 and y_2, and hence, the matrix $[S^{(1)}]$ is assembled to occupy the elements of the rows and columns of y_1 and y_2 of the matrix $[S]$, and the vector $\vec{Q}^{(1)}$ is assembled to occupy the elements of the rows y_1 and y_2 of the vector $\vec{Q}$. A similar procedure is used to assemble the matrices and vectors of other elements. (Note that the nodal degrees of freedom are y_2 and y_3, y_3 and y_4, and y_4 and y_5 for elements 2, 3, and 4, respectively. Then

$$[S] = \sum_{e=1}^{4} [S^{(e)}] = \frac{1}{24} \begin{array}{c} \begin{array}{ccccc} y_1 & y_2 & y_3 & y_4 & y_5 \end{array} \\ \left[\begin{array}{ccccc} 94 & -97 & 0 & 0 & 0 \\ -97 & 188 & -97 & 0 & 0 \\ 0 & -97 & 188 & -97 & 0 \\ 0 & 0 & -97 & 188 & -97 \\ 0 & 0 & 0 & -97 & 94 \end{array}\right] \end{array} \begin{array}{c} y_1 \\ y_2 \\ y_3 \\ y_4 \\ y_5 \end{array}, \tag{E1}$$

and

$$\vec{Q} = \sum_{e=1}^{4} \vec{Q}^{(e)} = \frac{1}{96} \begin{Bmatrix} 1 \\ 2+4 \\ 13+7 \\ 8+10 \\ 11 \end{Bmatrix} = \frac{1}{96} \begin{Bmatrix} 1 \\ 6 \\ 20 \\ 18 \\ 11 \end{Bmatrix} \begin{array}{c} y_1 \\ y_2 \\ y_3 \\ y_4 \\ y_5 \end{array}. \tag{E2}$$

Note that the various rows and columns of $[S]$ and $\vec{Q}$ are identified with the corresponding unknowns, $y_1, y_2, \ldots, y_5$. The boundary conditions $y_1 = 0$ and $y_5 = 0$ can be incorporated by deleting the rows and columns corresponding to y_1 and y_5 in

Eqs. (E1) and (E2). This results in the final system equations:

$$\frac{1}{24}\begin{bmatrix} 188 & -97 & 0 \\ -97 & 188 & -97 \\ 0 & -97 & 188 \end{bmatrix}\begin{Bmatrix} y_2 \\ y_3 \\ y_4 \end{Bmatrix} = \frac{1}{96}\begin{Bmatrix} 6 \\ 20 \\ 18 \end{Bmatrix}. \tag{E3}$$

The solution of Eqs. (E3) is given by

$$y_2 = 0.055497,\ y_3 = 0.092097,\ y_4 = 0.071454. \tag{E4}$$

The exact solution of the problem (defined by Eqs. (E1) and (E2) of Example 13.7) is given by

$$y(x) = \frac{\sin x}{\sin 1} - x. \tag{E5}$$

Equation (E5) can be used to find the exact solution at the nodal values by

$$(y_2)_{\text{exact}} = \frac{0.2474}{0.8415} - 0.25 = 0.0440,$$

$$(y_3)_{\text{exact}} = \frac{0.4794}{0.8415} - 0.50 = 0.0697, \tag{E6}$$

and

$$(y_4)_{\text{exact}} = \frac{0.6816}{0.8415} - 0.75 = 0.0600.$$

To improve the accuracy of the finite-element solution, either the number of elements or (and) the order of the interpolation polynomial can be increased. ◀

▶Example 13.9

Derive the finite-element equations for solving the parabolic partial differential equation

$$A\frac{\partial^2\phi(x,t)}{\partial x^2} = \frac{\partial\phi(x,t)}{\partial t};\ a \le x \le b;\ t \ge 0, \tag{E1}$$

with the boundary conditions

$$\phi(a,t) = f_0(t), \tag{E2}$$

$$\phi(b,t) = g_0(t), \tag{E3}$$

and the initial condition

$$\phi(x,0) = h_0(x), \tag{E4}$$

using Galerkin method.

Note: Physically, Eq. (E1) represents situations such as the transient heat conduction in a rod and the one-dimensional transient diffusion problem. In general, although f_0 and g_0 are functions of t and h_0 is a function of x, they are usually assumed as constants. The boundary conditions can also be specified in terms of $\frac{d\phi}{dx}$.

Solution

We divide the spatial (physical) domain (x) into a number of finite elements as shown in Fig. 13.16, and treat the time domain (t) using a finite-difference approach. An alternative approach is to treat x and t as coordinates of a two-dimensional space and use two-dimensional finite elements (such as triangles) for modeling. A linear interpolation model, Eq. (13.8), is used to approximate the spatial variation of the solution ϕ within a finite element as

$$\phi^{(e)}(x) = [N(x)]\vec{\phi}^{(e)} = [N_i(x) \;\; N_j(x)] \begin{Bmatrix} \phi_i \\ \phi_j \end{Bmatrix}, \tag{E5}$$

where the dependence of ϕ, ϕ_i, and ϕ_j on time t is not explicitly indicated for simplicity of notation. The residual $R^{(e)}$ for a typical element e, for the Galerkin method is defined as

$$R = \left(A\frac{\partial^2 \phi}{\partial x^2} - \frac{\partial \phi}{\partial t} \right)^{(e)}. \tag{E6}$$

The equations for element e are given by

$$\int_{x_i}^{x_j} R^{(e)}(x) N_i^{(e)}(x)\, dx = \int_{x_i}^{x_j} \left[A^{(e)} \frac{\partial^2 \phi^{(e)}(x)}{\partial x^2} - \frac{\partial \phi^{(e)}(x)}{\partial t} \right] N_i(x)\, dx = 0, \tag{E7}$$

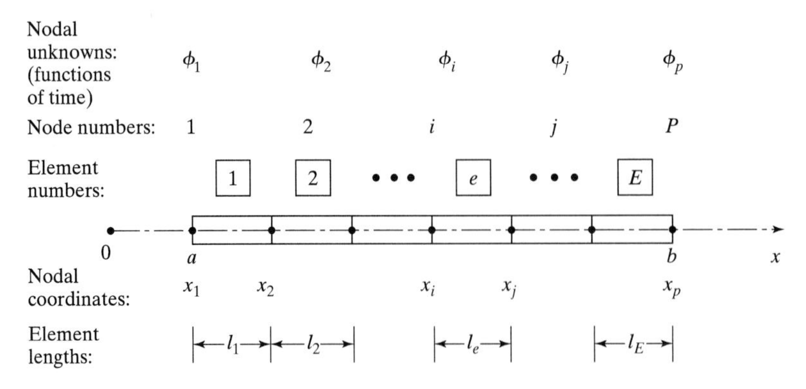

Figure 13.16 Finite element model of Example 13.9.

$$\int_{x_i}^{x_j} R^{(e)}(x) N_j^{(e)}(x)\, dx = \int_{x_i}^{x_j} \left[A^{(e)} \frac{\partial^2 \phi^{(e)}(x)}{\partial x^2} - \frac{\partial \phi^{(e)}(x)}{\partial t} \right] N_j(x)\, dx = 0, \quad \text{(E8)}$$

where $A^{(e)}$ represents the value of A for element e. By using the principle of integration by parts, we get

$$\int u\, dv = uv - \int v\, du, \quad \text{(E9)}$$

where

$$dv = A^{(e)} \frac{\partial^2 \phi^{(e)}(x)}{\partial x^2} dx; \; u = N_i(x), \quad \text{(E10)}$$

$$v = A^{(e)} \frac{\partial \phi^{(e)}(x)}{\partial x}; \; du = \frac{\partial N_i(x)}{\partial x}, \quad \text{(E11)}$$

and the first terms of Eqs. (E7) and (E8) can be expressed as

$$\int_{x_i}^{x_j} A^{(e)} \frac{\partial^2 \phi^{(e)}(x)}{\partial x^2} N_i(x)\, dx = A^{(e)} \frac{\partial \phi^{(e)}(x)}{\partial x} N_i(x) \Big|_{x_i}^{x_j} - \int_{x_i}^{x_j} A^{(e)} \frac{\partial \phi^{(e)}(x)}{\partial x} \frac{\partial N_i(x)}{\partial x} dx, \quad \text{(E12)}$$

and

$$\int_{x_i}^{x_j} A^{(e)} \frac{\partial^2 \phi^{(e)}(x)}{\partial x^2} N_j(x)\, dx = A^{(e)} \frac{\partial \phi^{(e)}(x)}{\partial x} N_j(x) \Big|_{x_i}^{x_j} - \int_{x_i}^{x_j} A^{(e)} \frac{\partial \phi^{(e)}(x)}{\partial x} \frac{\partial N_j(x)}{\partial x} dx. \quad \text{(E13)}$$

By substituting Eqs. (E12) and (E13) into Eqs. (E7) and (E8) and rearranging the terms, we obtain

$$\int_{x_i}^{x_j} A^{(e)} \frac{\partial \phi^{(e)}(x)}{\partial x} \frac{\partial N_i(x)}{\partial x} dx + \int_{x_i}^{x_j} \frac{\partial \phi^{(e)}(x)}{\partial t} N_i(x)\, dx = A^{(e)} \frac{\partial \phi^{(e)}(x)}{\partial x} N_i(x) \Big|_{x_i}^{x_j}, \quad \text{(E14)}$$

and

$$\int_{x_i}^{x_j} A^{(e)} \frac{\partial \phi^{(e)}(x)}{\partial x} \frac{\partial N_j(x)}{\partial x} dx + \int_{x_i}^{x_j} \frac{\partial \phi^{(e)}(x)}{\partial t} N_j(x)\, dx = A^{(e)} \frac{\partial \phi^{(e)}(x)}{\partial x} N_j(x) \Big|_{x_i}^{x_j}. \quad \text{(E15)}$$

Using Eq. (E5) with

$$\phi^{(e)}(x) = N_i(x)\phi_i + N_j(x)\phi_j \quad \text{(E16)}$$

and

$$\frac{\partial \phi^{(e)}(x)}{\partial x} = \frac{\partial N_i(x)}{\partial x}\phi_i + \frac{\partial N_j(x)}{\partial x}\phi_j, \quad \text{(E17)}$$

we can write Eqs. (E14) and (E15) as

$$\int_{x_i}^{x_j} A^{(e)} \frac{\partial N_i(x)}{\partial x} \left[\frac{\partial N_i(x)}{\partial x}\phi_i + \frac{\partial N_j(x)}{\partial x}\phi_j \right] dx + \int_{x_i}^{x_j} N_i(x) \left[N_i(x)\frac{\partial \phi_i}{\partial t} + N_j(x)\frac{\partial \phi_j}{\partial t} \right] dx = \left[A^{(e)} \frac{\partial \phi^{(e)}(x)}{\partial x} N_i(x) \right]_{x_i}^{x_j}, \quad \text{(E18)}$$

$$\int_{x_i}^{x_j} A^{(e)} \frac{\partial N_j(x)}{\partial x} \left[\frac{\partial N_i(x)}{\partial x}\phi_i + \frac{\partial N_j(x)}{\partial x}\phi_j \right] dx + \int_{x_i}^{x_j} N_j(x) \left[N_i(x)\frac{\partial \phi_i}{\partial t} + N_j(x)\frac{\partial \phi_j}{\partial t} \right] dx = \left[A^{(e)} \frac{\partial \phi^{(e)}(x)}{\partial x} N_j(x) \right]_{x_i}^{x_j}. \quad \text{(E19)}$$

Equations (E18) and (E19) can be expressed in matrix form as

$$\int_{x_i}^{x_j} A^{(e)} \begin{bmatrix} \left(\frac{\partial N_i}{\partial x}\right)^2 & \frac{\partial N_i}{\partial x}\frac{\partial N_j}{\partial x} \\ \frac{\partial N_i}{\partial x}\frac{\partial N_j}{\partial x} & \left(\frac{\partial N_j}{\partial x}\right)^2 \end{bmatrix} \begin{Bmatrix} \phi_i \\ \phi_j \end{Bmatrix} dx + \int_{x_i}^{x_j} \begin{bmatrix} N_i^2 & N_i N_j \\ N_i N_j & N_j^2 \end{bmatrix} \begin{Bmatrix} \frac{\partial \phi_i}{\partial t} \\ \frac{\partial \phi_j}{\partial t} \end{Bmatrix} dx = A^{(e)} \begin{Bmatrix} \frac{\partial \phi^{(e)}}{\partial x} N_i \\ \frac{\partial \phi^{(e)}}{\partial x} N_j \end{Bmatrix}_{x_i}^{x_j}. \quad \text{(E20)}$$

The various integrals appearing in Eq. (E20) can be evaluated, using Eqs. (13.9) and (13.10) and the relation $(x_j - x_i) = l_e$, as

$$\int_{x_i}^{x_j} \left(\frac{\partial N_i}{\partial x}\right)^2 dx = \int_{x_i}^{x_j} \left(-\frac{1}{l_e}\right)^2 dx = \frac{1}{l_e^2}(x_j - x_i) = \frac{1}{l_e}, \quad \text{(E21)}$$

$$\int_{x_i}^{x_j} \frac{\partial N_i}{\partial x}\frac{\partial N_j}{\partial x} dx = \int_{x_i}^{x_j} \left(-\frac{1}{l_e}\right)\left(\frac{1}{l_e}\right) dx = -\frac{1}{l_e}, \quad \text{(E22)}$$

$$\int_{x_i}^{x_j} N_i^2 \, dx = \int_{x_i}^{x_j} \frac{1}{l_e^2}\left(x_j - x\right)^2 dx = \frac{1}{l_e^2}\int_{x_i}^{x_j} (x_j^2 + x^2 - 2x_j x)\, dx$$
$$= \frac{1}{l_e^2}\left(x_j^2 x + \frac{1}{3}x^3 - x_j x^2\right)\Bigg|_{x_i}^{x_j} = \frac{l_e}{3}, \quad \text{(E23)}$$

$$\int_{x_i}^{x_j} N_i N_j \, dx = \int_{x_i}^{x_j} \frac{1}{l_e^2}(x_j - x)(x - x_i)\, dx = \frac{l_e}{6}, \quad \text{(E24)}$$

and

$$\int_{x_i}^{x_j} N_j^2 \, dx = \int_{x_i}^{x_j} \frac{1}{l_e^2}(x - x_i)^2 \, dx = \frac{l_e}{3}. \quad \text{(E25)}$$

The terms appearing on the right-hand side of Eq. (E20) can be evaluated as

$$A^{(e)} \frac{\partial \phi^{(e)}}{\partial x} N_i \Big|_{x_i}^{x_j} = -A^{(e)} \left. \frac{\partial \phi}{\partial x} \right|_i, \tag{E26}$$

and

$$A^{(e)} \frac{\partial \phi^{(e)}}{\partial x} N_j \Big|_{x_i}^{x_j} = A^{(e)} \left. \frac{\partial \phi}{\partial x} \right|_j, \tag{E27}$$

since N_i is equal to unity at x_i and zero at x_j, and N_j is equal to unity at x_j and zero at x_i. Note that the quantities $\frac{\partial \phi}{\partial x}|_i$ and $\frac{\partial \phi}{\partial x}|_j$ in Eqs. (E26) and (E27) will be nonzero only if boundary conditions are specified in terms of $\frac{\partial \phi}{\partial x}$ at nodes i and j. As stated earlier, the time derivatives of ϕ_i and ϕ_j are approximated using a backward-difference formula by

$$\left. \frac{\partial \phi_i}{\partial t} \right|_{t+\Delta t} = \dot{\phi}_i |_{t+\Delta t} = \frac{\phi_i(t+\Delta t) - \phi_i(t)}{\Delta t}, \tag{E28}$$

$$\left. \frac{\partial \phi_j}{\partial t} \right|_{t+\Delta t} = \dot{\phi}_j |_{t+\Delta t} = \frac{\phi_j(t+\Delta t) - \phi_j(t)}{\Delta t}, \tag{E29}$$

where Δt is a small time step. Substituting Eqs. (E28) and (E29) into Eq. (E20) yields the elemental equations:

$$\frac{A^{(e)}}{l_e} \begin{bmatrix} 1 & -1 \\ -1 & 1 \end{bmatrix} \begin{Bmatrix} \phi_i \\ \phi_j \end{Bmatrix}_{t+\Delta t} + \frac{l_e}{6\Delta t} \begin{bmatrix} 2 & 1 \\ 1 & 2 \end{bmatrix} \begin{Bmatrix} \phi_i \\ \phi_j \end{Bmatrix}_{t+\Delta t}$$
$$- \frac{l_e}{6\Delta t} \begin{bmatrix} 2 & 1 \\ 1 & 2 \end{bmatrix} \begin{Bmatrix} \phi_i \\ \phi_j \end{Bmatrix}_t = A^{(e)} \begin{Bmatrix} \left.\frac{\partial \phi}{\partial x}\right|_i \\ \left.\frac{\partial \phi}{\partial x}\right|_j \end{Bmatrix}_t. \tag{E30}$$

Equations (E30) can be rewritten in the more convenient form

$$\begin{bmatrix} 6\lambda_e + 2 & -6\lambda_e + 1 \\ -6\lambda_e + 1 & 6\lambda_e + 2 \end{bmatrix} \begin{Bmatrix} \phi_i \\ \phi_j \end{Bmatrix}_{t+\Delta t} = \begin{bmatrix} 2 & 1 \\ 1 & 2 \end{bmatrix} \begin{Bmatrix} \phi_i \\ \phi_j \end{Bmatrix}_t + 6\lambda_e l_e \begin{Bmatrix} \left.\frac{\partial \phi}{\partial x}\right|_i \\ \left.\frac{\partial \phi}{\partial x}\right|_j \end{Bmatrix}_t, \tag{E31}$$

$$\lambda_e = \frac{A^{(e)} \Delta t}{l_e^2}. \tag{E32}$$

Note that the time step Δt is the same for all elements, and in practical applications, $i = e$ and $j = e + 1$ for element e. ◀

▶Example 13.10

Derive the finite-element equations for solving the elliptic partial differential equation

$$\frac{\partial}{\partial x}\left(p(x, y)\frac{\partial \phi}{\partial x}\right) + \frac{\partial}{\partial y}\left(q(x, y)\frac{\partial \phi}{\partial y}\right) + r(x, y)\phi(x, y) = f(x, y) \qquad \text{(E1)}$$

with the boundary conditions

$$\phi(x, y) = g(x, y) \text{ on } S_1, \qquad \text{(E2)}$$

(Dirichlet boundary condition)

$$u(x, y)\frac{\partial \phi}{\partial x}n_1 + v(x, y)\frac{\partial \phi}{\partial y}n_2 + w(x, y)\phi(x, y) = h(x, y) \text{ on } S_2, \qquad \text{(E3)}$$

(mixed-boundary condition)

where n_1 and n_2 are the direction cosines of the outward normal to the boundary at point (x, y). Assume that the domain of the problem is the plane region shown in Fig. 13.17.

Solution

We shall use the variational approach to derive the finite-element equations. The functional I corresponding to the differential equation and the boundary condition on S_2, Eqs. (E1) and (E3), can be represented as [13.4]

$$I = \int\int_A \left[\frac{1}{2}p(x, y)\left(\frac{\partial \phi}{\partial x}\right)^2 + \frac{1}{2}q(x, y)\left(\frac{\partial \phi}{\partial y}\right)^2 - \frac{1}{2}r(x, y)\phi^2 + f(x, y)\phi\right] dA$$
$$+ \int_{S_2}\left[-h(x, y)\phi + \frac{1}{2}w(x, y)\phi^2\right] dS, \qquad \text{(E4)}$$

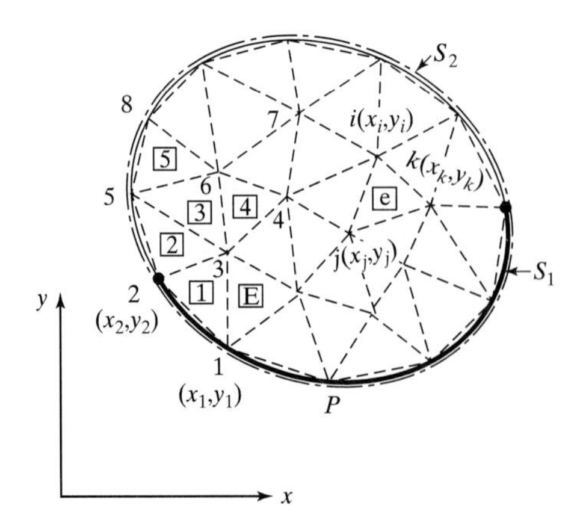

Figure 13.17 Solution domain of Eq. (E1).

where A is the total area of the solution region. Note that the boundary condition of Eq. (E2) will be incorporated at a later stage. The solution region is divided into triangular elements as shown in Fig. 13.17. The number, size, and orientation of the elements are chosen arbitrarily. A linear interpolation model is used to approximate the solution $\phi(x, y)$ within a typical finite element e so that

$$\phi^{(e)}(x, y) = N_i(x, y)\phi_i + N_j(x, y)\phi_j + N_k(x, y)\phi_k = [N(x, y)]\vec{\phi}^{(e)}, \tag{E5}$$

where

$$[N(x, y)] = \begin{bmatrix} N_i(x, y) & N_j(x, y) & N_k(x, y) \end{bmatrix} \tag{E6}$$

and

$$\vec{\phi}^{(e)} = \begin{Bmatrix} \phi_i \\ \phi_j \\ \phi_k \end{Bmatrix}, \tag{E7}$$

where i, j, and k are the nodes of the element e, and ϕ_i, ϕ_j, and ϕ_k are the nodal values of $\phi(x, y)$, which are treated as unknowns in the problem. (See Eq. (13.16)) By expressing the functional I as a summation of elemental contributions $I^{(e)}$ and substituting Eq. (E5) into Eq. (E4) yields

$$I = \sum_{e=1}^{E} I^{(e)}, \tag{E8}$$

where

$$\begin{aligned} I^{(e)} = \int\!\!\int_{A^{(e)}} \Bigg[& \frac{1}{2}p(x, y)\left(\frac{\partial \phi^{(e)}}{\partial x}\right)^2 + \frac{1}{2}q(x, y)\left(\frac{\partial \phi^{(e)}}{\partial y}\right)^2 \\ & - \frac{1}{2}r(x, y)(\phi^{(e)})^2 + f(x, y)\phi^{(e)} \Bigg] dA \\ & + \int\!\!\int_{S_2^{(e)}} \left[-h(x, y)\phi^{(e)} + \frac{1}{2}w(x, y)(\phi^{(e)})^2 \right] dS, \end{aligned} \tag{E9}$$

where $A^{(e)}$ is the area of the element e, and E is the number of finite elements. By substituting Eq. (E5) into Eq. (E9), we obtain

$$\begin{aligned} I^{(e)} = \int\!\!\int_{A^{(e)}} \Bigg[& \frac{1}{2}p(x, y)\vec{\phi}^{(e)T}\left[\frac{\partial N}{\partial x}\right]^T \left[\frac{\partial N}{\partial x}\right]\vec{\phi}^{(e)} \\ & + \frac{1}{2}q(x, y)\vec{\phi}^{(e)T}\left[\frac{\partial N}{\partial y}\right]^T \left[\frac{\partial N}{\partial y}\right]\vec{\phi}^{(e)} \Bigg] dA \end{aligned}$$

$$-\int\int_{A^{(e)}}\left[\frac{1}{2}r(x,y)\vec{\phi}^{(e)^T}[N]^T[N]\vec{\phi}^{(e)}+f(x,y)[N]\vec{\phi}^{(e)}\right]dA$$

$$+\int\int_{S_2^{(e)}}\left[-h(x,y)[N]\vec{\phi}^{(e)}+\frac{1}{2}w(x,y)\vec{\phi}^{(e)^T}[N]^T[N]\vec{\phi}^{(e)}\right]dS. \quad \text{(E10)}$$

The minimization of the functional I with respect to the nodal unknowns yields the conditions

$$\frac{\partial I}{\partial \phi_m}=\sum_{e=1}^{E}\frac{\partial I^{(e)}}{\partial \phi_m}=0,\ m=1,2,\ldots,P, \quad \text{(E11)}$$

where ϕ_m is a typical nodal unknown and P is the total number of nodes. Equation (E11) can be expressed in matrix notation as

$$\sum_{e=1}^{E}\frac{\partial I^{(e)}}{\partial \vec{\phi}^{(e)}}=\sum_{e=1}^{E}\left\{\int\int_{A^{(e)}}\left[p(x,y)\left[\frac{\partial N}{\partial x}\right]^T\left[\frac{\partial N}{\partial x}\right]\vec{\phi}^{(e)}\right.\right.$$

$$\left.\left.+q(x,y)\left[\frac{\partial N}{\partial y}\right]^T\left[\frac{\partial N}{\partial y}\right]\vec{\phi}^{(e)}\right]dA\right\}$$

$$-\sum_{e=1}^{E}\left\{\int\int_{A^{(e)}}\left[r(x,y)[N]^T[N]\vec{\phi}^{(e)}+f(x,y)[N]^T\right]dA\right\}$$

$$+\sum_{e=1}^{E}\left\{\int\int_{S_2^{(e)}}\left[-h(x,y)[N]^T+w(x,y)[N]^T[N]\vec{\phi}^{(e)}\right]dS\right\}=\vec{0}, \quad \text{(E12)}$$

which can be rewritten as

$$\sum_{e=1}^{E}[S^{(e)}]\vec{\phi}^{(e)}=\sum_{e=1}^{E}\vec{Q}^{(e)}, \quad \text{(E13)}$$

where the element characteristic matrix, $[S^{(e)}]$, is given by

$$\left[S^{(e)}\right]=\int\int_{A^{(e)}}\left[p(x,y)\left[\frac{\partial N}{\partial x}\right]^T\left[\frac{\partial N}{\partial x}\right]+q(x,y)\left[\frac{\partial N}{\partial y}\right]^T\left[\frac{\partial N}{\partial y}\right]\right.$$

$$\left.-r(x,y)[N]^T[N]\right]dA+\int\int_{S_2^{(e)}}w(x,y)[N]^T[N]\,dS \quad \text{(E14)}$$

and the element characteristic vector, $\vec{Q}^{(e)}$, is given by

$$\vec{Q}^{(e)}=\int\int_{A^{(e)}}f(x,y)[N]^T\,dA+\int\int_{S_2^{(e)}}h(x,y)[N]^T\,dS. \quad \text{(E15)}$$

Equations (13.16) through (13.19) give

$$[N(x, y)] = [\ N_i(x, y) \quad N_j(x, y) \quad N_k(x, y)\ ], \tag{E16}$$

$$\left[\frac{\partial N}{\partial x}\right] = \left[\ \frac{\partial N_i}{\partial x} \quad \frac{\partial N_j}{\partial x} \quad \frac{\partial N_k}{\partial x}\ \right], \tag{E17}$$

$$\left[\frac{\partial N}{\partial y}\right] = \left[\ \frac{\partial N_i}{\partial y} \quad \frac{\partial N_j}{\partial y} \quad \frac{\partial N_k}{\partial y}\ \right], \tag{E18}$$

$$\frac{\partial N_i}{\partial x} = \frac{1}{2\tilde{A}}(y_j - y_k), \tag{E19}$$

$$\frac{\partial N_j}{\partial x} = \frac{1}{2\tilde{A}}(y_k - y_i), \tag{E20}$$

$$\frac{\partial N_k}{\partial x} = \frac{1}{2\tilde{A}}(y_i - y_j), \tag{E21}$$

$$\frac{\partial N_i}{\partial y} = \frac{1}{2\tilde{A}}(x_k - x_j), \tag{E22}$$

$$\frac{\partial N_j}{\partial y} = \frac{1}{2\tilde{A}}(x_i - x_k), \tag{E23}$$

$$\frac{\partial N_k}{\partial y} = \frac{1}{2\tilde{A}}(x_j - x_i), \tag{E24}$$

where $\tilde{A} = A^{(e)}$ is given by Eq. (13.15). If the functions $p(x, y)$, $q(x, y)$, $r(x, y)$, $f(x, y)$, $h(x, y)$ and $w(x, y)$ are known, the integrations in Eqs. (E14) and (E15) can be performed. The system equations can be written, from Eq. (E13), as

$$[S]\vec{\phi} = \vec{Q}. \tag{E25}$$

The boundary conditions given by Eq. (E2) can be incorporated in Eq. (E25), and the resulting system equations can be solved to find the nodal unknowns, $\vec{\phi}$. ◀

▶Example 13.11

Derive the element characteristic matrix $[S^{(e)}]$ and the element characteristic vector $\vec{P}^{(e)}$ for the differential equation

$$\frac{\partial^2 \phi}{\partial x^2} + \frac{\partial^2 \phi}{\partial y^2} = f(x, y) = q_0 \tag{a}$$

using a linear triangular element.

Solution

Let i, j, and k denote the global nodes corresponding to the corners, 1, 2, and 3, respectively, of element e as shown in Fig. 13.18. From Eqs. (E14) through (E24) of

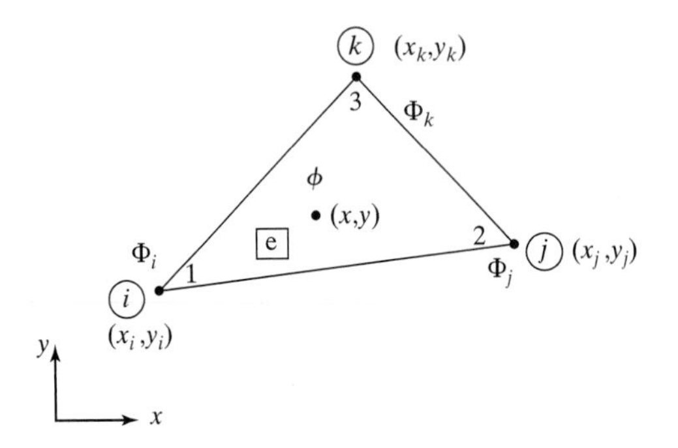

Figure 13.18 Typical triangular element.

Example 13.10, we obtain, with $p(x, y) = q(x, y) = 1$ and $r(x, y) = w(x, y) = 0$,

$$[S^{(e)}] = \frac{1}{4A^{(e)}} \begin{bmatrix} (b_i^2 + c_i^2) & (b_i b_j + c_i c_j) & (b_i b_j + c_i c_j) \\ \cdot & (b_j^2 + c_j^2) & (b_j b_k + c_j c_k) \\ \text{symmetric} & \cdot & (b_k^2 + c_k^2) \end{bmatrix}, \tag{b}$$

$$\vec{P}^{(e)} = -\int\int_{A^{(e)}} f(x, y)[N]^T \, dA. \tag{c}$$

Using $f(x, y) = q_0 = \text{constant}$, we see that Eq. (c) gives

$$\vec{P}^{(e)} = -\frac{q_0 A^{(e)}}{3} \begin{Bmatrix} 1 \\ 1 \\ 1 \end{Bmatrix}. \tag{d}$$

Here $A^{(e)} = \tilde{A}$ is given by Eq. (13.15) and b_i, b_j, b_k, c_i, c_j, and c_k are defined by Eqs. (13.17) through (13.19). ◀

13.8 Choice of Method

The application of the finite-element method for the solution of second-order ordinary differential equation and two-dimensional Poisson's equation are considered in this chapter. The method is also applicable for the solution of all types of linear and nonlinear differential equations. In fact, the method is extensively used for the solution of several practical engineering problems, including those related to stress analysis, heat transfer, fluid flow, electromagnetic fields, geomechanics, and aerodynamics. The method is especially convenient for solving problems involving arbitrary geometry. The method is applied to all types of equilibrium (or steady-state), eigenvalue, and propagation (or transient) problems. To illustrate the use of the finite-element method for the solution of parabolic and hyperbolic equations, consider the one-dimensional transient heat transfer equation

$$\frac{\partial T}{\partial t} = \frac{k}{\rho c}\left(\frac{\partial^2 T}{\partial x^2}\right), \quad 0 \le x \le L, \ 0 \le t \le \tau,$$

with initial condition

$$T(x, 0) = f(x) \tag{13.39}$$

and boundary conditions

$$T(0, t) = g(t)$$

$$\frac{\partial T}{\partial x}(0, t) = h(t),$$

where k is the thermal conductivity, c is the specific heat, ρ is the density, L is the length of the rod and τ is the time duration for finding the temperature distribution, $T(x, t)$. The problem indicated by Eq. (13.39) can be solved using two approaches. In the first approach, the time derivative is replaced by a finite-difference formula, for example, by a forward-difference formula, to obtain

$$\frac{T^{i+1} - T^i}{\Delta t} = \frac{k}{\rho c}\left(\frac{\partial^2 T}{\partial x^2}\right). \tag{13.40}$$

Equation (13.40) can be solved using the methods discussed in Section 13.7. In the second approach, the solution domain, $0 \le x \le L$ and $0 \le t \le \tau$, is considered as a rectangular region in (x, t)-space. The region is then divided into two-dimensional finite-elements, such as triangular elements, and the problem is solved as indicated in Section 13.7.

A number of general-purpose computer programs have been developed for the implementation of finite-element analysis. Some of the popular finite-element software packages include NASTRAN, ANSYS, ABAQUS, and ADINA. The finite-element method is implemented on a wide range of computers such as supercomputers, parallel computers, minicomputers, and personal computers.

13.9 Use of Software Packages

13.9.1 MATLAB

▶Example 13.12

Determine the solution of the second-order differential equation

$$a\frac{d^2\phi}{dx^2} + b\frac{d\phi}{dx} + c\phi = f(x), \ 0 \le x \le l, \tag{a}$$

with boundary conditions

$$\phi(0) = 0 \quad \text{and} \quad \phi(l) = 0 \tag{b}$$

for the following data: $a = 2, b = -3, c = 4, f(x) = d = 2.5, l = 1$.

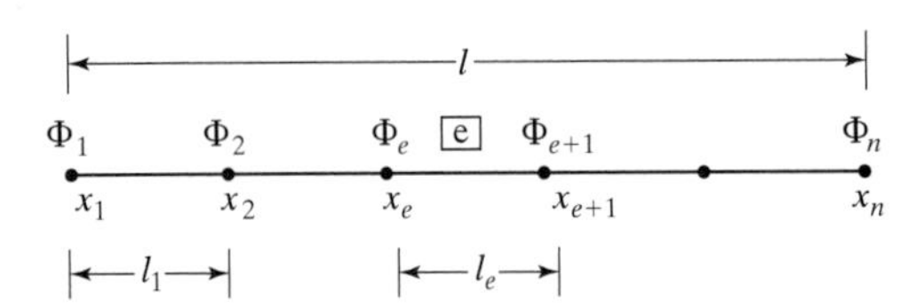

Figure 13.19 Finite-element idealization.

Solution

Finite-element equations

The finite-element idealization of the solution region is shown in Fig. 13.19. The element equations corresponding to Eq. (a) can be derived as (see Problem 13.30)

$$[K^{(e)}]\vec{\Phi}^{(e)} = \vec{P}^{(e)}, \tag{c}$$

where

$$[K^{(e)}] = -\frac{a}{l_e}\begin{bmatrix} 1 & -1 \\ -1 & 1 \end{bmatrix} + \frac{b}{2}\begin{bmatrix} -1 & 1 \\ -1 & 1 \end{bmatrix} + \frac{cl_e}{6}\begin{bmatrix} 2 & 1 \\ 1 & 2 \end{bmatrix}, \tag{d}$$

$$\vec{\Phi}^{(e)} = \begin{Bmatrix} \Phi_e \\ \Phi_{e+1} \end{Bmatrix}, \tag{e}$$

$$\vec{P}^{(e)} = \frac{dl_e}{2}\begin{Bmatrix} 1 \\ 1 \end{Bmatrix} \text{ if } f(x) = d \text{ is a constant,} \tag{f}$$

and

$$l_e = x_{e+1} - x_e \text{ is the length of element } e. \tag{g}$$

The generation of element matrices and vectors, the assembly of element matrices and vectors, the application of the boundary conditions of Eq. (b), and the solution of the resulting equations are accomplished through MATLAB programming. In the present case, five finite elements are used for the numerical solution.

The following input data are required:

CX(I) = x-coordinate of node I
A,B,C,D = coefficients of second-order differential equation ($f(x) = D$ is assumed)
XL = length of solution domain
NE = number of elements
PHI(1), PHI(NE+1) = boundary values of ϕ
EPS = convergence criterion

The following is the listing of the m.file and its output:

```
%P13_12.m
cx = [0.0, 0.2, 0.4, 0.6, 0.8, 1.0];
```

```
a = 2.0; b = -3.0; c = 4.0; d = 2.5; x1 = 1.0; ne = 5;
eps = 1.0e-6;
phi (1) = 0.0; phi (6) = 0.0;
nn = ne + 1;
nn2 = nn - 2;
nnp1 = nn2 + 1;
for i = 1: ne
   e1 (i) = cx(i + 1) - cx(i);
end
for i = 1: nn
   p(i) = 0.0;
   for j = 1: nn
      xk(i,j) = 0.0;
   end
end
for i = 1: ne
   xke(1,1) = -(a/e1(i)) - (b/2.0) + (c*e1(i)/3.0);
   xke(1,2) = (a/e1(i)) + (b/2.0) + (c*e1(i)/6.0);
   xke(2,1) = (a/e1(i)) - (b/2.0) + (c*e1(i)/6.0);
   xke(2,2) = -(a/e1(i)) + (b/2.0) + (c*e1(i)/3.0);
   pe(1) = d*e1(i)/2.0;
   pe(2) = pe(1);
   xk(i,i) = xk(i,i) + xke(1,1);
   xk(i, i+1) = xk(i, i+1) + xke(1,2);
   xk(i+1, i) = xk(i, i+1) + xke(2,1);
   xk(i+1, i+1) = xk(i+1, i+1) + xke(2,2);
   p(i) = p(i) + pe(1);
   p(i+1) = p(i+1) + pe(2);
end
for i = 1: nn2
   for j = 1: nn2
      xk2(i,j) = xk(i+1, j+1);
      p2(i) = p(i+1);
   end
end
for i = 1: nn2
   p2(i) = p2(i) - xk(i+1,1)*phi(1) - xk(i+1, nn)*phi(nn);
end

%Solution of the system of linear equations Ax=b
%using Gauss-seidel iterative algorithm

A = xk2; C = -A; b = p2; max = 50;
x = [0; 0; 0; 0];
```

```
i = 1;
while (i <= max)
   x1 = x;
   for j = 1: nn2
      C(j,j) = 0;
      x(j) = (C(j, :) *x + b(j))/A(j, j);
   end
   if norm(x1-x) <= eps
      break
   end
   i = i+1;
end
for i = 1: nn2
   phi (i+1) = x(i);
end
disp ('System stiffness matrix');
xk
disp ('System load vector');
p
disp ('Solution vector');
phi

% a = 2; b = -3; c = 4; d = 2.5; l = 1
>> P13_12
System stiffness matrix

xk =
   -8.2333    8.6333         0         0         0         0
   11.6333  -19.4667    8.6333         0         0         0
         0   11.6333  -19.4667    8.6333         0         0
         0         0   11.6333  -19.4667    8.6333         0
         0         0         0   11.6333  -19.4667    8.6333
         0         0         0         0   11.6333  -11.2333

System load vector

p =
    0.2500    0.5000    0.5000    0.5000    0.5000    0.2500

Solution vector

phi =
         0   -0.1008   -0.1694   -0.1882   -0.1381         0
```
◀

13.9.2 MAPLE

▶Example 13.13

Starting from the element characteristic matrix and characteristic vectors given by Eqs. (d) and (f) of Example 13.12, find the solution of the differential equation considered in Example 13.12 using MAPLE.

Solution

For the data given in Example 13.12, the characteristic matrix and characteristic vector of element e are given by

$$[K^{(e)}] = \begin{bmatrix} -8.23333333 & 8.63333333 \\ 11.63333333 & -11.23333333 \end{bmatrix}, e = 1, 2, \ldots, 5,$$

$$\vec{P}^{(e)} = \begin{Bmatrix} 0.25 \\ 0.25 \end{Bmatrix}, \quad e = 1, 2, \ldots, 5.$$

These element characteristic matrices and vectors are assembled and boundary conditions are applied (by deleting the rows and columns corresponding to zero values of Φ_i in assembled matrix and vector) to obtain the final equations as

$$[K]\vec{\Phi} = \vec{P}.$$

These equations are solved to find the desired solution, $\vec{\Phi}$, as

$$\vec{\Phi} = [K]^{-1}\vec{P}.$$

The following is the MAPLE program to implement the solution procedure:

```
P13_12.mws:a=2,b=-3,c=4,d=2.5,l=1

> with (linalg):

> KE := array([[-8.23333333, 8.63333333], [11.63333333,
  -11.23333333]]);
```

$$KE := \begin{bmatrix} -8.23333333 & 8.63333333 \\ 11.63333333 & -11.23333333 \end{bmatrix}$$

```
> Pe := array([[0.25], [0.25]]);
>
```

$$Pe := \begin{bmatrix} .25 \\ .25 \end{bmatrix}$$

```
> DOF := 6;
```

$$DOF := 6$$

```
> K := proc (Ke: :matrix, DOF, ii, jj)
> local i, j, KK;
> KK := matrix (DOF, DOF, 0);
> for i from 1 to DOF - 1 do
> for j from 1 to DOF - 1 do
> if j = i then
> KK[i,j] := KK[i,j] + Ke[1,1];
> KK[i,j+1] := KK[i,j+1] + Ke[1,2];
> KK[i+1,j] := KK[i+1,j] + Ke[2,1];
> KK[i+1,j+1] := KK[i+1,j+1] + Ke[2,2]; fi;
> od;
> od;
> KK[ii,jj];
> end:
>
> GK := array(1..4, 1..4): for i to 4 do for j to 4 do
> GK[i,j] := K(KE, DOF, i+1, j+1) od: od: print (GK);
```

$$\begin{bmatrix} -19.46666666 & 8.63333333 & 0 & 0 \\ 11.63333333 & -19.46666666 & 8.63333333 & 0 \\ 0 & 11.63333333 & -19.46666666 & 8.63333333 \\ 0 & 0 & 11.63333333 & -19.46666666 \end{bmatrix}$$

```
> P := proc (Pe, DOF, ii)
> local i, P;
> P :=matrix(DOF, 1, 0);
> for i from 1 to DOF - 1 do
> P[i,1] := P[i,1] + Pe[1,1];
> P[i+1,1] := P[i+1,1] + Pe[2,1];
> od;
> P[ii,1];
> end:
```

```
> 
> P(Pe, DOF, 3);
                                  .50
> 
> GP:=array(1..4, 1..1): for i to 4 do GP[i,1] := P(Pe, DOF,i+1)
> od: print (GP);
```

$$\begin{bmatrix} .50 \\ .50 \\ .50 \\ .50 \end{bmatrix}$$

```
> X: = linsolve(GK, GP);
```

$$x := \begin{bmatrix} -.1008004511 \\ -.1693724458 \\ -.1881627448 \\ -.1381315033 \end{bmatrix}$$

```
> 
```

◀

13.9.3 MATHCAD

▶Example 13.14

Solve the problem considered in Example 13.13 using MATHCAD.

Solution

By using the element characteristic matrices $[K^{(e)}]$ and vectors $\vec{P}^{(e)}$, the assembly of the global stiffness matrix $[GK]$ and load vector $\overrightarrow{PP}$, the derivation of the final assembled stiffness matrix $[K]$ and the load vector $\vec{P}$ (after incorporating the boundary conditions), and the solution of the equations are accomplished through a MATHCAD Program. The program and the results are shown as follows:

$$\text{Ke} := \begin{pmatrix} -8.23333333 & 8.63333333 \\ 11.63333333 & -11.23333333 \end{pmatrix}$$

$$\text{Pe} := \begin{pmatrix} 0.25 \\ 0.25 \end{pmatrix}$$

DOF := 6

$$KK(Ke,DOF) := \left| \begin{array}{l} \text{for } i \in 0..DOF-1 \\ \quad \text{for } j \in 0..DOF-1 \\ \qquad GK_{i,j} \leftarrow 0 \\ \text{for } i \in 0..DOF-2 \\ \quad \text{for } j \in 0..DOF-2 \\ \qquad \left| \begin{array}{l} GK_{i,j} \leftarrow GK_{i,j} + Ke_{0,0} \text{ if } i = j \\ GK_{i,j+1} \leftarrow GK_{i,j+1} + Ke_{0,1} \text{ if } i = j \\ GK_{i+1,j} \leftarrow GK_{i+1,j} + Ke_{1,0} \text{ if } i = j \\ GK_{i+1,j+1} \leftarrow GK_{i+1,j+1} + Ke_{1,1} \text{ if } i = j \end{array} \right. \\ GK \end{array} \right.$$

GK := KK(Ke,DOF)

$$GK = \begin{pmatrix} -8.233 & 8.633 & 0 & 0 & 0 & 0 \\ 11.633 & -19.467 & 8.633 & 0 & 0 & 0 \\ 0 & 11.633 & -19.467 & 8.633 & 0 & 0 \\ 0 & 0 & 11.633 & -19.467 & 8.633 & 0 \\ 0 & 0 & 0 & 11.633 & -19.467 & 8.633 \\ 0 & 0 & 0 & 0 & 11.633 & -11.233 \end{pmatrix}$$

$i := 0..DOF-3 \qquad j := 0..DOF-3$

$K_{i,j} := GK_{i+1,j+1}$

$$K = \begin{pmatrix} -19.467 & 8.633 & 0 & 0 \\ 11.633 & -19.467 & 8.633 & 0 \\ 0 & 11.633 & -19.467 & 8.633 \\ 0 & 0 & 11.633 & -19.467 \end{pmatrix}$$

$$PP(Pe,DOF) := \left| \begin{array}{l} \text{for } i \in 0..DOF-1 \\ \quad GP_i \leftarrow 0 \\ \text{for } i \in 0..DOF-2 \\ \quad \left| \begin{array}{l} GP_i \leftarrow GP_i + Pe_0 \\ GP_{i+1} \leftarrow GP_{i+1} + Pe_1 \end{array} \right. \\ GP \end{array} \right.$$

GP := PP(Pe,DOF)

$$GP = \begin{pmatrix} 0.25 \\ 0.5 \\ 0.5 \\ 0.5 \\ 0.5 \\ 0.25 \end{pmatrix}$$

i := 0..DOF − 3

$P_i := GP_{i+1}$

$$P = \begin{pmatrix} 0.5 \\ 0.5 \\ 0.5 \\ 0.5 \end{pmatrix}$$

$x := K^{-1} \cdot P$

$$x = \begin{pmatrix} -0.101 \\ -0.169 \\ -0.188 \\ -0.138 \end{pmatrix}$$

◀

13.10 Computer Programs

13.10.1 Fortran Programs

Program 13.1: Solution of one-dimensional problems

Example

Develop a computer program to solve Example 13.12.

Solution

The Fortran program developed is quite general and can be used for any general set of values of a, b, c, d, and l of the one-dimensional differential equation described in Example 13.12.

The following is the program output:

```
FINITE ELEMENT METHOD FOR 1-D PROBLEMS

DATA:
A = 2.00 B = -3.00 C = 4.00 D = 2.50
XL = 1.00 NE =    5

BOUNDARY CONDITIONS: PHI (1) = 0.00 PHI (6) = 0.00

SYSTEM STIFFNESS MATRIX

-0.8233E+01  0.8633E+01  0.0000E+00  0.0000E+00  0.0000E+00  0.0000E+00
 0.1163E+02 -0.1947E+02  0.8633E+01  0.0000E+00  0.0000E+00  0.0000E+00
 0.0000E+00  0.1163E+02 -0.1947E+02  0.8633E+01  0.0000E+00  0.0000E+00
 0.0000E+00  0.0000E+00  0.1163E+02 -0.1947E+02  0.8633E+01  0.0000E+00
 0.0000E+00  0.0000E+00  0.0000E+00  0.1163E+02 -0.1947E+02  0.8633E+01
 0.0000E+00  0.0000E+00  0.0000E+00  0.0000E+00  0.1163E+02 -0.1123E+02

SYSTEM LOAD VECTOR

0.2500E+00  0.5000E+00  0.5000E+00  0.5000E+00  0.5000E+00  0.2500E+00

SOLUTION VECTOR

0.0000E+00 -0.1008E+00 -0.1694E+00 -0.1882E+00 -0.1381E+00  0.0000E+00
```

◀

Program 13.2: Solution of two-dimensional problems

Example

Find the solution of the Poisson equation

$$\frac{\partial^2 \phi}{\partial x^2} + \frac{\partial^2 \phi}{\partial y^2} = 10000$$

over the domain $0 \le x \le 3$ and $0 \le y \le 3$ using the finite-element idealization shown in Fig. 13.20. The value of ϕ is specified as 200 over the boundary.

Solution

Finite-element equations

The element equations are given by $[K^{(e)}]\vec{\Phi}^{(e)} = \vec{P}^{(e)}$, with $[K^{(e)}]$ and $\vec{P}^{(e)}$ given by Eqs. (b) and (d) of Example 13.11, respectively. The assembly of element equations, the application of boundary conditions and the solution of the resulting equations are accomplished through the following computer Program. The program can be used for the solution of any Poisson or Laplace equation using any number of elements.

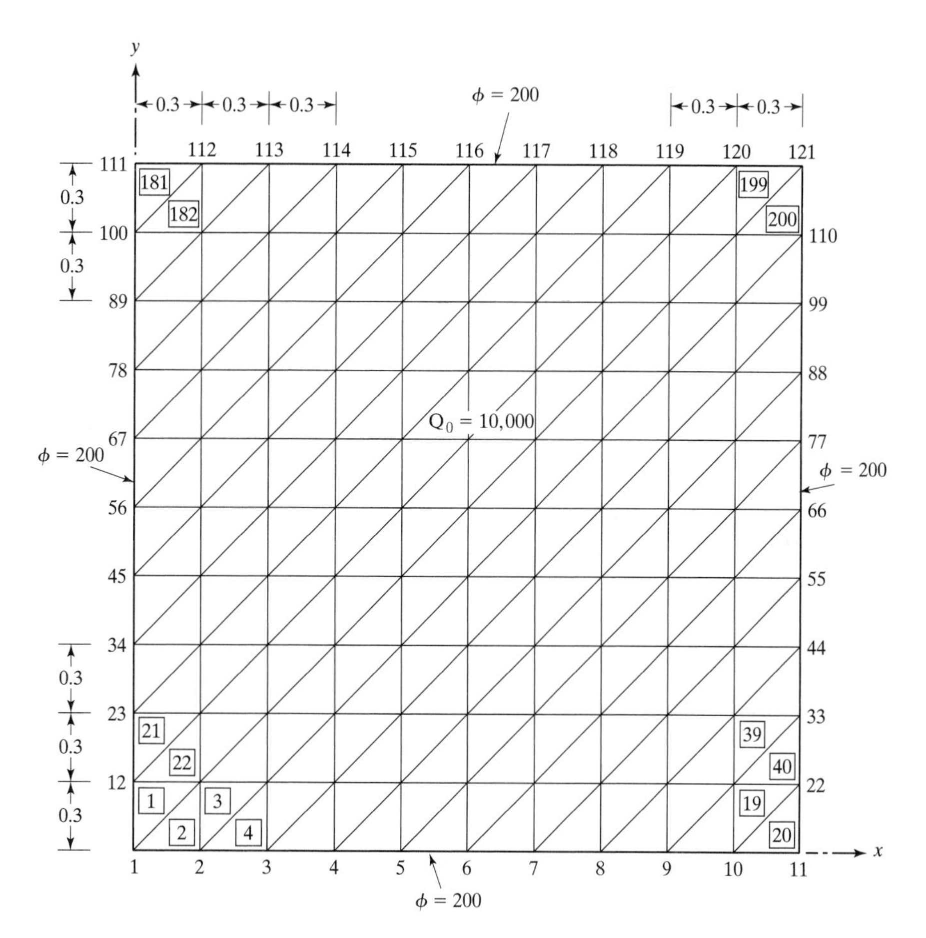

Figure 13.20 Finite element idealization.

The following input data are required:

NN = number of nodes
NE = number of elements
NFIX = number of degrees of freedom for which values are specified
NN1 = NN + 1
EPS = convergence criterion
IFIX(I) = I th degree of freedom whose value is specified
CONST(I) = specified value of I th degree of freedom
Q0 = value of F(X) = D = constant
XC(I), YC(I) = x, y coordinates of node I

With N = 121, NE = 200, NFIX = 40, EPS = 1.0×10^{-6}, IFIX (I) = 1, 2, 3, 4, 5, 6, 7, 8, 9, 10, 11, 12, 22, 23, 33, 34, 44, 45, 55, 56, 66, 67, 77, 78, 88, 89, 99, 100, 110, 111,

112, 113, 114, 115, 116, 117, 118, 119, 120, 121, CONST(I) = 200 for I = 1 to 40, and Q0 = 10000.0. Program 13.2 produces the following output:

```
FINITE ELEMENT METHOD FOR 2-D PROBLEMS

DATA:
NN = 121 NE = 200 NFIX = 40
Q0 = 0.1000E+05

SOLUTION OF POISSON EQUATION:
X-AXIS: HORIZONTAL, ROWS; Y-AXIS: VERTICAL, COLUMNS

200.0  200.0  200.0  200.0  200.0  200.0  200.0  200.0  200.0  200.0  200.0
200.0  328.1  406.3  454.0  480.0  488.3  480.0  454.0  406.3  328.1  200.0
200.0  406.3  542.9  629.6  677.7  693.1  677.7  629.6  542.9  406.3  200.0
200.0  454.0  629.6  743.9  808.1  828.8  808.1  743.9  629.6  454.0  200.0
200.0  480.0  677.7  808.1  882.1  906.0  882.1  808.1  677.7  480.0  200.0
200.0  488.3  693.1  828.8  906.0  931.0  906.0  828.8  693.1  488.3  200.0
200.0  480.0  677.7  808.1  882.1  906.0  882.1  808.1  677.7  480.0  200.0
200.0  454.0  629.6  743.9  808.1  828.8  808.1  743.9  629.6  454.0  200.0
200.0  406.3  542.9  629.6  677.7  693.1  677.7  629.6  542.9  406.3  200.0
200.0  328.1  406.3  454.0  480.0  488.3  480.0  454.0  406.3  328.1  200.0
200.0  200.0  200.0  200.0  200.0  200.0  200.0  200.0  200.0  200.0  200.0
```

13.10.2 C Program

Program 13.3: Solution of two-dimensional problems

Example

Consider Program 13.2 and write its *C* equivalent.

Solution

The following is the program's output:

```
Finite element analysis of two-dimensional problems

Data:
nn = 121 ne = 200 nfix =  40
q0 = 0.1000E+05

Solution of Poisson equation:
x-axis: horizontal, rows; y-axis: vertical, columns

200.0  200.0  200.0  200.0  200.0  200.0  200.0  200.0  200.0  200.0  200.0
200.0  328.1  406.3  454.0  480.0  488.3  480.0  454.0  406.3  328.1  200.0
200.0  406.3  542.9  629.6  677.7  693.1  677.7  629.6  542.9  406.3  200.0
200.0  454.0  629.6  743.9  808.1  828.8  808.1  743.9  629.6  454.0  200.0
200.0  480.0  677.7  808.1  882.1  906.0  882.1  808.1  677.7  480.0  200.0
200.0  488.3  693.1  828.8  906.0  931.0  906.0  828.8  693.1  488.3  200.0
200.0  480.0  677.7  808.1  882.1  906.0  882.1  808.1  677.7  480.0  200.0
```

```
200.0  454.0  629.6  743.9  808.1  828.8  808.1  743.9  629.6  454.0  200.0
200.0  406.3  542.9  629.6  677.7  693.1  677.7  629.6  542.9  406.3  200.0
200.0  328.1  406.3  454.0  480.0  488.3  480.0  454.0  406.3  328.1  200.0
200.0  200.0  200.0  200.0  200.0  200.0  200.0  200.0  200.0  200.0  200.0
```

◀

REFERENCES AND BIBLIOGRAPHY

13.1. S. S. Rao, *The Finite Element Method in Engineering*, 3d edition, Butterworth-Heinemann, Boston, 1999.

13.2. J. N. Reddy, *An Introduction to the Finite Element Method*, 2d edition, McGraw-Hill, New York, 1993.

13.3. Y. K. Liu, "Finite Element Modeling of the Head and Spine," *Mechanical Engineering*, Vol. 108, No. 1, January 1986, p. 60.

13.4. Y. W. Kwon and H. Bang, *The Finite Element Method Using* MATLAB, CRC Press, Boca Raton, FL, 1997.

13.5. G. R. Buchanan, *Schaum's Outline of Theory and Problems of Finite Element Analysis*, McGraw-Hill, New York, 1995.

REVIEW QUESTIONS

The following questions along with corresponding answers are available in an interactive format at the Companion Web site at http://www.prenhall.com/rao.

13.1. **Define** the following:

1. Simplex.
2. Bandwidth.
3. Degrees of freedom of a node.
4. Simplex element.
5. Complex element.
6. Multiplex element.
7. Residual.
8. Trial function.
9. Shape function.
10. Element characteristic matrix.

13.2. Answer **true or false** to each of the following:

1. Enforcing the boundary condition, $\Phi_i = 0$, in the finite-element method is equivalent to deleting the equation corresponding to the degree of freedom Φ_i.
2. The interpolation model for a simplex element is always linear.
3. The variational and the weighted residual finite element approaches can give different results for the same physical problem.
4. A shape function will have a value of unity at all nodes and zero at all other points of the element.
5. The finite-element method provides only an approximate solution.
6. The finite-element method permits the use of different material properties for different elements.
7. A given domain can be modeled in a variety of ways using finite elements.
8. The bandwidth can be reduced by suitably numbering the nodes of the finite-element mesh.
9. The finite-element method can be considered as a piecewise approximation procedure.
10. An interpolation function need not be a polynomial.
11. In the weighted residual approach, an integral function is minimized.

(*continued, page 1007*)

12. If the value of a combination of the field variable and its derivatives is specified, the resulting boundary condition is known as the mixed-boundary condition.

13.3. **Fill in the blanks** with suitable word(s):

1. In the finite-element method, a simple solution is assumed in each ______ .
2. To avoid numerical problems, all sides of three-dimensional elements should have the same ______ .
3. A node can have more than ______ degrees of freedom.
4. The overall accuracy of a finite-element solution depends on the approximating ______ .
5. Polynomial type of ______ functions are most widely used in the finite-element method.
6. A simplex element in three dimensions will have ______ nodes.
7. The shape function, $\frac{1}{l_e}(x_j - x)$, corresponds to node ______ in a one-dimensional element with nodes i and j.
8. The temperature distribution in a linear element is given by $\phi(x) = 80 - 16x$. The gradient of temperature inside the element is ______ .
9. The Galerkin method is a type of ______ method.
10. In general, the bandwidth can be reduced by numbering the nodes along the ______ dimension of the system.

13.4. **Give short answers:**

1. What is variational approach?
2. What is weighted residual approach?
3. How many nodal unknowns are required if the interpolation model, $\phi(x, y) = a_1 + a_2x + a_3y + a_4xy + a_5x^2 + a_6y^2$, is used for a triangular element?
4. How are the element characteristic matrices and vectors derived?
5. Name two types of weighted residual methods.
6. What is the role of trial functions in Galerkin method?
7. How is the residual defined in Galerkin method?
8. How can you improve the accuracy of a finite element solution?
9. What is the use of a space–time finite element?
10. What is a Dirichlet boundary condition?

13.5. Select the **most appropriate answer** out of the multiple choices given.

1. A complex element has
   (a) nonlinear or complex shape.
   (b) sides parallel to the coordinate axes.
   (c) nonlinear or complex interpolation polynomial.

(*continued, page 1008*)

2. The polynomial interpolation model, $\phi(x, y) = a_1 + a_2x + a_3y + a_4xy$, when used for a rectangular element with four corner nodes, represents a
   (a) linear model.
   (b) complex model.
   (c) nonlinear model.
3. A simplex element in two dimensions has the following characteristic:
   (a) triangle with three nodes.
   (b) triangle with six nodes.
   (c) rectangle with four nodes.
4. The area of a triangular element, given by

$$\tilde{A} = \frac{1}{2}\begin{vmatrix} 1 & 10 & 30 \\ 1 & 50 & 20 \\ 1 & 30 & 30 \end{vmatrix},$$

   is equal to
   (a) 200.
   (b) 100.
   (c) 300.
5. The (x, y)-coordinates of the three corner nodes 1, 2, and 3 of a linear triangular element are given by (2, 2), (10, 2), and (5, 10), respectively. The shape function, given by $\frac{1}{64}(-10 + 8x - 3y)$, corresponds to node.
   (a) 1.
   (b) 2.
   (c) 3.
6. In the variational method, the solution that minimizes the relevant functional satisfies the following:.
   (a) governing differential equation only.
   (b) boundary conditions only.
   (c) both the governing differential equation and the boundary conditions.
7. The partial differential equation

$$A\frac{\partial^2 \phi}{\partial x^2} = \frac{\partial \phi}{\partial t}, \quad a \le x \le b, \ t \ge 0,$$

   is a
   (a) parabolic equation.
   (b) hyperbolic equation.
   (c) elliptic equation.

(*continued, page 1009*)

8. Consider the partial differential equation

$$a_1 \frac{\partial^2 \phi}{\partial x^2} + a_2 \frac{\partial^2 \phi}{\partial y^2} + a_3 \phi = f(x, y),\ a_1,\ a_2,\ a_3 = \text{positive constants}$$

with suitable boundary conditions. This equation is called a
   (a) parabolic equation.
   (b) hyperbolic equation.
   (c) elliptic equation.

9. Consider the problem

$$\frac{d^2 y}{dx^2} + y + x = 0,\ 0 \le x \le 1,$$

with the boundary conditions $y(0) = y(1) = 0$. The exact solution of the problem is given by
   (a) $y(x) = \dfrac{\sin x}{\sin 1} - 1.$
   (b) $y(x) = \dfrac{\sin x}{\sin 1} - x.$
   (c) $y(x) = \dfrac{\sin 1}{\sin x} - 1.$

10. In a transient heat transfer problem, the nodal temperatures will be functions of
   (a) spatial variable(s) only.
   (b) time only.
   (c) both spatial variable(s) and time.

13.6. **Match the following:**

(1)	Approximate solution used in an element is called	(a)	better solution
(2)	Higher order polynomial models correspond to	(b)	simplex element
(3)	Lower order polynomial models correspond to	(c)	multiplex element
(4)	Linear model is used in	(d)	complex element
(5)	Midside nodes are used in	(e)	simple solution
(6)	Sides of the element are assumed parallel to coordinate axes in	(f)	interpolation function

PROBLEMS

Section 13.3

13.1. Number the nodes for the gear tooth shown in Fig. 13.21 to minimize the bandwidth of the resulting system matrix.

13.2. The finite-element modeling of a curved plate of rectangular cross section is shown in Fig. 13.22. (a) Determine the bandwidth of the corresponding system matrix when each node has three degrees of freedom. (b) Renumber the nodes for minimum bandwidth of the system matrix.

Section 13.4

13.3. A tapered bar is subjected to axial loading. The cross-sectional area of the bar is 100 mm^2 at $x = 20$ mm and 50 mm^2 at $x = 80$ mm. Assuming that the area of the bar varies linearly, express the area of cross section of the bar using a linear interpolation model. Also identify the shape functions.

13.4. A simplex element is used to approximate the temperature distribution in a straight fin (rod). If the temperatures at nodes i and j of an element are found to be 150° F and 50° F, respectively, determine the temperature distribution

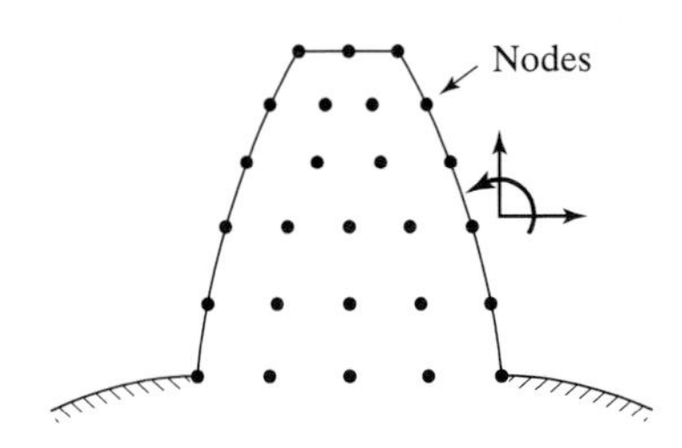

Figure 13.21 Gear tooth.

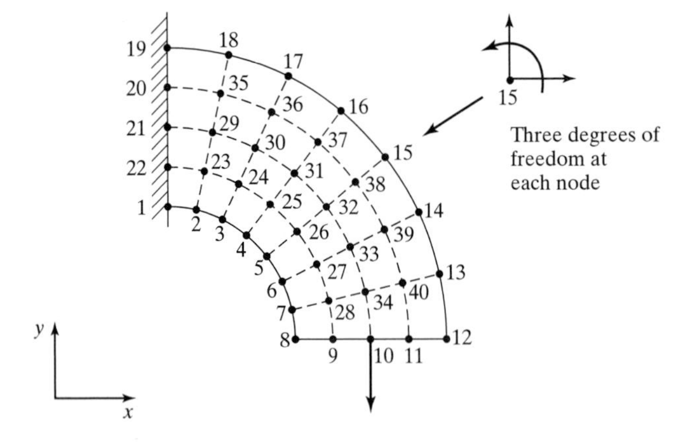

Figure 13.22 Curved plate.

in the element. The nodes i and j are located at distances of 10 in and 20 in from the origin, respectively. Also, find the temperature gradient inside the element.

13.5. The (x, y) coordinates of nodes i, j, and k of a two-dimensional simplex element, in inches, are (0, 10), (20, 30), and (40, 60), respectively. If the nodal temperatures are found to be $\phi_i = 110°$ F, $\phi_j = 90°$ F, and $\phi_k = 50°$ F, determine the following:
(a) shape functions corresponding to nodes i, j, and k.
(b) temperature distribution in the element.
(c) gradient of temperature inside the element.
(d) temperature at point P (30, 30).

13.6. Consider a three-dimensional simplex element as shown in Fig. 13.23. The field variable (solution), $\phi(x, y, z)$, is assumed to be

$$\phi(x, y, z) = a_1 + a_2 x + a_3 y + a_4 z,$$

where a_1, a_2, a_3, and a_4 are unknown constants. Express the solution in terms of the nodal unknowns ϕ_i, ϕ_j, ϕ_k, and ϕ_l and the corresponding shape functions $N_i(x, y, z)$, $N_j(x, y, z)$, $N_k(x, y, z)$, and $N_l(x, y, z)$.

13.7. A bar, subjected to an axial load, is modeled by one-dimensional elements with a quadratic interpolation model for the axial displacement. The end nodes, i and k, and the middle node j of an element are located at distances of 10 in, 20 in, and 15 in from the origin, respectively. If the axial displacements are found to be 0.009 in, 0.010 in, and 0.012 in at nodes i, j, and k, respectively, determine the following:
(a) shape functions.
(b) displacement distribution in the element.
(c) axial strain (rate of change of axial displacement in the axial direction) in the element.
(d) displacement at point P located at a distance of 18 in from the origin.

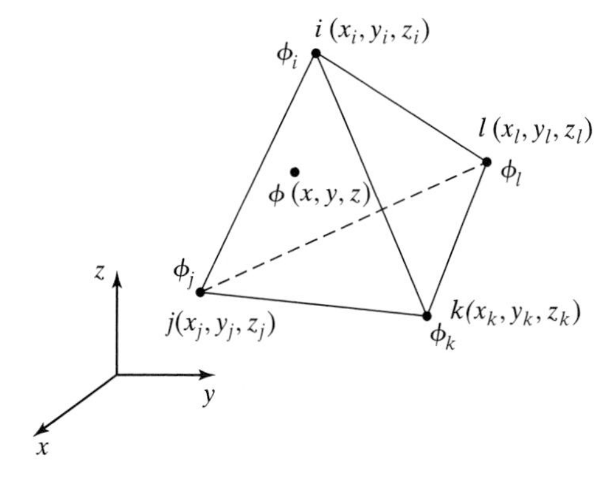

Figure 13.23 A three-dimensional simplex element.

Section 13.5

13.8. Consider the differential equation

$$\frac{d^2y}{dx^2} + 200x^2 = 0, \; 0 \le x \le 1, \tag{E1}$$

subject to the boundary conditions

$$y(0) = y(1) = 0. \tag{E2}$$

The functional I corresponding to the differential Equation (E1) is given by

$$I = \frac{1}{2}\int_0^1 \left\{ -\left(\frac{dy}{dx}\right)^2 + 400x^2 y \right\} dx. \tag{E3}$$

By dividing the solution domain ($x = 0$ to 1) into two finite elements of equal length with linear interpolation functions, derive the element characteristic matrices and vectors.

13.9. For the differential equation described in Problem 13.8, assume an approximate solution, which satisfies the boundary conditions, as

$$y(x) = a_1 x(1 - x^2), \tag{E4}$$

where a_1 is an unknown constant. Substitute this solution into Eq. (E3) of Problem 13.8, extremize I by setting $\frac{dI}{da_1} = 0$ (variational approach) and determine the value of a_1.

13.10. For the differential equation described in Problem 13.8, derive the element characteristic matrices and vectors using Galerkin approach. Divide the solution domain into two finite elements of equal length with linear interpolation functions.

Section 13.6

13.11. For the problem described in Problem 13.8, derive the system equations and find the solution.

13.12. For the differential equation described in Problem 13.8, assume an approximate solution, which satisfies the boundary conditions,

$$y(x) = a_1 x(1 - x^2), \tag{E1}$$

where a_1 is an unknown constant. Define the residual, R, as

$$R = \frac{d^2y}{dx^2} + 100x^2, \tag{E2}$$

substitute the solution, Eq. (E1), into Eq. (E2), equate the integral, $\int_0^1 R\,dx$, equal to zero (weighted residual approach), and determine the value of a_1. Compare the resulting solution with the one found in Problem 13.9.

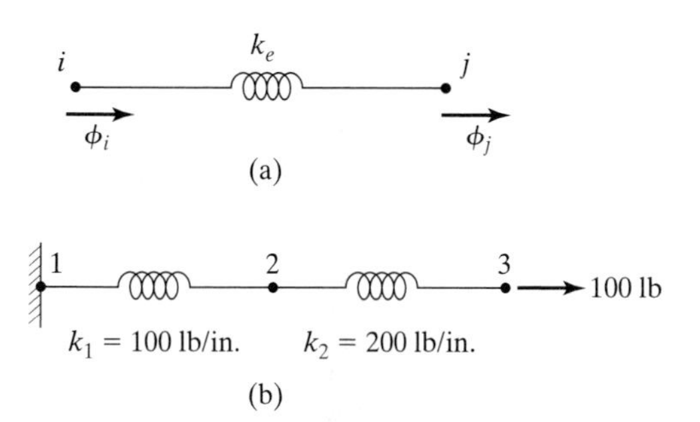

Figure 13.24 Springs in series.

Section 13.7

13.13. Consider the differential equation of Examples 13.7 and 13.8.
(a) derive the system of equations using two- and three-element idealizations.
(b) Find the solution of the problem corresponding to two- and three-element idealizations.

13.14. The stiffness matrix of a spring element, shown in Fig. 13.24(a), is given by

$$[k] = k_e \begin{bmatrix} 1 & -1 \\ -1 & 1 \end{bmatrix} \begin{matrix} \phi_i \\ \phi_j \end{matrix}, \quad \text{(columns } \phi_i, \phi_j\text{)}$$

where k_e is the stiffness of the spring. When two springs of stiffnesses $k_1 = 100$ lb/in and $k_2 = 200$ lb/in are connected in series with the left end (node 1) fixed and a load applied at the right end (node 3) as shown in Fig. 13.24(b), derive the system equations, apply the boundary conditions, and find the displacements of nodes 2 and 3.

Section 13.9

13.15. Use MATLAB to find the solution of the problem stated in Eqs. (a) and (b) of Example 13.12 for the following data:

$$a = 1, \; b = -3, \; c = 2, \; d = 1.0, \; l = 1.0.$$

(*Hint*: Change the data of line 3 in the program of Section 13.9.1.)

13.16. Find the solution of the Poisson equation

$$\frac{\partial^2 \phi}{dx^2} + \frac{\partial^2 \phi}{dy^2} = 5000$$

over the domain $0 \le x \le 3$ and $0 \le y \le 3$ using MATLAB. The finite-element grid is shown in Fig. 13.20, and the value of ϕ is specified as 100 over the boundary.

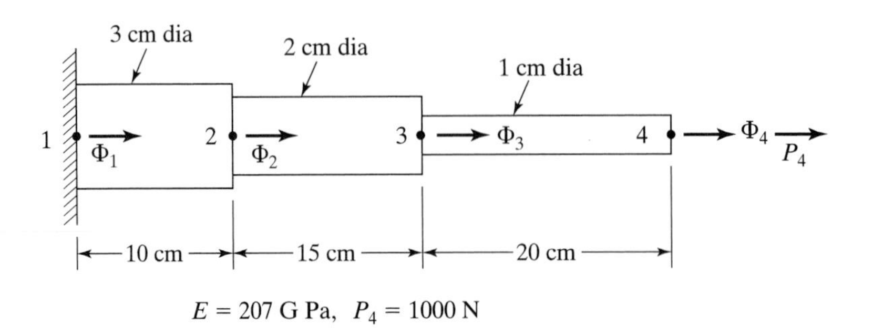

Figure 13.25 Stepped bar under axial load.

13.17. Solve the Poisson equation described in Problem 13.16 using MAPLE.

13.18. Solve the Poisson equation described in Problem 13.16 using MATHCAD.

Section 13.10

13.19. Write a computer program to find the displacements of nodes 1, 2, 3, and 4 of the stepped bar shown in Fig. 13.25 subjected to an axial load of 1000 N applied at node 4. The element equations are given by

$$[K^{(e)}]\vec{\Phi}^{(e)} = \vec{P}^{(e)},$$

where

$$[K^{(e)}] = \frac{A^{(e)}E^{(e)}}{l^{(e)}}\begin{bmatrix} 1 & -1 \\ -1 & 1 \end{bmatrix},$$

$$\vec{\Phi}^{(e)} = \begin{Bmatrix} \Phi_e \\ \Phi_{e+1} \end{Bmatrix}, \quad \vec{P}^{(e)} = \begin{Bmatrix} P_e \\ P_{e+1} \end{Bmatrix}.$$

$A^{(e)}$ is the cross-sectional area, $E^{(e)}$ is Young's modulus, $l^{(e)}$ is the length of element e, Φ_e is the displacement of node e, and P_e is the load applied at node e.

13.20. Use Program 13.1 to find the solution of the problem stated in Eqs. (a) and (b) of Program 13.1 for the following data:
(i) $a = 2, b = -3, c = 4, f(x) = 5.0, l = 1.$
(ii) $a = 1, b = 2, c = 3, f(x) = 4.0, l = 1.$

13.21. Use Program 13.2 to find the solution of the Poisson equation

$$\frac{\partial^2 \phi}{\partial x^2} + \frac{\partial^2 \phi}{\partial y^2} = 5000$$

over the domain $0 \leq x \leq 5$ and $0 \leq y \leq 5$ by specifying the value of ϕ as 100 over the entire boundary.

13.22. Use Program 13.3 to find the solution of the Poisson equation

$$\frac{\partial^2 \phi}{\partial x^2} + \frac{\partial^2 \phi}{\partial y^2} = 2000$$

over the domain $0 \leq x \leq 5$ and $0 \leq y \leq 5$ by specifying the value of ϕ as 200 over the boundary.

General

13.23. The fluid flow in a straight pipe is modeled using one-dimensional linear elements. If the locations of the nodes 1 and 2 of the element e from the origin are 20 cm and 28 cm, and the pressures are 80 psi and 40 psi, respectively, determine the following:

(a) pressure distribution in element e.

(b) pressure gradient in element e.

(c) pressure at a point P, located at a distance of 22 cm from the origin.

13.24. Three-dimensional simplex elements are used to find the pressure distribution in a fluid medium. The (x, y, z)-coordinates of nodes i, j, k, and l of an element are given by (3, 5, 3), (1, 1, 1), (5, 1, 1), and (3, 1, 7). Determine the shape functions $N_i(x, y, z)$, $N_j(x, y, z)$, $N_k(x, y, z)$, and $N_l(x, y, z)$.

13.25. The cubic interpolation model for a one-dimensional element is given by

$$\phi(x) = a_1 + a_2 x + a_3 x^2 + a_4 x^3,$$

where a_1, a_2, a_3, and a_4 are unknown constants. By using the four nodal values of the field variable shown in Fig. 13.26 as the unknowns, express the interpolation model in terms of ϕ_i, ϕ_j, ϕ_k, and ϕ_l, and identify the shape functions $N_i(x)$, $N_j(x)$, $N_k(x)$, and $N_l(x)$.

13.26. Consider the differential equation

$$\frac{d^4 y}{dx^4} + 4\frac{d^2 y}{dx^2} + 2y = 5, \quad 0 \leq x \leq 10, \tag{E1}$$

subject to the boundary conditions

$$y(0) = 0, \quad \frac{dy}{dx}(0) = 1, \quad y(10) = 10, \quad \frac{dy}{dx}(10) = 0. \tag{E2}$$

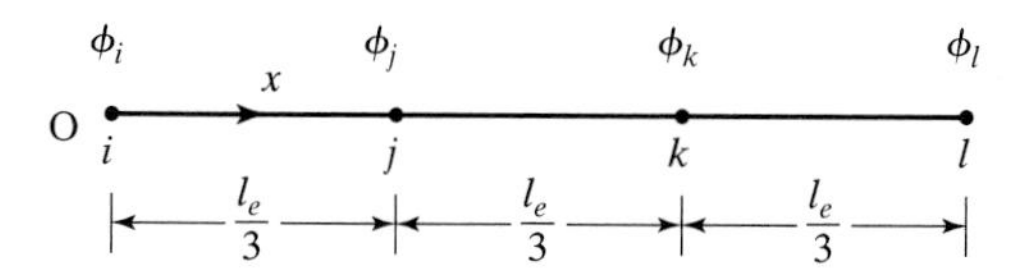

Figure 13.26 Cubic interpolation model.

The functional I corresponding to the differential Equation (E1) is given by

$$I = \frac{1}{2}\int_0^{10}\left\{\left(\frac{d^2y}{dx^2}\right)^2 - 4\left(\frac{dy}{dx}\right)^2 + 2y^2 - 10y\right\} dx. \tag{E3}$$

If the solution domain is to be divided into finite elements of equal length,

(a) determine the minimum order of the polynomial that can be used as the interpolation model.

(b) derive the element characteristic matrices and vectors for a cubic interpolation model.

(*Hint*: Use the ends of the element as nodes with y and $\frac{dy}{dx}$ as unknowns at each node.)

13.27. The system equations of a finite-element model, before applying the boundary conditions, are given by

$$10^6 \begin{bmatrix} 2 & -2 & 0 \\ -2 & 3 & -1 \\ 0 & -1 & 1 \end{bmatrix} \begin{Bmatrix} \phi_1 \\ \phi_2 \\ \phi_3 \end{Bmatrix} = \begin{Bmatrix} F_1 \\ 0 \\ 10 \end{Bmatrix},$$

(columns labelled ϕ_1, ϕ_2, ϕ_3)

where ϕ_1, ϕ_2, and ϕ_3 are the nodal values of the field variable and F_1 is the nodal reaction (unknown) at node 1. Incorporate the boundary condition $\phi_1 = 0$, and solve the resulting system of equations.

13.28. For a fluid flowing in a pipe, a linear interpolation model can be used for the variation of pressure. The characteristic matrix and vector of an element can be derived as (Fig. 13.27(a)):

$$[k] = c\begin{bmatrix} 1 & -1 \\ -1 & 1 \end{bmatrix}\begin{matrix} p_i \\ p_j \end{matrix} \quad \text{and } \vec{f} = \begin{Bmatrix} q_i \\ q_j \end{Bmatrix}\begin{matrix} p_i \\ p_j \end{matrix},$$

(columns labelled p_i, p_j)

where $c = \frac{\pi d^4}{128\mu l}$, d is the pipe diameter, μ is the viscosity of the fluid, l is the element length, q_i, and q_j are the flow rates of the fluid at nodes i and j, and p_i and p_j are the fluid pressures at nodes i and j.

For the four-segmented pipe network shown in Fig. 13.27(b), determine the fluid pressures at nodes 1 and 2. Data $Q = 3,456$ in^3/sec, d_i = diameter of pipe segment i (circular pipe) $= 4'', 3'', 2'', 3''$ for $i = 1, 2, 3, 4$, l_i = length of pipe segment $i = 200'', 300'', 400', 500''$ for $i = 1, 2, 3, 4$, $\mu = 10^{-7}$ lb$_f$-sec/in^2, and p_3 = pressure at node $3 = 0$.

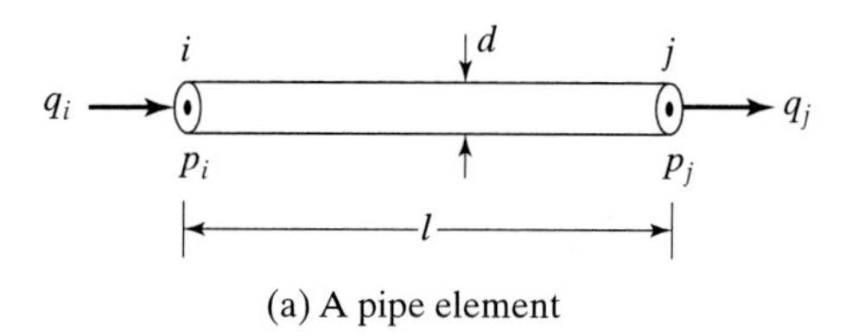

(a) A pipe element

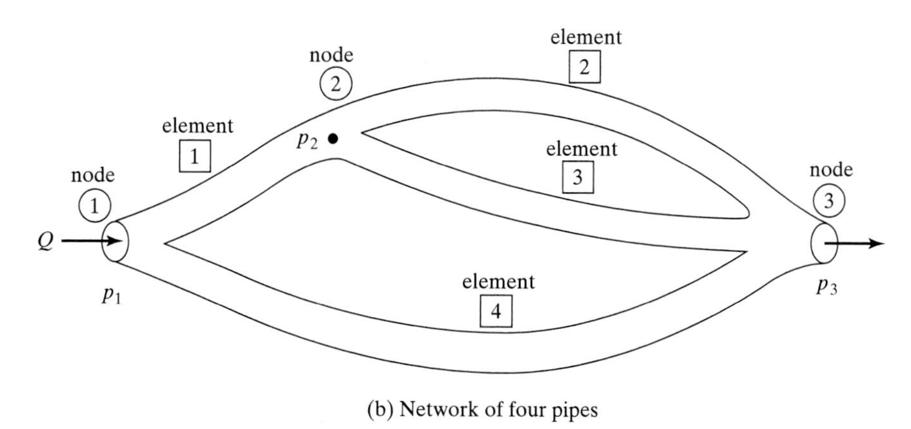

(b) Network of four pipes

Figure 13.27 Modeling of pipe network.

13.29. Consider the differential equation

$$\frac{d^2y}{dx^2} + 10y = \sin x \quad 0 \le x \le 1,$$

with the boundary conditions

$$y(0) = y(1) = 0.$$

Divide the solution region ($x = 0$ to 1) into E finite elements of equal length, use linear interpolation functions, and derive the elemental equations using a weighted residual approach.

13.30. Show that the element equations corresponding to the differential equation

$$a\frac{d^2\phi}{dx^2} + b\frac{d\phi}{dx} + c\phi = f(x), \; 0 \le x \le l,$$

are given by

$$[K^{(e)}]\vec{\Phi}^{(e)} = \vec{P}^{(e)},$$

where

$$[K^{(e)}] = -\frac{a}{l_e}\begin{bmatrix} 1 & -1 \\ -1 & 1 \end{bmatrix} + \frac{b}{2}\begin{bmatrix} -1 & 1 \\ -1 & 1 \end{bmatrix} + \frac{cl_e}{6}\begin{bmatrix} 2 & 1 \\ 1 & 2 \end{bmatrix},$$

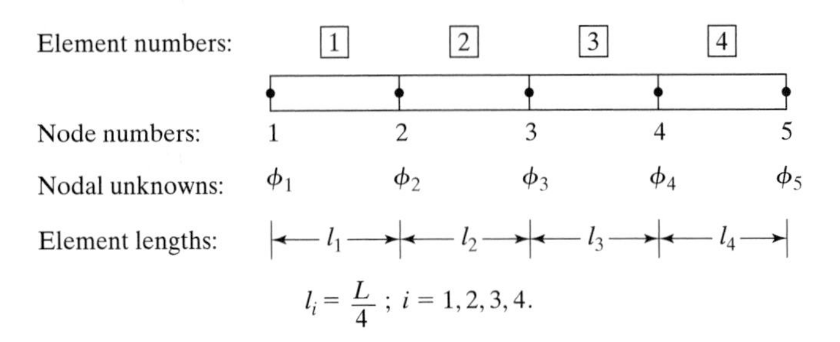

Figure 13.28 Modeling of uniform fin.

$$\vec{\Phi}^{(e)} = \begin{Bmatrix} \Phi_e \\ \Phi_{e+1} \end{Bmatrix},$$

$$\vec{P}^{(e)} = \frac{dl_e}{2} \begin{Bmatrix} 1 \\ 1 \end{Bmatrix} \text{ if } f(x) = d \text{ is a constant,}$$

and

$$l_e = x_{e+1} - x_e \text{ is the length of element } e.$$

PROJECTS

13.1. A uniform fin, shown in Fig. 13.28, is initially heated to a temperature of 150° C and then cooled. The governing equation is given by

$$\frac{k}{\rho c}\frac{\partial^2 \phi}{\partial x^2} = \frac{\partial \phi}{\partial t}, \quad 0 \le x \le 40 \text{ cm}, \ t \ge 0,$$

where k is the thermal conductivity, ρ is the density, and c is the specific heat of the material. At time $t = 100$ seconds, the temperatures of nodes 1 and 5 are known to be 80° C and 20° C, respectively. Assuming that $\frac{k}{\rho c} = 3.5$, determine the nodal temperatures of the fin at $t = 100$ seconds.

13.2. Find the solution of the Poisson equation considered in Section 13.8.2 by dividing the solution region into 25 (5 × 5) elements.

Appendix A

Basics of Fortran 90

1. Structure of a Fortran Program

The general form of a Fortran computer program includes the following items:

Heading
Specification part
Execution part
Subprogram part
End program statement

The *heading* marks the beginning of the program and gives the program a name. A heading is usually followed by comment lines about the purpose, algorithm used, input, output, and other relevant information about the program.

The following is an example of a heading (naming a program):

```
PROGRAM BESSEL-FUNCTION
```

The following is an example of a comment line:

```
!THIS PROGRAM COMPUTES BESSEL FUNCTIONS OF FIRST AND SECOND KIND
```

(A comment line starts with an exclamation mark (`!`) and can contain any number of characters up to the end of the line.)

In the *specification part*, the names and types of constants and variables used in the program are declared. The *execution part* contains statements that carry out the various steps of the algorithm that produces the results. When a computer program is to be developed for a large and complex problem, it is helpful to divide the problem into a number of simpler subproblems. Each subproblem is then coded to implement its own algorithm. The *subprogram part* contains the code for the solution of all subproblems. The *end program statement* marks the end of the program and stops execution.

2. Identifiers

Identifiers include program names and variables. Variables are names used to identify quantities that can change in value. Identifiers must begin with a letter and may be followed by up to 30 letters, digits, or underscores. The types of values (integer, real, complex, or logical) that each variable can have must be declared by placing a statement of the following form:

```
REAL::IAMOUNT,LIFE,TEMP_FINAL
INTEGER::TIME,QUA_PART,MONEY
```

3. `IMPLICIT NONE` Statement

The statement

```
IMPLICIT NONE
```

must appear at the beginning of the specification part of the program to cancel the implied naming convention. Without this statement, any undeclared variables whose names begin with I, J, K, L, M, or N will be considered as integers and others as real.

4. Use of Position

A line may have a maximum of 132 characters. A line may contain more than one statement, provided that the statements are separated by semicolons. If a line is to be continued to the next line, an ampersand (&) must be placed at the end of the first line. A maximum of 39 continuation lines are permitted. A comment line starts with an exclamation mark (!). Comment can also be attached to statements as follows:

```
INTEGER::IMAX!IMAX is the maximum number of iterations permitted
```

5. Statement Label

The statement label or number, if given, must precede the statement and be separated from it by at least one blank space. Statement labels must be integers in the range 1 through 99999.

6. Data Types

Fortran provides numeric, character, and logical data types. The numeric data can be integer, real, and complex types.

Consider the following examples:

Integer constants: 5, −40, 7562
Real constants: 3.201, 0.3201E+01, 0.42E−3

Complex constants: (2.1, 3.0), (3.0, –0.4)
(The first and second numbers in parenthesis denote the real and imaginary parts of the complex number, respectively.)

The following are character data: A set of acceptable characters—letters, digits, and punctuation marks—not included for computation is called a character (or string) constant. A string constant is enclosed within double quotation marks or apostrophes as

```
"ADAM","COMPUTED VALUE OF X"
'ADAM','COMPUTED VALUE OF X'
```

Note that blanks within the double quotations or apostrophes are also counted as character constants.

Logical data: If a quantity is defined as a logical variable, it can have two values: TRUE or FALSE.

7. Arrays

Arrays are to be declared through DIMENSION statements of the following form:

```
REAL,DIMENSION(3,4)::TEMP
```

(implies that the real variable TEMP has 3 rows and 4 columns)

```
INTEGER,DIMENSION(5:14)::ALFA
```

(implies that the one-dimensional array ALFA has subscripts ranging from 5 through 14)

```
LOGICAL,DIMENSION(4)::P
```

8. Assigning Values

Values can be assigned to variables in several ways:

```
INTEGER,DIMENSION(10)::A
A=(/3,6,9,12,15,18,21,24,27,30/)
A=(/(3*I,I=1,10)/)
A=(/3,6,(I,I=9,27,3),30/)

INTEGER,DIMENSION(10)::A=(/3,6,9,12,15,18,21,24,27,30/)

REAL,DIMENSION(3)::A=(/2.1,3.0,0.42),&
        B=(/8.5,3.91,7.5/),&
        C=(4.2,0.021,0.1E-3/)
```

9. Format Statements

A format statement defines the way a variable is to be read or written, as shown by the following examples:

```
PRINT*,IMAX,A1
```

(An asterisk is to be used when the precise form of the printout is not important. In this case, the format is determined by the type of variables to be printed; IMAX is an integer and A1 is a real number.)

In the statement

```
PRINT'(/,5X,I5,2X,E15.8)',IMAX,A1
```

the symbol / is used to leave a blank line. 5X denotes that the first 5 spaces are left blank, I5 indicates that the value of IMAX is printed as an integer within 5 spaces, 2X represents 2 blank spaces, and E15.8 denotes that the value of A1 is printed as a real number in exponential form using a total of 15 spaces, including 8 digits after the decimal. For example, if the value of IMAX is 5120 and that of A1 is 4251.327, they will be printed as follows:

```
......5120....0.42513270E+04
```

(a dot is used to denote a blank space).

```
PRINT 15,IMAX,A1
```

(Here the format is specified in a separate statement labeled 15.)

10. Read and Print Statements

These statements permit data to be read (entered as input) and printed (as output) during the execution of the program.

The following are examples:

```
READ*,IMAX,A1,A2
```

(Here the input data is to be separated by commas. For example, 5,6.8,3.1E-2 can be used to assign the values IMAX $= 5$, A1 $= 6.8$, and A2 $= 3.1 \times 10^{-2}$.)

```
      READ 10,A1,A2
      PRINT 10,A1,A2
   10 FORMAT(F12.3,2X,E14.6)
```

Here the values of A1 and A2 are read or printed in the format defined by the statement number 10.

11. Logical Operators

The logical operators are used to compare values of two expressions:

.EQ. (equal to)
.NE. (not equal to)
.LT. (less than)
.LE. (less than or equal to)
.AND. (both)
.OR. (either one)

12. DO Construct

The DO construct can be used for repetitive execution of one or more statements. The following example illustrates the use of a DO loop inside another DO loop (called nested DO loop):

```
DO I=1,IMAX
 DO J=1,JMAX
   FUN=I*J
   PRINT*,I,"",J,"",FUN
 END DO
END DO
```

13. IF Construct

An IF construct permits a programmer to specify the sequence of statements to be executed when a logical expression is true, but also to indicate an alternative sequence of statements to be executed when it is false.

For example,

```
IF(X.LE.0)THEN
  FUN=-X**2
 ELSE
  FUN=X**3
 END IF
```

14. Functions

Functions can be defined by either statements or subprograms in Fortran. The function statements are similar to formulas in mathematics. An example of a function statement is

```
Z=X**2-3.0*SIN(Y)
```

A function subprogram can be called into action by using a statement such as the following in the main program:

```
Z=FUN1(X,Y)
```

Then the function subprogram must be defined after the main program as follows:

```
  FUNCTION FUN1(X,Y)
! FUNCTION FUN1 WITH ARGUMENTS X AND Y
  FUN1=X**2-4.0*LOG(Y)
  END FUNCTION FUN1
```

Note that the definition of FUN1 can be more complex than indicated; it can include several statements or lines.

15. Library Functions (Examples)

```
Y=SQRT(X)              (Y=√(X))
I1=INT(X)              (I1=integer part of X)
IZ=MAX(I1,I2,...,IN)   (IZ=maximum out of I1,I2,...,IN)
Y=FLOAT(I1) or REAL(I1) (Y=real value equivalent of the integer I1)
Y=EXP(X)               (Y=e^X)
Y=SIN(X)               (Y=sinX)
Y=ABS(X)               (Y=|X|)
Y=LOG(X)               (Y=ln X)
Y=LOG10(X)             (Y=log10X)
Y=ASIN(X)              (Y=sin^-1X)
Y=MIN(X1,X2,...,XN)    (Y=minimum value of X1,X2,...,XN)
I1=NINT(X)             (I1=value of X rounded to nearest integer)
```

16. Subroutines

A subroutine is a subprogram that can be called into action by using a statement, such as the following, in the main program. It is similar to a function subprogram:

```
CALL SUB1(X1,X2,X3,Y)
```

Then the subroutine must be defined as follows, after the main program:

```
SUBROUTINE SUB1(Y1,Y2,Y3,Z)
---
---
Z=....
END SUBROUTINE SUB1
```

The number and type of arguments in the calling and subroutine statements must match; their names can be different. In the preceding example, the arguments `Y1, Y2, Y3`, and `Z` are assumed to be identical to `X1, X2, X3`, and `Y`, respectively. Also, `X1, X2`, and `X3` are assumed to be the input data to the subroutine, and `Y` is assumed to be the output quantity, computed and returned by the subroutine `SUB1`.

Appendix B

Basics of C Language

1. Comment

A comment is delimited by /* and */.

Example:
/* This is a comment.*/

2. Standard Headers

The following are some of the standard headers in C:

float.h (to describe floating-point types)
limits.h (to describe integer types)
math.h (to include mathematical functions)
stdio.h (to handle inputs and outputs)
stdlib.h (to include general utilities)
string.h (to handle arrays of characters)

3. void

The word *void* indicates that a function returns no value or expects no arguments.

Consider the statement

```
void fun (int x, int y)
```

Here, the function *fun* uses two parameters x and y, both of type *int*, and returns no value. In

```
float fun (void)
```

the function *fun* uses no parameter and returns a value of fun, of type *float*.

4. Constants

Different types of constants can be defined in C. The following are some examples:

Integer constants in decimal system. (Examples: 84, 3285)
Floating-point constants. (Examples: 0.2, 4.0e−2, 425.2e+3)
Character constants are characters enclosed in single quotes. (Examples: 'b', '#')

5. Data Types

Integer data types are *char, short int* (or *short*), *int*, and *long int* (or *long*).
Floating-point data types are *float, double*, and *long double*.

6. Arrays

An array is defined by writing the data type and the name followed by the size in brackets, as in

int $x[10]$
(This implies that $x[0], x[1], \ldots, x[9]$ are integers.)

7. Cast

A cast operator is used to convert an operand to a specific type as follows:

```
int i=5;
float z;
z=(float)i;
```

(This implies that the value of the integer `i` is converted to float and stored as `z`.)

8. do while

Example: *do* part1 *while* (condition);

Here, part1 is executed as long as condition is true, or nonzero. If part1 consists of a single statement, it is terminated with a semicolon. On the other hand, if part1 consists of multiple lines, it is enclosed in braces.

9. goto

goto is used to pass the control to a label.

Example: goto 10;

10. if

Example 1: *if* (condition) part1;

This implies that if condition is true, or nonzero, part1 is executed; otherwise, it skips to the statement immediately following part1.

Example 2: *if* (condition) part1 *else* part2;

This implies that part1 is executed if condition is true, or nonzero, and part2 is executed otherwise. In either case, the program skips to the statement immediately following part2.

11. Initialization

Variables are initialized as shown in the following examples:

```
int a=2;
int a[5]={2,16,7,21,33};
```

12. Label

A label is the first identifier on a line and is followed by a colon. A line that begins with a *label:* is a target of a *goto* statement.

13. return

The statement *return* implies that the function returns to its invoker without returning a value. The statement *return quantity*, or *return (quantity)*, implies that the function returns to its invoker the value of quantity.

14. Functions

A function is defined by a header and a body as the following shows:

```
double dotprod(double x[],double y[],int n);
{
int i;
double sum=0.0;
for(i=0;i<n;i++)
sum=sum+x[i]*y[i];
return(sum);
}
```

15. Library Functions

The following examples show some of the library functions in C and their use:

#include ⟨stdlib.h⟩
int abs (int a);
(returns the absolute value of the integer a)

```
#include <math.h>
double asin (double x);
```
(returns the arc sine of x in radians; value between $-\frac{\pi}{2}$ and $\frac{\pi}{2}$).

double exp (double x); (returns e^x)
double log (double x); (returns $\ln x$)
double log10 (double x); (returns $\log_{10} x$)
double pow (double x, double y); (returns x^y)

Appendix C

Basics of MAPLE

Several software packages, such as MACSYMA, DERIVE, Mathematica, and MAPLE, have been developed for symbolic mathematical computations. We describe the features of MAPLE in this section. As stated, MAPLE can be used for the symbolic, numeric, and graphical solution of mathematical problems. Although MAPLE sessions are to be started differently on different computers, the following features are applicable to all machines:

1. Syntax

(1) Upper- and lowercase letters are recognized differently.

(2) All expressions must end with a semicolon (;).

(3) Input lines always begin with the prompt symbol >, and output lines are always centered across the terminal screen.

(4) Spaces between symbols imply multiplication. (For example, $x\,y$ is assumed to be x multiplied by y.)

2. Algebraic Operations

(1) To perform simple arithmetic operations such as addition, subtraction, multiplication, or division of integers, fractions, polynomials, etc., the expression to be computed must be typed with a semicolon at the end as follows:

```
>1+5;
```
(to add the numbers 1 and 5)

```
>iquo(5,3);
```
(to truncate division of 5 and 3)

```
>abs(-8);
```
(to find the absolute value of -8)

```
>(x-y)^2;
```
(to find $(x - y)^2$; the operator ^ denotes exponentiation).

```
>expand(");
```
(the symbol " refers to the output of the immediately previous computation. In the present case, $(x - y)$^2 denotes the previous computation; hence, the output of the command will be as follows:

$$x^2+2xy+y^2$$

(2) The arguments of functions must be delineated with parentheses as denoted by the following examples:

```
>exp(4);
```
(to find the value of e^4)

```
>trunc(4.6);
```
(to truncate 4.6 to its integer part)

```
>arcsin(1.0);
```
(to find the value of $\sin^{-1} 1.0$)

```
>GAMMA(5);
```
(to evaluate the gamma function, $\Gamma(5)$)

```
>solve(x**2-2*x+1=0);
```
(to solve the equation $x^2 - 2x + 1 = 0$)

(3) Sets and lists must be delineated with curly brackets as follows:

```
>solve({a = b + 3, b = a + w},{a, b});
```
(to solve the system of equations, $a = b + 3$ and $b = a + w$, for the unknowns a and b)

(4) An expression can be substituted for another using the subs (substitute) command as follows:

```
>soln:=solve(4*x-y=x,x);
```

(5) The variable ranges involved in integration, counting, and plotting are indicated as $x = 0..4$ to mean that $0 \leq x \leq 4$. Another example is

```
>sum(1.0/(2*i-1),i=1..10);
```
(to find the value of $\sum_{i=1}^{10} \frac{1}{2i-1} = 2.133255530$)

3. MAPLE Operations

coeff() to collect like terms
evalf() to evaluate and find a numerical value
expand() to expand expression
simplify() to simplify expression
subs() to perform substitutions

4. Built-in Functions

abs() absolute value
sin() sine
arcsin() $\sin^{-1}$
exp() exponential
Im() imaginary part
Re() real part
sqrt() square root
ln() natural logarithm
erf() error function
dsolve() solve differential equation
fsolve() solve equation numerically
solve() solve equation symbolically

5. Calculus Functions

diff() find symbolic derivative
int() evaluate symbolic integral
laplace() take Laplace transform
invlaplace() take inverse Laplace transform
spline() find spline function
taycoef() determine coefficients of Taylor's expansion

6. Simple Examples

To find the numerical solution of algebraic equations, the following expressions are used:

```
>fsolve(expression);
```

```
>fsolve(set of equations, set of variables);
```

For example,

```
>fsolve(cos(x)=x);
```
gives .7390851332

and

```
>fsolve({2*x+3*y=8,3*x-y=1},{x,y});
```
gives x=1.0 and y=2.0

7. Plotting

To plot a graph, the following type of command can be used:

```
>plot(x^2+1,x=0..2);
```

(to plot the graph, $y = x^2 + 1$, in the range $0 \leq x \leq 2$).

Appendix D

Basics of MATLAB

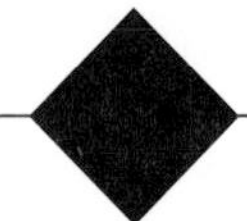

MATLAB, derived from MATrix LABoratory, is a software package that was originally developed in the late 1970s for the solution of scientific and engineering problems interactively. Originally, MATLAB was written in Fortran and based mainly on the packages LINPACK and EISPACK. Later, it incorporated tools for nonlinear equations, numerical integration, curve fitting, differential equations, optimization, and graphics. Currently, MATLAB is available in C and assembler languages. The software can be used to execute a single statement or a list of statements, called a script file. MATLAB provides excellent graphing and programming capabilities. It can also be used to solve many types of problems symbolically. Simple computations can be done by entering an instruction, similar to what we do on a calculator, at the prompt. The symbols to be used for the basic arithmetic operations of addition, subtraction, multiplication, division, and exponentiation are +, −, *, /, and ^, respectively. In any expression, the computations are performed from left to right, with exponentiation having the highest priority, followed by multiplication and division (with equal priority), and then addition and subtraction (with equal priority).

1. Variables

Variable names in MATLAB should start with a letter and can have a length of upto 31 characters in any combination of letters, digits, and underscores. Upper- and lowercase letters are treated separately. MATLAB treats all variables as matrices, although scalar quantities need not be given as arrays.

2. Arrays and Matrices

The name of a matrix must start with a letter and may be followed by any combination of letters or digits. The letters may be upper- or lowercase. Before performing arithmetic operations such as addition, subtraction, multiplication, and division on matrices, the matrices must be created using statements such as the following:

Row vector:

$$A = [1 \quad 2 \quad 3].$$

A row vector is treated as a 1-by-n matrix; its elements are enclosed in brackets and are separated by either spaces or commas.

Column vector:

$$A = \begin{matrix} [1 \\ 2 \\ 3] \end{matrix}, A = [1; \quad 2; \quad 3], \text{ or } A = [1 \quad 2 \quad 3]'.$$

A column vector is treated as a n-by-1 matrix. Its elements can be entered in different lines or in a single line using a semicolon to separate them or in a single line using a row vector with a prime on the right-side bracket (to denote the transpose).

To define the matrix

$$[A] = \begin{bmatrix} 1 & 2 & 3 \\ 4 & 5 & 6 \\ 7 & 8 & 9 \end{bmatrix},$$

the following specification can be used:

$$A = \begin{matrix} [1 & 2 & 3 \\ 4 & 5 & 6 \\ 7 & 8 & 9] \end{matrix}, \text{ or } A = [1 \quad 2 \quad 3; \quad 4 \quad 5 \quad 6; \quad 7 \quad 8 \quad 9].$$

3. Arrays with Special Structure

In some cases, the special structure of an array is used to specify the array in a simpler manner. For example, $A = 1:10$ denotes a row vector

$$A = [1 \quad 2 \quad 3 \quad 4 \quad 5 \quad 6 \quad 7 \quad 8 \quad 9 \quad 10]$$

and $A = 2:0.5:4$ represents the row vector

$$A = [2.5 \quad 3.0 \quad 3.5 \quad 4.0]$$

4. Special Matrices

Some of the special matrices are identified as follows:

$A = \text{eye}(3)$
implies an identity matrix of order 3:

$$A = \begin{bmatrix} 1 & 0 & 0 \\ 0 & 1 & 0 \\ 0 & 0 & 1 \end{bmatrix}.$$

$A = \text{ones}\ (3)$
implies a square matrix of order 3 with all elements equal to one:

$$A = \begin{bmatrix} 1 & 1 & 1 \\ 1 & 1 & 1 \\ 1 & 1 & 1 \end{bmatrix}.$$

$A = \text{zeros}\ (2, 3)$
implies a 2×3 matrix with all elements equal to zero:

$$A = \begin{bmatrix} 0 & 0 & 0 \\ 0 & 0 & 0 \end{bmatrix}.$$

5. Matrix Operations

To add the matrices $[A]$ and $[B]$ to get $[C]$, we use the following statement:

```
>C=A+B;
```

To solve a system of linear equations $[A]\vec{X} = \vec{B}$, we define the matrix A and the vector B and use the following statement:

```
> X = A\B
```

6. Functions in MATLAB

MATLAB has a large number of built-in functions, including the following:

Square root of x: sqrt(x)
Sine of x: sin(x)
Logarithm of x to base 10: log 10(x)
Gamma function of x: gamma(x)

7. M-files

MATLAB can be used in an interactive mode by typing each command from the keyboard. In this mode, MATLAB performs the operations much like an extended calculator. However, there are situations in which this mode of operation is inefficient. For example, if the same set of commands is to be repeated a number of times with different values of the input parameters, developing a MATLAB program will be quicker and efficient.

A MATLAB program consists of a sequence of MATLAB instructions written outside MATLAB and then executed in MATLAB as a single block of commands. Such a program is called a script file, or M-file. It is necessary to give a name to the script file. The name should end with `.m` (a dot followed by the letter `m`). A typical M-file (called `fibo.m`) is

```
file"fibo.m"

% m-file to compute Fibonacci numbers
f=[1 1];
i=1;
while f(i)+f(i+1)<1000
   f(i+2)=f(i)+f(i+1);
   i=i+1;
end
```

8. Plotting of Graphs

To plot a graph in MATLAB, we define a vector of values of the independent variable x (array x) and a vector of values of the dependent variable y corresponding to the values of x (array y). Then the $x-y$ graph can be plotted using the command:

```
plot(x,y)
```

As an example, the following commands can be used to plot the function $y = x^2 + 1$ in the range $0 \leq x \leq 3$:

```
x=0:0.2:3;
y=x^2+1;
plot(x,y);
hold on
x1=[0 3];
y1=[0 0];
plot(x1,y1);
grid on
hold off
```

Note that the first two lines are used to generate the arrays x and y (using increments of 0.2 for x); the third line plots the graph (using straight lines between the indicated points); the next six lines permit the plotting of x and y axes along with the setting up of the grid (using *grid on* command).

Appendix E

Basics of MATHCAD

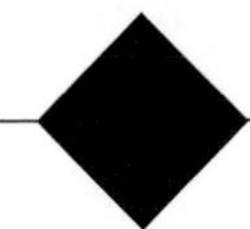

MATHCAD is a powerful engineering tool that combines the capabilities of a programming language and a spreadsheet. When equations are entered in MATHCAD, they appear on the screen in the same way we write them on a paper. For example, when we enter $(p - q)/(r + s)$ in MATHCAD, the expression is displayed as

$$\frac{p-q}{(r+s)}$$

on the screen.

1. MATHCAD Conventions

Italics denote scalar variable names, function names, and key words. **Bold** represents a menu command. It is also used to denote vectors and matrices.

MATHCAD *workspace:* When we start MATHCAD, a window appears on the screen. Different toolbars such as *Calculator, Graph, Matrix, Calculus, Math*, and *Programming* toolbars are available. Each button in the Math toolbar opens another toolbar of operators or symbols. The following are some of the buttons in the Math toolbar and the associated toolbars:

Calculator—Common arithmetic operators.

Graph—Various two- and three-dimensional plot types and graph tools.

Matrix—Matrix and vector operators.

Evaluation—Equal signs for evaluation and definition.

Calculus—Derivatives, integrals, limits, and iterated sums and products.

Symbolic—Symbolic keywords.

2. Calculations

Arithmetic Expression

Example

Compute the value of $25 - \frac{8}{104.5}$.

(1) Click the mouse anywhere in the workspace. A red crosshair appears.

(2) Type 25 − 8/104.5 =. When we type the equal sign, MATHCAD computes and displays the result as

$$25 - \frac{8}{104.5} = 24.923.$$

Equations

By defining variables and functions, we can link equations together and use intermediate results in further calculations.

Example

Compute the value of $\frac{x}{4}y^3$ when $x = 1.5$ and $y = 2.5$.

(1) At a red crosshair, type x followed by a colon:. (MATHCAD shows the colon as the definition symbol :=.) Type 1.5 to complete the definition of x.

(2) Press [↵] to move the crosshair to the next line. Type y : 2.5 and press [↵] again to bring the crosshair (cursor) to the next line.

(3) Type $x/4$[Space] ∗ y^3 and press the equal sign. MATHCAD computes and displays the result as shown here:

x:=1.5
y:=2.5
$\frac{x}{4} \cdot y^3 = 5.859$

Computation for a Range of Values

Example

Find the value of $\frac{x}{4}y^3$ when $y = 2.5$ and $x = 1.5, 1.6, \ldots, 1.9$.

(1) At a red crosshair, type x followed by a colon (:). Type 1.5,1.6;1.9 (1.5 indicates the first value of x, 1.6 the next value, a semicolon (;), denotes the range variable operator and 1.9 the last value of x).

(2) Press [↵] to move the crosshair to the next line. Type y : 2.5 and press [↵] again to bring the crosshair to the next line.

(3) Type $x/4$[Space]$*y$^3, press the equal sign, and press [↵]. MATHCAD computes and displays the result as follows:

```
x:=1.5,1.6..1.9
y:=2.5
```

$\frac{x}{4}\cdot y^3 =$

5.859
6.25
6.641
7.031
7.422

Creating a Matrix

Example

Define the matrix

$$\begin{pmatrix} 3 & 2 & 1 \\ 7 & 8 & 9 \end{pmatrix}.$$

(1) Click on [matrix icon] in the matrix toolbar to obtain the dialog box

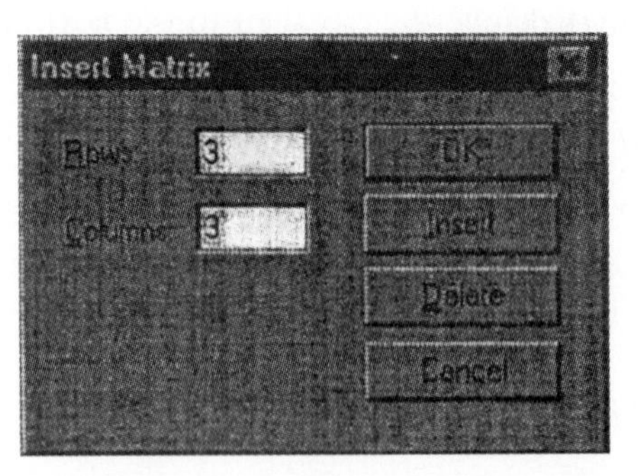

(2) Enter the number of rows and number of columns in the appropriate boxes. Click ok in the dialog box to display a matrix of placeholders:

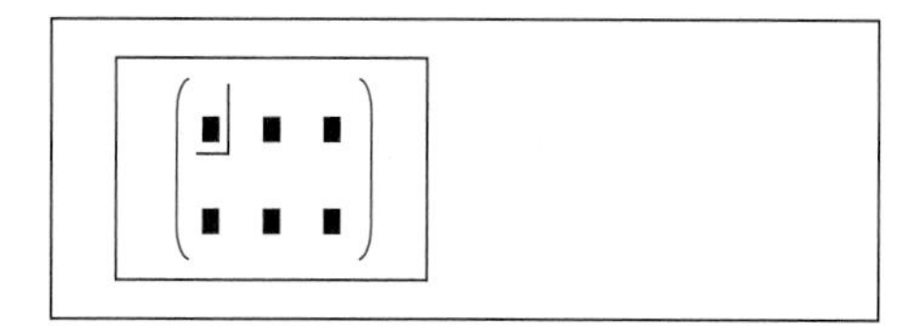

(3) Fill the placeholders by typing a number and pressing [Tab] to move to the next placeholder. Complete the definition of the matrix to obtain

$$\begin{pmatrix} 3 & 2 & 1 \\ 7 & 8 & 9 \end{pmatrix}.$$

Note that this matrix can be used in equations just like a number.

3. Some Built-in Functions in MATHCAD

(1) acos (z): gives $\cos^{-1}(z)$ in radians, where z is a real or complex number.

(2) cholesky (M) gives the lower triangular matrix L using Cholesky decomposition: $LL^T = M$, where M is a real, symmetric, positive-definite square matrix.

(3) cond2 (M) gives the condition number of the matrix M based on L_2—norm, where M is real or complex square matrix.

(4) cot (z) gives the cotangent of z, where z is in radians and not a multiple of π.

(5) dnorm (x, μ, σ) gives the normal probability density function corresponding to the real number x, mean value μ, and standard deviation σ.

(6) eigenvals (M) gives the vector of eigenvalues of the matrix M, where M is a real or complex square matrix.

(7) exp (z) gives the value of e^z, where z is a real or complex number.

(8) Im (z) gives the imaginary part of z, where z is a real or complex number.

(9) log (z) gives the value of $\log_{10} z$, where z is a real or complex nonzero number.

(10) tanh (z) gives the hyperbolic tangent of z, where z is a real or complex number.

(11) rank (M) gives the rank of the matrix M, where M is a real $m \times n$ matrix.

(12) root (*f(var), var*): gives a value of *var* at which the function *f(var)* is equal to zero.

Appendix F

Review of Matrix Algebra

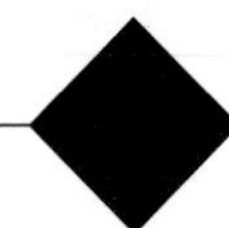

1. Definitions

Matrix: A matrix is a rectangular array of integer, real, or complex numbers. An array having m rows and n columns enclosed in brackets is called an m-by-n matrix. If $[A]$ is an $m \times n$ matrix, it is denoted as

$$[A] = \begin{bmatrix} a_{11} & a_{12} & \cdot & \cdot & \cdot & a_{1n} \\ a_{21} & a_{22} & \cdot & \cdot & \cdot & a_{2n} \\ \cdot & \cdot & \cdot & \cdot & \cdot & \cdot \\ \cdot & \cdot & \cdot & \cdot & \cdot & \cdot \\ \cdot & \cdot & \cdot & \cdot & \cdot & \cdot \\ a_{m1} & a_{m2} & \cdot & \cdot & \cdot & a_{mn} \end{bmatrix} \tag{1}$$

where the numbers a_{ij} are called the elements of the matrix. The first subscript i denotes the row and the second subscript j specifies the column in which the element a_{ij} appears.

Square matrix: If the number of rows (m) is equal to the number of columns (n), the matrix is called a square matrix of order n.

Column matrix: A matrix with m rows and 1 column is known as a column matrix or simply a column vector. A column vector $\vec{a}$ with m elements is denoted as

$$\vec{a} = \begin{Bmatrix} a_1 \\ a_2 \\ \cdot \\ \cdot \\ \cdot \\ a_m \end{Bmatrix}. \tag{2}$$

Row matrix: A matrix with 1 row and n columns is known as a row matrix or simply a row vector. A row vector $[b]$ with n elements is denoted as

$$[b] = [b_1 \, b_2 \ldots b_n]. \tag{3}$$

Diagonal matrix: A square matrix in which all the elements are zero, except those on the main diagonal is called a diagonal matrix. A diagonal matrix, $[A]$, of order n is denoted as

$$[A] = \begin{bmatrix} a_{11} & 0 & 0 & \cdot & \cdot & \cdot & 0 \\ 0 & a_{22} & 0 & \cdot & \cdot & \cdot & 0 \\ 0 & 0 & a_{33} & \cdot & \cdot & \cdot & 0 \\ \cdot & \cdot & \cdot & \cdot & \cdot & \cdot & \cdot \\ \cdot & \cdot & \cdot & \cdot & \cdot & \cdot & \cdot \\ \cdot & \cdot & \cdot & \cdot & \cdot & \cdot & \cdot \\ 0 & 0 & 0 & \cdot & \cdot & \cdot & a_{mn} \end{bmatrix}. \tag{4}$$

Identity matrix: If all the elements of a diagonal matrix are equal to 1, unity, then the matrix is known as an identity or unit matrix and is denoted as $[I]$.

Zero matrix: If all the elements of a matrix are equal to zero, then the matrix is called a zero or null matrix and is denoted as $[0]$.

Symmetric matrix: A square matrix for which the upper right half can be obtained by flipping the matrix about the main diagonal is called a symmetric matrix. Thus, for a symmetric matrix $[A] = [a_{ij}]$, $a_{ij} = a_{ji}$.

Transpose of a matrix: The transpose of an $m \times n$ matrix $[A]$ is defined as the $n \times m$ matrix obtained by interchanging the rows and columns of $[A]$, and is denoted as $[A]^T$. For example, if $[A]$ is given by

$$[A] = \begin{bmatrix} 2 & 4 & 5 \\ 4 & 1 & 8 \end{bmatrix}, \tag{5}$$

its transpose, $[A]^T$, is given by

$$[A]^T = \begin{bmatrix} 2 & 4 \\ 4 & 1 \\ 5 & 8 \end{bmatrix}. \tag{6}$$

It can be noted that the transpose of a column matrix (vector) is a row matrix (vector), and vice versa.

Trace: The trace of a square matrix is defined as the sum of the elements in the main diagonal. For example, the trace of the $n \times n$ matrix, $[A] = [a_{ij}]$, is given by

$$\text{Trace } [A] = a_{11} + a_{22} + \cdots + a_{nn}. \tag{7}$$

2. Determinant of a Matrix

If $[A]$ denotes a square matrix of order n, then the determinant of $[A]$, is denoted as

$$|[A]| = \begin{vmatrix} a_{11} & a_{12} & \cdot & \cdot & \cdot & a_{1n} \\ a_{21} & a_{22} & \cdot & \cdot & \cdot & a_{2n} \\ \cdot & \cdot & \cdot & \cdot & \cdot & \cdot \\ \cdot & \cdot & \cdot & \cdot & \cdot & \cdot \\ \cdot & \cdot & \cdot & \cdot & \cdot & \cdot \\ a_{n1} & a_{n2} & \cdot & \cdot & \cdot & a_{nn} \end{vmatrix}. \tag{8}$$

The value of a determinant can be determined in terms of its minors and cofactors. The *minor* of the element a_{ij} of the determinant $|[A]|$ of order n is a determinant of order $n-1$ obtained by deleting the row i and column j of the original determinant. The minor of a_{ij} is denoted as M_{ij}. The *cofactor* of the element a_{ij} of the determinant $|[A]|$ of order n is the minor of the element a_{ij}, with either a plus or a minus sign attached; it is defined as

$$\text{cofactor of } a_{ij} = \beta_{ij} = (-1)^{i+j} M_{ij}, \tag{9}$$

where M_{ij} is the minor of a_{ij}. For example, the cofactor of the element a_{32} of

$$|[A]| = \det[A] = \begin{vmatrix} a_{11} & a_{12} & a_{13} \\ a_{21} & a_{22} & a_{23} \\ a_{31} & a_{32} & a_{33} \end{vmatrix} \tag{10}$$

is given by

$$\beta_{32} = (-1)^5 M_{32} = -\begin{vmatrix} a_{11} & a_{13} \\ a_{21} & a_{23} \end{vmatrix}. \tag{11}$$

The value of a second-order determinant $|[A]|$ is defined as

$$\det[A] = \begin{vmatrix} a_{11} & a_{12} \\ a_{21} & a_{22} \end{vmatrix} = a_{11}a_{22} - a_{12}a_{21}. \tag{12}$$

The value of an nth-order determinant $|[A]|$ is defined as

$$\det[A] = \sum_{j=1}^{n} a_{ij}\beta_{ij}, \quad \text{for any specific row } i,$$

or

$$\det[A] = \sum_{i=1}^{n} a_{ij}\beta_{ij}, \quad \text{for any specific column } j. \tag{13}$$

Note: It can be seen from Eq. (13) that there are $2n$ different ways in which the determinant of a matrix can be computed. It can be shown that the number of arithmetic operations (multiplications, divisions, additions, or subtractions) required for the computation of a determinant of an $n \times n$ matrix is $O(n!)$. Thus, the number of arithmetic operations increases very rapidly with an increase in the size of the matrix.

Properties of determinants:

(1) The value of a determinant is not affected if rows (or columns) are written as columns (or rows) in the same order.

(2) If all the elements of a row (or a column) are zero, the value of the determinant is zero.

(3) If any two rows (or two columns) are interchanged, the value of the determinant is multiplied by -1.

(4) If all the elements of one row (or one column) are multiplied by the same constant a, the value of the new determinant is a times the value of the original determinant.

(5) If the corresponding elements of two rows (or two columns) of a determinant are proportional, the value of the determinant is zero. For example,

$$\det[A] = \begin{vmatrix} 4 & 6 & -8 \\ 3 & 4 & -6 \\ -2 & 2 & 4 \end{vmatrix} = 0.$$

3. Rank of a Matrix

For an $n \times n$ matrix $[A]$, consider all possible square submatrices that can be formed by deleting rows and columns. The rank of $[A]$ is then defined as the size of the highest order nonsingular submatrix. This implies that a square matrix of order n is nonsingular if and only if its rank is n.

4. Inverse Matrix

The inverse of a square matrix $[A]$ is written as $[A]^{-1}$ and is defined by the relationship

$$[A]^{-1}[A] = [A][A]^{-1} = [I], \tag{14}$$

where $[A]^{-1}[A]$, for example, denotes the product of the matrix $[A]^{-1}$ and $[A]$. The inverse matrix of $[A]$ can be determined by

$$\frac{\text{adjoint } [A]}{\det [A]}, \tag{15}$$

where adjoint $[A]$ is the adjoint matrix of $[A]$ and det $[A]$, the determinant of $[A]$, is assumed to be nonzero.

Adjoint Matrix

The adjoint matrix of a square matrix $[A] = [a_{ij}]$ is defined as the matrix obtained by replacing each element a_{ij} by its cofactor β_{ij} and then transposing. Thus,

$$\text{Adjoint }[A] = \begin{bmatrix} \beta_{11} & \beta_{12} & \cdot & \cdot & \cdot & \beta_{1n} \\ \beta_{12} & \beta_{22} & \cdot & \cdot & \cdot & \beta_{2n} \\ \cdot & \cdot & \cdot & \cdot & \cdot & \cdot \\ \cdot & \cdot & \cdot & \cdot & \cdot & \cdot \\ \cdot & \cdot & \cdot & \cdot & \cdot & \cdot \\ \beta_{n1} & \beta_{n2} & \cdot & \cdot & \cdot & \beta_{nn} \end{bmatrix}^T = \begin{bmatrix} \beta_{11} & \beta_{21} & \cdot & \cdot & \cdot & \beta_{n1} \\ \beta_{12} & \beta_{22} & \cdot & \cdot & \cdot & \beta_{n2} \\ \cdot & \cdot & \cdot & \cdot & \cdot & \cdot \\ \cdot & \cdot & \cdot & \cdot & \cdot & \cdot \\ \cdot & \cdot & \cdot & \cdot & \cdot & \cdot \\ \beta_{1n} & \beta_{2n} & \cdot & \cdot & \cdot & \beta_{nn} \end{bmatrix}. \tag{16}$$

▶Example

Find the adjoint matrix and the inverse of the matrix

$$[A] = \begin{bmatrix} 2 & 2 & 3 \\ 4 & 5 & 6 \\ 7 & 8 & 9 \end{bmatrix}.$$

Solution

The determinant of $[A]$ can be obtained by selecting the first column for expansion by

$$|[A]| = a_{11}\beta_{11} + a_{21}\beta_{21} + a_{31}\beta_{31}, \tag{E1}$$

where $a_{11} = 2$, $a_{21} = 4$, $a_{31} = 7$, $\beta_{11} = (-1)^{1+1}M_{11}$, $\beta_{21} = (-1)^{2+1}M_{21}$, $\beta_{31} = (-1)^{3+1}M_{31}$,

$$M_{11} = \begin{vmatrix} a_{22} & a_{23} \\ a_{32} & a_{33} \end{vmatrix} = (a_{22}a_{33} - a_{23}a_{32}) = (5 \times 9 - 6 \times 8) = -3,$$

$$M_{21} = \begin{vmatrix} a_{12} & a_{13} \\ a_{32} & a_{33} \end{vmatrix} = (a_{12}a_{33} - a_{13}a_{32}) = (2 \times 9 - 3 \times 8) = -6,$$

and

$$M_{31} = \begin{vmatrix} a_{12} & a_{13} \\ a_{22} & a_{23} \end{vmatrix} = (a_{12}a_{23} - a_{13}a_{22}) = (2 \times 6 - 3 \times 5) = -3.$$

Thus, Eq. (E1) gives

$$|[A]| = 2(-3) - 4(-6) + 7(-3) = -3.$$

The cofactors β_{12}, β_{22}, β_{32}, β_{13}, β_{23}, and β_{33} can be determined in a manner similar to those of β_{11}, β_{21}, and β_{31} and the adjoint matrix of $[A]$ can be obtained as

$$\text{adjoint } [A] = \begin{bmatrix} \beta_{11} & \beta_{21} & \beta_{31} \\ \beta_{12} & \beta_{22} & \beta_{32} \\ \beta_{13} & \beta_{23} & \beta_{33} \end{bmatrix} = \begin{bmatrix} -3 & 6 & -3 \\ 6 & -3 & 0 \\ -3 & -2 & -2 \end{bmatrix}. \tag{E2}$$

The inverse of the matrix $[A]$ is given by Eq. (15):

$$[A]^{-1} = \frac{1}{-3}\begin{bmatrix} -3 & 6 & -3 \\ 6 & -3 & 0 \\ -3 & -2 & 2 \end{bmatrix} = \begin{bmatrix} 1 & -2 & 1 \\ -2 & 1 & 0 \\ 1 & \frac{2}{3} & -\frac{2}{3} \end{bmatrix}.$$

◀

5. Basic Matrix Operations

Equality of matrices: Two matrices $[A]$ and $[B]$, having the same order, are equal if and only if $a_{ij} = b_{ij}$ for every i and j.

Addition and subtraction of matrices: The sum of the two matrices $[A]$ and $[B]$, having the same order, is given by the sum of the corresponding elements. Thus, if $[C] = [A] + [B] = [B] + [A]$, then $c_{ij} = a_{ij} + b_{ij}$ for every i and j. Similarly, the difference of two matrices $[A]$ and $[B]$ of the same order, is given by $[D] = [A] - [B]$ with $d_{ij} = a_{ij} - b_{ij}$ for every i and j.

Multiplication of matrices: The product of two matrices $[A]$ and $[B]$ is defined only if they are conformable (i.e., if the number of columns of $[A]$ is equal to the number of rows of $[B]$. If $[A]$ is of order $m \times n$ and $[B]$ is of order $n \times p$, then the product $[C] = [A][B]$ is of order $m \times p$ and is defined by $[C] = [c_{ij}]$, with

$$c_{ij} = \sum_{k=1}^{n} a_{ik}b_{kj}. \tag{17}$$

This means that c_{ij} is the quantity obtained by multiplying the ith row of $[A]$ and the jth column of $[B]$ and summing these products.

Other operations: If the matrices are conformable, the multiplication process is associative

$$([A][B])[C] = [A]([B][C]), \tag{18}$$

$$([A]+[B])[C] = [A][C]+[B][C]. \tag{19}$$

Note that $[A][B]$ is the premultiplication of $[B]$ by $[A]$ or the postmultiplication of $[A]$ by $[B]$. Also, the product $[A][B]$ is not necessarily equal to $[B][A]$. The transpose of a matrix product is given by the product of the transposes of the separate matrices in reverse order. Thus, if $[C]=[A][B]$, then

$$[C]^T = ([A][B])^T = [B]^T[A]^T. \tag{20}$$

The inverse of a matrix product is given by the product of the inverses of the separate matrices in reverse order. Thus, if $[C]=[A][B]$, then

$$[C]^{-1} = ([A][B])^{-1} = [B]^{-1}[A]^{-1}. \tag{21}$$

Appendix G

Statistical Tables

Table G.1 Standard Normal Distribution Function.

$$\text{Values of } \Phi(z) = \frac{1}{\sqrt{2\pi}} \int_{-\infty}^{z} e^{-x^2/2}\, dx$$

z	.00	.01	.02	.03	.04	.05	.06	.07	.08	.09
.0	.50000	.50399	.50798	.51197	.51595	.51994	.52392	.52790	.53188	.53586
.1	.53983	.54380	.54776	.55172	.55567	.55962	.56356	.56749	.57142	.57534
.2	.57926	.58317	.58706	.59095	.59483	.59871	.60257	.60642	.61026	.61409
.3	.61791	.62172	.62552	.62930	.63307	.63683	.64058	.64431	.64803	.65173
.4	.65542	.65910	.66276	.66640	.67003	.67364	.67724	.68082	.68439	.68793
.5	.69146	.69497	.69847	.70194	.70540	.70884	.71226	.71566	.71904	.72240
.6	.72575	.72907	.73237	.73565	.73891	.74215	.74537	.74857	.75175	.75490
.7	.75804	.76115	.76424	.76730	.77035	.77337	.77637	.77935	.78230	.78524
.8	.78814	.79103	.79389	.79673	.79955	.80234	.80510	.80785	.81057	.81327
.9	.81594	.81859	.82121	.82381	.82639	.82894	.83147	.83398	.83646	.83891
1.0	.84134	.84375	.84614	.84849	.85083	.85314	.85543	.85769	.85992	.86214
1.1	.86433	.86650	.86864	.87076	.87286	.87493	.87698	.87900	.88100	.88298
1.2	.88493	.88686	.88877	.89065	.89251	.89435	.89616	.89796	.89973	.90147
1.3	.90320	.90490	.90658	.90824	.90988	.91149	.91309	.91466	.91621	.91774
1.4	.91924	.92073	.92220	.92364	.92507	.92647	.92785	.92922	.93056	.93189
1.5	.93319	.93448	.93574	.93699	.93822	.93943	.94062	.94179	.94295	.94408
1.6	.94520	.94630	.94738	.94845	.94950	.95053	.95154	.95254	.95352	.95449
1.7	.95543	.95637	.95728	.95818	.95907	.95994	.96080	.96164	.96246	.96327
1.8	.96407	.96485	.96562	.96638	.96712	.96784	.96856	.96926	.96995	.97062
1.9	.97128	.97193	.97257	.97320	.97381	.97441	.97500	.97558	.97615	.97670

Table G.1 *(continued)*

z	.00	.01	.02	.03	.04	.05	.06	.07	.08	.09
2.0	.97725	.97778	.97831	.97882	.97932	.97982	.98030	.98077	.98124	.98169
2.1	.98214	.98257	.98300	.98341	.98382	.98422	.98461	.98500	.98537	.98574
2.2	.98610	.98645	.98679	.98713	.98745	.98778	.98809	.98840	.98870	.98899
2.3	.98928	.98956	.98983	$.9^20097$	$.9^20358$	$.9^20613$	$.9^20863$	$.9^21106$	$.9^21344$	$.9^21576$
2.4	$.9^21802$	$.9^22024$	$.9^22240$	$.9^22451$	$.9^22656$	$.9^22857$	$.9^23053$	$.9^23244$	$.9^23431$	$.9^23613$
2.5	$.9^23790$	$.9^23963$	$.9^24132$	$.9^24297$	$.9^24457$	$.9^24614$	$.9^24766$	$.9^24915$	$.9^25060$	$.9^25201$
3.0	$.9^28650$	$.9^28694$	$.9^28736$	$.9^28777$	$.9^28817$	$.9^28856$	$.9^28893$	$.9^28930$	$.9^28965$	$.9^28999$
3.5	$.9^37674$	$.9^37759$	$.9^37842$	$.9^37922$	$.9^37999$	$.9^38074$	$.9^38146$	$.9^38215$	$.9^38282$	$.9^38347$
4.0	$.9^46833$	$.9^46964$	$.9^47090$	$.9^47211$	$.9^47327$	$.9^47439$	$.9^47546$	$.9^47649$	$.9^47748$	$.9^47843$
4.5	$.9^56602$	$.9^56759$	$.9^56908$	$.9^57051$	$.9^57187$	$.9^57318$	$.9^57442$	$.9^57561$	$.9^57675$	$.9^57784$
4.9	$.9^65208$	$.9^65446$	$.9^65673$	$.9^65889$	$.9^66094$	$.9^66289$	$.9^66475$	$.9^66652$	$.9^66821$	$.9^66981$

Table G.2 Student's t Distribution.

Values of $t_{n,\alpha}$ in $P(X > t_{n,\alpha}) = \alpha$ (n = degrees of freedom)

n \ $\alpha = 0.4$		0.25	0.1	0.05	0.025	0.01	0.005	0.0025	0.001	0.0005
1	0.325	1.000	3.078	6.314	12.706	31.821	63.657	127.32	318.31	636.62
2	.289	0.816	1.886	2.920	4.303	6.965	9.925	14.089	22.327	31.598
3	.277	.765	1.638	2.353	3.182	4.541	5.841	7.453	10.214	12.924
4	.271	.741	1.533	2.132	2.776	3.747	4.604	5.598	7.173	8.610
5	0.267	0.727	1.476	2.015	2.571	3.365	4.032	4.773	5.893	6.869
6	.265	.718	1.440	1.943	2.447	3.143	3.707	4.317	5.208	5.959
7	.263	.711	1.415	1.895	2.365	2.998	3.499	4.029	4.785	5.408
8	.262	.706	1.397	1.860	2.306	2.896	3.355	3.833	4.501	5.041
9	.261	.703	1.383	1.833	2.262	2.821	3.250	3.690	4.297	4.781
10	0.260	0.700	1.372	1.812	2.228	2.764	3.169	3.581	4.144	4.587
11	.260	.697	1.363	1.796	2.201	2.718	3.106	3.497	4.025	4.437
12	.259	.695	1.356	1.782	2.179	2.681	3.055	3.428	3.930	4.318
13	.259	.694	1.350	1.771	2.160	2.650	3.012	3.372	3.852	4.221
14	.258	.692	1.345	1.761	2.145	2.624	2.977	3.326	3.787	4.140
15	0.258	0.691	1.341	1.753	2.131	2.602	2.947	3.286	3.733	4.073
16	.258	.690	1.337	1.746	2.120	2.583	2.921	3.252	3.686	4.015
17	.257	.689	1.333	1.740	2.110	2.567	2.898	3.222	3.646	3.965
18	.257	.688	1.330	1.734	2.101	2.552	2.878	3.197	3.610	3.922
19	.257	.688	1.328	1.729	2.093	2.539	2.861	3.174	3.579	3.883
20	0.257	0.687	1.325	1.725	2.086	2.528	2.845	3.153	3.552	3.850
21	.257	.686	1.323	1.721	2.080	2.518	2.831	3.135	3.527	3.819

Table G.2 *(continued)*

$n \backslash \alpha = 0.4$		0.25	0.1	0.05	0.025	0.01	0.005	0.0025	0.001	0.0005
22	.256	.686	1.321	1.717	2.074	2.508	2.819	3.119	3.505	3.792
23	.256	.685	1.319	1.714	2.069	2.500	2.807	3.104	3.485	3.767
24	.256	.685	1.318	1.711	2.064	2.492	2.797	3.091	3.467	3.745
25	0.256	0.684	1.316	1.708	2.060	2.485	2.787	3.078	3.450	3.725
26	.256	.684	1.315	1.706	2.056	2.479	2.779	3.067	3.435	3.707
27	.256	.684	1.314	1.703	2.052	2.473	2.771	3.057	3.421	3.690
28	.256	.683	1.313	1.701	2.048	2.467	2.763	3.047	3.408	3.674
29	.256	.683	1.311	1.699	2.045	2.462	2.756	3.038	3.396	3.659
30	0.256	0.683	1.310	1.697	2.042	2.457	2.750	3.030	3.385	3.646
40	.255	.681	1.303	1.684	2.021	2.423	2.704	2.971	3.307	3.551
60	.254	.679	1.296	1.671	2.000	2.390	2.660	2.915	3.232	3.460
120	.254	.677	1.289	1.658	1.980	2.358	2.617	2.860	3.160	3.373
∞	.253	.674	1.282	1.645	1.960	2.326	2.576	2.807	3.090	3.291

Table G.3 Chi Square Distribution.

Values of $\chi^2_{n,\alpha}$ in $P(X > \chi^2_{n,\alpha}) = \alpha$ (n = degrees of freedom)

$n \backslash \alpha$	0.995	0.990	0.975	0.950	0.050	0.025	0.010	0.005
1	392704.10^{-10}	157088.10^{-9}	982069.10^{-9}	393214.10^{-8}	3.84146	5.02389	6.63490	7.87944
2	0.0100251	0.0201007	0.0506356	0.102587	5.99146	7.37776	9.21034	10.5966
3	0.0717218	0.114832	0.215795	0.351846	7.81473	9.34840	11.3449	12.8382
4	0.206989	0.297109	0.484419	0.710723	9.48773	11.1433	13.2767	14.8603
5	0.411742	0.554298	0.831212	1.145476	11.0705	12.8325	15.0863	16.7496
6	0.675727	0.872090	1.23734	1.63538	12.5916	14.4494	16.8119	18.5476
7	0.989256	1.239043	1.68987	2.16735	14.0671	16.0128	18.4753	20.2777
8	1.34441	1.64650	2.17973	2.73264	15.5073	17.5345	20.0902	21.9550
9	1.73493	2.08790	2.70039	3.32511	16.9190	19.0228	21.6660	23.5894
10	2.15586	2.55821	3.24697	3.94030	18.3070	20.4832	23.2093	25.1882
11	2.60322	3.05348	3.81575	4.57481	19.6751	21.9200	24.7250	26.7568
12	3.07382	3.57057	4.40379	5.22603	21.0261	23.3367	26.2170	28.2995
13	3.56503	4.10692	5.00875	5.89186	22.3620	24.7356	27.6882	29.8195
14	4.07467	4.66043	5.62873	6.57063	23.6848	26.1189	29.1412	31.3194
15	4.60092	5.22935	6.26214	7.26094	24.9958	27.4884	30.5779	32.8013
16	5.14221	5.81221	6.09766	7.96165	26.2962	28.8454	31.9999	34.2672
17	5.69722	6.40776	7.56419	8.67176	27.5871	30.1910	33.4087	35.7185
18	6.26480	7.01491	8.23075	9.39046	28.8693	31.5264	34.8053	37.1565
19	6.84397	7.63273	8.90652	10.1170	30.1435	32.8523	36.1909	38.5823
20	7.43384	8.26040	9.59078	10.8508	31.4104	34.1696	37.5662	39.9968
21	8.03365	8.89720	10.28293	11.5913	32.6706	35.4789	38.9322	41.4011

Table G.3 (*continued*)

$n \backslash \alpha$	0.995	0.990	0.975	0.950	0.050	0.025	0.010	0.005
22	8.64272	9.54249	10.9823	12.3380	33.9244	36.7807	40.2894	42.7957
23	9.26043	10.19567	11.6886	13.0905	35.1725	38.0756	41.6384	44.1813
24	9.88623	10.8564	12.4012	13.8484	36.4150	39.3641	42.9798	45.5585
25	10.5197	11.5240	13.1197	14.6114	37.6525	40.6465	44.3141	46.9279
26	11.1602	12.1981	13.8439	15.3792	38.8851	41.9232	45.6417	48.2899
27	11.8076	12.8785	14.5734	16.1514	40.1133	43.1945	46.9629	49.6449
28	12.4613	13.5647	15.3079	16.9279	41.3371	44.4608	48.2782	50.9934
29	13.1211	14.2565	16.0471	17.7084	42.5570	45.7223	49.5879	52.3356
30	13.7867	14.9535	16.7908	18.4927	43.7730	46.9792	50.8922	53.6720
40	20.7065	22.1643	24.4330	26.5093	55.7585	59.3417	63.6907	66.7660
50	27.9907	29.7067	32.3574	34.7643	67.5048	71.4202	76.1539	79.4900
60	35.5345	37.4849	40.4817	43.1880	79.0819	83.2977	88.3794	91.9517
70	43.2752	45.4417	48.7576	51.7393	90.5312	95.0232	100.425	104.215
80	51.1719	53.5401	57.1532	60.3915	101.879	106.629	112.329	116.321
90	59.1963	61.7541	65.6466	69.1260	113.145	118.136	124.116	128.299
100	67.3276	70.0649	74.2219	77.9295	124.342	129.561	135.807	140.169

Index

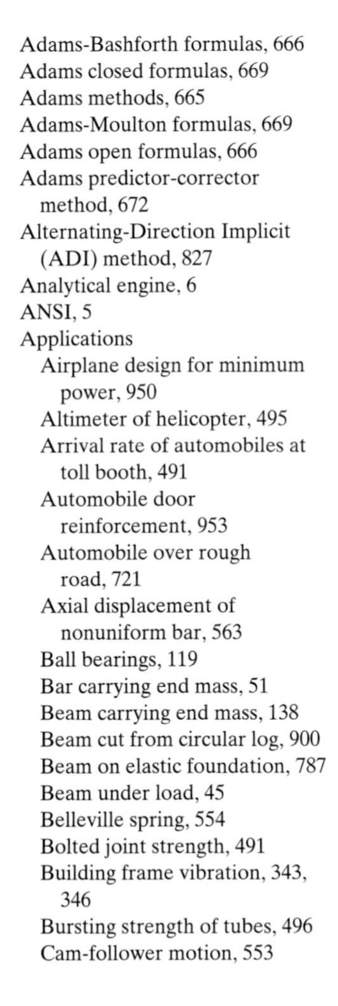